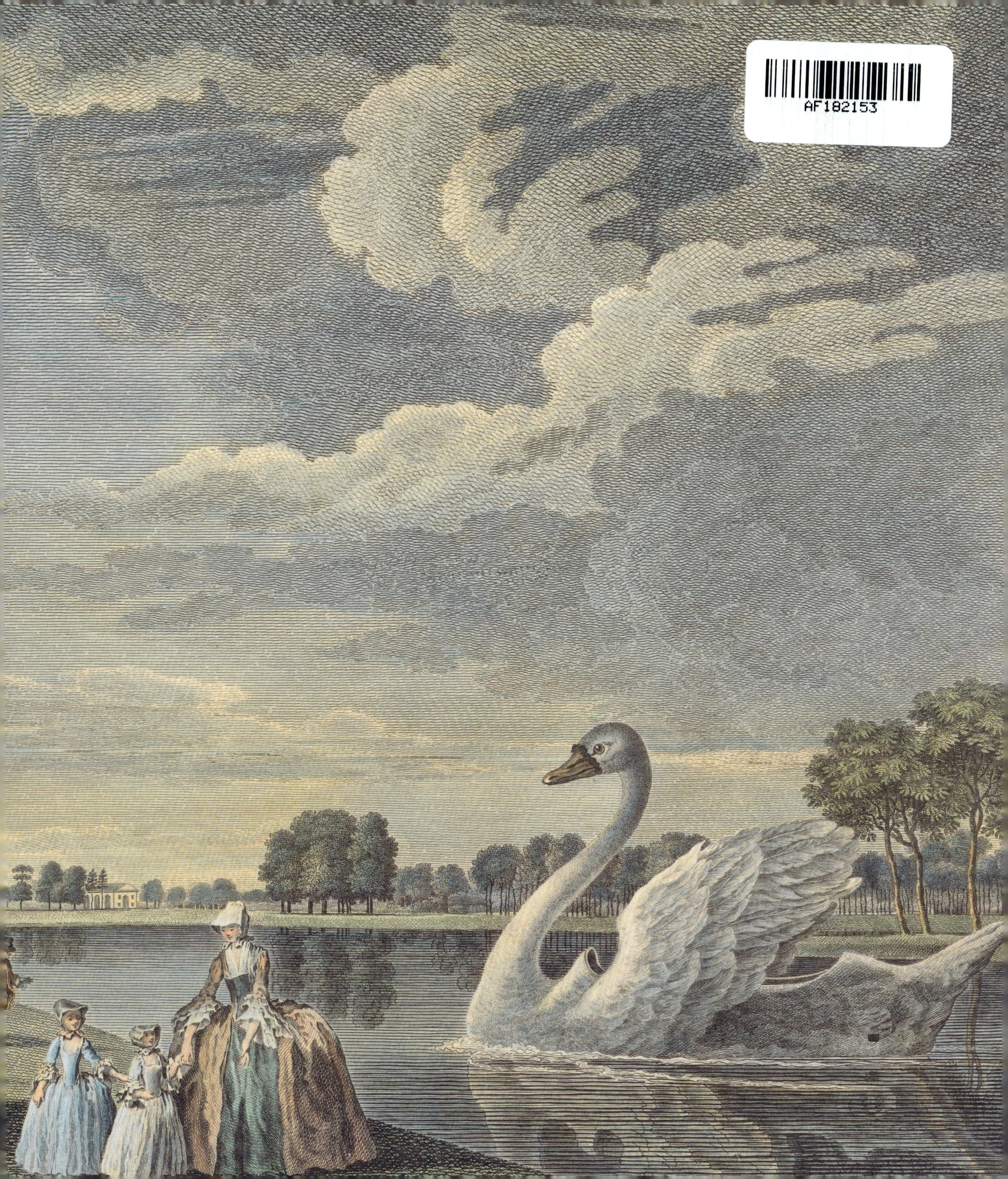

AF182153

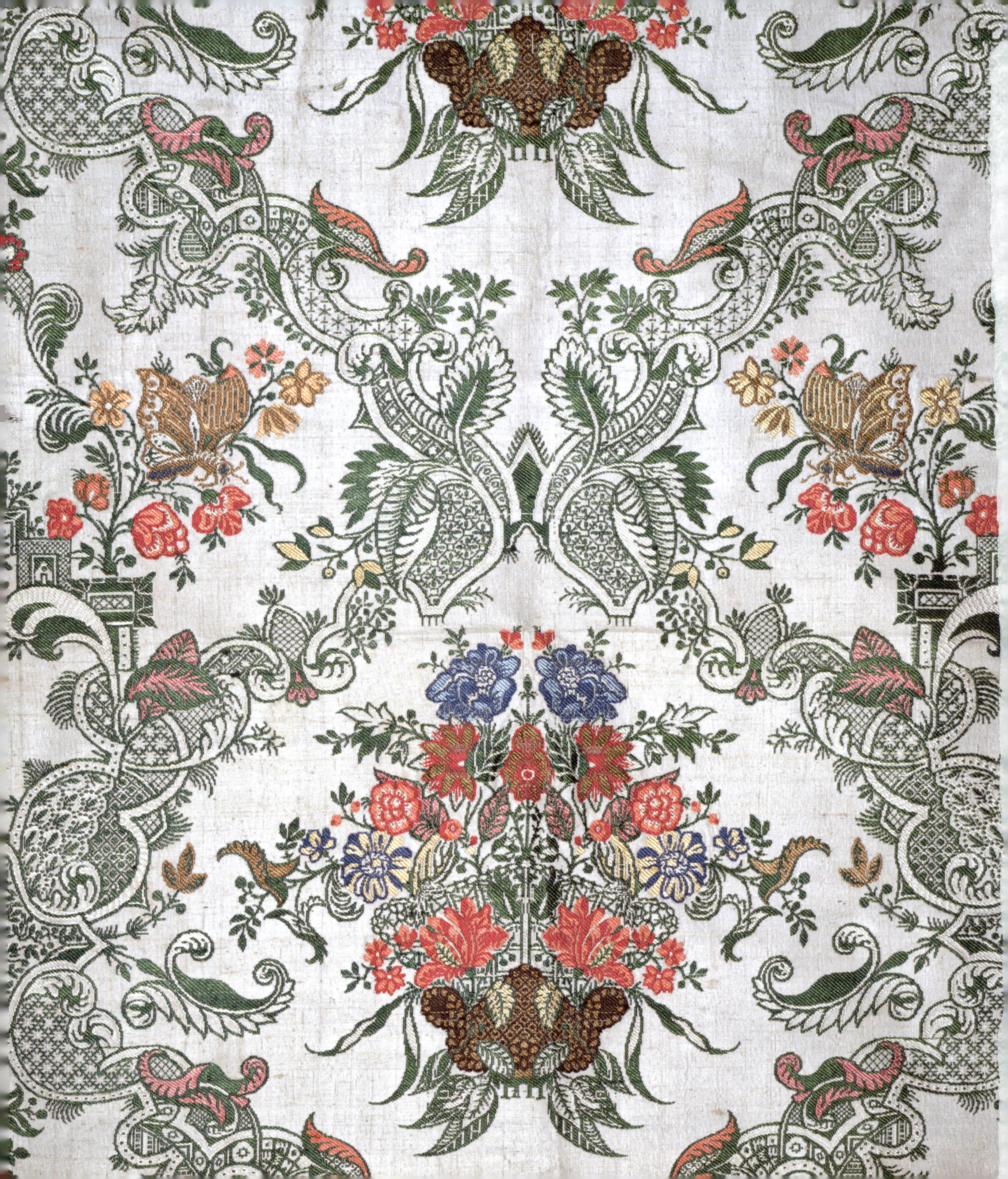

ENLIGHTENED PRINCESSES

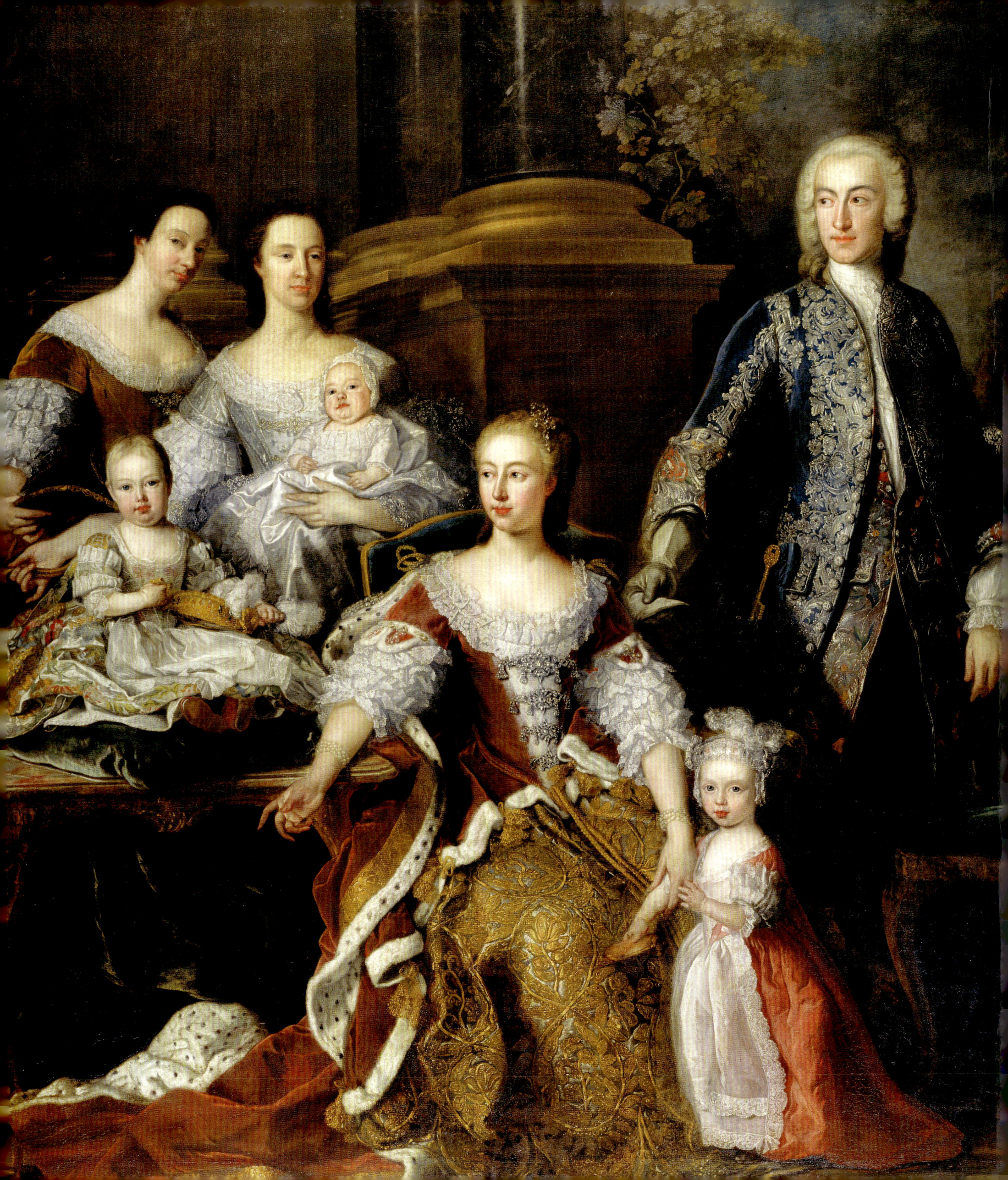

ENLIGHTENED PRINCESSES

CAROLINE, AUGUSTA, CHARLOTTE, AND THE SHAPING OF THE MODERN WORLD

EDITED BY

Joanna Marschner

with the assistance of David Bindman and Lisa L. Ford

WITH ESSAYS BY

Glenn Adamson • A. Cassandra Albinson • Robyn Asleson • Wolf Burchard
Donald Burrows • Lisa L. Ford • Todd Longstaffe-Gowan
Florence Grant • Tyler Griffith • Mark Hallett • Craig Ashley Hanson
Samantha Howard • Roisin Inglesby • Emma Jay • Berta Joncus
Kathryn Jones • Mark Laird • Ben Marsh • Joanna Marschner
Amy Meyers • Clarissa Campbell Orr • Frank Prochaska • Lee Prosser
Aileen Ribeiro • Joseph Roach • Jane Roberts • Cynthia E. Roman
Matthew Storey • John Styles

YALE CENTER FOR BRITISH ART, NEW HAVEN
HISTORIC ROYAL PALACES
YALE UNIVERSITY PRESS, NEW HAVEN AND LONDON

This publication accompanies the exhibition *Enlightened Princesses: Caroline, Augusta, Charlotte, and the Shaping of the Modern World*, co-organized by the Yale Center for British Art and Historic Royal Palaces, on view at the Yale Center for British Art, New Haven, from 2 February to 30 April 2017, and at Kensington Palace, London, from 22 June to 12 November 2017.

Exhibition curated by Joanna Marschner, with the assistance of Samantha Howard, and Amy Meyers, organizing curator at the Yale Center for British Art, with the assistance of Lisa L. Ford, Glenn Adamson, and Tyler Griffith.

Copyright © 2017 by Yale University

All rights reserved. This book may not be reproduced in whole or in part, including illustrations, in any form (beyond that copying permitted by Sections 107 and 108 of the U.S. Copyright Law and except by reviewers for the public press), without written permission from the publishers.

Library of Congress Cataloging-in-Publication Data

Names: Marschner, Joanna, editor. | Bindman, David, 1940- editor. | Ford, Lisa L., editor.
Title: Enlightened princesses : Caroline, Augusta, Charlotte, and the shaping of the modern world / edited by Joanna Marschner, David Bindman and Lisa L. Ford.
Description: New Haven : Yale University Press, 2017.
Identifiers: LCCN 2016029260 | ISBN 9780300217100
Subjects: LCSH: Caroline, Queen, consort of George II, King of Great Britain, 1683-1737—Art patronage. | Augusta, Princess of Wales, 1719-1772—Art patronage. | Charlotte, Queen, consort of George III, King of Great Britain, 1744-1818—Art patronage. | Arts, British—18th century. | Science—Great Britain—History—18th century. | Great Britain—Intellectual life—18th century.
Classification: LCC NX543 .E55 2017 | DDC 700.941/09033—dc23
LC record available at https://lccn.loc.gov/2016029260

A catalogue record for this book is available from the British Library.

Designed by Miko McGinty, Anjali Pala, and Claire Bidwell
Typeset in Adobe Caslon Pro by Tina Henderson
Printed and bound in Italy by Conti Tipocolor Srl

Endpapers: William Elliot, after William Woollett, *Kew: The White House, Orangery, and Temple of Arethusa* (detail), c. 1760, engraving, hand-coloured in watercolour, 14½ x 21 in. (36.8 x 53.3 cm). Yale Center for British Art, Paul Mellon Collection (cat. 19.15)
Half-title facing page: Unknown weavers (Spitalfields, London), dress fabric (detail), 1720–30, brocaded silk, 26⅝ x 20¼ in. (67.5 x 51.5 cm). Victoria and Albert Museum (cat. 28.04)
Frontispiece: Jean-Baptiste van Loo, *Augusta, Princess of Wales, with Members of her Family and Household*, 1739, oil on canvas, 98 x 89 in. (249 x 226 cm). Royal Collection Trust
p. v: Attributed to Princess Elizabeth, *Portlandia Grandiflora*, 1786–88, collage of coloured papers, with gouache and watercolour on paper. The Newport Garden Club, Newport, RI (fig. 6.12)
p. viii: Georg Dionysius Ehret, 'Magnolia', 1731, watercolour and gouache with gum arabic over pencil, 15 x 11½ in. (40.5 x 29.3 cm). Royal Collection Trust
p. x: Thomas Bowles, *The North Prospect of London Taken from the Bowling Green at Islington* (detail), 1752, hand-coloured engraving, 10¼ x 17⅛ in. (26 x 43.5 cm). Folger Shakespeare Library, Washington, DC (fig. 2.03)
p. xii: William Kent, *The Marriage of Henry V*, 1730–32, oil on canvas, 30 x 23⅞ in. (76.2 x 60.6 cm). Royal Collection Trust (cat. 12.06)
pp. 160–161: William Hogarth, *A Performance of 'The Indian Emperor or The Conquest of Mexico by the Spaniards'* (detail), 1732–35, oil on canvas, 51½ x 57¾ in. (130.8 x 146.7 cm). Private collection (fig. 7.02)
pp. 216–217: Johan Zoffany, *Queen Charlotte with Her Two Eldest Sons* (detail), c. 1765, oil on canvas, 44⅛ x 50½ in. (112.2 x 128.3 cm). Royal Collection Trust (fig. 1.15)
pp. 276–277: William Kent, 'A Sketch for the Queen's Library' (detail), c. 1736, pen and ink, and washes over pencil, on paper, 12⅞ x 17⅞ in. (32.7 x 45.3 cm). Sir John Soane's Museum, London (cat. 17.09)
pp. 322–323: William Woollett, after Joshua Kirby, *North Prospect of the Ruin, in the Gardens of Kew* (detail), 1763, etching, 12⅝ x 17¾ in. (32 x 45.2 cm). British Museum, London (fig. 20.01)
pp. 372–373: Benjamin West, *Princess Augusta, Princess Elizabeth, Prince Ernest, Prince Augustus, Prince Adolphus, and Princess Mary* (detail), 1776, oil on canvas, 66¼ x 71 in. (168.2 x 180.3 cm). Royal Collection Trust (cat. 23.18)
pp. 450–451: Chelsea (Lawrence Street Factory), 'Mecklenburg service', c. 1763, soft paste porcelain with Mazarin blue glaze, Royal Collection Trust (cat. 28.14)
p. 532: Eliza Lucas Pinckney, Letter to an unidentified person, p. 1 (detail), c. 1753, ink on paper, 11¾ x 14¼ in. (29.8cm x 36.2 cm). The National Society of The Colonial Dames of America in the State of South Carolina (cat. 30.10)

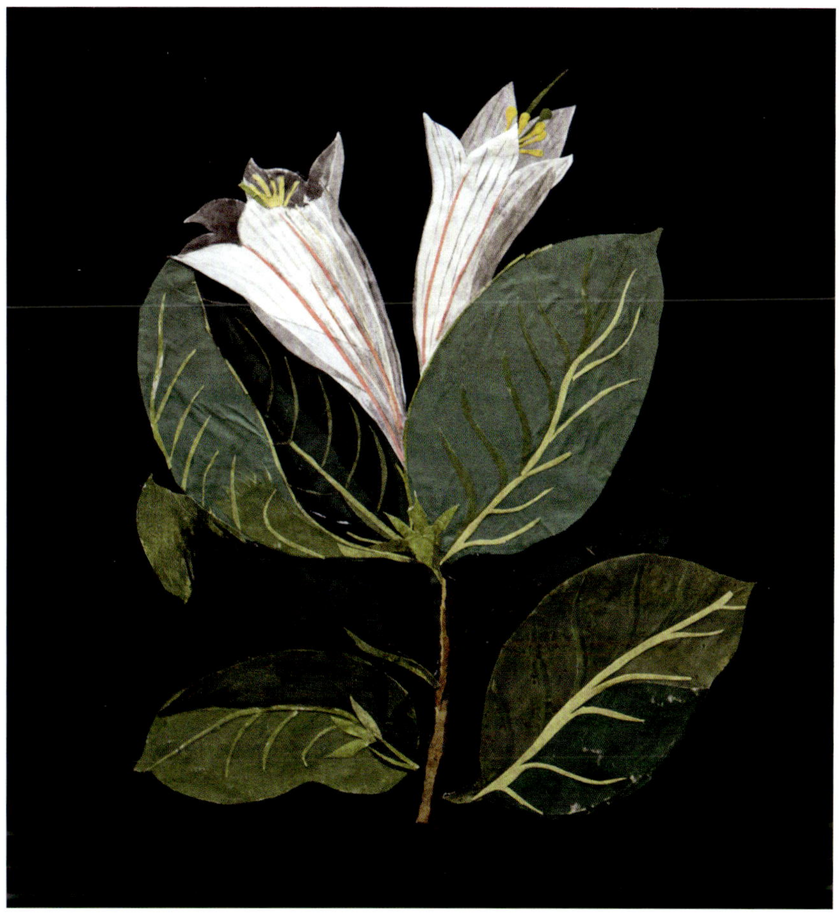

CONTENTS

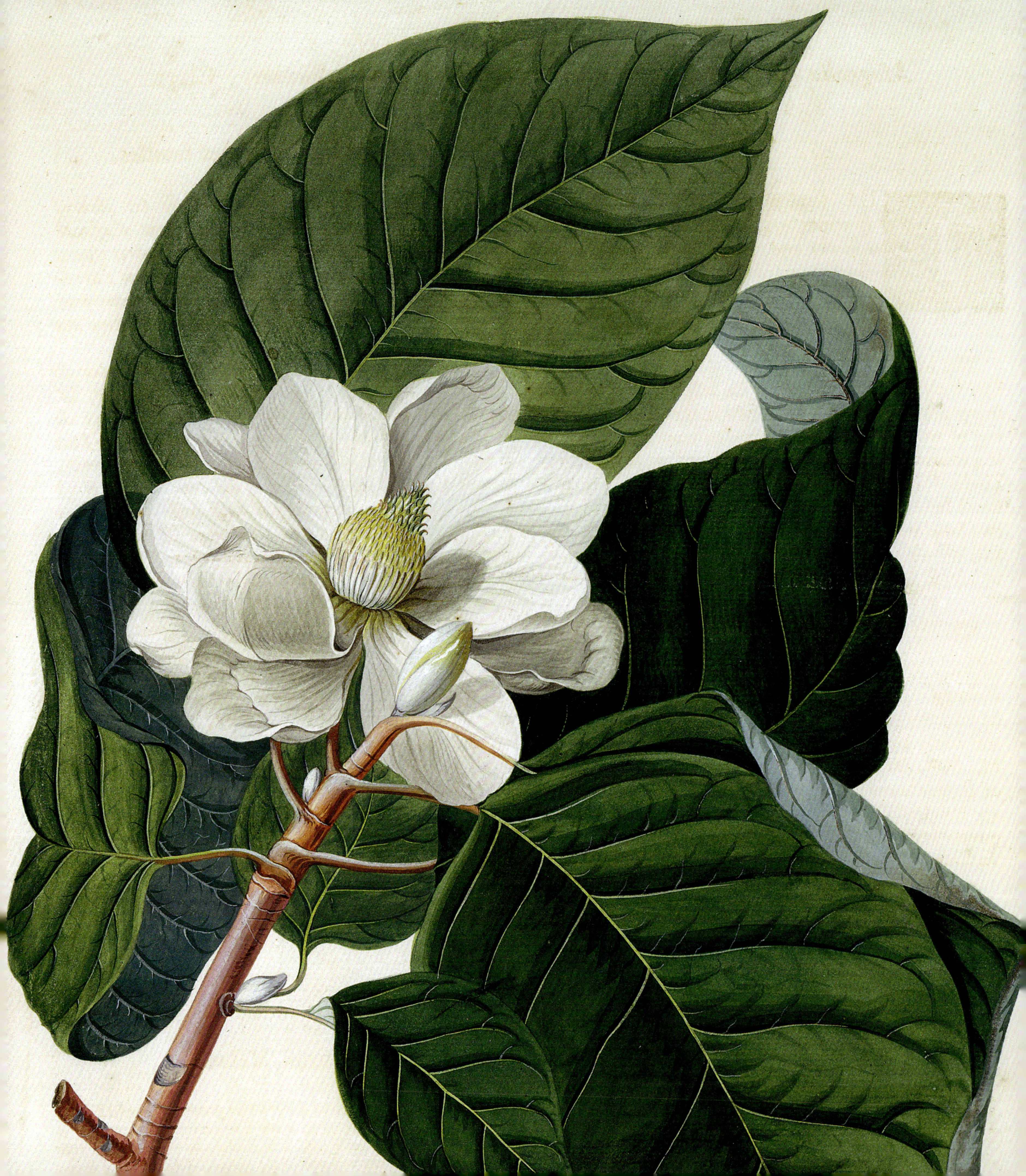

Foreword

Enlightened Princesses: Caroline, Augusta, Charlotte, and the Shaping of the Modern World* is the result of a collaborative project that has brought together the Yale Center for British Art (YCBA) and Historic Royal Palaces (HRP) in the planning of this important exhibition, along with its associated publication and programme. The exhibition, which will be mounted in 2017 at the YCBA and at Kensington Palace, explores the contributions made by Caroline of Ansbach (1683–1737), Augusta of Saxe-Gotha (1719–1772), and Charlotte of Mecklenburg-Strelitz (1744–1818) to the consolidation of the Hanoverian dynasty in its rule over Great Britain from the beginning of the eighteenth century through the first decades of the nineteenth. The project also examines the agency expressed by these three women in the shaping of British society more generally, along with their legacy, which has had a lasting impact on world culture into the modern age.

Caroline, Augusta, and Charlotte were three Protestant princesses who shared a common German ancestry. Caroline, the wife of the future George II, arrived in London in 1714, as the first Hanoverian king, George I, was crowned. She became Queen Consort after her husband succeeded his father in 1727. Augusta was married to Caroline's eldest son, Frederick, but never became queen as her husband died young. However, as mother of the next king she played a crucial part in the shaping of the monarchy for the later eighteenth century. Her son, George III, married Charlotte in 1761. She died in 1818, two years before her husband, who was by this time living at Windsor Castle, blind and deranged.

Much has been written about the Hanoverian kings regarding the embedding and defending of the new dynasty, the management of the volatile dynamic between Britain and Europe, the parts they played in the growth of the British Empire and, ultimately, the loss of Britain's thirteen colonies along the eastern seaboard of North America. However, the roles played by their wives—the women—has never been thoroughly evaluated. The research and debate engendered by this project, in the course of many conversations, a series of scholarly workshops, and an academic conference, have revealed the extent of their extraordinary contributions to the monarchy's success in their own time and to its longevity. Each princess became a significant patron of the arts and architecture. They championed new models of charity and philanthropy. They helped to create a sense of Britishness at home, as well as playing an important part in Britain's imperial agenda as it forged lucrative, but also highly problematic, relationships with the wider world.

The outcomes of their initiatives continue to have resonance in the world today. In the breadth of their ambition and their achievements, they provided a blueprint for royal women—and women of power and influence more broadly—right up to the present. For this, it is our intention to give to the princesses the attention they deserve, both in illuminating their many admirable accomplishments and in exposing the more questionable aspects of their endeavours, some of which were recognized in their own time and others of which have come into focus from today's moral vantage point.

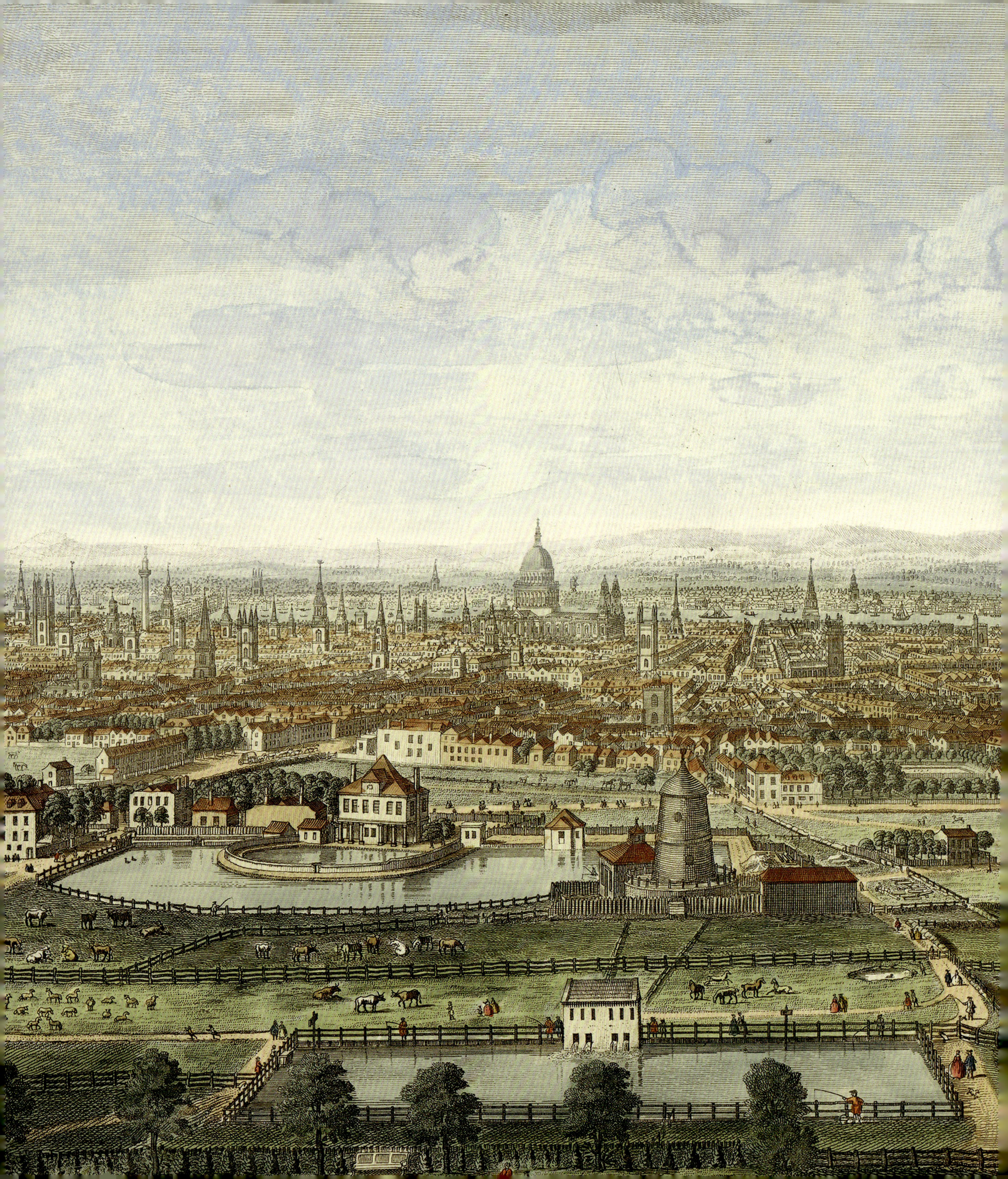

Enlightened Princesses has been guided brilliantly by the lead curator of the exhibition and editor of this volume, Joanna Marschner, Senior Curator at Kensington Palace, whose scholarly expertise and curatorial creativity underpin the project. Joanna's enthusiasm, dedication, diplomacy, and consideration have inspired the team, and staff members at the YCBA and HRP have learnt much from one another. Through this partnership, they have explored exciting new ways of working that will benefit both institutions long into the future.

Particular delight has been garnered from the collaborative examination of how intangible content can be curated most effectively. Research has revealed that the written record of the princesses' conversations with artists, writers, composers, architects, naturalists, physicians, tradespeople, and early industrialists has survived. There are manuals recording the choreographic notation of the dances they performed on formal occasions at court, and notes recording the scientific experiments set up to intrigue and amaze their visitors. The princesses' endeavours had an enduring, and sometimes transformative, impact on those who witnessed them. However, they have left little material legacy that can now conjure up these experiments and experiences. Together the team has assembled artists, dancers, performers, and even young engineers to recast these moments to create a new and powerful connection for a contemporary audience with our subject. For Historic Royal Palaces, the project has also played a part in achieving Independent Research Organisation status, formally recognizing its scholarly endeavour and reputation.

Echoing the princesses' curiosity and their clever nurturing of networks across Great Britain and the wider world, the project has brought together scholars from Britain, from many countries in Europe, and from the United States of America to shed light on the subject. The debate has proved so compelling that it has touched the agendas of many associated museums, galleries, and organizations, which have devised programmes to explore other aspects of the princesses' agency. We trust that the project will leave as one of its own legacies a lively research network, with a continuing exchange of information between and among museums and palaces, botanic gardens and artistic and scientific academies, and universities.

Many of these distinguished partners have been instrumental in the gathering of an extraordinarily rich array of artworks and artefacts for the exhibitions. The greatest number have been graciously lent by HM The Queen; these form the golden thread that makes the exhibition cohere.

Through this project we have discovered much about Caroline, Augusta, and Charlotte. Certainly there will be new discoveries and reappraisals relating to these women in the future; but we are gratified that following our work they can be properly recognized for the impact they made on life in Britain and in the wider world in their own time. Within the constraints of their diplomatic marriages and the restrictions of tradition and convention, they approached their role with boldness, confidence, and even bravery, seizing the opportunities presented by the age in which they lived.

Amy Meyers
Director
Yale Center for British Art

Michael Day CVO
Chief Executive
Historic Royal Palaces

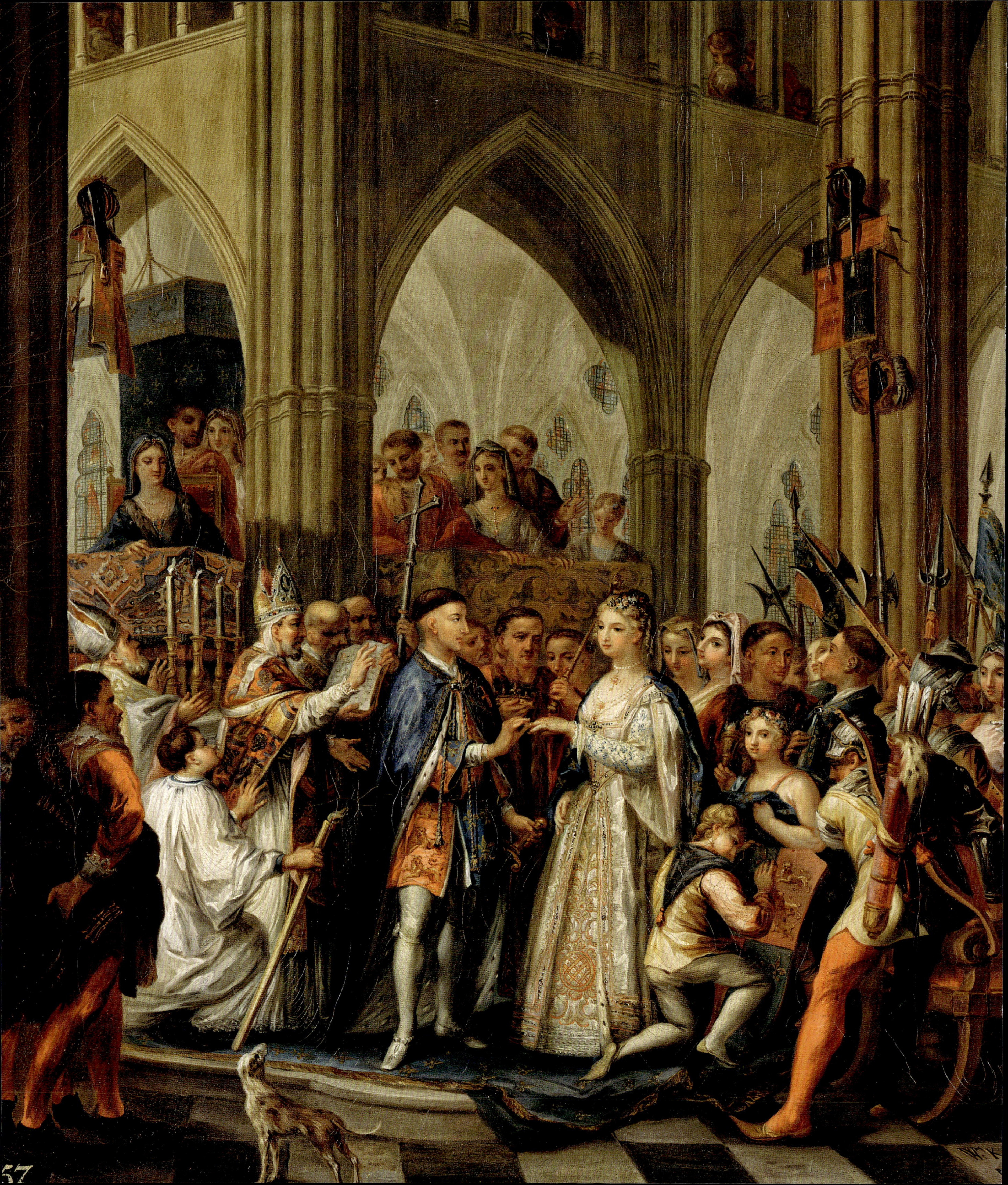

Acknowledgements

There are many people who have contributed in important ways to making *Enlightened Princesses: Caroline, Augusta, Charlotte, and the Shaping of the Modern World* such a rich and worthy exhibition at the Yale Center for British Art (YCBA) and Kensington Palace, and who have supported the making of this publication, which we hope will prove the enduring legacy of an exciting and rewarding collaborative research project.

We first must express our gratitude to HM The Queen, who has graciously and generously lent many key works to the exhibition. The Queen also has granted scholars working on the project access to the Royal Archives, which has ensured that we have been able to present much new information about the three princesses who sit at the heart of our study. We thank Jonathan Marsden, Director of the Royal Collection Trust and Surveyor of The Queen's Works of Art; Desmond Shawe-Taylor, Surveyor of The Queen's Pictures; Oliver Urquhart-Irvine, Royal Librarian; and their colleagues for their much appreciated encouragement and support for the project. Thanks also are due to the conservation staff and the Photographic Services team of Royal Collection Trust who have overseen with efficiency and courtesy many practical aspects surrounding this very important loan. We especially wish to thank Caroline de Guitaut and Hannah Belcher, who have managed the detailed planning which has ensured that fragile and precious artworks and artefacts have arrived in just the right places at just the right time.

We must extend our thanks to Her Majesty Queen Margrethe II of Denmark for permission to exhibit an important work from Rosenborg Castle, the Royal Collection of Denmark; to His Majesty King Willem-Alexander of the Netherlands for valuable materials from the Royal House Archive and the Royal Collection, The Netherlands; and to His Royal Highness The Prince of Hanover for a painting from Marienburg Castle, Hanover.

There are many other museums, art galleries, palaces, and private individuals in Great Britain, Denmark, Germany, and The Netherlands, as well as throughout the United States of America, who have contributed works from their collections to make the most splendid of exhibitions at both venues. We would like to express our thanks to Lord and Lady Baker; F. C. Shroeder from the Beinecke Rare Book & Manuscript Library, and Susan Gibbons from the Yale University Library, at Yale University; Richard Chesser, Tom Harper, Roly Keating, Tanya Kirk, and Gemma Renshaw, from the British Library; Hugo Chapman, Jill Cook, Jordina Diaz Ferrando, Hartwig Fischer, Neil MacGregor, Sheila O'Connell, Judy Rudoe, and Kim Sloan from the British Museum, London; David Knox and Jonathan Prown of the Chipstone Foundation; Tim Knox from the Fitzwilliam Museum, University of Cambridge; Stephanie Chapman, Colin Coleman, Alison Duke, and Caro Howell from the Foundling Museum, London; Mungo Campbell, Malcolm Chapman, David Gaimster, Stuart MacDonald, Elizabeth O'Neill, and Maggie Reilly from The Hunterian, University of Glasgow; Penny Johnson from the UK Government Art Collection; Nicole L. Bouché and Cynthia E. Roman from The Lewis Walpole Library, Yale University, Farmington, CT; Karin Dyko at Marienburg Castle, Hanover;

Catherine Hess, Kevin Salatino, and Laura Skandera Trombley from the Huntington Library, Art Collections, and Botanical Gardens, San Marino, CA; Carla Hayden and Rachel Waldron from the Library of Congress, Washington, DC; Thomas P. Campbell, Constance McPhee, and Melinda Watt from The Metropolitan Museum of Art, New York; Kathleen V. Jameson from the The Mint Museum, Charlotte, NC; Colin B. Bailey from The Morgan Library & Museum, New York; the Marquess of Bute, Lord Anthony Crichton-Stuart, Adam Ellis-Jones, and Alice Martin from Mount Stuart, Isle of Bute; Peter Crane and Tony Willis from Oak Spring Garden Library, Upperville, VA; Silke Ackerman and Lucy Blaxland from the Museum of the History of Science, University of Oxford; Paul Crouch, Michael Dixon, Jonathan Gregson, Hein Van Grouw, Andrea Hart, Charlie Jarvis, and Donney Nicholson from the Natural History Museum, London; Nicholas Cullinan, Lucy Peltz, and David McNeff from the National Portrait Gallery, London; Renee Marshall, Mrs William A. Marshall, and Scott Scholz of The National Society of the Colonial Dames of America in the State of South Carolina; Richard Broyd, Dame Helen Ghosh, and Fernanda Torrente from the National Trust, UK; Sir Moir Lockhead OBE from the National Trust for Scotland; Susan Fraser, LuEsther T. Mertz Library, and Gregory Long, from the New York Botanical Garden; Anthony W. Marx from the New York Public Library; Paul Anthony Madden of The Queen's College, University of Oxford; Michelé Green, Camilla Hampshire, and Shelley Tobin from the Royal Albert Memorial Museum & Art Gallery, Exeter; Fiona Ainsworth, Richard Deverell, Christopher Mills, Lynn Parker, Alan Paton, and Elizabeth Woodgyer from the Royal Botanic Gardens, Kew; Andreas Grinde, Jørgen Hein, and Jørgen Selmer from Rosenborg Castle, The Danish Royal Collection; Charlotte Eymael, and Dr Ph.C.B. Maarschalkerweerd from the Royal House Archive and the Royal Collection, The Netherlands; Charles Hind, Drawings and Archives Collections, British Architectural Library, and Jane Duncan, from the Royal Institute of British Architects; Julie Maxton, Keith Moore, and Paul Nurse from the Royal Society, London; Ian Blatchford, Alison Boyle, Jane Desborough, Jenni Fewery, Alex Johnson and Rebecca Rhodes from the Science Museum, London; Christopher Baker, Scottish National Portrait Gallery, Edinburgh, and John Leighton, from National Galleries of Scotland; Bruce Boucher and Helen Dorey from Sir John Soane's Museum, London; Nancy Davis, John Gray, and Margaret Grandine from the National Museum of American History, the Smithsonian Institution; the Stuart Collection; Neil Curtis from University Museums, University of Aberdeen; Julie Gardham, Special Collections, and Susan Ashworth, University of Glasgow Library; Clare Browne, Julius Bryant, Hannah Kaufman, Susan North, and Martin Roth from the Victoria and Albert Museum, London; Jon Benington from the Victoria Art Gallery, Bath; Rowan de Saulles and Marianne Templeton, Wellcome Library, London, and Jeremy Farrar, from the Wellcome Trust; Linda Eaton, Leslie Grigsby, Beth Parker Miller, Kathleen Orr, and David Roselle from Winterthur Museum, Garden and Library, Winterthur, DE; John Gallagher and Melissa Graffe from the Medical Historical Library, Harvey Cushing/John Hay Whitney Medical Library, Yale University; Ruthann McTyre from the Irving S. Gilmore Music Library, Sterling Memorial Library, Yale University; and L. Lynne Addison, Suzanne Boorsch, Pamela Franks, Laurence Kanter, and Jock Reynolds from the Yale University Art Gallery.

The curators could not have created an exhibition of this complexity without the help of many colleagues at the YCBA and Historic Royal Palaces (HRP). At the YCBA we would like to thank Richard Caspole, Soyeon Choi, Belene Day, Martina Droth, Kristin Dwyer, Amelia Ewan, Elisabeth Fairman, Gillian Forrester, Melissa Gold Fournier, Linda Friedlander, Matthew Hargraves, Betsy Kim, Francis Lapka, Chris Lotis, Anne Markowski, Beth Miller, Ronnie Rysz, Rebecca Sender, Maria Singer, Bernie Staggers, Jaime Ursic, Sarah Welcome, and Mary Regan-Yttre. Tyler Griffith and Jane Nowosadko have played a special role, in collaboration with Glenn Adamson, Senior Research Scholar, in executing the programme associated with the exhibition at the YCBA. At HRP we would like to thank Alexandra Attelsey, John Barnes, Tracy Borman, David Bransby, Jonathan Bridal, Colin Briggs, Adam Budhram, Michael Day, Rosanagh Fuller, Gina George, Rhiannon Goddard, Paul Gray, Georgina Grubb, Annie Heron, Wendy Hitchmough, Laura Hutchinson, Deirdre Murphy, Polly Putnam, Lee Prosser, Victoria Richards, Lisa Sensier, Deborah Shaw, Lea Violoni, Dan Wolfe, Natasha Woollard, and Lucy Worsley. A particularly important role has been played by the registrars for both organisations. At the YCBA we would like to thank Timothy Goodhue and Corey Myers. At HRP we would like to thank Rebecca Wallace. The exhibition design has been beautifully conceived at the YCBA by Stephen Saitas with the assistance of Kevin Derken, Rachel Hellerich, Richard Johnson, Abraham Omonte, Lyn Bell Rose, Greg Shea, and Dylan Vitale, working with Nathan Flis, A. Robin Hoffman, and Tyler Griffith. At Kensington Palace the exhibition has been designed with the greatest skill and elegance by the teams at Universal Design Studio and Bibliothèque, working with Dominique Driver and Daniel Green from HRP.

The curators would like to acknowledge an immense debt of gratitude to the Paul Mellon Centre for Studies in British Art, London, and especially appreciate the support and encouragement of Mark Hallett, Martin Postle, Sarah Ruddick, and Sarah Victoria Turner. We would also like to thank David Knox and Jonathan Prown of the Chipstone Foundation for their enthusiasm for and interest in the project. The Paul Mellon Centre and the Chipstone Foundation both underwrote a series of study days and a conference in the period leading up to the opening of the exhibition and the publication of this book, which allowed us to explore our subject with the benefit of the expertise of an extraordinary group of international scholars. Both ventures have been immeasurably enriched by such thoughtful and lively debate, and we thank, in addition to YCBA and HRP staff, all the guest participants: Malcolm Baker, Robert Blocker, Tracy Borman, Julius Bryant, James Bundy, Wolf Burchard, Donald Burrows, Francesco Casetti, Edward S. Cooke, Jr., Peter Crane, Michael Donaghue, Sebastian Edwards, Patricia Fara, Paul Freedman, John Goldfinch, Kate Goodridge, Andrew Hamingson, Craig Ashley Hanson, Rosemary Harden, Kate Heard, Sue Hill, Jason Holley, Kathryn James, Emma Jay, Martin Jean, Jonathon Jeffrey, Berta Joncus, Kathryn Jones, George Knight, Eric Larson, Todd Longstaffe-Gowan, Nicholas McGegan, Christopher Mills, Tristan Mostert, Jeanie O'Hare, Therese O'Malley, Clarissa Campbell Orr, Jonathan Prown, Jules Prown, William Purvis, Aileen Ribeiro, Joseph Roach, Jane Roberts, Cynthia E. Roman, Stephen Saitas, Vanessa Bezemer Sellers, Deborah Shaw, Desmond Shawe-Taylor, David K. Skelly, Kim Sloan, Michael Snodin, Robert Storr, Stephen Taylor, Andrew Thompson, Sonia Tomic, Katie Trumpener, Amanda Vickery, Laura Wexler, and Vincent Wilczyniski.

The legacy of these opportunities for in-depth discussion of the histories of the princesses, and the worlds they navigated, is a series of lively and thought-provoking essays in this book. For their interest, the excellent contributions, and the care and patience taken during the editorial stages of the book production, we would like to thank the authors: Glenn Adamson, A. Cassandra Albinson, Robyn Asleson, Wolf Burchard, Donald Burrows, Todd Longstaffe-Gowan, Florence Grant, Tyler Griffith, Craig Ashley Hanson, Samantha Howard, Roisin Inglesby, Emma Jay, Berta Joncus, Kathryn Jones, Mark Laird, Ben Marsh, Clarissa Campbell Orr, Frank Prochaska, Lee Prosser, Aileen Ribeiro, Joseph Roach, Jane Roberts, Cynthia E. Roman, Matthew Storey, and John Styles.

A tremendous debt of gratitude is owed to the team who have brought the book together. At the YCBA A. Robin Hoffman has created wonders with the help of Shaunee Cole and the team of designers, Miko McGinty, Anjali Pala, and Claire Bidwell. John Goldfinch must also be thanked for his kind editorial advice. It has been the greatest pleasure to work with Rose Bell, Liz Drury, Jane Horton, and most especially with Sally Salvesen, of Yale University Press.

Having early opportunities for discussion of the subject, and how this might be best presented to a public audience, allowed the exhibition teams to bring together a community of musicians, artists, dancers, architects, scientists, and engineers who have used their considerable talents to contribute to the richness of the exhibitions and the programmes that accompany them. We will be ever grateful to Michelle Addington, Michelle Aland, Andrej Bako, Ricardo Barros, the Battersea Arts Centre, Ailbhe Clyne, Emily Coates, Jane Cohan, Claire Davies, Lucinda Dawkins, Anna Dyson, Fred Fabre, Moira Goff, Sue Hill, Ruth Hogan, Kevin Hourigan, Michael Jean, Seth Kriebel, Edith Lalonger, Mercedes Kemp, Nicholas McGegan, David Micklem, Ailbhe Murphy, The Royal Academy of Music, Yinka Shonibare, RA MBE, Vincent Wilcynski, and Lawrence Wilen. Special thanks are due to Anne Daye and the Dolmetsch Historical Dance Society, which provided funds to allow some of the princesses' dances to be performed during the *Enlightened Princesses* conference.

There are many other individuals to whom we owe gratitude for contributing their research to enrich the project, for their help to the exhibition and publication team, or just for their kind support and interest. There are almost too many names to list here, and we are eternally grateful to them all. However, we would like to mention especially Tim Barringer, Ned Blackhawk, David Blight, Jill Campbell, Sir David Cannadine, Hazel Carby, Francesco Casetti, Linda Colley, Laura and James Duncan, Elizabeth Eger,

Gabriella Fenton, Meredith Gamer, Inderpal Grewal, James Harrington, John and Eileen Harris, Eleanor Hughes, Katia Johansen, Anna Kesson, Bianca du Mortier, Margaret Morton, Kirsten Aschengreen Piacenti, Jacqueline Riding, Joseph Roach, Sir Hugh Roberts, Trudie Rosa de Cavalho, Ivan Selin, Bill Sherman, Lisa Skogh, Paris Spies-Gans, William Tullett, Helen Watanabe-O'Kelly and Laura Wexler. There are a few people who we must draw out and thank for their exceptionally important contribution to the *Enlightened Princesses* project. As Curator of Paintings and Sculpture at the YCBA, and Acting Head of the department, A. Cassandra Albinson served as the organizing curator for the exhibition at the Center before accepting the position of Margaret S. Winthrop Curator of European Art at the Harvard Art Museums. Lisa L. Ford, Assistant Director for Research at the YCBA, then assumed a number of these curatorial duties for the exhibition after Cassandra's departure. A team of postdoctoral research associates also has supported the curators and editors in all parts of the venture. Without the enthusiasm and energy, as well as the scholarship, of Nina Amstutz, Florence Grant, Tyler Griffith, Peter Moore, and Samantha Howard we certainly would not have achieved our great ambition. We also appreciate the work of undergraduates Mary Edgington and Mary Reidy.

Finally, we would like to thank our families for their loving support and interest. We will be ever grateful for their kind patience as the princesses came to play such a large role in their lives as well as ours.

Joanna Marschner, David Bindman, and Lisa L. Ford
Editors

Joanna Marschner and Amy Meyers
Curators

FAMILY TREE

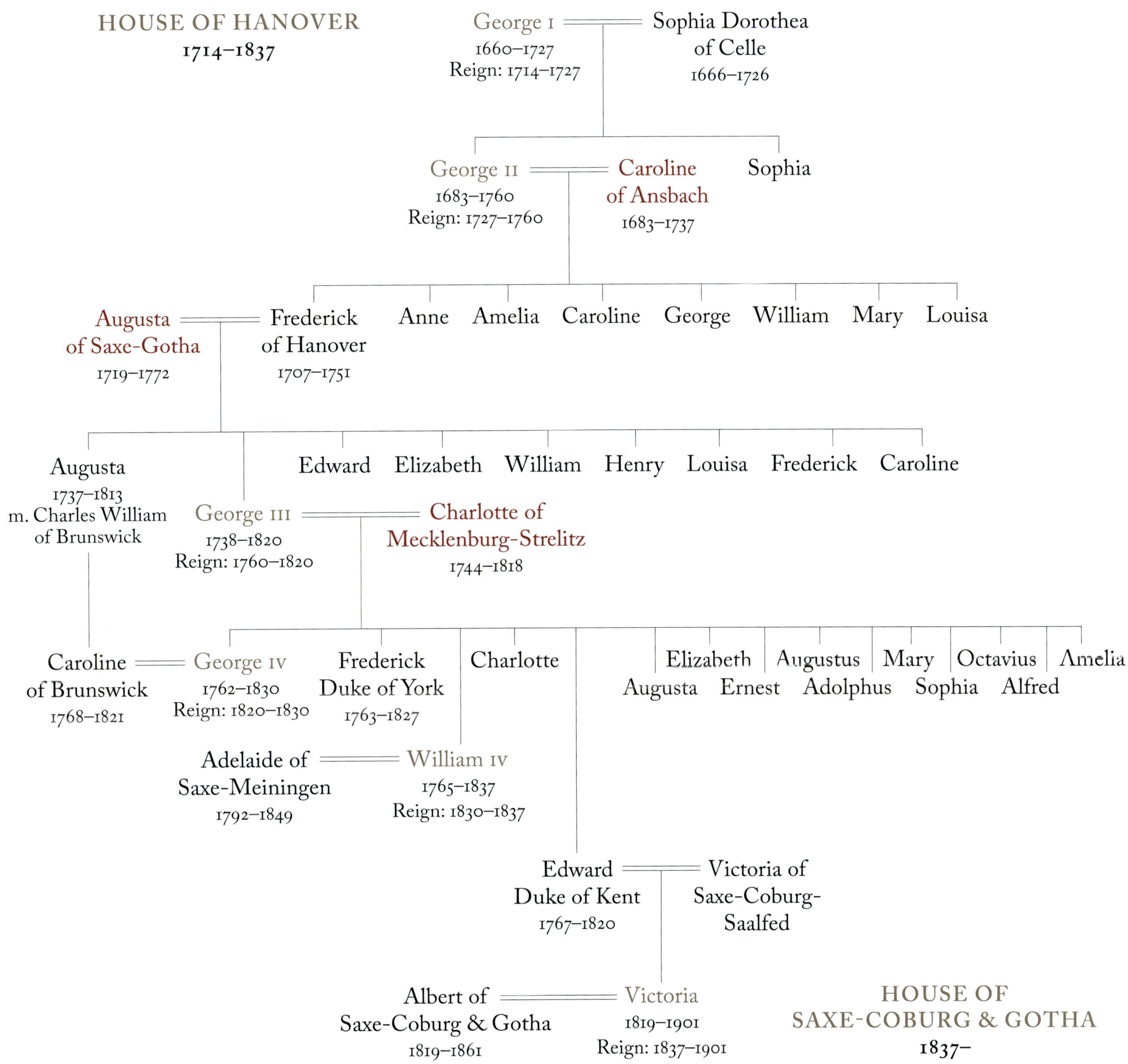

HOUSE OF HANOVER
1714–1837
George I
1660–1727
Reign: 1714–1727
Sophia Dorothea
of Celle
1666–1726
George II
1683–1760
Reign: 1727–1760
Caroline
of Ansbach
1683–1737
Sophia
Augusta
of Saxe-Gotha
1719–1772
Frederick
of Hanover
1707–1751
Anne
Amelia
Caroline
George
William
Mary
Louisa
Augusta
1737–1813
m. Charles William
of Brunswick
Edward
Elizabeth
William
Henry
Louisa
Frederick
Caroline
George III
1738–1820
Reign: 1760–1820
Charlotte of
Mecklenburg-Strelitz
1744–1818
Caroline
of Brunswick
1768–1821
George IV
1762–1830
Reign: 1820–1830
Frederick
Duke of York
1763–1827
Charlotte
Augusta
Elizabeth
Ernest
Augustus
Adolphus
Mary
Sophia
Octavius
Alfred
Amelia
Adelaide of
Saxe-Meiningen
1792–1849
William IV
1765–1837
Reign: 1830–1837
Edward
Duke of Kent
1767–1820
Victoria of
Saxe-Coburg-
Saalfed
Albert of
Saxe-Coburg & Gotha
1819–1861
Victoria
1819–1901
Reign: 1837–1901
HOUSE OF
SAXE-COBURG & GOTHA
1837–

Map of Northern Europe showing the Birthplaces of Caroline, Augusta, and Charlotte

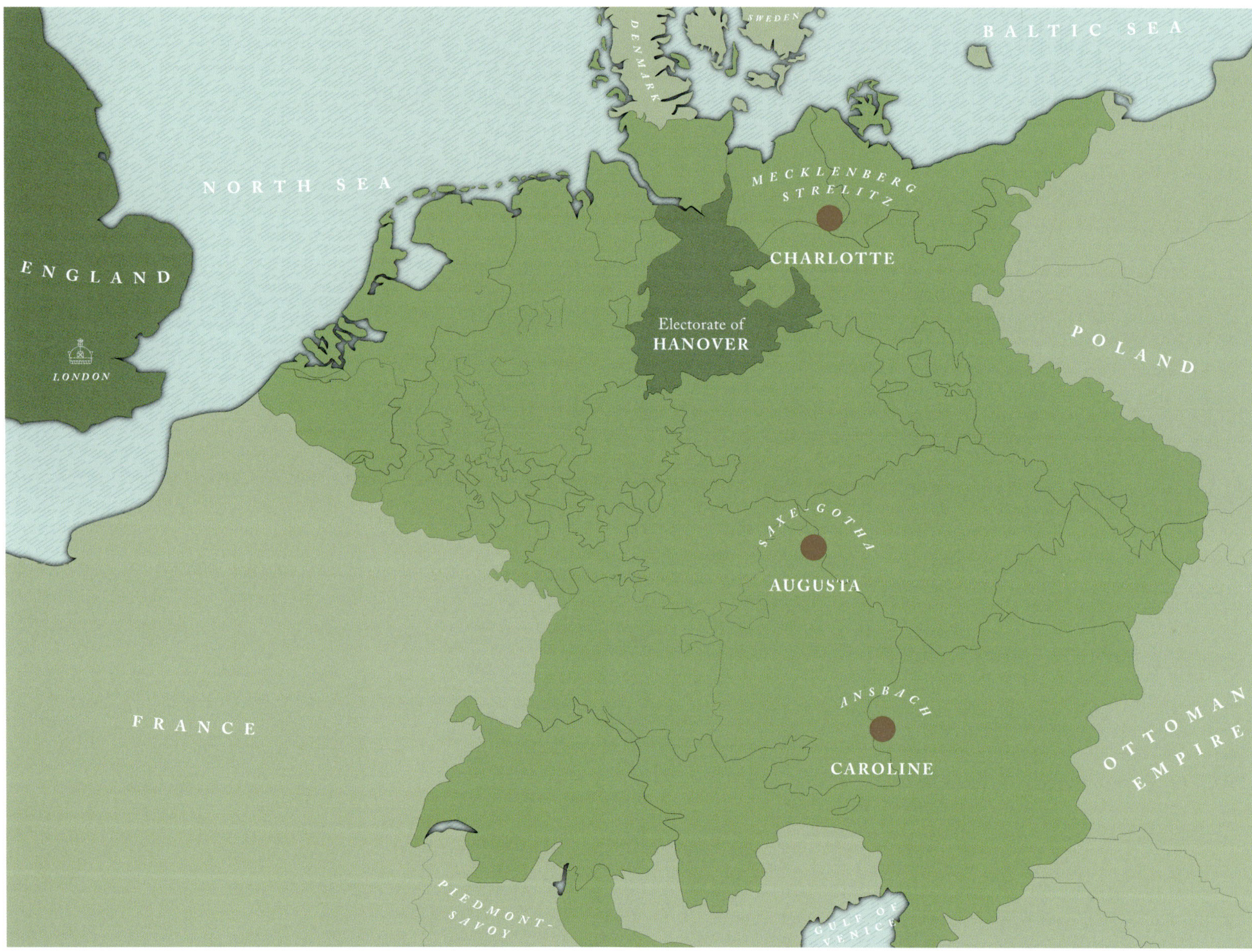

Map of the World showing the British Empire 1770

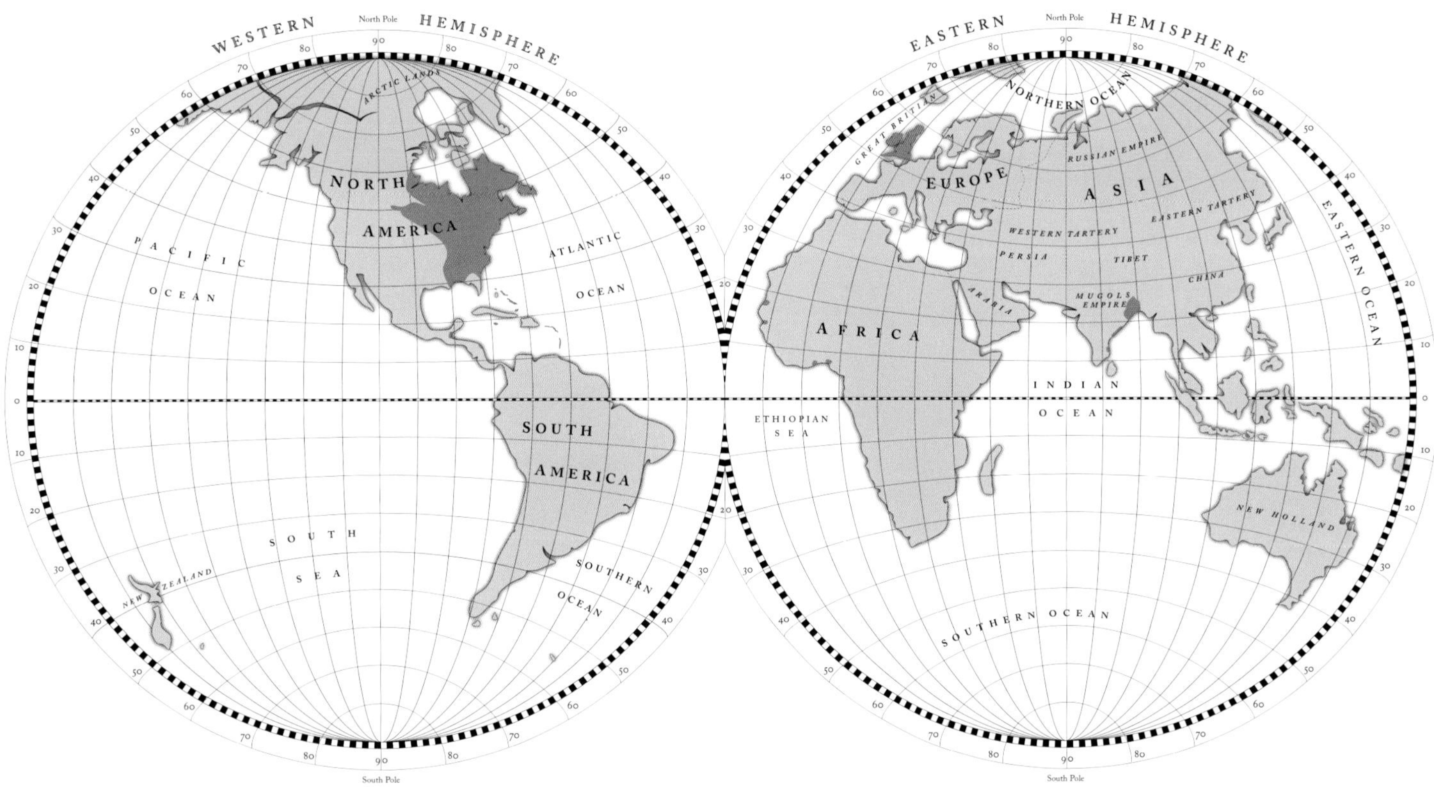

Map of the World showing the British Empire 1820

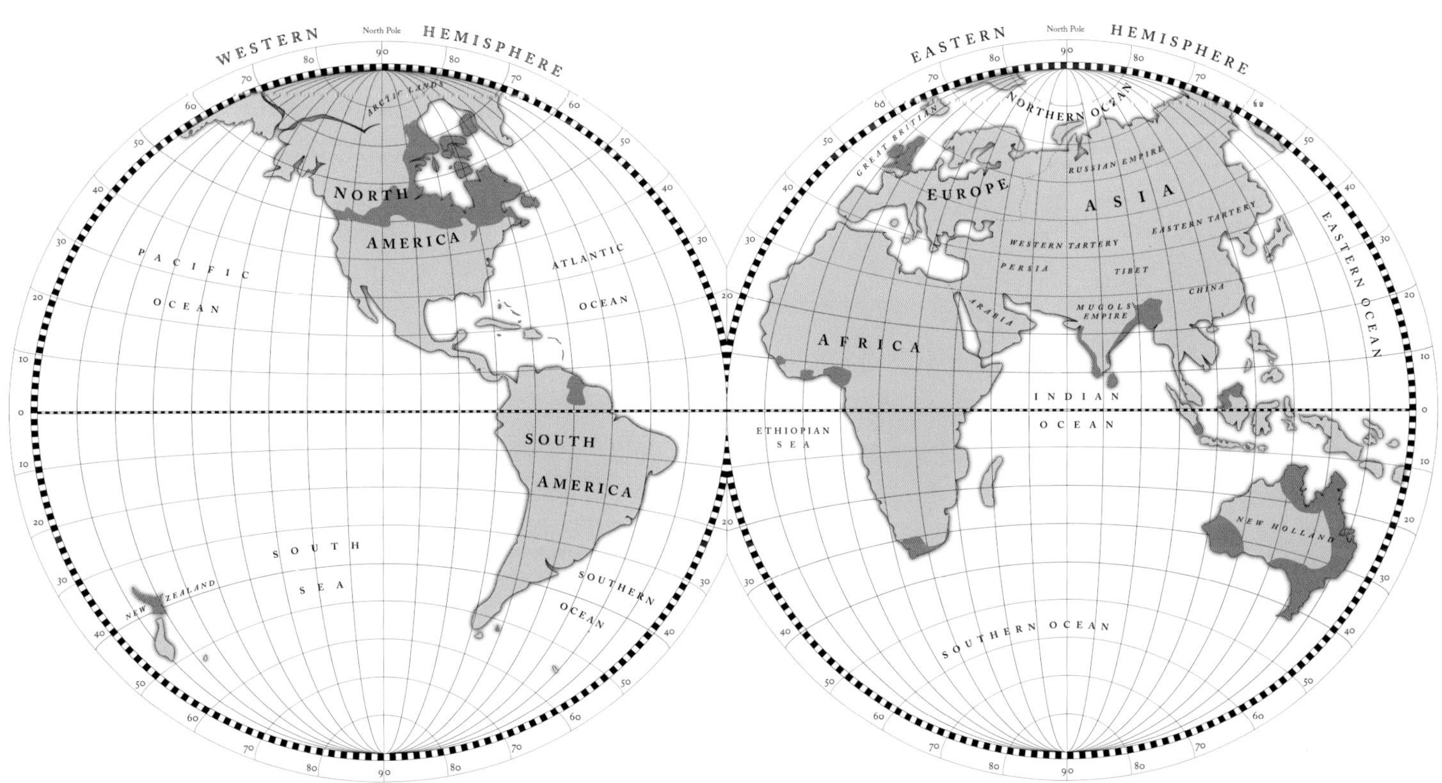

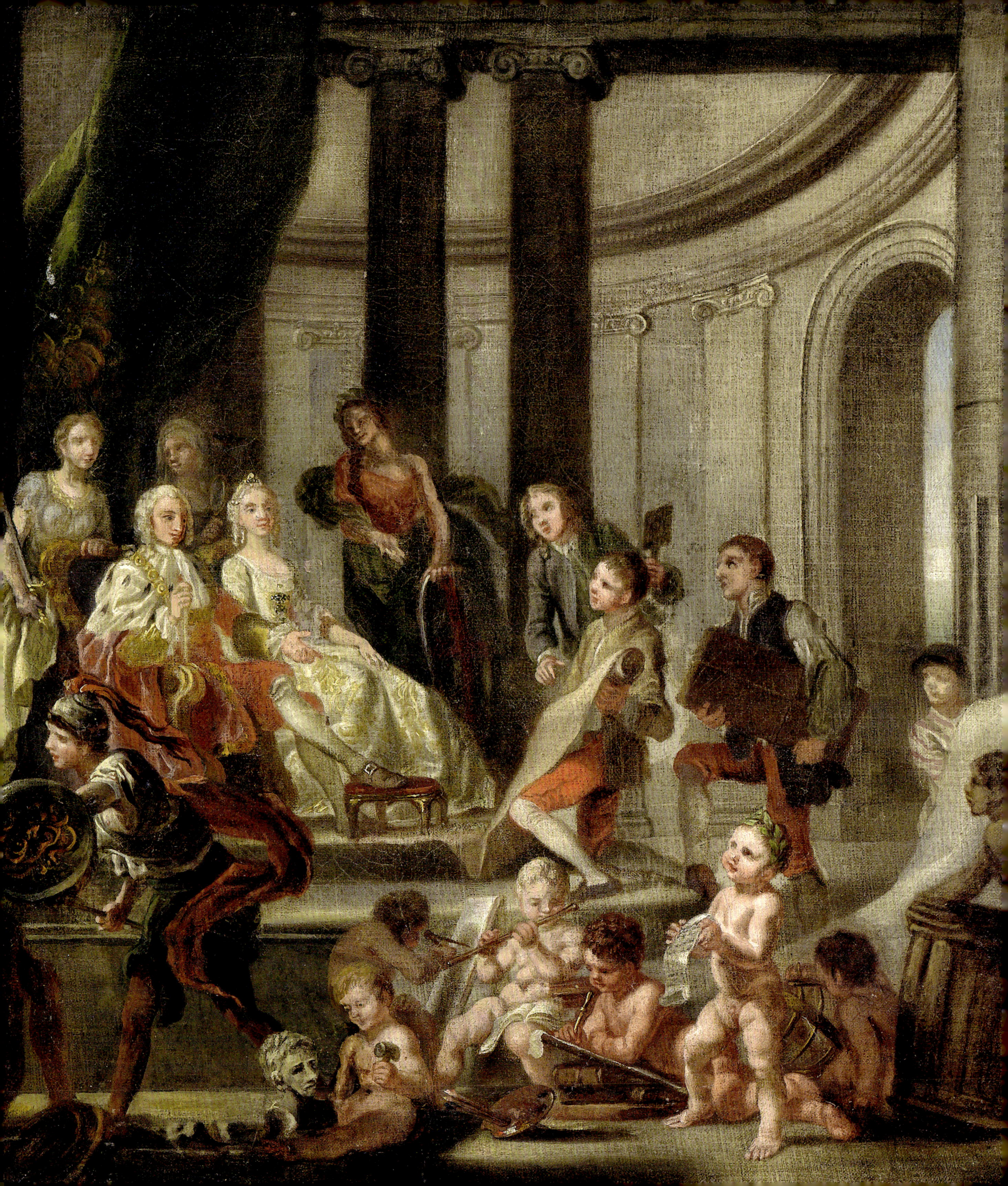

Enlightened Princesses: Caroline, Augusta, Charlotte, and the Shaping of the Modern World

Joanna Marschner

Caroline of Ansbach, Augusta of Saxe-Gotha, and Charlotte of Mecklenburg-Strelitz enjoyed variously the titles Princess of Wales, Queen Consort, Electress of Hanover, Regent, Princess Dowager, and Mother of the King. Of three different generations, their life histories straddle the long eighteenth century and are inextricably interwoven. Not only did they share roles but also a common German ancestry. Crucially, they all were Protestant. While their marriage choices had been dictated by others, as individuals they were curious about British culture and ideas, and this would lead to their making a contribution, directly and indirectly, to many aspects of life in their adopted country. Each Princess had a powerful influence on the one who followed, inheriting projects, plans, and debates, which they then shaped physically and philosophically, as their interest, taste, and sphere of influence allowed. Each embraced in turn the gifts and challenges of an extraordinarily dynamic century, before handing on the baton.

CAROLINE (1683–1737)

> The Queen talked with me at least half an hour upon my collection of printed Heads, Dr Couraye, the history of France, gardening, painting, flattery, and divers political and moral subjects.
>
> John Percival, Lord Egmont, December 1734[1]

Caroline of Ansbach (fig. 1.01) was the daughter of John Frederick, Margrave of Brandenburg-Ansbach (1654–1686), and his second wife, Eleanore of Saxe-Eisenach (1662–1696).[2] Following the death of her father, the family moved to Dresden on Eleanore's marriage to John George IV, Elector of Saxony, in 1692. This unhappy alliance ended two years later when John George died of smallpox. Caroline, her mother,

Detail of p. 24, fig. 1.14

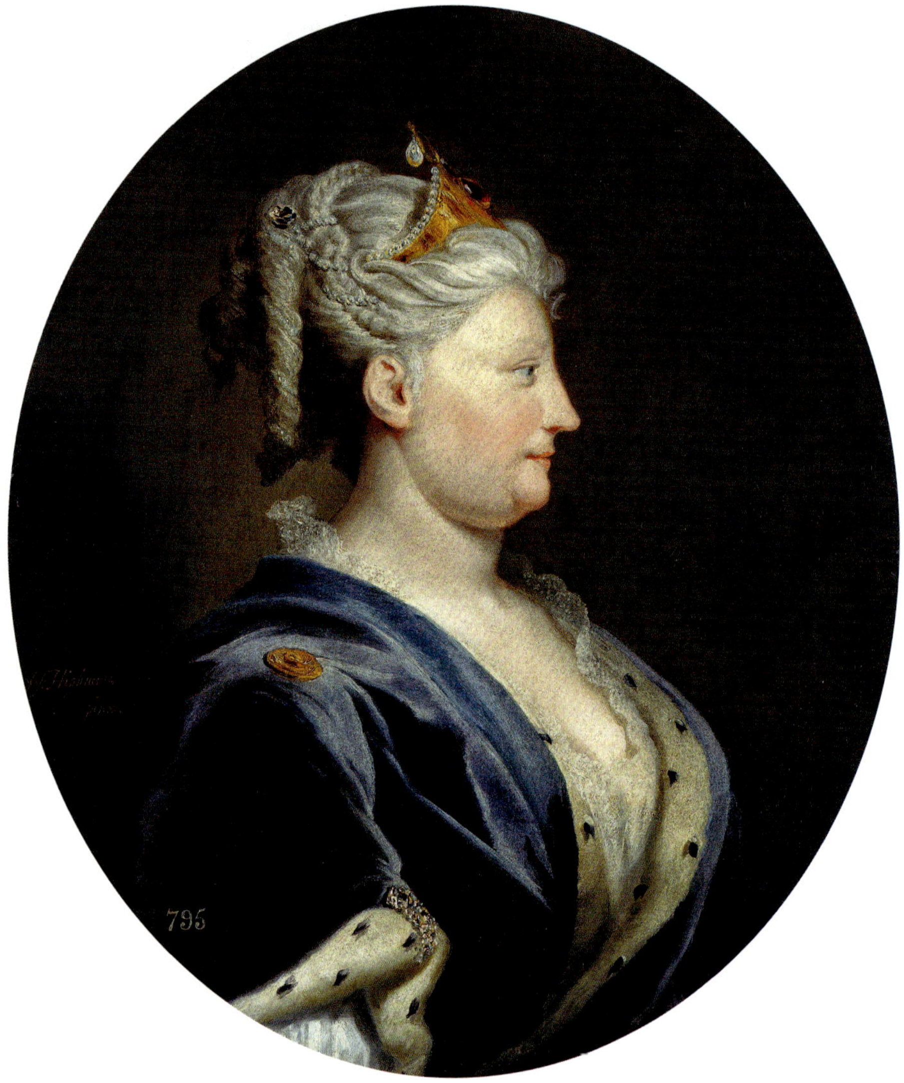

FIG. 1.01 Joseph Highmore, *Queen Caroline of Ansbach*, c. 1735, oil on canvas, 29⅝ x 24⅜ in. (75.1 x 61.8 cm). Royal Collection Trust (cat. 13.02)

and brother continued to live at Pretsch in Saxony until Eleanore's death in 1696. After a brief sojourn back in Ansbach, Caroline was despatched to live with new guardians: Frederick III, Elector of Brandenburg, first King in Prussia from 1701, one of her mother's kinsmen, and his wife Sophia Charlotte of Hanover. There is little evidence that Caroline received much formal education up to this time. Her spelling remained idiosyncratic throughout her life, and her handwriting was so poor, her husband would later claim she wrote 'like a cat'.[3]

Caroline joined Sophia Charlotte's household at Lietzenburg, the palace located a short distance from Berlin. It was there that she was introduced to natural philosophers Gottfried Wilhelm Leibniz and John Toland, writer and philosopher Pierre Bayle, and musician and composer George Frideric Handel, as well as other members of the lively intellectual circle encouraged by Sophia Charlotte. Caroline

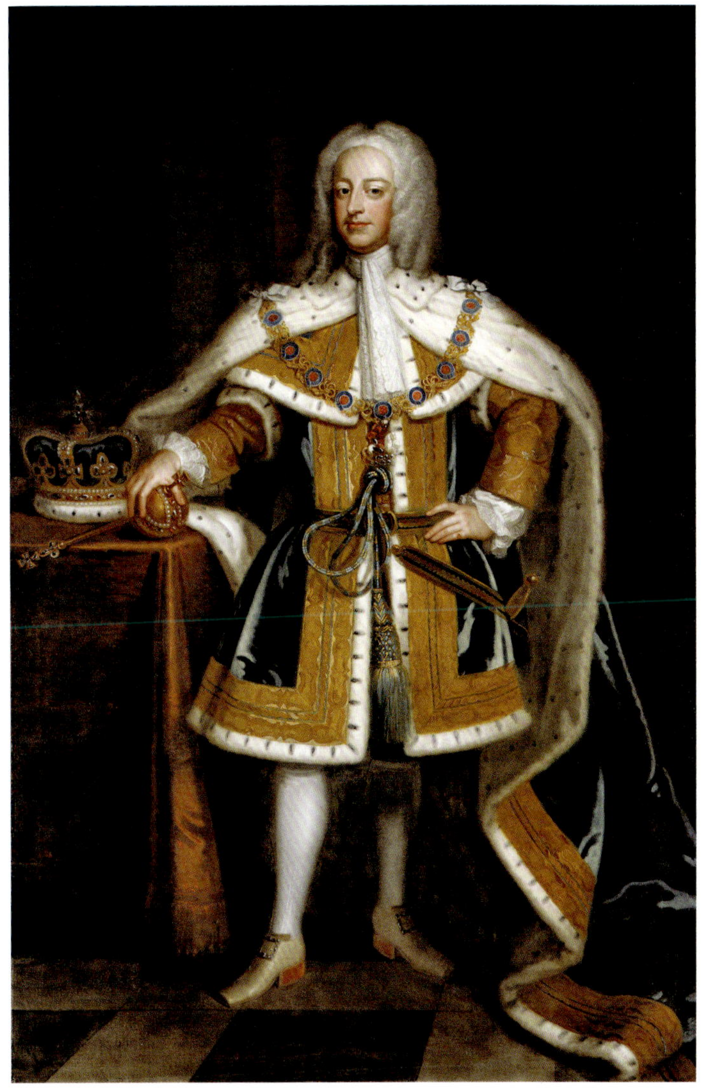

FIG. 1.02 Enoch Seeman the younger, *George II*, c. 1730, oil on canvas, 93¾ x 58¼ in. (238.3 x 148 cm). Royal Collection Trust

flourished in this environment: Sophia Charlotte declared Berlin became 'a desert' in her absence.[4]

In 1703 Caroline chose not to pursue her first marriage proposition from the Archduke Charles of Austria, subsequently Holy Roman Emperor Charles VI, after deciding she could not convert to Catholicism. This evidence of a stout allegiance to Protestantism would serve her well in the future. It made her especially attractive to George Augustus, electoral prince of Hanover, who paid her court in the same year (fig. 1.02).[5] The couple were married in 1705, in the chapel of the Leineschloss in Hanover. In 1707, Caroline gave birth to their first child, Frederick Louis, later Prince of Wales, and three daughters followed: Anne, Amelia, and Caroline Elizabeth.

At the Hanoverian court Caroline found a mentor in her husband's grandmother, the formidable dowager Electress Sophia, who encouraged her intellectual and literary interests. Lively conversations about history and philosophy continued with Leibniz, who had retreated to Hanover in 1705 after the death of Sophia Charlotte. Caroline embraced the exciting challenge of preparing herself for royal service: the Act of Settlement of 1701 had established Sophia and her heirs as successors to Queen Anne and the British crown. Caroline took English lessons and developed an interest in politics, welcoming any opportunity to converse with visitors from London.

Sophia would be thwarted in her ambition to become Queen. She died in Caroline's arms, in the garden at Herrenhausen, just a few weeks before Queen Anne's death in 1714. The throne passed to Sophia's son, George Louis, Elector of Hanover. George Augustus and Caroline, newly created Prince and Princess of Wales, followed George I to London, bringing their three little girls with them, and took up residence at St James's Palace. Their son, Frederick, was left in Hanover as token of the family commitment to the Electorate.

George I's court was sedate and secluded. Although he understood and spoke English, he preferred to conduct much official business with his ministers in French or Latin. He had left his estranged wife in Germany, and travelled with his mistress and half-sister, an arrangement his new compatriots found confusing. For the British he had little appeal beyond his Protestant faith. One of Caroline's first contributions to the smooth integration of the new regime would be to provide a strong female presence at court. Her position as wife of the heir to the throne was unambiguous, and she was mother to a clutch of healthy children, who were introduced into the heart of court ceremonial.

This situation, however, came to an abrupt end in 1717, when the always uneasy relationship between George Augustus and his father broke down completely. Caroline followed her husband on his banishment from court, despite having to leave her children in the King's charge. They set up home in Leicester House, and later acquired Richmond Lodge as a country retreat. Although the King contrived to allow Caroline daily visits to her daughters, even after a formal reconciliation between father and son in 1720, the children continued to live apart from their parents until the death of George I in June 1727. The births of three healthy children subsequently, William Augustus, Mary, and Louisa, provided a little solace.

In 1727 the new king, George II, and the Hanoverian regime still faced many challenges. The Jacobite threat remained serious, despite the fact that the Act of Union in 1707 joining Scotland with England and Wales under the Protestant faith and a common legislature, prevented a separate Scottish foreign policy, and lessened the chances of Stuart sympathizers establishing a rival power base. It is important to recall that it had been necessary to repulse expeditionary forces supporting James Stuart, the 'Old Pretender' in 1715, 1717, 1719, and 1720. It would not be until 1745 that the danger finally abated, after the defeat at Culloden of the rebellion led by James's son, Charles Edward Stuart, the 'Young Pretender'. The increasing prominence and influence of the First Minister inaugurated important shifts in the relationship between the monarch and Parliament. Powerful factions within the prosperous business community of London had an ambition to explore, and potentially exploit, the riches of the Americas, Africa, and Asia. The universities of Oxford and Cambridge and the learned societies of London had cornered the intellectual life of the nation and had their own connections with their counterparts in Europe. Coffee houses, assembly rooms, and pleasure gardens competed with the court as places of sociable gathering.

Caroline enjoyed just ten years as Queen Consort; she died at the age of fifty-four in 1737. During her years in London she worked strenuously to support her husband, to celebrate and promote the Hanoverian dynasty, and to build up the confidence of her new compatriots in the regime. She was lucky. Her guardians had ensured that she was aware of her dynastic capital as a Hohenzollern princess, as well as her good looks, to attract a suitor. Marriage to George Augustus, with its promise of the British throne, presented a particular and exciting challenge, and she was fortunate in growing to love her husband, and he her. He would nurse her devotedly though both smallpox and pneumonia. She must, in her turn, have given him support and good counsel—he clearly discussed politics with her, sought her advice, and trusted her judgment. During his four absences in Hanover in 1729, 1732, 1735, and 1736–7, he left her as regent entrusted with all 'matters domestic'.[6] Even when George II was in the country he allowed Caroline influence in areas of government, especially over ecclesiastical affairs. She would play the crucial role in the appointment of at least four, and possibly as many as eight, of the thirteen bishops consecrated between 1727 and 1737.[7] This trust allowed her to be bold and confident in pursuit of her goals. She pushed constantly against traditional assumptions that could have circumscribed her ambition, and her determination opened up opportunities for her successors to explore.

AUGUSTA (1719–1772)

> [She enquired] after Carolina as to the Government and constitution of it, whether the Laws were made by the Goverr and Council . . . whether we had Eathquaks, askd us concerning the Hurrican, what were the names of the Rivers between wch Charles town was situate wch she said she had forgot. Enquired concerning the Indians their colour, manners &c how many of them we had in our Interest of our houses, of what they were built, our wines, and from whence we had them, our eating and manner of dressing turtle, one of wch she was to have for dinner next day. She told me of the French settled among us . . . corrupting our Indians our manifactures . . . concerning our silk . . . how long the Province had been settled, how far it extended back, and many other questions.
>
> Eliza Lucas Pinckney to an unnamed correspondent, summer 1753[8]

Augusta of Saxe-Gotha (fig. 1.03), the wife of Frederick, Prince of Wales (fig. 1.04), the eldest son of George II and Caroline, was exhorted by an anonymous poet to learn by her mother-in-law's example. 'A second Caroline' was called for—'Be Caroline's bright pattern still in View' and 'O! May thy steps her royal paths pursue, still keep the bright original in view.'[9]

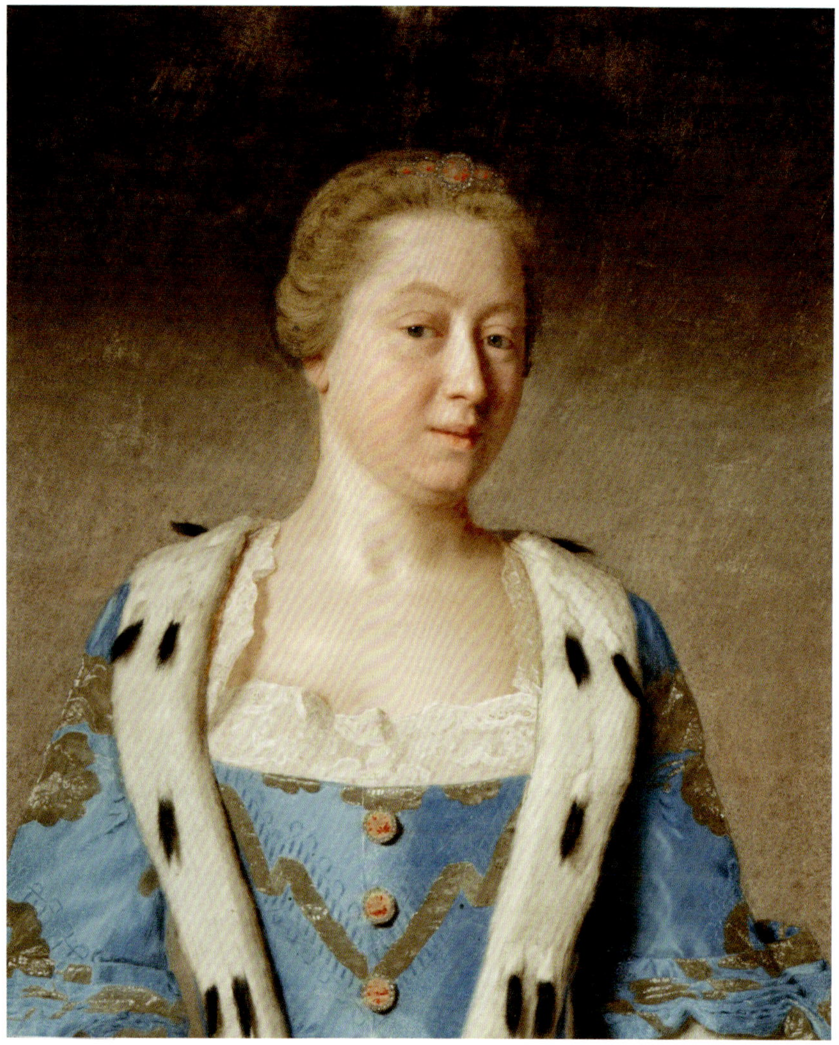

FIG. 1.03 Jean-Étienne Liotard, *Augusta, Princess of Wales*, 1754, pastel on paper, 25½ x 20¼ in. (64.8 x 51.4 cm). Royal Collection Trust (cat. 9.05)

Augusta was the thirteenth child of Frederick II, Duke of Saxe-Gotha-Altenburg, and his wife Magdalena Augusta of Anhalt-Zerbst. Saxe-Gotha was a tiny principality, and Augusta's marriage prospects had seemed inauspicious. However, George II was under pressure to find a bride for his rebellious son who had arrived in London from Hanover in 1728, estranged from his parents after many years of separation. His interest in his father's position had created an immediate tension; he was swiftly courted by the political opposition. Augusta, aged seventeen, was summoned to meet George II, who was passing the summer in his Electorate of Hanover, and despite the fact that her education had done little more than ensure she was a resolute Protestant and a competent harpsichord player, she impressed the King with her affability, good sense, and her Saxon connections. Sir Robert Walpole noticed, shrewdly, that despite her youth and naivety there was evidence that 'spoke strongly in favour of brains that had had but seventeen years to ripen'.[10] He cautioned Caroline to cultivate her

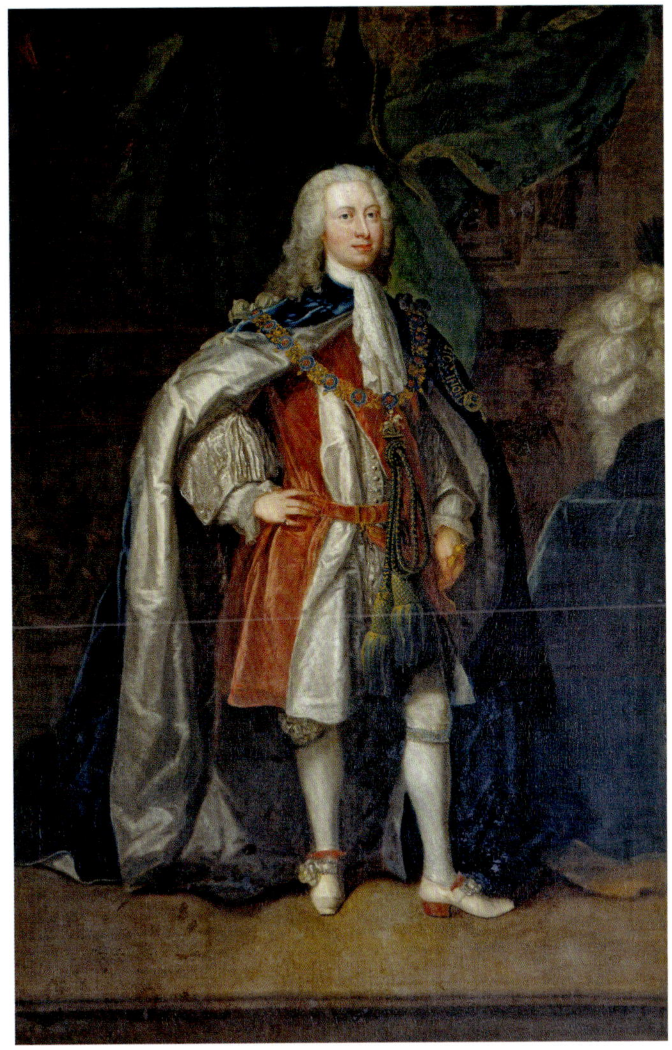

FIG. 1.04 Charles Philips, *Frederick, Prince of Wales*, 1737, oil on canvas, 94½ x 58⅞ in. (240.1 x 149.7 cm). Royal Collection Trust

carefully, despite the fact her parents had had so little appreciation of her destiny that they had failed even to arrange English lessons before her departure for London.

Following the wedding, which took place at St James's Palace in 1736, the relationship between monarch and heir did not improve. When Augusta went into labour with her first child at Hampton Court Palace the following year, her husband bundled her into a carriage in the depth of night for the uncomfortable journey back to St James's, determined that his child would be born in London and not under his parents' roof. Putting the life of mother and baby at risk, and causing this royal birth to be unwitnessed, were unforgivable actions. Once again, the heir to the throne was excluded from court by his father.

Augusta proved to be an ideal consort in this difficult situation. Her very innocence brought her immediate sympathy from George II, Caroline, and the court. She was a good mother: she and Frederick eventually had four sons and three daughters who survived infancy, and she is invariably depicted surrounded by her children. Their eldest daughter, Princess Augusta, was born in 1737, and a son George in the following year. They were followed by Edward, Elizabeth, William Henry, Henry, Louisa Anne, Frederick, and Caroline Matilda. As part of their campaign to rally the political opposition, the children would accompany their parents as they attended the theatre, entertainments, and visited manufactories, to the applause of the popular press.[11] This lively, comfortable, and wholesome image for this next generation of the royal family compared favourably to that of the court of George II, which by the middle of the century had lost much of its sparkle. The King in his later years lived quietly with his mistress and half-sister, travelling to Hanover whenever he could manage it.

Augusta's loyalty to her husband, and her decision to share his priorities, ensured they forged a strong, interdependent relationship. She proved a significant asset in Frederick's bid for popularity: she, in fact, learned English quickly, and in 1738 the *Gentleman's Magazine* noted that when she accompanied her husband to visit representatives of the commercial community in Bristol, she 'talked freely with the ladies in good English, which entirely won their hearts'.[12] George II seems to have respected her level-headedness, and sought her influence in managing the relationship with his son.[13] Frederick's advisers, George Bubb Dodington and John Percival (Perceval), second Earl of Egmont, were also in agreement that the best way of shaping Frederick's policy was through his wife.[14]

Augusta would never become a Queen Consort; Frederick, Prince of Wales, died unexpectedly in 1751. However, as mother of the future king who had yet to reach his majority, there was still potentially a role for her, as long as her son remained within her household. She immediately destroyed past evidence of her involvement in opposition politics and threw herself on the mercy of the King.[15] George II declared her the 'best of women', her income was protected, and he supported legislation that named her Princess Dowager and regent—albeit with the support of a council—in the event of his death before George, her eldest son, had reached the age of eighteen.[16]

Despite this victory, Augusta remained fearful of the King's choice of advisers to the Regency Council, and contemptuous of the moral and intellectual quality of those appointed to oversee the education of Prince George.[17] Her ambition was for him to claim the role anticipated for her husband—as the next 'Patriot King', imbued with her own strong Anglican Protestant morality. However, within the present arrangement he failed to thrive, and she sought to regain an element of control, isolating him from court society, and discreetly appointing her late husband's friend, John Stuart, third Earl of Bute, as his tutor in matters of politics and government.

Bute established an excellent rapport with his pupil, who grew in confidence and ability. However, when Prince George, on achieving his majority in 1756, appointed Bute his Groom of the Stole, the secrecy behind Bute's introduction into Augusta's household led to claims they had their own illicit relationship: this seemed the only explanation for the apparently unmerited ascent of a relatively unknown Scottish nobleman to the first rank of politicians. She was derided and lampooned by politicians, journalists, and satirists. Whether selflessly or unwittingly, her actions on behalf of her son demolished the confidence and trust others had in her abilities and good judgment. The access she, with her husband, had allowed to their family during past decades ensured that she was a figure of public fascination, and having presented herself as a model of a moral and peaceful domesticity, to be seen as having acted immorally was crushing. The critique extended to her relationship with Prince George. The *Oxford Magazine* wrote: 'Her name is Tyranny, and in a string/ She leads the shadow of an infant king.'[18]

Although it is thought unlikely that Bute and Augusta had an intimate relationship, such was the vociferousness of the critics that in her last years the Princess, who had been such a favourite of Londoners, rarely ventured beyond Carlton House and the White House at Kew, her royal homes. Her energies would be focused on her gardens. Even her considerable philanthropy would be conducted without publicity. Before Augusta, Princess Dowager, died in 1772, her son had only reluctantly allowed her a role in the new Regency Council, and would advise his new wife, Charlotte of Mecklenburg-Strelitz, to keep her own counsel. Charlotte would later describe to Elizabeth, Countess Harcourt:

> the dear king's great strictness, at my arrival in England, to prevent my making many acquaintances: for he always used to say, that in this Country, it was difficult to know how to draw a line, on account of the Politics of the Country; that there never could be kept up a Society without party, which was always dangerous for any woman to take part in, but particularly so for the Royal Family.[19]

CHARLOTTE (1744–1818)

> Happy England, were each female mind,
> To science more, and less to pomp inclin'd;
> If parents, by example, prudence taught,
> And from their QUEEN the flame of virtue caught!
> Skill'd in each art that serves to polish life,
> Behold in HER a scientifick wife![20]
>
> *The Ladies Poetical Magazine*, 1781

George III succeeded his grandfather in 1760, and married Charlotte of Mecklenburg-Strelitz in 1761 (figs 1.05, 1.06). She was born in Mirow, in the duchy of Mecklenburg-Strelitz in 1744, the second daughter of Duke Charles Louis and his wife Elizabeth of Saxe-Hildburghausen. Charlotte was educated by Mme Grabow, a cultivated gentlewoman and poet. She enjoyed a well-rounded education, and as well as art and music lessons, early on acquired a taste for serious reading—of theology in particular. She studied Latin and learned to speak French and Italian. From 1760 until her marriage she was a secular canoness at the imperial abbey of Hervoden, and as a younger daughter of a minor branch of an ancient family, she may have imagined she would remain unmarried and become an abbess. However, her destiny proved very different;

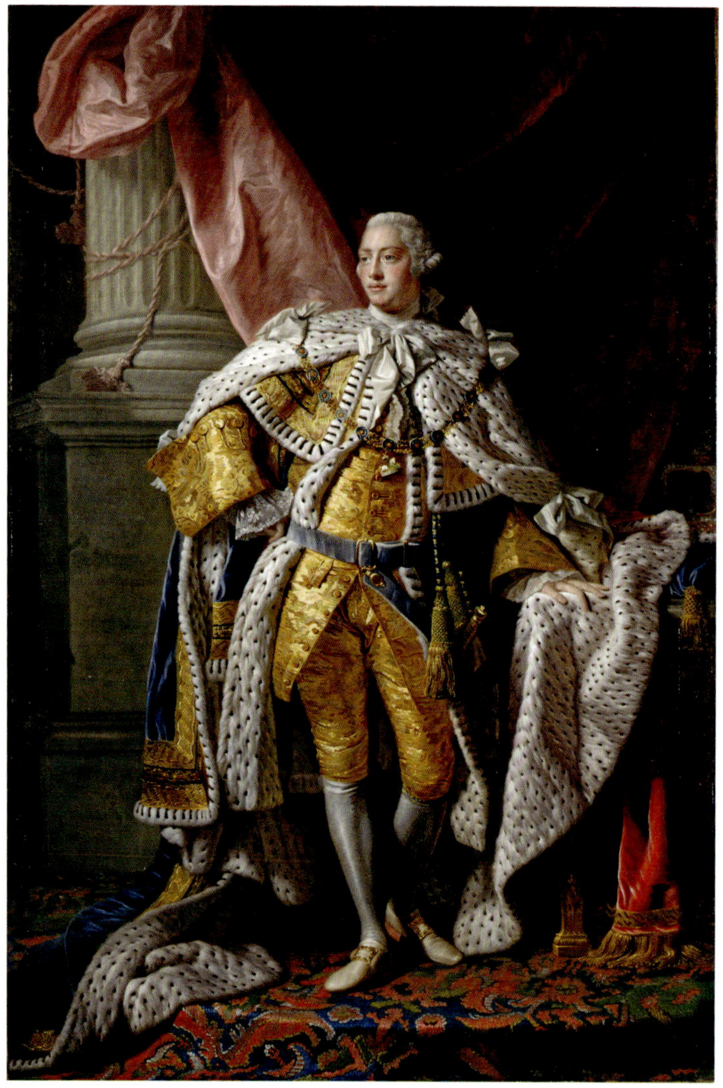

FIG. 1.05 Allan Ramsay, *George III*, 1761–62, oil on canvas, 98¼ x 64¼ in. (249.5 x 163.2 cm). Royal Collection Trust

she was chosen from a short list of candidates as bride for the King of England—she was young and healthy, of good character, of princely rank and Protestant.

Charlotte was fortunate in her marriage. An arranged match, it proved a perfect fit, and she enjoyed a happy life with her husband, who loved her dearly. In December 1785, Fanny Burney said of their relationship, 'their behaviour to each other speaks the most cordial confidence and happiness . . . in their different ways, and allowing for the difference of their characters, they left me equally charmed both with their behaviour to each other and to myself.'[21] They shared many interests—art, science, music, and theatre—as well as a deep religious faith. Like her immediate predecessors Charlotte was able to bear children without difficulty, swiftly fulfilling one of her principal functions, to secure an heir for the dynasty. She eventually gave birth to fifteen children—nine sons: George, Frederick, William, Edward, Ernest Augustus, Augustus Frederick,

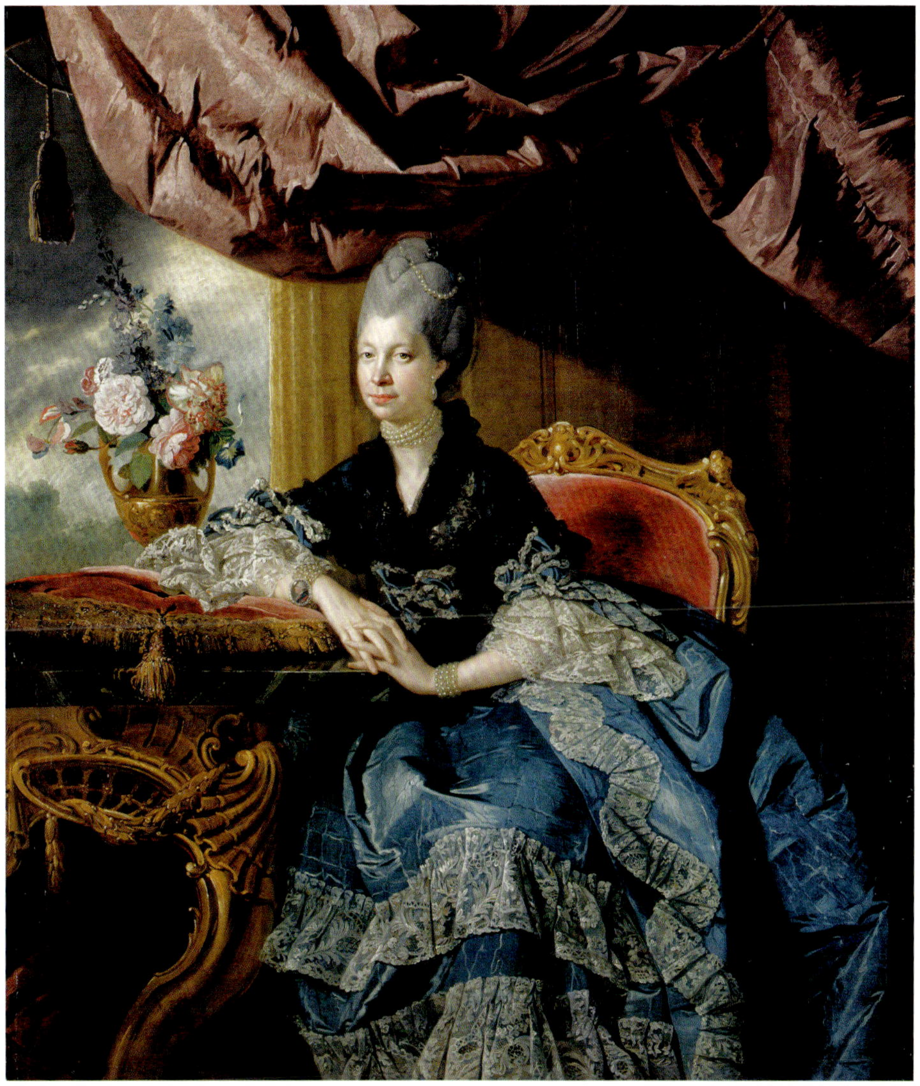

FIG. 1.06 Johan Zoffany, *Queen Charlotte*, 1771, oil on canvas, 64⅛ x 54 in. (162.9 x 137.2 cm). Royal Collection Trust (cat. 13.18)

Adolphus, Octavius, and Alfred, and six daughters: Charlotte, Augusta Sophia, Elizabeth, Mary, Sophia, and Amelia. Charlotte's evident delight in this large family contributed not just to the success of the marriage, but also to the creation of an image of a more informal stable domesticity for the royal family of the later eighteenth century.

Charlotte does not seem to have been her husband's confidante in affairs of state. Despite evidence in letters written to her brother Charles, later Charles II, Duke of Mecklenburg-Strelitz, that she was informed about the political system in her new homeland, and her maintenance of a library containing many books on British political and administrative history, she showed no ambition for a more active engagement.[22] In London her support for the regime was managed subtly through the Drawing Room over which she presided, and in the dispensing of court appointments in a society where most politicians were, or sought to be, courtiers, and knew there was advantage to be accrued from their wives' success in this arena. By these means she was able to support publicly her husband's dealings with his ministers, both positively and negatively.

Charlotte, who, like Caroline, was Electress of Hanover, seems to have been more inclined to exercise her political influence in the European arena. Unlike Caroline, she maintained a lively correspondence with members of her extended family in Germany. It was with the help of her brother Charles, who had been appointed by George III as military governor in Hanover in 1776, that she passed on personal requests concerning the royal homes there. She suggested to Charles how best he might manage public appointments and financial requests through the Hanoverian minister in London and the Hanoverian Regency Council. Although patriotically supporting Britain in the American War of Independence, she was equally engrossed in following events in Europe, monitoring the power struggles between the German states created by the Bavarian war of succession. Later she would not only correspond with her niece Louise, wife of Frederick William III of Prussia, but also would despatch her French reader, Jean-André Deluc, to visit her, in an effort to enlist support for Hanoverian interests during the French Revolutionary Wars.

In October 1788 George III had his first mental and physiological collapse. The episode was acutely distressing for Charlotte, and brought her an unasked-for prominence in political affairs. The terms of the government's Regency Bill in 1789, in seeking to prevent Prince George, the ambitious eldest son of George III and Charlotte, from consolidating a party around him, placed restrictions on the regent's power, and allowed Charlotte control of the King's person and household. This was interpreted by the political opposition as evidence of Charlotte's avarice, and following her energetic public support for her husband during his periods of remission from illness, Edmund Burke would claim that 'the power and dominance of the Queen' was finally being exercised.[23] Charlotte's relationship with her eldest son, which was badly damaged over the period, would not be fully healed until 1791. The King's illness recurred in 1801 and 1804, and after 1811 he was so mentally unstable, as well as rendered almost blind due to cataracts, that he would be confined, in seclusion, to his own apartment at Windsor Castle. The Regency Bill of 1811 restated Charlotte's charge of the King's person and household, assisted by a council. She used her agency over the next years to maintain sufficient continuity of procedure and personnel that, should her husband recover, he would be able to resume his duties swiftly. She supported her son by ensuring that the mechanisms of the court at St James's Palace were faithfully maintained, and provided him with sound advice about managing the public response to his own marriage break-up. Charlotte died in 1818, and was buried at Windsor where her husband continued to reside. The king outlived her by two years.

PRINCESS OF WALES, QUEEN CONSORT, PRINCESS DOWAGER, MOTHER OF THE KING

Discussion touching on the role of the consort begins to emerge in Renaissance texts such as Niccolò Machiavelli's *The Prince*, and Baldassare Castiglione's *Book of the Courtier*. In these texts, there is an appreciation that there are two facets of character that combine to make the successful ruler—the warrior hero, defending the honour of the realm, and the nurturer and encourager of its community. The latter role, which aligned to the responsibility women had for the care and championing of the family, comes increasingly to be vested in the consort—so often a woman. Their charge eventually extended well beyond the royal family, to include supporting the spiritual, social, and economic well-being of the nation at large. This distinction was neatly summarized by Hubert-François Gravelot when he drew up the design for the lottery medal for Henry Jernegan's silver cistern in 1735 (fig. 1.07). To complement the image of Britannia, surrounded with trophies, and the title, 'Both hands fill'd for Britain' with the sub-title 'George reigning', there is, on the obverse, the image of Caroline, tending little palm trees from a watering-can. While she may have been noted for her love of gardens and botany, the caption 'Growing arts adorn empire: Caroline protecting' indicates a greater ambition to associate the Queen with the nurturing of British interests at home and abroad.

However, for the eighteenth-century consort, achieving any freedom of action was complicated. As wife of the sovereign, or the heir to the throne, she was to be far more than helpmate to her husband; she was often his political confidante and principal supporter. She should also be fertile and successful in the nurturing of a large healthy family (fig. 1.08). For the British consort, it was essential that she was Protestant. When Caroline arrived in London in 1714 and assumed her position as the senior woman at court, it had been nearly forty years since there had been a Roman

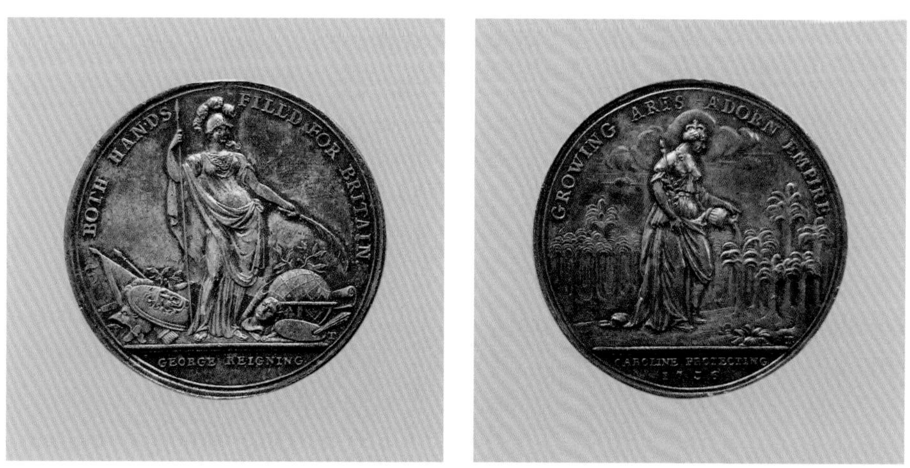

FIG. 1.07 John Sigismund Tanner, *Jernegan's Cistern Lottery Medal, Great Britain, George II and Caroline*, 1736, silver, 1½ in. (3.9 cm) diam. Yale University Art Gallery, Transfer from the Yale University Library, Numismatic Collection, 2001 (cat. 13.08)

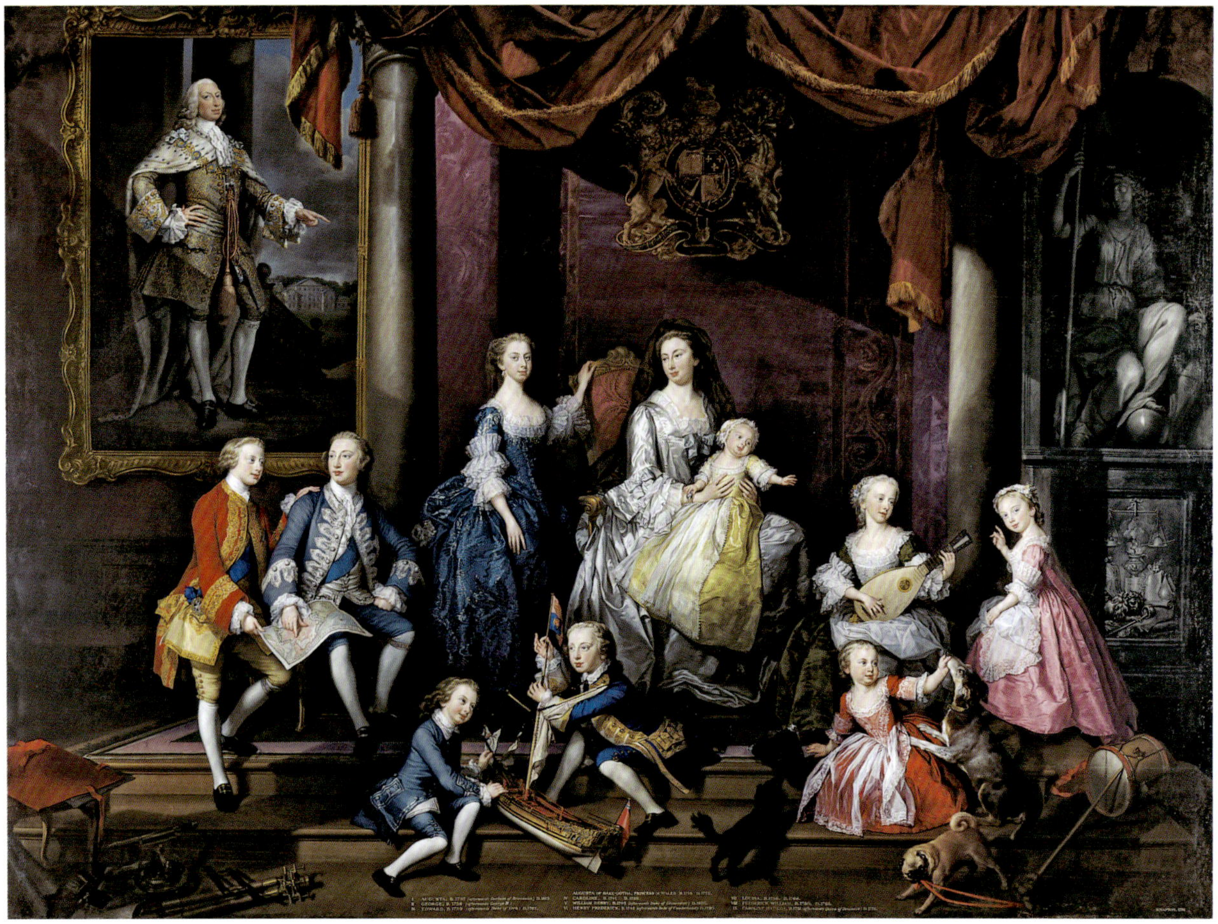

FIG. 1.08 George Knapton, *The Family of Frederick, Prince of Wales*, 1751, oil on canvas, 138¼ x 181½ in. (351 x 461 cm). Royal Collection Trust

Catholic consort. Indeed, it had been the confessional aspects of that marriage—between James II and Princess Mary Beatrice of Modena—that had contributed eventually to the terms of the Act of Settlement that had brought the Hanoverians to the throne.

There were many dangers the consort would be required to navigate. A consort who was too bold and independent could be seen as a political threat to the monarch, and the penalty for this was at best isolation, and at worst death. If the woman were barren, the situation was hardly better. Character, intelligence, and ambition were all important for achieving a workable equilibrium. It would help if the consort had access to money, and that the bond forged with her husband following their arranged marriage was good. In exploring the role of women at court, it is important to recognize that the court was a polycentric entity. The King's Household was but one of many—the heir, the siblings, the mistress and favourite, the dowager, and importantly the consort, had theirs too. The successful consort was able to influence and shape the court politically, socially, and physically, by her skilful negotiation between these rival power bases.

As marriages were arranged by the Hanoverian monarchs following the Act of Settlement, it was inevitable that they looked to the ruling families of German states for candidates. There was an ample supply of women deemed eligible because of their Protestant faith, who might in addition contribute through a marriage alliance to the furthering of family ambition in Europe. This German connection would prove enduring, if not invariably guaranteeing a successful match. Caroline, Augusta, and Charlotte were succeeded as consorts by Caroline of Brunswick, who failed to achieve a bond with her husband George IV, and Adelaide of Saxe-Meiningen, whose ambition was circumscribed by her husband William IV, one of the more retiring monarchs in British history. When Queen Victoria, as queen regnant, sought her consort, her eye fell upon her German kinsman, Albert of Saxe-Coburg-Gotha. His role and responsibilities, both devising and overseeing the educational programmes for the royal children and his championing of the arts and sciences, for instance, echo those of his female counterparts of previous centuries.[24]

In this world of opportunity, but also constraint, Caroline, Augusta, and Charlotte were all lucky in achieving a useful bond with their husbands. All had early success in bearing healthy children, which immediately conferred on them status and respect. In the sermon preached by William Talbot, then Bishop of Oxford, at the coronation of George I in 1714, Caroline was already identified as a 'Princess of the brightest Character' as a consequence of her 'hopeful issue'.[25] In 1818, the year of her death, Charlotte attended a reception at the Egyptian Hall at the Mansion House in the City of London in support of the London National Schools. For this occasion additional verses were added to the anthem 'God Save The King' expressly in her honour. Her maternal virtues were celebrated not only in terms of her own family and the dynasty, but also for the nation at large:

Then as the strain we raise,
Which our old Father's praise,
Still loves to breathe;
Mother of Britain, thou
Wilt not disdain the vow,
But let thy gentle brow
Grace the fair wreath.

Look where these little bands
Bless with their helpless hands
That brow serene!
Kindly approve, while we
Bid their pure infancy
Lisp forth its prayer for thee,
God bless the Queen.[26]

With this essential responsibility discharged, each Princess went on to build an affectionate bond with her husband. Despite evidence that Caroline had to endure George II's volatile temper, he was distraught when she died. Horace Walpole reported in a letter to Sir Horace Mann in August 1749, that the King 'has locked up half the palace [Kensington] since the Queen's death: so he does at St James's' and ordered that none of her possessions be moved.[27] When Walpole returned in 1758 he noticed that the logs laid in her bedroom hearth on the day she died were still in place.[28] At the King's request the Queen's coffin was designed to allow one side to be removed, ensuring that when the King was eventually laid alongside her in the vault he had had created under the Lady Chapel in Westminster Abbey, his coffin also having a removable portion, their remains could mingle. A token of Frederick's affection for Augusta is seen in the commissioning of the opera *Alfred*, with music by Thomas Arne and a libretto by David Mallet and James Thomson, to mark her birthday in 1740. It is best known for its finale, 'Rule, Britannia!' Fanny Burney speaks eloquently of the co-dependent relationship between George III and Charlotte: 'The King seems to admire as much as he enjoys her conversation, and to covet her participation in everything he either sees or hears.'[29] Although after 1804 Charlotte was so discomforted by her husband's irrational behaviour that the couple lived apart, they remained friends and continued to appear in public until 1810, when a recurrence of the King's illness was so severe he would eventually be sequestered at Windsor Castle. It is likely that Charlotte saw her husband for the last time in June 1812.

By gaining the trust and affection of their husbands the Princesses may have won a level of influence in the management of family politics, but they all went on to achieve varying degrees of more overtly political power too. Their skill and sensitivity at managing their fractious spouses ensured that the Princesses should be seen as their confidantes and advisers, and in the case of Caroline and Augusta the role was more akin to that of agent or partner, moving from passive support to more active involvement. Caroline, who numbered Voltaire as well as Leibniz among her admirers, was the most successful of all. She established such a rapport with Sir Robert Walpole, the First Minister, he would claim: 'Madam, without you I can do nothing . . . you, Madam, are the sole mover of this Court.'[30] Contemporary satirists were quick to identify her influence over her husband: 'You may strut, dapper George, but 'twill all be in vain;/ We know 'tis Queen Caroline, not you, that reign.'[31] During the riots which took place in London in reaction to the Excise Bill of 1733, it was Caroline's effigy, not the King's, which was burnt alongside that of Walpole.[32]

There were some who questioned Caroline's intelligence and criticized her for trespassing into intellectual debates and aspects of the national life they considered fell beyond her prerogative. But her position was unassailable, and it was only those who had nothing to lose who felt able to broadcast their views. Lady Mary Wortley Montagu, who had married out of her noble class, and therefore was denied court access, said she possessed no more than 'low cunning', and that she lacked 'having understanding enough to observe that falsehood in conversation, like red on the face, should be used very seldom and very sparingly'.[33] Alexander Pope, whose courtly connections were circumscribed by his Catholicism, cast her in *The Dunciad* as the 'Goddess of Dulness' for her promotion of writers he considered mediocre. The Scottish poet James Thomson, who aligned his fortunes to Frederick's party, satirized her as 'The World's Imperial Queen' who, proud of her ancestry and learning, sat surrounded by her collection of obscure antique relics, as sycophantic virtuosi flocked to her feet.[34]

Augusta, though fiercely loyal and supportive of her husband, managed skilfully and astutely to navigate a course between the rival power bases—Frederick's court and that of her father-in-law. On the one hand her presence in the room while Frederick and Lord Egmont debated the loyalty of various members of parliament, and the potential for their recruitment to the Prince's party, is recorded with no surprise, indicating that this was no unusual occurrence; on the other hand she had won the respect of George II, who despite the irreconcilable differences with his son, was prepared to deal with her to prevent his total alienation from his successors. When Frederick died so unexpectedly, she acted with an impressively disciplined incisiveness to protect her own and her family's position with the King and the government.

Charlotte may not have sought political prominence, but George III's bouts of illness brought her unexpected responsibilities. During the first shocking episode in 1788, on learning that medical reports were being sent to her son George, Prince of Wales, and that it was he who was overseeing the dissemination of information about the King's condition, she protested, insisting that 'nothing should be sent from the House without Her Majesty's concurrence', and subsequently she reviewed drafts of the text.[35] While it was suggested she found taking responsibility for 'the King's person' was 'much against her inclination', she expressed strong views about the choice of doctors brought in to treat her husband, and to an extent dictated their appointment.[36] The pathos of her situation—information about the King's illness was avidly reported by the press—brought her massive popular sympathy. On the King's recovery in 1789, one hundred and sixty loyal addresses were sent to Charlotte specifically: 'No female ever more justly deserved it . . . she is a pattern of domestic virtue which cannot be too much admired.'[37]

That the Princesses were able to elevate their sphere of political action from the familial to touch dynastic ambition, and even the politics of the state, was also as a consequence of changes in British society, which had begun to introduce opportunities not open to many of their European counterparts for women to engage more thoroughly with the wider community. Since the Interregnum in the seventeenth century, the traditional injunctions that women should be chaste, silent, and obedient had been undermined by a generation which wrote, preached, and even petitioned Parliament. Such female participation in the public sphere—even political sphere too, through the antics of the royal bedchamber—brought proto-feminist reference to 'the equality of women's merits and rights with the man' in 1669.[38] In 1676 in *The Gentleman's Companion*, William Ramsay, suggested women were wittier, and potentially better governors than men, and therefore should be educated to fulfil that potential.[39] Importantly, as Elisabeth Charlotte, Duchess of Orléans, would remind

Caroline, there was also a tradition of female rulers.[40] George I had succeeded Queen Anne, who in turn had succeeded her sister Mary II, who ruled jointly with her cousin William III. Queen Elizabeth I was revered as the bright star of the Tudor dynasty.

Into this new environment the Princesses brought their European experience. This included knowledge, to a greater and lesser extent, of female-generated *salon* culture, as well as of German approaches to philanthropy, grounded in Pietist philosophy. Each Princess was sufficiently confident, ambitious, and empowered to pursue projects and initiatives which brought European practice into British court culture. The British monarchy would be reshaped profoundly as a consequence.

By the mid-seventeenth century, well-born women were beginning to realize the deficiencies of their education and the concept of the *salon* emerged in Paris. This was an occasion when they could meet in small groups, often in the home, to discuss literature, theology, and new scientific discoveries. The meetings may not have had any official public or academic status, but attendance provided access into the world of letters, and allowed women a part in contemporary cultural and scientific as well as political debates. They provided an incentive for preparatory reading, and an opportunity to practise conversational skills, thereby ensuring a better understanding of the world for those women attending. Despite continuing British suspicion of female intellectual autonomy—clever, curious women made the 'least of wives', according to Thomas Gisborne, but fortunately were 'phenomena rarely seen in the median of Great Britain'—each Princess in turn created a network of connections, anchored in their Drawing Room, but breaking through the palace walls into elite society, and cultural and intellectual communities in London and the country.[41] Such circles had pivoted around the Electress Sophia Charlotte in Berlin and the Electress Sophia in Hanover, while a literary *salon* continued in Paris under the auspices of Sophia's niece, Elisabeth Charlotte—Liselotte—Duchess of Orléans, with whom Caroline corresponded until the latter's death in 1722. This family tradition of intellectuality and literary patronage provided a model of a princely freedom of opinion that Caroline sought to maintain. She was comfortable in this world, and was confident enough in her own dynastic capital to initiate her own conversations and debates, as Augusta and Charlotte would be, if to a lesser extent, in her wake.

But this activity would serve a function far greater than self-fulfilment. The most celebrated intellectual exchange over which Caroline presided was that between Leibniz and Sir Isaac Newton (fig. 1.09), and concerned their work on calculus. It was conducted through a series of letters, which passed between Leibniz and Dr Samuel Clarke, one of Caroline's protégés. Since 1711, Leibniz and Newton had been engaged in a fierce dispute over which of them should take credit for the discovery of 'the method of fluxions and fluents'. It would seem that Newton came by his calculation first, but before publishing his findings in any detail, Leibniz too had reached the same conclusion and circulated his proof in good faith. Caroline knew Leibniz from Berlin and Hanover, but realized, on her arrival in London, that Newton was venerated as a national hero, and if there was to be a new cultural energy at the heart of the court, Newton was too important not to be honoured. He was invited to her

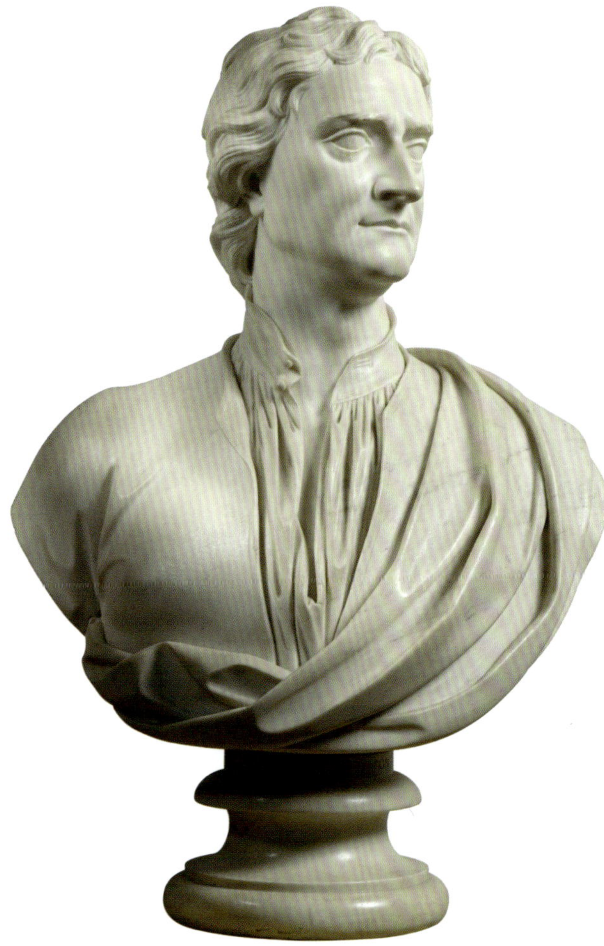

FIG. 1.09 Giovanni Battista Guelphi, *Sir Isaac Newton*, 1730–1, marble, 29⅛ x 20 x 11¾ in. (74 x 51 x 30 cm). Royal Collection Trust

Drawing-Room, where he conducted experiments to explore the effect of the vacuum, and 'with the colours' using prisms to refract light. Leibniz, who lobbied hard for a court appointment, never received the summons to London he craved. As Caroline built her circle and selected her heroes she would instead be celebrated for her 'particular Affection for this Country and the Natives of Great Britain': the *Gentleman's Magazine* noted, 'her own Leibniz is not allowed a place there.'[42] The *salon* brought her connection with her new compatriots, and through this built popular empathy for the fragile new dynasty.

Charlotte's scholarly inclination was at variance with Caroline's. She aligned herself with the Bluestockings and circles of female intellectual curiosity growing within British privileged classes, rather than looking to traditions of European royal intellectualism. If grounded in her personal taste, it also chimed with the public role she sought with her husband, that by following a simple exemplary way of life they might serve as a moral compass and conscience for the nation. This ambition was shaped by the philosophy of Jean-Jacques Rousseau, supreme advocate of rural simplicity and an

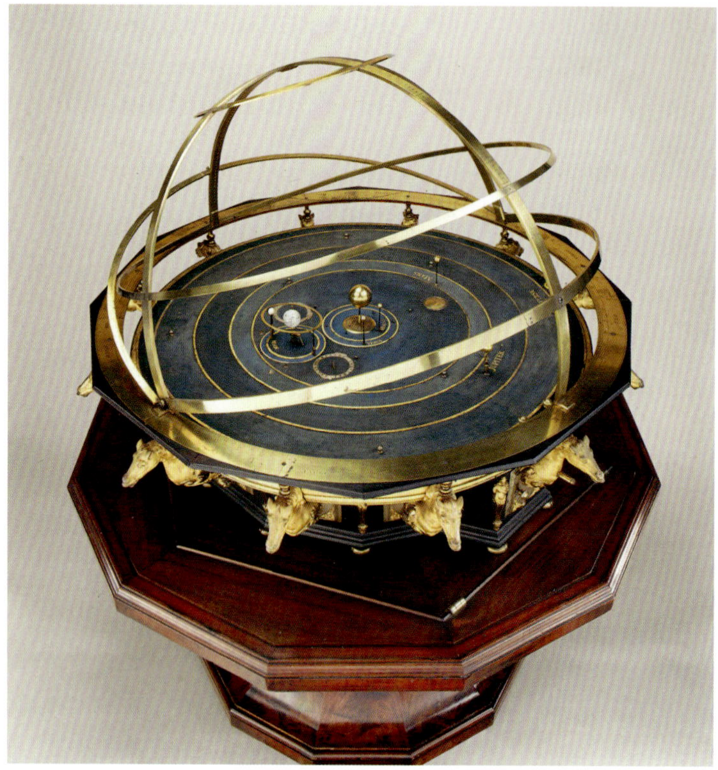

FIG. 1.10 Thomas Wright, Orrery, c.1740, wood (oak), brass gilt, and ivory, 6⅞ x 19½ x 17½ in. (17.5 x 49.5 x 44.5 cm). Science Museum, London

emotional religiosity.[43] If more informal in its composition, Charlotte's network of contacts also served the regime well. As the century progressed and Charles James Fox exhorted the community to 'Observe the magnificence of our metropolis—the extent of our empire—the immensity of our commerce and the opulence of our people', there was comfort in a monarchy which served as a solid, steady focus. George III and Charlotte travelled more extensively than their predecessors, lodging in the homes of the like-minded elite who formed their circle, as well as visiting hospitals and schools, and treating with industrialists.[44] This was a monarchy that was prepared to be seen, and to engage with ascendant middle England. It brought a level of popularity for the institution that ensured it did not have to weather the challenges levelled at the French royal family.

By drawing artistic, scientific, and literary communities into the palaces, and by travelling into the city to engage with them there, the Princesses had occasion to explore new ideas, philosophies, and practices—the opportunities that were presented by the age (fig. 1.10; see also cat. 16.09). While in 1716 Caroline delighted in the spectacle of science, as Newton set up experiments to refract light in her apartments, Augusta's Privy Purse accounts reveal that she too was interested in technologies involving light. She owned camera obscuras and magic lanterns, probably using them to enliven the educational programmes devised for her children. Charlotte followed in their footsteps, setting up large-scale back-lit transparencies as a contribution to the spectacle of royal celebration (fig. 1.11). The most splendid of all were devised by the Queen and her daughters for the Jubilee to commemorate the King's fiftieth anniversary in 1809. It was the first such occasion ever held, and a great success, serving in a small way to accelerate the adoption of political neutrality for the monarchy. It cut through party political and religious allegiances, was marked throughout the British Isles and the empire, and the royal family set right at its heart became a patriotic rallying point.

Caroline and Charlotte were each confronted with developing attitudes to the management of mental illness, and the rise of the sciences of psychology and psychiatry. In 1726 Caroline had coaxed George I to let her assume the charge of Peter the Wild Boy, a feral child adopted as a pet for the court after being found wandering in the woods near Herrenhausen.[45] Under the supervision of Dr John Arbuthnot, a series of experiments was conducted exploring Cartesian theories of nature or nurture: could Peter be taught the behaviours associated with polite human society, and would this give rise to the notions of responsibility, and build a conscience?[46] Caroline's death eventually brought an end to this programme, and Peter, who probably suffered from Pitt Hopkins Syndrome, which brings significant intellectual disability, was lodged with a farmer living near Berkhamsted in Hertfordshire.[47]

Charlotte had to navigate the private trauma of mental illness within her own family, which, because of the constitutional questions it occasioned, became a public

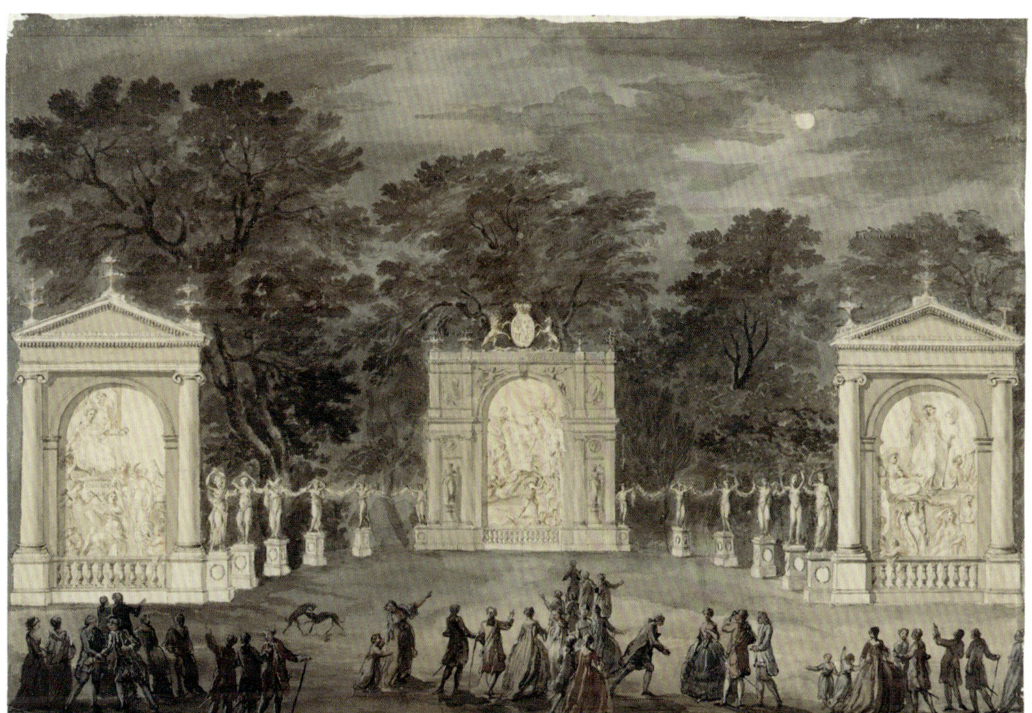

FIG. 1.11 Giovanni Battista Cipriani and Robert Adam, Design for *Illumination to Celebrate the Birthday of King George III*, 1763, watercolour and pencil on paper. Victoria and Albert Museum, London

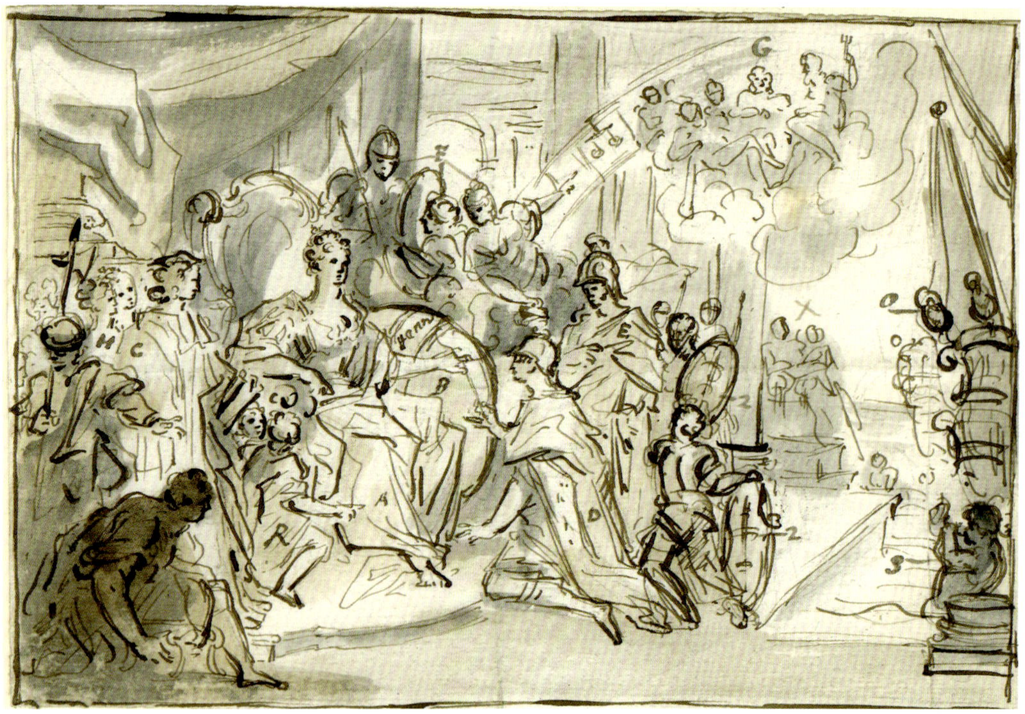

FIG. 1.12 William Kent, *Queen Caroline, as Britannia, Receiving the Gifts of the Earth*, 1730–37, pen and ink and wash over pencil on paper, 5¾ x 7⅞ in. (14.5 x 20.9 cm). British Museum, London (cat. 13.03)

event too. She negotiated with Dr Francis Willis and his brothers, and later Sir Henry Halford, practitioners at the forefront of the medicalization of insanity, as programmes of treatment were devised for her husband.[48] While Willis's approach to the King's care might only be seen as a mixture of morale boosting and psychological bullying, supplemented by more conventional medical practices such as blistering, experimental practices for merely managing mental illness gradually became the basis of a new science of psychiatry. The way in which symptoms were controlled became the means by which the patient might be cured.

An iconography of enlightenment developed around the imagery of the Princesses, representing virtues and qualities attributed to them, as well as perceptions of their role as the breadth of their interests touched the wider community. Of course, as was traditional, they were celebrated as mothers, for their care and nurturing of the family, and were depicted surrounded by their children, and tokens of their responsibility as educators; but other roles were increasingly identified and celebrated too (fig. 1.12). For an enterprise generated within the commercial community of London, Gravelot drew up his design for the Jernegan medal, which set up Caroline as the supporter of international trade, to balance George II as defender of the empire.[49] William Hopkins Craft, for the pair of enamel plaques made in 1773 to embellish an up-to-the-minute ormolu-mounted table clock with automaton movement, surrounded Charlotte with the attributes of agriculture (fig. 1.13). The sheaves of corn, the beehive, and lamb representing the rural pleasures and prosperity of Great Britain, are balanced by his

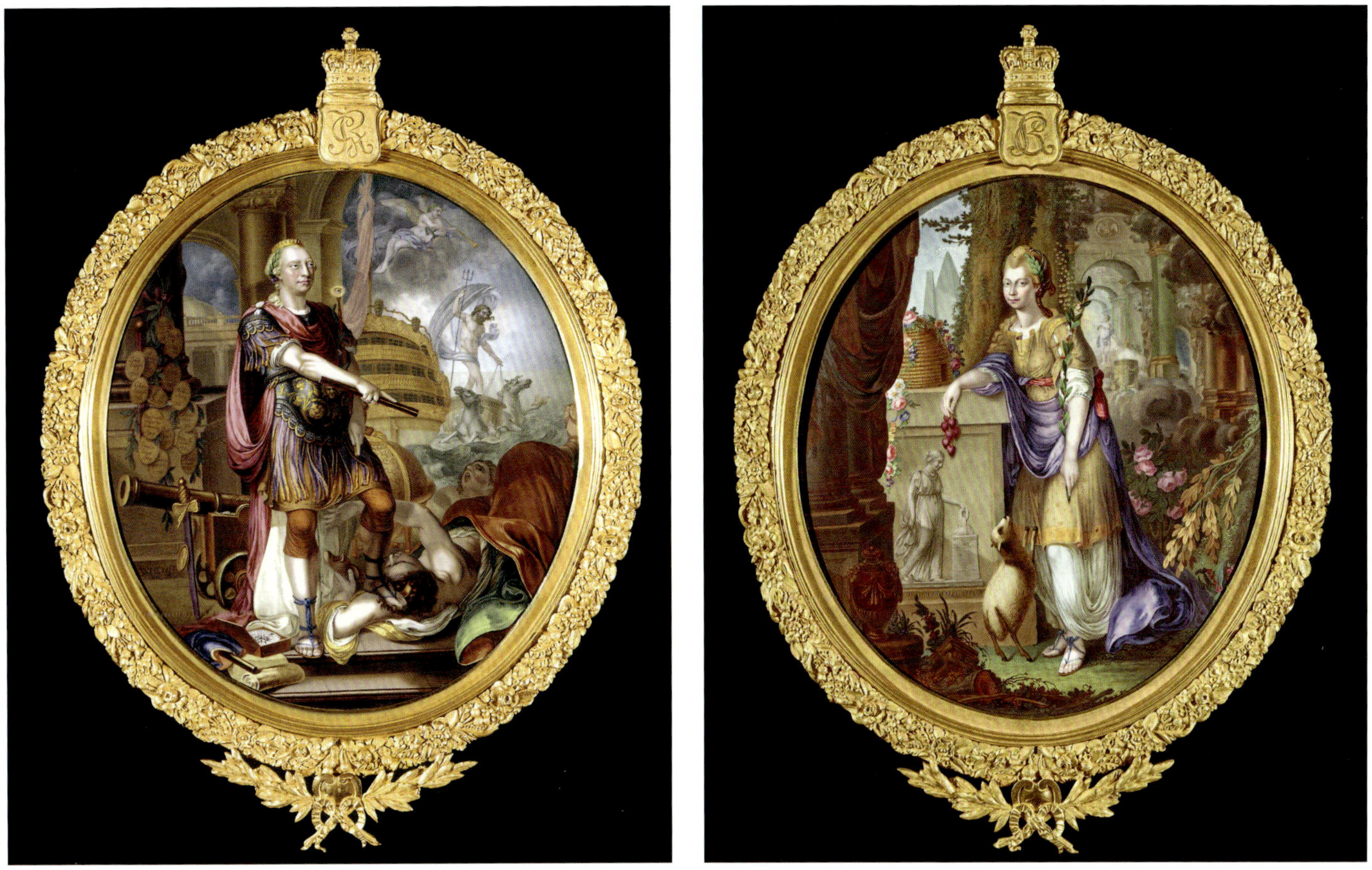

FIG. 1.13 William Hopkins Craft, *King George III as Lord of the Seas* and *Queen Charlotte with the Attributes of Agriculture*, 1773, enamel on metal. British Museum, London

image of George III as Lord of the Seas, looking to secure riches in the wider world.[50] Augusta may not have achieved the status of Queen Consort, but the interdependent relationship she successfully established with her husband was noted and recorded visually (fig. 1.14). Francis Hayman, in his oil sketch for one of the canvases in celebration of British virtues, intended as decoration for the Prince's Pavilion in the pleasure garden at Vauxhall, depicts Augusta seated alongside her husband, surrounded by the Muses, as artists present the plans for a new academy of the arts to them both.[51]

Each generation saw the image of monarchy drawn increasingly from the purely honorific into the present and everyday. For William Kent, Caroline was initially Minerva, the embodiment of wisdom, whose emblems had been inserted frequently by artists of earlier centuries into portraits of female clients, to honour their astuteness and creativity. Kent tucked them into the furnishing of the Queen's folly temples in the garden of Richmond Lodge, and the new decorations of the King's state apartments at Kensington Palace. But Caroline would later become Britannia, Britomart, Elizabeth I, and other incarnations which brought connections to new notions of Britishness.[52]

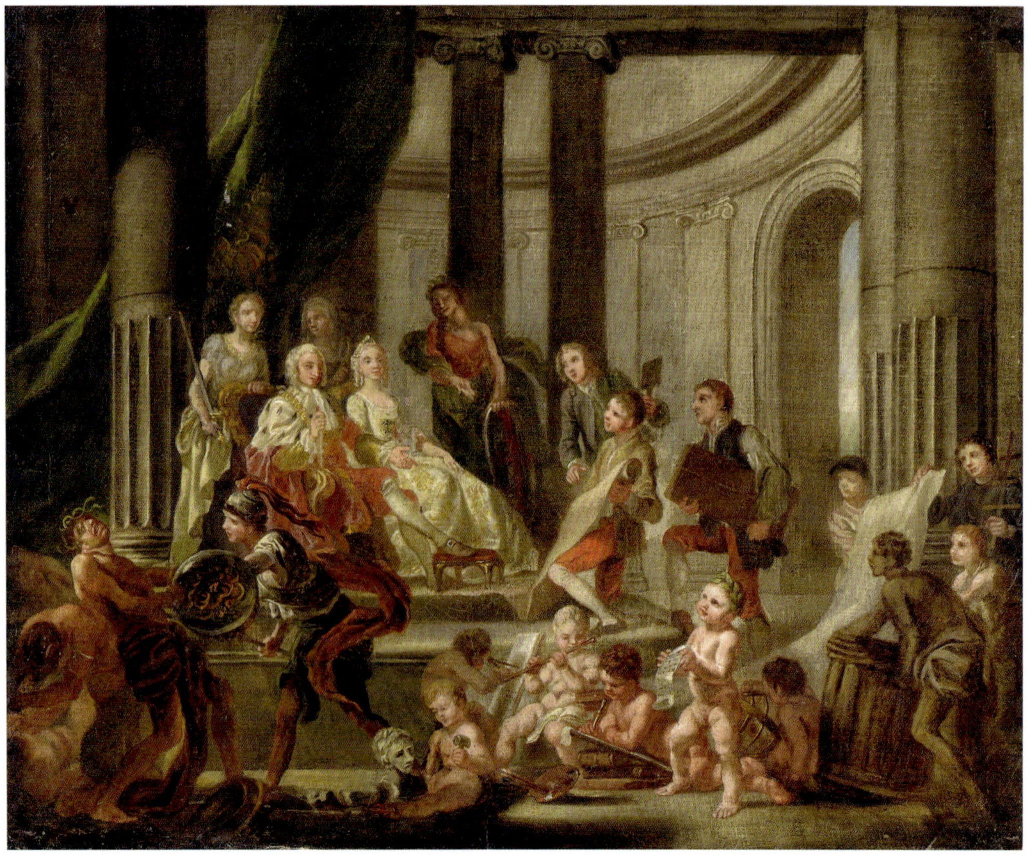

FIG. 1.14 Francis Hayman, *The Muses Paying Homage to Frederick Prince of Wales and Princess Augusta or The Artists Presenting a Plan for an Academy to Frederick Prince of Wales and Princess Augusta*, c. 1750–51, oil on canvas, 27¼ x 32¼ in. (69.2 x 81.8 cm). Royal Albert Memorial Museum, Exeter (cat. 13.14)

Confidence grew slowly: an early commission to William Hogarth in about 1732 for a conversation piece showing the family of George II and Caroline informally, at home in their apartments and garden, was cancelled after the production of two preparatory oil sketches.[53] Augusta was bolder and would be painted by Jean-Baptiste van Loo in 1739, with a flock of members of her household, in a pseudo-realistic royal interior.[54] A further shift occurred with the images supplied by Johan Zoffany for George III and Charlotte. In *Queen Charlotte with her Two Eldest Sons,* painted in about 1765, the Queen is seated at ease at her dressing table in Buckingham House, surrounded by the stuff of her life (fig. 1.15). The royal accounts confirm that the lace trimming the dressing table comes from Flanders, the carpet is Turkish, the silver-gilt toilet service is probably German, and framing the Queen are Chinese lacquer figures. Through the window a solitary flamingo can be seen strutting through the garden. Charlotte plays with her two infant children; George, the heir to the throne is dressed as Telemachus, his little brother Frederick as a Turk. It is an image of monarchy comfortable and confident as it engages with the riches of the wider contemporary world.[55]

Information about the extraordinary breadth of the Princesses' interests may be garnered from a rich vein of contemporary accounts. Letters and diaries written by the

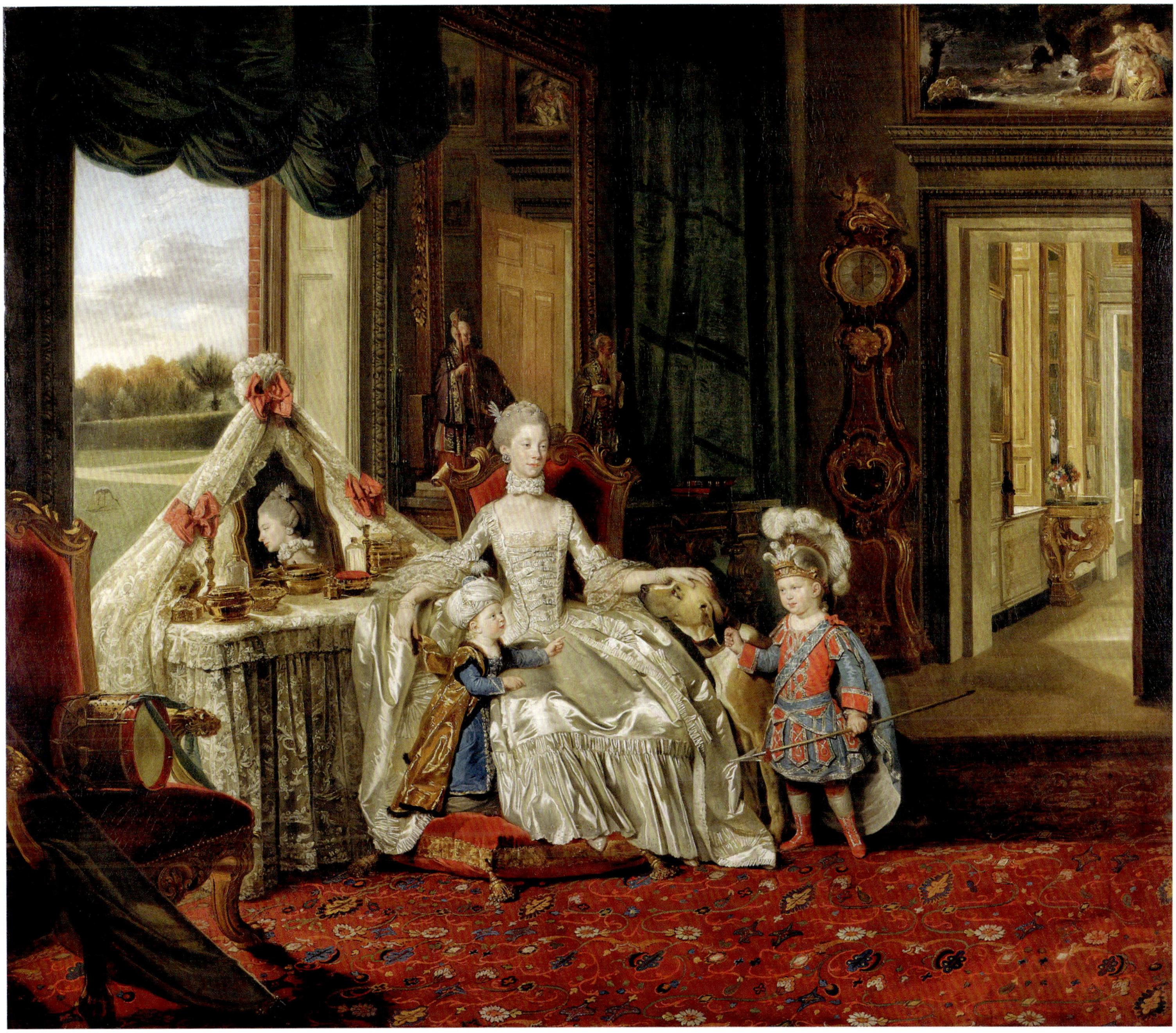

FIG. 1.15 Johan Zoffany, *Queen Charlotte with Her Two Eldest Sons*, c. 1765, oil on canvas, 44⅛ x 50½ in. (112.2 x 128.3 cm), Royal Collection Trust

Princesses themselves survive as well as those of their family members and their social circle. The Princesses' Privy Purse accounts, which chronicle their private expenditure, are especially revealing, recording payments for paintings commissioned or purchased, architects' fees for building and garden projects, bills for plants and birds, but also monies given to support organizations, charities, and individuals in whom the Princesses had a concern.[56] Inventories, lists, and sale catalogues were on occasion made of particular collections; Caroline's cabinet of curiosities, her most celebrated Picture Closet at Kensington Palace, and her library, which was carefully organized on Leibnizian principles, are carefully catalogued, for instance.[57] Following the deaths of both Augusta and Charlotte, sales of their personal effects were conducted at Christies' salerooms in Westminster.[58] The lots range from microscopes to medals and Mandarin dress, camera obscuras to clocks and cameos, harpsichords to hairpins, and paint boxes to porcelain. Charlotte's sale included her library; there were more than 4,500 lots of books. The Princesses' projects and initiatives were commented on by an enthusiastic press. For satirists they became habitual subjects.

Inventories, descriptions, and illustrations made by visitors to the royal homes, especially those made by the historian of British art and printmaker George Vertue, suggest how the collections were deployed. A series of watercolours made by Charles Wild, James Stephanoff, George Cattermole, and others, of the interiors of St James's Palace, Windsor Castle, Buckingham House, and Kensington Palace, among other royal homes, although they were not made until between 1815 and 1820, record the detail of the environments in which the Princesses conducted their lives, and reveal that some arrangements of furniture and paintings made by Caroline back in the 1730s were still in place.[59]

Over their lifetimes, Caroline, Augusta, and Charlotte worked creatively within the constraints that their position as consort laid upon them, to make many contributions that have shaped the British monarchy we know today. They left a material legacy, contributing through their architectural and gardening projects to the palaces and royal parks we can still enjoy. They brought substantial benefit to the royal collection by marshalling historic royal treasures, and overseeing their conservation, as well as adding new works. Their ambition touched the national good and imperial objective of the age, as they promoted industrial enterprise and championed projects designed to support colonial expansion, even if such policies have left political and social questions we must continue to address. They ensured the longevity of the dynasty, not only through their fertility, but in using their agency imaginatively to support and embed the Hanoverian regime, as it weathered political challenges in the first half of the eighteenth century. They were key to the identification of a practical role that the monarchy could assume within the organs of national government after the growth in ministerial authority, and the volume of public business created by protracted warfare, population growth, and national prosperity, circumscribed the King's room for manoeuvre as the eighteenth century progressed. As the monarch's powers declined, it gradually became easier to distinguish between monarch and minister, and to celebrate the former without owing allegiance to the latter. The Princesses' sense of social duty, grounded in their German experience, which had led them to develop and support a great range of religious, medical, and educational charitable initiatives, created a province in which the monarchy could be recast as the nation's moral conscience and heart. It was a role which was solidly patriotic but non-party political. The benevolent feminized 'Welfare Monarchy' that has flourished in Britain since the nineteenth century was the result. Importantly, they brought popularity to the monarchy. As the royal women managed their unpredictable spouses, their unruly families, and dealt with illness in an increasingly public arena, and as the press and souvenir industry reached ever-greater audiences, monarchy became soap opera. But this also brought a greatly increased interest in and loyalty for the institution. Ordinary lives were made to seem more important as it was evident that everyone grappled with the same concerns.

Thomas Paine claimed that monarchy, while it should be seen to be splendid as befitting its dignity, ought to be 'embodied . . . in persons', and each Princess variously through their *salons* and networks, their charitable work, and their engagement with the life of the nation, brought this for their generation.[60] They were colourful presences at court and had the romance and glamour, as well as irrationality and vulnerability, which brought popular devotion: this, Edmund Burke suggested, was what was essential to sustain a dynasty.[61]

NOTES

1. John Percival [Perceval] Earl of Egmont, *Manuscripts of the Earl of Egmont: Diary of Viscount Percival. Afterwards First Earl of Egmont*, 3 vols. (London: Historical Manuscripts Commission, 1920–23), 2: 138.
2. Caroline was baptized Wilhelmina Karoline. I would like to thank HM The Queen for graciously allowing me to consult the many rich sources of information held about the royal collections in the eighteenth century. I owe many thanks to the staff of the Royal Collection Trust, past and present, for their advice and support.
3. Royal Archives (RA) Add. MS. 28, no.52.
4. W. H. Wilkins, *Caroline the Illustrious: Queen Consort of George II*, 2 vols. (London: Longmans, Green & Co., 1901), 1: 18.
5. R. L. Arkell, *Caroline of Ansbach: George the Second's Queen* (Oxford: Oxford Univ. Press, 1939), 19.
6. Arkell, *Caroline of Ansbach*, 168. The National Archives (TNA) S.P.43. Regencies. 9, passim.
7. S. Taylor, 'Queen Caroline and the Church of England', in *Hanoverian Britain and Empire: Essays in Memory of Philip Lawson*, ed. S. Taylor, R. Connors, and C. Jones (Woodbridge: Boydell and Brewer, 1998), 92.
8. ELP 1015. Eliza Lucas Pinckney to unidentified person, 1753. The National Society of the Colonial Dames of America in the State of South Carolina (cat. 30.10).
9. *Gratulatio Academiae Oxeniensis in Nuptias . . . Frederici Principis Walliae* (Oxford, 1736), 51.
10. John Hervey, Baron Hervey, *Memoirs of the Reign of George the Second: From his Accession to the Death of Queen Caroline*, ed. J. W. Croker, 3 vols. (London, 1848), 2: 114–15.
11. *Gentleman's Magazine* 6 (1736): 231, describes a river trip taken by the Prince and Princess of Wales: having disembarked, the couple dined in public, 'the windows being thrown open to oblige the curiosity of the people'.
12. *Gentleman's Magazine* 8 (1749): 7.
13. As model of the ideal good mother and supportive wife, Augusta was careful to publicize her encounter with Isabella Gordon, Countess of Cromarty, who came to her with her young children to plead for her husband's life after the Jacobite rebellion in 1745. Calling her own children into the room she pointed out that should the rebellion have succeeded there would have been other innocent casualties—though she still lobbied successfully for Cromarty's life.

14. 18 July 1749. George Bubb Dodington, *The Political Journal of George Bubb Dodington*, ed. J. Carswell and L. Dralle (Oxford: Clarendon Press, 1965), 7.
15. John Percival [Perceval], Earl of Egmont, *Memoranda Book 20*, 20 March 1751, 198–99.
16. Horace Walpole noted 'the quiet inoffensive good sense of the Princess, who never said a foolish thing, or done a disobliging one since her arrival'. Horace Walpole, *Memoirs of King George II*, ed. J. Brooke, 3 vols. (New Haven and London: Yale Univ. Press, 1985), 1: 53.
17. William Augustus, Duke of Cumberland, the younger brother of Frederick, Prince of Wales, who was regarded by Frederick and Augusta as their greatest enemy, was a member of the Regency Council. Many of Augusta's actions between 1751 and 1756 were intended to prevent Prince George falling under his control.
18. From 'Kew Gardens', in *The Poetical Works of Thomas Chatterton. With Notices from his Life*, ed. C. Bonnycastle Willcox, 2 vols. (Cambridge, W. P. Grant, 1842), 1: 358.
19. Harcourt Papers 6, fols.109–10.
20. Quoted in Clarissa Campbell Orr, 'Queen Charlotte, "Scientific Queen"', in *Queenship in Britain, 1660–1837: Royal Patronage, Court Culture and Dynastic Politics*, ed. C. Campbell Orr (Manchester: Manchester Univ. Press, 2002), 236.
21. Frances Burney, *Diary and Letters of Madame d'Arblay*, edited by her Niece [Charlotte Barrett], 7 vols. (London: Henry Colburn, 1842–46), 2: 392.
22. Charlotte's library was sold on her death. *Catalogue of the Genuine Library, Prints, and Books of Prints, of an Illustrious Person, lately deceased, which will be sold by Auction . . . On Wednesday the 9th of June, 1819 and the following days, by Mr Christie . . .* (London: W. Bulmer, 1819).
23. Edmund Burke, *The Correspondence of Edmund Burke*, ed. T. W. Copeland and others, 10 vols. (Cambridge: Cambridge Univ. Press, 1958–78), 6: 2.
24. Two seminal works in the Royal Collection illustrate the dynamic between monarch and consort and attest to its non-gender specificity. In 1631–32 Sir Anthony van Dyck painted *Charles I and Henrietta Maria with their two eldest children, Prince Charles and Princess Mary* (RCIN 405353). Franz Xaver Winterhalter painted *The Royal Family* in 1846 (RCIN 405413) for Queen Victoria. In each painting the monarch and the heir to the throne confidently gaze out of the painting at the viewer. The consorts, Queen Henrietta Maria and Prince Albert, their arms encircling or gesturing to the younger offspring, are painted in profile gazing at their spouse and the heir whose education they supervise.
25. W. Talbot, *A Sermon Preach'd at the Coronation of King George In the Abbey Church of Westminster, October the 20th, 1714, by the Right Reverend Father in God, William Talbot, Lord-Bishop of Oxford* (London: J. Churchill, 1714).
26. *Morning Chronicle*, 30 April 1818.
27. Horace Walpole, Earl of Orford, *The Yale Edition of Horace Walpole's Correspondence*, ed. W. S. Lewis, Charles Bennett, and Andrew G. Hoover, 48 vols. (New Haven and London: Yale Univ. Press, 1937–83), 1: 279.
28. Horace Walpole, Earl of Orford, 'Walpole's Journals of Visits to Country Seats', ed. Paget Toynbee, *Walpole Society* 16 (1928): 16.
29. Burney, *Diary and Letters*, ed. Barratt, 2: 392.
30. John Hervey, Baron Hervey, *Lord Hervey's Memoirs: Edited from a Copy of the Original Manuscript in the Royal Archives at Windsor Castle*, ed. Romney Sedgwick (London: William Kimber, 1952), 112.
31. *The Craftsman*, 1727. Cited in Arkell, *Caroline of Ansbach*, 149.
32. Hervey, *Memoirs*, ed. Croker, 1: 203.
33. Lady Mary Wortley Montagu, *The Letters and Works of Lady Mary Wortley Montagu*, ed. Lord Wharncliffe, 2 vols. (London: R. Bentley, 1837), 1: 13.
34. C. Gerrard, '"The Castle of Indolence" and the Opposition to Walpole', *The Review of English Studies* 41, no. 161 (Feb. 1990): 54–64.
35. BM Add. MS. 41690-1. Willis Papers. 16 December 1788 and subsequent dates.
36. *Diary of Thomas, Earl of Ailesbury, Vol. 2: Great Britain*, Royal Commission on Historical Manuscripts, Fifteenth Report: Appendix 7 (London: HMSO, 1898), 297.
37. *The Times*, 18 April 1789. Quoted in L. Colley, 'The Apotheosis of George III: Loyalty, Royalty and the British Nation, 1760–1820', *Past and Present* 102, no. 1 (1984): 125.
38. A. Littleton, *A Sermon at the Funeral of the Rt. Hon. The Lady Jane . . .* (London, 1669), 19–20.
39. W. Ramsay, *The Gentleman's Companion* (1676), 10–13.
40. Elisabeth-Charlotte, Duchesse d'Orléans, *Fragments de Lettres Originales de Madame Charlotte-Elizabeth de Bavière, Veuve de Monsieur, Frère unique de Louis XIV, ecrites à S. A. S Monseigneur le Duc*

*Antoine-Ulric de B** W****, & à S. A. R. Madame la Princesse de Galles, Caroline, née Princesse d'Anspach, de 1715 à 1720*, 2 vols. (Hamburg: Maradan, 1778–88), 2: 4 (9 Jan. 1716).
41. Thomas Gisborne, *An Enquiry into the Duties of the Female Sex* (London: T. Cadell and W. Davies, 1797), 266.
42. *Gentleman's Magazine* 3 (Aug. 1733): 421.
43. R. A. Leigh, 'Rousseau's English Pension', in *Studies in Eighteenth-Century French Literature Presented to Robert Niklaus*, ed. J. H. Fox, M. H. Waddior, and D. A. Watts (Exeter: Exeter Univ. Press, 1975), 110–21.
44. W. Cobbett, *The Parliamentary History of England from the Earliest Period to 1803*, 36 vols. (London: Hansard, 1813), 17: 147; in Ronald Hyam and Martin Ged, *Reappraisals in British Imperial History* (London: Macmillan, 1975).
45. D. Defoe, *Mere Nature Delineated: Or A Body without a Soul. Being Observations Upon the Young Forester Lately bought to Town from Germany* (London, 1726). Peter died in 1785, and is buried in the churchyard at Berkhamsted.
46. J. Arbuthnot, *It Cannot Rain, but it Pours; or London Strow'd with Rarities* (London, 1726). James Burnett, Lord Monbdoddo, used Peter as a case study in his exploration of the evolution of language. See James Burnett, Lord Monbdoddo, *Of the Origin and Progress of Language*, 6 vols. (Edinburgh and London: J. Balfour and T. Cadell, 1773–92). 47. For further discussion of Peter the Wild Boy, see Lucy Worsley, *Courtiers: The Secret History of Kensington Palace* (London: Faber, 2010), 87–110.
48. [Richard Plantagenet Temple-Nugent-Brydges-Chandos-Grenville,] Duke of Buckingham and Chandos, ed., *Memoirs of the Court and Cabinets of George the Third: From Original Family Documents*, 4 vols. (London: Hurst and Blackett, 1853), 1: 47. W. W. Grenville to the Marquess of Buckingham, 10 Dec. 1788.
49. Hubert-François Gravelot, *Design for a Medal: Growing Arts Adorn Empire, Caroline Protecting*, c. 1736, pen and black ink over pencil, Yale Center for British Art, Paul Mellon Collection, B1975.4.444.
50. William Hopkins Craft, *Charlotte Surrounded by Attributes of Agriculture*, 1773, enamel on metal with ormolu frame, British Museum, London, 1891,0217.17.
51. Francis Hayman, *The Muses Paying Homage to Frederick Prince of Wales and Princess Augusta or The Artists Presenting a Plan for an Academy to Frederick Prince of Wales and Princess Augusta*, c. 1750–51, oil on canvas, Royal Albert Memorial Museum, Exeter, RAMM 392/1974.
52. William Kent (c. 1685–1748), *Queen Caroline, as Britannia, Receiving the gifts of the Earth*, 1730–37, pen and ink, and wash over pencil on paper, British Museum, London, 1927,0721.2.
53. William Hogarth (1697–1764), *The Family of George II*, c. 1731–32, oil on canvas, RCIN 401358; William Hogarth, *The Family of George II*, c. 1732, oil on canvas, National Gallery of Ireland, NGI 126.
54. Jean-Baptiste van Loo, *Augusta, Princess of Wales with Members of her Family and Household*, 1739, oil on canvas, RCIN 403502.
55. Johan Zoffany, *Queen Charlotte with her Two Eldest Sons*, c. 1765, oil on canvas. RCIN 400146.
56. Caroline's Privy Purse Accounts: RA Add. MS. 53993–54005, May 1730–March 1731; RA Add. MS. 54006–54014, May 1731–Sept. 1731; RA Add. MS. 54015–54023, Sept. 1731–1733. Augusta's Privy Purse Accounts: RA Add. MS. 55432–55652.
57. Caroline's library catalogue: British Library (BL) Add. MS. 11511. Caroline's inventory of curiosities: BL Add. MS. 20101, fol. 60.
58. Augusta's sale catalogue: *Catalogue of the Superbe Jewels, Trinkets, Plate, Gold and Silveer Medals, China . . . of a Person of Great Rank etc . . . which will be sold by Auction by Mr Christie, on Monday, Tuesday and Wednesday, the 1st, 2nd and 3rd of February, 1773.* Charlotte's sale catalogues: *A Catalogue of the First Part of a Magnificent Collection of Oriental Curiosities and Porcelain etc . . . Will be sold by Auction by Mr Christie, on Friday May 7, 1819, and Three following days. A Catalogue of the Remaining Part of a Valuable Collection of Curiosities, containing Carvings in Ivory, Trinkets, Coins . . . which will be sold by Auction by Mr Christie, on Monday, May 24, 1819, and Two following days. A Catalogue of Sundry Valuables Miscellaneous Effects . . . Which will be sold by Auction by Mr Christie . . . On Thursday August 26, 1819, and following Day.*
59. The watercolours were engraved and published as W. H. Pyne, *The History of the Royal Residences of Windsor Castle, St James's Palace, Carlton House, Kensington Palace, Hampton Court, Buckingham House and Frogmore*, 3 vols. (London: L. Harrison for A. Dry, 1816–19).
60. Edmund Burke, *Reflections on the Revolution in France by Edmund Burke & The Rights of Man by Thomas Payne* (New York: Anchor, 1973), 91.
61. Edmund Burke, *Reflections on the Revolution in France, and on the Proceedings in Certain Societies in London Relative to that Event* (London: Printed for J. Dodsley, 1790).

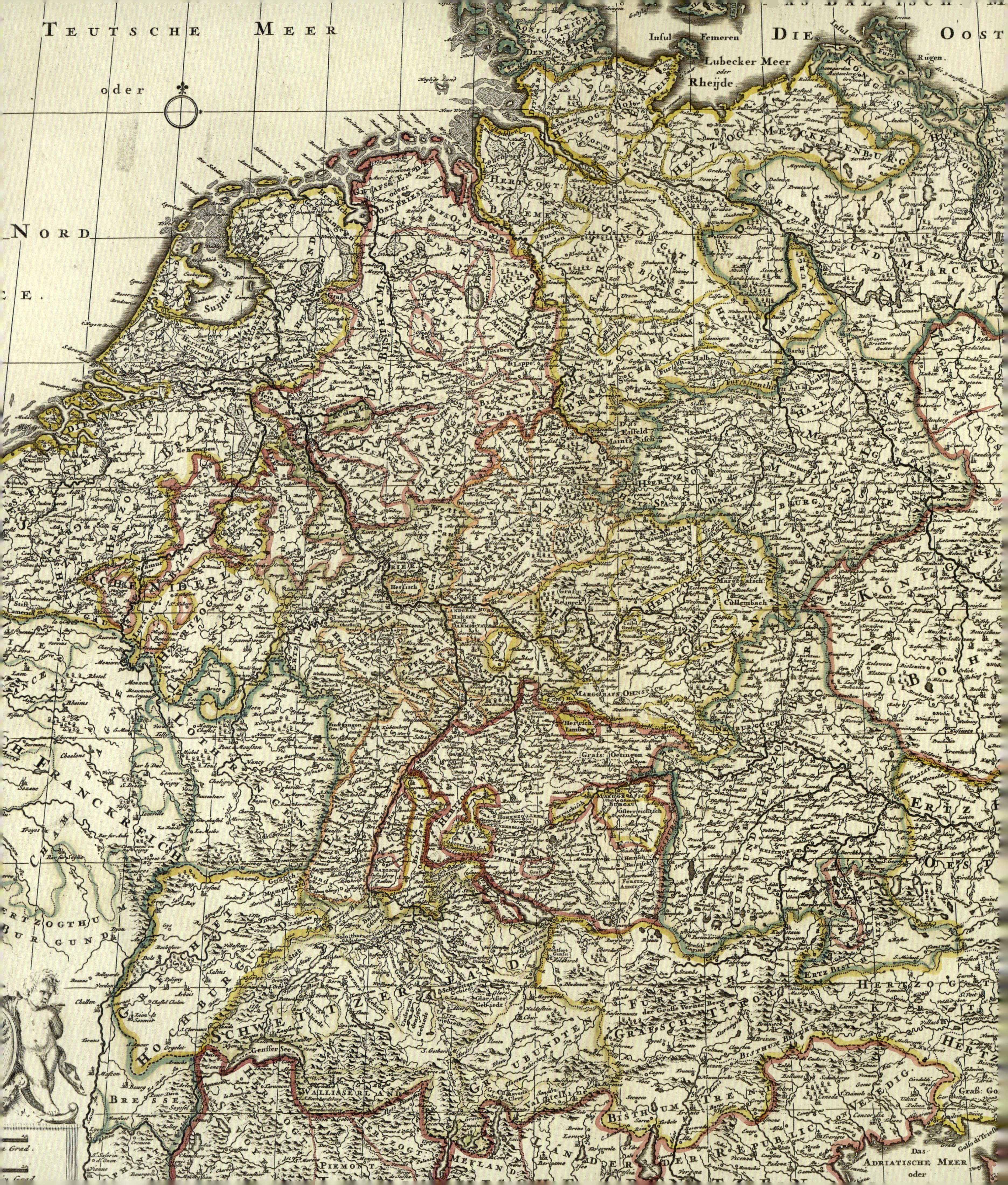
TEUTSCHE MEER
oder
NORD
DIE OOST
Insul Femeren
Lubecker Meer
oder
Rheyde
Rügen
FRANCKREICH
SCHWEITZERLAND
Das
ADRIATISCHE MEER
oder

Enlightened Princesses between Germany and Britain

John Styles

Among the ruling families of eighteenth-century Europe, it was not unusual for monarchs and their heirs to marry a German princess. Most famously, in 1770 the future Louis XVI of France married Marie-Antoinette, born Maria Antonia Josepha Johanna von Habsburg-Lothringen in 1755, daughter of the Holy Roman Emperor Francis I and his wife Maria Theresa, who ruled in Vienna and claimed authority as imperial overlords over most of Germany. If Marie-Antoinette's execution by guillotine during the French Revolution was unique among German princesses, her marriage to a king was not. German princesses married rulers of all the eighteenth-century monarchies of Europe, from Russia, Sweden, and Denmark in the north to Spain, Portugal, and Sardinia in the south. Europe was a continent where marriage alliances between ruling dynasties were diplomatically and militarily crucial. Germany was Europe's strategic crossroads. Its political and religious fragmentation ensured the availability of numerous eligible princesses.

In the eighteenth century Germany did not exist as a single state. Nevertheless, Germany was a recognized geographical entity, consisting of all or most of the German-speaking lands that comprised a huge swathe of territory in the middle of Europe (fig. 2.01). It stretched from the Alps to the North Sea, and from the valley of the river Rhine to the eastern shores of the Baltic. In the middle of the eighteenth century, its population was well in excess of fifteen million, more than twice that of Great Britain.[1] This territory, much of it nominally part of the Holy Roman Empire, was divided into hundreds of more or less independent principalities, duchies, bishoprics, cities, and counties. A few of these states were large—Brandenburg-Prussia at the start of the eighteenth century was larger than Scotland, Bavaria larger than Wales—but most were small, often tiny.[2] Yet many of their rulers, irrespective of their precise title, enjoyed the status of prince. It was a status they were entitled to pass on to all their children, both male and female, making them eligible to marry into monarchical dynasties both in the German lands and beyond.

Detail of p. 32, fig. 2.01

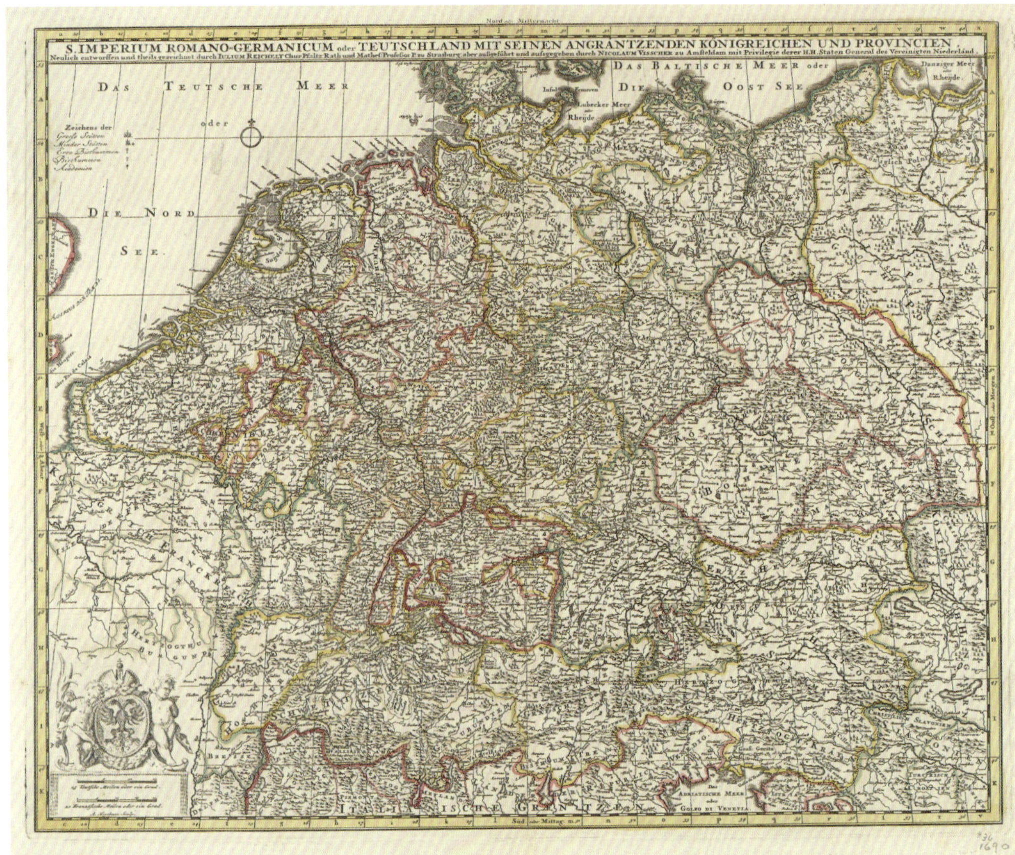

FIG. 2.01 A. Hogeboom, with engraver Julius Reichelt, *S. Imperium Romano-Germanicum, oder, Teutschland mit seinen angrantzenden Königreichen und Provincien* (Amsterdam: Nicolaes Visscher, 1690?), 20⅜ x 24⅜ in. (51.8 x 62 cm). Beinecke Rare Book & Manuscript Library, Yale University

Caroline of Ansbach, Augusta of Saxe-Gotha, and Charlotte of Mecklenburg-Strelitz, the three German princesses who are the subject of this book, married into the Brunswick-Lüneburg dynasty that ruled one of the larger German states—the Electorate of Hanover. In 1714 it became the reigning dynasty in the kingdoms of Great Britain and Ireland, which comprised one of the largest, wealthiest, and most powerful monarchies in Europe. London was to be their husbands' principal residence. Indeed, Charlotte's husband, George III, never set foot in Hanover. Once the three women had moved to Britain, they did not return to their native Germany. Their careers as royal consorts were pursued in a setting divorced from the one in which they had been raised. How did Britain, and the new life they embraced there, differ from their native Germany? How did their German background contribute to their conduct as royal wives in a new country?

The Thirty Years' War of 1618 to 1648 cast a long shadow over Germany during the later seventeenth and early eighteenth centuries. Partly a religious civil war between Germans, pitting Protestants against Catholics, partly a struggle between Germany's ruling dynasties over territorial authority and independence, the Thirty Years' War drew in many of the non-German powers of Europe, including Sweden,

France, Poland, Spain, and the Netherlands. The seemingly interminable conflict continued, with pauses, for year after year, fought principally on German soil with devastating results. By the time a compromise peace was achieved in 1648, Germany's population had declined by perhaps a third as a result of wartime plague, famine, and destruction. In many territories the economy was laid waste. Population growth and economic expansion were to be key priorities for German rulers for the next century.

The peace terms required all sides to accept the religious affiliations of the various German states as they had existed in the early years of the war. Hostility between Catholics and Protestants persisted, but henceforth it was to be managed within a broadly stable patchwork of hundreds of states, large and small, each with its specific religious allegiance. If the ruler of a state changed religion, its people were entitled to retain their faith. Nevertheless, with Germany divided between a predominantly Protestant north and a predominantly Catholic south, religion remained an extremely sensitive issue. Changes of princely faith threatened religious stability across Germany, so there was considerable pressure for Protestant rulers to marry Protestants, and Catholic rulers to marry Catholics. Other constraints on the sovereign power of rulers tended to diminish, allowing an extension of royal authority. In addition, rulers' territorial autonomy was affirmed by the peace treaty that ended the war, fixing for almost a century and a half the political fragmentation of Germany that rendered it such a prolific source of brides for the royal houses of Europe.

Like Germany, Britain endured a civil war in the mid-seventeenth century. It too was partly a war of religion, with different varieties of Protestantism pitted against each other and fear of a resurgent Roman Catholicism a powerful force in the background. The war also turned, as in the German lands, on the relationship between different states within the same territory, specifically the three kingdoms that shared the British Isles—England, Scotland, and Ireland. More prominent in the British conflict was the relationship between the ruler and the institutions—the parliaments—which represented at least some of the ruled. The war was fought out, with pauses, between 1642 and 1651. It resulted in the execution in 1649 of the reigning Stuart King, Charles I, and the establishment of a republic. The monarchy was restored in 1660 with the return of the exiled Charles II, the executed King's son, but many of the antagonisms that underlay the war remained unresolved.

Nevertheless, in the course of the next fifty years the British Isles saw a resolution of these conflicts. More radical than the peace treaty ending the Thirty Years' War in Germany, its terms explain the presence in Britain of the three German Protestant princesses. Charles II's successor, his brother, the openly Catholic James II of England and Ireland and James VII of Scotland, was deposed in 1688 in what its victorious supporters termed the 'Glorious Revolution'. James was toppled after a brief reign lasting less than four years by an invasion of Dutch troops. They were led by his son-in-law, the Protestant William of Orange, military leader of the Netherlands, who acceded to the throne jointly with his wife Mary. William's priority was to secure British support in his war with Louis XIV's France. In return for that support, he accepted a dramatic and, as it turned out, decisive tilting of the balance of constitutional authority in Britain away from the monarch towards Parliament. Henceforth, British monarchs ruled on terms dictated by their legislatures. Parliament became the ultimate source of political authority. The royal court remained important politically, ceremonially, and culturally, but it was no longer the focus of authority.

In 1660 Charles II had restored his father's brand of High-Church Protestantism as the doctrine of the Church of England, to wide acclaim. It was soon to find impressive physical expression in Christopher Wren's new London churches. The Puritans and other Protestants, like the Quakers, who could not stomach the new religious order, withdrew from the Church of England to establish their own dissenting congregations, which faced intermittent persecution. They comprised perhaps one-tenth of the population. Yet at the same time, the two kings who were the supreme rulers of the Church of England between 1660 and 1688, Charles II and James II and VII, either sympathized with the outlawed Catholic faith, or, in the case of James, professed it. Sons of a French Catholic princess, Charles and James had spent time in exile in Catholic Europe in the aftermath of the English Civil War. Both were married to Catholic queens. They were admirers of the absolutist style of Baroque Catholic monarchy practised by their cousin Louis XIV in France. James's authoritarian attempt to re-establish Catholicism in the mainstream of British life led directly to his forced abdication and return to exile.

In the aftermath of the 'Glorious Revolution', the assumption that all the king's subjects must profess the same faith was abandoned. Under William and his English Protestant wife, Mary, freedom of worship was extended to all Protestants. Britain became in effect a denominational society. This did not mean religious equality. A Protestant state church continued to exist, enjoying enormous political, economic, and organizational privileges. Other Protestant denominations had to finance their own clergy and premises. In England the state church was Anglican, in Scotland it was Presbyterian, a cynical but realistic accommodation to national differences in theology that marked a retreat from the unbending religious fanaticism of previous generations. Only Catholics remained subject to legal restrictions, one of which was critical for the future of the British monarchy. The Act of Settlement, passed by the English Parliament in 1701, barred anyone who was a Catholic, became a Catholic, or married a Catholic from the English and Irish thrones, and settled the royal succession on the German Protestant Electress Sophia of Hanover and her descendants. Although Sophia was a great-granddaughter of the first Stuart King of England, James I, many other members of the Stuart family had a superior claim. However, they were all Catholics. So when Sophia's son, the Elector of Hanover, became King of Great Britain and Ireland in 1714 as George I, it was not because he was the direct, hereditary heir to the throne, but because he fulfilled religious conditions laid down by a Parliament unwilling to countenance another Catholic monarch. 'A Protestant country can never have stable times under a Popish Prince', wrote one bishop in justification of the succession of the new German king, 'any more than a flock of sheep can have quiet when a wolf is their shepherd.'[3] George's heirs, and their consorts, were subject to the same religious conditions. By the time George became King, the Parliament that enacted those conditions included representatives from both England and Scotland. The union of the two countries in 1707 was the last of the major constitutional innovations that resolved the conflicts of the seventeenth century. The kingdoms of the British Isles continued to be ruled as a composite monarchy, but one that was much more effectively integrated than a century before.

These innovations were not, of course, universally accepted. Supporters of the legitimate royal line descended from the militantly Catholic James II—known as Jacobites—contested them for much of the eighteenth century. Exiled in 1688, James II died near Paris in 1701, but in 1714 his exiled son, also a committed Catholic, continued to claim the English throne as James III. The following year, in a response to the Hanoverian succession, he led his supporters in an unsuccessful attempt to seize back the throne from the new German king by means of a military campaign in Scotland and northern England. His son Charles Edward Stuart, later known as Bonnie Prince Charlie, was to attempt the same in 1745, with no greater success.

The presence in Britain of Princesses Caroline, Augusta, and Charlotte was therefore an outcome of the terms on which the British constitutional and religious conflicts of the seventeenth century were resolved. German and Protestant, they offered a marked contrast as British royal consorts from their seventeenth-century predecessors, of whom half had been Catholic, two British, and none German. Yet by insisting on a Protestant monarchy, the British Act of Settlement served to limit the pool of suitable royal consorts. It ruled out diplomatically ambitious, but religiously perilous, marriages like that of Charles I in 1625 to the French Catholic Princess Henrietta Maria. It is striking that Caroline, Augusta, and Charlotte all came from small, relatively insignificant German states. Their crucial qualification was their Protestantism; for a major European power like Britain, their diplomatic advantages as royal spouses were strictly limited.

The contrast between the minor German states where the Princesses were brought up and their new home in eighteenth-century London, the capital of Europe's leading commercial nation, was immense. The British Isles suffered widespread destruction in the civil wars of the mid-seventeenth century, but Britain's recovery was much faster than Germany's after its longer, more devastating Thirty Years' War. Britain's economic revival was assisted by a broader shift in the balance of Europe's economy during the sixteenth and seventeenth centuries away from the Mediterranean towards the north-western seaboard (fig. 2.02). A consequence of the rise of Atlantic commerce and global sea routes, the shift brought stagnation to Venice and Genoa and riches to London and Amsterdam, while bypassing much of Germany.

Eighteenth-century Britain was wealthy, although for some continentals its wealth did not always compensate for its other shortcomings. 'A cold, dark, dull, dirty country, where there is nothing but money', was how the poet James Thomson summed up French attitudes to England in 1732.[4] But it was not just money that excited French resentment. Britain was also newly powerful on the international stage. The Peace of Utrecht in 1713 brought to an end a long period of warfare between

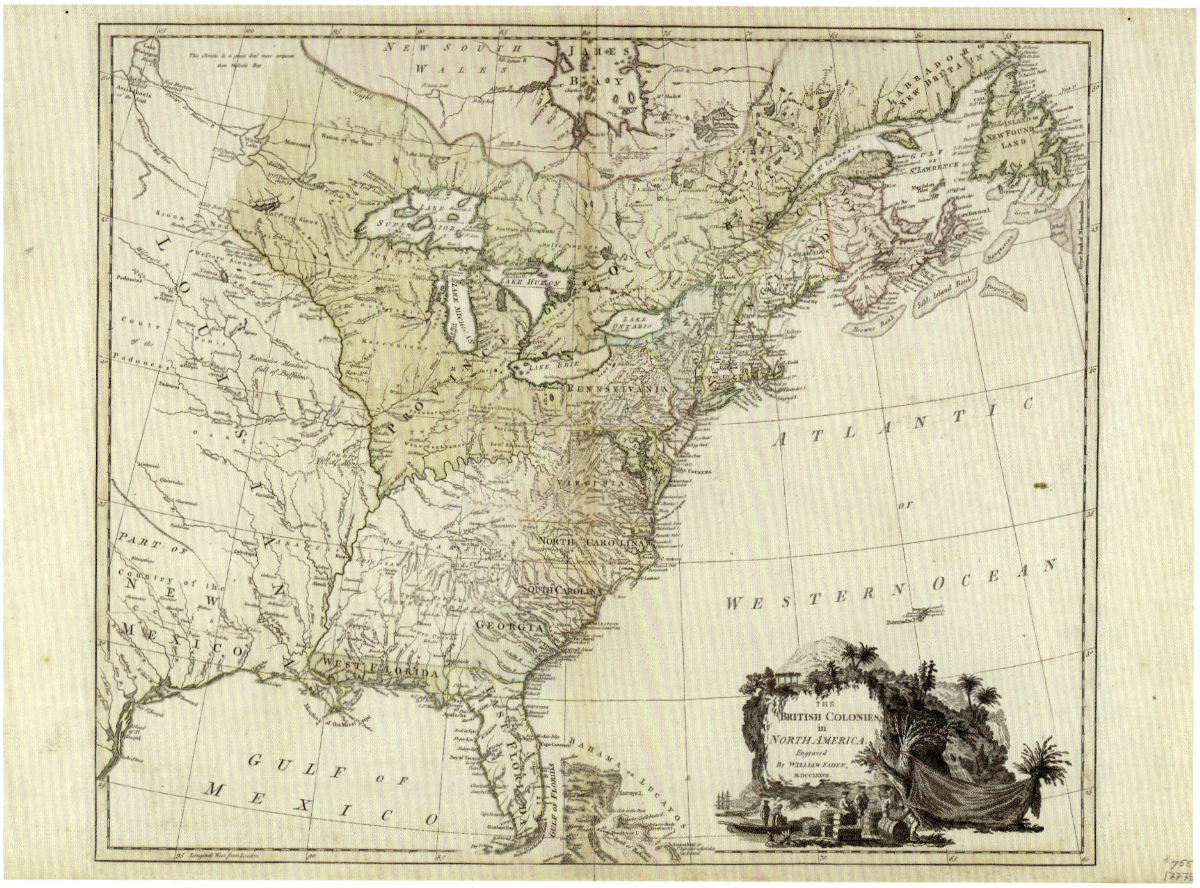

FIG. 2.02 William Faden, *The British Colonies in North America*, 1777, hand-coloured engraving, 20⅞ x 24¾ in. (53 x 63 cm). Beinecke Rare Book & Manuscript Library, Yale University

major European powers dating from 1689. These wars had arisen out of the ambitions of the French king, Louis XIV, for dominance in Europe. Britain emerged from the conflict immensely strengthened. Within the British Isles, the Act of Union between England and Scotland in 1707 brought about the political integration of the British mainland. Protestant dominance in still largely Catholic Ireland had been reinforced in the 1690s. In continental Europe, Britain's leadership of the coalitions that contained France made the country a major military and political power for the first time since the Middle Ages. In the extra-European world, its colonies in North America and the West Indies had been safeguarded and extended, and its trading links with India and China reinforced. Moreover, Britain was on the point of usurping the position of the Dutch as the leading European commercial power. This too prompted French resentment. 'The French were jealous of England', one visitor to Paris reported on his return in 1727, 'they said that we were for engrossing the whole commerce of the world.'[5]

Britain's self-image was that of a trading nation. Throughout the eighteenth century, no other major European economy was more heavily committed to international trade than the British. 'Our trade', announced Lord Carteret in 1739, 'is our chief support.'[6] Indeed, so dedicated were Britons to the cult of trade that the historian Thomas Mortimer could describe commerce in 1762 as 'the great idol of this nation, and to which she sacrifices every other consideration'.[7] The common seaman, notoriously truculent and uncouth, was fêted as the epitome of British courage and the mainstay of Britain's prosperity. The Royal Navy, which secured the trade routes, was the focus of national pride. Victorious admirals like Edward Vernon, George Brydges Rodney, first Baron Rodney, and Horatio, first Viscount Nelson were national heroes.

Trade followed the flag. Britain was at war for two-fifths of the period from 1714 to 1818, and was largely victorious, with the notable exception of the American War of Independence (1775–83). In all these wars, the French were her principal enemies, while trade and colonies figured prominently among the causes of conflict. In the eighteenth century, colonies offered unrivalled opportunities for the expansion of trade. They might serve as exclusive markets for manufactured exports, like Britain's North American colonies, with their fast-growing populations of European and African descent. They might serve as sources of exotic manufactures, like the muslins brought from Bengal by the English East India Company. They might serve as suppliers of industrial raw materials, like the indigo grown on the slave plantations of South Carolina to dye the textiles manufactured in Yorkshire and Lancashire, Norfolk and Somerset. Or they might serve as suppliers of tropical foodstuffs, like the British West Indian sugar islands of Barbados and Jamaica. The British colonies in the Americas, moreover, provided a seemingly endless stream of previously unknown plants and animals, to the delight and fascination of scientists and collectors.

It was the West Indies that were the jewels in the imperial crown, the object of the most intense colonial rivalry among the European powers. At the end of the eighteenth century, three-quarters of a million slaves from Africa toiled on sugar plantations in the British Caribbean.[8] They laboured to satisfy the insatiable appetite for sweetness among British consumers, and to generate great fortunes for their owners. Britain was the leading European slave-trading nation in the eighteenth century, followed closely by Portugal. It was only in 1807 that the British outlawed the slave trade, and later still, in 1833, that slavery itself was abolished in British territories. Meanwhile, thirteen North American colonies gained their independence in 1783 to form the United States, but thereafter Britain went on to conquer new colonial territories—not only in India, where the East India Company amassed a vast territorial empire after 1760, but also in Africa and Australia. By 1818, Britain exercised authority over perhaps a quarter of the world's population, some 200 million men and women.

The intensely commercial tenor of eighteenth-century British life extended to industry and agriculture. In retrospect, our view of eighteenth-century British manufacturing has been dominated by new steam engines and textile machinery. Brought together within the forbidding walls of the factory towards the end of the eighteenth century, they have come to define the term 'Industrial Revolution' for subsequent generations. These inventions made a critical contribution to transforming Britain into the leading industrial power in Europe. However, it is important to emphasize that a process of improvement and innovation in manufacturing was already well advanced by the mid-eighteenth century. A passion for invention extended to a far wider range of industries than those conventionally associated with the Industrial Revolution. Foreigners saw it as distinctively English. The Frenchman Jean-Bernard Le Blanc pointed out in 1747 that 'England has more than any other country of those machines so useful to the state, which really multiply men by lessening their work.'[9] A nation that at the start of the seventeenth century had exported mainly raw materials and semi-finished goods, was already in 1760, on the eve of the Industrial Revolution, one whose exports consisted principally of high-quality, finished manufactures—West Country woollen broadcloths, Spitalfields silks, Lancashire cottons, Staffordshire pots, Bristol glasswares, and a huge diversity of small metal goods, from humble nails to fine silverware. As the range and variety of manufactured products grew, imports came to consist less and less of manufactured goods and more and more of raw materials. The hunger for technical innovation extended to agriculture. Farmers and landowners experimented with crops, bred more productive animals, and tilled the soil in new ways. As a consequence, they succeeded in feeding the population of Great Britain without much recourse to food imports, even though the number of people more than doubled, from 6.5 million in 1700 to 14 million in 1820.

The single most spectacular manifestation of Britain's eighteenth-century commercial success was the size and dynamism of London (fig. 2.03). The city performed a triple role as the country's capital, its leading port, and its principal industrial centre. From London, goods were distributed across Great Britain, unhindered by the multiple customs barriers that impeded trade between the hundreds of German states. The country formed one large uninterrupted market, ringed by a fence of customs duties, designed to defend British manufactures against foreign competition. London was the arbiter of fashionable taste, the seat of the arts, and the crossroads of scientific and intellectual enquiry. In 1700 it was already the biggest city in western Europe, with over half a million people. No city in the German lands came close. Vienna, with just over 100,000 people, was the largest. A hundred years later, in 1800, London's population had swelled to almost a million. By then Vienna had 247,000 people, followed by Berlin, with 172,000.

By comparison with Britain, much of eighteenth-century Germany was commercially backward. When Sir Richard Arkwright, the British entrepreneur who invented the water-powered cotton spinning machine, died in 1792, his obituary in the *Gentleman's Magazine* noted he was 'immensely rich, and has left manufactories the income of which is greater than that of most German principalities, though derived from very opposite circumstance, those that promote the prosperity of a country'.[10] The British may have regarded the economic policies of German states as misguided, but German rulers did endeavour to develop the economies of their territories. Conscious that German manufactured goods were often inferior to those produced in the Netherlands, Britain, or France, they used subsidies and tax concessions, monopolies and state-run enterprises to encourage industries that replaced foreign imports—most famously luxury products such as porcelain, silks, tapestries, and mirrors. Yet when such state-dependent, mercantilist, beggar-my-neighbour policies were implemented

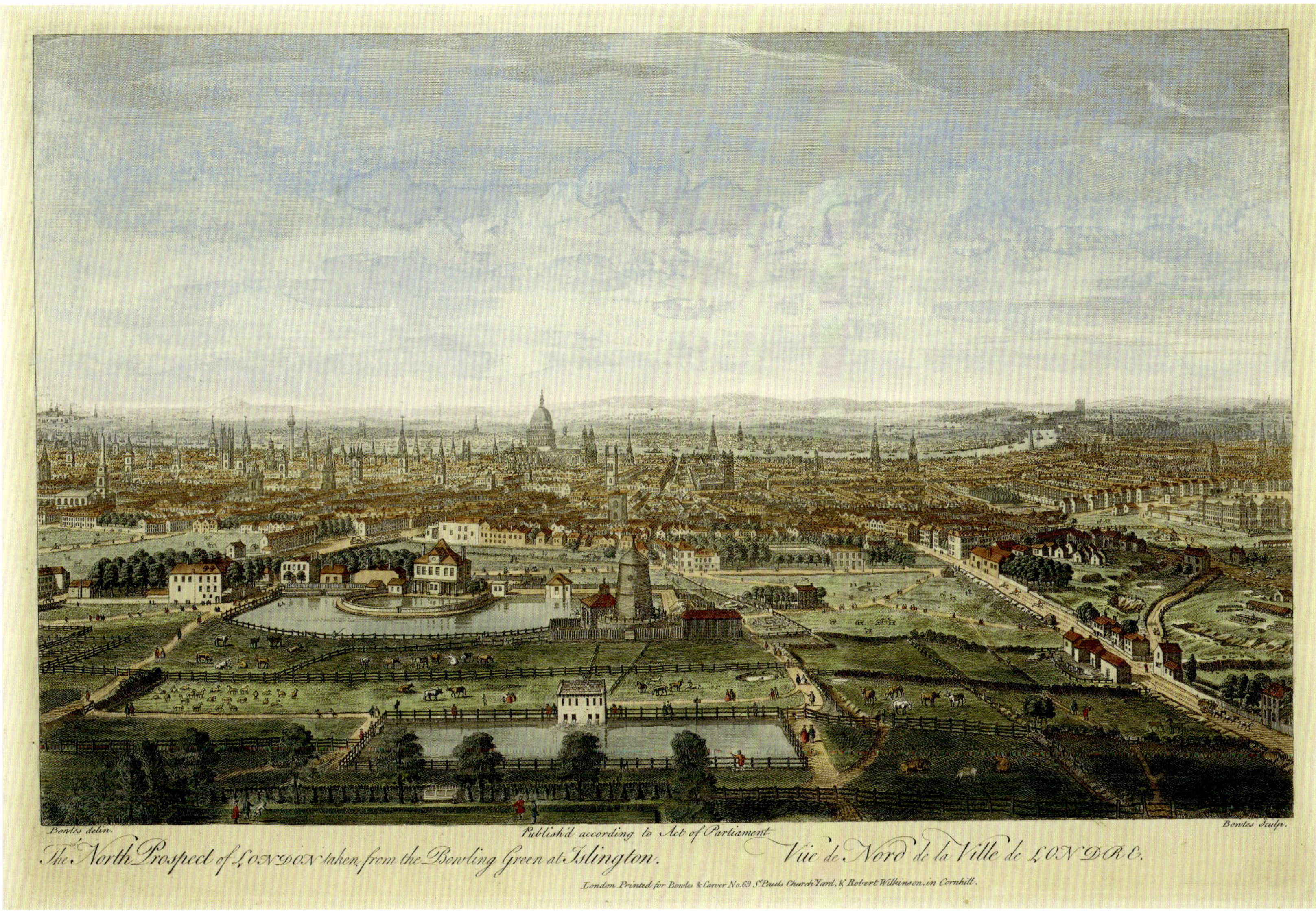

FIG. 2.03 Thomas Bowles, *The North Prospect of London Taken from the Bowling Green at Islington*, 1752, hand-coloured engraving, 10¼ x 17⅛ in. (26 x 43.5 cm). Folger Shakespeare Library, Washington, DC

by numerous small, independent states across Germany, they could be an impediment to commerce and to the growth of the German economy as a whole. Many German rulers saw little benefit in importing any foreign merchandise except as a source of customs revenue, charging duties even on goods that passed through their territories on the way to another destination.

Unlike Britain, the German states had little or no direct access to the exotic new goods from Asia and the Americas that flooded into Europe in the seventeenth and eighteenth centuries—chintzes from India, tea from China, tobacco from Virginia, sugar from Brazil or the West Indies. Imports of these goods through Hamburg and Bremen came via London, Amsterdam, or Nantes. Brandenburg-Prussia's attempts to enter the Atlantic slave trade and the Asian trade in tea and textiles ended in failure. In 1683 it established a settlement on the west coast of Africa, in modern-day Ghana, as a base for trading in gold and slaves, administered by an African Company based in Königsberg on the Baltic coast; but it was sold in 1721. Similarly, a Royal Prussian Asiatic Company, founded at Emden on the North Sea coast in 1750, was abandoned after only fifteen years.[11] Germany's main involvement in the Atlantic slave trade was as a supplier of linen textiles to the slave-trading nations, partly to exchange for slaves on the African coast, and partly to clothe plantation slaves in the Americas.

It was ironic, then, that many of the Germans who made these slave-trade textiles were themselves in a state of bondage.[12] In Germany east of the Elbe, a significant proportion of the rural population consisted of serfs, living in a form of hereditary, personal subjection to an overlord. Serfs were obliged to perform unpaid work for their lord and could be moved from one estate to another. Some lawyers even defined serfs and their children as the overlord's property, although, in contrast to slaves in the British Americas, their status was never racially defined, they were entitled to make contracts, and they could buy their freedom. Serfdom was especially prevalent in the duchy of Mecklenburg-Strelitz, the home of Princess Charlotte and economically one of the least developed of the German states. Most of the duchy's rural population was enserfed. In nearby Brandenburg-Prussia, where Princess Caroline spent part of her youth, serfdom was also widespread early in the eighteenth century.[13] Serfdom in these parts of eastern Germany was associated with huge rural estates and low crop yields, but throughout Germany agricultural output and agricultural techniques lagged behind those in Britain.

Of course, neither German princesses nor British royal consorts spent much time in the company of those who toiled in the fields, irrespective of whether they were serfs on the plains of northeastern Germany or lowly waged labourers in southeastern England. But contrasts between Germany and Britain also extended to life in the princely and royal courts where they resided.

The court was, in effect, the household of the monarch. It was the principal setting in which royal authority was performed. In monarchical systems organized around ruling dynasties, it was through dynastic households that royal power was exercised or delegated, royal patronage dispensed or refused, royal magnificence projected or withheld. In the eighteenth century, governing was not simply a matter of bureaucrats and institutions, ministers and parliaments, law books and tax collectors. It required the governing classes to cultivate personal relationships with their rulers and with their rulers' families. The key arena where this took place was the court. Attendance at court offered personal access to the monarch, opportunities to serve, and the possibility of patronage. Across eighteenth-century Europe, royal courts were crucial to the exercise of state power.[14]

The seventeenth and eighteenth centuries saw an elaboration of court life across Europe, driven by competition for prestige among its ruling dynasties. Royal courts increased in size, became less mobile than in earlier centuries, and were housed in ever-grander palaces. Leading these trends was the court of Louis XIV of France, the largest in Europe with some two thousand official posts. From 1682 it was housed in his new palace at Versailles, built in stages between 1661 and 1710 (fig. 2.04). Its location, twelve miles from Paris, provided an exclusivity unattainable at its predecessor in the heart of the city, the Louvre Palace. Huge, magnificent, and ruinously expensive, Versailles exercised an extraordinary hold on expectations among the crowned heads

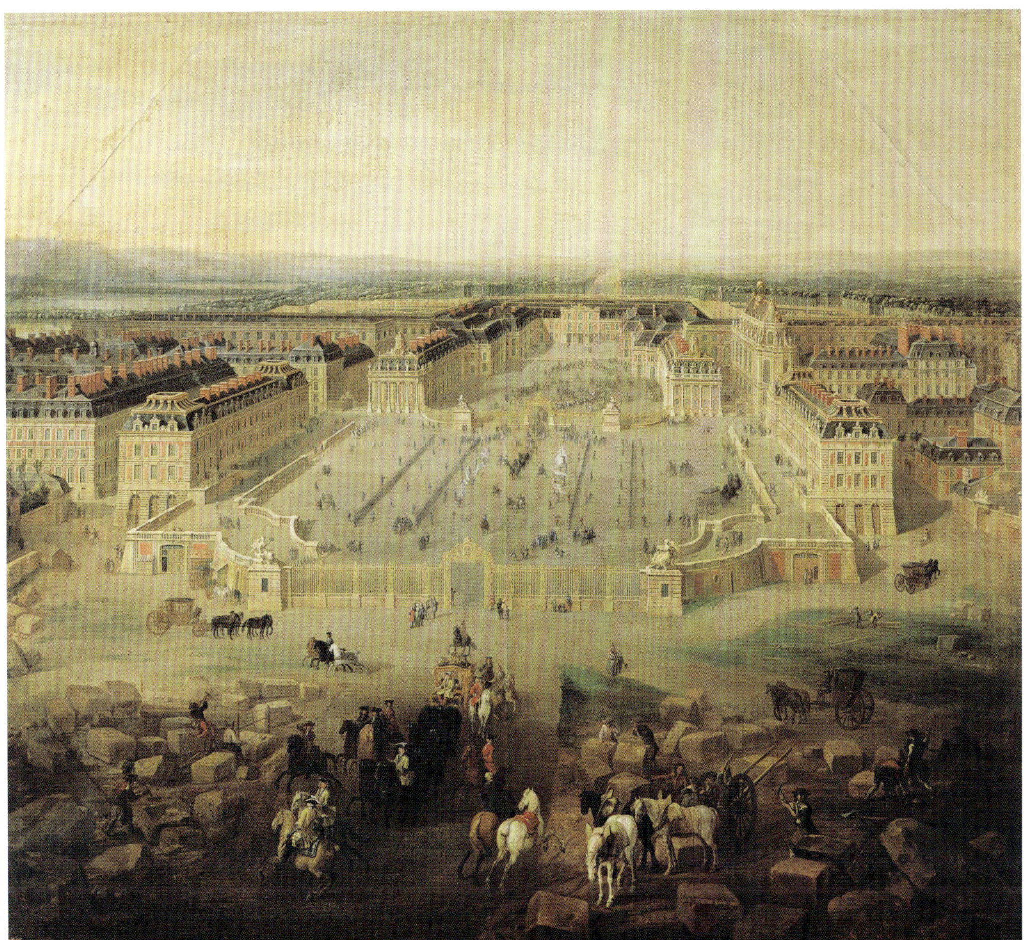

FIG. 2.04 Pierre-Denis Martin, *Le Château de Versailles as Seen from the Place d'Armes*, 1722, oil on canvas, 54¾ x 59 in. (139 x 150 cm). Château de Versailles, France

of Europe as to how a court should function. This was especially so in Germany. Louis XIV was famously an absolute monarch, who ruled without the restraint of an elected body representing his people. Most German rulers did not enjoy such untrammelled authority as Louis XIV, but their capacity to exercise personal power was substantial. Once peace returned after the Thirty Years' War, they competed with each other for prestige by expanding the personnel at their courts and building new palaces to house them. Indeed, the century after 1648 witnessed an astonishing surge of princely architectural endeavour. Most German rulers constructed entirely new palaces, or substantially rebuilt existing ones. Many copied features of Versailles on a lesser scale, at great cost to their states, which were smaller and poorer than France.

Hence, during the middle decades of the eighteenth century, Princess Charlotte's uncle and brother, who were successively dukes of Mecklenberg-Strelitz, oversaw the building of a palace on the site of a former royal hunting lodge at Neustrelitz (fig. 2.05). Despite their dukedom's poverty, it was accompanied, as at Versailles, by a new town laid out on a radial plan, which became their capital. At Gotha in central Germany, Duke Ernst I, Princess Augusta's great-grandfather, built a vast and architecturally pathbreaking Baroque palace, Schloss Friedenstein, at the end of the Thirty Years' War. Not content with one palace, his grandson, Augusta's father, constructed another, Schloss Friedrichsthal, next to it. Built in a French style between 1708 and 1711, it was partly financed by renting out the dukedom's troops as mercenaries. Princess Caroline had a peripatetic childhood, but as a young woman she lived with her guardians, the Elector and Electress of Brandenburg, mainly at Electress Sophia Charlotte's Lietzenburg palace outside Berlin. Lietzenburg, renamed Charlottenburg after Sophia Charlotte died in 1705, was yet another entirely new palace, built between 1695 and 1699. All these palaces were accompanied by extensive Baroque gardens laid out in the French manner.

The influence of France on German courts was not restricted to architecture and design. Its most striking manifestation was the use of French as the customary language at many German courts. It went hand-in-hand with the adoption of elements of French court ritual, although sometimes in what the English regarded as a distinctively formal German manner. Queen Charlotte's English biographer recorded that at the Neustrelitz palace in Mecklenberg-Strelitz, where she was brought up, 'much outward ceremony was observed, and splendour kept up, agreeable to the regimen of the German courts'.[15] As young women in Germany, Princesses Charlotte and Caroline learnt French, not English, although Princess Augusta learnt neither.

Frenchified ceremonial and splendour may have been common in the newly built Baroque palaces that housed eighteenth-century German courts, but their personnel remained relatively small in number. The royal court in London was larger, as befitted one of the richest monarchies in Europe, with an establishment numbering about a thousand for most of the eighteenth century. In the German lands, only the court at Vienna exceeded it in size, but neither rivalled the court at Versailles. Despite its size, expenditure on the British court was less than in the leading continental monarchies. The most obvious consequence was an absence of new, purpose-built royal palaces.

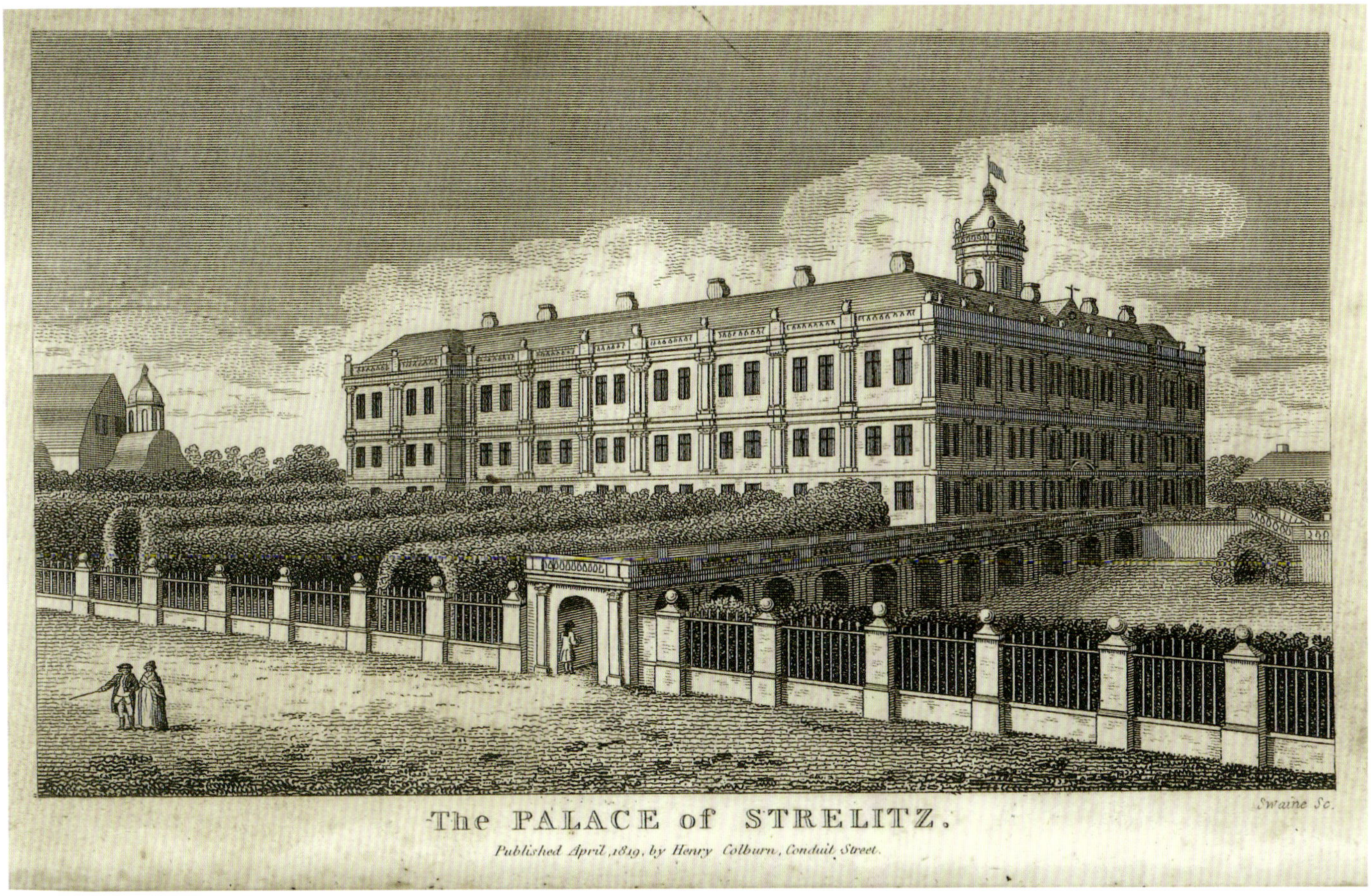

FIG. 2.05 John Watkins, 'The Palace of Strelitz', engraving, in *Memoirs of Her Most Excellent Majesty Sophia-Charlotte, Queen of Great Britain from Authentic Documents* (London: Printed for Henry Colburn, Public Library, Conduit-Street, Hanover-Square, 1819). Yale University Library

In the seventeenth century, the Stuart kings had remodelled their enormous Whitehall Palace in the heart of London. Most of it was then destroyed by two fires in the 1690s. It was never rebuilt. Subsequently St James's Palace became the principal royal residence in London, the venue for formal court occasions (fig. 2.06). Smaller than Whitehall Palace, it was nearly two hundred years old when George I ascended to the throne, and retained many of its Tudor features. In 1689 King William and Queen Mary bought what became Kensington Palace in the London suburbs and extended it over the next decade, but it remained relatively small and never replaced St James's as the official seat of the court. In the same decade, they also began rebuilding Hampton Court Palace, twelve miles south-west of London, but the work was left half-completed. Thereafter no new royal palaces were built or comprehensively rebuilt until the nineteenth century, although interiors were refurbished and updated. In addition, a number of buildings were purchased to turn into satellite royal residences, notably Buckingham House, bought in 1762 by George III as a private family residence for Queen Charlotte. Modest in scale, it was only in the nineteenth century that it was enlarged to become Buckingham Palace.

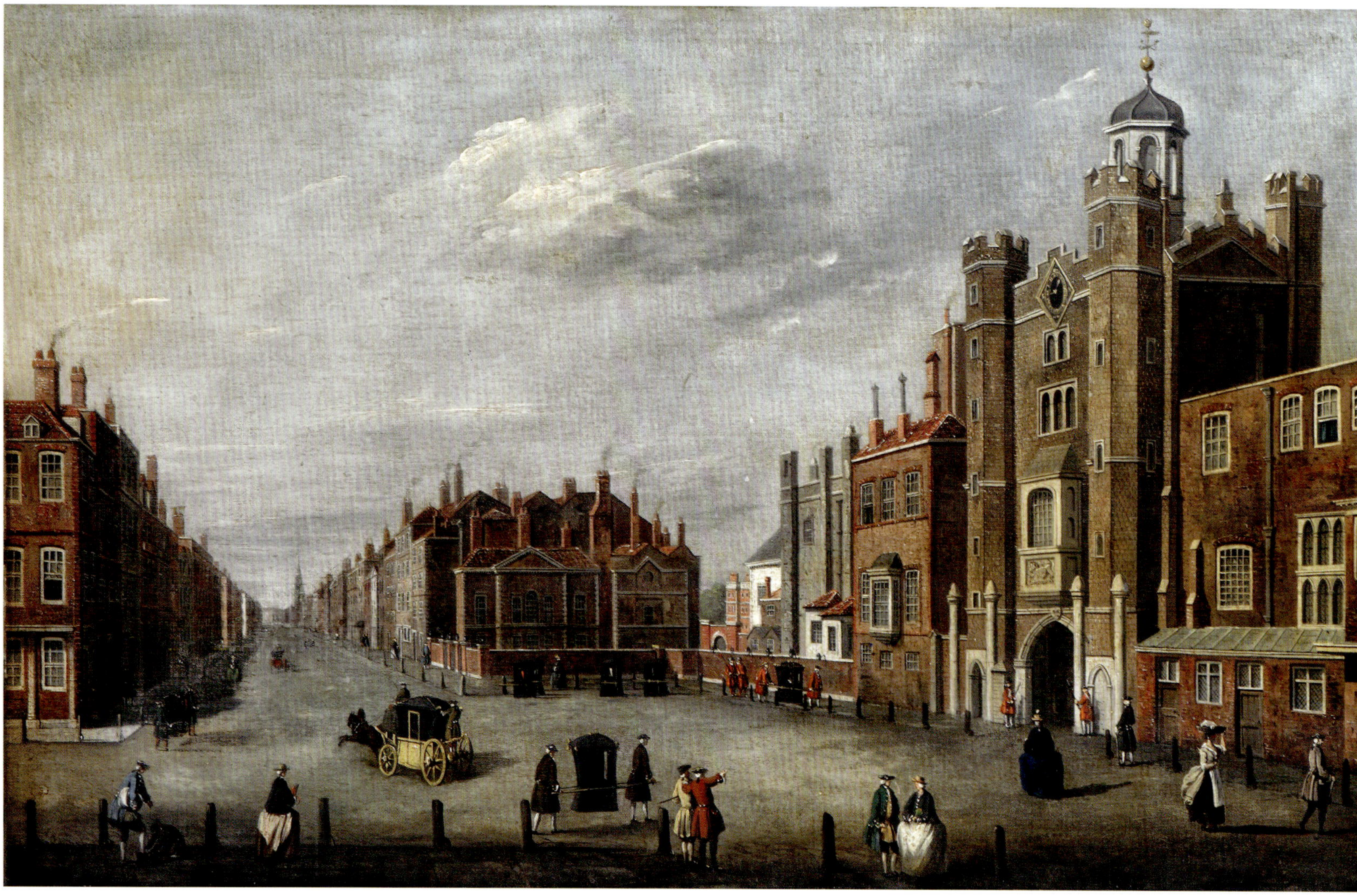

FIG. 2.06 William James, *St James's Palace with a View of Pall Mall*, 1700–53, oil on canvas, 21½ x 35½ in. (54.6 x 90.2 cm). National Trust

The absence of palace building did not mean the Hanoverian court in London lacked splendour and ostentation. César de Saussure, a Swiss visitor in 1725, considered that St James's Palace 'does not give you the impression from outside of being the residence of a great king', but he was impressed by its interior furnishings—velvet upholstery, beautiful tapestries, excellent paintings, and silver chandeliers (fig. 2.07).[16] Daniel Defoe, writing at the same time, agreed that 'the King's palace, tho' the receptacle of all the pomp and glory of Great Britain, is really mean', but he extolled

> the rich furniture within, I mean the living furniture, the glorious court of the King of Great Britain: the splendour of the nobility, the wealth and greatness of the attendants, the oeconomy of the house, and the real grandeur of the whole royal family, out-does all the courts of Europe, even that of France itself, as it is now managed since the death of Lewis the Great [Louis XIV].[17]

Defoe's eulogy to the glories of the British court under George I involved a degree of patriotic exaggeration, but formal court events at St James's did dazzle with a matchless profusion of silks and diamonds, testimony to the wealth of the English nobility.[18] If the court's setting in the heart of the metropolis was modestly archaic on the outside, its public rooms were furnished in a stately, opulent, and up-to-date manner. Nevertheless, court etiquette in London was less elaborate and punctilious than at Versailles. On their arrival in England, none of the three German Princesses were obliged to undergo the kind of ritual stripping and reclothing *à la française* that the young Marie-Antoinette endured when she arrived at the French border from Vienna in 1770.[19] Moreover, the Hanoverian monarchs cultivated a reputation for frugality, as contemporary satires suggest (figs. 2.08, 2.09). They dispensed with some of the more flamboyant aspects of Baroque kingship, such as a court theatre, relying instead on London's commercial suppliers of entertainment.[20] As we have seen, they also avoided major building projects. Appropriately for a country that was France's principal European rival, the standard language of the court in London was English. Compared to many German royal courts, at St James's the influence of France was muted.

As in all the royal courts of Europe, attendance at the British court was dominated by the nobility, but it was a nobility different from those of the German states and most other European countries. In the British Isles there were very few nobles. The British peerage created by the Act of Union with Scotland in 1707 numbered less than two hundred in the mid-eighteenth century. The separate Scottish and Irish nobilities added a few hundred more. They were a mere handful compared with the fifty thousand noble families in late eighteenth-century Germany. In Germany, where nobles were numerous, it was not unusual for them to be poor. In England, where nobles were few, almost all were fabulously wealthy, and becoming more so. The English nobility benefited not only from growing agricultural profitability after 1760, but also indirectly, through their ownership of urban and industrial property due to the expansion of trade and manufacturing. It was fortunate they were so wealthy, because they enjoyed fewer legally enshrined privileges than almost any other nobility in

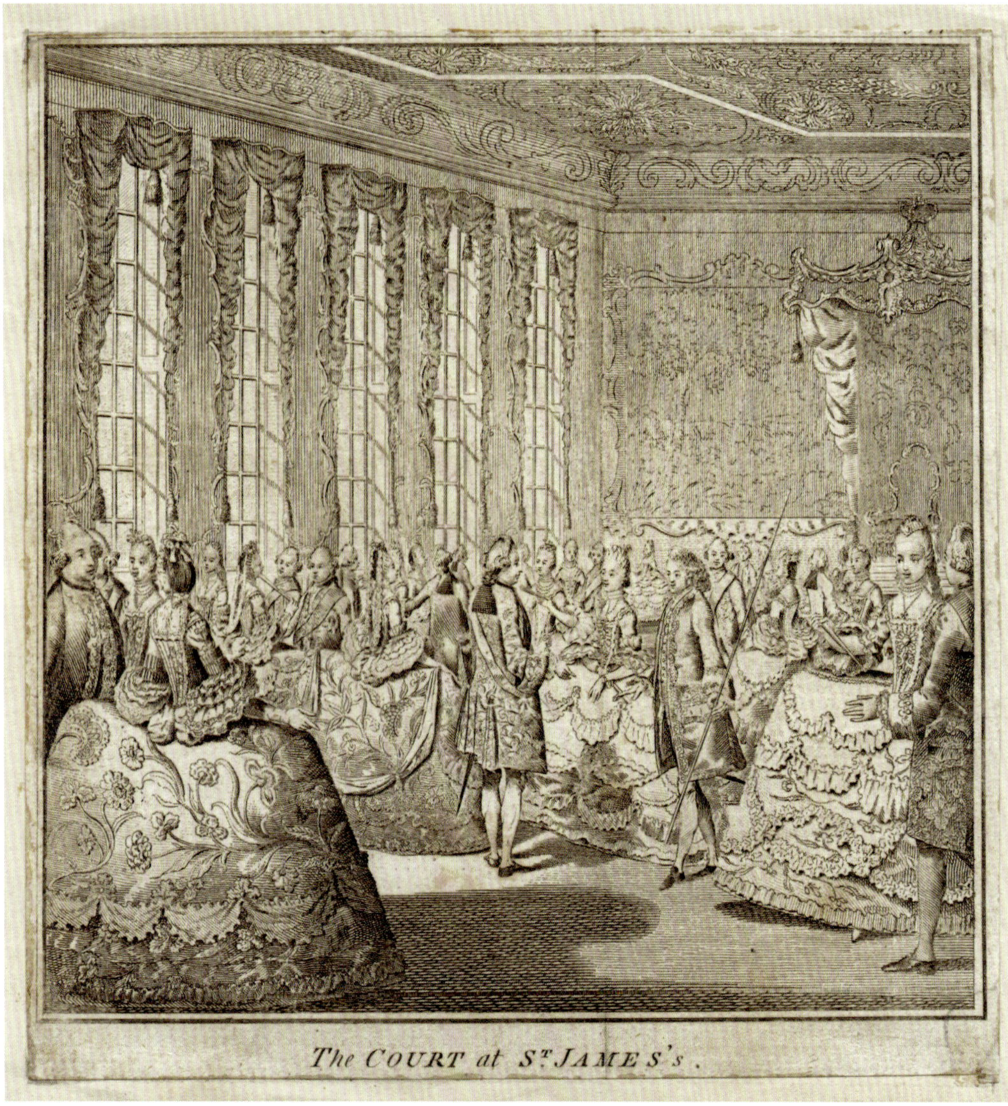

FIG. 2.07 Unknown engraver after unknown artist, *The Court at St James's*, 1766, etching with engraving on paper, 6¾ x 6¼ in. (17 x 16 cm). The Lewis Walpole Library, Yale University

Europe, apart from their politically crucial membership of the upper house of Parliament, the House of Lords.

In Germany, by contrast, nobles enjoyed a wide range of privileges denied to commoners, such as freedom from direct taxation, rights to own certain types of land, and, in some states, the exclusive right to wear expensive, fashionable fabrics. The precise terms of noble privilege varied from territory to territory, but the assignment privileges restricted to nobles were almost universal. Germany was a society of orders, in which people were ranked in a caste-like hierarchy of prestige, differentiated by legal and social regulations. Nobles stood at the pinnacle of this hierarchy and insisted on their separation from its lesser ranks. At many larger German courts, access to formal events was confined to nobles, although noble status was enjoyed by all their children, unlike England where male primogeniture applied to noble titles. At the theatre, nobles and non-nobles were often obliged to sit separately. In the Burgtheater in Vienna, a wooden barrier separated commoners from nobles throughout the eighteenth century.[21] In some territories nobles were excluded from trade by law, although nobles increasingly monopolized the senior ranks of their rulers' administrative bureaucracies. Marriages between noblemen and non-noble women were rare.

The royal court and elite culture the German Princesses encountered in Britain functioned differently, although the court remained an integral part of elite life in London. As in Germany, the court was dominated by the nobility, but the nobility's small numbers meant that those who attended court events included many members of the political and social elite who were persons of quality—often landowners—but not noble. Beyond the court, London's entertainment venues—coffee houses and pleasure gardens, theatres and concert halls—erected few formal barriers between nobles and wealthy commoners. Distinctions within such venues, and between them, turned principally on the ability to pay, not on whether a person bore a title. In these socially mixed venues, the different ranks had no choice but to meet each other, even if they did not necessarily mingle. Such encounters were regulated by the informal code of manners known as politeness. Politeness did not deny the legitimacy of social rank, but it required the different ranks to engage with each other in an agreeable manner. Hence the rule made in the early eighteenth century by Beau Nash, the

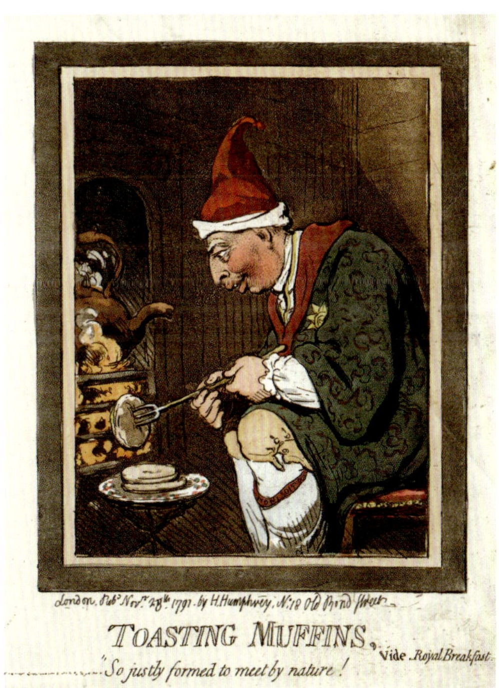

FIG. 2.08 James Gillray, *Toasting Muffins*, 1791, hand-coloured etching and aquatint on paper, 7½ x 5¾ in. (19 x 14.7 cm). The Lewis Walpole Library, Yale University

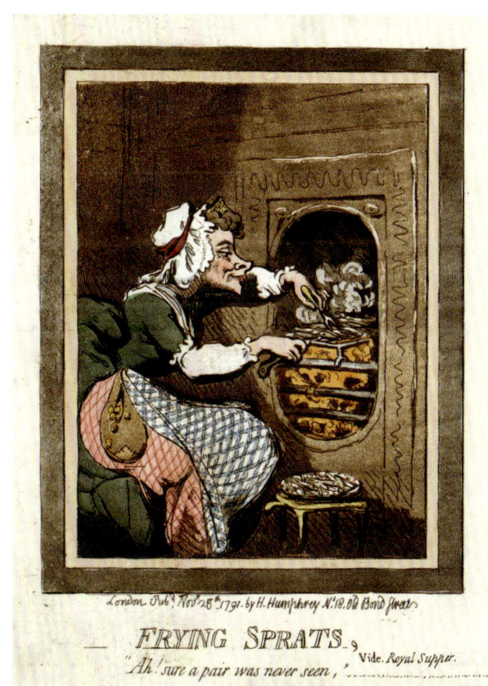

FIG. 2.09 James Gillray, *Frying Sprats*, 1791, hand-coloured etching and aquatint on paper, 7½ x 5¾ in (19 x 14.8 cm), on sheet 9½ x 13¾ in. (24 x 35 cm). The Lewis Walpole Library, Yale University

Master of Ceremonies at the resort city of Bath, that nobles should treat lesser visitors with the same courtesy they usually reserved for each other. 'That ease and open access', wrote the poet Oliver Goldsmith of Nash's rule, 'our gentry brought back to the metropolis, and thus the whole kingdom by degrees became more refined by lessons originally derived from him.'[22]

In most German states, the royal court was the single most important focus of political authority, but this was not the case in Britain. The 'Glorious Revolution' of 1688 established a mixed monarchical constitution, with a Parliament that represented at least some of the people and met regularly for several months each year. Both court and Parliament contributed to the exercise of power under this constitutional arrangement—ministers had to enjoy the confidence of the monarch and the legislature—but the terms on which the Hanoverian dynasty came to the throne meant Parliament had the upper hand. Parliament was a very public forum for the trenchant expression of political opinions opposed to those of the king and his ministers. Opposing views, in Parliament and beyond, were widely reported in a relatively free press, which was also prepared to critique the behaviour of the royal family. The members of that royal family had, therefore, to cultivate a public image in ways that were unknown in Germany, where the extensive newspaper press was wary of reporting domestic politics, at least within the boundaries of each state.[23] For the first half-century of Hanoverian rule, moreover, an additional incentive to cultivate the dynasty's image arose from the very real threat of a Jacobite restoration.

Freedom of the press and open public debate are at the heart of the way historians have come to understand the meaning of Enlightenment in eighteenth-century Britain. Traditionally the Enlightenment was treated as a predominantly French phenomenon, a philosophical movement that employed reason to critique the irrationalities of France's absolutist 'ancien régime' in the century before the French Revolution—especially its lack of freedom of speech and its intolerant enforcement of superstitious religious belief. Viewed in those terms, Britain did not need an Enlightenment, because it already enjoyed free public debate and religious toleration. Recent scholarship has taken a broader, European-wide view of the Enlightenment.[24] Now the term is used to describe the new ways of understanding and improving people's worldly lives that emerged in the eighteenth century, encouraged by corresponding advances in the understanding of nature made by thinkers such as Francis Bacon, René Descartes, and Isaac Newton. The British contribution was enormous, not merely because writers like John Locke, David Hume, and Adam Smith were read across Europe, but because Britain provided a model of how enlightened public engagement with these issues might be conducted, whether it was in popular debating clubs and reports in newspapers, or in meetings of scientific societies and erudite articles in learned journals. Nevertheless, royal influence and intervention remained important for the Enlightenment in Britain. The monarchy undertook a variety of initiatives which were enlightened in the broader sense, from the study of natural history to the promotion of inoculation.

The Enlightenment was different in the German-speaking lands than in Paris and London. Germany's contribution to Enlightenment thought was provided largely by officials—professors and bureaucrats—rather than the professional writers, wealthy intellectuals, and scholars with powerful patrons so prominent elsewhere. The governments of the many German states required numerous administrators and clergymen. To supply them, eighteenth-century Germany hosted some forty universities, dominated by faculties of theology, medicine, and law. Great Britain, by contrast, had only seven universities, five in Scotland and two in England. Not surprisingly, then, German Enlightenment thinking was often influenced by religious imperatives and focused on the question of how improvements in the administration of a state could increase the prosperity and well-being of its people. In the course of the eighteenth century this pragmatic approach developed into a specialist body of knowledge—a science of administration—that became the content of dedicated university courses, training future bureaucrats. The distrust of state and church found among Enlightenment thinkers elsewhere was notable by its absence.[25]

In Protestant Germany, the emergence of this science of administration was accompanied by the spread of the Pietist religious movement. It encouraged an intensely personal religious observance combined with the living of an active Christian life in the world. The fruits of its practical Christianity—schools, orphanages, dispensaries, work schemes for the poor—complemented the emphasis on social improvement developing in German Enlightenment thought at the same period. Pietism did not go unchallenged, however. It was not always welcomed by Protestants of different theological persuasions, or rulers unnerved by its criticisms of the vanities of courtly splendour.[26]

These various tendencies in the German Enlightenment came together in Brandenburg-Prussia at the start of the eighteenth century. This was the moment when the young Princess Caroline was living in the brand new Lietzenburg palace outside Berlin, where her guardian, Sophia Charlotte, Electress of Brandenburg, held court. The separate court her husband, the Elector Frederick III, maintained at Berlin imitated Louis XIV's Versailles in pomp, ceremony, and elegance. Yet Frederick also established a new university at Halle, a hundred miles south-west of Berlin, which was to become one of Germany's largest, and appointed to a professorship August Hermann Francke, the leading light of German Pietism. Frederick went on to champion the internationally influential orphanage Francke established at Halle, despite hostility from some of his courtiers and church officials. Meanwhile, his wife, the Electress, had a theatre built at the Lietzenburg palace to host Italian opera, supported new Academies of Arts and Sciences, and surrounded herself with philosophers and theologians, most notably Gottfried Wilhelm Leibniz, the outstanding figure in early German Enlightenment philosophy, whose work ranged across mathematics, the sciences, law, and theology. In this combination of Frenchified courtly elegance, enlightened learning, and pious good works, Brandenburg-Prussia at the start of the eighteenth century exemplified the German Enlightenment, but the same elements were reproduced to some degree in courts, large and small, across Protestant Germany.

Caroline of Ansbach, Augusta of Saxe-Gotha, and Charlotte of Mecklenburg-Strelitz were all brought up in this kind of enlightened German Protestant milieu.

It did not guarantee they would succeed in their role as British royal consorts. The marriage to the future George IV of their immediate successor, another German princess, Caroline of Brunswick, was to be a disaster, although her failure owed more to her husband's many shortcomings than to her own inadequacies. Britain, unlike their own tiny German states, was a major European and imperial power, wealthier and more socially diverse than any of the German territories, and without their caste-like social hierarchies. The role of the monarchy in Britain's mixed constitution was very different from its German equivalents; less powerful, constitutionally constrained by Parliament, and subject to an entirely unfamiliar degree of public scrutiny.

Yet these women's German upbringing endowed them with certain advantages as they began their new lives in Britain. German princesses were brought up in the expectation that dynastic marriages would be arranged for them. The fact that the three Princesses were both German and Protestant smoothed their entry into a dynasty that shared both those characteristics, although not without some hitches along the way. Augusta arrived in London knowing hardly a word of English, her mother having wrongly assumed that the court of a Hanoverian dynasty would be fluent in German, and as a Lutheran she was at first reluctant to receive Communion in the Church of England. An upbringing in the courts of Germany meant Caroline, Augusta, and Charlotte were genuinely enlightened princesses, well-prepared for London's cosmopolitan culture of art and architecture, science and collecting. Schooled in German Enlightenment piety, they were also well-equipped for the British Hanoverian emphasis on frugality, philanthropy, and patriotic service.

Britain, in its turn, offered Caroline, Augusta, and Charlotte one benefit entirely missing in their native Germany—a history of admired female rulers who reigned in their own right. In most of the monarchies of continental Europe, the Salic Law denied women the right to sit on the throne. In England, the Salic Law did not apply. Reigning queens had included Anne, immediately prior to the accession of George I; Mary II, who had ruled jointly with William of Orange; and most famously Elizabeth I, the last of the Tudors, revered as the quintessential Protestant monarch. Consequently queenship in Britain enjoyed a status it lacked in most of continental Europe. As one of Princess Caroline's German royal relatives, married to the younger brother of Louis XIV of France, reminded her, 'in England women can reign, but France has to be governed by men.'[27]

NOTES

The author would like to thank Isa Fleischmann-Heck, Hannah Greig, Joanna Marschner, Charlotte Mitchell, Luca Mola, Giorgio Riello, Michael Schaich and Klaus Weber for references and suggestions.

1. Ulrich Pfister and Georg Fertig, 'The Population History of Germany: Research Strategy and Preliminary Results', *MPIDR Working Paper WP 2010–035* (Rostock: MPIDR, 2010), 5.
2. Peter H. Wilson, *The Holy Roman Empire, 1495–1806* (Basingstoke: Palgrave Macmillan, 2011), 54–55.
3. Quoted in Linda Colley, *Britons: Forging the Nation, 1707–1837* (New Haven and London: Yale Univ. Press, 2005), 47.
4. Alan D. McKillop, *James Thomson (1700–1748): Letters and Documents* (Lawrence, KS: Univ. of Kansas Press, 1958), 83.

5. Quoted in Jeremy Black, *Natural and Necessary Enemies: Anglo-French Relations in the Eighteenth Century* (London: Duckworth, 1986), 140.
6. Quoted in Basil Williams, *Carteret and Newcastle: A Contrast in Contemporaries* (Cambridge: Cambridge Univ. Press, 1943), 119.
7. Thomas Mortimer, *A New History of England: From the Earliest Accounts of Britain to the Ratification of the Peace of Versailles, 1763*, 3 vols. (1766): 322.
8. B. W. Higman, 'Population and Labor in the British Caribbean in the Early Nineteenth Century', in *Long-Term Factors in American Economic Growth*, ed. Stanley L. Engerman and Robert E. Gallman (Chicago: Univ. of Chicago Press, 1986), 605.
9. Jean-Bernard Le Blanc, *Letters on the English and French Nations*, 2 vols. (London, 1747), 1:48.
10. *The Gentleman's Magazine* 62 (1792): 770.
11. John Gagliardo, *Germany under the Old Regime, 1600–1790* (Harlow: Longman, 1991), 145.
12. Anka Steffen and Klaus Weber, 'Spinning and Weaving for the Slave Trade: Proto-Industry in Eighteenth-Century Silesia', in *Slavery Hinterland: Transatlantic Slavery and Continental Europe, 1680–1850*, ed. Felix Brahm and Eve Rosenhaft (Woodbridge: Boydell & Brewer, 2016), 87–107.
13. Dieter K. Buse, *The Regions of Germany: A Reference Guide to History and Culture* (Westport, CT: Greenwood Press, 2005), 135; Markus Cerman, *Villagers and Lords in Eastern Europe, 1300–1800* (Basingstoke: Palgrave Macmillan, 2012), 11–17.
14. Jeroen Duindam, 'Rulers and Courts', in *The Oxford Handbook of Early Modern European History, 1350–1750*, Vol. 2: Cultures and Power, ed. Hamish Scott (Oxford: Oxford Univ. Press, 2015).
15. John Watkins, *Memoirs of Her Most Excellent Majesty Sophia-Charlotte, Queen of Great Britain from authentic documents* (London, 1819), 33.
16. Madame van Muyden, ed., *A Foreign View of England in the Reigns of George I and George II: The Letters of Monsieur César de Saussure to his Family* (London: John Murray, 1902), 40, 42.
17. Daniel Defoe, *A Tour through the Whole Island of Great Britain*, 2 vols. (1724–27; London: Dent, 1962), 1: 355.
18. Hannah Greig, 'Leading the Fashion: The Material Culture of London's Beau Monde', in *Gender, Taste and Material Culture in Britain and North America, 1700–1830*, ed. John Styles and Amanda Vickery (New Haven and London: Yale Univ. Press, 2007).
19. Caroline Weber, *Queen of Fashion: What Marie Antoinette Wore to the Revolution* (New York: Henry Holt, 2006), chapter 2.
20. Hannah Smith, *Georgian Monarchy: Politics and Culture, 1714–1760* (Cambridge: Cambridge Univ. Press, 2006), 64–81.
21. William Weber, 'The Cultural Authority of the Capital City', in *Opera and Society in Italy and France from Monteverdi to Bourdieu*, ed. Victoria Johnson, Jane F. Fulcher, and Thomas Ertman (Cambridge: Cambridge Univ. Press, 2007), 171.
22. Oliver Goldsmith, *The Works of Oliver Goldsmith*, ed. Arthur Friedman, 5 vols. (Oxford: Clarendon Press, 1966): 3:288–9. See also R. O. Bucholz, 'Going to Court in 1700: A Visitor's Guide', *The Court Historian* 5 (2000): 181–215, and Hannah Greig, '"All Together and All Distinct": Public Sociability and Social Exclusivity in London's Pleasure Gardens, ca. 1740–1800', *Journal of British Studies* 51 (2012): 50–75.
23. Eckhart Hellmuth and Wolfgang Piereth, 'Germany, 1760–1815', in *Press, Politics and the Public Sphere in Europe and North America, 1760–1820*, ed. Hannah Barker and Simon Burrows (Cambridge: Cambridge Univ. Press, 2002), 76–77.
24. John Robertson, 'Europe's Enlightenment', in *Oxford Handbook of Early Modern European History, 1350–1750*, Vol. 2: *Cultures and Power*, ed. Scott.
25. Gagliardo, *Germany under the Old Regime*, chapter 15; Erik S. Reinert, 'Johann Heinrich Gottlob von Justi—The Life and Times of an Economist Adventurer', in *The Beginnings of Political Economy: Johann Heinrich Gottlob von Justi*, ed. Jürgen Georg Backhaus (New York: Springer, 2009).
26. Douglas H. Shantz, *An Introduction to German Pietism: Protestant Renewal at the Dawn of Modern Europe* (Baltimore: Johns Hopkins Univ. Press, 2013), chapter 5.
27. Elisabeth-Charlotte, Duchesse d'Orléans to Caroline of Ansbach, 9 Jan. 1716, in Elisabeth-Charlotte, Duchesse de Orléans, *Fragmens de Lettres Originales de Made Charlotte-Elizabeth de Bavière, Veuve de Monsieur, Frère unique de Louis XIV*, 2 vols. (Hamburg: Maradan, 1788), 2: 4.

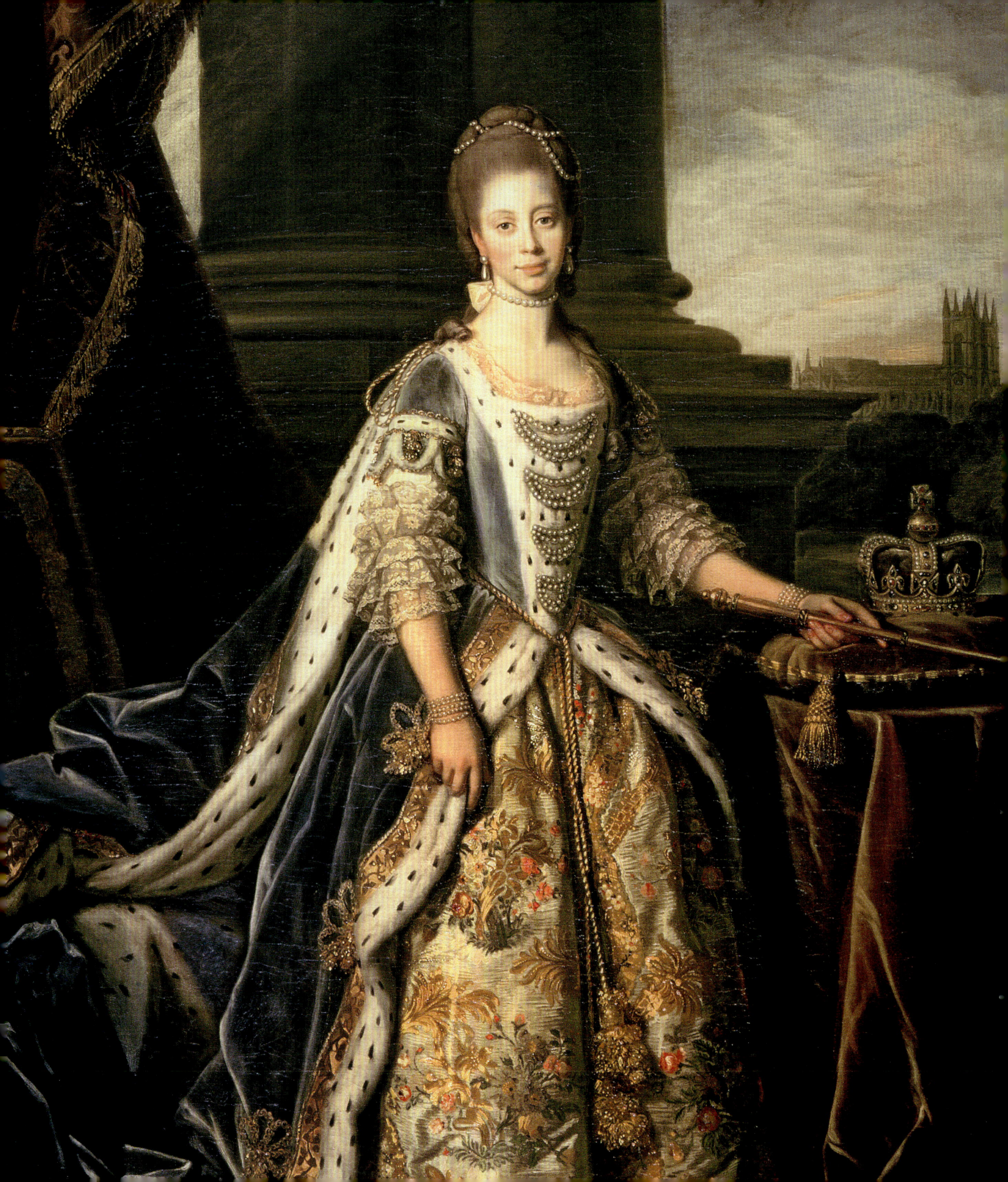

Partners in Power

Clarissa Campbell Orr

THE POWER OF DYNASTY

When the Hanoverian dynasty succeeded to the British throne in 1714, the monarchy had already had a female face since the turn of the century. Even before that, William III and his wife Mary II (reigned 1689–1702 and 1689–94) had been co-rulers, not king and consort. Queen Anne (reigned 1702–14) had been one of Britain's most successful rulers, inheriting a war coalition from her brother-in-law and cousin William of Orange, and forcing Louis XIV of France to accept the Act of Settlement of 1701, which nominated the Hanoverians as her successors. The war, which was concluded a few months before Anne's death, had included an astonishing sequence of British victories on land and sea, and demonstrated that Britain was now a European power to be reckoned with, despite her marginal island position and small population. No one could argue that a female sovereign was a liability, given that trade and the arts, particularly literature, had also flourished. But Queen Anne's tragic history of seventeen pregnancies, with the longest survivor, William, Duke of Gloucester, dying just after his eleventh birthday, had led to the 1701 Act of Settlement to ensure a Protestant monarchy. The Act determined that after her death the succession to the thrones of England, Scotland, and Ireland should pass to Sophia, Dowager Electress of Hanover (fig. 3.01), and her heirs. She, like Anne, had in common the same grandfather, James I of England and VI of Scotland. However, Sophia died only seven weeks before Anne, avoiding the accession of two ruling queens in a row. It was important to the first Georgians to insist that this Act of Settlement had only *regulated* the succession, by ruling out fifty-two Catholic claimants whose faith rendered their claims 'inconsistent with Protestant liberty'.[1] This interpretation meant that Parliament had not *chosen* or *elected* this new dynasty, but had merely *regulated* who should come next; and the Act still rested on the hereditary principle.

Detail of p. 71, fig. 3.10

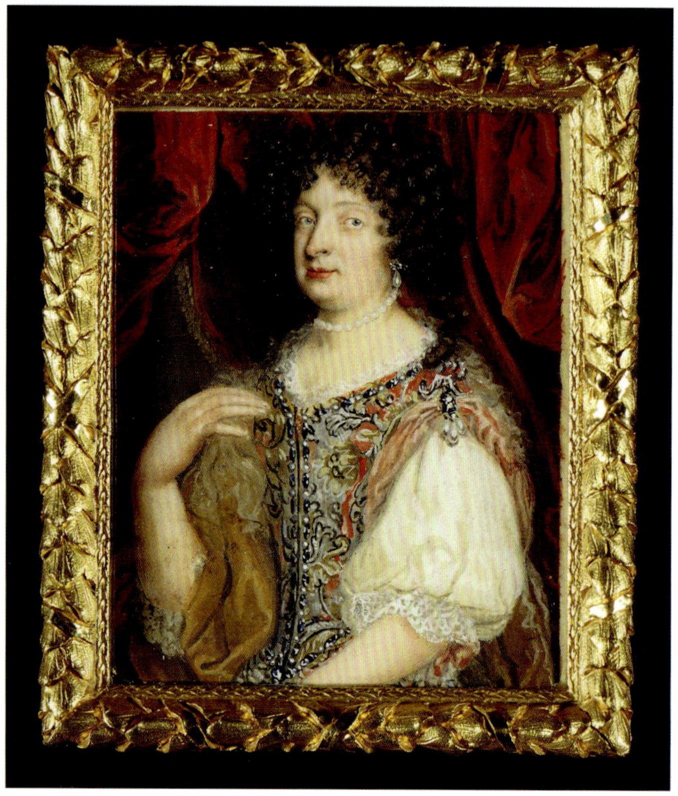

FIG. 3.01 Unknown artist, Danish, *Anna Sophia, Electress of Hanover*, c. 1676, watercolour on vellum, 4¾ x 2⅜ in. (7 x 5.9 cm). Royal Collection Trust

Nevertheless, the state of monarchy itself in Britain was relatively precarious. The pattern of instability since the Civil War of 1642 had resulted in the trial and beheading in 1649 of an anointed monarch, Charles I, followed by a republican interlude up to 1658 headed by a military leader, Oliver Cromwell, and his son, Richard. The restoration of Charles II in 1660 could not have been predicted with any confidence, nor what would be required for him to stay on the throne, given that a Parliament conscious of its power to challenge royal authority had also been restored. This uncertainty contrasted with continental-style absolutism, where a monarch could not be challenged effectively by any type of elected assembly. Charles II and his ministers proved adept at managing England's fractious political players, but the Catholic James II managed to alienate the two main groupings, Tories and Whigs, within only three years—1685 to 1688.

In 1688–89 James II abandoned the throne, sending ahead to France his young bride, the Catholic Mary of Modena, and newborn son, whom English loyalists, or Jacobites, considered the lawful future King James III. By 1701 William III had spent most of his reign fighting to defend his legitimacy, and when Louis XIV of France refused to accept the Act of Settlement, recognizing instead James 'III' as the next in line, he was preparing to fight again.

During Caroline of Ansbach and Augusta of Saxe-Gotha's time, Scotland challenged the monarchy as the last redoubt of Jacobite loyalty. But the 1745–46 rebellion saw a thorough military victory against them at Culloden, led by Caroline's second son, William, Duke of Cumberland. He was the only member of the Hanoverian

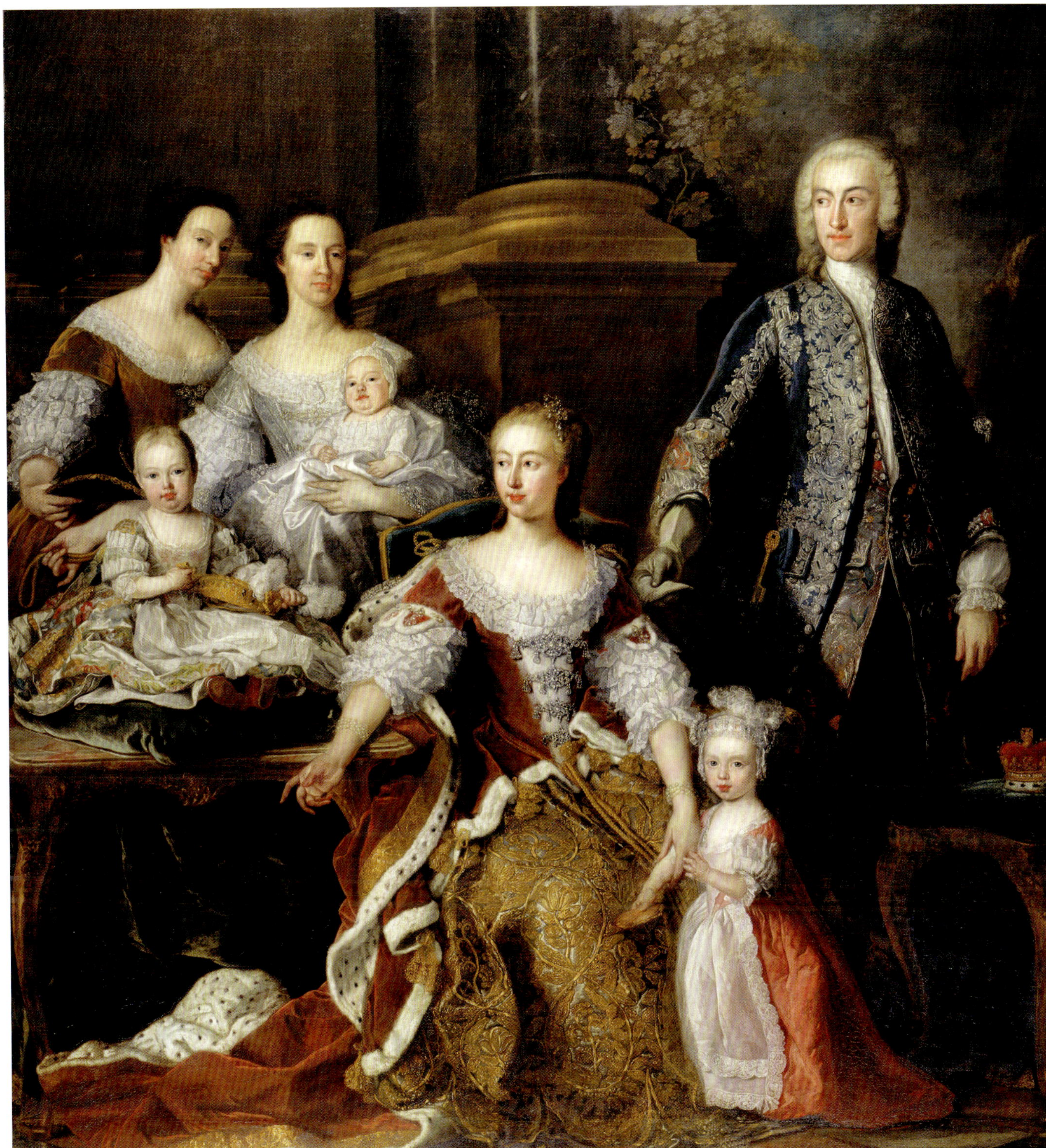

FIG. 3.03 Jean-Baptiste van Loo, *Augusta, Princess of Wales and Members of her Family and Household*, 1739, oil on canvas, 88 x 79⅛ in. (223.4 x 201 cm). Royal Collection Trust. Arabella Herbert is shown on the far left, grouped with Prince George (the future George III) and Prince Edward (later Duke of York).

only as a province to a despicable Electorate.'[3] The two governments were in fact conducted separately, with a German chancellery in London from which the King could direct his German business.

But the British were not anti-German in a twentieth-century sense. Germany was a geographical expression, not a unified state. In its nature the Holy Roman Empire of the German Nation was an anomaly, riddled with anomalies, and even contemporaries puzzled over it. It was a federation of sovereign states, numbering over three hundred, of three main types: principalities, ecclesiastical states, and city states. Principalities followed dynastic inheritance rules, which forbade female inheritance. The nine princes who elected the Emperor were the most important: they included Hanover, Saxony, Brandenburg, and Bavaria. Extra prestige accrued to the first three as each had also become linked to a kingdom: Hanover had a personal union of monarchies through inheritance with Great Britain, as did Brandenburg with Prussia; the Saxon Electors were twice chosen as Polish kings.[4]

The Princesses Caroline, Augusta, and Charlotte grew up familiar with this complex structure, for every part and function of which there was an historical rationale. In its last phase (1648–1806) it was still a useful entity, as it acted as a system of public law which maintained internal equilibrium and resolved conflicts through negotiation, not internal warfare.[5] Charlotte was to outlive its destruction in 1806, during the Napoleonic Wars; after that, several component parts achieved higher status as additional electors and the ecclesiastical states disappeared. When the Napoleonic era ended in 1815, Hanover, Saxony, and Bavaria became kingdoms in a new confederation of thirty-three states, while the Habsburgs were now transformed into hereditary Emperors of Austria. In this new structure, Charlotte was therefore a queen twice over, though, sadly for her, George III was not in sufficient health to act as Hanover's first king.[6] Her role as consort had weathered one of the most cataclysmic upheavals in all European history.

Although German princes were of equal status in legal terms, in practice some were more powerful than others, especially as their territories varied enormously in size and population. The powerful acted as patrons to their lesser 'brethren', for example by paying their military contribution to the Imperial army.[7] The nine Electors ranked highest of all, and the Hanoverians had only achieved this status in 1692 by dint of unifying several duchies, adopting primogeniture in inheritance to gain territorial and demographic weight, and by supporting the Emperor militarily and financially in his Turkish wars. Thereafter they were rivals with Brandenburg in exercising influence in northern Germany, especially when Mecklenburg-Schwerin, ruled by the senior branch of Charlotte's paternal dynasty, was temporarily put under an Imperial Commission after the deposition of its ruling duke, Karl Leopold, when Hanover gained supervision of it.

There were at least two strategies a lesser German prince could pursue to raise his profile in relation to others. One was through military prowess, gained by serving either in the Imperial army, almost continuously engaged along the Turkish frontier throughout the century, or in the larger princely armies of states such as

Brandenburg-Prussia. The other was by marriage into a more prestigious dynasty, either by their own marriages, or that of their daughters, sisters, nieces, or other relatives. George I had hoped his son could marry into the Swedish royal family, but when this fell through he permitted his son's own proclivity towards Caroline of Ansbach-Bayreuth, a small territory within the Bavarian (thus Catholic) orbit, which was nonetheless staunchly Protestant. As Caroline's guardian was Frederick I of Brandenburg-Prussia, she also counted as a client of this Protestant dynasty.[8]

Frederick, Prince of Wales's marriage into the Saxe-Gotha-Altenburg dynasty was one result of George II's desire to have a dynastic strategy independent of his father's, who was aiming for a double marriage between his granddaughter Amelia and his Prussian nephew, Crown Prince Frederick, and between his grandson Frederick and Princess Wilhelmine of Prussia. It was also a by-product of his dissatisfaction that Frederick, Prince of Wales, was meddling with the marriage of Caroline's Ansbach nephew. Frederick was summoned to England and had to kick his heels while his older sister Princess Anne took centre stage with her Dutch marriage.[9] There had never been marriages into the Saxe-Gotha-Altenburg dynasty before: the family was part of a group of lesser dukes who had once been powerful Electors, but had lost this title in the sixteenth century when they had sponsored the Lutheran Reformation. Though the patrimony was dispersed through seven sub-divisions, the Saxe-Gotha-Altenburg line was still wealthy and provided significant cultural leadership through having one of the largest court theatres in Germany, as well as maintaining important artistic and scientific collections.

Charlotte too was chosen from a 'new' dynasty. This time it was because George and his mother had already headed off a marriage within the older branch of the Brunswick dynasty, the Hanoverians being the younger branch. The dukes of Brunswick-Wolfenbüttel were military clients of Brandenburg-Prussia, and George III disapproved of the involvement of Britain, Brunswick-Wolfenbüttel, and other German allies in the Seven Years' War with France. His mother favoured another match with her own family, but the relevant princess was considered insufficiently devout. Charlotte's branch of the Mecklenburg family gained enough kudos to 'marry up' because of her maternal great-uncle, Joseph of Saxe-Hildburghausen, who had gained prestige in the Imperial army and had married an extremely rich heiress. This princess was niece to the even greater general in the employ of the Habsburgs, Prince Eugene of Savoy—the co-leader with John Churchill, first Duke of Marlborough, in Queen Anne's war with France. Charlotte's own dynasty was critical of Prussian recruitment tactics within its borders, so both from the British and the Mecklenburg standpoint her marriage was an anti-Prussian move, marking rapprochement with France after five years of war. Frederick II of Prussia professed not to care. He recognized that the Anglo-French rapprochement was irreversible and told his envoy, Count Karl-Wilhelm Finck von Finckenstein, who was dealing with British state correspondence, to send suitable congratulations on the intended marriage. When Finckenstein asked for more guidance on how he should word his message, Frederick replied with some irritation that the issue was a mere 'bagatelle' compared with impending military action and the nightmare that Austria and Russia might now link up.[10] The quid pro quo, however, was that George's sister Augusta was obliged to marry the Brunswick Crown Prince, Charles William Ferdinand, Duke of Brunswick-Wolfenbüttel, in recognition of his part in the original anti-French coalition. He was thus 'married up' to a king's daughter; but it never became a love-match.

The question that arises from the choice of relatively insignificant princesses as consorts is whether their German connections ever influenced policy. Caroline was chair of the Regency Council, which acted for the King when he visited Hanover in 1729, 1732, 1735, and 1736, but she did not alter any policy. Rather, as Joanna Marschner and Andrew Hanham have shown, she concentrated her energies during her decade as Queen in 'anglicising' the dynasty. She brought continental flair to her reorganization of the royal collections, and in her various acts of patronage, highlighted Hanoverian links with German ancestors who had English connections, such as the English Plantagenet Queen Matilda, wife of the Brunswick Emperor, Henry the Lion; or English affinities with German culture, as in her support for the Anglo-Saxon scholar Elizabeth Elstob. Elstob admired the Anglo-Saxon Church and wrote a grammar of Anglo-Saxon which she dedicated to Caroline.[11] Augusta, of course, never had an electoral role, as her husband predeceased his father.

In a long residence of fifty-eight years, Charlotte helped to keep the British royal family in tune with Germany.[12] Two of her brothers were in the Hanoverian service, and three of her sons were educated at Göttingen University in Germany. Her second son, Frederick, was made Prince-Bishop of Osnabrück by his father, becoming a German Imperial ruler in his own right.

The year 1778 saw Charlotte balancing her different patriotic loyalties most conspicuously. She supported the army regiments preparing to travel to America, after the French and Spanish Bourbons joined the American side, and sent charitable aid to Hanoverian regiments scattered at sea who were replacing British soldiers defending Minorca. But in a real act of intervention in the German Empire, she drew George's attention to the Bavarian succession problem, when Frederick II of Prussia rallied some of the lesser princes into a league to prevent Habsburg over-reach.[13] In 1785 she was the direct conduit to her relatives in Mecklenburg-Schwerin when George III decided to join a subsequent princely league, designed for a similar purpose. None of these actions conflicted with British foreign policy, or required British money; but it does show that she helped George maintain his electoral role, as well as supporting his royal one. Though she kept out of British domestic politics, she clearly felt it was within her legitimate remit to keep track of events within the German Empire.

Later, in 1806, when Hanover was occupied by the French, she encouraged her niece Louise of Mecklenburg-Strelitz, the Prussian queen, to urge the Prussians to abandon neutrality and fight the French. During the Congress of Vienna which concluded the wars, she encouraged her son George in his capacity as the Prince Regent (he succeeded his father, as King George IV in 1820) to advance the claims of Mecklenburg-Strelitz to retain its identity while remaining in the Prussian orbit and gaining the status of Grand Duchy.[14] Again, none of these initiatives put her at odds with British foreign policy, but arguably, as she was the German-born one of the royal couple and the King was too ill to intervene, she had a more up-to-date cognizance of German developments than did the King or her eldest son. Her roles as British and Hanoverian consort required different knowledge and tactics, but she succeeded in juggling both of them.[15]

When contemporaries of the Princesses referred to 'the Empire' they invariably meant the Holy Roman Empire, not the British Empire as it existed in the eighteenth century. Among the latter's earliest acquisitions in South Asia was Bombay (modern Mumbai), which was part of the dowry of Charles II's Portuguese wife, Catherine of Braganza. Britain by no means controlled all of India: its legal status was as revenue collector of the Mughal Empire, in its Bengal province. All Indian affairs were mediated through the East India Company, a trading monopoly. So the two Princesses who became queens were never Indian empresses. They were, though, queens of the Thirteen Colonies in North America, although in Caroline's era, Maryland, Pennsylvania, and large portions of the Carolinas were still owned by their proprietors.

THE POWER OF RELIGION

The Princesses were eligible as brides only because they could represent the Protestant faith unequivocally.[16] In Great Britain Catholicism was associated with continental absolutism, and the threat that church lands sold off by Henry VIII in the 1530s might be restituted if a Catholic monarch succeeded. Caroline scored especially highly since she had come under pressure from the Wolfenbüttel line to convert to Catholicism in order to marry the Emperor's second son, Karl, who was the claimant to the Spanish crown. When his older brother Joseph died, Karl became Emperor, so her status might have been higher still. Caroline's steadfast determination to remain Protestant made her well liked in Britain.[17] Augusta also had the religious asset of descent from Martin Luther's most significant patron, Frederick III, Elector of Saxony.

By the 1760s the key religious issue was to keep the monarchy aligned with observant Christianity itself rather than to head off the Catholic Jacobites—the bigger danger was considered by then to be scepticism and free-thinking. It was because Frederica of Saxe-Gotha was considered to be too much influenced by Voltairean philosophy that she had been rejected as George III's bride; her mother was Frederick II's main confidante after his sister Wilhelmina died in 1758. George and Charlotte both positioned themselves against the materialist strands of the French Enlightenment and sponsored thinkers who could counteract modern philosophy, such as the Scottish academic James Beattie, whose critique of the scepticism of David Hume was considered by the former's supporters to be definitive (fig. 3.04). Charlotte's readers, Jean-André Deluc and Marie Elisabeth Boué, Madame de la Fite, shared this critique Once the French republicans 'dechristianised' France in the early 1790s, the couple almost felt on crusade to defend religion *tout court*.[18] George planned to give this

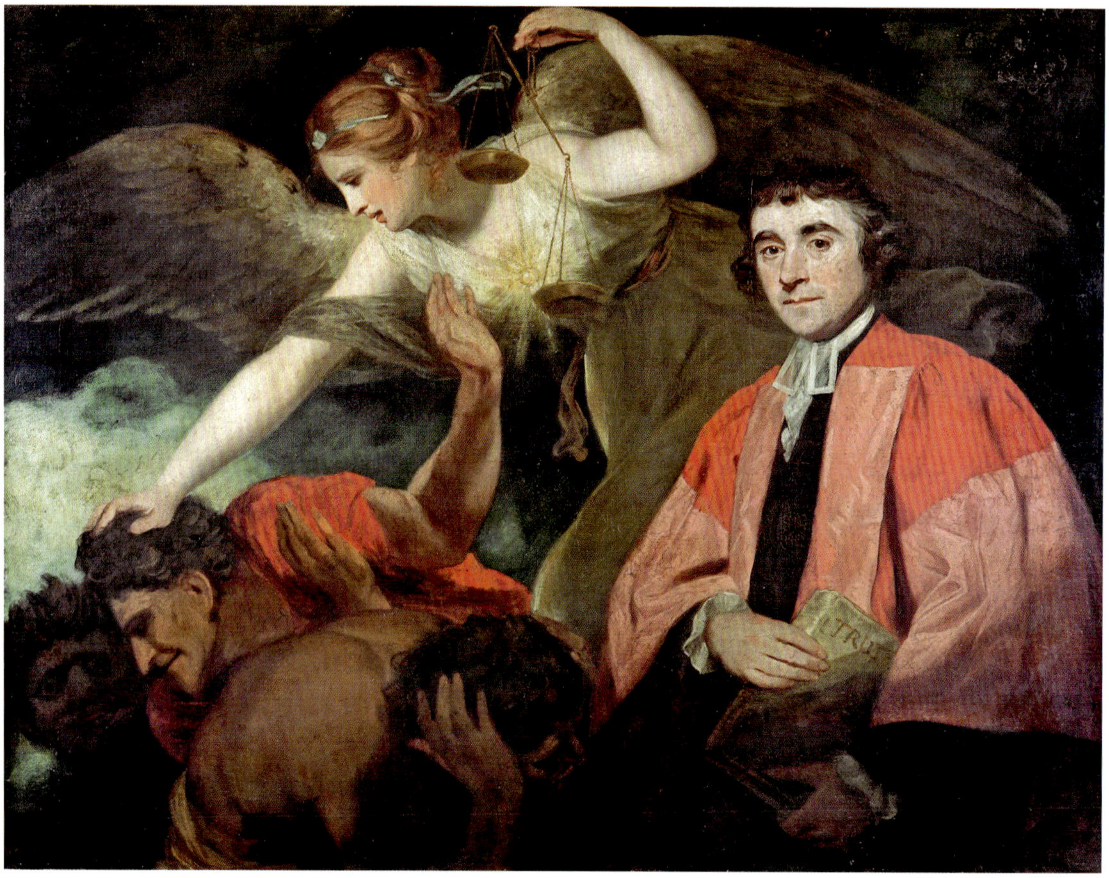

FIG. 3.04 Sir Joshua Reynolds, *The Triumph of Truth: Dr James Beattie*, 1773, oil on canvas, 48 x 61 in. (122 x 155 cm). University of Aberdeen (cat. 16.28)

religious stance visible embodiment in the Chapel of Revealed Religion at Windsor, to be painted by the American Quaker Benjamin West.[19] Left unfinished when illness overwhelmed the King, his son as Prince Regent later used the space to create the Waterloo Chamber to celebrate victory over Napoleon.

Though the defence of revealed religion varied in intensity between the three Princesses, all would have agreed to the conventional belief in studying natural science as God's handiwork. Francis Bacon had articulated the view that the Book of God's Word, or Holy Scripture, was paralleled by the Book of God's World, or Nature. What mattered more than their beliefs was that the Princesses should be dynastically Protestant.[20] But each had a personal commitment to piety as well. Caroline sponsored lectures by Samuel Clarke in the Boyle lecture series, founded to demonstrate the truth of the Christian religion, which advocated a concord between Newtonian science and Providential theology. The King also allowed her to take an interest in Anglican church appointments and she may have influenced over half of the decisions to appoint bishops.[21] Augusta was more concerned with religious morality than theology, and brought up George III to eschew masculine libertinism. Charlotte took no more interest in ecclesiastical appointments than she did in political ones, but by her era the increasing trend was to place the monarchy above politics as the head of voluntary organizations. She concentrated her own efforts on charities concerned with women's health, as well as education. Her last public appearance was at a prize-giving for the National Schools.[22] At a personal level, she and the King were close friends of Richard Hurd, Bishop of Worcester, who had been preceptor of their two eldest sons, and whose portrait was in her room at Kew Palace.[23]

In the French court, a '*parti dévot*' clustered around Louis XV's queen Marie Leszczyńska and her unmarried daughters. There was no equivalent 'devout party' as such in British political culture. The admittance to court of men and women whose morals were perceived to be lax was different under Caroline and Charlotte. In general there was more elasticity for both sexes at George II's court; George and Charlotte tried to draw stricter moral lines. Charlotte's Lady of the Bedchamber, Elizabeth Gunning, sixth Duchess of Hamilton, stuck loyally by her daughter, Elizabeth, twelfth Countess of Derby, after she eloped with John Sackville, third Duke of Dorset. But Charlotte wrote to the King:

> Her indelicacy goes a great way, and I am of opinion with my humblest advice your Majesty would now do better to let the D[uke] of Manchester [the Queen's Chamberlain] acquaint her that her daughter should not come to court, & be ill received, as letting her come will open the door to others, for though she is not divorced, she has stood a public trial. I declare I shall neither speak to mother or daughter, and if she leaves my family I shall but get rid of an impertinent person who has always behaved disrespectful [sic] to me.[24]

Charlotte used the term 'family' in the then current sense of 'Household'. When Gertrude Russell, Duchess of Bedford, a powerful political hostess, asked of the King in 1775 that her rakish nephew, Richard Fitzpatrick, be given a post in the Royal Household, thus cementing her husband's political following to Lord North's ministry, the King's vehement reply was, 'I will not allow a gamester in my family.'[25]

THE POWER OF THE CONSORT'S HOUSEHOLD

Legally and financially, royal brides were not mere appendages to their husbands: they brought a dowry with them of a greater or lesser amount, and had their own household with parallel offices to the King's. These covered at least fifteen different categories and included women *and* men from the top to near the bottom of English society. Key posts included a Lord Chamberlain and Vice-Chamberlain, who regulated ceremonies and access; a Master of the Horse, in charge of transport and stabling; a Groom of the Stole, who was head of the attendants of the Princess—her Ladies and Women of the Bedchamber and the Maids of Honour; plus Gentleman Ushers and Pages. Every household also needed professional services—a Treasurer, Secretary, Attorney-General, Physician, and Chaplain. Augusta, for example, appointed as her

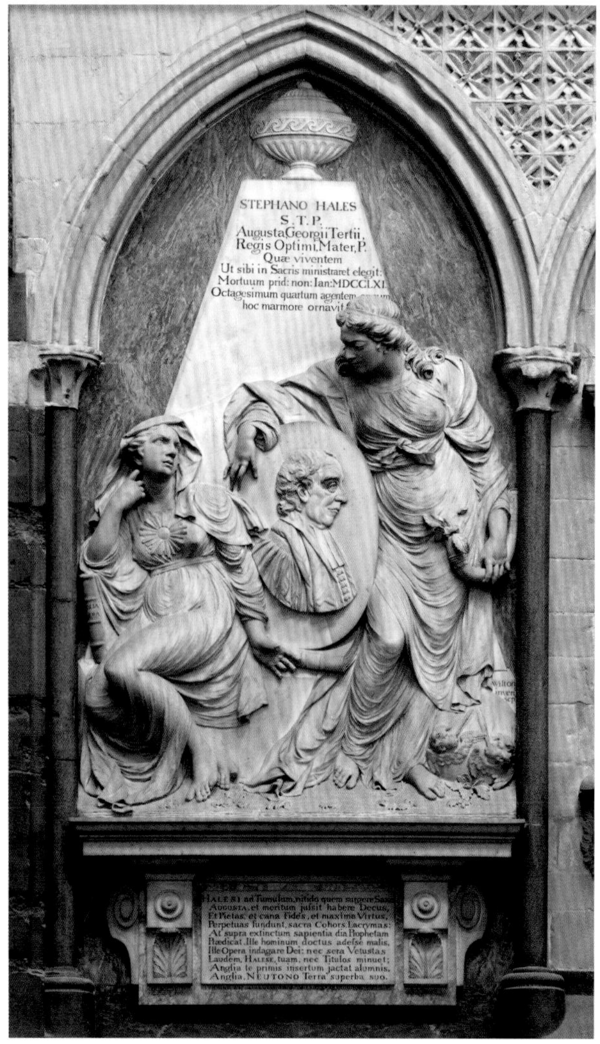

FIG. 3.05 Joseph Wilton, *Monument to Stephen Hales* (commissioned by Augusta). Westminster Abbey, London

Chaplain the widely regarded Reverend Stephen Hales (figs. 3.05 and 3.06), whose work on plant physiology was seen as an important extension of Newtonian science. He taught botany to the future George III, and was also a noted philanthropist who, *inter alia*, improved ventilation in ships and in prisons.[26]

The money to finance Royal Households was voted by Parliament, who thus gained power of scrutiny over it; or, in the case of a Prince and Princess of Wales, money came from the revenue of the Duchy of Cornwall, which normally accrued to the Prince. The money was therefore to cover all the expenditure of a consort's household, including salaries for the governors and governesses looking after her children, and the board, lodging, and clothing of the children themselves. Household personnel extended right down to humbler roles such as footmen, wet-nurses, and seamstresses, so consorts were substantial employers, and their employees were both salaried and entitled to a pension. For a role of high status, such as Lady of the Bedchamber, an

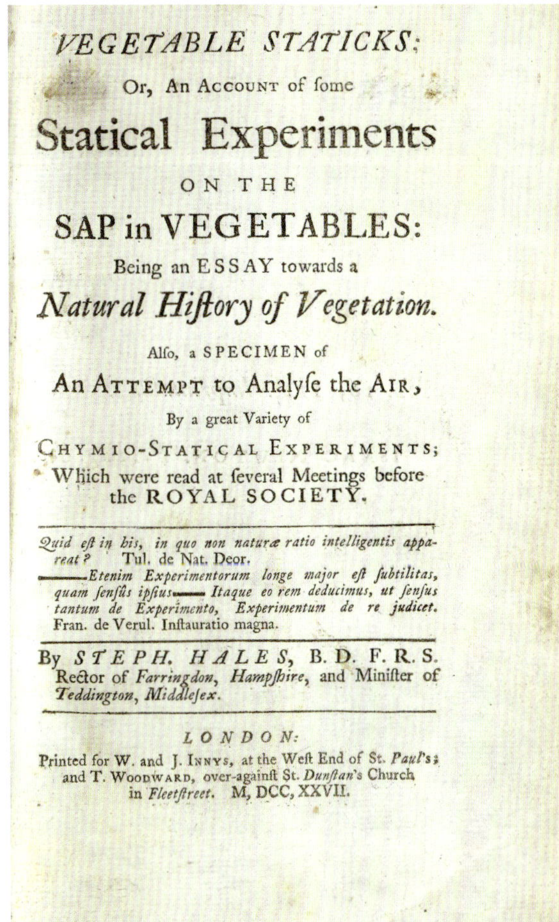

VEGETABLE STATICKS:
Or, An ACCOUNT of ſome
Statical Experiments
ON THE
SAP in VEGETABLES:
Being an ESSAY towards a
Natural Hiſtory of Vegetation.
Alſo, a SPECIMEN of
An ATTEMPT to Analyſe the AIR,
By a great Variety of
CHYMIO-STATICAL EXPERIMENTS;
Which were read at ſeveral Meetings before
the ROYAL SOCIETY.

Quid eſt in his, in quo non naturæ ratio intelligentis appareat? Tul. de Nat. Deor.
——*Etenim Experimentorum longe major eſt ſubtilitas, quam ſenſûs ipſius*—— *Itaque eo rem deducimus, ut ſenſus tantum de Experimento, Experimentum de re judicet.* Fran. de Verul. Inſtauratio magna.

By *STEPH. HALES*, B. D. F. R. S.
Rector of *Farringdon*, *Hampſhire*, and Miniſter of *Teddington*, *Middleſex*.

LONDON:
Printed for W. and J. INNYS, at the Weſt End of St. *Paul's*; and T. WOODWARD, over-againſt St. *Dunſtan's* Church in *Fleetſtreet*. M, DCC, XXVII.

FIG. 3.06 Stephen Hales, *Vegetable Staticks: Or, An Account of some Statical Experiments on the Sap in Vegetables: Being an Essay towards a Natural History of Vegetation* (London: Printed for W. and J. Innys and T. Woodward, 1727), title page. Royal Botanic Gardens, Kew, Library, Art & Archives Collections (cat. 16.22)

extra perk was the payment of the dowry of their daughters; it was also the expectation of the Maids of Honour when they married that they would have a dowry. Such a role was attractive to less well-off gentry families.[27] Sometimes husband and wife teams were appointed: Thomas Fermor, first Earl of Pomfret, was Master of the Horse to Caroline, and his wife, Henrietta, was a Lady of the Bedchamber.

A consort thus had money at her command and an economic base to attract supporters and traditions of service from families who were courtiers rather than aspiring politicians; though for men the line between the two roles was not clear-cut.[28] Potentially, a consort could rally her own supporters in opposition to her husband or her father-in-law, for what always mattered to courtiers was the degree and type of access they had to the monarch. Facilitating this access was the main way a consort might be able to help ministers to get appointed, or, more commonly, to dispose in turn of constant demands for their patronage: regimental or church vacancies, administrative

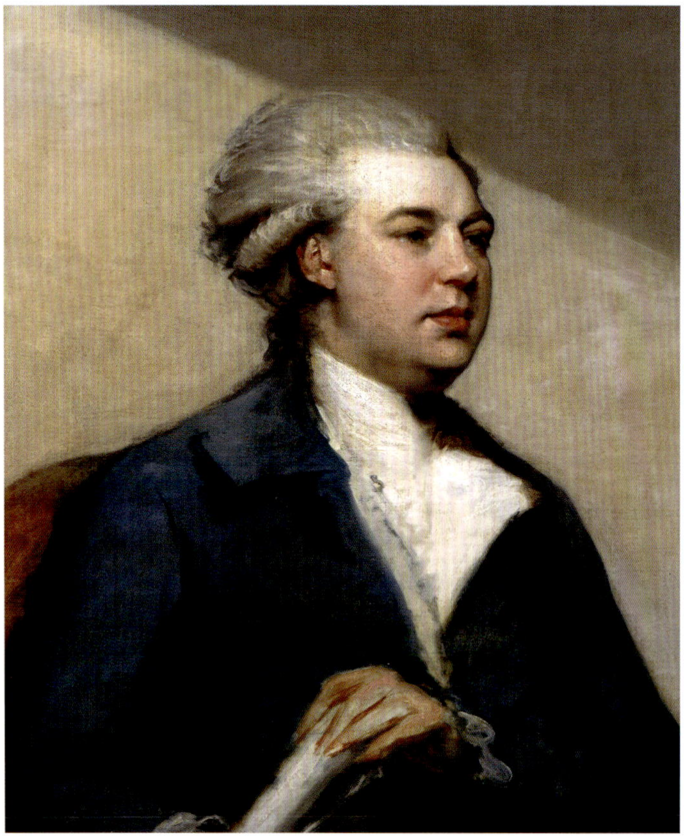

FIG. 3.07 John Russell, RA, *Portrait of Frederick North, 2nd Earl of Guilford PM*, 1760s, oil on canvas, 31 x 27 in. (78.7 x 68.6 cm). Philip Mould, Historical Portraits

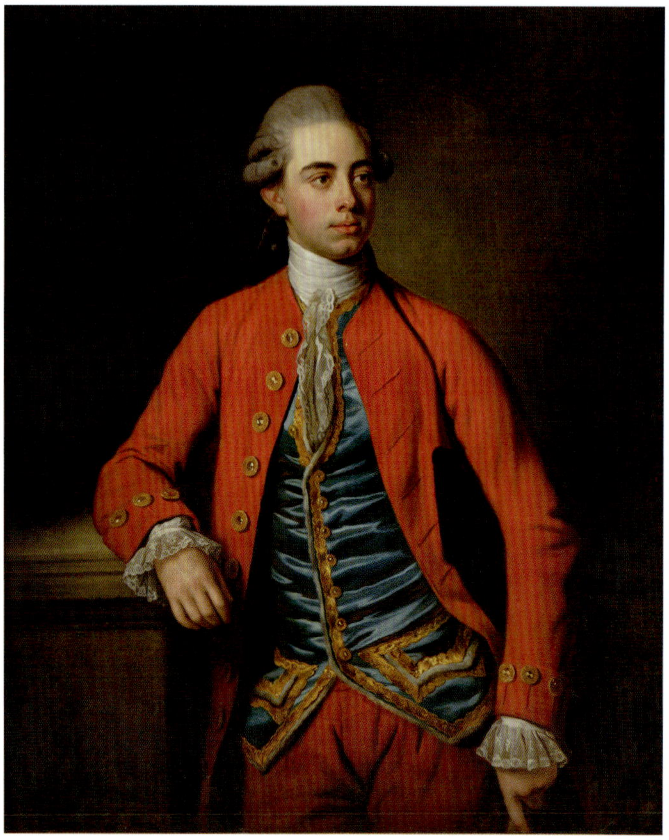

FIG. 3.08 Nathaniel Dance-Holland, *George North, 3rd Earl of Guilford*, undated, oil on canvas, 50 x 40 in. (127 x 101 cm). Philip Mould, Historical Portraits

posts—for instance in the Treasury—or diplomatic roles. Consorts and political wives hardly ever helped engineer the ministerial appointment in the first instance, but once their husbands were ministers, the wives were often approached to gain the minister's ear to dole out patronage at the next level down. The imagined power of Queen Caroline was exceptional, but at the time, and during the nineteenth century, it was widely believed that Sir Robert Walpole survived the transition from the reign of George I to George II because of his friendship with Queen Caroline. Recent scholars believe this is oversimplified, but there can be no question of Caroline's acute political intelligence and the importance of this element in her marriage. One biographer, Stephen Taylor, has concluded: 'it is difficult to deny that Caroline was the most powerful queen consort of the Hanoverian period. Indeed . . . she probably exercised more influence over English government than any queen since Elizabeth I.'[29]

In reality the three consorts were not so much embroiled directly in party conflict, but they were implicated in the generational politics inherent in the Westminster parliamentary system when sons rebelled against their father's ministers and rallied a group of disaffected politicians, hoping for an alternative future. Charlotte maintained a united front with the King concerning their eldest son's indebtedness, but Caroline and Augusta both co-operated with their husbands when they were in opposition to the King. Augusta facilitated access to their home for Frederick's parliamentary 'fixers', George Bubb Dodington and John Percival, first Earl of Egmont. Egmont was present when opposition strategy was discussed, and ensured that papers revealing the extent of this opposition were destroyed the moment he died.[30]

THE POWER OF SOCIABILITY

The main role of the Princesses in the English parliamentary monarchy was, therefore, by social means to help cement to the crown the political and financial elites and their relatives and clients. Once she became the King's consort, a queen presided alongside the king over the most important Drawing-Room in the four nations, the royal Drawing-Room, usually held at St James's Palace on Sundays. It was the main arena for both sexes to display loyalty to the Hanoverians; Jacobites excluded themselves or were formally excluded. Once former Jacobites changed their loyalties, in some cases converting from Catholicism, it was often easier for them to achieve a place at court than a political role; it gave them an extra reason to adhere to the Hanoverians. Many families served Royal Households across the generations (figs 3.07, 3.08). In contrast, loyalist families travelling abroad, or sending a son on his grand tour, would attend a Drawing-Room before departure to show they were not leaving for the Jacobite Pretender's court in Rome, and they would signify their return by another visit. Other occasions co-hosted by King and consort were the monarch's birthdays, when it was customary to wear new clothes as well as jewels, and dancing as well as refreshments were included.

Historians generally regard the Hanoverian era as a period of greater political stability and civility after the intense 'rage of party' characteristic of the later Stuart era up to Queen Anne's death. No one wanted a return to the resolution of political differences through civil war; gradually the notion of a 'loyal' opposition emerged. As the main social arena where elites and their protégés met, the court was the counterpart to evolving political customs at Westminster. Politicians might posture in the Lords and Commons, but in the Drawing Room they had to meet, mingle, and be polite to ally and foe alike. This also applied to worship at the Chapel Royal, in St Stephen's Chapel in Westminster, or parish churches in the evolving West End, such as St George's, Hanover Square.[31] Of course, politeness at court might be expressed through gritted teeth and with an angry heart well concealed by a smile and a bow. Bitter disappointment or disapproval might still break through, and George II seldom disguised his feelings when he disapproved of someone by 'rumping'—that is, turning his back towards them. Decades later George III used the same technique against Admiral Nelson, turning his back on the naval hero at a Drawing Room to talk to an obscure general, to show his moral disapproval of Nelson's openly living with Emma, Lady Hamilton.[32]

THE POWER OF FEMININITY

Female fertility was at the heart of dynastic power. Dynastic marriages demonstrated male maturity: the King as symbolic father of his people. For all three princely partners, marriage was associated with the male milestone of attaining an independent household with attendant financial means, and demonstrating their power to father offspring, thus maintaining the dynasty. Being a successful mother biologically was the most obviously important role for a consort. In 1714 Caroline was already a known quantity in this respect. Apart from her son Frederick, left behind with his great-uncle to represent the dynasty, she brought with her three daughters, Anne, Amelia, and Caroline, and was to have three more surviving children: the soldier Prince William, Duke of Cumberland, plus Louisa and Mary, who married respectively the Danish king Frederick V, and the Landgrave Frederick II of Hesse-Kassel. But her first pregnancy on English soil demonstrated the internal politics of the dynasty in dramatic form. In 1717 her husband was aligned with parliamentary critics of his father, who were splitting the Whig party. To demonstrate his independence from his father, the future George II took issue with his father's choice of godparents for the new baby prince, George William. The ensuing argument led to his exclusion from court, and Caroline followed him. Legally her children were wards of the ruler, so while her husband and father-in-law were at odds, the newborn and his siblings were under George I's direct control. Caroline was eventually allowed to see her children, and as royal children were invariably wet-nursed her separation from the baby did not risk its life. When the baby died in February 1718, the autopsy showed it had a malformation, and would never have survived. Meanwhile the Prince and Princess attracted courtiers to them by their entertaining at Richmond.

Princess Augusta experienced a similar situation. Marriage had given Frederick a degree of emancipation from paternal control; and it was paralleled by his move towards political opposition. He was determined that his wife should give birth at St James's Palace, the dynasty's London 'Headquarters', so she was rushed there after her first labour had commenced in 1737. The child was a girl, named after her mother. The split between father and son which followed his defiant and perhaps reckless action persisted for three years, and George II made it clear that courtiers and politicians supporting the Prince were not welcome at his court. But Augusta and Frederick were amply fertile, producing a boy at the second try, and seven other children.

Augusta became the political confidante of her husband in a second phase of opposition. She was present on occasion when he discussed political business with his friends and supporters, Bubb Dodington, and Lord Egmont, and they sought her intercession to shape Frederick's policy. Swiftly and astutely she concealed all traces of Frederick's schemes as soon as her husband died.[33] As a dowager, despairing of her son's lack of self-confidence, she turned to her late husband's friend, John Stuart, the third Earl of Bute, in the hope that the apparent dullard would be galvanized by such a mentor. Augusta had chosen well: the sustained personal attention inspired young George to concentrate on preparing to be king. But this was at the cost of Augusta's

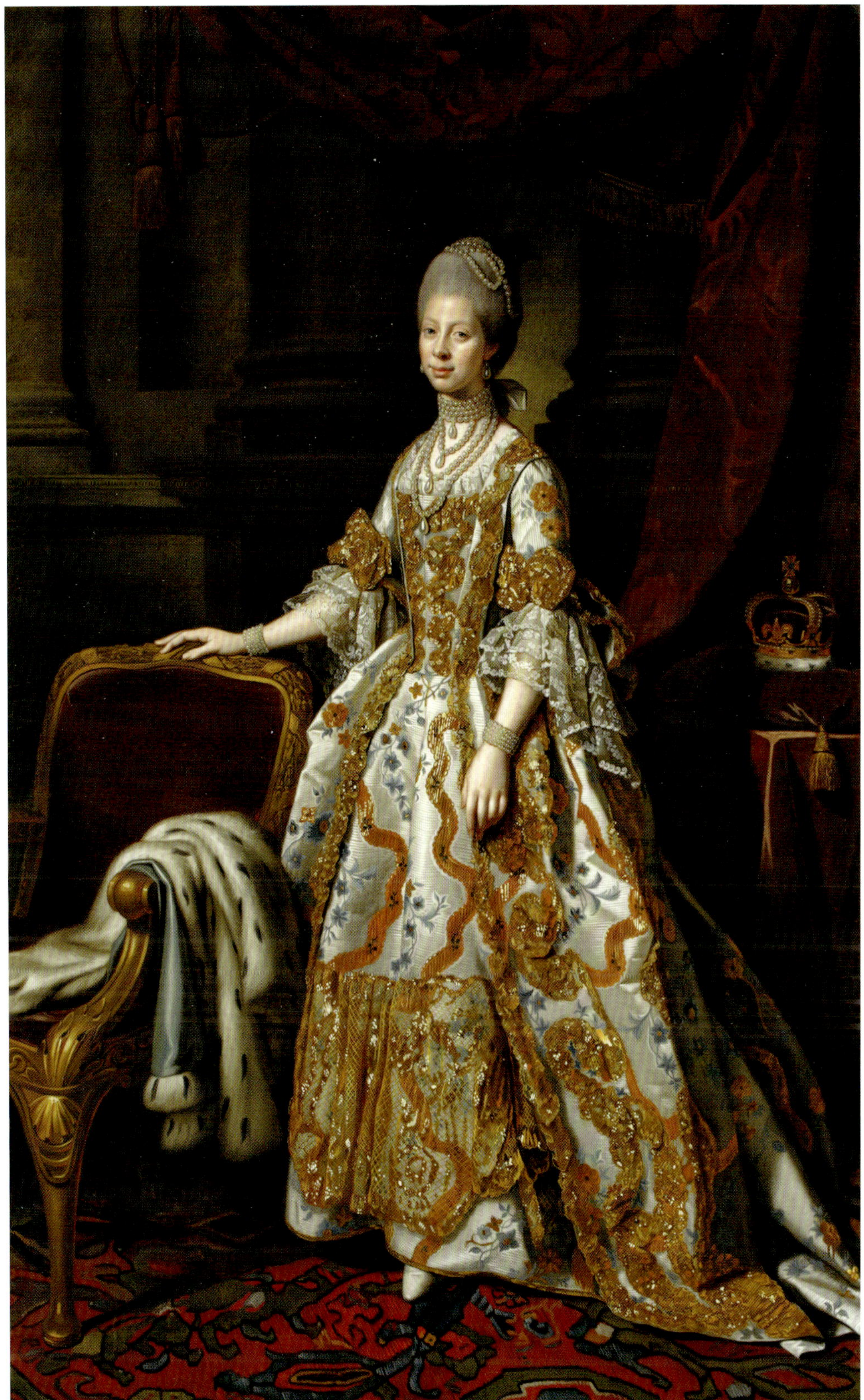

FIG. 3.09 Nathaniel Dance-Holland, *Queen Charlotte*, 1769, oil on canvas, 100 x 59 in. (254 x 150 cm). National Trust, Uppark

reputation: her continuing friendship with Bute made her a political liability and the subject of lubricious cartoons suggesting he was her lover, so unintentionally she gave the crown an air of disrepute (see fig 3.02).

The significance of George's personal pathway through marriage and fatherhood within a very short time has been underplayed by historians, who have tended to focus only on the evolution of George III's early statecraft, while the details of Charlotte's own development have been overlooked. Historians have concentrated on George III's first ten years as King and the political difficulties he had in creating stable ministries.[34] He was, however, only twenty-two years old, and had been nowhere in England other than Kew and St James's, and a few other London locations. Unlike aristocrats of his age he had neither been to university nor on the customary grand tour to Paris, Rome, or Vienna, let alone to Hanover.[35] He was certainly over-dependent on Bute, making him one of the two Secretaries of State, and treating him as his Prime Minister, whereas previously Bute only held a Royal Household position as Groom of the Stole, and had no political base in the House of Lords. Nevertheless, what has been virtually ignored is the degree to which his early troubles took place at a time when the King was passing rapidly through significant milestones in the male life cycle.[36] George became King in October 1760; in September 1761 he met his chosen bride for the first time, married her immediately, was crowned, and eleven months later they had become parents of a male heir. Instead of flailing helplessly, he acquired a satisfactory marriage: the Princess from a remote part of Germany gave him self-confidence and an emotional focus apart from Lord Bute.

Charlotte accordingly identified herself with her husband's interests and put them ahead of her family. This can be illustrated by her behaviour when George was considering how best to award places in the Order of the Garter (the highest royal chivalric honour, solely in the sovereign's gift). He wanted to reward Bute, as well as his own younger brother Edward, Duke of York, and his new brother-in-law, Adolphus Frederick IV, Duke of Mecklenburg-Strelitz. Bute wrote to the King urging him to consider the two royal appointments ahead of his own claims, but George decided to appoint Bute first, followed by Adolphus and then his own brother. Charlotte's response to this sequence of events was reported by the King to Bute:

> my own conduct to her and to her only *made her esteem herself as belonging to me, & to me only, that she thought it very probable that it might be thought too much for a brother and brother-in-law to have it at once*, & therefore expressed satisfaction at my deferring it, thinking it more for my interest which she from inclination as well as duty prefers to every other consideration, on this I told her the very handsome part my Dearest Friend had acted this morning, & then produced his letter, as it related solely to this transaction, she read it attentively & was pleased with the manner of it, and said I could with honour give it to no-one whilst the man I had ever titled my friend was without, and desired me when I should think it proper to congratulate Him in Her name.[37]

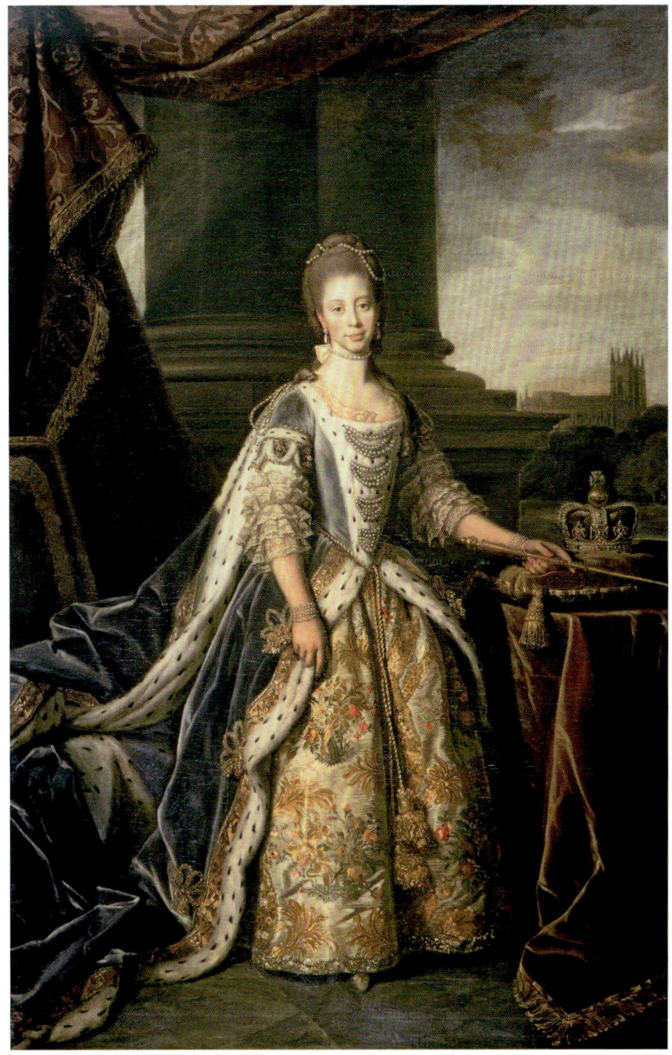

FIG. 3.10 Nathaniel Dance-Holland, *Queen Charlotte, Wife of George III*, 1773–74, oil on canvas, 94½ x 57⅞ in. (240 x 147 cm). The State Hermitage Museum, St Petersburg

Charlotte wrote to thank Bute for 'making interest' on behalf of her brother: 'You will receive this acknowledgement as a mark of my particular esteem for your person, and believe that no Body one in this world can value more that person who is the King's friend than I do.' At the same time Charlotte agreed to a ministerial suggestion that she exemplify royal generosity by making a charitable donation to a hospital when the Peace terms with France were presented to Parliament. It was hoped this would offset the risk of riots over them.[38] There was a lot for her to learn in a short time about her role, which was to be social and philanthropic rather than political, and this remained the essence of her position to the end.

The couple between them commissioned three pictures celebrating the Queen as wife with their first two sons: one from Bute's protégé Allan Ramsay, and two by a German painter very much to the Queen's liking, Johan Zoffany.[39] Their successful personal and political adjustment to the roles of King, consort, and continuers of the dynasty was very important to them.

Charlotte's role as companion and friend of the King did not mean that she was a political confidante as well. George made it clear to Charlotte early on that her role was to be the open-minded hostess of all shades of opinion at court, and to discourage any attempt to make her the figurehead of a Queen's party or other faction. As she recollected later to her friend Lady Elizabeth Harcourt:

> I am most truly sensible of the dear king's strictness, at my arrival in England, to prevent my making many acquaintances . . . : that there never could be society without party, which was always dangerous for any woman to take part in, but particularly so for the Royal Family . . . but he allowed me to be Civil to all.[40]

In the future, there would have to be extreme self-discipline on her part, which Nathaniel Dance-Holland captured in 1770 in his portrait of her in court dress, weighed down by her robes (fig. 3.10). The year before he had painted a fresher, more spontaneous woman, unencumbered by ceremony (fig. 3.09). As she grew older she made a clear distinction between what she could say as Queen and what she could say securely in privacy. Lady Harcourt said: 'I remember once saying to her, 'I should like to tell you something, but pray promise never to let the Queen know it'. She laughed, and said, 'oh no, *she* can have no business with what passes between us in our private, unreserved conversation.'[41]

NATURE OR NURTURE?

A consort's power might seem to rest primarily on her ability to reproduce, but was there any way that a Hanoverian consort might want to move it away from biology to reflect merit and capacity? The later seventeenth century inaugurated debate about the relative influence of nature and nurture in relation to women's capabilities, and these discussions began in court circles. Two of the earliest articulations of feminism *avant la lettre* were Mary Astell's *A Serious Proposal to the Ladies* of 1694, which advocated a college for ladies as a kind of spiritual and educational retreat before they went on to the marriage market, and Judith Drake's sprightly *An Essay in Defence of the Female Sex*, published in 1696. Both were dedicated to Princess Anne, then heiress presumptive to the throne, who was about to become a monarch as well as a dynastic mother. Caroline as Queen was celebrated as a person of intellectual ability, and she surrounded herself with intelligent and cultured Ladies of the Bedchamber, such as Henrietta Louisa Fermor, Countess of Pomfret, who was also a Lady in Waiting to Princess Augusta on her marriage. In 1739 a series of dialogues entitled *Woman not inferior to Man: OR A short and modest Vindication of the natural Right of the FAIR-SEX to a perfect Equality of Power, Dignity, and Esteem, with the Men* was anonymously published by 'Sophia', that is, Wisdom. Her identity remains unknown but she may have been part of the extended network of Lady Fermor.[42] Its chief arguments are that it is only inadequate education that has created the appearance of unequal abilities in women, and therefore, if this were corrected, women would be fit for political office and even military leadership, as well as attainment in the sciences and professions. The female types portrayed include Novilia, who claims to know what the king said in bed to his wife the previous night.

The dating of this book might explain the robust reply given to Anne, Caroline's eldest daughter, happily married to Willem IV of Orange, by her unwed sister Amelia. She had been considered several times as royal consort to Prussian, Swedish, and Danish kings—and in the latter case was gallingly superseded by her younger and prettier sister Louisa. She wrote to Anne:

> I sometimes believe you think that unmarried women have no places in heaven for you think nobody can have the least happiness without being tied from morning to night to a creature which may tire one's life out and plague one incessantly with their good advice or commands whilst one is full as wise as they . . . They . . . might make a figure in life if they had a more manly education, we poor women have great disadvantages without making them still worse and putting oneself into a Jackanapes power. I am glad you are happy . . . but those things don't happen often.[43]

As the century developed it was evident that women were becoming ever more prominent as part of the general cultural conversation and as patrons and cultural producers. They were consistently gaining more scope to be the early years' instructors of their children, and using this position to obtain a better education for themselves. Queens Consort had a unique role, but this did not mean they were hermetically sealed off from other women, and they inevitably absorbed or even intensified the new developments taking place, and which seeped into their cognizance through other women in court service. Caroline took a conspicuous interest in educating her children. Richard Samuel's imagined group portrait of modern Muses in 1776 (now in the National Portrait Gallery, London) included the real-life figures of the painter Angelica Kauffmann and the Bluestocking hostess Mrs Elizabeth Montagu.[44]

Several women associated with the circle had links to Queen Charlotte, including Elizabeth Carter, to whom the Queen lent books, and Mary Hamilton, who was sub-governess to the princesses. The Queen was also recognized as a supporter of women authors. Maria Edgeworth argued that enlightened and educated women were now an unstoppable force, making a liberally minded fictional father in 1795 say 'more knowledge and literature are expected from [the female sex] in society.'[45] In 1820 Caroline of Brunswick, the estranged wife of George IV, a German princess herself, who conspicuously failed to establish herself as part of the royal family, became a figurehead for female wrongs precisely because his treatment of her now seemed alarmingly retrograde.[46] It could be argued that in a quiet way the Princesses had advanced women's status by reflecting and supporting the cultural claims being made by non-royal women throughout the century.

NOTES

1. This was the wording that ruled out Catholic succession in the 1689 Bill of Rights.
2. Christine Gerrard, 'Queens-in-waiting: Caroline of Anspach and Augusta of Saxe-Gotha as Princesses of Wales', in *Queenship in Britain, 1660–1837: Royal Patronage, Court Culture and Dynastic Politics*, ed. Clarissa Campbell Orr (Manchester: Manchester Univ. Press, 2002), 143–61.
3. Cited by Brendan Simms in 'Pitt and Hanover', in *The Hanoverian Dimension in British History, 1714–1837*, ed. Brendan Simms and Torsten Riotte (Cambridge: Cambridge Univ. Press, 2007), 32–33.
4. Prussia was outside the boundaries of the Holy Roman Empire despite being predominantly German speaking. Poland was an elective monarchy.
5. Out of many sources on the Holy Roman Empire, see Peter H. Wilson, *Absolutism in Central Europe* (London: Routledge, 2000).
6. Her fifth son, Ernest, led the liberation of Hanover from French occupation, and her seventh son, Adolphus, acted as Governor until the death of his brothers George IV and William IV, her first and third sons, when Ernest became King. As a surviving son of George III he took precedence over his niece Victoria.
7. Peter H. Wilson, 'Prussia's Relations with the Holy Roman Empire, 1740–1786', *Historical Journal* 51, no. 2 (June 2008): 337–71.
8. Andrew C. Thompson, *George II: King and Elector* (New Haven and London: Yale Univ. Press, 2011), 27–31.
9. Thompson, *George II*, 82, 109–10, 112–13, 116–17.
10. *Politische Corresponenz Friedrichs des Grossen*, 46 vols. (Berlin: Alexander Duncker, 1893), 20: 548, 559, 565.
11. Joanna Marschner, *Queen Caroline: Cultural Politics at the Early Eighteenth-Century Court* (New Haven and London: Yale Univ. Press, 2014); Mechthild Gretsch, 'Elstob, Elizabeth (1683–1756)', *Oxford Dictionary of National Biography* (Oxford: Oxford Univ. Press, 2004; online ed., Oct. 2007), http://www.oxforddnb.com/view/article/8761, accessed 13 March 2014; Elizabeth Elstob, *Rudiments of Grammar for the English-Saxon Tongue* (London: W. Bowyer, 1715), 4; Andrew Hanham, 'Caroline of Brandenburg-Ansbach and the 'Anglicisation' of the House of Hanover', in *Queenship in Europe, 1660–1815: The Role of the Consort*, ed. Clarissa Campbell Orr (Cambridge: Cambridge Univ. Press, 2004), 276–99.
12. Clarissa Campbell Orr, 'Life and Culture at Court in England and Hanover: An Anglo-German Comparison', in *Hannover-Coburg Gotha-Windsor: Problems and Perspectives of a Comparative Anglo German Dynastic History from the 18th to the 20th Century*, ed. Frank-Lothar Kroll and Martin Muncke, *Prince Albert Studies*, Vol. 32 (Berlin: Duncker & Humblot, 2015).
13. Clarissa Campbell Orr, 'Dynastic Perspectives', in *Hanoverian Dimension*, ed. Simms and Riotte, 241–46.
14. Clarissa Campbell Orr, 'Anglo-German Kinship Networks in 1832: Dynastic Survival and Adaptation', in *Royal Kinship: Anglo-German Family Networks, 1815–1914*, ed. Karina Urbach (Munich: K. G. Saur, Prince Albert Foundation Publications, 2008), 45–74; see 65–66.
15. Campbell Orr, 'Anglo-German Kinship Networks', 60–61.
16. Andrew Thompson, 'The Confessional Dimension', in *The Hanoverian Dimension*, ed. Simms and Riotte, 161–82.
17. Hanham, 'Caroline of Brandenburg-Ansbach'.
18. Clarissa Campbell Orr, 'The Late Hanoverian Monarchy and the Christian Enlightenment', in *Monarchy and Religion: The Transformation of Royal Culture in Eighteenth-Century Europe*, ed. Michael Schaich (Oxford: Oxford Univ. Press, 2007), 317–42.
19. David Watkin, *The Architect King: George III and the Culture of the Enlightenment* (London: Royal Collection Publications, 2004), 132–36.
20. Thompson, 'Confessional Dimension'.
21. Stephen Taylor, 'Queen Caroline and the Church of England', in *Hanoverian Britain and Empire: Essays in Memory of Philip Lawson*, ed. S. Taylor, R. Connors, and C. Jones (Woodbridge: Boydell and Brewer, 1998), 82–101.
22. Olwen Hedley, *Queen Charlotte* (London: John Murray, 1975), 290. This was an Anglican charity.
23. G. M. Ditchfield and Sarah Brewer, 'Hurd, Richard (1720–1808)', *Oxford Dictionary of National Biography* (Oxford: Oxford Univ. Press, 2004; online ed., Jan. 2008), http://www.oxforddnb.com/view/article/14249, accessed 4 March 2016.

24. A. Aspinall, ed., *The Later Correspondence of George III, 1783–1810*, 5 vols. (Cambridge: Cambridge Univ. Press, 1962–70), 5: 654.
25. Cited in *House of Commons, 1754–1790*, 'Hon. Richard Fitzpatrick, 1748–1813', 2: 433–35, from Sir John Fortescue, ed., *The Correspondence of King George III from 1760 to December 1783*, 6 vols. (London: Macmillan, 1927–28), 3: 190.
26. D. G. C. Allen, 'Hales, Stephen (1677–1761)', *Oxford Dictionary of National Biography* (Oxford: Oxford Univ. Press, 2004; online ed., Oct. 2006), http://www.oxforddnb.com/view/article/11915, accessed 18 Oct. 2012.
27. See Frances Harris, "The Honourable Sisterhood': Queen Anne's Maids of Honour', *The British Library Journal* 19, no. 2 (Autumn 1993): 181–98.
28. Clarissa Campbell Orr, 'New Perspectives on Hanoverian History', *The Historical Journal* 52, no. 2 (2009): 513–29.
29. Stephen Taylor, 'Caroline [Princess Caroline of Brandenburg-Ansbach] (1683–1737)', *Oxford Dictionary of National Biography* (Oxford: Oxford Univ. Press, 2004; online ed.), http://www.oxforddnb.com/index/101004720/Caroline, accessed 1 March 2016.
30. John Percival, Earl of Egmont 'Memoranda Book' (20 March 1751), 198–99. In A. N. Newman, ed., 'Leicester House Politics, 1750–60', from the Papers of John, Second Earl of Egmont, in *Camden Miscellany*, Vol. 23, Camden 4th Series Vol. 7 (London: Royal Historical Society, 1969).
31. This mix of political opposition tempered by parliamentary and religious custom persists. This essay was written during the 2015 British General Election. The morning after, two party leaders out of three resigned after a conspicuous defeat, yet at 3:00 pm the same day they stood alongside the victorious Prime Minister at the National Cenotaph in Whitehall for the commemoration of the war dead of VE Day in 1945.
32. Thompson, *George II*, 159; Christopher Hibbert, *George III: A Personal History* (London: Penguin, 1998), 81.
33. J. L. Bullion, '"To play what game she pleased without observation": Princess Augusta and the Political Drama of Succession, 1736–1756', in *Queenship in Britain*, ed. Campbell Orr, 207–35.
34. See Jeremy Black, *George III: America's Last King* (New Haven and London: Yale Univ. Press, 2006); Herbert Butterfield, *George III and the Historians* (London: Collins, 1957); Hibbert, *George III*; and John Cannon, 'George III (1738–1820)', *Oxford Dictionary of National Biography* (Oxford: Oxford Univ. Press, 2004; online ed., May 2013), http://www.oxforddnb.com/view/article/10540, accessed 29 May 2016.
35. There is a slight possibility he might have made an incognito visit to Scotland when still Prince of Wales.
36. Jeroen Duindam, *Dynasties: A Global History of Power, 1300–1800* (Cambridge: Cambridge Univ. Press), 2015.
37. My italics. The Bute Archive at Mount Stuart, BU/108/97. I am very grateful to Alice Martin, Head of Collections at Mount Stuart, for facilitating my visit to Mount Stuart in May 2015, and to Lynsey Nairn, Archivist, and her team for every assistance in accessing material and discussing George III's progression to manhood with me.
38. The Bute Archive at Mount Stuart, BU/114, Queen Charlotte to Lord Bute, 2 May 1762; Romney Sedgwick, ed., *Letters from George III to Lord Bute, 1756–1766* (London: Macmillan, 1939), 171, note 1.
39. Clarissa Campbell Orr, 'Six Courts and Four Empires: Zoffany as Courtier', in *Johan Zoffany RA: Society Observed*, ed. Martin Postle (New Haven and London: Yale Univ. Press, 2011), 75–99.
40. Edward William Harcourt, ed., *The Harcourt Papers*, 14 vols. (Oxford: Privately printed, 1880–95), 6: 109–10.
41. Harcourt, *Harcourt Papers*, 4: 79.
42. Karen O'Brien, *Women and Enlightenment in Eighteenth-Century Britain* (Cambridge: Cambridge Univ. Press, 2009), 17–18.
43. Veronica Baker-Smith, *Royal Discord: The Family of George II* (London: Athena Press, 2008), 109.
44. Richard Samuel, *Portraits in the Characters of the Nine Muses in the Temple of Apollo*, oil on canvas, 1776, 52 x 61 in. (132.1 x 154.9 cm). National Portrait Gallery, London, NPG 4905.
45. Maria Edgeworth, *Letters for Literary Ladies*, ed. Claire Connolly (London: Dent, 1993), 19.
46. There is a large literature on this: for excellent analysis of many facets, see Anna Clark, *Scandal: The Sexual Politics of the British Constitution* (Princeton, NJ, and Oxford: Princeton Univ. Press, 2004).

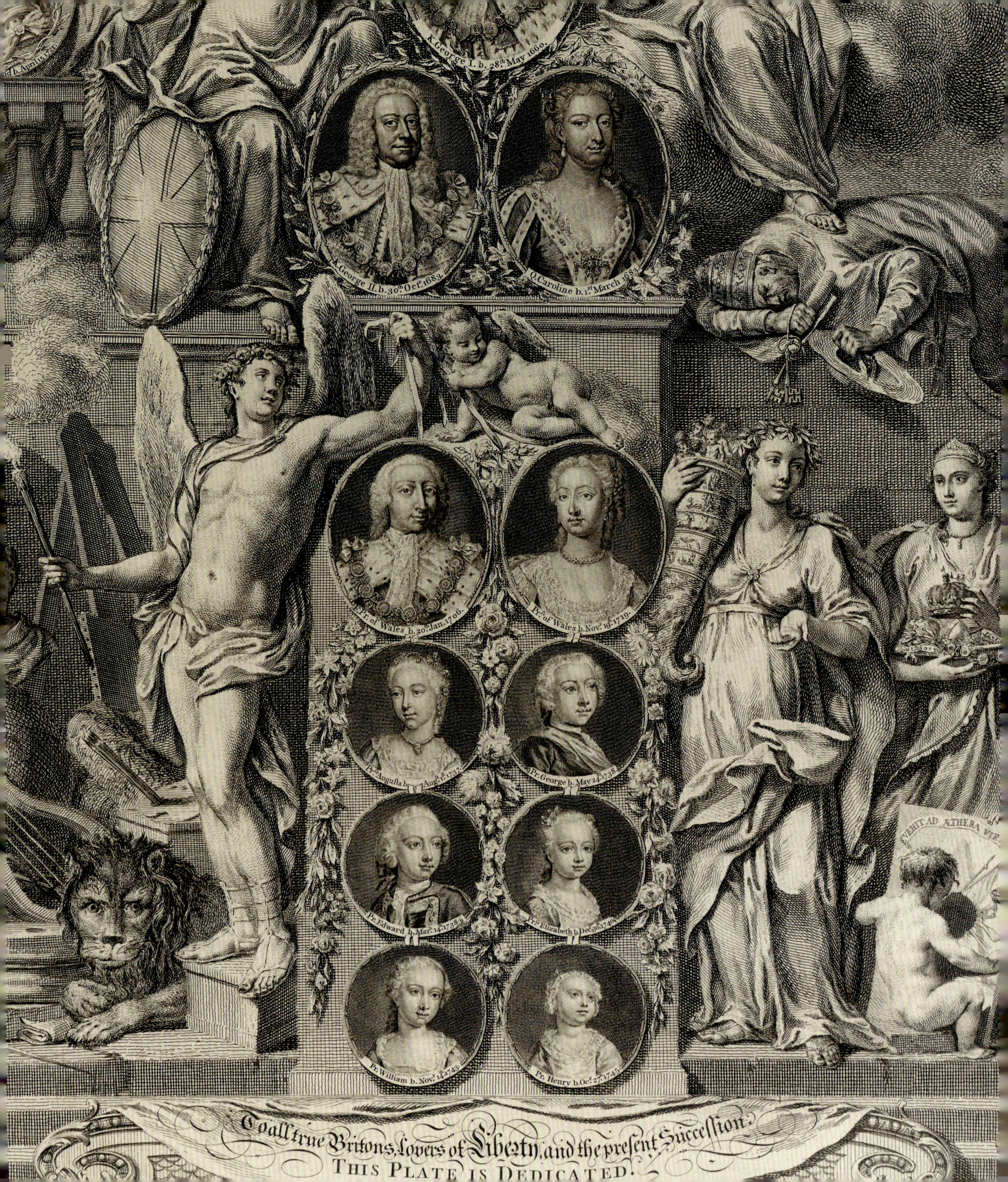
To all true Britons, Lovers of Liberty, and the present Succession,
THIS PLATE IS DEDICATED.

Cornucopia: Royal Female Portraiture and the Imperatives of Reproduction

A. Cassandra Albinson and Mark Hallett

In May 1748, a lengthy newspaper advertisement in London's *General Evening Post* alerted readers to the publication of 'A PRINT of the ROYAL FAMILY, from King George I to the present Time; adorned with Emblematic Figures'.[1] This impressive engraving (fig. 4.01), executed on what the advertisement described as a 'large SHEET-PLATE', had been designed by Samuel Wale and engraved by Johann Sebastian Müller. As the notice went on to declare, 'on the upper Part of the Picture sit Britannia and Liberty, holding the British Imperial Crown over the Heads of his late and present Majesty [that is, George I and George II] and Queen Caroline.' Meanwhile, 'over the heads of the Prince and Princess of Wales [Prince Frederick and Princess Augusta of Saxe-Gotha], and their present Royal Issue, are represented Cupid and Hymen', holding what are described as the 'sacred Bands' of matrimony. In the print, the 'happy Union' between Frederick and Augusta is denoted by the figures of Fruitfulness and Honour, the latter of whom 'stands ready to confer Diadems on the Princely Offspring'. In the far left background of the image, we can see the figure of Minerva leaning against a shield adorned with the profile of George II's youngest son, William Augustus, Duke of Cumberland, who had recently played a central role in putting down a Jacobite revolt; he, in turn, is shown being presented with a crown of laurel by 'the Figure of Glory'. Elsewhere, the sprawling figure of 'Popish Superstition' is depicted being trampled under Liberty's feet; near them, as the advertisement confirms, 'is the Figure of *Justice*, which denotes the Equity of the Government under this august Line'. Meanwhile, the symbols in the lower part of the engraving—which include a globe, an easel, a palette, paintbrushes, and a lyre—'allude to Arts and Sciences, which have flourish'd in these auspicious Reigns'.

Wale and Müller's engraving, which deploys a host of long-standing genealogical and commemorative conventions, offers a visually complex and patriotically forthright defence of the Hanoverian dynasty in the aftermath of the recent Jacobite invasion.[2] At the same time it offers a suggestive starting point for thinking afresh about the

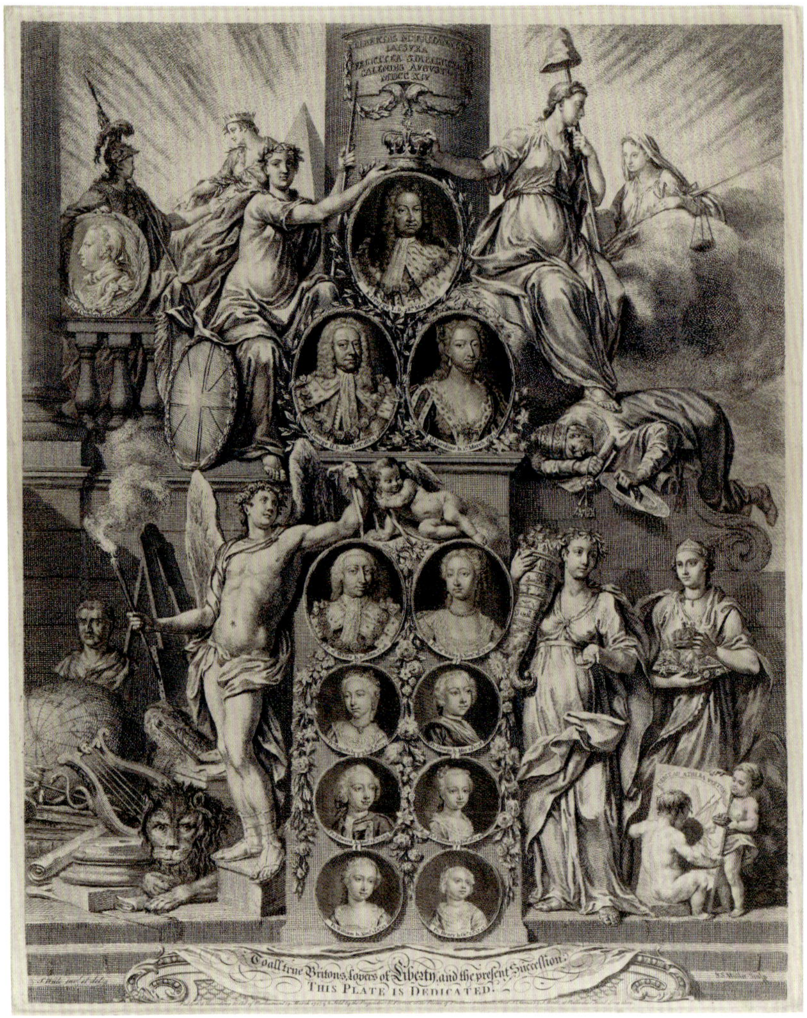

FIG. 4.01 Johann Sebastian Müller, engraver, after Samuel Wale, *Emblematic Image of the Hanoverian Succession*, 1748, etching and engraving on paper, 20¾ x 16½ in (52.8 x 41.8 cm). British Museum, London

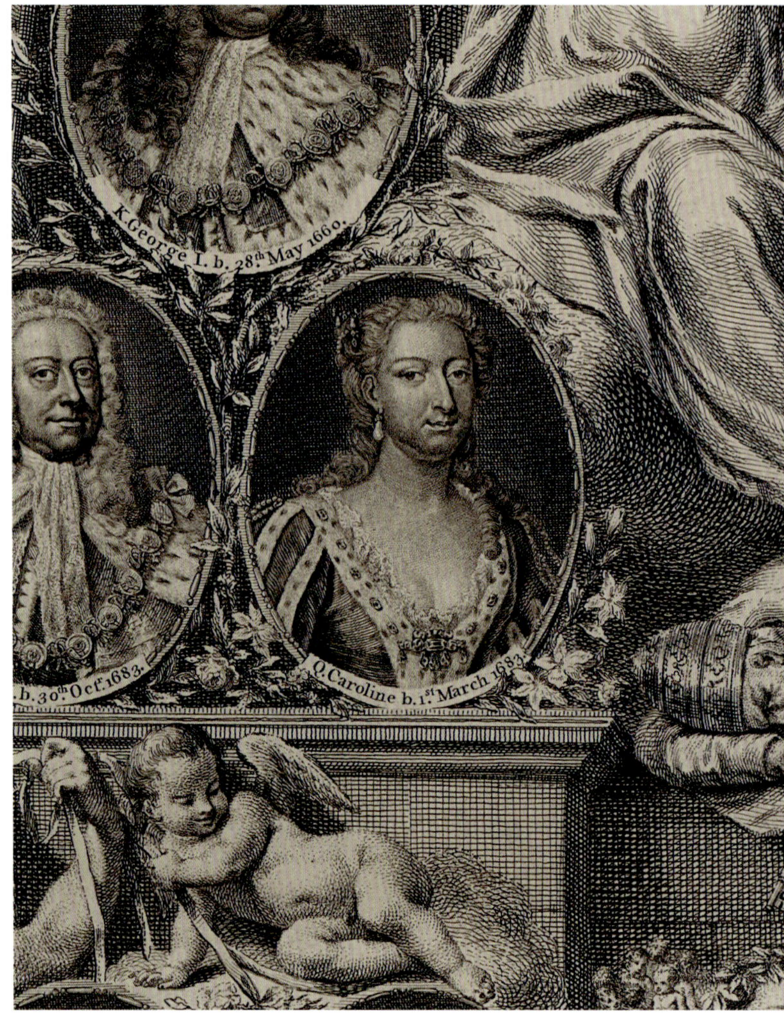

FIG. 4.02 Johann Sebastian Müller, engraver, after Samuel Wale, *Emblematic Image of the Hanoverian Succession* (detail), 1748. British Museum, London

workings of female royal portraiture in this period and, in particular, about the relationship of such portraiture to two different forms of reproduction. It is immediately evident that the portraits of such figures as Caroline of Ansbach, Augusta of Saxe-Gotha, and later Charlotte of Mecklenburg-Strelitz, were enmeshed in complex processes of pictorial reproduction. An inspection of the portrait of Caroline that is incorporated into the group of roundels shown hanging in front of the depicted column (fig. 4.02) reveals that in all likelihood, Wale based this portrait on that found in a print issued by the distinguished mezzotint engraver John Faber the younger in 1739 (fig. 4.03). It is evident that Wale's roundel offers an exact, if truncated, duplicate of this portrait, showing Caroline facing in the same direction. Faber's own printed image, which depicts 'Her Most Excellent Majesty Queen Caroline', itself reproduced a painted portrait of the Queen executed by John Vanderbank the younger in 1736, now in the collection of Goodwood House, Sussex (fig. 4.04). These three different

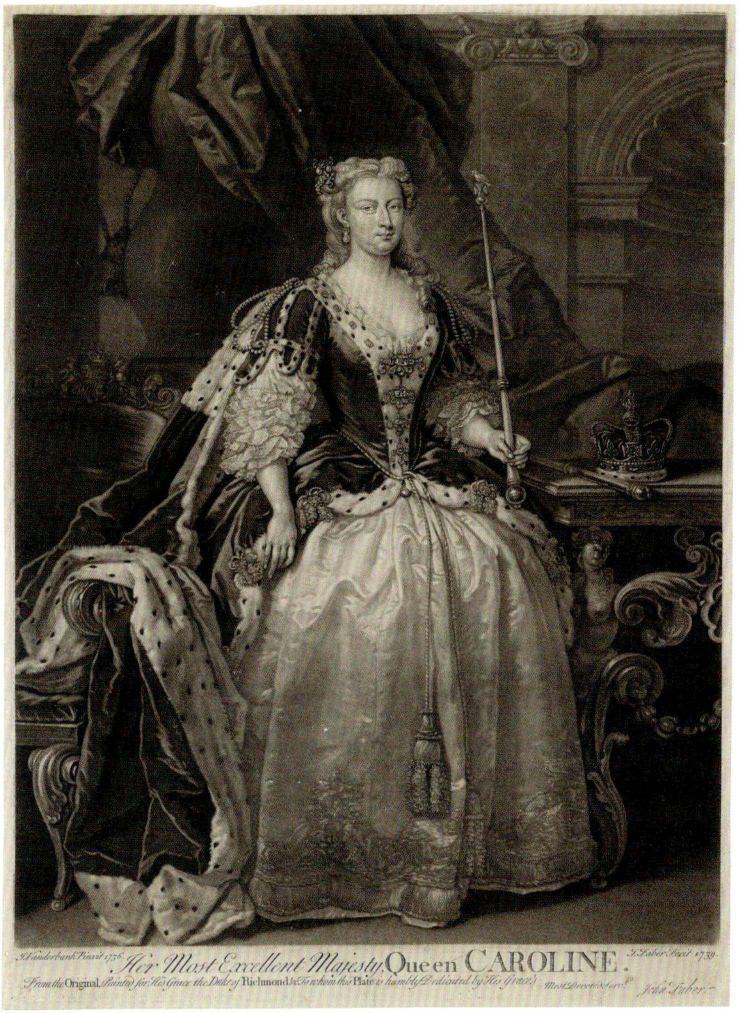

FIG. 4.03 John Faber the younger, *Her Most Excellent Majesty Queen Caroline*, 1739, mezzotint, 18¾ x 13¾ in. (47.5 x 34.9 cm). Royal Collection Trust

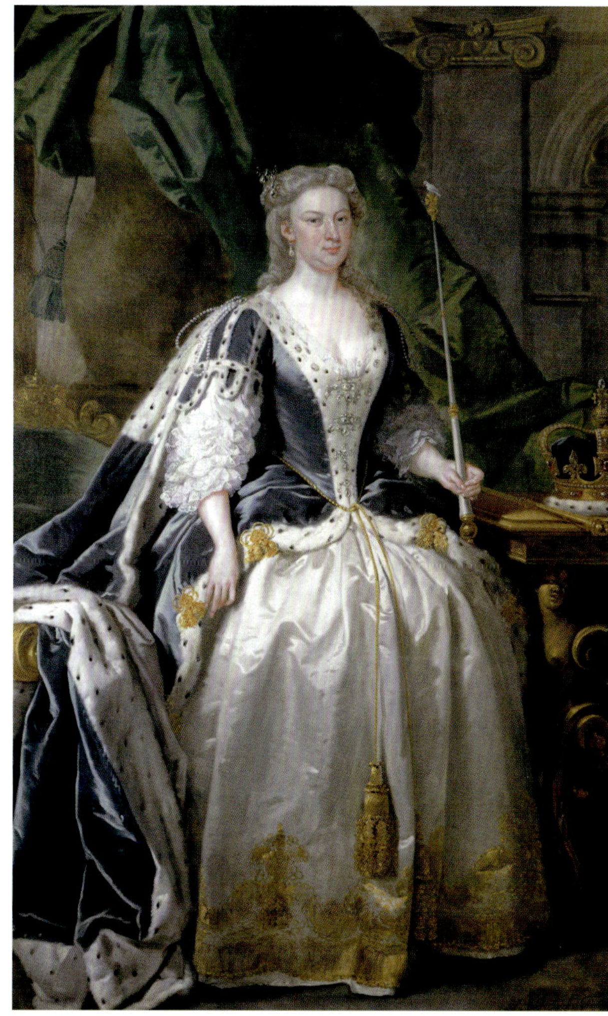

FIG. 4.04 John Vanderbank the younger, *Queen Caroline*, 1736, oil on canvas. Goodwood House Collection, Duke of Richmond and Gordon

iterations of Caroline's image, which mobilized her likeness across a variety of pictorial settings, alert us to the fact that the female royal portraits of the period were not only self-sufficient works of art, but also images that were continually reactivated and reused across a variety of media.

A second point of interest within Wale and Müller's image is its emphasis on generational reproduction. For the Hanoverian kings, reproduction in the form of numerous royal children with a claim to the throne became the cornerstone of a dynastic strategy for staying in power. The house of Hanover had to outlast any challenges from the Stuarts and to avoid the problem of its predecessors, William and Mary and Queen Anne, who had failed to produce any royal heirs, putting the whole succession in peril. The print of 'the ROYAL FAMILY, from King George I to the present Time' places a special emphasis on generational succession, and on the 'present royal issue', Frederick and Augusta's six children, whose faces smile out at us from the lower part of the print. This emphasis on fertility finds allegorical expression in the figure of 'Fruitfulness', who carries a cornucopia that issues forth the miniaturized bodies and faces of numerous infants (see fig. 4.01 and p. 76). The inclusion of this figure and her overflowing cornucopia is a telling one: though the engraving incorporates portraits of kings and princes and is filled with masculine symbolism of shields, swords, and monumental public architecture, it is more powerfully defined by a feminized imagery of procreation and proliferation, articulated through the numerous women and children, both real and allegorical, who crowd its spaces.

This essay explores female royal portraits of the Georgian period through the prism of both pictorial and bodily reproduction. There seem to be good reasons for doing so. As has already started to become apparent, the appeal and meanings of such portraits were shaped by the dynamics of graphic reproduction, as in Wale and Müller's print, or through a parallel form of visual recycling, in which a commissioned oil painting would become the basis for scores of painted copies. At the same time, these women's portraits were continually and increasingly embedded in narratives of royal bodily reproduction. From representations of Caroline in the first half of the century to those of Charlotte and her family at the end of the century, royal female portraiture was ever-more closely aligned with a burgeoning mythology of Hanoverian fertility. The continual recycling of these women's portraits through prints and paintings meant that this narrative of royal fertility, together with its associated storylines of maternal care and domestic tenderness, was disseminated to all corners of British dominion, at home and overseas. In this process, the British art world itself served as a kind of cornucopia, spilling forth thousands of images of royal children and their devoted mothers.

CAROLINE

In 1716, two years after her arrival in England and her proclamation as Princess of Wales, Caroline of Ansbach was painted by Sir Godfrey Kneller, long established as the leading portraitist to the British court.[3] His full-length depiction of the Princess (fig. 4.05) grafts her youthful likeness onto the schematic pictorial apparatus of the state portrait, and bedecks her with the regalia of her new position—ermine-lined robes, an elaborately decorated chair, and a coronet that perches on a nearby table. Built into the production of such images was the expectation that they would obey certain laws of sartorial, gestural, and pictorial decorum; make a powerful visual impact on the spectator; and operate in harmony with one or more pictorial companions, in this case Kneller's depiction of her husband, George, Prince of Wales (fig. 4.06). Kneller excelled in producing state portraits that met these various criteria and that, in a different way from the lush imagery of his predecessor Sir Peter Lely, managed to communicate both a fitting degree of royal splendour and the dignified forms of courtly restraint associated with the post-Restoration-period monarchy. Central to this effect was the fact that Kneller's works of this kind typically subordinated

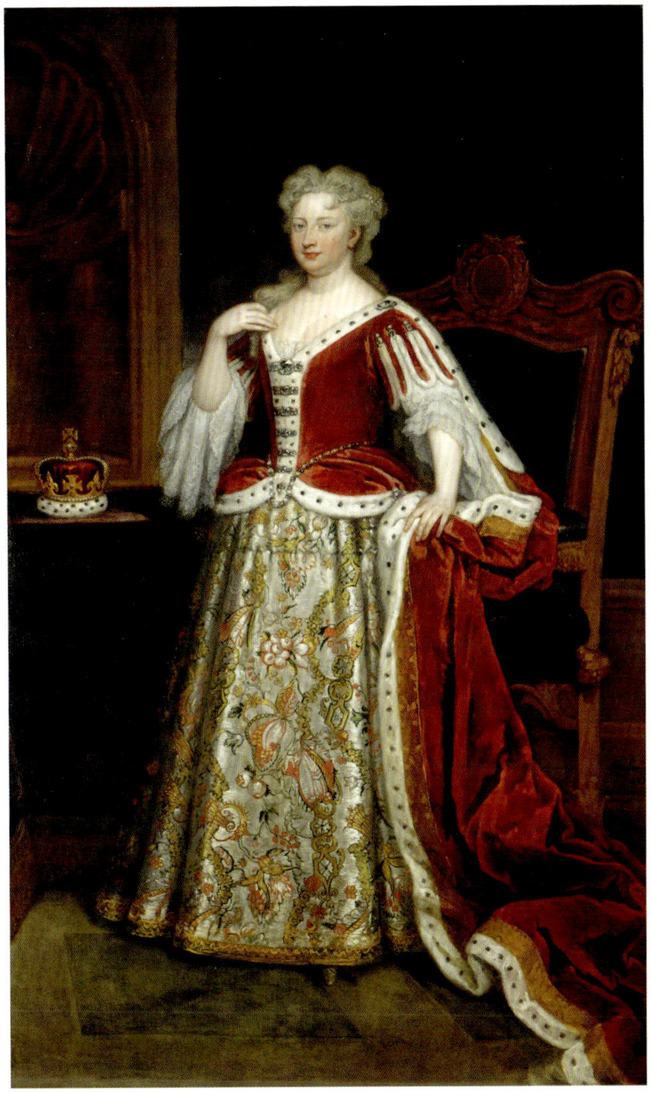

FIG. 4.05 Sir Godfrey Kneller, *Queen Caroline of Ansbach, when Princess of Wales*, 1716, oil on canvas, 94½ x 55¾ in. (240 x 141.6 cm). Royal Collection Trust

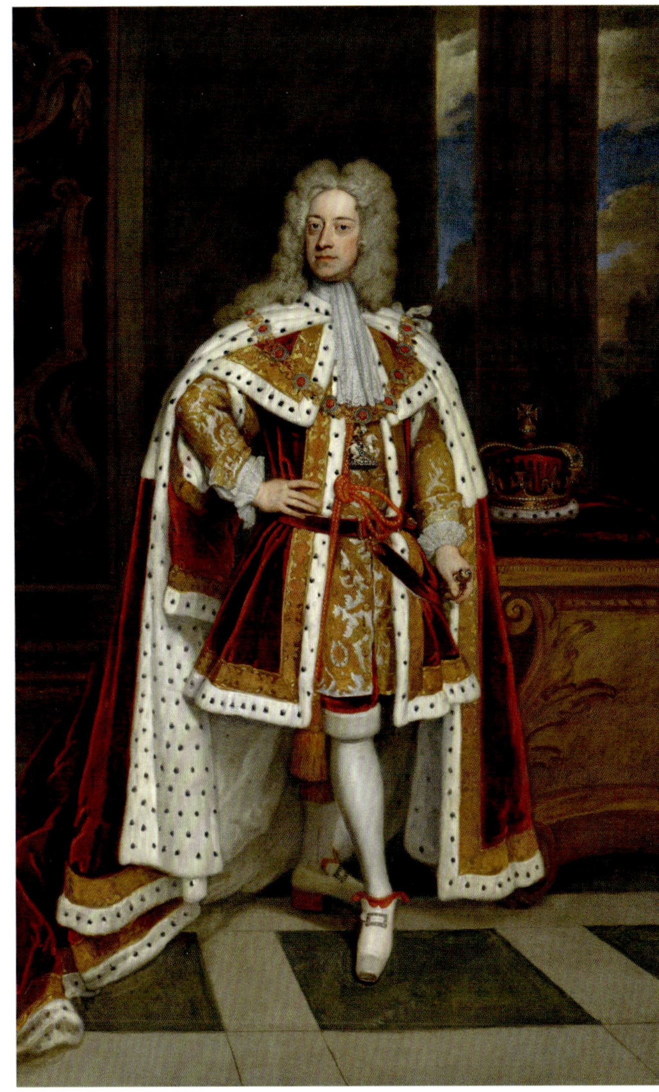

FIG. 4.06 Sir Godfrey Kneller, *George II when Prince of Wales*, 1716, oil on canvas, 94⅛ x 58⅜ in. (239.2 x 148.4 cm). Royal Collection Trust

rich areas of colour and pattern—particularly in the depiction of dress—to a taut, pared-down, and sharply delineated imagery of the head and shoulders, the bodily silhouette, and the pictorial setting, all of which gave his pictures the 'remarkable simplicity' lauded after his death by André Rouquet.[4]

This combination of pictorial qualities—in particular, their clear outlines and uncluttered settings—also ensured that Kneller's state portraits met another crucial expectation for such images: an openness to painted and graphic reproduction. His portrait of Caroline in 1716 was no exception, and became the prototype for an enormous number of paintings and prints, in which the artist's original template was continually revised and reformulated. As a demonstration of this, three engraved versions of Kneller's image produced by the distinguished printmaker John Smith might be

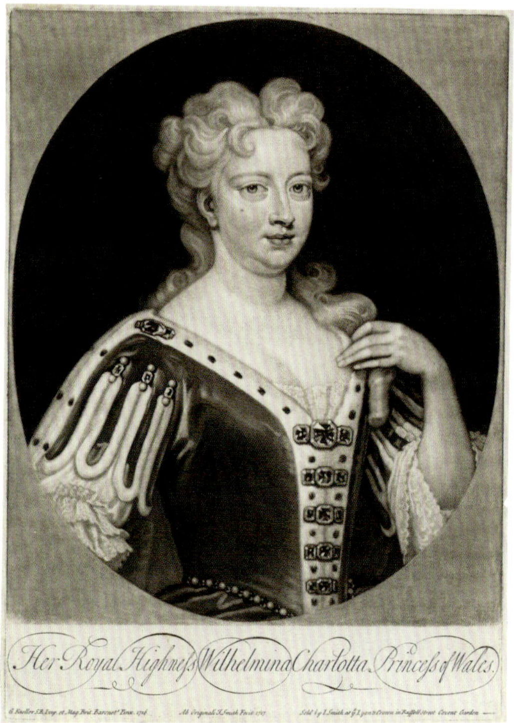

FIG. 4.07 John Smith after Sir Godfrey Kneller, *Her Royal Highness Wilhelmina Charlotta, Princess of Wales* [Caroline], 1717, mezzotint, 11¾ x 9⅞ in. (30 x 25 cm). Royal Collection Trust

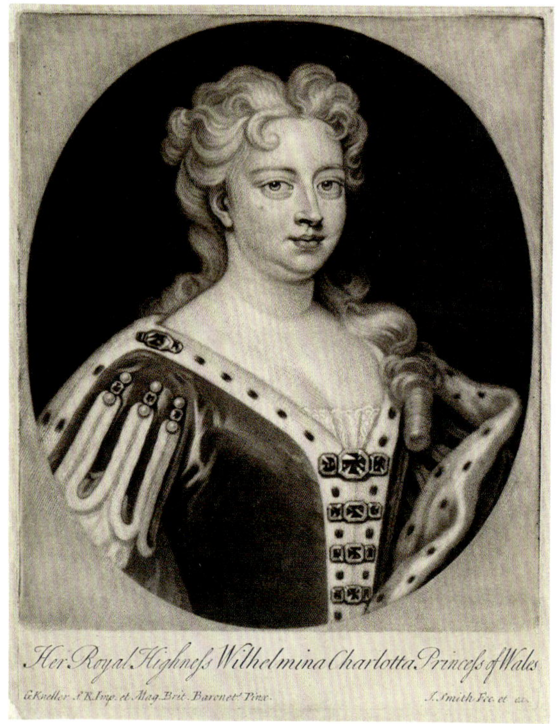

FIG. 4.08 John Smith after Sir Godfrey Kneller, *Her Royal Highness Wilhelmina Charlotta, Princess of Wales* [Caroline], c. 1716–27, mezzotint, 7 x 5⅞ in. (17.8 x 15 cm). Royal Collection Trust

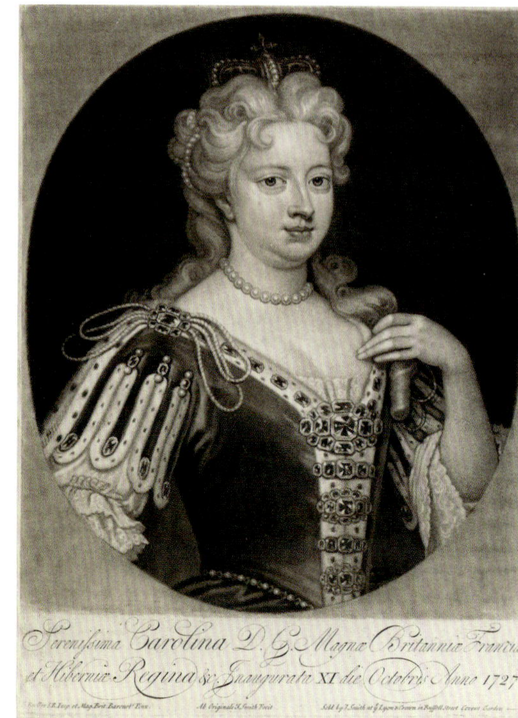

FIG. 4.09 John Smith after Sir Godfrey Kneller, *Serenissima Caroline D. G. Magnae Britanniae Franciae et Hiberniae Regina &c.*, c. 1727–43, mezzotint, 11¾ x 9⅞ in. (29.7 x 25 cm). Royal Collection Trust

cited.[5] The first is a print published in 1717 (fig. 4.07), which takes the form of an oval bust adapted from the full-length original. Smith exploits the schematic simplicity of Kneller's portrait to create a stark and arresting two-toned image, in which Caroline's pale unadorned face, neck, hand, and forearm, and her long blond hair, are played off against a background of almost unrelieved darkness. The work's focus on the Princess's smiling face and bare shoulders, the twists and turns of her hair, and the delicate touch of her fingers on her skin, would give it a particular appeal in the marketplace; the image emphasizes Caroline's youthful attractiveness as much as her royal status. Intriguingly, Smith produced a second print after Kneller's portrait that offers a subtly different address to the viewer, in which the Princess's entire left arm, and the gesture she makes with her fingers, is erased and replaced by a swirl of ermine-lined cloak (fig. 4.08). The image is thereby made heavier in feel and more decorous in its approach. On Caroline's accession to the role of Queen in 1727, Smith issued a third version with a fresh inscription (fig. 4.09). She has acquired not only a new title, but also the knots and twists of pearls that adorn her hair, neck, shoulders, and arms. The ermine that fringes her robe is more thickly dotted and bejewelled, and the coronet that had sat on a nearby table in Kneller's portrait, is now placed on top of her head. In this process, Smith generates an image of Caroline that reuses the painter's decade-old prototype whilst also offering a highly topical representation of the newly crowned Queen.

If Smith's three images confirm the ways in which the individual royal female portrait was subject to constant pictorial recycling, another reworking of Kneller's depiction of Caroline shows the early emergence of the narratives of Hanoverian female fecundity. In 1716, the poet Joseph Addison published a poem entitled *To Her Royal Highness the Princess of Wales*, which was paired with a poem addressed *To Sir Godfrey Kneller, on his Picture of the King*.[6] The poem dedicated to Caroline, who had now given birth to four children and whose marriage was celebrated in the period as an example of the virtues of conjugal love, proceeds almost immediately to celebrate her fertility.[7] Addison sees this fertility as heralding a new era of continuity and peace within a nation that had long been fractured by the uncertainties and failures of royal succession:

> No longer shall the widow'd Land bemoan
> A broken Lineage, and a doubtful Throne;
> But boast her Royal Progeny's Increase,
> And count the Pledges of her future Peace.
> O Born to strengthen and to grace our Isle![8]

Addison lauds the beauties and charms of the 'fair' Princess's three daughters, Anne, Amelia, and Caroline, and imagines them attracting the attentions of a series of love-struck, high-born, but ultimately unsuccessful suitors:

> While you fair PRINCESS, in your Off-spring smile
> Supplying Charms to the succeeding Age,
> Each Heav'nly Daughter's Triumphs we presage;
> Already see th' Illustrious Youths complain,
> And pity Monarchs doom'd to sigh in vain.[9]

The poet then turns to Caroline's eldest son, Frederick, and imagines him as a handsome and desirable youth being enveloped in the embrace of a female 'Albion':

> Thou too, the Darling of our fond Desires,
> Whom *Albion*, opening wide her Arms, requires,
> With manly Valour and attractive Air
> Shalt quell the Fierce, and captivate the Fair,
> O *England's* younger Hope! In whom conspire
> The Mother's Sweetness, and the Father's Fire![10]

The presence of this healthy and desirable brood of royal children presages an era in which writers, 'Pleas'd with the Prospect of successive Reigns', can at last turn from the discourses of rebellion and strife to focus on preoccupations of a gentler and more soothing kind, including those of 'filial love':

To milder Sounds each Muse shall tune the Lyre,
And Gratitude, and Faith to Kings inspire,
And Filial Love; bid impious Discord cease,
And sooth the madding Factions into Peace.[11]

Sometime after 1717, the engraver and print-publisher John Clark produced a pair of prints that reproduced Addison's poems to Caroline and Kneller and that were illustrated with elaborate headpieces.[12] The print featuring verses devoted to the Princess shows her portrait, deriving from that by Kneller, surrounded by three figures in a classical architectural setting (fig. 4.10). A seated figure on the left who rests the caduceus against her shoulder and probably represents Peace, gestures towards Caroline's portrait. She is paired with the voluptuous figure of Fruitfulness on the right, holding an overflowing cornucopia whilst being offered gifts by a helmeted companion.

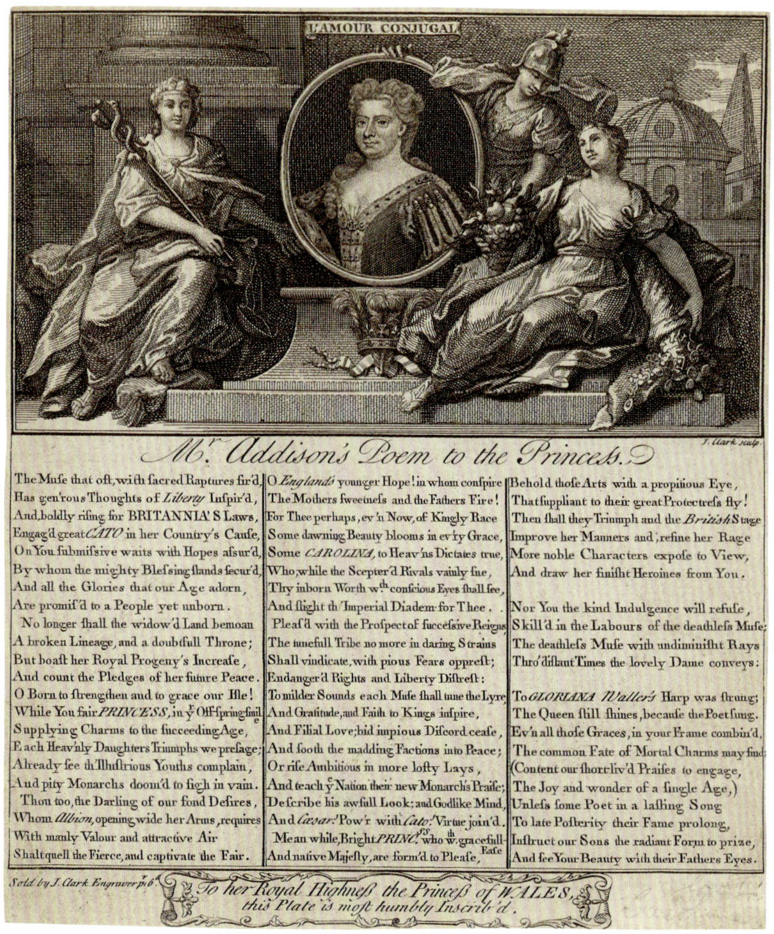

FIG. 4.10 John Clark, after Sir Godfrey Kneller, Bt, *Caroline Wilhelmina of Brandenburg-Ansbach when Princess of Wales* with *Mr. Addison's Poem to the Princess*, c. 1714–27, line engraving and etching, 7⅞ x 6½ in. (20 x 16.6 cm), National Portrait Gallery, London, Purchased with help from the Friends of the National Libraries and the Pilgrim Trust, 1966

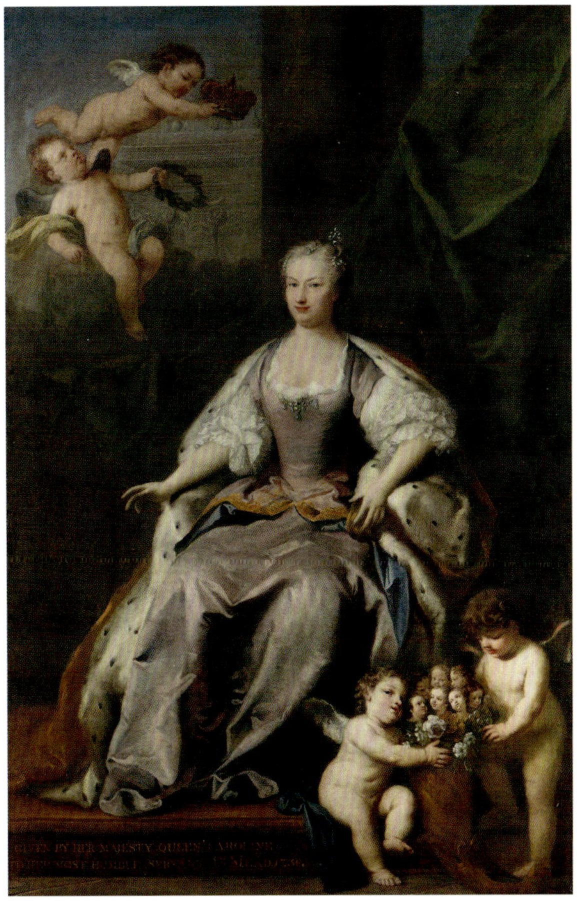

FIG. 4.11 Jacopo Amigoni, *Caroline Wilhelmina of Brandenburg-Ansbach*, 1735, oil on canvas, 96⅛ x 59⅞ in. (244 x 152 cm). National Portrait Gallery, London

Clothed in loose-fitting drapery that clings to her outstretched body, Fruitfulness sprawls across the plinth that supports Caroline's portrait, gazing up not only at her male companion but also at the engraved inscription—'L'AMOUR CONJUGAL'—that surmounts the portrait, alluding to the conjugal love that Caroline famously enjoyed with her husband. Here, then, the Princess is pointedly aligned with a figure who not only symbolizes the fecundity celebrated by Addison, but who, in her languorous body language and her invocation of the pleasures of the marriage bed, gives this fecundity a strikingly sensual form.

Caroline's association with fertility and with her ever-growing number of 'royal progeny' was rendered even more explicit in a portrait she commissioned from the Italian artist Jacopo Amigoni in 1735, and that she gave to the royal physician and art-lover Dr Richard Mead.[13] Though Caroline was fifty-two years old at the time the portrait was painted, her face is youthful and her figure slim. She is accompanied by four putti, two of whom carry a crown of laurels and a crown of state, as if bringing them from the more public realms that operate beyond the painting's fictional setting; the other two putti are shown operating in a more private and intimate zone of the

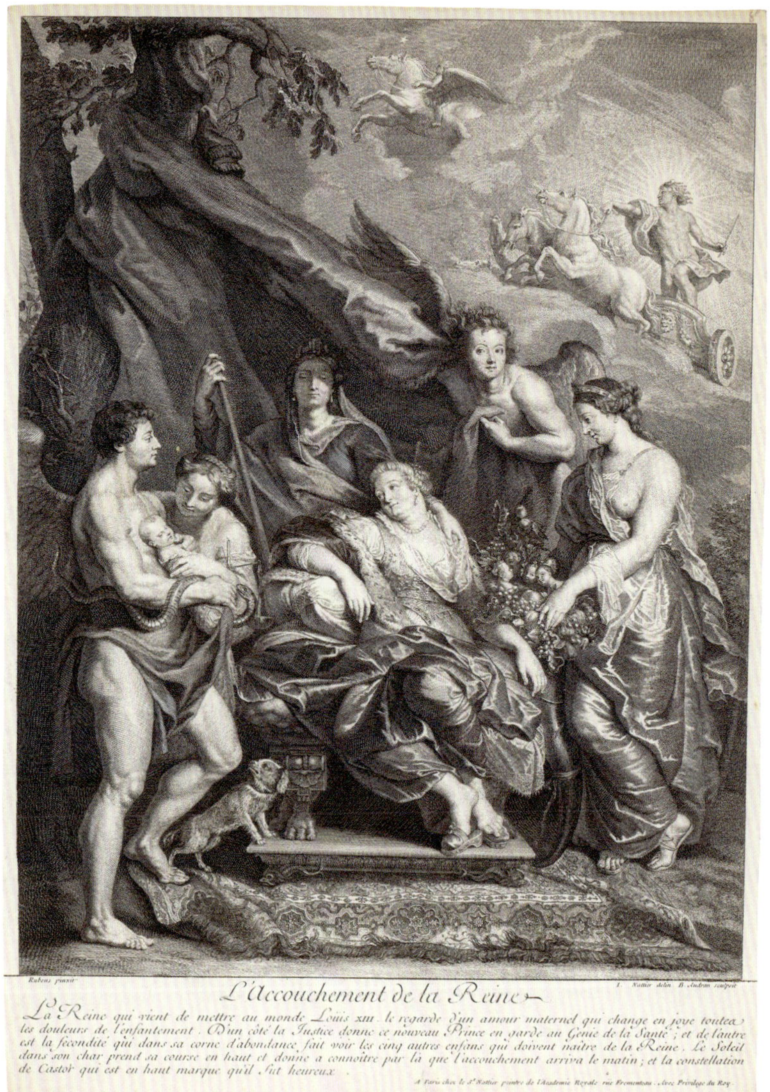

FIG. 4.12 Benoît Audran I after Jean-Baptiste Nattier after Sir Peter Paul Rubens, *L'Accouchement de la Reine*, 1710, engraving. Fine Arts Museums of San Francisco, Achenbach Foundation for Graphic Arts

image, tending to another cornucopia, this time one that overflows with the miniaturized heads of the Queen's surviving children, now seven in number (fig. 4.11).

Amigoni's portrait—like the later image by Wale and Müller—draws upon a canonical image of regal fertility: Rubens's celebrated allegorical portrait, *The Birth of the Dauphin at Fontainbleau*, produced in the early 1620s as part of his great cycle of paintings depicting the life of the French queen Marie de Medici. In this work, which was reproduced as an engraving in 1710 (fig. 4.12), the queen is shown soon after childbirth, sitting exhaustedly but happily on a richly canopied throne, gazing at her newly born son, cradled in the arms of the figure of Good Health. On her left stands the personification of Conjugal Fecundity, again holding a cornucopia issuing forth a stream of infants' heads, this time symbolizing the five further children that Marie would bear during her time as queen. Amigoni's portrait, in engaging with Rubens's famous precedent in such a thoroughgoing way—a reference that a connoisseur like Mead would have spotted in an instant—asked its viewers to equate Caroline with another great female founder of a royal dynasty, Marie de Medici. At the same time, the allusion to Rubens's example might have alerted those same viewers to the *absence* in Amigoni's portrait of an equivalent figure of Conjugal Fecundity standing between the Queen and the cornucopia that lies at her feet; from this perspective, indeed, Caroline herself can be seen to assume this form of allegorical identity. More accurately, perhaps, the Italian artist's portrait is one that depicts her engaging with a number of different personae: that of Conjugal Fecundity; that of a Queen (symbolized by the crown and laurel wreath that are about to be lowered onto her head); and that of a proud and healthy mother, juxtaposed with the collection of portrait miniatures depicting her burgeoning brood of children.

AUGUSTA

In the spring of 1736, Caroline's eldest son Frederick married Princess Augusta of Saxe-Gotha, who was also to play an active part in the pictorial mythologies of Hanoverian fertility. Like her mother-in-law, she was represented as Princess of Wales in a magnificent full-length state portrait, painted in 1742 by Jean-Baptiste van Loo (fig. 4.13). Some three years earlier, however, another portrait by the same artist had activated a parallel pictorial role for Augusta, that of an attentive mother operating at the heart of a busy Royal Household (fig. 4.14).[14] In this innovative group portrait, which draws upon the fashionable forms of the conversation piece, the main compositional focus is Augusta herself, who is pictured sitting at the painting's centre, tenderly holding the arm of her eldest daughter, Princess Augusta. Towards her right, we find the seated figure of her eldest son Prince George, the future George III. Behind him stand two members of the Royal Household—Arabella Herbert, who was Governess to Princess Augusta, and, to her left, holding the infant figure of Prince Edward in her arms, Lady Jane Hamilton, Mistress of the Robes. The standing man on the right is Sir William Irby, who had previously been a page to George I and George II, and was now Vice-Chamberlain to Augusta herself. He proffers her a letter with his gloved hand.

In this picture, the regalia which will continue to be found in the later state portrait of Augusta by Van Loo, is subordinated to a celebration of her authoritative role in the management of a particular branch of the Royal Household and the education and upbringing of her children. Yet the artist also ensures that our eyes soon turn away from Augusta herself, and towards the body of the royal heir. The Princess and her attendants' attention is drawn relentlessly towards the left-hand side of the canvas and the figure of the future king. Though Augusta provides comfort and contact for her daughter with one hand, her other hand, her eyes, face, and upper body move towards her eldest son. These gestures to her right mean that she ignores the message held by Irby, who also casts his eyes towards the Prince. Lady Jane Hamilton, though she faces

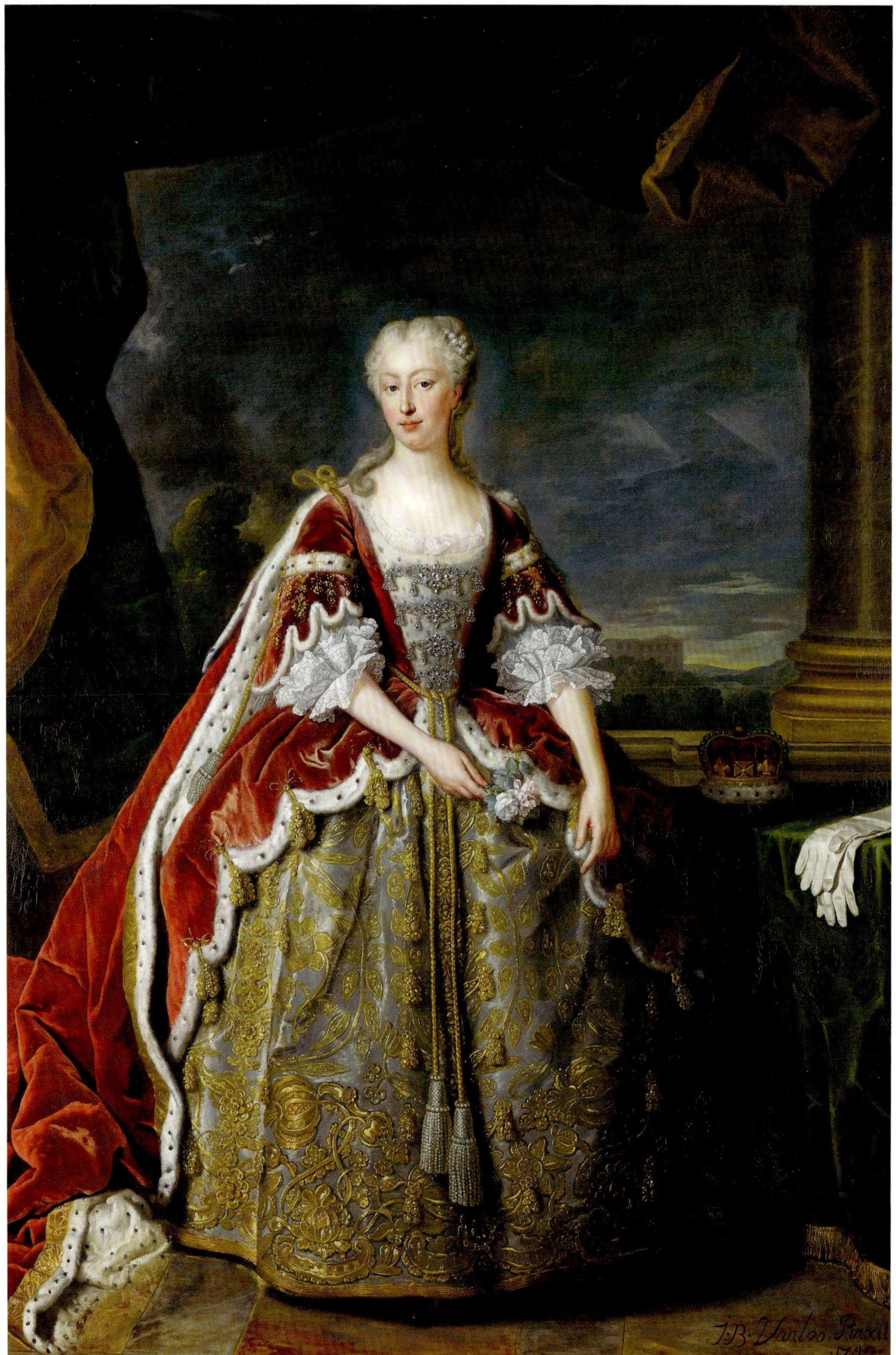

FIG. 4.13 Jean-Baptiste van Loo, *Augusta, Princess of Wales*, 1742, oil on canvas, 94½ x 62 in. (240 x 157.5 cm). Royal Collection Trust

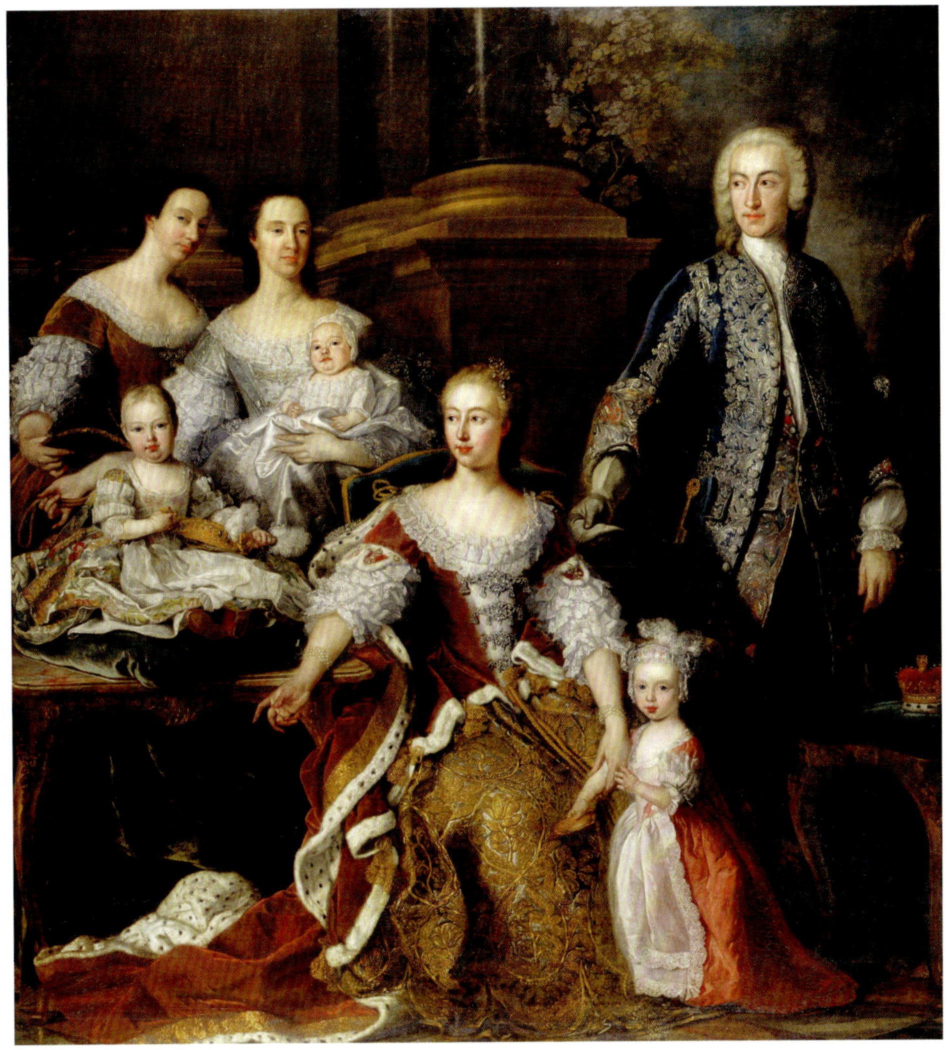

FIG. 4.14 Jean-Baptiste van Loo, *Augusta, Princess of Wales, with Members of her Family and Household*, 1739, oil on canvas, 98 x 89 in. (249 x 226 cm). Royal Collection Trust

Augusta, stretches her right hand behind George, playing with a cord or strap behind his back; Mrs Herbert holds the Prince's hat in her right hand.

George, despite his seemingly unobtrusive location at the picture's left edge, is thus placed at the heart of a centripetal swirl of gestures and looks that pull the viewer's attention away from his mother and towards him. We, in turn, are invited to appreciate the maternal, selfless, and animated gaze of Augusta herself, who subordinates her own pictorial as well as familial authority to her son. Here it is worth noting that Caroline had died the previous year, so Augusta was now the most senior woman at court. Her power and prestige stemmed directly from her role in bringing up her son, the future king. In France, the tutelage of the royal children did not belong to the queen; the governess was of such power that she was almost a second mother figure. In Great Britain, however, queens oversaw their children's education; it was their main arena of authority within the Royal Household.[15]

The pictorial turn towards the Princess's children found expression in a number of other royal portraits that were reproduced as engravings in this period. In 1738 John Faber published a mezzotint after Charles Philips's tender portrait of the Princess's eldest daughter, Augusta, as a young baby (fig. 4.15). In the early 1740s Henry Overton published a print after Enoch Seeman the younger's portrait of Princess Augusta and her two brothers, in which she assumes the role of looking after George and Edward, and stands over them with a gesture that suggests both a playful assumption of authority and sisterly pride (fig. 4.16). In 1744 John Faber produced a print after a humorous fancy painting by Philippe Mercier that identified the picture's three protagonists as Princess Augusta and her two brothers, who are pictured training a puppy to assume the pose of a soldier (fig. 4.17). Finally, a life-sized portrait of Frederick and Augusta's six eldest children, painted by the visiting Swiss artist Barthélémy Du Pan in 1746, was turned into an engraving—again by Faber—soon afterwards (fig. 4.18).[16] It depicts the couple's offspring playing in a pastoral setting, happily performing a variety of roles including, in Princess Augusta's case, that of a protective older sister, and in George's case, that of a soldier firing an arrow from a bow. Such pictures demonstrate the ways in which the fertility of such royal consorts as Augusta could be promoted not only through their own portraiture, but also through the proliferating imagery of their children.

Augusta's identification with these children and with the role of a mother and guardian became more pointed after the sudden death of her husband in the spring of 1751, an event that generated an outpouring of commemorative poetry which also acknowledged the heavy familial responsibilities that Augusta now inherited. Drawing upon such classical predecessors as Cornelia, the Roman matron famous for her devotion to her children, poets wrote of a future in which the widowed Princess's own devotion to her offspring—and to her sons in particular—would be essential to ensure the continuation of the Hanoverian line. One writer trusted that the new heir to the throne, Prince George, would thrive under 'the royal mother's watchful eye', while another wrote that 'AUGUSTA's prudent hand shall still sustain / With all Cornelia's care her filial train.'[17] Poets also wrote of the need for Augusta to suspend her grief in order to take on full responsibility for her children's education and development. An Oxford poet, William Dobson, enjoined her to 'Cease, fair Augusta, cease to grieve', and to 'view that lovely, numerous line, / Sweet-blending Frederick's charms with thine.' Even as Augusta laments the passing of her late husband, he suggests that she must 'Begin th' important charge, / Which the wise monarch's tend'rest care / Commits to thee: their youth to rear, / Their op'ning minds enlarge.'[18]

In 1751, the same year as this kind of commemorative poetry was being written and circulated, the Princess Dowager and her children were depicted in a monumental portrait by the artist George Knapton. The deceased Prince of Wales appears in a portrait on the wall, pointing down to Augusta, who wears a mourning veil but appears otherwise composed, and a group of nine children. Caroline Matilda, who was born some four months after her father's death, is pictured in her mother's arms (fig. 4.19). Knapton's work provides a reassuring image of royal stability and succession in the

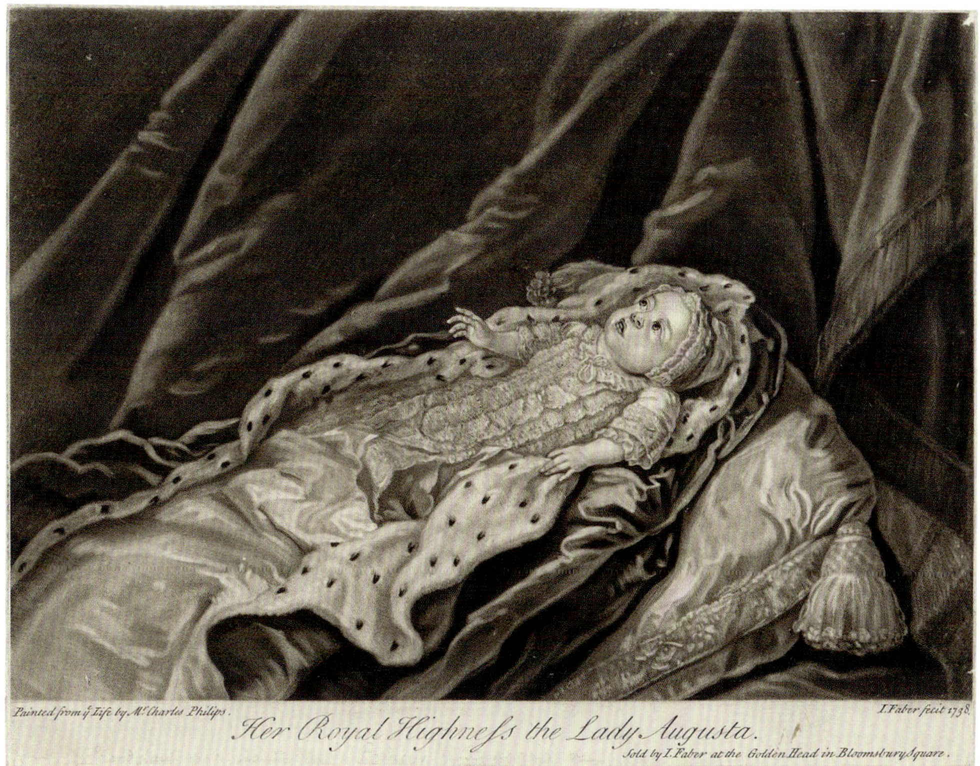

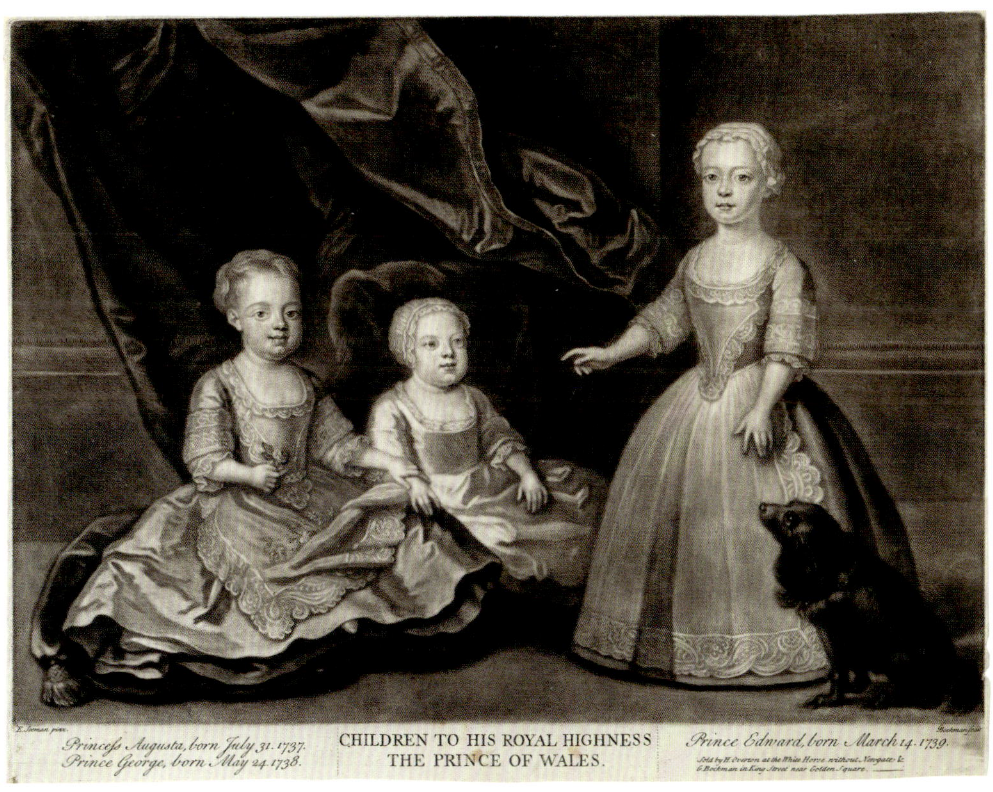

FIG. 4.15 John Faber the younger, engraver, after Charles Philips, *Her Royal Highness the Lady Augusta*, 1738, mezzotint, 10⅝ x 11⅜ in. (27 x 29 cm). Royal Collection Trust

FIG. 4.16 Gerhard Bockman, engraver, after Enoch Seeman the younger, *Children to His Highness the Prince of Wales,* c. 1739–51, mezzotint, 9⅝ x 13¼ in. (24.3 x 33.6 cm). Royal Collection Trust

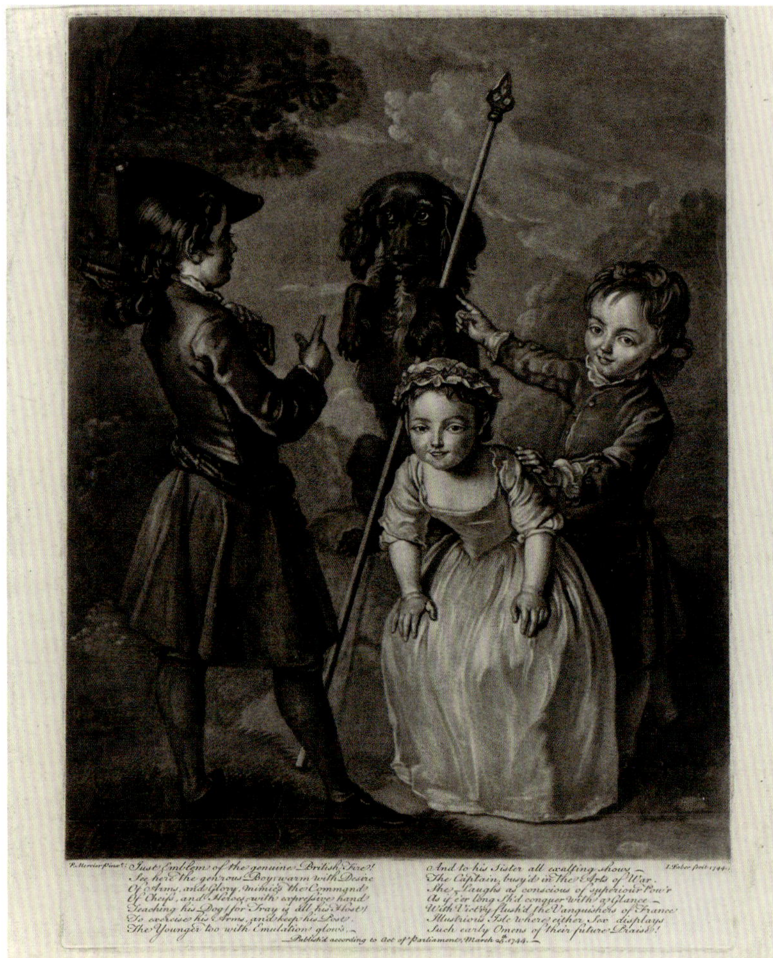

FIG. 4.17 John Faber the younger, engraver, after Philippe Mercier, *Playing Soldier (Children of Frederick, Prince of Wales)*, 1744, mezzotint, 13⅞ x 9⅞ in (35.3 cm x 25.cm). British Museum, London

wake of Frederick's death. Desmond Shawe-Taylor has suggested that this painting provides an allegory of constitutional monarchy, balanced by Frederick and the future George III on the one side and by the British constitution, represented by a statue of Britannia standing in the shadows, on the other.[19] This insight alerts us to the way in which the picture is poised between the languages of portraiture and allegory. With its schematic composition and orchestrated symbolism, as well as its carefully distributed portraits, the painting is not so far from the 'Emblematic' image by Wale and Müller with which this essay began. In this instance, however, the newly widowed Princess, raised on a throne-like chair, takes centre stage, and is placed precisely halfway between the portrait of her late husband and the figure of Britannia. She thus acquires a quasi-allegorical status in her own right: as represented by Knapton, she is a figure who is both an individually portrayed Princess Dowager surrounded by her children *and* someone who embodies that more abstract ideal of maternal plenty with which the Hanoverians were increasingly being associated.

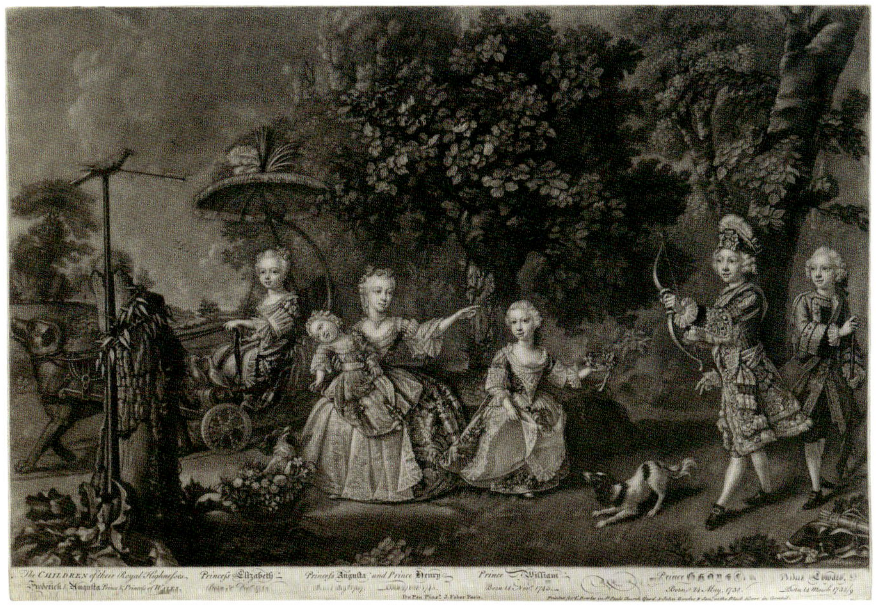

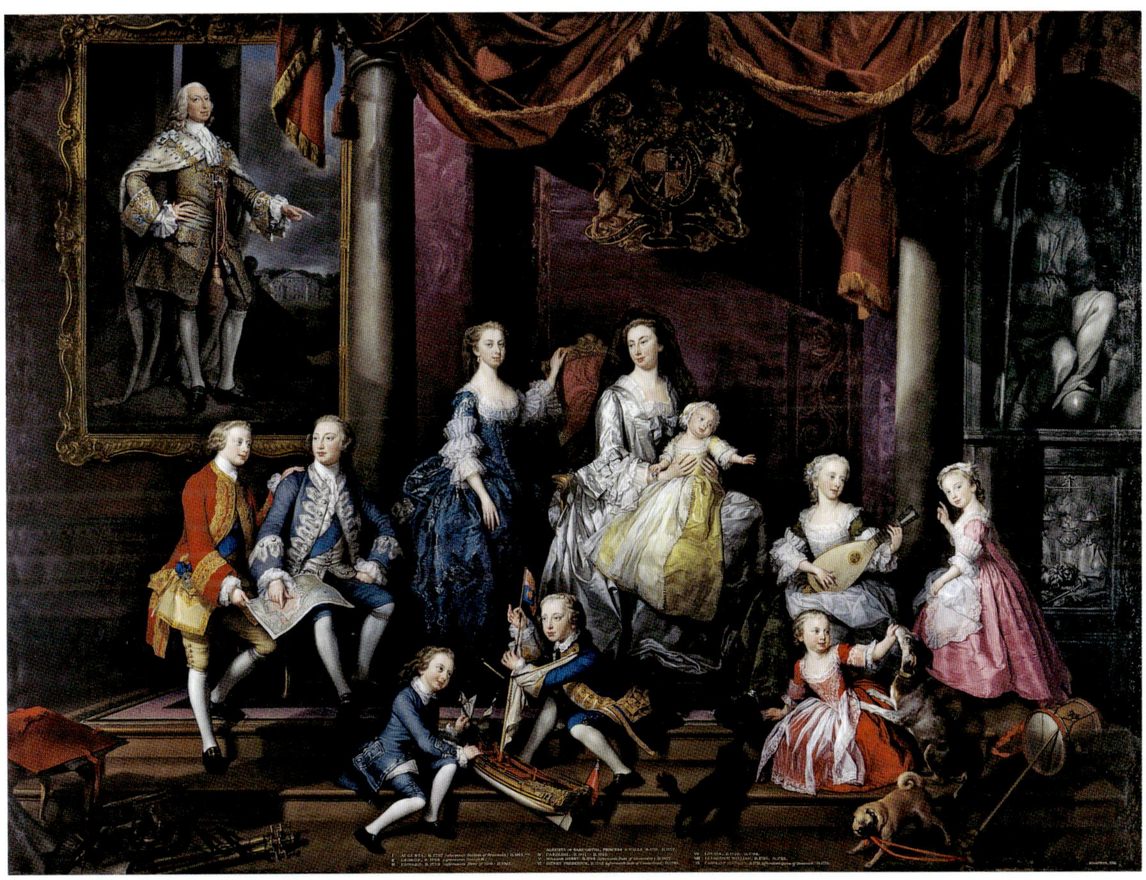

FIG. 4.18 John Faber the younger, engraver, after Barthélémy Du Pan, *The Children of their Royal Highnesses Frederick & Augusta Prince and Princess of Wales*, c. 1745–56, mezzotint, 13 x 19¾ in. (33 x 50.2 cm). Royal Collection Trust

FIG. 4.19 George Knapton, *The Family of Frederick, Prince of Wales*, 1751, oil on canvas, 138¼ x 181½ in. (351 x 461 cm). Royal Collection Trust

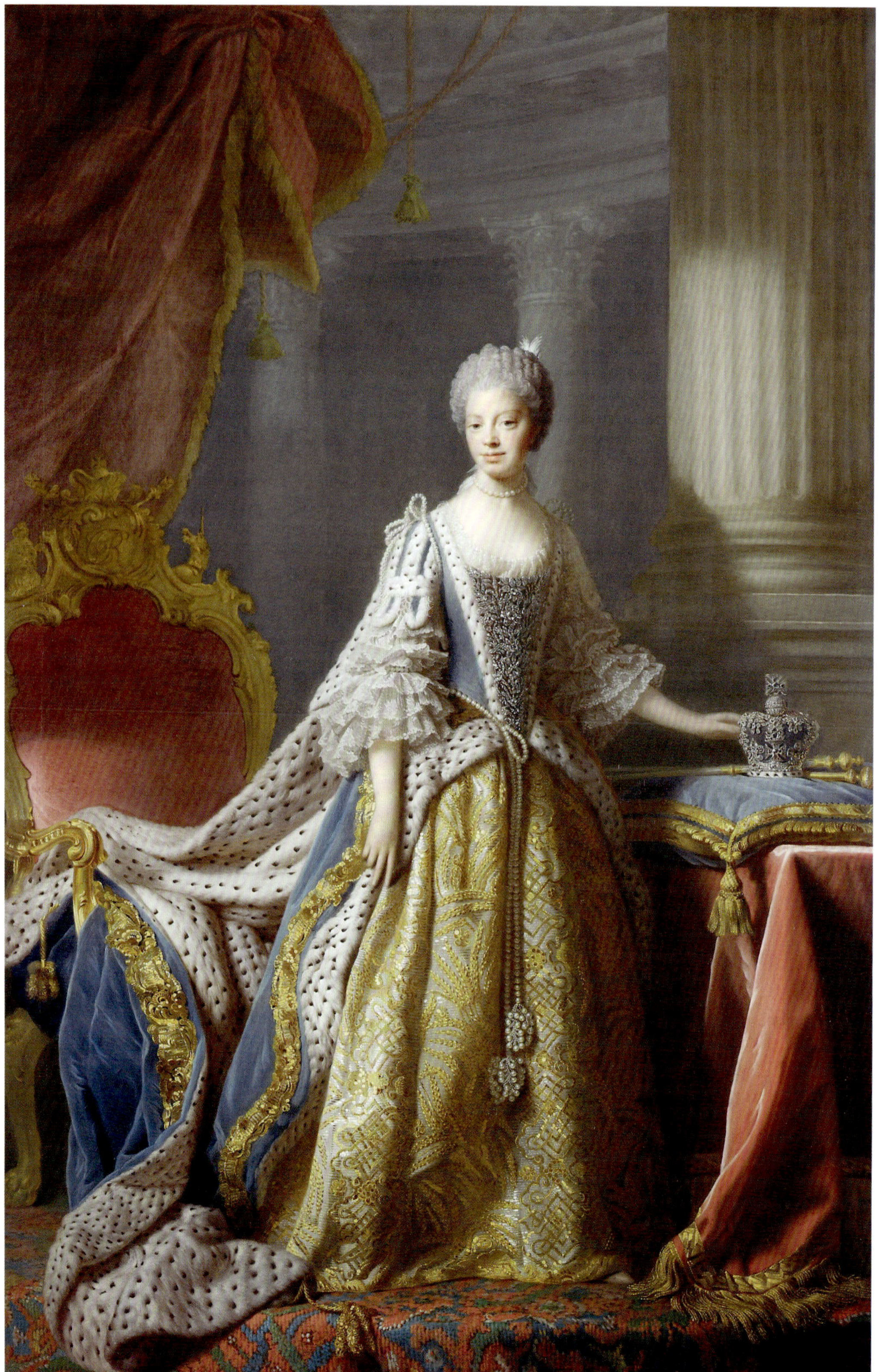

FIG. 4.20 Allan Ramsay, *Queen Charlotte*, c. 1760–1, oil on canvas, 98 x 63½ in. (248.9 x 161.3 cm). Royal Collection Trust

CHARLOTTE

In 1760, Prince George acceded to the throne. In the following year, he married Princess Charlotte of Mecklenburg-Strelitz, who, after being crowned Queen, was depicted in her coronation robes by the Scottish artist and newly appointed court painter Allan Ramsay (fig. 4.20).[20] A slightly revised version of this work became the model for the official coronation portrait of the Queen, which, with the artist's companion portrait of the King, Ramsay was regularly required to reproduce.[21] As John Ingamells has written, 'the replication of the state portraits of the King and Queen was to concern Ramsay for the rest of his life.'[22] The originals hung at St James's Palace, and a number of pairs were painted for other royal residences, including Buckingham House and Carlton House; most, however, were painted for the Lord Chamberlain's office, for distribution to the homes of the nobility, to ambassadors' residences, and to the houses of provincial and colonial governors. In all, more than a hundred pairs of these coronation portraits were ordered and painted. An early nineteenth-century memoir by the artist Joseph Moser recalled the dramatic impact of such commissions on the artist's Soho-Square studio: 'the ardour with which these beloved objects were sought for by distant corporations and transmarine colonies was astonishing; the painter with all the assistance he could procure could by no means satisfy the diurnal demands that were made in Soho-Square upon his talents and industry.'[23]

This intense process of pictorial cloning, which placed an enormous pressure on Ramsay's practice, played a crucial role within the visual dynamics of the Hanoverian monarchy. More particularly, the artist's portraits of this type provided a means through which George and Charlotte, in the form of their pictorial surrogates, retained a visual presence throughout their newly expanded imperial territories. Like their more humble and ubiquitous relation, the coin, these works provided a familiar and authoritative visual currency of the monarchy that, in this painted form, helped bind together a highly disparate collection of public spaces—courtly, aristocratic, civic, and imperial—into a single symbolic realm. The state portrait of a Hanoverian consort was thus always a collective portrait, mobilized by the process of pictorial reproduction and structured around the simultaneous projection of the same royal image within a multitude of interiors: to properly understand the workings and power of Ramsay's coronation portrait of Charlotte, we thus need to imagine its image of the Queen being broadcast from scores of walls, day after day, to thousands of viewers across the globe.

This, however, was not the only way in which Charlotte's image was being crafted in oil paint during this period. Speedily meeting the expectations demanded of all the Hanoverian consorts, she had given birth to two sons, Prince George and Prince Frederick, in 1762 and 1763 respectively; soon afterwards, as in the case of Caroline and Augusta, she began being pictured together with her young children. Thus in 1764, the visiting German artist Johan Zoffany began a painting that depicted Charlotte, George, and Frederick in one of the lushly appointed interiors of Buckingham House (fig. 4.21).[24] Though, as recent scholars have noted, the painting is clogged with symbolism, what is of particular note is its strikingly informal and domesticated

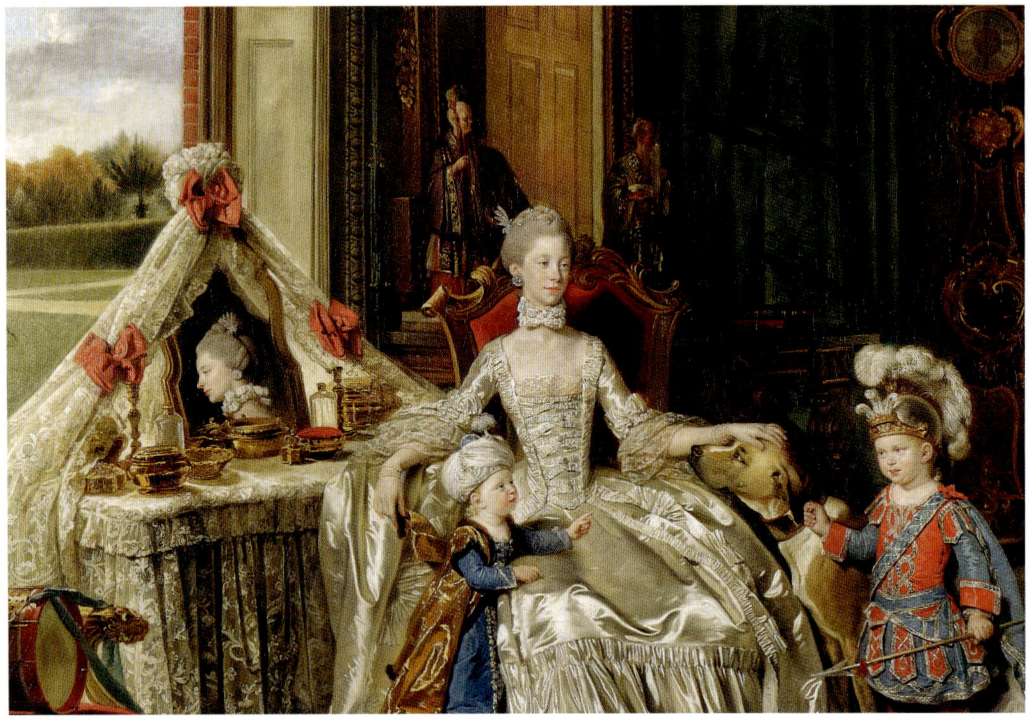

FIG. 4.21 Johan Zoffany, *Queen Charlotte with her Two Eldest Sons* (detail), c. 1765, oil on canvas, 44⅛ x 50⅜ in. (112 x 128 cm). Royal Collection Trust (see p. 25, fig. 1.15)

representation of royal motherhood. The Queen is seated in a relaxed and graceful fashion at her dressing table, being pleasurably distracted by the playfully theatrical appearance of her two young sons: as Mary Webster has written, 'the action of the figures is simple and delightful: the two little-boys, in fancy dress, have come into their mother's dressing-room to show themselves to her in their costumes.'[25] Reinforcing this sense of having captured a private moment within the everyday life of the Royal Household, and powerfully amplifying his portrait's claims to verisimilitude, Zoffany offers us a hyper-alert pictorial inventory of the room's luxurious contents: its elegant chairs, richly decorated carpets, the patterned table cover, and shining mirrors; even the tiny gleaming jars and boxes that make up the Queen's silver-gilt toilet-service. Here, we seem to have travelled a long way from the grandly classical spaces and weighty regal iconography of the conventional state portrait, and even from the imagery of such pictures as those of Van Loo and Knapton, both of whom placed their narratives of motherly affection and childish play within abstracted and obviously fictionalized environments. In Zoffany's image, which he completed in 1765, the royal female portrait, and the imagery of Hanoverian fertility, is not only situated indoors, but aligned with the daily narratives and exquisitely modelled accessories of courtly domestic life.

A second depiction of Charlotte by Allan Ramsay was produced in the same year, and also shows her with her two sons Prince George and Prince Frederick (fig. 4.22).[26] As a physical object, this painting is a kind of collage: the heads of all three members

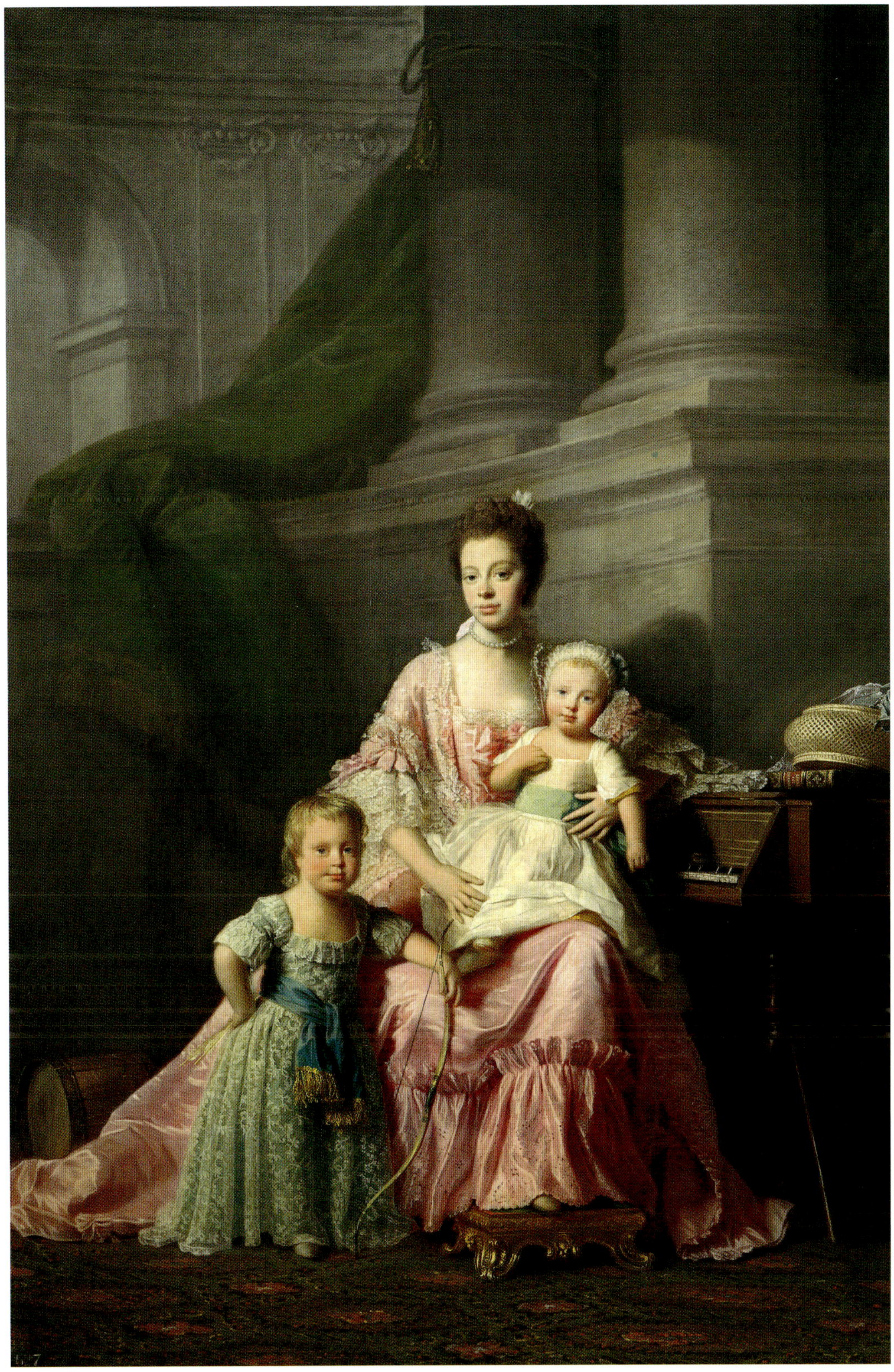

FIG. 4.22 Allan Ramsay, *Queen Charlotte with her Two Eldest Sons*, c. 1764–69, oil on canvas, 97¼ x 65 in. (247.8 x 165 cm). Royal Collection Trust

of the royal family have been painted on separate pieces of canvas, then sewn into the larger rectangle of canvas that supports the portrait as a whole. But this work is also a collage in another sense, for it stitches together, and seeks to resolve, precisely those two kinds of royal portraiture we have just been examining. A single, extended line snakes across the centre of the painting, bounding the figures and furniture that dominate the foreground. It runs along the contents of the needlework basket on the far right of the picture, up and along Charlotte's left arm, Frederick's head, the Queen's necklace and her discreetly dressed hair, and then down along her cheek, shoulder, and right sleeve, along George's shining blonde head and jutting right arm, along the swathe of pink dress at the painting's lower left edge, and then, finally, around the abandoned toy drum in the shadows. Above this line, the picture is dominated by the structures and symbolism of the grandest kind of full-length state portrait—the huge classical columns that tower over the picture's human subjects, resting on a monumental, curving classical plinth; the great swathes of deep-green drapery, fringed with gold tassels; and the looming arched gateway, decorated with pilasters and stretching out and above the painting's upper edge. Below this line, meanwhile, the painting is characterized by the kinds of imagery associated with Zoffany's picture. The Queen, we are asked to imagine, has been playing at her harpsichord, an activity that the painting suggests she regularly intermixes with other accomplishments such as needlework, the appreciation of art (a folio of drawings lies against the harpsichord's legs), and serious forms of reading that enable her to hone her role as a responsible royal mother: a copy of Locke's *Some Thoughts on Education* is tucked under the needlework basket. Again, however, she is shown as happily giving up these activities to cradle her youngest son and to entertain George, the young heir to the throne, who, just as in Zoffany's work, is given an infantile accessory of war—this time a bow—to play with, and thereby to broadcast his burgeoning masculine prowess. Completing the imagery of ease and elegance that dominates the lower half of Ramsay's painting, the tip of a royal shoe is seen resting on a carved footstool, itself placed on the richly patterned carpet that spans the picture's foreground.

Here, then, we see the artist constructing an image of Hanoverian fertility and tender motherhood that yokes together the long-standing iconography of the state portrait and an emergent pictorial language of royal domesticity. Subtle forms of handling and chiaroscuro mask the potential violence of such a conjunction, reconciling the upper and lower levels of the composition. Even so, the silhouette whose boundaries we have just been tracing is, if not a pictorial fault line, eloquent of the stresses involved in trying to find a way of promoting an elevated, historically rooted, and highly 'public' identity for the Queen while also endorsing her as an affectionate modern mother, engaged in the more private rituals of family life. Whatever these strains, the painting seems to have been appreciated as a successful exercise in royal portraiture: there survive several studio replicas, though nowhere near the number generated by the same artist's coronation picture.

Somewhat surprisingly, neither of Ramsay's two ambitious portraits of the Queen seem to have been turned into an engraving. The forms of reproduction to which they were subjected, though highly intensive in the former case, stayed

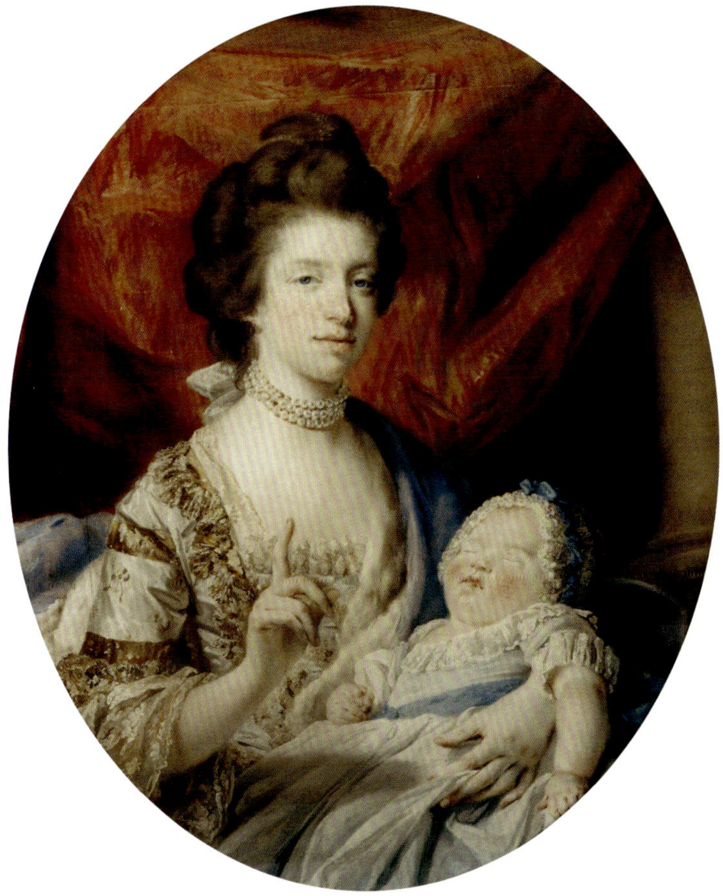

FIG. 4.23 Francis Cotes, *Queen Charlotte with Charlotte, Princess Royal*, 1767, pastel on vellum, 36⅝ x 30⅞ in. (93 x 78.5 cm). Royal Collection Trust

within the realm of painting, perhaps in order to guarantee their potency within the elaborate interiors for which they were produced. Similarly, Zoffany's more domestic portrait of the Queen and her children, painted as a commission from the King, was also not translated into a graphic image. However, a range of other representations of Charlotte and her children reached a far wider audience, both through engraved reproduction and through the means of the annual exhibitions of contemporary painting that had begun taking place in London since 1760. An especially striking example of such work, and one that again prompted a series of pictorial revisions, is Francis Cotes's pastel portrait of Charlotte with her fourth child, Princess Charlotte Augusta Matilda (fig. 4.23).[27] This modestly scaled picture, which was displayed at the Society of Artists' exhibition of 1767, offered visitors to the display not only another advertisement of the Queen's fertility, but also an especially powerful sense of being in her presence, thanks to its conceit of showing Charlotte as if looking out at the viewer and, through the gesture she makes with her right hand, enjoining us to be silent, so as not to awake her sleeping child.[28]

Through such means, Cotes produced an image that—for a poetic critic writing in a contemporary pamphlet, at least—saw Charlotte's identity as an 'Empress' being thoroughly, if temporarily, absorbed into that of a tender mother and loyal 'friend', someone who, loving her country in the same kinds of ways as she loves her children and husband, in turn provides a 'balm' for the kingdom itself:

> How on thy canvas, Cotes, with joy is seen,
> The tend'rest Mother, and the mildest Queen;
> Who can her dignity with her meekness blend,
> And lose awhile the Empress in the friend;
> Who makes humility her highest boast,
> For then the pious Queen commands the most.
> The joy of Britain in her bosom lies,
> What inexpressive sweetness in her eyes!
> Maternal fondness and maternal grace,
> Breathe in her air, and beam upon her face.
> With boundless charity from heaven that springs,
> The balm of Kingdoms, and the crown of Kings![29]

Cotes's pastel seems also to have functioned as a preparatory study for a full-length painting that placed 'mildness' and 'meekness' within a far more grandiose pictorial tradition.[30] This portrait in turn was reproduced as an engraving by William Wynne Ryland in 1770 (fig. 4.24). Here, maternal tenderness and fashionable ease are again harnessed to the weighty symbolism of the state portrait. The familiar devices of the column, the drapery, and the crown-bedecked table are reintroduced, to present a kind of internal frame, enclosing the Queen's body and reasserting her regal authority even in the face of her seemingly intimate interaction with the spectator.

The new public exhibitions that featured such works as Cotes's pastel, themselves introduced a new logic of pictorial reproduction and adaptation into royal portraiture, and into British painting more generally. This was one that operated not through the changes that were made to the individual image, but through the forms of repetition and recycling that characterized the annual hang at these exhibitions.[31] For, year after year, visitors to these displays would encounter—typically hanging in the same, central position on the exhibition room's main walls—one or more royal portraits; and often these were of Charlotte and her children. Here, then, was another space in which the imagery of a Hanoverian Queen and her royal offspring was constantly being reproduced and revised, if in a far looser manner than in the case of those graphic or painted translations of royal portraits we have hitherto been looking at.

As an example of this phenomenon, we can turn to a pair of Royal Academy exhibition submissions by Benjamin West, who enjoyed a great deal of court patronage in the 1770s and 1780s, and who regularly submitted royal portraits to the annual displays. In 1777, a decade after Cotes had displayed his portrait of the Queen with her sleeping child, West exhibited a portrait at the Royal Academy of Charlotte with the same daughter, which was quickly translated into a mezzotint (fig. 4.25).[32] Charlotte, the Princess Royal, has now grown up to become an upright young girl, shown

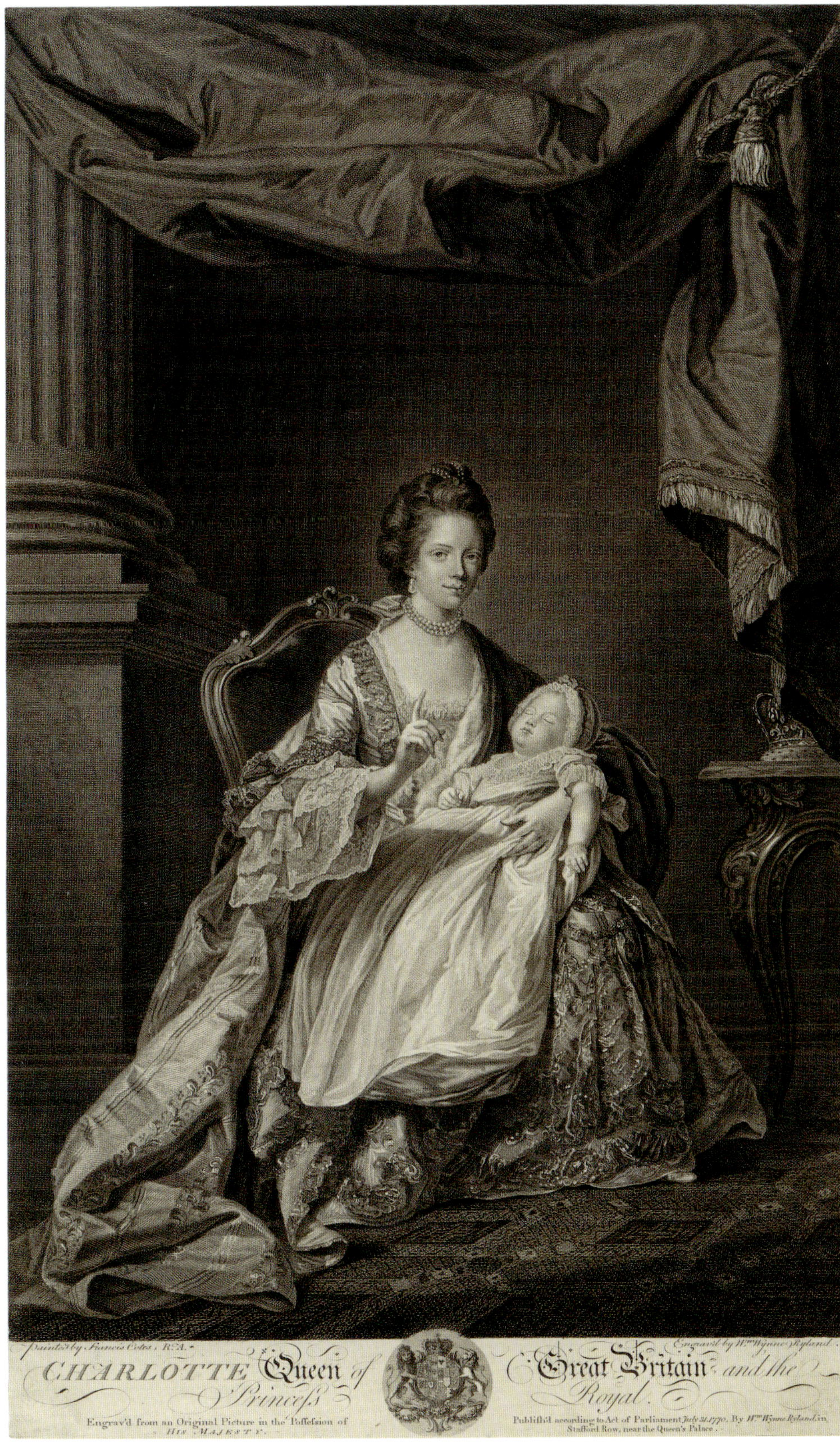

FIG. 4.24 William Wynne Ryland, engraver, after Francis Cotes, *Charlotte, Queen of Great Britain and the Princess Royal*, 1770, engraving, 21⅛ x 13¼ in. (53.8 x 33.8 cm). Royal Collection Trust

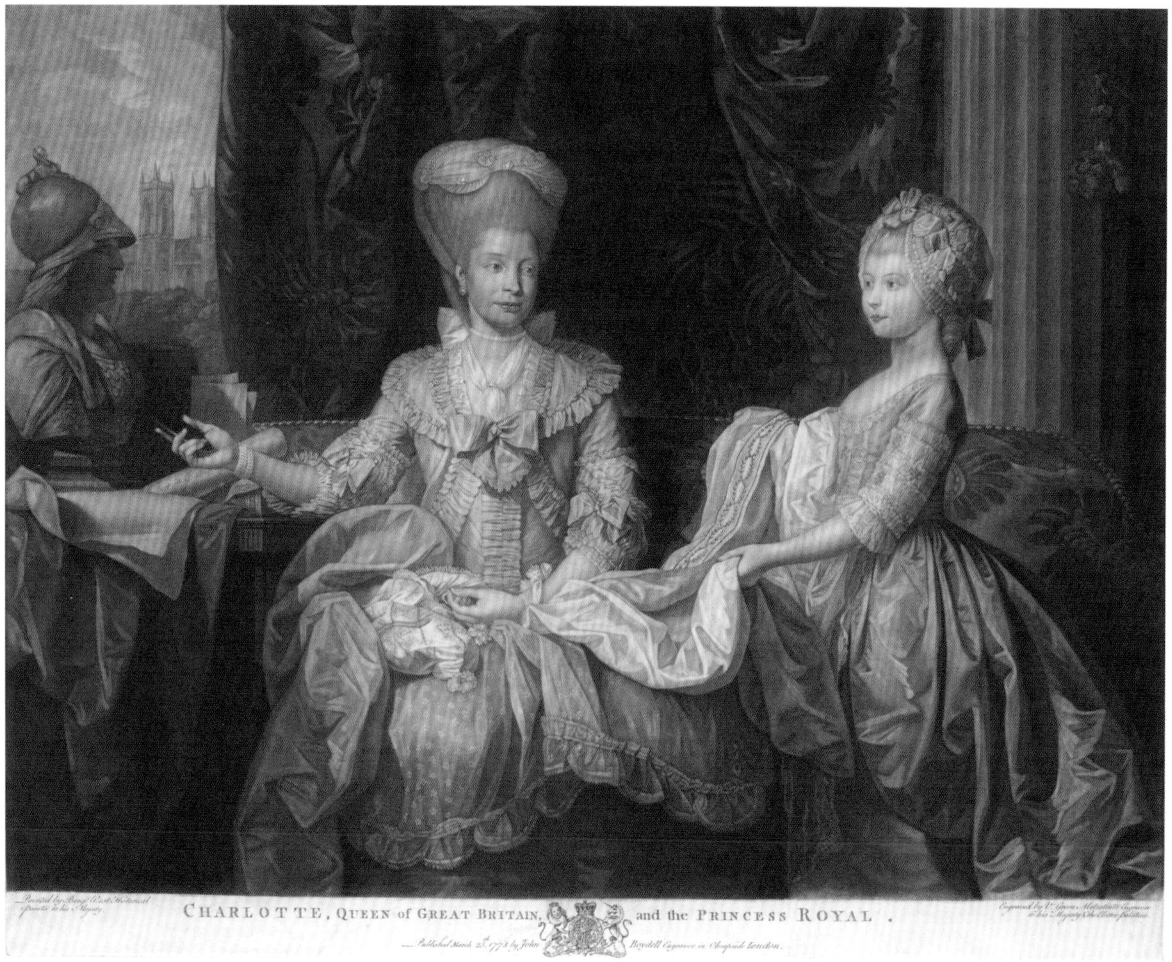

FIG. 4.25 Valentine Green, engraver, after Benjamin West, *Queen Charlotte, Queen Of Great Britain and The Princess Royal*, 1778, mezzotint, 18⅝ x 22¾ in. (47.4 x 57.8 cm). UK Government Art Collection (cat. 13.20)

listening to her parent discoursing on the virtues symbolized by the classical bust of Athena that stands on a nearby table, while assisting her mother in her needlework. The pictorial narratives of queenship and maternity being articulated on the walls of the annual exhibitions were replenished by a second royal portrait submitted to the Academy by the artist, this time in 1780 (fig. 4.26).[33] In this ambitious full-length, the standing figure of the Queen—her head a near mirror image of that depicted in West's earlier portrait—is juxtaposed with the regalia of state, while her children, now thirteen in number, are depicted in the background, gathered together on a lawn in front of Windsor Castle and the Queen's Lodge.

This painting provides a final example of how artists of the period sought to combine the conventions of state portraiture with a new, more domesticated imagery of royal family life. Here, indeed—to reuse the metaphor suggested by that patched-up painting by Ramsay looked at earlier in this essay—it is as if a rectangular and relatively informal conversation piece-sized group portrait of the royal children has been stitched into the canvas carrying a state portrait of their mother. The Queen's

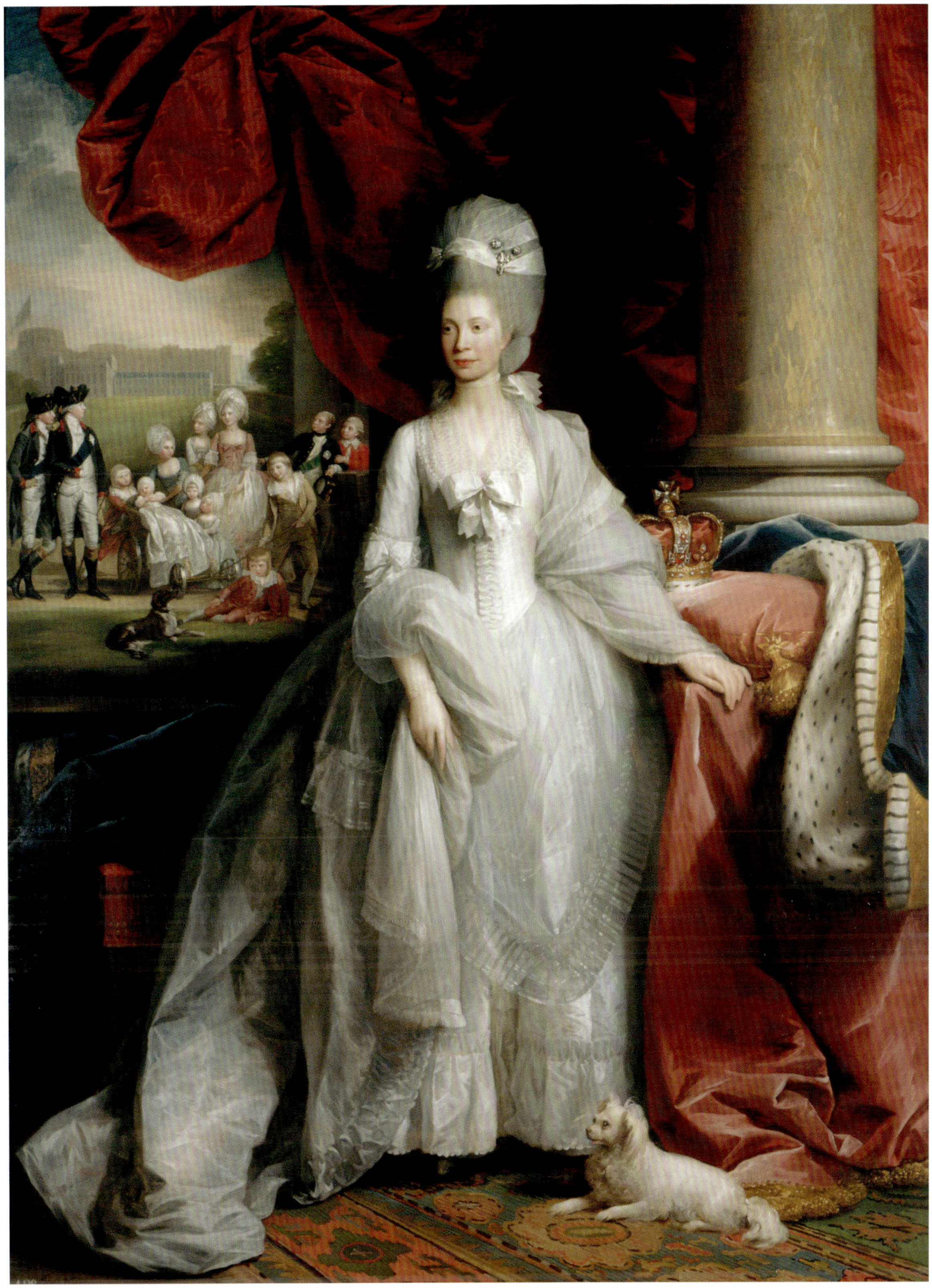

FIG. 4.26 Benjamin West, *Queen Charlotte*, 1779, oil on canvas, 101 x 71½ in. (256.5 x 181.6 cm). Royal Collection Trust

gaze, which seems simultaneously to look out of the canvas—as if across to the companion state portrait of George III—*and* to encompass the children gathered some distance behind her, captures in its dualities and ambiguities the demands being made not only on royal female portraiture in this period, but on the Hanoverian Queens and Princesses themselves.

CONCLUSION

Portraiture is always in some senses a discourse of reproduction, in that it reproduces the face and figure of an individual or individuals. Furthermore, it is a genre that has proven especially ripe for pictorial reproduction. In the Georgian period, this particular category of the arts was also closely tied to the issue of bodily reproduction and familial continuity. For eighteenth-century royal women, the birth of children, and especially the production of a healthy male heir, was of critical importance. The pressure that these circumstances placed on the Hanoverian royal consorts of the period was intense, and the ways in which these same circumstances shaped the character, circulation, and display of their self-presentation were far-reaching. The portraits that resulted, which were disseminated and revised across a multiplicity of sites and media, placed a premium on emphasizing the fruitfulness of the Hanoverian line, and on promoting women such as Caroline, Augusta, and Charlotte as ideal patriotic mothers, whose tender care of their many offspring served national as well as dynastic interests.

This was certainly the case in one of the most famous, if also most unusual, royal portraits of the period—Thomas Gainsborough's remarkable group portrait, completed in 1783, of George III, Queen Charlotte, and their thirteen children (fig. 4.27).[34] This work, which the artist exhibited in that year's summer exhibition at the Royal Academy, is another pictorial composite, this time made up of small, individually painted canvases, all of which have been separately framed and then grouped together in three rows that, running left to right, and one above the other, take us from the portrait of the uniformed, ruddy-faced king on the upper left to the pale, delicately painted face of Prince Alfred on the lower right. Though this work is often seen as unusual, even odd, in relation to Gainsborough's wider practice as a portraitist, it offers a fitting culmination to the story we have been telling. The portraits of children, placed in identical frames, flow forth from those of their parents, and from each other. Gainsborough's royal portrait, then, is all about reproduction, both bodily and pictorial. Furthermore, if those roundels were rearranged, as they so easily could be, George and Charlotte could be placed side-by-side at the top of a vertical column of images, with their children placed in a succession of pairs beneath them. Looking down such a column, our eyes would finally rest on the solitary figure of Prince Alfred, who can be imagined waiting for yet another sibling to join him, thereby ensuring a nice symmetry to the structure as a whole. Rearranging the images in this way, of course, again takes us right back to the image by Wale and Müller with which

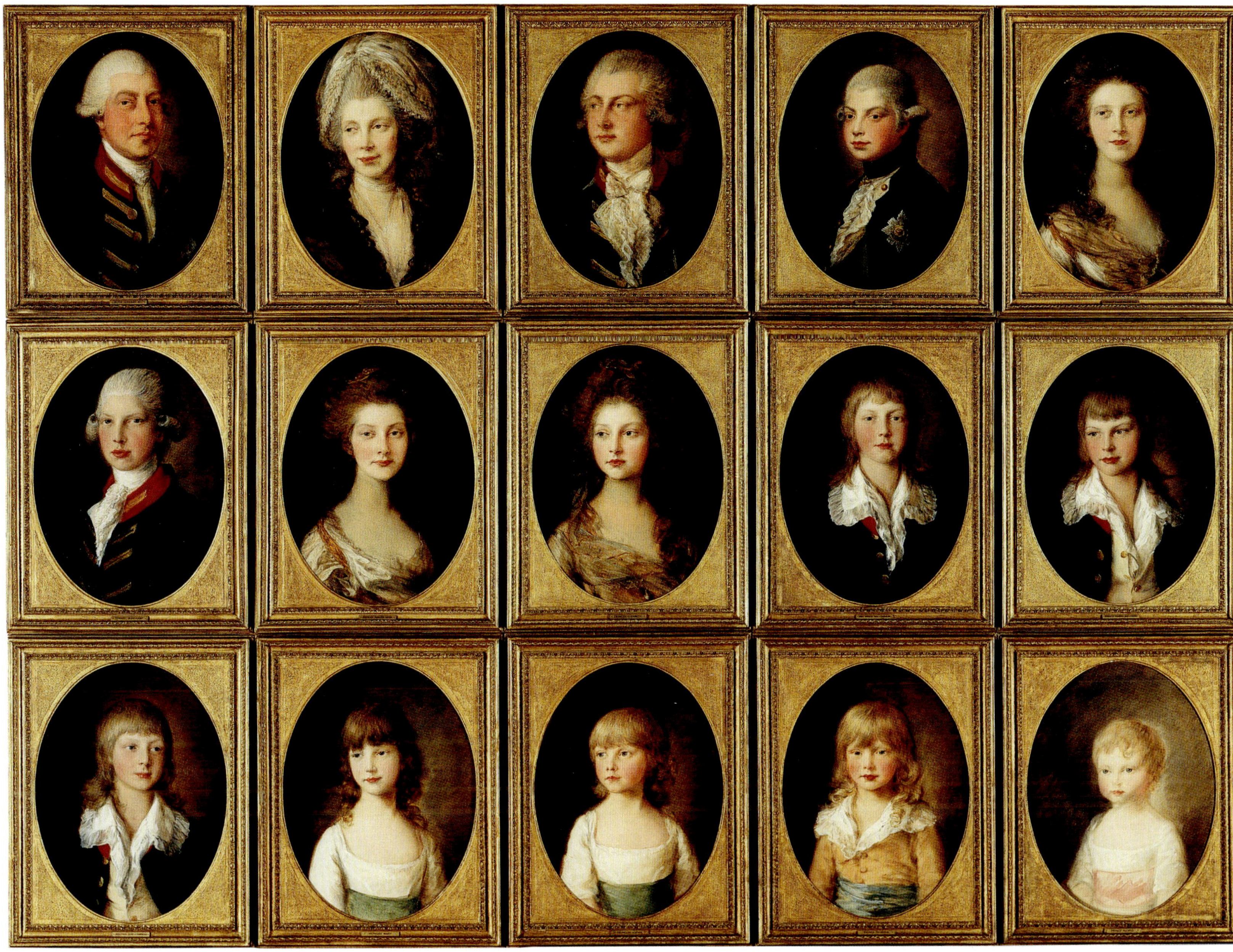

FIG. 4.27 Thomas Gainsborough, *Miniatures of George III, Queen Charlotte, and Their Thirteen Children*, c. 1783, oil on canvas. Royal Collection Trust

we began. In doing so, it helps us to understand that Gainsborough's group portrait too is a child of its forebears, born out of the fertile modern traditions of Hanoverian royal portraiture.

NOTES

1. *General Evening Post*, 19–21 May 1748. For more on this print, see Kate Heard's catalogue entry in Desmond Shawe-Taylor, ed., *The First Georgians: Art and Monarchy, 1714–1760* (London: Royal Collection Trust, 2014).

2. The print's reliance on the pictorial conventions used to represent the royal family tree is especially evident; for a suggestive graphic precedent to Wale and Mueller's work, see Renald Elstack's print, *The most happy Unions contracted betwixt the Princes of the Blood Royall of theis towe Famous Kingdomes of England & Scotland*, published by John Speed in 1603.

3. Kneller's portraiture, royal and otherwise, cries out for further research; for a survey of his career, together with a useful catalogue of his works, see J. Douglas Stewart, *Sir Godfrey Kneller and the English Baroque Portrait* (Oxford: Oxford Univ. Press, 1983); for Lely's royal portraiture, particularly of women, see Catharine MacLeod and Julia Marciari Alexander, *Painted Ladies: Women at the Court of Charles II* (London: National Portrait Gallery, 2001).

4. André Rouquet, *The Present State of the Arts in England* (London, 1755), 34.

5. For an especially stimulating and informative introduction to Smith's work and to the mezzotint portraiture of this period, see the National Portrait Gallery online research project *Early History of Mezzotint*, http://www.npg.org.uk/research/programmes/early-history-of-mezzotint/john-smith-mezzotint-printmaker-biography.php.

6. Anon. [Joseph Addison], *To Her Royal Highness the Princess of Wales, with the Tragedy of Cato, Nov. 1714, To Sir Godfrey Kneller, on his Picture of the King* (London, 1716).

7. For a later reference to their virtuous conjugal bliss, see the anonymously published *The Conjugal Lovers* (London, 1724), which is 'Most Humbly Dedicated to Their Royal Highnesses the Prince and Princess of Wales', and the dedication to which suggests their role as 'bright and illustrious' exemplars of 'Conjugal Love' (iii).

8. Addison, *To her Royal Highness*, 2.

9. Addison, *To her Royal Highness*, 2–3.

10. Addison, *To her Royal Highness*, 3.

11. Addison, *To her Royal Highness*, 3.

12. The engraving incorporating Addison's poem to Kneller is accessible online at the Royal Collection website. See https://www.royalcollection.org.uk/collection/603683/george-i.

13. For more on this portrait, and on the portraiture of Caroline more generally, see Joanna Marschner, *Queen Caroline: Cultural Politics at the Early Eighteenth-Century Court* (New Haven and London: Yale Univ. Press, 2014), especially 112–15.

14. This painting is reproduced and discussed in Elizabeth Einberg's Tate Gallery exhibition catalogue, *Manners and Morals: Hogarth and British Painting, 1700–1760* (London: Tate Gallery, 1987), 120–21.

15. See Thomas E. Kaiser, "Scandal in the Royal Nursery: Marie-Antoinette and the 'Gouvernantes des Enfants de France'", *Historical Reflections/Réflexions Historiques*, vol. 32, n. 2 (Summer 2006): 403–420.

16. For Du Pan's painting, see Desmond Shawe-Taylor's entry in Shawe-Taylor, ed., *First Georgians*, 366–67.

17. Anon., 'Upon George, Prince of Wales', *The London Magazine* (1751): 229; Allen Bathurst, 'To his Royal Highness George Prince of Wales', in Anon., *The English Poems collected from the Oxford and Cambridge verses on the Death of his Royal Highness Frederick Prince of Wales* (Edinburgh, 1752), 16.

18. William Dobson, 'To her Royal Highness The Princess Dowager of Wales', in Anon., *English Poems*, 31–32.

19. See Desmond Shawe-Taylor, 'Ruling a Free Nation', in *First Georgians*, ed. Shawe-Taylor, especially 22–25.

20. For a good introduction to this picture, see the entry in Alastair Smart, *Allan Ramsay: A Complete Catalogue of his Paintings*, ed. John Ingamells (New Haven and London: Yale Univ. Press, 1999), 88–90.

21. For an introduction to those copies and a detailed listing of a large number of examples, see Smart, ed. Ingamells, *Allan Ramsay*, 111–21.
22. See Smart, ed. Ingamells, *Allan Ramsay*, 112.
23. Quoted in Smart, ed. Ingamells, *Allan Ramsay*, 112.
24. For this painting, see Jane Roberts, ed., *George III and Queen Charlotte: Patronage, Collecting and Court Taste* (London: The Royal Collection, 2009), 26–27; Mary Webster, *Johan Zoffany, 1733–1810* (New Haven and London: Yale Univ. Press, 2011), 135–38; Clarissa Campbell Orr's entry on the painting in Martin Postle, ed., *Johan Zoffany RA: Society Observed* (New Haven and London: Yale Univ. Press, 2011), 204; and Marcia Pointon, *Hanging the Head: Portraiture and Social Formation in Eighteenth-Century England* (New Haven and London: Yale Univ. Press, 1997), 162–68.
25. Webster, *Johan Zoffany*, 137.
26. See Smart, ed. Ingamells, *Allan Ramsay*, 90–91; see also the discussion of this image in David H. Solkin, *Art in Britain, 1660–1815* (New Haven and London: Yale Univ. Press, 2015), 194–95.
27. For this painting, see Edward Mead Johnson, *Francis Cotes* (Oxford: Phaidon, 1976), 37–38, 83, and Roberts, ed., *George III and Queen Charlotte*, 27–28.
28. As David Solkin has recently noted, the iconography of Cotes's painting follows on very closely from that found in Ramsay's 1764–65 portrait of Charlotte and her two sons. Solkin, *Art in Britain*, 194–95.
29. Anon., 'Le Pour & la Contre. Being a poetical Display of the Merit and Demerit of the Capital Paintings, exhibited at Spring Gardens', quoted in *London Chronicle*, 5–7 May 1767.
30. For this work, see Mead Johnson, *Francis Cotes*, 85.
31. For this phenomenon, and its impact on artists' submissions to the annual exhibitions, see Mark Hallett, *Reynolds: Portraiture in Action* (New Haven and London: Yale Univ. Press, 2014), especially 343–81.
32. For this picture, see Helmut von Erffa and Allen Staley, *The Paintings of Benjamin West* (New Haven and London: Yale Univ. Press, 1986), 473.
33. For this picture, see Von Erffa and Staley, *Paintings of Benjamin West*, 469–71.
34. For a good introduction to this work in the context of the Academy's annual exhibitions, see Michael Rosenthal, *The Art of Thomas Gainsborough: 'A Little Business for the Eye'* (New Haven and London: Yale Univ. Press, 1999), 104–5.

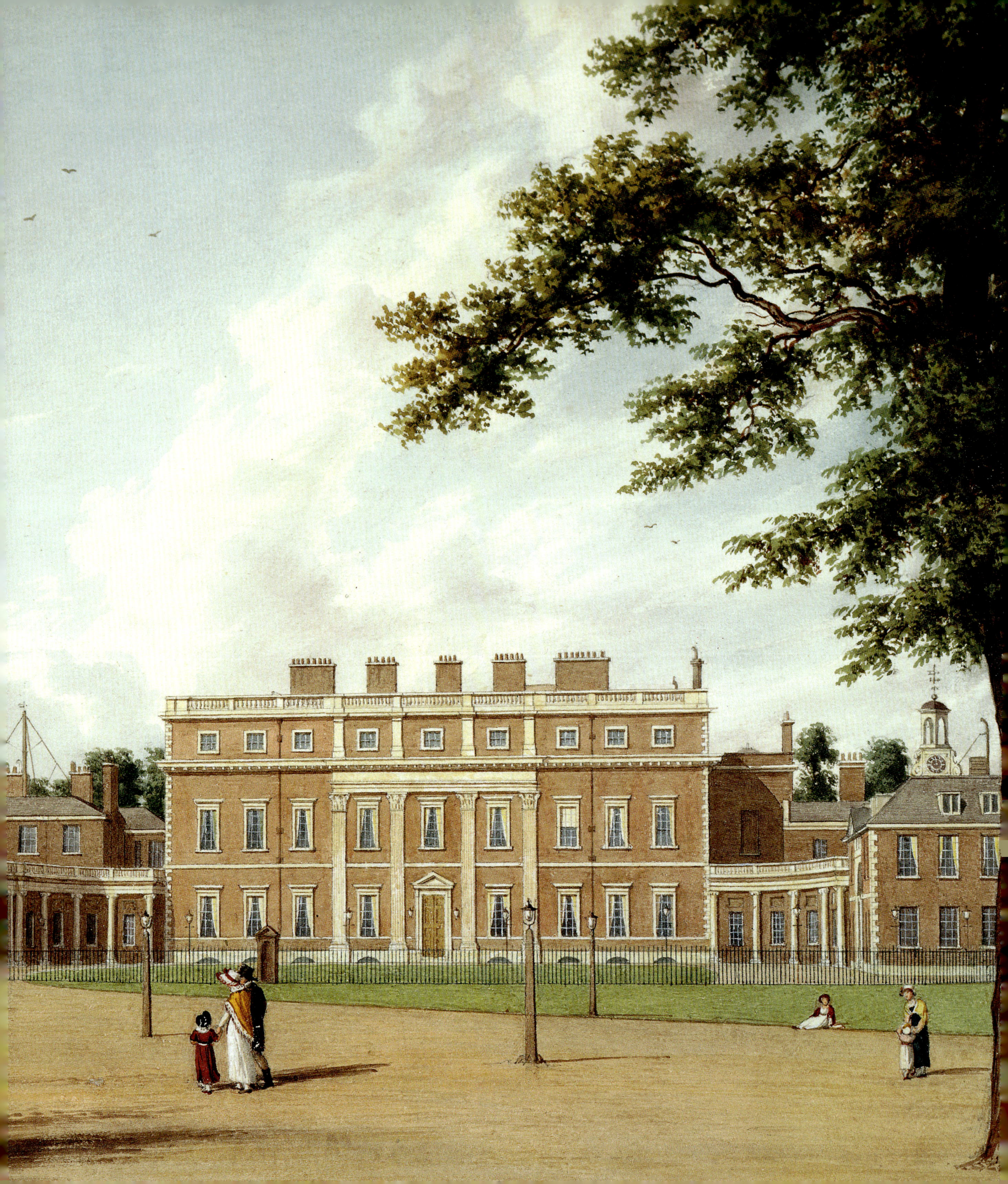

The German Princesses as British Patrons

Wolf Burchard and Kathryn Jones

The first Hanoverian monarchs, George I and George II, have repeatedly—but not always justifiably—been described as philistines, but this can hardly be said of Caroline of Ansbach, Augusta of Saxe-Gotha, and Charlotte of Mecklenburg-Strelitz.[1] All three women were enthusiastic patrons of the arts. All three were born, raised, and educated in Germany, and were therefore foreigners when they first set foot in Britain. However, unlike the Archduchess Marie-Antoinette of Austria, who was obliged to discard her roots to become entirely French when she married Louis XVI, Caroline, Augusta, and Charlotte always retained a German role, for eighteenth-century British Queen Consorts were simultaneously Electresses of Hanover, while Princesses of Wales were *Kurprinzessinnen*, that is, Electoral Princesses.

CAROLINE

In spite of overall similarities, the three Princesses were very different individuals with different upbringings, operating under different circumstances. Caroline was born in the tiny margravedom of Ansbach. Her early childhood was overshadowed by the consecutive decease of her father, stepfather, and mother between 1686 and 1696. Aged thirteen, the orphan Princess was taken into the care of her Hohenzollern relative Sophia Charlotte of Hanover, wife of Frederick III, Elector of Brandenburg and sister of George I. Sophia Charlotte and her mother, Sophia, Electress of Hanover, oversaw Caroline's unusually thorough intellectual upbringing. At Schloss Lietzenburg outside Berlin (now known as Schloss Charlottenburg), Sophia Charlotte surrounded herself and Caroline with learned men and women, including the renowned polymath of the era, Gottfried Wilhelm Leibniz.[2] Leibniz entertained a dynamic scholarly dialogue with all three women—Sophia, Sophia Charlotte, and Caroline—which the latter pursued in writing once she moved to England.

Detail of p. 123, fig. 5.21

At the Hohenzollern court in Berlin, Frederick III's desire to become King of Prussia meant that he was determined to showcase not only interests in the natural sciences, history, philosophy, and poetry, but also in the visual arts and architecture. The court was divided between Frederick's lavish household at the *Stadtschloss* in Berlin and his wife's *Musenhof*—the court of muses—at Lietzenburg. The distinct contrast between the politically determined and pomp-loving king-to-be and his piously intellectual consort is incisively summarized in what is probably an apocryphal anecdote, recorded by their grandson Frederick the Great. On her deathbed, Sophia Charlotte is reported to have said: 'I am leaving now to satisfy my curiosity about the origins of those things that Leibniz was never able to explain to me . . . and to the king, my husband, I give the spectacle of a funeral, which presents him with yet another opportunity to display his magnificence.'[3] That tension between Frederick's artistic and architectural aspirations, which aimed to reinforce his political potency, and Sophia Charlotte's genuine scientific and philosophical curiosity, may have laid the foundation for Caroline's understanding of the importance of royal patronage.

Upon her marriage to the future King George II in 1705, Caroline moved to Hanover and continued to be under the direct influence of the Electress Sophia. Like the Hohenzollern, the Hanoverians were an ambitious dynasty seeking to make the decisive jump from being electors to kings. As their efforts came to fruition in 1714 with the accession of George I, Caroline had to assume the new role of Princess of Wales and move from Hanover to St James's Palace, where her apartments were refitted in the latest fashion.[4] Her enjoyment of her new living quarters did not last, however, as the King expelled her and her husband from court following a dispute over the choice of godparents to their short-lived son George William. Until their own accession in 1727, the Prince and Princess of Wales established a vibrant home at Leicester House, contrasting with the stuffy entourage of the shy George I.[5] Although the move to non-royal premises could have prompted Caroline to form a fine collection of pictures to furnish her family's new home, the financial constraints imposed on the Prince of Wales's household forbade almost any kind of artistic luxury.[6]

For Caroline, the most fertile period of patronage and collecting was the decade between George II's accession in 1727 and her death in 1737. Her unlimited access to the royal collection enabled her to pursue personal interests as well as shape the public perception of the Hanoverian dynasty. The Queen had a particular affection for Kensington Palace, which served her as suburban *maison de plaisance* in the same way that Lietzenburg had done for Sophia Charlotte of Brandenburg and Herrenhausen for the Electress Sophia of Hanover, although—contrary to these fashionable retreats of Caroline's childhood guardians—Kensington's plain red-brick exterior hardly advertised the modernity of its interiors. Adorned on a tight budget, the Palace's state apartments, largely the result of a scheme carried out by William Kent in the 1720s, bore the most up-to-date decor the royal family had to offer at the time.[7] The disparity between its dull exterior and its more fashionable (albeit rather controversial) internal appearance, illustrates how the early Hanoverians sought acceptance amongst their new British subjects. While they required appropriate decorum to be taken seriously by their courtiers, they did not want to offend the outside world by exhibiting signs of ostentation and overspending.

Caroline's Picture Closet and cabinet rooms at Kensington Palace reflect particularly well her understanding of the values and the importance of art and historical legacies. She used several spaces to display her varied collections.[8] Her private apartments included the so-called 'Little Library' as well as the two cabinet rooms, which gradually became a shrine of royal history akin to German cabinets of curiosities or the Italian *studiolo*.[9] Her small Picture Closet, tucked away between the Great Drawing Room and the King's Gallery, contained more than four hundred items by the time she died, all meticulously recorded in George Vertue's catalogue and by line drawings published as engravings around 1755.[10] The room was densely hung, and Caroline's fascination for the sixteenth century came to the fore in her display of a large group of Holbein drawings depicting members of Henry VIII's court. Apparently unearthed by the Queen herself in a forgotten desk at Kensington in 1727, the drawings were taken out of their albums, individually mounted, and hung in her rooms at Richmond Lodge, where John Percival, first Earl of Egmont, admired them in 1735 before they were eventually moved to Kensington Palace.[11] A unique visual record of Tudor history, the Holbein portraits allowed the Queen to immerse herself in one of the most defining periods in the history of the kingdom over which her husband now reigned.

In addition to these very fine drawings, the closet contained an array of wax and ivory portraits as well as numerous miniatures of European monarchs and princes. Prior to the Queen's involvement, the few miniatures that had remained in the royal collection after the Civil War, just over fifty, were dispersed over several residences. Not only did Caroline have them all gathered together and reorganized in her closet at Kensington, but she also made significant additions to this collection.[12] These comprised an intriguing group of forty-nine portraits of Hanoverian forebears, mostly sixteenth-century dukes of Brunswick and their consorts, dating from the 1590s, of which another, apparently very similar set of fifty-two miniatures was kept in Caroline's closet at Herrenhausen.[13] The Kensington set was extended by yet another eight miniatures depicting fifteenth-century Brunswick ancestors, which may well have been commissioned by the Queen herself and painted by an English miniaturist to match the original German group (figs 5.01–12).[14]

The Picture Closet at Kensington also featured a small selection of old master paintings, such as Lucretia 'sitting and stabbing herself', then attributed to Annibale Carracci; *The Seven Children of the King and Queen of Bohemia*, a copy after Cornelis van Poelenburgh (a reference to the dynastic link between the Hanoverians and the Stuarts); two portraits then thought to be of Hans Holbein and his wife; and, most importantly, a large portrait of Queen Elizabeth I as a young woman.[15] Caroline, rather like Electress Sophia, had a great historical affinity for the Elizabethan era: Elizabeth I clearly constituted a powerful role model for her. In December 1713, about eight months prior to George I's accession to the throne, Leibniz sent Caroline a letter from Vienna proclaiming his loyalty: 'I pray God to grant you the same measure of

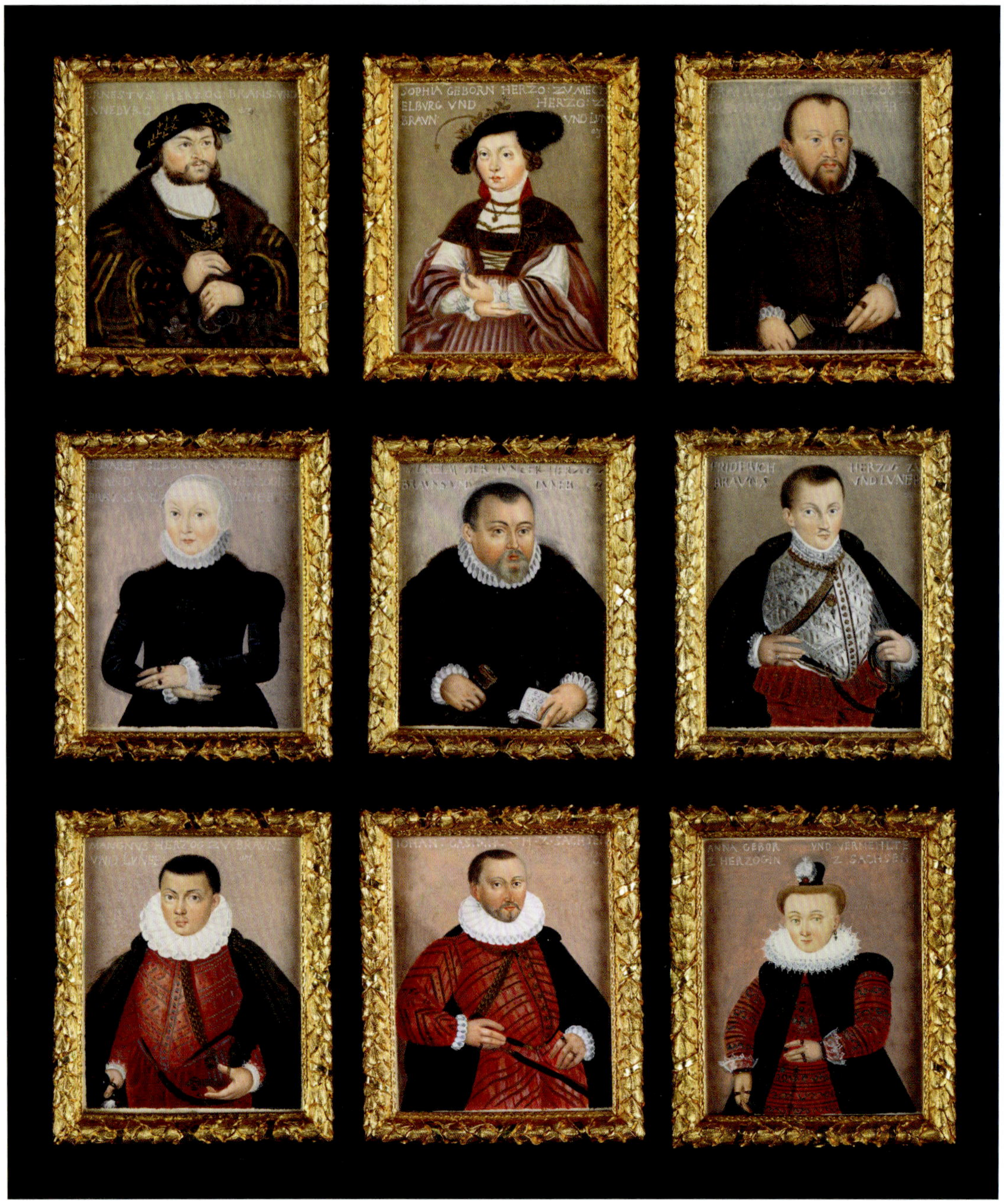

FIGS. 5.01–5.09 Brunswick-Lüneburg Court miniaturist, c. 1595, watercolour on vellum. Royal Collection Trust

TOP

FIGS. 5.01 *Ernest of Celle, Duke of Brunswick-Lüneburg*, 2¾ x 2¼ in. (6.9 x 5.6 cm)

FIG. 5.02 *Sophia, Duchess of Brunswick-Lüneburg*, 2¾ x 2¼ in. (6.9 x 5.6 cm)

FIG. 5.03 *Francis Otto, Duke of Brunswick-Lüneburg*, 2¾ x 2⅛ in. (7 x 5.5 cm)

CENTRE

FIG. 5.04 *Elizabeth Magdalene, Duchess of Brunswick-Lüneburg*, 2¾ x 2⅛ in. (7 x 5.5 cm)

FIG. 5.05 *William the Younger, Duke of Brunswick-Lüneburg*, 2¾ x 2⅛ in. (7 x 5.5 cm)

FIG. 5.06 *Frederick II, Duke of Brunswick-Lüneburg*, 2¾ x 2⅛ in. (7 x 5.5 cm)

BOTTOM

FIG. 5.07 *Magnus, Duke of Brunswick-Lüneburg*, 2¾ x 2⅛ in. (7 x 5.5 cm)

FIG. 5.08 *John Casimir, Duke of Saxe-Coburg*, 2¾ x 2⅛ in. (7 x 5.5 cm)

FIG. 5.09 *Anna, wife of Johann Casimir, Duke of Saxe-Coburg*, 2⅝ x 2⅛ in. (6.8 x 5.4 cm)

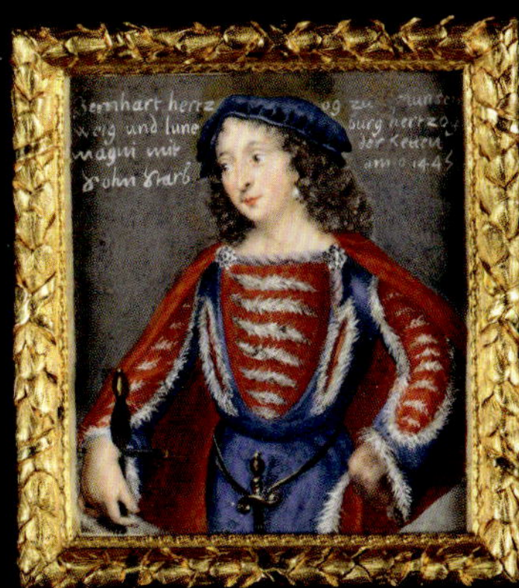

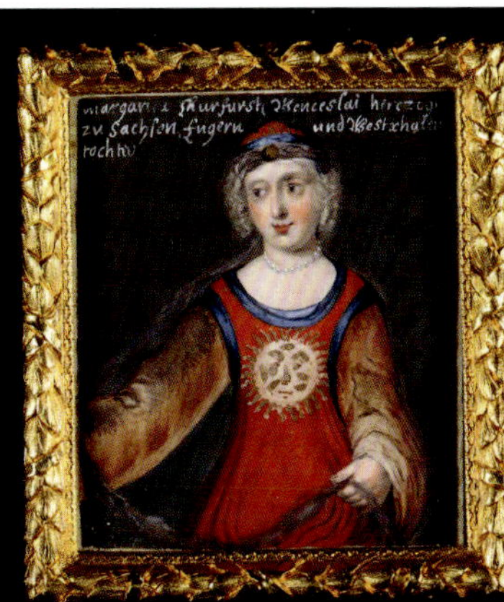

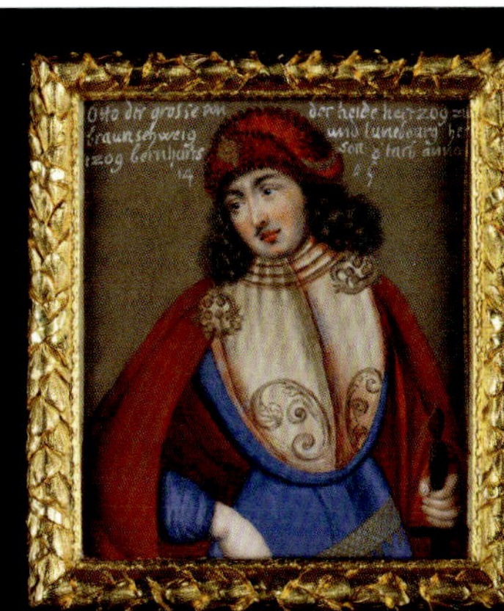

FIG. 5.10 German School, *Bernard I, Duke of Brunswick- Lüneburg*, 1600–1735, watercolour on vellum, 3¼ x 2¾ in. (8.1 x 7.0 cm). Royal Collection Trust (cat. 10.10)
FIG. 5.11 German School, *Margareta, Duchess of Brunswick-Lüneburg*, 1600–1735, watercolour on vellum, 3¼ x 2⅞ in. (8.1 x 7.2 cm). Royal Collection Trust (cat. 10.11)
FIG. 5.12 Court miniaturist, *Otto the Great, Duke of Brunswick Luneburg*, c. 1600–1720, watercolour on vellum, 2⅝ x 2¼ in. (6.8 x 5.8 cm). Royal Collection Trust

years as the Electress [Sophia] enjoys, and the same good health. And I pray also that you may one day enjoy the title of Queen of England so well worn by Queen Elizabeth, which you so highly merit.' Caroline soon replied:

> I accept the comparison which you draw, though all too flattering, between me and Queen Elizabeth as a good omen. Like Elizabeth, the Electress's rights are denied by a jealous sister with a bad temper [Queen Anne], and she will never be sure of the English crown until her accession to the throne.[16]

References to Elizabeth I pervaded Caroline's collecting and commissions. Her cabinet of curiosities contained at least two sixteenth-century cameos showing the profile of 'the Virgin Queen',[17] and the Little Library contained a finely illuminated genealogical volume featuring the pedigree of all the Kings of England, drawn up, according to Horace Walpole, for Elizabeth I.[18] Probably the most peculiar outcome of her patronage was 'Merlin's Cave', a Gothic folly designed by William Kent for Richmond, exhibiting the Queen's understanding of Britain's historical past. Staging a tableau of six life-sized wax figures, amongst which were Queen Elizabeth and Merlin himself, it illustrated one of the wizard's prophecies heralding the return of the ancient kings of England—as supposedly performed by the Tudors, of which the Hanoverians claimed to be the rightful heirs (fig. 5.13).[19]

Caroline's thorough engagement with the historical past of Britain's ruling dynasties, as well as its nation as a whole, is particularly pronounced in her imposing book collection.[20] Over the course of her life the Queen acquired a vast library, mostly scattered over several rooms between her apartments and those of the King at St James's Palace. When, in 1735, George II was reluctant to return from Hanover

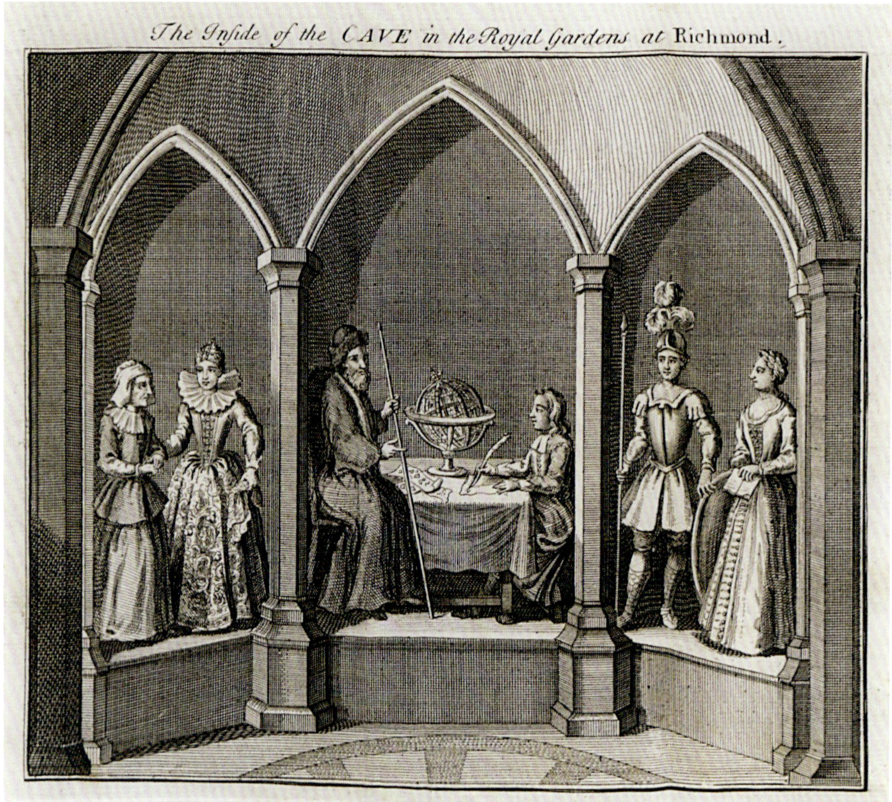

FIG. 5.13 Edmund Curll, 'The Inside of the Cave in the Royal Gardens at Richmond', in *The Rarities of Richmond: being exact descriptions of the Royal Hermitage and Merlin's Cave. With his life and prophecies . . .* (London: E. Curll and J. Read, 1736); (cat. 12.07)

because he did not want to leave his German mistress, Amalie von Wallmoden, behind, Caroline suggested that her books should be removed from those rooms in order for the mistress to move in.[21] This accommodating measure gave Caroline the opportunity to ask William Kent to design her a purpose-built library with enough room to house her ever-growing collection of books. Kent supplied at least four different proposals for the Queen's Library, and his initial designs were exceedingly more ambitious than the building which was constructed.[22] The remarkable interior he first imagined featured a cross vault upheld by four pairs of columns subdividing the library into three bays, as well as large Venetian windows on one side, which would have been echoed by gently curved niches bearing bookshelves on the opposite side (fig. 5.14).[23] John Cornforth noted that had the library been erected according to the architect's initial intentions, 'it would have been the most complex architectural interior designed by Kent and unrivalled in the first half of the century.'[24]

The much more modest space eventually built was a single-storey pavilion containing a double-cube room generously lit by large windows overlooking Green Park (fig. 5.15).[25] Its neo-Palladian architectural vocabulary is further testimony to the Queen's desire to present herself as a British consort and patron. Caroline was well acquainted with the stylistic alternatives of French and German neo-classicism and Rococo, and could have easily commissioned a room more in line with continental

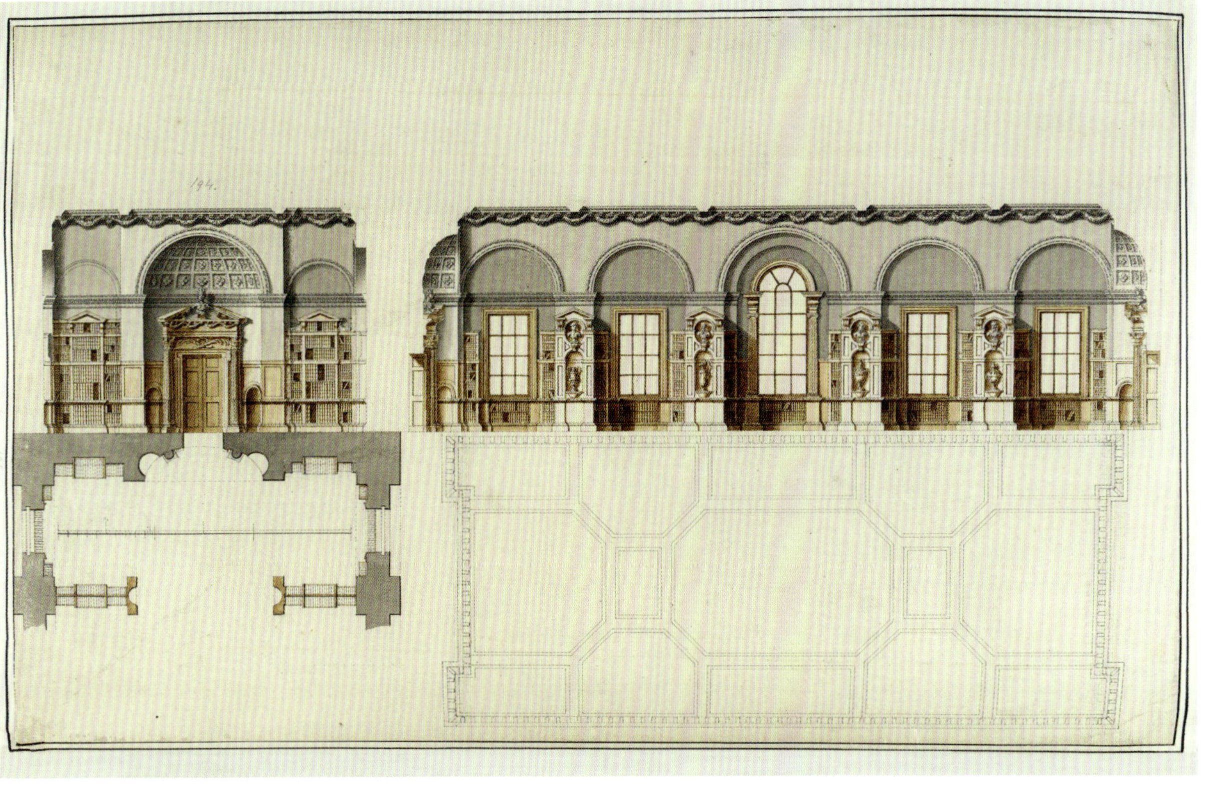

FIG. 5.14 William Kent, *A Sketch for the Queen's Library*, c. 1736, pen and ink, and washes over pencil on paper, 12⅞ x 17⅞ in. (32.7 x 45.3 cm). Sir John Soane's Museum, London, SM Vol. 147/197

fashions.[26] Instead, the library was manifestly a British interior. The brackets between the bookshelves held terracotta busts of Plantagenet and Tudor kings and queens—including, once again, Elizabeth I—that Caroline had commissioned from Michael Rysbrack.[27] Originally, Kent sought to create a visual echo to the busts by inserting roundels into the coving comprising two-dimensional portraits of British worthies, such as Newton and Shakespeare. Only two years earlier, in 1733, Kent had designed *The Temple of British Worthies* for Richard Temple, Viscount Cobham, at Stowe—an almost semi-circular structure with niches bearing busts of medieval monarchs as well as poets, scientists, and theologians.[28] Following the unexpected death of the Queen, the scheme was apparently never carried out, and in the event, George II discontinued the commission for further terracotta busts, so that the set was never complete.

The library at St James's Palace was to be Caroline's last commission. The Queen was taken ill as she inspected the building site in October 1737 and died shortly afterwards.[29] We therefore do not know how she would have organized and arranged her books in her new library, or who would have had access to this sanctuary of learning, located far away from the state rooms, in unusual proximity to the stables. The library was pulled down in 1825, and its only architectural survivals are a pair of chimney-pieces designed by Kent and moved to the state apartments at St James's Palace by George IV in 1822 (fig. 5.16). Intriguingly, these marble vestiges bear small heraldic

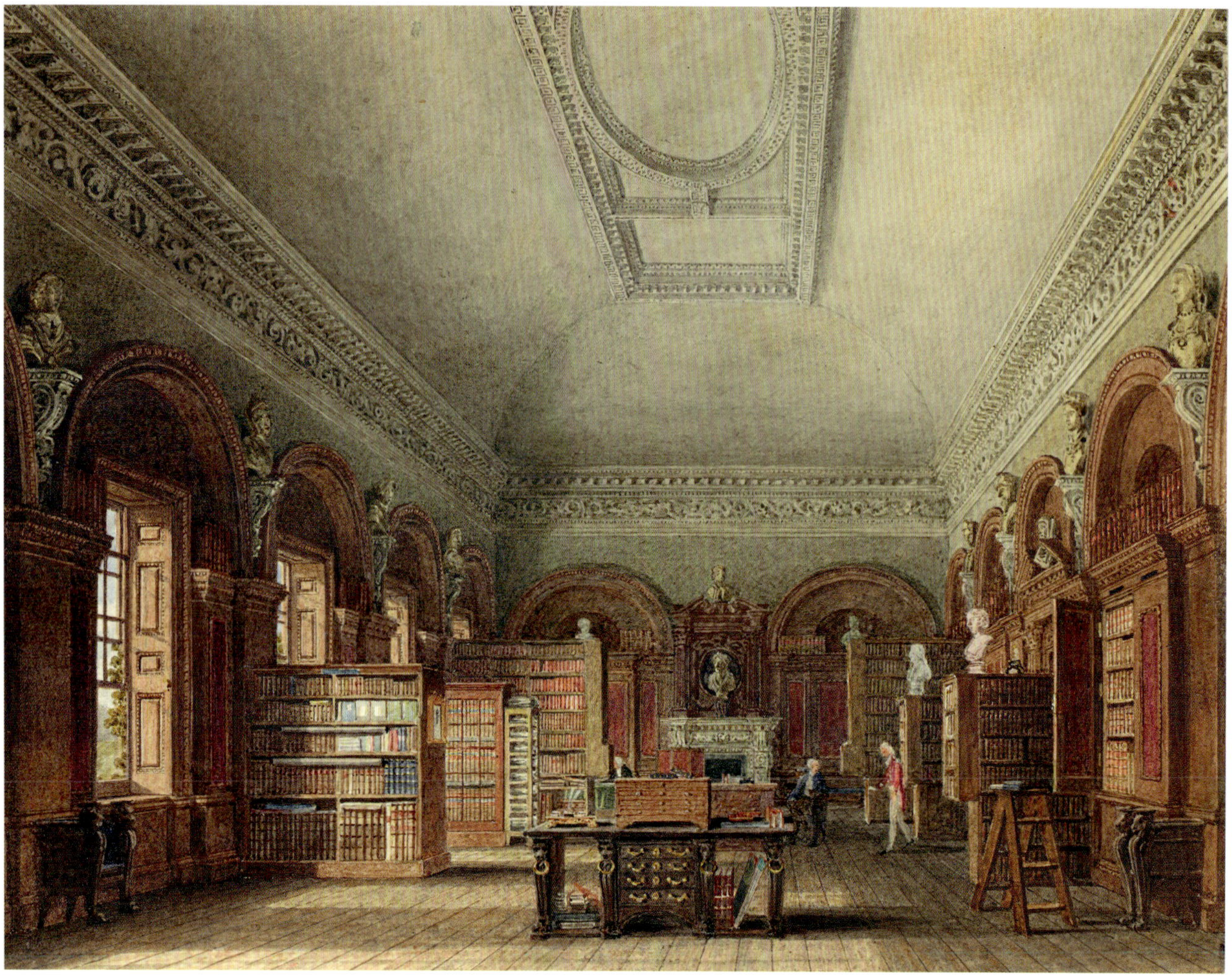

FIG. 5.15 Charles Wild, *St James's Palace: The Queen's Library*, c. 1819, pen and ink with watercolour and gouache on paper, 7⅞ x 9⅞ in. (20 x 25.1 cm). Royal Collection Trust (cat. 17.10)

shields taken exclusively from the Brandenburg-Ansbach arms (no English lion or Saxon steed can be found), which suggest that in spite of her fascination with, and dedication to, British culture and history—exhibited in the library's overall decoration—it was Caroline's intellectual education at the court of Brandenburg that made her into the representative of the early Enlightenment that she became.[30]

AUGUSTA

Following Caroline's death, George II's daughter-in-law, Augusta, Princess of Wales, became the most senior woman of the realm, and yet to this day her patronage remains something of an enigma.[31] Certainly during the fifteen-year period of her conjugal life, her commissions and collecting are difficult to disentangle from that

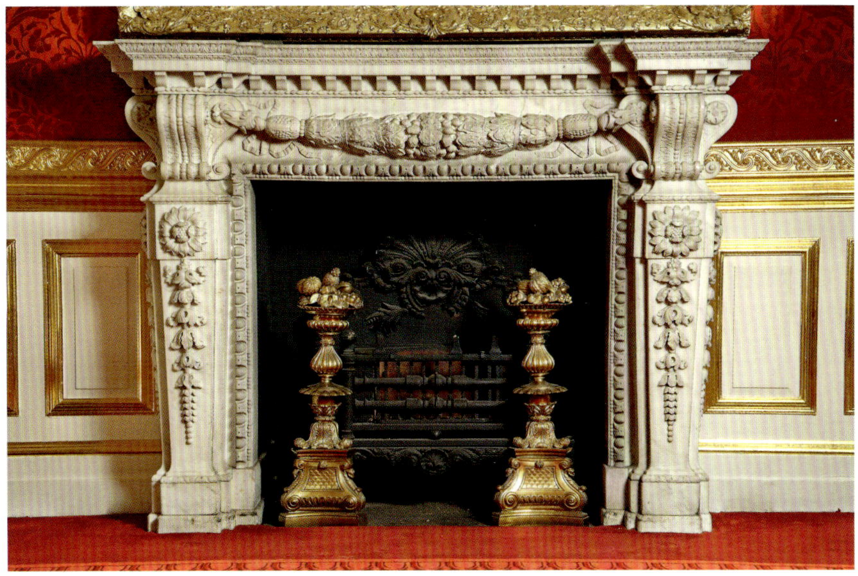

FIG. 5.16 William Kent, designer, Chimneypieces in the State Rooms at St James's Palace. Royal Collection Trust

of her art-loving husband Frederick. The Prince of Wales was a keen patron and collector—rivalled only by Charles I and George IV—to whom the royal collection owes some of its greatest masterpieces.[32] His dismissal from St James's Palace in 1737 led him to take up residence at Carlton House, Norfolk House, and eventually Leicester House—homes for which he was in need of pictures and furnishings.[33] While Augusta's early education may not have been quite as scholarly as Caroline's, the new Princess of Wales was no stranger to the implications of princely patronage. Augusta was born in the economically thriving city of Gotha in Saxony; her parents were Frederick II, Duke of Saxe-Gotha-Altenburg, and Magdalena Augusta of Anhalt-Zerbst. Through her mother, Augusta was second cousin to Catherine the Great of Russia, the most powerful patron and collector of the eighteenth century, who wrote to the Princess of Wales in the autumn of 1760 to remind her of the blood ties between the crowns of Britain and Russia.[34]

Frederick's engagement to Augusta was relatively unexpected. The Princess had not been prepared to marry the heir to the British throne. She spoke almost no English, and even after the engagement, her mother considered it superfluous for her to learn the language, for she assumed that twenty years after the arrival of the Hanoverians, Britons would be speaking German.[35] When Augusta married Frederick, Prince of Wales, in the spring of 1736, a little pamphlet was issued, which stressed the ancientness of her origins and the unrivalled Protestantism of the Saxon dynasty from which she descended.[36] With regards to her father, the pamphlet read:

> Frederick II . . . was a steddy [*sic*] Protestant . . . His People ador'd him; and indeed he treated them as a Father rather than a Sovereign. The Peasants, his

Subjects, are said to be the richest in all Germany . . . His Court was, of all the Courts of Saxony next to Dresden, the most numerous and magnificent.[37]

Augusta was raised at Schloss Friedenstein, a vast Baroque palace erected in the immediate aftermath of the Thirty Years' War (fig. 5.17).[38] Duke Frederick II had a real passion for architecture. His favoured architect Wolf Christoph Zorn von Plobsheim built him churches, schools, and Schloss Friedrichsthal, a suburban 'retreat', in fact only a stone's throw away from Friedenstein. The ducal collection and library grew significantly during his reign. He acquired three important collections of books and in 1712 went to great lengths to purchase the numismatic collection of Prince Anton Günther of Schwarzburg-Arnstadt for 100,000 thalers. The Duke took great pride in his internationally renowned collection of ancient and modern coins and medals, for which he had a room lavishly refitted at Schloss Friedenstein. Trompe-l'oeil frescoes adorned the walls and ceilings with pilasters and balustrades open to the skies. The coins and medals were kept in elaborate cabinets aligned along the walls, alternating with imposing gilt busts of Roman emperors.[39]

As Princess of Wales and as her father's daughter, Augusta understood the educational value of historic coins and medals, which led her to accumulate an important numismatic collection in her own right. In this she offered a parallel to Caroline, whose cabinets in the library at Kensington contained around one thousand medals.[40] Augusta's collection was kept at Somerset House and is mentioned in a memorandum from Lord North, when the house was cleared after her death.[41] While some of these pieces were sold at Christie's, others may have been absorbed into George III'

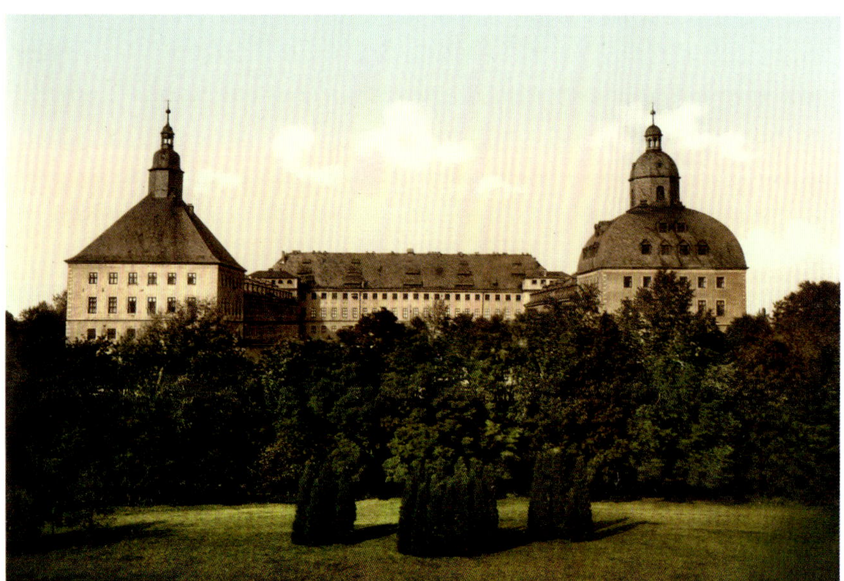

FIG. 5.17 Schloss Friedenstein (exterior), Gotha, Thuringia, Germany, c. 1890–1900, photochrom photomechanical print. Library of Congress, Washington, DC, Photochrom Print Collection

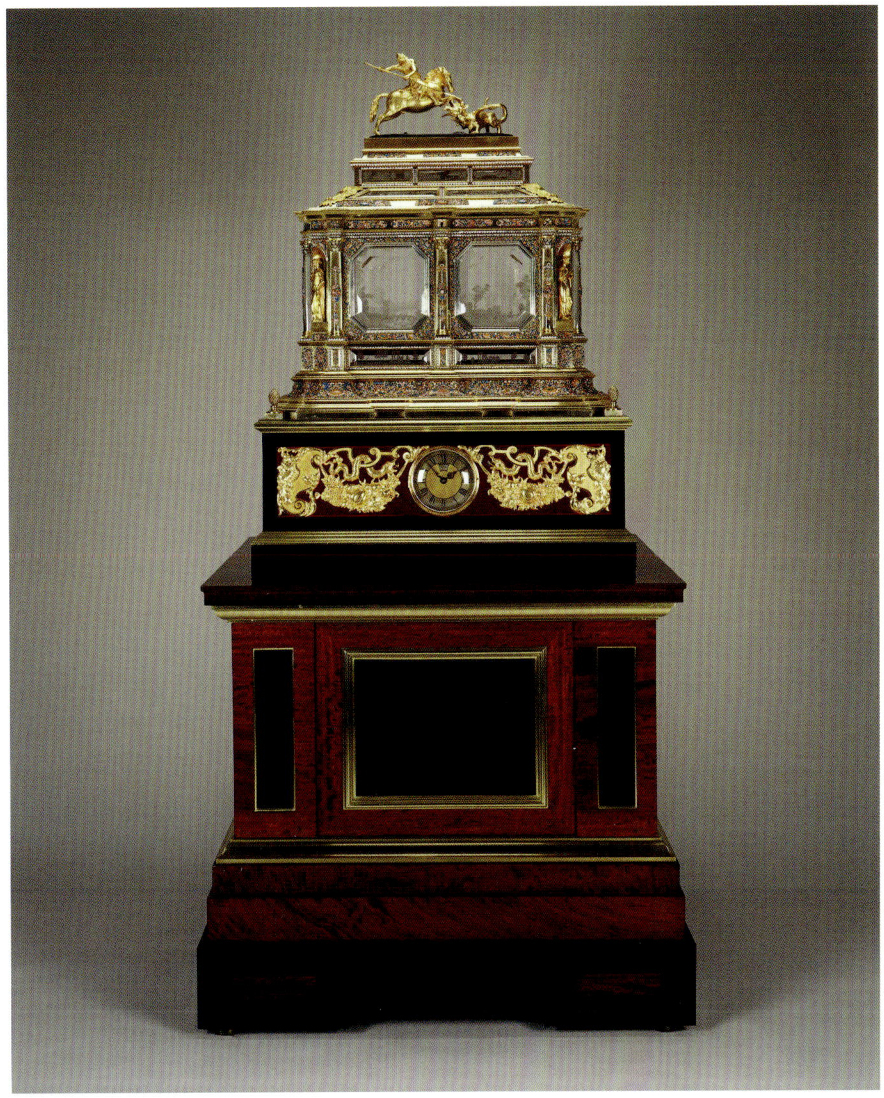

FIG. 5.18 Melchior Baumgartner, *The Temple and Oracle of Apollo* (casket), 1664, rock crystal, bronze, enamel, silver gilt; and Charles Clay, clock with organ, c. 1740, mahogany, ebony, bronze, 44½ x 35¾ x 29⅞ in. (113 x 91 x 76 cm). Royal Collection Trust

collection, later presented to the British Museum by George IV.[42] Like Caroline, Augusta appears to have created something of a *Wunderkammer*, although its contents are difficult to assess and only glimpses are offered in her accounts. In 1739, for example, she purchased an amber figure for 15 guineas, and a few months later a bill from Mr Natter was submitted for wax figures; Japan boxes appear with regularity in the accounts as well as items from Paul Daniel Chevenix the toy-man, Taylor who supplied ceramics, and a certain Desnoyer who sold Meissen porcelain to the Princess in 1737.[43] A silver-gilt cup and cover hallmarked by Paul de Lamerie is one of the few items of plate to have returned to the royal collection after Augusta's death, and appears in the inventories as having belonged to her, rather than to Frederick.[44] That Augusta may have been aware of the categorizations of the traditional *Wunderkammer*, as laid

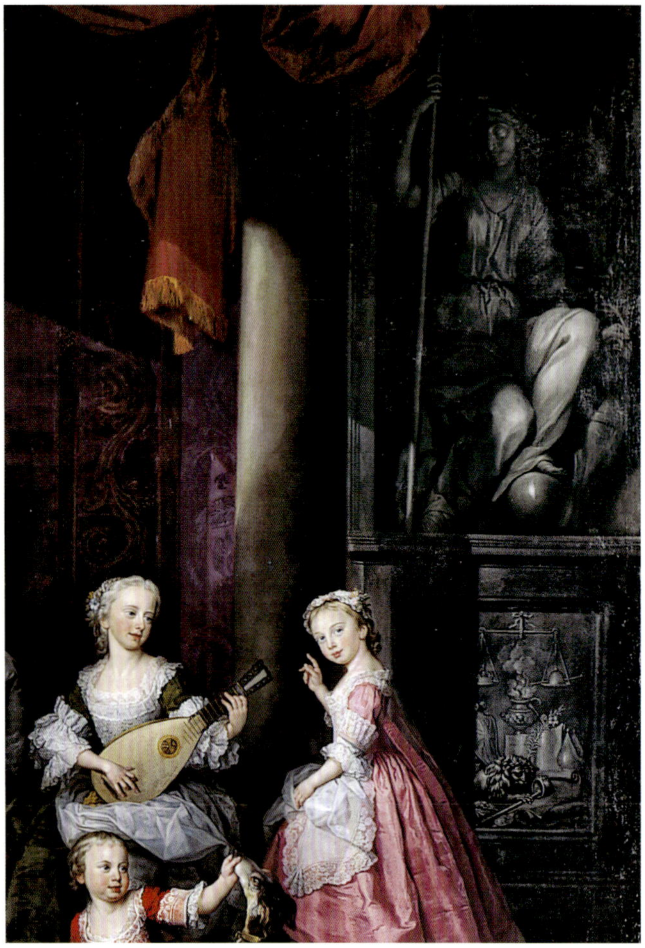

FIG. 5.19 George Knapton, *The Family of Frederick, Prince of Wales* (detail), 1751, oil on canvas. Royal Collection Trust (see p. 14, fig. 1.08)

down by Samuel Quiccheberg in the sixteenth century, may be too much of a stretch, but in a nod to the scientific she certainly owned a camera obscura and she was trained at the turning lathe.[45] Perhaps the work which might most closely be associated with the type of object to appear in a contemporary *Wunderkammer* is the organ clock, the work of the 'ingenious Mr Clay' (fig. 5.18).[46] This remarkable piece of furniture incorporates an engraved rock crystal and enamel-mounted casket by the Augsburg craftsman Melchior Baumgartner. Its composition, so characteristic of southern Germany, suggests the work would have appealed strongly to Augusta, who acquired it from Charles Clay's widow in 1759.[47]

Princess Augusta's most significant pictorial commission was a huge family portrait by George Knapton, painted in the immediate aftermath of Frederick's death in 1751. It shows the Princess Dowager enthroned between two columns, flanked by a state portrait of her late consort on the left and a statue of Britannia on the right, and surrounded by her nine children. The allegorical bas-relief on the plinth of Britannia represents the British constitution by depicting a British lion guarding Magna Carta

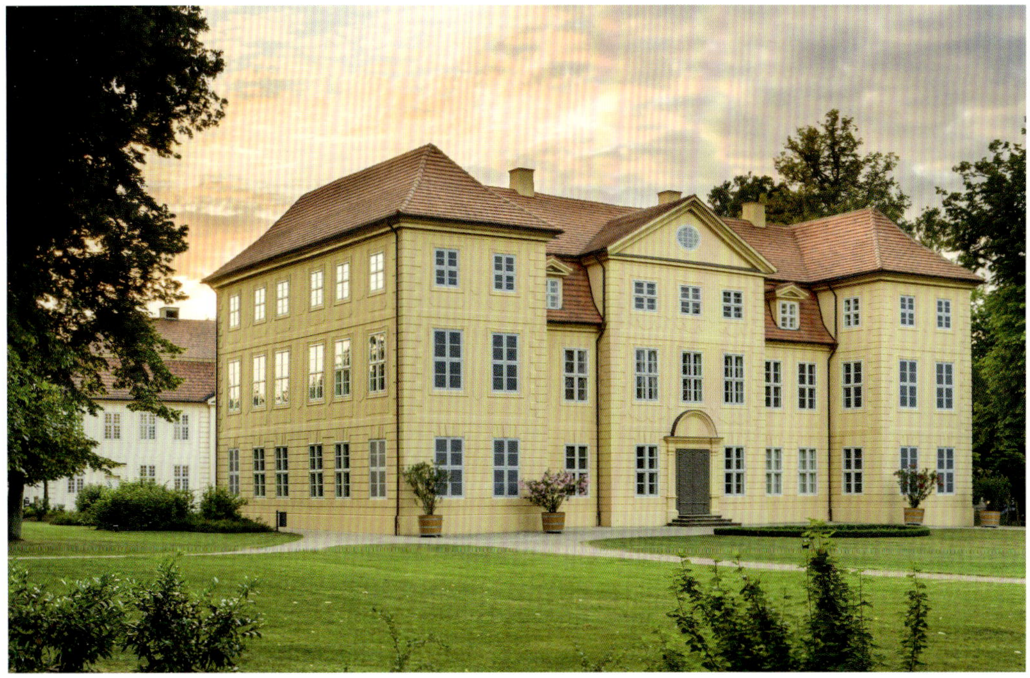

FIG. 5.20 Schloss Mirow, Germany. Staatliche Schlösser und Gärten Mecklenburg-Vorpommern

and the Act of Settlement, as well as a perfectly balanced set of scales holding a crown on one side and a cap of liberty on the other (fig. 5.19). As Desmond Shawe-Taylor has shown, the entire family portrait therefore 'reads as a pair of scales with the Prince and the constitution held in balance. The Princess sits enthroned on the fulcrum, as a guardian of these principles rather than claimant to the throne, and supervises her children.'[48] The picture not only illustrates how Augusta perceived her own role in securing the future of the Hanoverian dynasty, but also her—and presumably her late husband's—understanding of Britain's constitutional values, contrasting with the absolutism performed in most other parts of Europe.

CHARLOTTE

Augusta and Frederick's eldest son, George III, succeeded his grandfather George II in October 1760. Although he was the first Hanoverian ruler actually born and educated in Britain, his subjects continued to be well aware of his Germanic roots and those of his future consort. Charlotte of Mecklenburg-Strelitz arrived in London in September 1761, and the royal couple was married that same day, with the coronation following two weeks later. With regard to the wedding night, the contemporary novelist Sarah Scott could not help but note that '[The queen] and the king were in bed, and that all night after her journey and so long a voyage. Nothing but a German constitution could have undergone it.'[49] Queen Charlotte's background departed significantly from that of Caroline and Augusta. Not only was her father, Duke

Charles Louis of Mecklenburg-Strelitz, a minor prince, he was not even a ruling one.[50] Charlotte grew up in the utmost simplicity. She spent most of her sheltered childhood under the care of her mother Elizabeth Albertine of Saxe-Hildburghausen at Schloss Mirow, a tiny manor house on the southeast border of her uncle's dukedom (fig. 5.20). Mirow was just the type of picturesque palace that was the literary inspiration for the princely courts in Grimms' fairy tales, or maybe even the home of the German Baron Thunder-ten-tronckh in Voltaire's *Candide* (1759). Largely destroyed by fire in 1742, the palace was rebuilt between 1749 and 1752 to the designs of Christoph Julius Löwe. To decorate her suite of rooms, Charlotte's mother, the then Dowager Duchess, employed Johann Melchior Kambly, Frederick the Great's most accomplished sculptor and ébéniste. Kambly created a suite of intimate interiors in the late Rococo fashion, for which a set of elevation drawings by the sculptor Simon Gehle survives at Waddesdon Manor.[51]

We are relatively well informed about life at Mirow through the incorrigible pen of Frederick the Great, who saw much of Charlotte's family as he sojourned at his nearby hunting lodge, Schloss Rheinsberg. The then Crown Prince of Prussia mockingly called his neighbours *Miroquois*, hinting at their poverty, provincial attitudes, and overall lack of culture. He discerned that while they were kind, jolly, and entertaining, conversations always remained notably superficial.[52] Nonetheless, Charlotte's parents were well aware of the importance of a suitable upbringing—in the absence of any fortune, all they could give their children were their titles and a good education.[53] To this end, they employed the botanist and theologian Gottlob Burchard Genzmer as their tutor, who was also an old university friend of the pioneering Hellenist and founding father of modern archaeology, Johann Joachim Winckelmann, and the Swedish botanist, physician, and zoologist Carl Linnaeus.[54]

For Charlotte, the contrast between Mirow and London must have been dramatic. Nonetheless, the royal headquarters were comparatively humble. St James's Palace was George III's official residence, as had been the case for his predecessors George I and George II—a palace that was subject to mockery throughout the eighteenth century. In 1761, at the very beginning of George III's reign, the bookseller Robert Dodsley had criticized St James's for being a 'low, plain and mean building with no single beauty on the outside to recommend it. It is at once the contempt of foreign nations, and the disgrace of the British. It is therefore to be hoped that the erecting of so necessary a structure would not be much longer neglected.'[55] Although a passionate amateur architect, George III refrained from complying with Dodsley's vision of building a new palace more appropriate to his rank. Instead, he bought Buckingham House on the other side of Green Park as a private home for his new Queen, officially transferred to her in 1775 by Act of Parliament, in exchange for Somerset House, which had served as a dower house to previous queens.[56]

From 1762 and over a period of fifteen years, the building—which later became Buckingham Palace—underwent considerable transformation under the supervision of Sir William Chambers.[57] Rather than embellish the overall appearance of the house, Chambers toned it down: the façade was simplified by removing two angle

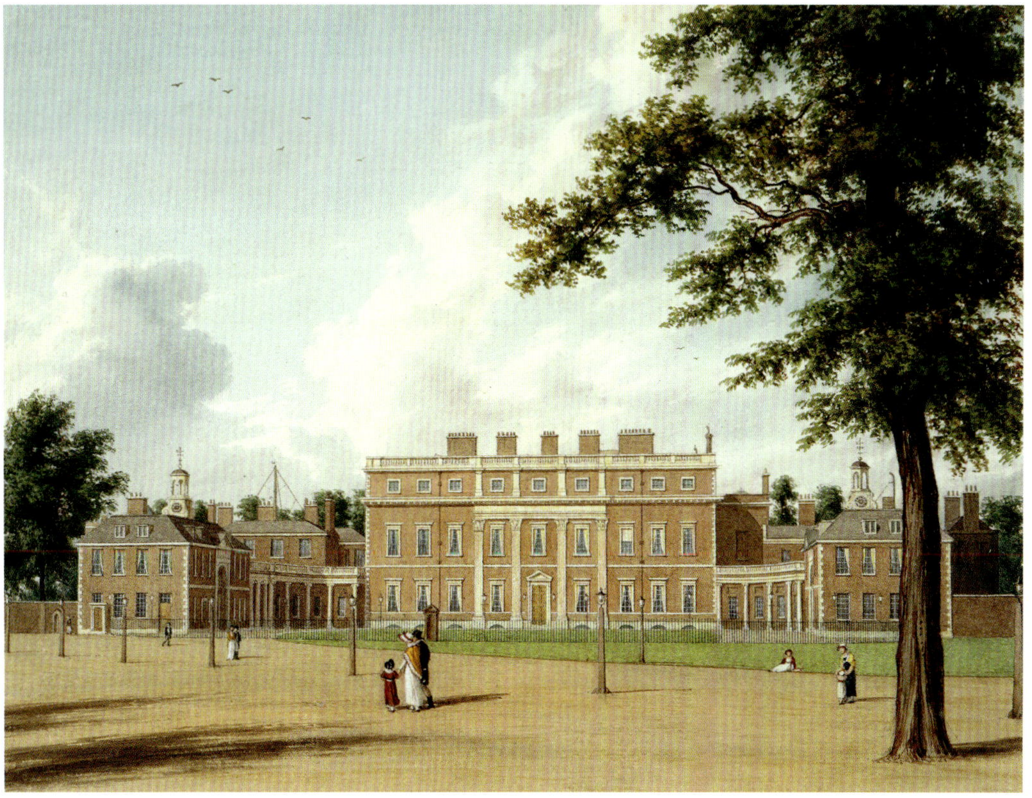

FIG. 5.21 William Westall, *Buckingham House: The East (Entrance) Front*, 1819, watercolour and gouache over pencil, 8½ x 10⅝ in. (21.8 x 27.0 cm). Royal Collection Trust

pilasters, parts of the cornice, as well as the sculptures towering on the balustrade; he further withdrew the fountain and the large Baroque gates, replacing them by a less intimidating iron grille (fig. 5.21).[58] Formality was replaced by domesticity. However, the Queen's House, as it was first known, was not an exclusively private retreat. Charlotte held court here, and she was less averse to artistic extravagance than her husband. Her lavish first-floor suite of rooms contrasted with the comparatively austere ground-floor apartments of the King.[59] The largest and most important room was the Queen's Saloon, which later functioned as a presence chamber or throne room (fig. 5.22). Chambers replaced Louis Laguerre's opulent Baroque murals with a much more restrained neo-classical decoration of grisailles and arabesques.[60] Initially, the room was hung with the famous Raphael cartoons, brought over from Hampton Court in 1763 and transferred to Windsor Castle in 1787–88.[61] These jewels of Charles I's otherwise widely dispersed art collection occupied a unique status amongst George III's and Queen Charlotte's contemporaries. Most notably, Sir Joshua Reynolds, President of the Royal Academy from its establishment in 1768, considered the cartoons as well as Michelangelo and Raphael's frescoes at the Sistine Chapel, 'The principal works of modern art'.[62] Installing the cartoons in the most prestigious space at the Queen's House was a clear sign that the royal couple understood their supreme artistic value.

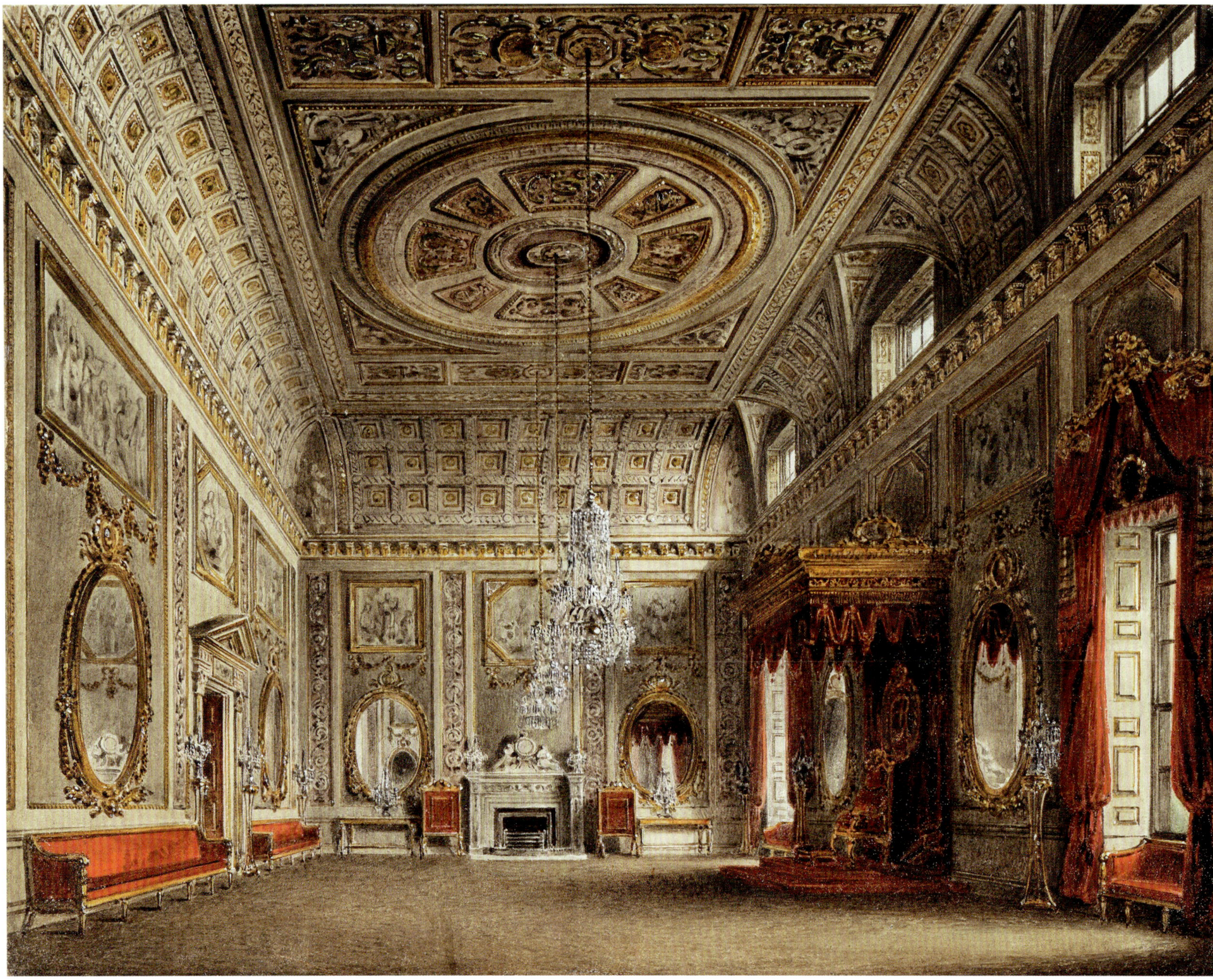

FIG. 5.22 James Stephanoff, *Buckingham House: The Saloon*, 1818, watercolour and gouache over pencil on paper, 8¼ x 9¾ in. (21.0 x 25.1 cm). Royal Collection Trust

One thread of continuity with Charlotte's predecessors may be teased out of her collections. Like both Caroline and Augusta she sought to create a 'theatre of wisdom' or *Wunderkammer* at Buckingham House, encapsulating in microcosm elements of the whole world. No inventory of these objects survives, but elements of it appear in the sale of the Queen's collections held over a series of days at Christie's after her death.[63] The sale included cups formed from coconuts, rhinoceros horns, and agate as well as a Baroque pearl transformed into a figure of Harlequin and a rock crystal bust of Charles I. Alongside these was a piece of furniture described as a wardrobe, fitted out with sliding shelves and cabinets which contained Charlotte's collection of 'naturalia', particularly shells, ores, minerals, exotic birds, and frames containing preserved butterflies. Revealingly, a group of works of art from Buckingham House later passed to her son George, then Prince Regent, at Carlton House.[64] Among this group were objects which could be traced back to Caroline's *Wunderkammer*, including a stuffed humming bird, vases of rock crystal, branches of red and white coral, a distinctive vase with chimera handles and unicorn feet, a crimson glass vase, two agate goblets, and a group of daggers. Most distinctive of all is a marble sculpture of a 'shock dog' (a Pomeranian), which Walpole claimed had come from the closet of Charles I and was the work of Gian Lorenzo Bernini.[65] This object can clearly be picked out of the inventories of Caroline's closets at Kensington.

The collecting and commissioning of art as well as the rearranging of the royal collection was one of the passions that George III and Queen Charlotte shared.[66] Their appreciation of old master paintings, contemporary works, and the decorative arts led them to make crucial acquisitions, including Joseph 'Consul' Smith's vast collection of books, as well as paintings and drawings by Canaletto, Sebastiano Ricci, Francesco Zuccarelli, Pietro Longhi, and others[67]—a collection that would, at least to a degree, be made accessible to polite society.[68] George III was also a great admirer of the work of Anthony van Dyck; he therefore reacquired the accomplished group portrait of Charles I's five eldest children for Buckingham House.[69] Indeed, such was George and Charlotte's taste for the Flemish master that they ordered a family portrait from Johan Zoffany depicting the royal family sporting Caroline costumes.[70]

As opposed to the early Hanoverians, who exhibited an aversion to having their likenesses captured on canvas, George and Charlotte commissioned numerous portraits of themselves and the royal children. Allan Ramsay, Reynolds, Thomas Gainsborough, Benjamin West, Sir William Beechey—each had had his own take on the royal image. The German-born Zoffany, who had arrived in London in the autumn of 1760, only a year before Charlotte, was an early favourite of the royal couple.[71] He may have been introduced to the King and Queen by John Stuart, third Earl of Bute, who had had his children portrayed by the artist around 1763–64.[72] George and Charlotte commissioned no less than twenty-one portraits and conversation pieces from Zoffany, of which seventeen survive in the Royal Collection.

Zoffany was nominated to the Royal Academy in 1769, but fell out of favour with his royal patrons following his failure 'to paint a perspective view of the Tribuna' at the Galeria degli Uffizi according to Queen Charlotte's vision (fig. 5.23).[73] In 1772 the

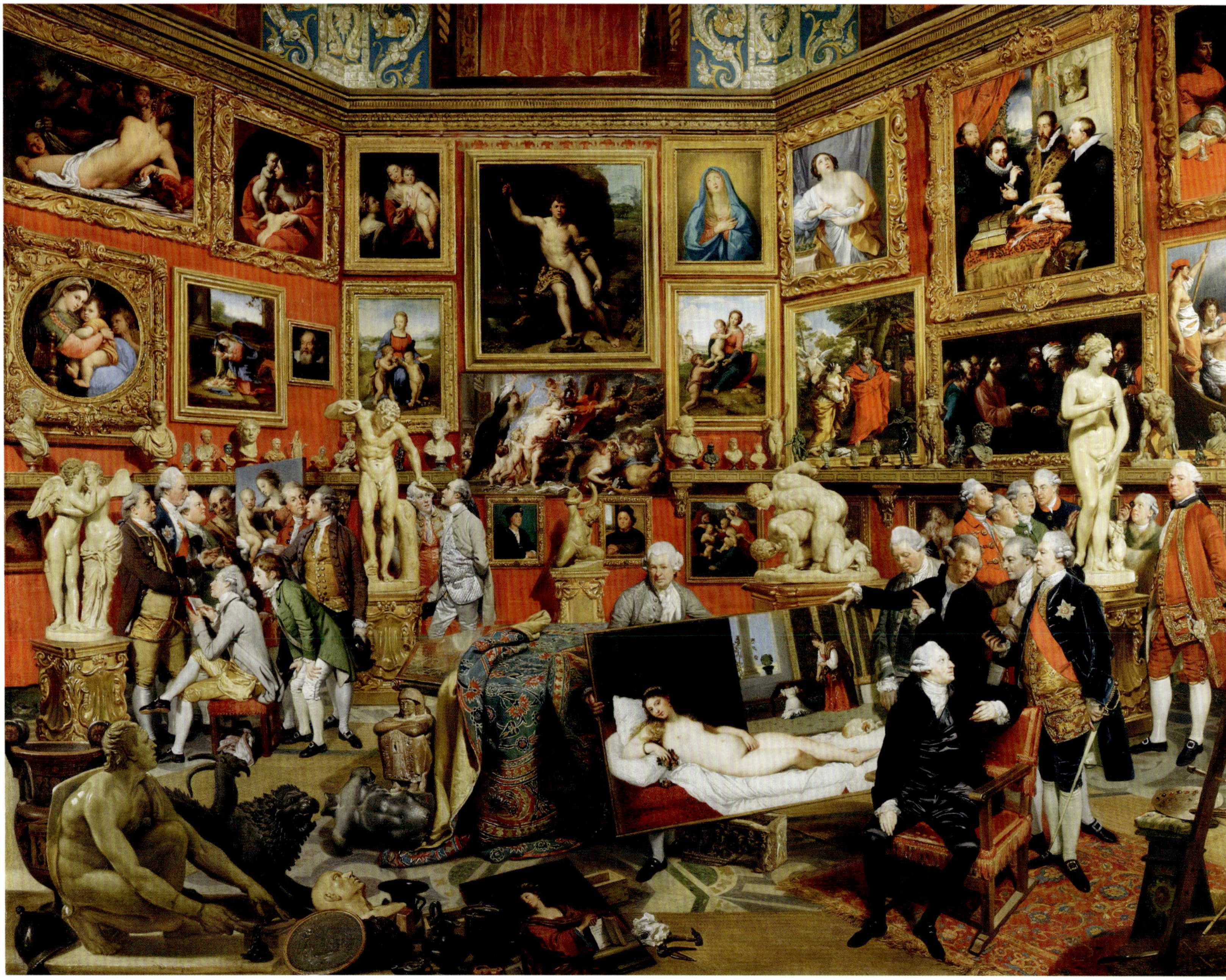

FIG. 5.23 Johan Zoffany, *The Tribuna of the Uffizi*, 1772–77, oil on canvas, 48⅝ x 61 in. (123.5 x 155 cm). Royal Collection Trust (cat. 9.11)

artist had been sent to Florence to capture this iconic space with all its details—its pictures, frames, and sculptures.[74] In 1778 the finished canvas was brought back to London, where it met with Charlotte's disapproval: 'The Queen wd. not suffer the picture to be placed in any of her apartments', Joseph Farington noted in his diary in December 1804.[75] Today, Zoffany's *Tribuna* is considered a key summation of late eighteenth-century grand tours, representing British gentlemen immersing themselves in the fine arts of the Italian Renaissance. Charlotte's younger brother George Augustus also went on a grand tour (paid for by George III) travelling from Leiden to London and from Paris to Pisa, Florence, and Rome, from where he visited many other Italian cities.[76] In Rome, he was looked after by Winckelmann. The tutor and his student quickly grew extremely fond of each other, as Winckelmann recorded: 'The Prince of Mecklenburg-Strelitz won't leave the house without me. . . . He is a friend, son, student, crony. With him, I am returning to my past youth.'[77]

Although fascinated by art and architecture, the great masters of the Italian Renaissance and antiquarianism, as King and Queen, George III and Charlotte were hindered from undertaking their own grand tour. Had they ever been Prince and Princess of Wales together, they might have been allowed to travel across Europe rather like the Grand Duke Paul and his consort Marie Feodorovna had done in 1782, fourteen years before their accession to the throne of Russia. George III never even set foot in Hanover, although he planned to do so repeatedly. However, in the 1780s and 1790s, the King and Queen sent all their sons, except for George IV, to be representatives of the dynasty and to be educated either in Hanover or to attend the University of Göttingen for a few years.[78] Throughout their lives, George and Charlotte endorsed their close family ties to Hanover, Mecklenburg-Strelitz, and other German principalities, even if more privately than publicly. Born, raised, and educated in Britain, George III could be much more relaxed than his predecessors with regard to his German roots. Nevertheless, the artistic patronage demonstrated by George and Charlotte showed a pointed bias towards British craftsmen. Although Charlotte acquired some historic pieces of German silver or items of French porcelain, the significant new commissions of the reign were awarded to British manufactories. From the Chelsea factory in 1763 came the so-called Mecklenburg-Strelitz service, a gift to Charlotte's brother Adolphus Frederick IV.[79] Both the Wedgwood and the Worcester factories named designs after the Queen, in order to curry favour and to win patronage from members of the court. Josiah Wedgwood developed Queen's Ware—an earthenware with cream-coloured body—after the completion of a tea service for Charlotte, supplied in 1766, which gained him the royal warrant.[80] When George III visited the Worcester factory in 1788, he suggested that the lily pattern being developed might be renamed 'Royal Lily' in honour of Queen Charlotte.[81]

Through her skilful support of the arts, Caroline was key to establishing the Hanoverian dynasty as the rightful ruling house of Britain. Augusta, as illustrated in Knapton's portrait, saw herself as a securer of the dynasty and its understanding of Britain as a constitutional monarchy. Charlotte, while reflecting on Caroline's patronage, may have had a less politically motivated attitude to the visual arts. Her influence on her sons' collections cannot be overstated. Like Augusta before her, Charlotte held great sway over her sons. Both Frederick, Duke of York, and Augustus, Duke of Sussex, amassed significant collections of their own—Frederick commissioned English and French silver and porcelain; Augustus compiled a vast library and an important silver collection—and it appears that it was from Charlotte that George IV developed his own love of Sèvres porcelain and items of plate from the continent as well as his taste for the exotic.

NOTES

1. For the patronage of George I and George II, see most recently Desmond Shawe-Taylor, ed., *The First Georgians: Art and Monarchy, 1714–1760* (London: Royal Collection Trust, 2014); Wolf Burchard, 'Face to Face with the First Georgians', *The Georgian* 1 (2014): 4–8; and Jarl Kremeier, 'George I and George II as Patrons of Architecture in Hanover', *The Georgian Group Journal* 22 (2014): 1–20, with further references.
2. George MacDonald Ross, 'Leibniz und Sophie Charlotte', in *Sophie Charlotte und ihr Schloß: Ein Musenhof des Barock in Brandenburg-Preußen*, ed. Gerd Bartoschek and others (Munich: Prestel, 1999), 95–105.
3. 'Ich gehe jetzt, meine Neugier befriedigen über die Urgründe der Dinge, die mir Leibniz nie hat erklären können . . . und dem König meinem Gemahl bereite ich das Schauspiel eines Leichenbegängnisses, welches ihm neue Gelengeheit gibt, seine Pracht darzuthun.' See Karl August Varnhagen von Ense, *Leben der Königin von Preußen Sophie Charlotte* (Berlin, 1837), 232, previously cited in Veronica Biermann, '"Ma chère Pelnits" Henriette Charlotte von Pöllnitz (um 1670–1722), "Erstes Kammerfräulein" Sophie Charlottes', in *Sophie Charlotte und ihr Schloß*, ed. Bartoschek and others, 80.
4. Rufus Bird, 'The Furniture and Furnishing of St James's Palace, 1714–1715', *Furniture History* 50 (2014): 147–203.
5. Wolf Burchard, 'Houses, Palaces and Gardens: The First Georgians and Architecture', in *First Georgians*, ed. Shawe-Taylor, 56.
6. Hannah Smith, 'The Political Economy of the Early Georgian Court', in *Georgian Monarchy: Politics and Culture, 1714–1760* (Cambridge: Cambridge Univ. Press, 2006), 64–73.
7. Steven Brindle, 'Royal Commissions', in *William Kent: Designing Georgian Britain*, ed. Susan Weber (New Haven and London: Yale Univ. Press, 2013), 271–77, with further references.
8. Kathryn Jones and Desmond Shawe-Taylor, '"An Amiable Philosopher on the Throne": Queen Caroline and the Encouragement of Learning', in *First Georgians*, ed. Shawe-Taylor, 287.
9. George Vertue, *The Collections of Pictures, Paintings etc at Kensington, Hampton Court and in the Castle of Windsor* (London, 1750), MS, Royal Collection, Surveyor's Office (OM 30).
10. For a discussion of the contents of Queen Caroline's Picture Closet, see Joanna Marschner, *Queen Caroline: Cultural Politics at the Early Eighteenth-Century Court* (New Haven and London: Yale Univ. Press, 2014), 97–101. Horace Walpole's inventory was published in W. Bathoe, *A Catalogue of Pictures Belonging to King James the Second (copied from a ms. In the Library of the Earl of Oxford): To which is Appended a Catalogue of the Pictures, Drawings, Limnings, Enamels, Models in Wax and the Ivory Carvings etc. at Kensington in Queen Caroline's Closet, next the State Bedchamber* (London, 1758).
11. Shawe-Taylor, ed., *First Georgians*, 274.
12. Vanessa Remington, '"Curious Limnings': Portrait Miniatures in the Early Georgian Period', in *First Georgians*, ed. Shawe-Taylor, 320–21.
13. For the Kensington miniatures, see Graham Reynolds, *The Sixteenth and Seventeenth-Century Miniatures in the Collection of Her Majesty The Queen* (London: Royal Collection Trust, 1999), 177–202, nos. 181–229, and Shawe-Taylor, ed., *First Georgians*, 329, nos. 195–98. The Herrenhausen group appears to have become part of the Welfenmuseum in 1861. Their subsequent fate is unknown. We are grateful to Bernd Wedekind for sharing this information with us. While his research on the sixteenth-century Brunswick miniatures has not yet been published, his entries are accessible at the archives of the Bomann Museum in Celle.
14. Reynolds, *Sixteenth and Seventeenth-Century Miniatures*, 208–12, nos. 208–46, and Shawe-Taylor, ed., *First Georgians*, 330, nos. 199–202.

15. The so-called self-portrait of Holbein and that of his wife were given to Queen Caroline by Horace Walpole and are now attributed to Hans Brosamer; see Kate Heard and Lucy Whitaker, *The Northern Renaissance: Dürer to Holbein* (London: Royal Collection Trust, 2011), 138–41, nos. 60 and 61. The portrait of Elizabeth I is now attributed to William Scrots.
16. W. H. Wilkins, *Caroline the Illustrious, Queen-Consort of George II and sometime Queen Regent: A Study of her Life and Time*, 2 vols. (New York and Bombay, 1901), 1: 102–4, as cited in Shawe-Taylor, ed., *First Georgians*, 272.
17. Royal Collection inventory numbers (hereafter RCIN) 65186 and 65187. See Kirsten Aschengreen Piacenti and John Boardman, *Ancient and Modern Gems and Jewels in the Collection of Her Majesty the Queen* (London: Royal Collection, 2008), nos. 230–31, and Kathryn Jones in *First Georgians*, ed. Shawe-Taylor, 293, nos. 160–61.
18. RCIN 1112557.
19. Judith Colton, 'Merlin's Cave and Queen Caroline: Garden Art as Political Propaganda', *Eighteenth-Century Studies* 10, no. 1 (1976): 1–20.
20. Marschner, *Queen Caroline*, 121–47.
21. George II felt particularly responsible for the well-being of Amalie von Wallmoden and therefore had his German court architect Johann Paul Heumann build her a dower house close to the Leineschloss in Hanover. See Bernd Adam, 'Hannoversche Adelspalais des Barock und Rokoko', in *Stand und Repräsentation, Kultur- und Sozialgeschichte des hannoverschen Adels vom siebzehnten bis zum neunzehnten Jahrhundert*, ed. Silke Lesemann and Annette von Stieglitz (Hanover: Verlag für Regionalgeschichte, 2004), 22–27.
22. Marschner, *Queen Caroline*, 121 ff.; Brindle, 'Royal Commissions', 294–97; Katharine Eustace, 'The Politics of the Past: Stowe and the Development of the Historical Portrait Bust', *Apollo* 148 (July 1998): 31–40; and Wolf Burchard, 'St James's Palace: George II's and Queen Caroline's Principal London Residence', *The Court Historian* 16, no. 2 (Dec. 2011): 188 ff. All four with further references.
23. William Kent's designs are preserved at the Sir John Soane's Museum in Fauntleroy's edition of Thomas Pennant's 1790 *Account of London* (147/192–198).
24. John Cornforth, *Early Georgian Interiors* (New Haven and London: Yale Univ. Press, 2004), 174.
25. Shawe-Taylor, *First Georgians*, 78, no. 29, and Brindle, 'Royal Commissions', 294–97, with further references.
26. Precisely at the time when Caroline was building her library, her nephew modernized the palace at Ansbach, creating new suites of Rococo interiors. A set of designs showing the alterations was sent to the Queen in 1736 and still survives at the British Library; see Heinz Braun, 'Ansbacher Spätbarock: Ein Beitrag zur Stilgeschichte der Baukunst im Fürstentum Brandenburg-Onolzbach, 1695–1791', *Jahrbuch für fränkische Landesforschung* 16 (1956): 465, and Josef Maier, 'Residenzschloß Ansbach: Gestalt und Ausstattung im Wandel der Zeit', *Jahrbuch des Historischen Vereins für Mittelfranken* 100 (2005): 176, 532.
27. Marschner, *Queen Caroline*, 86 ff.; Joanna Marschner, 'Michael Rysbrack's Sculpture Series for Queen Caroline's Library at St James's Palace', in *Burning Bright: Essays in Honour of David Bindman*, ed. Diana Dethloff, Tessa Murdoch, and Kim Sloan, with Caroline Elam (London: UCL Press, 2015), 27–35; Jonathan Marsden in *First Georgians*, ed. Shawe-Taylor, 284–85, nos. 151–52; and Burchard, 'St James's Palace', 189. All four with further references.
28. Eustace, 'Politics of the Past', 37; Burchard, 'St James's Palace', 189–90.
29. Howard Colvin, *The History of the King's Works* (London: HMSO, 1963–82), 5: 242.
30. Burchard, 'St James's Palace', 191.
31. John Harris, 'George III's Parents: Frederick and Augusta', in *The Wisdom of George the Third*, ed. Jonathan Marsden (London: Royal Collection Trust, 2005), 21 and 23–25.
32. Oliver Millar, 'Frederick, Prince of Wales', in *The Tudor, Stuart and Early Georgian Pictures in the Collection of Her Majesty The Queen*, 2 vols. (London: Phaidon, 1963), 1: 28–30, and Frances Vivian, *A Life of Frederick, Prince of Wales, 1707–1751: A Connoisseur of the Arts*, ed. Roger White (Lewiston, NY, and Lampeter: Edwin Mellen Press, 2006).
33. Burchard, 'Houses, Palaces and Gardens', 56–57.
34. The Royal Archives (hereafter RA) preserves two letters from Catherine the Great to Princess Augusta: one in which she reminds her cousin of the blood ties with her mother (RA Geo/Main 242 [19 Nov. (OS) 1760]), and another in which she wishes closer relations between Great Britain and Russia (RA Geo/Main 29 [13 Oct. (OS?) 1762]).

35. Andrew Thompson, *George II: King and Elector* (New Haven and London: Yale Univ. Press, 2011), 113.
36. *Memoirs of the Ancestors of Her Royal Highness Augusta, Princess of Wales, and of Saxe-Gotha. Particularly Frederick the Wise, John the Good and the Constant and John Frederick the Magnanimous, Protectors of the Great Luther* (London, 1736).
37. *Memoirs of the Ancestors of Her Royal Highness Augusta, Princess of Wales*, 74–75.
38. Marc Rohrmüller, 'Schloss Friedenstein: Architektur, Distribution, Ausstattung', in *Ernst der Fromme (1601–1675). Bauherr und Sammler*, ed. Juliane Ricarda Brandsch (Gotha: Schlossmuseum, 2001), 11–20; Marc Rohrmüller, 'Ernst der Fromme und die Höfische Pracht', in *Ernst der Fromme (1601–1675): Staatsmann und Reformer*, ed. Roswitha Jacobsen and Hans-Jörg Ruge (Buch: Wissenschaftliche Beiträge und Katalog zur Ausstellung, 2002), 159–66; Jörn Tillmann Rieckhof Nackas, *Schloss Friedenstein in Gotha: Entwürfe für die Residenz Ernsts des Frommen (1601–1675)* (Berlin: Freie Univ., Magisterarbeit, 2010).
39. Wolfgang Steguweit and Martin Hoernes, 'Gothas Gold und Silber: Der Münzschatz Herzog Friedrichs II. von Sachsen-Gotha-Altenburg kehrt zurück nach Friedenstein', *Arsprototo* 4 (2011): 38–41.
40. In 1763, Horace Walpole appended a hand-written list to his copy of Vertue's catalogue of the Picture Closet at Kensington. The list is now in the Royal Collection (RCIN 1112557). Walpole makes note of Queen Caroline's medal collection housed in several drawers, amongst which was a complete series of Roman Emperors.
41. British Library, North (Sheffield Park) papers, Add. MS. 61860, vol. I.
42. *Catalogue of the superb Jewels, Trinkets, Plate, Gold, and Silver Medals, China &c. of a Person of great Rank, dec.* [Augusta, Princess of Wales], Christie's London, Wednesday, 3 Feb. 1773, 11–12; Andrew Burnett, '"The King Loves Medals": The Study of Coins in Europe and Britain', in *Enlightenment: Discovering the World in the Eighteenth Century*, ed. Kim Sloan (London: British Museum, 2004), 122–31.
43. All of these items appear in Augusta's Privy Purse accounts, RA Geo Add. MS. 17/65/1.
44. RCIN 51016, recorded in an inventory of plate in 1832 as having belonged to the Princess Dowager of Wales.
45. RA Geo Add. MS. 55540, bill from P. Ramsden, 2 July 1768, for altering a glass frame for a camera obscura: 5 s.; RA Geo Add. MS. 55539 for mention of the lathe.
46. RCIN 30037.
47. Hugh Roberts, 'So Beautiful a Style', *Country Life*, 23 Nov. 1995, 58–59, and Rufus Bird in *First Georgians*, ed. Shawe-Taylor, 218–19, no. 120, with further references.
48. Shawe-Taylor, ed., *First Georgians*, 22.
49. Sarah Scott, in John Doran, *A Lady of the last Century (Mrs Elizabeth Montagu)* (London, 1873), 97–98, as cited in Friederike Drinkuth, *Queen Charlotte: A Princess from Mecklenburg-Strelitz Ascends the Throne of England* (Schwerin: Thomas Helms, 2011), 33. Taking advantage of public interest in the new Queen, Sarah Scott further published [Mrs G. L. Scott], *The History of Mecklenburgh; from the First Settlement of the Vandals in that Country, to the Present Time* (London, 1762), telling the story of Charlotte's origins.
50. Marcus Köhler, 'The Courts of Hanover and Strelitz', in *Wisdom of George the Third*, ed. Marsden, 60–81.
51. For the Dowager Duchess's art collection at Schloss Mirow, see Friedrike Drinkuth, 'Zur Identifikation und Bedeutung einer Aussteuerliste von Herzogin Elisabeth Albertine zu Mecklenburg-Strelitz', *Mecklenburgische Jahrbücher* 129 (2014): 265–82, and Alastair Laing, 'German Drawings', in *The James A. de Rothschild Bequest at Waddesdon Manor: Drawings for Architecture, Design and Ornament*, gen. ed. Geoffrey de Bellaigue, 2 vols. (Aylesbury: Waddesdon Manor, 2006), 2: 614–77. For Gehle drawings, see nos. 465a–g (for Schloss Neustrelitz), inv. nos. 266/1–6, accession nos. 1648, 1650–54 respectively (pp. 618–22), and 466a–g (for Schloss Mirow), inv. nos. 227/1–7, accession nos. 1655–62 respectively (pp. 624–29).
52. For correspondences between Frederick the Great and his father Frederick William I (autumn 1736), see Johann D. E. Preuss, *Œuvres de Frédéric le Grand*, Vol. 27/3 (Berlin, 1856), 120 ff., as referred to in Drinkuth, *Queen Charlotte*, 7.
53. Köhler, 'Courts of Hanover and Strelitz', 65.
54. Drinkuth, *Queen Charlotte*, 10 ff.
55. Robert Dodsley, *London and its Environs Described*, 6 vols. (London, 1761), 3: 214.
56. David Watkin, 'Buckingham House', in *The Royal Interiors of Regency England* (London and Melbourne: J. M. Dent, 1984), 76.

57. Colvin, *History of the King's Works*, 5: 134–38; Jane Roberts, 'Sir William Chambers and George III', in *Sir William Chambers: Architect to George III*, ed. John Harris and Michael Snodin (New Haven and London: Yale Univ. Press, 1996), 46–49; David Watkin, 'Buckingham House', in *The Architect King: George III and the Culture of the Enlightenment* (London: Royal Collection Publications, 2004), 81–99: all three with further references.
58. John Harris, 'The Queen's House', in John Harris, Geoffrey de Bellaigue, and Oliver Millar, *Buckingham Palace* (London: Thomas Nelson, 1968), 24.
59. Harris, 'The Queen's House', 27.
60. Jane Roberts, 'Buckingham House: The Saloon', in *George III and Queen Charlotte: Patronage, Collecting and Court Taste*, ed. Jane Roberts (London: Royal Collection, 2004), 118, no. 111; Gervase Jackson-Stops, '"A Noble Simplicity": Pyne's Views of Buckingham House', *Apollo* 138 (Aug. 1993): 45–47; Watkin, 'Buckingham House' (1984), 82–83.
61. John Shearman, *Raphael's Cartoons in the Collection of Her Majesty The Queen, and the Tapestries for the Sistine Chapel* (London: Phaidon, 1972), and Mark Evans, 'The "afterlife" of the Cartoons, 1600–1865', in *Raphael: Cartoons and Tapestries for the Sistine Chapel*, ed. Mark Evans, Clare Browne, and Arnold Nesselrath (London: V&A, 2010), 60–61.
62. Sir Joshua Reynolds, *Discourses on Art*, ed. Robert R. Wark (San Marino, CA: Huntington Library, 1959), 81; also cited in Evans, 'The "afterlife" of the Cartoons', 60.
63. Christie's, London, May and June 1819. See especially 17 May, lots 25, 30; 18 May, lot 8; 19 May, lots 76, 81; 25 May, lots 64–80; and 26 May, lots 81–89.
64. RCIN 1112775: Benjamin Jutsham, Receipts, 1816–29, vol. II, f.124, 19 Jan. 1821.
65. RCIN 1112557.
66. Christopher Lloyd, 'King, Queen and Family', in *George III & Queen Charlotte*, ed. Roberts, 16 ff.
67. Royal Collection Trust, *A King's Purchase: King George III and the Collection of Consul Smith* (London: Queen's Gallery, 1993).
68. Watkin, 'Buckingham House' (2004), 88.
69. Lloyd, 'King, Queen and Family', 179–81, no. 155.
70. Lloyd, 'King, Queen and Family', 30–31, no. 7.
71. Martin Postle, 'Johan Zoffany: An Artist Abroad', in *Johan Zoffany RA: Society Observed*, ed. Martin Postle (New Haven and London: Yale Univ. Press, 2011), 20 ff.
72. Lloyd, 'King, Queen and Family', 27.
73. Letter from Horace Mann to Horace Walpole dated 25 Aug. 1772, as quoted in Desmond Shawe-Taylor, *The Conversation Piece: Scenes of Fashionable Life* (London: Royal Collection Trust, 2009), 130, no. 25.
74. Sir Oliver Millar, *The Later Georgian Pictures in the Collection of Her Majesty the Queen*, 2 vols. (London: Phaidon, 1969), 1: 154–57, no. 1211; Shawe-Taylor, *The Conversation Piece*, 130–37, no. 25; and Postle, 'Johan Zoffany', 230–32, no. 53, with further references
75. Shawe-Taylor, *The Conversation Piece*, 133.
76. Helmut Borth, 'Im Schatten der königlichen Schwester: Georg August (1748–1785), Herzog zu Mecklenburg', in *Herzoghaus Mecklenburg-Strelitz: Von gekrönten Häuptern, blaublütigen Kuckuckskindern und der Mirower Fürstengruft* (Friedland: Steffen, 2014), 81–84.
77. As quoted in Borth, 'Im Schatten der königlichen Schwester', 83: 'Der Prinz von Mecklenburg-Strelitz will ohne mich nicht aus dem Hause gehen. . . . Er ist mir ein Freund, Sohn, Schüler, Spießgeselle. Ich kehre mit ihm zu meiner verflossenen Jugend zurück.'
78. Frederick, Duke of York and Bishop of Osnabrück, lived in Hanover from 1781–83; William, Duke of Clarence (future William IV), from 1783–86; Edward, Duke of Kent from 1786–87; and Augustus, Duke of Sussex in 1790. Ernest Augustus, Duke of Cumberland (later King of Hanover) and Adolphus Frederick, Duke of Cambridge, briefly lived in Hanover following their studies at the University of Göttingen; see Georg Schnath, *Das Leineschloss: Kloster, Fürstensitz, Landtagsgebäude* (Hanover: Hahnsche Buchhandlung, 1962), 105–6.
79. RCIN 57011–57029. The service returned to the Royal Collection in 1947.
80. Jane Roberts, ed., *George III & Queen Charlotte: Patronage, Collecting and Court Taste* (London: Royal Collection Trust, 2004), 307.
81. Roberts, ed., *George III & Queen Charlotte*, 310.

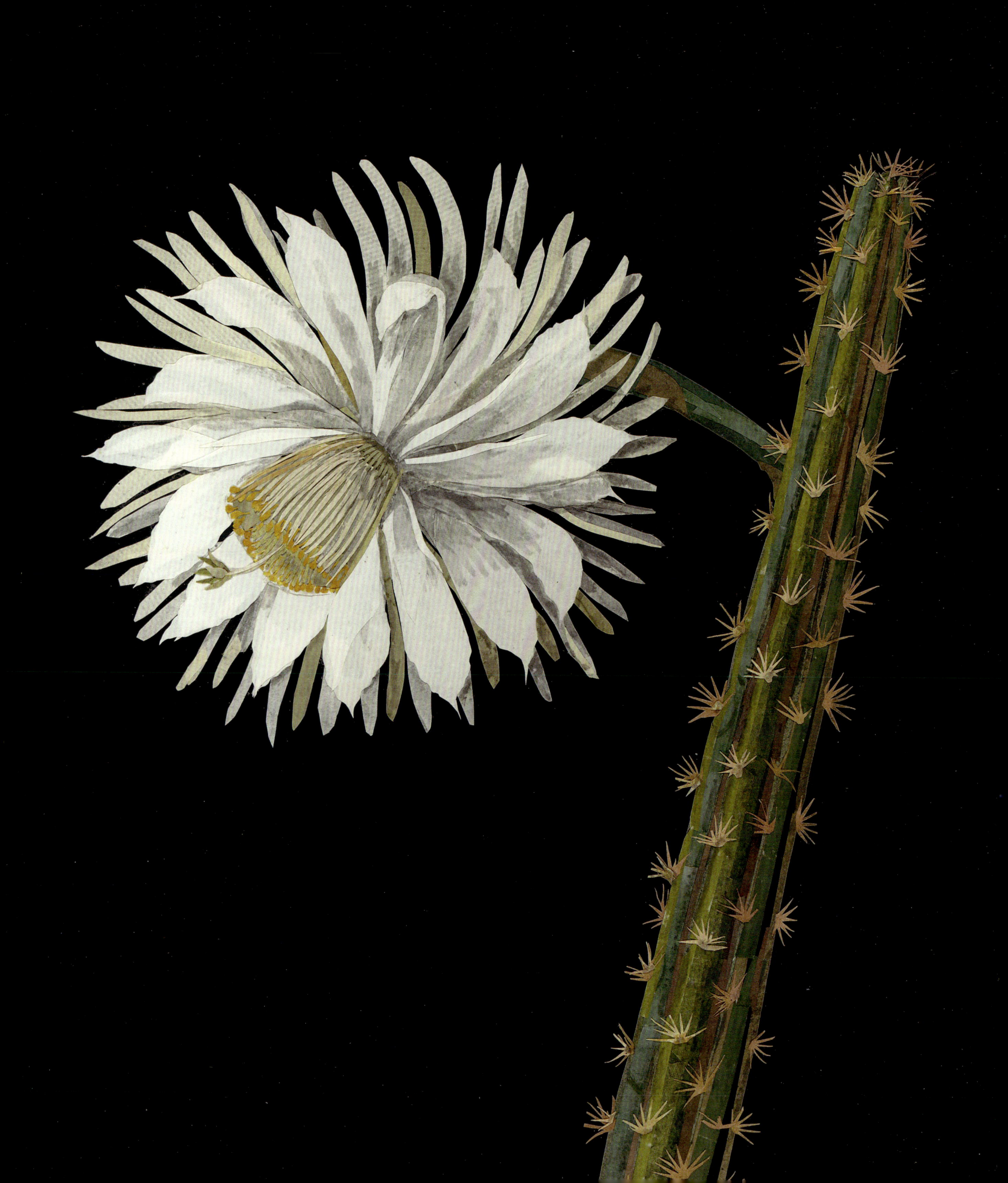

An August Domain: Gardening and Collecting in the Princesses' Kew

Mark Laird

How significant was the role of Caroline of Ansbach, Augusta of Saxe-Gotha, and Charlotte of Mecklenburg-Strelitz in shaping the modern landscape, and what is their importance as collector–naturalists? In this essay I will examine their influence beyond the eighteenth century, and beyond the traditional history of men as landscapers, focusing on Kew and its development from royal estates to scientific collecting ground.[1] For the landscape historian it remains an intriguing task to give new emphasis to the women involved, including Mary Granville, Mrs Delany, with her collages of Kew specimens, and Charlotte's daughter Princess Elizabeth's accomplishments under the sway of Delany, Mary Moser, and Margaret Meen at Windsor and Frogmore, as well as at Kew. Thus Kew stands for several seats of cultured creativity which, at a peak around 1800, owed much to Caroline, Augusta, and Charlotte.

The stretch of the River Thames near Richmond had long been associated with the court, intellectual life, and fashion. So it was not surprising that a royal red-brick house, leased as Ormonde Lodge by James Butler, second Duke of Ormonde, until his exile in 1715, came to the attention of George, Prince of Wales, and his wife Caroline as they looked for an Arcadia away from St James's Palace after their banishment by George I. In 1718 they moved in, and for almost twenty years the park was subject to improvements under Caroline. Her Richmond Gardens is significant for two remarkable buildings of the 1730s: 'Merlin's Cave', and a hermitage containing busts of contemporary worthies, both designed by William Kent.[2]

In 2013 Stephen Bending, contributing to a new gendered history of the landscape garden, drew attention to the lasting impact that Caroline's 'masculine national heritage' had on women as well as men.[3] His work prompts a revision to the conventional history of Kew. By focusing on the gardening and collecting interests of Caroline, Augusta, and Charlotte instead of only considering John Stuart, third Earl of Bute, Sir Joseph Banks, William Chambers, and Lancelot 'Capability' Brown, along with botanists and gardeners such as Sir John Hill and William Aiton, or taxonomists

Detail of p. 366, cat. 21.09

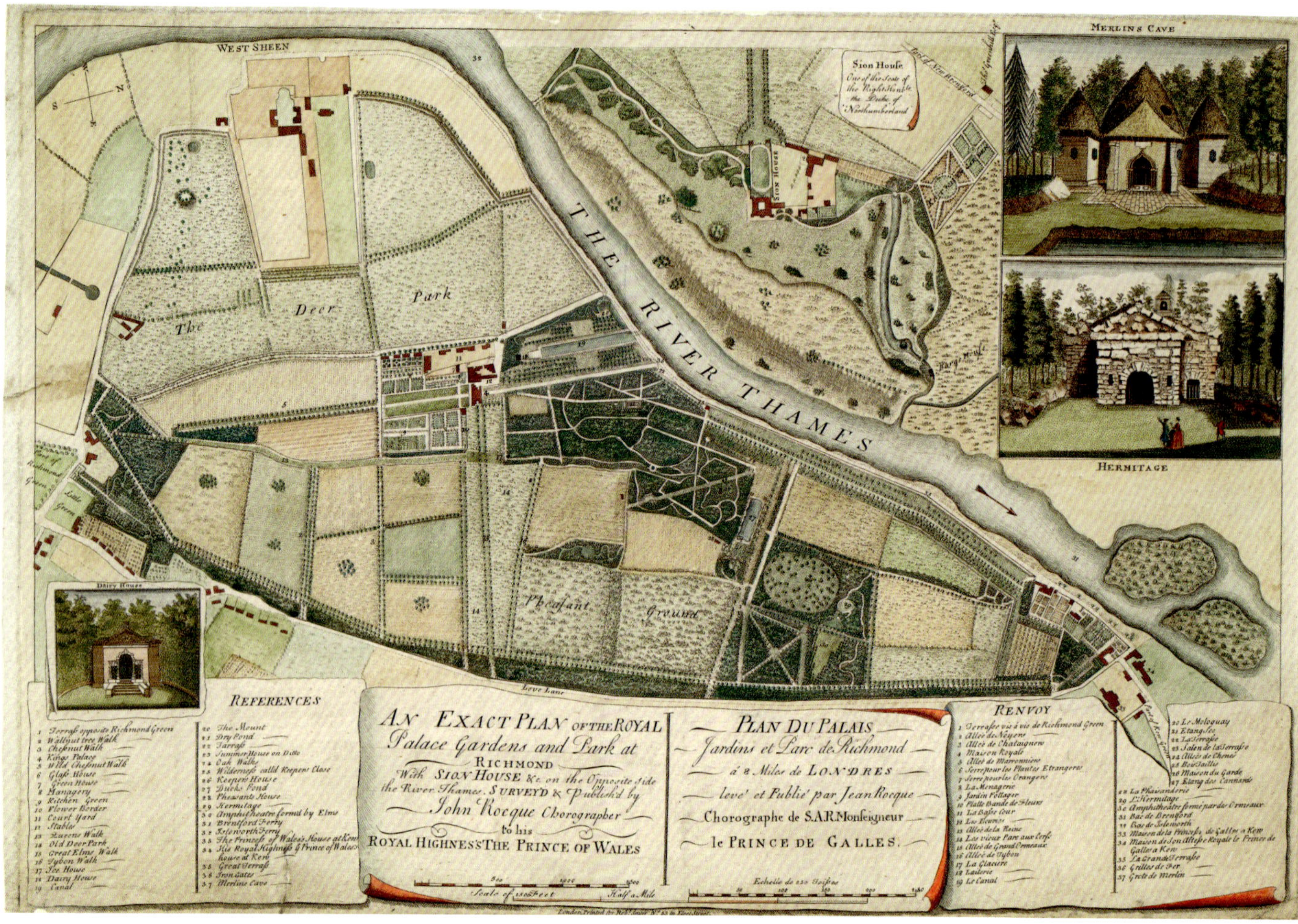

FIG. 6.01 John Rocque, *Plan of the House, Gardens, Park & Hermitage of Their Majesties at Richmond: and of Their R.H. the Prince of Wales & the Princess Royal of Kew*, 1736, hand-coloured etching and engraving on paper, 24⅜ x 37¾ in. (62 x 96 cm). Royal Botanic Gardens, Kew, Library, Art & Archives Collections

and plant collectors such as Daniel Solander and Francis Masson, the masculine view of eighteenth-century Kew can be challenged.[4]

Kent's two iconic buildings in Caroline's Richmond Gardens are often reproduced in isolation, but they should be seen among the twenty-one features of John Rocque's plan of 1736 (fig. 6.01), dedicated to the Duke of Richmond and furnished with a variety of views. The Hermitage (17), which is adjacent to the Amphitheatre (18), is surrounded by the Forest Oval and Diagonal Wilderness. Merlin's Cave (14), at the east end of the 'Ducks Pond' (13), is ensconced in a large wilderness of irregular as well as regular paths. Visitors began their tour at the 'Palace' (6) and were then directed to the Dairy (7), overlooking a canal, where they could pause to admire its display of delicate china bowls and jars.

Alternative circuits then beckoned. One led to the Tuscan Temple on a Mount (9) at the far end of a Thames-side walk—a walk enjoyed on the 'new Terrass' (16), as views by Bernard Lens III (fig. 6.02), Marco Ricci, and the elusive P. Brookes, an engraver working for John Boydell, amply show. Constructed between 1729 and 1734, the grassed terrace, lined by an inner row of elms, attracted fashionable crowds, especially on Sunday evenings in summer. The plan and supporting views make Charles Bridgeman's protean layout as important as Kent's architectural contributions.

Before Bridgeman left his imprint, Richmond's royal history had begun with James I, who used it as a hunting lodge. In the time of George I, Caroline, then Princess of Wales, brought the renamed Richmond Lodge to the attention of Alexander Pope; he arrived there to discuss her garden ideas in September 1719. While he would poke fun at the warring gardener 'Criticks' who joined him on that occasion in a letter to his friend Lord Bathurst, he pointed the way to what was emerging as a new synthesis: a 1725 'Bridgmannick Theatre' in his garden at Twickenham.[5] Bridgeman's 'workmen' at Richmond evidently undertook the work for Pope. Pope and Bridgeman were thus leading England towards the style soon to be associated with Kent at Carlton House. A plan dated pre-1727 and attributed to Bridgeman shows the Richmond house and gardens occupying about three hundred and eighty acres: eleven acres of formal gardens, twenty-two acres of woodland with sinuous paths, but with much

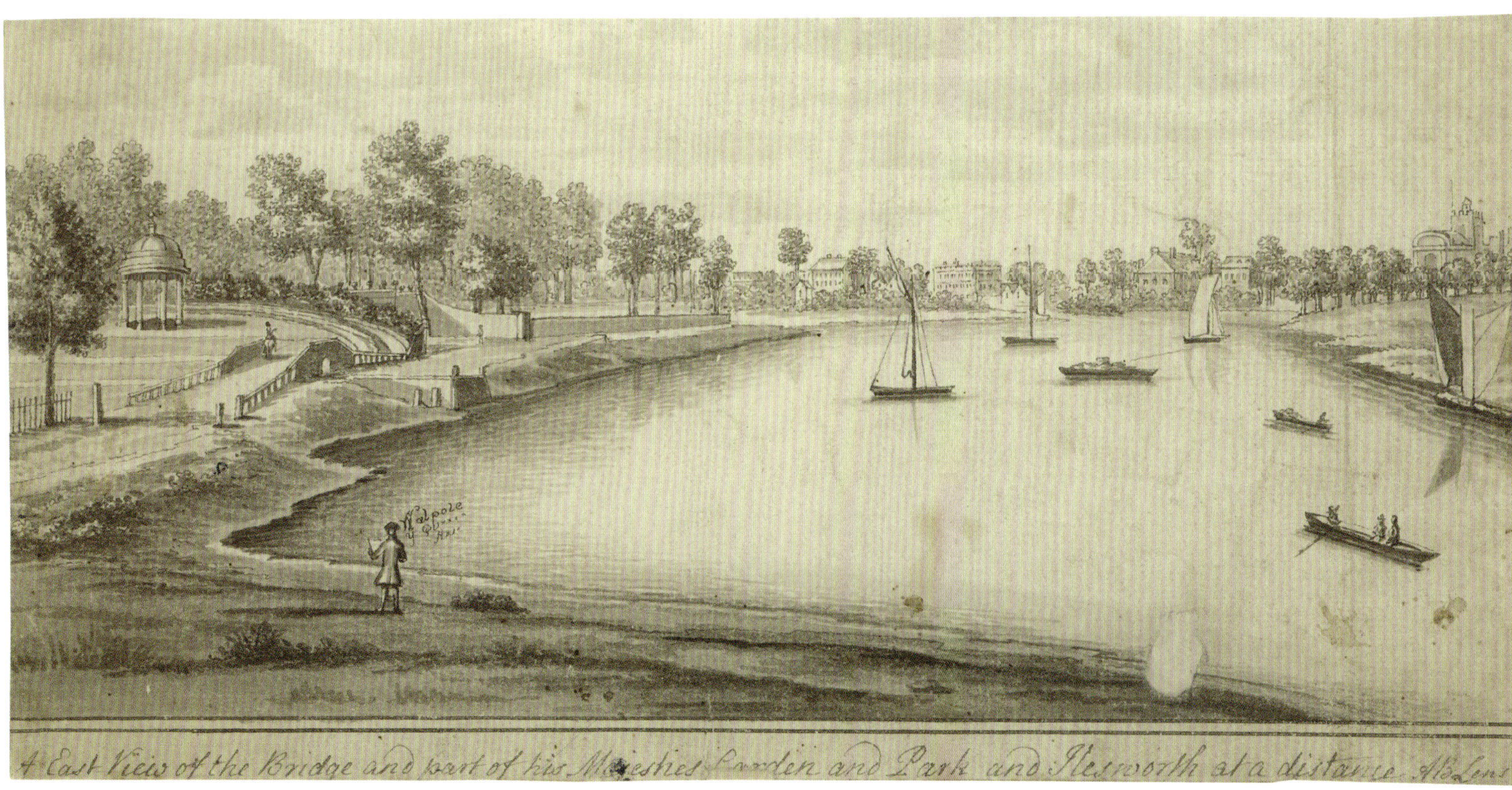

FIG. 6.02 Bernard Lens III, *A East View of the Bridge and part of his Majesties Garden and Park*, undated, graphite on paper, 6⅛ x 12¼ in. (15.7 x 31.2 cm), London Borough of Richmond upon Thames

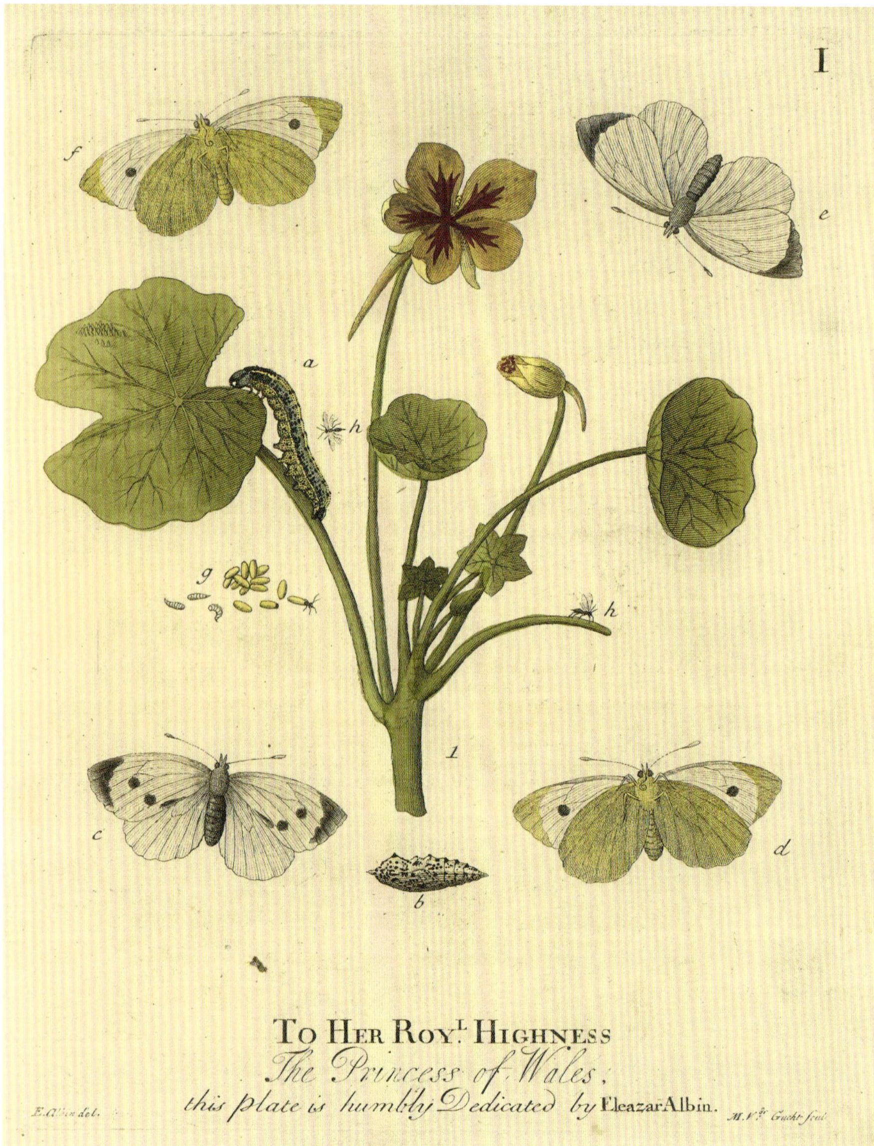

FIG. 6.03 Unknown engraver (possibly Gerard Vandergucht) after Eleazar Albin, artist, '*Papilio albus vulgaris major*: the Great White Butterfly', 1720, hand-coloured engraving, in Eleazar Albin, *Natural History of English Insects* . . . (London: Printed for the author, and sold by William and John Innys at the West End of St Pauls, 1720), plate 1. Beinecke Rare Book & Manuscript Library, Yale University (cat. 29.47)

of the remaining acreage still virgin fields.[6] It was the expanse of meadows and cornfields that made a tour of Richmond so memorable.

The expanded references in Rocque's 1754 plan make clear that these eleven-acre formal gardens to the south of the Palace featured a Glass House, a Green House, and a Flower Border near the Kitchen Green. There was also a 'Menagery'. Little is known of Caroline's plant and animal collections, but her significance as a patron of horticulture and natural history can be inferred from dedicated publications and from her library. On the subscribers' page at the front of Robert Furber's *Twelve Months of Flowers* (1730), the auricula 'Potters Queen Caroline' is given a strategic position at the top—a counterweight to the dedication to her son, Frederick, Prince of Wales, who had been out of favour with his parents. The catalogue of her library of around 1740 includes two works by Batty Langley, Richard Bradley's *The History of Succulent Plants* (1716), and many titles covering natural history, including two copies of Eleazar Albin's *Natural History of English Insects* (1720). This was dedicated to her as Princess of Wales. The plate of the 'Cabbage White', *Pieris brassicae* (fig. 6.03) was also 'humbly dedicated to Her Roy.l Highness The Princess of Wales'. As Queen, she appeared under her 'august name Carolina' in the dedication of Mark Catesby's first volume of *The Natural History of Carolina, Florida and the Bahama Islands* (1731).

In May 1729, several British newspapers published the announcement that Catesby, introduced by John Carteret, second Earl Granville, presented that first volume to 'her Majesty', who had just been appointed the Queen Regent while George II was in his German dominions.[7] Carteret still held his position as a Lord Proprietor of the Carolina Colony, even after the administration was overthrown in 1719. He facilitated the marriage of Frances Thynne, one of Caroline's circle of women, to Algernon Seymour, seventh Duke of Somerset, styled Earl of Hertford. Lady Hertford's gardening at Percy Lodge, later called Richings Park, in Buckinghamshire, proved as notable as that of the banished Henrietta St John, Lady Luxborough at Barrells in Warwickshire.[8]

When Caroline died in November 1737, the first *Magnolia grandiflora* to flower in London had just been painted by Georg Dionysius Ehret, leading to its illustration as plate 61 in the second volume of Catesby's *Natural History* in 1743 (fig. 6.04). It was also presented as Catesby's engraving (after Ehret) in Christopher Gray's *Catalogue of American Trees and Shrubs*. While the legacy of Caroline's landscaping is apparent around Kensington Palace today, the acclimatization of *Catalpa* and *Magnolia* from the Carolinas constitutes another enduring gift to modern gardening. The 1720s and 1730s saw the first generation of climate trials; a second came at Kew in the 1750s and 1760s.[9]

Catesby dedicated the second volume of his *Natural History* to Augusta. The record of her Privy Purse expenditure provides details about what engaged her during Caroline's final years. Now newly transcribed, this allows the landscape historian to revisit the significance of William Kent's assistance to Frederick, Prince of Wales, in the layout at Carlton House and in the first stages of the design of the garden at the White House at Kew. The estate located adjacent to Richmond Lodge, with its extensive gardens developed by Queen Caroline, had been similarly acquired as a rural retreat in 1731, but from 1737 was used with greater frequency. It is clear that it was in the landscaping at Carlton House that a new 'culture of horticulture' was born, for by August 1736, purchases of flowers were appearing in great numbers in Robert Furber's nursery bills.[10]

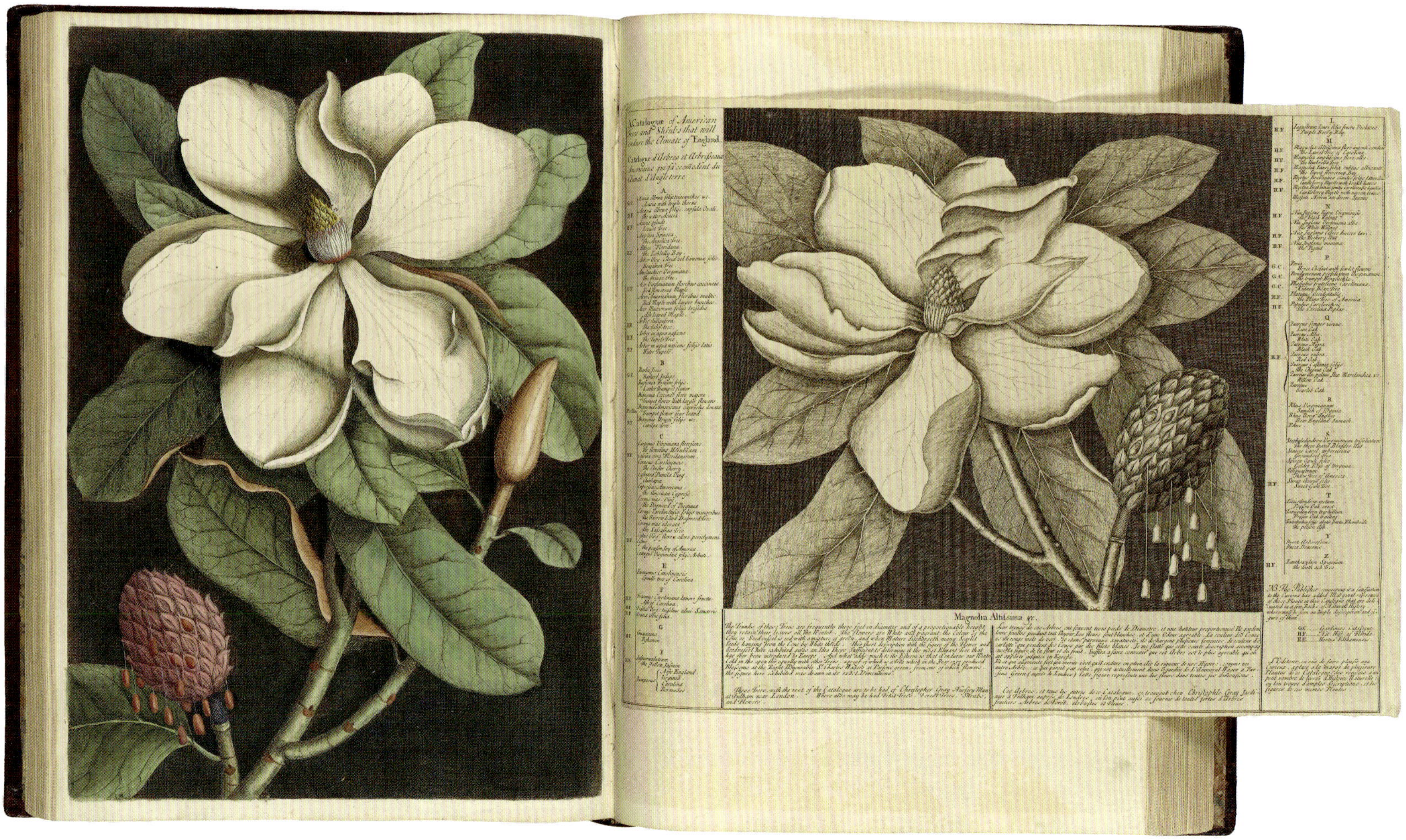

FIG. 6.04 Left: Georg Dionysius Ehret, artist and engraver, 'Magnolia altissima', 1737–38, hand-coloured etching, in Mark Catesby, *The Natural History of Carolina, Florida, and the Bahama Islands . . .* (London: Published by the Author, 1731–43; Appendix 1747), vol. 2, plate 61; right: Mark Catesby, artist and engraver, *A Catalogue of American Trees and Shrubs that will endure the Climate of England*, c. 1737, etching, published by Christopher Gray, Fulham. Oak Spring Garden Library, Upperville, VA (cat. 29.17)

Other evidence from the Privy Purse accounts suggests that a significant amount was spent on the development of Augusta's 'culture of birds'. The financial entries confirm that during her first year of marriage, Augusta took an interest in songsters that became a fully-fledged passion by the 1760s, when she kept native songbirds, such as the linnet and chaffinch, in her Aviary Flower Garden at Kew.[11] Recent information has also come to light which shows that when Eliza Lucas Pinckney, a plantation owner from South Carolina, visited England in 1753 together with her husband and young daughter, they brought for Augusta three American birds: a *Spinus tristis* (American goldfinch), a *Passerina ciris* (painted finch, or painted bunting), and a *Passerina cyanea* (blue linnet, or indigo bunting). These gifts indicate that Augusta's interests were widely known.[12]

The Privy Purse accounts itemize the purchase of many birds from July 1736 to December 1737, sometimes for Kew but probably also for the other estates maintained by the couple, including Carlton House:

1736/7	
February:	To a paroquet £ – 10. –
	To 4 canary birds £2.2. –
	To Smith's Bill for loive cupps for ye bird cage £1.12. –
	Marshall's bill for a bird cage £29.15. –
1737	
March:	To a Magpye at Kew £2.2. –
	To a boy with bird's nets £ – 5. –
	Went's bill for a Virginia nightingale £2.12.6
June:	for a poor woman for a bird and a cage £1.1. –
	Went's bill for birds £3.3. –
August:	Horsley ye bird cagemakers bill £6.1.0
December:	Went's bill for Virginia Nightengales £2.2.[13]

The Virginia nightingale was the North American *Cardinalis cardinalis* (cardinal). Like the exotic canary, it was by nature a songbird.[14] The accounts continue in the same direction over the following seasons. Amongst the items purchased are ten canaries, for a staggering price of around £45—a similar amount to that paid for over four hundred and fifty large standard horse chestnut trees in the Prince's household accounts for 1734–35—and a 'white robin' for 1 guinea.

The accounts for 1740–43 are less revealing about birds but more significant for their references to Augusta's contribution to the running of her husband's property of Cliveden in Buckinghamshire, and to the expenses involved in visiting Lord Cobham's Stowe, also in Buckinghamshire, and Charles Hamilton's Painshill in Surrey. They record the payment for one copy of the second volume of Mark Catesby's *Natural History*, which was dedicated to Augusta.

Given that Part 10, concluding Catesby's *Natural History*, was presented to the Royal Society on 15 December 1743, the same year in which the list of subscribers, which records Augusta's interest, was published, it can be assumed that her payment of 'Mr Keatsby's subscription' in 1743 refers to his book. At 22 guineas for a complete set, it was one of the most expensive publications of the eighteenth century. If this assumption is correct, the full set probably came with the addition of 'plants and flowers' supplied by Catesby.[15] The expenditure of £50 contrasts sharply with the miserable 1 guinea paid to 'weeding women' at Stowe—which was even less than that received by the poorly paid male gardeners, who themselves earned less than servants. Taken together, the accounts suggest that Augusta was becoming familiar with avant-garde landscaping, as well as with the most lavish and up-to-date publication on North America's natural productions. A later set of Privy Purse accounts supports the idea that Augusta was directly responsible for paying John Dillman, who since his appointment by her husband in 1732 had maintained the gardens of all their London properties, as well as Durdans near Epsom. Only Cliveden fell to the care of gardener John Morris.[16] After the Prince's death in 1751, Augusta would retain Dillman as her gardener up to his retirement in 1753. Coming from Dillenburg in Germany, he had earlier served Lady Elizabeth St André, daughter of Algernon Capell, second Earl of Essex, at Kew before Frederick's acquisition of her estate there.

With Bridgeman's death in 1738, a new era of gardening began. William Kent became the presiding genius, with Dillman the steady hand. Yet by the 1750s, after the deaths of Kent in 1748 and the Prince of Wales in 1751, the impact of Lancelot 'Capability' Brown's new radical approach to the management of the landscape was increasingly evident after his appointment by George III to supervise new works at Richmond.

Augusta lived on in the White House, Kew, as Princess Dowager of Wales until her death in 1772, though in her final years Carlton House was more regularly used. She would appoint as her chaplain Stephen Hales, who encouraged her interest in natural history; she would also expand Frederick's Kew collections. Just before his death, Augusta appreciated the gesture that Griffith Hughes made in his dedication of two plates of 1750 to the couple: the 'Cabbage-tree, Palma Maxima' (*Roystonea oleracea*) to Frederick, and the 'Coco-Nut tree, Palma Nucifera' (*Cocos nucifera*) to herself.

The garden of the White House appears on John Rocque's plan of Richmond of 1734 and subsequent editions, notably those of 1736 (see fig. 6.01) and 1748. Unlike the radically remodelled White House, the 'Kentian garden' appears striking only for Kent's garden seat, which William Chambers illustrated in his 1763 volume on Kew. The lawns with statues and flanking groves are regular and rather cramped. Indeed, a later 1754 edition of the Rocque plan is notable for illustrating how the parkland surrounding Hugh Percy, first Duke of Northumberland's, Syon House on the west side of the River Thames had already advanced into the style characterized by trees planted in clumps and belt associated with 'Capability' Brown, while Richmond remained stuck in the earlier tradition. Hence it was not until the remodelling in the 1750s and

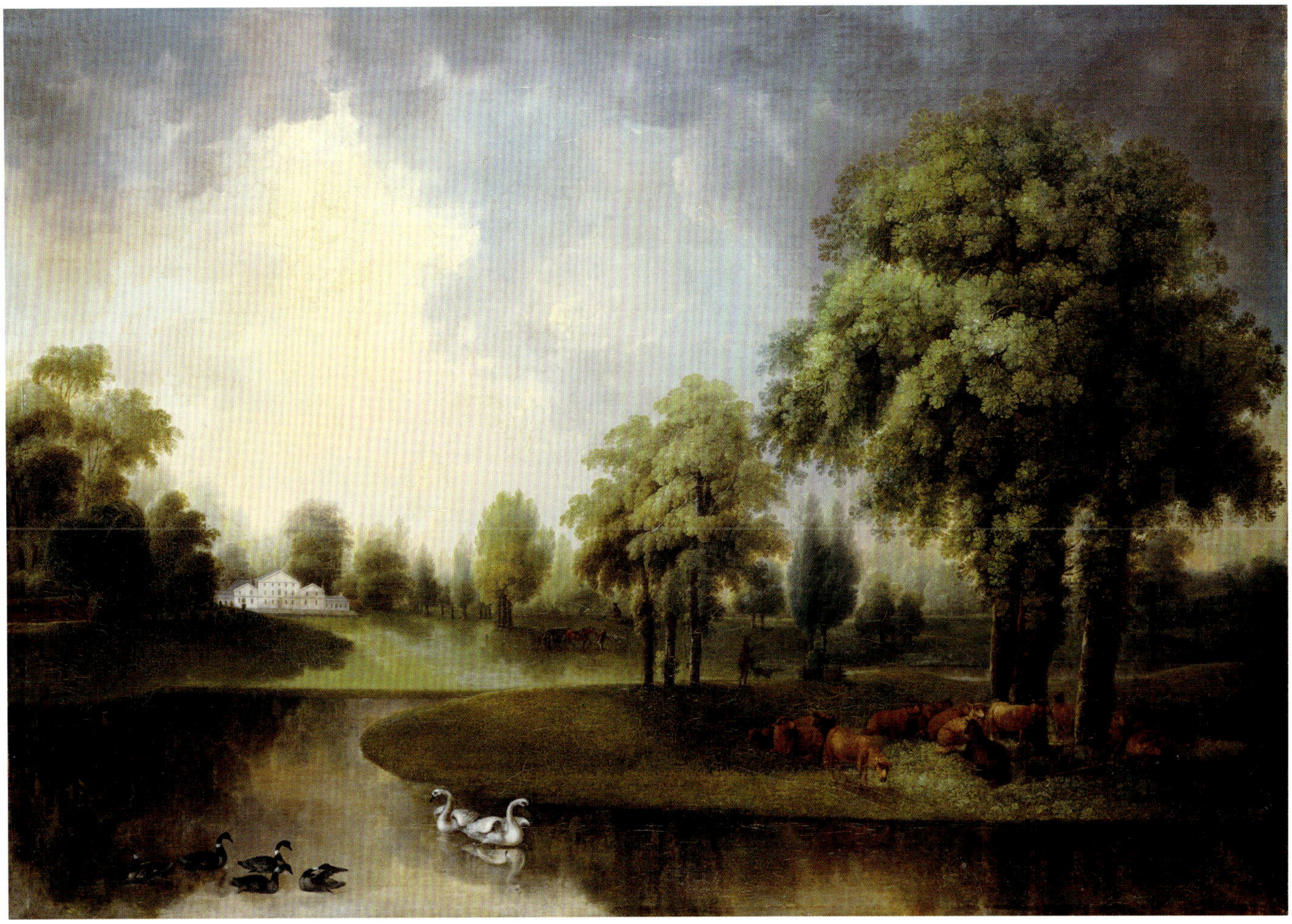

FIG. 6.05 Johann Jakob Schalch, *The White House at Kew*, c. 1759–60, oil on canvas, 29½ x 40¾ in. (75 x 103.7 cm). Royal Collection Trust

1760s that the White House acquired an expansive setting. It had been only after 1740 that the 'Pheasant Ground' fell under royal control and the estate was consolidated. There was also a challenge to Augusta's right to reside at the White House in 1759, only resolved when the lease was renegotiated for another twenty-one years, and she was able to complete the building of her 'Eden' in the years immediately following.

Kent's architectural iconography conceived for the garden surrounding Richmond Lodge sits in marked contrast to the iconographical construct which underpinned the design of the new garden of the White House, just adjacent, developed under the auspices of Augusta and Lord Bute and Chambers. In the course of converting the 'Desart' into an 'Eden', an impressive array of exotic structures, of varying ephemerality, including a mosque, Alhambra, and a pagoda were introduced. Johann Jakob Schalch's two paintings of the White House of around 1759–60 are amongst the most evocative representations of Augusta's 'earthly paradise'. One view of Kew's lake is by moonlight. The mood is magical and other-worldly (figs 6.05–06).

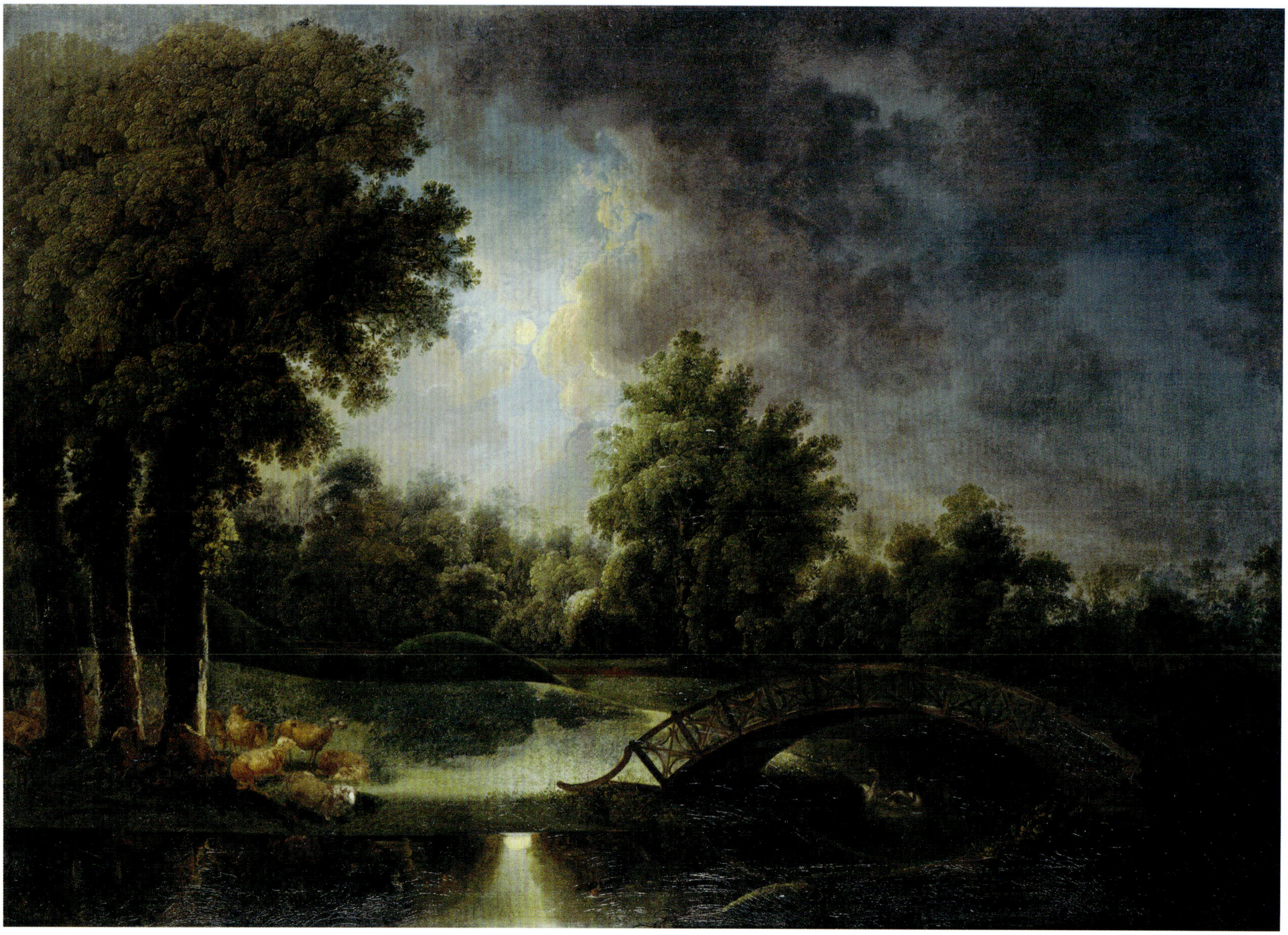

FIG. 6.06 Johann Jakob Schalch, *The Gardens at Kew by Moonlight*, c. 1760, oil on canvas, 30 x 41 in. (76.1 x 104.2 cm). Royal Collection Trust (cat. 18.06)

George Ritso's poem of 1763, *Kew Gardens*, which was dedicated to the 'Dowager Princess of Wales', contains a literary evocation of that 'earthly paradise'. As moonlight gives way to daylight, a choir of local native birds joins the nightingale (*Luscinia megarhynchos*), the classical Philomela:[17]

> The feather'd Choir, that all the Night had slept,
> By Heav'n's indulgent Care securely kept;
> From their warm Wings their nestled Heads remove,
> And charm with mattin Song the listning Grove . . .
> The mounting Lark expands her agile Wings;
> And makes the Welkin vocal, as she sings.
> The whistling Blackbird, and melodious Thrush
> Solace the Bough, or warble in the Bush.
> Shrill goldfinches begin their charming Notes;
> And louder Linnets strain their little Throats:
> Along the wafting Air the melting Music floats.
> The chirping Chaffinch, in the spangled Thorn,
> With sportive Sparrows, greets the buxom Morn.

Local linnets and chaffinches thus fly in and out at liberty, while their captive country cousins are housed inside the Aviary.[18] Ritso makes something of this conceit of 'Bretheren' before turning to the Chinese peacock pheasant. The real encomium, however, is saved for the Princess Dowager:

> Nature 'form'd thee, surely from a perfect Plan
> Superior to thy Sex, a softer Man.
> Forth from her Palace comes the Royal Dame,
> *August* in Person, as *August* in Name.
> Her cheerful Presence brightens all the Scene
> The Flow'rs look fairer and the Grass more green.

Augusta's passion for plant collecting, which began as this garden took shape, is confirmed by the lists of plants which were delivered by plant collectors and nurserymen in the late 1760s.[19] In 1768 alone, her gardener William Aiton's bill for shipping and transportation amounted to £90; in 1769 one bill came to £10 18s. 6d. In the final months of 1768, pots alone cost £16 8s. In spring 1769, the cost of orange tubs amounted to £15.

It was scarcely surprising that Augusta wanted some pictorial record of her collections. Having just returned from Newfoundland, Joseph Banks met Sydney Parkinson at the Vineyard Nursery in Hammersmith and persuaded him to draw plants both there and at Kew. Parkinson produced a fine set of twelve paintings of plants from these gardens between 1767 and 1768. His portrait of the diminutive North American mayflower, *Epigaea repens*, compares favourably with Ehret's portrait of the

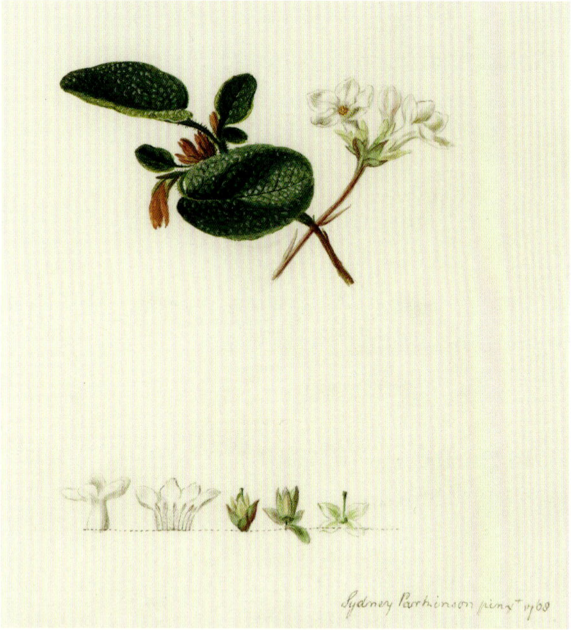

FIG. 6.07 Sydney Parkinson, *'Epigaea repens'*, 1768, pencil and watercolour on paper. Natural History Museum, London, Botany Library Collection

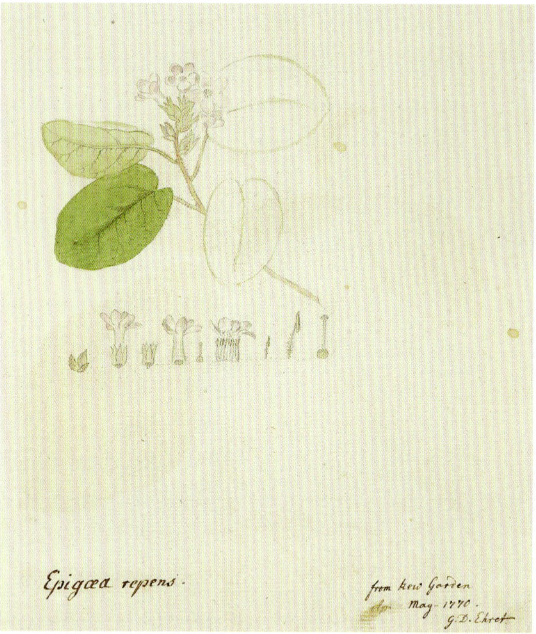

FIG. 6.08 Georg Dionysius Ehret, *'Epigaea repens': from Kew Garden*, sketch 107, May 1770, watercolour, graphite, and ink on paper. Natural History Museum, London, The Georg Ehret Drawings Collection

same species, drawn at Kew in 1770 (figs 6.07 and 6.08); and his painting of an orchid on a tree is especially vivid.[20] John Hill's *Hortus Kewensis* of 1768—listing over two thousand seven hundred species of herbaceous plants—contains twenty of his engravings, while his *Twenty-five New Plants, rais'd in the Royal Garden at Kew* of 1773 represented the latest acquisitions made before Augusta's death in February 1772.[21]

Dr John Hope of Edinburgh, who visited Kew amongst other English gardens in 1766, was attentive to the techniques for dealing with the acclimatization of 'Gardenia augusta', *G. jasminoides* (fig. 6.09), along with many North American plants that required soil remediation, mulching, and winter covering and coppicing. It is these consequential acclimatization trials under Lord Bute, John Hill, and William Aiton that made Augusta's Kew the pioneering rival to Chelsea Physic Garden, thereby anticipating Joseph Banks's promotion of Kew's scientific prowess in the early 1770s.

While all this was happening in Augusta's Kew, Richmond Lodge was back in use again as a summer retreat for her son George III and his bride Charlotte of Mecklenburg-Strelitz. Caroline's original garden was soon to disappear, as in 1764, once 'Capability' Brown's plan of 10 December was approved, the re-landscaping commenced. By 1766, lead salvaged after the demolition of Merlin's Cave was being sold off to a plumber, Mr Devall. Only the Hermitage survived into the next century. The greatest destruction of all was of the Terrace—'a boast of the country' and the 'admiration of all foreigners'—which was replaced by sweeping lawns, in a move applauded by Uvedale Price and Arthur Young.[22] The survey by Peter Burrell and Thomas

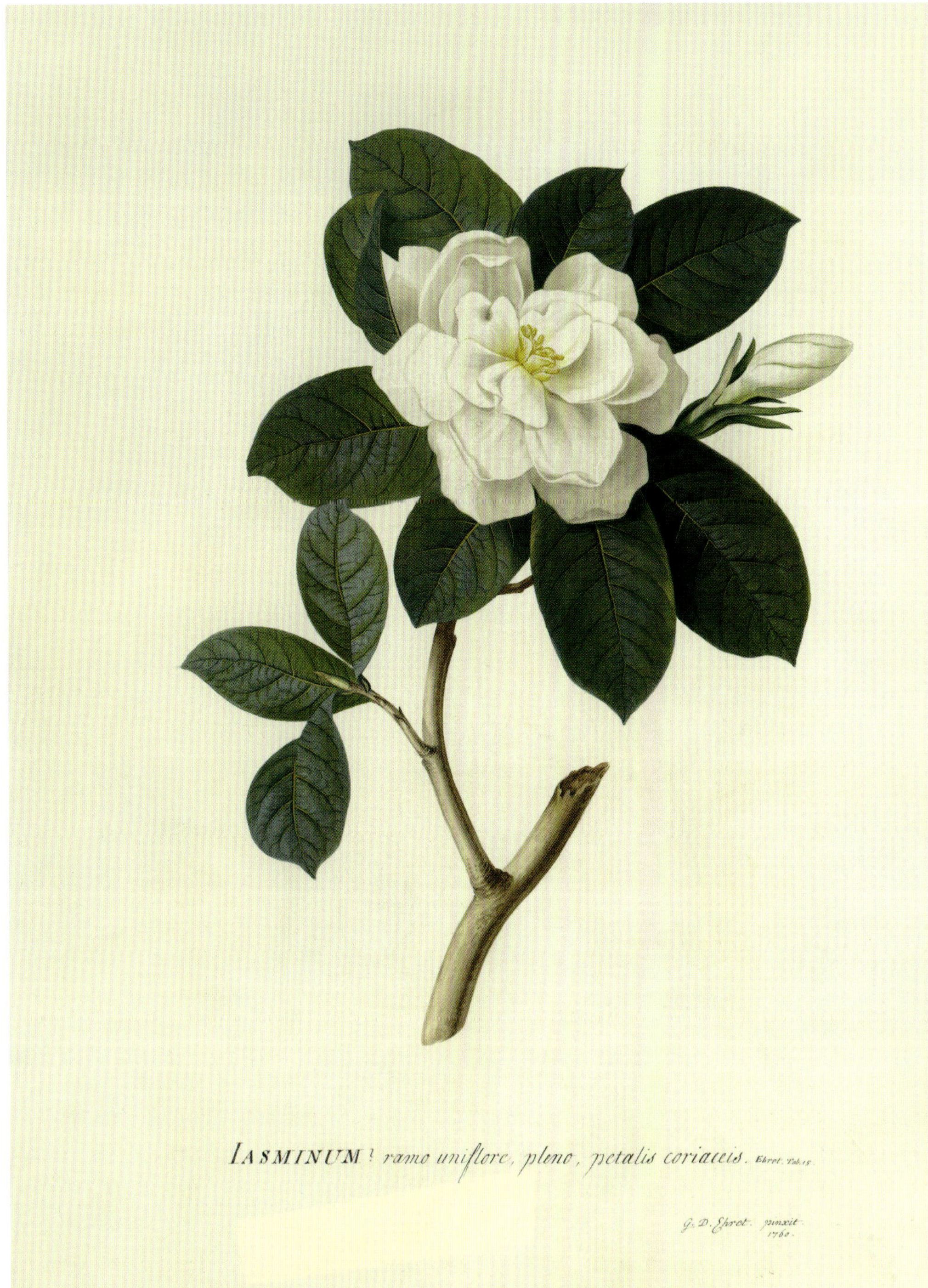

FIG. 6.09 Georg Dionysius Ehret, '*Jasminum ramo uniflore pleno petalis coriaieis*' [Cape Jasmine], 1760, gouache on vellum, 19⅝ x 13¾ in. (50 x 35 cm). Royal Botanic Gardens, Kew, Library, Art & Archives Collections (cat. 29.21)

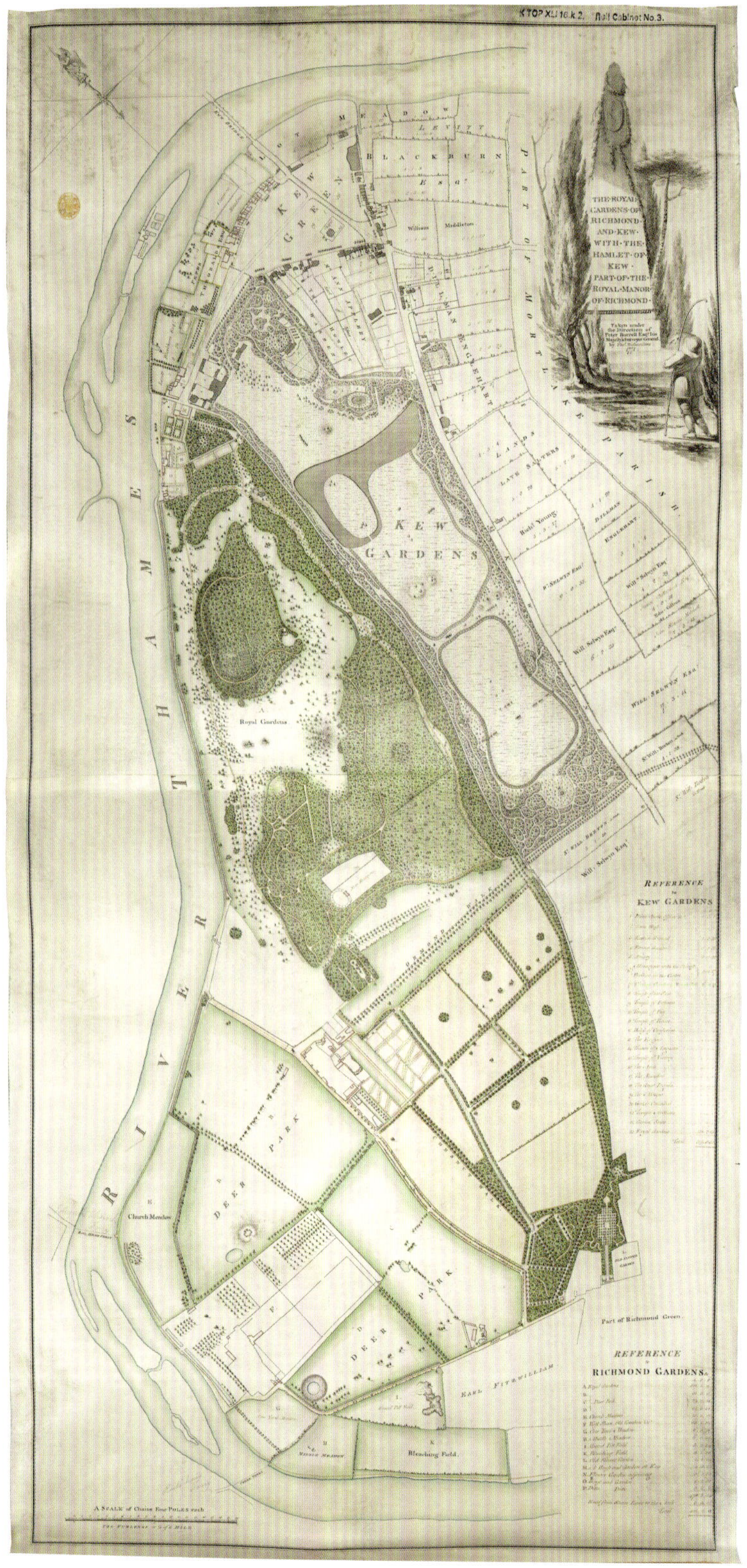

FIG. 6.10 Peter Burrell and Thomas Richardson, *Map of the Royal Gardens of Richmond and Kew with the Hamlet of Kew. Part of the Royal Manor of Richmond, taken under the Direction of Peter Burrell, his Majesty's surveyor General*, 1771, ink on vellum, 46⅝ x 21 in. (118.5 x 53.3 cm). British Library, Maps and Drawings Collection

Richardson of 1771 (fig. 6.10) provides evidence of the conversion, and is the first indication of how the Richmond Gardens (or 'Royal Gardens') and 'Kew Gardens' were edging towards becoming an eventual unity. Also apparent is the 'New Menagerie' that would come to be associated with the Queen's Cottage, a cottage orné conceived by Charlotte in the furthest reaches of Richmond Gardens.

With the death of Augusta, Charlotte installed her own family in the White House. Soon after, Richmond Lodge would be demolished, leaving her elder children, the Prince of Wales and Prince Frederick, lodged in the Dutch House, adjacent to the White House, which served as their schoolroom. Meanwhile, a structure known as 'Queen Caroline's House' served as a nursery for their brother Prince Ernest and other younger siblings. From 1775, George III spent about three summer months of every year at Kew. Once a week, the children accompanied their parents to Richmond Gardens to play with pets and inspect the new additions in the menagerie.

While Caroline had kept her menagerie at Richmond Lodge for 'deer and wild beasts', with tigers still housed in the 'Royal Paddock Garden' in 1740, Augusta concentrated her efforts on the creation of an Aviary at the White House with its collection of pheasants and other domestic and exotic species in cages encircling an ornamental fish pond.[23] The 'New Menagerie', which was created in the later eighteenth century in the Richmond Lodge grounds on the footprint of Caroline's menagerie, would reiterate the elliptical form of Augusta's Aviary and also eschewed rapacious creatures. The Queen's Cottage, which probably had its origins in the former menagerie keeper's accommodation, looked to the west towards an enclosure for exotic birds and pacific beasts, including cattle from Algeria and India and a 'hog like a porcupine in skin, with navel on its back'. George Stubbs's portrait of a Nilgai (*Boselaphus tragocamelus),* dated 1769, conveys the impressive stature of this quadruped. The same artist's *Kongouro from New Holland (Kangaroo)* (fig. 6.11) records the most prized possession of the collection—a marsupial from the newly explored Australia, then called New Holland. Before the conversion of the kangaroo paddock into a flower garden after 1806, the population had grown close to twenty.[24]

The *London Magazine* for 1774 was among the first to refer to the converted single-storey menagerie house as the 'Queen's Cottage'—a change occurring around 1771/72. Meanwhile, the King initiated the unification of the White House, Kew, estate with Richmond Gardens, which took place between 1773 and 1774: a full merger would only be completed legally in 1802.

In August 1771, Banks and Solander, just returned from the South Seas, were presented to George III at Kew. Although it was Augusta's physician, Sir John Pringle, who first persuaded George III to dispatch a plant collector to the Cape of Good Hope in 1772, Banks's first-hand experience of a month at the Cape on the return of the *Endeavour* probably made it a *fait accompli*. Francis Masson was dispatched, becoming Kew's first official plant collector, gathering many species between 1772 and 1774 including the very significant *Strelitzia reginae.*

The close relationship that Banks had to the King led inevitably to the development of a 'masculine' history of 'His Majesty's Garden'. Yet when Mary Delany came

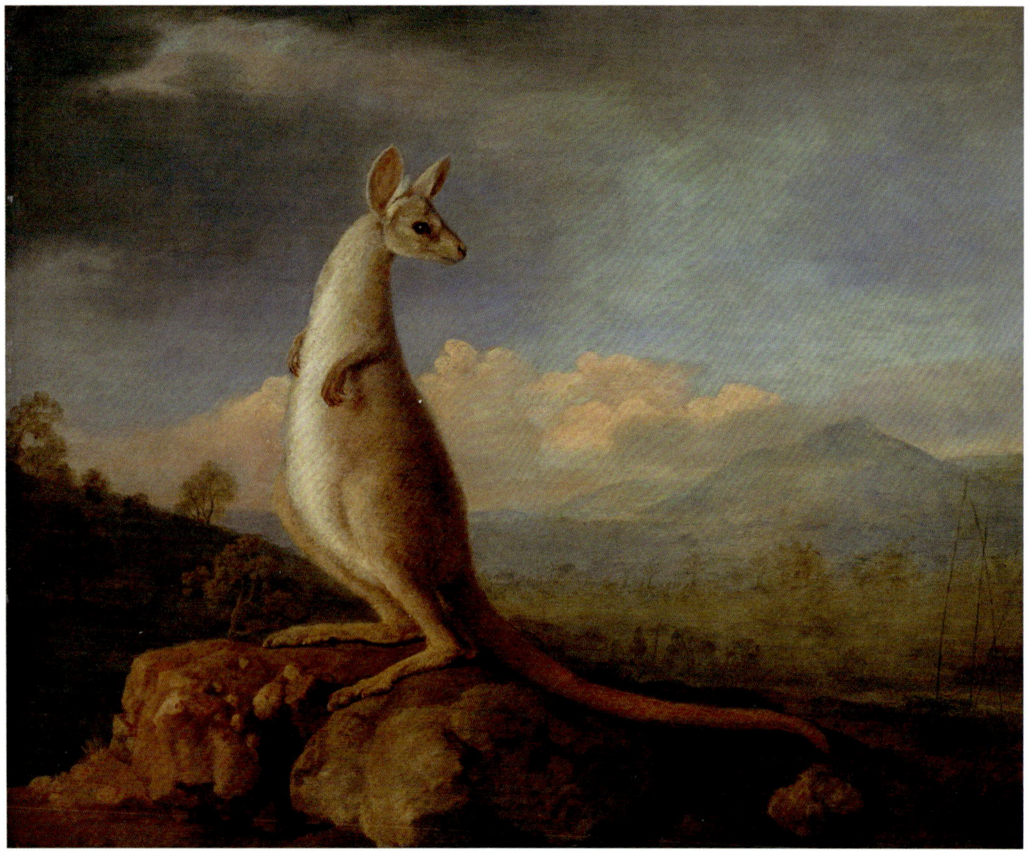

FIG. 6.11 George Stubbs, *The Kongouro from New Holland (Kangaroo)*, 1772, beeswax on mahogany panel, 23⅞ x 28 x⅜ in (60.5 cm x 71.5 cm x 1.0 cm). National Maritime Museum, Greenwich

to seek Kew specimens for her collages after 1776, it is notable that the authority came from Charlotte. Indeed, in a letter of October 1777, Delany wrote of being '*plentifully* supplied' from 'the Queen's garden at Kew'.[25] Her 'Dianthus Caryophyllus' drawn at Bulstrode, the country estate of Margaret Cavendish Bentinck, Duchess of Portland—friend, admirer, and protector of Delany—was from 'Pink or Carnation' stock supplied by Charlotte in August 1778. The 'Cactus Grandiflorus' (compare *Harrisia gracilis*) of 12 August 1778 was cut on 'the Day the King & Queen & the Royal family were at Bulstrode' (p. 366, cat. 21.09). The collage 'Amaryllis Regis' (*Hippeastrum vittatum*) of 16 March 1781 came from a plant supplied by Charlotte to Delany at St James's Palace. And, when Charlotte visited Bulstrode on 'Saturday 20 Octr 1781', Delany undertook the collage 'Conyza inuloides' (*Allogopappus canariensis*).

In 1781, Charlotte cemented the bond by giving Mrs Delany a needlework pocket-book containing exquisite versions of the tools employed in collage making. Although Delany ceased her work towards the end of 1782—just short of the one thousand collages she aspired to as her 'Flora Delanica'—she did not abandon the nurturing of other accomplished women, including Charlotte's artistically inclined daughters. She could well have played a role in overseeing some of their works in what became an album of over forty botanical collages, now in the Newport Garden Club, Rhode Island.[26] During Delany's final years spent in Windsor, Charlotte encouraged her daughters in their artistic initiatives and Princess Elizabeth in particular emerged as the most versatile artist (fig. 6.12).

The links to Delany and Kew are evident from the inscriptions on some of her collages. On the reverse of the image of a variegated-leaf pelargonium is written 'Kew 1786'. The 'China Aster' is recorded as having been completed on 3 September 1786, following another collage on 2 September, and several works continued to be worked on through October 1786 before there was a break. The collage of spindle has written in graphite on the verso: 'Copied from one of Mrs. Delany's sent to me at Kew November': significantly, the date November 1786 coincides with Princess Elizabeth's recuperation at Kew when she was well enough to see Delany, following an illness at Windsor in August and September that year. Collage making resumed from January

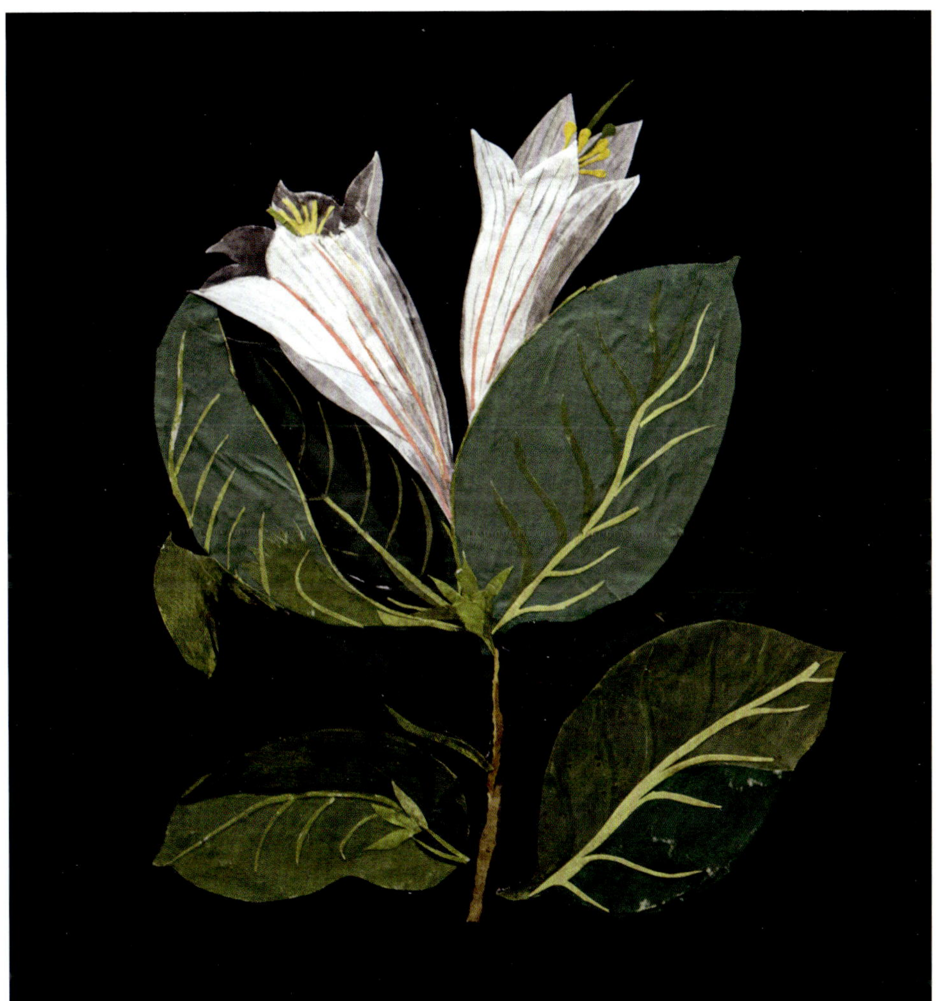

FIG. 6.12 Attributed to Princess Elizabeth, *Portlandia Grandiflora*, 1786–88, collage of coloured papers, with gouache and watercolour on paper. The Newport Garden Club, Newport, RI

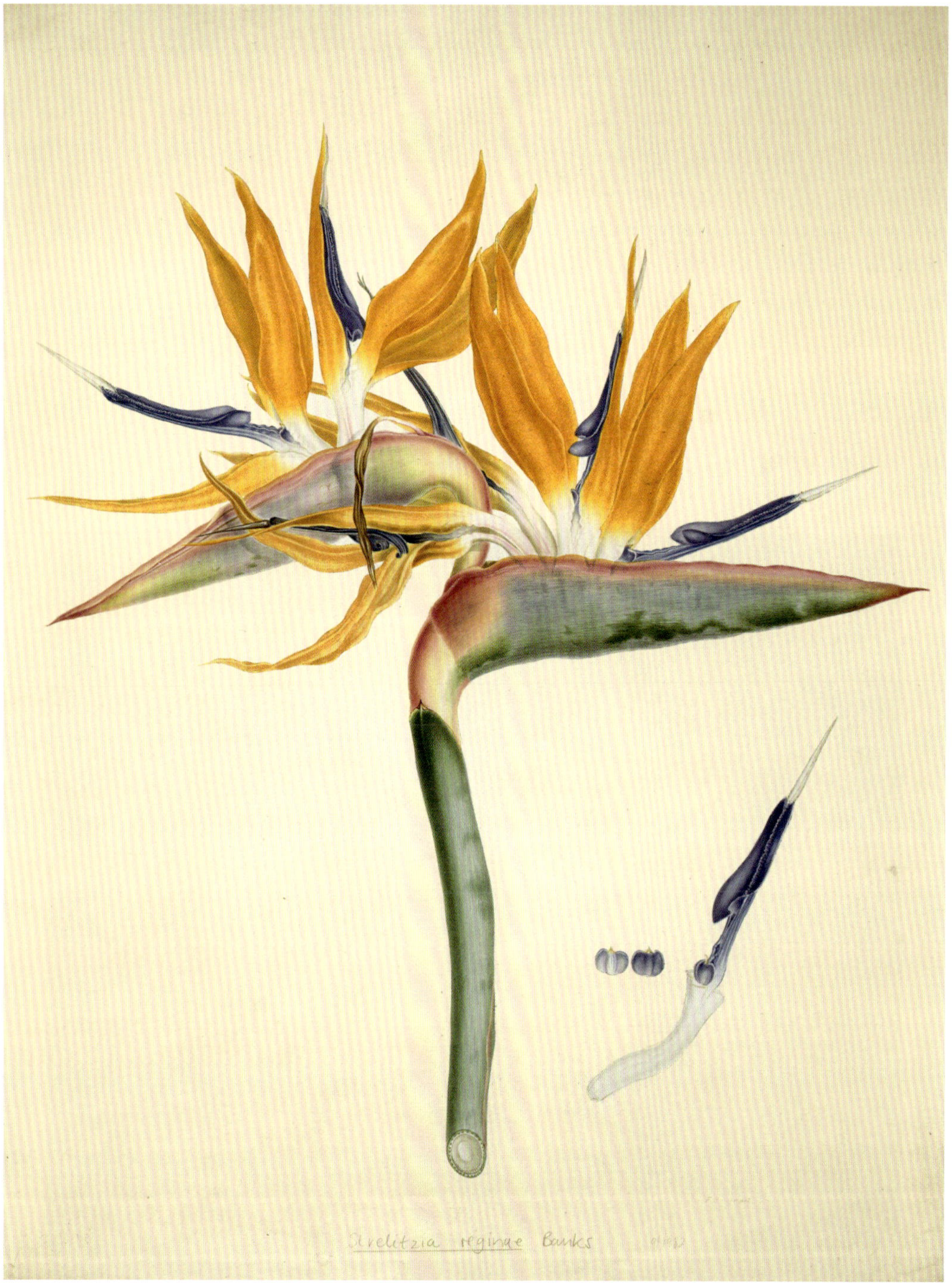

FIG. 6.13 Francis (Franz Andreas) Bauer, '*Strelitzia Reginae* Banks', design 144 for *Strelitzia Depicta*, 1812, watercolour on paper, 22⅛ x 16⅛ in. (56.2 x 41 cm). Natural History Museum Library and Archives, London (cat. 20.08)

to March 1787, and the final dated collage is of lily of the valley (*Convallaria majalis*), completed on 'Sunday, February 17, 1788'. This was a flower beloved by Delany. She was by then in decline, having already alluded to being 'Very much out of order' early in 1786.[27] It was after a visit to Kew in March 1788 that Mrs Delany caught a chill that developed into pneumonia. She died on 15 April, just a month short of her eighty-eighth birthday. Many other interesting connections and borrowings are indicated in the annotations to the works. The *Portlandia* of the album was inscribed on the verso: 'Began at Windsor from one of Mrs. Delany's and finished at the Queen House from memory January 13th 1787.'[28]

The authorship of the forty-four surviving works—of varied execution and artistic merit—remains uncertain, but it is reasonable to attribute the *Portlandia* to Princess Elizabeth when it is compared with those that have inscriptions all in the same hand. Her paper cutting at Frogmore is documented in the 1790s and, as the most artistically adept daughter, she would work in many different media, including painting fans. By 1792, she had produced a number of large flower paintings, including a fine 'Flower piece with bird's nest' after a still life by Margaret Meen.[29]

Meen, active from 1775 onwards, and teaching flower and insect painting in London, completed her volume *Exotic Plants from the Royal Gardens at Kew* in 1790. It was dedicated to Charlotte. Fittingly, it features a portrait of *Strelitzia reginae*. When this extraordinary 'Bird of Paradise' flower was introduced to Kew from the Cape in 1773, Banks and Aiton agreed that the botanical name would come from Charlotte's natal home of Mecklenburg-Strelitz. James Sowerby's depiction of *Strelitzia reginae*, made in 1787, was engraved for use in William Aiton's *Hortus Kewensis*, published in 1789; in 1807, Robert Thornton published *Strelitzia reginae* in his *Temple of Flora*; and Francis (Franz Andreas) Bauer, resident artist at Kew from 1790 to 1840, produced the ultimate study of the genus *Strelitzia*, including *S. reginae* in 1812 (fig. 6.13).[30]

Two charming notecards, now in the Fitzwilliam Museum, show Bauer shifting from science to decoration. Both are dated 30 March 1814. The first, signed by 'Jos: Banks', with his seal and his flower, *Banksia*, makes much of the *Strelitzia* association, but uses myrtle and forget-me-not, to signify affection for and remembrance of 'ELIZA', presumably Princess Elizabeth (fig. 6.14). The second, with roses, thistles, shamrock, forget-me-nots, and heartsease in the surround, was signed 'S. S. Banks' for Joseph's sister, Sarah Sophia Banks. This decorative use of floral illustration was where Princess Elizabeth excelled—demonstrated by a pair of flower vases of 1812, for example, or in the decoration of the Drawing Room on the first floor of the Queen's Cottage, dated to 1805. A tented room, the walls were painted with trompe l'oeil bamboos, nasturtium, and convolvulus.

In contrast to this surviving decorative interior, the conversion of the 'kangaroo paddock' attached to the Queen's Cottage into a 'flower garden' in December 1806 is a lost project from Charlotte's time. The 'Plan of Kew Gardens and the environs' of around 1785 does not include the area around the Queen's Cottage and, sadly, no later plan appears to document the conversion. The 1785 plan, however, in its representation of Augusta's gardens around a dozen years on from her death, serves as a document of

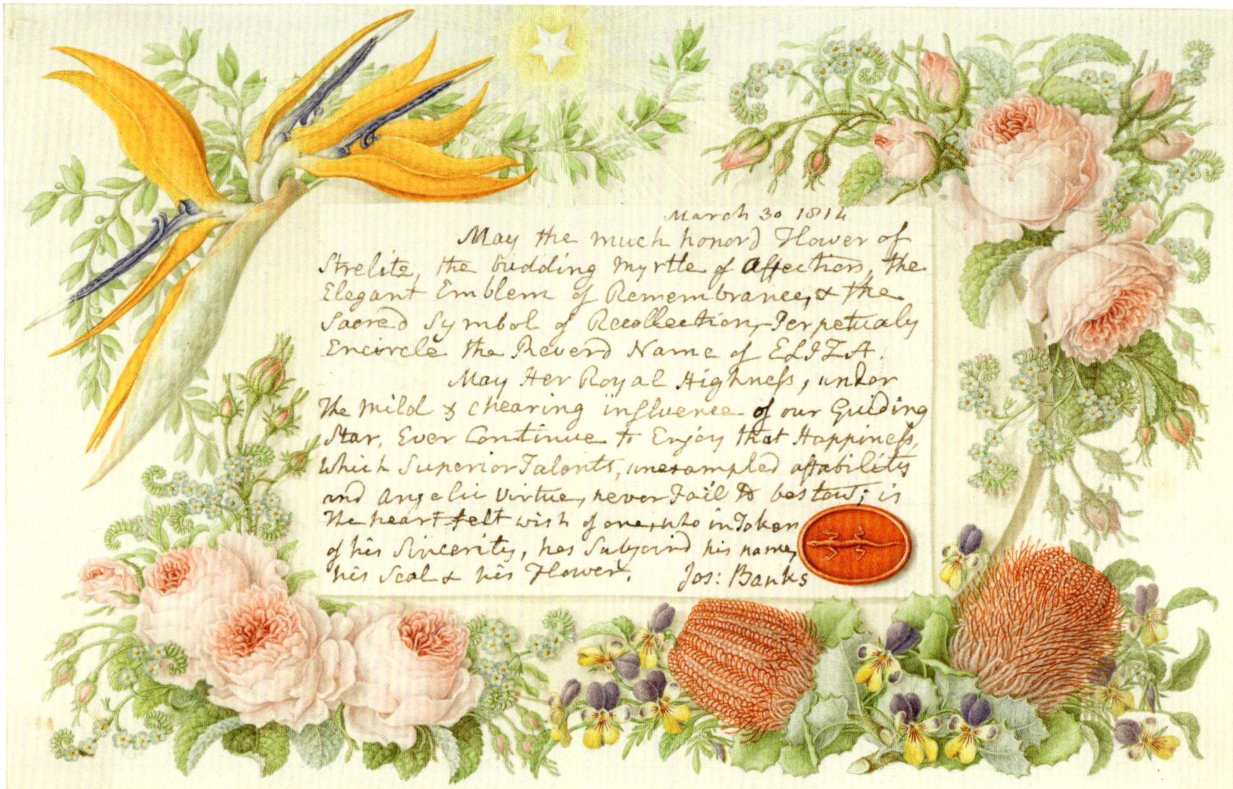

FIG. 6.14 Francis (Franz Andreas) Bauer, Notecard with a surround of mixed flowers with a central hand-inscribed message by Joseph Banks in rectangular central reserve, 30 March 1814, watercolour and pencil on paper, 5⅜ x 8¼ in. (13.5 x 20.8 cm). Fitzwilliam Museum, Cambridge

disappearance—in this instance of her elliptical Aviary enclosure and the ha-has that encircled the lawns—as well as of survival and continuity in regard to exotic buildings and plantings. The Franz Joseph Manskirch view of the White House, engraved by H. Schutz in 1798, captures the mature landscape, appearing as a Brownian scene, seemingly devoid of flowers or botanical enclosures. The same artist's view of the Alhambra and Pagoda, also of 1798, likewise depicts mature trees and plantations, without the shrubs and specimen woody species that had graced Augusta's pleasure ground. These views appear to consolidate the 'masculine' view of Kew, far from the image of it as a flowery, 'feminine' domain.

Yet flowers and flowering shrubs continued to assume significance in the Princesses' Kew. During the 1790s, Margaret Meen gave lessons in flower painting to Queen Charlotte's daughters Charlotte, the Princess Royal, and Princess Elizabeth. With their teacher, they meticulously coloured Francis Bauer's *Erica* engravings, and copied plates in botanical works. Robert Thornton, in his *New Illustration of the Sexual System of Linnaeus* (1799–1807), wrote of the diverse plants of Kew becoming subjects of their work, accomplished 'with a grace and skill which reflect on these personages the highest honour'.[31] But it was not the scientific brilliance of Bauer that was transmitted, so much as the floral brilliance of Mary Moser's images of bunches of flowers produced between 1792 and 1797. Princess Elizabeth, given an opportunity to apply her decorative skills at Frogmore, decorated the narrow room called the 'Cross Gallery' with panels that incorporated her cut-paper silhouettes and paintings of flower garlands in oil and tempera. These were in the manner of Mary Moser and Lady Diana Beauclerk's work at Little Marble Hill.[32]

What evidence is there, then, that Charlotte and the princesses engaged in genuine botanizing at Kew, Windsor, and Frogmore? The role of Jean-Jacques Rousseau as a botanical mentor to women, and as the Duchess of Portland's self-styled 'herboriste' between 1767 and 1773, is one important dimension explored elsewhere.[33] The appointment of Jean-André Deluc, the Genevan savant who knew Rousseau, as the Queen's Reader in French in 1774, led Charlotte to geology through his letters of 1778 which touch on this subject. Yet Rousseau's thinking was kept at arm's length. Indeed, while he eventually accepted a pension from George III it was on the condition that it was not publicly known. The titles on natural history or philosophy, horticulture, farming, and gardening, which are listed in Charlotte's library on its sale in 1819, are wide-ranging and impressive. They point to a familiarity with Linnaean science and to diverse natural systems or taxonomies. For example, Lord Bute's *Botanical Tables* of 1785, priced at £111 16s., was paramount among the titles. It emerged from the work of his protégé, John Hill, whose twenty-six volumes of *The Vegetable System* were published between 1759 and 1775. Seven of the twelve sets Bute published went to women, including Charlotte and Catherine II, Empress of Russia; as Bute suggested in his introduction to the work, it was 'composed solely for the amusement of the Fair Sex'.[34]

Yet surrounding herself with learned works did not mean that Charlotte had mastery of such learning. Although James Edward Smith, President of the newly formed Linnean Society of London, gave 'regular lectures (or rather conversations) in Zoology & Botany' to Charlotte and the princesses, there is little evidence of this instructional role bearing fruit beyond Banks's remark that Charlotte 'studies botany intensively & really reads with perseverance elementary books'.[35] Smith spoke from his notes, which Charlotte copied and then read aloud at the conclusion of each session to check her accuracy. According to one source, James Edward Smith was chosen to 'arrange the queen's herbarium' in 1791, but fell out of favour at court because his book spoke too highly of Rousseau.[36] In 1796 he 'retired to Norwich', visiting London only occasionally thereafter. Yet there are many specimens in the Smith Herbarium collected at Kew from 1781 onwards, and into the 1790s and 1800s, which suggest that a connection was maintained.

With some diffidence Charlotte told Lord Bute, who had presented her with the copy of *Botanical Tables*, that she was about to compile a 'Herbal from impressions on black paper',[37] thus preserving actual plants rather than using cut-paper images mounted on black. That 'Herbal' suggests a type of learning modelled on Mary Delany's collages. Although the 'Herbal' has not been located, a 'Delany-style' collage on a black ground

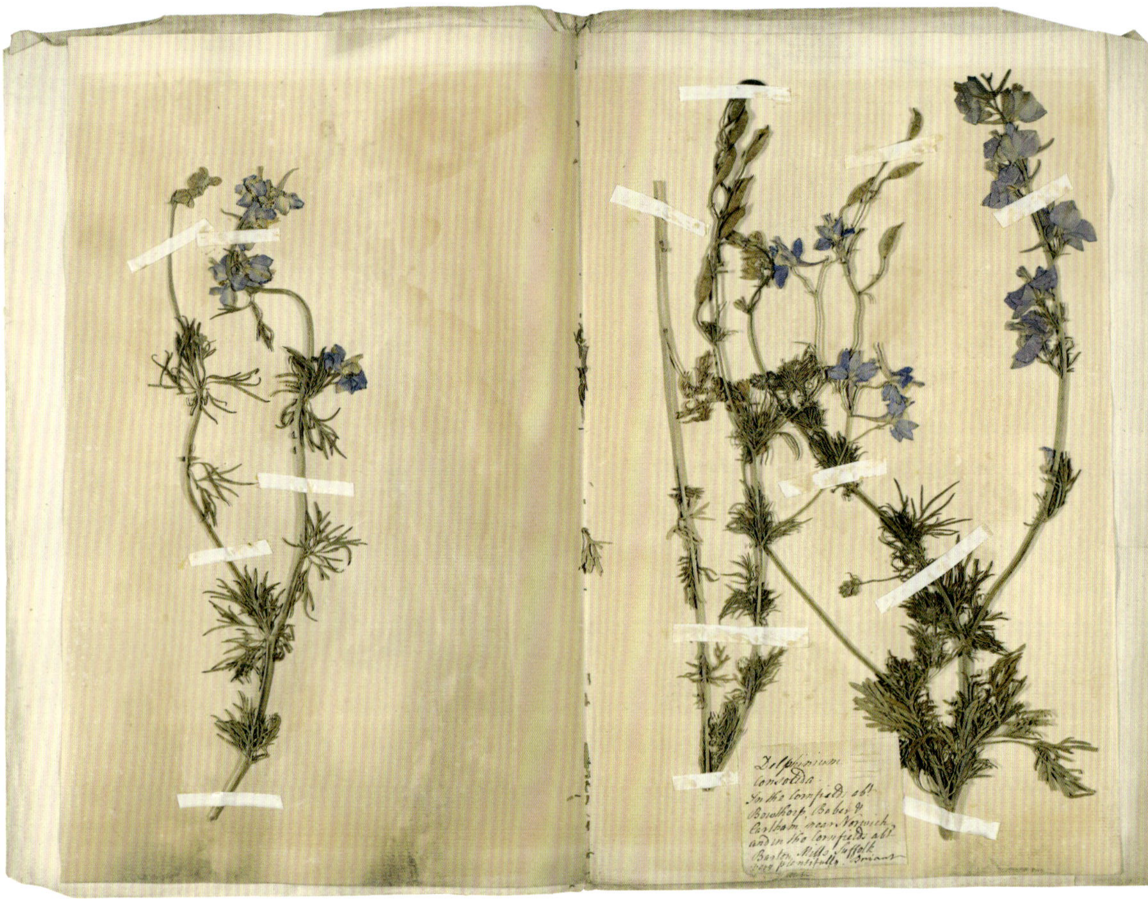

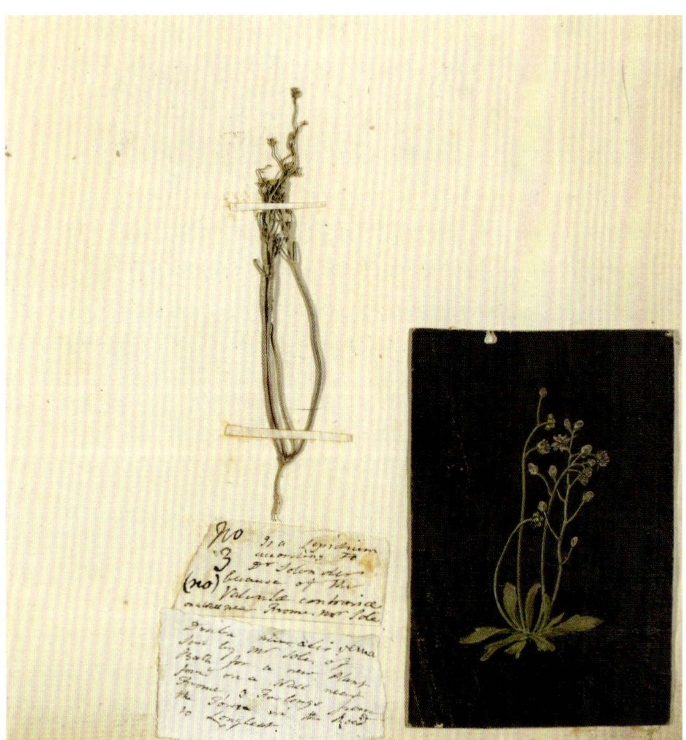

FIG. 6.15A John Lightfoot, Herbarium sheet with *Delphinium*, late 18th century, botanical specimen on paper, 16 x 12⅞ in. (40.5 x 32.7 cm). John Lightfoot Herbarium, Royal Botanic Gardens, Kew, Library, Art & Archives Collections (cat. 22.05)

FIG. 6.15B John Lightfoot, Herbarium sheet with *Lepidium*, late 18th century, botanical specimen with collage of coloured papers and drawing, 16¼ x 13 in. (41.4 x 33 cm). John Lightfoot Herbarium, Royal Botanic Gardens, Kew, Library, Art & Archives Collections (cat. 22.05) has been found next to a single specimen in the Lightfoot Herbarium, an assemblage of pressed flower specimens mounted on paper, which George III bought for Charlotte as a gift for 100 guineas in 1788 (fig. 6.15). Originally housed in twenty-four mahogany cabinets at Frogmore, and now in the collection of the Royal Botanic Gardens, Kew, this collection deserves more scholarly attention, not least to answer one question: was this unique collage done at Bulstrode under the eyes of Delany and John Lightfoot; or was it perhaps added later, that is, after Delany's death in 1788, as a royal tribute to her genre of botanical collage?

The Reverend John Lightfoot was chaplain and botanist to the Duchess of Portland, widowed since 1761, and he remained at Bulstrode during Delany's own extended widowhood. As well as publishing *Flora Scotica* in 1777, he had of course corresponded with many naturalists, taken part in botanical excursions, and amassed a collection of dried plants including algae. While, on his death, the main part of this collection was first acquired by the royal family, the seaweeds ended up separated from the rest of the herbarium, and are now kept at the Natural History Museum, London. This algal herbarium, along with Lightfoot's life and work as a whole, has been the subject of several studies.[38] His work with the Duchess's shells has also received scholarly treatment.[39] Yet more of the Bulstrode story needs to be explored; for, with the auctioning of the Duchess's collections in 1786 following her death the previous year, a group of Ehret botanical studies ended up in the collections of the Royal Botanic Gardens, Kew, including his *Parnassia palustris* among beautiful portraits of English moisture-loving plants, and there will be more to discover elsewhere.

In this way, studies of British flora were transferred from the Duchess's Bulstrode to the Queen's domain.[40] Specimens of English cranesbills—*Geranium columbinum*, *G. pusillum*, *G. sylvaticum*—include labels that refer respectively to Norfolk, the botanist William Hudson, Kew, Dr Bryant (who was likely Charles Bryant, d. 1799, Beadle, Court of Guardians, Norwich collecting for Lightfoot), as well as to the Oxford Physic Garden and to Ingleborough in the Yorkshire Dales. The specimen of *Linum radiola*, syn. *Radiola linoides* was from 'Penzance'.

Lightfoot's English geraniums were collected at a critical moment for systematic study, coinciding with Delany's cranesbills and their relatives which are recorded in thirty-six collages, the largest single group in 'Flora Delanica', and including *Geranium sylvaticum* of the Lightfoot Herbarium. Under the generic name *Geranium*, Delany included species from southern Africa now placed in the genus *Pelargonium*. This latter genus was first distinguished and named by the French botanist Charles Louis L'Héritier de Brutelle, and first published as such in Aiton's *Hortus Kewensis* of 1789.

L'Héritier played an important role as catalyst in the genesis of *Hortus Kewensis*, the benchmark work on plants cultivated at Kew during Charlotte's lifetime. It provides the names and provenance of some 5,600 species. Its genesis can be traced back to 1773, when William Aiton, just a year after Augusta's death and doubtless under the sway of Joseph Banks, began compiling a descriptive plant list of 3,400 species. Aiton adopted the same title for his work that John Hill had used when listing the plants grown in Augusta's garden. Banks got wind of the fact that L'Héritier, who had been

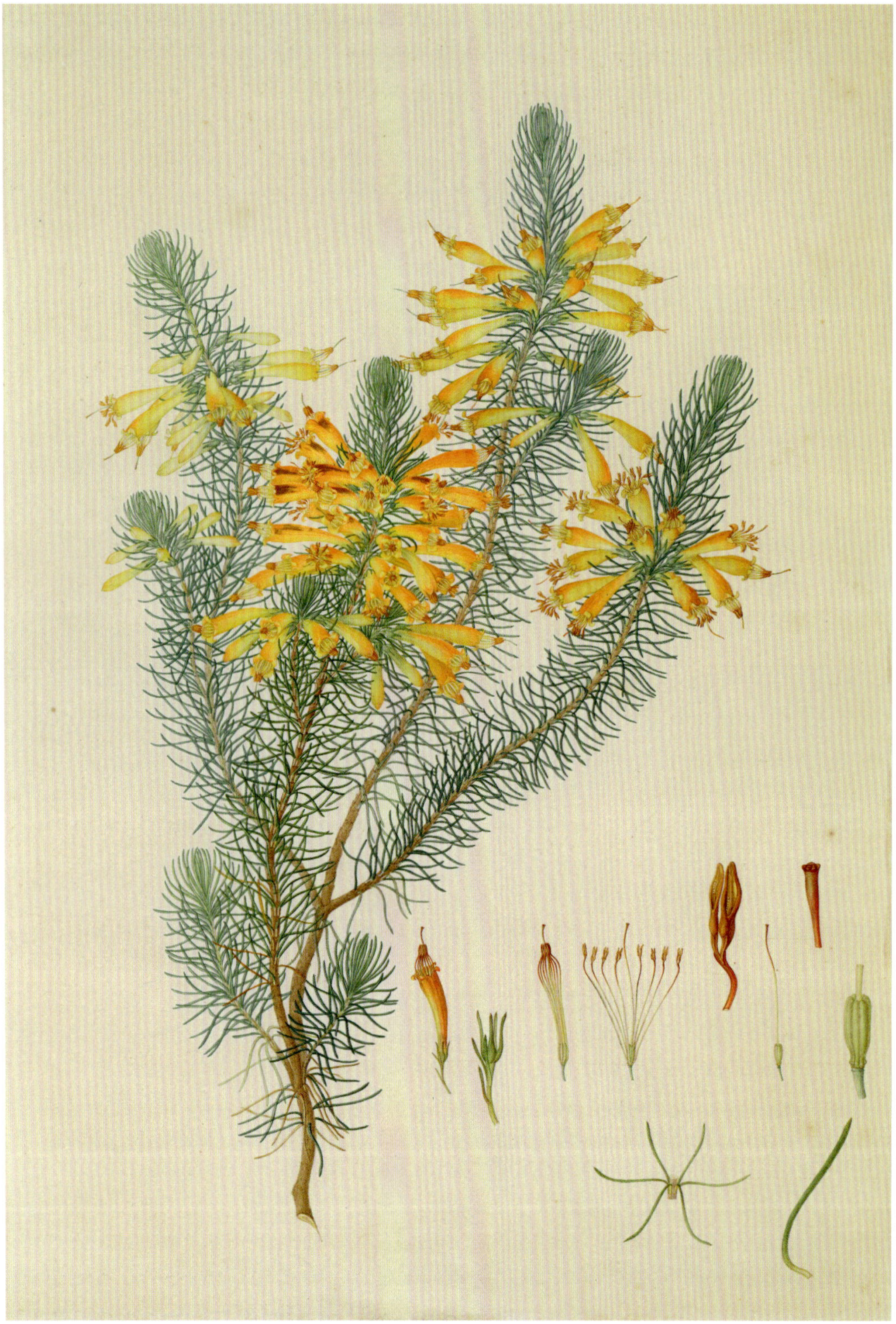

FIG. 6.16 Francis (Franz Andreas) Bauer, '*Erica grandiflora*', c. 1803, watercolour on paper. Natural History Museum, London, Bauer Collection

working at Kew, appeared on the brink of beating Aiton with a serial publication, *Stirpes Novae*. Thus he indicated to James Edward Smith that he would ensure the Kew gardener moved fast so as to 'little molest L'Heritier'.[41] In practice, Jonas Dryander, the Swedish botanist editing Aiton's lists, sustained a cordial relationship with the French botanist, submitting proofs to him in 1788, while receiving back some proofs of L'Héritier's *Sertum Anglicum*, along with comments and corrections. The appearance of the first part of *Sertum Anglicum* in January 1789, with its fine drawings of two *Lightfootia* species amid engravings of early drawings by Pierre Joseph Redouté, also prodded Banks and Aiton into very prompt action.

When documenting how the three-volume *Hortus Kewensis* of 1789 became the five volumes of 1810–13, itemizing eleven thousand species, it is important to mention the role that Charlotte played. The 'KEW Records Book, 1793–1809', for example, shows how her 'British Herbarium' was being augmented in 1799, and how Charlotte and the princesses were involved in having Kew plants sent to Frogmore. According to Banks, this was only after 'H[er] Majesty' sought assurance of their being at Kew.[42] That assurance also allowed, for example, the dispatch of plants to Grace, Lady Neale 'by Command of her R.H. Princess Augusta', Charlotte's second daughter, including 'Erica persoluta', 'E. ventricosa', 'E. baccans', 'E. Massoni', and 'E. grandiflora'.[43]

In 1790 Francis Bauer set up house at Kew, and, until his death in 1840, he produced one of the finest works of botanical art. His *Delineations of Exotick Plants cultivated in the Royal Gardens at Kew* first appeared in 1796, and depicted ten of Francis Masson's Cape heaths. Ten further illustrations followed in 1797, and ten more in 1803. *Erica grandiflora*, introduced by Masson from South Africa in 1785, is among those listed in Princess Augusta's 'Command' order for Lady Neale. Bauer's original painting of *E. grandiflora* in the Natural History Museum has a luminous colouring that adds to the virtuosity of his three-dimensional modelling (fig. 6.16). The lasting impact that *Erica* species have had on gardening in the British Isles is evident in the founding of the Heather Society in 1963. Thus helping to shape the modern world of horticulture, the combined resources, will power, tenacity, and talent of Banks, Masson, and Bauer should be celebrated today—just as much as Banks's contribution under George III to economic botany, which receives a prominent paragraph in the 'Historical Description' (forming part of the citation for the Royal Botanic Gardens, Kew, as a UNESCO World Heritage Site).

In light of the emphasis of UNESCO's 'masculine' inscription, the works of female accomplishment associated with Kew, and the objects and flora and fauna linked to those circles of creativity and culture fostered by Caroline, Augusta, and Charlotte, could be deemed the forgotten 'intangible heritage'. The discussion of many of those objects in this volume begins to set the record straight: the modern world of horticulture owes much to the resources and vision of these three forward-looking Princesses. They helped make Kew what it is today: an august domain.

NOTES

1. Ray Desmond, *Kew: The History of the Royal Botanic Gardens* (London: Harvill Press with The Royal Botanic Gardens, Kew, 1995).
2. For a good account of these, see Joanna Marschner, 'Queen Caroline of Anspach and the European Princely Museum Tradition', in *Queenship in Britain, 1660–1837: Royal Patronage, Court Culture and Dynastic Politics*, ed. Clarissa Campbell Orr (Manchester: Manchester Univ. Press, 2002), 133–37, and Joanna Marschner, *Queen Caroline: Cultural Politics at the Early Eighteenth-Century Court* (New Haven and London: Yale Univ. Press, 2014), 71–84. See also, in this volume, Lee Prosser, 'Architectural Patronage', and Todd Longstaffe-Gowan, 'Hankering after Horticulture'.
3. Stephen Bending, *Green Retreats: Women, Gardens and Eighteenth-Century Culture* (Cambridge: Cambridge Univ. Press, 2013), 222–23.
4. http://whc.unesco.org/en/list/1084, accessed 20 July 2015.
5. See Desmond, *Kew*, 2–3: 'Several Criticks were of several opinions: One declar'd he would not have too much Art in it, for my notion (said he) of gardening is, that it is only sweeping Nature; Another told them that Gravel walks were not of a good taste, for all of the finest abroad were of loose sand: a third advis'd peremptorily there should not be one Lyme-tree in the whole plantation; a fourth . . .'. Letter from Alexander Pope to Lord Bathurst, 13 Sept. 1719, quoted in Alexander Pope, *The Works of Alexander Pope*, ed. J. W. Croker, intro. and notes W. Elwin and W. J. Courthope, 10 vols. (London: John Murray, 1871–89), Vol. 8: *Correspondence III*, (1872), 327–29.
6. The National Archives, Kew, MR 696.
7. See E. Charles Nelson, '"The Truly Honest, Ingenious, and Modest Mr Mark Catesby, F.R.S.": Documenting his Life (1682/83–1749)', in *The Curious Mister Catesby*, ed. E. Charles Nelson and David J. Elliott (Athens, GA: Univ. of Georgia Press, 2015), 14.
8. See Clarissa Campbell Orr, 'Mrs. Delany & the Court', in *Mrs. Delany and her Circle*, ed. Mark Laird and Alicia Weisberg-Roberts (New Haven and London: Yale Univ. Press, 2009), 42–44; see also Clarissa Campbell Orr, 'Mary Delany and Queen Charlotte: The Botanizing Court', in *Hanover and England: A Garden and Personal Union: German and British Garden Culture Between 1714 and Today*, ed. Marcus Köhler and Joachim Wolschke-Bulmahn (Hanover: Druckerei Hartmann, 2014).
9. See Mark Laird, 'From Callicarpa to Catalpa: The Impact of Mark Catesby's Plant Introductions on English Gardens of the Eighteenth Century', in *Empire's Nature: Mark Catesby's New World Vision*, ed. Amy R. W. Meyers and Margaret Beck Pritchard (Chapel Hill, NC, and London: Univ. of North Carolina Press, 1998), 184–227; Mark Laird, 'Mark Catesby's Plant Introductions and English Gardens of the Eighteenth Century', in *Curious Mister Catesby*, chap. 19; Mark Laird, *A Natural History of English Gardening, 1650–1800* (New Haven and London: Yale Univ. Press, 2015), chaps. 3 and 5.
10. Laird, *Natural History*, chap. 3.
11. For the Kew Aviary, see Laird, *Natural History*, chap. 5.
12. Campbell Orr, 'Mary Delany and Queen Charlotte'.
13. The Royal Archives, RA Add. MS. 17/65/1.
14. Mark Laird, 'Lilac and Nightingale: A Heritage of Scent and Sound at Horace Walpole's Strawberry Hill', in *Sound and Scent in the Garden*, ed. Dede Fairchild Ruggles (Washington, DC: Dumbarton Oaks, 2017), 193–215.
15. For Catesby's supply of plants with a subscription, see Laird, *Natural History*, 146.
16. Augusta's Privy Purse, RA Geo Add. MS. 17/65/2.
17. I am grateful to Florence Grant for her transcription of Queen Charlotte's copy in the Pierpont Morgan Library, New York: George Ritso, *Kew Gardens. A Poem, humbly inscribed to Her Royal Highness the Princess Dowager of Wales* ([London] Printed by M. Lewis and sold by R. and J. Dodsley, 1763).
18. See Laird, *Natural History*, chap. 5.
19. See again Laird, *Natural History*, 255–57.
20. Botany Library, Natural History Museum, Ehret large folio 107.
21. For background information, see Desmond, *Kew*, chap. 3. For Ehret and Kew collections, see Gerta Calmann, *Ehret: Flower Painter Extraordinary* (Boston, MA: New York Graphic Society/Little Brown, 1977); for acclimatization, see Laird, *Natural History*, chap. 5.
22. Desmond, *Kew*, 64–66.
23. Desmond, *Kew*, 74.
24. Desmond, *Kew*, 75.
25. See Laird and Weisberg-Roberts, eds., *Delany*, chap. 7, 163.

26. I am grateful to Patricia H. Fernandez, President, Newport Garden Club, Pam Russell, Archives and History Chair of the Club, and Mary Daniels, formerly of the GSD, Harvard University, for making it possible for me to view the album in the Redwood Library, Newport, RI, in April 2015. I am indebted to Jane Roberts for scholarly advice on all aspects of the Royal Archives.
27. Laird, *Natural History*, chap. 6, 323.
28. See also, in this volume, Jane Roberts, 'Frogmore and the Princesses as Artists', 375–77.
29. For this background, see Kim Sloan, '"*A Noble Art": Amateur Artists and Drawing Masters, c. 1600–1800* (London: British Museum Press, 2000), 76 and 236–37.
30. See William Aiton, *Hortus Kewensis, or, A catalogue of the plants cultivated in the Royal Botanic Garden at Kew* (London, 1789), and Robert Thornton, *Temple of Flora,* Part 3 of *New Illustration of the Sexual System of Linnaeus* (1799–1807).
31. See Desmond, *Kew*, 80.
32. Sloan, '*Noble Art*', 76.
33. See Clarissa Campbell Orr, 'Queen Charlotte, "Scientific Queen"', in Campbell Orr, *Queenship*, 236–40 passim, 247–58 passim.
34. See Laird, *Natural History*, chap. 7, 379.
35. Cited in Desmond, *Kew*, 79.
36. See G. S. Boulger, *rev.* Margot Walker, 'Smith, Sir James Edward (1759–1828)', *Oxford Dictionary of National Biography* (Oxford: Oxford Univ. Press, 2004; online ed.), http://www.oxforddnb.com/view/article/25825, accessed 20 July 2015.
37. Desmond, *Kew*, 79.
38. Jean K. Bowden, *John Lightfoot, his Work and Travels: With a Biographical Introduction and a Catalogue of the Lightfoot Herbarium* (Kew: Bentham-Moxon Trust, Royal Botanic Garden; Pittsburgh: Hunt Institute for Botanical Documentation, Carnegie Mellon Univ., 1989); Jenny A. Bryant, Linda M. Irvine, and Emma Ruffle, 'Insights into the Life and Work of Rev. John Lightfoot (1735–1788), with Particular Reference to his Algal Herbarium and its Conservation', *The Linnean* 28, no. 1 (April 2012): 26–43.
39. Beth Fowkes Tobin, *The Duchess's Shells: Natural History Collecting in the Age of Cook's Voyages* (New Haven and London: Yale Univ. Press, 2014).
40. Laird, *Natural History*, 296.
41. Cited in Desmond, *Kew*, 105.
42. Desmond, *Kew*, 79.
43. 'KEW Records Book, 1793–1809' (MS), listed on page 52. I am grateful to Florence Grant for the transcription.

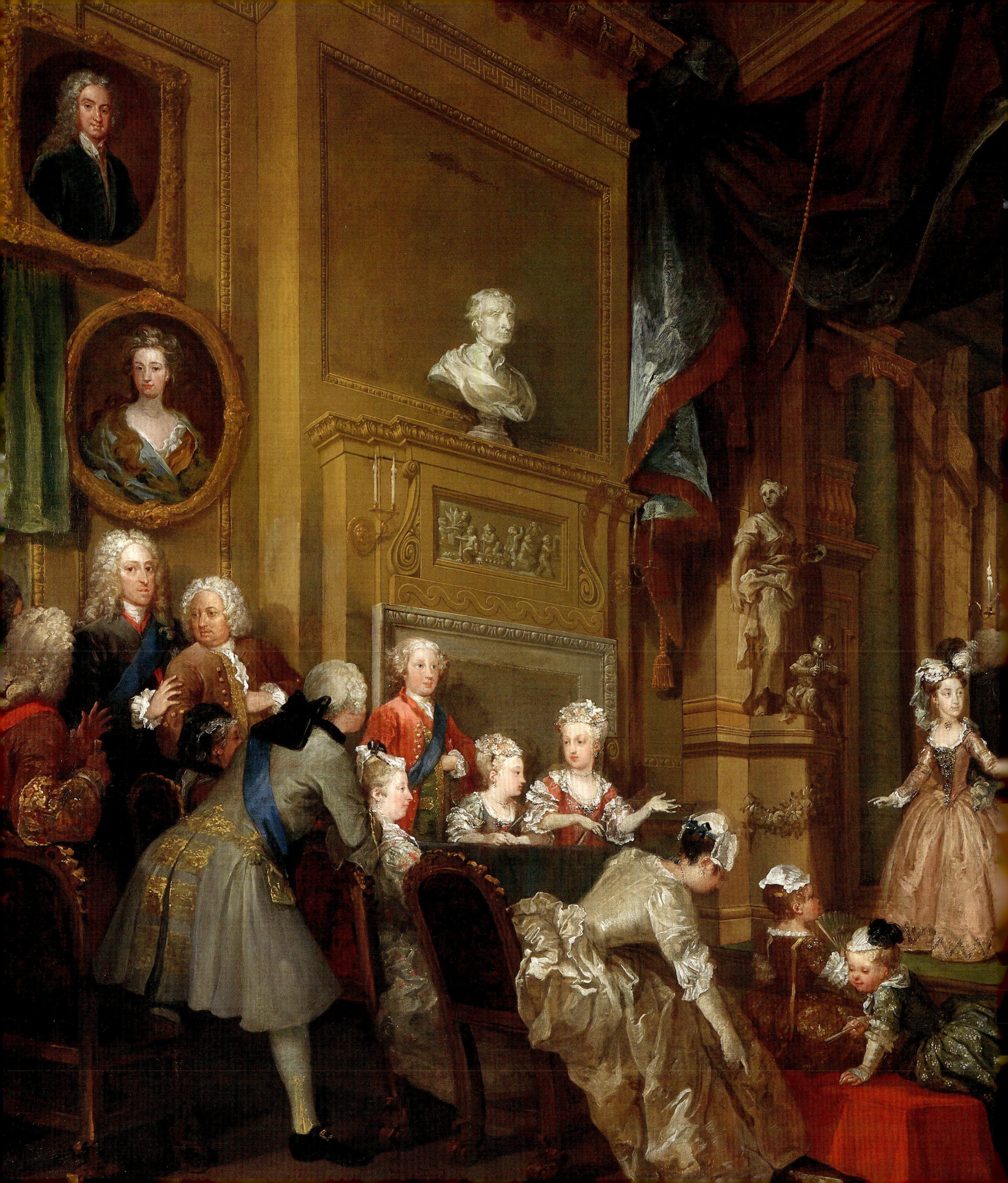

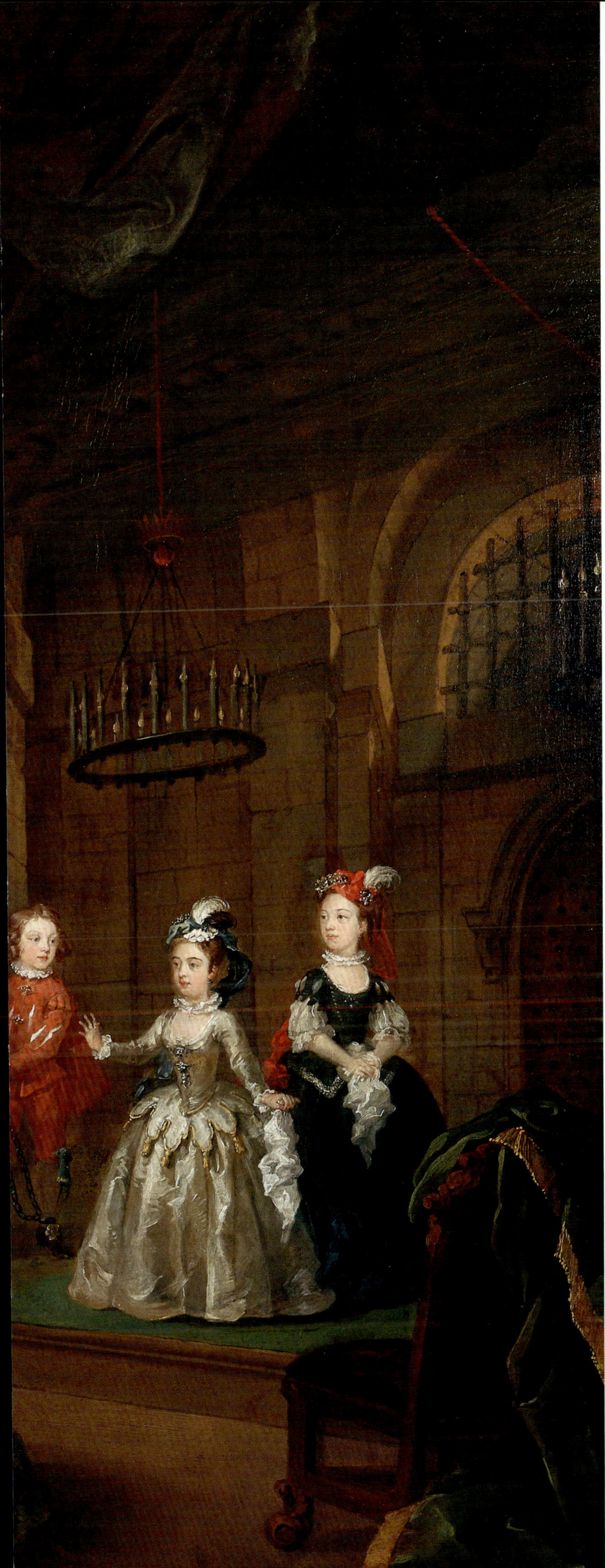

Court as a Stage

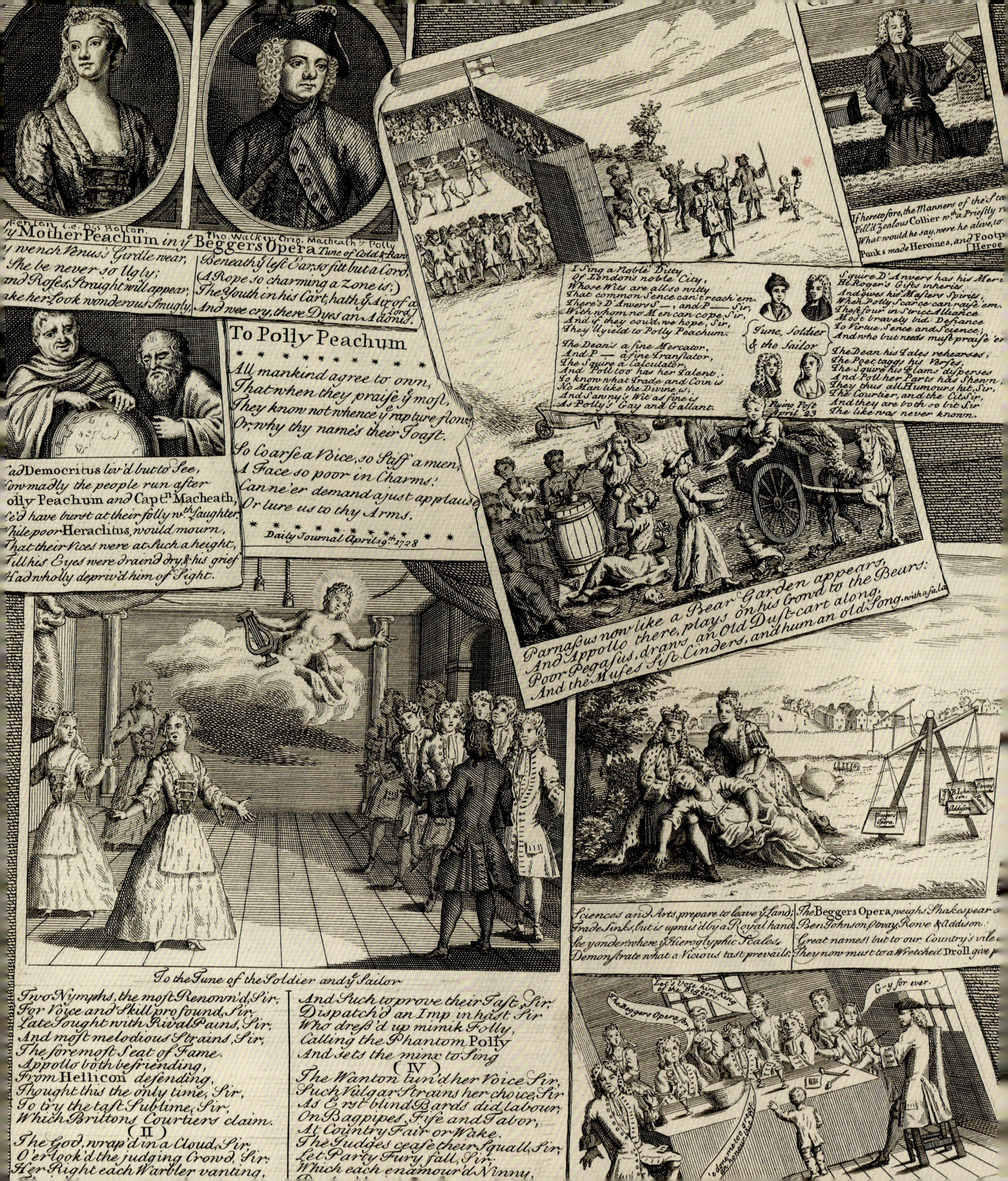
Fenton i.e. D.ss Bolton.
y Mother Peachum in ye Beggers Opera
Tho. Walker Orig. Macheath & Polly
Tune of Cold & Ran
y wench Venus's Girdle wear,
She be never so Ugly;
and Roses, Straight will appear,
ake her Look wonderous Smugly.
Beneath ye left Ear, so fitt but a Cord,
(A Rope so charming a Zone is;)
The Youth in his Cart, hath ye Air of a Lord
And wee cry, there Dyes an Adonis.
If heretofore, the Manners of the St
Fill'd Zealous Collier wth a Priestly r
What would he say, were he alive, to
Punks made Heroines, and Footp [Heroe
I Sing a Noble Ditty
Of London's noble City,
Whose Wits are all so witty
That common Sence can't reach'em.
There's D'Anvers S—, and P— Sir,
With whom no Men can cope Sir,
And if they cou'd, we hope, Sir,
They'll yield to Polly Peachum.
The Dean's a fine Mercator,
And P— a fine Translator,
The Squire a Calculator,
And Poll too has her Talent.
To know what Trade and Coin is
No Man like the Divine is;
And Sawny's Wit as fine is
As Polly's Gay and Gallant.
Tune, Soldier & the Sailor
Flying Post April 23
Squire D'Anvers has his Meri
He Roger's Gifts inherits
And gives his Masters Spirits,
When Polly scarce can raise'em.
These four in Strict Alliance
Most bravely bid Defiance
To Virtue, Sence and Science;
And who but needs must praise 'em
The Dean his Tales rehearses,
The Poet taggs his Verses,
The Squire his Flams disperses
And Poll her Parts has Shewn,
They thus all Humours hit, Sir,
The Courtier, and the Cit, Sir,
And they are both so bit Sir
The like was never known.
To Polly Peachum
All mankind agree to own,
That when they praise ye most,
They know not whence ye rapture flows,
Or, why thy name's their Toast.
So Coarse a Voice, so Stiff a mien,
A Face so poor in Charms!
Can ne'er demand a just applause
Or lure us to thy Arms.
Daily Journal April 19th 1728
Had Democritus liv'd but to See,
How madly the people run after
Polly Peachum and Captn Macheath,
He'd have burst at their folly wth Laughter.
While poor Heraclitus, would mourn,
That their Vices were at such a height,
Till his Eyes were drain'd dry, & his grief
Had wholly depriv'd him of Sight.
Parnassus now like a Bear Garden appears,
And Appollo there, plays on his Crowd to the Bears:
Poor Pegasus, draws an Old Dust-cart along,
And the Muses Sift Cinders, and hum an old Song. with a fa. la
Sciences and Arts, prepare to leave ye Land;
Trade Sinks, but is uprais'd by a Royal hand.
See yonder where ye Hieroglyphic Scales,
Demonstrate what a Vicious tast prevails:
The Beggers Opera, weighs Shakespear a
Ben Johnson, Otway, Rowe & Addison.
Great names! but to our Country's vile
They now must to a Wretched Droll give p
Let's Vote him King of the Beggers.
G-y for ever.
The Beggers Opera
To the Tune of the Soldier and ye Sailor
Two Nymphs, the most Renown'd, Sir,
For Voice and Skill profound, Sir,
Late Sought with Rival Pains, Sir,
And most melodious Strains, Sir,
The foremost Seat of Fame.
Appollo both befriending,
From Hellicon desending,
Thought this the only time, Sir,
To try the tast Sublime, Sir,
Which Britons Courtiers claim.
(II)
The God, wrap'd in a Cloud, Sir,
O'erlook'd the judging Crowd, Sir:
Her Right each Warbler vanting,
And Such to prove their Tast, Sir,
Dispatch'd an Imp in hast Sir
Who dress'd up mimik Folly,
Calling the Phantom Polly
And sets the minx to Sing
(IV)
The Wanton tun'd her Voice Sir,
Such Vulgar Strains her choice, Sir
As Erst blind Bards did labour,
On Bagpipes, Fife and Tabor,
At Country Fair or Wake.
The Judges cease their Squall, Sir,
Let Party Fury fall, Sir,
Which each enamour'd Ninny,

Public and Private Stages

Joseph Roach

In the absence of a written constitution, the theatre is especially useful to a well-ordered commonwealth. Civil society needs such broadly accessible community rituals through which to assert its coherence and continuity: in order to be remembered, that which has not been written down must be repeatedly acted out. After the writer and politician Richard Steele attended the burial of the great tragedian Thomas Betterton in Westminster Abbey, for example, he commended in particular the propriety of the actor's deportment on the stage as exemplary for all Britons in the conduct of national life: 'There is no Human Invention so aptly calculated for the forming a Free-born People as that of the Theatre.'[1] In eighteenth-century Britain, as in ancient Athens, sovereign authorities not only sponsored theatrical performances on officially sanctioned stages, producing the works of their greatest dramatists, but they also mirrored those theatrical actions on the fractious stage of mixed government, enacting the conflict of checks and balances among tyranny, oligarchy, and democracy. While Betterton himself concluded that the theatre of his time was a 'Mimic State', succeeding generations of courtiers and politicians under the Hanoverian monarchs returned the compliment with the sincerest form of flattery.[2]

In that spirit, understanding the court as a stage, it is possible see how the roles of the three Princesses—Caroline of Ansbach, Augusta of Saxe-Gotha, and Charlotte of Mecklenburg-Strelitz—were scripted as performances, but only in part by black-letter law, more often by custom, and sometimes by self-inventing improvisation. Duty required them to perform private virtues as well as public acts for the general good throughout their lives, and each Princess in her own way obliged, though not in every instance happily. Each played a featured female role in the staging of a continuously reconstructed monarchy between the reigns of two queens, Anne and Victoria, who ruled as sovereigns in their own right; and each contributed, however distantly, to the long-running performance of the current one. In so doing, each of the three Princesses played more than a supernumerary part in the creation of the most important political document of the modern world never to have been written.

Astringently surveying Hanoverian polity over the long term in *The English Constitution* in 1867, Walter Bagehot stresses the importance of putting on a good show. He plays off the symbolic role of what he calls the 'visible form' of government, which consists of a few glamorous people doing interesting things, against the 'efficient form', which consists of many practical people doing boring things. The first form is charismatic; the second, bureaucratic; but each has its role to play in exercising the power of an unwritten constitution. 'The apparent rulers of the English nation', Bagehot writes:

> are like the most imposing personages of a splendid procession: it is by them that the mob are influenced; it is they whom the spectators cheer. The real rulers are secreted in second-class carriages; no one cares for them or asks about them, but they are obeyed implicitly and unconsciously by reason of the splendour of those who eclipsed and preceded them.[3]

Detail of p. 169, cat. 7.07

Bagehot here describes the practical scheme of limited monarchy and aristocratic preferment that had evolved under the Hanoverians, and his account illuminates the contributions of eighteenth-century royals to the consolidation of those institutions.

In order to lead the splendid procession, Caroline, Augusta, and Charlotte did many interesting things, but they also pioneered modern ways in which to do them. That included performing visible government in a burgeoning culture of print-world publicity. As Princess of Wales, Caroline showed herself in public with her husband at the theatre and opera and visited the City on Lord Mayor's day, making news. As Queen Consort, she encouraged the correspondence between Gottfried Wilhelm Leibniz and Samuel Clarke on Sir Isaac Newton's doctrines, commissioned innovative garden designs in the new 'natural' style, and revived the custom of dining in public on Sundays. For her part, Augusta played on patriotic sympathies when she accompanied Frederick, Prince of Wales, to popular fairs and parades, wore woollens in solidarity with the weavers of Spitalfields, and championed the English-speaking theatre over the Italian opera. Charlotte became an accomplished amateur botanist, patronized Wolfgang Amadeus Mozart and Johann Christian Bach, and commissioned extraordinary illuminated decorations to mark royal celebrations, as well as generously supporting worthy charities (fig. 7.01). Treating the court as a stage, the Princesses learned the publicist's art of how to command attention without seeming to seek it. Caroline, like an industrious housewife, rescued a collection of precious Hans Holbein sketches from the drawer of a forgotten old bureau and displayed them in Richmond Lodge and later Kensington Palace. Augusta and Charlotte, following Caroline as patronesses of the apparently unobtrusive art of landscaping, both contributed to the development of the gardens at Kew, to which, in 1762, was added the Pagoda, a fifty-metres-tall exclamation mark to their horticultural iteration of visible government.

Like other members of the royal family in Georgian Britain, however, the Princesses sometimes attempted to assert their roles in ways that were not merely symbolic. Oedipal friction and dynastic anticipation typically put the Princes of Wales in opposition to their respective kings and fathers. Caught in the middle, their partners had to take sides or, like Augusta, serve as go-betweens, risking the penalty of exile from the Royal Household that included their own children. For her part, Caroline supported her husband in his early struggle to court the allegiance of his future English subjects by subverting the policies of the stubbornly Germanic George I. But then she squandered the diminishing capital of her popularity by forming a behind-the-scenes alliance with Sir Robert Walpole, the widely despised precursor to ministerial practitioners of Bagehot's 'efficient form' of government. In turn, Augusta assisted Frederick in advancing the designs of the Leicester House faction against the policies of George II and the preferences of Caroline. When Frederick died unexpectedly in 1751 while still Prince of Wales, she skilfully negotiated with her father-in-law to retain charge of raising her son as heir to the throne. To Charlotte fell the responsibility of taking care of the King's person and the Royal Household during the regencies occasioned by George III's madness. But the blunt force of efficient government caused all three Princesses to fall back on the 'visible form' of their unwritten powers, which obligated them to do interesting things and to be seen doing them.

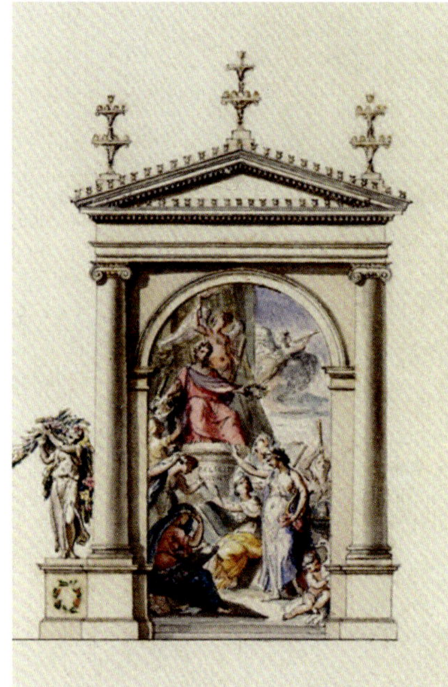

FIG. 7.01 Robert Adam, *General Design of a Transparent Illumination, Ordered by the Queen and Executed in Honour of his Majesty's Birth Day* (detail: Britannia directing Painting, Sculpture and Architecture), 1763, pencil, pen and watercolour on paper, 18⅞ x 49¼ in. (48 x 125 cm). Royal Collection Trust

The gardens that Caroline had planted at Kensington and Richmond certainly demonstrated her showmanship. If pervasive influence is taken as the criterion, her part in encouraging, though she did not start it, the development of the English garden may be her most significant contribution to the modern world: the idea of assisting, rather than replacing, Nature in creating a garden has shaped the styles of landscaping in most country estates, urban parks, suburban towns, and cemeteries down to the present day. Such designs feature architecture as well as horticulture, appealing to memory and imagination as well as sensuous appreciation. A generation after Sir John Vanbrugh unsuccessfully beseeched Sarah Churchill, Duchess of Marlborough, to save the ruins of 'Rosamund's Bower' as a romantically evocative architectural feature in the grounds of Blenheim Palace in Oxfordshire, Caroline's architect William Kent created 'Merlin's Cave' at Richmond. Adjacent to the rustic 'Hermitage', which featured marble busts of Newton, John Locke, and other worthies of Enlightenment science, Merlin's Cave ironically counterbalanced the Hermitage sculptures with waxwork figures of Merlin, Merlin's secretary, 'Henry VIII's Queen', Queen Elizabeth, and 'a Witch'. The combined programme of the Hermitage and Merlin's Cave shows Caroline's desire to project her affinity for Englishness, celebrating its idiosyncrasies by contrasting the deep-rooted magic of national folklore with the triumphs of modern science, putting the regime in even-handed touch with both. George II, less interested once crowned than Caroline in connecting with his British subjects, blustered about the expense of this folly. But the public came in droves nonetheless when the garden opened for tours.

The satirists, however, poisoned the tips of their darts, by which Caroline had already been stung, and they targeted her garden architecture and much more besides with their fusillades, showing the downside of high visibility when the royal court functions as a public stage. As part of the controversy over the runaway hit of John Gay's *The Beggar's Opera* (1728), both of the two opposing sides let fly at the Queen Consort. Most of the criticism of the regime came from those taking Gay's side, but a satirical print hostile to *The Beggar's Opera*, *The Stage Medley Representing the Polite Taste of the Town* (cat. 7.07), took aim at the monarchy as well. Whereas Gay specialized in subtly shifting satirical digs at multiple victims whose identities sometimes overlap, sometimes diverge, no special effort at ferreting out innuendo is required by the depiction of the members of the royal family in *The Stage Medley*. Deploring the success of *The Beggar's Opera* as indicative of the general degeneracy of the nation, the print shows the personification of Trade falling lifelessly into the feeble arms of George II and Caroline.

Nothing that the moralists said, however, diminished the ballad opera's success, which came in part at the expense of the government. William Hogarth's iconic *Scene from The Beggar's Opera* shows the audience of recognizable notables taking delight in Gay's send-up of the Italian opera which George and Caroline prominently patronized, of Robert Walpole, who was known sarcastically as 'the Queen's Minister', and of contemporary courtiers generally—who get lampooned as sycophants, whoremongers, and thieves. Macheath, the highwayman hero of *The Beggar's Opera*, swearing his eternal fidelity to his beloved Polly, asks rhetorically: 'Is there any force that could tear me from thee?', to which he answers himself satirically: 'You might sooner tear a pension out of the hands of a courtier.'[4] Frederick, Prince of Wales, who aligned himself with 'The Country Party', conspicuously opposed the Italian opera as foreign and, with Augusta at his side, publicly championed the English-language drama, making theatre-going a potentially subversive performance in itself.[5] Sharing the satirical stance of *The Beggar's Opera*, they symbolically positioned themselves among the onstage audience famously depicted by Hogarth.[6] Gay's ballad opera gave politically minded spectators a lot to think about. It dramatized the conspiratorial relationship between the corrupt Mr Peachum, who stands in for Walpole in several satirical jibes, and a Lady Macbeth-like Mrs Peachum, who commands him regally to perpetrate his nefarious deeds. Urging him to arrange the judicial murder of Macheath, she speaks in the modern, Orwellian language that those in power use to express their preferences obliquely, inciting their agents to action while preserving deniability: 'The thing, husband, must and shall be done. For the sake of intelligence we must take other measures.'[7] Himself a master of innuendo, Gay lets the insinuation linger. How many of the wits in attendance at the premiere of *The Beggar's Opera* might have heard Mrs Peachum's use of the first-person plural pronoun as the royal 'We'? No one will ever know for certain, but the correct answer is probably on the high side of the scale between nobody and everybody. When anti-government incendiaries burnt Walpole's effigy, they also torched the image of Caroline, not George, alongside that of the hated minister.[8]

Before his untimely death, Frederick, an amateur playwright himself, and Augusta staged private theatricals at home. They cast courtiers and sometimes their children in featured roles. These intimate but well-publicized domestic performances likely resembled the production that Hogarth depicted of John Dryden's *The Indian Emperor* with ten-year-olds cast in the leading roles, which John Conduitt, Master of the Royal Mint, and his wife put on in their home in 1731, with George and Caroline's children joining the audience (fig. 7.02).[9] The precise politics of the Conduitts' choice of Dryden's play are unclear, but Hogarth's conversation piece captures the poignancy of the children gravely acting Dryden's heroic roles, mastering adult passions in the cause of dynastic and imperial supremacy. Frederick and Augusta's private staging of Joseph Addison's *Cato*, the most inexhaustible allegory of British and American patriotism in the eighteenth century, was even more pointed. Frederick himself drew out the political meanings by writing a new prologue and epilogue. But even *Cato* paled in partisanship next to James Thomson and David Mallet's *Masque of Alfred*, set to music by Thomas Arne, which Frederick commissioned and produced at Cliveden House in 1740 to honour Augusta on her birthday. A paean to English Liberty under benevolent monarchy, the plot of the masque shows Alfred the Great driven from the throne by an invasion led by Ivar the Dane. Alfred wanders incognito through the forests of Somersetshire until he is able to rally a band of loyal retainers to defeat Ivar, reunite with his loving queen,

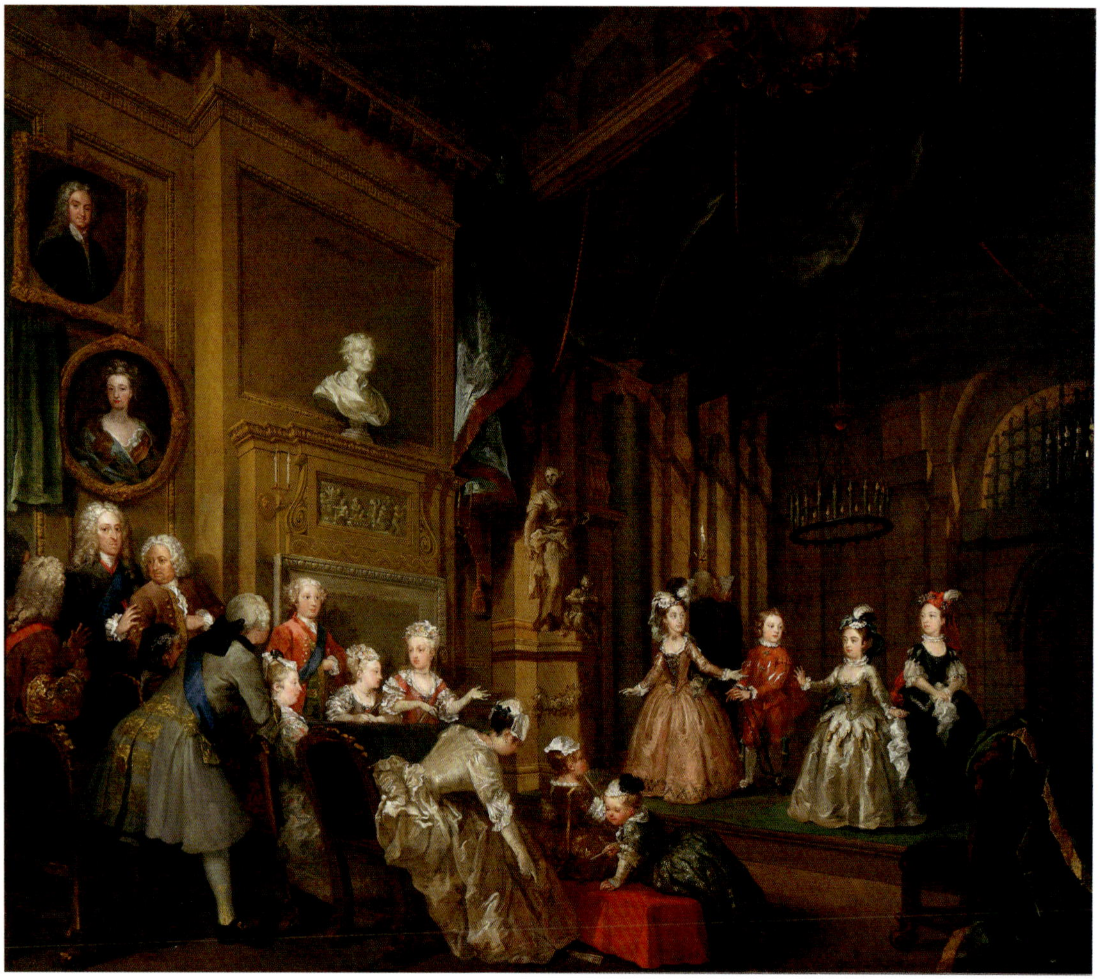

FIG. 7.02 William Hogarth, *A Performance of 'The Indian Emperor or The Conquest of Mexico by the Spaniards'*, 1732–35, oil on canvas, 51½ x 57¾ in. (130.8 x 146.7 cm). Private collection

and resume his lawful reign, banishing tyranny from the Anglo-Saxon kingdom.[10] This tendentious oracle, alas, misspoke. Like the shocking onset of George III's madness, Frederick's sudden death at the age of forty-four highlights the fragility of dynastic succession and sheds light on the prominence of children in the private theatricals, in the painted conversation pieces of the time, and in the unwritten but unalterable script that the three Princesses performed. Fertility presented them with a fundamental obligation of governing visibly: producing a legitimate heir—and as many spares as possible. This volume is a record of their many other performances in that constitutional genre, whereby they made the court their stage.

NOTES

1. Richard Steele, *The Tatler*, ed. Donald F. Bond, 3 vols. (Oxford: Oxford Univ. Press, 1987), 2: 423 (No. 167, Thursday, 4 May 1710).
2. Charles Gildon, *The Life of Thomas Betterton, the Late Eminent Tragedian* (London: Printed for Robert Gosling, 1710), 10.
3. Walter Bagehot, *The English Constitution*, ed. R. H. S. Crossman (1867; Ithaca, NY: Cornell Univ. Press, 1966), 249.
4. John Gay, *The Beggar's Opera*, ed. Edgar V. Roberts (1728; Lincoln, NE: Univ. of Nebraska Press, 1969), 28.
5. John L. Bullion, 'To play what game she pleased without observation: Princess Augusta and the Political Drama of Succession', in *Queenship in Britain, 1660–1837: Royal Patronage, Court Culture and Dynastic Politics*, ed. Clarissa Campbell Orr (Manchester: Manchester Univ. Press, 2002), 210.
6. Romita Ray, '*Veluti in speculum*: Spectatorship in Hogarth's *The Beggar's Opera*', in *'Among the Whores and Thieves': William Hogarth and The Beggar's Opera*, ed. David Bindman and Scott Wilcox (New Haven and London: Yale Univ. Press, 1997), 35–40.
7. Gay, *The Beggar's Opera*, 26.
8. Baron John Hervey, *Memoirs of the Reign of King George the Second: From His Accession to the Death of Queen Caroline*, ed. J. W. Croker, 2 vols. (London: John Murray, 1848), 1: 203. *Fog's Weekly Journal*, 14 April 1733.
9. Ronald Paulson, *Hogarth*, Vol. 2: *High Art and Low, 1732–1750* (New Brunswick, NJ: Rutgers Univ. Press, 1992), 1–3.
10. *The Masque of Alfred* was revived in the colonies as the first official production of the College of Philadelphia to celebrate the Christmas holiday of 1756–57. See Jason Shaffer, *Performing Patriotism: National Identity in the Colonial and Revolutionary American Theater* (Philadelphia, PA: Univ. of Pennsylvania Press, 2007), 110–13.

7.03

STEPHEN DEMAINBRAY (1710–1782)
Camera obscura
Before 1753
Oak, brass, and glass
7⅝ x 8¼ x 16⅞ in. (19.5 x 21 x 43 cm)
Science Museum, London, 1927–1139

In 1768 the natural philosopher and astronomer Stephen Demainbray became the first superintendent of the Observatory in Kew Gardens where George III's and Charlotte's collections of 'philiosophical instruments' were kept. His own lecturing apparatus accompanied him to Kew and was merged with the royal collection. The catalogue compiled of Charlotte's artefacts lists two camera obscuras, one of which had formerly belonged to Demainbray. Augusta's Privy Purse accounts contain references to payments made for the repair of a camera obscura, and two were included in the Christie's sale of her effects (1–3 February 1773). In the wake of Sir Isaac Newton's experiments involving the refraction of light undertaken for Caroline in 1716, scientific experiments, especially those involving light, contributed to the spectacle of court occasion. As well as providing entertainment, camera obscuras were used in the educational programmes devised for the royal children.

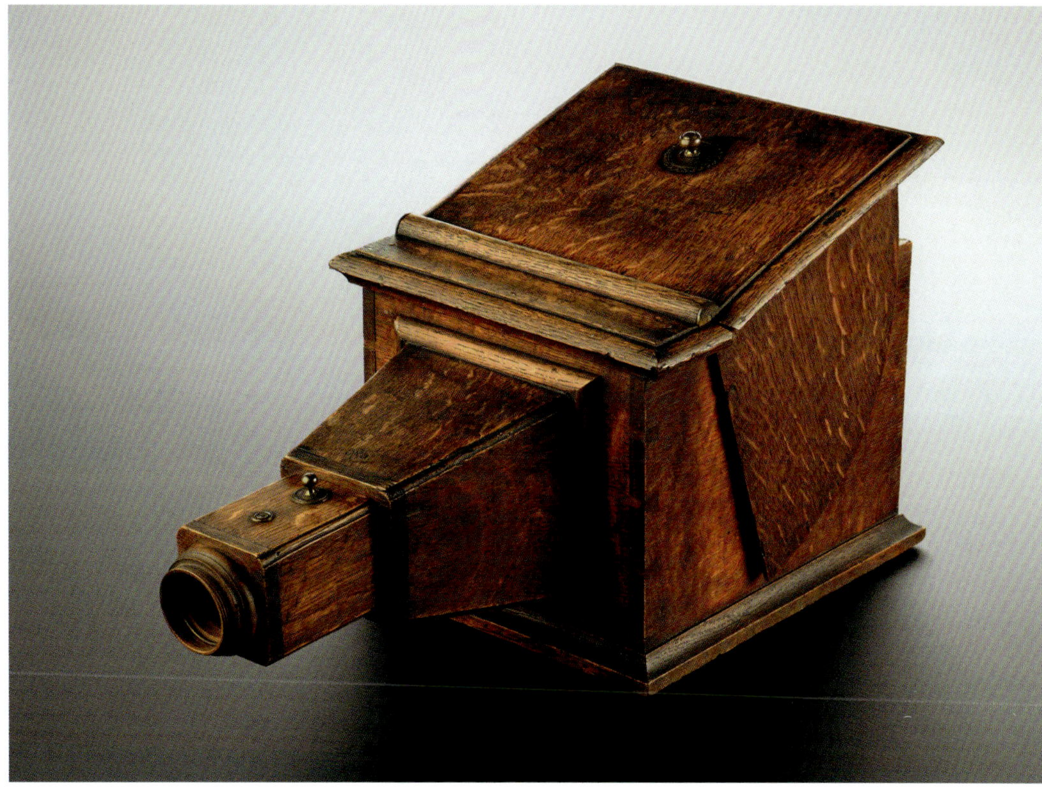

7.03

7.04†

J. MERIGOT (active 1772–1791), engraver, after **MATTHEW COTES WYATT** (1777–1862)
A Jubilee at Frogmore
c. 1809
Etching and aquatint
14¾ x 22¼ in. (37.3 x 56.3 cm)
Royal Collection Trust, RCIN 700895

The royal Jubilee of 1809, the first of its kind, celebrated the beginning of George III's fiftieth year as king, in more than six hundred locations in the British Isles and in its colonies. The fête held at Frogmore, Charlotte's favourite residence, and designed by her and her daughters, was attended by nearly 1,200 people from the local community. There were fireworks and illuminations, an 'elegant supper', and an elaborate water pageant on the lake, celebrating the triumph of Britannia. The temple on the island, a temporary structure made to house a large back-lit projection, was designed by Princess Elizabeth, and executed by the artist James Wyatt. The celebration was recorded in a watercolour by James Wyatt's son, Matthew Cotes Wyatt.

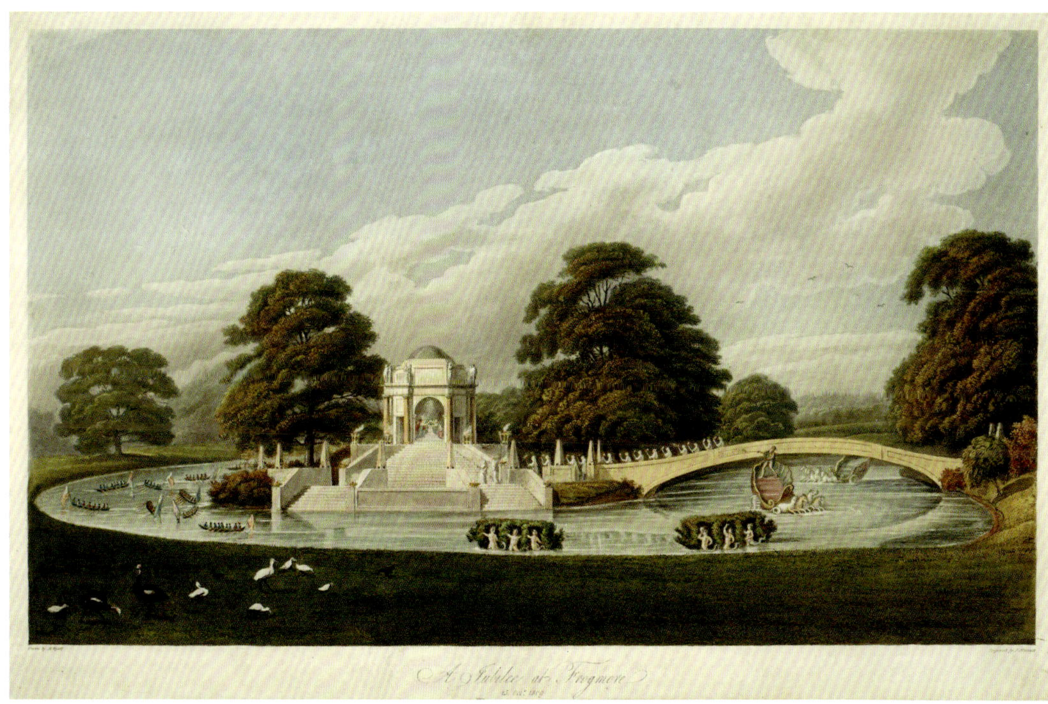

7.04

* = YCBA only
† = HRP only

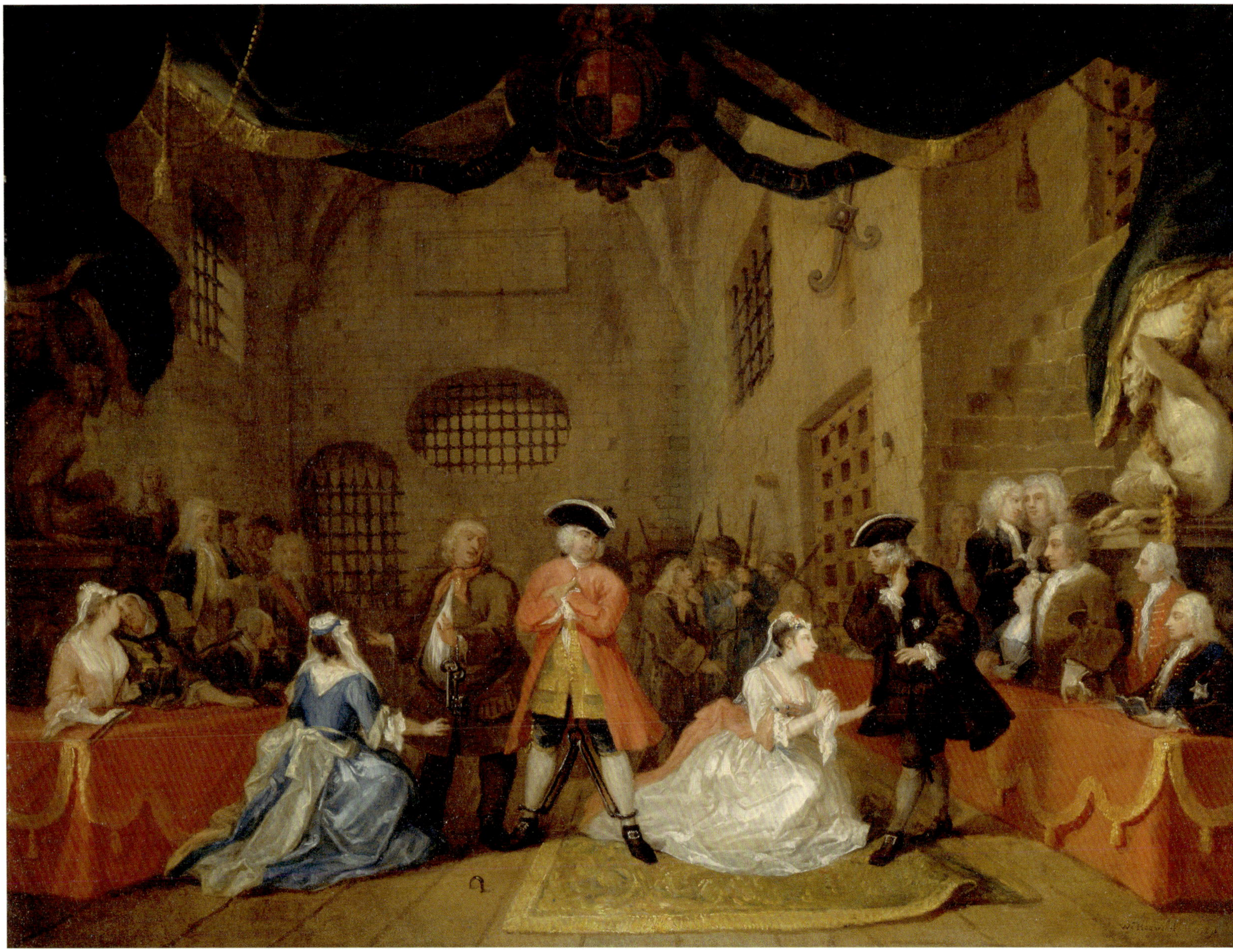

7.05

THEATRICAL PERFORMANCE

7.05*

WILLIAM HOGARTH (1697–1764)
The Beggar's Opera
1729
Oil on canvas
23 ¼ x 30 in. (59.1 x 76.2 cm)
Yale Center for British Art, Paul Mellon Collection, B1981.25.349

John Gay's *Beggar's Opera* was first produced by John Rich at the Lincoln's Inn Fields Theatre in January 1728. It was an unprecedented success, running for sixty-two performances in its first season. With its English ballad interludes and its setting in London's criminal underworld, the play challenged the vogue for Italian opera that Hogarth had satirized in his earliest engravings. Hogarth's painting depicts the climax of the play, set in Newgate Prison. Macheath, a gentlemanly highwayman bigamist, stands in chains at centre stage as his two wives, Polly Peachum and Lucy Lockit, appeal against his conviction. Hogarth painted five versions of the scene; this is the last and marks the culmination of his rapid development as a painter.

Gay sought Caroline's patronage for many years, dedicating his play *The Captives* to her in 1724, and *Fables* to her youngest son, William Augustus, in 1725. In 1727 she offered him the post of Gentleman Usher to the infant Princess Louisa, which he turned down as too demeaning. Political difference would lead to the cancelling of a commission to William Hogarth of a group portrait of the royal family in about 1731–32. Two preparatory oil sketches survive: one in the Royal Collection (RCIN 401358), the other in the National Gallery of Ireland (NGI. 126).

7.06*

ROBERT DODD (1748–1815), engraver, after
WILLIAM HOGARTH (1697–1764)
'The Indian Emperor' or 'The Conquest of Mexico'
1792
Etching and engraving on paper
17¹¹⁄₁₆ x 21½ in. (44.9 x 54.6 cm)
Yale Center for British Art, Gift of David Bindman, B2000.2.1

John Dryden's play *The Indian Emperour, or the Conquest of Mexico by the Spaniards, being the Sequel of The Indian Queen*, first performed in the spring of 1665, told the inherently political story of Hernán Cortés and the expansion of European power to North America. This performance, in which the actors are children, took place in the home of John Conduitt and his wife Catherine. The occasion was recorded by William Hogath in about 1732–35. The host and his wife are modestly represented as portraits, and their most distinguished family connection, Sir Isaac Newton, appears as a portrait bust. George II and Caroline's younger children take their place in the audience. The prompt, who can be seen in the shadows at the back of the image, was the natural philosopher Dr John Desaguliers, one of Caroline's protégés. The performance was such a success that it was repeated before the King and Queen at St James's Palace on 27 April 1732.

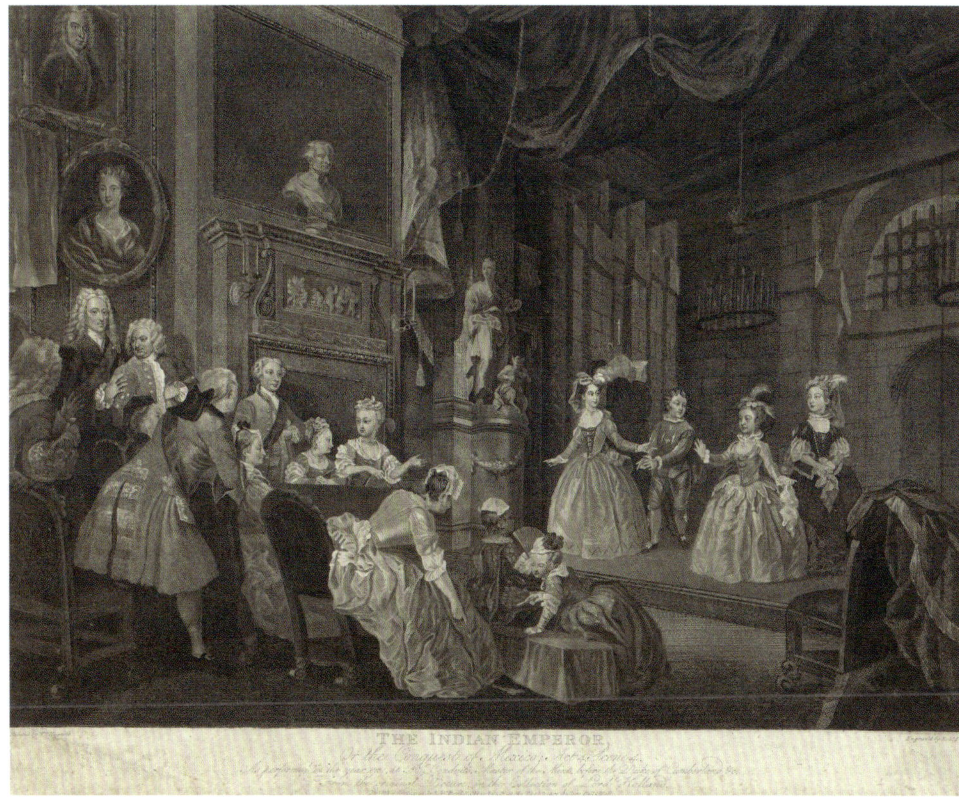

7.06

7. 07*

UNKNOWN ENGRAVER
The Stage Medley representing the Polite Tast [sic] of the Town & the Matchless Merits of Poet Gay Polly Peachum & Capt. Macheath
1728
Etching on paper
14⅛ x 10¼ in. (36 x 26.1 cm)
British Museum, 1868,0808.3527

This satire on Gay's *Beggar's Opera* is made up of a medley of scenes from the opera. They are presented as if cut from newspapers and journals of the day. Some of the journals are identified by name: 'To the Tune of the Soldier and ye Sailor' is identified with the 'Daily Journal April 19th, 1728', and 'To Polly Peacham' with the 'Daily Journal April 10 1798'. Identifications of the actors are added in pen and ink by Edward Hawkins, from whom this satire was acquired by the British Museum: 'Gay / Miss Fenton i.e. Dss Rolton / Tho. Walker Orig. Macheath & Polly / Apl. 1728'.

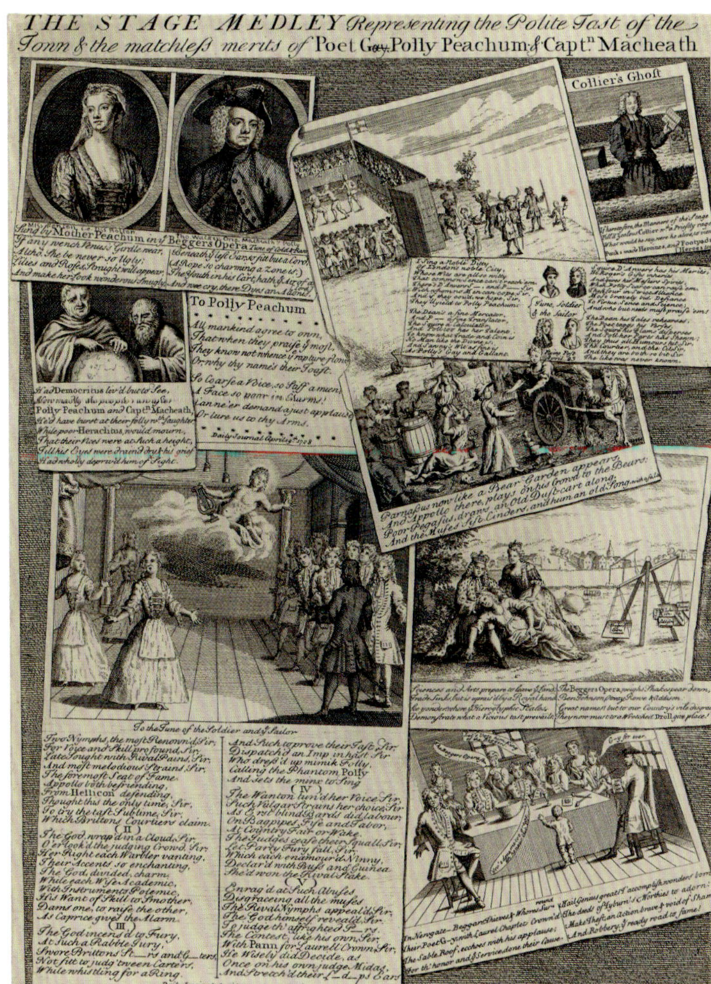

7.07

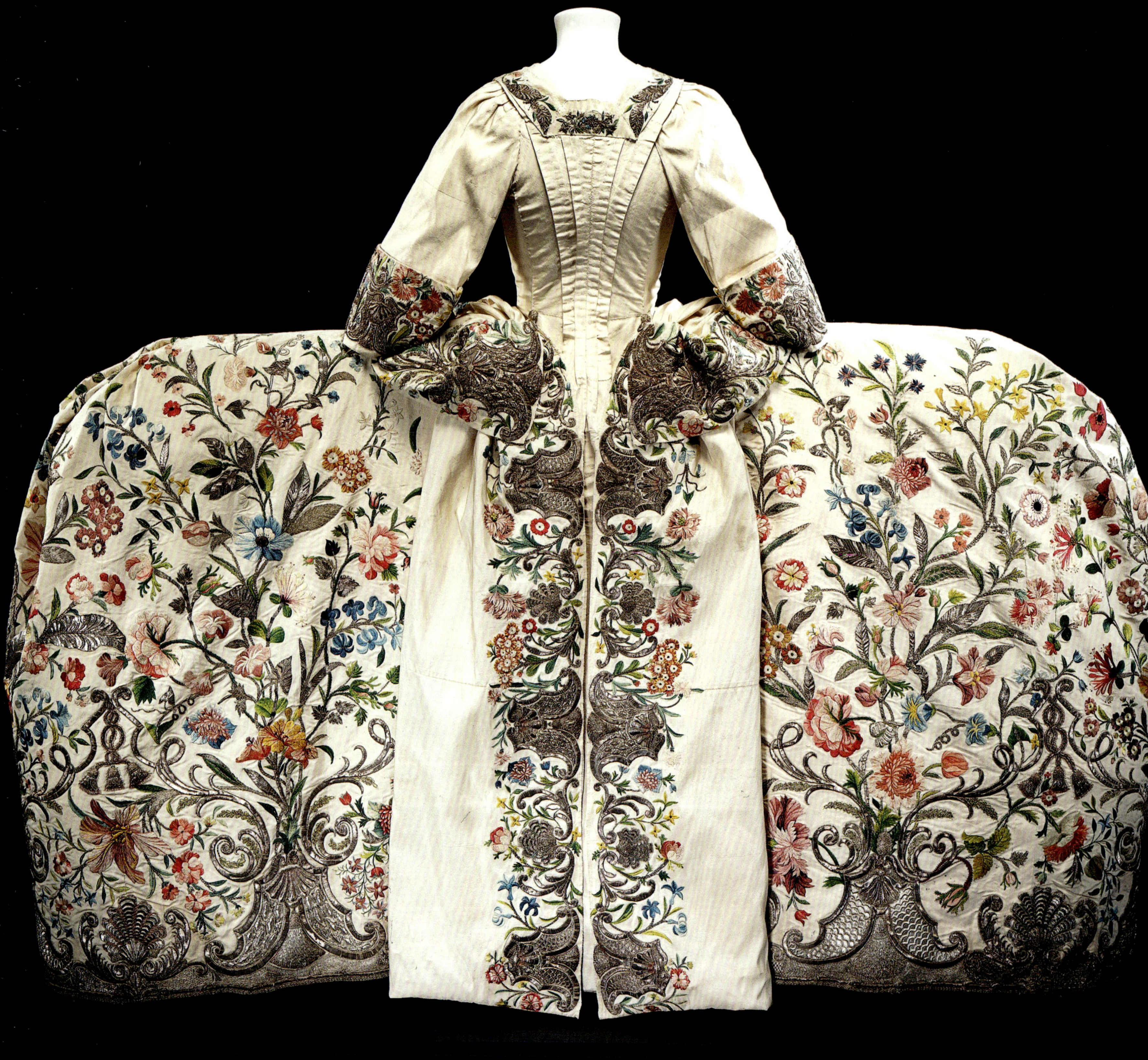

Being at Court

Aileen Ribeiro

If the past is a foreign country, then royal courts with their hierarchies, tedium, and stiff and elaborate clothing might seem especially alien in the twenty-first century. In *Sartor Resartus* (1835) Thomas Carlyle suggests the absurdity of reverence for the clothing of 'pompous Ceremonials . . . Royal Drawing-rooms, Levees, Couchees', and wonders 'whether to laugh or to weep'.[1] But in eighteenth-century London, being at court—the centre of power, influence, and fashion—was essential for the ruling elite, even in a constitutional monarchy. The court was not just a physical place where the monarch and the royal family met the nobility, courtiers, and the political caste, but it also encompassed a code of behaviour and costume specific to the occasion, signifying exclusive membership of a charmed circle. Performances of monarchy (formal meetings of the ruler and court emerged during the reign of Charles II) had become more regular and strictly organized by the early eighteenth century. They included Levees—men-only occasions attended by the nobility and ministers of the crown—and Drawing-Rooms, where both men and women appeared, and which were generally regarded as more brilliant occasions, especially those which celebrated royal birthdays and culminated in a ball.[2] Courts provided opportunities for political business (therefore regular attendance was important), and for the presentation of young scions of the aristocracy to continue the tradition of physical allegiance to the ruler and for them to receive acknowledgment in their turn. If the King spoke more than once to an individual at a Levee or a Drawing-Room it was regarded as a sign of royal favour, but conversation was usually limited to a few words, small talk, smiles, and courtly gestures such as bows and curtsies; everyone stood in a circle waiting to be approached by the monarch or his consort. For it was at court that acts of seeing and being seen took place. As Pierre Bourdieu remarks: 'a class is defined as much by its *being-perceived* as by its *being*.'[3] And one way in which preferential access to the monarch at court could be signalled was via distinctive dress. The writer and politician Richard Steele considered that society being 'form'd of different Ranks', should be 'distinguished by Habits, that all proper Duty or Respect might attend their Appearance'.[4]

Clothing is more than just a thing in itself; it also reflects national traditions and social customs. It is evidence of how a society regards itself and the importance it attaches to status and ceremony, which themselves underline political and social systems. As fashion follows power, it is unsurprising that in Europe it was in France, under Louis XIV, that the first court dress for women, the *grand habit*, was initiated. This evolved from the formal fashion of the 1670s into a style comprising a rich and decorated, heavily boned, off-the-shoulder, and pointed bodice with short sleeves (*corps de robe*), skirt, and a train (*bas de robe*), pulled back at the sides with jewelled clasps. Such a dress was worn with hoops (which first appeared in the early eighteenth century), and became a pompous and unyielding costume that lasted until the French Revolution, having been adopted by most European monarchies. In England this costume, known as a 'stiff-bodied gown', and rarely depicted in art, was worn only on the grandest of royal occasions such as royal marriages, as it was in 1736 by Augusta of

Detail of p. 175, cat. 8.06

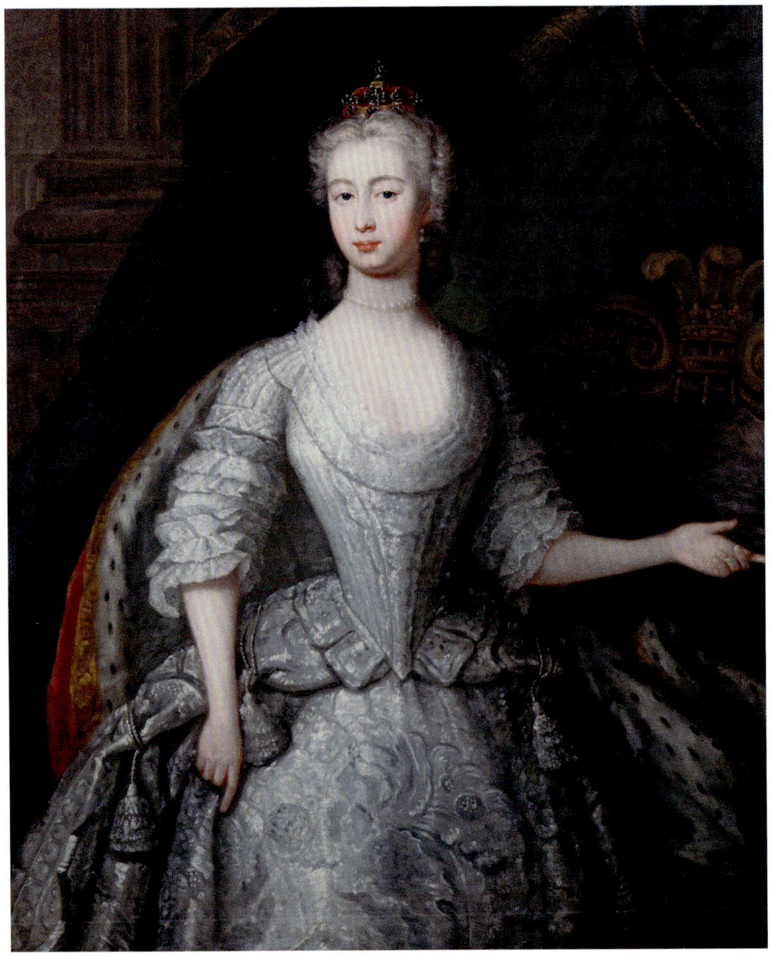

FIG. 8.01 Charles Philips (1708–1748/49), *Augusta of Saxe-Gotha, Princess of Wales*, c. 1736, oil on canvas, 50 ¼ x 40 in. (127.6 x 101 cm). National Portrait Gallery, London

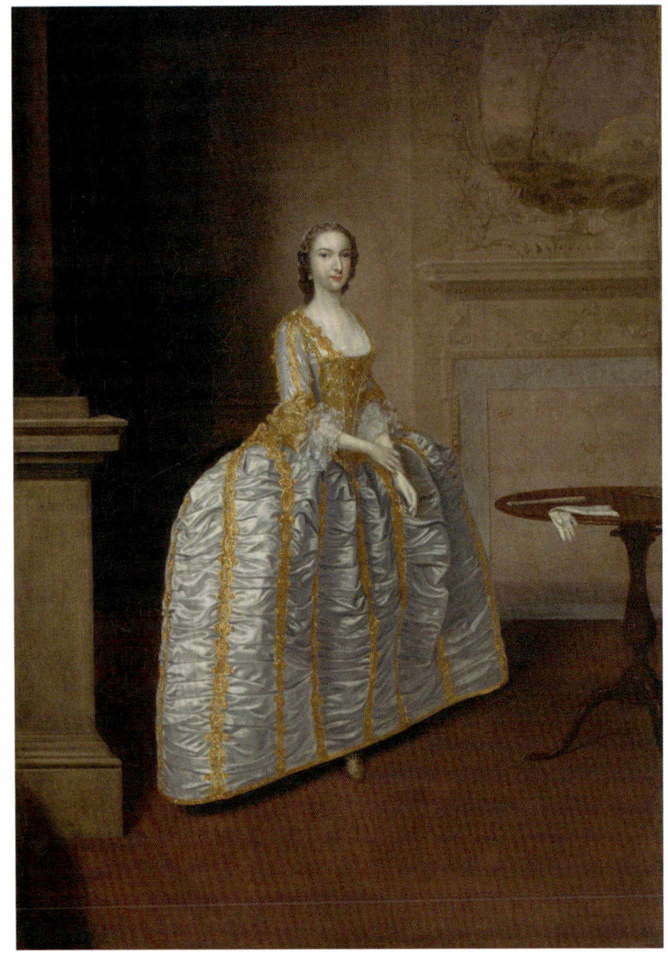

FIG. 8.02 Arthur Devis, *Alicia Maria Carpenter, Countess of Egremont*, 1745, oil on canvas, 24⅛ x 16⅝ in. (61.3 x 42.2 cm). Fine Arts Museums of San Francisco, Palace of the Legion of Honor, Mildred Anna Williams Collection

Saxe-Gotha for her marriage to Frederick, Prince of Wales (fig. 8.01). In the portrait made by Charles Philips to mark the occasion, she is shown in cloth of silver, a back-fastening bodice with lace ruffles (*manchettes*) attached to the sleeves, a skirt worn over the bell-shaped hoop of the 1730s, and a train attached to the waist. This is clothing which was literally stiff as well as being 'stiff-bodied'; that is, it had a fiercely boned bodice and a hooped skirt, which was difficult to control or move in. Any physical awkwardness in appearance (which was the case with Augusta) was only exacerbated by this cumbersome costume.[5]

In the early Hanoverian period, men at court wore clothing in the current fashionable style but made of the richest fabrics, along with the insignia of knightly orders. Women's court dress was the formal mantua, originally a loose gown with looped-up back drapery, which by the reign of George I comprised an open robe with a stylized train, a decorated stomacher at the bodice front, and a hooped skirt, all of rich silks such as brocades, damasks, and embroideries. Eagle-eyed observers might notice when a dress was not new, nor sufficiently expensive, and to keep up appearances less wealthy women borrowed lace (for the requisite triple-sleeve ruffles and lappets—pendants worn in the headdress) and expensive jewellery, as Mary Granville, Mrs Delany sometimes did. In her lengthy descriptions of court dress, 'rich but not showy' and 'genteel' were her terms of approval. But she could occasionally be waspish about embroidery that was too vulgar, noting in 1739 that one woman's over-elaborate design was 'much properer for a stucco staircase than the apparel of a lady'.[6] By the 1730s, as Delany records, court dress might be shown off to relatives and friends, and Charles Philips's *A Tea Party at Lord Harrington's House, St James's* (1730) shows a rare example of court dress mingling with less formal everyday clothes. In theory, those attending court were encouraged to wear 'British materials' and not 'French stuffs', but this seems often to have been ignored.[7] Taste was the guiding rule for the selection of fabrics, and there is no real evidence that either patriotic feeling or loyalty to the crown determined the choices that were made.

As the century progressed, the train became narrower, pushed to the centre back, and—as the hoop became much larger—consequently far less visible, as in the portrait of 1745 by Arthur Devis showing the sixteen-year-old Alicia Maria Carpenter at her first court presentation in a ruched white silk dress trimmed with gold lace (fig. 8.02). Such a dress required the wearer's expert handling and perfect deportment, accomplishments taught by a dancing master, who would also give lessons on the elegant gestures of the hands and arms and how to deploy a fan with grace.

From the 1760s, as the mantua became increasingly remote from fashion, another 'full dress', the *sacque* or *robe à la française*, with double pleats falling from the shoulders at the back, and with elaborate decoration, was accepted as court wear on less formal occasions; Charlotte of Mecklenburg-Strelitz can be seen in this dress, in a portrait by Nathaniel Dance made in about 1769 (fig. 8.03). The Queen wears a beautiful Lyons silk trimmed with gold lace and net. When by the 1780s the *sacque* in its turn was no longer fashionable, it was adopted for the more formal court occasions and worn over a large hoop—as we can see in Thomas Gainsborough's portrait of Charlotte of 1781 (fig. 8.04). This shows her dressed in a shimmering *sacque* of white gauze spangled in gold over white silk. This style of dress—'too antiquated for any other place than court', claimed the *Fashionable Magazine* in 1786—was the type Sarah Siddons reluctantly had to wear when reading to the royal princesses in 1785: 'a Sacque or Negligee with a hoop, treble ruffles and Lappets, in which costume I felt not at ease'.[8] Similar reservations about this dress were expressed by Frances (Fanny) Burney when employed in the Queen's household the following year, but she had to accept that at St James's Palace, although she 'had nothing to do with the Drawing Room', it was 'the Etiquette of my place'.[9] Yet while the *sacque* could on rare occasions be worn outside court, the mantua was, as the American Abigail Adams remarked, 'a Court dress you cannot make use of anywhere else'. She complained about having to wear at a Drawing-Room in 1785 what the *Magazine a la Mode* had called in 1777 a 'Robe de Cour but more properly the *Royal Robe*, because her Majesty generally appears in this dress',[10] and asked her mantua-maker (dress-maker) to 'let my dress be elegant but plain as I could possibly appear with decency.' The resulting dress was still elaborate, however, made of white silk trimmed with white crape, lilac ribbon, and imitation needle lace.[11]

As clothing in general became simpler in style and plainer in fabrics—a trend encouraged by the politics of equality promoted by the French Revolution—the absurdity of the vast-hooped Royal Robe became more pronounced, with the increasingly high waistline of the 1790s making the dress, even when made of

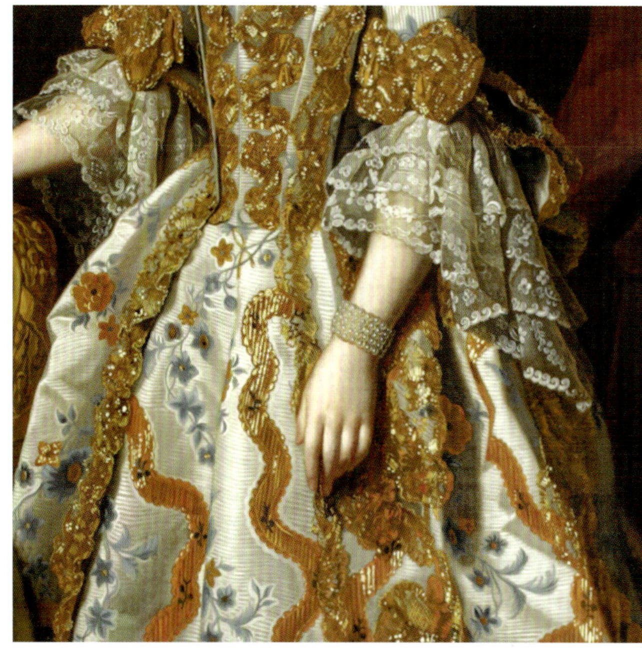

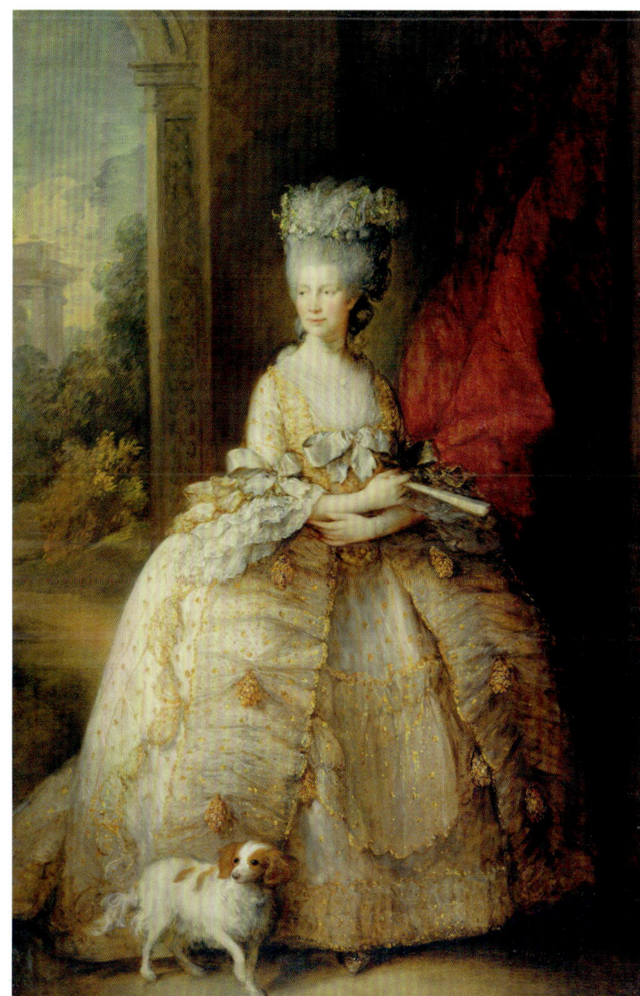

FIG. 8.03 Sir Nathaniel Dance-Holland, *Queen Charlotte* (detail), 1769, oil on canvas, 100 x 59 in. (254 x 150 cm). National Trust, Uppark (see p. 69, fig. 3.09)

FIG. 8.04 Thomas Gainsborough, *Queen Charlotte*, 1781, oil on canvas, 94 x 62½ in. (238.8 x 158.7 cm). Royal Collection Trust

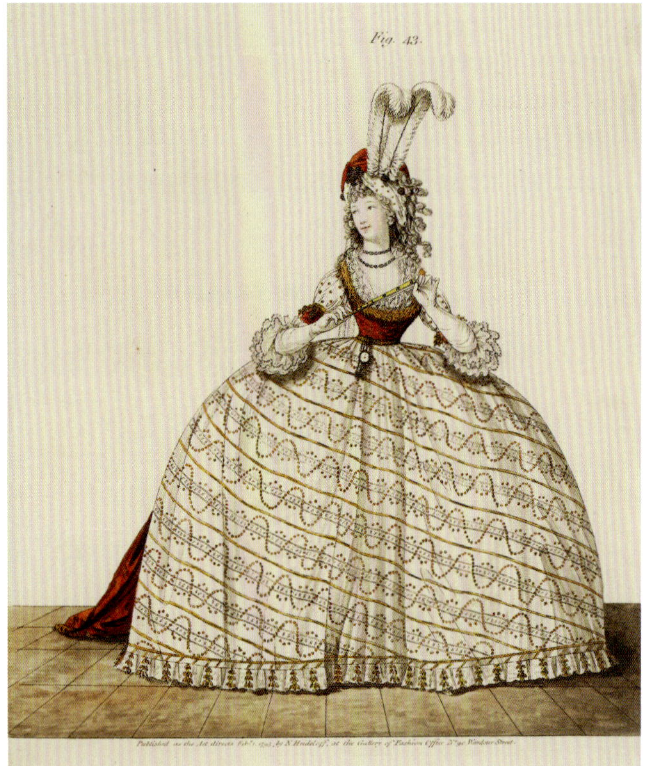

FIG. 8.05 Nicolaus van Heideloff, 'Court Dress, made by Madame Beauvais, Milliner to her Majesty. February 1795', hand-coloured aquatint engraving, from *The Gallery of Fashion* (London: 1794–95), vol. 1, plate XLIII, 9⅝ x 7¾ in. (24.9 x 19.7 cm). Yale Center for British Art, Paul Mellon Collection

fashionable fabrics, look out of proportion. This can be seen in a design for a court dress of white silk crape 'embroidered in gold foil and spangles' with a bodice and train of red velvet, made 'by Madame Beauvais, Milliner to her Majesty', to be worn on the Queen's birthday in January 1795 (fig. 8.05). Such designs, although beautifully illustrated, were certainly bizarre; and Niklaus von Heideloff, publisher of the fashionable journal *The Gallery of Fashion*, had to insist that the dresses were 'not *imaginary*, but *really existing* ones'.[12] The archaic and unchanging nature of this court dress, only abandoned after Charlotte's death, was out of step with the rapidly changing fashion for clothes which, by the end of the eighteenth century, were increasingly made from non-elite fabrics such as cotton and woollen cloth, reflecting the political and social changes taking place in Great Britain.

NOTES

1. Thomas Carlyle, *Sartor Resartus* (1835; London: J. M. Dent, 1984), 45. By 'Couchees' he refers to the formal undressing of the King and Queen when they retired at night, a custom introduced at the court of Louis XIV.
2. Drawing-Rooms were also held by Queens Consort, the Prince of Wales, and other senior members of the royal family.
3. Pierre Bourdieu, *Distinction: A Social Critique of the Judgement of Taste*, trans. Richard Nice (Cambridge, MA: Harvard Univ. Press, 1984), 483.
4. Richard Steele, *The Spectator*, no. 478, 8 Sept. 1712.
5. Lord Hervey remarked in 1736 on Augusta's 'ill-made' figure, commenting that 'in spite of all the finery of jewels and brocade', she could achieve only 'an ordinary air'. See John Hervey, *Lord Hervey's Memoirs*, ed. Romney Sedgwick (London: B. T. Batsford, 1963), 117.
6. Lady Llanover, ed., *The Autobiography and Correspondence of Mary Granville, Mrs Delany*, 6 vols. (London: Richard Bentley, 1861–62), 2: 28.
7. *The Magazine a la Mode or Fashionable Miscellany. Particularly adapted to the Use of People of Fashion of Both Sexes* (London: J. Wenham, 1777), 4. The wearing of 'French stuffs' was generally disapproved of throughout the eighteenth century, but especially at times when the country was at war with France. Such was the case when at Princess Augusta's birthday celebration on 19 Nov. 1748, the Prince of Wales declared himself 'greatly displeased' to see 'the nobility and gentry at Leicester-House [in] . . . French stuff' (*The Gentleman's Magazine*, Nov. 1748, 521). Members of the royal family generally embraced this ambition, wearing dress made from textiles sourced locally or from within the territories overseas in which the British had an interest. In 1735, Caroline wore a dress made of silk grown within a trading and colonization initiative in the American colony of Georgia, with the intention of promoting the venture.
8. William van Lennep, ed., *Reminiscences of Sarah Kemble Siddons, 1773–1785* (Cambridge, MA: Harvard Univ. Press, 1942), 21.
9. Peter Sabor, ed., *Court Journals and Letters of Frances Burney*, Vol. 1: *1786* (Oxford: Clarendon Press, 2011), 50.
10. *The Magazine a la Mode* (1777), 4.
11. Charles Francis Adams, ed., *Letters of Mrs Adams, the Wife of John Adams*, 4th ed. (Boston: Wilkins, Carter & Co., 1848), 255.
12. Unpaginated introduction to Niklaus von Heideloff's *Gallery of Fashion*, Vol. 2: *1795–96*.

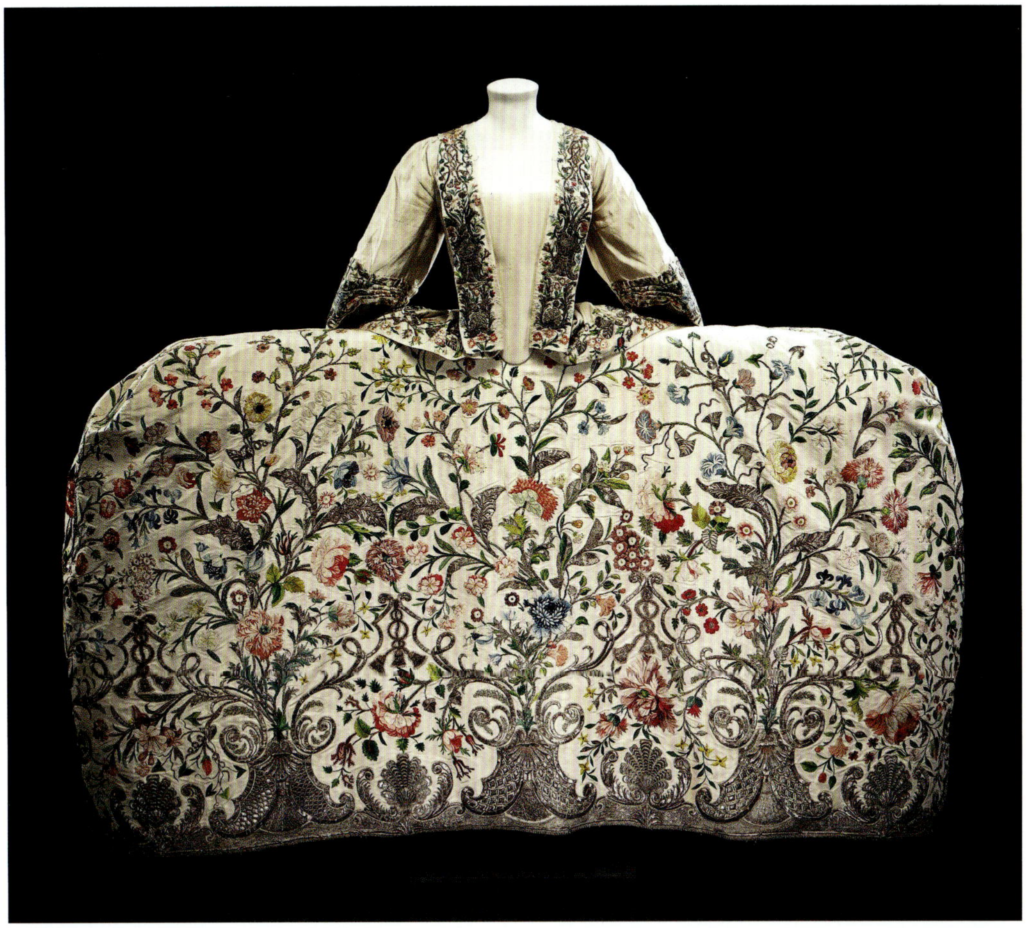

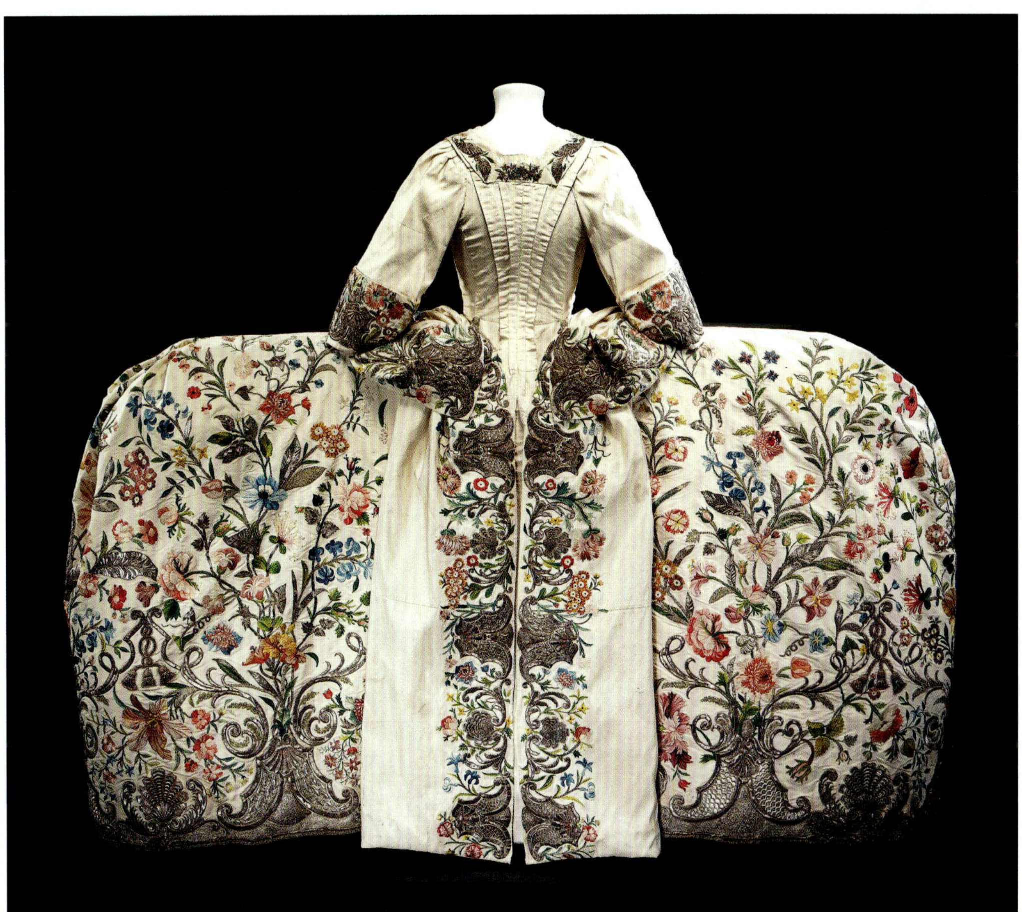

8.06

8.06*

UNKNOWN MAKER, British
Court mantua, comprising robe, petticoat, and stomacher
1740–45
Silk embroidered with coloured silk and silver thread
66⅞ x 70⅞ x 31½ (170 x 180 x 80 cm)
Victoria and Albert Museum, T.260&A–1969

This mantua is associated with Isabella Courtenay who married Dr John Andrew in Exeter Cathedral on 14 May 1744. As was customary, Mrs Andrew probably wore the ensemble for her presentation at court after her marriage. Mary Granville, Mrs Delany, describes dresses worn at Norfolk House, the court of Frederick, Prince of Wales, and Princess Augusta in February 1741. The Duchess of Queensbury's clothes pleased her best. Her mantua of white silk was embroidered with a design of rolling hills covered with thickets of trees and flowers, 'nastersians, ivy, honeysuckles, periwinkles, convolvuluses', and 'many of the leaves finished with gold'.

* = YCBA only
† = HRP only

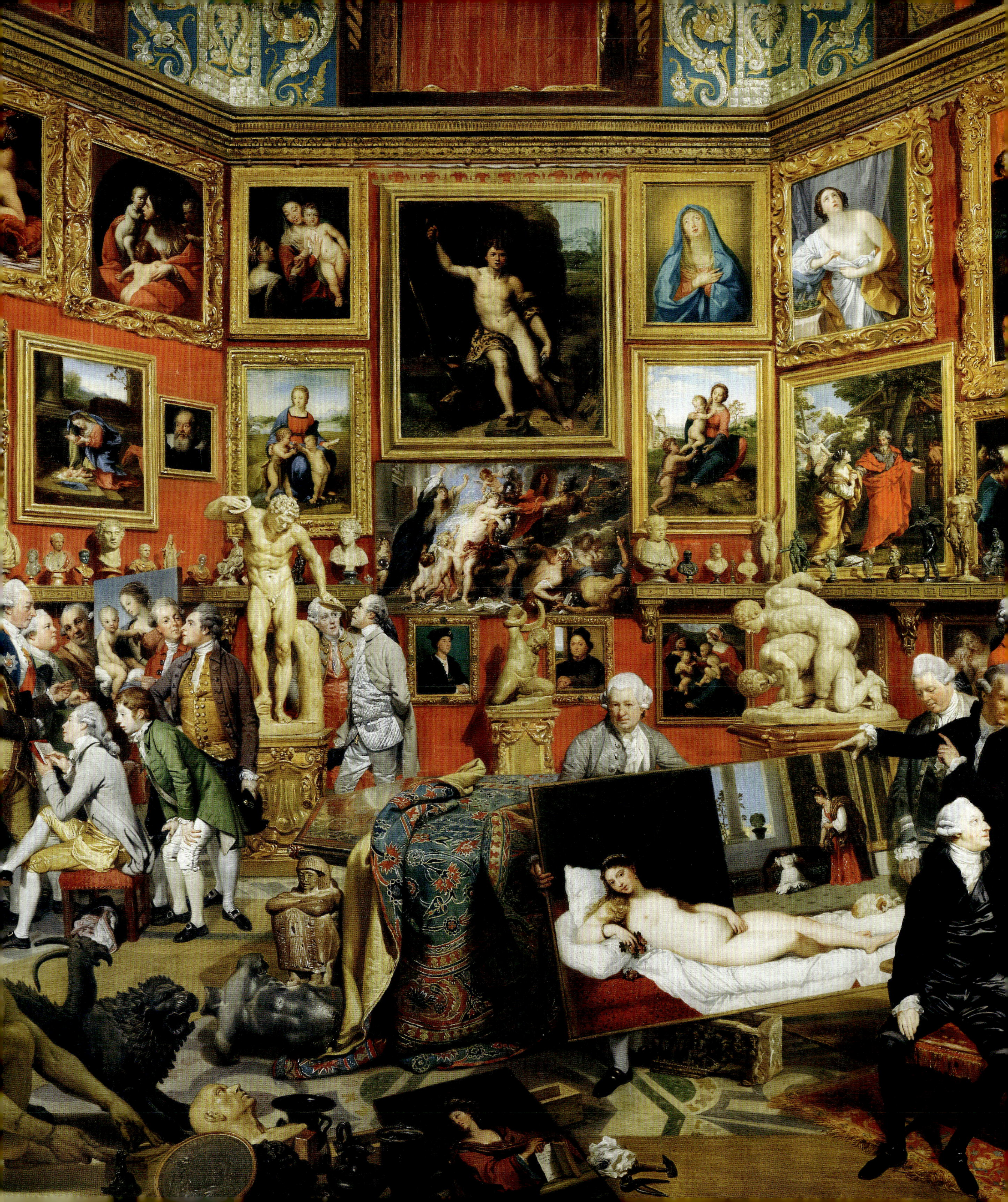

Rediscovery, Collecting, and Display

Robyn Asleson

Painting played a key role in the court life of the royal Princesses Caroline of Ansbach, Augusta of Saxe-Gotha, and Charlotte of Mecklenburg-Strelitz, often serving as the means by which each aligned herself with a specific set of associations, both aesthetic and personal. Paintings commissioned and collected by the three women reflect their individual tastes and interests, but they also attest to the predominant concerns of their courts, which varied from generation to generation.

Caroline had experienced the way that court life in Germany was played out against a backdrop of ancestral portraits that expressed the continuity of hereditary authority. Finding no equivalent in England, she located and displayed historical royal portraits to enhance the Hanoverian dynasty's aura of legitimacy and to assert its 'Englishness'.[1] Her efforts to promote representations of England's past monarchs reflect her longstanding fascination with genealogy as well as her sense of the lessons to be learned from the past. 'There is . . . a satisfaction to see the portraits of eminent persons dead and gone', she remarked in 1735: 'Their examples may be of use to those who are living to excite them to like actions.'[2]

Caroline first began to assemble a collection of royal portraits at her country retreat, Richmond Lodge. In 1724 the engraver and antiquary George Vertue had observed 'at Richmond in the princess [*sic*] house' a portrait of Henry VIII's elder daughter Mary I, a half-length copy of the iconic three-quarter-length painting of 1554 by the Netherlandish artist Anthonis Mor van Dashorst (fig. 9.01).[3] Caroline's version, enriched by a flourish of intricately patterned red and gold cloth in the foreground, may have formed part of the 'set of kings' that once hung in Culford, the Suffolk country seat of Charles, first Earl of Cornwallis. Caroline reportedly 'begged' Cornwallis for the fifteen paintings on panel, which filled gaps in the dynastic display she had begun to assemble.[4] Although the paintings possessed little artistic distinction, their quaint medieval style would have appealed to her antiquarian taste.

At a time when Stuart loyalists were challenging the legitimacy of Hanoverian rule, the portrait of Queen Mary might have provided a useful object lesson on the perils of Catholic rule. Other portraits in the collection served a more exemplary function, beginning with the early Plantagenet ruler Edward III, whose venerable appearance is enhanced by a bejewelled crown and ermine robe (fig. 9.02). Edward's defeat of the French at Crécy in 1346 had made him the archetypal warrior-king and the subject of poems and sermons exhorting Hanoverian rulers to equivalent glory.[5] The series of monarchs included an unbroken chain of Tudor portraits tracing the dynasty from the union of Henry VII and Elizabeth of York, who are represented holding the symbolic roses of Lancaster and York respectively. Their marriage concluded the battle over the succession during the Wars of the Roses and introduced a new royal line that resulted in the golden eras of Henry VIII and Elizabeth I, furnishing a hopeful precedent for the Hanoverians.

Parallels between past and present are also implicit in three small history paintings made for Caroline by William Kent, perhaps better known as an architect and designer.[6] Kent's pictures illustrate

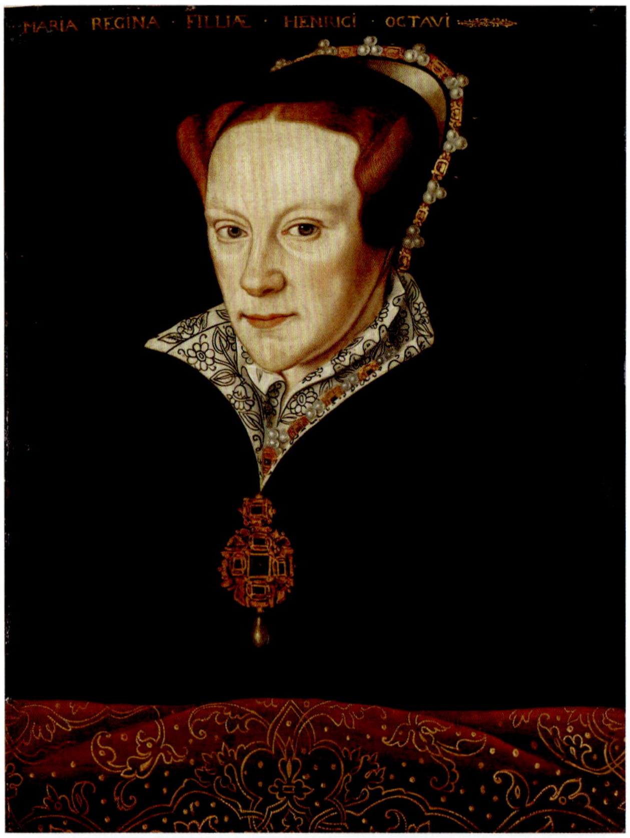

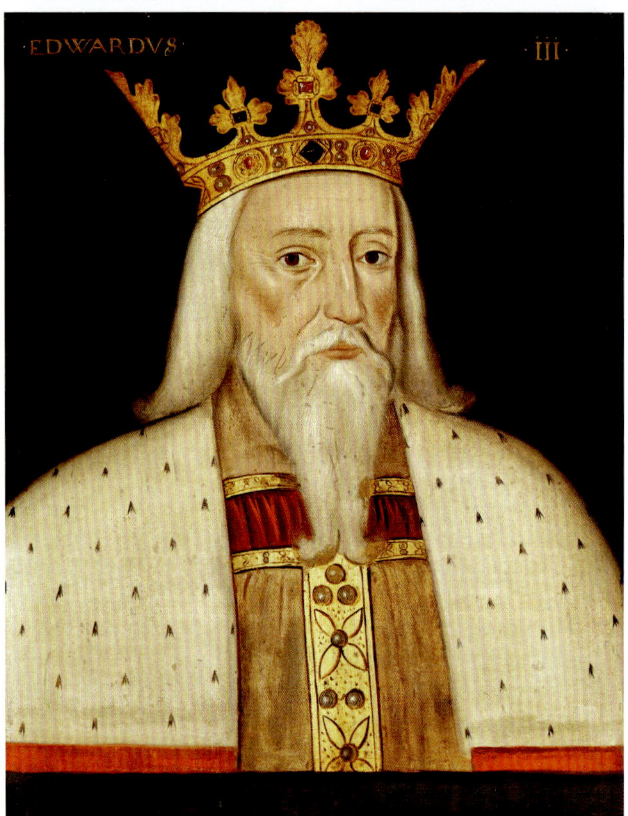

FIG. 9.01 Unknown painter, after Anthonis Mor van Dashorst, *Mary I*, oil on wood panel, 22⅝ x 17⅛ in. (57.5 x 43.5 cm). Royal Collection Trust

FIG. 9.02 Attributed to British School, *Edward III*, c. 1600, oil on wood panel, 22½ x 17⅜ in. (57.2 x 44 cm). Royal Collection Trust

events from the reign of Henry V, a popular Prince of Wales who had succeeded his less-favoured father and gone on to become a triumphant warrior-king, conquering the French at Agincourt in 1415. Not only did the similarities in their circumstances make Henry a flattering prototype for George II, but he had also been the unwitting initiator of the Tudor dynasty through his marriage to Catherine of Valois, grandmother of Henry VII by her second husband, Owen Tudor. Kent's painting *The Marriage of Henry V* of 1730–31 commemorates the king's dynastic importance, emphasized by the profusion of heraldic devices, most notably the Plantagenet coat of arms (see p.224, cat. 12.06). Kent may have modelled Henry's profile on a portrait in Caroline's royal collection.

The presentation of historical portraits at Richmond was a trial run for Caroline's more extensive displays at Kensington Palace, where reminders of hereditary authority provided a backdrop to the political theatre of the court. The dynastic portraiture displays the Queen created at St James's Palace, Hampton Court Palace, and Windsor Castle provide further evidence of her fascination with royal history and genealogy.[7]

Caroline's estranged eldest son, Frederick, Prince of Wales, established a number of royal residences in and around London where he and his wife Augusta presided over a rival court. Like his mother, Frederick used art to 'anglicize' his image, seeking the acceptance of England's aristocratic elite through his activities as a connoisseur and collector of old master paintings and other valuable works of art.[8] Augusta, on the other hand, found it advantageous to avoid association with this traditionally masculine field of endeavour, adhering rather to the submissive, maternal, and domestic image that she had developed as an implicit critique of her mother-in-law's assertive personality.[9] While Frederick's dressing room at Leicester House featured some of his most highly prized and valuable paintings, Augusta's contained relatively minor works, including nineteen gouache copies after old master paintings—principally Italian and Flemish landscapes and classical subjects—made in the late 1740s by the French artist Joseph Goupy.[10] The sole surviving example, a reduced copy of *Belisarius Receiving Charity* made in 1746–47 (fig. 9.03), focuses on the theatrical gestures of the figures, with little of the rich colour and chiaroscuro of the original, then thought to be by Van Dyck, which had been purchased in Paris by Richard Boyle, third Earl of Burlington, during his Grand Tour in 1715.[11]

While deferring to her husband in matters of connoisseurship, Augusta took the lead in commissioning family portraits that asserted her maternal authority over the domestic sphere. Following Frederick's untimely death in 1751, Augusta turned to portraiture as a means of keeping his memory alive, while also emphasizing her assumption of responsibility for their children—particularly their eldest son, George, heir to the throne. Although

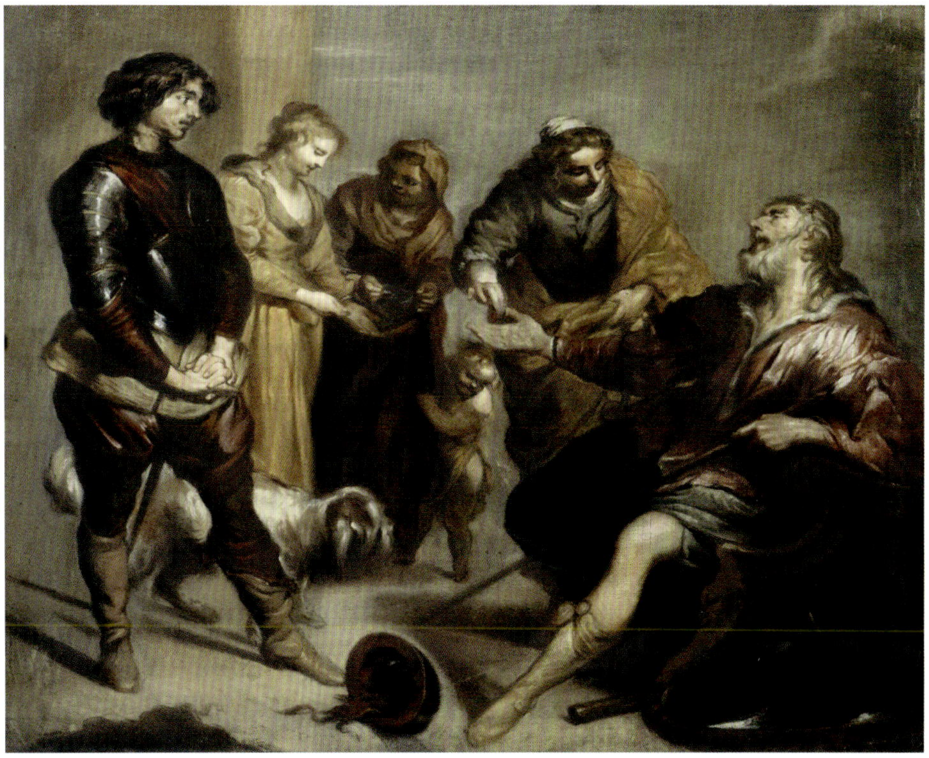

FIG. 9.03 Joseph Goupy, *Belisarius Receiving Charity*, 1746–47, pastel on canvas, 21⅞ x 28 in. (55.7 x 71.2 cm), Royal Collection Trust

many of the paintings were intended for semi-public spaces (as indicated by their grand scale and idealized imagery), others were meant for more private viewing. In 1754 Augusta commissioned a series of intimate portraits of family members from Jean-Étienne Liotard, a Swiss painter fashionable amongst an international royal and aristocratic clientele. Underscoring the conjugal affiliation that was the source of her royal standing and domestic authority, Augusta had Liotard create a pair of portraits of herself and her deceased husband executed in pastel on paper, and slightly smaller pastels on vellum representing their children.

Liotard represented Augusta with unsparing frankness, exaggerating neither youth nor beauty (cat. 9.05). The wariness of her sidelong glance contrasts with the wide-eyed innocence of her five-year-old daughter, Princess Louisa Anne, whose diminutive stature and delicacy are emphasized by the inclusion of an oversized chair (cat. 9.09). Each of the other children is characterized with insightful individuality. The precocious dignity and poignant vulnerability of fourteen-year-old Princess Elizabeth Caroline suggests the transitional nature of adolescence (cat. 9.07). Her younger brother, Henry Frederick, Duke of Cumberland, also appears to straddle the border between youth and adulthood, turning from the childish occupation of building a house of cards—a traditional symbol of the vanity of worldly pursuits—to fix the viewer with a sombre gaze.

Following a long residence in Constantinople, Liotard had continued to trade on his exotic cachet after returning to Europe by sporting a long beard and Turkish dress, which he publicized through self-portraits distributed to potential clients such as Augusta. He also reproduced some of the royal pastels as miniatures to be given as mementos to family members and worn on the body, as shown in the miniature of Augusta's eldest daughter, also called Augusta, who wears a brooch containing a miniature of her father, Frederick (cat. 9.10).[12] Liotard's jewel-like miniature of George, Prince of Wales, and the pastel he created around the same time, convey an impression of dignity and self-possession befitting the boy's destiny as King of England (cat. 9.06).

After ascending to the throne in 1761, George III and his wife Charlotte used family portraiture to construct a public image of harmonious, informal domesticity. The Queen commissioned many of these portraits, patronizing the leading male painters of the day. Over time, however, she became more invested in encouraging the work of women artists.[13] Most notable are the elaborate botanical decorations that Charlotte commissioned from the Royal Academician Mary Moser for her country retreat Frogmore. The estate soon became the site of elaborately orchestrated royal fêtes and public occasions open to ticket holders, providing a wider audience for Charlotte's enthusiasm for botany and patronage of women artists.[14]

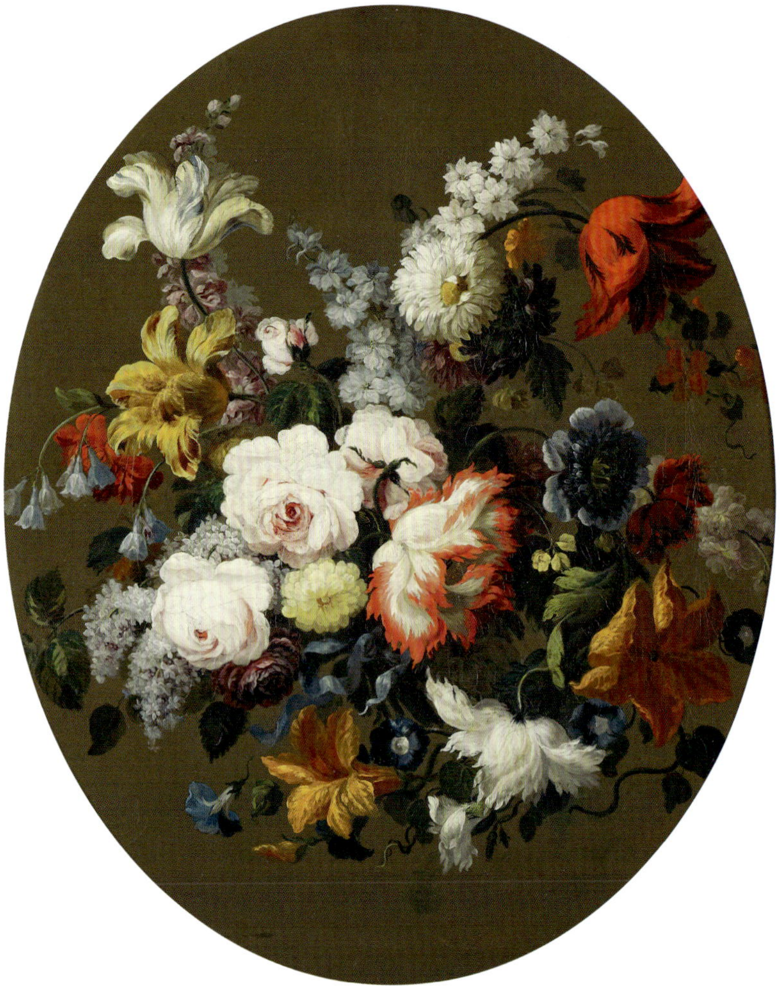

FIG. 9.04 Mary Moser, *A Bunch of Flowers*, c. 1792–97, oil on canvas, 33⅝ x 30¾ in. (85.5 x 78 cm). Royal Collection Trust

As a complement to the extensive gardens she developed at Frogmore, Charlotte employed a number of botanical artists to decorate the interior of her house.[15] The south pavilion, constructed in the mid-1790s by James Wyatt, became known as 'Miss Moser's Room'. It featured the artist's ceiling painting of floral garlands set against a trompe-l'oeil sky, simulating the effect of an open-air trellis to connect the interior space with the gardens beyond. Moser elaborated the floral motif through a number of paintings set into the walls. In a pair of oval canvases, she represented an encyclopedic array of spring and summer flowers that had long flourished in English gardens, such as roses, peonies, tulips, lilies, poppies, fritillaries, delphiniums, and bluebells (fig. 9.04). Absent are the more exotic specimens of recent importation prized by English gardeners, such as *Strelitzia reginae*, named in Charlotte's honour. Moser's restriction to established English flowers suggests a patriotic motive, and reflects Charlotte's original intentions for the gardens at Frogmore, as she described them in a letter of February 1791 to her son Augustus: 'My chief Plants are to be Natives of England and all such foreign ones as will thrive in our Soil.'[16]

Like her predecessors Caroline and Augusta, Charlotte was herself a foreign importation, who nevertheless found the means of thriving on English soil. All three Princesses did so, in part, through their active participation in the collection and patronage of paintings that shaped their public identities, while also contributing to the 'Englishness' of their courts.

NOTES

1. Joanna Marschner, *Queen Caroline: Cultural Politics at the Early Eighteenth-Century Court* (New Haven and London: Yale Univ. Press, 2014), 88, 90, 93–94, 186; Catherine Tite, *Portraiture, Dynasty and Power: Art Patronage in Hanoverian Britain, 1714–1759* (Amherst, NY: Cambria Press, 2010), 23, 63–64.
2. See Lord Egmont's diary entry for 21 April 1735, in *Manuscripts of the Earl of Egmont: Diary of The First Earl of Egmont (Viscount Percival)*, Historical Manuscripts Commission 2 (London: His Majesty's Stationery Office, 1923), 171.
3. 'Vertue's Note Book A.b. (British Museum Add. MS23069)', *The Walpole Society* 18 (1929): 133.
4. James Granger, *A Supplement, Consisting of Corrections and Large Additions, to a Biographical History of England* (London: T. Davies and others, 1774), 6.
5. Hannah Smith, *Georgian Monarchy: Politics and Culture, 1714–1760* (Cambridge: Cambridge Univ. Press, 2006), 53.
6. Catherine Arbuthnott, 'Kent and Italy'; Steven Brindle, 'Kent the Painter'; Julius Bryant, 'Kent and the Gothic Revival', in *William Kent: Designing Georgian Britain*, ed. Susan Weber (New Haven and London: Yale Univ. Press, 2013), 78, 139, 249; Desmond Shawe Taylor, ed., *The First Georgians: Art and Monarchy, 1714–1760* (London: Royal Collection Trust, 2014), 283.
7. Marschner, *Queen Caroline*, 79, 104–6, 106, 182; Tite, *Portraiture, Dynasty and Power*, 35.
8. Kimerly Rorschach, 'Frederick, Prince of Wales (1707–1751) as a Patron of the Visual Arts: Princely Patriotism and Political Propaganda' (PhD diss., Yale University, 1985), 265–66, 268–69 and passim.
9. John L. Bullion, '"To play what game she pleased without observation": Princess Augusta and the Political Drama of Succession, 1736–1756', in *Queenship in Britain, 1660–1837: Royal Patronage, Court Culture and Dynastic Politics*, ed. Clarissa Campbell Orr (Manchester: Manchester Univ. Press, 2002), 212, 214, 21.
10. Rorschach, 'Frederick, Prince of Wales as a Patron of the Visual Arts', 239–40, 261.
11. The painting was then in the collection of Lord Burlington and attributed to Anthony van Dyck. It is currently attributed to Luciano Borzone and hangs in the Music Room at Chatsworth.
12. N. S. Trivas, 'Liotard's Portraits of Frederick Lewis, Prince of Wales, and His Family', *Burlington Magazine* 68, no. 396 (March 1936): 118.
13. Heidi A. Strobel, *The Artistic Matronage of Queen Charlotte (1744–1818): How a Queen Promoted Both Art and Female Artists in English Society* (Lewiston, NY: The Edwin Mellen Press, 2011).
14. Marcia Pointon, *Strategies for Showing: Women, Possession, and Representation in English Visual Culture, 1665–1800* (Oxford: Oxford Univ. Press, 1997), 160–65.
15. Strobel, *Artistic Matronage*, 213–17.
16. Queen Charlotte to Prince Augustus, 25 March 1791; quoted in *The Later Correspondence of George III, 1783–1810*, ed. A. Aspinall, 5 vols. (Cambridge: Cambridge Univ. Press, 1962–70), 1: no. 665, quoting RA. 47907–8.

9.06†

JEAN-ÉTIENNE LIOTARD (1702–1789)
George, Prince of Wales
1754
Pastel on vellum
16 x 11¾ in. (40.6 x 29.8 cm)
Royal Collection Trust, RCIN 400897

9.07†

JEAN-ÉTIENNE LIOTARD (1702–1789)
Princess Elizabeth Caroline
1754
Pastel on vellum
15½ x 11⅞ in. (39.4 x 30.2 cm)
Royal Collection Trust, RCIN 400902

9.08†

JEAN-ÉTIENNE LIOTARD (1702–1789)
Henry Frederick, Duke of Cumberland
1754
Pastel on vellum
15¾ x 12 in. (40 x 30.5 cm)
Royal Collection Trust, RCIN 400895

9.09†

JEAN-ÉTIENNE LIOTARD (1702–1789)
Princess Louisa Anne
1754
Pastel on vellum
15¾ x 12 in. (40 x 30.5 cm)
Royal Collection Trust, RCIN 400900

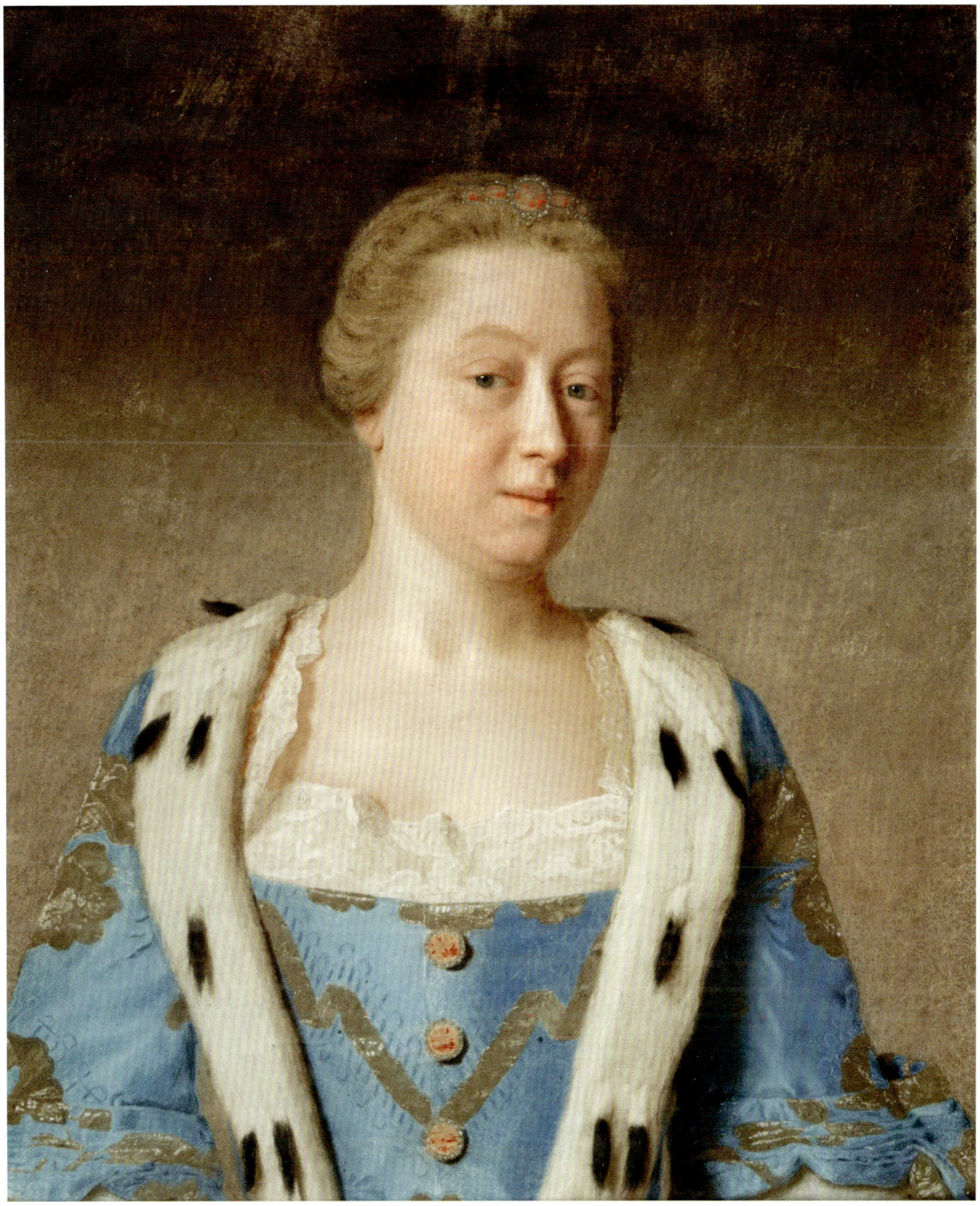

9.05

9.05†

JEAN-ÉTIENNE LIOTARD (1702–1789)
Augusta, Princess of Wales
1754
Pastel on paper
25½ x 20 ¼ in. (64.8 x 51.4 cm)
Royal Collection Trust, RCIN 400892

Jean-Étienne Liotard was introduced to Augusta by either John Stuart, third Earl of Bute, or Prince William, Duke of Cumberland, in 1753, and was commisioned by her to make a series of portraits of her family. Her own portrait and that of her late husband, Frederick, Prince of Wales, are likely to have been the first completed. George Bubb Dodington noted on 14 February 1754 that the Princess was sitting for the artist. The set was completed by slightly smaller portraits of all but her youngest child. The pastels provide candid informal likenesses of each family member, with the exception of Prince Frederick's portait, which must have been worked up from an earlier painting. Liotard was paid 240 guineas from Augusta's Privy Purse on 15 August 1755 (RA Add. MS. 55448).

* = YCBA only
† = HRP only

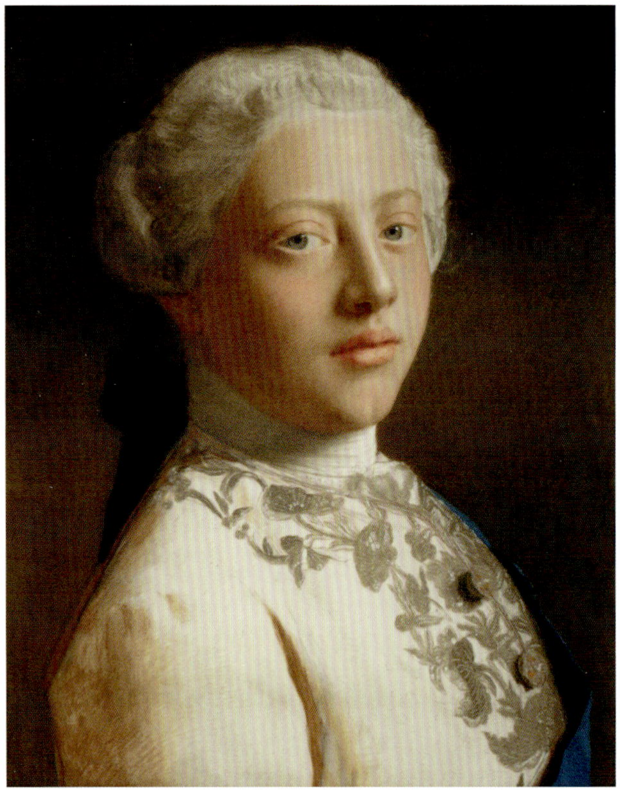

9.06

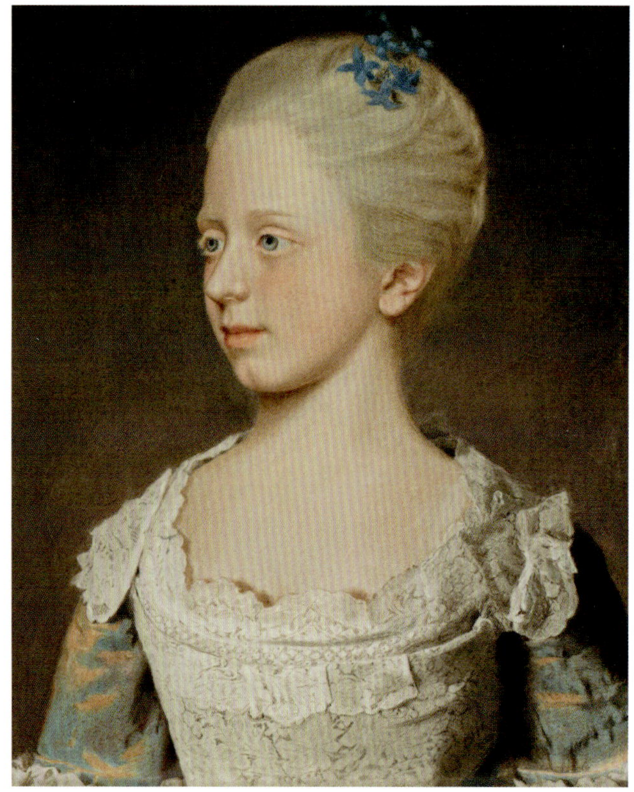

9.07

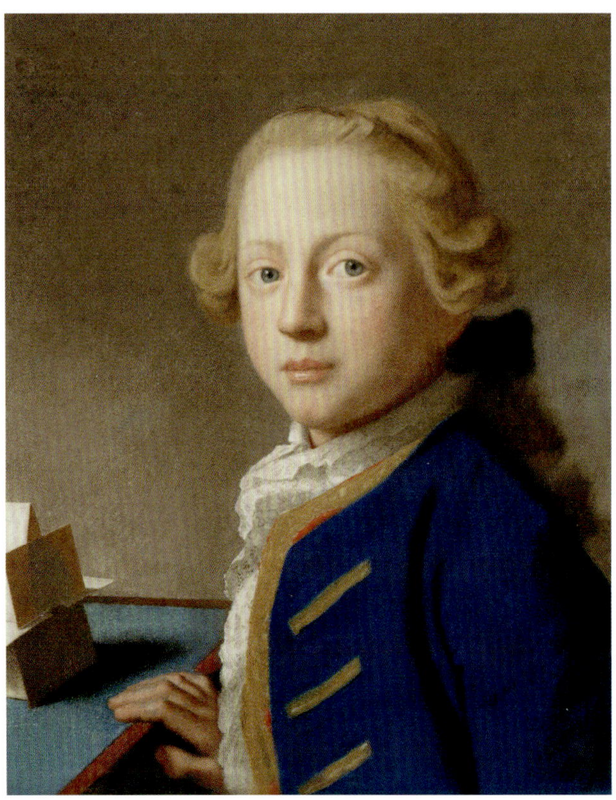

9.08

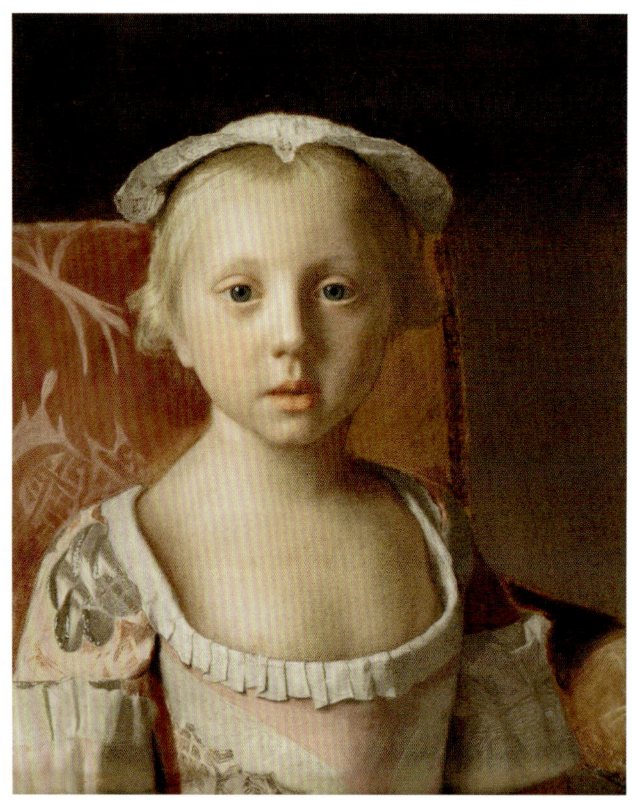

9.09

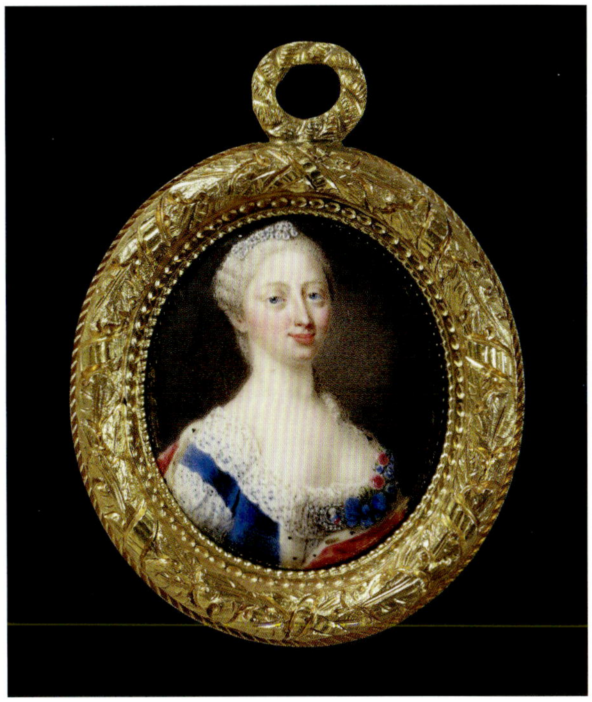

9.10

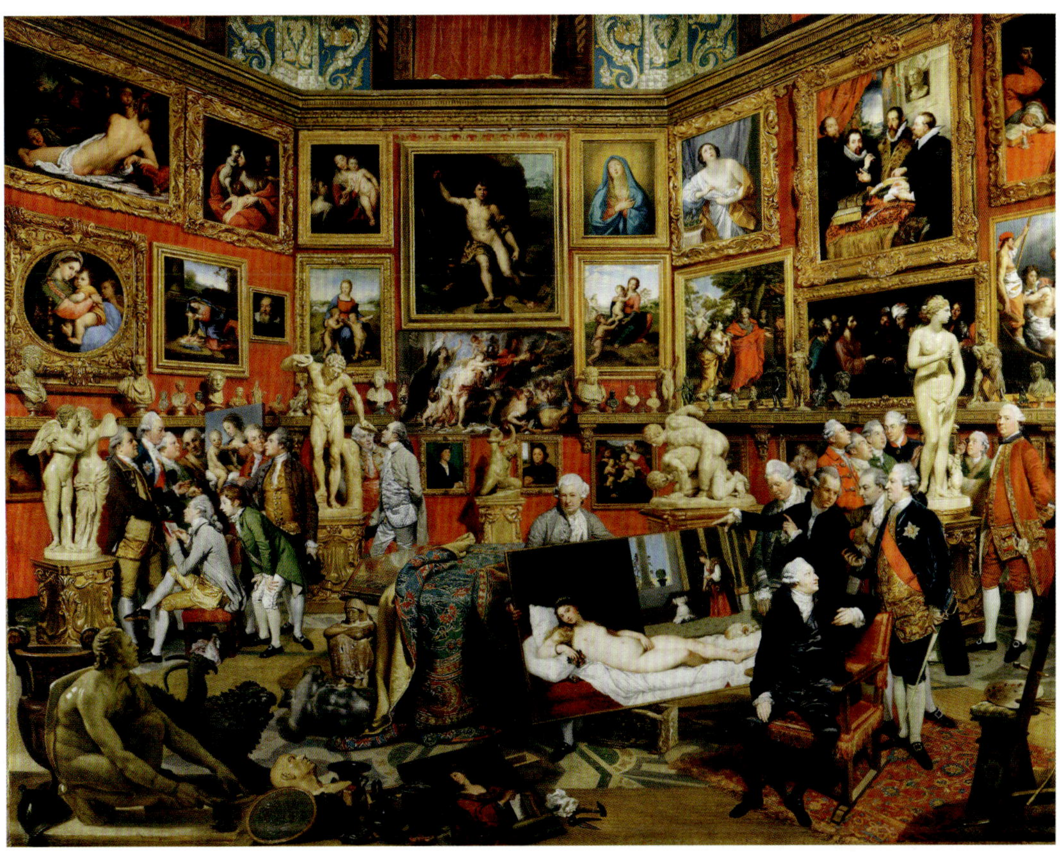

9.11

9.10*

JEAN-ÉTIENNE LIOTARD (1702–1789)
Princess Augusta, Duchess of Brunswick
1758
Enamel
1½ x 1¼ in. (3.6 x 3.1 cm)
Royal Collection Trust, RCIN 421846

Portrait miniatures of Princess Augusta and George, Prince of Wales, the eldest children of Frederick, Prince of Wales, and Princess Augusta, after their portraits in pastel made by Liotard in 1754, were probably commissioned from the artist by their mother. Such works subsequently circulated as gifts. Prince George would present a copy of his miniature to his aunt Anne, Princess of Orange (Stichting Historische Verzamelingen van het Huis Oranje-Nassau, Apeldoorn).

9.11†

JOHAN ZOFFANY (1733–1810)
The Tribuna of the Uffizi
1772–77
Oil on canvas
48⅝ x 61 in. (123.5 x 155 cm)
Royal Collection Trust, RCIN 406983

In 1772 Johan Zoffany left London with a commission from Charlotte to paint the 'Florence Gallery'. She provided letters of introduction and £300 living expenses. He did not return until 1779, when the painting had been completed. The 'Florence Gallery', The Tribuna, an octagonal room designed by Bernardo Buontalenti for Francesco de'Medici, is located in the Uffizi, and in the eighteenth century it contained the most important paintings and antiquities of the Medici collection. Zoffany was able to negotiate for additional works to be transferred temporarily from the Pitti Palace to enrich the arrangements. Charlotte was not pleased with the image, because Zoffany had chosen to include many portraits of dilettantes and collectors looking at the art, which was originally supposed to have been the focus of the painting. She had to be coaxed to pay the remainder of the artist's fee. The painting hung first at Kew Palace and later in the Upper Library at Buckingham House.

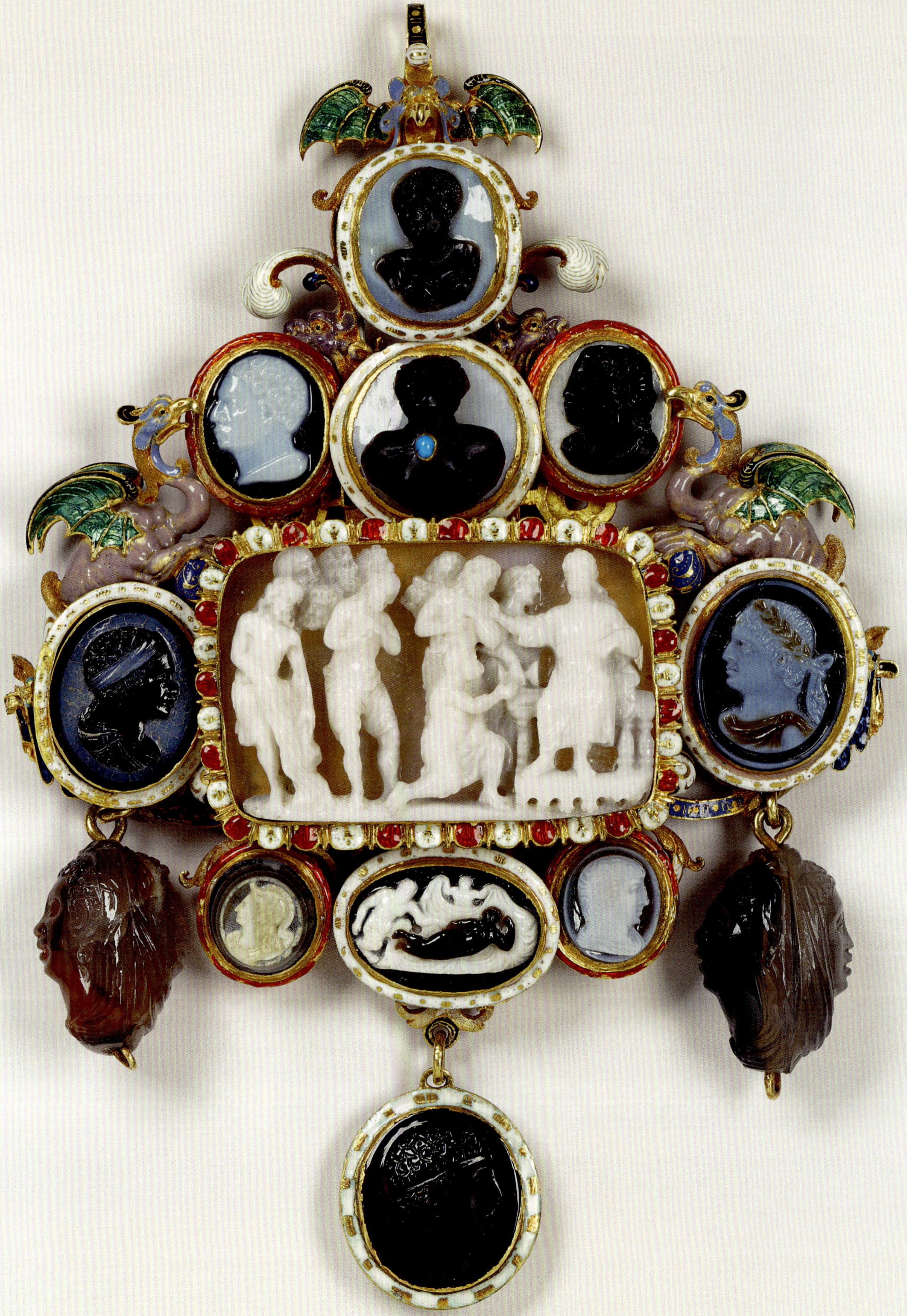

Adorning the Cabinet

Kathryn Jones

Something of Caroline of Ansbach's interests can be captured in a study of the closet rooms at Kensington Palace. The Picture Closet, which was catalogued by George Vertue in 1743, with accompanying plans showing the dense hang of paintings, drawings, miniatures, and wax and ivory carvings, reflected her interests in history, genealogy, and collecting. The catalogue was said to have been drawn up 'at the Queen's command', although by the time Vertue's plan was printed Caroline had been dead for seven years.[1] The Picture Closet lay at the heart of the state apartments, close to the King's Gallery and State Bedchamber. Here Caroline could display her particular love of genealogy, which reflected the princely virtue of collecting and enhanced the credentials of the Hanoverian family as descendants of the Guelph dynasty and the ruling houses of Britain. Her interest was encouraged by the scholar Gottfried Wilhelm Leibniz, who confirmed the place of the Hanoverians within the Holy Roman Empire by connecting the Guelphs with the Italian Este family.[2] As a young bride in Hanover, Caroline had carried out her own researches into European pedigrees. She had previously spent time in Dresden at the court of her stepfather, the Elector John George IV, where the collections were housed in the renowned *Kunstkammer*, the Grünes Gewölbe (Green Vaults). At Lietzenburg, later called Charlottenburg, the court of Sophia Charlotte of Hanover, the wife of her guardian, Frederick III, first King in Prussia, she was also aware of the collections, and was probably familiar with the *Kunstkammer* at Salzdahlum, belonging to the Hanoverian family connection Anton Ulrich of Brunswick-Wolfenbüttel. It was with a fierce desire to establish her own place among these great dynasties that Caroline set about establishing a collection of portraits at Kensington.

Her interests were reflected in the selections of miniatures, often framed together in groups of eight or nine, which appeared in the Closet. Eighty-eight miniatures of her Hanoverian forebears were mounted in frames and interspersed with drawings by Hans Holbein the younger, landscapes, and still life paintings. Of these, forty-nine were painted in Brunswick-Lüneburg by an unidentified artist who created the entire set, representing the Dukes of Brunswick and their consorts, in the years between 1593 and 1597 (cats. 10.04–10.09). The works in the group follow a common format, the portraits produced in rectangular frames of similar size, with the name of the sitter inscribed at the top of each image. A further group of eight miniatures extended the series, stretching the dynasty back to the fifteenth century (cats. 10.10–10.13). The fanciful rendition of the sitters and the inscriptions in a romanticizing Gothic script were probably done by an eighteenth-century artist at the request of Caroline herself, emulating the sixteenth-century style of the Brunswick-Lüneburg court miniaturist (cats. 10.04–10.09).[3]

A further group of miniatures representing the family of Frederick III of Denmark, his consort Sophia Amalia, and their children was also placed in the Closet by Caroline, again for their connections with the Brunswick-Lüneburg dynasty. They appear to have been created in the Danish court around 1676 (cats. 10.14–10.17).

Pride of place in the Closet must have been given to two miniatures of George I's parents, the Elector Ernest Augustus of

See p. 199, cat. 10.29

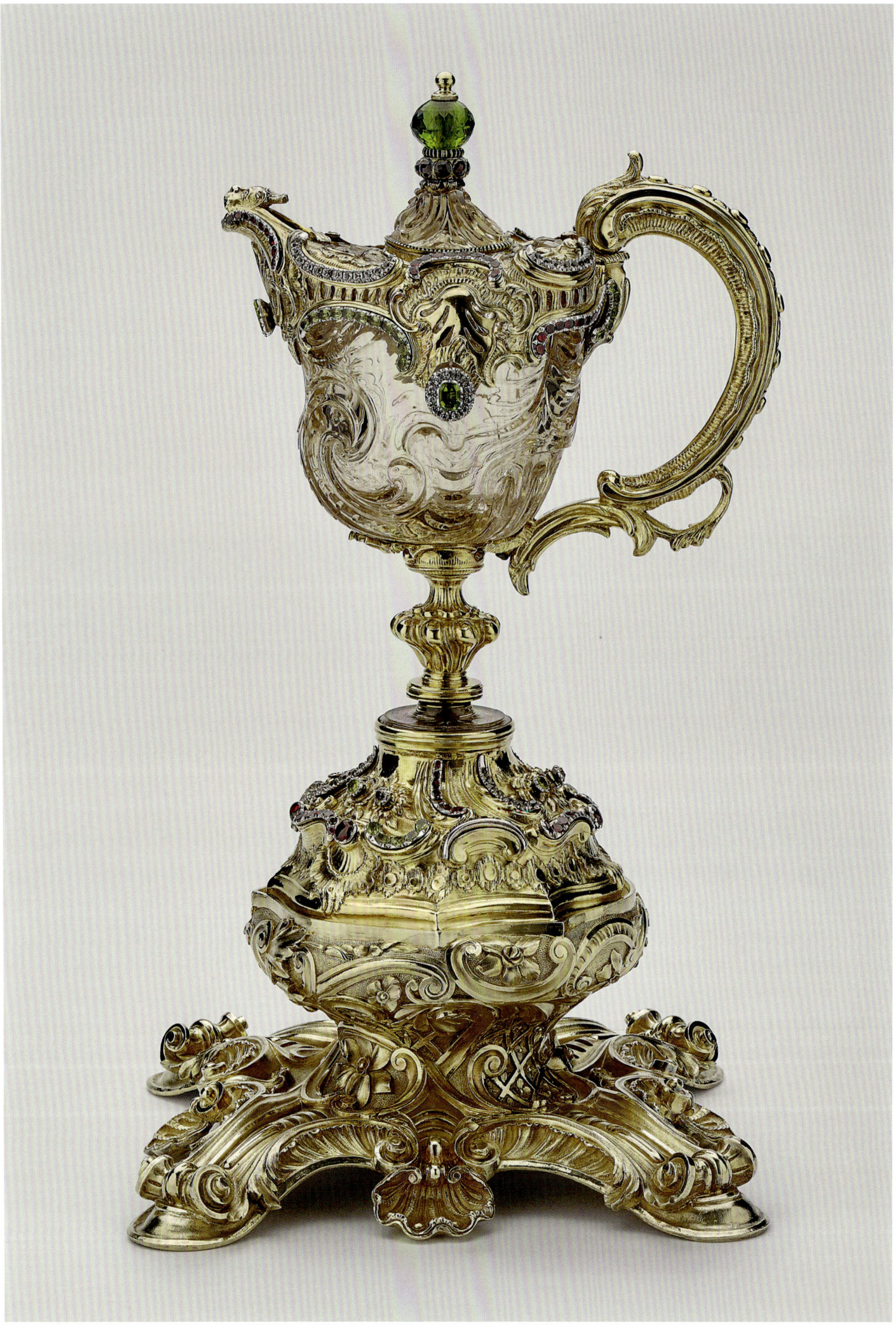

FIG. 10.01 Ewer: probably Viennese, c. 1730, rock crystal; later mount by Philip Rundell, 1819–20, silver and gem stones, 11⅞ in. (30 cm). Royal Collection Trust

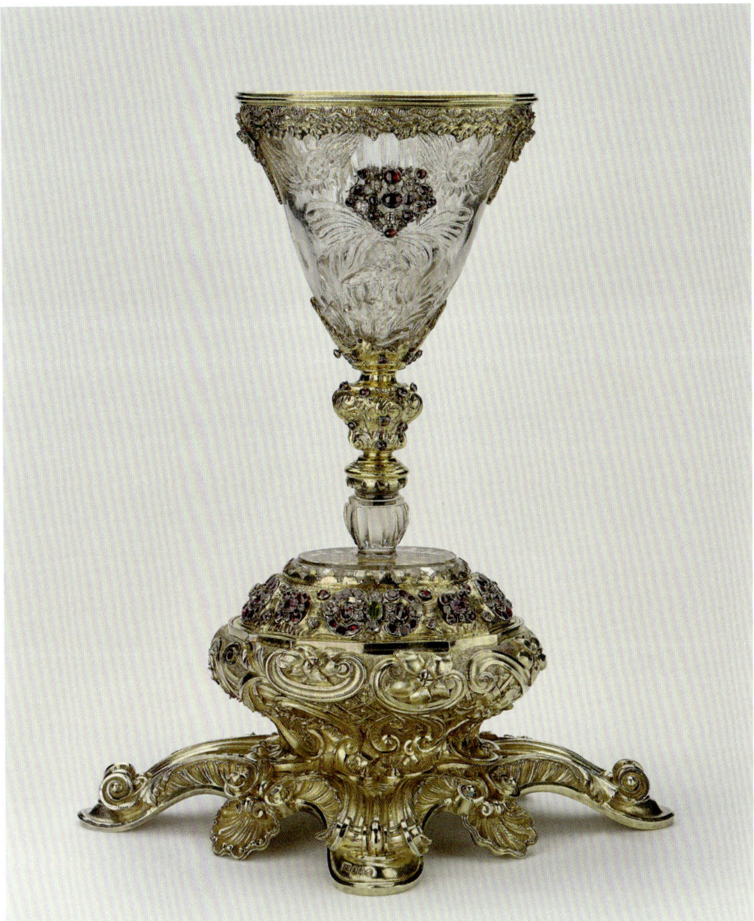

FIG. 10.02 Chalice: cup, probably Viennese, c. 1730, rock crystal; later mount by Philip Rundell, 1819–20, silver and gem stones, 12⅛ x 9⅝ x 8⅝ in. (30.8 x 24.2 x 21.9 cm). Royal Collection Trust

Hanover, depicted as a Roman commander, and his consort Sophia, by the French artist Jean Michelin, who worked in the court in Hanover for almost twenty years (cats. 10.19–10.20). The Electress was known personally to Caroline, as she spent much time at the court of her daughter Sophia Charlotte, where Caroline had lived as a child, and was later her grandmother-in-law. Caroline added to the group hanging in the Closet a set of enamel miniatures of her own Brandenburg-Ansbach family by Christian Frederick Zincke (cats. 10.21–10.23), a Dresden-born artist who worked in England under the first court enamellist, Charles Boit, whom Zincke succeeded.

The setting of these images in the Picture Closet was not entirely a public one—the room was reserved for the more private visitor to the court—but these works nevertheless placed an emphasis on pedigree and connection that served as a backdrop to the day-to-day life of the Queen. Caroline created similar *Ahnengalerie*, or ancestral portrait series, in her decorative schemes at Hampton Court and the library at St James's Palace, but neither was as extensive as the Kensington hang.

If the Picture Closet was semi-public, the other spaces at Kensington used to house Caroline's collections were more private still. Among them were the Little Library and two cabinets off it, which contained the *Wunderkammer* or *Schatzkammer* of the Queen. These spaces broadly followed the model of the ideal princely closet as laid down by Francis Bacon and Samuel Quiccheberg in previous centuries, and included works of *naturalia* (corals, shells, bezoars, unicorn horns, and rock crystal); *ethnographica* (Turkish and Indian weapons, Chinese porcelain); *scientifica* (a portable brass sundial), and, above all, *artificialia* or man-made objects. The latter included turned ivory, glass, and mounted hardstones and silver. A 1763 inventory made by Horace Walpole includes rock crystal works, which may possibly be identified through their later progress into Charlotte of Mecklenburg-Strelitz's Closet (figs 10.01, 10.02).[4] The item described by Walpole as 'a ewer flaw'd with chimera spout', together with another chalice, were noted again when they were transferred from Charlotte's collections at Buckingham House to Carlton House in the reign of George IV.[5]

Caroline was particularly proud of her group of around one thousand medals, including a full set of Roman Emperors. She also owned a prodigious collection of cameos and intaglios, catalogued sometime after her death by her dresser Mrs Purcell, and later recorded by Walpole before they were sent to George III.[6] Many of the gemstones reflected her interest in family lines and are largely the work of French or Italian carvers. A number of the cameos reflect the Queen's fascination with the sixteenth and early seventeenth centuries, including portraits of Henry VIII, Edward VI, Philip II of Spain, Elizabeth I, and James VI and I (cats. 10.25–10.27). These sat alongside other works including one depicting Henry IV of France and an image of the Emperor Leopold carved in amethyst.

The cameo double portrait of Henry VIII and Edward VI, a purely imagined scene based on the works of Hans Holbein, was formerly catalogued as an original sixteenth-century work, but it must almost certainly have been an eighteenth-century commission, possibly from Caroline herself, although no record of it survives in her accounts (cat. 10.24). The work was clearly intended to be handled: the stone is cut in the rear in such a way that when held before a light, the features of the faces from the front have greater depth and subtlety. Another cameo of Henry VIII alone, which was in the Closet, follows the same format, with the back of the stone hollowed out to give depth to the portrait from the front. The portrait is based on the Swiss-born medallist Jean Dassier's engraving for a coin struck in 1731. Both these works and perhaps the small bust of Hercules, which although probably originally intended to be mounted on a piece of furniture is beautifully cut to be turned in the hand, suggest that Caroline's collection was meant to be studied, handled, and to serve as an educational tool.

Some of the gemstones may have come from the Stuart collections. At least one of a number of images of Philip II was recorded in the collections of Charles I, and an impression of the cameo of James I and VI was created in wax by Elias Ashmole in 1660.[7]

Walpole's commentary on the gemstone collection does not rate many of the works very highly except for two: the large images of Elizabeth I, which Walpole attributed to Valerio Belli, known in the eighteenth century as Vicentino. The attribution seems to have been due to John Evelyn, who claimed to have seen an image 'exceedingly like' the Elizabeth in the closet of Charles II, and known to have been the work of Vicentino, though Belli was dead before Elizabeth became queen.[8] Walpole retained the attribution, however, perhaps because Belli was one of the few stone carvers to appear in Giorgio Vasari's *Lives*. Elizabeth appears to have been something of a role model for Caroline.[9] A ring with an image of Elizabeth was also listed as being in the Closet, and Vertue recorded a 'portrait of Elizabeth in Chinese dress'.[10]

Some of the gemstones collected by the Queen were not portraits, but were intended to be appreciated as works of art: the exquisite Adoration of the Magi, carved in almost painterly fashion in places, is a particularly fine example (cat. 10.28). Caroline's love of gemstones was reflected in her correspondence with Elisabeth Charlotte of the Palatinate, known as Liselotte. Between 1715 and 1722, engraved stones are mentioned in at least one letter.[11] Liselotte may have been responsible for the gift of the cameo of Joseph with his brothers, surrounded by twelve further stones and mounted in an elaborate gold and enamel tabernacle-style frame (cat. 10.29). The framing of this work may have been influenced by the goldsmith Johann Melchior Dinglinger at the court of Augustus the Strong in Dresden, and it suggests a renewed interest in gemstones as works of art. A similarly opulent work is a *commesso* pendant formed from a cameo of an unknown woman created in amethyst, carnelian, ruby, garnet, emerald, and gold, with an elaborately enamelled back, which was recorded in Caroline's collections at Kensington. Little is known of its provenance. It may have been created in the French court in the sixteenth century, and may have appealed to Caroline for its associations with Dido, Queen of Carthage, celebrated for her ambition, astuteness, bravery, and steadfastness since classical antiquity.[12]

NOTES

1. George Vertue, *Catalogue of the Pictures, Drawings, Limnings, Enamels, Models in Wax & the Ivory Carvings &c., at Kensington in Queen Caroline's Closet, next the State Bedchamber, Taken at Queen Caroline's Command by Mr Vertue in September 1743* (RCIN 1112557), 13.
2. N. Gädecke, 'Die Rolle des Historikers: Gottfried Wilhelm Leibniz und der Aufstieg des Welfenhauses', in *Hannover und die englische Thronfolge*, ed. H. Barmeyer (Bielefeld: Verlag für Regionalgeschichte, 2005), 157 ff.; G. van der Heuvel, 'Theorie ohne Praxis: Leibniz Rolle in der Politik des Hauses Hannover', in *Ehrgeiz, Luxus und Fortune* (Hanover: Historisches Museum, 2001), 85–87.
3. Desmond Shawe Taylor, ed., *The first Georgians: Art and Monarchy, 1714–1760* (London: Royal Collection Trust, 2014), 330, cat. nos. 199–202.
4. Horace Walpole's hand-written list was appended to his copy of Vertue's catalogue of the Picture Closet in 1763. The manuscript is now in the Royal Collection (RCIN 1112557).
5. RCIN 1112775: Benjamin Jutsham, Receipts, 1816–29, vol. 2, fol. 124, 19 Jan. 1821. For the ewer and the chalice, see RCIN 50551 and RCIN 50552 respectively.
6. 6 BM Add. MS. 20101, f. 60: 'The inventory of the Curiositys & Medals in the Cabinet in His Majesty's Library of wch Mrs Pursell had the original'. The inventory must have been drawn up before 1755, when Mrs Purcell died.
7. Oxford MS. Ashmole 1138; see M. Henig, Appendix II, in K. Aschengreen Piacenti and J. Boardman, *Ancient and Modern Gems and Jewels in the Collection of Her Majesty The Queen* (London: Royal Collection, 2008), 269–81, especially 272, no. 12.
8. John Evelyn, *Sculptura*, 2nd ed. (London, 1755), 62.
9. See in this volume, Clarissa Campbell Orr, 'Partners in Power', 53–75.
10. Royal Archives: GEO Add. MS. 16.
11. S. Paas and A. Alexander, eds., *Liselotte von der Pfalz: Madame am Höfe des Sonnenkönigs* (Heidelberg: HVA, 1996), 167–71.
12. Aschengreen Piacenti and Boardman, *Ancient and Modern Gems and Jewels*, 182, no. 279.

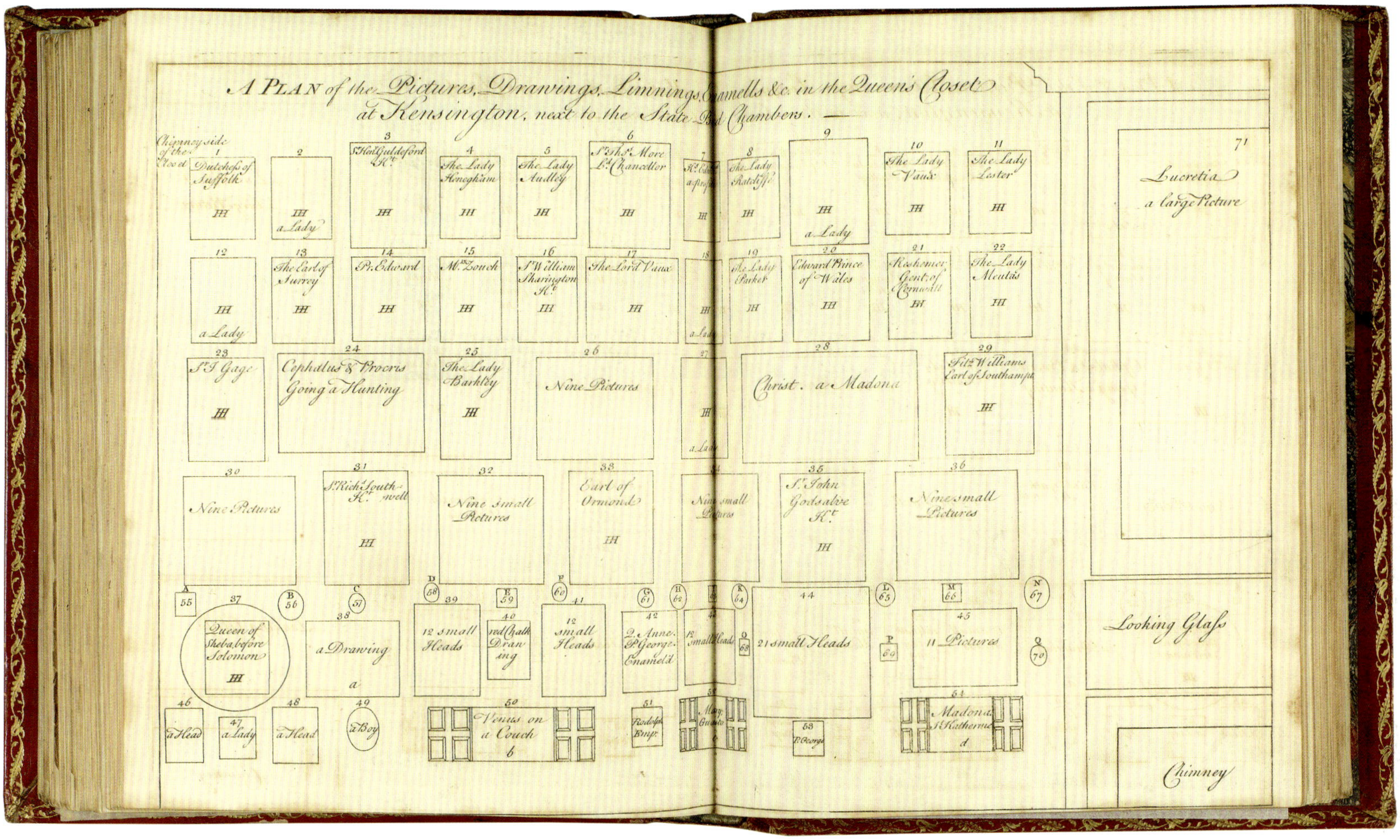

10.03

CARING AND THE 'GREATEST STORE OF THE ENGLISH'

10.03*

GEORGE VERTUE (1684–1756), artist and editor, and **HORACE WALPOLE**, fourth Earl of Orford (1717–1797), editor
'A Plan of the Pictures, Drawings, Limnings, Enamells &c. in the Queen's Closet at Kensington, next to the State Bed Chambers'
In *A catalogue of the collection of pictures, &c. belonging to King James the Second; to which is added, a catalogue of the pictures and drawings in the closet of the late Queen Caroline, With their exact Measures; And also of the Principal Pictures in the Palace at Kensington*
1758
Printed for William Bathoe, London
9⅞ in. (25 cm)
The Lewis Walpole Library, Yale University, 27 7, Copy 2
Exhibited: 'Chimney wall', unpaginated appendix

By August 1735 it is evident that Caroline had filled a small room adjacent to the King's Drawing Room at Kensington Palace with a great number of family portraits, portrait miniatures, as well as the Holbein drawings and other works relating to the Tudor dynasty. George Vertue was granted a warrant from the Lord Chamberlain in 1743 which gave him extended access to Queen Caroline's Closet and permission to 'copy or draw after those heads by Holbein' and to complete a 'scheme to show the nature, magnitude and disposition of them [the pictures] on the four sides of the room', together with a listing of all the contents. He spent two or three days each week completing the task, working from a room provided by the palace housekeeper, Henry Lowman. The original drawings survive in the British Museum (BL Add. MS. 15752) It was published by George Bickham in 1755 in *Apelles Britannicus*, the first proto-guidebook to the palace. In 1758 it appeared in a collection of royal inventories published by William Bathoe.

* = YCBA only
† = HRP only

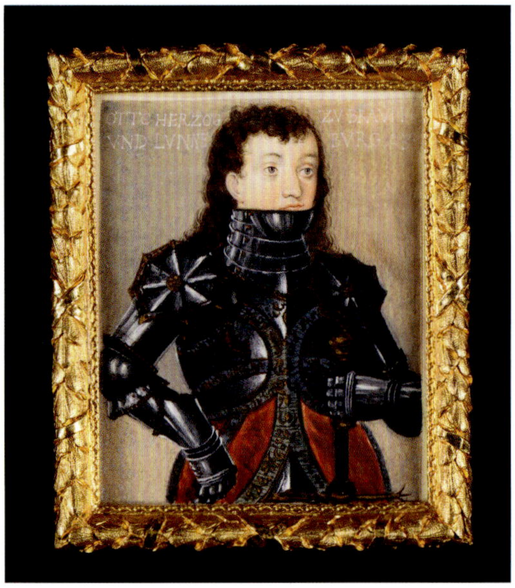

10.04

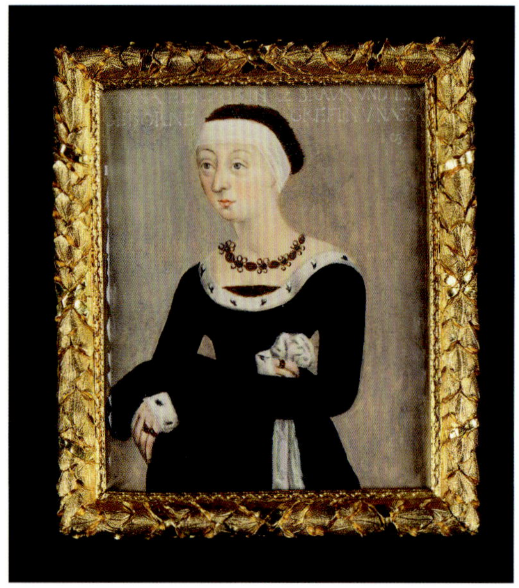

10.05

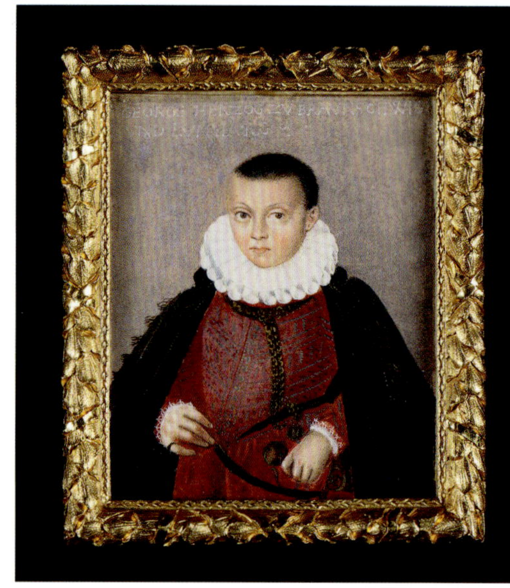

10.06

10.04*

BRUNSWICK-LÜNEBURG COURT MINIATURIST (active c. 1595)
Otto II, the Magnanimous, Duke of Brunswick-Lüneburg (1439–1471)
c. 1595
Watercolour on vellum
3¼ x 2⅝ in. (8.3 x 6.8 cm)
Royal Collection Trust, RCIN 420431

10.05*

BRUNSWICK-LÜNEBURG COURT MINIATURIST (active c. 1595)
Anna, Duchess of Brunswick-Lüneburg (1440/41–1514)
c. 1595
Watercolour on vellum
3¼ x 2½ in. (8.1 x 6.3 cm)
Royal Collection Trust, RCIN 420432

10.06*

BRUNSWICK-LÜNEBURG COURT MINIATURIST (active c. 1595)
George, Duke of Brunswick-Lüneburg-Kalenberg (1582–1641)
c. 1595
Watercolour on vellum
3¼ x 2⅝ in. (8.1 x 6.7 cm)
Royal Collection Trust, RCIN 420445

10.07*

BRUNSWICK-LÜNEBURG COURT MINIATURIST (active c. 1595)
George Frederick, Margrave of Brandenburg (1539–1603)
c. 1595
Watercolour on vellum
3¼ x 2⅝ in. (8.3 x 6.7 cm)
Royal Collection Trust, RCIN 420477

10.08*

BRUNSWICK-LÜNEBURG COURT MINIATURIST (active c. 1595)
John George, Elector of Brandenburg (1525–1598)
c. 1595
Watercolour on vellum
3¼ x 2¾ in. (8.2 x 6.8 cm)
Royal Collection Trust, RCIN 420471

10.09*

BRUNSWICK-LÜNEBURG COURT MINIATURIST (active c. 1595)
Anna Catherina, Queen of Denmark (1575–1612)
c. 1595
Watercolour on vellum
3¼ x 2½ in. (8.2 x 6.2 cm)
Royal Collection Trust, RCIN 420482

Six portrait miniatures from a set of forty-nine acquired by George I, George II, or Caroline, and first noted in 1743 by George Vertue as part of the display in Queen Caroline's Closet at Kensington Palace. This substantial series is all from the same, unidentified painter who worked mainly at the Brunswick-Lüneburg court, but also visited Brandenburg and other principalities. This, and similar collections of portrait miniatures displayed by Caroline, enabled her to create an ancestral lineage linking the Hanoverian dynasty to various European courts, and the Tudor–Stuart line of descent.

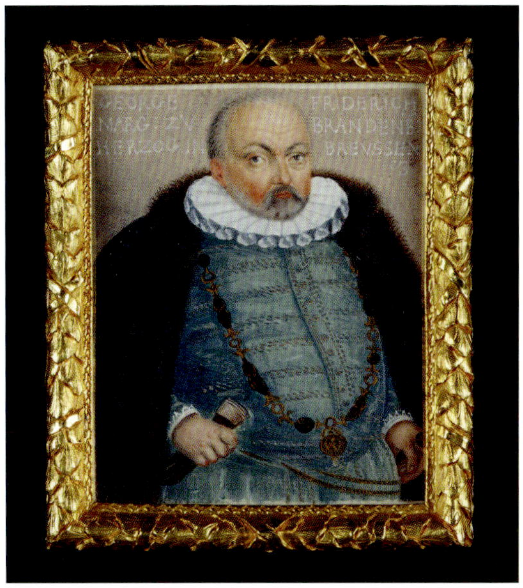

10.07

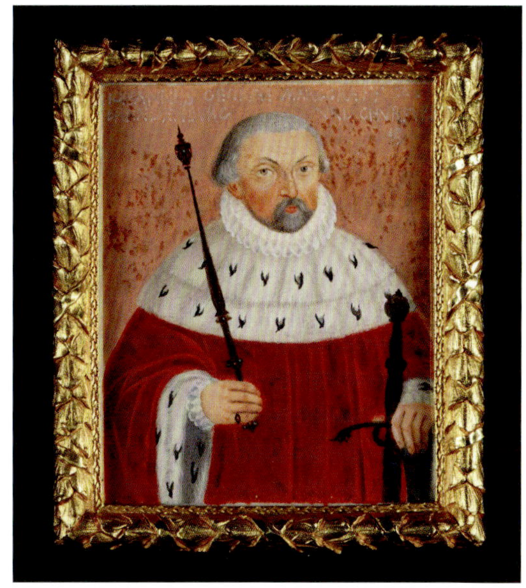

10.08

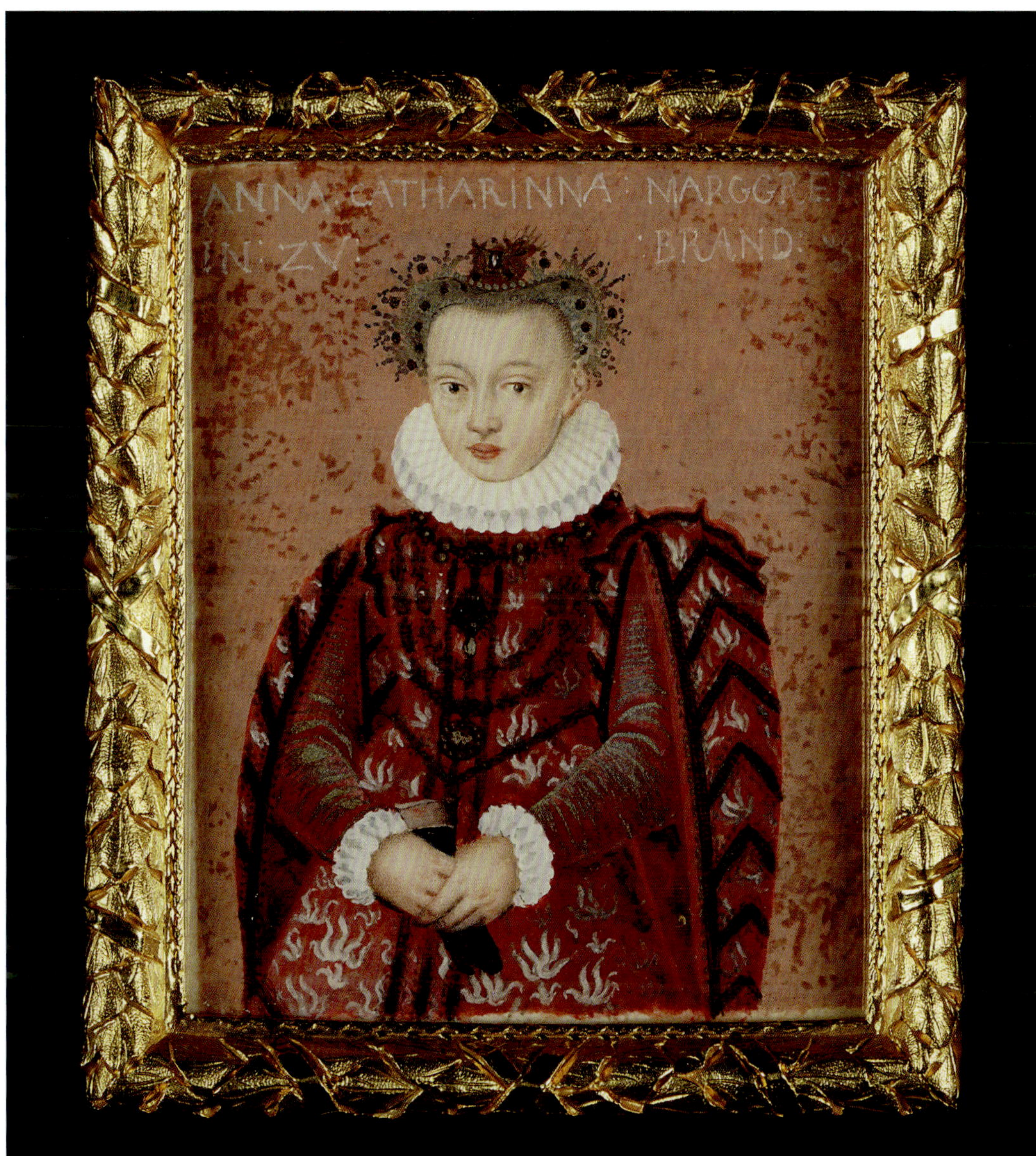

10.09

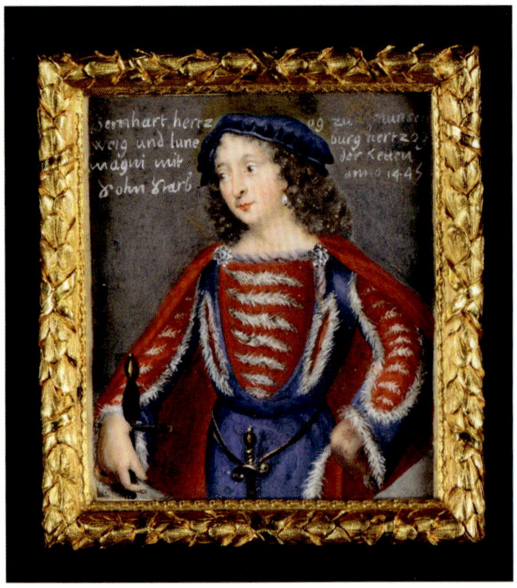

10.10

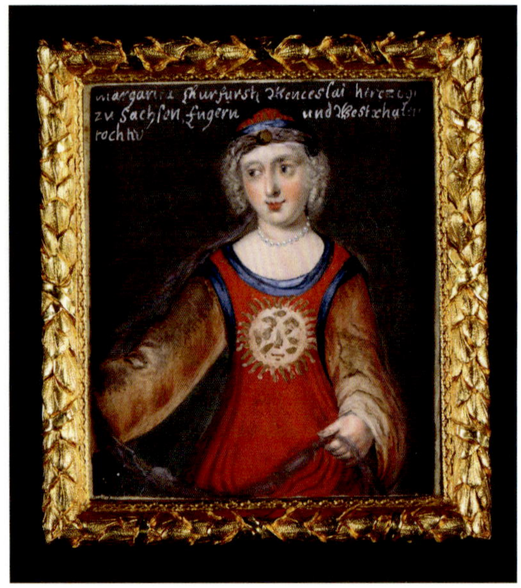

10.11

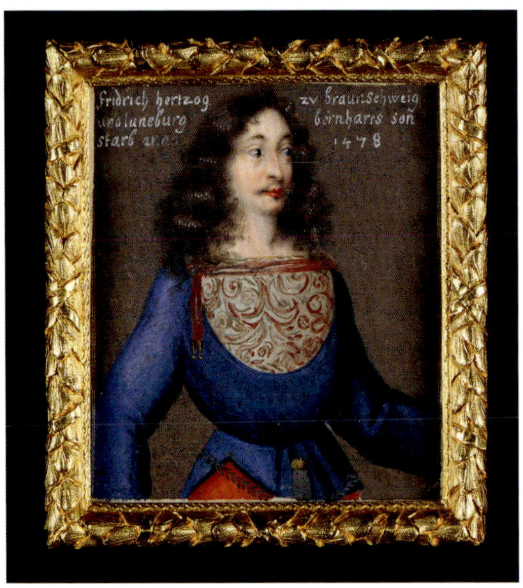

10.12

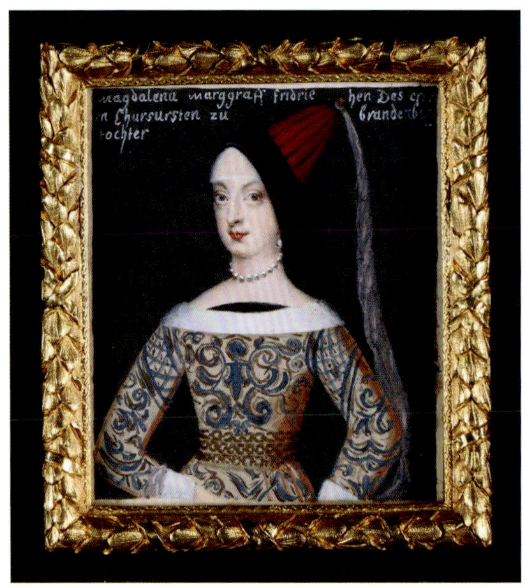

10.13

10.10*

GERMAN SCHOOL
17th century
Bernard I, Duke of Brunswick-Lüneburg (d. 1434)
1600–1735
Watercolour on vellum
3¼ x 2¾ in. (8.1 x 7.0 cm)
Royal Collection Trust, RCIN 420423

10.11*

GERMAN SCHOOL, 17th century
Margareta, Duchess of Brunswick-Lüneburg (d. 1418)
1600–1735
Watercolour on vellum
3¼ x 2⅞ in. (8.2 x 7.2 cm)
Royal Collection Trust, RCIN 420424

10.12*

GERMAN SCHOOL, 17th century
Frederick the Pious, Duke of Brunswick-Lüneburg (d. 1478)
1600–1735
Watercolour on vellum
3¼ x 2⅞ in. (8.2 x 7.1 cm)
Royal Collection Trust, RCIN 420427

10.13*

GERMAN SCHOOL, 17th century
Magdalena, wife of Frederick the Pious, Duke of Brunswick-Lüneburg (d. 1454)
1600–1735
Watercolour on vellum
3¼ x 2⅞ in. (8.2 x 7.3 cm)
Royal Collection Trust, RCIN 420428

Among the series of dynastic miniatures installed in Caroline's Closet at Kensington Palace were eight romanticized portraits of fifteenth-century members of her Brunswick-Lüneburg family, from which this is a sample. Painted in the seventeenth or early eighteenth century, by an unknown German artist, their purpose was to extend the line of ancestry for the Hanoverian family further into the past. The Gothic script used to identify the subjects was intended to enhance the antiquarian impression. They were first noted in 1743 by George Vertue as part of the display in Queen Caroline's Closet at Kensington Palace.

10.14*

DANISH SCHOOL, 17th century
Frederick III, King of Denmark (1609–1670)
c. 1676
Watercolour on vellum
3⅜ x 2⅞ in. (8.4 x 7.2 cm)
Royal Collection Trust, RCIN 420479

10.15*

DANISH SCHOOL, 17th century
Sophia Amalia, Queen of Denmark (1628–1685)
c. 1676
Watercolour on vellum
3⅜ x 2⅞ in. (8.5 x 7.1 cm)
Royal Collection Trust, RCIN 420480

10.16*

DANISH SCHOOL, 17th century
Anna Sophia, Electress of Saxony (1647–1717)
c. 1676
Watercolour on vellum laid on card with a gessoed back
3⅜ x 2⅞ in. (8.4 x 7.1 cm)
Royal Collection Trust, RCIN 420555

10.17*

DANISH SCHOOL, 17th century
Frederica Amalia, Princess of Denmark (1649–1704)
Inscribed 1676
Watercolour on vellum
3¼ x 2⅝ in. (8.2 x 6.5 cm)
Royal Collection Trust, RCIN 420487

In Queen Caroline's Closet at Kensington Palace there was a series of seven miniatures depicting Frederick, King of Denmark, his wife, and four of their six children, as well as Charlotte Amalie of Hesse Kassel, consort to Christian V of Denmark. Three of them are inscribed with the date 1676, which may be the date of production for the series. Some are related to, or copied from, oil paintings of the sitters. These seven miniatures were first noted in the Closet in 1743 by George Vertue.

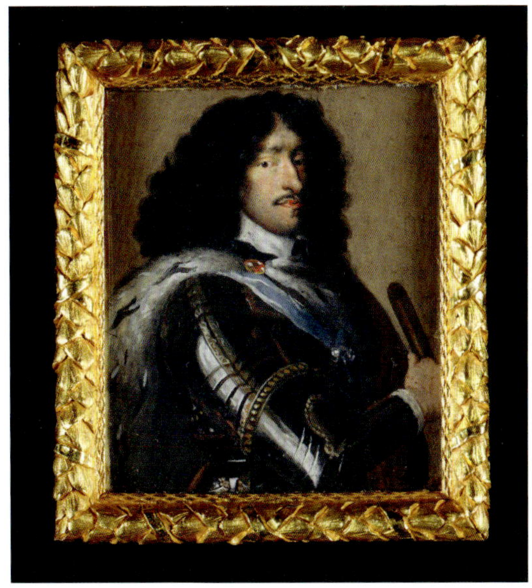

10.14

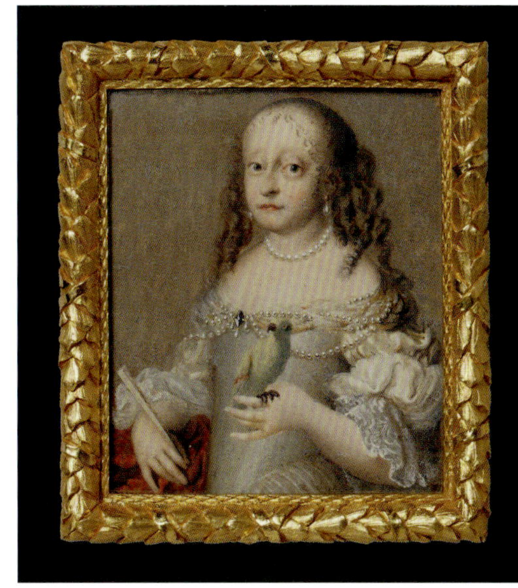

10.15

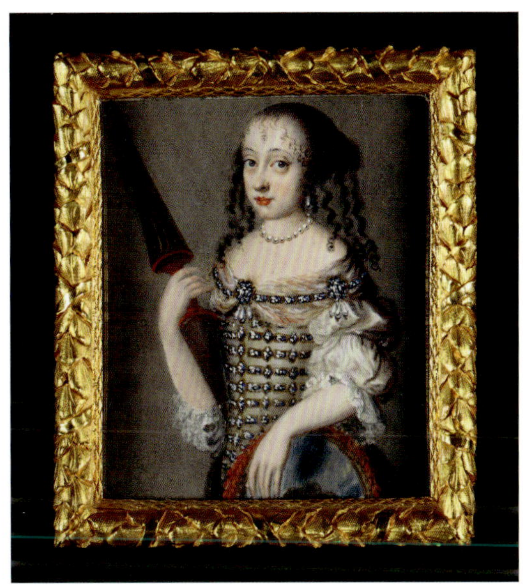

10.16

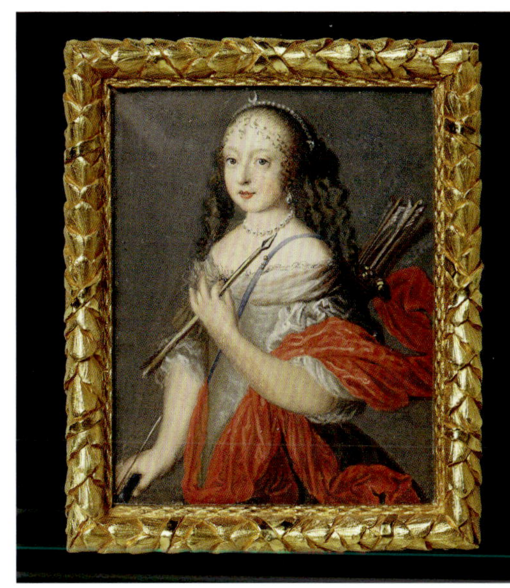

10.17

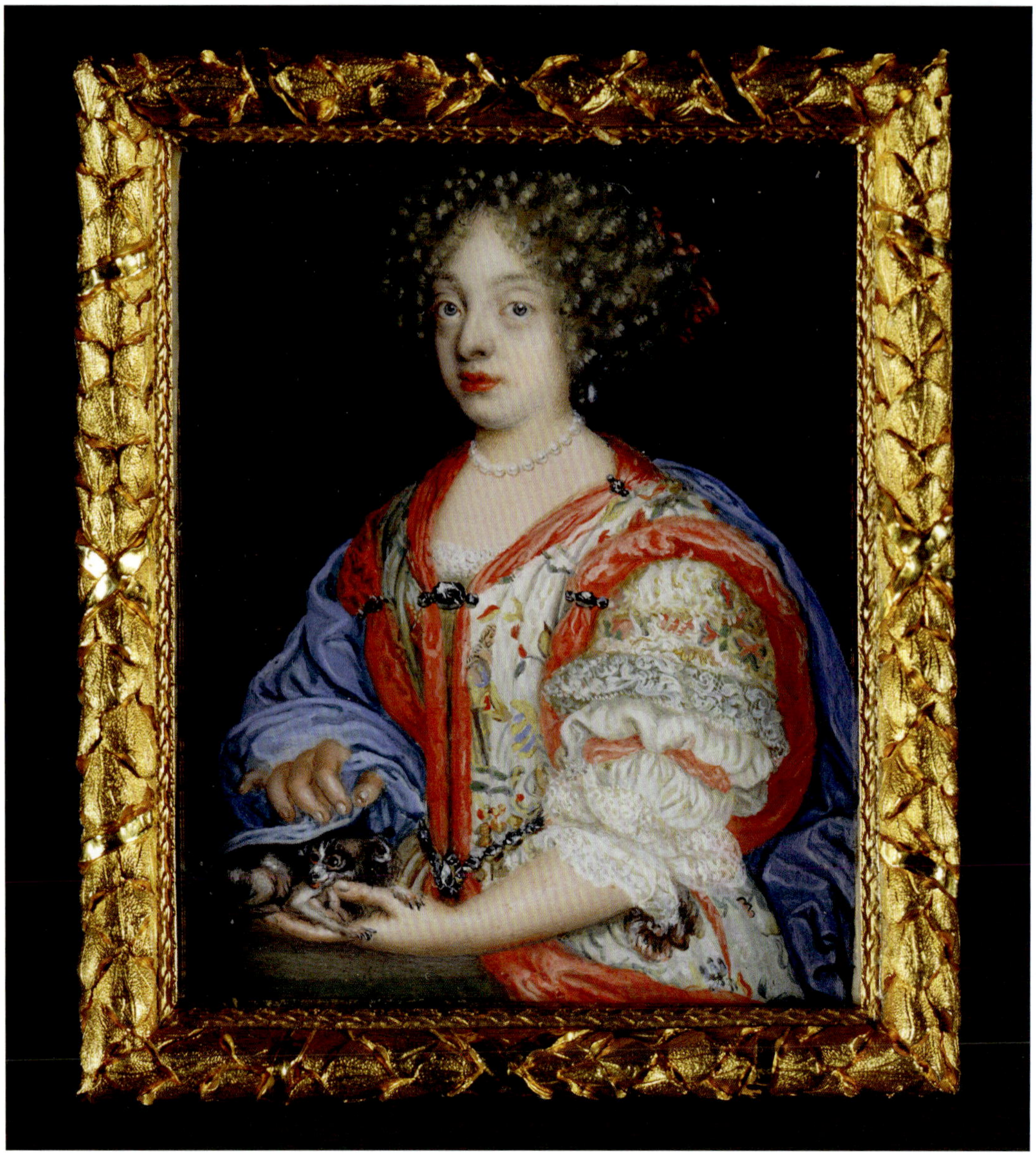

10.18

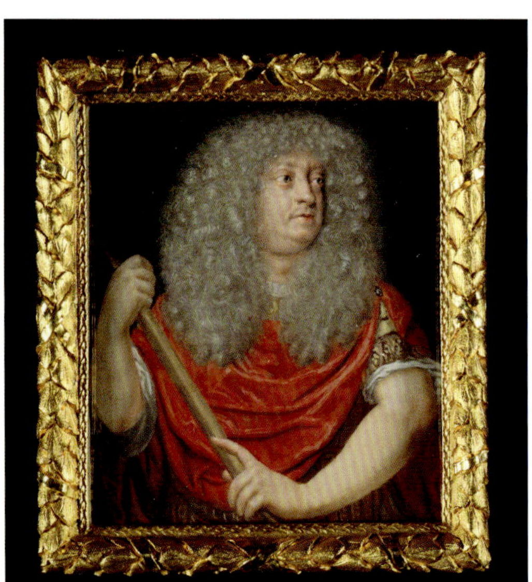

10.19

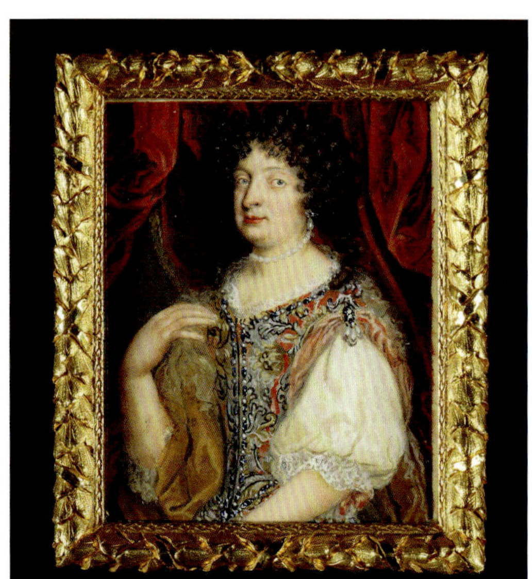

10.20

10.18*

GERMAN SCHOOL, 17th century
Benedicta Henrietta, Duchess of Brunswick-Lüneburg (1652–1730)
c. 1680
Watercolour on vellum
3¼ x 2¾ in. (8 x 6.8 cm)
Royal Collection Trust, RCIN 420461

This is one of eighty-eight small-scale portraits of members of German princely families and others, recorded by George Vertue in Queen Caroline's Closet at Kensington Palace in 1743. Benedicta Henrietta, Duchess of Brunswick-Lüneburg, was both sister-in-law and niece to the Electress Sophia. She had a stronger claim to the British throne than Sophia, but came from a Catholic family, and bore no male children. The miniature is related to a later full-length oil painting of the Duchess in Schloss Marienburg.

10.19*

JEAN MICHELIN (1629–1696)
Ernest Augustus, Elector of Hanover (1629–1698)
c. 1670
Watercolour on vellum laid on card
3¼ x 2⅞ in. (8.2 x 7.1 cm)
Royal Collection Trust, RCIN 420624

10.20*

JEAN MICHELIN (1629–1696)
Sophia, Electress of Hanover (1630–1714)
c. 1670
Watercolour on vellum laid on card
3¼ x 2¾ in. (8.3 x 6.9 cm)
Royal Collection Trust, RCIN 420627

The French miniature painter Jean Michelin worked for the Hanoverian court from 1668 to 1688. There were four portrait miniatures of recent members of the Hanoverian family by the artist in Queen Caroline's Closet, according to Vertue's 1743 inventory, including that of Ernest Augustus, Elector of Hanover, and his wife Sophia, grand-daughter of King James I of England and VI of Scotland. The Electress Sophia's claim to the English throne was ratified by the Act of Settlement in 1701.

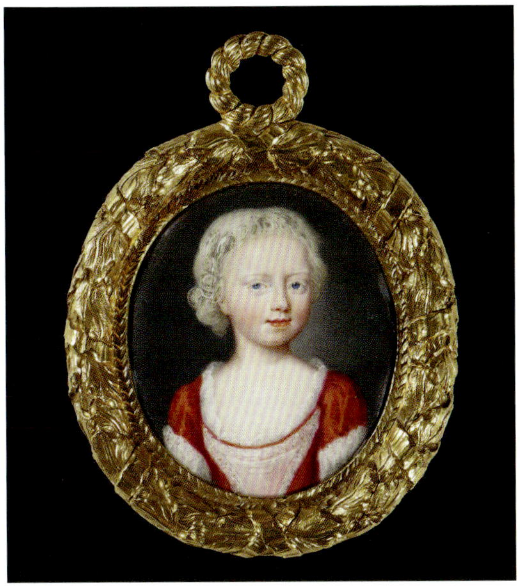

10.21

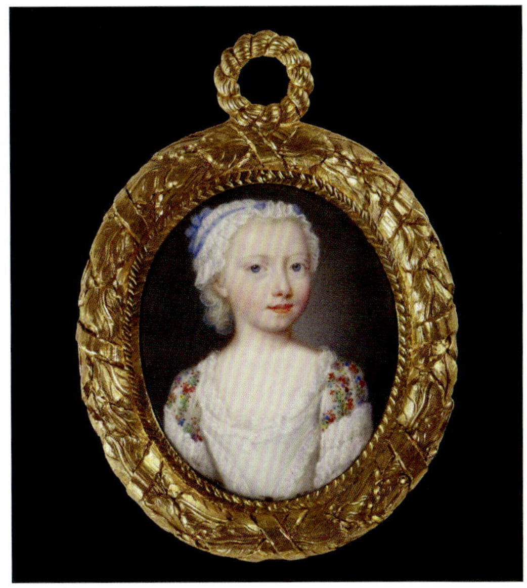

10.22

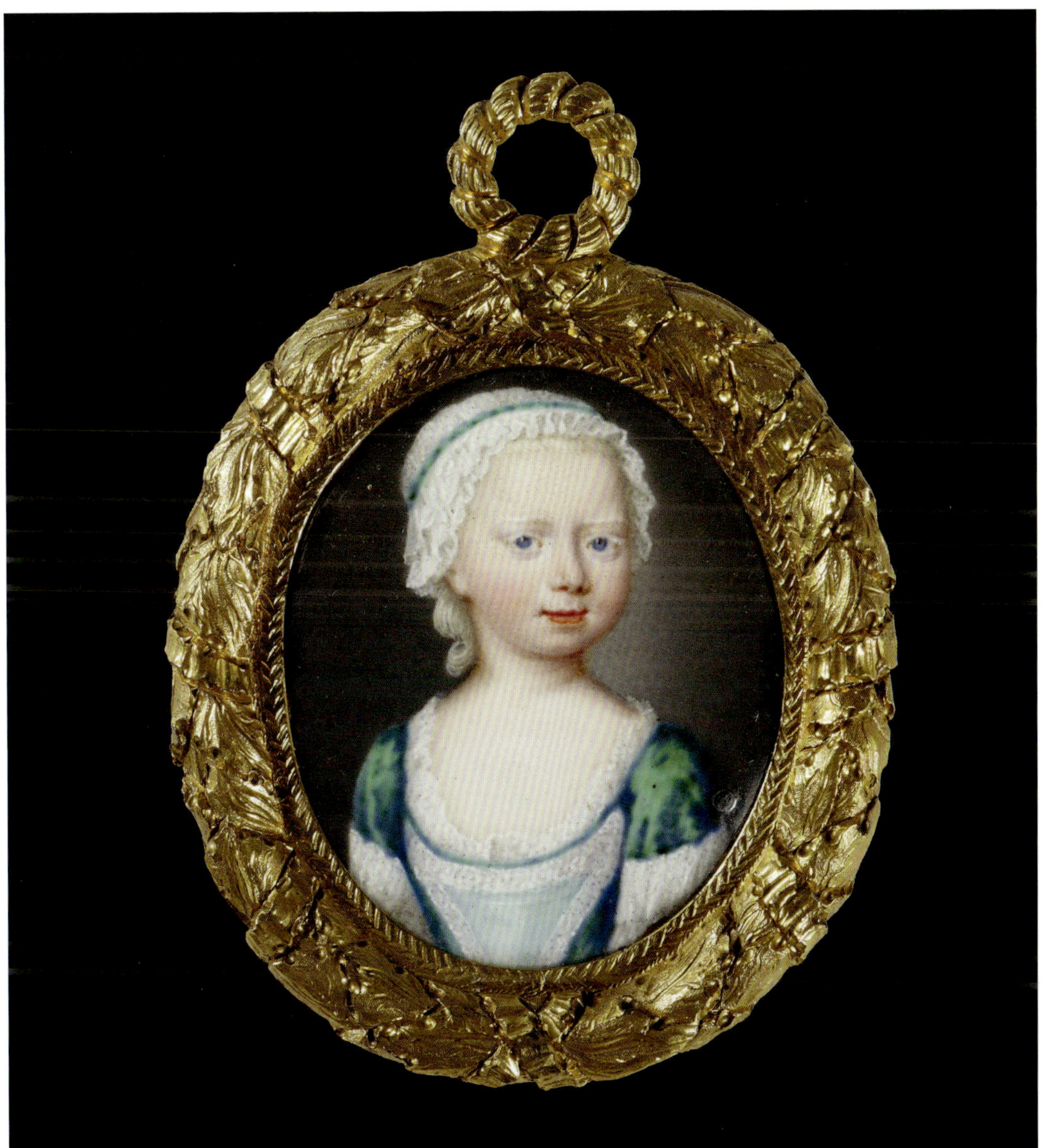

10.23

10.21*

CHRISTIAN FREDERICK ZINCKE
(1683/84–1767)
Frederick, Prince of Wales, as a Child
c. 1714
Enamel
1¾ x 1⅝ in. (4.4 x 4.1 cm)
Royal Collection Trust, RCIN 421804

10.22*

CHRISTIAN FREDERICK ZINCKE
(1683/4–1767)
Anne, Princess Royal, later Princess of Orange, as a Child
c. 1709–14
Enamel
1¾ x 1⅝ in. (4.5 x 4.0 cm)
Royal Collection Trust, RCIN 421805

10.23*

CHRISTIAN FREDERICK ZINCKE
(1683/84–1767)
Princess Amelia as a Child
c. 1714
Enamel
1⅞ x 1⅝ in. (4.6 x 4.0 cm)
Royal Collection Trust, RCIN 421806

The portrait miniatures of the three eldest children of George II and Caroline, as Prince and Princess of Wales, were probably painted at the time of George I's coronation in October 1714. Christian Frederick Zincke was born in Dresden, the son of a goldsmith. He settled in England in 1706 and studied enamel painting under Charles Boit. Boit had been Enamel Painter to the Court until Queen Anne died in 1714 when he moved to France. Zincke took over much of his work and worked extensively for members of the royal family; possibly because he spoke German he was well liked. Vertue recorded that when Zincke made portraits of George II and Caroline in 1732, the Queen asked him to make the King look youthful, while George II instructed him to ensure that Caroline looked no more than twenty-eight years old. In February 1732 Zincke was appointed Cabinet Painter to Frederick, Prince of Wales, and Augusta's Privy Purse accounts for 1738 reveal the artist undertook private commissions for her too. Vertue's drawings and listing of the contents of Caroline's Picture Closet reveals that it contained portraits of her children.

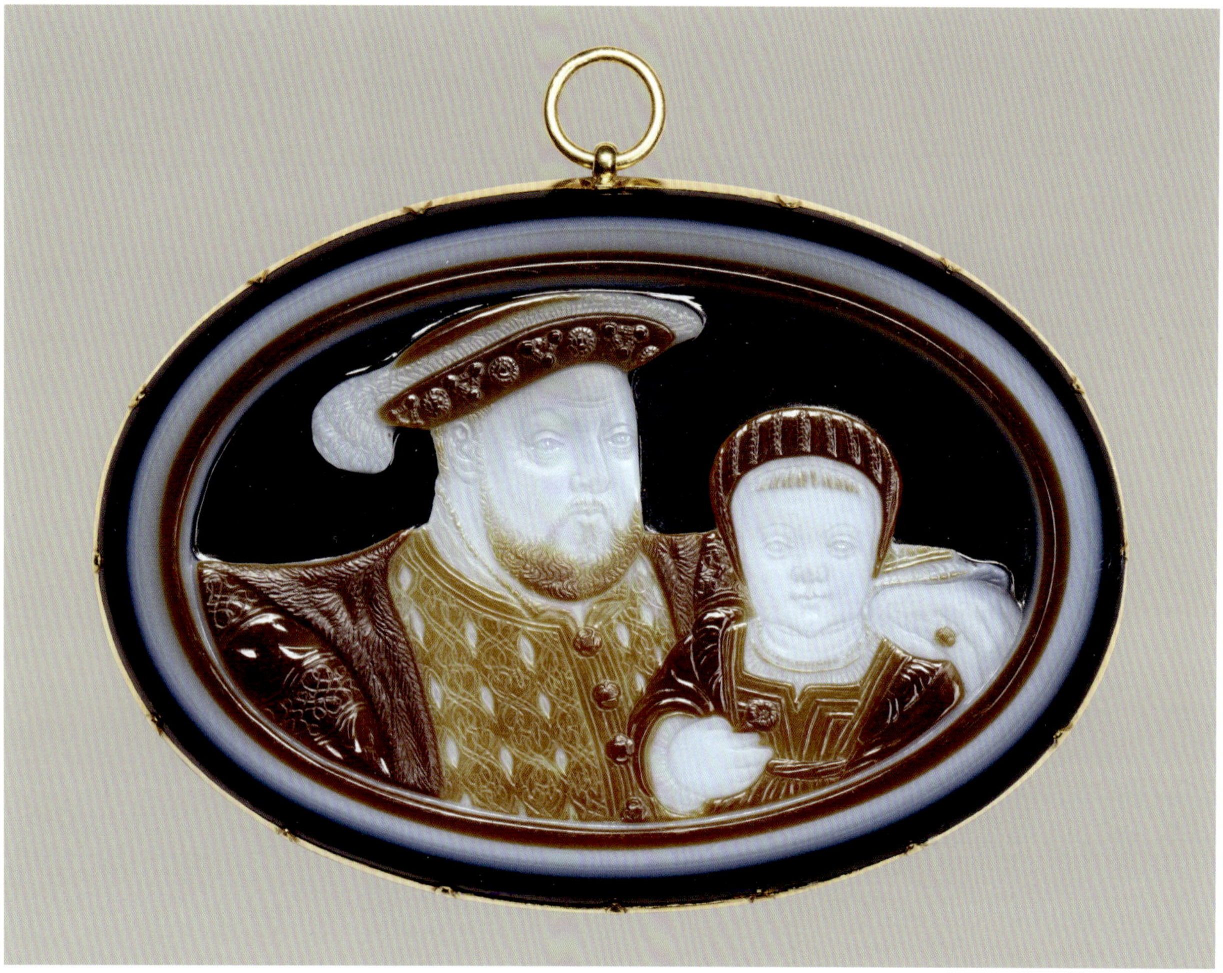

10.24

CAROLINE'S CABINET OF CURIOSITIES

10.24*

UNKNOWN JEWELLER, after HANS HOLBEIN THE YOUNGER (1497/98–1543)
Henry VIII and Prince Edward
18th century
Sardonyx and agate cameo in gold mount
2¼ x 3⅛ in. (5.7 x 7.8 cm)
Royal Collection Trust, RCIN 65189

This cameo of Henry VIII and his son Edward, with another in the Royal Collection showing Henry VIII alone, is carved in very low relief that provides an unusual, almost stereoscopic effect when they are held up to the light. The cameo of the King with his son has so many similarities with a gem from the collection of Lord Burlington (Chatsworth House) it is probable that they come from the same workshop. The sources for the images resemble typical images of the subjects by Hans Holbein, but the gems themselves are likely to be early eigtheenth-century productions inspired by these. This cameo was mentioned in the inventory made of Caroline's collection of curiosities by her laundress, Margaret Purcell, some time between 1737 and Mrs Purcell's death in 1755 (BM Add. MS. 20101, fol. 60). It is subsequently noted in Horace Walpole's annotations to his copy of W. A. Bathoe's volume of royal inventories, now in the Royal Collection, together in his tidy list of rarities from Caroline's collections, which were sent from Kensington Palace to Buckingham House on 22 March 1764.

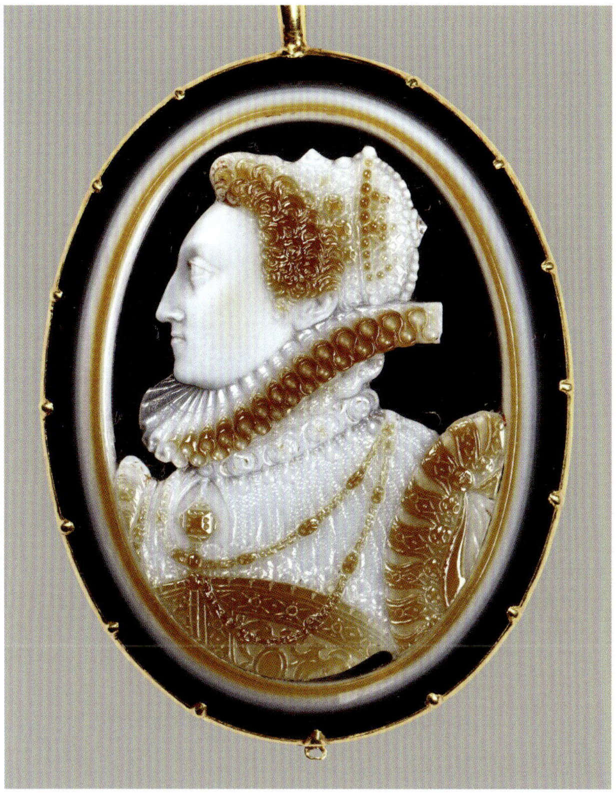

10.25

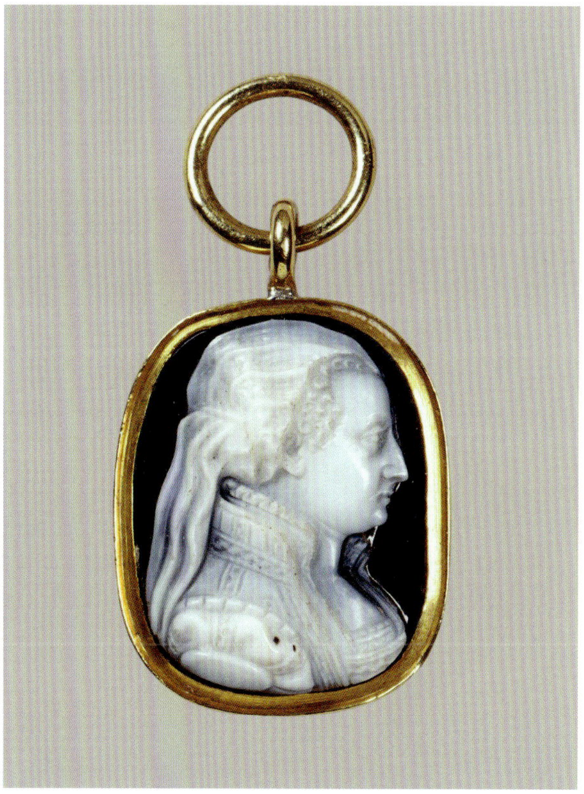

10.26

10.25*

UNKNOWN JEWELLER, English
Elizabeth I
Cameo: c. 1575–85; mount: early 18th century
Sardonyx and agate cameo in gold mount
2⅝ x 2¼ in. (6.7 x 5.5 cm)
Royal Collection Trust, RCIN 65186

The greater part of Caroline's collection of curiosities comprised material collected by her English royal predecessors. King James I on his accession in 1603 had inherited an impressive collection of virtuoso metalwork and curiosities commissioned and collected by Henry VIII and Elizabeth I. The lists of Caroline's curiosities made by Horace Walpole in 1763 and 1764 (RA Geo Add. MS. 16) include several cameos of Elizabeth I, described variously as '2 of Eliz by Valerio Vincentino, the larger good', 'Queen Elizabeth small by Valerio' and 'Large cameo of Q. Elizabeth in profile by ditto, set in gold', this last probably the example shown here. Elizabeth I was an important inspiration for Caroline. Her imagery was built into the Queen's folly temples at Richmond between 1730 and 1735, and she was selected as one of the subjects for the series of busts illustrating key figures in the royal lineage, commissioned from Michael Rysbrack in 1736, which were intended as decoration for her new library at St James's Palace.

10.26*

UNKNOWN JEWELLER, Italian
Mary, Queen of Scots
Cameo: c. 1560; mount: late 18th century
Agate and onyx cameo in gold mount
3/4 x⅝ in. (1.9 x 1.5 cm)
Royal Collection Trust, RCIN 65198

This cameo of a widow was traditionally described as Mary, Queen of Scots, but the identity is controversial. It may have been the 'very fine head of Mary Queen of Scots, profile with a veil' or 'A most beautiful head of Mary Queen of Scots, profile, cameo on onyx, I believe by Valerio Viventino or a better hand', mentioned by Horace Walpole in his lists of curiosities (c. 1764) from Caroline's Cabinet at Kensington Palace, and it was among those objects which moved from the Kensington collection under his auspices to join other collections made by George III at Buckingham House. A jewel associated with Queen Mary, a pivotal figure linking the Tudor and Stuart dynasties, would be seen as very valuable in a collection drawn up to celebrate the distinction and pedigree of the English monarchy.

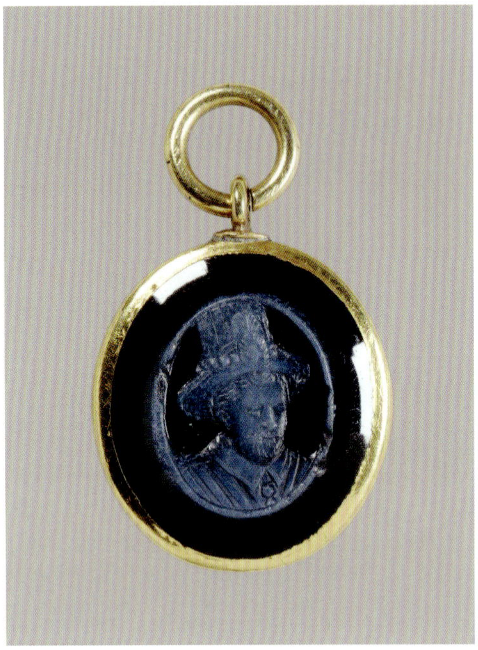

10.27

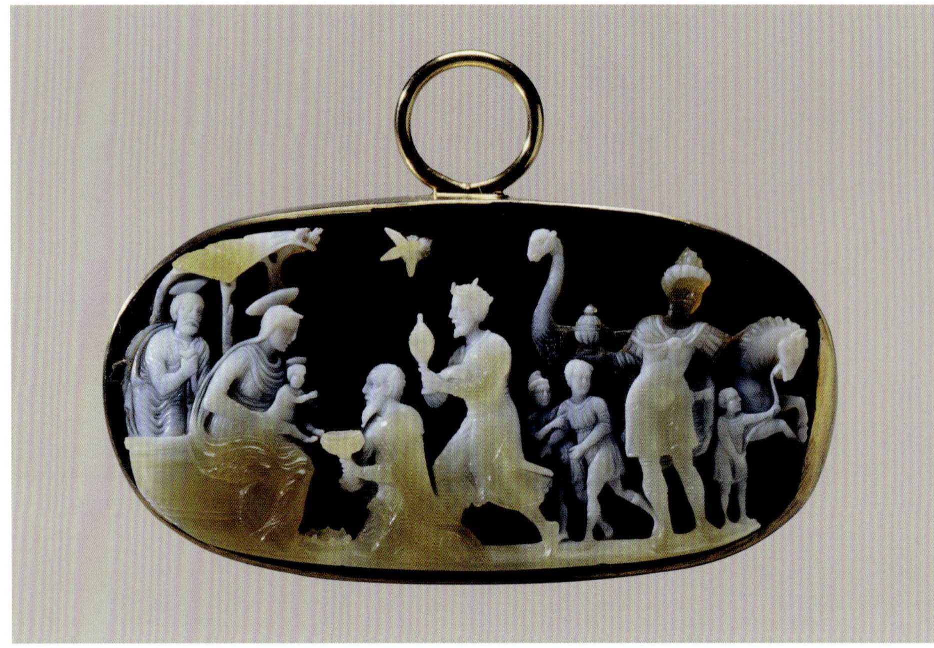

10.28

10.27*

UNKNOWN JEWELLER, possibly French
James I
Cameo: c. 1590; mount: 18th century
Sardonyx and agate cameo in gold mount
3/4 x⅝ in. (1.9 x 1.7 cm)
Royal Collection Trust, RCIN 65191

This cameo was already in royal hands by June 1660, when Elias Ashmole took wax impressions of objects from Charles II's collection. At that time it was mounted in a ring; it was remounted some time in the later eighteenth century. In his list of curiosities Walpole described the gem as 'King James full face with his hat on, cameo', and noted that it was one of a number of cameos commemorating the first Stuart monarch he ensured were transferred from the Kensington collection for George III (RA Geo. Add. MS. 16).

10.28*

UNKNOWN JEWELLER, northern Italy
The Adoration of the Magi
Cameo: 16th century; mount: c. 1700
Agate and sardonyx cameo in gold mount
1⅜ x 2⅝ x⅝ in. (3.4 x 6.5 x 1.5 cm)
Royal Collection Trust, RCIN 65175

This outstanding North Italian, sixteenth-century cameo is by an unknown craftsman, though the names Matteo dal Nassaro or Dominicus Romanus are tentatively suggested. The scene, deeply undercut, depicts the Adoration of the Magi. A masterpiece of virtuoso carving, a sense of space and perspective has been created as the pale camel's head and neck fade into the darker layer, behind the group of pages. The work serves to show Caroline's exquisite taste, perhaps nurtured at the courts of Saxony and Brandenburg where she spent her early years, and where there were distinguished collections of gems, cameos, and the finest metalwork. The cameo was first recorded in an inventory drawn up by Margaret Purcell sometime between 1737 and 1755, where it is described as 'An Onex of the Adoration of the Kings'.

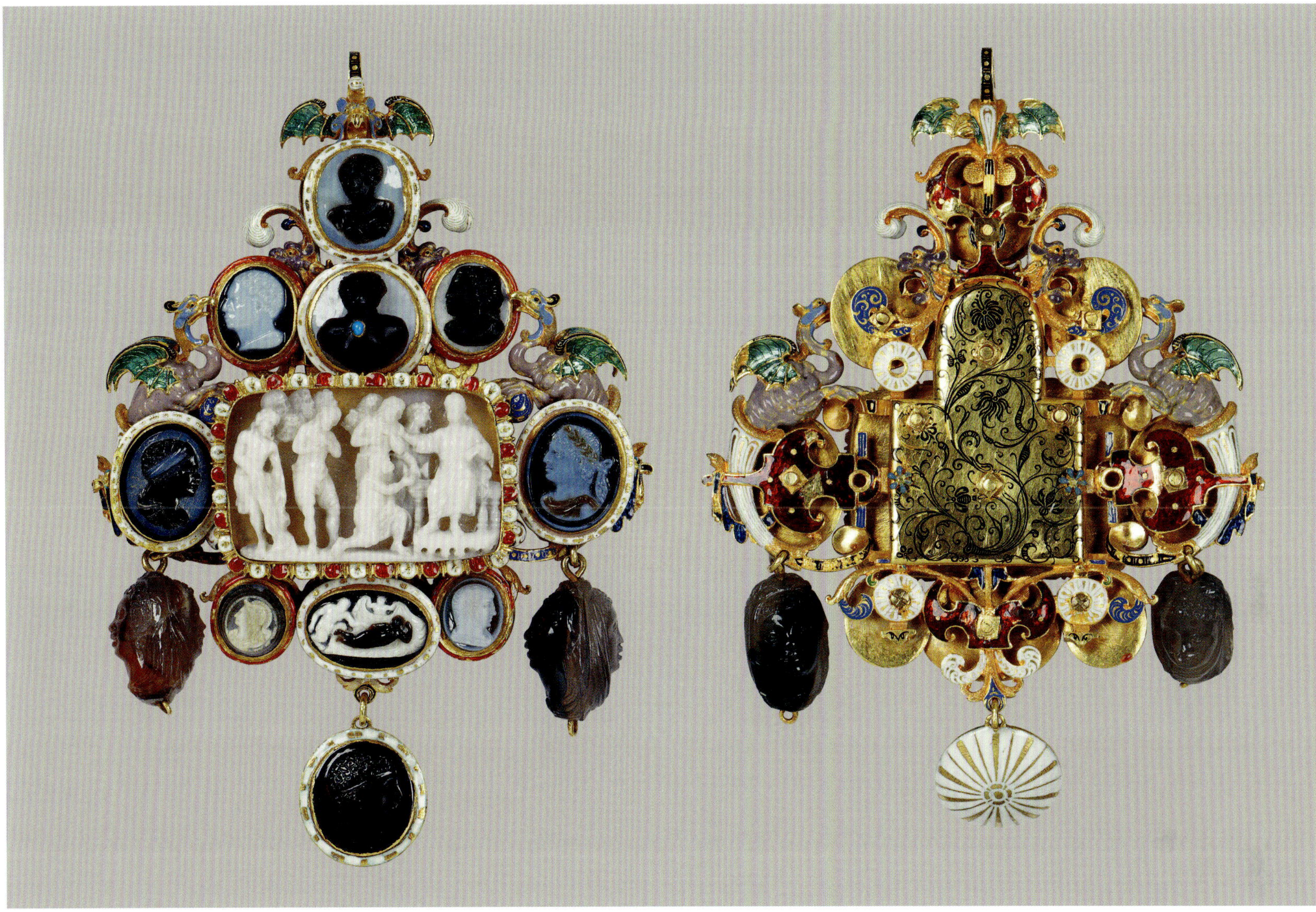

10.29

10.29*

UNKNOWN JEWELLER, French
Pendant with thirteen cameos
Cameos: early 17th century; mounts: 18th century
Gold, émail en ronde bosse, émail champlevé, sardonyx, turquoise, black and white onyx, sard, brown and grey agate, black and grey agate, and niccolo
5⅛ x 3½ in. (13 x 8.9 cm)
Royal Collection Trust, RCIN 65256

The jewel is built around an early seventeenth-century gold openwork pendant decorated with enamel. The setting of the original central gem suggests it may have been the work of Jean Vangrol, the Flemish jeweller and valet de chambre to Marie de Médicis, who worked in Paris until his death in 1640, or a craftsman in his circle. The central element was replaced, probably in the eighteenth century, with a cameo possibly by Francesco Tortorino, a Milanese gem-engraver, showing Joseph and his brothers. It was at this time that the additional antique and contemporary cameos were added. The jewel was part of Caroline's collection, and may have been a gift from her kinswoman Elisabeth Charlotte of the Palatinate, Duchess of Orléans—Liselotte—with whom Charlotte conducted an extensive correspondence between 1715 and 1722. Liselotte, who had amassed a fine collection of gems, discussed with her half-sister her desire to find French 'engraved stones' for Caroline. It seems likely this is the piece mentioned by Margaret Purcell in her inventory of Caroline's collection of curiosities housed at Kensington Palace, as 'A hanging Jewell of onexes containing 12 heads and a piece of figures in the middle' (BM Add. MS. 20101, fol. 60). Horace Walpole descibes it as a 'Large cameo set round with several others' when it was sent to George III.

10.30*

UNKNOWN SCULPTOR, Northern Italy
Hercules
Bust: second half 16th century; socle: late 19th/ early 20th century
Saxon agate, mounted on gilt metal plinth
2¼ x 1½ x 1¼ in. (5.7 x 3.8 x 3.0 cm)
Royal Collection Trust, RCIN 65740

This gemstone bust of the warrior and hero Hercules with a lion skin draped over his head is cut in the round. Carvings of this kind were produced in the ducal workshops in Florence by Milanese craftsmen for the decoration of fine furniture. Margaret Purcell lists this among other small-scale statuary carved from wood, coral, ivory, precious and semi-precious stone in her inventory of Caroline's collection of curiosities made between 1737 and 1755. This is presumed to be the artefact noted by Horace Walpole in June 1763, and also perhaps the 'Bust of Hercules in agate' recorded among items sent by Walpole from Kensington Palace to George III in March 1764.

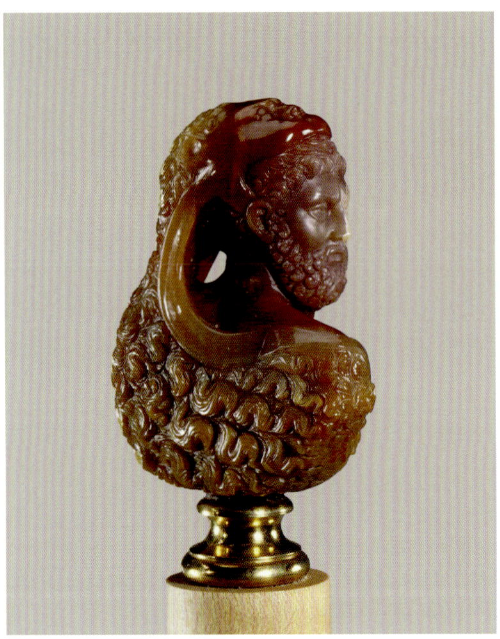

10.30

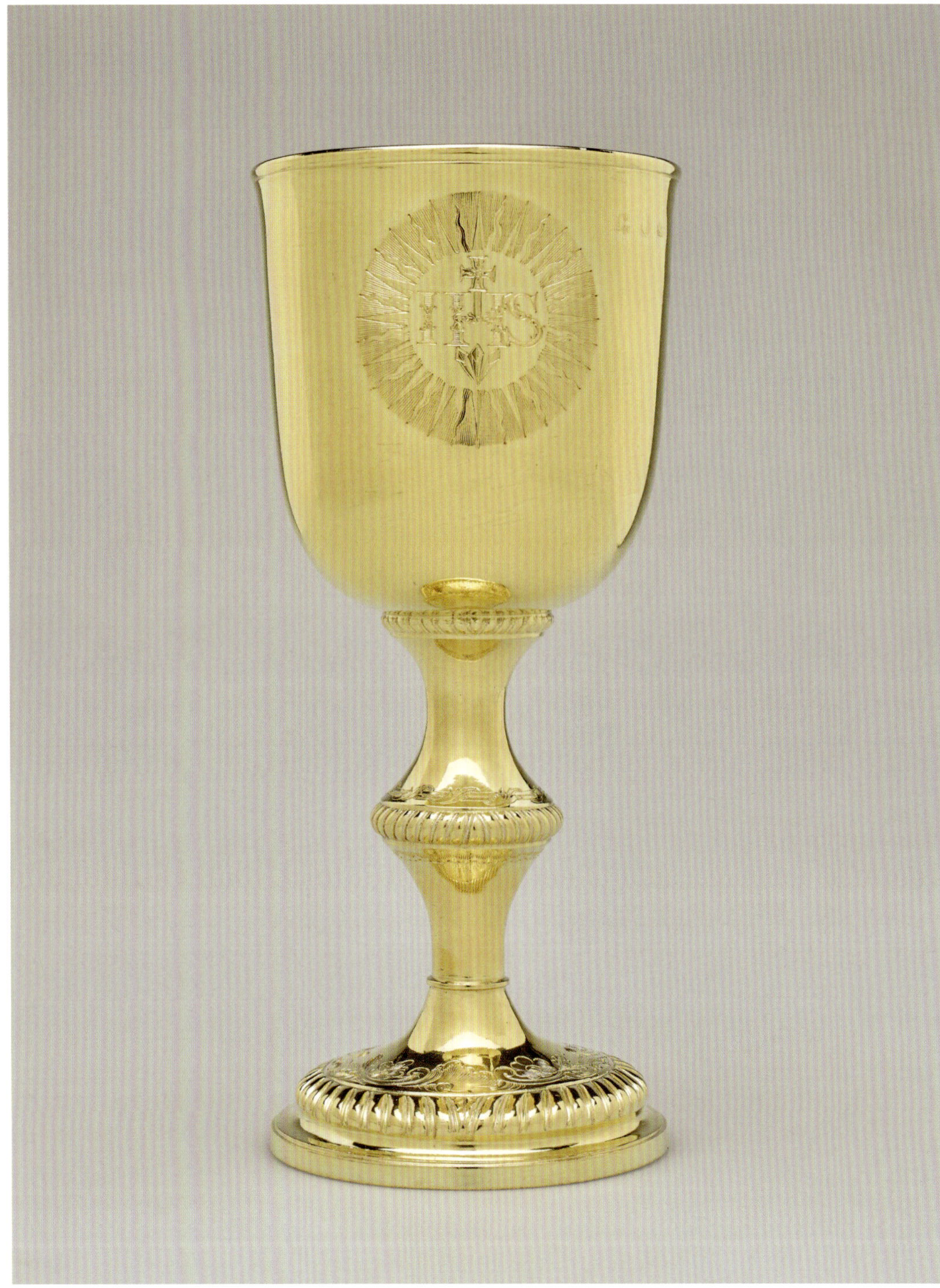

10.31

COMMUNION PLATE

10.31

THOMAS HEMING (1722/23–1801), goldsmith
Chalice
1760–61
Silver gilt
8¼ x 3½ x 3½ in. (21 x 9 x 9 cm)
Royal Collection Trust, RCIN 48467

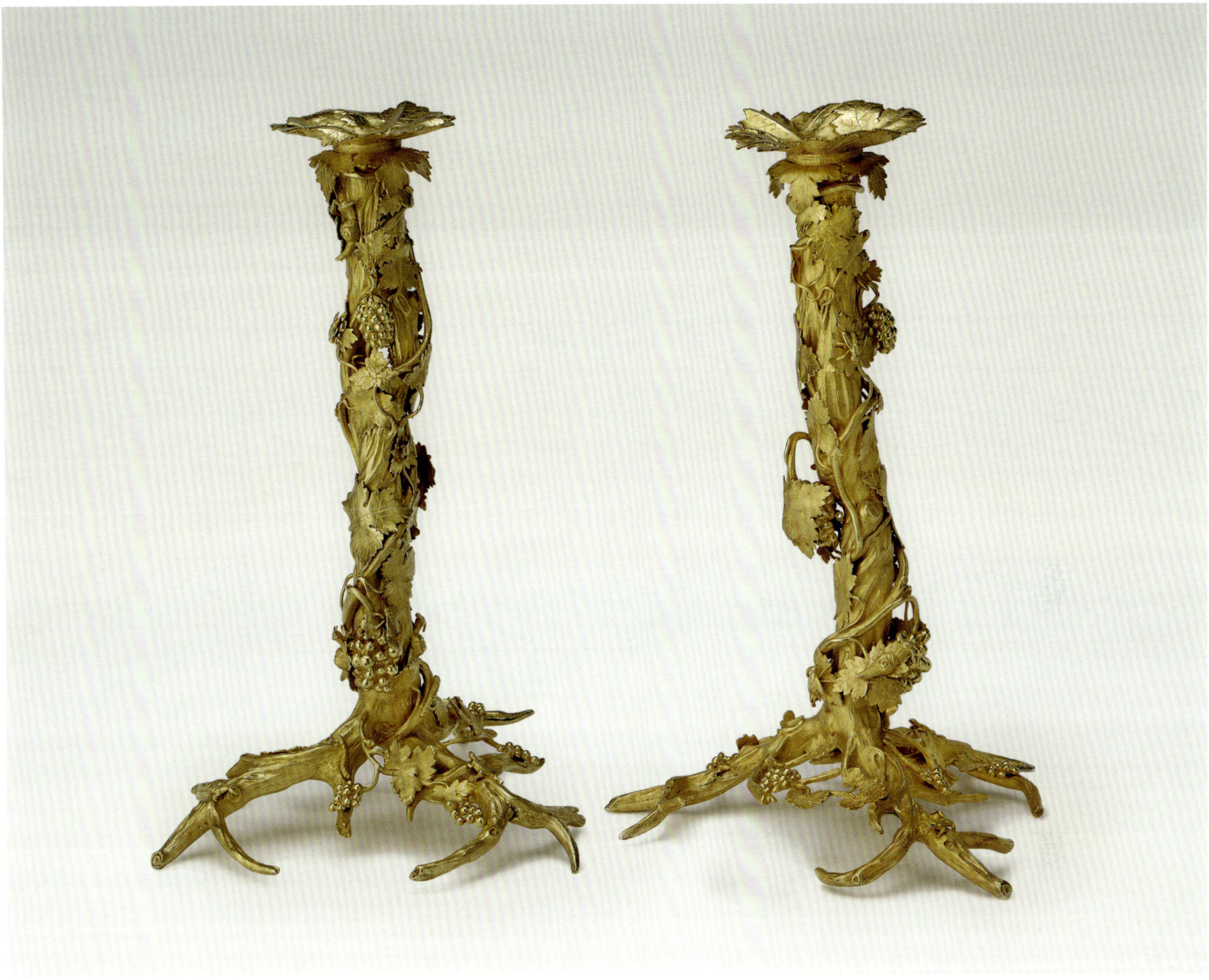

10.32

10.32

THOMAS HEMING (1722/23–1801), goldsmith
Pair of candlesticks
1757–58
Silver gilt
14¾ x 9⅜ x 9 in. (37.3 x 23.7 x 23 cm)
Royal Collection Trust, RCIN 51619

For all the Princesses their Protestant faith had been an essential qualification in the brokering of their marriages into the Hanoverian family. It was important that this was observed publicly, but each of the Princesses held their beliefs with sincere conviction. Visitors to Hampton Court today can enter Caroline's oratory, within her private apartments. It was important in their new roles that communion was taken according to the Anglican rite. Frederick admitted to Caroline that he found Augusta weeping, and that she 'talked of her conscience' when he insisted she attend the Lutheran Chapel. Once convinced that her failure to attend the services would imperil her status as Consort she agreed to put her scruples aside. For the private oratories communion plate would be required. The bowl of the chalice is engraved with the IHS monogram in glory, and with a coronet and AP monogram. This is for Augusta, Princess of Wales. One leaf on each candlestick is also engraved with an AP monogram and coronet.

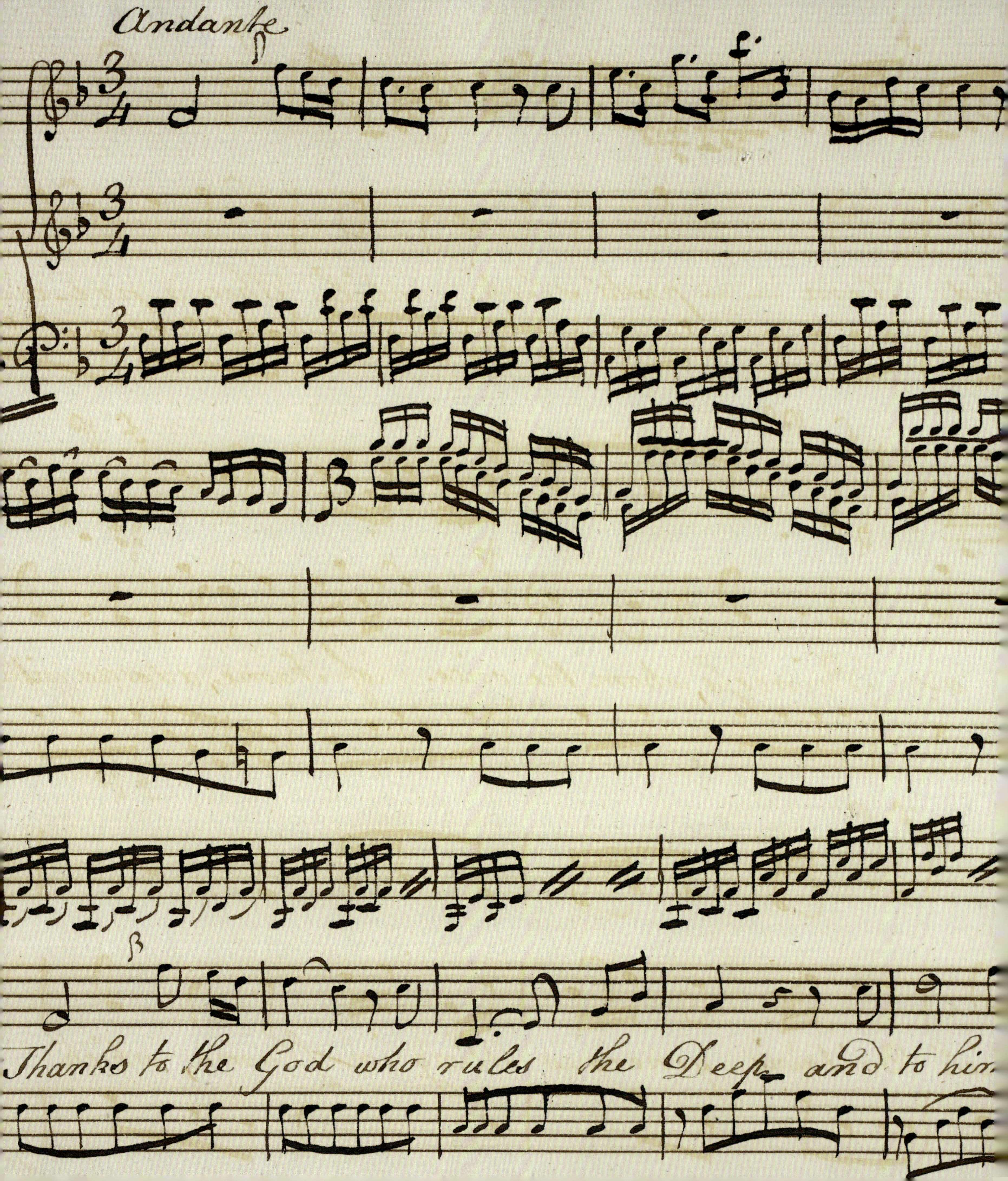
Andante
Thanks to the God who rules the Deep and to him

Music and Representation

Donald Burrows and Berta Joncus

'A love of music to distraction runs through our family', wrote Charlotte of Mecklenburg-Strelitz's eldest daughter in 1788.[1] Caroline of Ansbach, Augusta of Saxe-Gotha, and Charlotte all shared this love, which crystallized in Charlotte's patronage of the young Wolfgang Amadeus Mozart. In the early eighteenth century, the thirteen-year-old Caroline received a rigorous musical education from her guardian Sophia Charlotte, Electress of Brandenburg-Prussia, later first Queen of Prussia, who enlisted her keyboard master, Johann Anton Coberg, to teach her.[2] According to standard pedagogy, fifty years later Charlotte too would have learned to play, sing, extemporize, and harmonize at sight. As well as being a passion, music was also a powerful medium through which the Princesses' tastes, political allegiances, and propitious conduct could be broadcast. Their musical activities took place in companies domestic, select, and public. Within these three overlapping spheres, each Princess cultivated music in their own way as part of their self-representation.

Mentored by Sophia Charlotte, Caroline came to appreciate 'learned'—contrapuntally complex—cantatas, especially those of Agostino Steffani, whose manuscripts Sophia Charlotte bequeathed to Caroline. After marrying and moving to Hanover in 1710, Caroline immediately recognized the genius of one visiting virtuoso, George Frideric Handel (fig. 7.01). In deference to Caroline, Handel during his visit wrote 'for her . . . practice' five duets, modelled after, and even quoting from, Steffani.[3] Handel's appointment as Kapellmeister to the Hanoverian court, which changed the face of British music, was indebted to Caroline's taste.

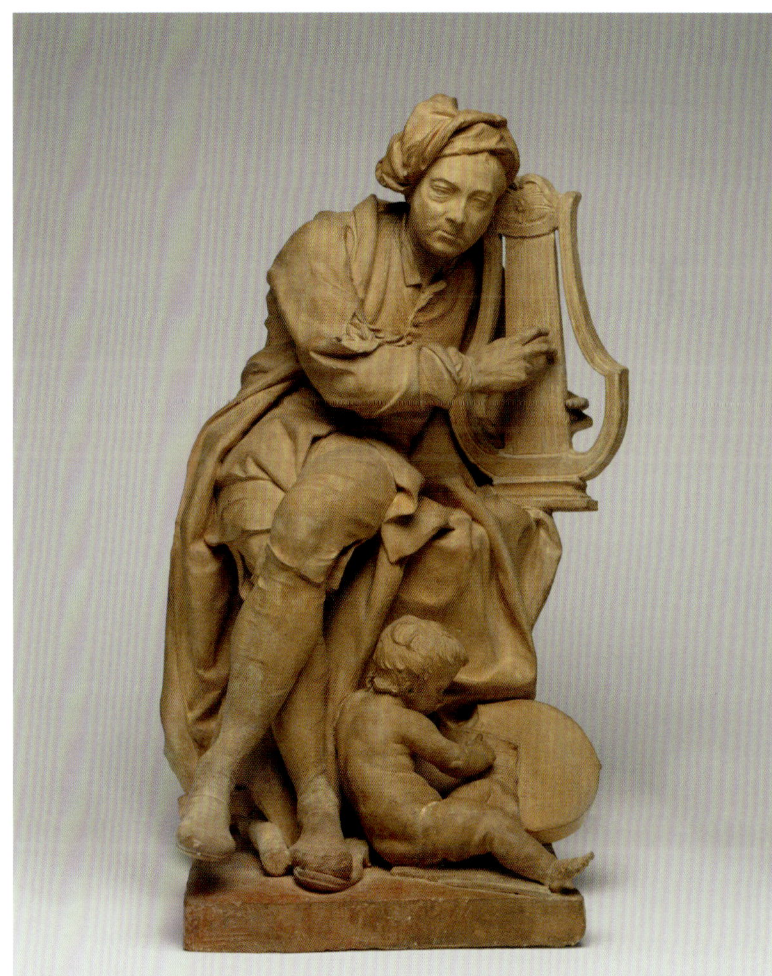

FIG. 11.01 Louis François Roubiliac, *George Frideric Handel*, before 1738, terracotta, 18½ x 10⅝ x 14⅛ in. (47 x 27 x 36 cm). Fitzwilliam Museum, Cambridge

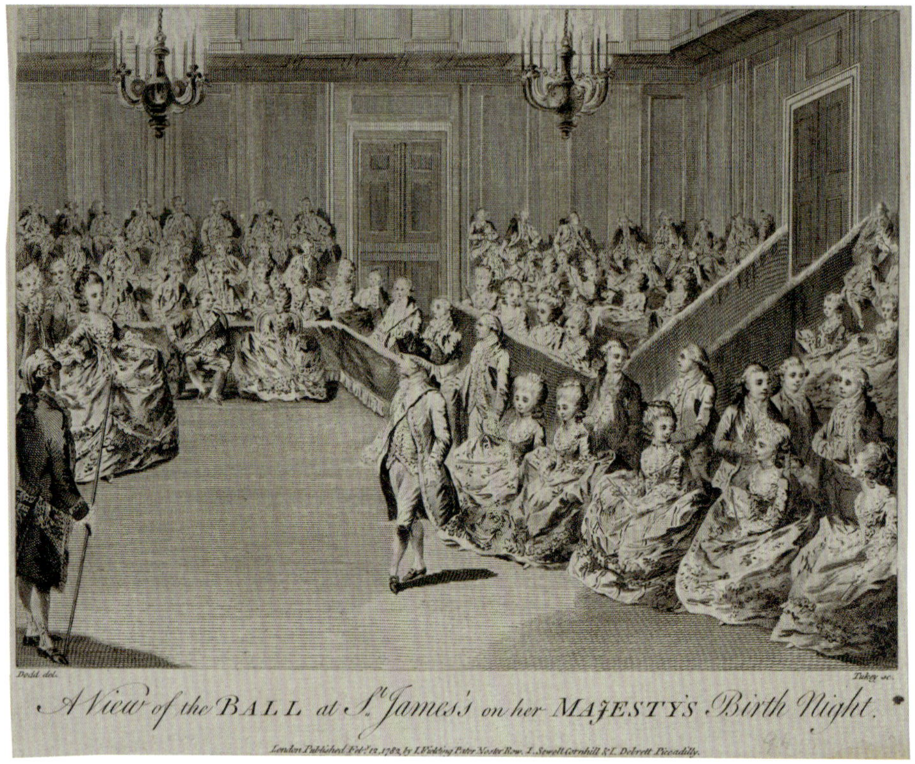

FIG. 11.02 Unknown engraver after Daniel Dodd, *A View of the Ball at St James's on Her Majesty's Birth Night, 1781*, 1782, engraving with etching on paper, 7¼ x 8⅝ in. (18.4 x 21.8 cm). Royal Collection Trust

In 1714, Caroline moved with the Hanoverian court to London, and for Britain a new era of princess-led musical connoisseurship began. It started with the education of the royal children. Caroline's seven-year-old eldest son, Frederick, required by order of King George I to stay in Hanover as token of the family commitment to the Electorate, was left under the instruction of 'Denoyé', undoubtedly a member of the Desnoyer family who were celebrated dance masters at the court of Louis XIV in France. When the elder Desnoyer died in 1721, his son George Desnoyer was appointed to provide Frederick's dance lessons.[4] In London, Caroline organized instruction for her other children because, in the later words of Philip Dormer Stanhope, fourth Earl of Chesterfield, 'to dance well, is absolutely necessary, in order to sit, stand, and walk well.'[5] According to the dancing master John Essex, Caroline 'retriv'd' the 'Grandure & Gayety of the Court' by holding court balls, which had lapsed under Queen Anne.[6] From 1715 she held, among other assemblies, birthday celebrations with dances named after and often led by royal family members. Printed choreographies allowed those outside the court to learn these dances and their tunes. Generations later, Charlotte still cultivated the tradition revitalized by Caroline, presiding over birthday balls that opened with *la belle danse* performed by members of the royal family (fig. 11.02).

In 1717, George I expelled Caroline and George, Prince of Wales, from court, and assumed responsibility for the education of the couple's older children. However, the King's quarrel was with his son, and after the death of one of the children while in his charge, he allowed Caroline limited access to her family. She probably had a hand in the 1717 appointment of Handel as music master to the three young princesses at St James's Palace. Tutored by Handel, Caroline's eldest daughter, Princess Anne, outstripped her mother's musicianship. According to Handel, Anne was the 'Flower of all Crown Princesses' who alone could move him 'to any teaching'.[7] Philippe Mercier's 1733 painting of Caroline's elder children, *'The Music Party': Frederick, Prince of Wales with his Three Eldest Sisters* (fig. 11.03), represents a family united in harmony, in the sense that they are portrayed playing together using the musical skills encouraged by their mother. But Mercier's image belies the antagonism that existed between Caroline's eldest son, Frederick, Prince of Wales, and the rest of the family after he joined them in London in 1728. Rebelling against his tight-fisted father, from 1733 Frederick supported the Opera of the Nobility company, formed to rival the opera company of Handel, the royal music master.[8] Backing her husband, Caroline encouraged London's opera patrons to favour Handel, and stepped up the frequency of her private concerts, apparently to counter

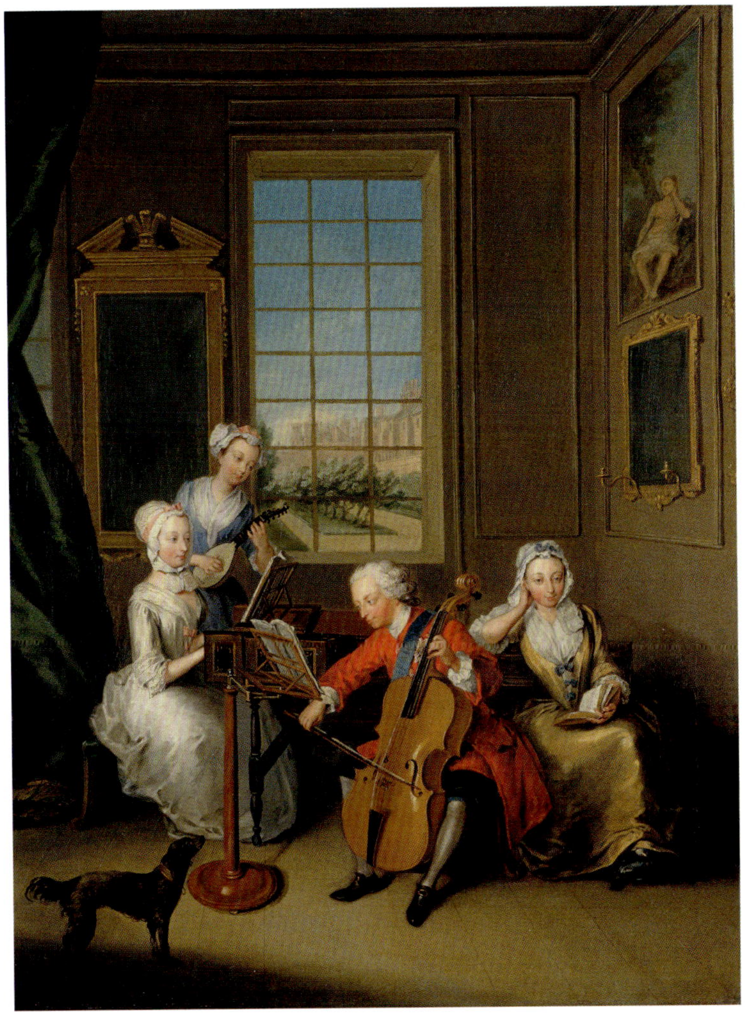

FIG. 11.03 Philippe Mercier, *'The Music Party': Frederick, Prince of Wales with his Three Eldest Sisters*, 1733, oil on canvas, 31¼ x 22¾ in. (79.4 x 57.8 cm). Royal Collection Trust (cat. 23.06)

Frederick's—which Handel was eventually forbidden from directing.[9] Driving the Nobility Opera was its patrons' desire to hear star singers. Caroline deftly co-opted many of these for her private concerts: after a 1735 concert with the Nobility Opera stars Francesca Cuzzoni, castrati Carlo Maria Michelangelo Nicola Broschi, called Farinelli, and Francesco Bernardi Senesino, Caroline organized first a concert with Handel's vocalists, castrato Giovanni Carestini and Anna Maria Strada del Po, then another bringing together the singers from the rival companies.[10] That Caroline was more conciliatory towards Frederick during the King's sojourn in Hanover in the summer of 1736, when she let him command two Handel operas the following season, suggests that the hostility towards her son was, at least in the instance of their musical patronage, put on to oblige her husband.

Frederick's wife Augusta excelled in the performance of docility. Thrust into her husband's family conflicts from 1736, she swiftly learned to win favour by appearing innocent.[11] Her musical patronage underscored her submission: from 1736 she employed the dance master George Desnoyer (the son of Frederick's original dance master) who had taken over Frederick's tuition in Hanover in 1722, then later followed him to London.[12] In addition she paid for the music master Giuseppe Sammartini—a composer-oboist and player in the Nobility Opera band—and took keyboard lessons, although her sister-in-law Princess Amelia sneeringly reported that because Augusta 'doth not advance at all with her harpsichord she is already grown quite tir'd of it'.[13] In 1751 she appointed a second Italian music master, Pietro Domenico Paradies, to tutor Prince George.[14] The engagement of these Italian music masters represented Frederick's taste, which broke with the Hanoverian tradition of employing Germans. As well as employing Handel, Caroline had secured the services of Carl Friedrich Weideman to provide musical tuition for her younger children.[15]

Music was also crucial to Frederick's oppositional patriotic agenda. Famously he commissioned the masque *Alfred*, with music by Thomas Arne, and with its chorus, 'Rule Britannia!'. It is less well known that during the evening in 1740 when it was performed, a second masque, *The Judgement of Paris*, a resetting by Giuseppe Sammartini of William Congreve's famous libretto of 1701, was also put on. Singing the title role himself, Frederick enacted yielding to Venus after she promises him the 'matchless fair' Helen, who, 'were she for the Crown contending', would be 'Queen'—surely an allusion to the watching Augusta.[16] To connect Augusta with Britain, Sammartini echoed Henry Purcell's air 'Fairest Isle, all Isles excelling' when setting similar words ('Fairest Shee [*sic*] all Nymphs transcending'). The production likely also featured Desnoyer, who had led the 1733 Drury Lane pantomime *The Judgement of Paris*, and who, since 1736, had been part of Augusta and Frederick's inner circle.[17] As a celebrity, for the public Desnoyer embodied royal taste for this generation, as Handel had for Caroline. By commanding such celebrities, the rival royal patrons showed their power. In 1738, when Handel was banned by George II from any participation in Frederick's concerts, Frederick forbade Desnoyer from teaching his younger siblings.[18]

After Frederick's early death in 1751, Augusta shadowed the musical taste of the Scottish John Stuart, third Earl of Bute. Augusta sought Bute's favour because she thought him uniquely fit, in the absence of Frederick, to train her son to be king. On his recommendation, from the late 1750s she championed the German-born composer John Christopher Smith (Johann Christoph Schmidt), who had succeeded his father, another John Christopher Smith, as Handel's copyist and amanuensis. As well as taking keyboard lessons from Smith the younger, Augusta hosted performances of his neo-Handelian oratorios at Carlton House.[19]

Through Augusta, Smith's collection of Handel autographs was eventually incorporated into the Royal Music Library. Her later acquisitions through the Scottish publisher Robert Bremner—perhaps another link to Bute—show that Augusta kept pace with the fashion for instrumental chamber music.

Augusta's interest in instrumental music mirrored that of her daughter-in-law Charlotte, who married George III in September 1761. Charlotte was a virtuoso, having studied rigorously for years with Johann Linike, a violinist, composer, and veteran of Handel's opera band. Excellence in the making of music, to which George III was 'extremly [*sic*] partial', had reportedly been one of the King's criteria when choosing a consort.[20] Shortly after her wedding Charlotte instituted regular music concerts at which the royal family, including Augusta, performed to a select company—though the King played only in private with Charlotte. The new music master, the composer and keyboard player Joseph Kelway, was tasked with creating the Queen's Band of exclusively English players.

Critical of their skills, Charlotte replaced Kelway in 1762 with Johann Christian Bach, whom she summoned from Milan, and also imported twelve German players for the Band. Bach was a friend of the virtuoso Carl Friedrich Abel who had earlier arrived from Dresden, and after his London début on 5 April 1759 had been appointed to the King's Band on the recommendation of Prince Edward, Duke of York and Albany.[21] Charlotte would make Abel a member of her Chamber Band, which gave 'private quartet parties' twice a week. In 1779 Mary Granville, Mrs Delany described one such party as 'two hours' of the 'most delightful entertainment . . . particularly the first fiddle by Cramer, Abel on the Viol de Gambo (tho' I don't like the instrument) and a new hautboy'.[22] In 1766 an organ was built in the morning room of the Queen's House, the present Buckingham Palace, and musical evenings with the Queen's Band were further enriched by performances from travelling virtuosi and prodigies. In 1764 the eight-year-old Mozart performed at the invitation of Charlotte, who sang while he accompanied her, and to whom he dedicated six sonatas.[23] Even after the King's ill-health had made concerts of the Queen's Band much less frequent, Charlotte hosted Joseph Haydn in 1795.[24]

Music was a crucial representative medium for Caroline, Augusta, and Charlotte, each of whom became progressively more active as a performer, more public as a patron, and more invested in instrumental music than their predecessor. Successively, they helped shape British music throughout the eighteenth century.

NOTES

1. John Brooke, *King George III* (London: Constable, 1972), 302; cited also in Hans Joachim Marx, '"A Love of Music to Distraction": Musik im Leben der englischen Königin Charlotte (1744–1818)', *Archiv für Musikwissenschaft* 71, no. 1 (2014): 2.
2. Colin Timms, *Polymath of the Baroque: Agostino Steffani and his Music* (Oxford and New York: Oxford Univ. Press, 2003), 46.
3. Colin Timms, 'Steffani's Influence on Handel's Chamber Duets', in *Handel Tercentenary Collection*, ed. Stanley Sadie and Anthony Hicks (Basingstoke: Macmillan Press, 1987), 222–45. The quotation is from John Mainwaring's 1760 Handel biography.
4. Moira Goff, 'Desnoyer, Charmer of the Georgian Age', *Historical Dance* 4, no. 2 (2012): 3.
5. Earl of Chesterfield to his son, 16 May 1751; cited in Goff, 'Desnoyer', 6.
6. Dedication 'To Her Royal Highness', from John Essex, *For the Further Improvement of Dancing: A Treatise of Choreography, or the Art of Dancing Country Dances after a new Character* (London, 1715).
7. Jacob Wilhelm Lustig, *Inleiding tot de muzykkunde*, 2nd ed. (1751; Groningen: H. Vechnerus, 1771), 172.
8. John Hervey, *Lord Hervey's Memoirs*, ed. Romney Sedgwick, rev. ed. (1952; London: B. T. Batsford, 1963), 42–43; Sir John Buckworth to the Earl of Essex, 13 May 1736, BL Add. MS. 37735, ff. 177–78.
9. Newburgh Hamilton to the Earl of Strafford, 3 Jan. 1738, BL Add. MS. 31142, f. 202.
10. Queen Caroline to Princess Anne, 18 Feb. 1735, Koninklijk Huisarchief, The Hague KH A30/430, and Countess of Burlington to Earl of Burlington, June 1736, BL Add. MS. 75358.
11. John L. Bullion, '"*To play what game she pleased without observation*": *Princess Augusta* and the Political Drama of Succession, 1736–1756', in *Queenship in Britain, 1660–1837: Royal Patronage, Court Culture and Dynastic Politics*, ed. Clarissa Campbell Orr (Manchester: Manchester Univ. Press, 2002), 207–35.
12. Goff, 'Desnoyer', 6.
13. Princess Amelia to Princess Anne, 21 Dec. 1736, Koninklijk Huisarchief, The Hague KH A30–430.
14. The National Archives, Establishment Book 30, 22.
15. Joanna Marschner, *Queen Caroline: Cultural Politics at the Early Eighteenth-Century Court* (New Haven and London: Yale Univ. Press, 2014), 18.
16. The air's verses ran: 'Fairest Shee [*sic*] all Nymphs transcending/ That the sun himself has seen/ Were she for the Crown contending/ Thou would'st own her, beauties [*sic*] Queen.' BL Add. 17860, ff. 109–14. Lady Catherine Hanmer as Venus sang. Performing the other roles were Lady Falconberg (Juno), Lady Middlesex (Pallas), and Prince Lobkowitz (Mercury). Mariateresa Dellaborra, '*The Judgement of Paris* (1740), pastorale di Giuseppe Sammartini', *Rivista Italiana di Musicologia* 48 (2013): 43.
17. Goff, 'Desnoyer', 4–6.
18. See note 10 above.
19. William Coxe, *Anecdotes of George Frederick Handel and John Christopher Smith* (London: W. Bulmer and Co., 1799), 52–54.
20. Walley C. Oulton, *Authentic and Impartial Memoirs of Her late Majesty, Charlotte, Queen of Great Britain and Ireland* (London: J. Robins and Co., 1819), 10; cited in Marx, 'A Love of Music to Distraction', 4.
21. Walter Knape, *Karl Friedrich Abel: Leben und Werk eines frühklassischen Komponisten* (Bremen: Schünemann Univ. Press, 1973), 52.
22. Letter of 29 Oct. 1779, quoted in Simon McVeigh, *Concert Life in London from Mozart to Haydn* (Cambridge: Cambridge Univ. Press, 2005), 51.
23. Describing his son's performance at court on 19 May 1764, Leopold Mozart wrote: 'He played so splendidly on the King's organ that they all value his organ-playing more highly than his clavier-playing. Then he accompanied the Queen in an aria which she sang.' Quoted in McVeigh, *Concert Life in London*, 51.
24. Marx, '"A Love of Music to Distraction"', 9; McVeigh, *Concert Life in London*, 52.

CAROLINE AND THE MUSICAL LIFE OF THE COURT

11.04

THOMAS HUDSON (1701–1779)
George Frideric Handel
c. 1756
Oil on canvas
30⅛ x 25⅛ in. (76.4 x 63.7 cm)
Royal Collection Trust, RCIN 400960

In 1717, Handel was appointed music tutor to Princesses Anne, Amelia, and Caroline, the elder daughters of George and Caroline. The princesses, especially Anne, were gifted musicians, and their association with the composer continued until his death in 1759. Handel composed anthems for the Coronation of George II and Caroline in 1727, for the wedding of Frederick, Prince of Wales, and Augusta in 1736, and for the funeral of Caroline in 1737, as well as for many other royal celebrations. His creative development, which saw him move from composing elaborate Italianate opera, to producing oratorios sung exclusively in English, played a part in the anglicization of the monarchy and the building of a sense of Britishness for the nation.

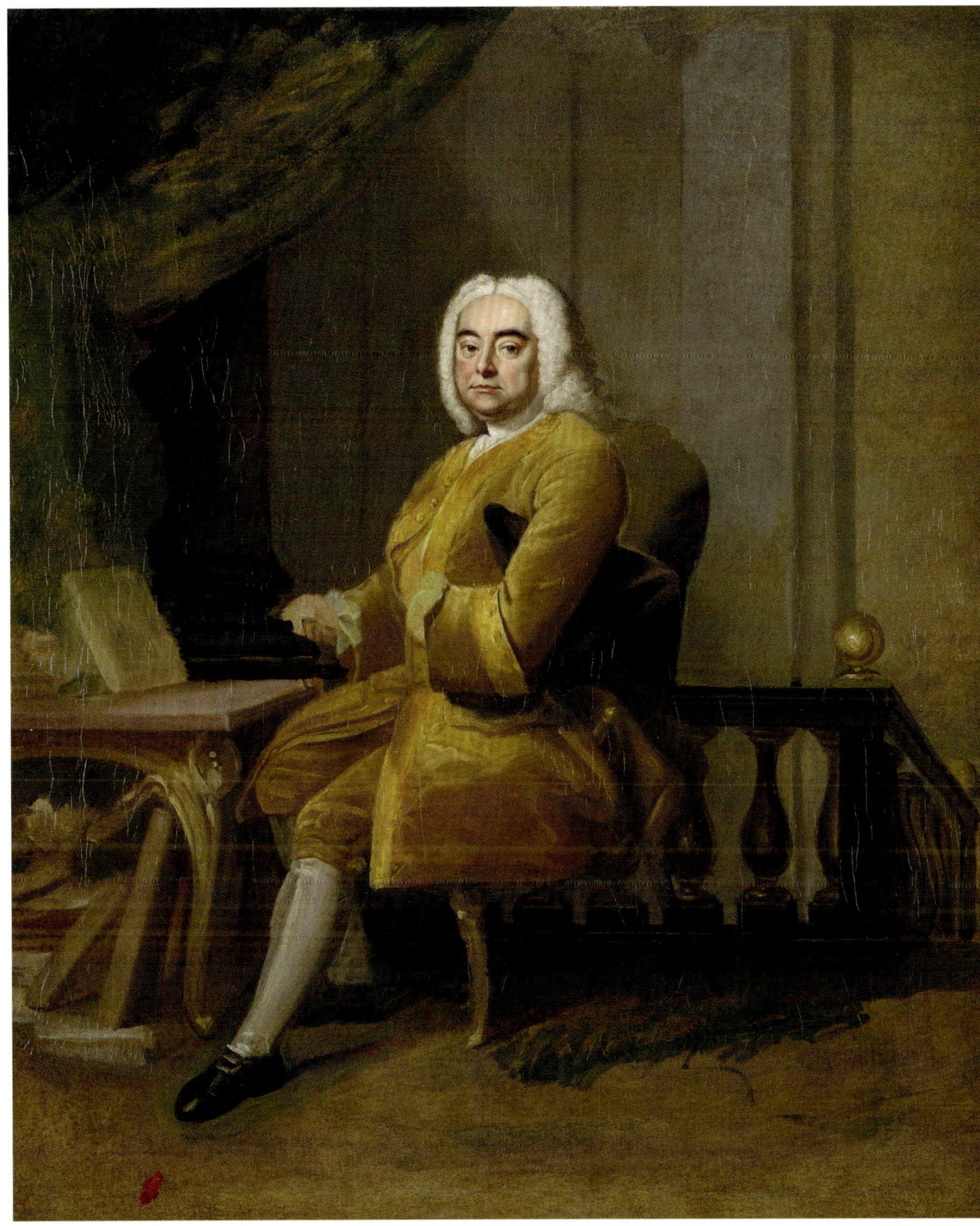

11.04

* = YCBA only
† = HRP only

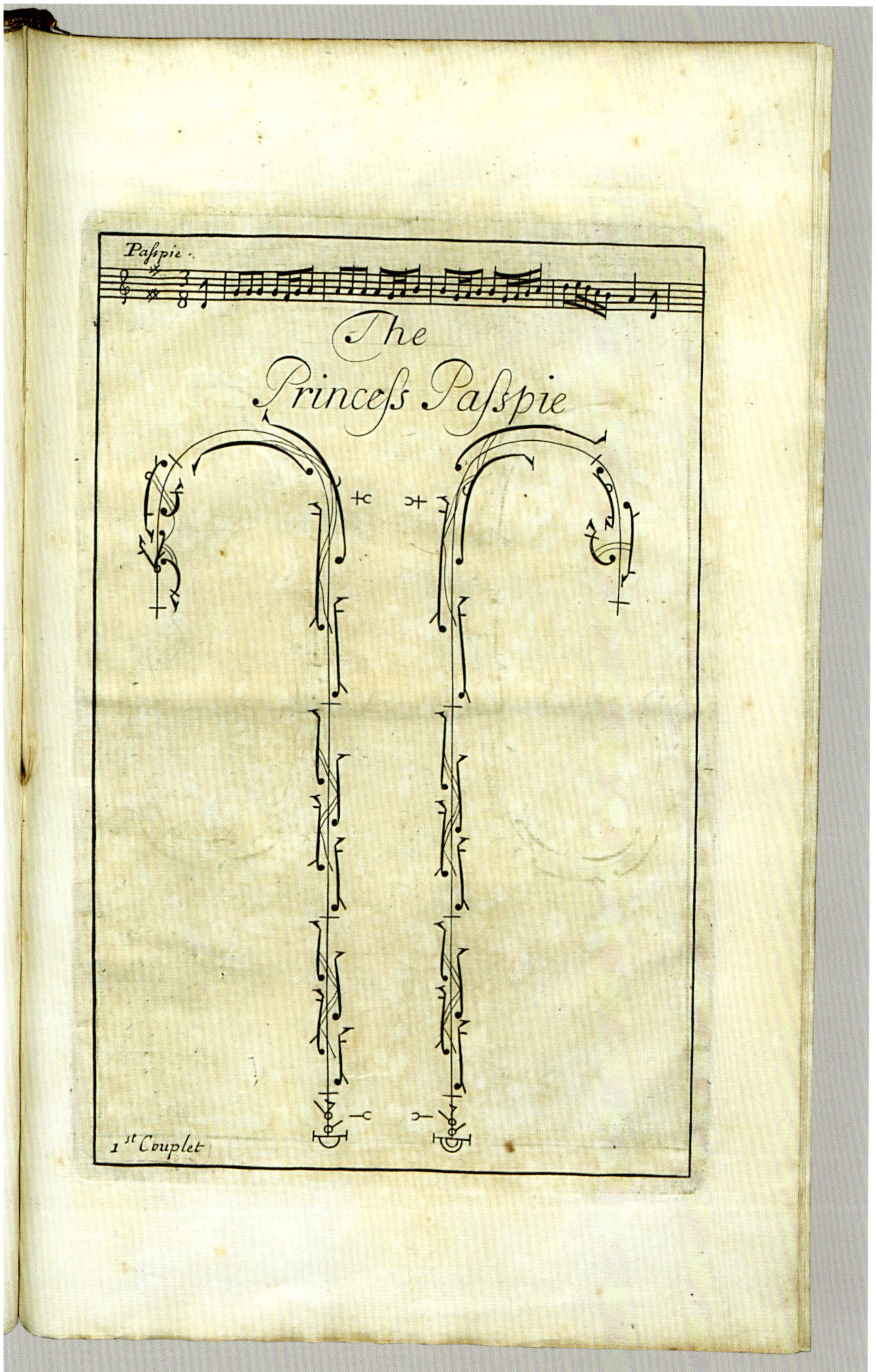

11.05

11.05

RAOUL AUGER FEUILLET (1659/60–1710), author, with **JOHN ESSEX** (1680–1744), translator, author, and editor
For the Further Improvement of Dancing. A Treatise of Chorography, or the art of dancing country dances after a new character . . . with a Collection of Country Dances, and a New French Dance Call'd the Princess's Passpied
c. 1715
John Walsh, John Hare, [etc], London
British Library, 60.h.28
* Exhibited: *The Princess Passpie*: 1st figure (illustrated), and title page
† Exhibited: 2nd figure; 3rd figure

John Essex, dancer, choreographer, and teacher was involved in several publications which explained the dance notation of the French dance masters Raoul Auger Feuillet and Pierre Rameau to a British audience. *For the Further Improvement of Dancing* (1710) included the music and choreographic notation for country dances. In 1715 a second edition dedicated to Caroline as Princess of Wales was printed containing four additional country dances by Essex to supplement the original ten, and 'The Princess's Passepied', named for Caroline, Essex's only surving ball dance. Caroline ensured country dances took their place in the ceremonial at court. A dancing lawn was laid out in front of one of her folly temples at Richmond, ensuring picturesque and patriotic entertainments were part of the experience of her garden. This handsomely bound volume from George III's library may be the one presented to Caroline by Essex.

11.06

ANTHONY L'ABBÉ (1666/7–1753), choreographer and dancing-master, and **EDMUND PEMBERTON** (d. c. 1735), editor
The Princess Ann's Chacone. A New Dance for his Majties Birth-day
1715–19
Edmund Pemberton, London
British Library, Music Collections h.801.b.
* Exhibited: *The Princess Royale. A New Dance for his Majestys Birth Day* (London: Edmund Pemberton, 1715)
† Exhibited: *The Princess Amelia. A New Dance for his Majties Birth-day* (London: Edmund Pemberton, 1718) and *The Princess Ann's Chacone. A New Dance for his Majties Birth-day* (London: Edmund Pemberton, 1719); (illustrated)

Anthony L'Abbé, a dancer at the Académie Royale de Musique in Paris, was coaxed to come to London in 1698 by Thomas Betterton, manager of Lincoln's Inn Fields Theatre. L'Abbé remained in London for the next thirty years, and became one of the most highly respected dancing-masters of his day. The dedication to Princess Anne of his dance 'The Princess Royale' in 1715 confirms that she was already his pupil; that dance was the first of a series of annual ball-dances, many of which were named after members of the royal family. By 1720, he had been formally appointed dancing master to the three young princesses Anne, Amelia, and Caroline, granddaughters of George I. He held this post until at least 1737, at a salary which rose to £240 in comparison to the £200 paid to Handel as Princess Anne's music master. Thirteen engraved notations of the annual ball-dances created by L'Abbé during this period survive in the British Library, together with one theatrical solo devised in 1711, all published by the dancing master Edmund Pemberton. The young princesses danced for the company at court, as demonstration of the liveliness and sustainability of the new Hanoverian regime. It was reported widely that they were seen dancing at court following their innoculation against smallpox, as evidence of the safety of the process, playing their own part in Caroline's promotion of the medical practice.

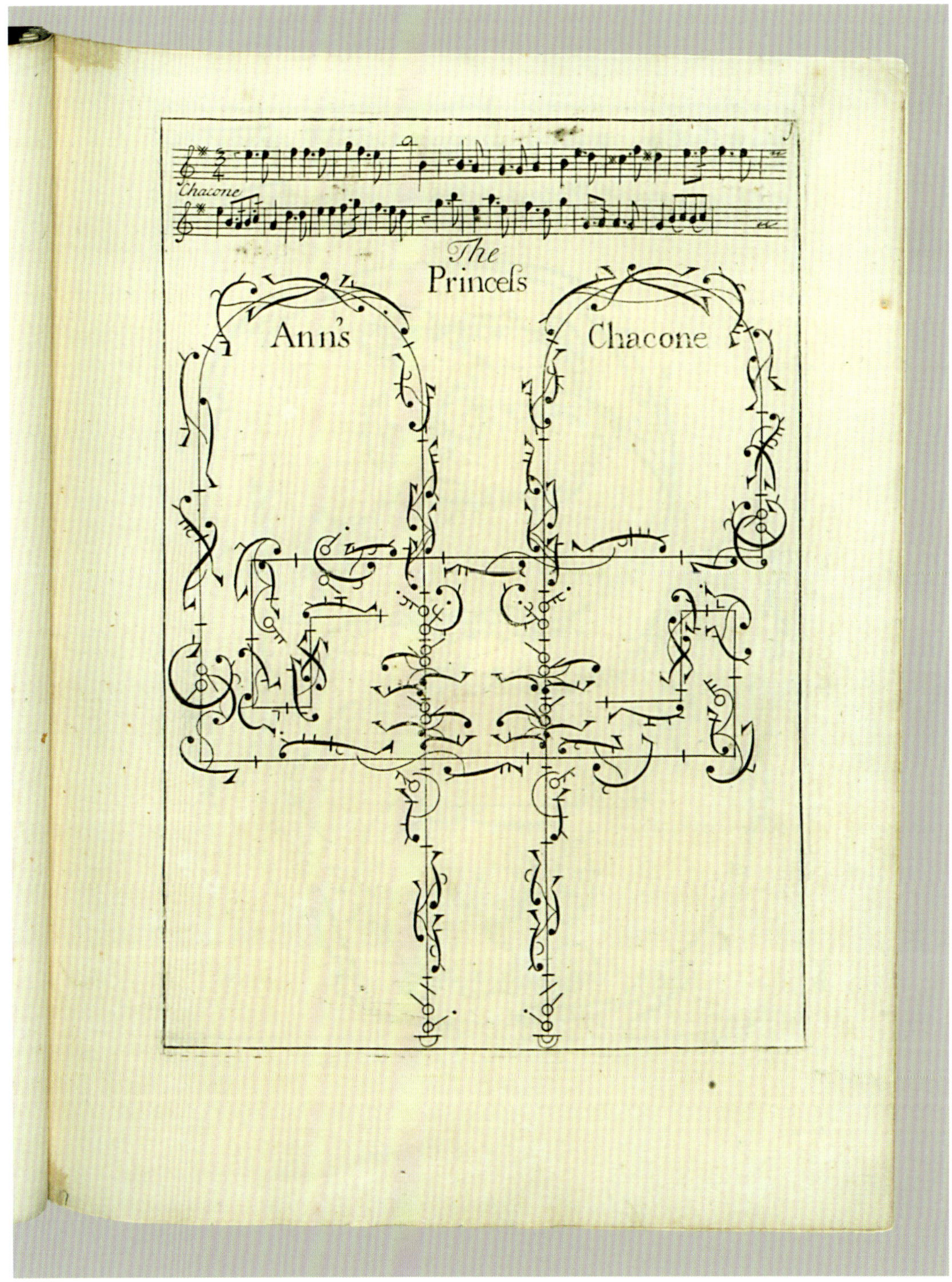

11.06

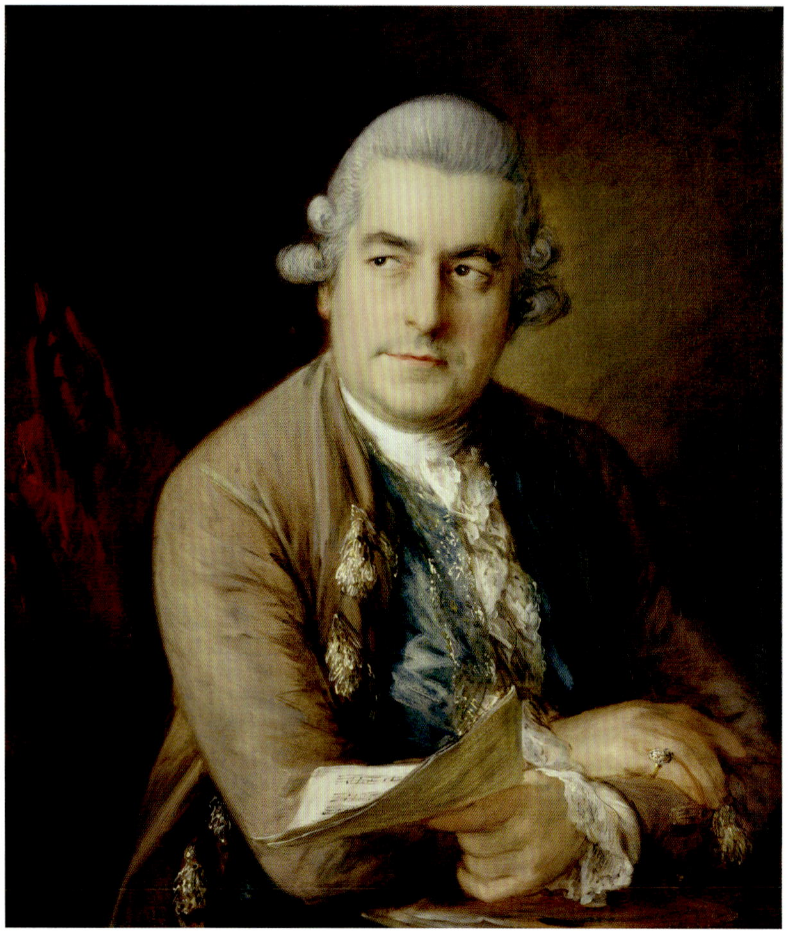

11.07

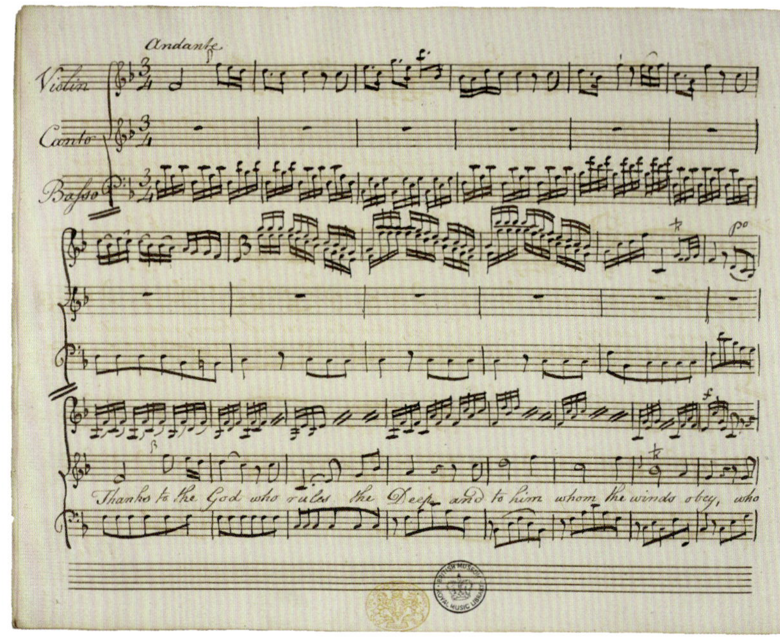

11.08

CHARLOTTE AND THE MUSICAL LIFE OF THE COURT

11.07*

THOMAS GAINSBOROUGH (1727–1788)
Johann Christian Bach
1776
Oil on canvas
29¾ x 24⅜ in. (75.5 x 62 cm)
National Portrait Gallery, London, NPG 5557

Johann Christian Bach, son of Johann Sebastian Bach, arrived in London from Milan in 1762, and after the considerable success with his operas *Orione* and *Zanaida* at the King's Theatre, was appointed singing teacher, and Master of The Queen's Band to Charlotte in 1763. Charlotte was an accomplished harpsichord player. He organized her chamber music concerts, which took place several times each week. When Bach died in considerable debt in 1782, she settled the bills on his estate and provided a pension for his widow, Cecilia.

11.08†

JOHANN CHRISTIAN BACH (1735–1782), composer, and JOHN LOCKMAN (1698–1771), author
'Ode on the auspicious Arrival and Nuptials of Her present most gracious Majesty Queen Charlotte, by John Lockman. Set to music by Mr. Bach'
18th century
Manuscript
9½ in. (24 cm)
British Library, London, R.M. 22.a.17
Exhibited: ff. 1v–2r ('Thanks to the God who rules the deep and to him whom winds obey'); (illustrated)

This presentation volume of music for solo voice with a three-part chorus and accompaniment for violin and bass was written to celebrate the arrival of Charlotte of Mecklenburg-Strelitz in England. Handwritten annotation to the endpapers by Frederick Nicolay, Keeper of the Queen's Music Library, confirms that the manuscript is from the Queen's private collection.

11.09*

THOMAS GAINSBOROUGH (1727–1788)
Carl Friedrich Abel
1777
Oil on canvas
88⅝ x 59½ in. (225 x 151 cm)
Huntington Library, Art Collections and Botanical Gardens, San Marino, 25.19

Köthen-born Carl Friedrich Abel, a pupil of Johann Sebastian Bach and celebrated viola de gamba player, travelled to London in about 1758–59. From 1764 until his death in 1787 he joined his compatriots Johann Christian Bach, Johann Christian Fischer, and Carl Friederich Weideman, to perform with them, and to compose for private music-making and for The Queen's Band. Thomas Gainsborough's portrait of his great friend Abel, who gave him music lessons in return for paintings, was purchased by Charlotte after the sale of Abel's effects on his death. This painting was later sold at Christie's as lot 93 in the auction of the Queen's effects on 24–26 May 1819. A second portrait of the musician, by Alexandre-Auguste Robineau (RCIN 403540), was also acquired by his royal patrons.

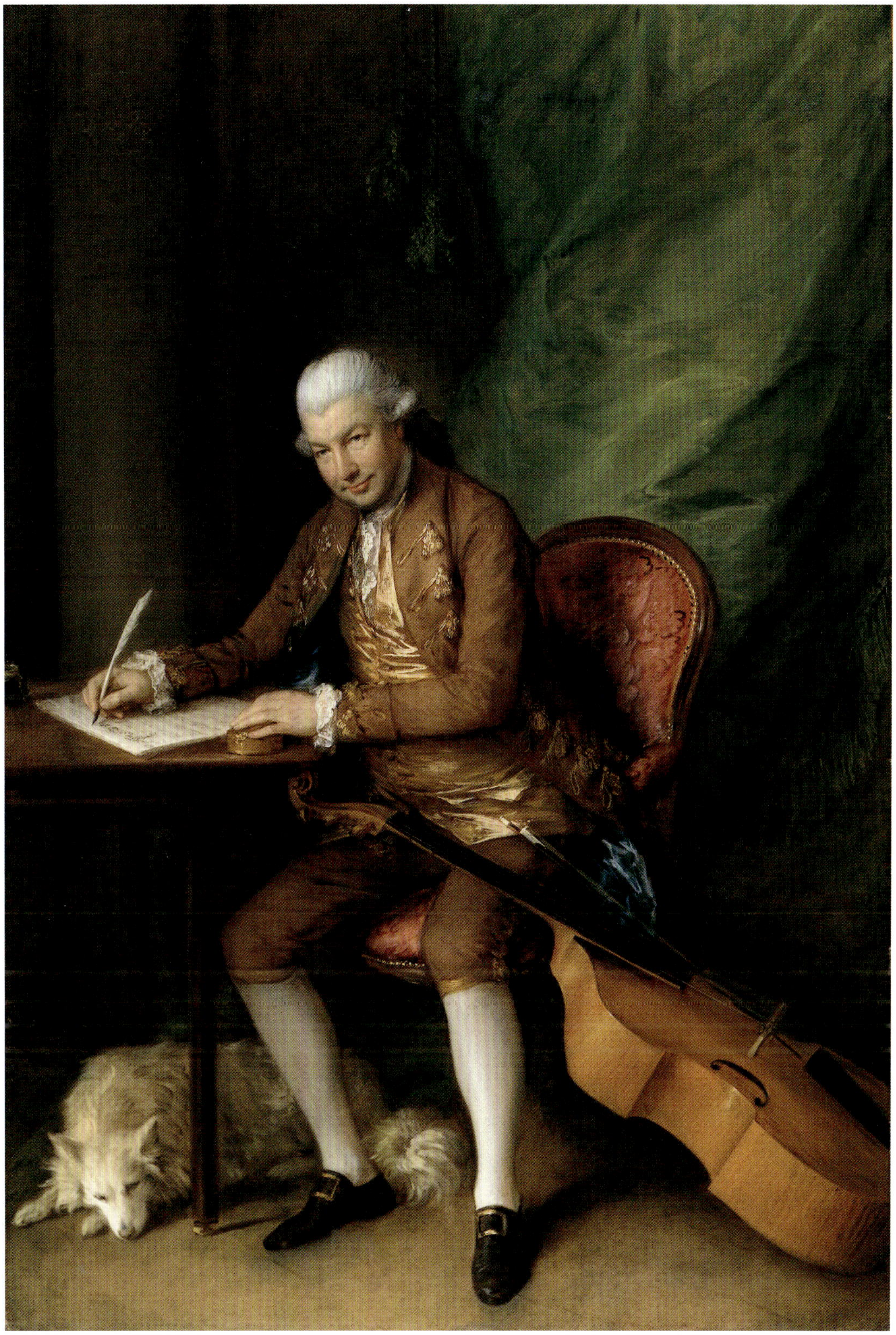

11.09

II.10*

JOHANN CHRISTIAN BACH (1735–1782) and **CARL FRIEDRICH ABEL** (1723–1787)
Six sonates pour le clavecin: accompagnées d'un violon ou flute traversière et d'un violoncelle: oeuvre II
c. 1765
Printed by Welcker, London
Beinecke Rare Book & Manuscript Library, Yale University, OSB MSS 146, Box 837
Exhibited: 'Six Sonatas dedicated to Charlotte', title page (illustrated)

This volume is inscribed, in the hand of Frederick Nicolay, Keeper of the Queen's Music Library, 'This volume belongs to the Queen, 1788.' It contains Bach's Opus 2, dedicated to Augusta, and Abel's Opus 5, dedicated to Charlotte. Nicolay's parents had arrived at court in the train of Augusta of Saxe-Gotha in 1736. As a young man, Frederick served as Dancing Master to George III as Prince of Wales, and from 1762 served as Page of the Backstairs to Charlotte, as well as playing violin in The Queen's Band of Musicians, and The Queen's Chamber Band. It was Nicolay, as guardian of the Queen's music, who wrote on the manuscript listing of Queen Caroline's music collection that all the duplicates should be sent to Hanover. This princely collection entered the holdings of the Beinecke Library in 2008.

II.11†

JOHANN CHRISTIAN BACH (1735–1782), **CARL FRIEDRICH ABEL** (1723–1787), and **ANTONÍN KAMMEL** (1730–1784)
Six Sonatas for two Violins and a Violoncello, with a Thorough Bass for the Harpsichord (3 parts)
1777
John Welcker, London
British Library, Music Collections Hirsch III.34
Exhibited: pp. 2–3 (opening of Bach, Sonata 1 for violin), and pp. 8–9 (opening of Abel, Sonata IV for violin; illustrated)

The Queen's Band, under the leadership of Johann Christian Bach, employed some of the leading instrumentalists of the day, including the viola de gamba player and composer C. F. Abel, the oboist J. C. Fischer and the violinist Antonín Kammel. J. C. Bach was in attendance at St James's Palace three times a week. Among his many dedications to his royal pupil, Queen Charlotte, was an organ concerto: its finale comprised a set of variations on 'God Save the King'. Bach's and Abel's compositions for The Queen's Band were numerous, and under the Queen's patronage their work was performed in a series of concerts in London between 1765 and 1781.

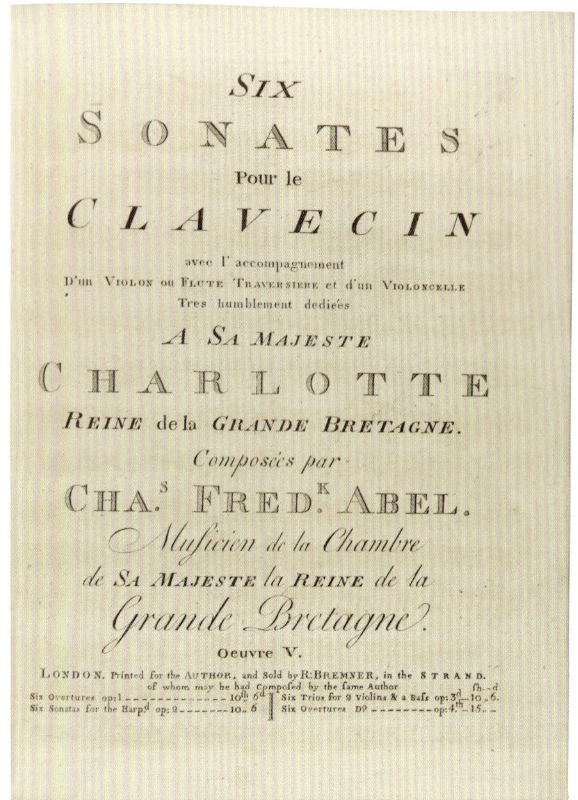

II.10

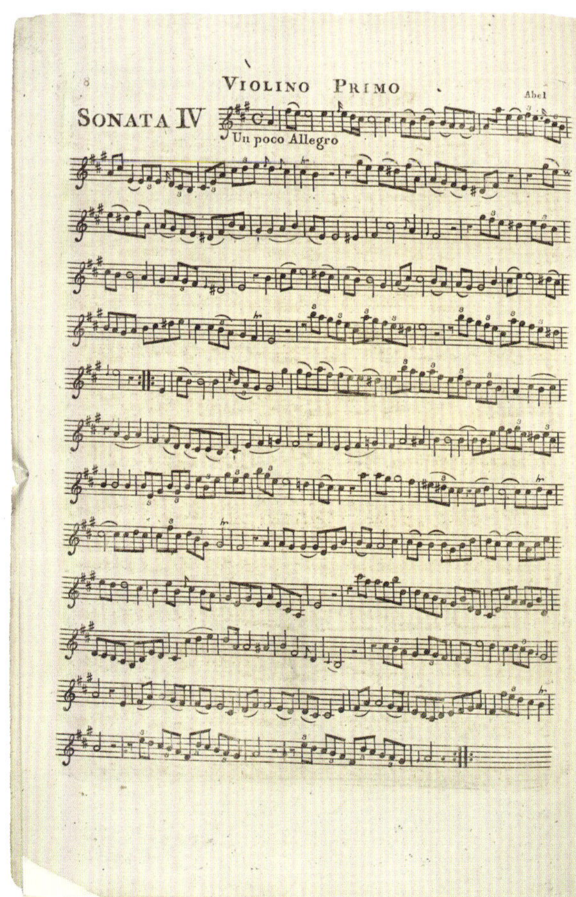

II.11

II.12†

WOLFGANG AMADEUS MOZART (1756–1791), with **LEOPOLD MOZART** (1719–1787) and **UNKNOWN ENGRAVER**
Six Sonates pour le clavecin. Composed for violin (flute), cello and harpsichord
1765
Printed for the author, London
British Library, Music Collections R.M. 11.f.5
Exhibited: Violin (flute) f. 1 (Sonata 1: allegro; illustrated) Violin (flute) ff. 6–7 (Sonata III: allegro; Sonata IV: allegro)

II.13*

WOLFGANG AMADEUS MOZART (1756–1791), composer, with **UNKNOWN ENGRAVER**
Six Sonates pour le clavecin
2nd edition, unbound
1765
Printed for the author and sold at his lodgings, London
Irving S. Gilmore Music Library, Yale University, M241 M939 K.10–15 1765+
Exhibited: Dedicatory epistle (illustrated)

The set of six sonatas (K10–15) for harpsichord, violin (or flute), and cello, was commissioned by Queen Charlotte on 25 October 1764, after the eight-year-old Wolfgang Amadeus Mozart had perfomed with his sister, Maria Anna, on three occasions at the Queen's House. They were dedicated to her on 18 January 1765. The Queen's copy, now in the British Library, has a manuscript violin part in the hand of Leopold Mozart. Before the Mozart family left London in July 1765, Leopold arranged to have the works set up and printed in Soho as 'Opus III', and he managed their sale from the lodgings they had taken in St Martin's in the Fields. The printed edition has a fulsome dedication to Charlotte which includes details of the young Mozart's encounter with the royal family.

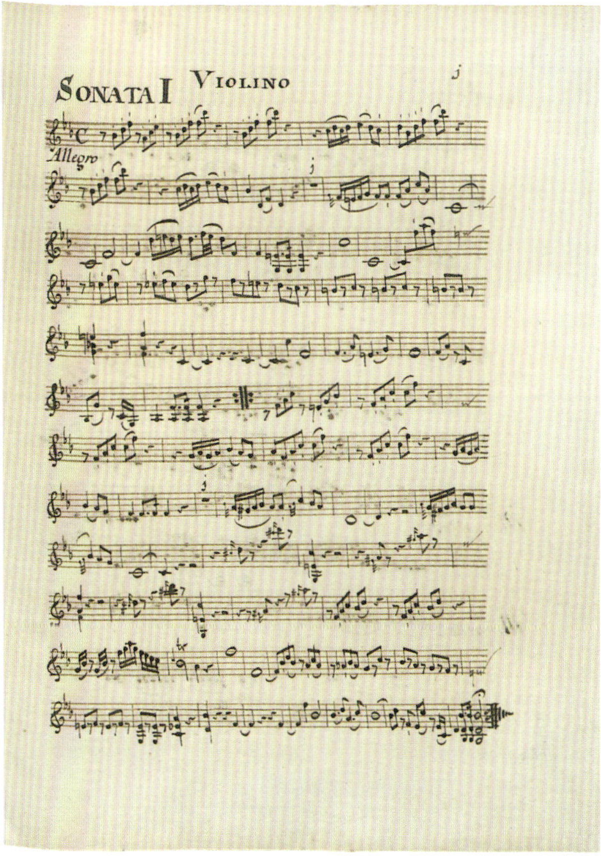

II.12

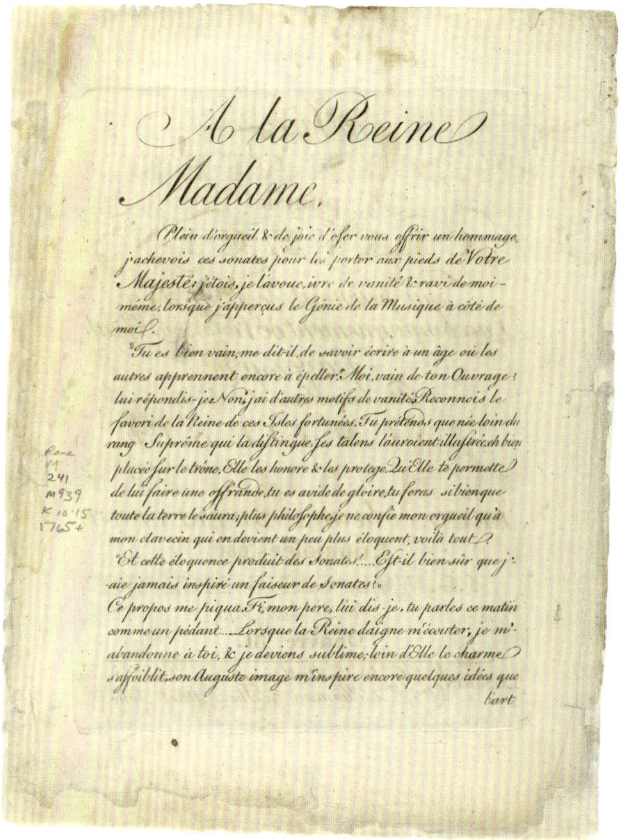

A la Reine

Madame,

Plein d'orgueil & de joie d'oser vous offrir un hommage, j'achevois ces sonates pour les porter aux pieds de Votre Majesté; j'étois, je l'avoue, ivre de vanité & ravi de moi-même, lorsque j'apperçus le Génie de la Musique à côté de moi.

"Tu es bien vain, me dit-il, de savoir écrire à un âge où les autres apprennent encore à épeller." Moi, vain de ton Ouvrage! lui répondis-je. Non, j'ai d'autres motifs de vanité. Reconnois le favori de la Reine de ces Isles fortunées. Tu prétends que née loin du rang Suprême qui la distingue, ses talens l'auroient illustrée. eh bien placée sur le trône, Elle les honore & les protege. Qu'Elle te permette de lui faire une offrande, tu es avide de gloire, tu feras si bien que toute la terre le saura; plus philosophe, je ne confie mon orgueil qu'à mon clavecin qui en devient un peu plus éloquent, voilà tout.

"Et cette éloquence produit des Sonates!.... Est-il bien sûr que j'aie jamais inspiré un faiseur de Sonates?"

Ce propos me piqua. Fi, mon pere, lui dis-je, tu parles ce matin comme un pédant... Lorsque la Reine daigne m'écouter, je m'abandonne à toi, & je deviens sublime; loin d'Elle le charme s'affoiblit, son Auguste image m'inspire encore quelques idées que
l'art

II.13

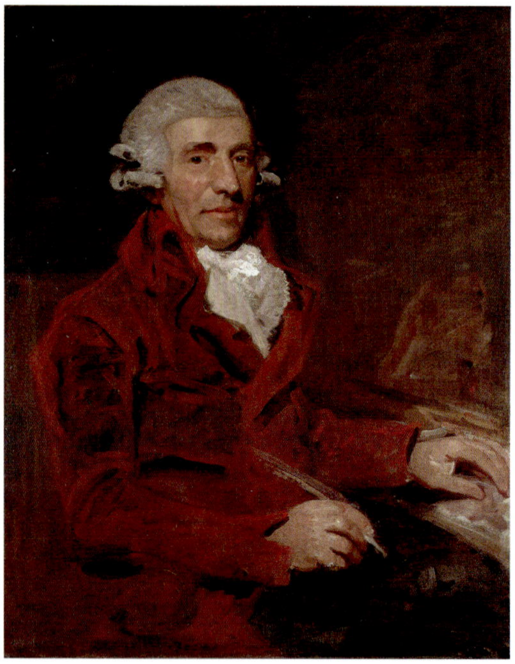

II.14

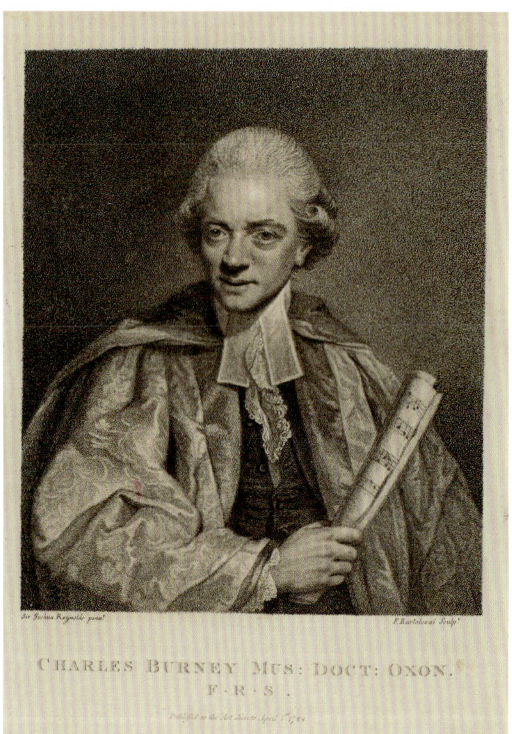

II.15

II.14*

JOHN HOPPNER (1758–1810)
Franz Joseph Haydn
1791–92
Oil on canvas
36¼ x 28⅛ in. (92.1 x 71.5 cm)
Royal Collection Trust, RCIN 406987

Joseph Haydn travelled to London to conduct a series of concerts for the German impresario Johann Peter Salomon in 1791–92. He returned in 1794–95 and in 1795, after being introduced to George III and Charlotte by George, Prince of Wales, played for them at a series of concerts. He was offered, but declined accommodation at Windsor Castle. Prince George (later George IV) commissioned this portrait from John Hoppner, his Principal Portrait Painter, during Haydn's first London period.

II.15*

CHARLES BURNEY (1726–1814)
A General History of Music from the Earliest Ages to the Present Period (4 vols, 1776–89)
1776
Printed for the author, London
The Lewis Walpole Library, Yale University, A 53 B 93 776
Exhibited: Frontispiece (Francesco Bartolozzi, 1728–1815, engraver, after Sir Joshua Reynolds, 1723–1792, 'Charles Burney Mus: Doct: Oxon., F.R.S.')

II.16*

SIR JOSHUA REYNOLDS (1723–1792)
Charles Burney
1781
Oil on canvas
29½ x 24 in. (74.9 x 61 cm)
National Portrait Gallery, London, NPG 3884

Organist and musicologist Charles Burney, the father of Frances (Fanny) Burney, later Madame d'Arblay, the novelist and diarist, was the author of the monumental *A General History of Music*. Burney had ambitions at Court and following the dedication of his book to Queen Charlotte, his daughter was appointed as Second Keeper of the Robes, a post she held between 1786 and 1790. Burney's dealings with the king and queen brought him more disappointment than satisfaction. His attempt to write a brief account of the 1784 Handel commemoration festival was complicated by the king's attempts to interfere; he had been passed over as Master of The King's Band in 1779 in favour of John Stanley, and again on Stanley's death in 1786 by William Parsons. It seems likely that Fanny's appointment was intended in some way to make up for this latter snub.

In 1770 Charles Burney undertook a tour of France and Italy to collect material for what later became *A General History of Music* (published separately as *The Present State of Music in France and Italy* in May 1771). A further volume, covering Germany and the Netherlands, followed in 1773. These works established Burney as a leading musicologist and saw his election to the Royal Society. Burney dedicated his subsequent four-volume *A General History of Music from the Earliest Ages to the Present Period* to Queen Charlotte.

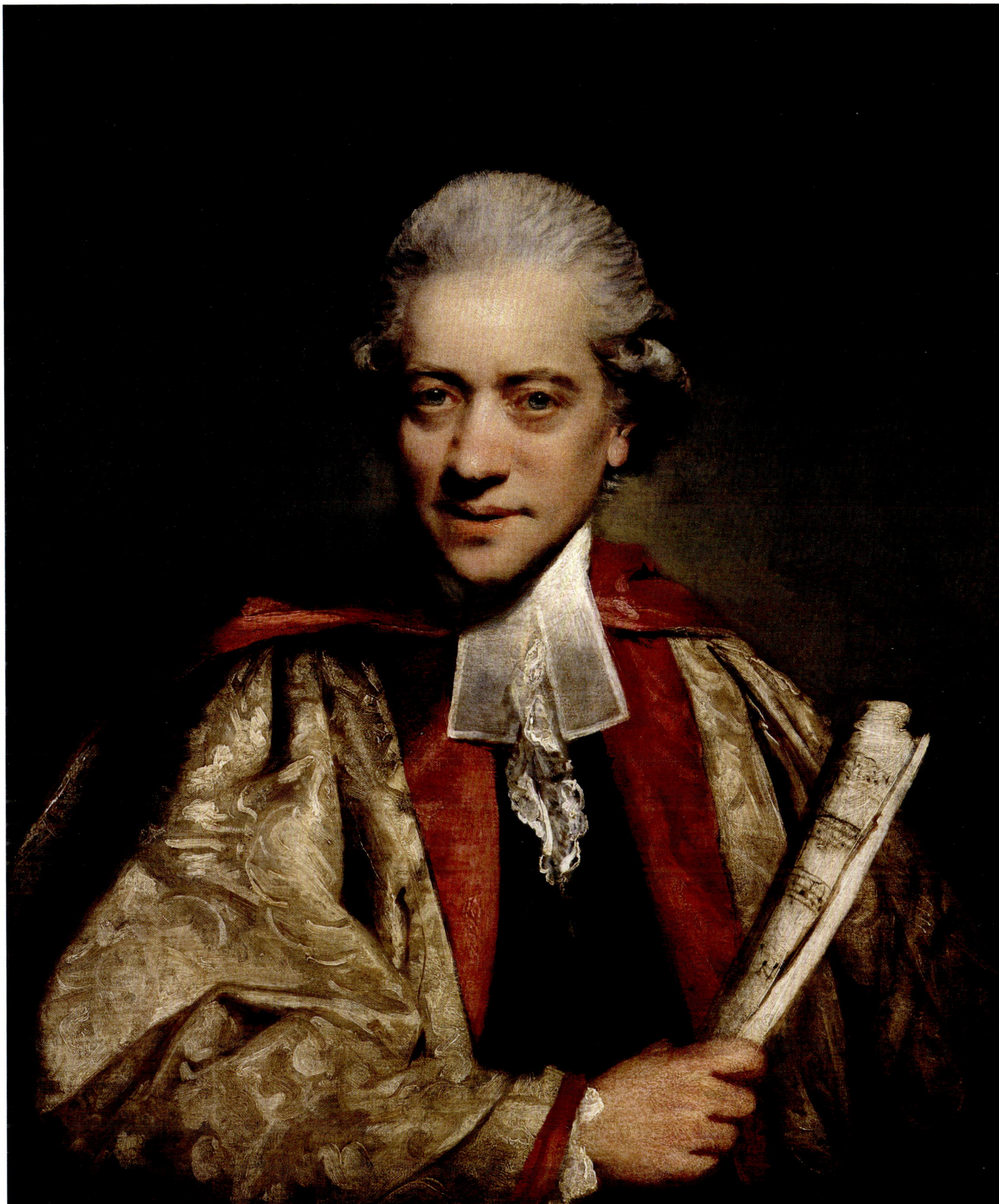

II.16

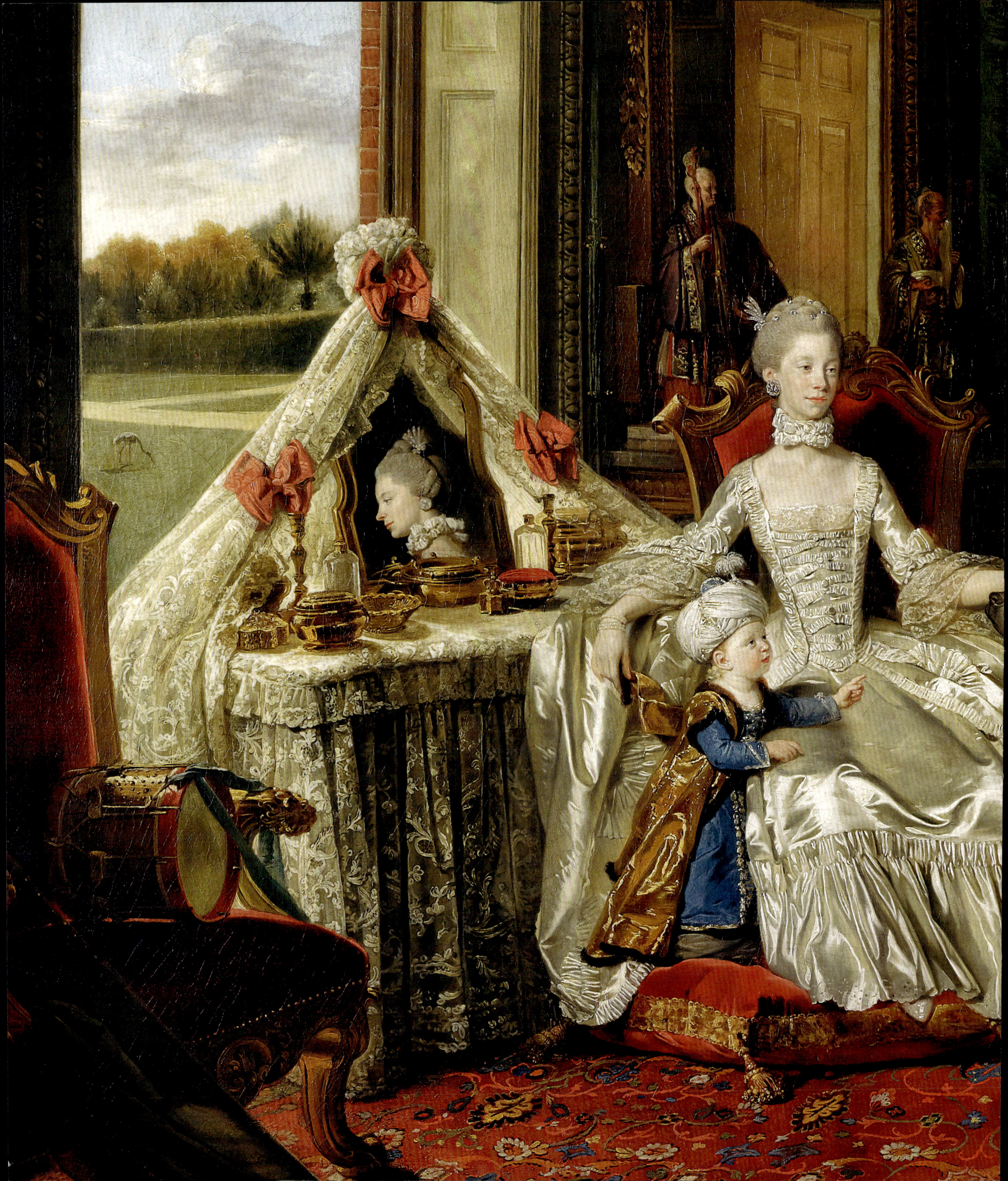

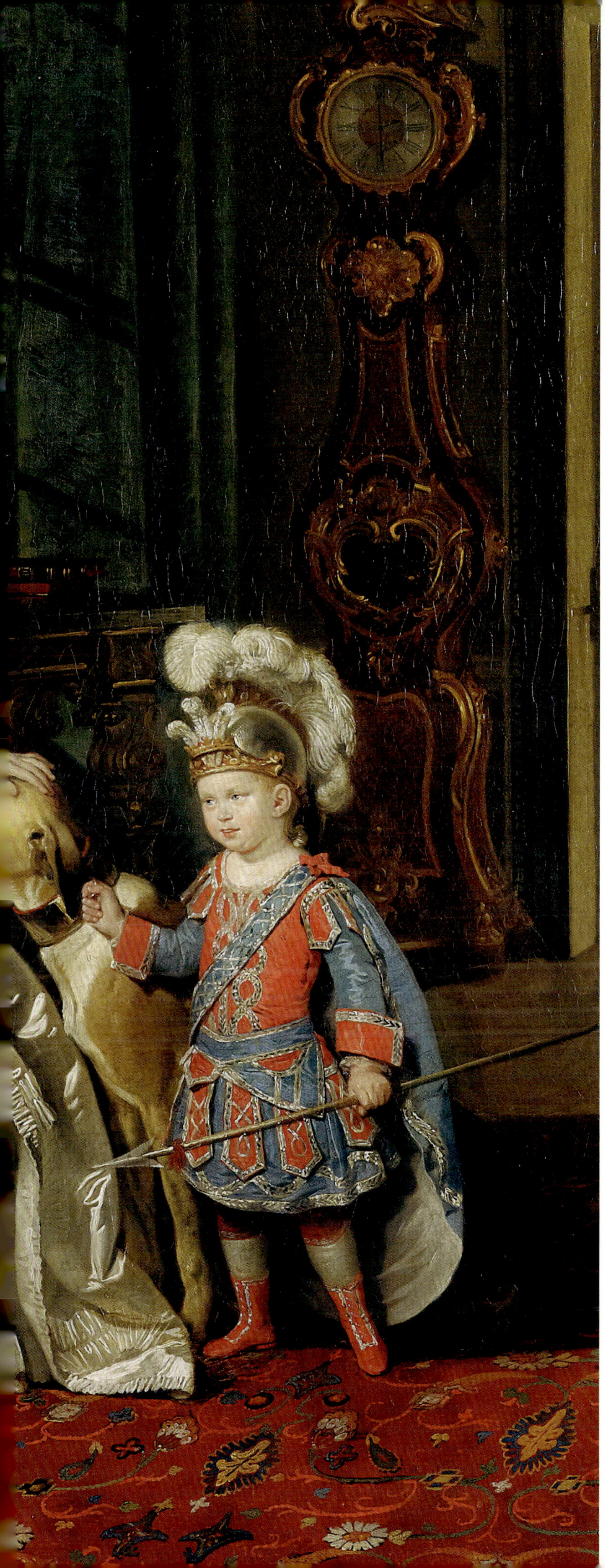

The Image of Monarchy

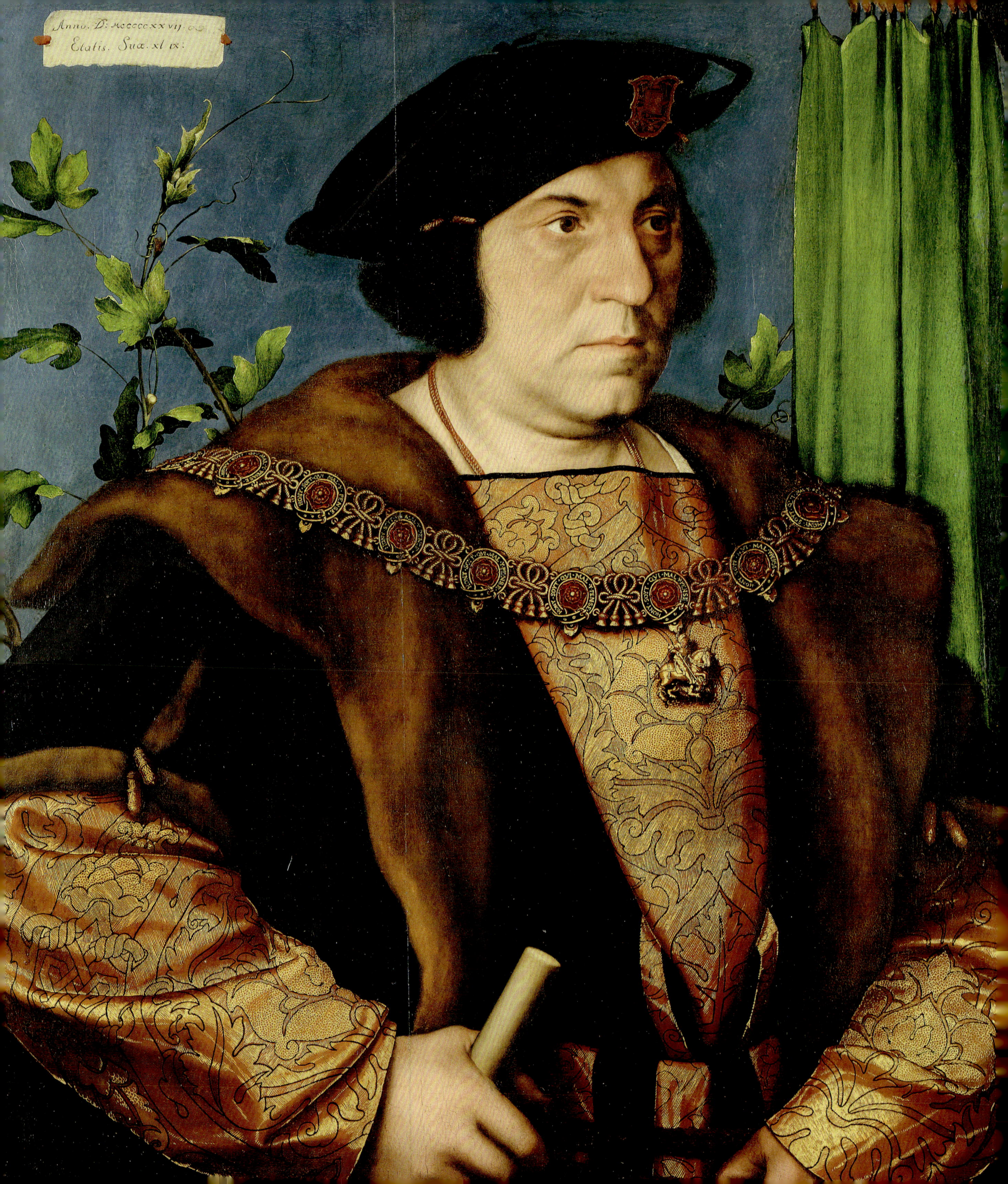
Anno. D: MCCCCCXXVII.
Etatis. Suæ. xl ix:

Using Britain's Past

Lisa L. Ford

The extraordinary dynastic changes to the British throne from the fifteenth through to the eighteenth centuries were brought about by the ultimate necessity of having a continuous blood line; the Bill of Rights of 1689 and the Act of Settlement of 1701 added the further necessity of a Protestant succession even where Catholics might have a better claim in terms of inheritance. Dynastic founders in support of their claims often emphasized connections to an ancient and mythical British past, as well as to more recent dynasties. The first Tudor monarch, Henry VII, could claim descent through Edward III, but he also invoked connections through his Welsh blood to ancient or mythical British rulers, by carrying the banner of the Red Dragon of Wales at Bosworth Field and using the red dragon as a heraldic supporter on the royal coat of arms. He claimed descent from Cadwalader, sometimes called the last King of the Britons, and named his son and heir Arthur, after the legendary British king.[1] James I, the first Stuart king and son of Mary, Queen of Scots, built his mother a grand tomb and placed it within Henry VII's chapel at Westminster in the aisle directly opposite that of Elizabeth I, the inscription of which marked her descent and his from Henry VII and Edward IV.[2]

When the Hanoverians came to the throne it was through the descent of George I from James I, which linked the new dynasty not only to ancient associations, but to the Protestant Stuarts as well. However, it was not George I who embraced the task of connecting the new dynasty to the British past but Caroline of Ansbach, particularly on her elevation to Queen Consort of England in 1727. Caroline's interest in collecting and through her art and architectural commissions evoking those connections has already been eloquently detailed by Joanna Marschner, and discussed by Wolf Burchard and Kathryn Jones, in 'The German Princesses as British Patrons'.[3] Her scope was wide and thorough: Merlin's Cave at Richmond touched on the ancient Arthurian claims and recent Tudor ancestry, while the Hermitage honoured more modern British 'worthies' such as Robert Boyle and Isaac Newton. At Kensington Palace her Picture Closet was thickly hung with portraits and objects delineating Hanoverian–British ancestry, and featured prominently the drawings by Hans Holbein of the younger of members of the Tudor royal family, courtiers, churchmen, and intellectuals (fig. 12.01), thus paying homage to both Henry VIII's court and one of the most renowned court artists of Germanic background, while in the Privy Chamber, Holbein's masterly portrait of Sir Henry Guildford extended the Tudor theme (p. 218). The busts she commissioned from Michael Rysbrack for the library at St James's Palace offered a line of great British royal ancestors through the ages, including King Alfred, lauded for his wisdom and staunch support of Anglo-Saxon liberties; Edward III, progenitor of both the Lancastrian and Yorkist dynasties; the Black Prince, hero of the field at Crécy and Poitiers (cat. 12.08); Catherine of Valois, ancestress of the Tudors; Henry VII, who claimed both ancient British and Lancastrian inheritance, and was the exemplar of a successful dynastic shift; Elizabeth of York, the Yorkist link in the union of the two dynasties; and Edward VI and Elizabeth I, the monarchs most closely

Hans Holbein the younger, *Sir Henry Guildford*, 1527, oil on oak panel, 32½ x 26¼ in. (82.7 x 66.4 cm). Royal Collection Trust

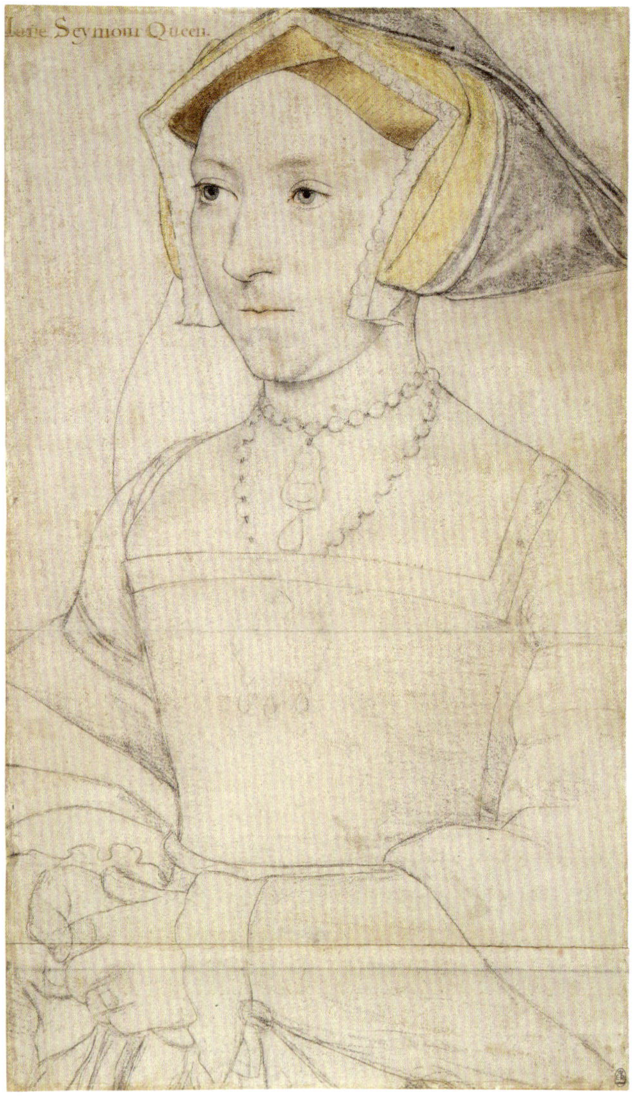

FIG. 12.01 Hans Holbein the younger, *Queen Jane Seymour*, 1536–37, black and coloured chalks, pen and ink, and metalpoint, on paper, 19½ x 11 ¼ in. (50 x 28.5 cm). Royal Collection Trust

associated with the Protestant settlement (fig. 12.02).[4] Henry V also enjoyed a prominent place as the subject of another Rysbrack bust, and of three paintings depicting major moments in his rule, produced for Caroline by William Kent (fig. 12.03, and cat. 12.06). Elizabeth I held a special place in Caroline's heart, and her collecting and display of images of that renowned Tudor queen, whom she held to be an exemplar for herself, included two cameos (fig. 12.04, and cat. 10.25) and a painting of Elizabeth as princess, which were among the items in the Closet, as well as the Rysbrack bust, and a waxwork image in Merlin's Cave.

The Holbein drawings were a particularly notable addition to Caroline's cabinet display.[5] Described as 'perhaps the most remarkable visual record of a royal court in European Art', when the Hanoverians succeeded to the throne in 1714, the drawings were lying forgotten in a bureau at Kensington Palace, where they were rediscovered by Caroline.[6] She was fortunate they were still among the royal possessions, as their ownership in the years between Holbein's tenure as court artist and the Hanoverian accession is eccentric. This history, with some debate on details, is oft-rehearsed; the most definitive version is that they may have been in Holbein's studio on his death in 1543, and were subsequently purchased or acquired either by Henry VIII or Edward VI in the form of 'a great booke of Pictures doone by Haunce Holbyn'. This 'booke' had been acquired by the Lord Chamberlain, Henry Fitzalan, twelfth Earl of Arundel by the time of the young king's death.[7] From there, the book passed into the possession of John, first Baron Lumley, Arundel's son-in-law, on the Earl's death. It likely came back into royal hands when Henry, Prince of Wales, acquired Lumley's books on that Lord's death in 1609, and passed to Charles I on Prince Henry's death, but was then bartered by Charles to Philip Herbert, fourth Earl of Pembroke in exchange for his Raphael of St George and the Dragon.[8] Pembroke gave the drawings to Thomas Howard, Earl of Arundel, who allowed Wenceslaus Hollar to engrave some of them. Finally, they returned into royal hands by 1675, probably purchased by Charles II, who may have been advised to do so by Peter Lely.[9]

Though Holbein was an artist and craftsman in many media, producing designs for stained glass; jewellery and goldsmith's work; painting schemes for temporary constructions; illuminating books and designing title pages; and painting religious and allegorical scenes, the works in England for which he is best known are portraits. Indeed, though the painting itself is no longer extant, Holbein's Whitehall Palace mural showing Henry VII, Henry VIII, Elizabeth of York, and Jane Seymour, set the dominant pattern for Henry VIII's portraiture and representations that are still reproduced today on everything from scholarly volumes to tea towels.

As well as serving as a record of Henry's court, the drawings offer intriguing insights into Holbein's techniques and artistic practices. Though portraits are not extant for all the sitters represented by the drawings, either due to subsequent loss of the works, or lack of a commission, those which are extant track his changes from drawing to canvas. Susan Foister observed how the pendant portraits of Sir Henry Guildford and Lady Guildford show adjustments to the shape of Sir Henry's face which may be indicative of the patron's wishes rather than of Holbein's finesse, and that Lady Guildford's expression in her portrait is markedly less animated than in the *ad vivum* drawing.[10] Many of them are annotated with notes on the colour and materials of the sitter's clothing, presumably as an aide-mémoire in working up the final portrait.

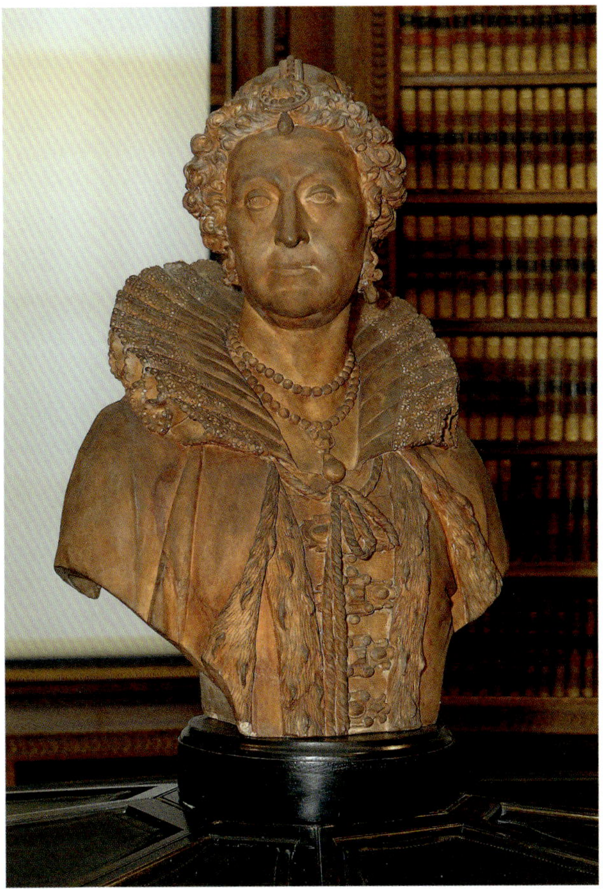

FIG. 12.02 Michael Rysbrack, *Elizabeth I*, c. 1736, terracotta, 25¾ x 17⅞ x 11½ in. (65.5 x 45.5 x 29 cm). Royal Collection Trust

The drawings of Lady Lister and Sir Thomas LeStrange (cats. 12.09–12.10) have rarely been on display. Lady Lister's precise identity still engenders debate, and no known portrait after this drawing has been found. Lady Lister's identity is complicated by conflicting references to her in modern assessments. K. T. Parker identified her as 'Jane, daughter of Ralph Shirley, widow of Sir John Dawtrey'.[11] However, information from Tudor House in Southampton, Dawtrey's home, refers instead to his widow as Isabel, daughter of Ralph Shirley, and his first wife as Jane William, and further claims that Isabel was a widow for ten years before marrying Lister, and bringing Tudor House to the marriage, as well as being 'an important merchant in her own right', dealing in wool and buying and selling millstones.[12] The identification of Sir Thomas Strange (or LeStrange) is supported by at least one existing portrait of him which closely follows this drawing, though the portrait is no longer attributed to Holbein, based on comparisons to other Holbein portraits and dendrochronological analysis.[13]

When mounted and displayed on the walls of Caroline's Closet at Kensington, the Holbein drawings must have seemed like a watchful, if archaic, crowd of onlookers, who formed, with their domination of the wall space, a ready-made sixteenth-century court interspersed among the other representations of royal relatives and courtiers past and present. Mingled with paintings representing her Hanoverian antecedents, and Tudor/Stuart monarchs, they stood as a major addition to Caroline's line of heritage.

Such zeal to connect the Tudors and Hanoverians on Caroline's part did not go unnoticed, though not all her efforts were universally admired, as satirists mocked her attempts to play up the Tudor connection in Merlin's Cave, and many were confused by its complicated message and unidentifiable figures.[14] The Hermitage was more to the general taste, and one poet claimed, of its sculptural figures, 'The Bustos rear'd by her command, Thro' ev'ry age shall speak her praise.'[15] However, attitudes changed and in George III's reign, the Holbein drawings were removed from the walls of Caroline's Closet at Kensington and remounted into volumes; Merlin's Cave was dismantled as part of the improving plans of Lancelot 'Capability' Brown at Kew in 1766; and the Hermitage was abandoned to gradual ruin. The Rysbrack terracotta models for the St James's busts were mounted in the

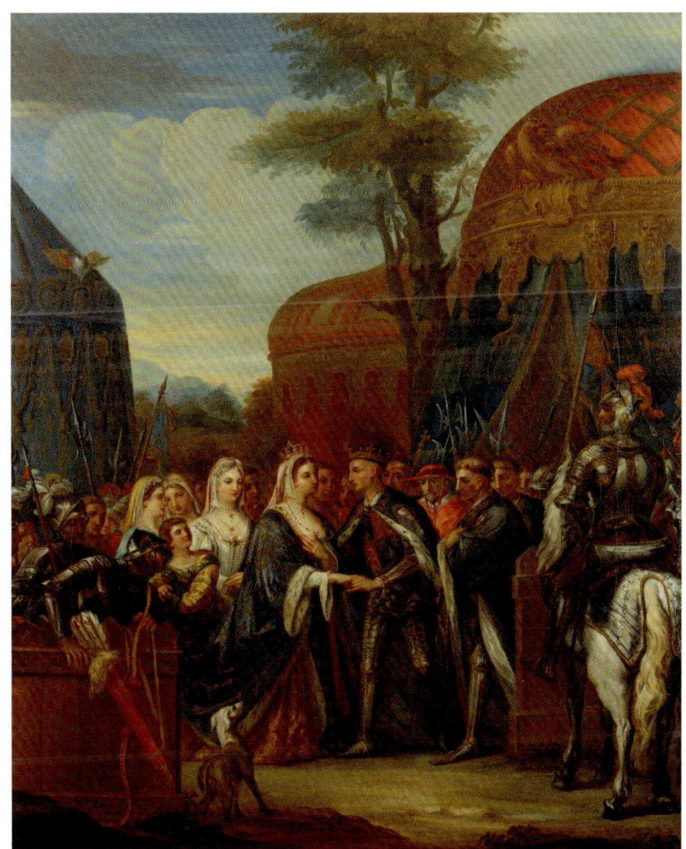

FIG. 12.03 William Kent, *The Meeting between Henry V and the Queen of France*, c. 1728–30, oil on canvas, 30 x 23⅞ in. (76.2 x 60.5 cm). Royal Collection Trust

Queen's Library after her death, and remained there until 1825. On Caroline's death, she herself joined her beloved Tudors, as George II provided the final link to the Tudor dynasty and the British past by creating a new vault in the Henry VII chapel with a grand sarcophagus where he and Caroline were both interred.[16]

Her daughter-in-law, Augusta of Saxe-Gotha, Princess of Wales, inherited the mantle of consort, but her collecting and creative pursuits were more redolent of the British present and connections to empire, than a hearkening back to the glories of the British past. Augusta's husband, Frederick, Prince of Wales, displayed busts of the Black Prince and King Alfred at Carlton House, presenting himself as an exemplar of British patriotism, and Augusta contributed to this with her commission to George Knapton of the posthumous portrait of her family, whose iconography celebrates Frederick's championship of these British ideals (see p. 93, fig. 4.19).[17] But after Frederick's death Augusta's particular passions were invested in the continuing development of the estate at Kew, where the plant collecting and cultivation of the gardens so avidly pursued by herself and John Stuart, third Earl of Bute, reflected an engagement with a wider world. The inclusion of a folly, in the form of a Gothic cathedral, alongside the Pagoda, Alhambra, Mosque, and other buildings celebrating exotic architectural styles from other cultures, is probably best read within the contemporary fashion for the 'Gothick'.

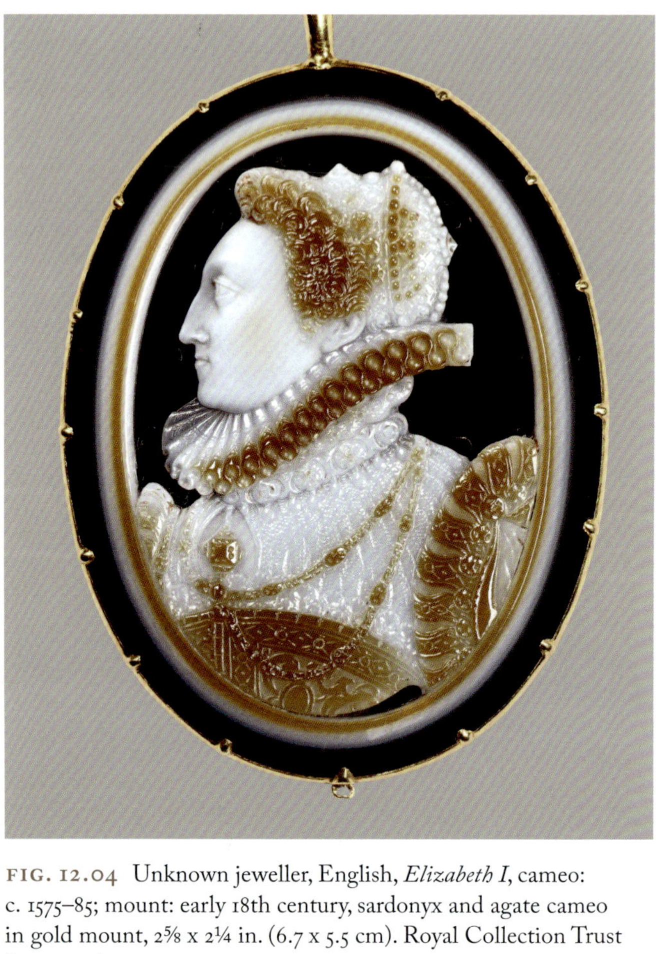

FIG. 12.04 Unknown jeweller, English, *Elizabeth I*, cameo: c. 1575–85; mount: early 18th century, sardonyx and agate cameo in gold mount, 2⅝ x 2¼ in. (6.7 x 5.5 cm). Royal Collection Trust (cat. 10.25)

Both Caroline and Augusta also demonstrated a familiarity with the work of those engaged in exploring the ancient British past. One figure who caught the attention of both was William Stukeley, a physician and clergyman, and a passionate antiquary, who was a driving force behind the refounding of the Society of Antiquaries. Stukeley, who is credited as a pioneer in British field archaeology, avidly pursued his interest in excavation and the examination and recording of ancient sites and objects on his travels through the British Isles, and wrote valuable treatises on the structure and possible significance of the stone circles at Stonehenge and Avebury. Caroline's architects for Merlin's Cave both were familiar with Stukeley and his works, and Caroline owned a copy of *Itinerarium Curiosum*, Stukeley's magisterial survey of ancient sites and objects in Great Britain.[18] Stukeley gradually focused his interest on pre-Roman Britain, and developed elaborate theories on the religion and culture of the Druids, even adopting the 'Druidic' name Chynodonax.[19] His circle of friends included Sir Hans Sloane, Richard Mead, Sir Isaac Newton, and Stephen Hale, and he wrote in his diary of a meeting with Augusta in which they discussed his book on Stonehenge and the purpose of what he identified as Druidic 'instruments' which were found in her garden at Kew. Stukeley dedicated his publication *Palaeographia Sacra*, of 1763, to Augusta, referring to her as 'Veleda, Archdruidess of Kew', and expressing his gratitude for her interest in the objects excavated there (fig. 12.05).[20]

Charlotte of Mecklenburg-Strelitz connected to the British past most directly through her production of 'Grangerized' books and written catalogues, and her print collecting, which enabled her to illustrate those catalogues lavishly. In a letter of February 1772 to her brother Charles, she wrote: 'I love being occupied, both in itself, and as a means of being sustained in disagreeable moments and above all against the temptations of a corrupt world.'[21] That desire to be occupied led her to transcribe an account of the King of Denmark's visit to England in 1606, and the advice of James II to his son, as well as to collaborate on a history of the English kings, transcribed by her daughter, Mary, and illustrated by her with prints. But the private nature of the volumes suggests that their creation was not an attempt to promote British history but simply a way of finding interesting subjects with which to occupy herself in the years of living quietly at her beloved Frogmore.[22] Indeed, William Westall told the diarist Joseph Farington that when Charlotte expressed her interest in transcribing the Stuart documents and Westall offered to do it for her, to save her the 'laborious undertaking', she replied: 'I am

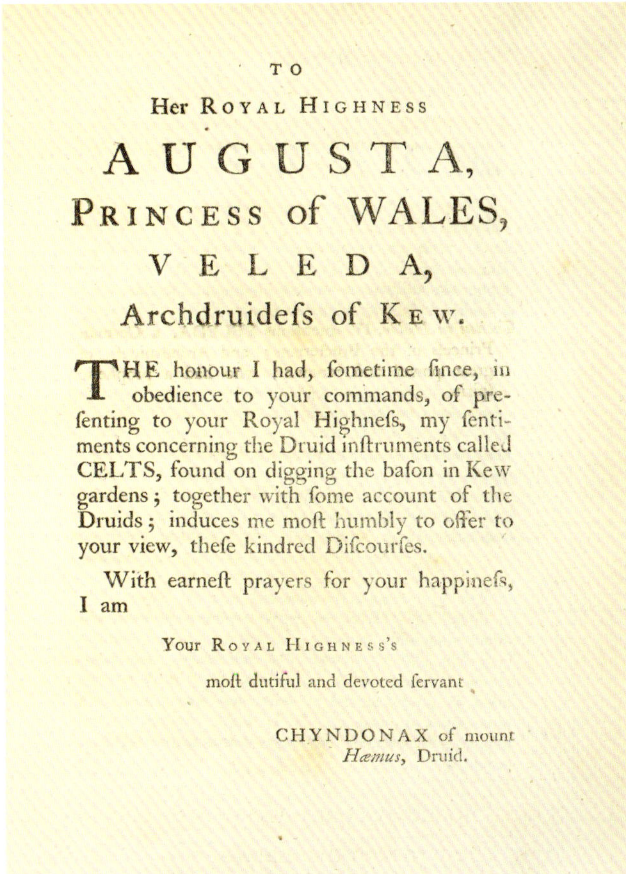

TO

Her ROYAL HIGHNESS

AUGUSTA,

PRINCESS of WALES,

VELEDA,

Archdruideſs of KEW.

THE honour I had, ſometime ſince, in obedience to your commands, of preſenting to your Royal Highneſs, my ſentiments concerning the Druid inſtruments called CELTS, found on digging the baſon in Kew gardens; together with ſome account of the Druids; induces me moſt humbly to offer to your view, theſe kindred Diſcourſes.

With earneſt prayers for your happineſs, I am

Your ROYAL HIGHNESS's

moſt dutiful and devoted ſervant

CHYNDONAX of mount *Hæmus*, Druid.

FIG. 12.05 William Stukeley, *Palæographia Sacra. Or Discourses on Sacred Subjects* (London, 1763), dedication page. British Library, London

accustomed to such work; I have 400 pages of extracts which I have from various works.'[23]

Caroline remains the consort who, in her collecting, commissioning, and display of British monarchs, worthies, and curious figures through artworks and architectural schemes, most assiduously strove to establish the connections of the Hanoverians to the British past. But in fairness to the subsequent royal consorts, Caroline's zeal in so thoroughly laying out those connections left little ground for Augusta and Charlotte to cover.[24]

NOTES

1. *Bernard André: The Life of Henry VII*, trans. Daniel Hobbins (New York: Italica Press, 2011), 6–8; *The Anglica Historia of Polydore Vergil, A.D. 1485–1537*, ed. and trans. Denys Hay (London: Royal Historical Society, 1950), 5; Sydney Anglo, *Images of Tudor Kingship* (London: Seaby, 1992), 56–59.
2. Peter Sherlock, 'The Monuments of Elizabeth Tudor and Mary Stuart: King James and the Manipulation of Memory', *Journal of British Studies* 46, no. 2 (April 2007): 267–68. See also http://www.westminster-abbey.org/our-history/royals/mary-queen-of-scots.
3. The best overall discussion is in Joanna Marschner, *Queen Caroline: Cultural Politics at the Early Eighteenth-Century Court* (New Haven and London: Yale Univ. Press, 2014).
4. Joanna Marschner, 'Michael Rysbrack's Sculpture Series for Queen Caroline's Library at St James Palace', in *Burning Bright: Essays in Honour of David Bindman*, ed. Diana Dethloff, Tessa Murdoch, and Kim Sloan, with Caroline Elam (London: UCL Press, 1015), 27–36, for an extended discussion of the busts and their meanings.
5. Martin Clayton, *Holbein to Hockney: Drawings from the Royal Collection* (London: Royal Collection Publications, 2004), 9.
6. *Holbein and the Court of Henry VIII: The Queen's Gallery, Buckingham Palace, 1978–1979* (London: The Queen's Gallery, c. 1978), 7.
7. *Holbein and the Court of Henry VIII*, 11.
8. It has been suggested that it may have been the third Earl, William Herbert, who made this exchange with the king. http://www.nga.gov/content/ngaweb/Collection/art-object-page.28.html. See Susan Foister, *Holbein and England* (New Haven and London: Yale Univ. Press, 2004), 23.
9. Jane Roberts, *Drawings by Holbein from the Court of Henry VIII* (New York: Johnson Reprint Corporation, 1987), 21–23, comprises the best summary of histories of the drawings. See also Clayton, *Holbein to Hockney*, 10–14, and *Holbein and the Court of Henry VIII*, 11–12.
10. Susan Foister, *Holbein in England* (London: Tate Publishing, 2006), 26–27.
11. K. T. Parker, *The Drawings of Hans Holbein in the Collection of His Majesty the King at Windsor Castle* (London: Phaidon Press; New York: Oxford Univ. Press, 1945), 41.
12. *Tudor House and Garden: 900 Years of History in the Heart of Southampton's Old Town* (Southampton City Council, 2013), 8.
13. I am grateful to Nancy E. Edwards, Curator of European Art/Head of Academic Services at the Kimbell Art Museum, Fort Worth, for this information.
14. Judith Colton, 'Merlin's Cave and Queen Caroline: Garden Art as Political Propaganda', *Eighteenth-Century Studies* 10, no. 1 (Autumn 1976): 1–20. See, in this volume, Cynthia Roman, 'Disrespect: Undermining Royal Image' and Todd Longstaffe-Gowan, 'Hankering after Horticulture'.
15. *Gentleman's Magazine: or, Monthly Intelligencer*, 1 July 1733, 369.
16. Tim Tatton-Brown and Richard Mortimer, eds., *Westminster Abbey: The Lady Chapel of Henry VII* (Woodbridge, Suffolk, and Rochester, NY: Boydell Press, 2003), 324–25.
17. *Deliciae Britannicae; or, the curiosities of Kensington, Hampton Court, and Windsor Castle, delineated; with occasional reflctions* [*sic*] (London, [1742]), 172–73. See, in this volume, Wolf Burchard and Kathryn Jones, 'The German Princesses as British Patrons', 109–131.
18. Marschner, *Queen Caroline*, 61–62.
19. David Boyd Haycock, 'Stukeley, William (1687–1765)', *Oxford Dictionary of National Biography* (Oxford: Oxford Univ. Press, 2004; online ed.), http://www.oxforddnb.com/view/article/26743, accessed 23 July 2016.
20. A commentary by Stukeley between the title page and preface of *Palaeographica Sacra* states: 'Tacitus in Histor. IV, mentions VELEDA, a German Princess of the Bructurians; and Archdruidess, of great power and renown; who had a prophetic spirit.' *The Family Memoirs of the Rev. William Stukeley, M.D. and the Antiquarian and other Correspondence of William Stukeley, Roger & Samuel Gale etc*, ed. W. C. Lukis, 3 vols. (Durham: Surtees Society, 1882–87). For the discussion of the excavation at the White House, see vol. 3 (1887), 210–12. William Stukeley, *Palaeographia Sacra, or Discourse on Sacred Subjects* (London: Richard Hett, 1763).
21. Clarissa Campbell Orr, 'Charlotte (1744–1818)', *Oxford Dictionary of National Biography* (Oxford: Oxford Univ. Press, 2004; online ed., May 2009), http://www.oxforddnb.com/view/article/5162, accessed 10 Aug 2015.
22. See, in this volume, Jane Roberts, 'Frogmore and the Princesses as Artists', 375–77.
23. Joseph Farington, R.A., *The Farington Diary*, ed. James Greig (London: Hutchinson & Co. [1922–28]), 11–15, April 1813, 4430.
24. I am indebted to Joanna Marschner for this observation.

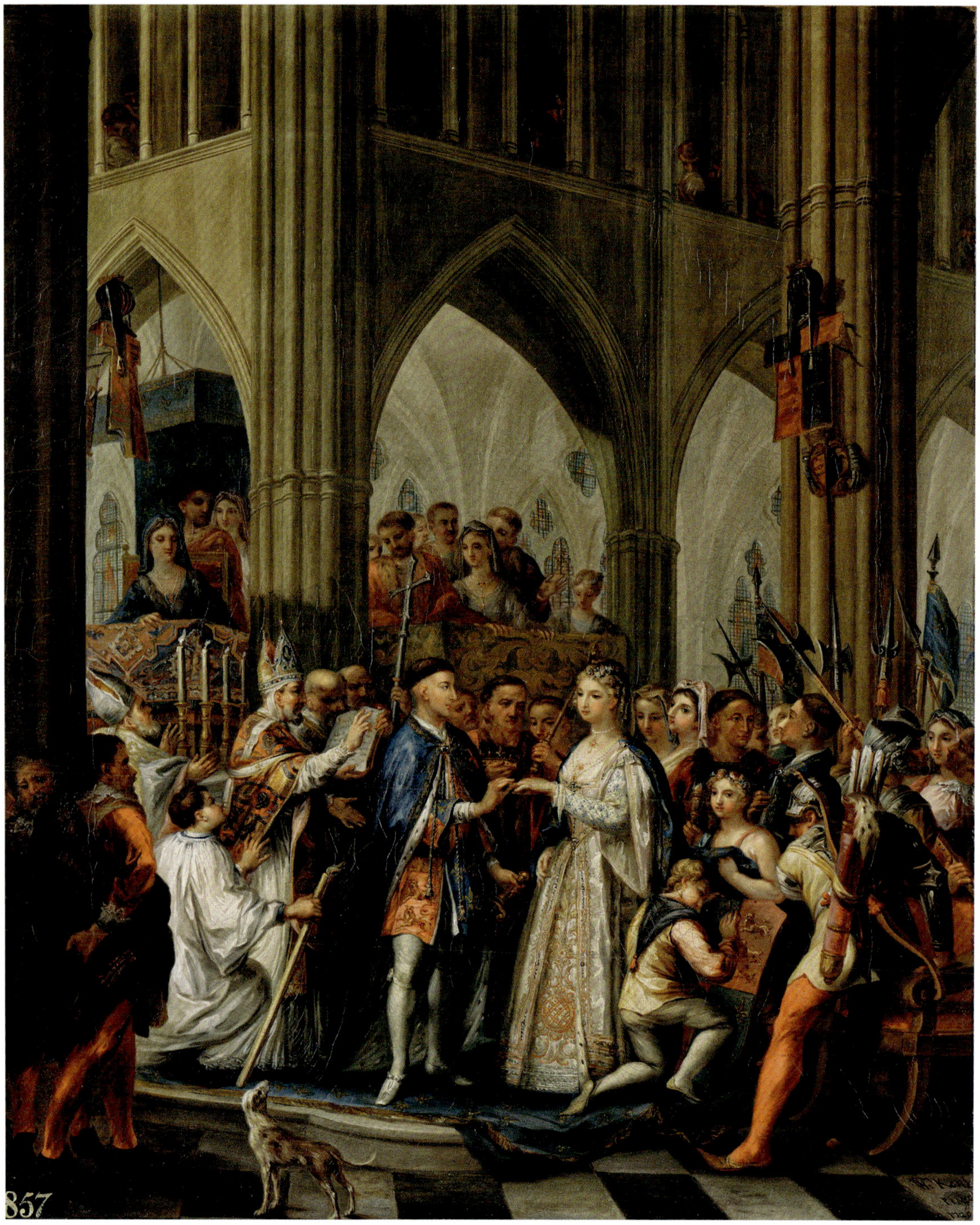

12.06

12.06

WILLIAM KENT (c. 1685–1748)
The Marriage of Henry V
1730–32
Oil on canvas
30 x 23⅞ in. (76.2 x 60.6 cm)
Royal Collection Trust, RCIN 402900

One of a set of three scenes from the life of Henry V that Caroline commissioned from William Kent. Her Privy Purse accounts record payments to Kent 'for pictures' in 1730 and 1731, which may relate to this series and suggest the private nature of the order. By 1758 Horace Walpole noted that they were hanging in the room that had served as her Dressing Room at St James's Palace. The other two paintings are *The Battle of Agincourt* and *The meeting between Henry V and the Queen of France*. Kent's first royal commission was to decorate Kensington Palace, which he gained in 1722 thanks to the support of Lord Burlington. He was appointed Master Carpenter in 1726, and would fulfill an important role in the achieving of Caroline's ambition for the garden at Richmond Lodge. In 1739 he succeeded Charles Jervas as Principal Painter to George II. Henry V continued to be celebrated as a royal hero by Caroline. He was included among the subjects for a series of portrait busts commissioned from Michael Rysbrack as decoration for her library.

* = YCBA only
† = HRP only

12.07[†]

EDMUND CURLL (c. 1675–1747)
'The Inside of the Cave in the Royal Gardens at Richmond'
In *The Rarities of Richmond: being exact descriptions of the Royal Hermitage and Merlin's Cave. With his life and prophecies . . .*
1736
E. Curll and J. Read, London
British Library, 8630.f.31.(1.)

Caroline installed a series of six life-sized waxworks in Merlin's Cave, one of a series of folly temples in the garden of Richmond Lodge. They represented figures associated with the ancient mythical history of the British nation. While their identities were hard even for contemporaries to establish, it is likely that the group comprised Merlin and his secretary; Queen Elizabeth and Elizabeth of York; Britomart and her nurse Glauce—taking inspiration from Edmund Spenser's *Faerie Queene* or possibly from Ludovico Ariosto's *Orlando Furioso*. They had been sculpted by Mrs Mary Salmon, who ran a popular waxworks attraction in Aldgate in London, and were modelled after members of Caroline's household staff. Caroline's appropriation of Elizabeth I as a model may have had a political as well as a personal resonance. In addition to being a female monarch, she was a champion of Protestantism, and the epitome of Englishness, as well as romantic, and just a little exotic.

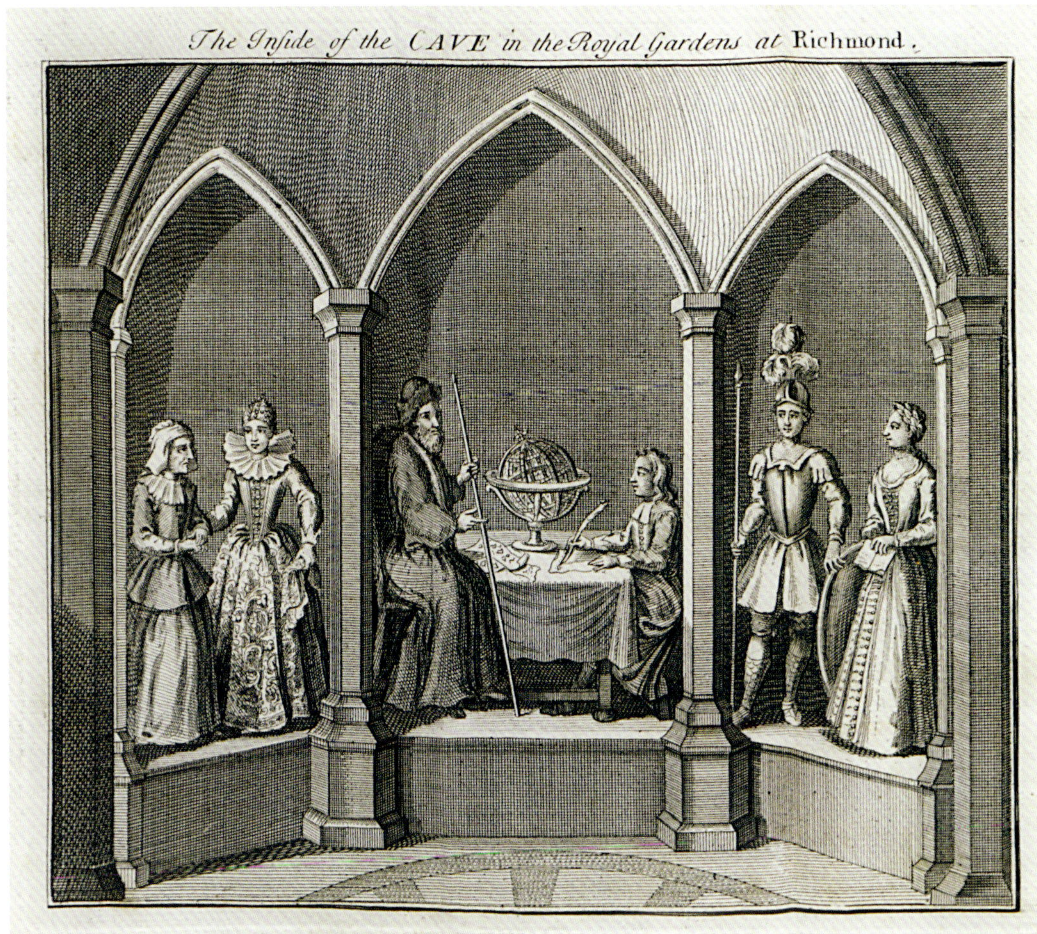

12.07

12.08*

MICHAEL RYSBRACK (1684–1770)
Edward, Prince of Wales, the Black Prince
c. 1737
Terracotta
25⅜ x 20⅛ x 9⅞ in. (64.5 x 51 x 25 cm)
Royal Collection Trust, RCIN 37067

The first reference to Caroline commissioning a series of busts of English monarchs from Michael Rysbrack is in 1735. The *Gentlemen's Magazine* suggests that they were intended initially as decoration for one of the folly temples in the garden at Richmond. The series was never completed. When Caroline died in 1737, George II asked Rysbrack to return the terracotta modellos which had been finished and these were installed on wall brackets in Caroline's recently built library, designed by William Kent, at St James's Palace. There were at least eleven busts in the series, but only three: Edward, Prince of Wales, the Black Prince; Edward VI, and Elizabeth I survive. The remainder, which included King Alfred, Henry V and his wife Catherine of Valois, and Henry VII and his wife Elizabeth of York, were removed to Windsor Castle after the library was demolished in 1825, and were irretrievably damaged when the shelf on which they were stored collapsed in the early twentieth century.

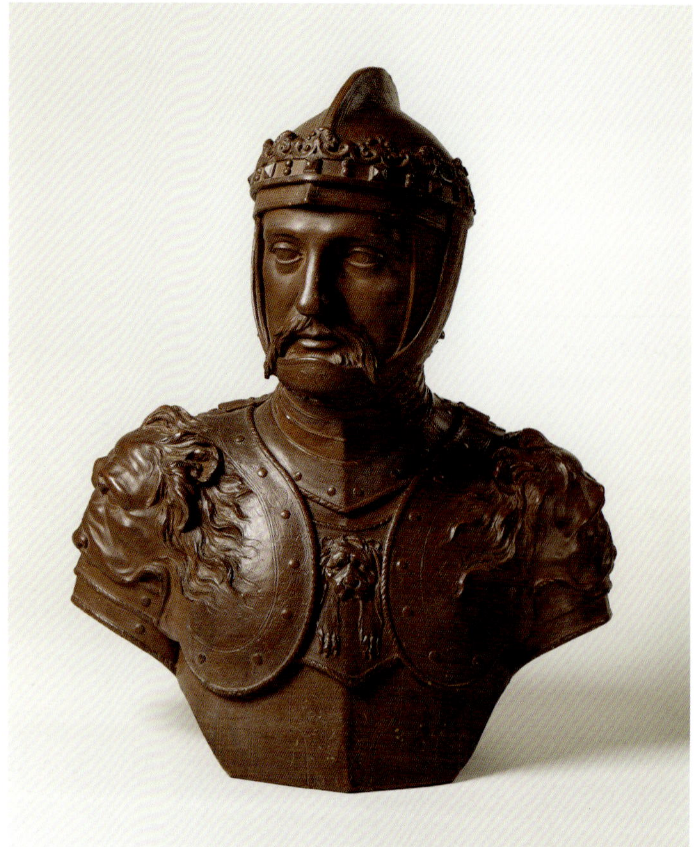

12.08

12.09

HANS HOLBEIN THE YOUNGER
(1497/98–1543)
The Lady Lister
1532–43
Black and coloured chalks, pen and ink and brush and ink on pale pink prepared paper
11½ x 8 ¼ in. (29 x 21 cm)
Royal Collection Trust, RCIN 912219

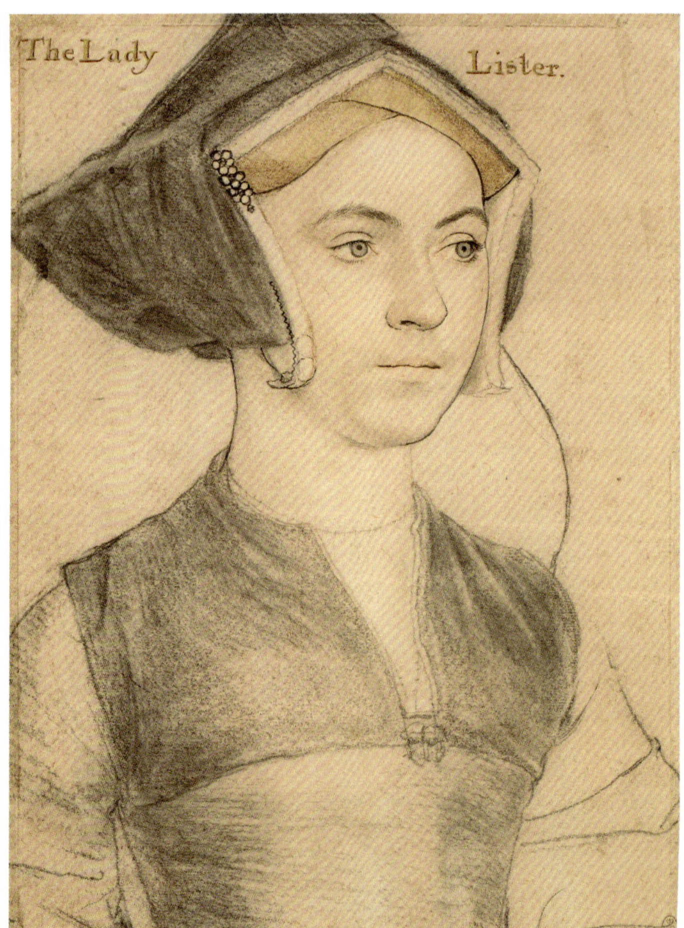

12.09

12.10

HANS HOLBEIN THE YOUNGER
(1497/98–1543)
Sir Thomas Lestrange
c. 1536
Black and coloured chalks, pen and ink, and metalpoint on pale pink prepared paper
9⅝ x 8¼ in. (24.3 x 21 cm)
Royal Collection Trust, RCIN 912244

In 1728 Caroline located a portfolio of works on paper in a bureau in the King's Closet at Kensington Palace. The collection included works by Leonardo da Vinci, Raphael, Michelangelo, and also Hans Holbein the younger. The works were carefully listed before the Holbein drawings were removed to Richmond Lodge. In about 1735 the collection returned to Kensington Palace, where it formed the neucleus of the hang in Queen Caroline's Picture Closet. The portrait drawing of Thomas Lestrange, a senior courtier, holding the post of Squire of the Body in the early years of Henry VIII's reign, was a study for a painting. It is close to a painting of Lestrange at the Kimbell Art Museum, Fort Worth, Texas. The precise identity of Lady Lister is harder to establish. Encouraged by John Percival, first Earl of Egmont, Caroline allowed George Vertue to make engraved copies of the drawings in 1736, after conducting experiments to ensure that the originals would not be damaged by the process.

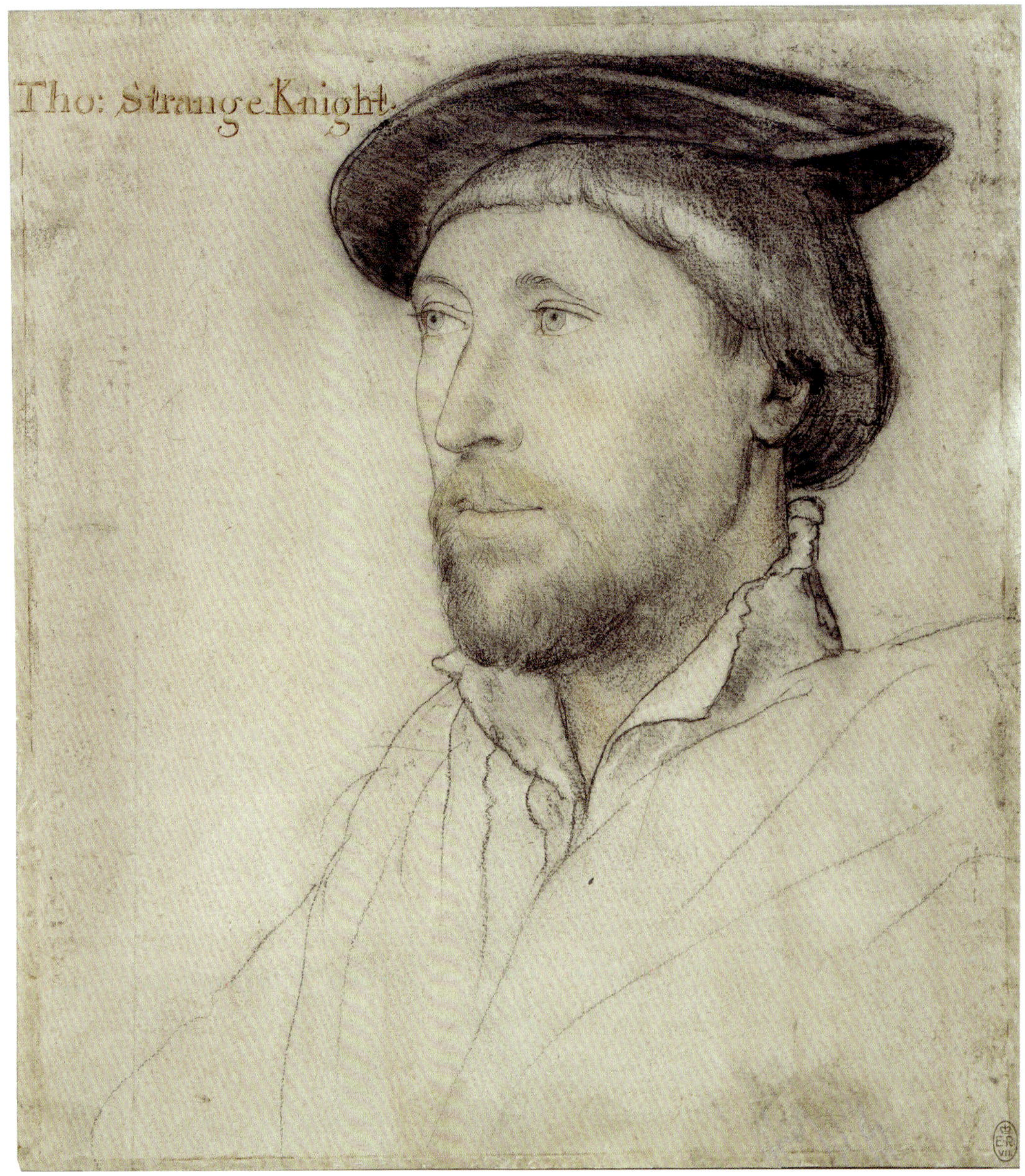

12.10

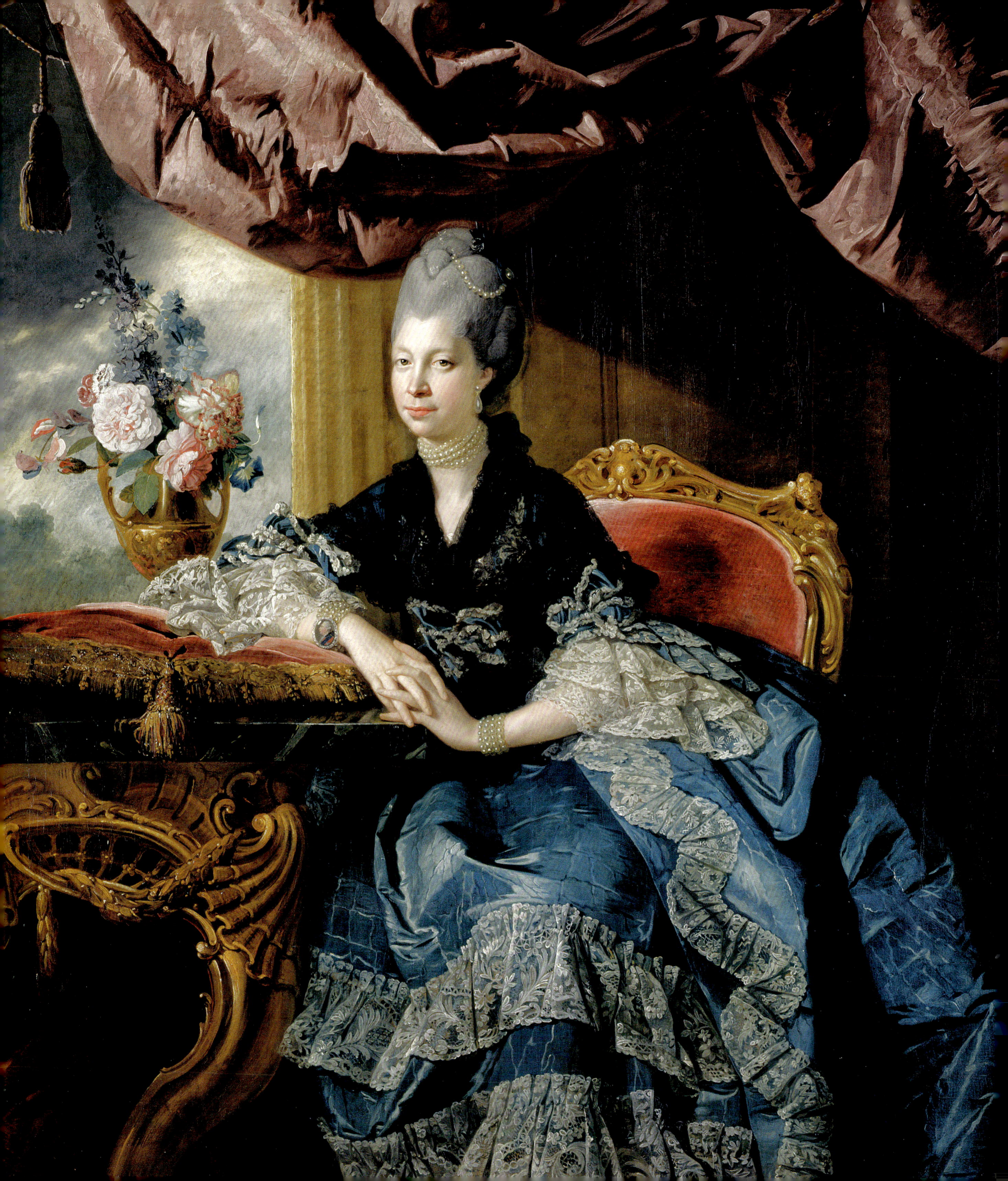

Portraits of the Princesses

Matthew Storey

Formal portraits were one of the few ways available to eighteenth-century monarchs to present themselves as they wished to be seen by their subjects. Portraits could be displayed in public places, but could also be reproduced in the form of engravings, widely distributed and often copied. The effect of the repetition of familiar poses, costume, and allegorical imagery in different versions of portraits of Caroline of Ansbach, Augusta of Saxe-Gotha, and Charlotte of Mecklenburg-Strelitz, often paired with their husbands, helped their own political agenda. They promoted the continuity of the Hanoverian dynasty with that of their Stuart predecessors, and present a positive view of a Protestant dynasty as fertile, secure, and British.

Caroline, the most learned of the three Princesses, placed herself and the Hanoverian dynasty within a long tradition of British history, going back to the Middle Ages. Michael Rysbrack's posthumous marble bust of Caroline had been commissioned by her for the St James's Palace library between 1736 and 1737.[1] It was displayed in the company of a marble bust of her husband and a pantheon of British kings and queens in terracotta.[2] Significantly, the group included Elizabeth I, to whom Caroline, as another Protestant British queen, had been compared by Gottfried Wilhelm Leibniz.[3]

The grandest and most revealing portrait of Caroline was made by Jacopo Amigoni in 1735 (see p. 85, fig. 4.11) and given to the royal physician Dr Richard Mead. It exemplifies all the major themes of dynastic image making, celebrating Caroline's fertility and success in perpetuating the dynasty into the future by including in the image a cornucopia filled with small children.[4] Caroline was also the subject of allegorical representation, seen in its grandest form in the vast mural of the Hanoverian succession in the Upper Hall of the Royal Hospital at Greenwich, painted by Sir James Thornhill between 1718 and 1725. Caroline is here depicted as Prudence, accompanied by her children, demonstrating again her central role as mother of a new dynasty.[5] Much smaller in scale, but equally potent, William Kent's sketch of *Queen Caroline as Britannia receiving the Gifts of the Earth* of around 1730–40 shows Caroline enthroned at the centre of power and patronage, attended by Mars and Minerva (cat. 13.03). Her influence extends to the heavens, which are signified by the signs of the zodiac, and to the seas, represented by the stern of a ship on the right.

Medals were another medium produced to ensure the wide circulation of the royal image. They were collected and easily carried as a sign of family or political allegiances. Some were produced to commemorate particular occasions and moments of national significance, such as the example by John Croker made to mark Caroline's coronation (cat. 13.04). In one case a medal was produced in riposte to another. In 1731 a Jacobite medal was issued in Italy depicting the two sons of James Francis Edward Stuart, 'The Old Pretender', championing these determined claimants to the British throne. In response, in 1732 George II commissioned a medal from Croker and John Sigismund Tanner showing his own family of seven children, restating that the Hanoverian family were well established and fit to serve the

Detail of p. 246, cat. 13.18

nation into further generations (cat. 13.05).[6] Henry Jernegan commissioned Hubert-François Gravelot in 1736 to design a medal to be given to ticket holders participating in a lottery for a large silver cistern, or wine cooler, to raise funds for the building of Westminster Bridge (cat. 13.08). Gravelot's design presents an interesting commentary on the respective roles of the king and queen. The obverse has the motto 'GEORGE REIGNING'. On the reverse it shows an image of Caroline tending young palm trees, representing the knowledge, arts, and industry needed for the encouragement of Britain's imperial ambition.[7] She is accompanied by the motto, 'GROWING ARTS ADORN EMPIRE, CAROLINE PROTECTING'.

While many of the images of Caroline mentioned above were initiated by her, or on her behalf, the profile portrait of Caroline by Joseph Highmore of about 1735 may have been painted speculatively, without a sitting from the Queen (cat. 13.02). George Vertue records that on one occasion a request for a sitting had been denied, but Highmore's skill in creating a likeness from memory is mentioned in his obituary in connection with a portrait of Caroline.

Augusta, as the wife, and later widow, of Frederick, Prince of Wales, was under similar pressures to Caroline to provide a large successful family in the interest of perpetuating the dynasty, and to celebrate the pedigree of their family. The position of the Hanoverian dynasty became more secure after the defeat of the Jacobite supporters at the Battle of Culloden in 1745. The portrait of Augusta by the studio of Jean-Baptiste van Loo, produced in about 1742, shows an awareness of a more contemporary taste characteristic of Frederick and Augusta's patronage (cat. 13.10).[8] The Frenchness of the painting is revealed in the lighter tone, the smoothness of the paintwork, and the highlighting of the costume and jewellery. However, the choice to be depicted in ceremonial dress is still conventional, representing the Princess as a symbol of continuity and linking the portrait to royal models from past centuries.

Another formal portrait of Augusta in state robes by Jeremiah Davison was a lasting memorial of a successful event (cat. 13.09). It was given to the city of Bath, together with a portrait of Frederick, as a mark of thanks for the hospitality the couple had received after they had stayed in the city for six weeks in late 1738, as part of Frederick's tour of cities in the south west of England to garner support for the royal family, and to celebrate local industry.[9] It was noted that at a feast in nearby Bristol, Augusta 'talked freely with the Ladies in good English, which entirely won their Hearts'.[10] The success of this venture, which was followed by the royal couple pursuing other initiatives to promote domestic industries in their bid to court popularity, would inform the allegory of *The Muses Paying Homage to Frederick, Prince of Wales and Princess Augusta* of about 1750–51 by Francis Hayman, a sketch for a painting intended for the great pleasure grounds in London's Vauxhall Gardens (cat. 13.14). In it the royal couple are presented with a plan for a royal academy for artists, though Frederick died before he could become its patron.[11] The allegory also celebrates a range of other British activities and industries including textile manufacture, suggesting the breadth of Frederick and Augusta's interests, influence, and patronage.[12]

Although portraits of the three Princesses usually served an official function, some, like the portrait of Caroline by Amigoni of 1735, and the 1769 portrait of Augusta by Allan Ramsay, were given as personal gifts.[13] Caroline gave her portrait by Amigoni to her doctor as a mark of esteem. The Ramsay was given to the Brunswick court, into which Augusta's eldest daughter, also named Augusta, had recently married, as a token of dynastic affiliation. Many royal portraits were also commissioned by clients outside the royal family as a mark of respect, loyalty, or affection. Ramsay was commissioned by John Stuart, third Earl of Bute, to make a full-length portrait of Augusta, probably as a token of their close friendship, though five versions of the portrait are known, suggesting a broader audience for the image.[14] The outdoor setting suggests their shared enthusiasm for gardening, developed through their work together on the gardens at Kew.[15]

Other portraits suggest more personal encounters, such as a pastel portrait of Augusta commissioned in 1754, after her husband's death, from Jean-Étienne Liotard, who was known for his simple and direct portraits (see cat. 9.05). The Princess's head is slightly tilted with her lips parted as if about to speak. She commissioned the portrait together with those of her eight elder children, and they all show a similar remarkable intimacy (cats. 9.06–9.10).[16]

The number of portraits of Charlotte far exceeds those of Caroline and Augusta, not only because of the length of her reign, but because of a more systematic policy of using portraiture for political ends. Also, George III and Charlotte found congenial painters in Ramsay and Johan Zoffany. Ramsay's coronation portraits of George III and Charlotte of 1761–62 were the most widely disseminated official portraits of the eighteenth century. British ambassadors and governors of British colonies were entitled to a pair of the portraits, paid for by the Lord Chamberlain, whose accounts list payments for eighty-six pairs of portraits, with a further six just of the King.[17] One notable version of Charlotte was supplied to Montagu Wilmot, the Governor of Nova Scotia.[18] Those not entitled to the portraits could directly commission their own, so the number produced was even higher.

While Ramsay was Principal Painter to George III and Charlotte, Zoffany's great skill as a painter of conversation pieces lead him to be indelibly associated with the depiction of royal family life. Through his intimate and relatively informal portraits, that included their children, they were able to project an image of themselves as happily and productively married, for example in *Queen Charlotte with her Two Eldest Sons* of about 1765 (p. 25, fig. 1.15). These paintings are almost unique in royal portraiture for their informality and charm. However, Zoffany was a versatile artist and the pair of portraits of the King and Queen painted in 1771, and circulated widely in mezzotint, are necessarily more formal. George's active stance and military accessories are set against Charlotte's alert but more contemplative pose. Her portrait is remarkable for her regal dignity but also for the minute depiction of her expensive dress and accessories. She wears a portrait miniature of George on a bracelet given to her by him, demonstrating her fidelity to her husband.

Charlotte was also the subject of allegorical representation. In a 1772 print by Thomas Burke after Angelica Kauffmann, *Queen Charlotte raising the Genius of the Fine Arts*, she is shown gently rousing the sleeping spirit with a maternal touch and regal bearing (cat. 13.21). Through her deeds the arts would be crowned with success. The winged genius is in fact a portrait of her eldest son, the future George IV.[19] The image combines to celebrate Charlotte as mother, queen, muse, and patron, her fertility linked to her encouragement of the arts. The enamel of *Queen Charlotte Surrounded by the Attributes of Agriculture* of 1773 shows Charlotte as the goddess Ceres, surrounded by imagery of abundance, her fertility associated with the health and happiness of Britain. A celebration of feminine attributes, it contrasts with the paired enamel of George III as Lord of the Seas, a battleship behind him; he is dressed in Roman armour, trampling America underfoot.

The portraits of the three Princesses in different ways address the insecurities of the Hanoverian dynasty in the eighteenth century, for the successive dramas of the royal succession had a European-wide impact. The advent of the Hanoverians in 1714 had caused widespread consternation. The Act of Settlement's insistence that the English monarch had to be Protestant had brought George I to the throne despite his extremely weak dynastic connection; there were a great many Catholic family members with stronger claims. In response, the Scottish supporters of these rival Jacobite claimants rose up the following year with much support in England, and it was only in 1745 that such threats finally receded. Added to this was a particular anxiety over royal fertility, caused by Queen Anne's frequent miscarriages, and a vague sense that the Hanoverians were foreign usurpers. Portraits and other artworks commissioned by the royal family had to address these concerns, deliberately associating the Hanoverians with portrait formulae associated with the Stuarts and earlier royal dynasties. A particular emphasis was placed on the ability of the royal consorts in the eighteenth century to produce children. Their portraiture also had to convey a sense that the Princesses were a benevolent presence in public life, encouraging trade and the arts, and supporting the welfare of the nation. In different ways all these attributes define royal portraiture in the eighteenth century.

NOTES

1. Desmond Shawe-Taylor, ed., *The First Georgians: Art and Monarchy, 1714–1760*, (London: Royal Collection Trust, 2014), 34–37; cat. nos. 5 and 6.
2. Shawe-Taylor, ed., *First Georgians*, 78; cat. no. 29.
3. W. H. Wilkins, *Caroline the Illustrious: Queen-Consort of George II, and sometime Queen-Regent: A Study of her Life and Time*, 2 vols. (New York and Bombay: Longmans, 1901), 1: 102–4, as cited in Shawe-Taylor, ed. *The First Georgians*, 272. For a more detailed discussion of this portrait, see, in this volume, Lisa L. Ford, 'Using Britain's Past', 219–223.
4. This agenda is explored in more detail in this volume in Cassandra Albinson and Mark Hallett, 'Cornucopia: Portraiture and the Imperatives of Reproduction,' 77–107.
5. James Thornhill, *An Explanation of the Painting in the Royal Hospital at Greenwich*, (London, 1730), 18–23.
6. Shawe-Taylor, ed., *The First Georgians*, 47; cat. no. 15.
7. Shawe-Taylor, ed., *The First Georgians*, 317; cat. no. 188.
8. For an account of Van Loo's influence on British painting, see David H. Solkin, *Art in Britain, 1660–1815* (New Haven and London: Yale Univ. Press, 2015), 125–26.
9. For an account of the visit, see Frances Vivian, *A Life of Frederick, Prince of Wales, 1707–1751: A Connoisseur of the Arts*, ed. Roger White (Lewiston, NY, and Lampeter: Edwin Mellen Press, 2006), 272–86.
10. *Gentleman's Magazine* 8 (1738): 603.
11. Brian Allen, *Francis Hayman* (New Haven and London: Yale Univ. Press, 1987), 64–66, and 122; cat. no. 48.
12. Kimerly Rorschach, 'Frederick Prince of Wales (1707–51), As Collector and Patron', *The Walpole Society* 55 (1989/90): 1–76 (14).
13. The portrait by Jacopo Amigoni is discussed in more detail in this volume in Cassandra Albinson and Mark Hallett, 'Cornucopia: Portraiture and the Imperatives of Reproduction', 77–107, and, Craig Ashley Hanson, 'Medical Practice', 427–33.
14. Alastair Smart, *Allan Ramsay: A Complete Catalogue of his Paintings*, ed. John Ingamells (New Haven and London: Yale Univ. Press, 1999), 74–75; cat. no. 21.
15. Oliver Millar suggested that the urn may be the one now on the East Terrace at Windsor, but that the setting was intended to be Kew. However, the urn appears in two other portraits by Ramsay, so the garden setting is more likely to be general than specific. See Millar, ed., *Later Georgian Pictures in the Collection of Her Majesty the Queen*, 2 vols. (London: Phaidon, 1969), 1: 97; cat. no. 1002, and Smart, *Allan Ramsay*, 74–75; cat. no. 21.
16. For a discussion of this series, see Jane Roberts, ed., *George III & Queen Charlotte: Patronage, Collecting and Court Taste* (London: Royal Collection, 2004), 23; cat. no. 2, and Shawe-Taylor, ed., *First Georgians*, 368–69; cat. nos. 242–44.
17. Jacob Simon, 'Frame Studies II: Allan Ramsay and Picture Frames', *The Burlington Magazine* 136 (July 1994): 444–55.
18. Smart, *Allan Ramsay*, 115; cat. nos. 192aw.
19. Angela Rosenthal, *Angelica Kauffman: Art and Sensibility* (New Haven and London: Yale Univ. Press, 2006), 165–66.

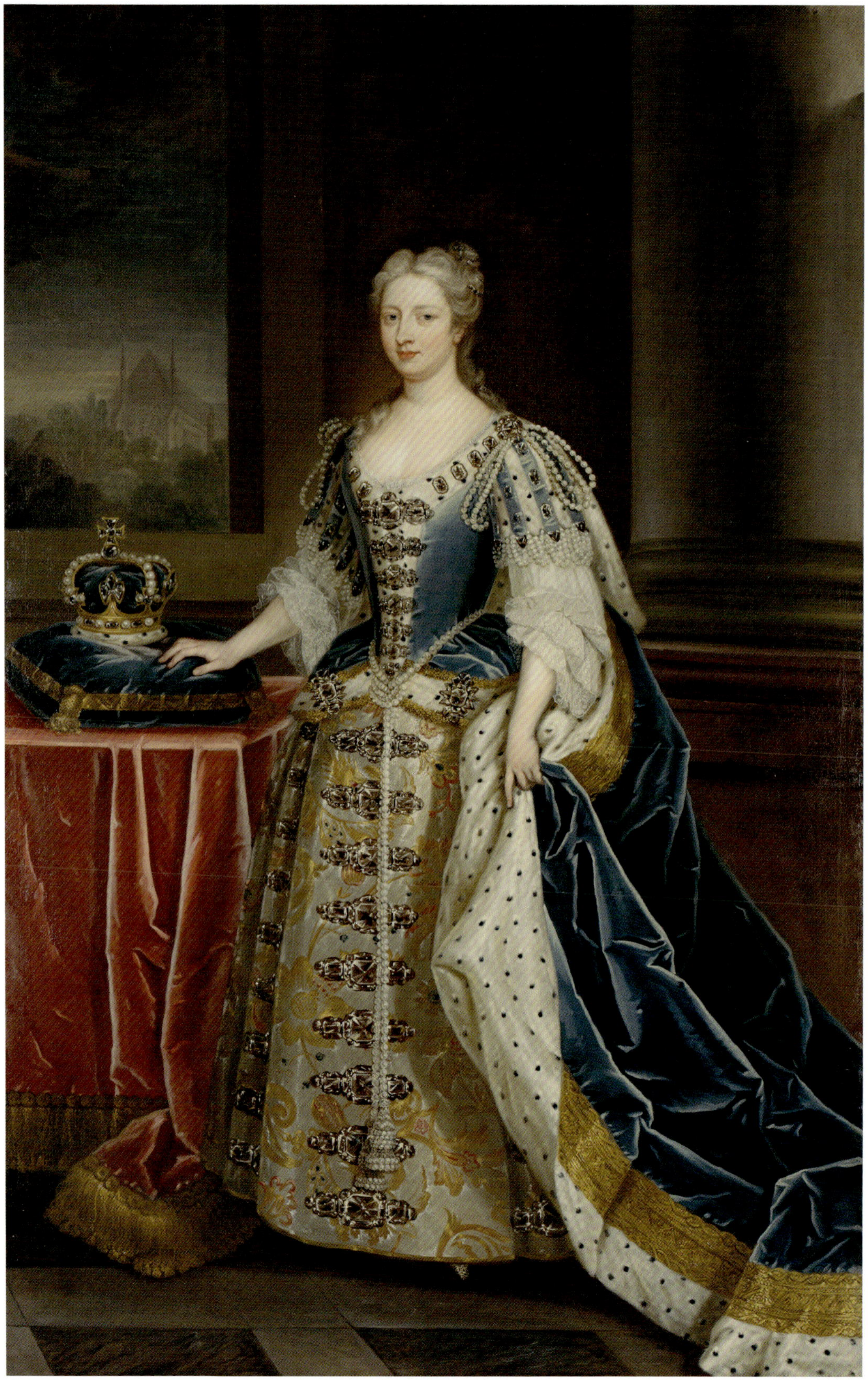

13.01

IMAGES OF CAROLINE

13.01*

CHARLES JERVAS (c. 1675–1739), after **ENOCH SEEMAN THE YOUNGER** (c. 1694–1744)
Caroline Wilhelmina of Brandenburg-Ansbach (1683–1737) Consort of King George II
c. 1727
Oil on canvas
94⅞ x 60⅜ in. (241 x 153.5 cm)
UK Government Art Collection, GAC 0/83

Charles Jervas found a patron in Sir Robert Walpole, who engineered his appointment in 1723 as King's Painter to George I, a post that was re-confirmed on the accession of George II. Jervas would make several copies of this portrait of Caroline, based on the coronation portrait Enoch Seeman, an artist of Dutch extraction, had been commissioned to produce in 1727, together with a companion portrait of George II. Depicted in their coronation robes they served as official portraits for the new regime. The original versions of these are in the Royal Collection, where they hung within significant dynastic picture groups at Windsor Castle and Kensington Palace. Several decades after Jervas's death Horace Walpole dismissed his style as 'a light, flimsy kind of fan painting as large as the life'. More recently he has been recognized as an accomplished portrait painter and a translator of considerable ability.

13.02

JOSEPH HIGHMORE (1692–1780)
Queen Caroline of Ansbach
c. 1735
Oil on canvas
29⅝ x 24⅜ in. (75.1 x 61.8 cm)
Royal Collection Trust, RCIN 406035

Joseph Highmore was the nephew of Thomas Highmore, Serjeant Painter to William III. After attending Sir Godfrey Kneller's drawing academy and anatomical lectures given by William Cheselden, his first commission for the Crown was awarded in about 1725, when he painted knights of the Order of the Bath. This portrait of Caroline was made in about 1735 shortly after he had returned from a study tour of northern Europe. It is uncertain whether the artist was granted a sitting by the Queen. George Vertue records that on one occasion access was denied. However, Highmore's obituary emphasized his ability to 'take a likeness by memory as well as by sitting'. The work was engraved by John Tinney.

* = YCBA only
† = HRP only

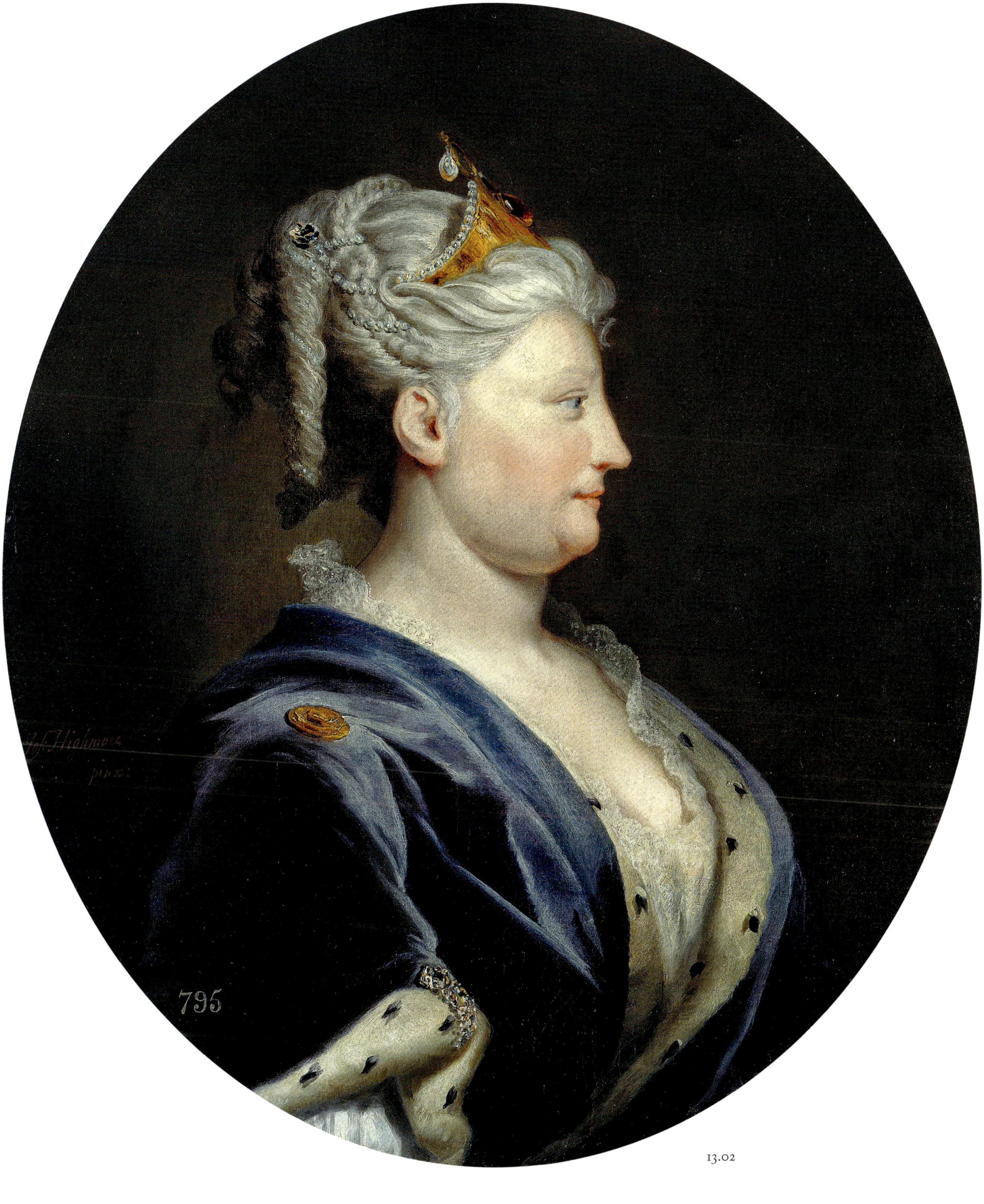

13.02

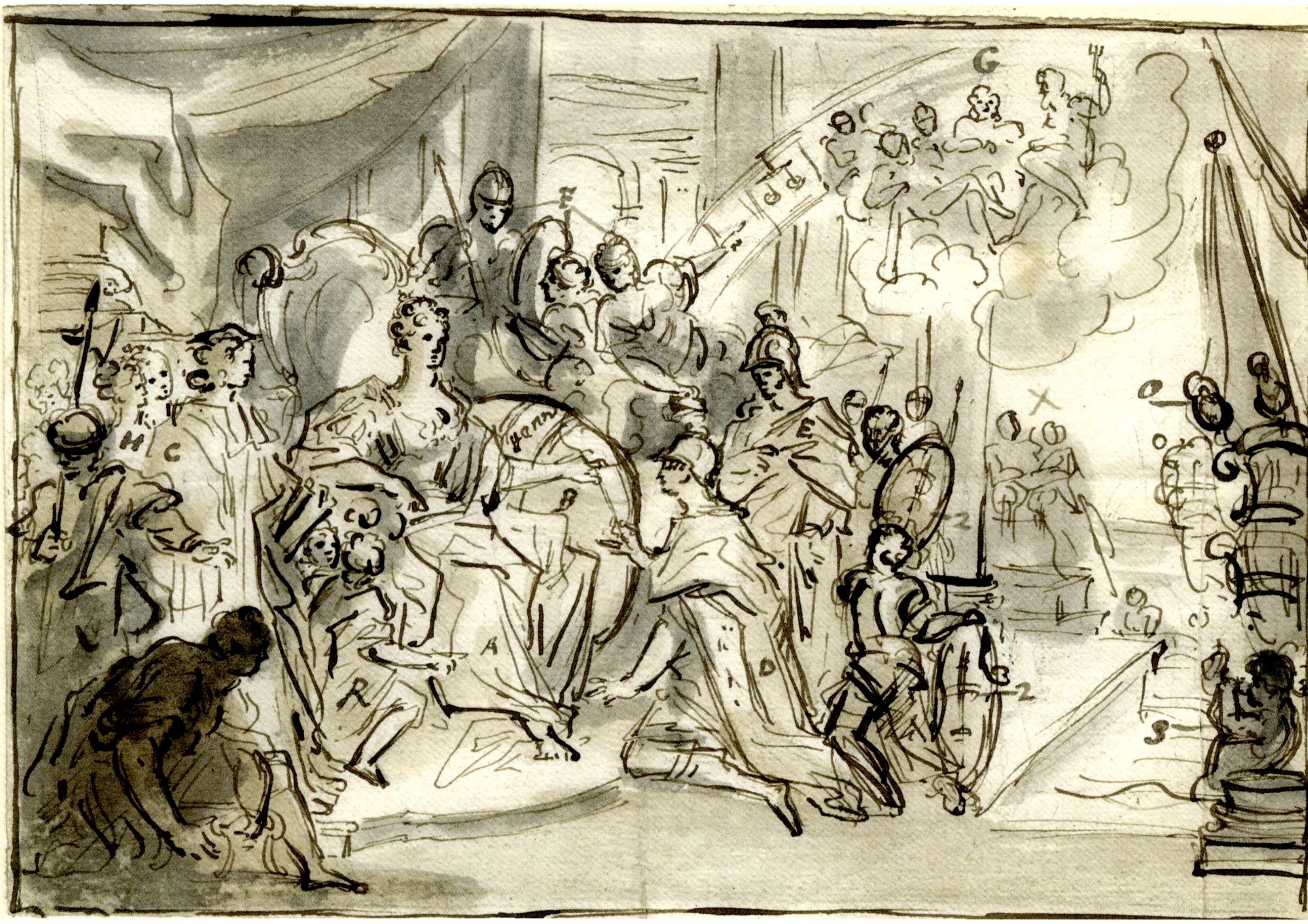

13.03

13.03

WILLIAM KENT (c. 1685–1748)
Queen Caroline, as Britannia, Receiving the Gifts of the Earth
1730–37
Pen and ink, and wash over pencil on paper
5¾ x 7⅞ in. (14.5 x 20.9 cm)
British Museum, 1927,0721.2

Caroline probably first met William Kent while he was working on new decorations for the King's State Apartments at Kensington Palace, where work had started in 1722. Kent's appointment, ousting the King's Serjeant Painter James Thornhill, was officially credited to Thomas Coke, the King's Vice-Chamberlain, but Richard Boyle, third Earl of Burlington, Kent's patron and admirer, almost certainly played a part. In this drawing Kent shows Caroline enthroned as Britannia, being presented with the gifts of an imperial nation, in honour of her interest and encouragement of international trade and colonial expansion. A ship stands as a symbol of Britain's maritime prowess.

13.04*

JOHN CROKER (1670–1741)
Medal commemorating the Coronation of Queen Caroline
1727
Silver
1⅜ in. (3.5 cm) diameter
Yale University Art Gallery, Transfer from the Yale University Library, Numismatic Collection, 2001, 2001.87.26146

13.05*

JOHN CROKER (1670–1741) and **JOHN SIGISMUND TANNER** (c. 1704–1775)
Medal Commemorating the Royal Family
1732
Silver
2¾ in. (6.9 cm) diameter
Royal Collection Trust, RCIN 443233

John Croker, formerly Johann Crocker, an engraver of coins and medals, was born in Dresden, the son of a woodcarver and cabinet-maker to the electoral court of Saxony. After apprenticeship to a goldsmith and jeweller in Dresden, he arrived in England in 1691. In 1697 he was appointed assistant to Henry Harris (1633/34–1704), chief engraver at the Royal Mint. When Harris died in 1705, Croker was appointed to succeed him. From 1702 to 1732 Croker was constantly engaged in medal engraving. His designs commemorated events rather than persons and were always struck, not cast. George Vertue noted that 'towards his latter end of life he was much assisted in his business of the Mint by Mr Tanner'; this was John Sigismund Tanner, apprenticed in 1729, who was to succeed him.

The medal commemorating Caroline's coronation shows her portrait in profile on the obverse. The reverse shows her in full length, gesturing to a veiled figure holding a book on her right (perhaps a personification of religious faith), and with Britannia on her left. The medal depicting George II, Caroline, and their seven children was a direct riposte to another produced in the preceding year in Italy that showed the sons of James Stuart, the 'Old Pretender' to the British throne. Their robust and numerous family was one of the means by which the couple sought to establish the dynastic legitimacy of the Hanoverians.

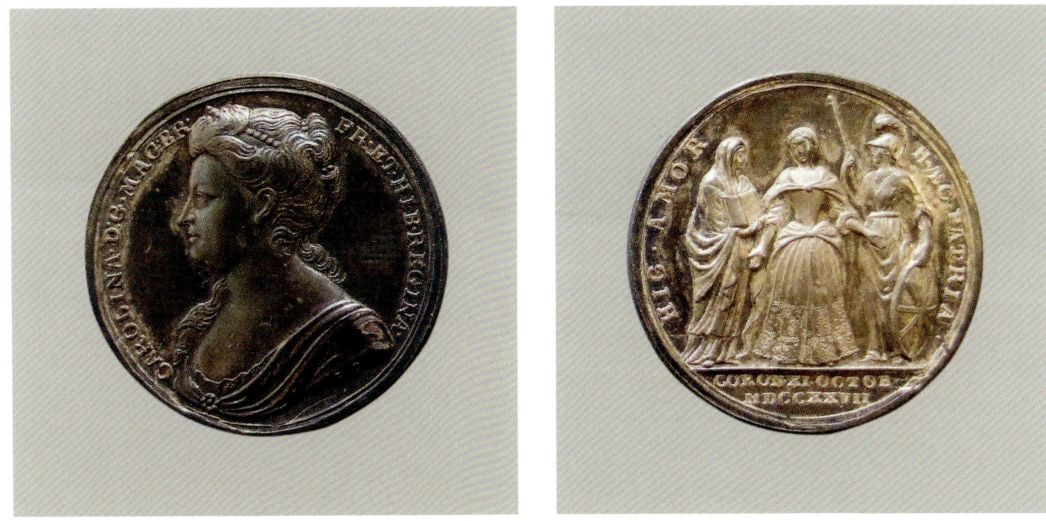

13.04

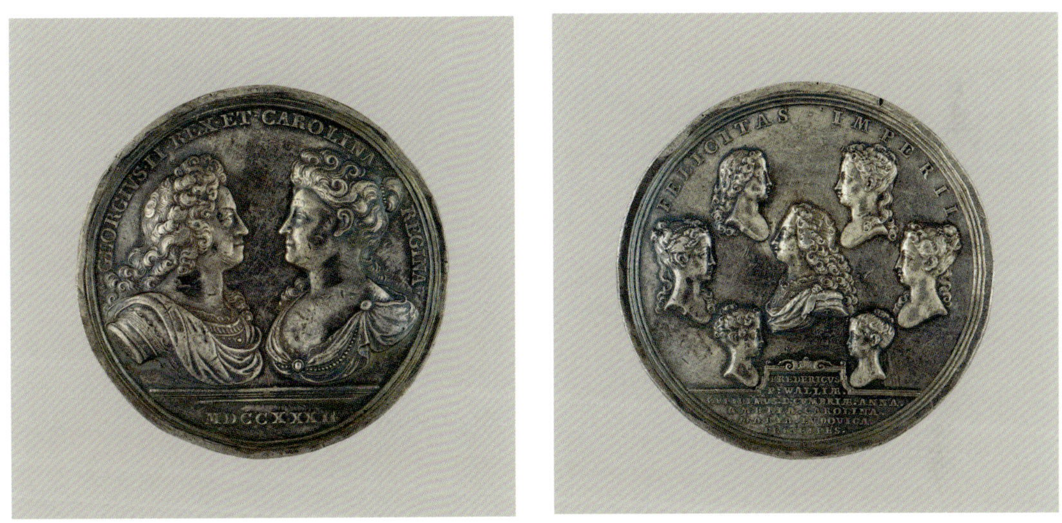

13.05

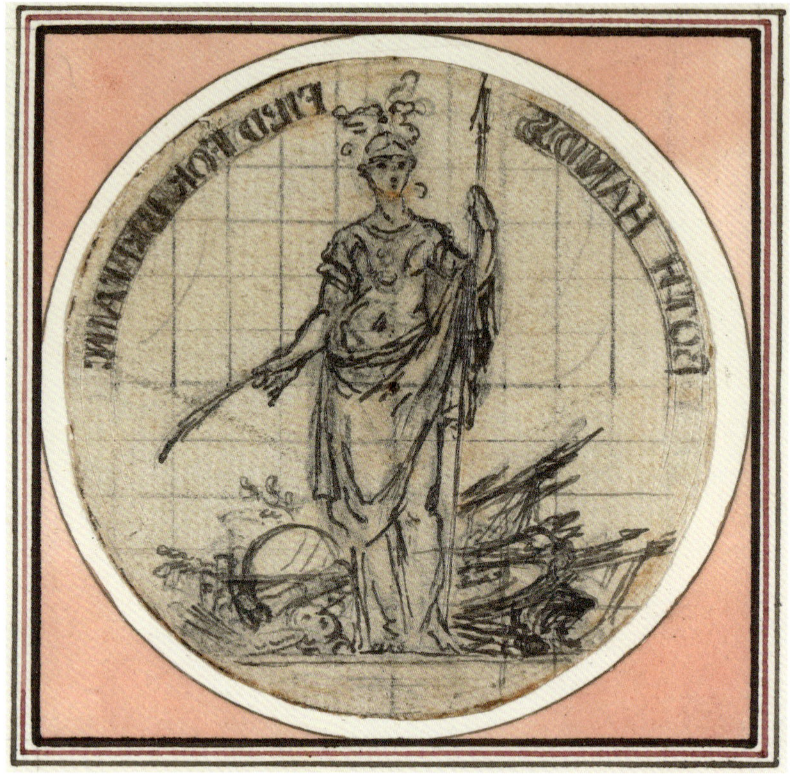

13.06

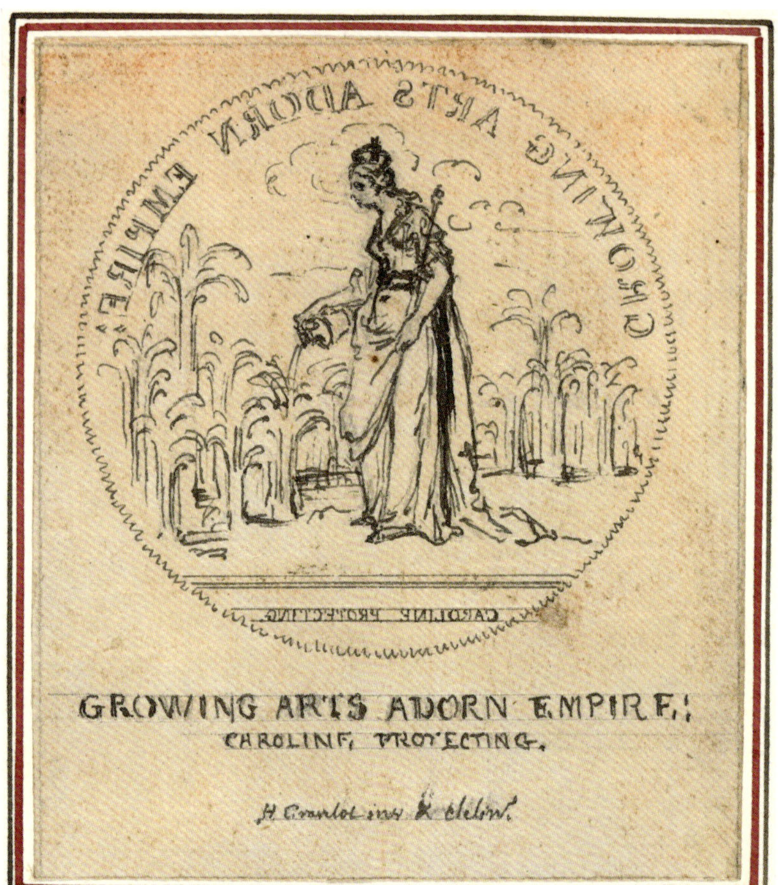

13.07

13.06*

HUBERT-FRANÇOIS GRAVELOT
(1699–1773)
Design for a Medal: Both Hands Fill'd for Britain
c. 1736
Pen and ink, pencil, and chalk with gold leaf, watercolour, and pen on paper
4⅛ x 4⅛ inches (10.5 x 10.5 cm)
Yale Center for British Art, Paul Mellon Collection, B1975.4.442

13.07*

HUBERT-FRANÇOIS GRAVELOT
(1699–1773)
Design for a Medal: Growing Arts Adorn Empire, Caroline Protecting
c. 1736
Pen and ink over pencil, with gold leaf, watercolour, and pen on paper
4⅞ x 4⅜ in. (12.4 x 11.1 cm)
Yale Center for British Art, Paul Mellon Collection, B1975.4.444

French engraver, book illustrator, designer and drawing master Hubert-François Gravelot emigrated to London in 1732, where he quickly became a central figure in the introduction of the Rococo style into Britain, through book illustration and designs for metalwork, furniture, and other decorative wares. In about 1736 Gravelot was asked by Henry Jernegan to devise a design for a lottery medal (13.08). The design shows George II on the obverse, Caroline on the reverse. The captions 'Both hands fill'd for Britain: George reigning' and 'Growing arts adorn empire: Caroline protecting', and the manner in which the subjects are depicted, reflect contemporary attitudes to their respective roles. George II is shown as defender and champion; Caroline is shown as nurturer, tending little palm trees, an allusion to her promotion of trading initiatives internationally, as well as her love of the natural sciences.

13.08*

JOHN SIGISMUND TANNER
(c. 1704–1775)
Jernegan's Cistern Lottery Medal, Great Britain, George II and Caroline
1736
Silver
1½ in. (3.9 cm) diameter
Yale University Art Gallery, Transfer from the Yale University Library, Numismatic Collection, 2001, 2001.87.26154

Henry Jernegan (Jerningham) of Russell Street, a London goldsmith-banker, commissioned the Jernegan Cistern in 1734–35 for a client, Littleton Poyntz Meynell, who wanted to own the largest silver wine cistern ever made. The sculptor Michael Rysbrack modelled the bacchanalian scenes on the bowl, the crouching panthers beneath, and the satyr handles. The project took four years to complete and the vessel eventually weighed around 227 kilos (8,000 ozs). The leading silversmith whose mark is struck on the cistern was the German immigrant Charles Kändler, probably Carl Rudolf Kaendler, elder brother of the noted Meissen porcelain modeller. When asked by Henry Jernegan to pay the final bill for the cistern, Meynell refused, and in 1737 Jernegan offered the cistern as a lottery prize in order to cover his costs. Each person who purchased a ticket was given Jernegan's lottery medal with images celebrating George II and Caroline. Designed by Hubert-François Gravelot, the medals were struck by John Sigismund Tanner, chief engraver at the Royal Mint. More than 30,000 tickets were sold. The winner of the lottery, Major William Battine of East Marden, Sussex, sold the cistern to the Grand Duchess Anna Leopoldovna of Russia in 1738. The cistern is now in the collection of the State Hermitage Museum in St Petersburg.

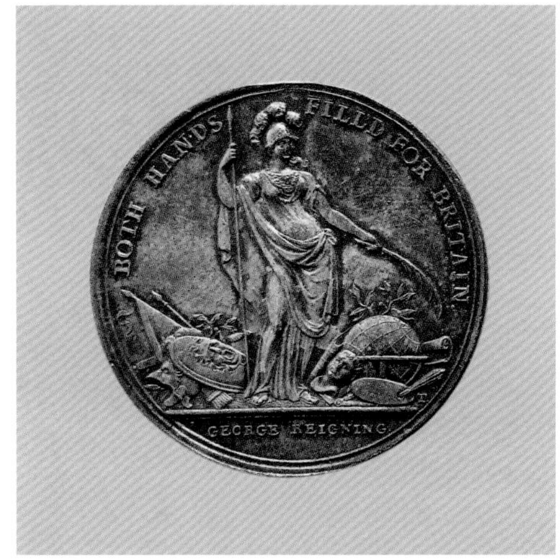

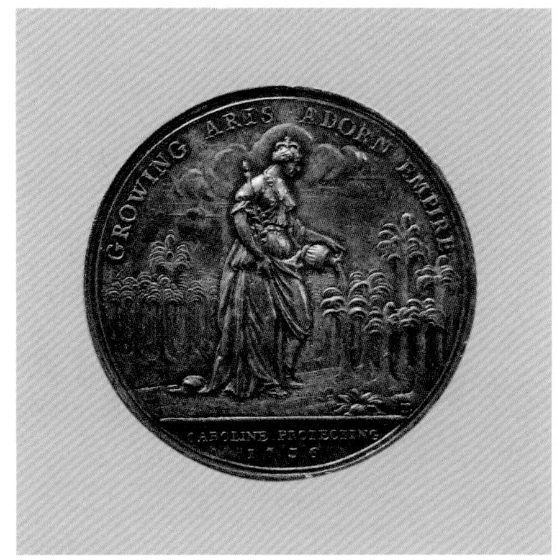

13.08

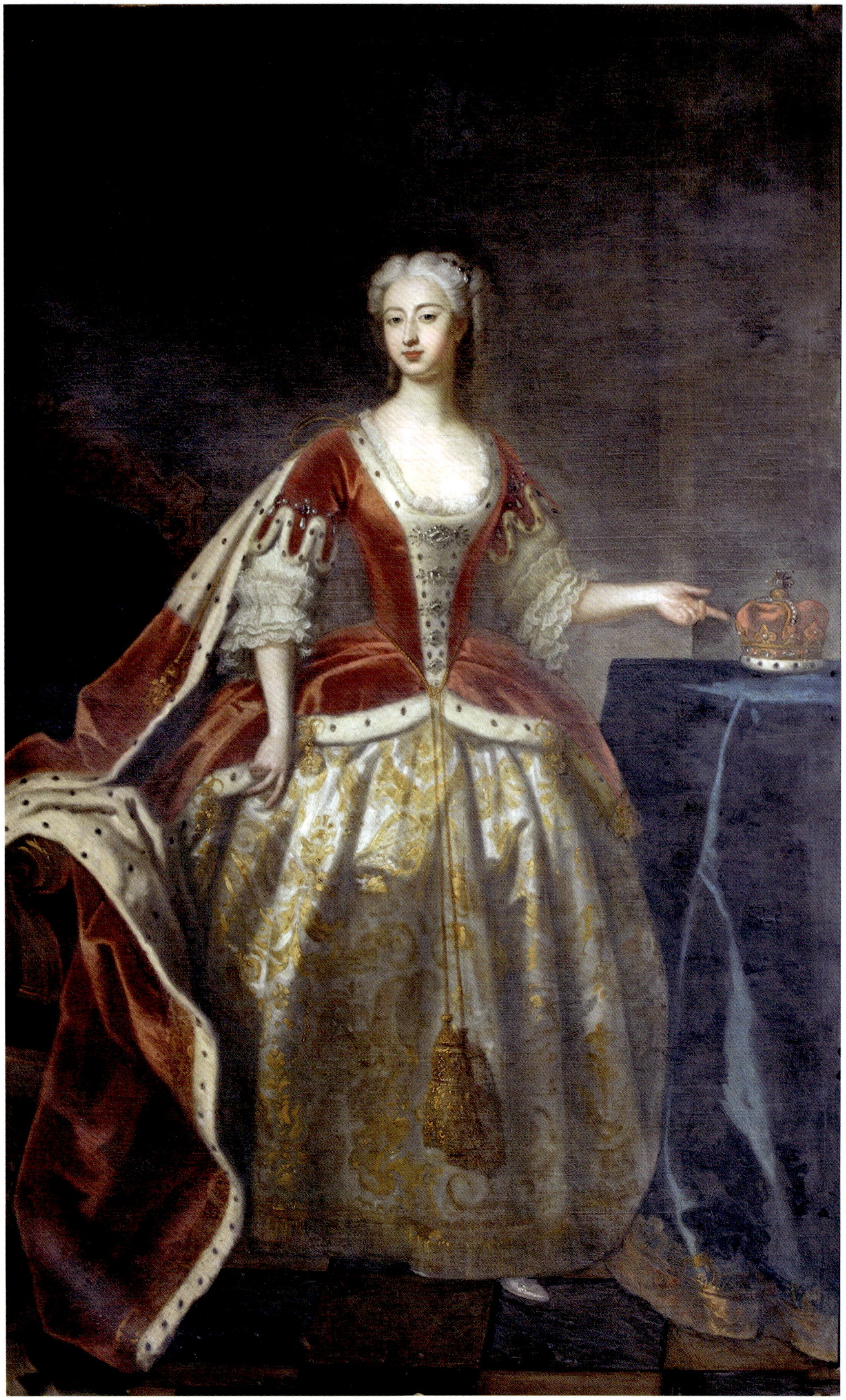

13.09

IMAGES OF AUGUSTA

13.09*

JEREMIAH DAVISON (c. 1695–1745)
Augusta of Saxe-Gotha, Princess of Wales
c. 1738
Oil on canvas
100 x 59⅞ in. (254 x 152 cm)
Victoria Art Gallery, Bath, BATVG: P: 1984.7

Jeremiah Davison was born in England of Scottish parents. He was first known as a copyist, and indeed George Vertue encountered 'Davison, an ingenious young painter' at Kensington Palace and Hampton Court Palace, where he had been given permission to work from paintings by Titian, and Van Dyck's portrait of *Charles I on Horseback*. He would later set up his own studio with the assistance of the well-known drapery painter Joseph van Aken. Frederick, Prince of Wales, sat to him in 1730, though this portrait has been lost. Augusta and Frederick visited the city of Bath in 1738, after which residents erected an obelisk in their honour in Queen Square. This portrait, with a companion piece of Frederick, Prince of Wales, was a subsequent gift from the couple to the Corporation of Bath.

13.10*

UNKNOWN PAINTER, after **JEAN-BAPTISTE VAN LOO** (1684–1745)
Princess Augusta of Saxe-Gotha Altenburg, Princess of Wales
1740–45
Oil on canvas
29½ x 24½ in. (74.9 x 62.2 cm)
National Trust, Hartwell House, Buckinghamshire, NT 1548064

Jean-Baptiste van Loo arrived in London in 1737. His early commissions there were for the playwright Colley Cibber, the First Minister Sir Robert Walpole, and Prince Frederick, Prince of Wales, and Princess Augusta. The prime version of this portrait, which was balanced by one of Prince Frederick, was painted as a present for William Pulteney, Earl of Bath (1684–1764). Each cost 60 guineas.

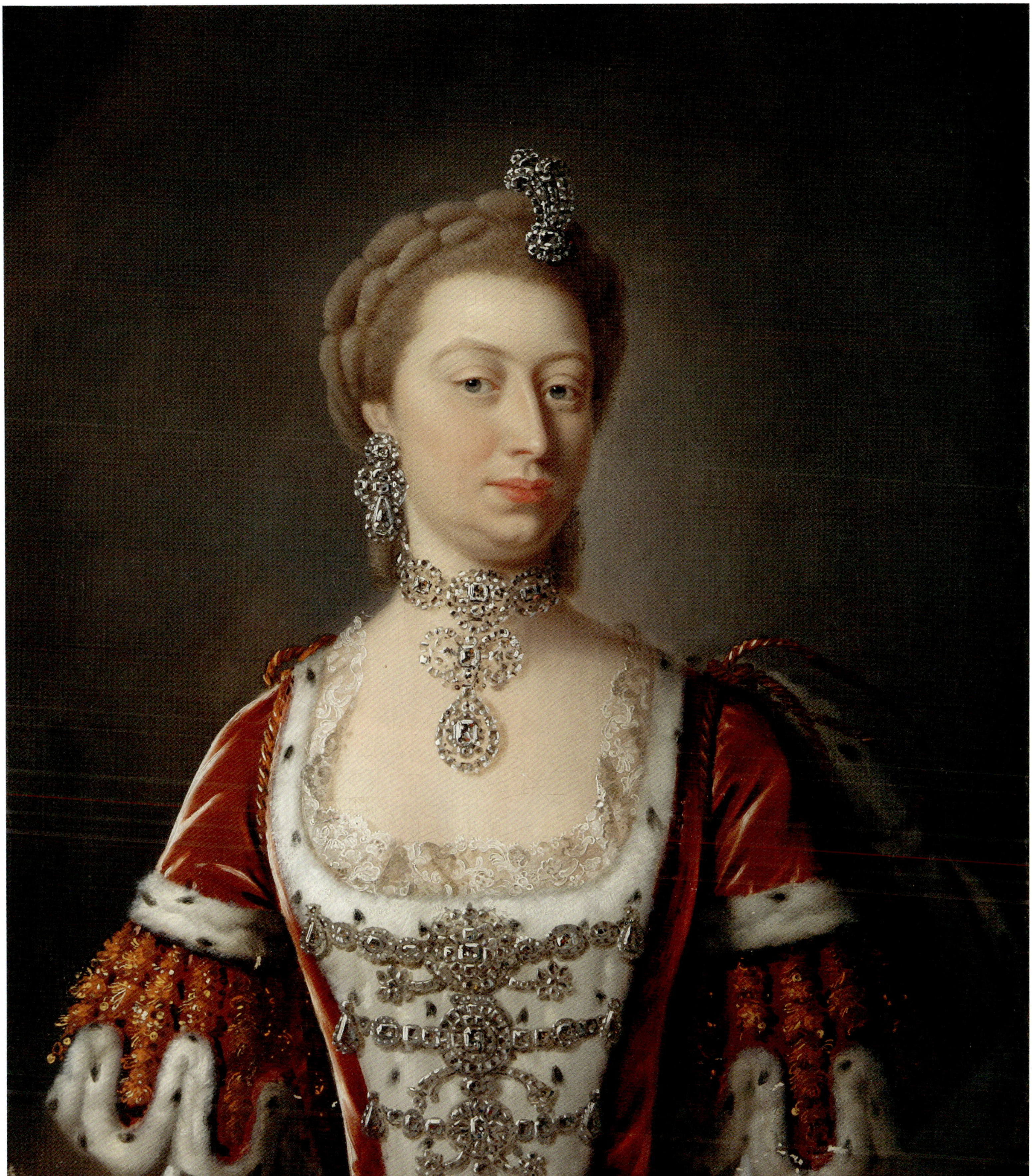

13.10

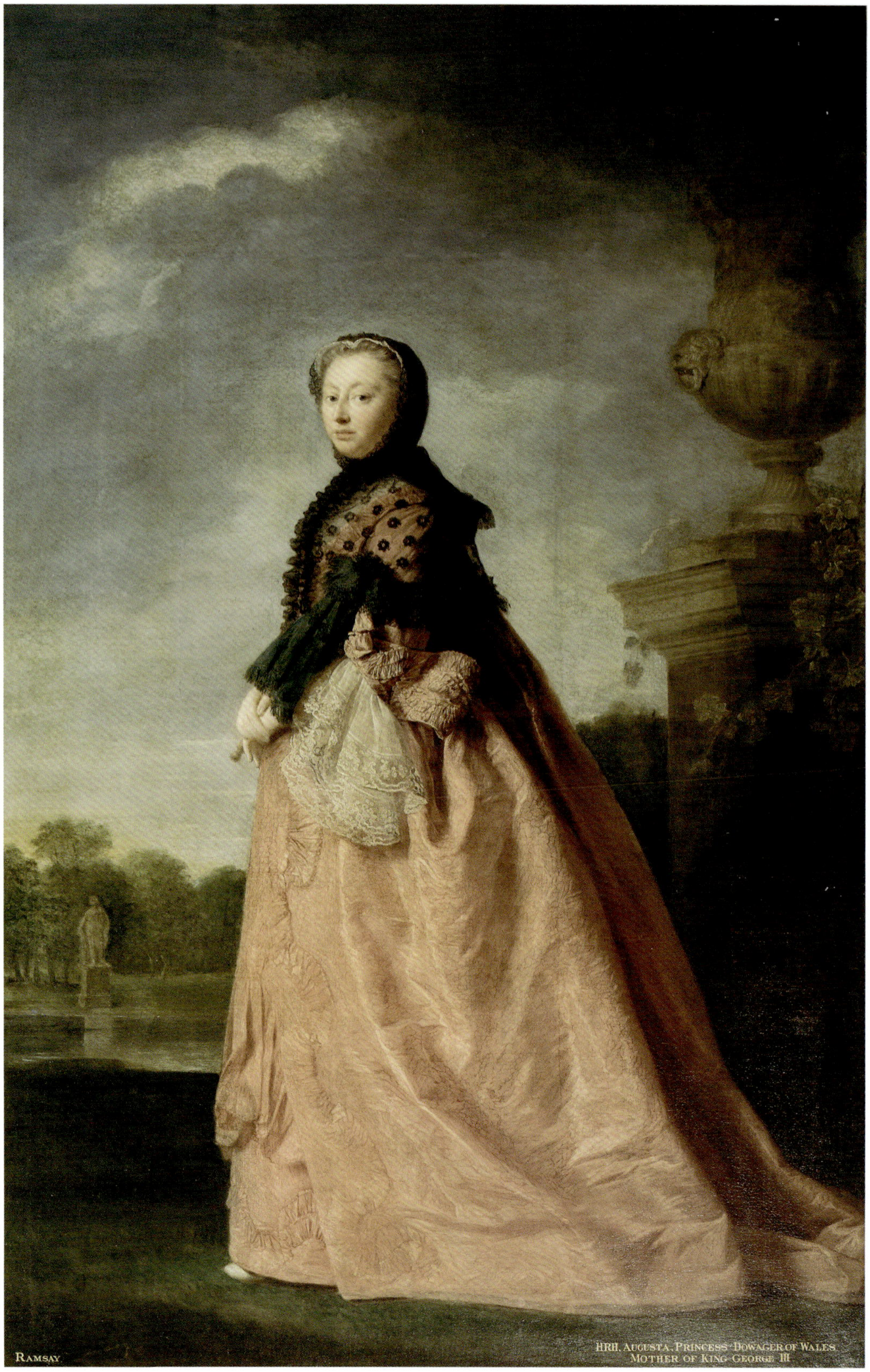

13.11

13.11*

ALLAN RAMSAY (1713–1784)
Augusta, Princess of Wales
1764
Oil on canvas
119 x 80 in. (302.3 x 203.2 cm)
Mount Stuart Trust, Mount Stuart

In or about 1764, Allan Ramsay began to paint the last in a series of grand manner portraits for John Stuart, third Earl of Bute. Ramsay had already made portraits of the Earl and his son, Lord Mountstuart, and his royal pupil, George, Prince of Wales. The set was completed with this portrait of Augusta, the young prince's mother, who had earlier contrived to appoint Bute as moral tutor to her son. Ramsay's familiarity with George III brought his appointment as one of 'His Majesty's Painters in Ordinary' on the accession. The choice of a Scotsman caused tensions within the artist community in London. Reynolds in particular continued to make strenuous efforts to attract royal favour, not least by painting one of Charlotte's bridesmaids in her wedding finery. For Lord Bute the commission of the portrait of Augusta was in recogition of their friendship and their many shared interests.

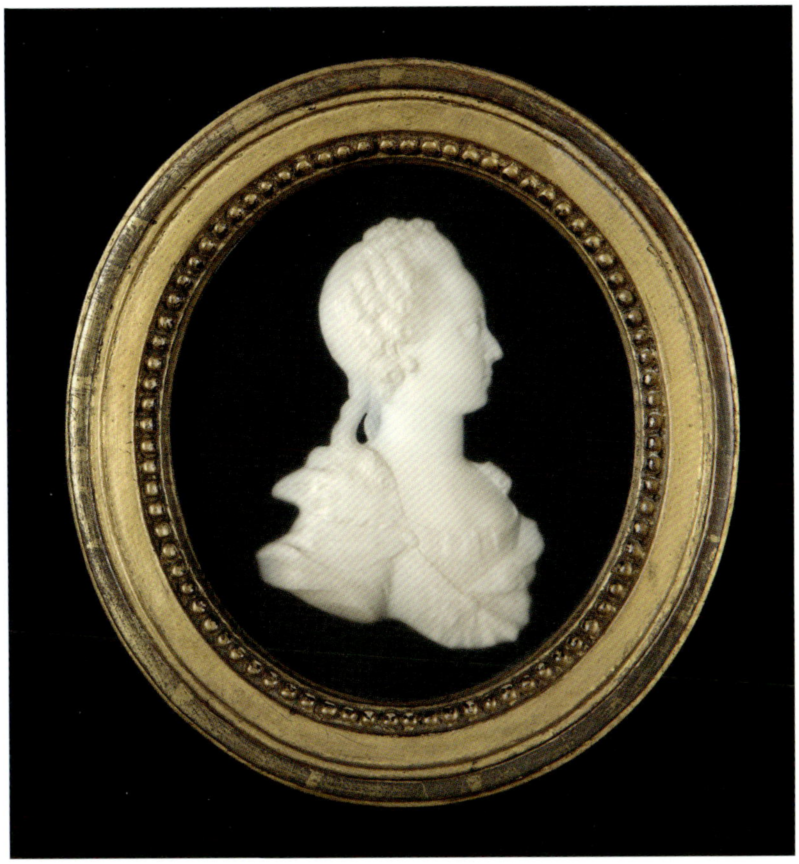

13.12

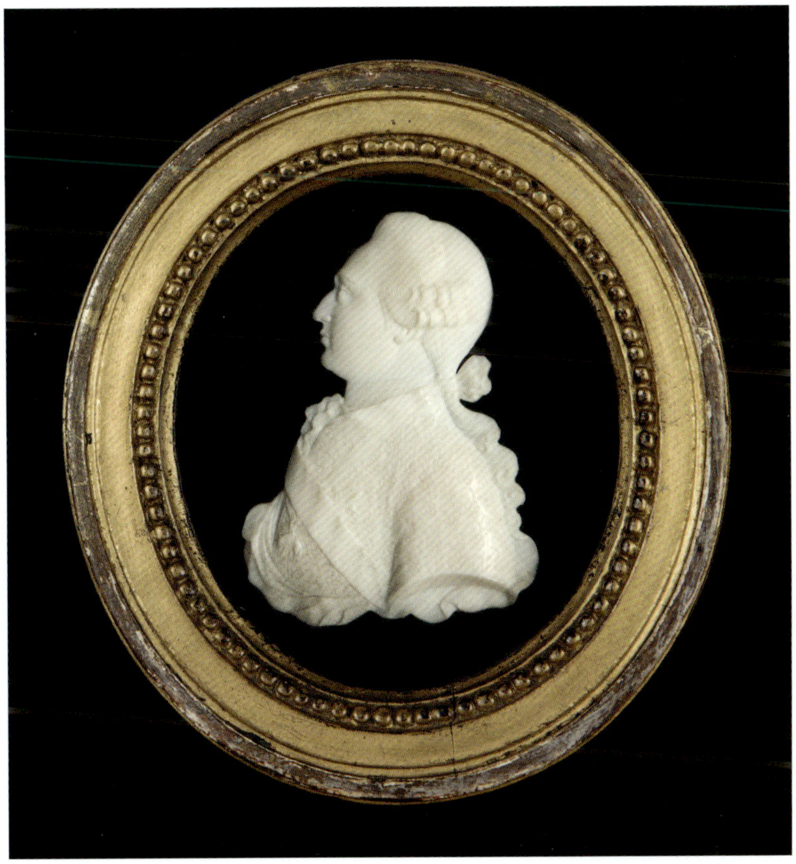

13.13

13.12, 13.13

ISAAC GOSSET (1713–1799)
Augusta, Princess of Wales and *Frederick, Prince of Wales*
18th century
Wax on glass
Each 7⅛ x 5⅞ x 1 ¼ in. (18 x 15 x 3 cm)
Lady Baker, née Muir

With his elder brother Jacob Gosset (1701–1788), Isaac began his career in the Gosset family workshop as a maker of picture frames. He would continue to accept commissions for frames in later life: the work was prestigious and profitable. Gosset frames were supplied to William Hogarth and Thomas Gainsborough, and the latter painted Isaac's portrait. His reputation rests, however, on the wax cameo portraits he produced chiefly for the adornment of gentlemen's cabinets. He was noted for his rapid execution of remarkable and exquisite likenesses of his sitters in profile. Isaac Gosset made wax portraits of George II, Prince Frederick and Augusta, and George III and Charlotte. His images of members of the royal family were used by Josiah Wedgwood as the basis by for portraits in ceramic (see also cat. 28.12). This pair of waxes was acquired by the Reverend John Muir, Minister from 1816 to 1865 at St Vigeans, Angus, Scotland.

13.14

FRANCIS HAYMAN (c. 1708–1776)
The Muses Paying Homage to Frederick Prince of Wales and Princess Augusta (The Artists Presenting a Plan for an Academy to Frederick Prince of Wales and Princess Augusta)
c. 1750–51
Oil on canvas
27¼ x 32¼ in. (69.2 x 81.8 cm)
Royal Albert Memorial Museum, Exeter, RAMM 392/1974

Francis Hayman, painter, engraver, and book illustrator, was probably born in Exeter. He would be apprenticed in 1718 to Robert Browne, a history painter in London. Much of his early work was providing decorations for the theatre. A commission from Jonathan Tyers, the entrepreneurial proprietor of the pleasure gardens at Vauxhall, to paint portraits of his family, led to a commission to paint large-scale decorations there for supper boxes. More than fifty of Hayman's works are recorded in contemporary guidebooks to Vauxhall Gardens. Frederick, Prince of Wales, had first visited the pleasure gardens in 1732, confirming their fashionable status. Over the following decade Tyers used Frederick and Augusta to court publicity and they were the inspiration for much of the decoration for the garden's buildings. Augusta, who frequently accompanied her husband as he travelled within London and elsewhere, is presented as equal partner, to be serenaded by the muses and honoured by artists.

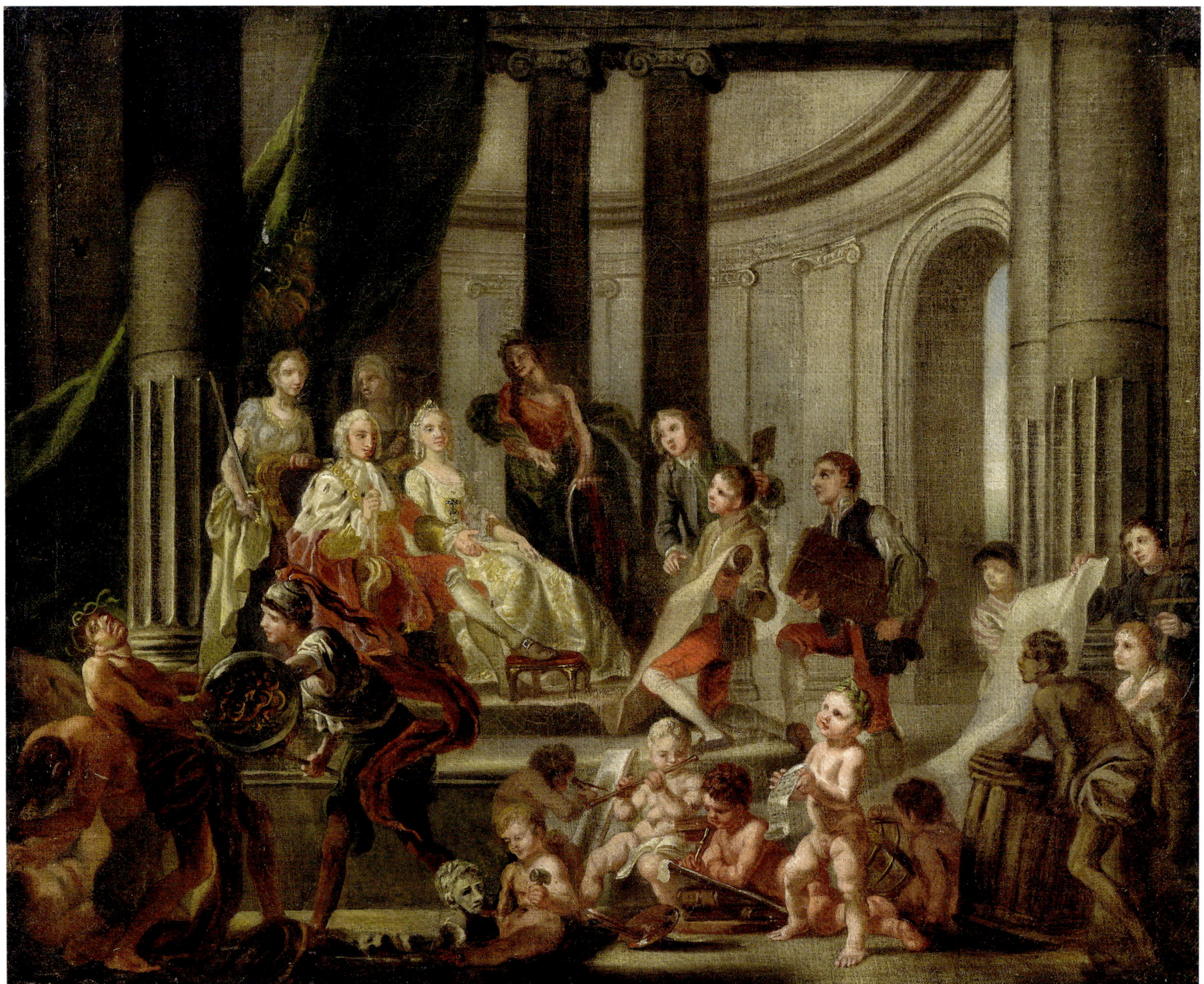

13.14

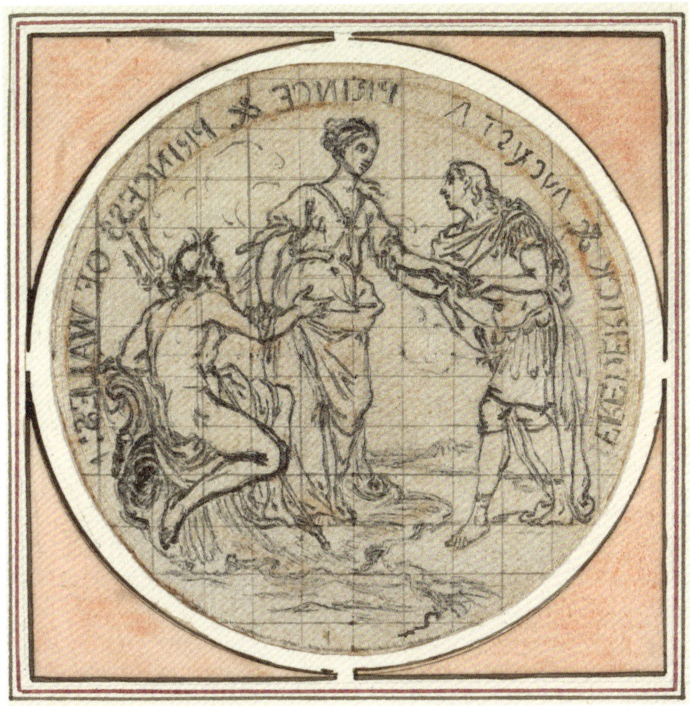

13.15

13.15*

HUBERT-FRANÇOIS GRAVELOT
(1699–1773)
Design for a Medal: Frederick and Augusta, Prince and Princess of Wales
Undated
Pen and ink over pencil with chalk, with gold leaf, watercolour, and pen on paper
4⅛ x 4⅛ inches (10.5 x 10.5 cm)
Yale Center for British Art, Paul Mellon Collection, B1975.4.440

It is probable that Hubert-François Gravelot worked up this design to commemorate the marriage of Frederick, Prince of Wales, and Augusta of Saxe-Gotha in 1736. It shows Augusta being greeted by the Prince, dressed in heroic Roman dress. Neptune, the classical god of the sea, looks on, and signifies Augusta's status as a German princess from across the seas. No medal made to this design has been identified, so Gravelot's design may have been produced speculatively.

IMAGES OF CHARLOTTE

13.16*

STUDIO OF ALLAN RAMSAY (1713–1784)
Queen Charlotte; Princess Sophia Charlotte of Mecklenburg-Strelitz, 1744–1818. Queen of George III
c. 1763
Oil on canvas
98¼ x 64¼ in. (249.5 x 163.3 cm)
Scottish National Gallery, Edinburgh, PG 217

On the accession of George III in October 1760, Allan Ramsay's first commission from the young king was to paint a full-length portrait to be sent to Hanover. Following the coronation, which took place the next September, the artist was commanded to revise the image to show the King in coronation robes, as well as painting a new pair of pendant portraits showing him and Charlotte in their coronation finery. This last pair of portraits proved to be the model for many copies which were sent to family members in the German states, and ambassadors and governors of British provinces thoughout the world. Ramsay finished the prime version of the portrait of Charlotte in March 1762. According to Allen Cunningham, the Queen's crown and regalia included in the image were sent to Ramsay's house in Soho Square in order that they should be rendered accurately. However, Horace Walpole remarked on seeing the image for the first time: 'It is much flattered, and the hair is vastly too light.'

13.17*

HENRY ROBERT MORLAND
(c. 1717–1797)
Charlotte Sophia, Queen Consort of George III
1780–90
Oil on canvas
48⅞ x 39 in. (124 x 99 cm)
The Queen's College, University of Oxford, Accession No. 70

Commissioned as a gift to Queen's College, Oxford, from John Thomas, Chaplain-in-Ordinary to the Queen, the portrait marks Charlotte's donation of £1,000 for the rebuilding of parts of the college damaged by fire in 1778. She is shown holding architectural plans of the library. The Queen's College, a constituent college of the University of Oxford, was founded in 1341 by Robert de Eglesfield in honour of Queen Philippa of Hainault, wife of King Edward III of England. In later generations other female members of the royal family took an interest, and on occasion provided funds in its support. Caroline gave £2,000 to a building fund and this generosity was marked by the inclusion of her statue by Sir Henry Cheere in the 'open temple' by Nicholas Hawksmoor over the main gate.

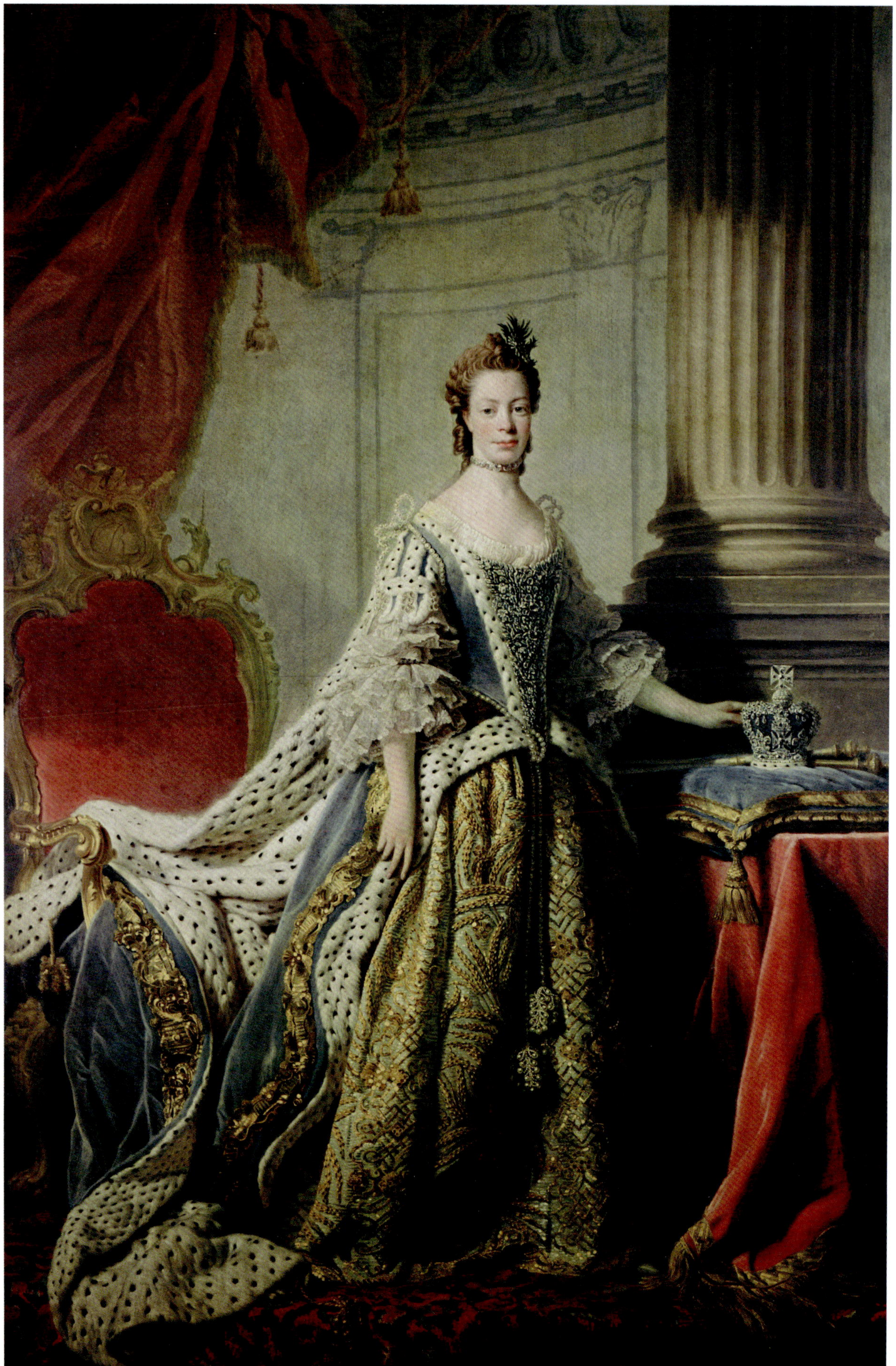

13.16

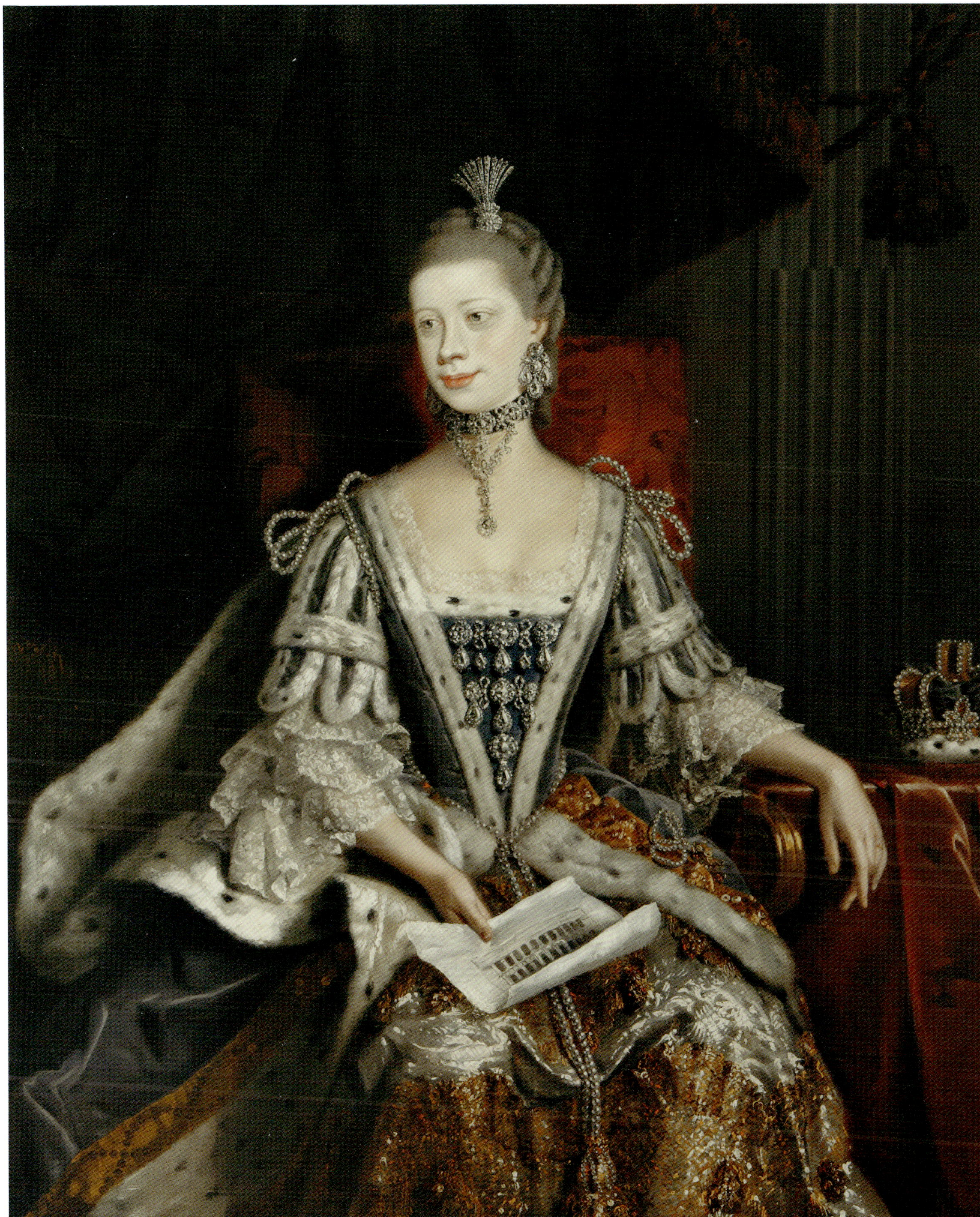

13.17

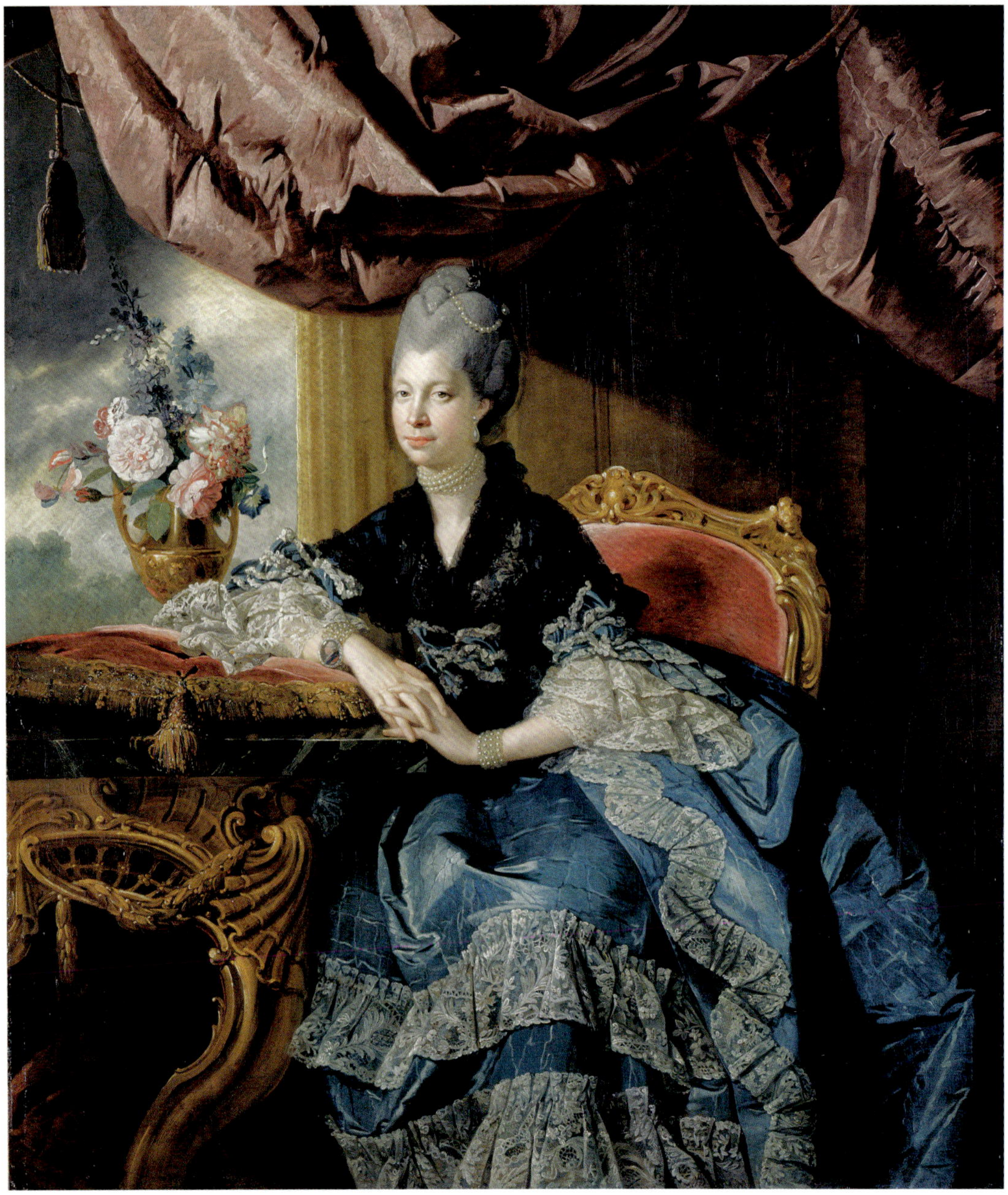

13.18

13.18

JOHAN ZOFFANY (1733–1810)
Queen Charlotte
1771
Oil on canvas
64⅛ x 54 in. (162.9 x 137.2 cm)
Royal Collection Trust, RCIN 405071

The German artist Johan Zoffany arrived in London in 1760 and soon established a reputation for informal conversation pieces in which accurate and lively portraits were set in surroundings showing the sitters' taste and circumstances. John Stuart, third Earl of Bute, commissioned portraits of his children from Zoffany around 1763–64, and probably introduced the artist to the George III and Charlotte. The King nominated Zoffany to the new Royal Academy in 1769. He continued to receive royal commissions for a further ten years, but after the unfavourable reception of *The Tribuna* he was supplanted by other, younger artists. Zoffany was probably commissioned by George III or Charlotte to produce this informal portrait. It shows the Queen wearing fashionable dress and pearl jewellery that had been a wedding present from her husband. The bracelet on her right wrist bears a miniature portrait of King George by Jeremiah Meyer. The vase of flowers may allude to Charlotte's interest in botany.

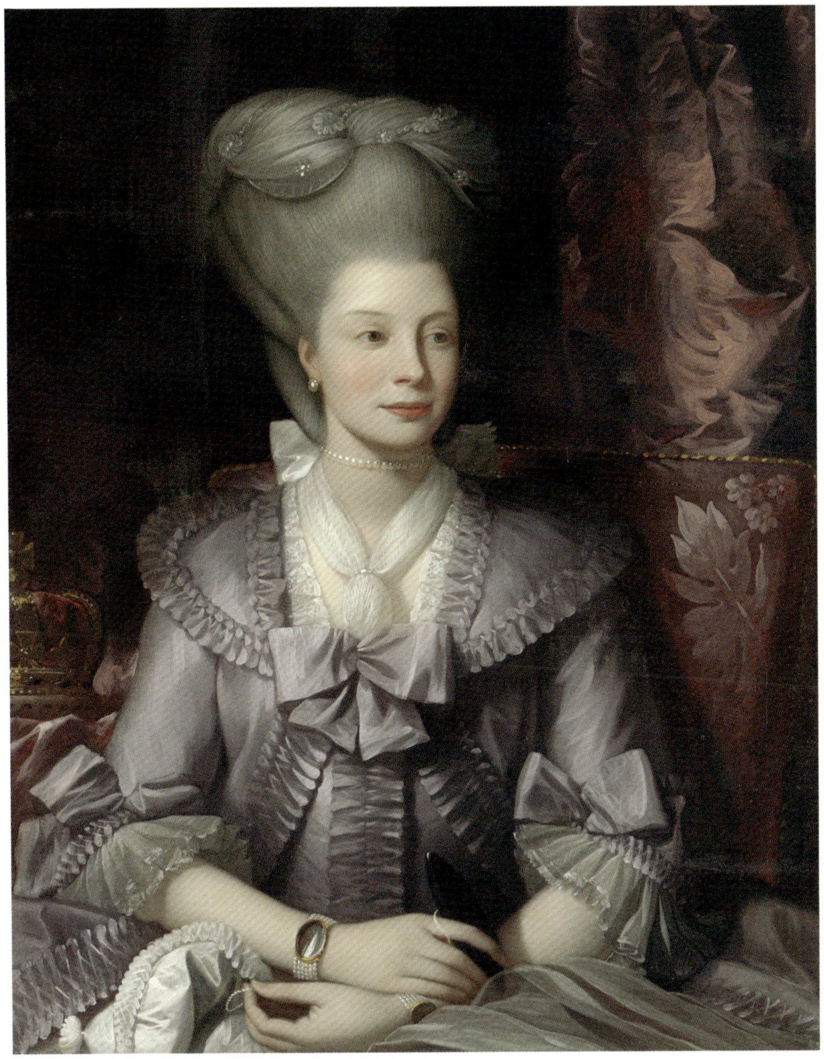

13.19

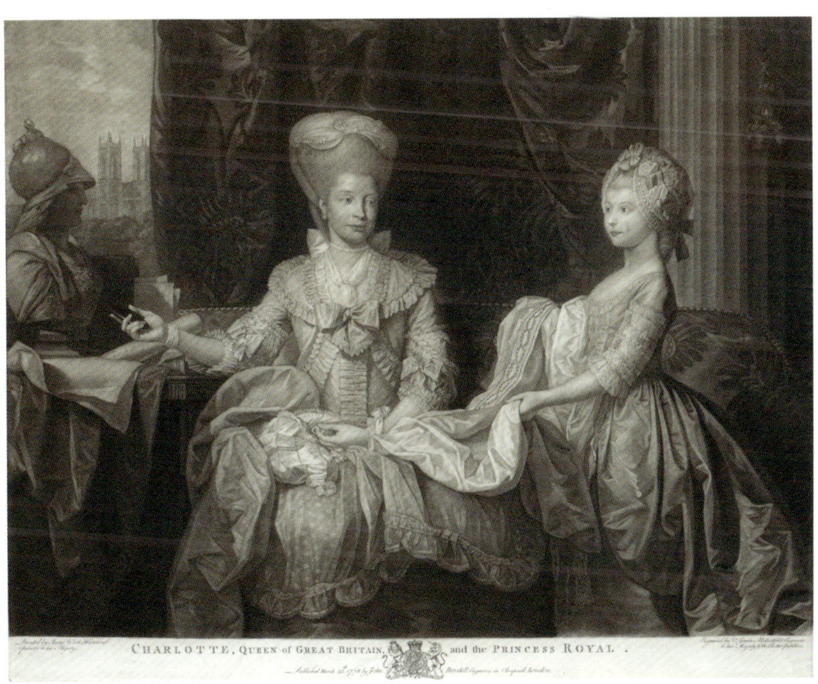

13.20

13.19*

BENJAMIN WEST (1738–1820)
Queen Charlotte
1777
Oil on canvas
36⅛ x 28 in. (91.8 x 71.1 cm)
Yale Center for British Art, Paul Mellon Collection, B1977.14.115

American-born Benjamin West received his first commission from George III in 1768, and he subsequently produced many works for the King and Queen. Between 1776 and 1778 George commissioned five double or group portraits of his family to hang together in the King's Closet at St James's Palace. This painting shows Queen Charlotte in the same dress and holding the fabric and tatting shuttle that appear in the double portrait West painted of her and her eldest daughter Charlotte, Princess Royal, which was exhibited at the Royal Academy in 1776 (Royal Collection Trust, RCIN 404573). A preparatory study also survives showing the Queen alone (Royal Collection Trust, RCIN 452460). This portrait may have been worked up from the study.

13.20*

VALENTINE GREEN (1739–1813), engraver, after BENJAMIN WEST (1738–1820)
Queen Charlotte, Queen of Great Britain and the Princess Royal
1778
Mezzotint
18⅝ x 22¾ in. (47.4 x 57.8 cm)
Published by John Boydell (1719–1804),
UK Government Art Collection, GAC 10048

This mezzotint by Green reproduces a portrait commissioned from West by George III and exhibited at the Royal Academy in 1777. The painting was part of a set of five double or group portraits of the royal family, intended to hang together in the King's Closet at St James's Palace. It is an affectionate image showing Charlotte and her eldest daughter working together industriously on a piece of embroidery. It casts Charlotte as a nurturer and educator, as well as alluding to her concern to support British textile industries. A bust of Minerva, the goddess of wisdom, presides over the scene.

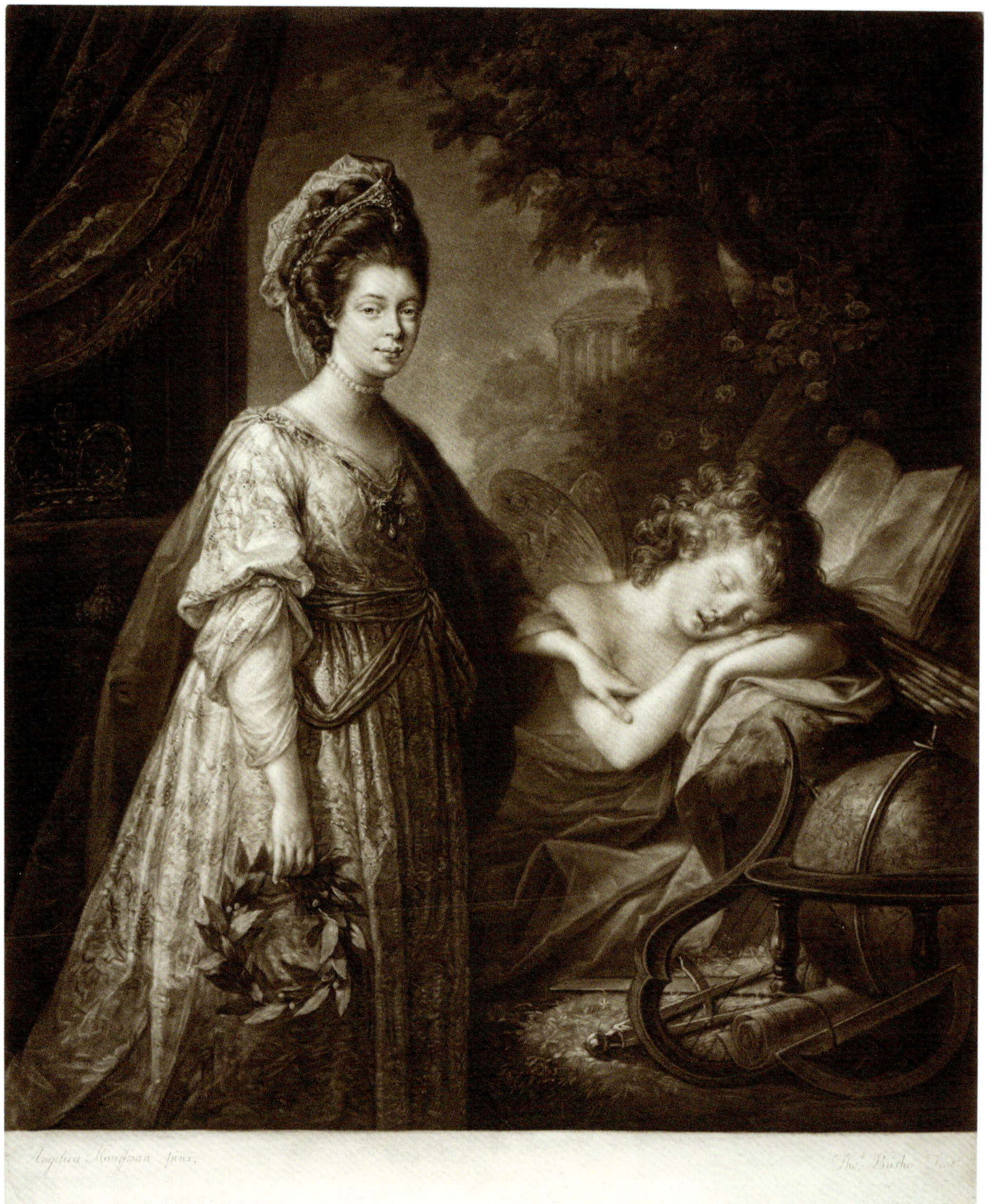

13.21

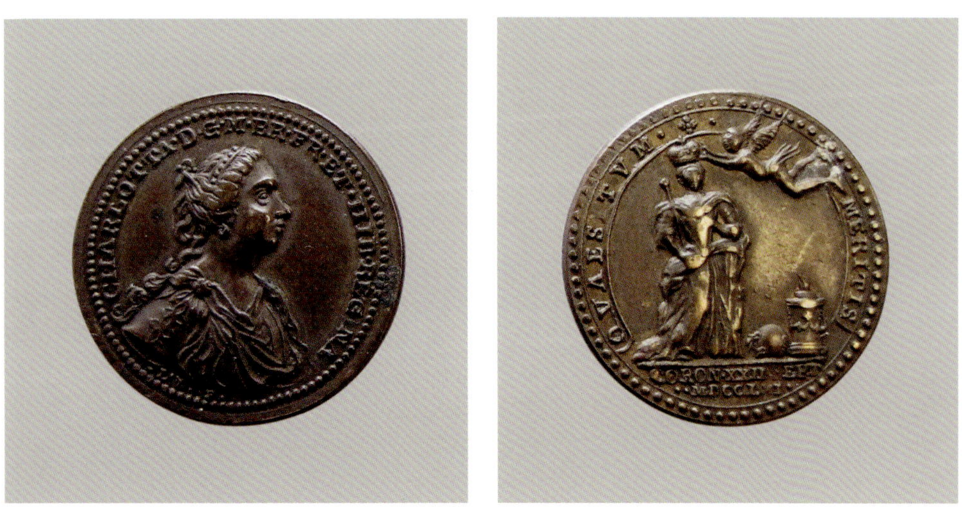

13.22

13.21

THOMAS BURKE (1749–1815), engraver, after
ANGELICA KAUFFMANN (1741–1807)
Queen Charlotte Raising the Genius of the Fine Arts
1772
Mezzotint on paper
18⅝ x 14⅞ in. (47.2 x 37.8 cm)
Published by William Wynne Ryland, London
Royal Collection Trust, RCIN 604619

Angelica Kauffmann was one of only two women who were founder members of the Royal Academy. The other was Mary Moser. Kauffmann subsequently chose subjects for her work which saw women as heroes, drawing attention to her unique status as a female history painter, and perhaps to attract female patrons. She received support from Queen Charlotte, who was commemorated in this image engraved by Thomas Burke. Kauffmann was later able to claim the Empress Catherine of Russia, Maria Theresa of Austria and her daughter Queen Carolina of Naples among her patrons.

13.22*

JOHANN LORENZ NATTER (1705–1763)
Coronation of Queen Charlotte
1761
Bronze
1⅜ in. (3.4 cm) diameter
Yale University Art Gallery, Transfer from the Yale University Library, Numismatic Collection, 2001, Gift of Irving Dillaye-Vann, B.A. 1897, 2001.87.9

In 1741 or earlier, Johann Lorenz Natter, from Biberach in Swabia, came to England to work as a medallist and gem engraver, bringing with him a collection of antique gems and sulphur casts from Italy. He became a Fellow of the Society of Antiquaries of London in 1755, and of the Royal Society in 1757 after publishing his 'Treatise on the Ancient Method of Engraving on Precious Stones compared with the Modern' in 1754. According to Rogers Ruding, cleric and numismatist, he was serving as engraver or assistant engraver at the Royal Mint at the beginning of the reign of George III. This medal commemorating the coronation of Queen Charlotte shows her portrait in profile on the obverse. On the reverse she is depicted standing full length, being crowned by a winged figure.

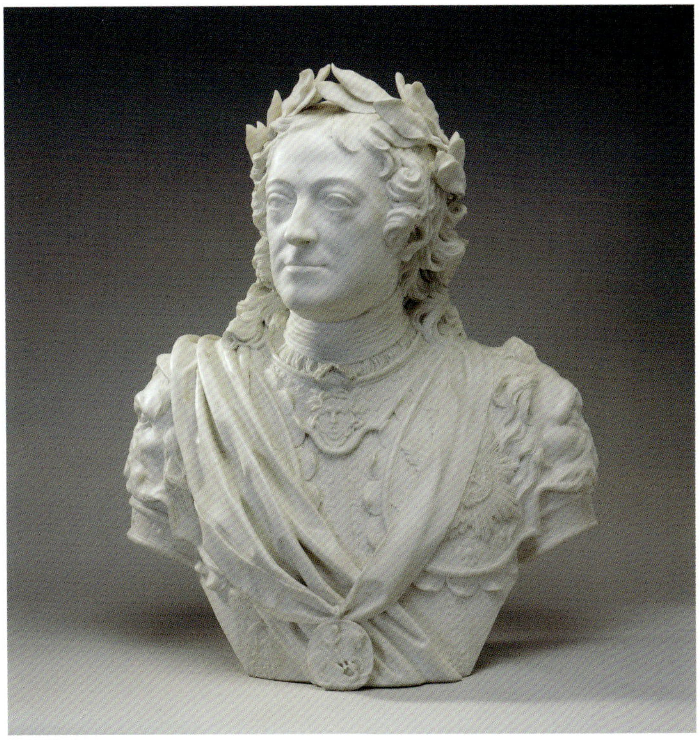

13.23

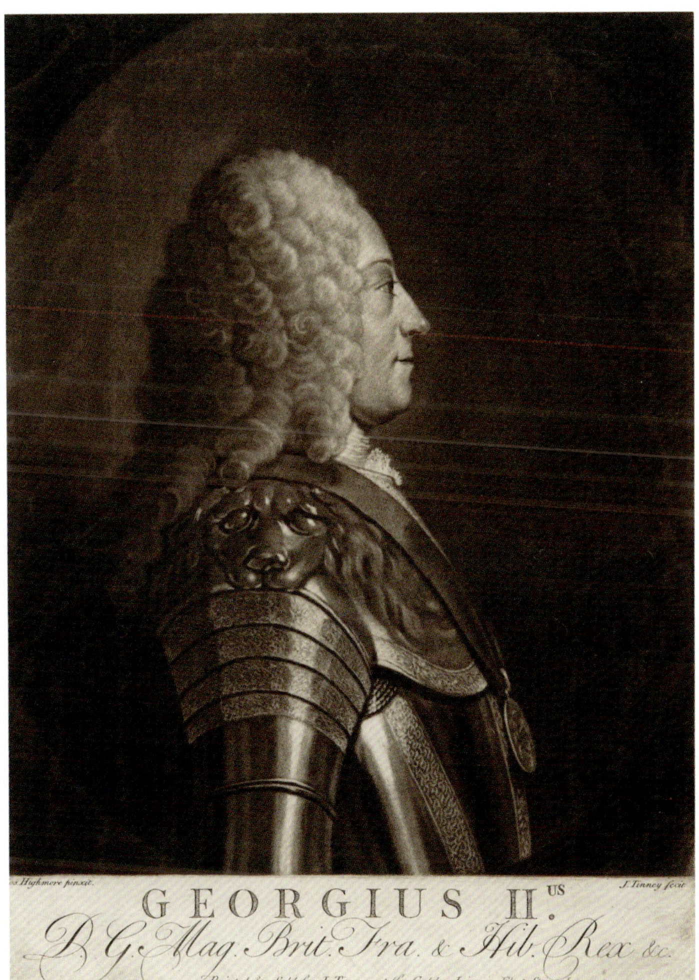

13.24

PRINCES, KINGS, AND HUSBANDS

13.23

MICHAEL RYSBRACK (1684–1770)
George II
1739
Marble
27½ x 21⅞ x 12¾ in. (69.8 x 55.5 x 32.5 cm)
Royal Collection Trust, RCIN 31322

The bust of George II and its pair of Caroline, were installed on brackets to preside over the chimney-pieces at each end of Caroline's library at St James's Palace. Caroline seems to have placed the commission with Rysbrack shortly before her death in 1737. The busts were first produced in terracotta, and then copied in marble. George Vertue, in his diaries, provides valuable information about the process. In 1738 he noted that 'the KING . . . sat to [Rysbrack] at Kensington twice. to have his picture modelled in Clay. the likeness much approvd on—and with a good Air.—also a Moddel of the Queen vastly like. Tho' not done from the life.' The resulting terracotta models are signed and dated 1738 (Royal Collection Trust RCIN 1411–1412). In 1739 Vertue recorded that 'two Marble Bustos the one of his present Majesty from a Model done from the life by Mr Rysbrack—and another busto of the late Majesty Q. Caroline, both were erected in the New Library at St. James, Green Park'.

13.24*

JOHN TINNEY (d. 1761), engraver, after
JOSEPH HIGHMORE (1692–1780)
Georgius II
1730–39
Engraving on paper
13⅞ x 9¾ in. (35.1 x 24.9 cm)
Historic Royal Palaces, HRP3000928

In about 1735 Joseph Highmore completed a portrait of George II in strict profile, to balance his portrait of Caroline. The work was lost in a fire in 1824, but had previously been engraved by John Tinney. The original is also recorded by Charles Wild (1781–1835) in his watercolour of the interior of the West Ante Room at Carlton House made in about 1818 (Royal Collection Trust, RCIN 922175). This was published as one of the illustrations to William Henry Pyne's three-volume *History of the Royal Residences of Windsor Castle, St James's Palace, Carlton House, Kensington Palace, Hampton Court, Buckingham House and Frogmore*.

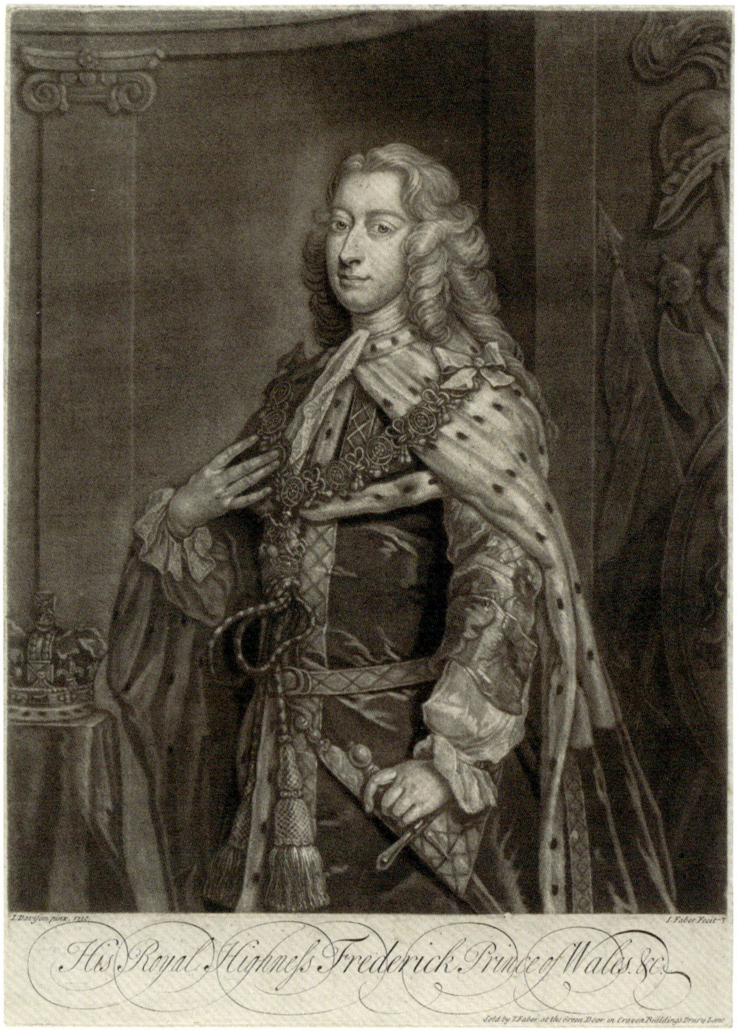

13.25

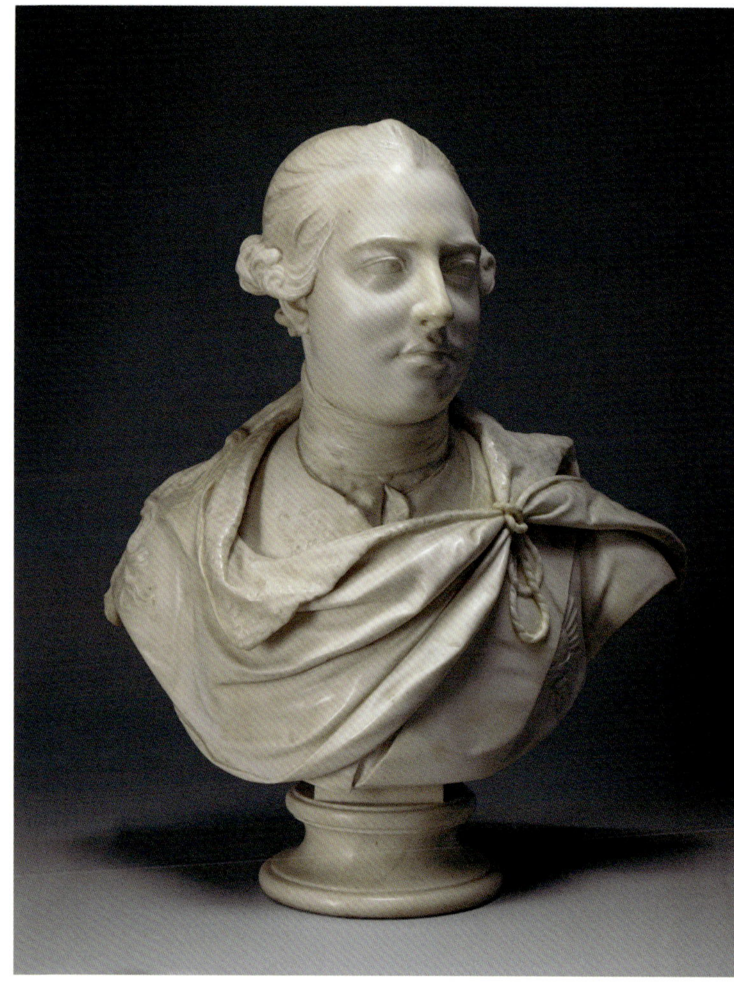

13.26

13.25*

JOHN FABER THE YOUNGER (c. 1695–1756), engraver, after **JEREMIAH DAVISON** (c. 1695–1745)
His Royal Highness Frederick, Prince of Wales
After 1730
Mezzotint on paper
14 1/16 x 10 in. (35.7 x 25.4 cm)
Published by John Faber, London
Yale Center for British Art, Gift of John Hay Whitney, Yale BA 1926, Yale MA (HON.) 1956, Transfer from the Yale University Library and the Yale University Art Gallery, B1994.4.174

Frederick, Prince of Wales, sat for Jeremiah Davison in 1730. This portrait is now lost, but it is known through the engraving of the work made by John Faber. Davison would be commissioned by the Prince to make a second full-length portrait, together with a balancing work showing Augusta. These were presented to the City of Bath following a visit made by the royal couple in 1738 (see cat. 13.09).

13.26*

JOHN NOST III (active 1749–d. 1780)
George III
1764
Marble
25 x 19¼ x 10 in. (63.5 x 48.9 x 25.4 cm)
Yale Center for British Art, Paul Mellon Collection, B1977.14.22

John Nost the younger was apprenticed to Henry Scheemakers in 1726. From 1750 onwards Nost served as an instructor in the modelling classes at the Dublin Society Schools and took on apprentices, including John Crawley and Patrick Cunningham. Among the first works Nost produced in Dublin was a statue of George II, placed in an arched niche over the door of the Weavers' Hall in the Coombe. Nost made a series of busts of George III. This example and another in the British Museum were completed in 1764. In 1765 the Earl of Northumberland commissioned him to produce an idealized statue of George III as a Roman general, now in the City Hall in Dublin.

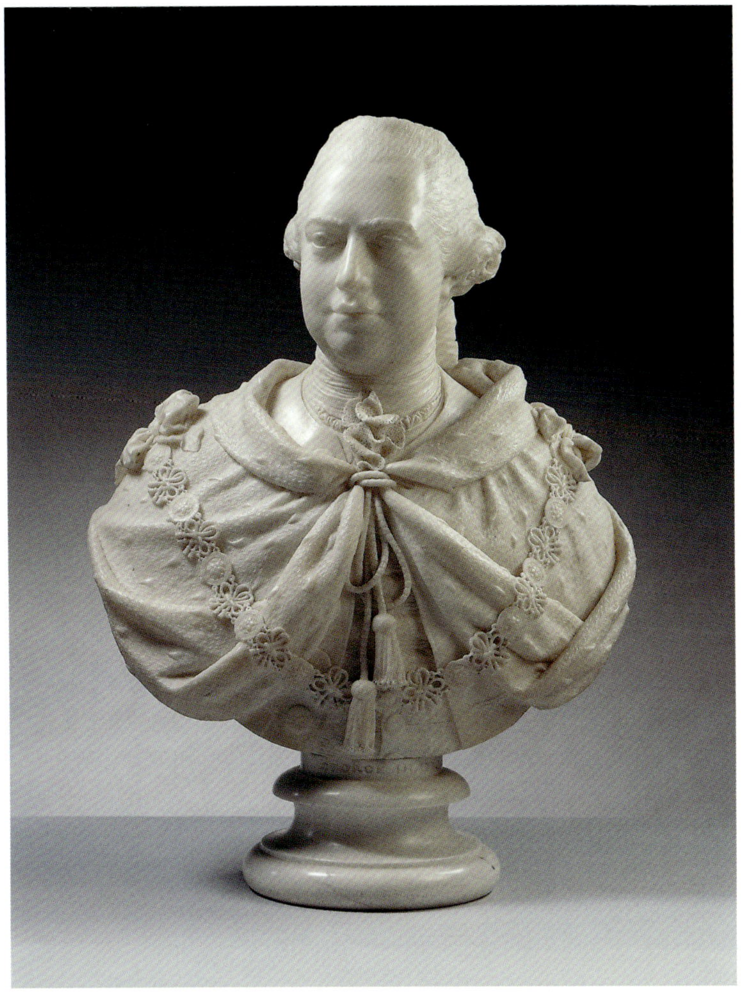

13.27

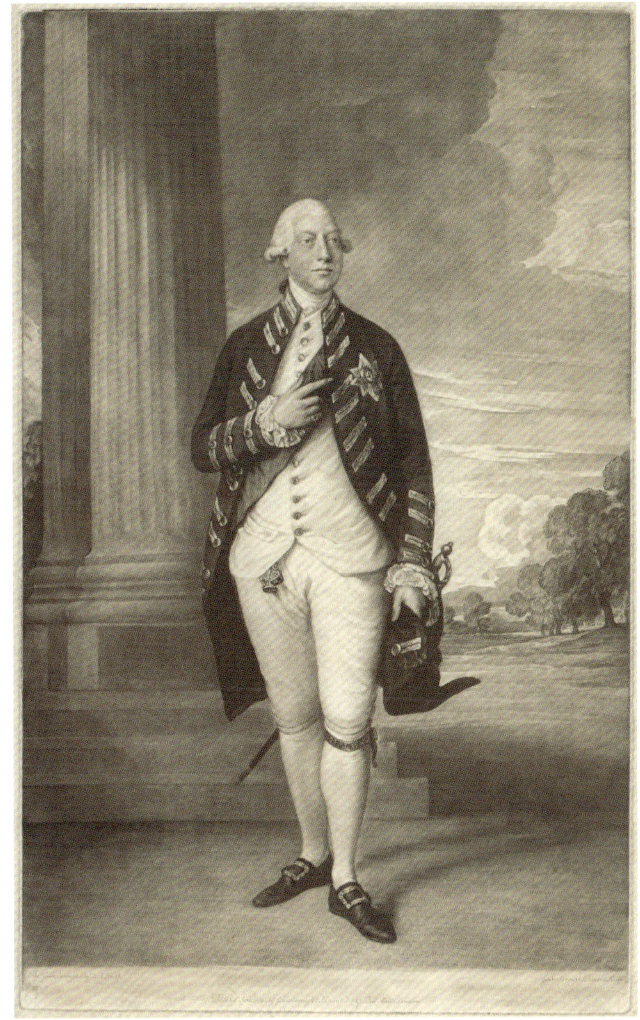

13.28

13.27†

JOHN BACON (1740–1799)
George III
1775
Marble
31⅛ x 21¼ x 12¼ in. (79 x 54 x 31 cm)
Royal Collection Trust, RCIN 31610

The prime version of the portrait bust of George III was commissioned by William Markham, Dean of Christ Church, Oxford, in 1770, showing the King wearing robes of state, and the insignia of the Order of the Garter. Markham was appointed governor to George, Prince of Wales, the following year. The bust came to serve as a favoured state image, and would be widely reproduced. This work was commissioned by the King as a gift for his eldest son in about 1775. Charlotte owned her own version. John Bacon had entered the Royal Academy Schools in the foundation year of 1768, and was awarded the Academy's first gold medal for sculpture in 1769.

13.28*

GAINSBOROUGH DUPONT (1754–1797), engraver, after THOMAS GAINSBOROUGH (1727–1788)
George III
1790
Mezzotint on paper
24½ x 14¾ in. (62.2 x 37.5 cm)
Yale Center for British Art, Paul Mellon Fund, B1970.3.368

Gainsborough Dupont was the nephew and pupil of Thomas Gainsborough and took over Gainsborough's studio in 1788. He painted portraits and landscapes in a style similar to that of his uncle, and from 1779 onwards he made a series of mezzotints after Gainsborough's portraits, including this one of George III. The original, together with a pendant of Charlotte, had been commissioned by the King, and when completed had been exhibited at the Royal Academy in 1781, before being displayed at Buckingham House.

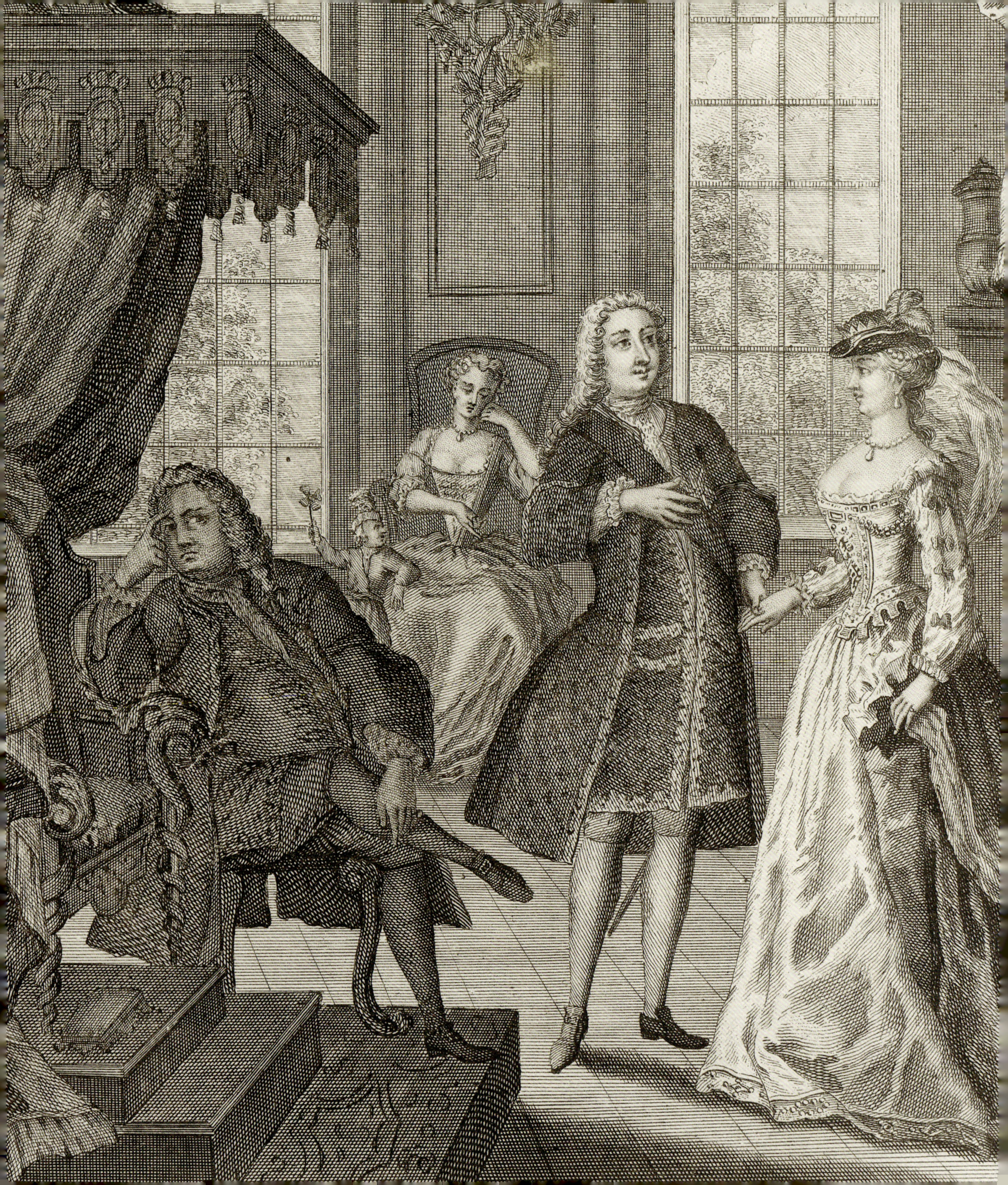

Disrespect: Undermining the Royal Image

Cynthia E. Roman

Visual satirists were no respecters of persons, whether politicians, nobility, kings, or queens. And the protocols of monarchy afforded little immunity against the rude, derisive wit and fantastical imagination of graphic satirists in England.[1] In fact, the contemporary and posthumous reputation of the Hanoverian consorts is indelibly inflected by irreverent satires that were produced outside royal control and in contrast to celebratory portraiture. The Princesses appeared in graphic satires with increasing prominence in inverse proportion to their waning political influence. Satires of Caroline of Ansbach numbered only a handful; Augusta of Saxe-Gotha was featured in many more; but Charlotte of Mecklenburg-Strelitz would be a perennial target of some of the most vicious caricatures of the century.

Anonymous and amateur satires of Caroline and Augusta largely lampooned their political influence rather than their personal character, in keeping with the trend of political satire prints in the earlier eighteenth century for emblematic strategies. A pivotal shift in the 'progress' of satire is generally credited to the aristocratic amateur George Townshend, who began to deploy the genteel tradition of caricatured portraits—the physiognomic distortion of face and body to reveal character flaws—in the service of interested political activism.[2] While Townshend's innovative example did not inspire immediate followers, by the end of the century the emblematic satire gave way to increasingly venomous personalized caricatures, most notable the fabulously malicious representations by James Gillray of Charlotte as a deformed and witch-like crone, the personification of greed, exploiting her position as consort for all it was worth. The extremity of her characterization is, however, due less to a changed attitude towards royalty than to major developments in the language of visual satire, fostered by increasing professionalization of the genre. Further, the facility of the etching process at the hands of skilful artists allowed for quick drafting and timely publication of politically topical events.

One of the first visual satires to target Caroline, *The Festival of the Golden Rump* (c. 1737), is strikingly brazen about her role as queen (cat. 14.04).[3] The obscenely scatological representation of the English court derives from an anonymously written text published in a new oppositional journal, *Common Sense*, in which a dream narrative parodies the servile rituals of George II's court. The cloven-footed, farting satyr represents George II; the high priestess with clyster-pipe Caroline; and the chief magician with a huge stomach Robert Walpole, the Prime Minister. Caroline administers the clyster, because she was believed to manage the King's temper.[4] The representation of Caroline's person is not demeaning in itself, rather the narrative context and bawdy props suggest her undue power and influence over the King.

Similar strategies broadly apply to satires on Augusta. From the moment of her betrothal and marriage to Frederick, Prince of Wales, in 1736, Augusta attracted the attention of graphic satirists. *View here Three different States in real Life—The Pimp the Miss forsaken and the Wife* (see p. 252, and cat. 14.06) mocks the marriage of Frederick and Augusta in a scene filled with innuendos about domestic transgressions and legitimate royal

Detail of p. 257, cat. 14.06

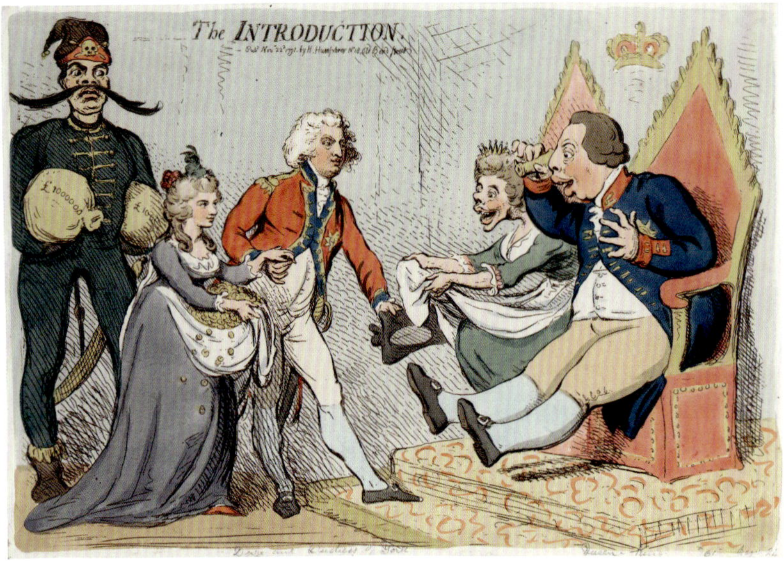

FIG. 14.01 James Gillray, *The Introduction*, 1791, hand-coloured etching on paper, 9⅞ x 13⅞ in. (25 x 35.2 cm). The Lewis Walpole Library, Yale University

issue. The representation of Augusta is in itself not demeaning at all, though the wider implications of the scene, which implicitly compares her to Henry VIII's unfortunate consort Anne Boleyn, is decidedly so.

Augusta's growing influence at court attracted a number of satires that clustered around particular events, personalities, and alliances. Her engagement of John Stuart, third Earl of Bute, to tutor her son, the future George III, lead to rumours that the two were lovers. In *Tempora Mutantur* (c. 1762), the Princess introduces Bute, wearing Scottish plaid, to her son the King who sits on his throne at the far right, his blindfold signifying that he is led unwittingly by his mother (cat. 15.02). Folly stands nearby holding a child's rattle inscribed 'Peace'. An illicit liaison is suggested but the figure of Augusta is not caricatured.

By the 1780s and 1790s graphic satire entered a new professional phase in which the anonymity of the artist was no longer the norm, and caricature became an almost universal mode. Charles James Fox, Charlotte, and her own husband, George III, were the principal victims of caricature that enabled them to be reduced to a small number of attributes that were repeated endlessly by different artists. Gillray's renderings of Charlotte as an uncouth, ugly, and unstylish Queen who shared the King's greed and miserliness, probably did more to define her public character than her official portraits; indeed, in some cases the latter may have been responses to the caricatured image rather than provoking it.[5] A particularly demeaning print, Gillray's *The Introduction* (fig. 14.01) parodies the royal couple's response to the betrothal of the Duke of York to Princess Frederica, the eldest daughter of the King of Prussia. The depiction of the Queen with bulging eyes, open mouth, and large lips, thirsting after a generous dowry, characterizes her as pure avarice. Her tiny crown and plain dress, as the wife of 'Farmer George', make her supremely unregal and ridiculous. The 1791 print, *From the Originals at Windsor* by Henry Kingsbury, likewise mocks George and Charlotte's greed in a parody of the famous painting of misers by Quinten Metsys that survives in many versions (see cat. 15.08). A bribery scandal in 1786 around the Warren Hastings trials led to accusations that the King and Queen had disreputably received jewels, undermining the moral authority of the crown. The royal couple's greed for diamonds became a recurring theme in caricature, as in the anonymous print, *The Queen of Hearts cover'd with Diamonds* (c. 1786; see cat. 28.22).[6]

George Moutard Woodward and Richard Newton probably derived their malicious characterization of Charlotte from Gillray. They continued to show her as an absurd foreign consort well matched to her equally ridiculous husband. Woodward's *The*

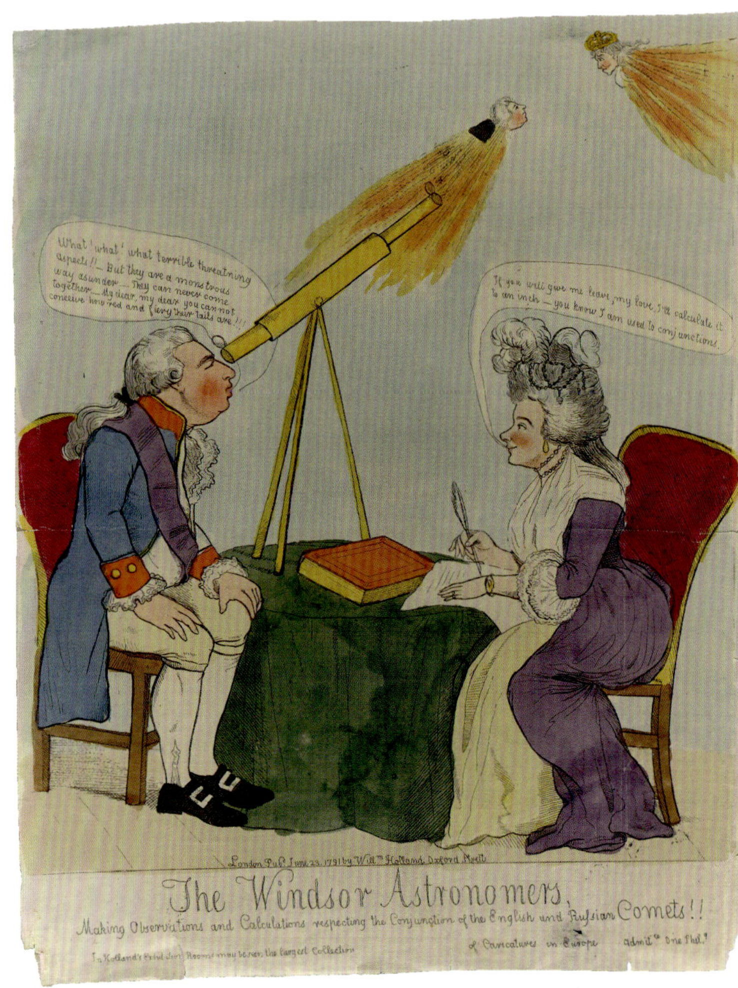

FIG. 14.02 G. M. Woodward, *The Windsor Astronomers, Making Observations and Calculations Respecting the Conjunction of the English and Russian Comets*, 1791, hand-coloured etching on paper, 17⅝ x 13½ in. (44.8 x 34.2 cm). British Museum, London

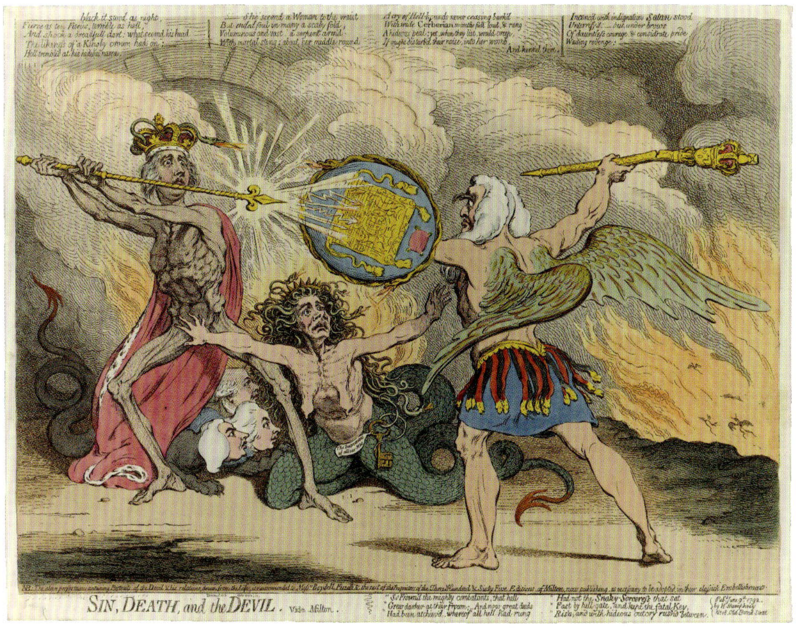

FIG. 14.03 James Gillray, *Sin, Death and the Devil: Vide Milton*, 1792, hand-coloured etching, 15 ¼ x 15⅞ in. (38.8 x 40.3 cm). British Museum, London

Windsor Astronomers, Making Observations and Calculations Respecting the Conjunction of the English and Russian Comets (fig. 14.02) presents Pitt's Russian policy as an imminent collision between two national leaders in the form of comets observed by the royal couple through a telescope, mocking also their amateur interest in astronomy. In 1792, the brilliant Richard Newton—still in his teens—invented a cruel scene of the royal couple as a comic Hector and Andromache, reflecting growing alarm in England over rumours of military preparations for war with France.[7]

Gillray's *Anti-Saccharrites—or John Bull and his Family leaving off the use of Sugar* of 1792 (cat. 15.09) shows George and Charlotte with their family, who suffer miserably from being deprived of sugar in order to contribute to the campaign against the slave trade. This places Charlotte at the centre of debates around slavery and empire. Though Gillray's inscription below the print 'respectfully' exhorts families in Great Britain to follow her noble example, the image of the grossly ugly, meagrely dressed, and emaciated Queen is anything but respectful: her exemplary gesture in supporting abolition by boycotting sugar is brutally caricatured as royal miserliness. Even more offensive is Gillray's *Sin, Death, and the Devil: Vide Milton* (fig. 14.03), also of 1792, in which Charlotte is portrayed as half serpent, half aging hag, with withered breasts and Medusa-like locks of hair. She is placed at the centre of a bitter conflict between William Pitt, then prime minister, and Edward Thurlow, first Baron Thurlow, rendered here in a parody of the Satan, Sin, and Death episode from Milton's *Paradise Lost*. Pitt is cast as Death, Thurlow as the Devil, and the Queen as Sin; in a supremely bawdy gesture, Charlotte's outstretched hand conceals or perhaps grasps the minister's groin.[8]

Continuing from the eighteenth century into the early nineteenth, satirists continued to find Charlotte an irresistible subject for caricature in her role as royal mother and adviser to her wayward son, the Prince of Wales, during his regency, just as Augusta had been targeted earlier in relation to her own son George III. An anonymous print of the Prince of Wales, *The Queen's [Ass]* of 1787, shows him attired in stripes, suggesting a comparison with a painting hanging on the wall behind him of the zebra that once belonged to Charlotte.

The ample satirical attention paid to the three Princesses ironically confirms the effectiveness of their efforts to establish influence and to legitimate dynasty. The brutality of some of the caricatures was the other side of their celebrity, as it was for such politicians as Charles James Fox and William Pitt, and also kings themselves, especially George III. Though visual satires of Caroline and Augusta disrespectfully placed them in compromising positions, it was Charlotte who received the full force of caricature from the most brilliant practitioners of the genre. It was her particular misfortune to be a public figure in an age when caricature was at its most unrestrained.

NOTES

1. On the Hanoverians in satire, see Sheila O'Connell, 'Attacks on the House of Hanover in Satirical Prints', in *Loyal Subversion: Caricatures from the Personal Union between England and Hanover (1714–1837)*, ed. Anorthe Kremers and Elisabeth Reich (Göttingen and Bristol, CT: Vandenhoeck & Ruprecht, 2014), 35–51.
2. Diana Donald, *The Age of Caricature: Satirical Prints in the Reign of George III* (New Haven and London: Yale Univ. Press, 1996), 46–50; Amelia Rauser, *Character Unmasked: Irony, Authenticity, and Individualism in Eighteenth-Century English Prints* (Newark, NJ: Univ. of Delaware Press, 2008), 46–50.
3. Any study of British satire depends on the magisterial catalogue of the collections in the British Museum. See F. G. Stephens and M. D. George, *Catalogue of Political and Personal Satires in the British Museum*, 12 vols. (London: British Museum, 1870–1954). See also the British Museum Catalogue Online, http://www.britishmuseum.org/research/collection_online/search.aspx
4. Mark Hallett, *The Spectacle of Difference: Graphic Satire in the Age of Hogarth* (New Haven and London: Yale Univ. Press, 1999), 134–38; O'Connell, 'Attacks on the House of Hanover', 37–38.
5. On Gillray's images of royalty, see Richard Godfrey, *James Gillray: The Art of Caricature* (London: Tate Publishing, 2001), 167–85.
6. Marcia Pointon, *Brilliant Effects: A Cultural History of Gem Stones and Jewellery* (New Haven and London: Yale Univ. Press, 2009), 193–97.
7. On Newton's royal satires of 1792, see David Alexander, *Richard Newton and English Caricature in the 1790s* (Manchester: The Whitworth Art Gallery and Manchester Univ. Press, 1998), 29–31.
8. See Ian Haywood, 'Milton's Monsters: James Gillray, Sin, Death and the Devil (1792)', in *Romanticism and Caricature* (Cambridge: Cambridge Univ. Press, 2013), 12–32.

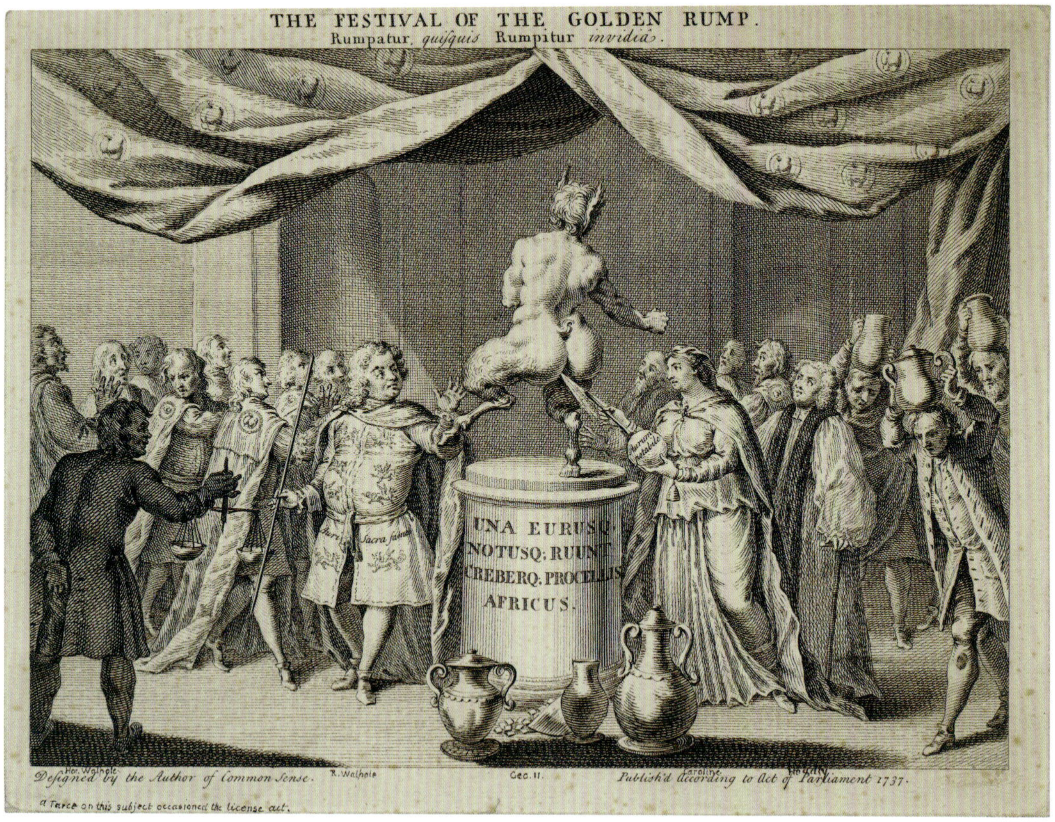

14.04

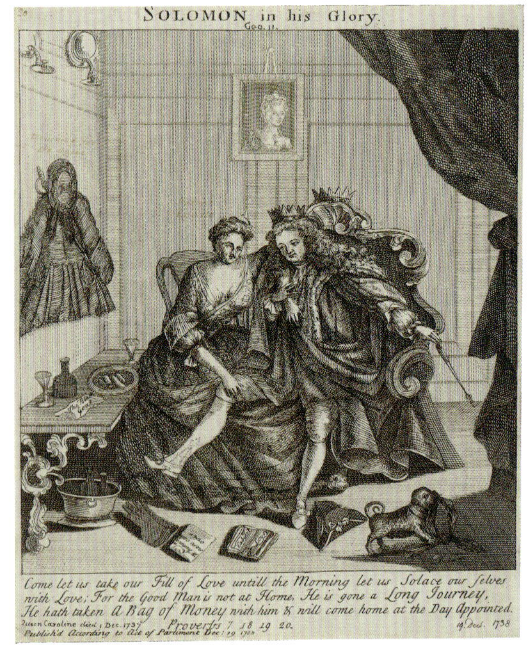

14.05

FAMILY POLITICS AND POWER

14.04

UNKNOWN ENGRAVER, possibly **GERARD VANDERGUCHT** (1696–1776)
The Festival of the Golden Rump. Rumpatur, quisquis Rumpitur invidiâ
1737
Etching and engraving on paper
8⅛ x 10½ in. (20.8 x 26.8 cm)
Published 'by the Author of Common Sense [*sic*]', London
* The Lewis Walpole Library, Yale University, 737.05.07.01 (not illustrated)
† Department of Prints and Drawings, British Museum, Cc,3.173 (illustrated)

This satire decries the power and influence that Queen Caroline and prime minister Robert Walpole were perceived to exert over her husband, George II. It is based on a 'vision' published in *Commonsense, or the Englishman's Journal* in the same year. The King appears as a satyr, standing on an altar and breaking wind; Caroline approaches to administer an enema.

14.05*

UNKNOWN ENGRAVER
Solomon in his Glory
c. 1738
Etching on paper
10¼ x 8 in. (26 x 20.4 cm)
British Museum, 1868,0612.3596

This satire from the year after Queen Caroline's death shows George II dallying with his mistress, Amelie Walmoden, rather than mourning his wife. On the pier table to the left is a paper inscribed 'The Black Goat', and scattered across the floor are Lord Rochester's book *Cabinet of Love*, and another paper with the words 'A Whimnam new co[me] over'. The King's mourning clothes hang from a hook on the wall, and a little dog toys with his mourning band. A portrait of Caroline presides over the scene.

14.06*

UNKNOWN ENGRAVER
View here Three different States in real Life—The Pimp The Miss forsaken and the Wife
1735–36
Etching on paper
13 x 8 in. (33.1 x 20.2 cm)
British Museum, 1858,0417.506

This satire on Frederick, Prince of Wales was made at the time of his marriage to Augusta of Saxe-Gotha, and is based in reverse on Hogarth's *Henry the Eighth and Anne Boleyn*. The scene is set in a palace where the Prince leads his bride, dressed in Van Dyck costume, towards the empty throne, beside which sits a melancholy Robert Walpole (in the place of Hogarth's Wolsey). In the background the Prince's mistress Anne Vane, who takes the place of Katherine of Aragon, is accompanied by her young son, Cornwall Fitz-Frederick Vane, wearing a hat with the Prince of Wales's feathers and holding a toy windmill in allusion to James Stuart, the Old Pretender.

* = YCBA only
† = HRP only

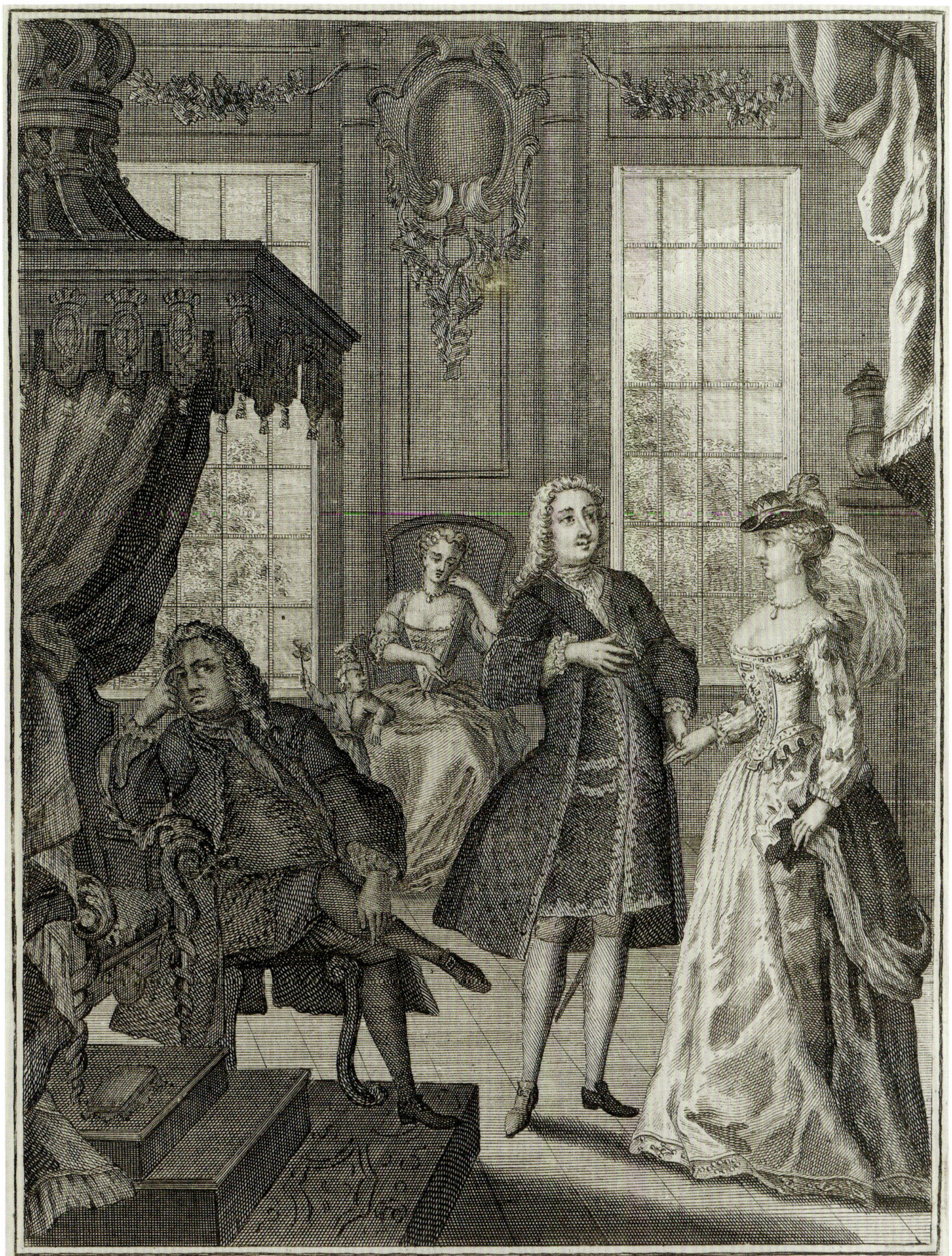

View here Three different States in real Life
The Pimp the Miss forsaken and the Wife
The Happy Pair with Mutual Transport Smile
And by Fond Looks each others care beguile
Backwards behold the Effects of Lawless Love
In silent Grief each heedless Maid reprove
She feels the pangs of Scorn, her Lovers hate

Mourns her Undoing & grows wise too late
The useful Pimp that necessary Tool
A Blundering Sycophant a Polish'd Fool
That Foe to Sacred Bands resigns his Post
And Grieves to see his wonted Honours lost
He & the Lady both in Secret pine
And Fret to See this *MODISH COUPLE Join.

Price 6d

14.06

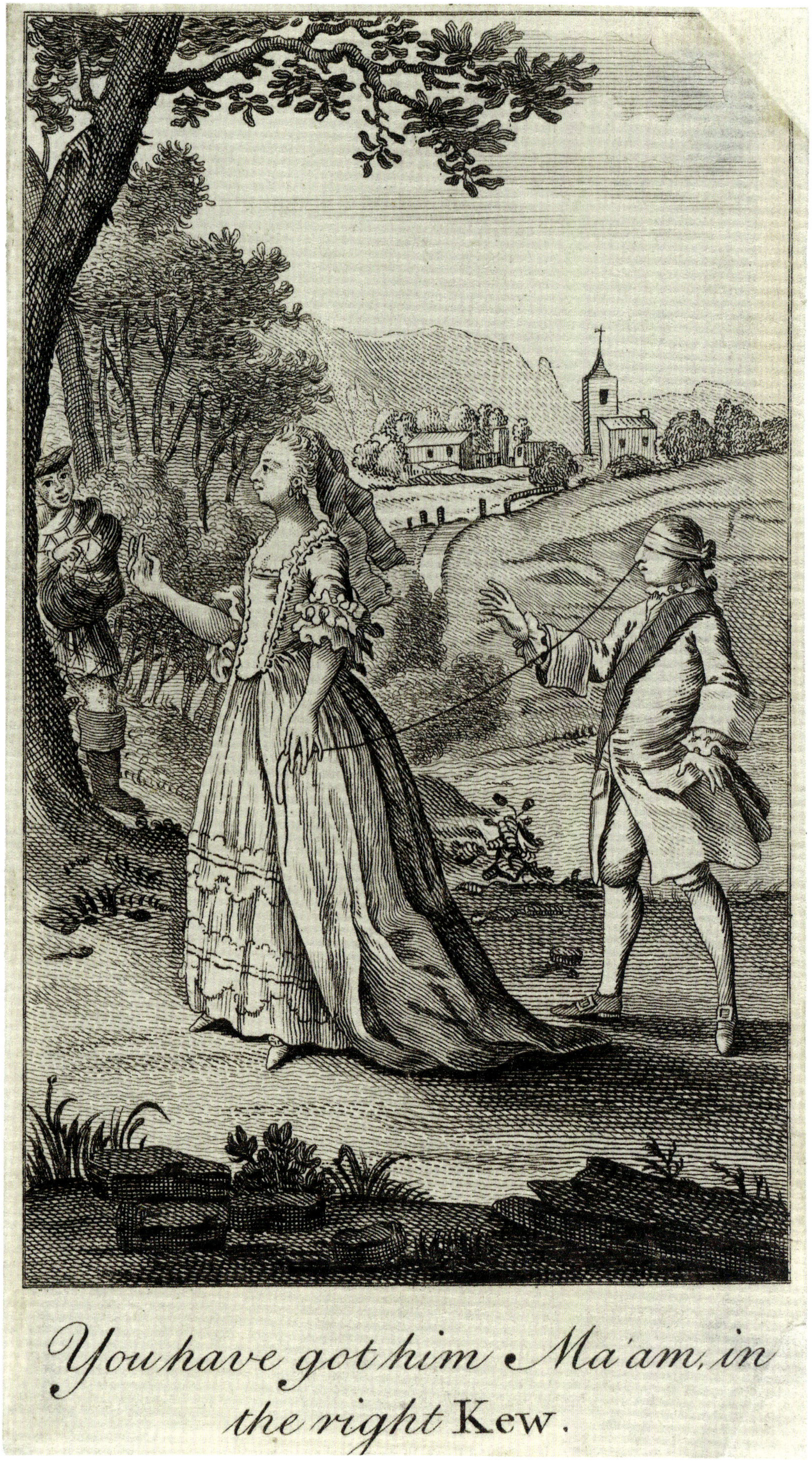

14.07

14.07†

UNKNOWN ENGRAVER
You have got him Ma'am, in the right Kew
1768
Etching and engraving on paper
4⅝ in x 2¾ in (11.6 cm x 7 cm)
British Museum, 1868,0808.13297

Like other satires on Princess Augusta, this example uses Kew gardens as a synecdoche for her supposed adultery with John Stuart, third Earl of Bute, and the pair's perceived influence over her son George III. Here, Augusta leads the blindfolded King by a string through his nose, and beckons towards Bute, who lurks behind a tree in the middle distance.

AUGUSTA, MOTHER OF THE KING, AND POLITICAL NEGOTIATION

14.08†

UNKNOWN ENGRAVER
The Royal Dupe
1770
Etching and engraving on paper
6⅞ x 4½ in. (17.4 x 11.3 cm)
Engraved for the *Political Register*
British Museum, 1868,0808.9894

The Royal Dupe is a caricature on the theme of Augusta's alleged affair with the Earl of Bute and his perceived sinister influence over her son. This image shows her holding the sleeping King in her arms while Bute grasps his sceptre. The politician Henry Fox picks the royal pocket, while scenes of murder (from *Hamlet*) and royal adultery (between Queen Isabella and the Earl of March) adorn the walls.

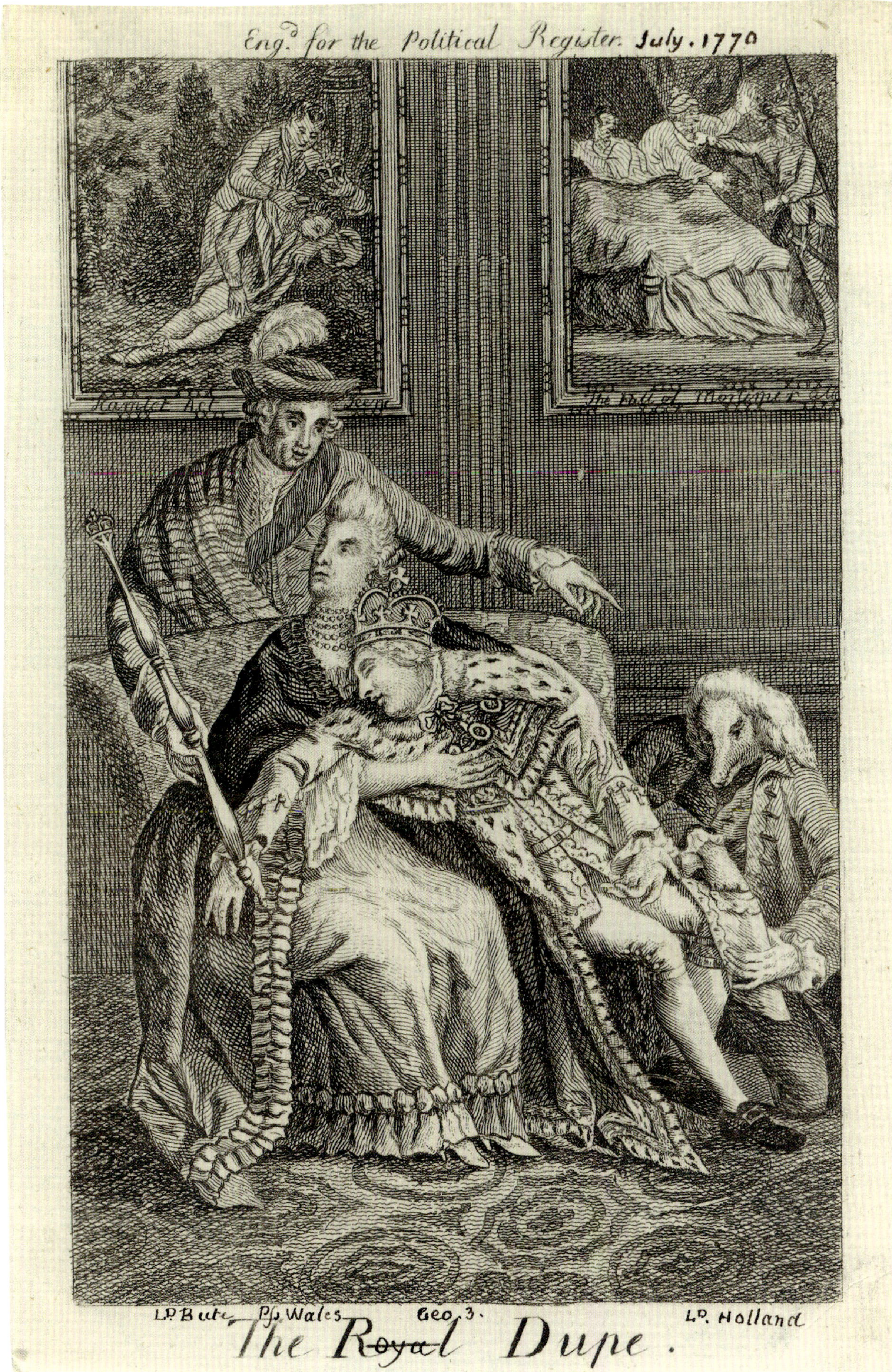
Eng.d for the Political Register. July. 1770
Ld. Bute
Ps. Wales
Geo. 3.
Ld. Holland
The Royal Dupe.

14.08

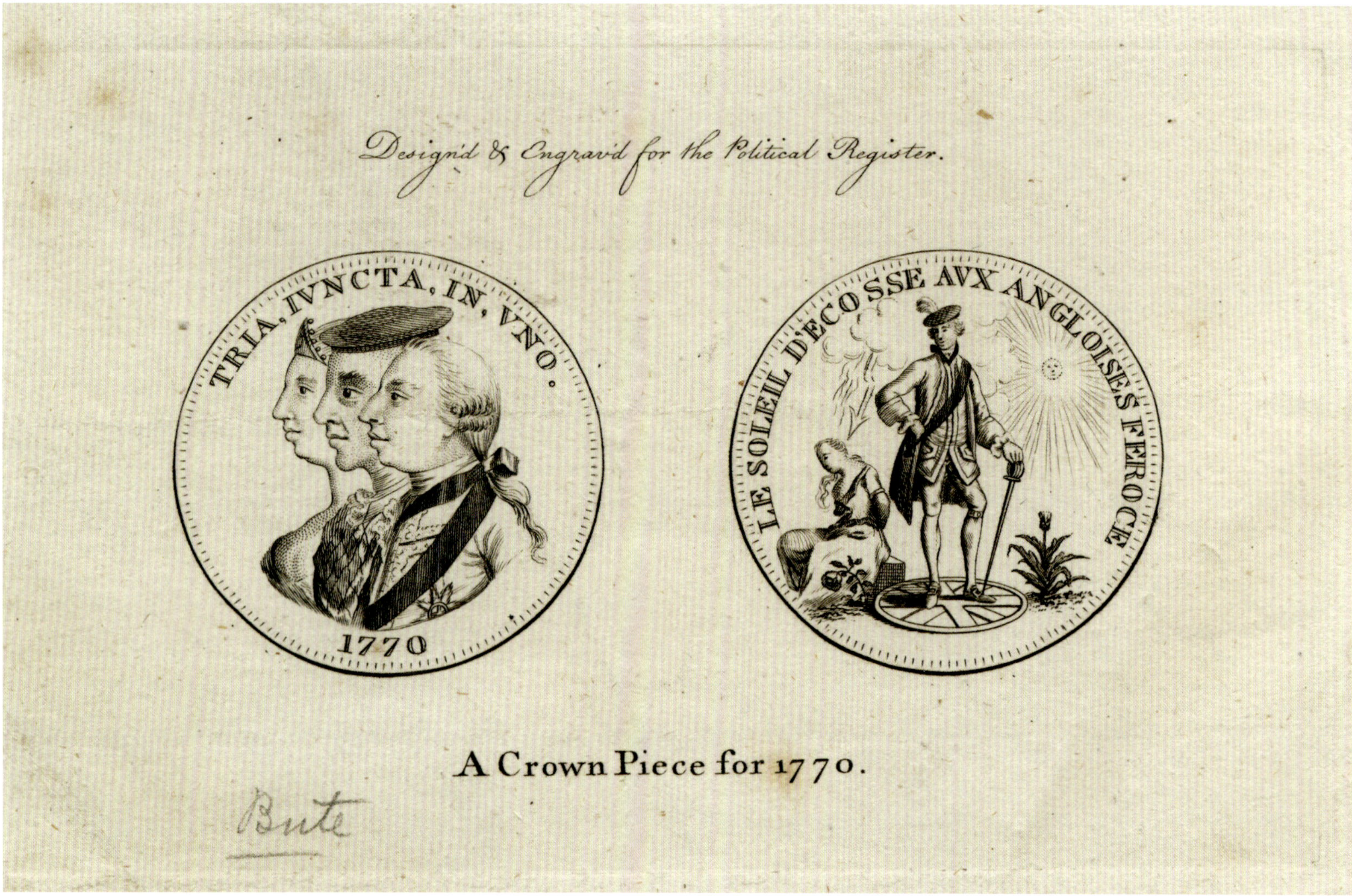

14.09

14.09*

UNKNOWN ENGRAVER
Crown piece for 1770
1770
Etching on paper
4¾ x 7⅞ in. (12 x 20 cm)
Designed and engraved for the *Political Register*
The Lewis Walpole Library, Yale University, CT
770.01.00.06

This satirical design plays with familiar conventions for medallions and coins to represent Princess Augusta, the Earl of Bute, and George III as a perverse ruling family. On the reverse, Bute stands on the shield of Britannia, who sits captive, while an English rose which grows on one side of him wilts and a Scottish thistle on the other side of him thrives.

14.10†

UNKNOWN ENGRAVER
Nero Fiddling, Rome Burning, Pompaja & Agrippina Smiling
May 1770
Etching and engraving on paper
5⅝ x 3⅞ in. (14.4 x 9.8 cm)
Engraved for the *Oxford Magazine*
British Museum, 1871,0610.9889

This satire deploys a classical theme to warn of the supposedly destructive effects of Princess Augusta's power over her son, George III. The text refers to George's strong rebuke of the address and petition for constitutional rights presented by the politician William Beckford; George, as Nero, imagines that the spectacle of London ablaze will teach his subjects a lesson, while Queen Charlotte (Nero's mistress, 'Pompaja') simply enjoys the music. Augusta, in the character of Nero's mother, Agrippina, urges on her son but chides him for neglecting to seize the city's wealth before the conflagration.

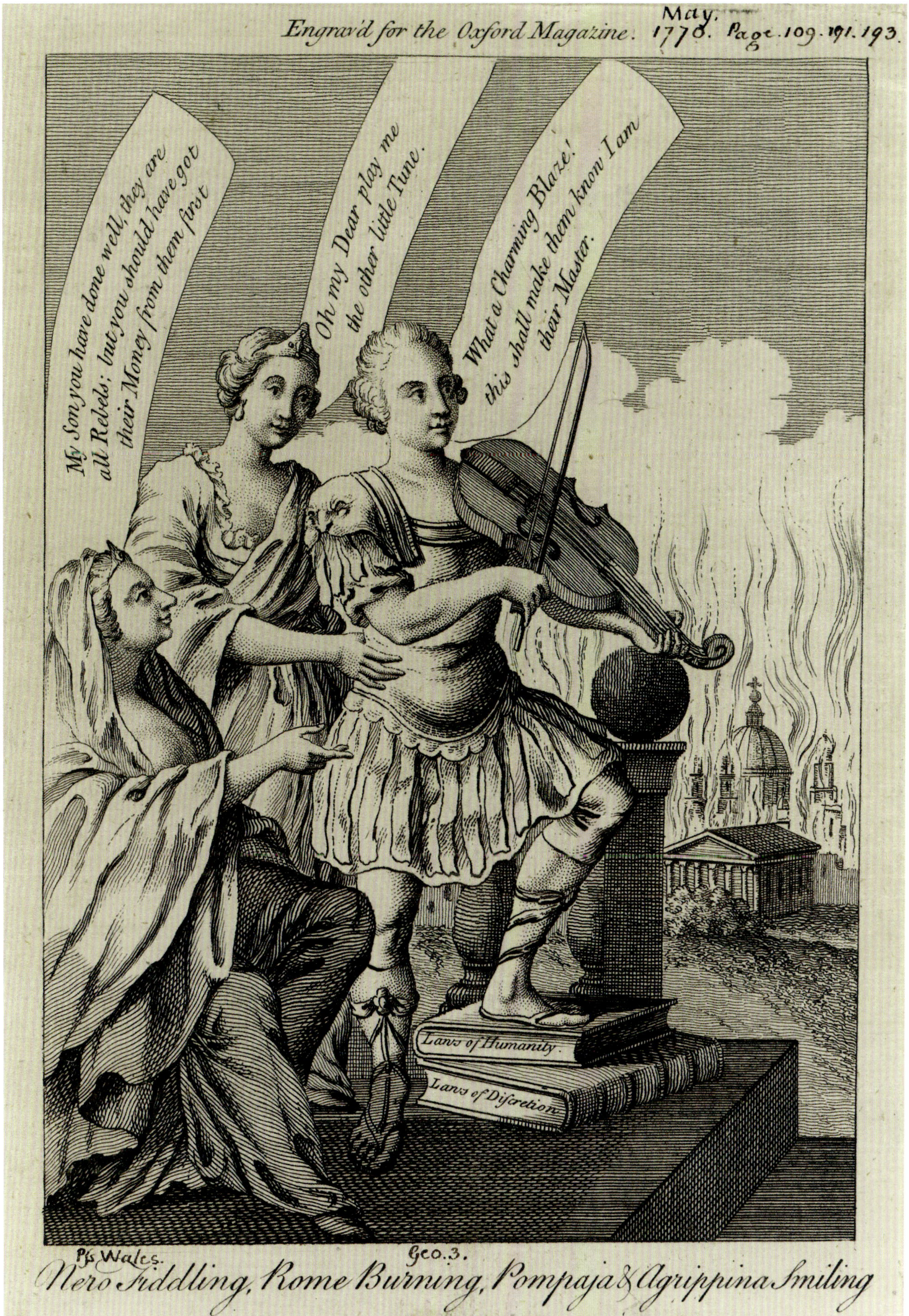
Engrav'd for the Oxford Magazine. May. 1770. Page. 109. 191. 193.
My Son you have done well, they are all Rebels; but you should have got their Money from them first
Oh my Dear play me the other little Tune.
What a Charming Blaze! this shall make them know I am their Master.
Laws of Humanity.
Laws of Discretion
Ps Wales.
Geo. 3.
Nero Fiddling, Rome Burning, Pompaja & Agrippina Smiling

14.10

CHARLOTTE—AN UNWELCOME INHERITANCE

14.11*

THOMAS ROWLANDSON (1756–1827)
The Prospect Before Us / Pitt and His Allies in His Fight for the Regency
1788
Pen and ink, and wash over pencil on paper
8¹¹⁄₁₆ x 12⅛ in. (22.1 x 30.8 cm)
Yale Center for British Art, Paul Mellon Collection, B1975.4.1378

Rowlandson's drawing is a response to debates over the form of a future regency that arose during George III's period of mental illness in 1788. Queen Charlotte, preceded by Mrs Schwellenberg, her German Woman of the Bedchamber and friend, advances towards the House of Lords, William Pitt restraining her with leading-reins. Half a crown is suspended over her head; the remaining half is given to Pitt. In his left hand is a paper inscribed 'I think myself as much entitled to be Regent as the Prince of Wales'. The Queen tramples on the Prince of Wales's feathers and coronet, tagged with the inscription 'My Sons Right'. Mrs Schwellenberg urges the Queen to secure her jewels. The most prominent spectator is Warren Hastings, wearing a jewelled turban with European dress; he lifts up his arm, exclaiming, 'My Diamonds will now befriend me. Huzza.' A man wearing blue and buff says, 'he never meddled with a Petticoat before.'

14.12

JAMES GILLRAY (1757–1815)
Wierd Sisters; Ministers of Darkness; Minions of the Moon
1791
Hand-coloured etching and aquatint on paper
9⅞ x 13¾ in. (25 x 34.9 cm)
Published by Hannah Humphrey (c. 1745–1818), London
* Yale Center for British Art, Paul Mellon Collection, B1981.25.853 (not illustrated)
9⅞ x 13⅞ in. (25 x 35.1 cm)
† British Museum, 1851,0901.570 (illustrated)

A satire exploring the potential shift in royal roles following the first illness of George III takes its inspiration from Henry Fuseli's well-known picture illustrating *Macbeth*, Act 1, scene 3. The three witches are Tory politicians Henry Dundas, William Pitt, and Edward Thurlow, who gaze with apprehension at the moon. This has the Queen's smiling profile in a brightly lit waxing section, and the head of George III in the shadowy waning portion. Beneath the title is inscribed: 'They should be Women!—and yet their beards forbid us to interpret,—that they are so.' Above the image is the caption: 'To H: Fuzelli Esqr this attempt in the Caricatura-Sublime, is respectfully dedicated.' The print is dated 23 December 1791.

14.13*

RICHARD NEWTON (1777–1789), printmaker
Arming in the Defence of the French Princes, or, The Parting of Hector and Andromache
1792
Hand-coloured etching on paper
14⅝ x 16⅛ in. (37 x 41 cm)
Published by William Holland (active 1782–1817), London
The Lewis Walpole Library, Yale University, 792.05.08.01++

In this image, Newton pokes gentle fun at George III and Queen Charlotte, depicting them as a simple-minded but affectionate couple. George, dressed in the armour of Henry VIII, prepares for the fight against the army of the Republic in the French Revolutionary Wars, promising Charlotte, who weeps into a handkerchief, to 'frighten the rascals into good manners'. William Pitt, also wearing armour, carries a shield with the motto 'Ways and Means'.

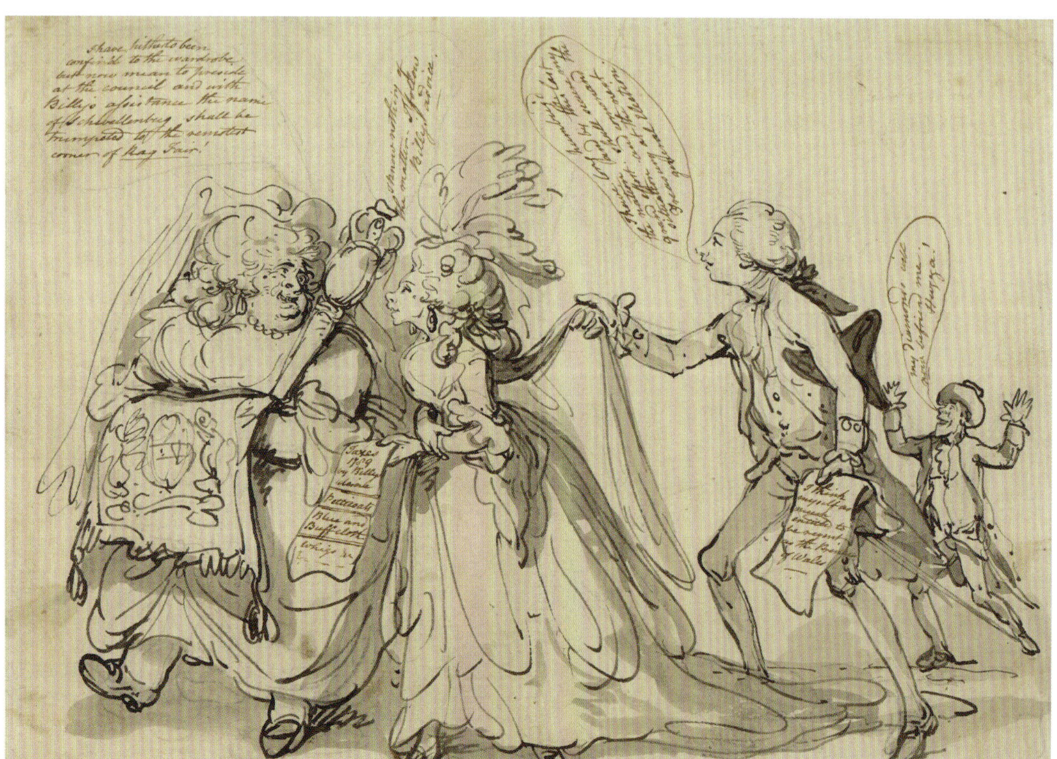

14.11

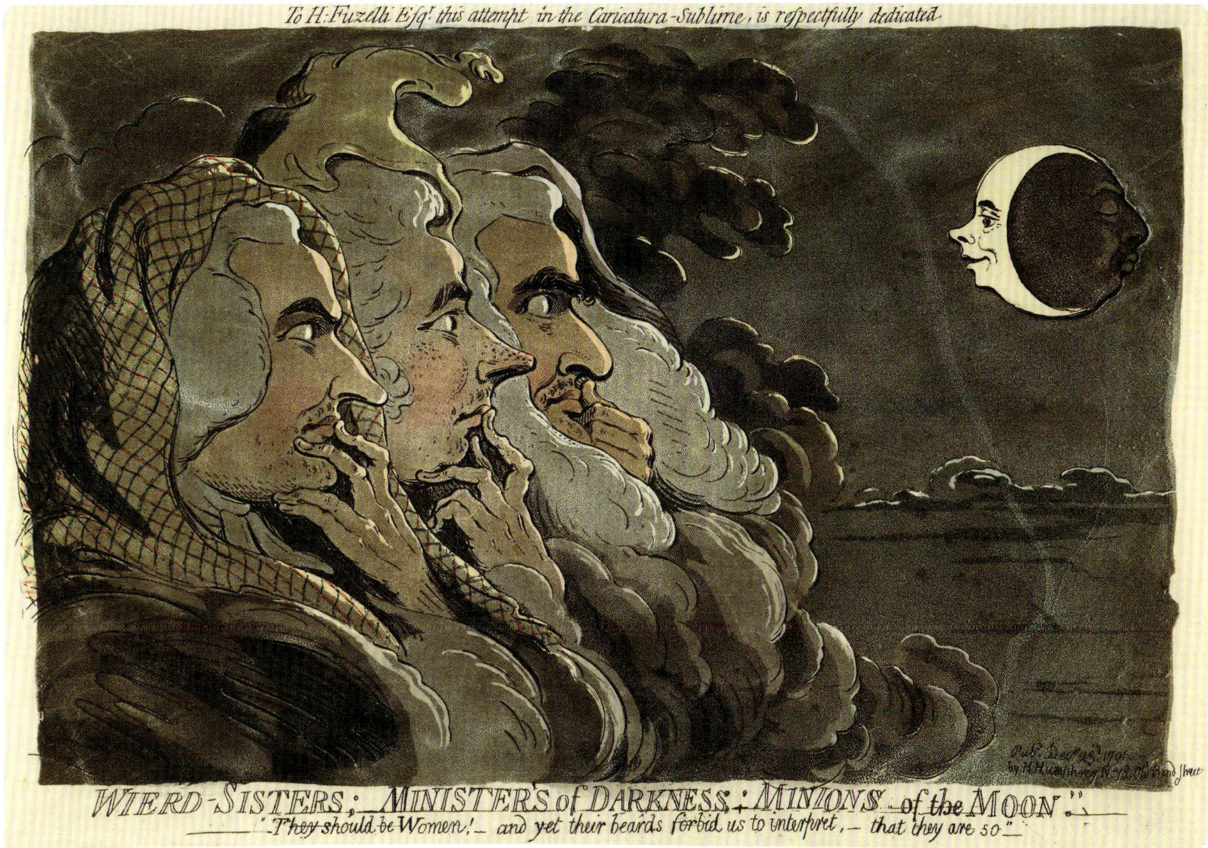
To H: Fuzelli Esq. this attempt in the Caricatura-Sublime, is respectfully dedicated.
WIERD-SISTERS; MINISTERS of DARKNESS; MINIONS of the MOON."
"They should be Women! – and yet their beards forbid us to interpret, – that they are so"

14.12

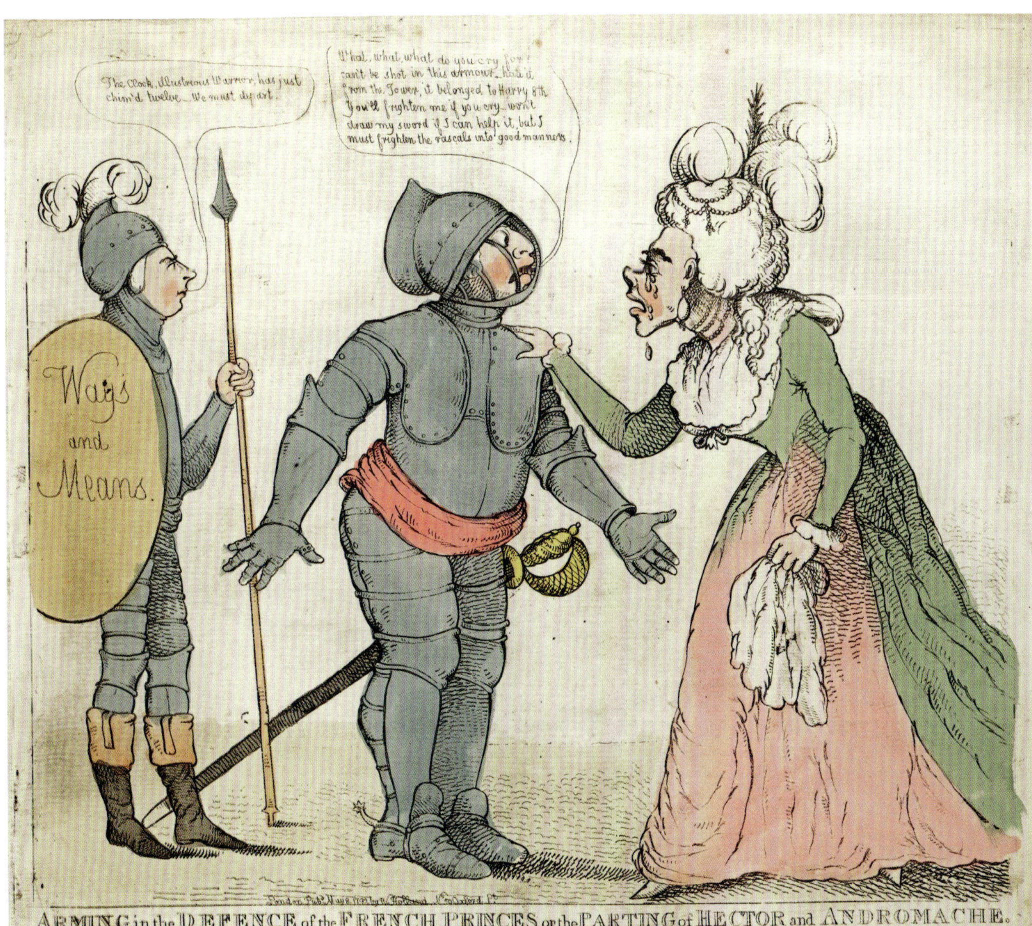
The Clock, Illustrious Warrior, has just chim'd twelve – We must depart.
What, what, what do you cry for? can't be shot in this armour – had it from the Tower, it belonged to Harry 8th You'll frighten me if you cry – won't draw my sword if I can help it, but I must frighten the rascals into good manners.
Wags and Means.
ARMING in the DEFENCE of the FRENCH PRINCES or the PARTING of HECTOR and ANDROMACHE.

14.13

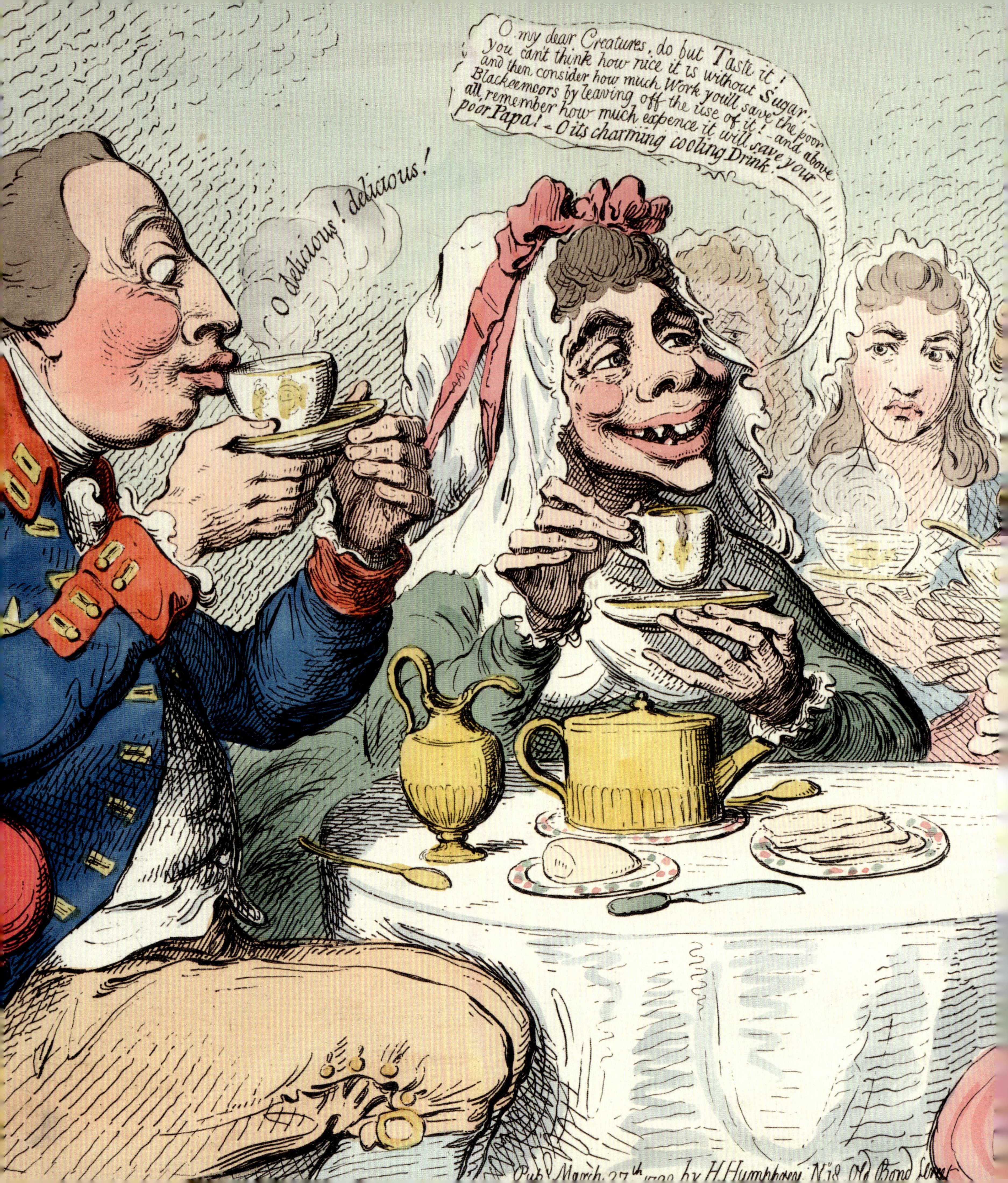
O delicious! delicious!
O my dear Creatures, do but Taste it! you can't think how nice it is without Sugar: and then consider how much Work you'll save the poor Blackeemoors by leaving off the use of it! — and above all, remember how much expence it will save your poor Papa! — O its charming cooling Drink!

Visual Satire and Caricature

Florence Grant

The accelerating rise of graphic satire from the 1730s onwards exposed Caroline of Ansbach, Augusta of Saxe-Gotha, and Charlotte of Mecklenburg-Strelitz to increasing volumes of public comment and derision, usually focused on the disruptive political and moral effects of female power. The stability of dynastic rule rested primarily on the provision of legitimate heirs, and thus was vulnerable to the complex relationships within large and often unruly families. Satirists throughout the century asserted that the intimate life of the royal family was uncomfortably bound up with the political life of the nation.

Graphic satires on the Princesses present a mixture of progressive and conservative attitudes, operating in a vigorous public sphere in which political opposition to the monarchy could slide easily into personal attack and ridicule. Satirists used gendered insults and stereotypes in their criticisms of Caroline, Augusta, and Charlotte and of the state of the royal family as a whole.[1] Of the three Princesses, Caroline exerted the most direct political power, taking charge of domestic affairs during George II's absences in Hanover and guiding numerous church appointments.[2] The genre of graphic satire was just emerging during her lifetime, and only a handful of surviving examples relate directly to her. *The Festival of the Golden Rump* (c. 1737) and *Solomon in His Glory* (c. 1738) are linked through the negative stereotype of the weak man ruled by women (cats. 14.04 and 14.05).[3] Supposedly dominated by Caroline during her lifetime and abandoned to mistresses after her death, George II was portrayed as incapable of real authority.

A densely allusive satire on Augusta's marriage to Frederick, Prince of Wales, *View here Three Different States in Real Life—The Pimp the Miss Forsaken and the Wife* (c. 1735–36), exposed the Prince's domestic arrangements as well as Caroline's attempts to identify the Hanoverians with the legacy of England's Tudor monarchs (cat. 14.06).[4] But satirists largely ignored Augusta until after the death of her husband in 1751 and the succession of her son, George III, in 1760. After that the stereotype of the licentious widow placed her at the centre of a series of unpopular political developments.[5] Her supposed adultery with John Stuart, third Earl of Bute, aroused fear and hostility because both she and her 'lover' were presumed to have wielded undue influence over her weak-minded son.[6]

That Bute was a Scot made matters worse. At the succession of George III, English memories of the Jacobite uprising of 1745 were still raw. Two satires, *The Caledonians Arrival, in Money-Land* (1762) and *Scotch Paradice: a View of the Bute[eye]full Garden of Edenborough* (c. 1763), were evidence of the fear that Bute's patronage would bring a flood of his fellow countrymen to empty the national treasury and corrupt the government (cats. 15.03 and 20.06). Augusta plays a central role, helping Bute distribute bags of money to his tattered followers, and reaching, Eve-like, to accept a tainted apple. The gardens at Kew, the site of Bute and Augusta's closest collaboration, set the scene for *Scotch Paradice*, as well as *You have got him Ma'am in the right Kew* (1768), in which Augusta leads her blindfolded son by the nose while Bute lurks behind a nearby tree (cat. 14.07), and *A Curious*

Detail of p. 274, cat. 15.09

Animal that was seen in the Gardens near Richmond (1767), in which Bute, half-man and half-goat, cavorts before the pagoda designed for Augusta by William Chambers. Augusta also appeared in images in which acts and policies of Bute's administration were being attacked.[7] Her presence hinted that there were disordered relations of sexuality and power between her, Bute, and George III. In *Tempora Mutantur* (c. 1762), a criticism of the terms that Bute negotiated for the end of the Seven Years' War, Augusta appears regally dressed and powerful, presenting Bute to a blindfolded and childish King. In contrast, *The seizure, or give the Devil his Due* (c. 1763) shows her indecently posed, dishevelled, humiliated, and helpless to intervene as the Devil drags Bute off to Hell for proposing a tax on cider.[8]

If satires on Augusta attributed her son's deficiencies to a childish dependency on his mother, Charlotte's image was also drawn into controversies surrounding George III's incapacity to rule. In 1788 he suffered a bout of mental illness that prompted intense debate over the powers to be granted to the Prince of Wales as regent, should the King fail to recover. The Prince's supporters demanded full powers, while the Prime Minister, William Pitt wished to impose limitations.[9] Charlotte was represented as a supporter of Pitt and the limited regency; two satires from this period present opposing views of her role: one infantile, the other hinting at sinister power. In *The Prospect before Us: Pitt and His Allies Fight for the Regency* (1788) she exclaims, 'I know nothing of the matter. I follow Billy's advice', as Pitt holds the train of her gown in the manner of a parent teaching a child to walk. In James Gillray's *Wierd Sisters; Ministers of Darkness; Minions of the Moon* (1791), a travesty of the witches in *Macbeth*, Pitt; the politician Henry Dundas, Viscount Melville; and Lord Chancellor Edward Thurlow, first Baron Thurlow, gaze at a crescent moon whose illuminated face bears Charlotte's profile, with that of George III in darkness behind her (cat. 14.12). The artist deplored the prospect of her growing power, as her husband fell into mental disorder and decline.[10]

George III and Charlotte's relative seclusion, the mutual affection of their married life, and the distance she maintained from national politics left Charlotte less open to the kind of attack that Augusta had suffered. However, later in the century as a new emphasis was placed on the role of mothers in the moral formation of their children, avarice and parsimony were attributed to her in graphic satire, and were used to implicate her in a current scandal.[11] *The Queen of Hearts cover'd with Diamonds* (c. 1786) refers to the gift of a large diamond delivered by Warren Hastings, which brought accusations of bribery during his trial for corruption following his resignation from the post of governor-general of Bengal (see cat. 28.22).[12]

In *Anti-Saccharrites,—or John Bull and his Family leaving off the use of Sugar* (1792), Charlotte, George III, and their daughters take part in efforts to curtail the slave trade by boycotting sugar, here motivated not by any strong moral sentiment but by penny pinching (cat. 15.09).[13] In her later life, Charlotte was increasingly portrayed in satirical prints as a witch-like figure with a tyrannical hold over her adult daughters. Her supposed miserliness was transformed into a failing of maternal character that in turn distorted the moral formation of her spendthrift, womanizing sons and unmarried daughters.[14] The print *Sales by Auction!—or Provident Children disposing of their deceased Mother's Effects for the Benefit of the Creditors!!* (1819) drove the point home after her death (cat. 15.10). The heirs crowded on the left appear generally incapable; the auction lots consist of ragged clothes; and behind the auctioneer there is a painting of a woman weighing coins in a scale.

Cultural anxiety about women's power made Caroline, Augusta, and Charlotte particularly vulnerable to attack as the genre of graphic satire expanded over the course of the eighteenth century. For every virtue or positive action the Princesses might cultivate, satirists found a grotesque inversion. The target in reality was not principally the three Princesses, but the problematic interplay of royal family dynamics and national politics for which their images were made to stand.

NOTES

1. For misogyny in eighteenth-century satirical prints, see Vic Gatrell, *City of Laughter: Sex and Satire in Eighteenth-Century London* (London: Atlantic Books, 2006), 376–587; for radical critiques of the royal family's sexuality, see 505–20.
2. See, in this volume, Joanna Marschner, 'Enlightened Princesses: Caroline, Augusta, Charlotte and the Shaping of the Modern World', 1–51
3. Laura Gowing, *Gender Relations in Early Modern England* (Harlow: Pearson, 2012), 33–34, 61.
4. See, in this volume, Cynthia E. Roman, 'Disrespect: Undermining the Royal Image', 253–55
5. For chastity and women's public reputations, see Gowing, *Gender Relations*, 62–63.
6. Kenneth Baker, *George III: A Life in Caricature* (New York: Thames & Hudson, 2007), 18.
7. George III appointed Lord Bute a member of the Privy Council in October 1760, and a member of the Cabinet and Secretary of State in March 1761. Bute became first lord of the Treasury in May 1762 and acted as head of government until his resignation in April 1763. See Karl Wolfgang Schweizer, 'Stuart, John, third earl of Bute (1713–1792)', *Oxford Dictionary of National Biography* (Oxford: Oxford Univ. Press, 2004; online ed., Oct. 2009), http://www.oxforddnb.com /view/article/26716, accessed 1 May 2016.
8. Baker, *George III*, 24–27.
9. Olwen Hedley, *Queen Charlotte* (London: John Murray, 1975), 153–72.
10. Baker, *George III*, 115.
11. Roy Porter, *Enlightenment: Britain and the Creation of the Modern World* (London and New York: Penguin, 2000), 327–30.
12. Baker, *George III*, 118–19. The caricature is also discussed in Roman, 'Disrespect', 354.
13. Baker, *George III*, 124. This image is also discussed in Roman, 'Disrespect', 355.
14. Baker, *George III*, 182–83, 194–95, 196–97.

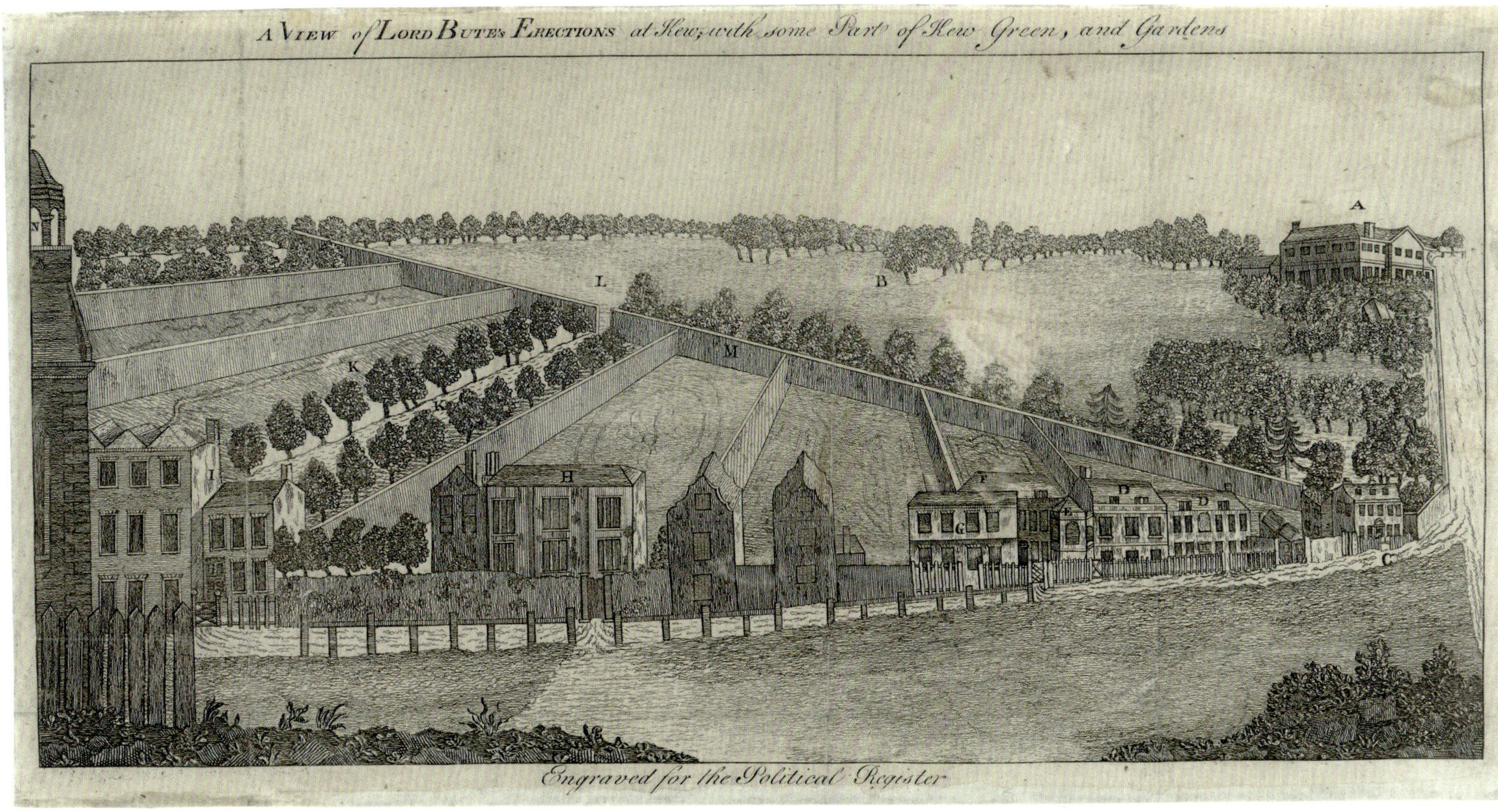

15.01

ROYAL POWER COMPROMISED

15.01*

UNKNOWN ENGRAVER
A View of Lord Bute's Erections at Kew c. 1767
Engraving and etching on paper
8 x 14¾ in. (20.3 x 37.6 cm)
Engraved for the *Political Register*
Royal Botanic Gardens, Kew, Library, Art & Archives Collections

This satire takes the form of a panoramic view of Kew Green, with the church of St Anne in the foreground. The house used by Lord Bute which lay to one side of the green is clearly marked, as is the house in which Princess Augusta lived. Annotations suggest to the viewer just how easy it would have been for the couple to have met, unobserved, by making their way through the adjoining gardens. The title sits above the image. The identification of the potential assignation locations is noted in a separate key published on the following page of the publication, titled 'Explanation of the Print'.

Explanation of the Print. 9

A The P. D. of Wales's houfe.
B Kew gardens.
C A houfe belonging to lord Bofton, chamberlain to the P. D. of Wales.
D A houfe now inhabited by Gen. Græme, fecretary to the queen, but late belonging to Sir Harry Erfkine.
E A houfe inhabited by Mr. Pennel Hawkins, one of his Majefty's furgeons.
F Mr. Howlett's.
G A houfe for the P. D. of Wales's fervants, commonly called *Hell Houfe.*
H The houfe in which lord Bute's family refides.
I A houfe built for lord Bute *to ftudy in*, where none of his family refides.
K A double row of trees, fhading a gravel walk that leads down from lord Bute's houfe to L.
L A door in the P. D. of Wales's garden-wall, which communicates with lord Bute's garden.
M No communication between Kew gardens, and the houfe in which lord Bute's family refides.
N The cupola of Kew-Green chapel, from whence this view was taken.

In a feries of letters, publifhed fome time ago in the pub- ic prints, figned Anti Sejanus, a charge was brought againft the favourite, of building palaces, villas, &c. which occa- fioned an anfwer from his beft informed advocate, in the fol- lowing words:
" The only villa that the earl of Bute hath had within the " compafs of three years, is a fmall old houfe upon Kew-green, " with a new detached brick building on one fide of the gar- " den, ufed by him in part for a library, and in part occupied " y a under fervant o e royal family." * *See a letter figned* "] *'n the Public A er of Monday, Auguft* 26, 1765.

* = YCBA only
† = HRP only

15.02*

UNKNOWN ENGRAVER

Tempora Mutantur
1762
Etching and engraving on paper
8¼ x 10⅜ in. (20.9 x 26.3 cm)
Historic Royal Palaces, HRP3006282

This satire from the early years of George III's reign explores political moves towards the concluding of the Seven Years' War promoted by Lord Bute and opposed by William Pitt. George III sits blindfolded on a throne, and is amused by Folly with a child's rattle as Princess Augusta introduces Lord Bute. Pitt is crowned by Fame, but looks to a weeping figure of Britannia. Minerva gestures towards ships of the navy with spiders' webs in their rigging, and with brooms at the masts indicating that they are for sale. The Union flag is falling into the sea; Admiral Hawke (identified by the hawk on his padlocked sword), turns away wringing his hands. The British lion is chained and muzzled at his feet, and the French birds crow triumphantly.

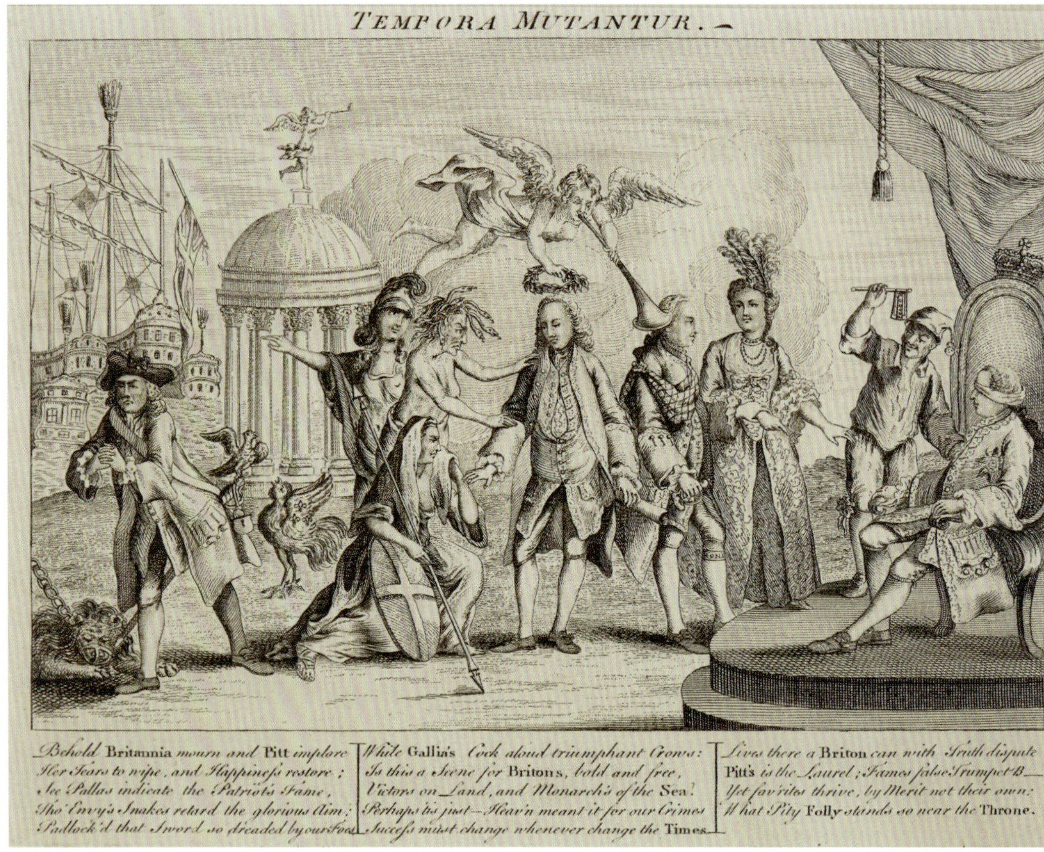

15.02

15.03*

UNKNOWN ENGRAVER

The Caledonians Arrival in Moneyland
1763
Etching on paper
4¾ x 7½ in. (12 x 19 cm)
Published by Edward Sumpter (active 1763–1787), London
The Lewis Walpole Library, Yale University, 762.05.00.17

With the memory of the Jacobite rebellion of 1745 still fresh, English hostility towards the Scots fuelled satirical attacks on Princess Augusta because of her presumed affair with the Scotsman, the Earl of Bute, and a perception that her favour would stoke further Scottish ambition in England. In this example, Augusta and Bute, the in-coming First Minister, are shown distributing sacks of money to Bute's cronies.

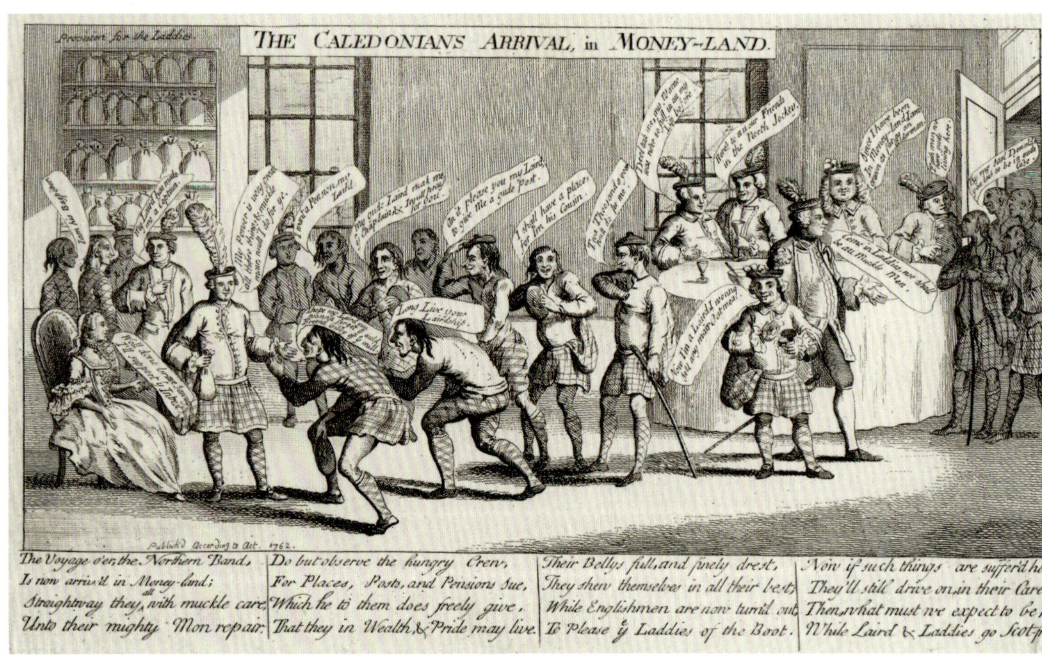

15.03

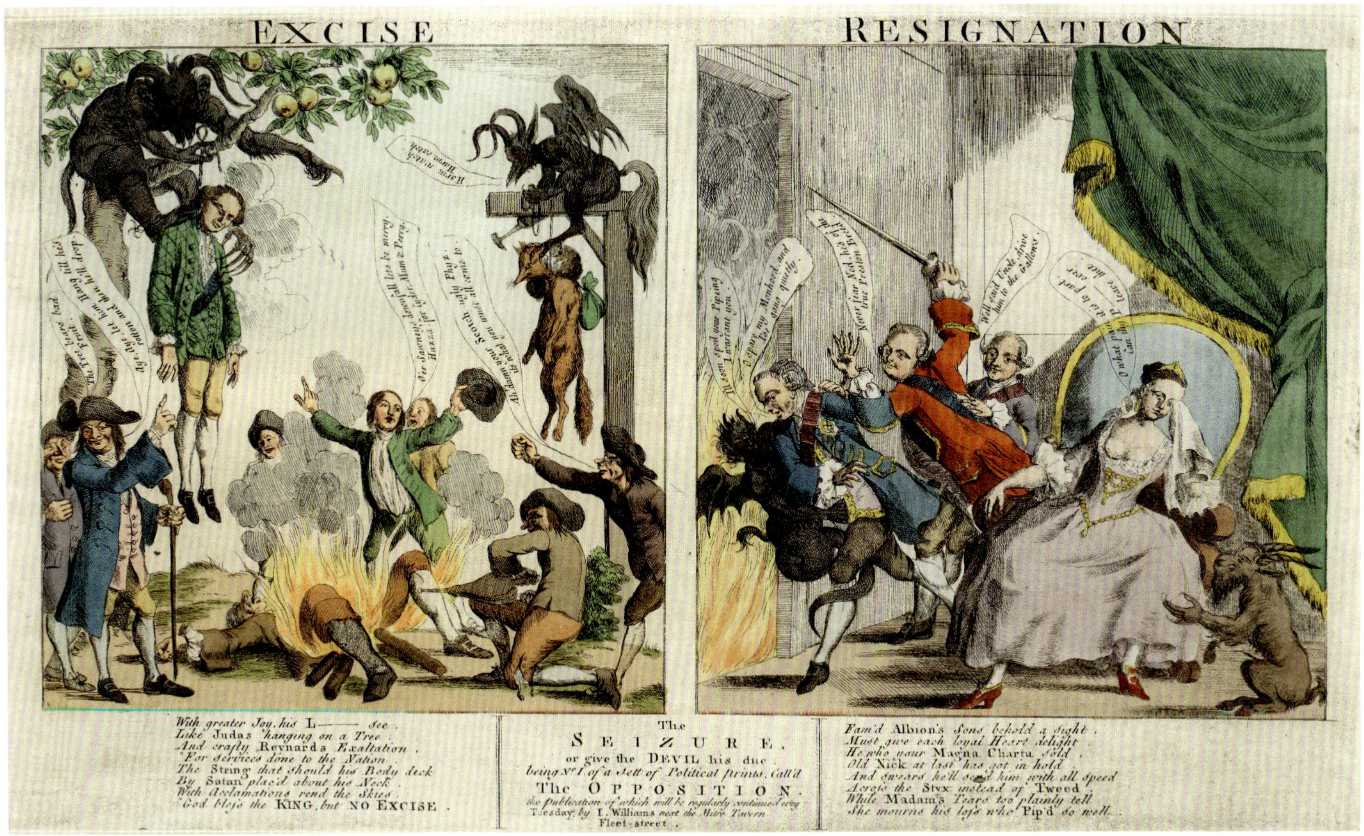

15.04

15.04*

UNKNOWN ENGRAVER, after **JEFFERYES HAMETT O'NEALE** (1734–1801)
Excise and Resignation: The Seizure, or give the Devil his due / The Opposition
1763
Engraving
15½ x 19½ (39.5 x 49.5 cm)
Published by John Williams
Historic Royal Palaces, HRP3006286

This two-part satire was published on the resignation of Lord Bute as First Minister in April 1763. On the left, a scene of execution, headed 'Excise', shows Bute, identified by his plaid waistcoat, hanging from an apple tree (referring to the Cider Bill), with a devil sitting on the branch above him wearing a large boot (a pun on Bute) on one leg, the other ending in a claw; two men below exult. In the centre a man (evidently another representation of Bute as he wears a large boot on one leg) lies in a fire as his detractors fan the flames. On the right is a scene set in a grand interior, captioned 'Resignation', in which Princess Augusta sprawls weeping in a chair as a devil drags Bute towards a door through which flames and smoke can be seen. William, Duke of Cumberland, and Augusta's son, Edward, look on with pleasure.

MAKING FUN OF THE QUEEN

15.05

HENRY HOWARD (active c. 1760–1762)
The Queen's Ass: A New Humorous Allegorical Song
1762
Broadside and etching
13¾ x 8⅞ in. (34.8 cm x 22.4 cm)
Published by the author, London
British Museum, 1868,0808.4198

This is one of many satires inspired by the arrival of one of the earliest zebras in Britain in 1762 as a gift to Queen Charlotte from the governor of South Africa. A bawdy song by Henry Howard is accompanied by an etching of the animal. The title is a rude pun on the zebra's common name, derived from its similarity to a striped donkey. This print was made for *The Political Register*, a popular journal.

15.06*

JOHN JONES (c. 1745–1797), printmaker
The Asses of Great Britain
1762
Engraved broadside with etching
12¼ x 8¼ in. (31 x 21 cm)
Published by John Williams, London
The Lewis Walpole Library, Yale University, 762.08.12.01

This caricature was published in 1762 in response to *The Queen's Ass* (cat. 15.05). The rude ballad 'Fartinando', which was to be sung to the tune of 'The Ass in the Chaplet', appears beneath an image in which the Queen's zebra has knocked Henry Howard, author of the *Queen's Ass* song, to the ground. Journalists Arthur Murphy and Tobias Smollett raise their hands in surprise. Bute looks on, dressed in tartan, riding a tamed British lion. He is accompanied by a parody of a Jewish businessman sitting in the stocks, and the cleric George Whitfield, who gazes into a mirror which reflects the image of an ass. In the background are the writer Henry Fielding, and the three Cherokee chiefs who were brought to London in 1762. Charles Churchill, poet and satirist, wielding a stick, chases off Lord Bute's supporters.

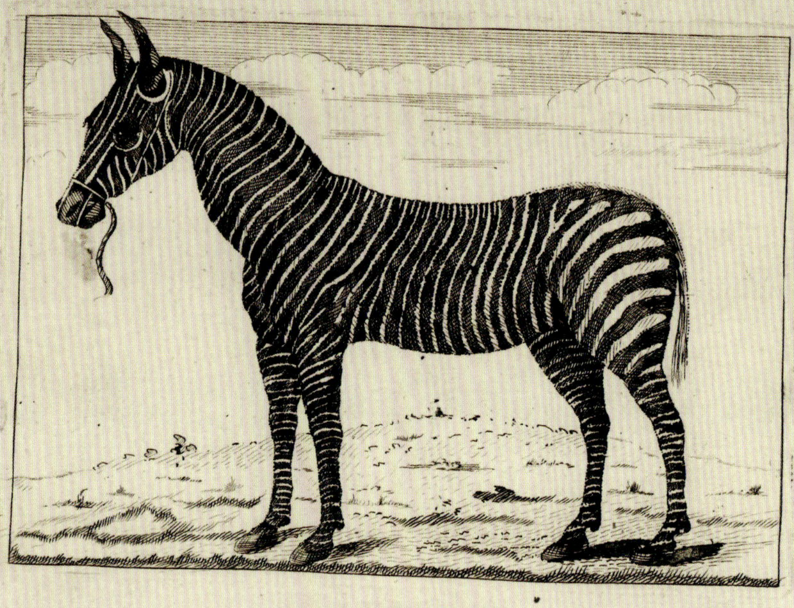

The QUEEN's ASS.

A

NEW HUMOROUS ALLEGORICAL SONG.

Honi ſoit qui mal y penſe.

By H. HOWARD.

To the Tune of, *Stick a Pin there.*

I.

YE Bucks and ye Jemmies who amble the Park,
Whoſe Hearts and whoſe Heads are as lightſome as Cork,
Through *Buckingham-Gate*, as to *Chelſea* you paſs,
Without Fee or Reward, you may ſee the Q———'s A——.
See the Q———'s A—: See the Q———'s A---,
Without Fee or Reward, &c.

II.

A Sight ſuch as this ſurely never was ſeen;
Who the Deuce would not gaze at the A--- of a Q——n?
What Proſpect ſo charming!—What Scene can ſurpaſs
The delicate Sight of her M———'s A---?

III.

Though ſqueamiſh old Prudes with Invective and Spleen,
May turn up their Noſes, and cenſure the Q———n;
Crying out,—" 'Tis a Shame, that her Q———ſhip, alas,
" Should take ſuch a Pride—in expoſing her A---."

IV.

Let them rail if they will; yet I'll bett Ten to One,
Not a Prude of them all but would alter her Tone,
Provided that Fortune, ſo kind to each Laſs,
Had beſtow'd *ſuch* an A--- as her M———'s A---.

V.

The Fribbles cry out, " 'Tis a Sin and a Shame
" To ſuffer a Sight with ſo *filthy* a Name:"
Though they rail, yet will each take a Peep thro' his Glaſs,
For who wou'd not peep at her M———'s A---?

VI.

From *M**kl**b***h St**l**z*, a Place of Renown,
This good-natur'd P--n--ſs came here for a Crown;
And now in Return to the Folks as they paſs,
She kindly repays them—by ſhewing her A---.

VII.

Ye Gods! I with Pleaſure cou'd gaze Day and Night,
At ſo charming, ſo pretty, ſo curious a Sight:
In truth, I muſt own—nay, I ſwear by the Maſs,
I could kiſs (if no Treaſon)---her M———'s A---.

VIII.

But this for a Subject, though loyal, I fear
Would be look'd on by ſome Folks as coming *too* near;
Then in Prudence my Paſſion I'll ſtifle, alas!
Content but to gaze on her M———'s A---.

IX.

Reſign'd to my Fate, thus to gaze and no more,
In vain for Poſſeſſion I ſigh and implore;
But Scripture informs us that all Fleſh is Graſs,
And *ſuch*, I preſume, is her M———'s A---.

X.

Since then there is no mighty Diff'rence between
The A--- of a Subject, and that of a Q---n,
Let each Lad full of Glee take his Bottle and Glaſs,
And drink the Q———'s Health----not forgetting her A---.
Not forgetting her A---: Not forgetting her A---,
And drink the Q———'s Health, &c.

Sold by the AUTHOR, oppoſite the Union Coffee-Houſe, in the Strand, near Temple-Bar; and by all the Print and Pamphlet-ſellers.

(PRICE SIXPENCE.)

N. B. In a few Days will be publiſhed, *A Song on the CHEROKEES*, with a Head-Piece.

3 Chiefs arrived 21 June 1762

1762

15.05

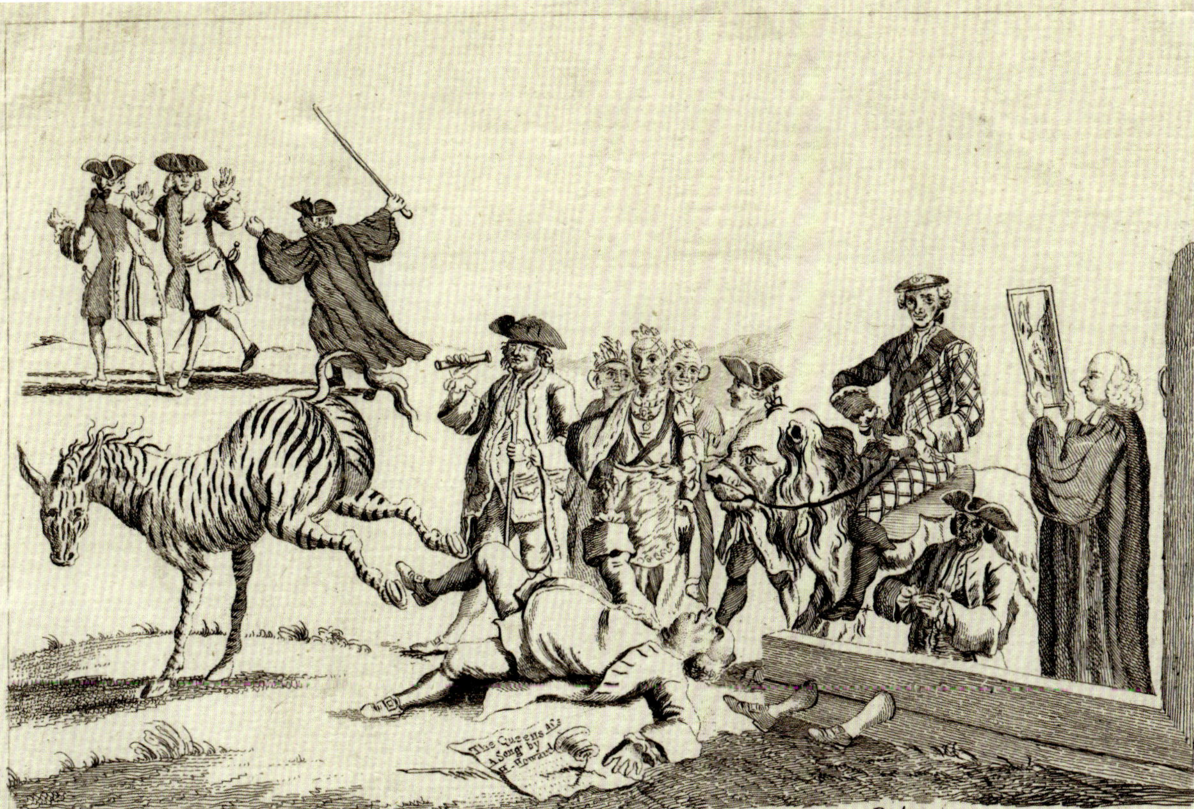

J. Jones delin et sculp^t

THE

ASSES OF GREAT BRITAIN,

An Answer to

HARRY H_______D'S ASS

by FART-INANDO

A Modern Political Ass-trologer.

Tune the Ass in the Chaplet!

Permit me good People (a Whimsical Bard)
And Snarl not ye Critical Class
If once I presume without fee or Reward
To prove that each BRITON'S an ASS.

First view HARRY H____D that Scribling fat Wight
With Forehead well cover'd with Brass
A Dinner is wanting then sits down to Write
And to the whole TOWN shows his A__.

At the best POST in BRITAIN see SAWNEY now PLAC'D
Who thought it wou'd er'e come to Pass
When the LYON shou'd thus be so vilely disgrac'd
And led by the Nose like an Ass.

An ASS wee are told found a LYON'S rough Hide
And fain for Grim LEO wou'd Pass
But when like the BRITON to frighten he try'd
His Braying discover'd the ASS.

The AUDITOR also attempted to roar
In Billingsgate Wit did surpass
The NORTH BRITON came a good Cudgel he bore
And Smartly Corrected the ASS.

Old SHYLOCK the JEW who in CHANGE ALLEY Strives
The Wealth of the Land to Amass
While into your Pocketts he openly Dives
Of each BULL & BEAR makes an ASS.

Let Sly canting SQUINTUM that sanctify'd Prig
But once take a peep in the Glass
Instead of a Saint with the Spirit grown Big
He'll there see the form of an ASS.

When M_RE sally'd forth the fair Sex to relieve
Like QUIXOTE or Sir HUDIBRASS
That FANNY was Scratching as Truth did believe
But now finds himself a Dull ASS.

Blind JUSTICE who owes the sad loss of his sight
To some unkind Love-inflam'd Lass
May boast he can plainly Discern wrong from Right
JACK CATCH will soon prove him an ASS.

But now to conclude Sirs I think it high Time
This Sing Song Satyrical Farce
And if you dont kindly Encourage his Rhyme
The AUTHOR will look like an ASS.

Publish'd According to Act of Parliament by J. Williams next the Mitre Tavern, Fleet Street London.

BMC 3941

1763

15.06

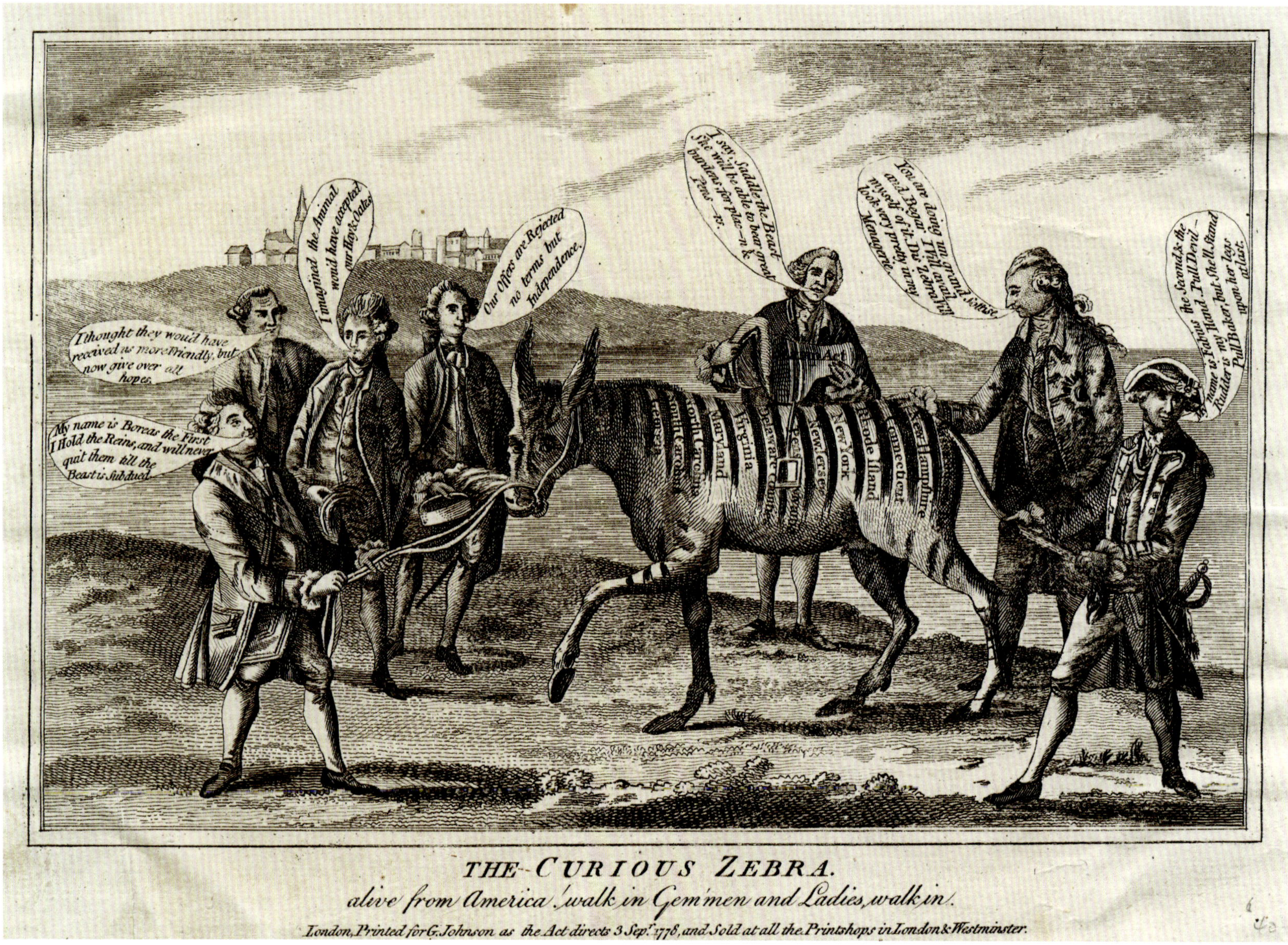
I thought they would have received us more Friendly, but now give over all hopes.
My name is Boreas the First I Hold the Reins, and will never quit them till the Beast is Subdued.
I imagined the Animal would have accepted our Hay & Oates
Our Offers are Rejected no terms but Independence.
I say, Saddle the Beast She will be able to bear great burdens for plac—n & Pens—s.
You are doing um grand sense and Begar I will avail myself of it. Dis Zebra will look very pretty in my Menagerie.
My name is Fabius the Second & the Rudder is my Hand. Pull Devil—Pull Baker, but She'll Stand upon her legs at last.
South Carolina
North Carolina
Maryland
Virginia
Delaware Counties
New Jersey
New York
Rhode Island
Connecticut
New Hampshire
THE CURIOUS ZEBRA.
alive from America! walk in Gem'men and Ladies, walk in!
London, Printed for G. Johnson as the Act directs 3 Sept. 1778, and Sold at all the Printshops in London & Westminster.

15.07

15.07*

UNKNOWN ENGRAVER

The Curious Zebra: Alive from America!

1778

Etching and engraving with stipple engraving

8¼ x 11 in. (21 x 28 cm)

Published by G. Johnson, London

The Lewis Walpole Library, Yale University, 778.09.03.01

This satire acknowledges the public fascination with the royal family's menagerie, and Queen Charlotte's African zebra in particular, by characterizing the newly independent American colonies as such an exotic beast, now set apart from British sensibilities and control. The zebra's stripes are labelled with the names of the first thirteen colonies, and the animal is being treated in different ways by four men. George Grenville attempts to place a saddle on its back, Lord North holds it by a halter, a Frenchman and George Washington, wearing military uniform, pull the zebra back by its tail. Frederick Howard, Lord Carlisle, and Governors William Eden and George Johnstone, who had been sent to treat with the Americans after the War of Independence, coax the zebra with hay and oats.

PUBLIC PERCEPTIONS: ROYAL PARSIMONY

15.08†

UNKNOWN ENGRAVER

From the Originals at Windsor

1791

Hand-coloured etching

17½ x 14⅛ in. (44.4 x 36 cm)

Published by Jacob Dowse, London

British Museum, 1868,0808.10332

This punning satire on George III and Queen Charlotte shows them in the character of two misers after a sixteenth-century painting in the royal collection. This work by a follower of Marinus van Reymerswaele (RCIN 405707) is thought to have been in its turn after a now-lost painting by Jan van Eyck of a banker and his client. With their love of family, modest lifestyle, and avoidance of any elaborate court ceremonial, George III and Charlotte were a frequent target of satirists who aligned these characteristics merely with parsimony.

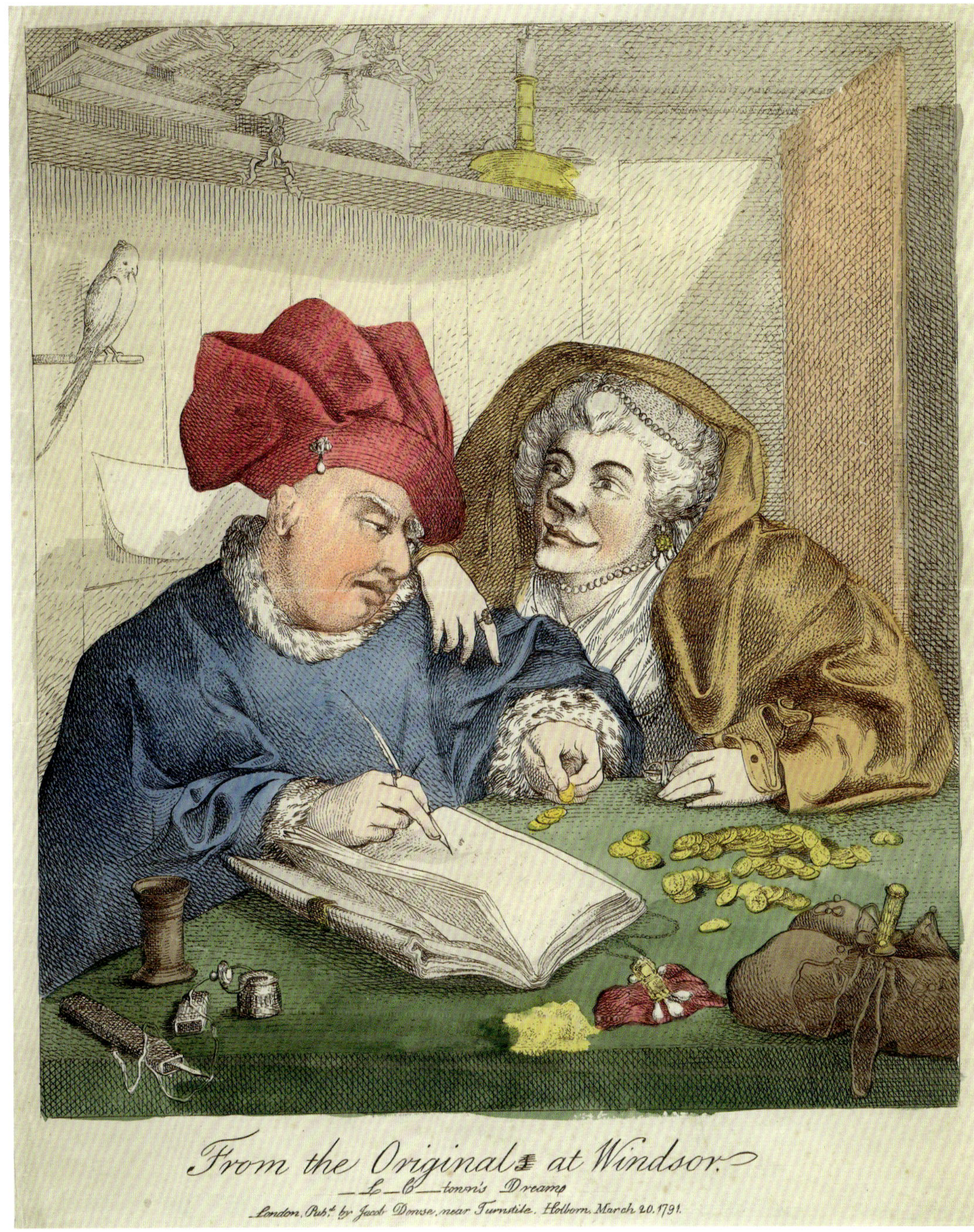

15.08

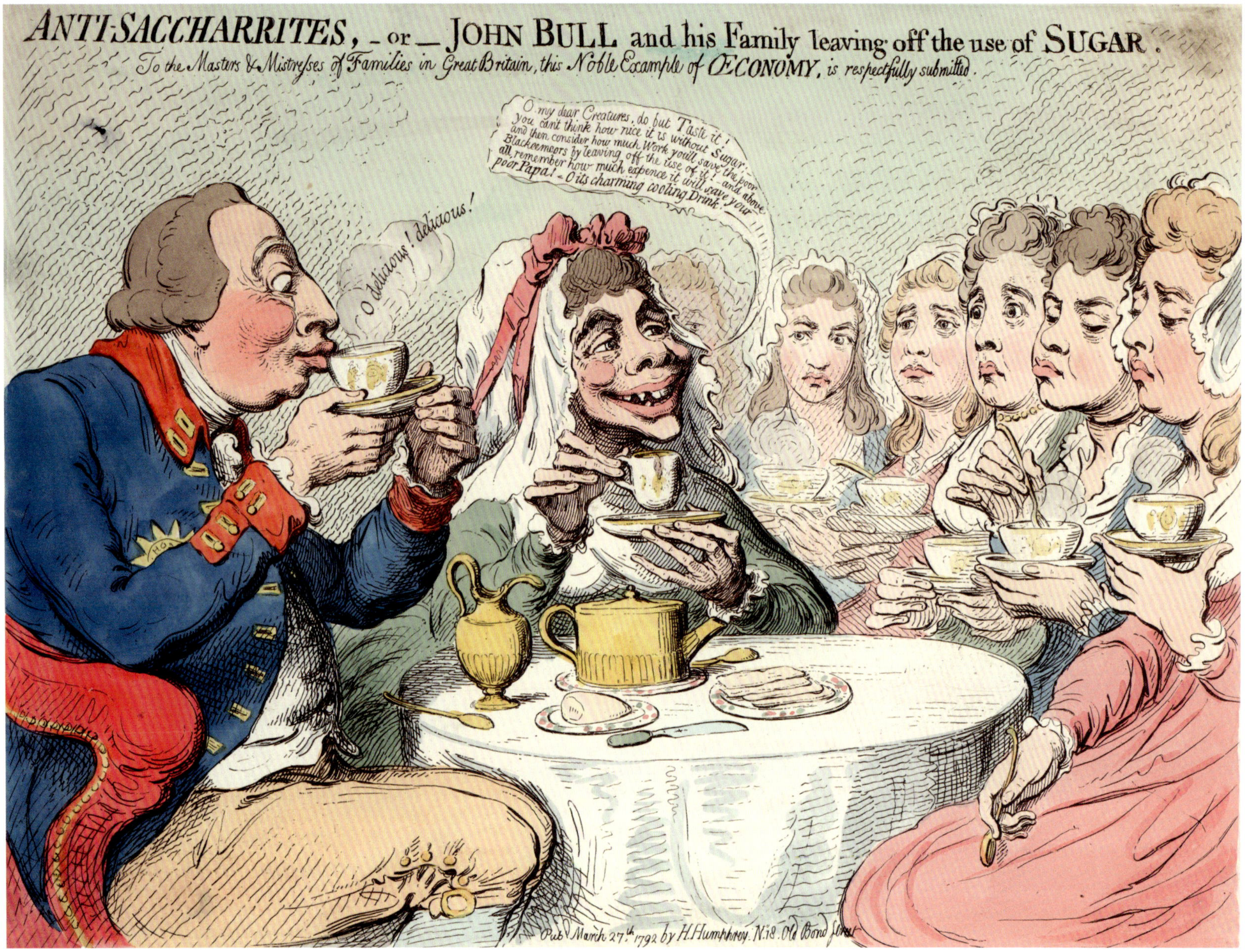

15.09

15.09

JAMES GILLRAY (1757–1815)
Anti-Saccharrites—or John Bull and his Family leaving off the use of Sugar
1792
Hand-coloured etching on paper
12½ x 15¾ in. (31.3 x 39.9 cm)
Published by Hannah Humphrey (c. 1745–1818), London
* The Lewis Walpole Library, Yale University, 792.3.27.01, Impression 1 (illustrated)
12½ x 16⅛ (32 x 40.9 cm)
† Historic Royal Palaces, HRP30063276; LB046 (not illustrated)

In satirical prints Queen Charlotte was increasingly characterized as a miser and Gillray's depictions of her seem to betray real antipathy on his part. Here, with her husband George III she cajoles her six pouting daughters to drink tea without sugar. The caption reads 'O my dear Creatures, do but Taste it! You can't think how nice it is without Sugar', while the inscription continues 'consider how much Work you'll save the poor Blackeemoors by leaving off the use of it!', suggesting that this gesture was motivated by the King and Queen's desire to align themselves with anti-slavery initiatives. The artist concludes his commentary with the words 'above all, remember how much expence it will save your poor Papa!'. This and the title given to the work, 'To the Masters & Mistresses of Families in Great Britain, this Noble Example of Œconomy, is respectfully submitted', suggest that the work was intended as a critique of royal frugality.

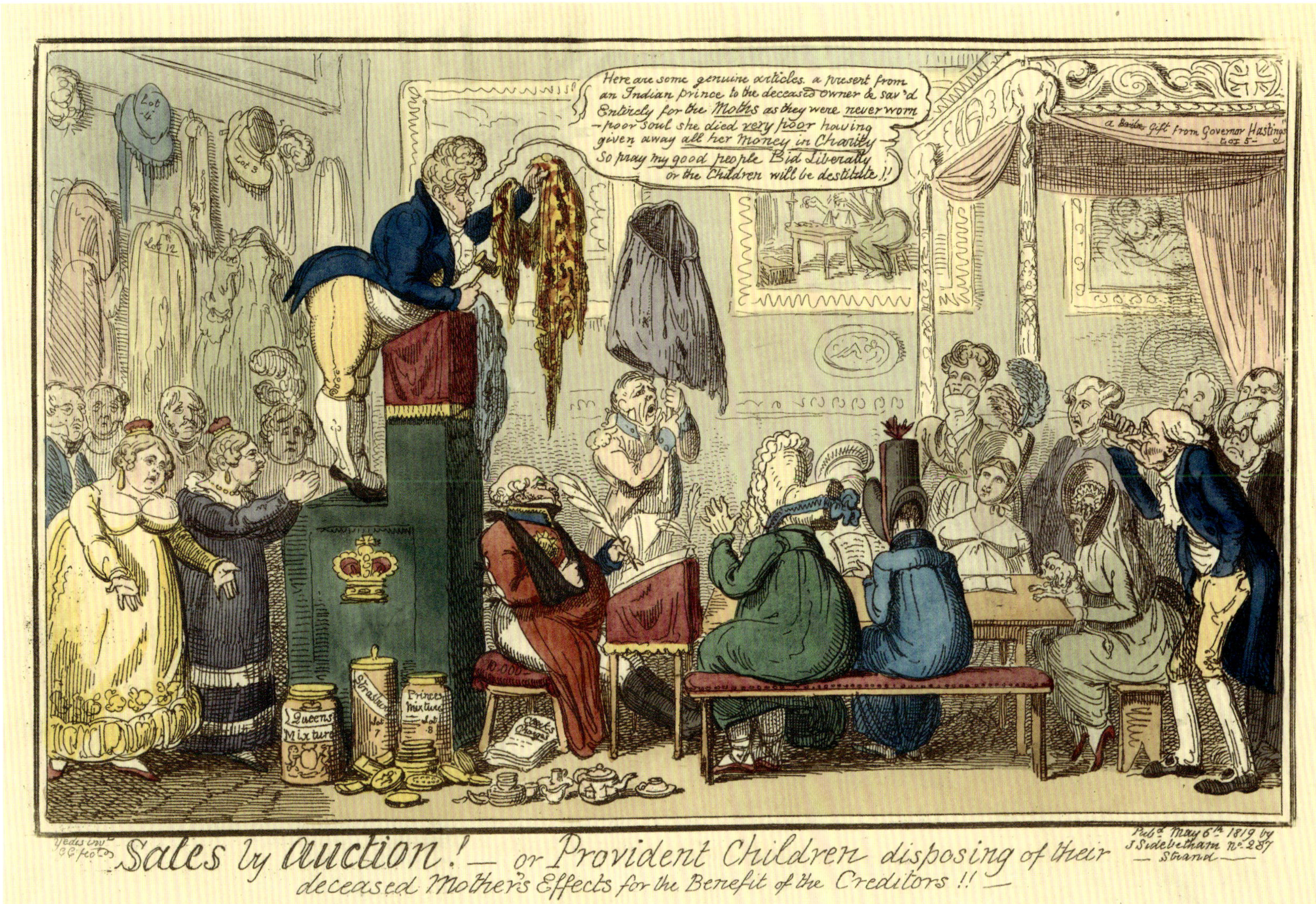

15.10

15.10*

GEORGE CRUIKSHANK (1792–1878), printmaker, after **UNKNOWN ARTIST**
Sales by Auction!—or Provident Children disposing of their deceased Mother's Effects for the Benefit of the Creditors!!
1819
Etching with hand-colouring in watercolour on paper
12⅜ x 16⅝ in. (31.2 x 42 cm)
Published by J. Sidebotham (active 1802–c. 1829), London
Yale Center for British Art, Paul Mellon Collection, B1977.14.1240

This etching by Cruikshank, which reprises the theme of Queen Charlotte's miserliness, imagines the scene at the sale of her well-worn effects after her death. The sale, held over several days in May 1819, was for the benefit of the Queen's unmarried daughters. In the caricature the Queen's eldest son, the Prince Regent, holds up a tattered shawl, declaring: 'Here are some genuine articles from an Indian Prince to the deceased owner & sav'd Entirely for the Moths as they were never worn—poor soul she died very poor having given away all her money in Charity So pray my good people Bid Liberally or the Children will be destitute!!' Charlotte's association with the discredited Warren Hastings continued, and the ivory bed depicted has the inscription 'Bribe' scored through and replaced with the word 'Gift'.

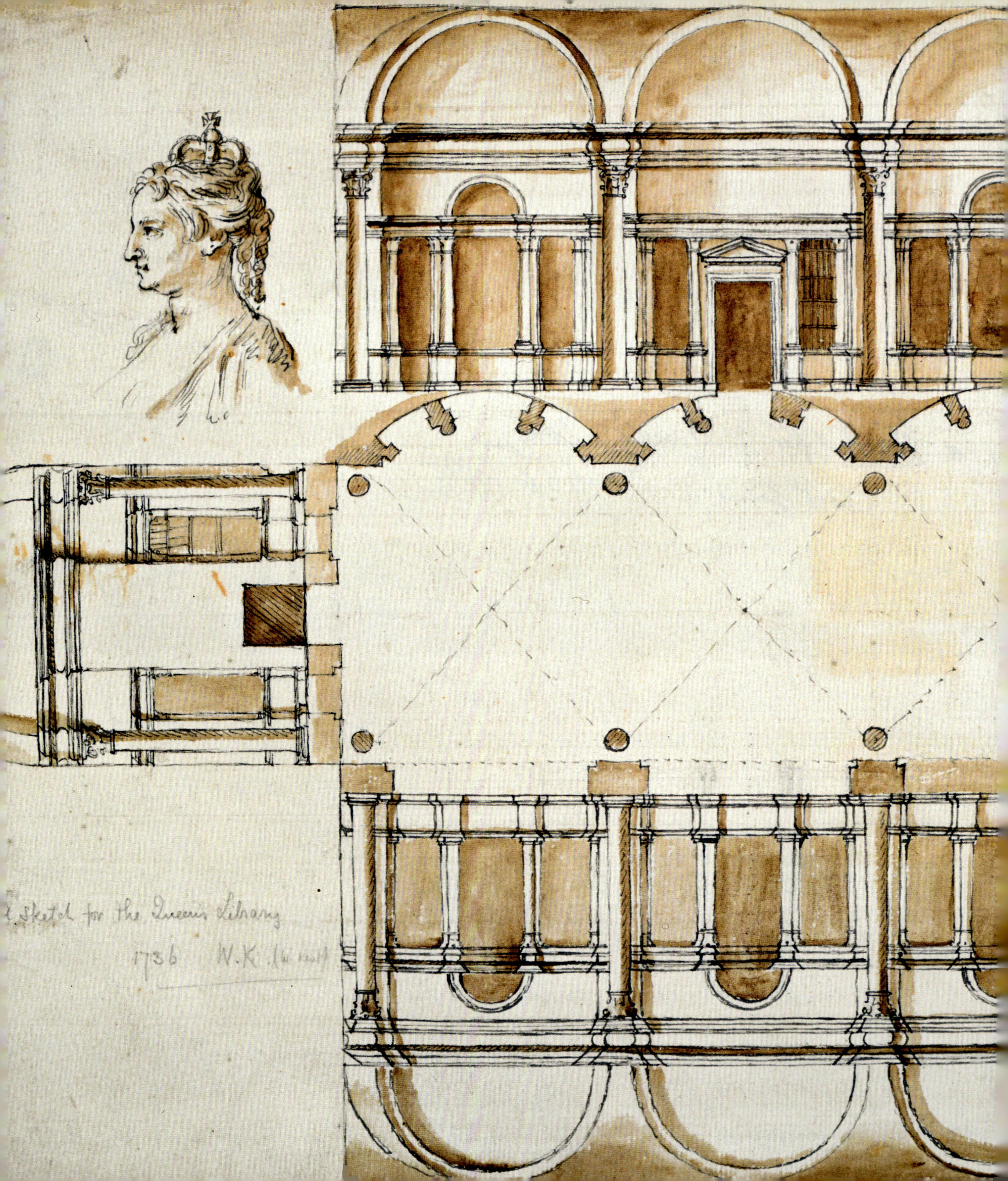
Sketch for the Queens Library
1736 W.K.

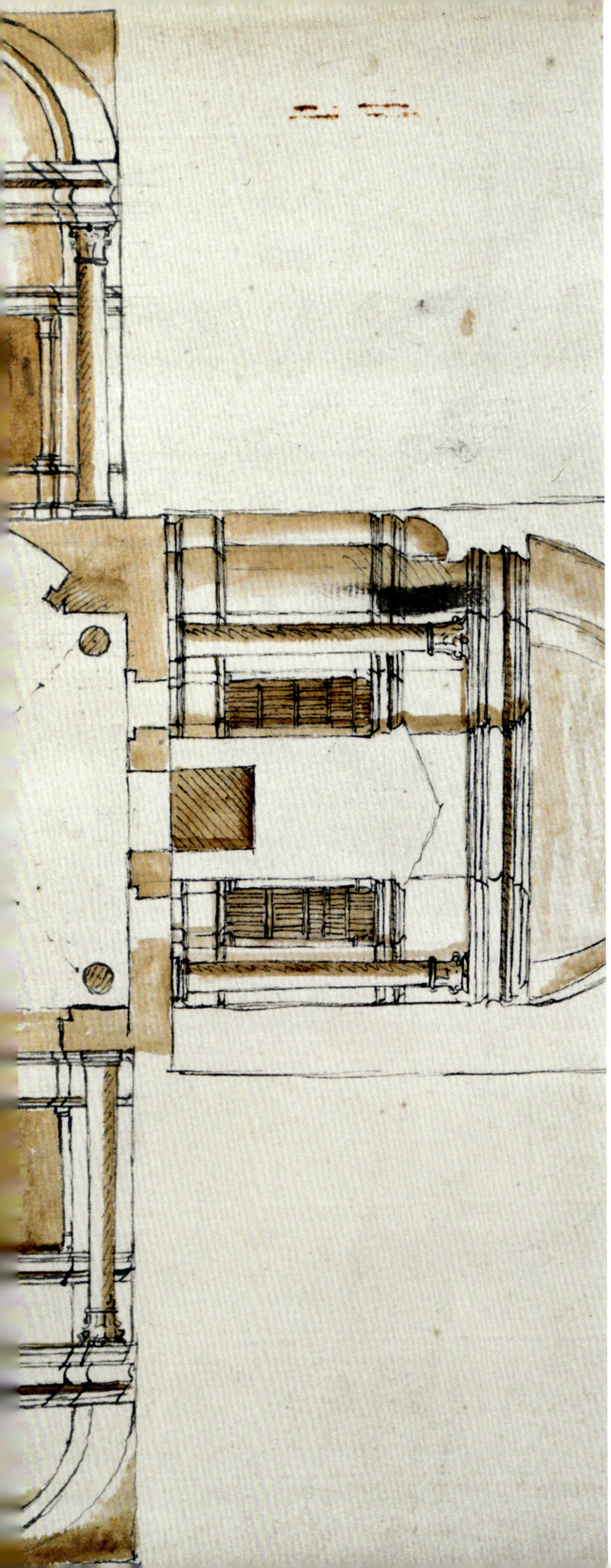

Intellectual Life

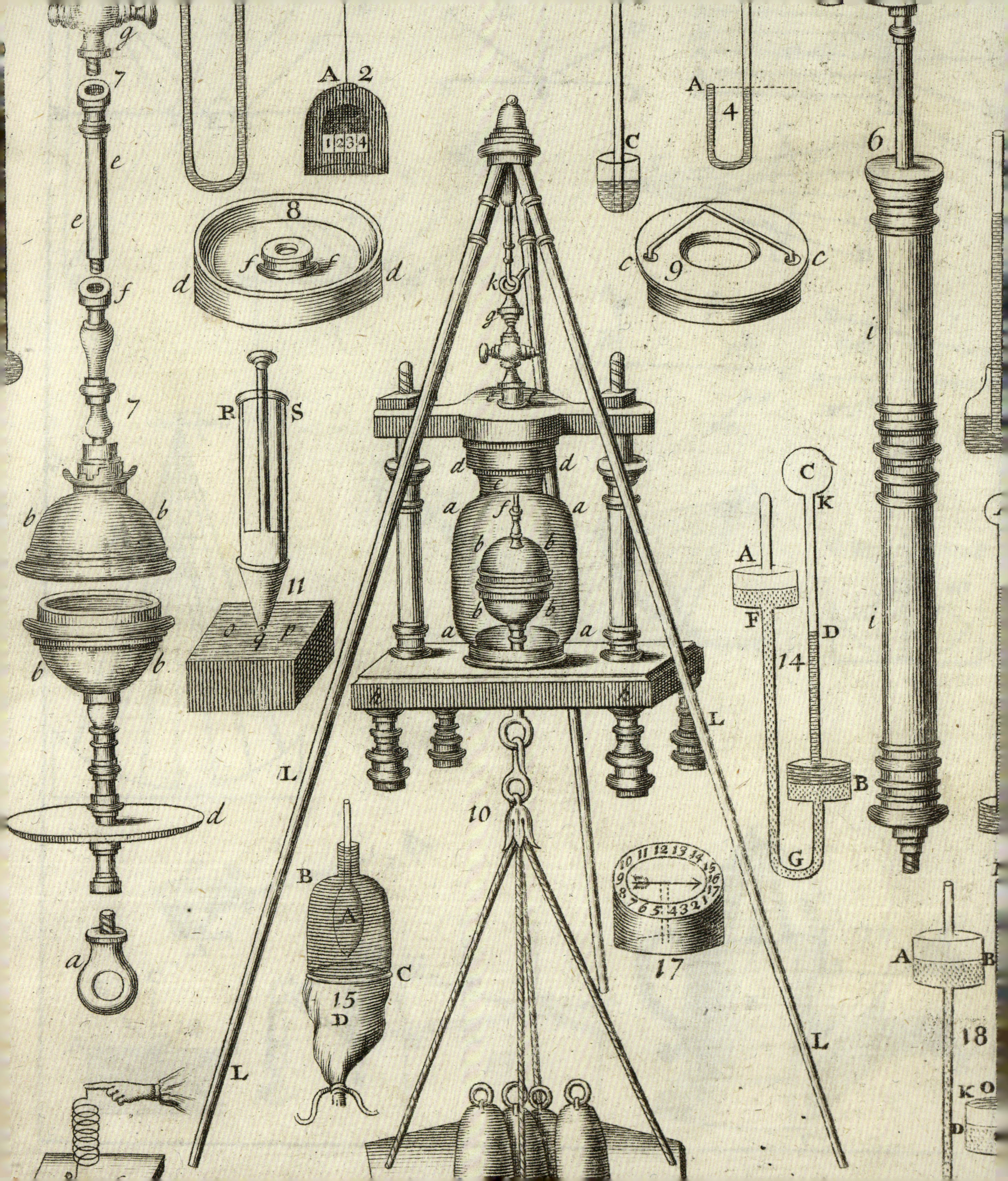

2
1234
4
6
7
8
9
10
11
14
15
17
18
R S
o q p
A B C D
C K D F G
L
a b d e f g
c
i
9 10 11 12 13 14 15 16 17
8 7 6 5 4 3 2 1

Conversations, Knowledge, and Influence

Clarissa Campbell Orr

For Caroline of Ansbach, Augusta of Saxe-Gotha, and Charlotte of Mecklenburg-Strelitz, their most significant conversations were about theology and natural philosophy, which meant both the broad intellectual frameworks through which the created world was understood, as well as the empirical research which these fostered. Caroline was revered for her refusal to convert to Roman Catholicism, a precondition of marrying into the powerful Habsburg dynasty, after discussing theological differences with Gottfried Wilhelm Leibniz.[1] As Electors, the Hanoverians were leaders of the Protestant grouping in the German Empire, but as British kings and queens consort, they were required to adhere to the Church of England.

Leibniz articulated a view of nature in which matter in motion was conceptualized as the displacement of tiny particles (monads), which he discussed with Sophia, the Dowager Electress of Hanover, and with Caroline, when she was still at the Hanoverian court (see p. 54, fig. 3.01). On her arrival in Great Britain, Caroline was baptized as an Anglican, discarding Leibniz's philosophical views and adopting the natural philosophy broadly associated with Sir Isaac Newton (cat. 16.04). He and his followers thought of matter as formed of atoms on which a variety of forces acted across empty space—gravity and chemical affinity among them. Their laws of operation could be expressed mathematically, as in the law of gravity.[2] It was argued that empirical research would eventually provide answers to all kinds of inquiry. Caroline commissioned the Reverend Samuel Clarke to publish his correspondence with Leibniz on these contrasting philosophies (see cat. 16.06). Her acknowledgment of the part Clarke played in the achieving of this ambition, as well as admiration of his own philosophy, led to her commissioning his portrait of about 1729/30, probably from Charles Jervas (fig. 16.01).

Newton's followers developed a 'physico-theology' which united religion with science, and invoked the Deity's providential supervision of the irregularities of the solar system as a parallel to His disposal of historical events, such as the 1688 'Glorious Revolution' which had ended Catholic rule and introduced parliamentary monarchy. Among the early Newtonians was the chemist Robert Boyle, discoverer of Boyle's law (cat. 16.10). His will founded a series of lectures whose purpose was to defend the truths of Christianity against free-thinking, atheism, and scepticism. Dr Samuel Clarke had given two series of these lectures in Queen Anne's reign. The modern-minded cleric William Wollaston, who delineated a form of rational religion which downplayed the role of scripture or revelation to an extent unacceptable to many Anglicans, was nonetheless included amongst the sculptural series Caroline commissioned from Giovanni Battista Guelphi for her pantheon, the Hermitage, at Richmond (cat. 16.12).[3]

Newton's unification of astronomy and mechanics to describe the solar system was popularly explained by itinerant lecturers using models of the solar system known as 'orreries' after their patron, Charles Boyle, fourth Earl of Orrery, a distant relation of the famous chemist (cat. 16.09). Caroline commissioned her own example from Thomas Wright, a notable maker of scientific instruments, and had it set up in the middle of her gallery, a

Detail of p. 287, cat. 16.08

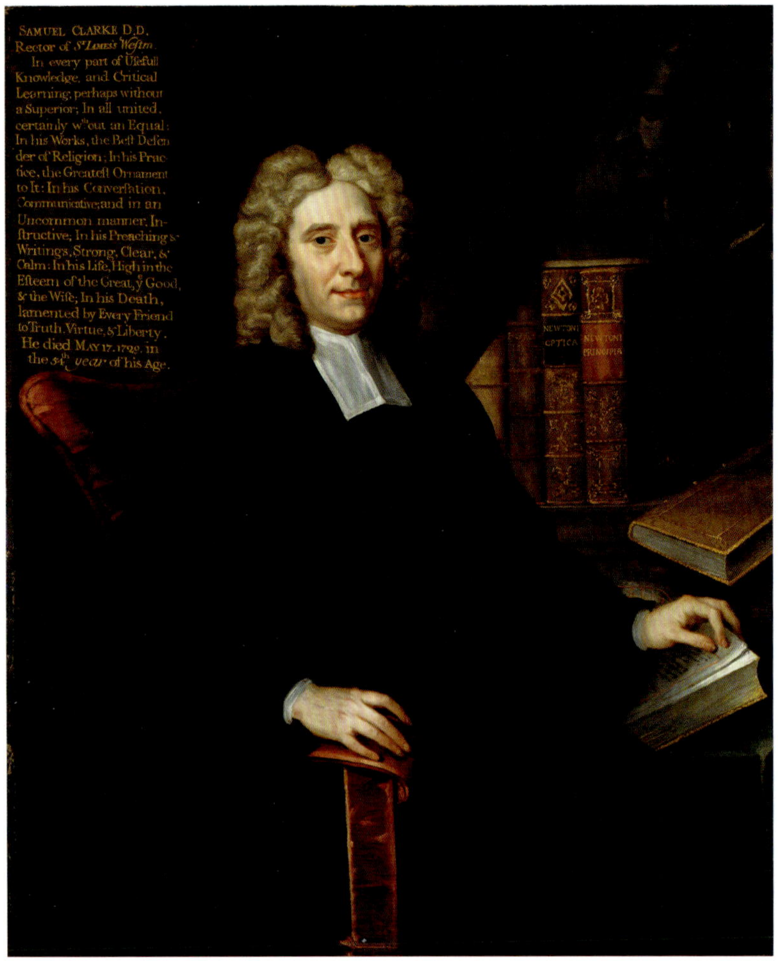

FIG. 16.01 Charles Jervas, *Samuel Clarke,* c. 1729–30, oil on canvas, 50½ x 40½ in. (128.2 x 103 cm). Royal Collection Trust (cat. 16.05)

senior reception room, in Kensington Palace. Astronomy attracted widespread attention in royal and aristocratic circles in Caroline's lifetime; a French Protestant cleric, John Theophilus Desaguliers, was one of its most successful popularizers (cat. 16.07). He had first given lectures at the Royal Society on this subject in George I's reign; George II and Caroline both attended his lectures on occasion and subscribed to his *Lectures of Experimental Philosophy* (cat. 16.08).[4] This was also the era of great collectors such as the royal physician Dr Hans Sloane, who succeeded Newton as President of the Royal Society (fig. 16.02). Their extensive and varied collections combined attributes of the older 'cabinets of curiosity' with a more modern, systematic method of classifying and displaying the natural world. Sloane's collection was open to selected visitors, while not being a public institution in his lifetime. Caroline visited it with her children to interest them in the new sciences of men as well as of nature.[5]

Caroline was also familiar with some of the literary developments of Queen Anne's reign. One of the most important was the new genre of the periodical magazine, published several times a week and featuring short essays on society and culture. The most significant were *The Tatler*, founded by Richard Steele and Joseph Addison in 1709, and their subsequent collaboration *The Spectator* (1711–12). They remained in wide circulation through collected editions; Caroline had copies in her library at St James's Palace, and Charlotte was still reading *The Spectator* decades later (cat. 17.12). The magazines were part of the Whig party's response to a more modern, socially mobile society, and they had reached out to the Hanoverian dynasty even before its succession in 1714. Addison was part of the delegation which travelled to Hanover to present the Order of the Garter to the future George I in 1706. By 1717 Addison was Secretary of State for the Southern Department and less active as a writer, though Caroline would later be influenced by his ideas on garden design.

By 1714, Addison had also broken with Jonathan Swift, who had followed Tory views since 1710. As Dean of St Patrick's Cathedral, Dublin, Swift was the second kingdom's greatest living patriot and a leading critic of English financial policy, which he believed hindered Irish prosperity. *Gulliver's Travels* of 1726, his satire on politics in the broader sense, was interpreted as an allegory of specific royal and political figures, Caroline included. She found it impossible to draw Swift into her orbit, though he would send her gifts of Irish textiles which she enthusiastically promoted within the court (cat. 17.15).[6]

Alexander Pope was similarly immune to court flattery, though he did participate in Caroline's conference on gardening and garden design philosophy in 1719, and she later contributed materials to enable the completion of his own garden at Twickenham.[7] As a Catholic he was an outsider, even though when Tory politicians were in power from 1710–14 he was close to some of their leading political figures and literary supporters. They saw him as poet of an Arcadian age of peace, prosperity, and taste that would emerge after the end of the wars with France and Spain. Anne's death in August 1714, soon after the Peace of Utrecht in 1713, cancelled these hopes, and Pope was to make himself free of patronage with the huge commercial success of his translations of the Greek Homeric classics (cat. 17.13). One of his subscribers was Caroline, as Princess of Wales. However, in his mock-heroic epic *The Dunciad*, published in its first edition in 1728, he skewered the figures of little real talent who succeeded in the new literary marketplace; and Caroline, by now the Queen, was ridiculed as Queen of the Dunces (cat. 17.14).

Caroline and the many intelligent female readers who flourished in her household and at court, however, benefited from deluxe translations of the classics. One example is the anthology of 1717, edited by Sir Samuel Garth, of translations by many hands, Addison's included, of Ovid's *Metamorphoses*, which was

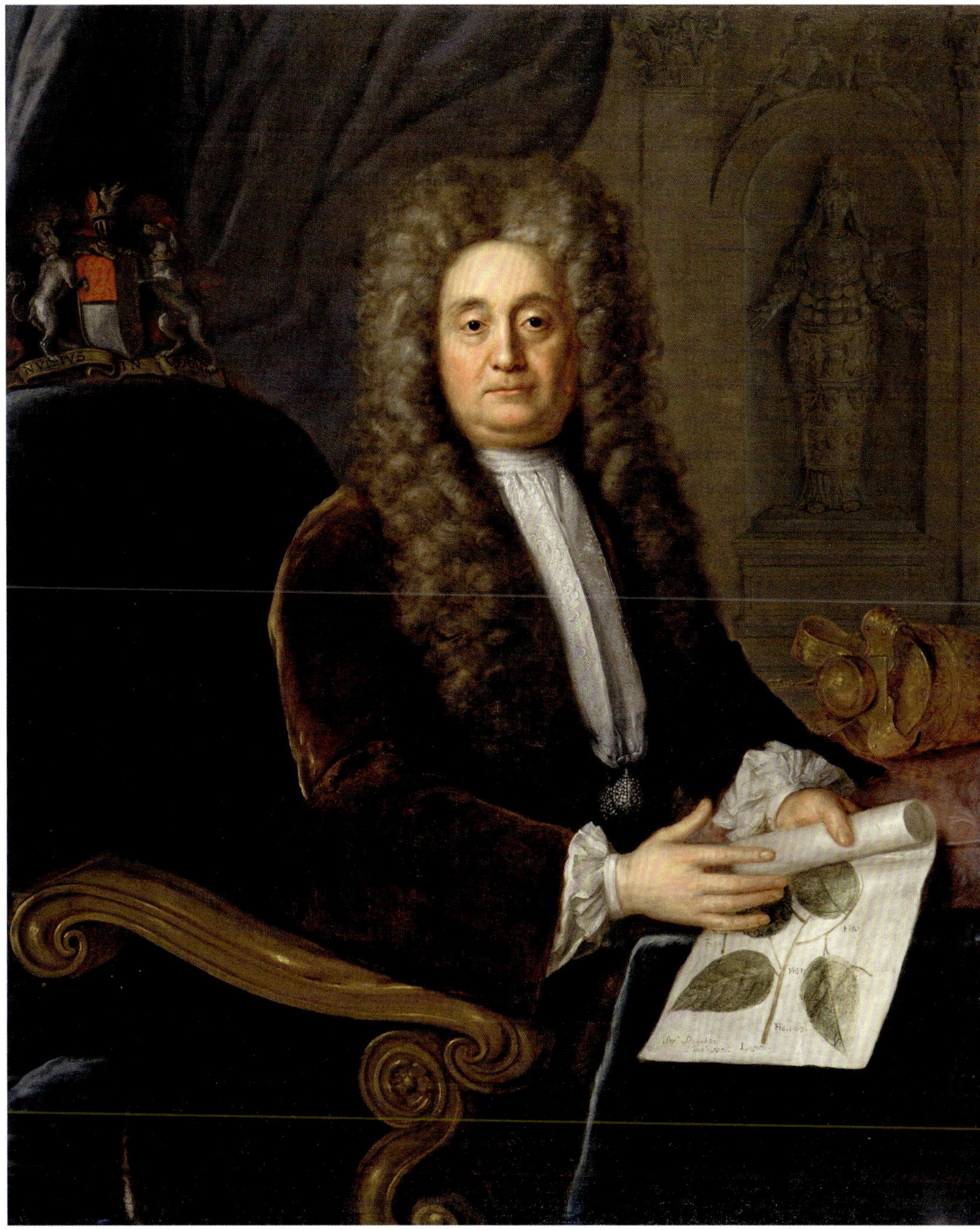

FIG. 16.02 Stephen Slaughter, *Sir Hans Sloane, Bt*, 1736, oil on canvas, 49½ x 39¾ in. (125.7 x 101 cm). National Portrait Gallery, London (cat. 29.14)

dedicated to Caroline as Princess of Wales. Each book within this collection had its own dedicatee, including some of Caroline's Ladies of the Bedchamber. The work made Latin myths available to wider audiences, including those who lacked a classical education. In 1715, the Anglo-Saxon scholar Elizabeth Elstob dedicated her *Rudiments of Grammar for the English-Saxon Tongue* to Caroline, pointing out, somewhat implausibly, that Hanoverians were descended from the original Anglo-Saxons. In 1730, Caroline pensioned the now-impoverished scholar through the aegis of Mary Pendarves, later Mary Delany.[8]

As to directly political conversations, it used to be thought that Sir Robert Walpole was able to extend his hold on ministerial power after the death of his first patron, George I, because of Caroline's influence over her husband behind the scenes. While Caroline undoubtedly enjoyed conversing with Walpole, and he was flattering in his assessment of her diplomatic skills, it is argued that Walpole's hard work in attending on the King when his civil list was being discussed in the House of Commons was also the key, though this is not to deny the effective complementary powers exercised by the couple.

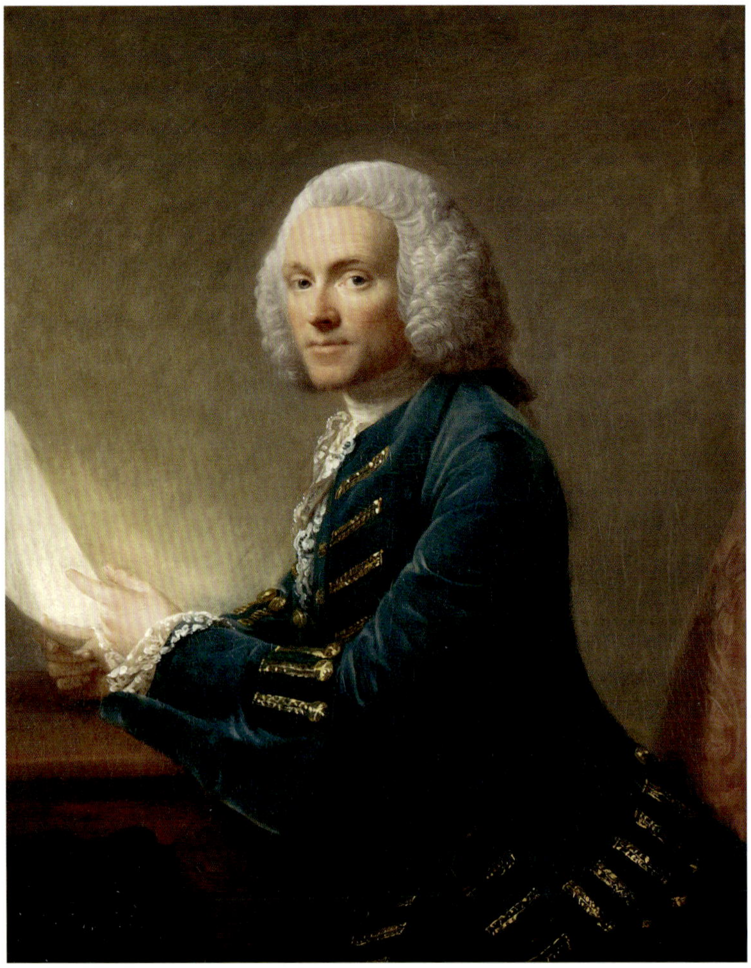

FIG. 16.03 Allan Ramsay, *Dr William Hunter*, 1764–65, oil on canvas, 37⅞ x 29½ in. (96 x 75 cm). Hunterian Museum and Art Gallery, University of Glasgow (cat. 26.09)

Augusta showed her loyalty to Anglicanism in the commissioning of communion plate for her private oratory, even if she was habitually late at worship: she was visible if unpunctual.[9] In 1760 she appointed Stephen Hales as her household chaplain; he was then regarded as the greatest British scientist after Newton (cats. 16.21–16.22). His expertise was in plant physiology, but he was also famous for practical inventions; the air pump he developed would make a significant contribution to the better ventilation of prisons, as his monument at Westminster Abbey, built with funds contributed by the Princess, spells out (see p. 64, fig. 3.05).[10]

As a widow, Augusta continued to develop plans originally conceived by her husband, Frederick, Prince of Wales, for the gardens surrounding the White House at Kew. In this she was assisted and encouraged by the knowledgable amateur botanist and Scottish peer, John Stuart, third Earl of Bute (cats. 16.17–16.18). Another significant botanist in Augusta's circle was John Hill, who catalogued the range of plants there.[11] Bute's visits to her residence attracted rumours of a clandestine love affair, but the real reason for them was that she was worried that her son, the future George III, seemed slow to learn. The personal attention the young Prince received from Bute, his 'dearest friend' and mentor, who became his unofficial tutor, was rewarded with his pupil's increased competence. When George was given his own household he continued to share it with his mother—Bute and Augusta argued he was too young to marry—and Bute was appointed Groom of the Stole, a Royal Household, not a government, post.

However, when George I inherited the throne and Bute was given a *political* role as Secretary of State in 1761, the King was suspected of over-promoting a favourite. This was particularly offensive in a monarchical culture where the king had to share power with his ministers in Parliament. Given a free press, cartoons implicating Bute as the beneficiary of 'Petticoat Power' grew ever more lubricious (see cats. 14.08 and 20.06), but attempts by the government to prosecute John Wilkes, the radical journalist and politician, and the King's and Bute's leading critic, for libel in 1763, and to hinder his taking a seat once elected as a Member of Parliament, raised serious constitutional issues as well as causing riots.

When George married Charlotte in 1761, he stressed to her the importance of not interfering—or even seeming to interfere—in Britain's system of parliamentary rule: that is, she should avoid her mother-in-law's example. Soon Bute backed away from power by resigning, and Augusta concentrated on her botanical studies. After her death when part of her estate was put up for auction, the list of lots for sale revealed her taste for exquisite decorative objects, reflecting the interests of her husband, but also included scientific instruments, books, and a sizeable collection of medals.

Charlotte also liked botany, but did not feel free to pursue it until her mother-in-law had died. She was probably aware of the botanical ideas of Jean-Jacques Rousseau some time before they were published in English translation (cat. 16.26), since on his visit to England in 1766 he became friendly with the great patron of natural history, Margaret Cavendish Bentinck, Duchess of Portland, styling himself her 'herboriste'. The Duchess's home was near Windsor Castle, and there was a lot of informal sociability between the two households. The young King had been persuaded by David Hume and his circle to give Rousseau a pension, despite the fact that Rousseau was Europe's foremost critic of court society and of modernity, and notwithstanding Hume's scepticism was as alien to George and Charlotte as was Rousseau's wayward Deism.[12] It was thought at the time that the Scottish poet and philosopher James Beattie had effectively refuted Hume's philosophy (cat. 16.28). Beattie would be sponsored at court by the bluestocking hostess Mrs Elizabeth Montagu and the devout peer

William Legge, second Earl of Dartmouth, and when he visited London he lodged with John and Anne Hunter, the brother and sister-in-law of the Queen's gynaecologist William Hunter (fig. 16.03). As her physician she employed Sir John Pringle, who coupled the practice of medicine with the editing of Newton's *Principia Mathematica*.

Charlotte shared many other interests with her husband, including astronomy. A list of the 'Philosophical Instruments' owned by the Queen rivals in its length and variety a parallel listing of the scientific collections owned by the King (cat. 16.24). Pringle, as President of the Royal Society, made a gift of a watch to the Queen in recognition of her interest. The court diaries of the writer Frances Burney demonstrate how much the theological purpose of all these scientific studies—to show how science authenticated revealed religion—percolated throughout the court. Other shared interests of Charlotte and George were in the fine arts and architecture: the young couple took lessons in architectural drawing from Sir William Chambers (see cat. 19.12), who had designed many of the buildings at Kew, and Joshua Kirby, a specialist in perspective drawing.[13]

The eighteenth century was an era of both terrestrial and celestial exploration, and Sir Joseph Banks (cat. 16.29), later President of the Royal Society, had travelled on Captain James Cook's first voyage, which also mapped the newly 'discovered' Botany Bay in Australia. He, his wife, and his sister all became personal friends of Charlotte and her daughters.[14]

Despite her care in keeping out of politics, Charlotte was not immune from scandal: Warren Hastings and his wife Marian had given her expensive Indian ivory furniture (see cat. 28.20) and jewels, which implicated her in his trial for misconduct and corruption as the first Governor of Bengal (see cat. 28.21).[15] William Pitt the younger proved an able Prime Minister from 1783, but because his appointment created rumours that the King was trying to increase royal power, Charlotte's rapport with him during the King's illness in 1788–89 prompted James Gillray to caricature her as a power behind the throne.

NOTES

1. These discussions took place in 1703. See Andrew C. Thompson, *George II: King and Elector* (New Haven and London: Yale Univ. to Press, 2011), 27–31.
2 *A Collection of Papers, Which passed between the late learned Mr. Leibnitz, and Dr. Clarke, in the Years 1715 and 1716: Relating to the Principles of Natural Philosophy and Religion* (London: James Knapton, 1717).
3. British Library (BL) Add. MS. 11511, which lists Caroline's books, includes the following titles by Sir Isaac Newton: *Traité d'Optique sur les Réflexions, Réfractions, Inflexions et les Couleurs de la Lumière*, and *Recueil de Diverse Pièces, sur la Philosophie, la Réligion Naturelle, l'Histoire, Les Mathématiques, &c. Par Mrs. [sic] Leibnitz, Clarke, Newton, & autres Autheurs célèbres*. The list also contains the following titles by Samuel Clarke: *A Discourse Concerning the Being and Attributes of God, The Obligations of Natural Religion, and the Truth and Certainty of the Christian Revelation*, and *A Collection of Papers which Pass'd between the Late Learned Mr Leibnitz and Dr Clarke, in the Years 1715 and 1716, Relating to the Principles of Natural Philosophy and Religion*. William Wollaston, a disciple of Clarke, was author of *The Religion of Nature Delineated*. Henriette-Charlotte von Pöllnitz reported to Leibniz that Caroline read Locke's work, such as his *Essay Concerning Human Understanding*. Caroline's library also contained works by Boyle.
4. Desagulier performed an experiment in front of George I and Caroline in 1717, to demonstrate that in an evacuated cylinder, a guinea, a piece of paper, and a feather would fall at the same velocity. See J. T. Desaguliers, 'An Account of an Experiment to Prove an Interspers'd Vacuum; Or to Shew that all Places are not Equally Full', *Philosophical Transactions of the Royal Society* 30 (1717–19): 717–20.
5. *The Family Memoirs of the Rev. William Stukeley M.D. and the Antiquarian and other Correspondence of William Stukeley, Roger & Samuel Gale etc*, ed. W. C. Lukis, 3 vols. (Durham: Surtees Society, 1882–87), 73: 76, 80, 73: 73.
6. John Gay, *John Gay: Dramatic Works*, ed. John Fuller, 2 vols. (Oxford: Oxford Univ. Press, 1983), I: 344. For a discussion of Caroline's promotion of Irish textiles procured with the help of Swift, see Jonathan Swift, *The Correspondence of Jonathan Swift D. D.*, ed. F. Elrington Ball, 6 vols. (London: G. Bell and Sons, 1910–14), 3: 354, 357, 360, 364, 371, 374; and Alexander Pope, *The Works of Alexander Pope*, Vol. 8: *Correspondence III*, ed. J. W. Croker, intro. and notes by W. Elwin and W. J. Courthope (London: John Murray, 1872), 90.
7. Letter from Alexander Pope to Lord Bathurst, 13 Sept. 1719, quoted in *Works of Alexander Pope*, ed. Croker, 8: 327–29. For information about the relationship between the development of Caroline's Garden at Richmond and Pope's garden at Twickenham, see Peter Martin, *Pursuing Innocent Pleasures: The Gardening World of Alexander Pope* (Hamden, CT, and Oxford: Archon Books, 1984), and Joanna Marschner, *Queen Caroline: Cultural Politics at the Early Eighteenth-Century Court* (New Haven and London: Yale Univ. Press, 2014), 50–1.
8. Lady Llanover, ed., *The Autobiography and Correspondence of Mary Granville, Mrs Delany*, 6 vols. (London: Richard Bentley, 1860–61), 1: 264.
9. John L. Bullion, '"To play what game she pleased without observation": Princess Augusta and the Political Drama of Succession, 1736–1756', in *Queenship in Britain, 1660–1837: Royal Patronage, Court Culture and Dynastic Politics*, ed. Clarissa Campbell Orr (Manchester: Manchester Univ. Press, 2002), 207–35, 217.
10. Stephen Hales's most significant publications include *Statical Essays, containing Vegetable Staticks* (London: W. and J. Innys, 1728), and *A description of ventilators: whereby great quantities of fresh air may with ease be conveyed into mines, gaols, hospitals* (London: W. Innys, R. Manby, and T. Woodward, 1743).
11. John Hill's most significant publications include *The British Herbal: an history of plants and trees, natives of Britain, cultivated for use, or raised for beauty* (London: T. Osborne, J. Shipton and others, 1756).
12. R. A. Leigh, 'Rousseau's English Pension', in *Studies in Eighteenth Century French Literature presented to Robert Niklaus*, ed. John Howard Fox, Mark H. Waddicor, and Derek A. Watts (Exeter: Exeter Univ. Press, 1975), 109–22.
13. The royal family visited annually after the Royal Academy was founded in 1768.
14. John Gascoigne, *Joseph Banks and the English Enlightenment: Useful Knowledge and Polite Culture* (Cambridge: Cambridge Univ. Press, 1994).
15. P. J. Marshall, *The Impeachment of Warren Hastings* (Oxford: Oxford Univ. Press, 1965).

CAROLINE'S SALON

16.04

JOHN VANDERBANK THE YOUNGER (1694–1739)
Isaac Newton
1726
Oil on canvas
50 x 40 in. (127 x 101.6 cm)
Royal Society, London, RS.9255

In 1716, after inviting Sir Isaac Newton to her Drawing Room, Caroline wrote to Leibniz reporting, 'Tomorrow we see the experiments with the colours, and the one I have seen to prove the vacuum has nearly converted me.' She was conscious that Newton was so much venerated by his compatriots that the regime would benefit from his being drawn into the court circle. He would later provide tuition in mathematics to Prince William Augustus. As well as commissioning a portrait bust of the philosopher from Giovanni Battista Guelphi, she also commissioned or acquired a painted portrait probably by Charles Jervas, after an earlier portrait made in 1689 by Sir Godfrey Kneller, which took its place in the group of images of contemporary heroes she marshalled at Windsor Castle. Isaac Newton was elected a Fellow of the Royal Society in 1672; he served as its President from 1703 to 1727. Martin Folkes, Newton's successor as President presented a picture of himself by William Hogarth and one of Francis Bacon to the Society, in addition to this portrait of Newton by John Vanderbank.

16.05

CHARLES JERVAS (c. 1675–1739)
Samuel Clarke
c. 1729–30
Oil on canvas
50½ x 40½ in. (128.2 x 103 cm)
Royal Collection Trust, RCIN 403011

After the death of John Locke, the Reverend Doctor Samuel Clarke was seen as one of England's foremost theologians and metaphysicians. He held the living of St James's, Piccadilly in London, and was appointed Chaplain in Ordinary to Queen Anne. In 1714, knowing of Caroline's intellectual interests by repute, he hastened to St James's Palace to present her with copies of his books. Caroline was impressed and he became a frequent member of her salon. In the 1720s Caroline began to assemble painted images of her contemporary heroes, including Dr Clarke. She hung them in her Dressing Room at Windsor Castle. On Clarke's death Caroline commissioned Alexander Pope to write his epitaph, and this was inscribed on the painting.

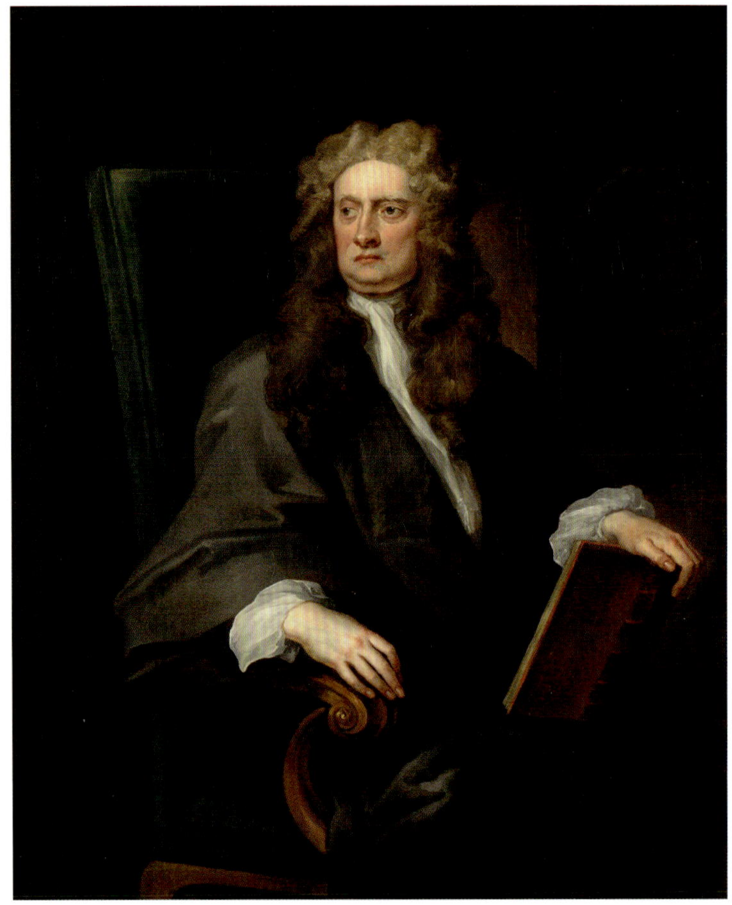

16.04

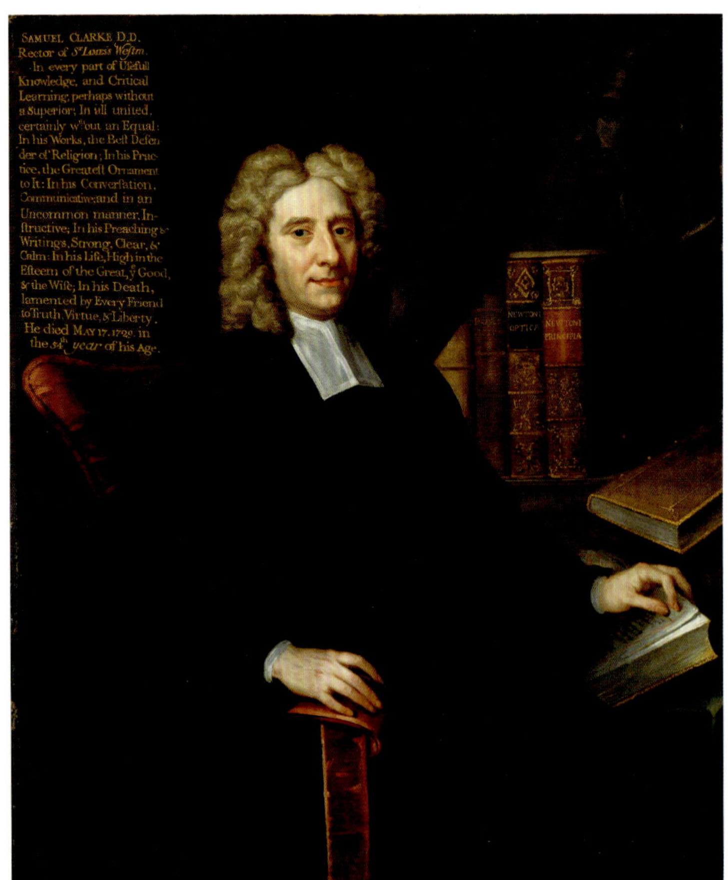

16.05

* = YCBA only
† = HRP only

16.06*

SAMUEL CLARKE (1675–1729), with GOTTFRIED WILHELM LEIBNIZ (1646–1716)
A Collection of Papers, Which passed between the late learned Mr. Leibnitz and Dr. Clarke in the Years 1715 and 1716, Relating to the Principles of Natural Philosophy and Religion, With an Appendix, to which are added, Letters to Dr. Clarke concerning liberty and necessity . . . Also, Remarks upon a book, entituled, A Philosophical Enquiry Concerning Human Liberty
1717
7½ in. (19 cm)
Printed for James Knapton, London
Beinecke Rare Book & Manuscript Library, Yale University, 1987 415
Exhibited: Title page (illustrated)

Between 1715 and 1716 Caroline presided over a debate between Sir Isaac Newton and Gottfried Wilhelm Leibniz concerning Newton's work on calculus, or as he termed it, 'the motion of fluxions and fluent'. At the heart of the exchange was the question of who should be given credit for its discovery. Newton claimed his work had commenced in 1666, but he delayed publishing his proofs in their final form until 1704. Leibniz, though he had had sight of some of Newton's work, reached his conclusions by a different method, and published his work in good faith in 1684. Caroline had first met Leibniz when she resided at Lietzenburg as the ward of Elector Frederick III, later first King of Prussia. He was also a member of the Dowager Electress Sophia of Hanover's intellectual circle, of which Caroline was later a part. The debate was conducted through letters which passed between Leibniz and Dr Samuel Clarke.

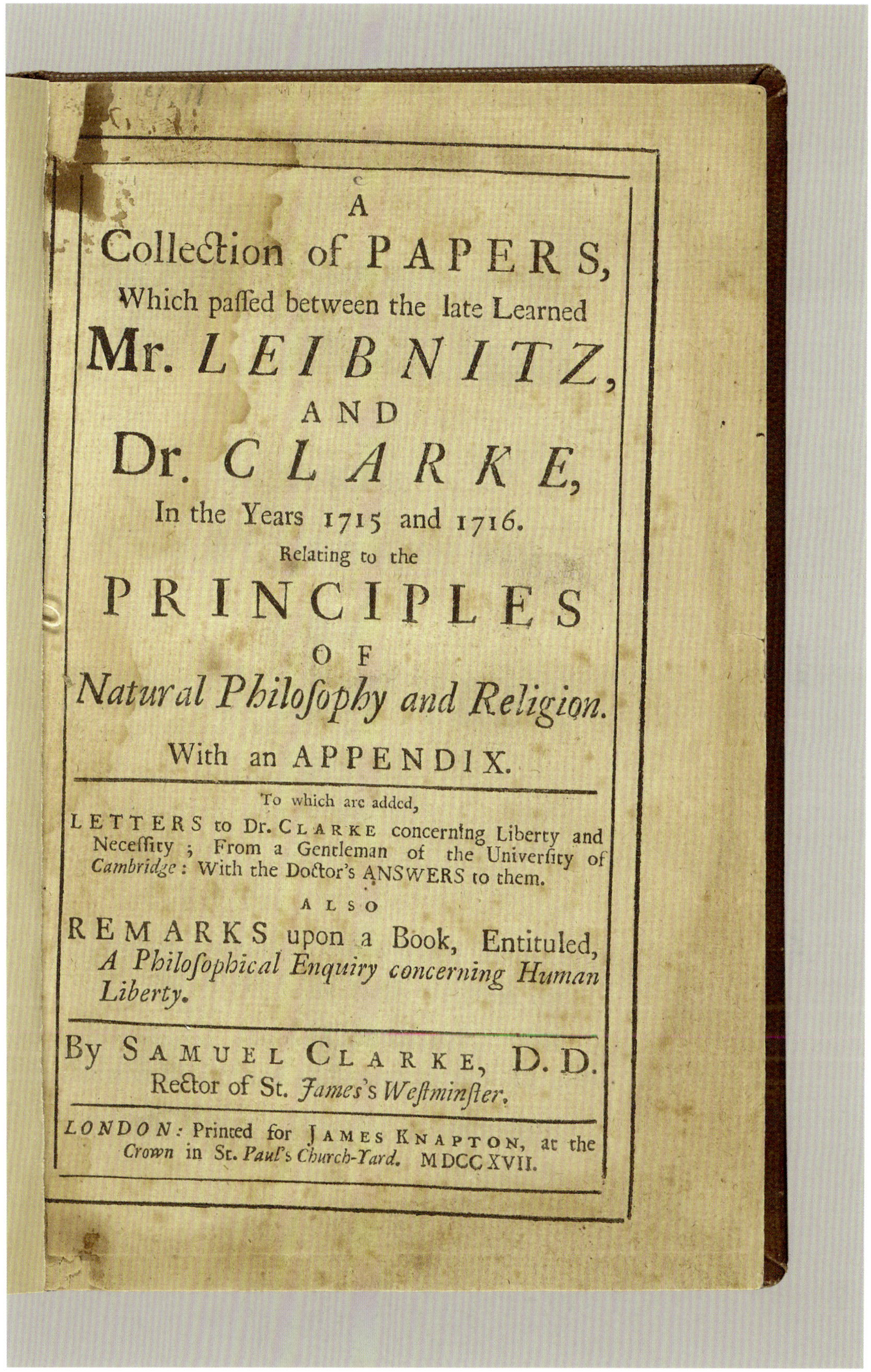
A
Collection of PAPERS,
Which paſſed between the late Learned
Mr. *LEIBNITZ*,
AND
Dr. *CLARKE*,
In the Years 1715 and 1716.
Relating to the
PRINCIPLES
OF
Natural Philoſophy and Religion.
With an APPENDIX.

To which are added,
LETTERS to Dr. CLARKE concerning Liberty and Neceſſity; From a Gentleman of the Univerſity of *Cambridge*: With the Doctor's ANSWERS to them.
ALSO
REMARKS upon a Book, Entituled, *A Philoſophical Enquiry concerning Human Liberty.*

By SAMUEL CLARKE, D.D.
Rector of St. *James*'s *Weſtminſter*.

LONDON: Printed for JAMES KNAPTON, at the *Crown* in St. *Paul*'s *Church-Yard*. MDCCXVII.

16.06

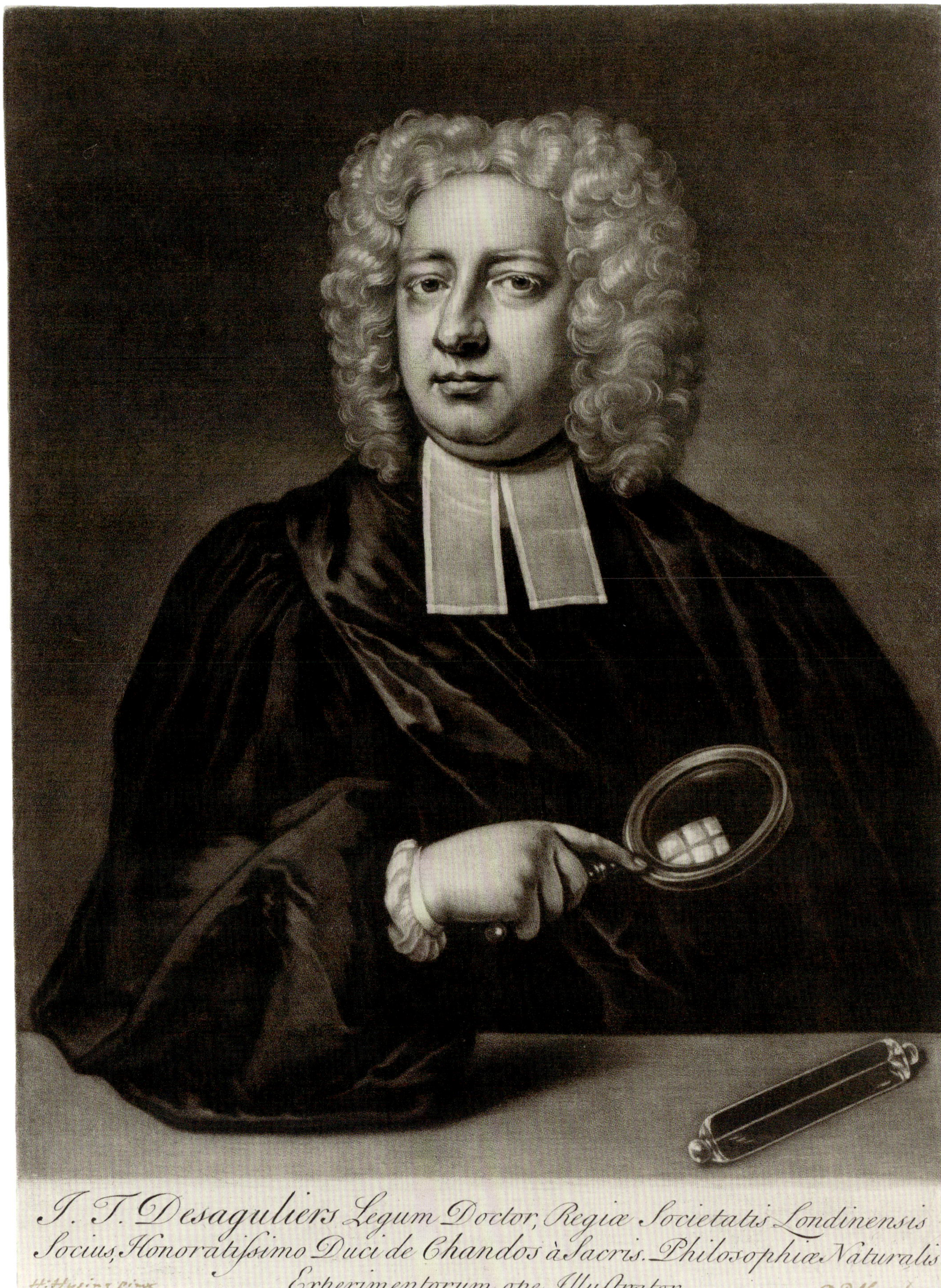
J. T. Desaguliers Legum Doctor, Regiæ Societatis Londinensis
Socius, Honoratissimo Duci de Chandos à Sacris. Philosophiæ Naturalis
Experimentorum ope Illustrator.

16.07

16.07*

PETER PELHAM (c. 1697–1751), engraver, after **HANS HYSING** (1678–1753)
J. T. Desaguliers
1725
Mezzotint
13⅝ x 9⅞ in. (34.5 x 25.2 cm)
Published by John Bowles (1701?–1779), London
British Museum, Q,1.63

John Theophilus Desaguliers was an early promoter of Isaac Newton's natural philosophy. In 1714 Sir Isaac Newton, President of the Royal Society, invited Desaguliers to replace Francis Hauksbee as demonstrator at the Society's weekly meetings; he was soon thereafter made a Fellow of the Royal Society. In 1717 Desaguliers lodged at Hampton Court and gave lectures in French on scientific subjects to King George I. One experiment, at which Caroline was also present, demonstrated that in an evacuated cylinder a guinea coin, a piece of paper, and a feather would fall at exactly the same velocity. He continued to lecture and conduct experiments at court for George II, Caroline, and members of Caroline's household. In *The Newtonian System of the World, the Best Model of Government: An Allegorical Poem*, written to celebrate the accession in 1727, he was complimentary about the astronomical competence of the royal family, but especially of Caroline.

16.08*

JOHN THEOPHILUS DESAGULIERS (1683–1744)
Lectures of Experimental Philosophy. Wherin the principles of mechanicks, hydrostaticks, and opticks, are demonstrated and explained . . . by a great number of curious experiments . . . Illustrated with . . . copper plates . . . To which is added, a description of Mr. Rowley's machine, called the orrery . . .
1719
Printed for W. Mears, London
Medical History Library, Yale University, 17th cent Boyle 366a
Exhibited: Unknown engraver after unknown artist, 'Experimental Demonstrations', plate 6

By the time of his death John Theophilus Desaguliers had given over 140 courses of some twenty lectures each on mechanics, hydrostatics, pneumatics, optics, and astronomy. He kept his lectures up to date, published notes for his audience, and designed his own apparatus, including a renowned planetarium to demonstrate the solar system, and a machine to explain tidal motion. He was invited to give lectures to George II, Caroline, and their children at court.

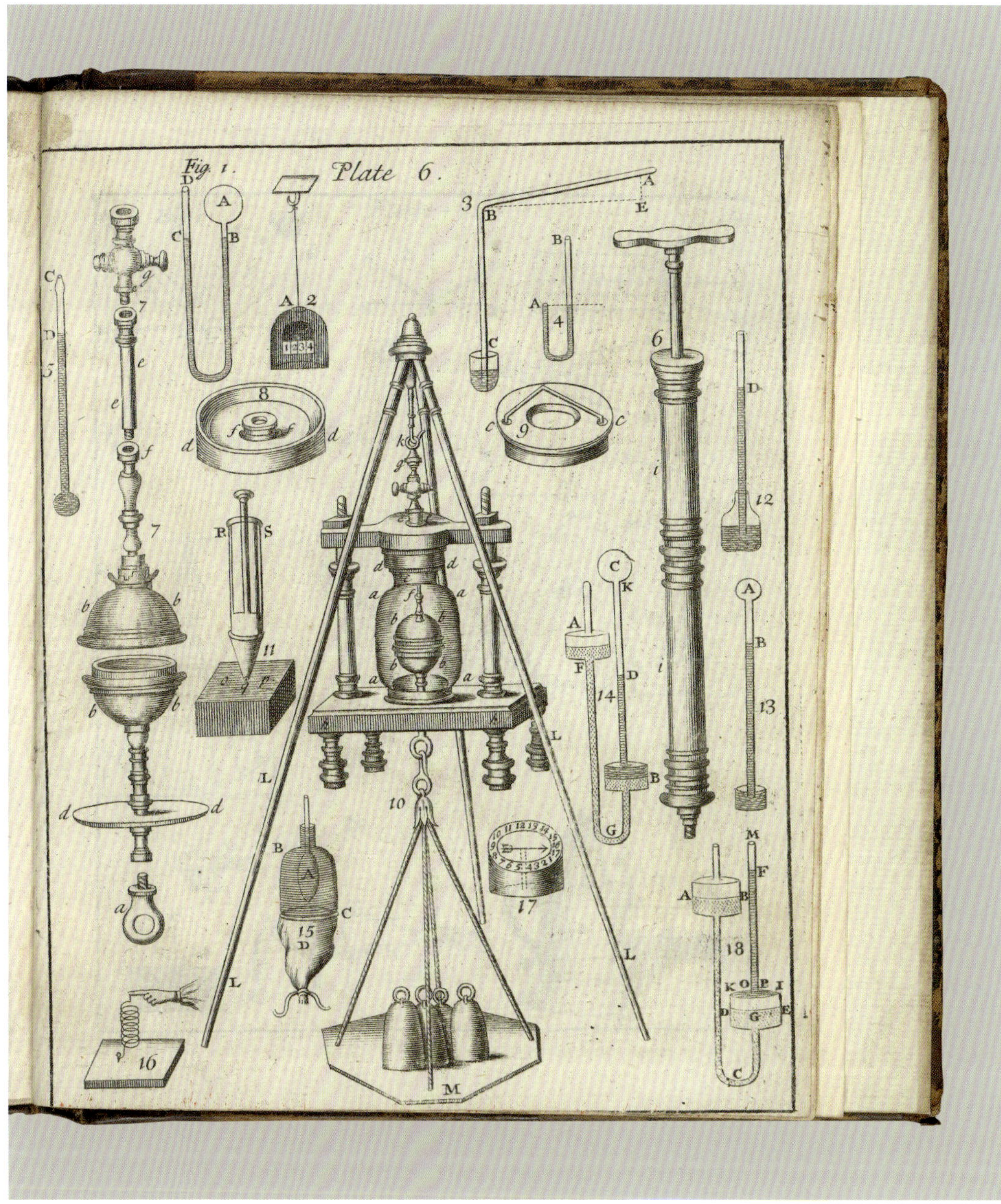

16.08

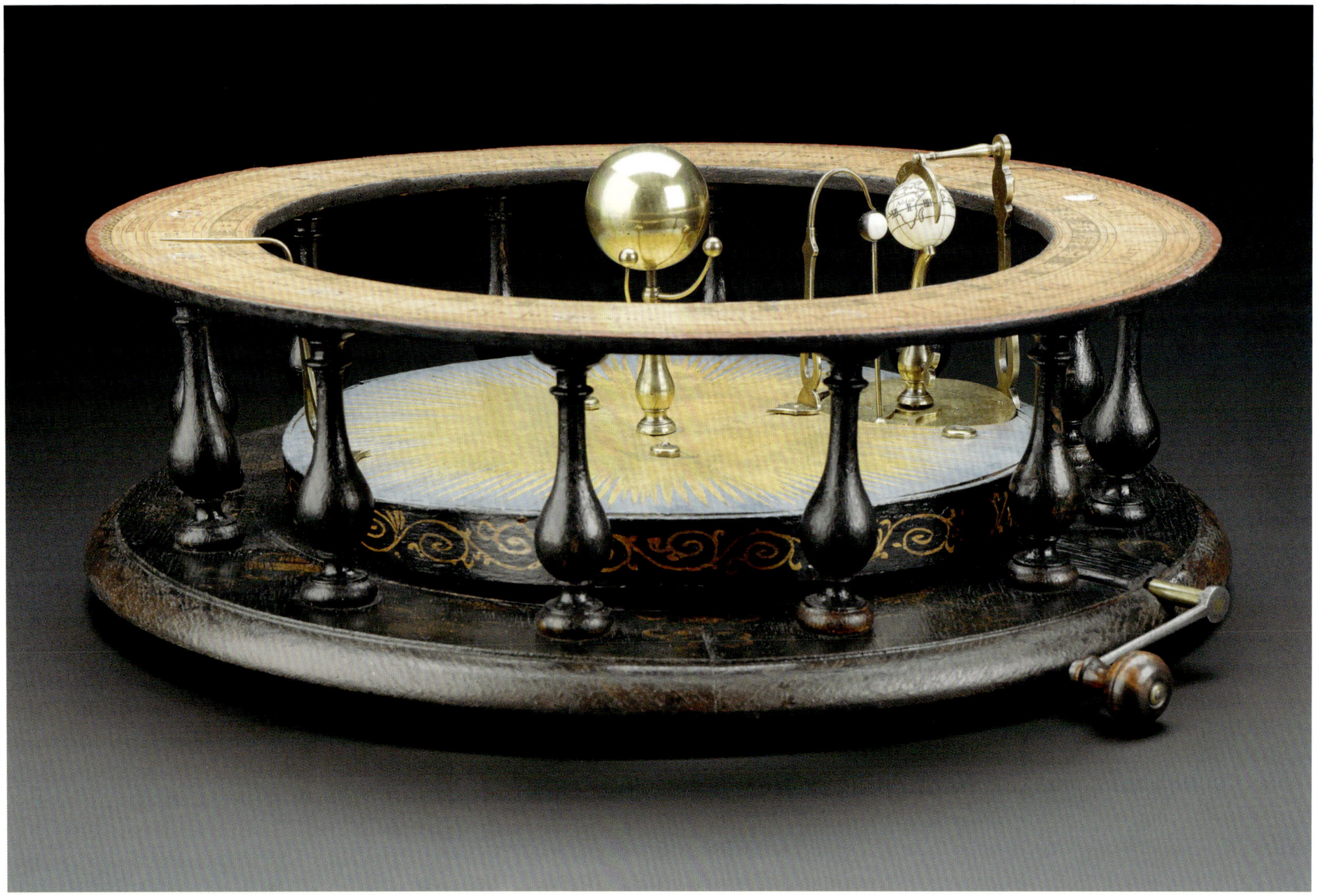

16.09

16.09

THOMAS WRIGHT (1711–1786), maker
Orrery
c. 1740
Wood (oak), brass gilt, and ivory
6⅞ x 19½ x 17½ in. (17.5 x 49.5 x 44.5 cm)
Science Museum, London, 1927 1414

In 1733, the instrument maker Thomas Wright was paid £336 'for making new all the Machinery and Wheel-Work to perform the Motions of all the planets & Co to the Great Orrery in Her Majesty's Gallery at Kensington'. Wright may have in fact reconditioned and modified an orrery made earlier by John Rowley. The dial plate is in the form of concentric rings, painted blue. Each carries a planet which moves on rollers when the handle is turned. The armillary hemisphere rests on the calendar ring which is engraved with the signs of the zodiac. The calendar ring is supported on short pillars, which are mounted on gilded horses' heads. By contrast to the grand production supplied to Caroline, this small orrery, also made by Thomas Wright, was designed to appeal to the taste and pockets of 'Ladies and Gentlemen rather than noblemen or Princes'. It is operated like Caroline's Great Orrery by turning a crank handle, but lacks the elaborate super-structure and costly ornaments.

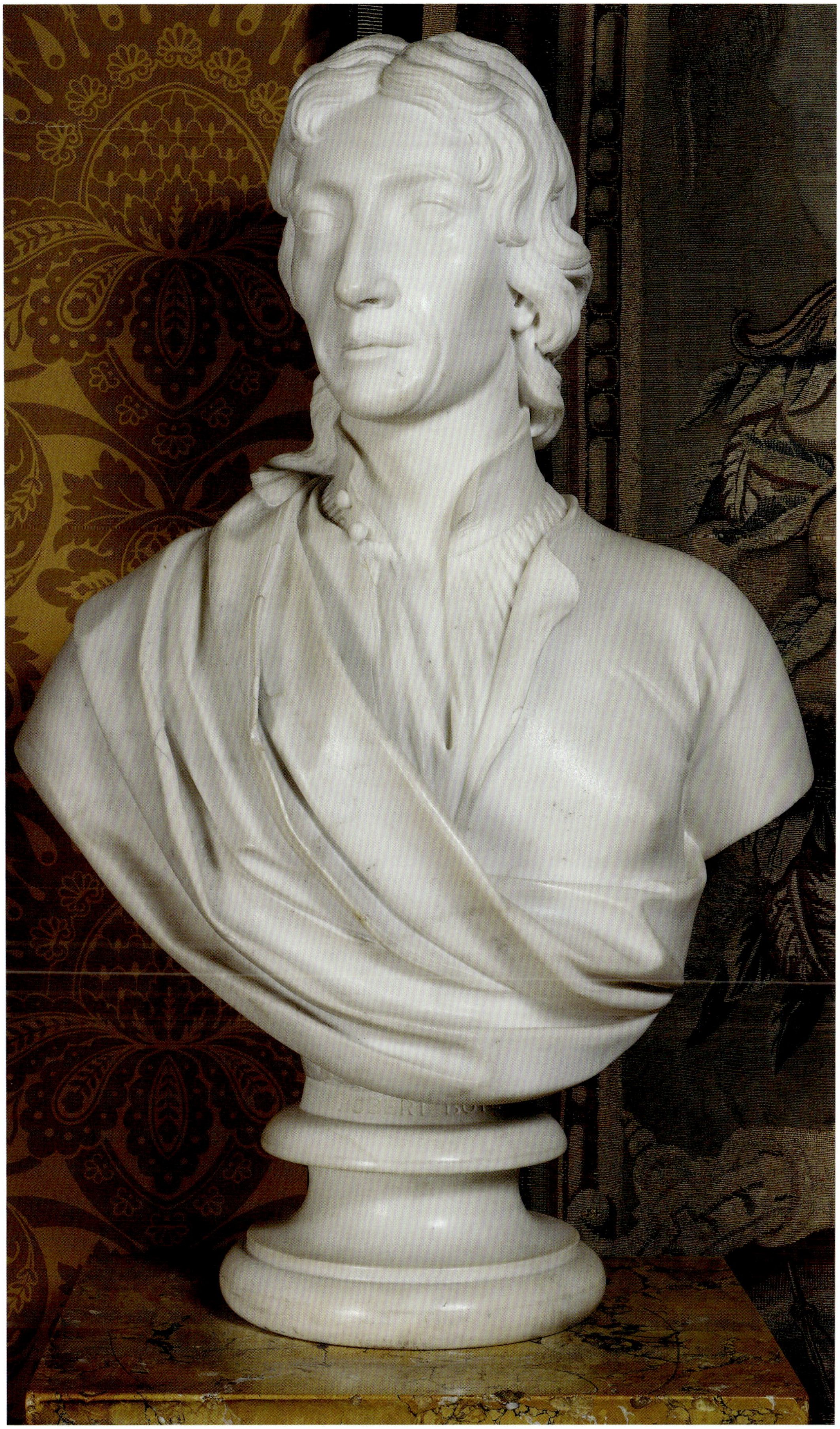

16.10

CAROLINE'S PANTHEON

16.10

GIOVANNI BATTISTA GUELPHI
(1688–1736)
Sir Robert Boyle
c. 1733
Marble
29½ x 19¼ x 9½ in. (75 x 49 x 24 cm)
Royal Collection Trust, RCIN 1393

In 1731 there is the first reference to a set of portrait busts being commissioned by Caroline, from the Italian sculptor Giovanni Battista Guelphi, to celebrate contemporary natural philosophers and theologians. The choice of artist is likely to have been influenced by Robert Boyle, third Earl of Burlington, who had brought Guelphi from Rome to London. There were eventually five busts in the series, and they would be installed in the Hermitage, a folly temple built to a design by William Kent in the garden of Richmond Lodge. As chemist, physicist, and philosopher, Boyle embodied the marriage between natural philosophy and Newtonian science, and presided over the pantheon. His bust was set on a plinth in an apse to anchor the arrangement, with a corona of golden rays arranged behind him.

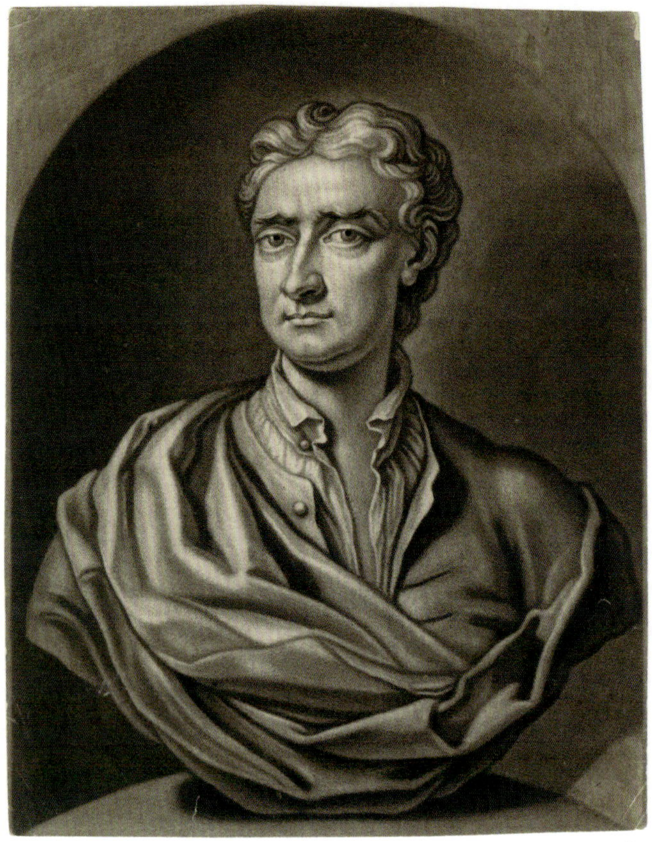

16.11

The set of portrait busts of contemporary heroes commissioned by Caroline from Giovanni Battista Guelphi as decoration for the Hermitage in Richmond Gardens included the philosphers Sir Isaac Newton and John Locke, the theologians Dr Samuel Clarke and William Wollaston, as well as Sir Robert Boyle. Caroline had drawn Newton, Clarke, and Wollaston into her salon discussions. Locke and Boyle had died before her arrival in London, but were still seen as lions of the age. The sculptural series was much admired. A poetry competition was set up by the *Gentleman's Magazine* in 1732 in its honour. The series was recorded in a set of engravings made by John Faber. For Caroline it brought repect; she was commended for taking 'statuary into her protection'. It was also noted that 'her own Leibniz is not allowed a place there'.

16.11*

JOHN FABER THE YOUNGER (c. 1695–1756), engraver, after GIOVANNI BATTISTA GUELPHI (1688–1736), sculptor
Sir Isaac Newton
c. 1736
Mezzotint on paper
12¼ x 9½ in. (31.2 x 24.2 cm)
National Portrait Gallery, London, NPG D3726

16.12*

JOHN FABER THE YOUNGER (c. 1695–1756), engraver, after GIOVANNI BATTISTA GUELPHI (1688–1736), sculptor
Mr William Wollaston
c. 1736
Mezzotint on paper
14 x 10 in. (35.6 x 25.4 cm)
Published by Thomas Bowles (1689/90?–1767) and John Bowles (1701?–1779), London
Royal Collection Trust, RCIN 664136

16.13*

JOHN FABER THE YOUNGER (c. 1695–1756), engraver, after GIOVANNI BATTISTA GUELPHI (1688–1736), sculptor
John Locke
c. 1736
Mezzotint on paper
14 x 10 in. (35.6 x 25.4 cm)
Published by Thomas Bowles (1689/90?–1767) and John Bowles (1701?–1779), London
Royal Collection Trust, RCIN 658185

16.14*

JOHN FABER THE YOUNGER (c. 1695–1756), engraver, after GIOVANNI BATTISTA GUELPHI (1688–1736), sculptor
Samuel Clarke D.D.
c. 1736
Mezzotint on paper
5½ x 64¼ in. (35.3 x 25.3 cm)
Published by Thomas Bowles (1689/90?–1767) and John Bowles (1701?–1779), London
Royal Collection Trust, RCIN 652618

16.15*

JOHN SIMON (c. 1675–1751), engraver and publisher, after HANS HYSING (1678–1753)
The Right honble Sir Robert Walpole
1727
Mezzotint on paper
12⅞ x 9¾ in. (32.8 x 25 cm)
Published by John Simon
Tipped into *A Description of the Villa of Horace Walpole* (extra-illustrated copy)
Strawberry Hill: Printed by Thomas Kirgate, 1784
The Lewis Walpole Library, Yale University, Folio 33 30, Copy 11

Robert Walpole, first Earl of Orford, a Whig from the gentry class, rose to become de facto the first Prime Minister of Great Britain. After the accession of George I, he was appointed to a series of evermore influential government offices. The resignation of Lord Sunderland and the death of Lord Stanhope in 1721 left Walpole as the most important figure in the administration, and he was appointed First Lord of the Treasury, Chancellor of the Exchequer, and Leader of the House of Commons. When George I died, Walpole's position was threatened, but, on Caroline's advice, George II agreed to keep him in office. Walpole and Caroline appear to have become great political allies; it seems that the First Minister took her advice on the managing and influencing of her volatile husband. He would claim, 'Madam, without you I can do nothing . . . . You, Madam are the sole mover of this Court.'

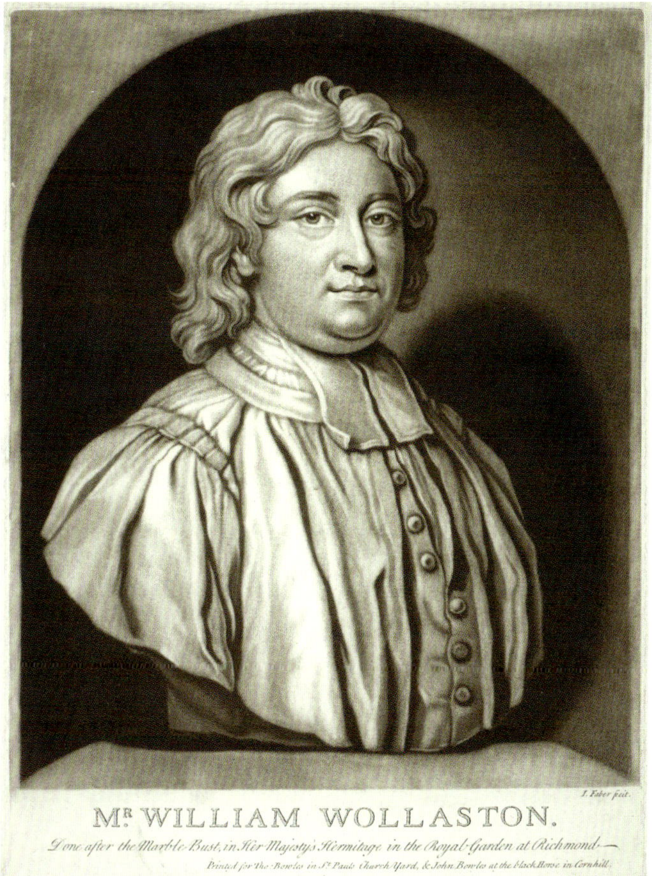

16.12

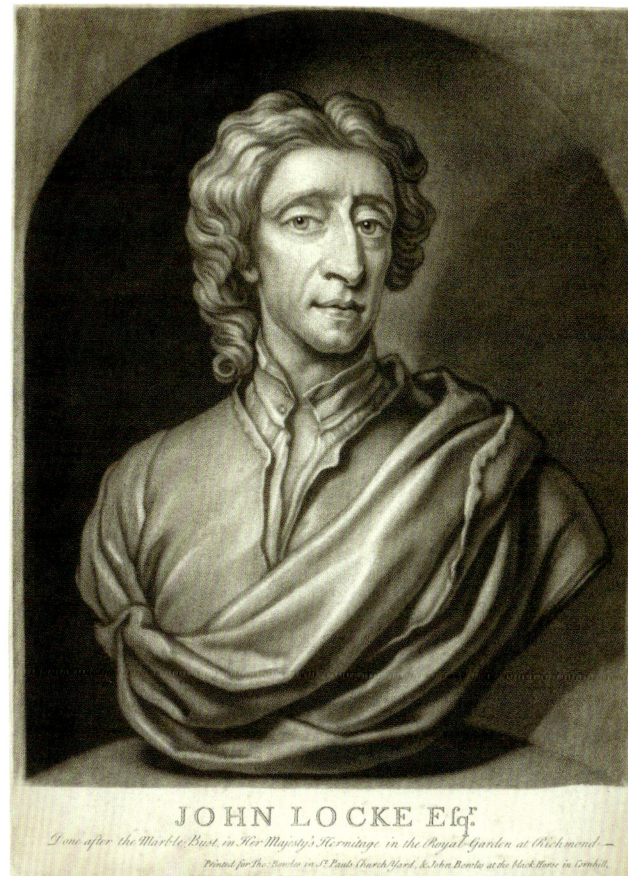

16.13

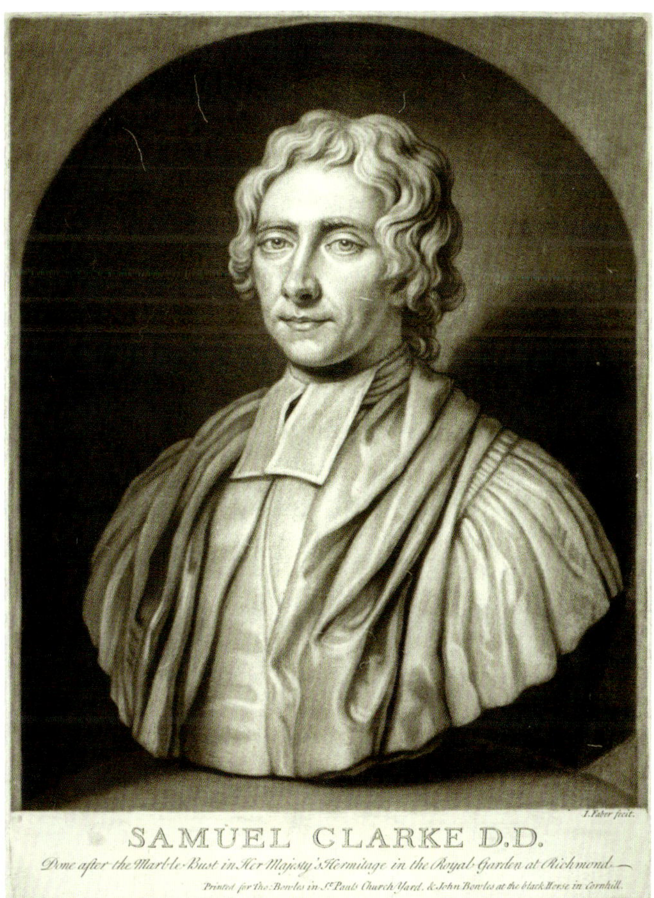

16.14

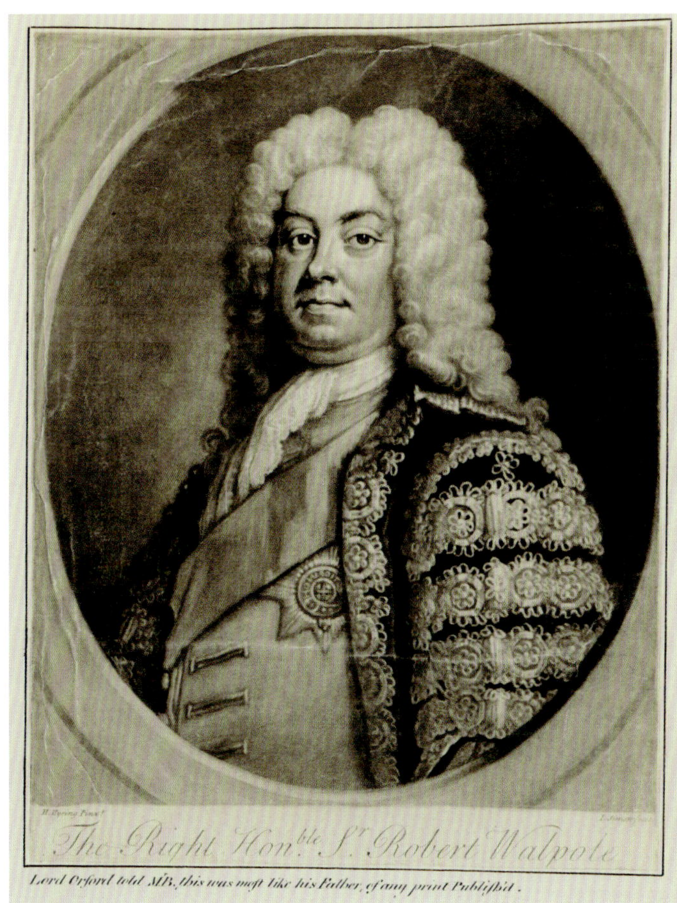

16.15

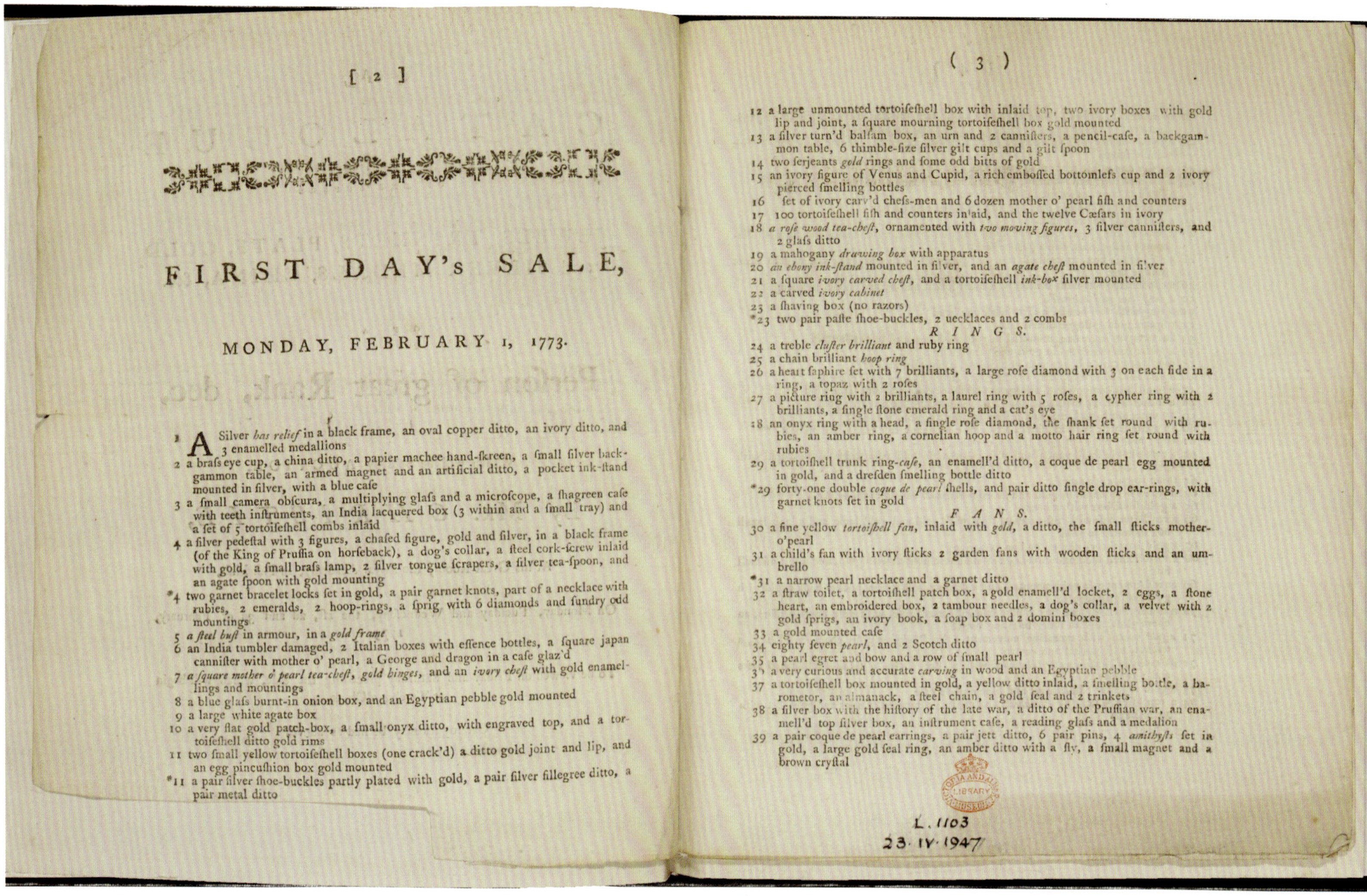

[2]

FIRST DAY's SALE,

MONDAY, FEBRUARY 1, 1773.

1 A Silver *bas relief* in a black frame, an oval copper ditto, an ivory ditto, and 3 enamelled medallions
2 a braſs eye cup, a china ditto, a papier machee hand-ſkreen, a ſmall ſilver back-gammon table, an armed magnet and an artificial ditto, a pocket ink-ſtand mounted in ſilver, with a blue caſe
3 a ſmall camera obſcura, a multiplying glaſs and a microſcope, a ſhagreen caſe with teeth inſtruments, an India lacquered box (3 within and a ſmall tray) and a ſet of 5 tortoiſeſhell combs inlaid
4 a ſilver pedeſtal with 3 figures, a chaſed figure, gold and ſilver, in a black frame (of the King of Pruſſia on horſeback), a dog's collar, a ſteel cork-ſcrew inlaid with gold, a ſmall braſs lamp, 2 ſilver tongue ſcrapers, a ſilver tea-ſpoon, and an agate ſpoon with gold mounting
*4 two garnet bracelet locks ſet in gold, a pair garnet knots, part of a necklace with rubies, 2 emeralds, 2 hoop-rings, a ſprig with 6 diamonds and ſundry odd mountings
5 *a ſteel buſt* in armour, in a *gold frame*
6 an India tumbler damaged, 2 Italian boxes with eſſence bottles, a ſquare japan canniſter with mother o' pearl, a George and dragon in a caſe glaz'd
7 *a ſquare mother o' pearl tea-cheſt, gold hinges,* and an *ivory cheſt* with gold enamellings and mountings
8 a blue glaſs burnt-in onion box, and an Egyptian pebble gold mounted
9 a large white agate box
10 a very flat gold patch-box, a ſmall onyx ditto, with engraved top, and a tortoiſeſhell ditto gold rims
11 two ſmall yellow tortoiſeſhell boxes (one crack'd) a ditto gold joint and lip, and an egg pincuſhion box gold mounted
*11 a pair ſilver ſhoe-buckles partly plated with gold, a pair ſilver fillegree ditto, a pair metal ditto

(3)

12 a large unmounted tortoiſeſhell box with inlaid top, two ivory boxes with gold lip and joint, a ſquare mourning tortoiſeſhell box gold mounted
13 a ſilver turn'd balſam box, an urn and 2 canniſters, a pencil-caſe, a backgammon table, 6 thimble-ſize ſilver gilt cups and a gilt ſpoon
14 two ſerjeants *gold* rings and ſome odd bitts of gold
15 an ivory figure of Venus and Cupid, a rich emboſſed bottomleſs cup and 2 ivory pierced ſmelling bottles
16 ſet of ivory carv'd cheſs-men and 6 dozen mother o' pearl fiſh and counters
17 100 tortoiſeſhell fiſh and counters inlaid, and the twelve Cæſars in ivory
18 *a roſe wood tea-cheſt*, ornamented with *two moving figures*, 3 ſilver canniſters, and 2 glaſs ditto
19 a mahogany *drawing box* with apparatus
20 *an ebony ink-ſtand* mounted in ſilver, and an *agate cheſt* mounted in ſilver
21 a ſquare *ivory carved cheſt*, and a tortoiſeſhell *ink-box* ſilver mounted
22 a carved *ivory cabinet*
23 a ſhaving box (no razors)
*23 two pair paſte ſhoe-buckles, 2 uecklaces and 2 combs

R I N G S.

24 a treble *cluſter brilliant* and ruby ring
25 a chain brilliant *hoop ring*
26 a heart ſaphire ſet with 7 brilliants, a large roſe diamond with 3 on each ſide in a ring, a topaz with 2 roſes
27 a picture ring with 2 brilliants, a laurel ring with 5 roſes, a cypher ring with 2 brilliants, a ſingle ſtone emerald ring and a cat's eye
28 an onyx ring with a head, a ſingle roſe diamond, the ſhank ſet round with rubies, an amber ring, a cornelian hoop and a motto hair ring ſet round with rubies
29 a tortoiſhell trunk ring-*caſe*, an enamell'd ditto, a coque de pearl egg mounted in gold, and a dreſden ſmelling bottle ditto
*29 forty-one double *coque de pearl* ſhells, and pair ditto ſingle drop ear-rings, with garnet knots ſet in gold

F A N S.

30 a fine yellow *tortoiſhell fan*, inlaid with *gold*, a ditto, the ſmall ſticks mother-o'pearl
31 a child's fan with ivory ſticks 2 garden fans with wooden ſticks and an umbrello
*31 a narrow pearl necklace and a garnet ditto
32 a ſtraw toilet, a tortoiſhell patch box, a gold enamell'd locket, 2 eggs, a ſtone heart, an embroidered box, 2 tambour needles, a dog's collar, a velvet with 2 gold ſprigs, an ivory book, a ſoap box and 2 domini boxes
33 a gold mounted caſe
34 eighty ſeven *pearl*, and 2 Scotch ditto
35 a pearl egret and bow and a row of ſmall pearl
36 a very curious and accurate *carving* in wood and an Egyptian pebble
37 a tortoiſeſhell box mounted in gold, a yellow ditto inlaid, a ſmelling bottle, a barometor, an almanack, a ſteel chain, a gold ſeal and 2 trinkets
38 a ſilver box with the hiſtory of the late war, a ditto of the Pruſſian war, an enamell'd top ſilver box, an inſtrument caſe, a reading glaſs and a medalion
39 a pair coque de pearl earrings, a pair jett ditto, 6 pair pins, 4 *amithyſts* ſet in gold, a large gold ſeal ring, an amber ditto with a fly, a ſmall magnet and a brown cryſtal

LIBRARY
L. 1103
23 · IV · 1947

16.16

AUGUSTA AND HER ADVISERS

16.16

JAMES CHRISTIE (1730–1803)
A Catalogue of the Superb Jewels, Trinkets, Plate, Gold and Silver Medals, China, &c. of a Person of Great Rank
1773
Inscribed in pencil: 'Princess of Wales.—Widow of Frederick Lewis Prince of Wales & Mother of George III'
8¾ in. (22.5 cm)
Christie's, London
Victoria and Albert Museum, National Art Library, 23.XX.17730201
Exhibited: pp. 2–3 (illustrated)

Following the death of Augusta, Princess Dowager of Wales, in 1772, an auction was held of her effects at Christie's auction rooms in Pall Mall over three days in February 1773. The catalogue runs to many pages, and serves to show the breadth of Augusta's interests. As well as many lots of plate and ceramics are long lists of jewellery, clocks, and watches. There are boxes and small-scale statuary made from precious metals, tortoiseshell, and ivory. Some of the trinkets are noted as being Indian. There are microscopes and camera obscuras and 'a mahogany drawing box with apparatus'. Perhaps the most unusual lot was 'Queen Elizabeth coronation gloves'.

16.17*

ALLAN RAMSAY (1713–1784)
John Stuart, 3rd Earl of Bute
1758
Oil on canvas
94⅛ x 57⅞ in. (239 x 147 cm)
National Trust for Scotland, 65.468

16.18

WILLIAM WYNNE RYLAND (1732–83), engraver, after ALLAN RAMSAY (1713–84)
John, 3rd Earl of Bute, first Lord of the Treasury
1763
Engraving with etching
23⅝ x 15⅝ in. (60 x 39.6 cm)
Published by W. W. Ryland, London
Royal Collection Trust, RCIN 662399

When Frederick, Prince of Wales, died unexpectedly in 1751, Augusta had swiftly to renegotiate her position with her father-in-law, George II. She was determined to retain charge of the education of her eldest son, George. Despite achieving success in this, the prince failed to flourish academically or exhibit the values or virtues she and her late husband had envisaged for the next 'Patriot King'. To remedy this she isolated him from court society, and discreetly appointed John Stuart, third Earl of Bute, as his tutor in matters of politics and government. When Prince George, on achieving his majority in 1756, appointed Bute his Groom of the Stole, the secrecy behind Bute's introduction into Augusta's household led to claims they had their own illicit relationship: this seemed the only explanation for the apparently unmerited ascent of a relatively unknown Scottish nobleman to the first rank of politicians. She was subsequently much derided and lampooned by politicians, journalists, and satirists. Whether selflessly or unwittingly, her actions on behalf of her son demolished the confidence and trust others had in her abilities and good judgment. This portrait of Bute was commissioned by Prince George, the future George III, and presented to Bute's son, John Stuart, Lord Mountstuart. Bute subsequently commissioned a portrait of George III as Prince of Wales for Augusta, and he finally commissioned a portrait of Augusta. This complicated series of gifts and commissions reflects Bute's singularly influential position in the royal household. Ramsay's portrait of Bute was engraved by William Wynne Ryland a short time after it was completed.

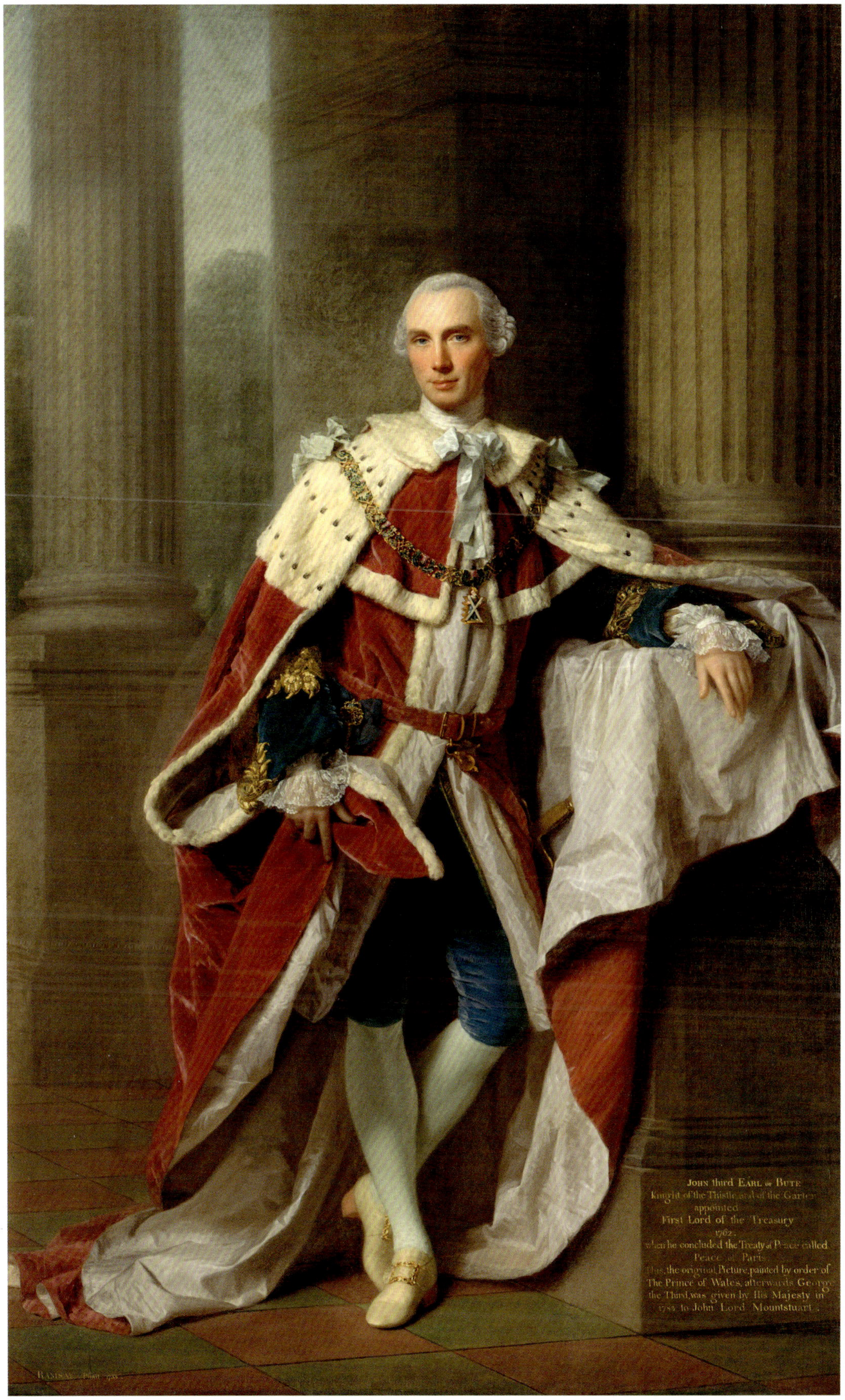

16.17

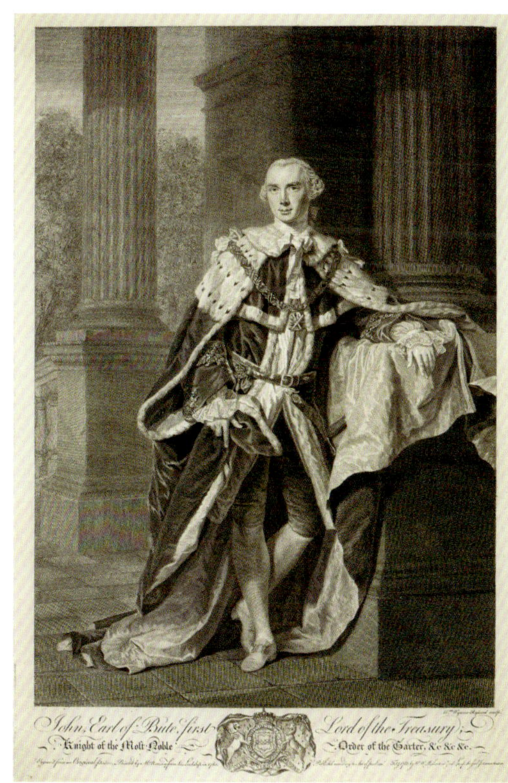

16.18

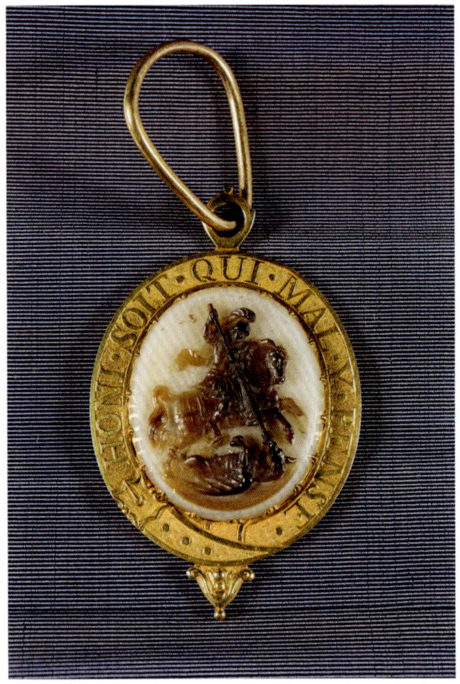

16.19

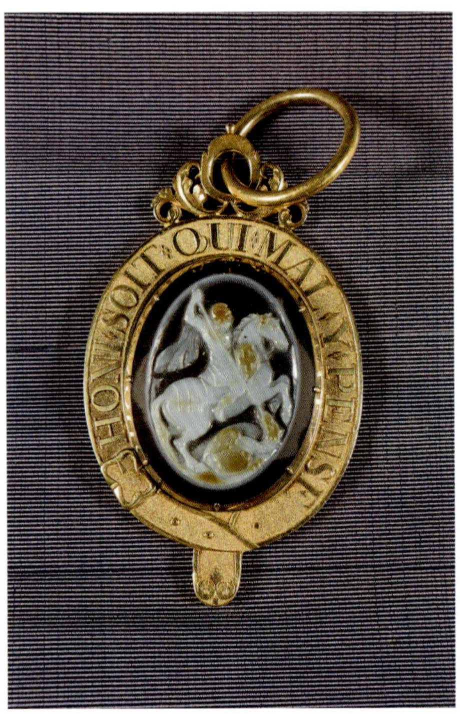

16.20

16.19*

UNKNOWN MAKER
The Most Noble Order of the Garter
Lesser George with sash
Gold, with onyx cameo; sash: silk grosgrain
Cameo: 3⅞ x 1⅝ in. (9.9 x 4.3 cm)
c. 1750 with 17th-century cameo
Mount Stuart Trust, Mount Stuart; inventory no. b02678
Provenance: Given by George III to John Stuart, third Earl of Bute

16.20*

UNKNOWN MAKER
The Most Noble Order of the Garter
Lesser George with sash
Gold, with onyx cameo and blue enamel; sash: silk grosgrain
Cameo: 3½ x 1¾ in. (9 x 4.4 cm)
c. 1760
Mount Stuart Trust, Mount Stuart, inventory no. b02678
Provenance: Given by Augusta, Princess Dowager of Wales, to John Stuart, third Earl of Bute

John Stuart, third Earl of Bute, was created a Knight of the Most Noble Order of the Garter by his former pupil, King George III, in 1762, at the same time that the young King conferred the same distinguished honour on his brother Edward. Lord Bute's son, the first Marquess of Bute, recorded in 1792 that the onyx badge was given to his father by the King, who said he had selected this jewel 'knowing it has belonged to the Stuart King's of England'. As a token of his affection the King went on to present Bute with the Garter, a key element of the insignia of the Order, from his own knee. Augusta, Princess Dowager of Wales, presented Bute with another Lesser George, this one also set with an onyx cameo, but with blue enamel decoration to the reverse. The jewel was suspended from a broad silk grosgain ribbon that was worn as a sash asymmetrically across the chest, as the habitual signal that the honour had been conferred. The alacrity of Bute's appointment created political tensions not only within the noble and political community, but also within the royal family.

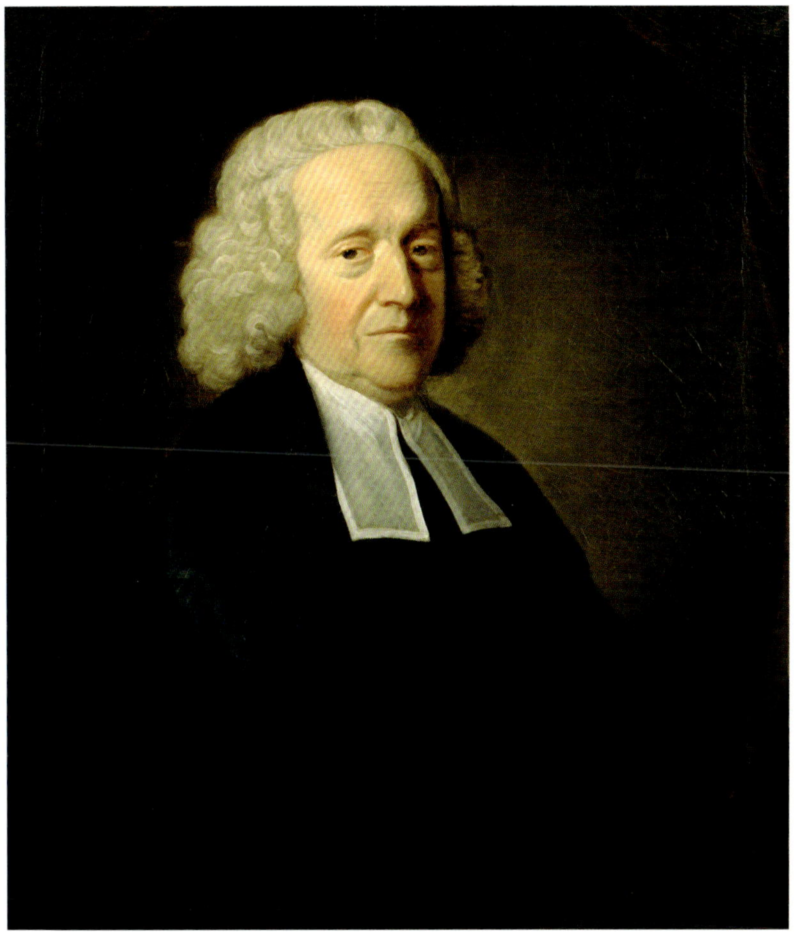

16.21

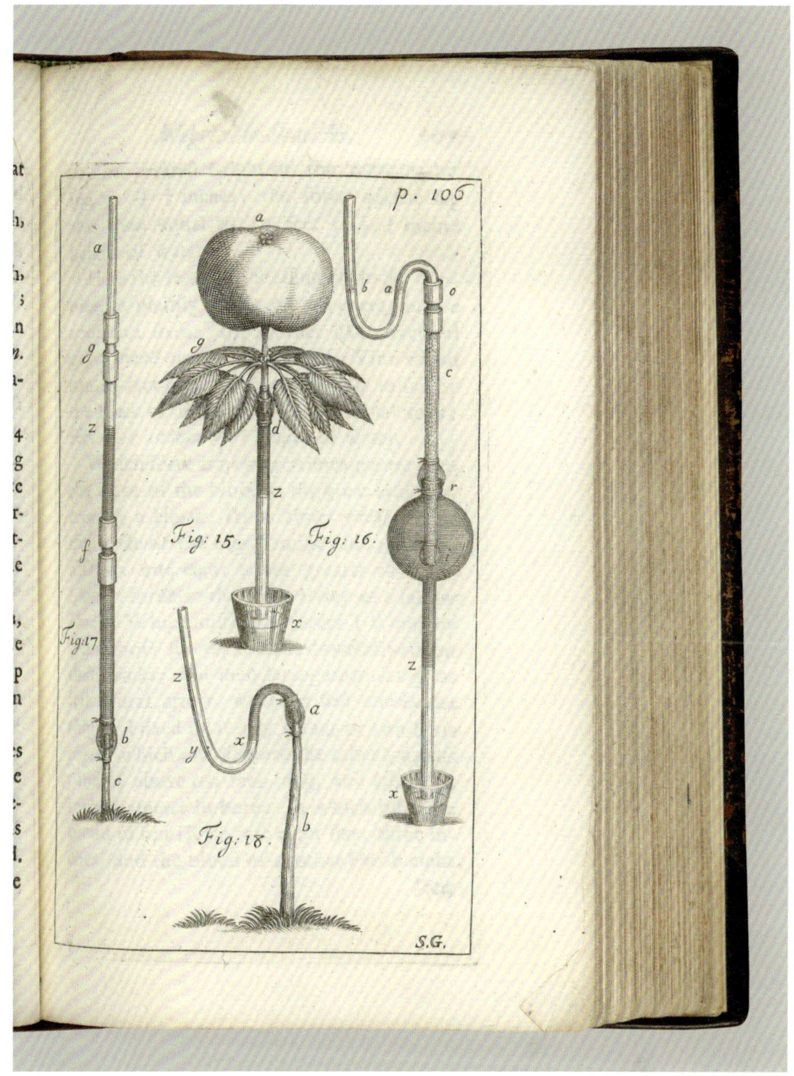

16.22

16.21

STUDIO OF THOMAS HUDSON
(1701–1779)
Stephen Hales
c. 1759
Oil on canvas
30⅛ x 25 in. (76.5 x 63.5 cm)
National Portrait Gallery, London, NPG 1861

16.22†

STEPHEN HALES (1677–1761)
Vegetable Staticks: Or, An Account of some Statical Experiments on the Sap in Vegetables: Being an Essay towards a Natural History of Vegetation
1727
7⅞ in. (20 cm)
Printed for W. and J. Innys and T. Woodward, London
† Royal Botanic Gardens, Kew, Library, Art & Archives Collections, PRE-LINN-A
Exhibited: p. 42, with facing plate (not illustrated)
* Medical Historical Library, Yale University, 18th cent 41
Exhibited: 'S. G.', engraver, 'Figures 15–18', p. 106 facing plate (illustrated)

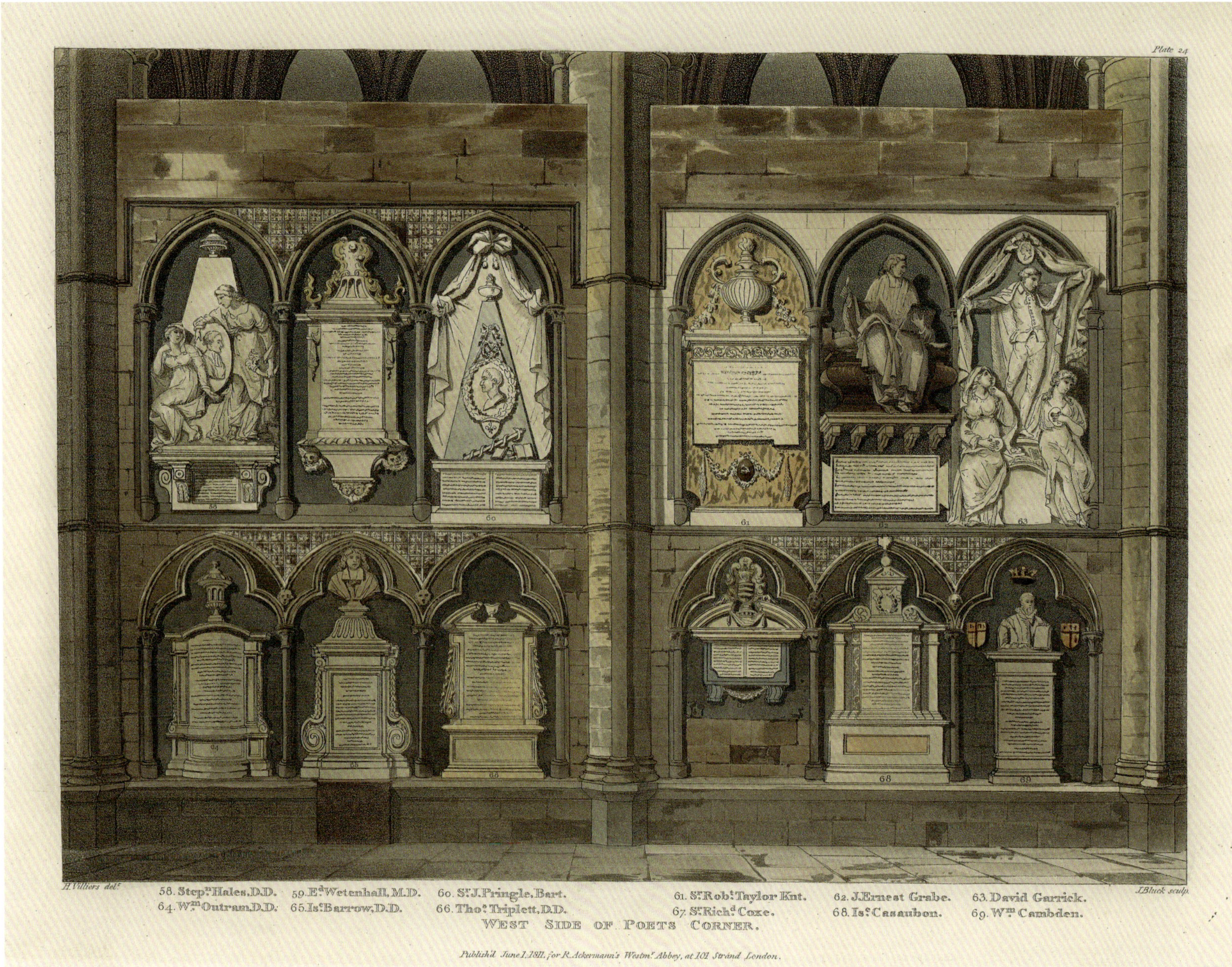

16.23

16.23*

JOHN BLUCK (active 1791–1819), after **JEAN-FRANÇOIS-MARIE HUET-VILLIERS** (1772–1813)
'West Side of Poets Corner'
In William Combe, *History of the Abbey Church of St Peter's Westminster*, vol. 2, plate 24
1812
14⅛ in. (36 cm)
Printed for R. Ackermann, by L. Harrison and J. C. Leigh, London
Yale Center for British Art, Paul Mellon Collection, Folio A/D/77 Copy 1

Stephen Hales was elected a Fellow of the Royal Society in 1718. His experiments came to notice when he read a paper to the society on the effect of the sun's rays on the sap of plants. His book on the same subject, the *Vegetable Staticks* (by 'staticks' he meant functional equilibrium, the measured balance of input and output), was published in 1727. His work for the colony of Georgia brought him into contact with native American peoples, eminent politicians, religious zealots, and sea captains. Created a doctor of divinity by Oxford University after 1733, the peak of his ecclesiastical career was his appointment as chaplain to Augusta, Princess Dowager of Wales, in 1751. Hales introduced Augusta to natural history, guiding her and her children's botanical studies. His monument in Westminster Abbey, paid for by the Princess, was carved by Joseph Wilton. It shows the figures of Religion and Botanical Science supporting a medallion portrait of the dead philosopher, and a Latin inscription linking him to Newton. The monument can be seen in the upper tier, far left, in the engraving of the arrangement of memorials on the west side of Poets' Corner in Westminster Abbey.

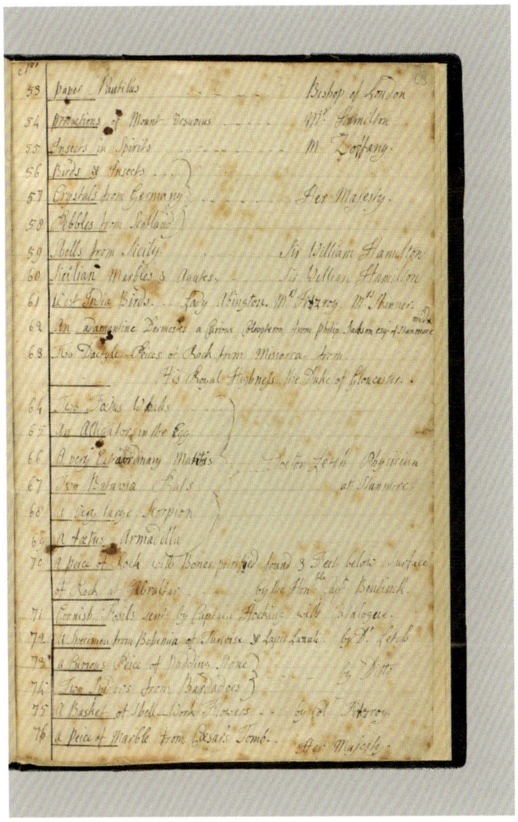

53 Paper Nautilus — Bishop of London
54 Productions of Mount Vesuvius — Wm. Hamilton
55 Insects in Spirits — Mr Zoffany
56 Birds & Insects
57 Crystals from Germany — Her Majesty
58 Pebbles from Scotland
59 Shells from Sicily — Sir William Hamilton
60 Sicilian Marbles & Agates — Sir William Hamilton
61 West India Birds — Lady Abingdon. Mrs Fitzroy, Mrs Skinner.
62 An adamantine Dermestes a curious Coleopteron from Philip Jackson esq. of Stanmore
63 Two Dactyle Pieces of Rock from Minorca from His Royal Highness the Duke of Gloucester.
64 Two Fœtus Whales
65 An Alligator in the Egg
66 A very Extraordinary Matitis — Doctor Letch Physician
67 Two Batavia Rats — at Stanmore
68 A very large Scorpion
69 A Fœtus Armadillo
70 A Peice of Rock with Bones petrified found 3 Feet below Surface of Rock at Gibraltar — by the Honble Capt Bentinck.
71 Cornish Fosils sent by Captain Hocking with Catalogue.
72 A Specimen from Bohemia of Turcoise & Lapis Lazuli — by Dr Letch
73 A Curious Peice of Pudding Stone — by Ditto
74 Two [illegible] from Barbadoes
75 A Basket of Shell Work Flowers — by Col Fitzroy
76 A Peice of Marble from Cæsar's Tomb — Her Majesty

16.24

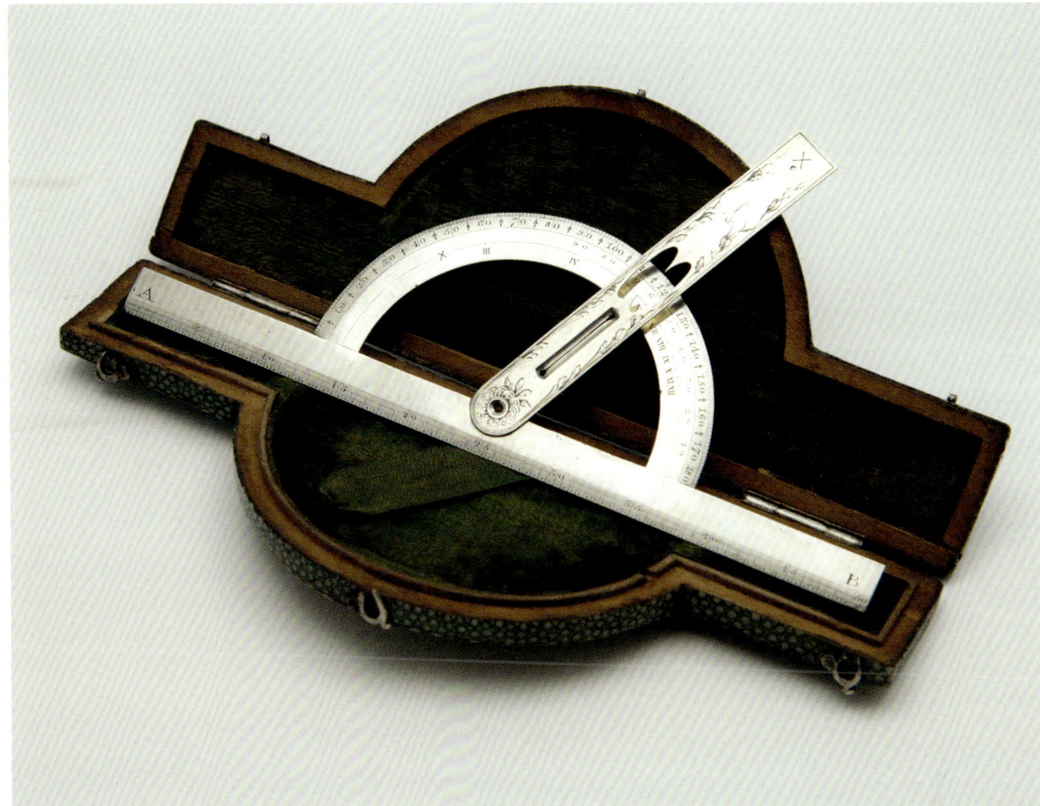

16.25

CHARLOTTE AND HER NETWORKS

16.24

Royal Observatory, Richmond
'Catalogue of the Apparatus of Philosophical Instruments, Her Majesty has deposited in the Royal Observatory at Richmond'
Manuscript
13 x 8¼ x 1½ in. (33 x 21 x 4 cm)
Science Museum, London, MS/0348
* Exhibited: items 275–305
† Exhibited: items 96–127, and items 128–68 (illustrated)

This manuscript catalogues scientific instruments owned by Queen Charlotte; an introductory note states that it was, 'at the time of her death, in the possession of Mary Anne Rigaud, sister of the Late Stephen Peter Rigaud, Savilian Professor of Astronomy and Radcliffe Observer at Oxford, and Observer at Richmond, to their late Majesties George III, George IV, and William IV.' The note was written, and the catalogue was subsequently presented to the 'Kew Observatory of the British Association for the Advancement of Science', by Stephen Rigaud's son in 1855. The main body of the catalogue is divided into sections which variously list equipment used for astronomy; for mechanical experiments; for optical experiments and entertainments; and for hydraulics and measuring. There are sections for models—anatomical and mechanical. At the end of the volume there are several pages listing gifts sent to the Queen for her scientific collections. This includes Cornish minerals and gems from William Borlase; and biological specimens including 'An Alligator in the Egg', 'A Very Extraordinary Matitsis' and 'Two Batavia Rats' from a Dr Letch, Physician at Stanmore; as well as 'A lizzard in spirits' and 'A dried cat' from the Countess of Effingham.

16.25

JOSHUA KIRBY (1716–1774), designer, and **GEORGE ADAMS** (1750–1795), maker
Architect's protractor
1765
Silver in a fitted case of wood, covered in green rayskin, and lined with dark green velvet
Length of base-bar: 6 in. (15.2 cm); radius of arc: 1⅝ in. (4.0 cm); radius of arm: 3 in. (7.6 cm)
Museum of the History of Science, University of Oxford, inventory no. 23312

This protractor was designed by Joshua Kirby and made for Queen Charlotte by George Adams, mathematical instrument maker to George III. Kirby had been tutor in perspective drawing to George when Prince of Wales. The Queen's list of scientific instruments is evidence of the breadth of her interests, with many entries relating to the equipment needed for practical engagement in scientific pursuits. While no architectural drawings associated with the Queen are known to survive, there are many examples of her drawing and works in watercolour.

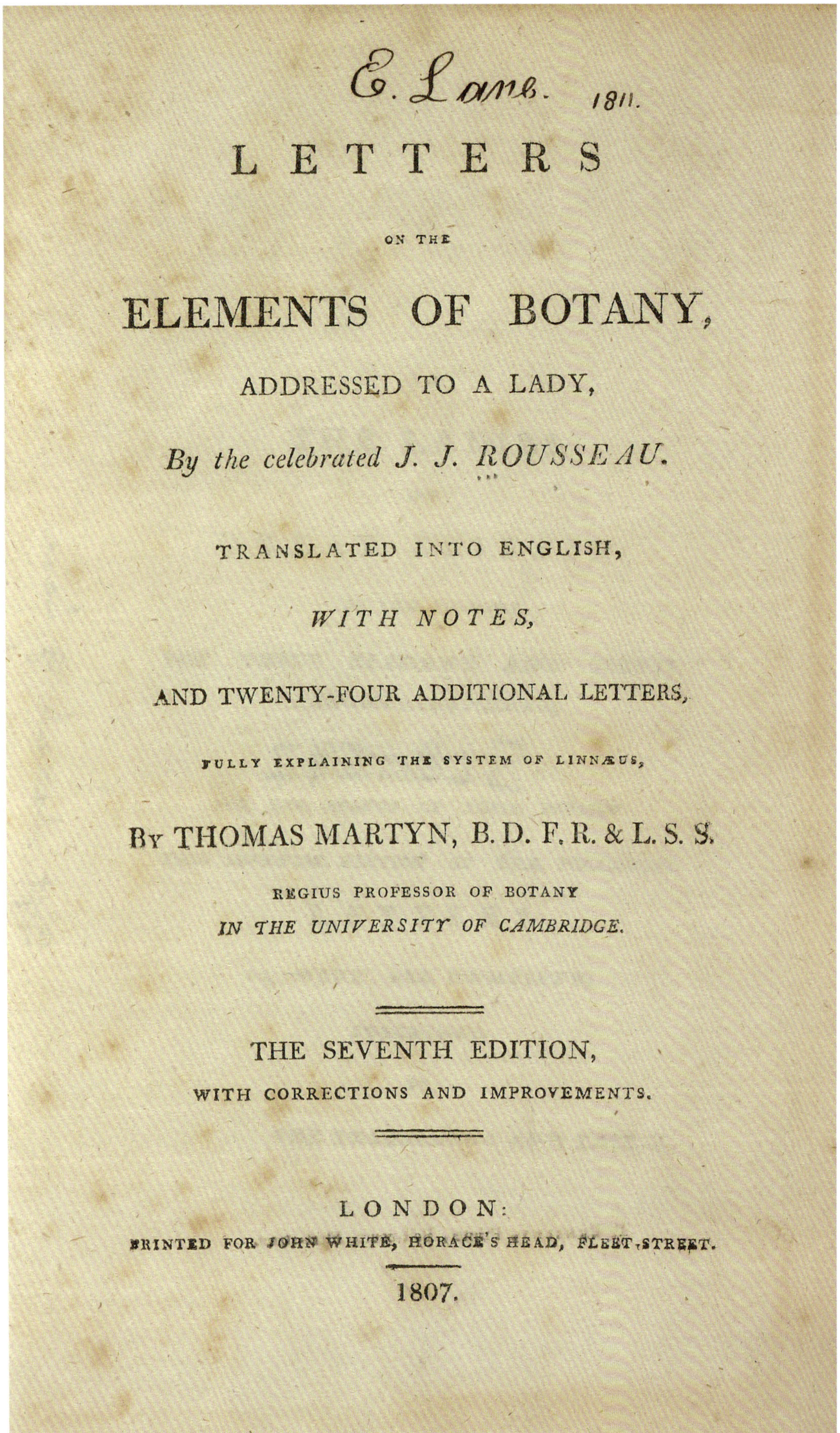

E. Lane. 1811.

LETTERS

ON THE

ELEMENTS OF BOTANY,

ADDRESSED TO A LADY,

By the celebrated J. J. ROUSSEAU.

TRANSLATED INTO ENGLISH,

WITH NOTES,

AND TWENTY-FOUR ADDITIONAL LETTERS,

FULLY EXPLAINING THE SYSTEM OF LINNÆUS,

BY THOMAS MARTYN, B. D. F. R. & L. S. S.

REGIUS PROFESSOR OF BOTANY

IN THE UNIVERSITY OF CAMBRIDGE.

THE SEVENTH EDITION,

WITH CORRECTIONS AND IMPROVEMENTS.

LONDON:

PRINTED FOR JOHN WHITE, HORACE'S HEAD, FLEET-STREET.

1807.

16.26

16.26*

JEAN-JACQUES ROUSSEAU (1712–1778),
with THOMAS MARTYN (1735–1825),
translator and editor
Letters on the Elements of Botany: Addressed to a Lady . . . Translated into English, with notes, and twenty-four additional letters, fully explaining the system of Linnaeus, 7th edition
1807
9 in. (23 cm)
Printed for J. White, London
Beinecke Rare Book & Manuscript Library,
Yale University, Si8r 647
Exhibited: Title page

Rousseau had been declared guilty of blasphemy in France for the theories expressed in his writing, and to avoid further confrontation in 1766 he travelled with David Hume to England where he made contact with the Duchess of Portland, George Simon, and Lord and Lady Harcourt, who were among Charlotte's dearest friends. Rousseau would later style himself 'L'Herboriste' to the Duchess of Portland. His reputation flew before him: when he attended the theatre shortly after his arrival in London, Hume observed that the King and Queen were more concerned to see Rousseau than the performance. George III offered him a pension, despite the divergence between his philosophy and the King's own stoutly maintained Christian orthodoxy. Charlotte's library at Frogmore included a copy of Rousseau's *Letters on the Elements of Botany.*

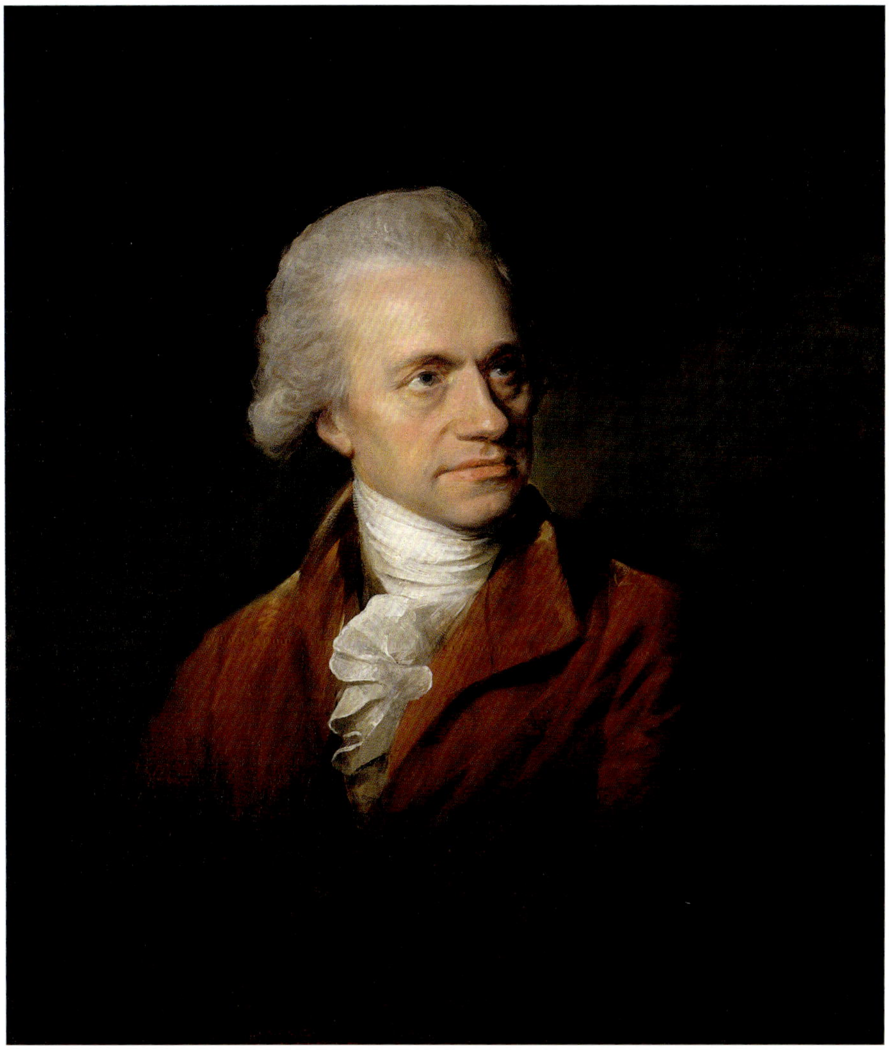

16.27

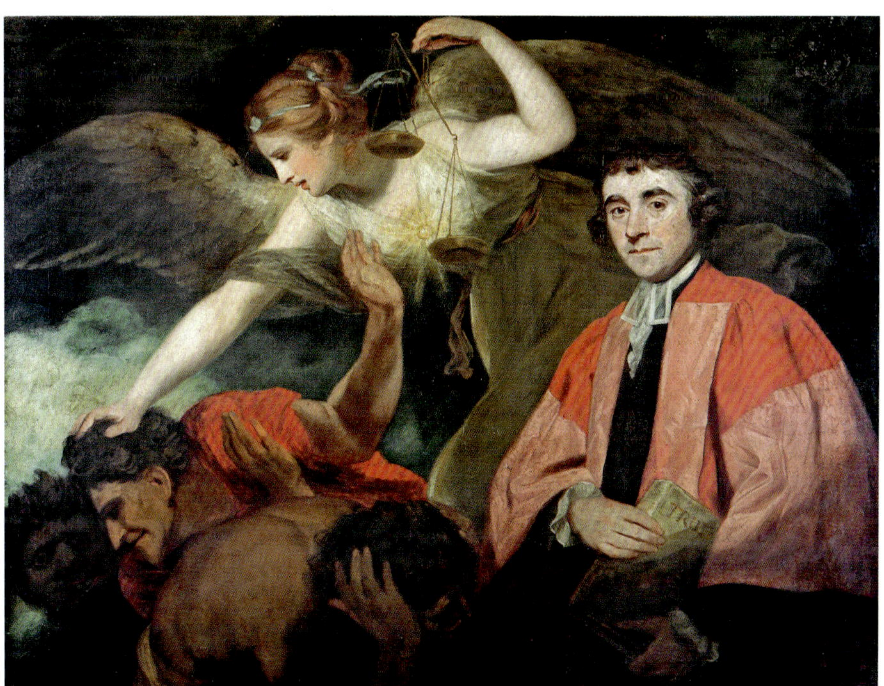

16.28

16.27*

LEMUEL FRANCIS ABBOTT
(c. 1760–1802)
Sir William Herschel
1785
Oil on canvas
30 x 25 in. (76.2 x 63.5 cm)
National Portrait Gallery, London, NPG 98

Frederick William Herschel, born in Hanover, travelled to England to follow a careeer as a musician. It was through friendships with the Reverend John Michell of Thornhill and the English Astronomer Royal Nevil Maskelyne that his interest in astronomy was encouraged. When Herschel discovered what he suspected was a new planet, in 1781, Maskelyne told Sir Joseph Banks, then President of the Royal Society, that the 'musician of Bath' had made a discovery which had 'much merit', and this brought a meeting with George III in May 1782. In honour of the King, Herschel named the planet Georgium Sidus (the Georgian Planet), and was rewarded with an appointment as the King's Astronomer, with a salary of £200 per annum. He successfully petitioned Queen Charlotte for an appointment and salary for his sister, Caroline, 'by way of encouraging a female astronomer'. She spent many hours polishing the mirrors of high performance telescopes so that the amount of light captured was maximized. For her work as William's assistant, she was granted £50 annually from the King. In the nineteenth century Georgium Sidus was renamed Uranus. Herschel sat for this portrait on the encouragement of his friend, the naturalist Sir William Watson. It was said by Caroline Herschel to be a good likeness.

16.28*

SIR JOSHUA REYNOLDS (1723–1792)
The Triumph of Truth: Dr James Beattie
1773
Oil on canvas
48 x 61 in. (122 x 155 cm)
University of Aberdeen, ABDUA 30003

The Scottish moral philosopher James Beattie, co-founder of the Royal Society of Edinburgh in 1783, opposed the scepticism of Hume and others. He was granted an audience with George III and Queen Charlotte in 1773, which led to the granting of an annual royal pension of £200. In the following year, John Majendie, Charlotte's English tutor, presented copies of the fifth edition of Beattie's *Essay on the Nature and Immutability of Truth* to the royal couple. Beattie sat for Reynolds at the time of his royal meeting.

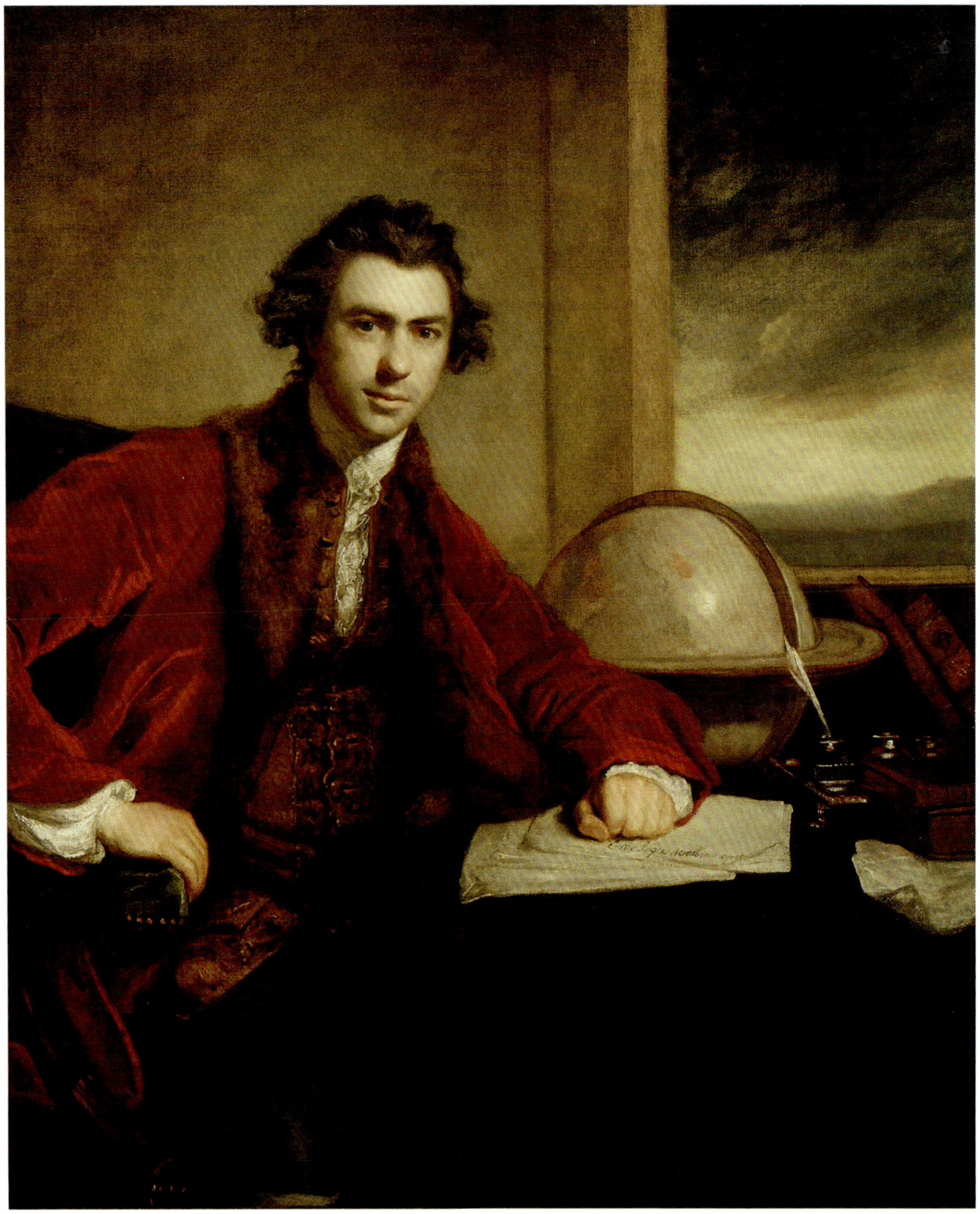

16.29

16.29

SIR JOSHUA REYNOLDS (1723–1792)
Sir Joseph Banks, Bt
1771–73
Oil on canvas
50 x 40 in. (127 x 101.5 cm)
National Portrait Gallery, London, Purchased with help from the National Heritage Memorial Fund, the Art Fund and the Pilgrim Trust, 1986, NPG 5868

Joseph Banks, Fellow of the Royal Society since 1768 and friend of the Swedish botanist Daniel Solander, was invited on the recommendation of the Society to join *H.M.S. Endeavour* on an expedition to the Pacific led by Lieutenant James Cook. For nearly three years Banks and his associates collected natural history specimens, made drawings, compiled vocabularies, and kept journals as they voyaged to South America, Tahiti, New Zealand, the eastern coast of Australia, New Guinea, and the Cape of Good Hope. They discovered about 1,400 new species and, on their return to England in 1771, the reception given to Banks overshadowed that of Cook. For the rest of his life, he was the supreme organizer, patron, and government adviser on all matters connected with science. He became President of the Royal Society in 1778 and held the position until his death. As the collection of plants and trees in the royal garden at Kew was augmented with species from the many parts of the world in which Britain had colonial interests, Banks became its quasi-director, providing access for scholars and collectors, and encouraging debate. He engaged in many discussions with Charlotte, who, he claimed a little dismissively, 'studies botany intensely & reads really with perseverance Elementary books'. Banks sat to Reynolds in 1772, and again after his voyage to Iceland in 1773.

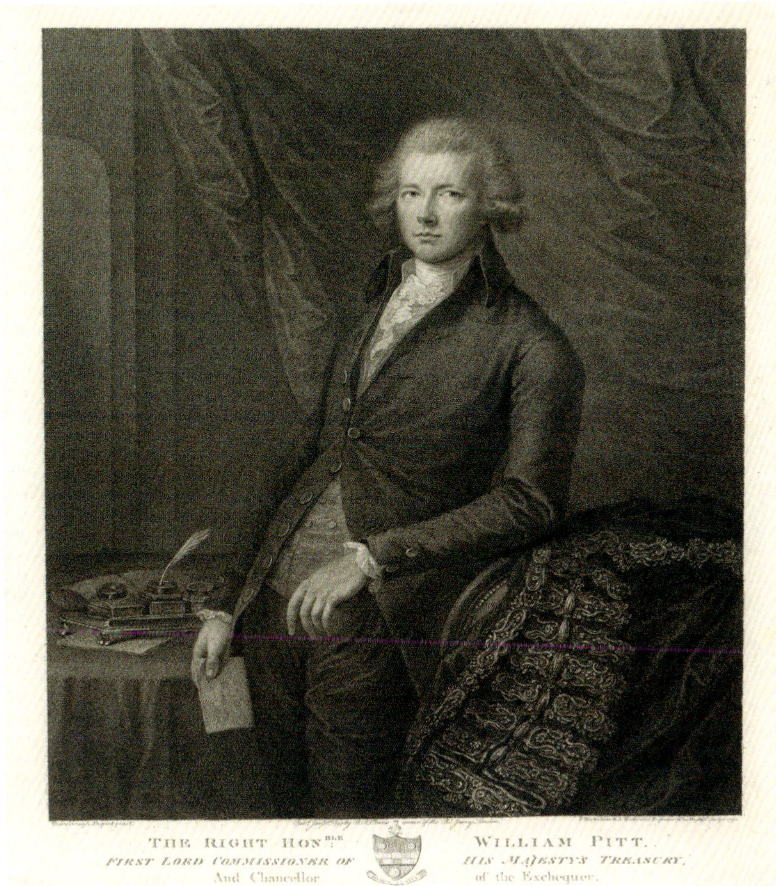

16.30

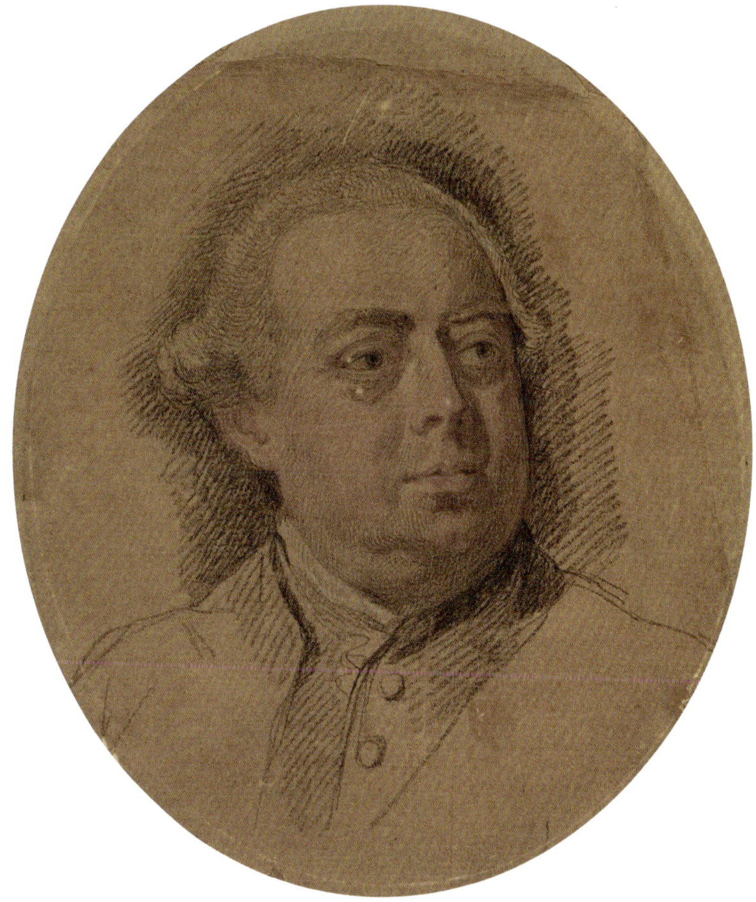

16.31

16.30*

FRANCESCO BARTOLOZZI (1728–1815), engraver, after **GAINSBOROUGH DUPONT** (1754–1797)
The Right Honourable William Pitt
1799
Engraving on paper
20¼ x 15¼ in. (51.4 x 38.7 cm)
Published by Benjamin Beale Evans (c. 1765–1824), London
Yale Center for British Art, Paul Mellon Collection, B1977.14.19580

In 1783 William Pitt the younger became the youngest Prime Minister at the age of twenty-four. He left office in 1801, but was Prime Minister again from 1804 until his death in 1806. He is best known for leading Britain in the wars against France and Napoleon. Pitt was an outstanding administrator who worked for efficiency and reform, bringing in a new generation of outstanding bureaucrats. He raised taxes to pay for the French wars and cracked down on radicalism. To meet the threat of Irish support for France, he engineered the Acts of Union in 1800 and tried (but failed) to get Catholic Emancipation as part of the Union. During George III's first illness in 1788, William Pitt served as a great support to Charlotte as she sought to maintain a level of influence in the care of her husband, in opposition to the Prince of Wales and his supporter Charles James Fox.

16.31*

FRANCESCO BARTOLOZZI (1728–1815)
Frederick, Lord North
1764–1802
Chalk on paper prepared with wash
5⅜ x 4½ in. (13.7 x 11.4 cm)
The Metropolitan Museum of Art, Morris K. Jesup Fund, 1955, 55.106.8

Frederick North, second Earl of Guilford, more often known by his courtesy title, Lord North, which he used from 1752 until 1790, was the son of Francis North, first Earl of Guilford, friend of George III, and Treasurer to Queen Charlotte. In 1754 he was elected at the age of twenty-two as the Member of Parliament for Banbury. In December 1767, he succeeded Charles Townshend as Chancellor of the Exchequer and with the resignation of the secretary of state Henry Seymour Conway in early 1768, became Leader of the Commons as well. In 1770, North became Prime Minister and had immediate success during the Falklands Crisis. Most of North's government would be focused on the growing problems with the American colonies and, later on, supporting British interests during the American War of Independence, which broke out in 1775. North was forced out of office by a motion of no confidence in 1782 on account of the British defeat at Yorktown the year before. In April 1783 he returned to power as Home Secretary in an unlikely coalition with the radical Whig leader Charles James Fox, under the nominal leadership of the Duke of Portland. King George III, who detested the radical and republican Fox, never forgave this supposed betrayal. One of the major achievements of the coalition was the signing of the Treaty of Paris which formally ended the American War of Independence.

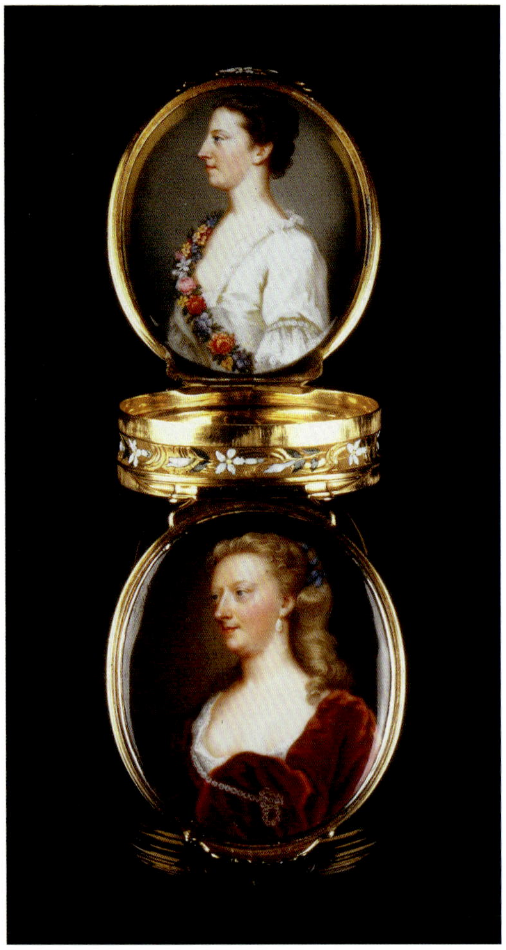

16.32

16.32*

CHRISTIAN FREDERICK ZINCKE
(1683/5–1767)
'Friendship' Box
1740
Gold and enamel
2 x 1⅝ x 1 in. (5.1 x 4.1 x 2.5 cm)
The Stuart Collection

This gold box with multiple portraits was commissioned by Margaret Cavendish Bentinck, Duchess of Portland. It commemorates a network of female friendship and intellectual activity among the Duchess herself; Elizabeth Montagu; the writer and artist Mary Granville, Mrs Delany (formerly Pendarves); and Mary Howard, Lady Andover. Queen Charlotte, as she created her network of sociability, looked to models such as this, formed within the privileged classes during the second half of the eighteenth century. Zincke's tiny miniature portraits are revealed sequentially as the box is opened to allow each woman her moment of appreciation, and making the piece more akin to a jewel than a functional receptacle.

16.33*

ANTONIO POGGI (active 1776–1801), print seller, and **PIETRO ANTONIO MARTINI** (1738–1797), engraver, after **JOHANN HEINRICH RAMBERG** (1763–1840)
Fan with leaf showing King George III and his family visiting The Royal Academy Exhibition in the Summer of 1788
c. 1790
Engraved, hand-coloured paper with carved and pierced ivory sticks and guards
Open: 8⅞ x 15⅛ x 1⅛ in. (22.5 x 38.4 x 3 cm)
Victoria and Albert Museum, Given by Dagmar and Gladys Farrant in memory of Arthur and Maud Loscombe Wallis, T.56–1933

16.34*

PIETRO ANTONIO MARTINI (1738–1797), after **JOHANN HEINRICH RAMBERG** (1763–1840)
Portraits of their majesties, George III and Queen Charlotte, and the Royal Family visiting the Exhibition of the Royal Academy, 1788
1790
Colour engraving on vellum
Image: 9⅞ x 19 in. (25.1 x 48.2 cm)
Sheet: 9½ x 20⅛ in. (24.2 x 51 cm)
Victoria and Albert Museum, London, E.771–1925

William Hogarth's success, in 1746, in establishing the Foundling Hospital as an early venue for contemporary artists in Britain to show their work, was followed by the Society of Artists of Great Britain and the Free Society of Artists. These stimulated the ambition to create a society to support the arts, through education and exhibition. The prominent architect Sir William Chambers used his connections with George III to gain royal patronage and financial support for the Royal Academy of Arts, which was founded through a personal act of King George III on 10 December 1768. Joshua Reynolds was made its first President. Founder members included Francis Cotes, Nathaniel Dance-Holland, Thomas Gainsborough, Francis Hayman, Thomas Sandby, Paul Sandby, Benjamin West, and Richard Wilson. There were two women members, Angelica Kauffmann and Mary Moser. William Hoare and Johan Zoffany were added to this list later by the King, and were known as nominated members.

The Royal Academy was initially housed in cramped quarters in Pall Mall, though the King allowed its library and teaching department temporary accommodation in Old Somerset House, then a royal palace. In 1780 it was installed in purpose-built apartments in the first completed wing of New Somerset House in the Strand, designed by Chambers, the Academy's first treasurer. The first Royal Academy exhibition of contemporary art, open to all artists, was held in 1769. George III, Charlotte, and their family were habitual visitors. The event, held annually, became such a popular place of gathering that modestly priced souvenirs were produced for visitors, including printed paper fans. The Royal Academy Summer Exhibition has continued to be staged every year without interruption to the present day.

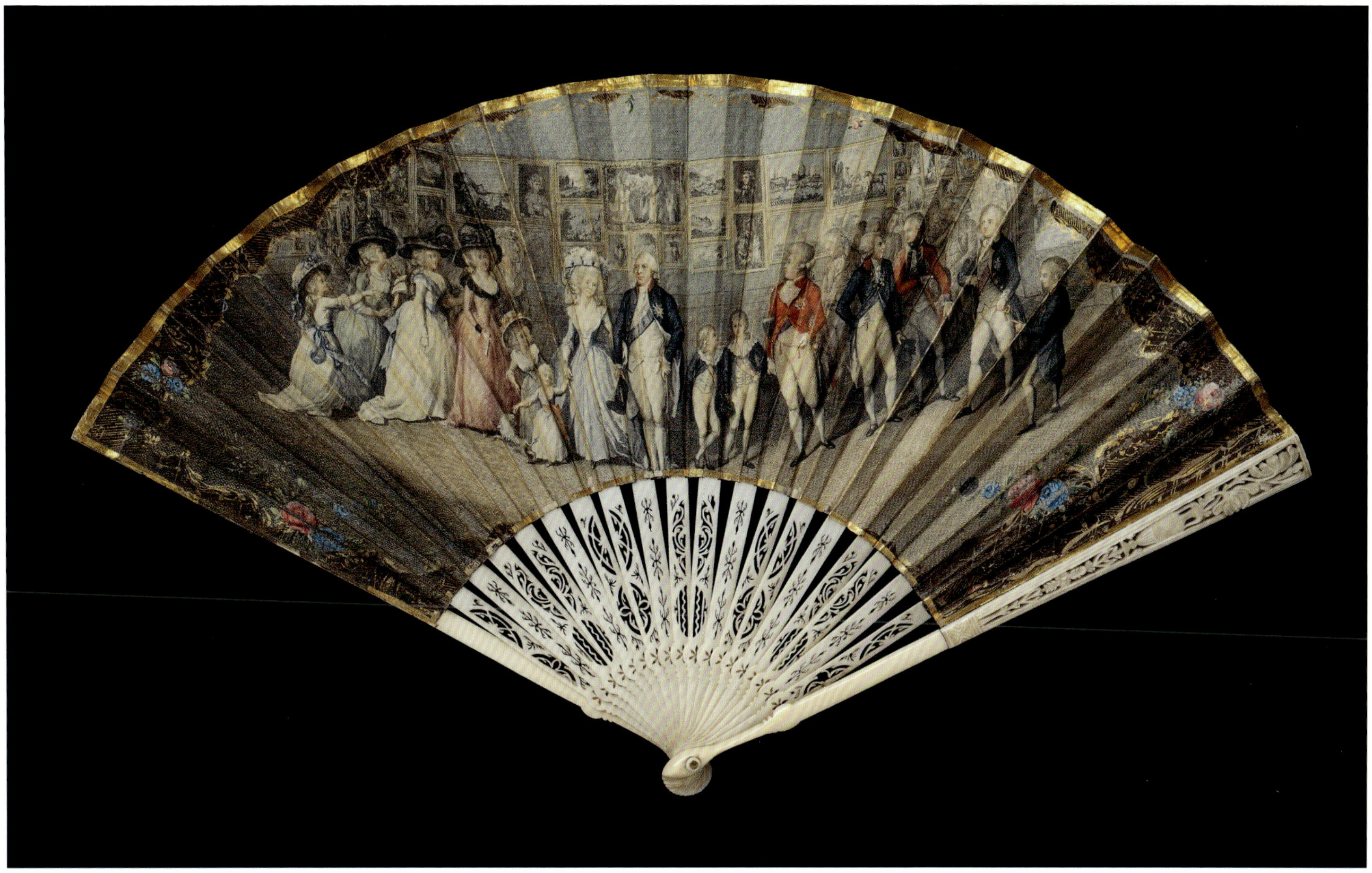

16.33

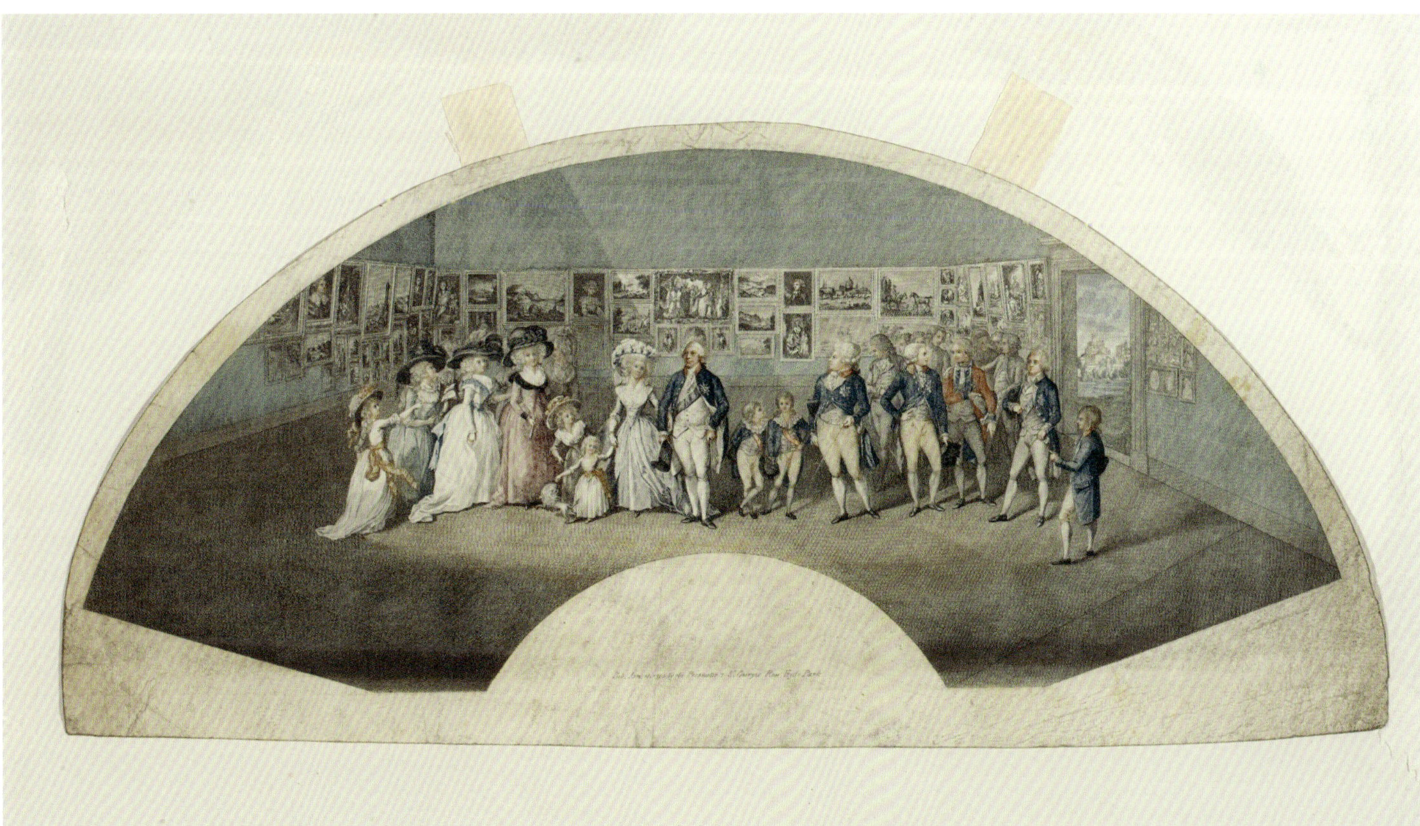

16.34

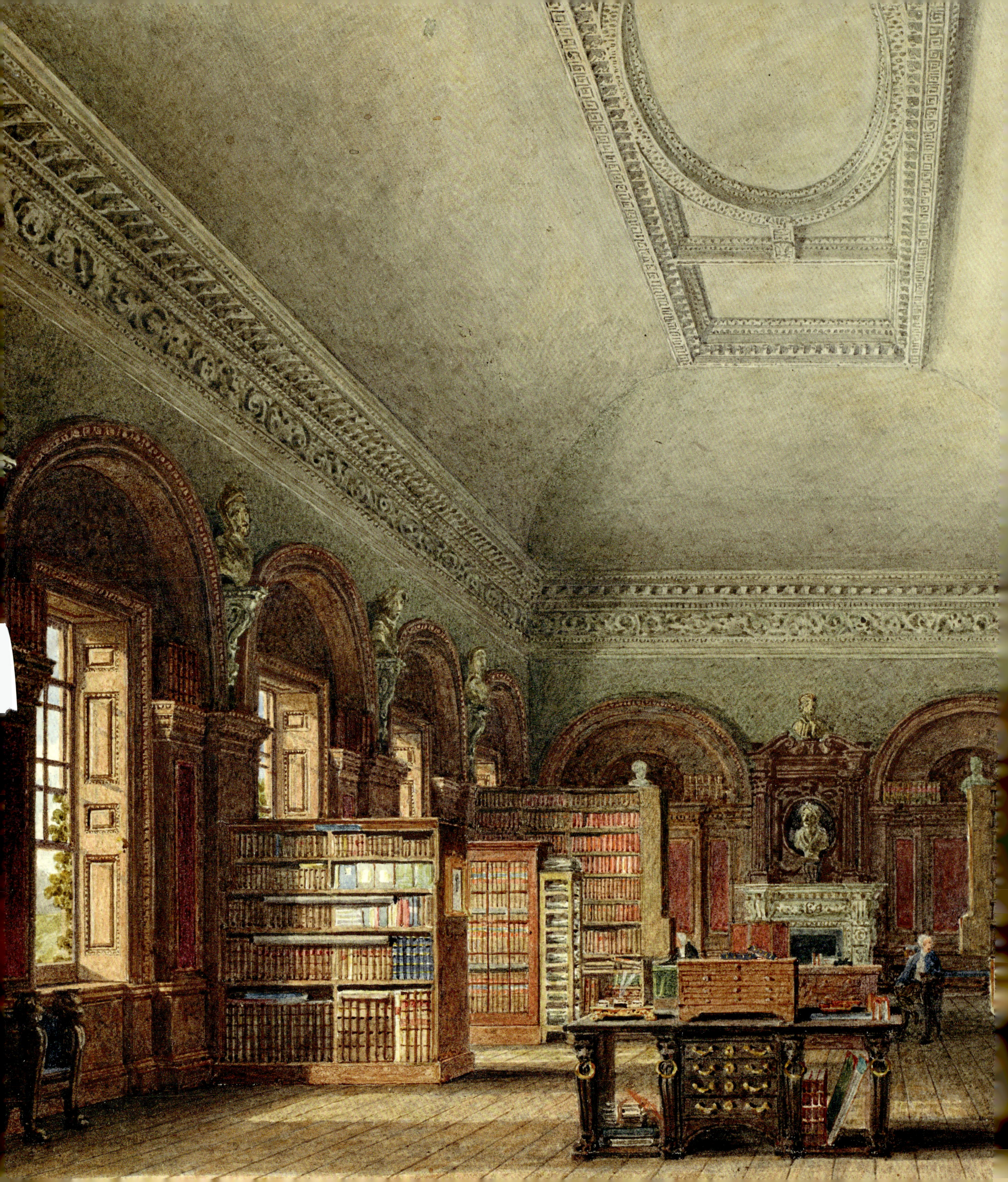

Libraries and their Contents

Emma Jay

Between the Hanoverian accession in 1714 and the death of Charlotte of Mecklenburg-Strelitz in 1818, book collecting burgeoned at the British court. So many members of the royal family collected books that over the hundred-year period there were not one, but many royal libraries in the different palaces, from St James's Palace and Kensington to Richmond, Windsor, and Kew.[1] Caroline of Ansbach and Charlotte created substantial libraries at St James's and Frogmore respectively (p. 304 and fig. 17.01). Augusta of Saxe-Gotha also collected books, although she seems to have had a much smaller collection than the other two. All three were strongly committed to supporting their children's education, and different generations of royal children went on to collect books. The most striking example is Augusta's eldest son, George III, who formed an immense scholarly collection at Buckingham House (now known as the King's Library, and kept at the British Library) and incorporated into it some of the books from earlier royal libraries.[2]

We can associate the royal family's deepening interest in book collecting with other developments in the century such as the growth of popular literacy, the flourishing of print culture, and the use of books and objects to bring together knowledge, discoveries, and ideas about the world.[3] Caroline, Augusta, and Charlotte's impact on the history of royal book collecting reflects the increasing participation of women in learning and literary culture, whether as readers, patrons, collectors, educators, or writers. Female book collectors in royal circles included Henrietta Louisa Fermor, Countess of Pomfret, and Margaret Cavendish Bentinck, Duchess of Portland. There were also strong traditions of women's book collecting at German courts: for instance, Augusta's sister-in-law, Louisa Dorothea, Duchess of Saxe-Gotha, who had keen literary interests and corresponded with Voltaire, kept a library which became part of the main ducal library at Schloss Friedenstein in Gotha.[4]

Caroline's library of three thousand volumes at St James's Palace was the product of many years of collecting.[5] She acquired books that helped her pursue her wide-ranging intellectual and cultural interests. Her reputation for learning encouraged writers to dedicate their works to her, for instance Samuel Clarke's 1717 *Collection of Papers*, a scholarly exchange of letters with Gottfried Leibniz in which Caroline herself had been involved (cat. 16.06). The library was more than just a personal collection, as it effectively became the central library for the royal family in Britain, reflecting the wider context of educational, social, political, and religious activities at court. Many of the books were handsome folios or presentation editions with fine bindings, and the library building, designed by William Kent, created a sense of prestige, elegance, and royal tradition. St James's was not the only place where Caroline kept books, as she had smaller collections elsewhere. Books were integral to the meaning of her two grottoes at Richmond, the Hermitage and Merlin's Cave (fig. 17.02). The small library in Merlin's Cave contained copies of works by the philosophers who were commemorated in sculpture in the Hermitage, for example William Wollaston's *The Religion of Nature Delineated* of 1722.[6] Its keeper was Caroline's protégé, Stephen Duck, whose

Detail of p. 312, cat. 17.10

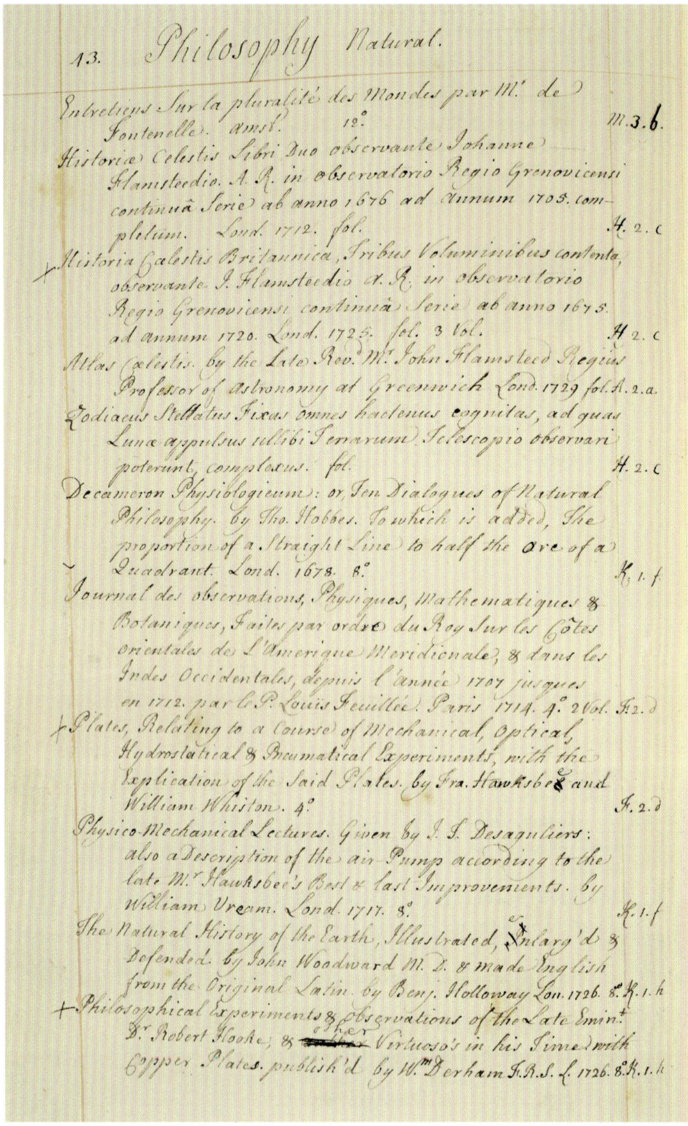

13. Philosophy Natural.

Entretiens sur la pluralité des Mondes par Mr. de Fontenelle. Amst. 12°. M.3.b.

Historiæ Celestis Libri duo observante Johanne Flamsteedio. A.R. in observatorio Regio Grenovicensi continuâ Serie ab anno 1676 ad Annum 1705. completum. Lond. 1712. fol. H.2.c

Historia Cælestis Britannica, Tribus Voluminibus contenta, observante J. Flamsteedio A.R. in observatorio Regio Grenovicensi continua Serie ab anno 1675 ad annum 1720. Lond. 1725. fol. 3 Vol. H.2.c

Atlas Cœlestis. by the Late Rev. Mr. John Flamsteed Regius Professor of Astronomy at Greenwich Lond. 1729 fol. H.2.a.

Zodiacus Stellatus Fixas omnes hactenus cognitas, ad quas Lunæ appulsus ubibi Terrarum Telescopio observari poterunt, complexus. fol. H.2.c

Decameron Physiologicum: or Ten Dialogues of Natural Philosophy. by Tho. Hobbes. To which is added, The proportion of a Straight Line to half the arc of a Quadrant. Lond. 1678. 8°. K.1.f

Journal des observations, Physiques, Mathematiques & Botaniques, Faites par ordre du Roy Sur les Côtes orientales de l'Amerique Meridionale, & dans les Indes Occidentales, depuis l'année 1707 jusques en 1712. par le P. Louis Feuillée. Paris 1714. 4° 2 Vol. H.2.d

Plates, Relating to a Course of Mechanical, Optical, Hydrostatical & Pneumatical Experiments, with the Explication of the said Plates. by Fra. Hauksbee and William Whiston. 4°. H.2.d

Physico-Mechanical Lectures. Given by J. T. Desaguliers: also a Description of the Air Pump according to the late Mr Hauksbee's Best & last Improvements. by William Vream. Lond. 1717. 8°. K.1.f

The Natural History of the Earth, Illustrated, Inlarg'd & Defended. by John Woodward M.D. & made English from the Original Latin by Benj. Holloway Lon. 1726. 8°. K.1.h

Philosophical Experiments & Observations of the Late Emint. Dr. Robert Hooke, & other Virtuoso's in his Time with Copper Plates. publish'd by W. Derham F.R.S. L. 1726. 8°. K.1.h

FIG. 17.01 'A Catalogue of the Royal Library of Her Majesty Queen Caroline Distributed into Faculties, 1743', 1743, bound manuscript, 15¼ (38.7 cm). Royal Collection Trust (cat. 17.07)

transformation from Wiltshire farm labourer into court poet had been brought about by education and reading as well as the patronage of powerful people.

When Caroline died in 1737, George II took charge of her books. This should remind us that he had literary interests of his own, keeping a private library at Hanover, as George I and the Electress Sophia had done before him. He was responsible for the larger Hanoverian court library where Leibniz had been librarian, and in 1757 he donated to the British Museum another royal library, now known as the Old Royal Library, which contained books owned by monarchs going back to Edward IV, and had been kept out of use for many years. The 1760s were a time of change, as the new King and Queen established several libraries in their British residences. George III added some of Caroline's books to his collections at Buckingham House and Windsor, but what proportion of her library he kept is not known.

Augusta's library has proved hard to trace. Questions remain about its scope, the extent to which it was distinct from that of her husband Frederick, Prince of Wales, and where her surviving books can be found. Evidence of ownership is a general difficulty in the study of royal libraries, as books do not always bear distinguishing marks such as ciphers, bookplates, or coats of arms. However, some King's Library items with bindings apparently done for Frederick could have been owned or co-owned by Augusta. James Thomson, one of the Patriot Whig writers who looked to Frederick for patronage, dedicated two plays to Augusta, *Agamemnon* (1738) and *Edward and Eleonora* (1739); the latter was banned by the Lord Chamberlain for its political implications.

Augusta's accounts show some payments for books in French and English: for instance, in 1769 she bought the second volume of L'Abbé Banier's French translation of Ovid's *Metamorphoses*.[7] The development of her garden at Kew gave rise to publications including John Hill's *Hortus Kewensis* (1768) and William Chambers's *Plans, Elevations, Sections, and Perspective Views of the Gardens and Buildings at Kew* (1763), both of which were dedicated to her.

Charlotte's Frogmore library was much more modest than George III's Buckingham House library—a personal collection first and foremost. It contained about 4,500 books, and was a place of retreat and seclusion, used primarily by the Queen and her unmarried daughters. Charlotte kept a printing press, but this was for private purposes, with no more than sixty copies of each publication being printed.[8] She shared Augusta's interest in botany, and one of the works dedicated to her was John Stuart, third Earl of Bute's *Botanical Tables* (1784). It is difficult to tell whether she adopted any books that had belonged to Augusta or Caroline, because the Frogmore collection was sold to settle her debts in 1819, a few months after her death.[9] Some items escaped the sale: the music manuscripts, among them works by her music master, J. C. Bach, were never kept at Frogmore and now form part of the Royal Music Collection at the British Library. Some of her books can be found in the Royal Library at Windsor, having been bought back by the royal family, and some exist in other collections. A volume of chamber sonatas by J. C. Bach and others, identified as Charlotte's by Frederick Nicolay, her music librarian, became part of her son Ernest Augustus, King of Hanover's collection. Along with other music books in the collection (see fig. 17.03, cat. 23.24) it was acquired by the Beinecke Rare Book & Manuscript Library at Yale University in 2008.[10]

Despite some key differences, there were strong similarities between Charlotte's library at Frogmore and Caroline's library at St James's. The two queens had several interests in common,

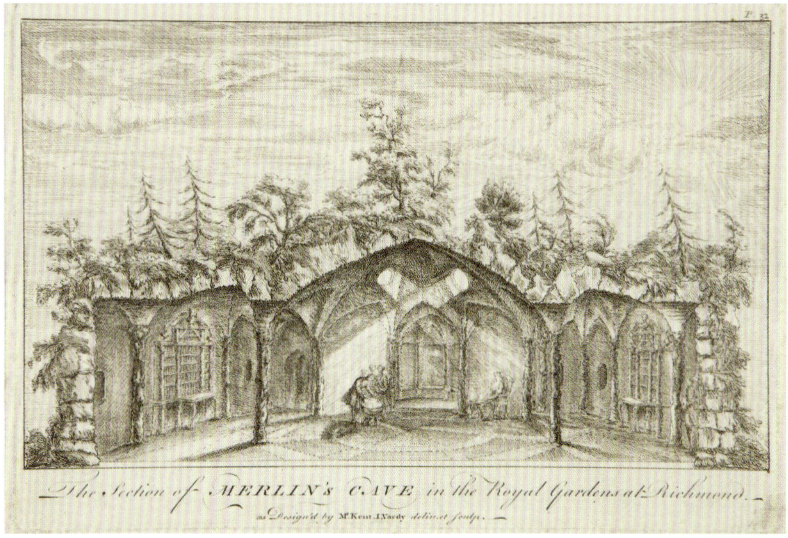

FIG. 17.02 John Vardy, after William Kent, 'A Section of Merlin's Cave in the Royal Gardens at Richmond', in *Some Designs of Mr Inigo Jones and Mr Wm Kent* ([London?]: J. Vardy, 1744), plate 33. Historic Royal Palaces (cat. 17.08)

particularly religion, and between them they owned a vast number of books on the subject. Both libraries were predominantly modern, in that they mostly contained books published in their owners' lifetimes. Indeed, Charlotte often had later editions of the same works. Both were influenced by contemporary trends in reading and publishing: for instance, Charlotte had a high proportion of books in smaller formats like duodecimo, which became popular in the later eighteenth century.[11] Both libraries covered a broad sweep of subjects, and both were dependent on the international book market and the circulation of ideas across Europe. Both were multilingual, although the fact that books in English constituted over 50 percent of Charlotte's collection, as opposed to only 35 percent of Caroline's, indicates the Hanoverians' progress in negotiating a British identity in the hundred years after 1714. Finally, both libraries were a product of, and did much to inspire, the increasingly close involvement in book collecting amongst the royal family and the wider court, a development in which women had an important part to play.

THE PRINCESSES' BOOKS

A library is in one sense a physical space, and Caroline and Charlotte created spaces to house and display their books at St James's Palace and Frogmore respectively. Caroline's library at St James's was an expression of taste as well as learning: she commissioned it from architect and designer William Kent, who was sought after by many in the fashionable world (see cat. 19.06). Kent produced a building in the Palladian style, decorating one of his designs with a pencil sketch of the Queen herself.[12] An early nineteenth-century watercolour by Charles Wild shows a spacious and orderly room with light streaming in from the windows overlooking Green Park.[13] The library's decoration indicated Caroline's desire to promote her dynastic connections to the British royal family. The sculptor Michael Rysbrack created a set of terracotta busts of past kings and queens including Alfred, Elizabeth I, and Edward, Prince of Wales, the so-called 'Black Prince', who was co-opted into eighteenth-century political discourse as an example of Gothic virtue (cat. 12.07).

Caroline and Charlotte's libraries were not hermetically sealed, but connected to the wider world in various ways. Both collections were products of contemporary publishing and the book trade. As Caroline's catalogues reveal, she mostly collected modern printed books in a variety of languages, particularly French, taking advantage of the growing market for books published in France.[14] Her books were 'Distributed into Faculties' (arranged by subject), from divinity and philosophy to plays and operas. The broad scope points to the influence of so-called 'universal' libraries in contemporary Germany, which brought together books on every subject and in every language. The Queen's own interests also shaped the library: for instance, she had a copy of *Poems on Several Occasions* (1736) by the 'Thresher Poet' Stephen Duck, to whom she gave a pension, a home, and a role tending her small library in Merlin's Cave. In George Bickham's engraving for the folio edition, based on a portrait by Sir James Thornhill, Duck sits with quill pen in hand, the earnest

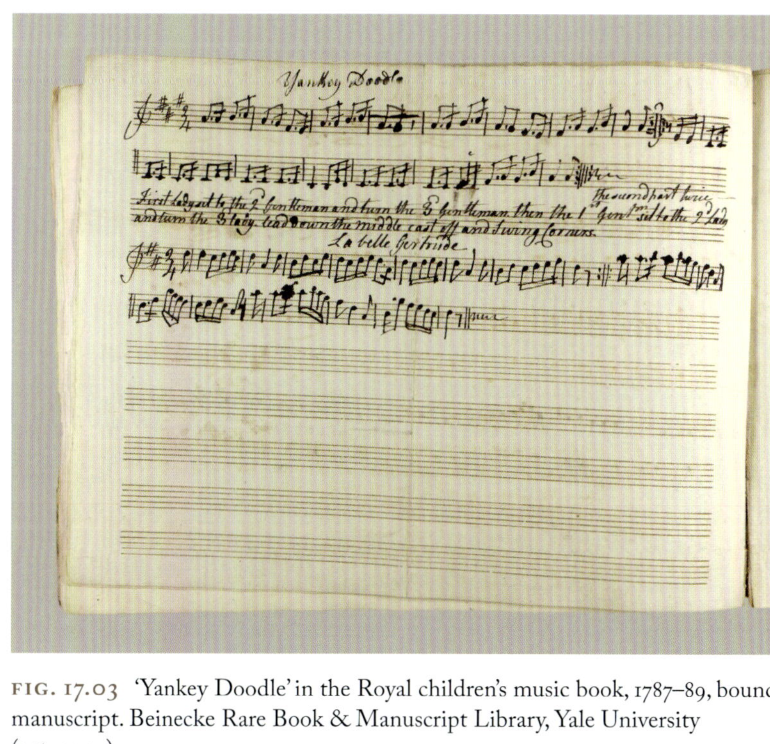

FIG. 17.03 'Yankey Doodle' in the Royal children's music book, 1787–89, bound manuscript. Beinecke Rare Book & Manuscript Library, Yale University (cat. 23.24)

expression on his face suggesting a search for poetic inspiration (fig. 17.04). However, in this period traditional literary patronage coexisted with increasingly commercial forms of publishing and a more independent role for authors. Duck's *Poems* were both dedicated to his benefactress Caroline and published by subscription, a popular method of publishing through which subscribers covered the costs of production in advance.

The institution of the court played a large role in the development of royal literary patronage. Access to royalty was mediated through the Royal Household, and writers needed allies if they were to gain success. Many of the writers who dedicated books to members of the royal family were hoping for places at court, in some cases with a sense of ambivalence about the dependence inherent in court service. After John Gay dedicated his *Fables* of 1726 to Prince William (cat. 17.16), Caroline offered him the post of Gentleman Usher to the youngest princess, Louisa, but he declined it as inappropriate; arguably her intention had been not to insult him but to acknowledge that he composed for children.[15] Caroline and Gay had been on good terms while she was Princess of Wales, and he and Alexander Pope were friends of her Woman of the Bedchamber, Henrietta Howard. Jonathan Swift also struck up a friendship with Mrs Howard when staying in England in 1726, and paid several visits to Caroline, but his views of both women turned sour. Swift and Gay's feelings of betrayal were political as well as a matter of wounded pride, as the opposition's hopes of a broader administration embracing Tories and Whigs were dashed after the coronation of 1727.

Pope, a Tory and a Roman Catholic, could never have been a courtier under the Hanoverians. Obliged by the penal laws to live at some distance from London, he settled in Twickenham and fashioned himself as a poet of opposition to the 'Court'—a term implying Sir Robert Walpole's government as well as the royal court—even if his life was not quite as separate from Caroline's as he liked to imply. *The Dunciad* of 1728 was, among other things, a satirical response to the coronation of George II, and Pope made his political antipathy to the Hanoverians clearer by reworking and expanding the poem over a number of years, finally producing a four-book version in 1743 in which the similarities between the goddess Dulness, Queen of the Dunces, and Caroline were impossible to ignore (see cat. 17.14).[16] *The Dunciad* both satirized and drew inspiration from the proliferation of print and the commercial book trade, mocking dozens of authors who wrote for money (most of them Whigs). The presence in the Queen's library of two copies of the 'Variorum' edition of 1729, a mock-scholarly version with footnotes, prompts questions about whether she noticed its references to herself, and if so, whether she viewed these with a sense of humour.

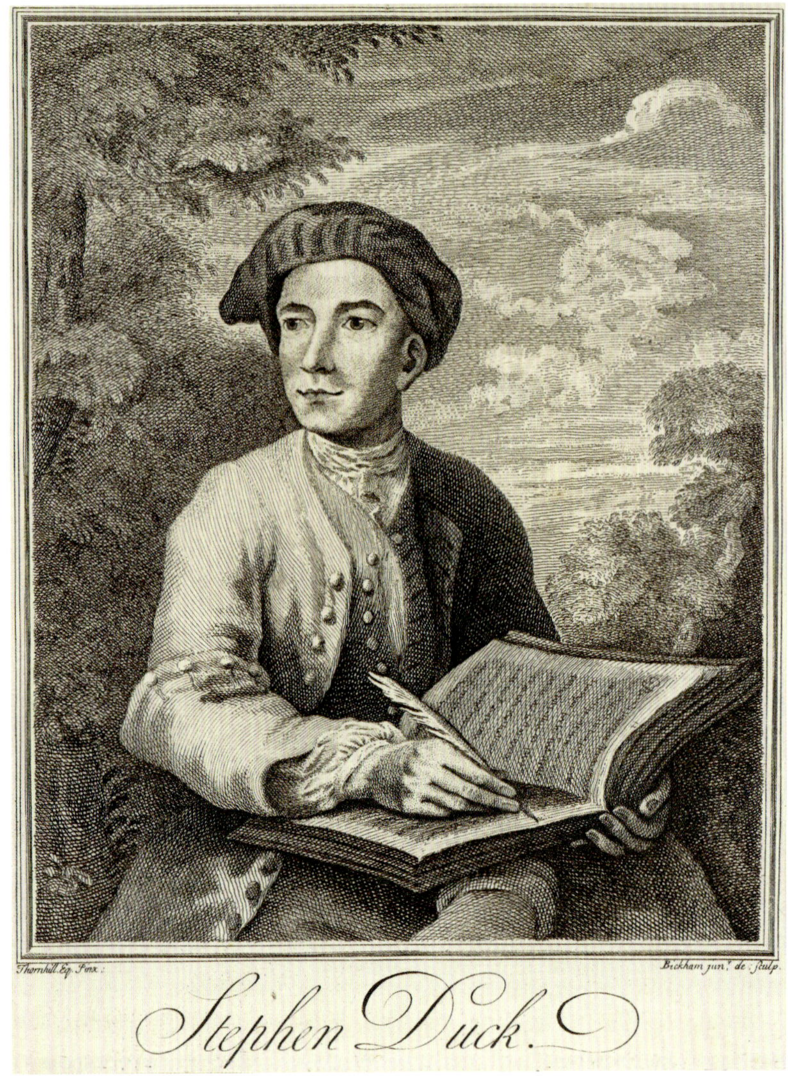

FIG. 17.04 George Bickham the Younger, after Sir James Thornhill, *Stephen Duck*, c. 1736, etching, 8½ x 6 in. (21.5 x 15.4 cm), Royal Collection Trust (cat. 17.17)

For other writers, Caroline's known interest in literature provided an opportunity of regaining favour after a period of exile. When the former Jacobite George Granville, Lord Lansdowne, published his *Genuine Works* in 1732 (incorporating a poem to Mary of Modena, James II's Catholic wife), he presented a copy to Caroline, scribbling a dedicatory verse inside it (fig. 17.05).[17] Granville uses the Tory language of divine right in describing the two parties' loyalty to the Queen: 'To Tory-Doctrine, Even Whigs resign, / And in your Person, own a Right-Divine.' Yet although it is reminiscent of early modern manuscript culture, this poem was later published in Robert Dodsley's *Collection of Poems by Several Hands* of 1748, a best-selling and much-reprinted anthology. Indeed, Charlotte had a copy of the 1782 edition in her library.[18] Perhaps Dodsley's *Collection* also had a more personal meaning for Charlotte: she knew Granville's niece, the botanical artist Mary Granville, Mrs Delany, and provided her with a house at

Windsor. The court was a network of connections, with the same families interacting with the royal family and its household across different generations.

Charlotte formed her library on similar principles to Caroline's, acquiring books on a range of subjects in English, French, German, and Italian that mostly dated from the 1760s onwards. Like Caroline, she was enormously interested in religion, and the theology section of her library was by far the biggest, making up 18 percent of the total. Her exquisite folding reading stand contained a marquetry panel decorated with her crowned cipher. She enjoyed being read to and employed two readers, Jean-André Deluc and Marie Elisabeth la Fite.[19] Even in the secluded setting of Frogmore, Charlotte's reading was to some extent a sociable activity, and her choice of reading matter was informed by recommendations from her officers and friends. Book collecting enabled her to maintain an interest in the circulation of ideas across Europe and stay in touch with her relatives and friends in German courts. Later eighteenth-century literary developments influenced her collection, such as female authorship. Her second Keeper of the Robes, the novelist Frances Burney, also acted as an unofficial reader and discussed books with her. Burney recommended Ellis Cornelia Knight's tale *Dinarbas* of 1790, which was dedicated to the Queen, for the princesses to read.[20] Knight served as Charlotte's companion in the years 1805 to 1813. They spent mornings together at Frogmore and, according to Knight's autobiography, Charlotte gave her a key to the library and allowed her to borrow any books she liked.[21] Knight was to cause offence, however, by moving out of Charlotte's service and into the position of companion to the Queen's granddaughter, the young Princess Charlotte.

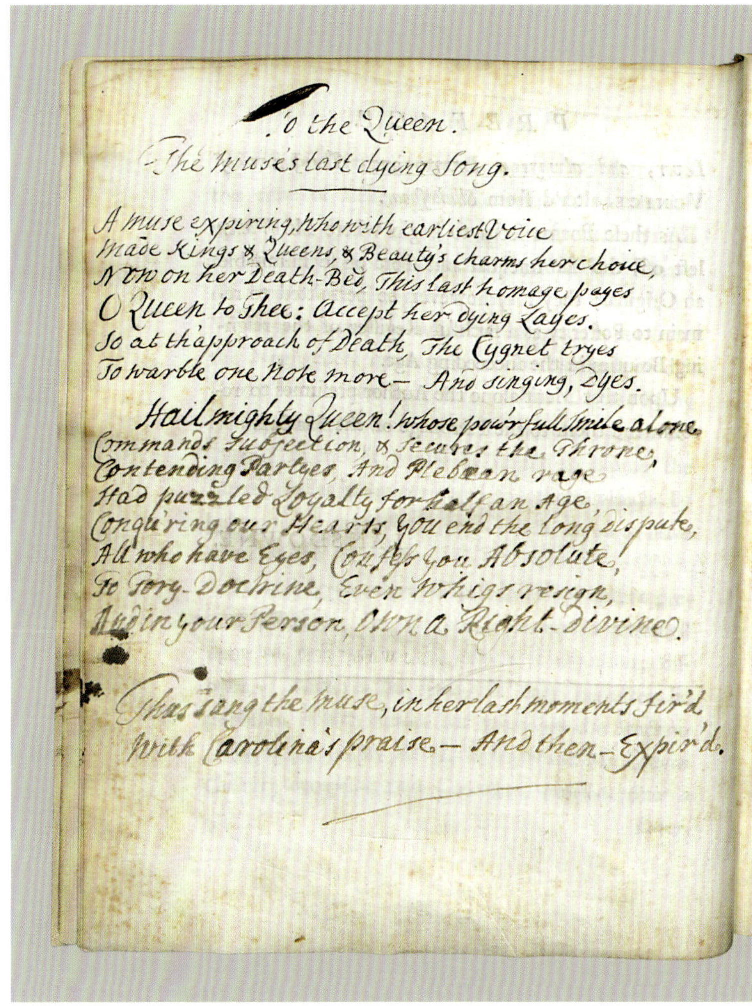

To the Queen.
The Muse's last dying Song.

A Muse expiring, who with earliest Voice
Made Kings & Queens, & Beauty's charms her choice,
Now on her Death-Bed, This last homage payes
O Queen to Thee: Accept her dying Layes.
So at th'approach of Death, The Cygnet tryes
To warble one Note more— And singing, Dyes.

Hail mighty Queen! whose pow'rfull Smile alone
Commands Subjection, & secures the Throne,
Contending Parties, And Plebean rage
Had puzzled Loyalty for half an Age,
Conqu'ring our Hearts, You end the long dispute,
All who have Eyes, Confess you Absolute,
To Tory Doctrine, Even Whigs resign,
And in your Person, Own a Right Divine.

Thus sang the Muse, in her last moments fir'd
With Carolina's praise— And then—Expir'd.

FIG. 17.05 George Granville, Baron Lansdowne, *The Genuine Works in Verse and Prose*, 2 vols. (London: J. Tonson and L. Gilliver, 1732), unpaginated dedication. British Library, London

Unlike Caroline, Charlotte became closely involved in the physical production of books. Her librarian Edward Harding was also a printer, and he set up the printing press at Frogmore.[22] Among the works issued by the Frogmore Press was Knight's *Translations from the German in Prose and Verse* of 1812. Another was a set of short histories of different countries, printed on cards and displayed in boxes (fig. 17.06). Either the Queen or Knight wrote the text of these 'Chronological Abridgements', and the Queen presented them to children, presumably for educative purposes. She was industrious, taking much time and care to produce a 'Copy of a Manuscript in the Prince Regent's Possession, found in the Stuart Papers' of 1812 (cat. 17.20). This handwritten transcription of a notebook written by James II is illustrated with prints taken from other books, demonstrating Charlotte's fondness for this contemporary practice of extra-illustration.[23]

There were many connections between Caroline and Charlotte's libraries and the development of reading, publishing, and book collecting in royal circles and beyond. How ironic, therefore, that the dispersal of Charlotte's library after her death was by auction, a very typical method of book purchasing in the eighteenth century—indeed, the auction was widely used as a building-block for new collections. Caroline had bought books at auction, and her mentor and friend Leibniz had done the same, looking at book sale catalogues to spot items or whole collections that would be worth buying. Charlotte's wish had been for her books and other personal effects to be divided between her four younger daughters, but her debts needed to be repaid, and sadly the Frogmore library was sold at Christie's in 1819, raising about £4,500. The copy of the sale catalogue from the King's Library of George III, now at the British Library, is marked up with the names of buyers and prices.[24] Through this sale, Charlotte's collection was returned to the literary marketplace from which it had been created.

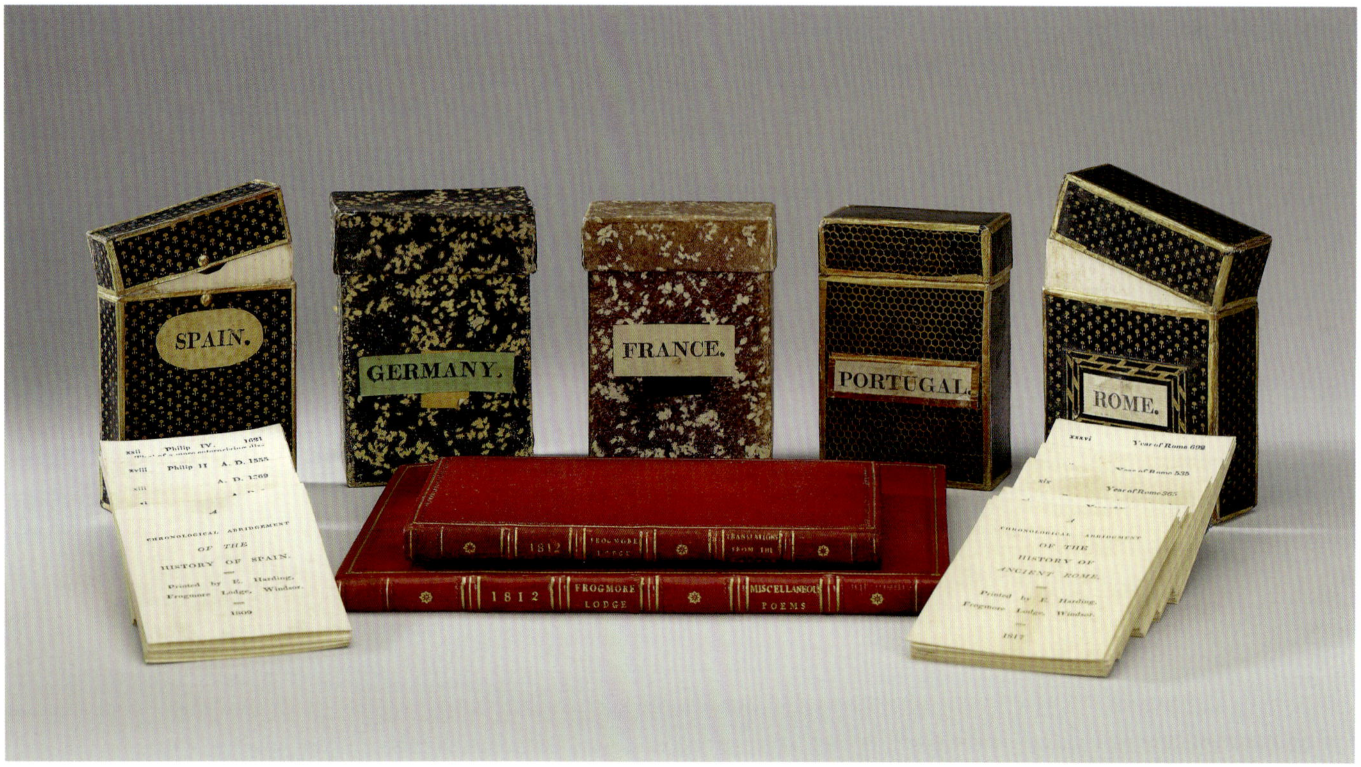

FIG. 17.06 Frogmore Press, 'Chronological Abridgements' of the histories of Spain, Germany, France, Portugal, and Rome (1809–1817); *Translations from the German* (1812); and *Miscellaneous Poems* (1812). Royal Collection Trust

NOTES

1. See Clarissa Campbell Orr, 'Lost Royal Libraries and Hanoverian Court Culture', in *Lost Libraries: The Destruction of Great Book Collections Since Antiquity*, ed. James Raven (Basingstoke: Palgrave Macmillan, 2004), 163–80.
2. See John Goldfinch, 'Royal Libraries in the King's Library', in *1000 Years of Royal Books and Manuscripts*, ed. Kathleen Doyle and Scot McKendrick (London: British Library, 2013), 213–36.
3. See Kim Sloan with Andrew Burnett, eds., *Enlightenment: Discovering the World in the Eighteenth Century* (London: The British Museum, 2003).
4. Bernhard Fabian, gen. ed., *Handbuch der Historischen Buchbestände in Deutschland*, Vol. 19: *Thüringen A–G*, ed. Friedhilde Krause (Hildesheim: Georg Olms, 1998), 242–45.
5. Emma Jay, 'Queen Caroline's Library and its European Contexts', *Book History* 9 (2006): 31–55; Emma Jay, 'Court Patronage Reconsidered: The English Literature in Queen Caroline's Library', *Library and Information History* 30, no. 2 (2014): 75–89.
6. British Library C.120.h.6 (6), fol. [16]r. The publication date of this copy of the Wollaston is not specified; the work first appeared in 1722, but there were several later editions.
7. Royal Archives (RA) GEO/MAIN/55616.
8. Jane Roberts, 'Edward Harding and Queen Charlotte', in *Burning Bright: Essays in Honour of David Bindman*, ed. Diana Dethloff, Tessa Murdoch, and Kim Sloan, with Caroline Elam (London: UCL Press, 2015), 146–59.
9. See *A Catalogue of the Genuine Library, Prints, and Books of Prints, of an Illustrious Personage, Lately Deceased* (London: W. Bulmer, 1819).
10. Beinecke Rare Book & Manuscript Library, Yale University, Hanover Royal Music Archive, James Marshall and Marie-Louise Osborn Collection, OSB MSS 146, Series II. Music Collections: Bound Collections, Box 837, Bound collection (1788, undated).
11. For later trends, see William St Clair, *The Reading Nation in the Romantic Period* (Cambridge: Cambridge Univ. Press, 2004).
12. Kent's designs are in an extra-illustrated copy of Thomas Pennant's *Some Account of London*, 4th ed. (London: Robert Faulder, 1805), Sir John Soane's Museum, vol. 147/197. See cat. 17.09.
13. See W. H. Pyne, *The History of the Royal Residences of Windsor Castle, St. James's Palace, Carlton House, Kensington Palace, Hampton Court, Buckingham House, and Frogmore*, 3 vols. (London: A. Dry, 1819), 3: plate facing page 60.
14. There are two surviving versions of the catalogue of Caroline's library after it was installed in Kent's building at St James's: British Library Add. MS. 11511 [1741]; and Royal Library, Windsor, RCIN 1028932.a, 1743–[1760].
15. For Swift's account of this episode, see *The Correspondence of Jonathan Swift*, ed. Harold Williams, 5 vols. (Oxford: Clarendon Press, 1963–65), 4: 97–101.
16. See Alexander Pope, *The Dunciad in Four Books*, ed. Valerie Rumbold, 2nd ed. (Harlow: Longman, 2009).
17. Caroline's copy is in the British Library, shelfmark 77.k.9–10.
18. See *A Catalogue of the Genuine Library*, 66.
19. Clarissa Campbell Orr, 'Charlotte of Mecklenburg-Strelitz, Queen of Great Britain and Electress of Hanover: Northern Dynasties and the Northern Republic of Letters', in *Queenship in Europe, 1660–1815: The Role of the Consort*, ed. Clarissa Campbell Orr (Cambridge: Cambridge Univ. Press, 2004), 368–402.
20. Campbell Orr, 'Lost Royal Libraries', 175–76.
21. *The Autobiography of Miss Knight, Lady Companion to Princess Charlotte*, ed. Roger Fulford (London: William Kimber, 1960), 83.
22. Roberts, 'Edward Harding and Queen Charlotte', 152.
23. On this practice, later known as 'Grangerizing' after the clergyman James Granger, author of the *Biographical History of England* (1769), see Roberts, 'Edward Harding and Queen Charlotte', 146–48.
24. British Library, shelfmark 123.f.16.

17.07

FRANCIS SAY (d. 1748) and **STAFF OF THE ROYAL HOUSEHOLD**
'A Catalogue of the Royal Library of Her Majesty Queen Caroline Distributed into Faculties, 1743'
1743
Manuscript, bound
15 ¼ (38.7 cm)
Royal Collection Trust, RCIN 1028932.a
Exhibited: ff. 44–45

There are several catalogues of Caroline's books, including a listing of her collection of music, but this is the most compendious. It survives in two versions, one owned by the Royal Collection Trust, the other in the British Library. They were likely written by professional scribes working under the auspices of Francis Say, appointed Caroline's Librarian in 1737. They contain about three thousand titles, carefully arranged by subject, which follow a classification hierarchy based on that championed by Leibniz.

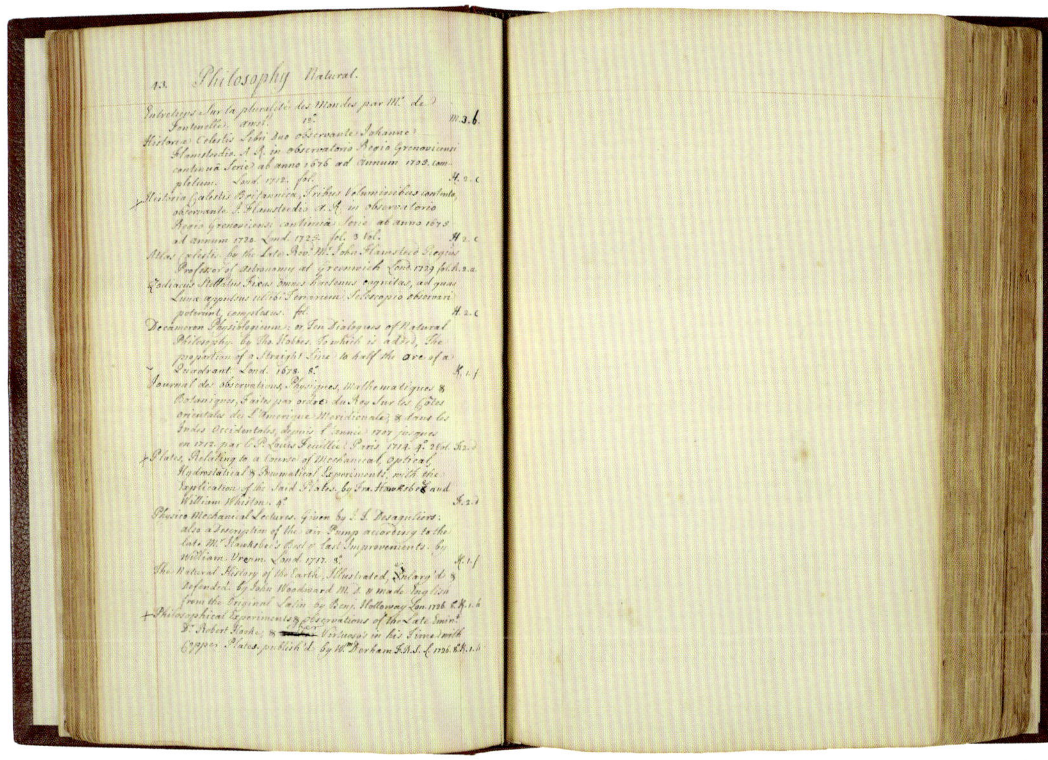

17.07

17.08†

JOHN VARDY (1718–1765), engraver, after **WILLIAM KENT** (c. 1685–1748)
'A Section of Merlin's Cave in the Royal Gardens at Richmond'
In *Some Designs of Mr Inigo Jones and Mr Wm Kent*, plate 33
1744
J. Vardy, [London?]
10½ in. (26 cm)
Historic Royal Palaces, HRP07788

Within Merlin's Cave, one of the folly temples in the garden of Richmond Lodge, Caroline established a small library in about 1735 for the use of the Thresher-poet, Stephen Duck. He would serve as keeper of the Cave and quasi-hermit. Drawn from collections of books possibly scattered over other royal homes, the volumes were bound in white vellum 'in the neat Plainness of Quakerism'. Authors represented included Alexander Pope, Jonathan Swift, Samuel Clarke, and others in the Queen's acquaintence, as well as classic texts from the Renaissance and antiquity.

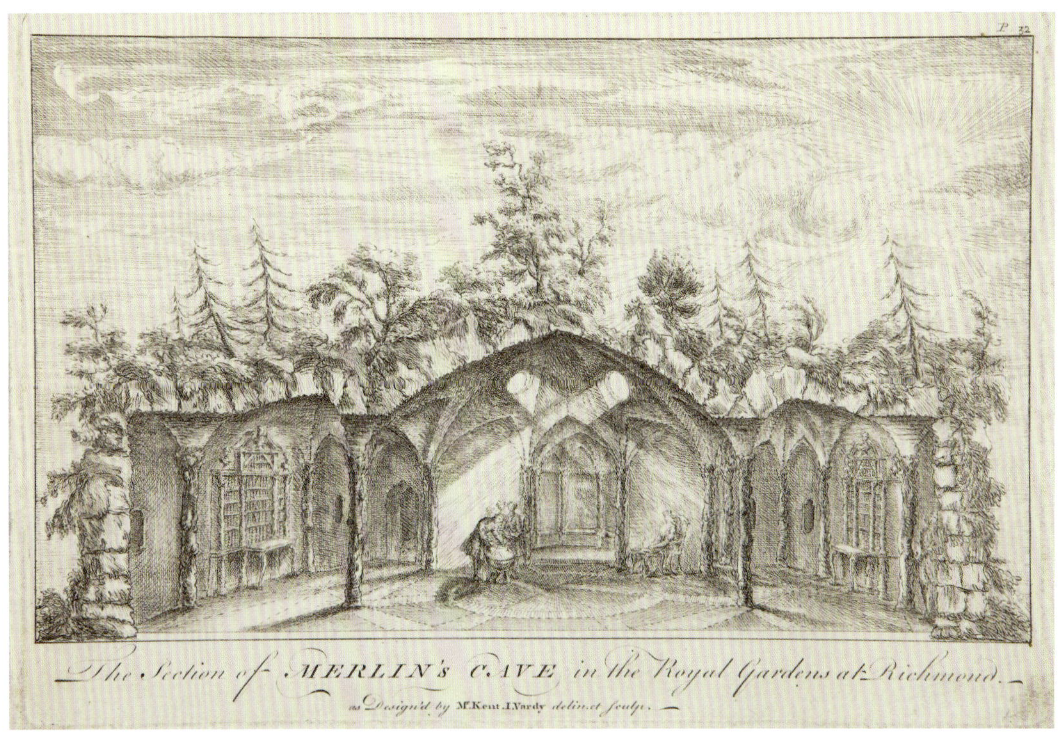

17.08

* = YCBA only
† = HRP only

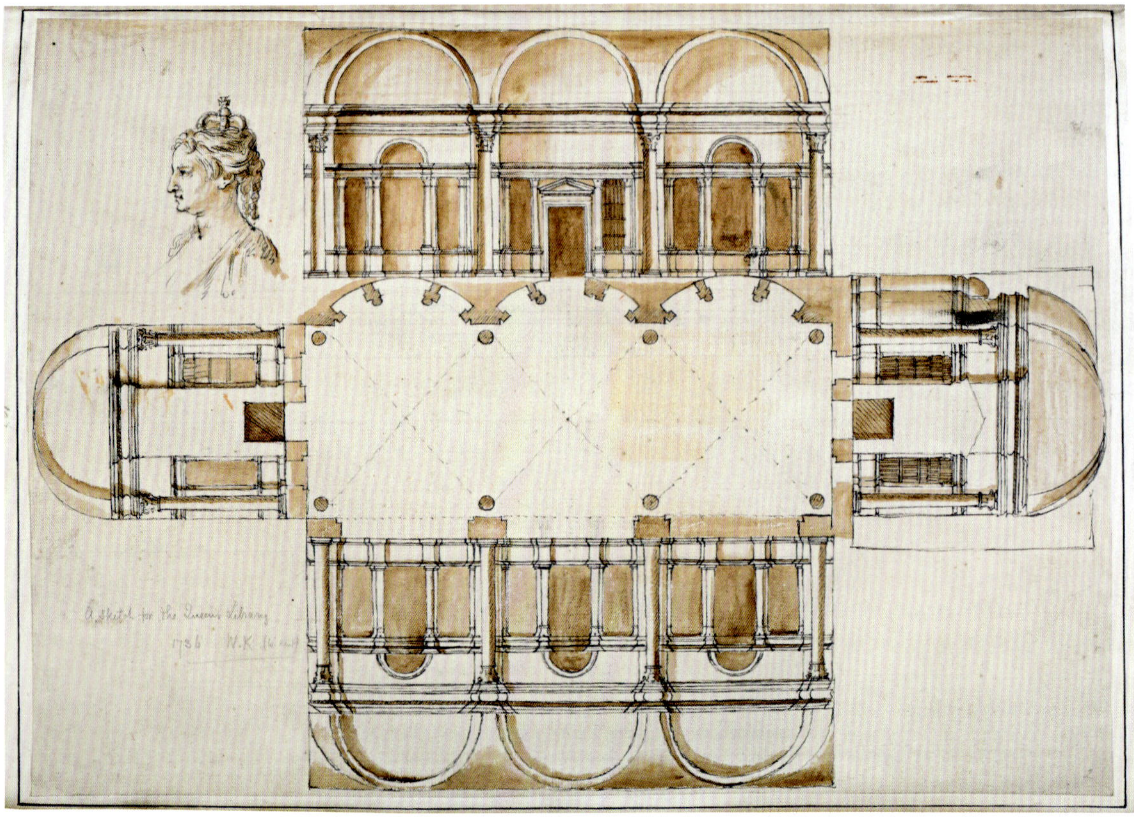

17.09

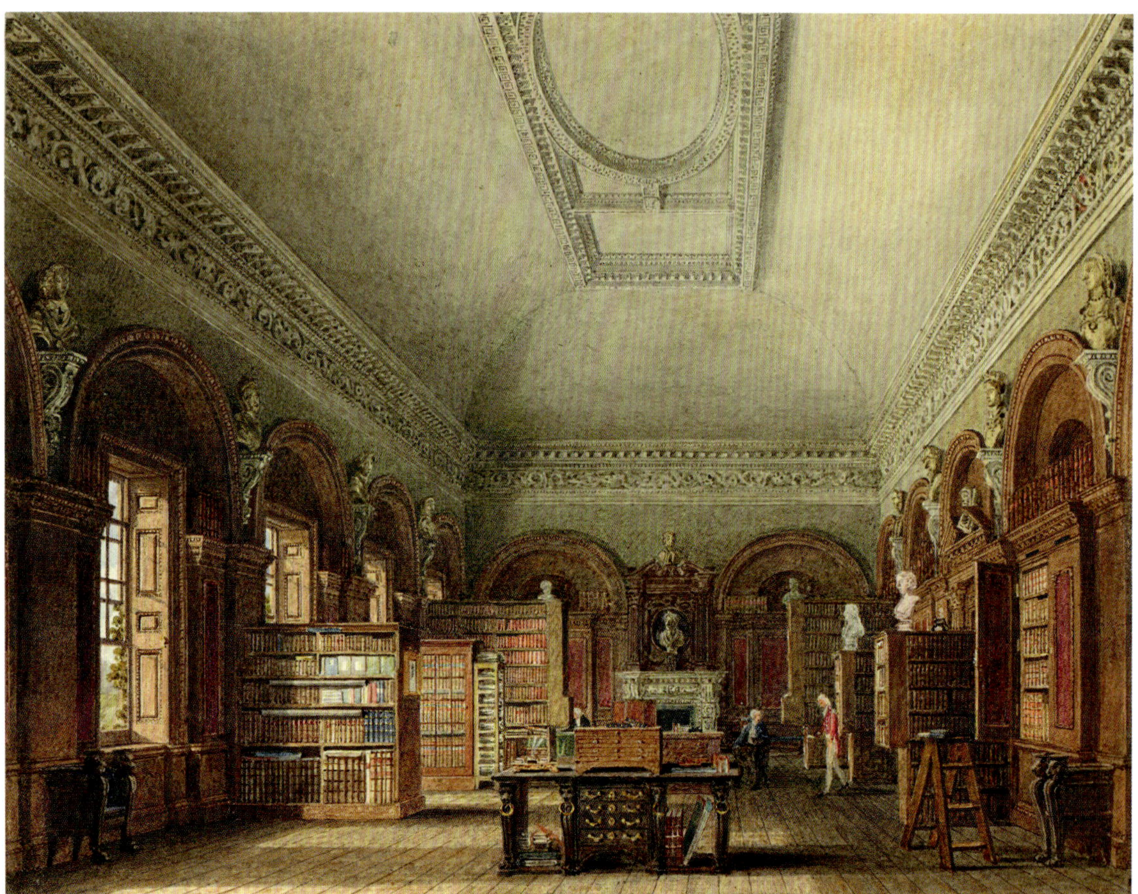

17.10

17.09*

WILLIAM KENT (c. 1685–1748)
A Sketch for the Queen's Library
c. 1736
Pen and ink, and washes over pencil
12⅞ x 17⅞ in. (32.7 x 45.3 cm)
Sir John Soane's Museum, London, SM Vol. 147/197

Draft plan and laid-out elevations prepared for Queen Caroline's library at St James's Palace, with a sketch portrait of the Queen included in the margin. There was evidently much debate about the final design details for Caroline's new library at St James's Palace. Six drawings survive in Sir John Soane's Museum which provide a series of alternative layouts for the bookcases and the detail of the chimney-pieces, cornice, and ceiling. Contemporary images and the building records suggest that the eventual scheme took elements from several of the proposals. The building works were supervised for the Queen by Andrews Jelfe, who had earlier worked on the Hermitage and Merlin's Cave. The cabinet-maker Benjamin Goodison made the mahogany bookcases, with doors of brass mesh.

17.10

CHARLES WILD (1781–1835)
St James's Palace: The Queen's Library
c. 1819
Pen and ink with watercolour and gouache
7⅞ x 9⅞ in. (20 x 25.1 cm)
Royal Collection Trust, RCIN 922168

In 1737, the building of the library designed by William Kent for the stable yard at St James's Palace was finished, and Caroline's extensive collection of books could be installed. With windows overlooking Green Park there had been every expectation that this would provide a congenial setting for reading and conversation. However, it was while supervising the final fitting out that Caroline was taken ill, and she died before the work was finally complete. George II, however, ensured that the books were installed, and that Michael Rysbrack's terracotta busts of kings and queens were set on brackets above the bookshelves. The arrangements can be seen in this watercolour made by Charles Wild in 1819. The library was demolished in 1825.

CAROLINE AND LITERATURE

17.11

JACOBUS HOUBRAKEN (1698–1780), engraver, after **SIR GODFREY KNELLER** (1646–1723)
Joseph Addison
1748
Engraving
19⅞ x 14¼ in. (50.5 x 36.2 cm)
Yale Center for British Art, In Memory of John V. McCarthy, B2002.3.1

17.12*

JOSEPH ADDISON (1672–1719), editor, with **RICHARD STEELE** (1672–1729), editor
The Spectator, Vol. 1, No. 1
1 March 1711
13 x 11 in. (33 x 28 cm)
Printed for Sam Buckley, London
Beinecke Rare Book & Manuscript Library, Yale University, Z17 0221
Exhibited: Title page

Caroline's scheme for a new garden for Richmond Lodge was influenced by the ideas of Joseph Addison, the Whig journalist philosopher, who with the Earl of Shaftesbury and William Temple advocated new ways of expressing humankind's connection with nature. Caroline's garden followed Addison's radical prescription that the variety of nature should be celebrated and that even honest farmland had its place in the garden. It was chiefly through Richard Steele that Addison was able to devote himself to periodical literature, for which he is best known. Steele had founded *The Tatler* on 12 April 1709 and Addison contributed his first essay on 20 May. After the closure of *The Tatler*, Addison and Steele launched *The Spectator* on 1 March 1711, embarking on one of the most triumphant literary projects of the age. Caroline's library included *The Spectator*, as well as other works by Steele and Addison.

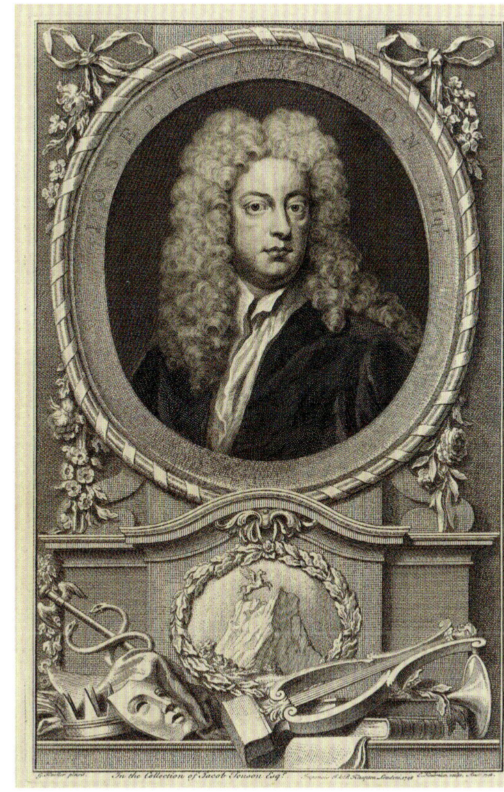

17.11

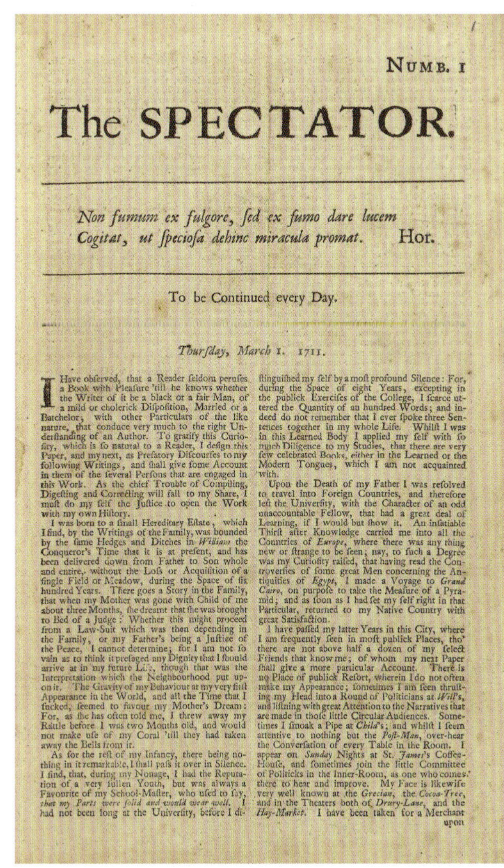

NUMB. 1

The SPECTATOR.

Non fumum ex fulgore, ſed ex fumo dare lucem
Cogitat, ut ſpecioſa dehinc miracula promat. Hor.

To be Continued every Day.

Thurſday, March 1. 1711.

I Have obſerved, that a Reader ſeldom peruſes a Book with Pleaſure 'till he knows whether the Writer of it be a black or a fair Man, of a mild or cholerick Diſpoſition, Married or a Batchelor, with other Particulars of the like nature, that conduce very much to the right Underſtanding of an Author. To gratify this Curioſity, which is ſo natural to a Reader, I deſign this Paper, and my next, as Prefatory Diſcourſes to my following Writings, and ſhall give ſome Account in them of the ſeveral Perſons that are engaged in this Work. As the chief Trouble of Compiling, Digeſting and Correcting will fall to my Share, I muſt do my ſelf the Juſtice to open the Work with my own Hiſtory.

I was born to a ſmall Hereditary Eſtate, which I find, by the Writings of the Family, was bounded by the ſame Hedges and Ditches in *William* the Conqueror's Time that it is at preſent, and has been delivered down from Father to Son whole and entire, without the Loſs or Acquiſition of a ſingle Field or Meadow, during the Space of ſix hundred Years. There goes a Story in the Family, that when my Mother was gone with Child of me about three Months, ſhe dreamt that ſhe was brought to Bed of a Judge: Whether this might proceed from a Law-Suit which was then depending in the Family, or my Father's being a Juſtice of the Peace, I cannot determine; for I am not ſo vain as to think it preſaged any Dignity that I ſhould arrive at in my future Life, though that was the Interpretation which the Neighbourhood put upon it. The Gravity of my Behaviour at my very firſt Appearance in the World, and all the Time that I ſucked, ſeemed to favour my Mother's Dream: For, as ſhe has often told me, I threw away my Rattle before I was two Months old, and would not make uſe of my Coral 'till they had taken away the Bells from it.

As for the reſt of my Infancy, there being nothing in it remarkable, I ſhall paſs it over in Silence. I find, that, during my Nonage, I had the Reputation of a very ſullen Youth, but was always a Favourite of my School-Maſter, who uſed to ſay, *that my Parts were ſolid and would wear well.* I had not been long at the Univerſity, before I diſtinguiſhed my ſelf by a moſt profound Silence: For, during the Space of eight Years, excepting in the publick Exerciſes of the College, I ſcarce uttered the Quantity of an hundred Words; and indeed do not remember that I ever ſpoke three Sentences together in my whole Life. Whilſt I was in this Learned Body I applied my ſelf with ſo much Diligence to my Studies, that there are very few celebrated Books, either in the Learned or the Modern Tongues, which I am not acquainted with.

Upon the Death of my Father I was reſolved to travel into Foreign Countries, and therefore left the Univerſity, with the Character of an odd unaccountable Fellow, that had a great deal of Learning, if I would but ſhow it. An inſatiable Thirſt after Knowledge carried me into all the Countries of *Europe*, where there was any thing new or ſtrange to be ſeen; nay, to ſuch a Degree was my Curioſity raiſed, that having read the Controverſies of ſome great Men concerning the Antiquities of *Egypt*, I made a Voyage to *Grand Cairo*, on purpoſe to take the Meaſure of a Pyramid; and as ſoon as I had ſet my ſelf right in that Particular, returned to my Native Country with great Satisfaction.

I have paſſed my latter Years in this City, where I am frequently ſeen in moſt publick Places, tho' there are not above half a dozen of my ſelect Friends that know me; of whom my next Paper ſhall give a more particular Account. There is no Place of publick Reſort, wherein I do not often make my Appearance; ſometimes I am ſeen thruſting my Head into a Round of Politicians at *Will's*, and liſtning with great Attention to the Narratives that are made in thoſe little Circular Audiences. Sometimes I ſmoak a Pipe at *Child's*; and whilſt I ſeem attentive to nothing but the *Poſt-Man*, over-hear the Converſation of every Table in the Room. I appear on *Sunday* Nights at St. *James's* Coffee-Houſe, and ſometimes join the little Committee of Politicks in the Inner-Room, as one who comes there to hear and improve. My Face is likewiſe very well known at the *Grecian*, the *Cocoa-Tree*, and in the Theaters both of *Drury-Lane*, and the *Hay-Market*. I have been taken for a Merchant upon

17.12

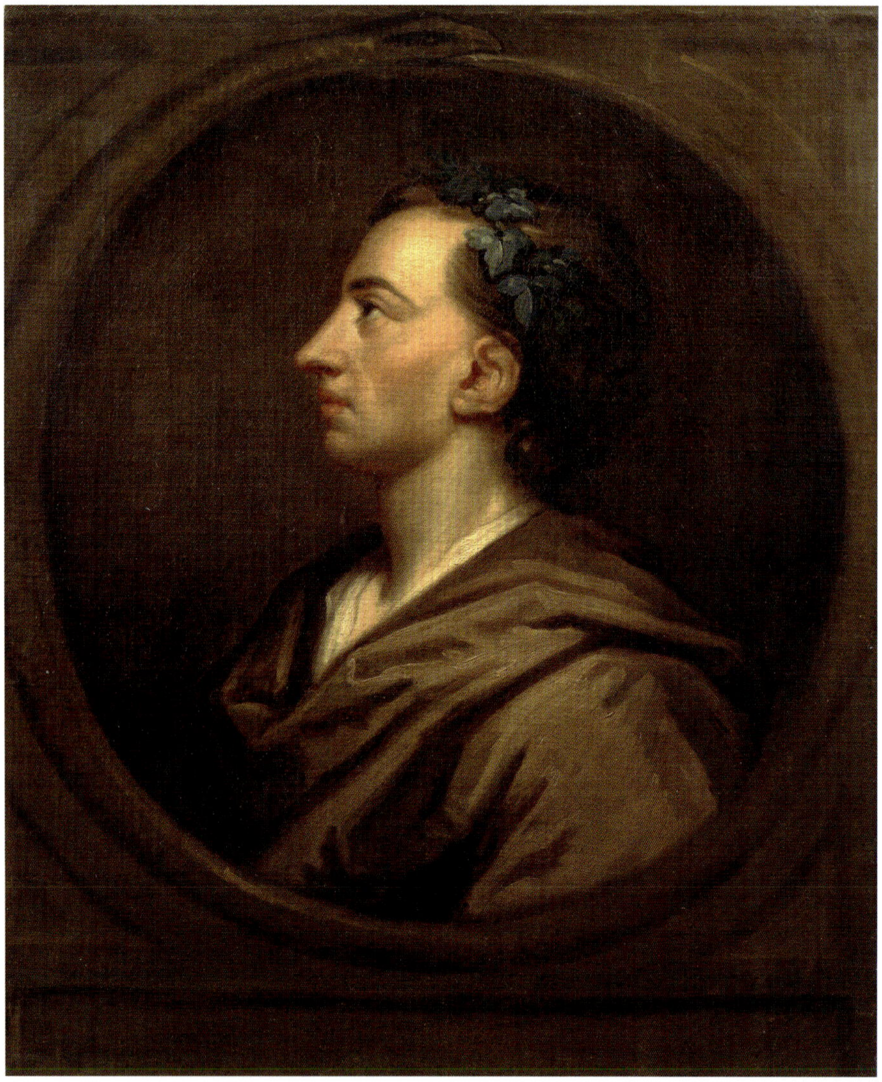

17.13

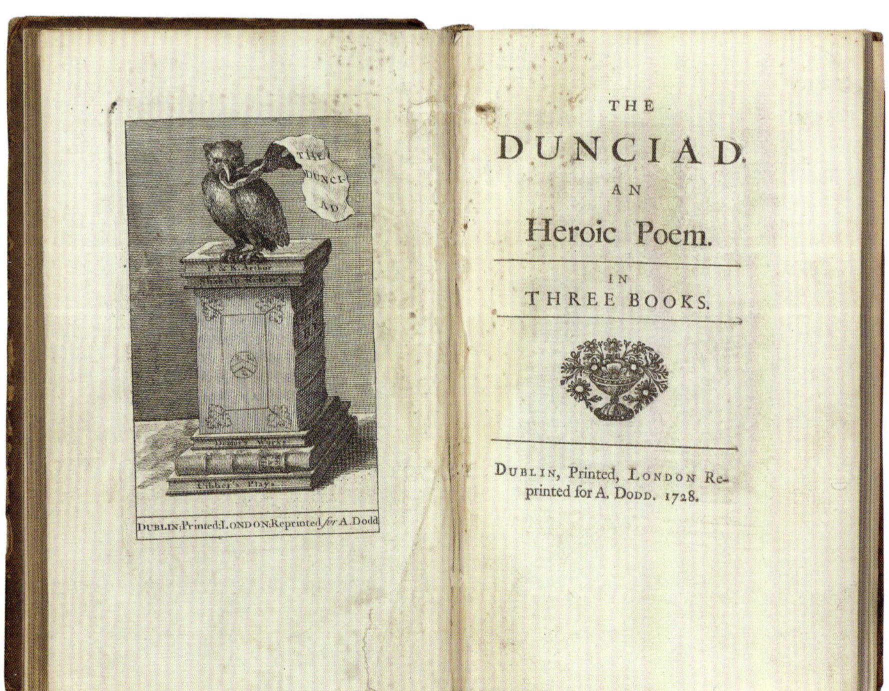

17.14

17.13*

STUDIO OF SIR GODFREY KNELLER
(1646–1723)
Alexander Pope Profile, Crowned with Ivy
c. 1721
Oil on canvas
28¼ x 23 in. (71.8 x 58.4 cm)
Yale Center for British Art, Gift of Margaret Wimsatt in memory of William Kurtz Wimsatt Jr, B1991.33

17.14*

ALEXANDER POPE (1688–1744)
The Dunciad. An Heroic Poem. In Three Books.
1728
Reprinted for Anne Dodd, London
Beinecke Rare Book & Manuscript Library, Yale Collection of American Literature, ik P810 728b
Exhibited: Title page

Queen Caroline's library included this contemporary satire in which she was targeted for ridicule. As a Roman Catholic, Alexander Pope was barred from office at Court, but Caroline seems nonetheless to have drawn the philosopher-writer into her debates. A description of her gardening conference held in 1719, at which Pope was evidently present, is found in one of his letters to his patron Lord Bathurst. In the development of his own garden at Twickenham he received building materials left over from the works on the royal garden at Richmond, possibly though the auspices of Charles Bridgeman, the gardener working both for him and for Caroline. Any early admiration for the Queen dissipated once the gift of offices and favours to some of his friends proved unforthcoming. In the *Dunciad* he cast her as the 'Queen of Dulness' for her promotion of those he thought mediocre.

17.15

CHARLES JERVAS (c. 1675–1739)
Jonathan Swift
c. 1718
Oil on canvas
48½ x 38¼ in. (123.2 x 97.2 cm)
National Portrait Gallery, London, NPG 278

Though known for his biting wit, often directed at departments of government and the monarchy, Caroline was determined to meet the Irish writer Jonathan Swift at her Drawing-Room. He received his invitation though Henrietta Howard, later Lady Suffolk, Caroline's Woman of the Bedchamber, who explained that Caroline 'loved to see odd persons, and having sent for a wild boy from Germany, had curiosity to see a wild Dean from Ireland'. It took many requests before Swift agreed. He sent Caroline Irish textiles, which she and her daughters wore and promoted at court. According to Swift, this resulted in them becoming the toast of the Irish weavers. Initially he praised Caroline as a 'freethinker', but matters changed after his friend John Gay felt himself humiliated when the royal office he had sought resulted in the offer of an appointment as Gentleman Usher to an infant princess, and he became deeply critical.

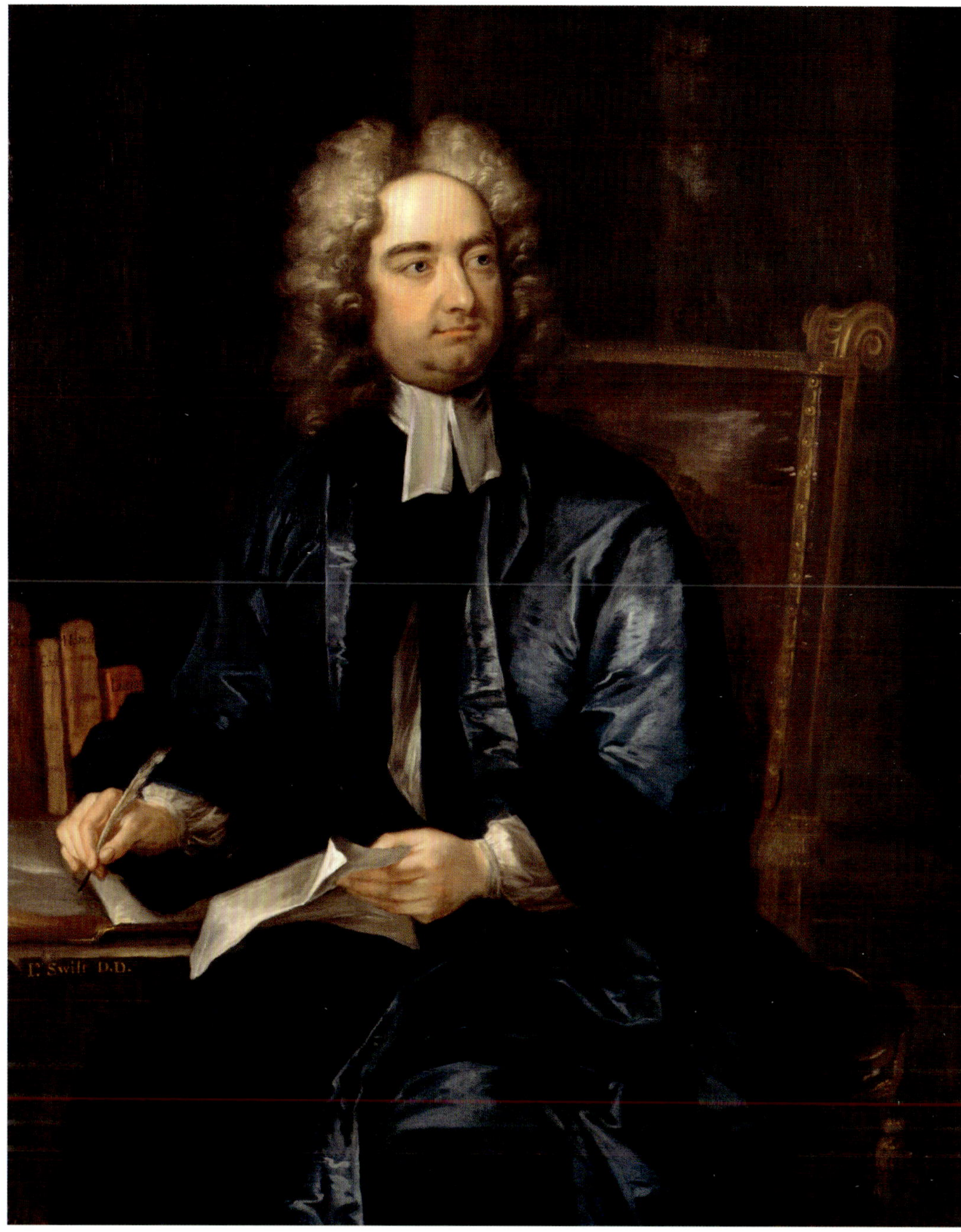

17.15

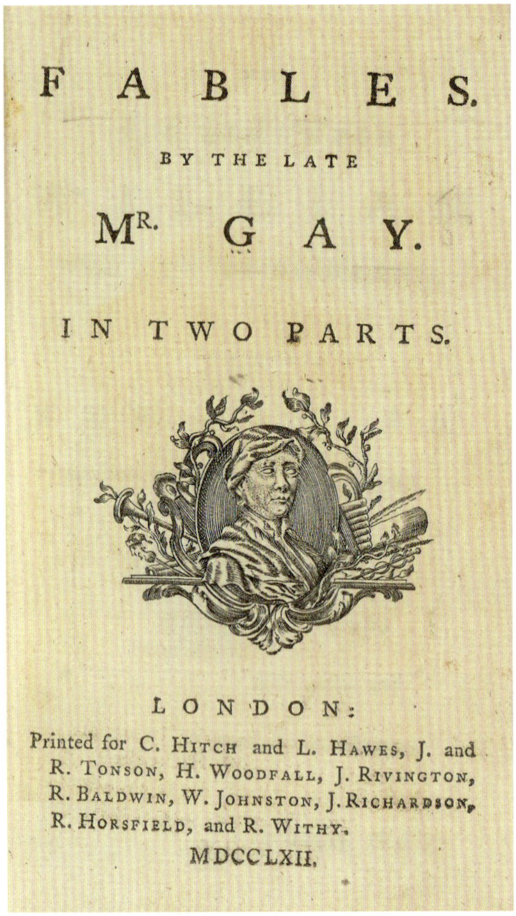
FABLES.

BY THE LATE

MR. GAY.

IN TWO PARTS.

LONDON:
Printed for C. HITCH and L. HAWES, J. and
R. TONSON, H. WOODFALL, J. RIVINGTON,
R. BALDWIN, W. JOHNSTON, J. RICHARDSON,
R. HORSFIELD, and R. WITHY.
MDCCLXII.

17.16

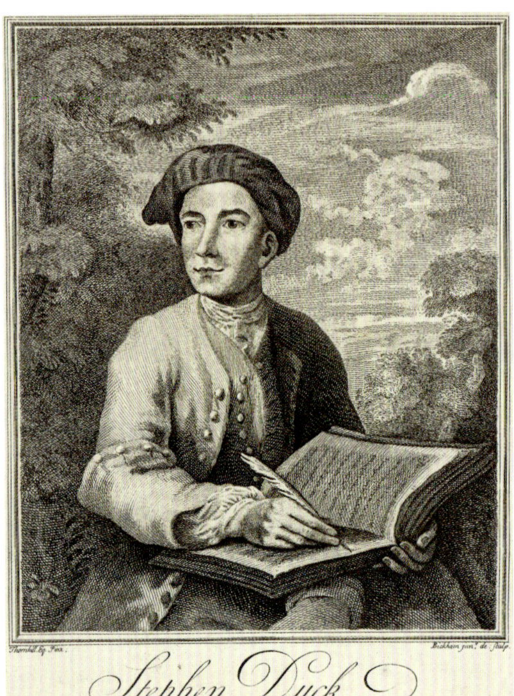

17.17

17.16*

JOHN GAY (1685–1732)
Fables by the late Mr. Gay
1762
Printed for C. Hitch and L. Hawes, J. and R. Tonson, H. Woodfall, J. Rivington, R. Baldwin, W. Johnston, J. Richardson, R. Horsfield, and R. Withy, London
Beinecke Rare Book & Manuscript Library, Yale University, ik G252 727hdc
Exhibited: Title page

In 1727 John Gay wrote *Fifty-one Fables in Verse* and dedicated it to six-year-old Prince William, later the Duke of Cumberland. For many years Gay had been assiduous in his attendance on Caroline, seeking an appointment with a salary to provide him with a little financial stability. He was offered the situation of Gentleman Usher to the Princess Louisa, Prince William's two-year-old sister. Whether the offer was meant as a snub to the ambitious author is hard to establish. It may be that having written a book for children, Caroline thought this post and its salary would be appropriate. However, Gay turned down the offer as too demeaning. He would be congratulated by his friends Alexander Pope and Jonathan Swift for not allowing himself to be caught in the royal thrall. Gay's *Fables* could be found on the shelves of Caroline's Library.

17.17

GEORGE BICKHAM THE YOUNGER (c. 1706–1771), engraver, after SIR JAMES THORNHILL (1675–1734)
Stephen Duck
c. 1736
Etching on paper
8½ x 6 in. (21.5 x 15.4 cm)
Royal Collection Trust, RCIN 653851

Stephen Duck was an agricultural labourer from Wiltshire, whose natural talent for poetry brought him to the attention of Dr Alured Clarke, Caroline's Clerk of the Chamber. After the publication of his works 'The Thresher's Labour' and 'The Shunammite' in 1730, Clarke brought him to London and introduced him to Caroline. She, impressed and interested by his work, appointed him initially Yeoman of the Guard and Keeper of Duck Island in St James's Park in 1733. In 1735 he was made hermit-interpreter for the folly temples Caroline had constructed within the garden at Richmond Lodge. A small specialist library was established there for his education and improvement. Duck remained 'Librarian of the Queen's Select Library' until he took holy orders, and was appointed chaplain to the Brigade of Dragoon Guards, and moved to a living at Byfleet. He committed suicide in 1756, by walking into the River Thames. The engraving of Stephen Duck, after a work by Sir James Thornhill, was included as a frontispiece for his *Poems on Several Occasions* (1736).

CHARLOTTE'S LIBRARIES

17.18*

JAMES CHRISTIE (1730–1803)
A Catalogue of the Genuine Library, Prints, and Books of Prints, of an Illustrious Personage, Lately Deceased
Annotated in pencil: 'Charlotte Q. Consort of K. George III'
1819
British Library, 123.f16
Exhibited: pp. 108–109, lots 3306–62

Queen Charlotte's library was sold at auction over twenty-three days between 9 and 23 June 1819. It was just one of six sales of the Queen's effects, which took place at Christie's auction house in Westminster during the spring and summer that year. The beneficiaries were to be Charlotte's unmarried daughters. For the sale of the library the books had been divided by subject. The auction of the theology section alone took four days. There were in total 4,515 lots.

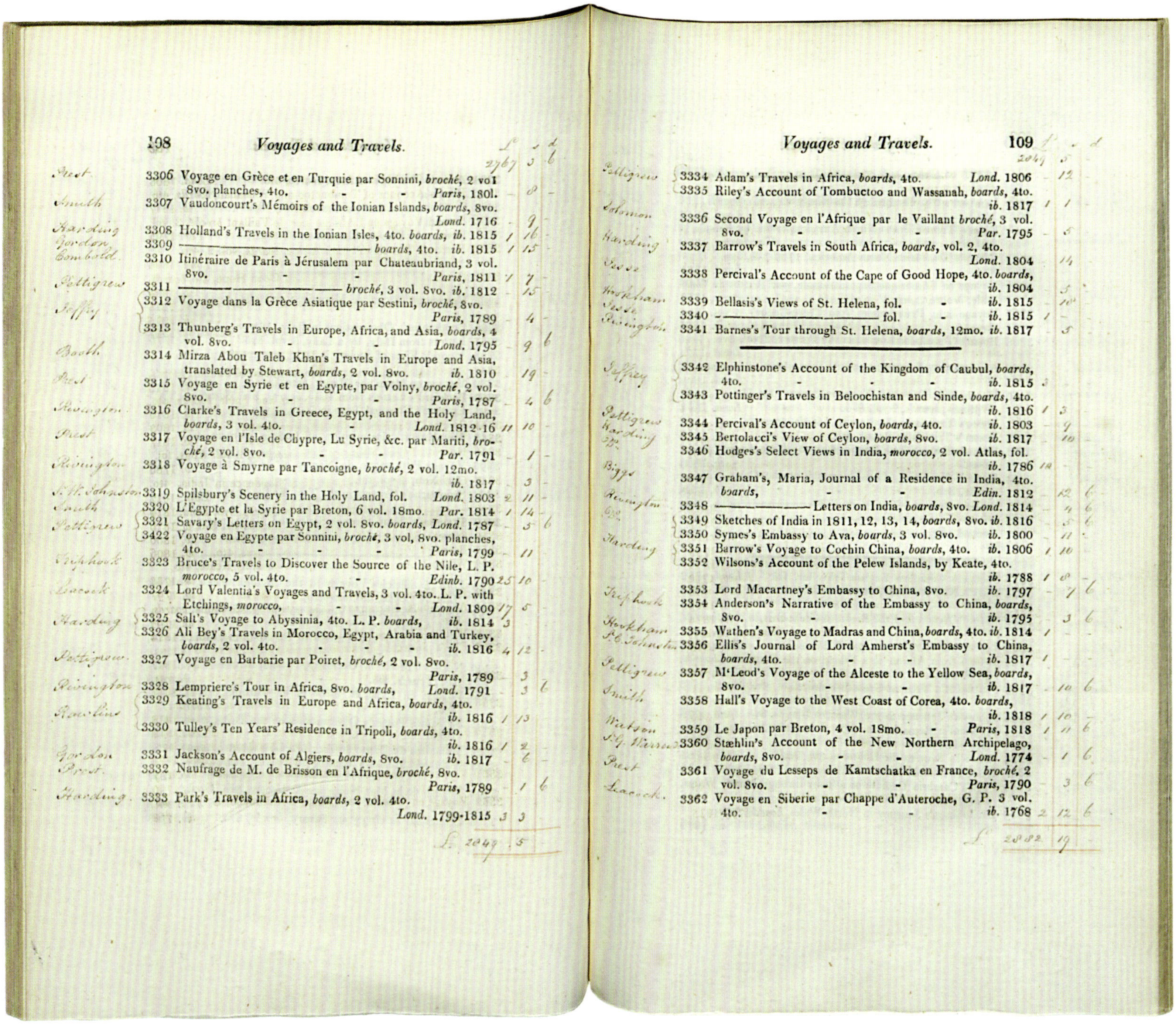
108 *Voyages and Travels.*

3306 Voyage en Grèce et en Turquie par Sonnini, *broché*, 2 vol 8vo. planches, 4to. - - *Paris*, 1801.
3307 Vaudoncourt's Mémoirs of the Ionian Islands, *boards*, 8vo. *Lond.* 1716
3308 Holland's Travels in the Ionian Isles, 4to. *boards*, *ib.* 1815
3309 ——————— *boards*, 4to. *ib.* 1815
3310 Itinéraire de Paris à Jérusalem par Chateaubriand, 3 vol. 8vo. - - *Paris*, 1811
3311 ——————— *broché*, 3 vol. 8vo. *ib.* 1812
3312 Voyage dans la Grèce Asiatique par Sestini, *broché*, 8vo. *Paris*, 1789
3313 Thunberg's Travels in Europe, Africa, and Asia, *boards*, 4 vol. 8vo. - - *Lond.* 1795
3314 Mirza Abou Taleb Khan's Travels in Europe and Asia, translated by Stewart, *boards*, 2 vol. 8vo. *ib.* 1810
3315 Voyage en Syrie et en Egypte, par Volny, *broché*, 2 vol. 8vo. - - *Paris*, 1787
3316 Clarke's Travels in Greece, Egypt, and the Holy Land, *boards*, 3 vol. 4to. - *Lond.* 1812-16
3317 Voyage en l'Isle de Chypre, Lu Syrie, &c. par Mariti, *broché*, 2 vol. 8vo. - - *Par.* 1791
3318 Voyage à Smyrne par Tancoigne, *broché*, 2 vol. 12mo. *ib.* 1817
3319 Spilsbury's Scenery in the Holy Land, fol. *Lond.* 1803
3320 L'Egypte et la Syrie par Breton, 6 vol. 18mo. *Par.* 1814
3321 Savary's Letters on Egypt, 2 vol. 8vo. *boards*, *Lond.* 1787
3422 Voyage en Egypte par Sonnini, *broché*, 3 vol, 8vo. planches, 4to. - - - *Paris*, 1799
3323 Bruce's Travels to Discover the Source of the Nile, L. P. *morocco*, 5 vol. 4to. - *Edinb.* 1790
3324 Lord Valentia's Voyages and Travels, 3 vol. 4to. L. P. with Etchings, *morocco*, - - *Lond.* 1809
3325 Salt's Voyage to Abyssinia, 4to. L. P. *boards*, *ib.* 1814
3326 Ali Bey's Travels in Morocco, Egypt, Arabia and Turkey, *boards*, 2 vol. 4to. - - - *ib.* 1816
3327 Voyage en Barbarie par Poiret, *broché*, 2 vol. 8vo. *Paris*, 1789
3328 Lempriere's Tour in Africa, 8vo. *boards*, *Lond.* 1791
3329 Keating's Travels in Europe and Africa, *boards*, 4to. *ib.* 1816
3330 Tulley's Ten Years' Residence in Tripoli, *boards*, 4to. *ib.* 1816
3331 Jackson's Account of Algiers, *boards*, 8vo. *ib.* 1817
3332 Naufrage de M. de Brisson en l'Afrique, *broché*, 8vo. *Paris*, 1789
3333 Park's Travels in Africa, *boards*, 2 vol. 4to. *Lond.* 1799-1815

Voyages and Travels. 109

3334 Adam's Travels in Africa, *boards*, 4to. *Lond.* 1806
3335 Riley's Account of Tombuctoo and Wassanah, *boards*, 4to. *ib.* 1817
3336 Second Voyage en l'Afrique par le Vaillant *broché*, 3 vol. 8vo. - - - *Par.* 1795
3337 Barrow's Travels in South Africa, *boards*, vol. 2, 4to. *Lond.* 1804
3338 Percival's Account of the Cape of Good Hope, 4to. *boards*, *ib.* 1804
3339 Bellasis's Views of St. Helena, fol. - *ib.* 1815
3340 ——————— fol. - *ib.* 1815
3341 Barnes's Tour through St. Helena, *boards*, 12mo. *ib.* 1817

3342 Elphinstone's Account of the Kingdom of Caubul, *boards*, 4to. - - - *ib.* 1815
3343 Pottinger's Travels in Beloochistan and Sinde, *boards*, 4to. *ib.* 1816
3344 Percival's Account of Ceylon, *boards*, 4to. *ib.* 1803
3345 Bertolacci's View of Ceylon, *boards*, 8vo. *ib.* 1817
3346 Hodges's Select Views in India, *morocco*, 2 vol. Atlas, fol. *ib.* 1786
3347 Graham's, Maria, Journal of a Residence in India, 4to. *boards*, - - *Edin.* 1812
3348 ——————— Letters on India, *boards*, 8vo. *Lond.* 1814
3349 Sketches of India in 1811, 12, 13, 14, *boards*, 8vo. *ib.* 1816
3350 Symes's Embassy to Ava, *boards*, 3 vol. 8vo. *ib.* 1800
3351 Barrow's Voyage to Cochin China, *boards*, 4to. *ib.* 1806
3352 Wilsons's Account of the Pelew Islands, by Keate, 4to. *ib.* 1788
3353 Lord Macartney's Embassy to China, 8vo. *ib.* 1797
3354 Anderson's Narrative of the Embassy to China, *boards*, 8vo. - - *ib.* 1795
3355 Wathen's Voyage to Madras and China, *boards*, 4to. *ib.* 1814
3356 Ellis's Journal of Lord Amherst's Embassy to China, *boards*, 4to. - - *ib.* 1817
3357 M'Leod's Voyage of the Alceste to the Yellow Sea, *boards*, 8vo. - - *ib.* 1817
3358 Hall's Voyage to the West Coast of Corea, 4to. *boards*, *ib.* 1818
3359 Le Japon par Breton, 4 vol. 18mo. - *Paris*, 1818
3360 Stæhlin's Account of the New Northern Archipelago, *boards*, 8vo. - - *Lond.* 1774
3361 Voyage du Lesseps de Kamtschatka en France, *broché*, 2 vol. 8vo. - - *Paris*, 1790
3362 Voyage en Siberie par Chappe d'Auteroche, G. P. 3 vol. 4to. - - *ib.* 1768

17.18

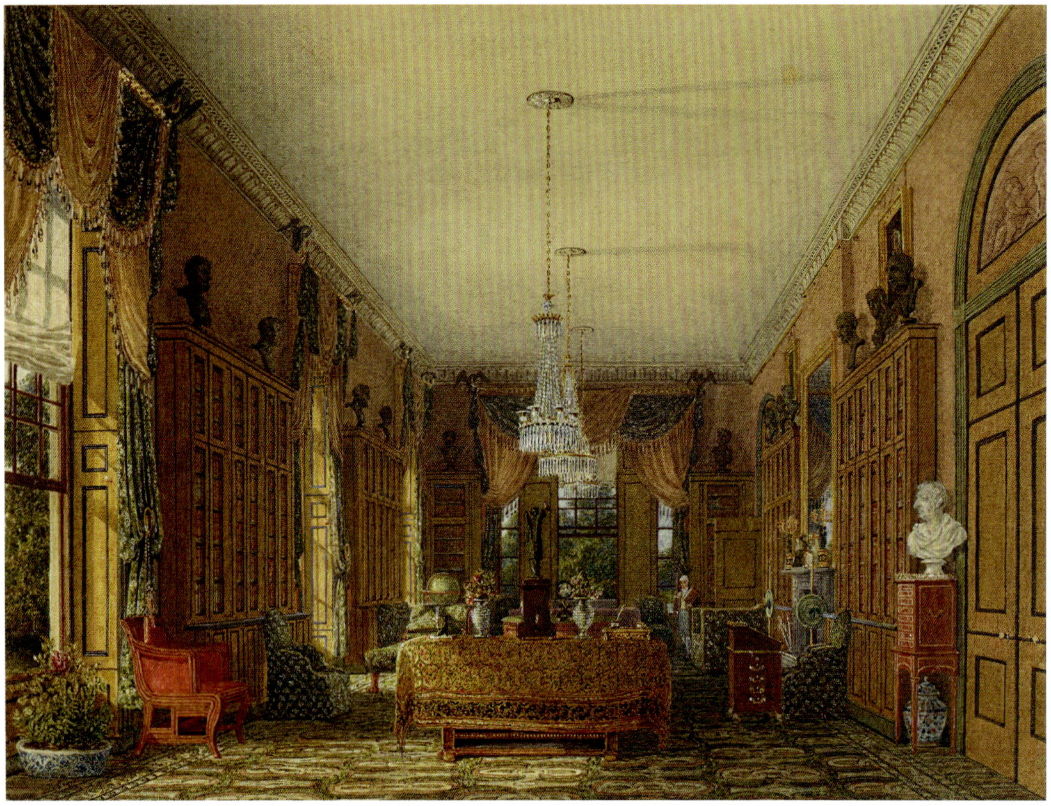

17.19

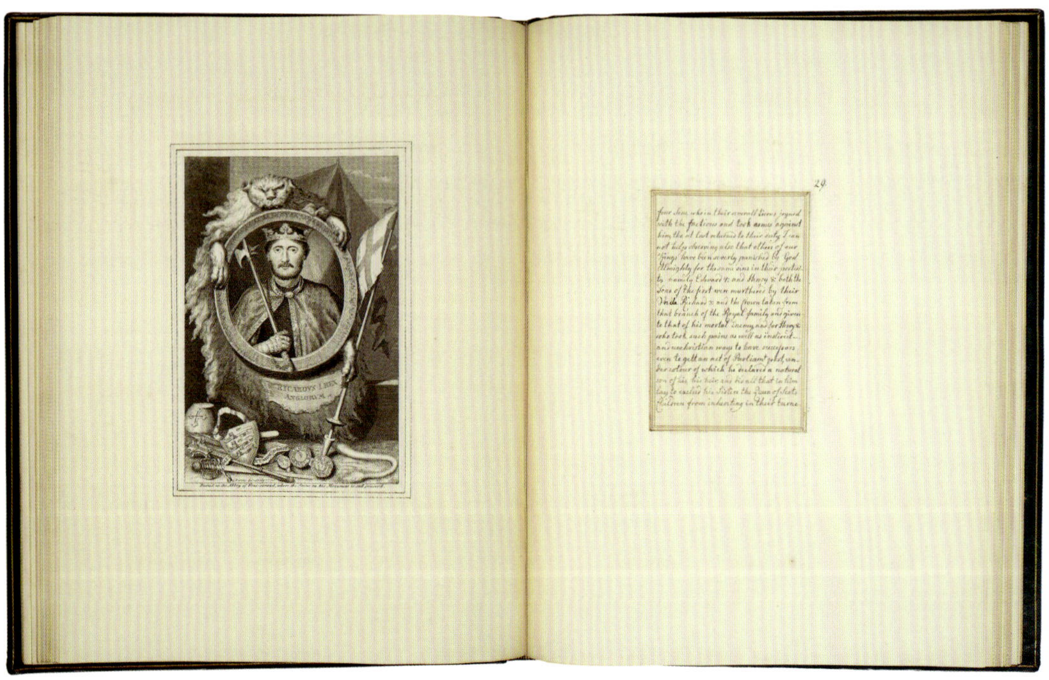

17.20

17.19

CHARLES WILD (1781–1835)
Frogmore House: The Queen's Library
c. 1817
Watercolour and gouache over pencil on paper
7⅞ x 10⅜ in. (20 x 26.1 cm)
Royal Collection Trust, RCIN 922120

Queen Caroline's library at Frogmore housed her botanical collections, printing press, and a substantial portion of her book collection, including the herbarium of the Reverend John Lightfoot, acquired by George III in 1788 as a gift for the Queen. This record of the library's interior was prepared for one of the plates in William Henry Pyne's multi-volume *The History of the Royal Residences of Windsor Castle, St James's Palace, Carlton House, and Frogmore* (London: Printed for A. Dry, 1816–19).

17.20*

QUEEN CHARLOTTE (1744–1818), copyist, after JAMES II (1633–1701), author
'Copy of a manuscript in the Prince Regents possession, found in the Stuart papers / transcribed by Her Majesty Queen Charlotte in her own handwriting. anno domini 1812'
1812
Manuscript, bound with engraved prints
22½ in. (57 cm)
Royal Collection Trust, RCIN 1046711
Exhibited: p. 29, 'Ricardus I. Rex Anglorum: Buried in the Abbey of Font-evraud, where the Statue on his Monument is still preserv'd.'

In this folio volume Queen Charlotte copied out a small notebook of advice written in exile by James II for his son. The original notebook had been brought from the Continent with other papers belonging to the Stuarts, and Charlotte illustrated her transcription with portrait prints, mounted together with the text. Charlotte, like Caroline before her, had a fascination with genealogy, and this led to the making of a collection of printed portraits, or 'heads'. During the eighteenth century this had become a fashion extending even to those of moderate means, allowing them to feel they were contributing in a small way to the contemporary debate about nationalism. The practice of pasting such imagery into volumes came to be known as 'grangerizing', after the Reverend James Granger (1723–1776), whose *Biographical History of England* (1769) included blank pages for illustrations to be supplied by the reader.

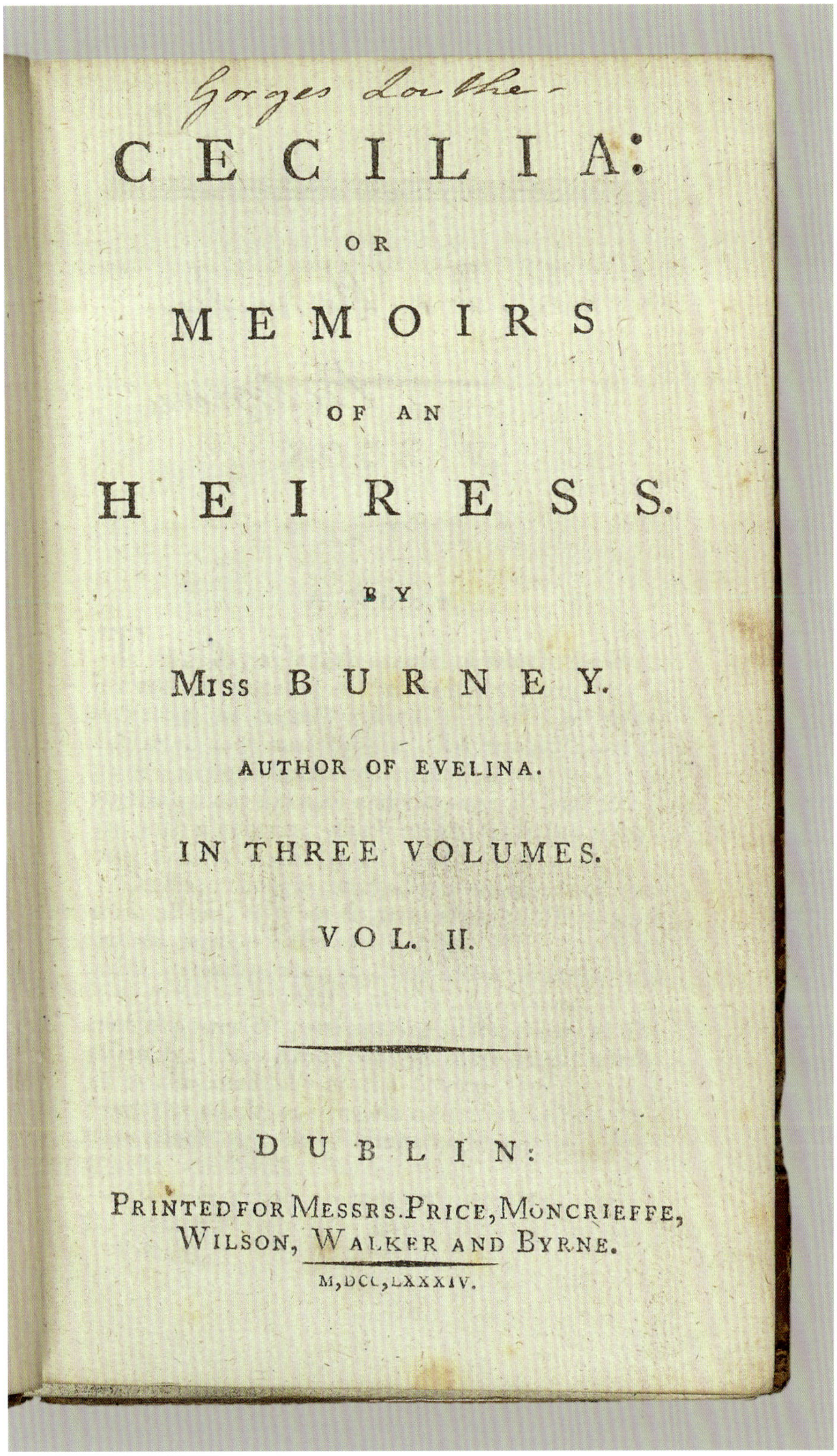
Gorges Lowther

CECILIA:

OR

MEMOIRS

OF AN

HEIRESS.

BY

MISS BURNEY.

AUTHOR OF EVELINA.

IN THREE VOLUMES.

VOL. II.

DUBLIN:

PRINTED FOR MESSRS. PRICE, MONCRIEFFE, WILSON, WALKER AND BYRNE.

M,DCC,LXXXIV.

17.21

QUEEN CHARLOTTE AND HER LITERARY INTERESTS

17.21*

FANNY [FRANCES] BURNEY

(1752–1840)
Cecilia: or, Memoirs of an Heiress. By Miss Burney. Author of Evelina (5 vols)
1782
7 in. (18 cm)
T. Payne and son and T. Cadell, London
Beinecke Rare Book & Manuscript Library, Yale University, Z78 9
Exhibited: Title page (illustrated), and vol. 5, p. 379

Frances—Fanny—Burney was the daughter of the ambitious musicologist Charles Burney. She was self-educated but the spirited journal-diaries she wrote from a young age caught the attention of Samuel Crisp, who encouraged her to write her first novel, *Evelina or the History of a Young Lady's Entrance into the World*, published anonymously in 1778, without her father's knowledge or permission. It was a critical success and brought her interest from Dr Johnson, Edmund Burke, and Hester Thrale. In 1782 she published *Cecilia: or Memoirs of an Heiress*. In 1785, thanks to her association with Mary Granville, Mrs Delany, Fanny travelled to the court of King George III and Queen Charlotte, where the Queen offered her the post of 'Keeper of the Robes', with a salary of £200 per annum. Unmarried at thirty-four, she decided that improved social status and income might allow her greater freedom to write. She developed a warm relationship with the Queen and princesses that lasted into her later years. However, frustrated at the lack of time available for writing and aggravated by court politics, Fanny left her post in 1790. The journal-diary written over these years provides a vivid account of life at court, and the familial dynamics within the royal family. The catalogue of Charlotte's library shows that she owned copies of all Fanny Burney's works.

17.22*

JOSEPH NOLLEKENS (1737–1823)
Samuel Johnson
1777
Plaster
24 x 14 x 11 in. (61 x 35.6 x 27.9 cm)
Yale Center for British Art, Paul Mellon Fund, B1990.17

17.23*

JOHN JONES (c. 1745–1797), engraver, after **SIR JOSHUA REYNOLDS** (1723–1792)
James Boswell of Auchinleck, Esq.
1786
Mezzotint
17⅞ x 12¼ in. (45.4 x 31.1 cm)
Published by J. Jones, London
Yale Center for British Art, Paul Mellon Fund, B1970.3.78

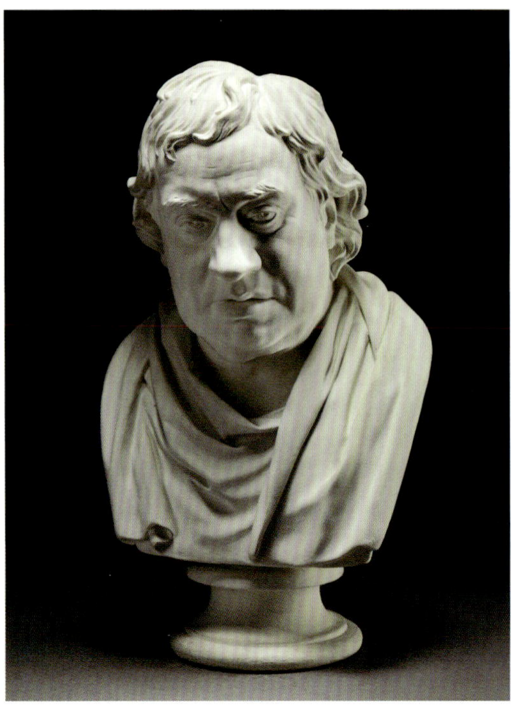

17.22

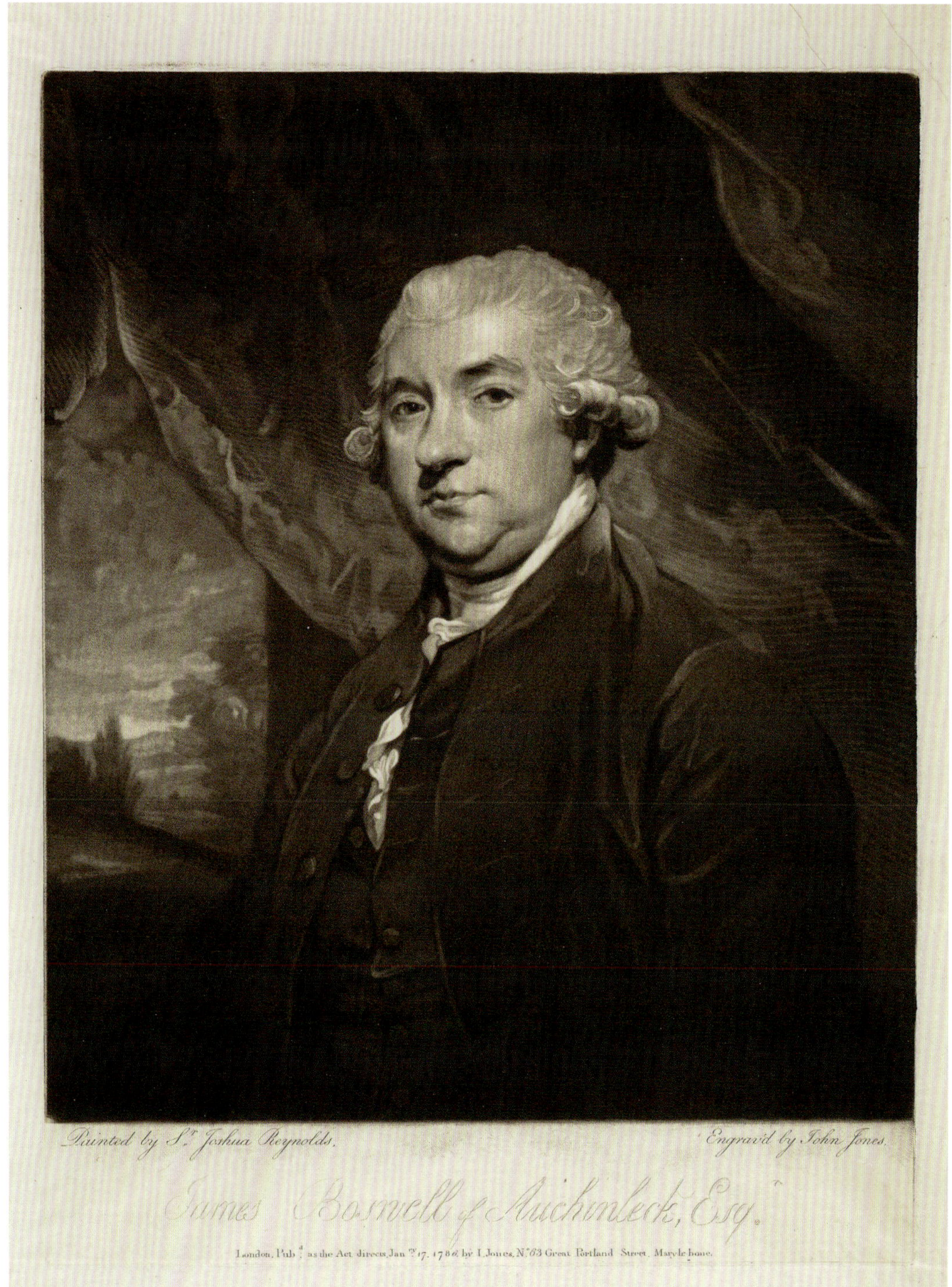

17.23

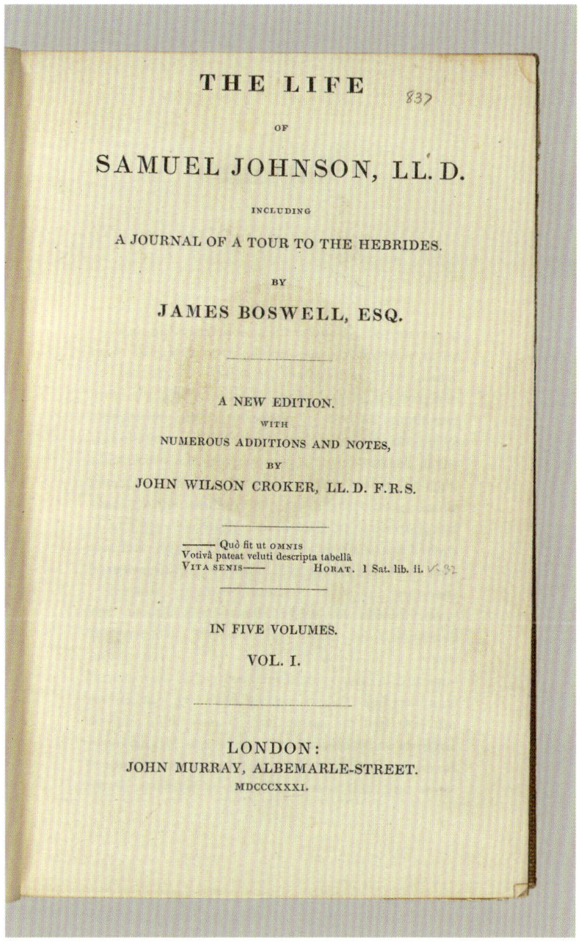

THE LIFE

OF

SAMUEL JOHNSON, LL. D.

INCLUDING

A JOURNAL OF A TOUR TO THE HEBRIDES.

BY

JAMES BOSWELL, ESQ.

A NEW EDITION.

WITH

NUMEROUS ADDITIONS AND NOTES,

BY

JOHN WILSON CROKER, LL. D. F.R.S.

—— Quò fit ut OMNIS
Votivâ pateat veluti descripta tabellâ
VITA SENIS—— HORAT. 1 Sat. lib. ii.

IN FIVE VOLUMES.

VOL. I.

LONDON:
JOHN MURRAY, ALBEMARLE-STREET.
MDCCCXXXI.

17.24

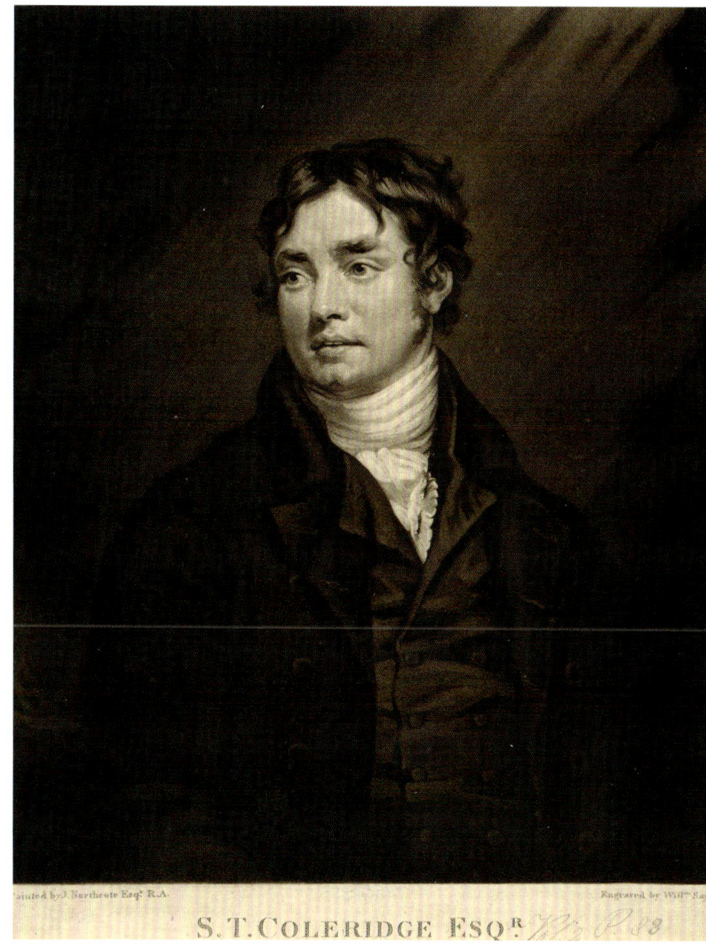

17.25

17.24*

JAMES BOSWELL (1740–1795)
The Life of Samuel Johnson, LL.D. (2 vols)
1791
12 ¼ in. (31 cm)
Printed by Henry Baldwin for Charles Dilly, London
Beinecke Rare Book & Manuscript Library, Yale University, BEIN Im J637 +W791 copy 1
Exhibited: Title page (illustrated) with frontispiece by James Heath (1757–1834), engraver, after Sir Joshua Reynolds (1723–1792)

James Boswell's biography of Dr Samuel Johnson was a popular and critical success when first published. Johnson was a key player in literary circles in London from the mid-eighteenth century, revered for his *Dictionary of the English Language.* He had encouraged Fanny Burney in her literary ambition; Burney leaves a compelling record of Charlotte and her own reading habits, and their discussions which touched on Boswell's work:

'Miss Burney, have you heard that Boswell is going to publish a life of your friend Dr Johnson?'

'No, ma'am.'

'I tell you as I heard. I don't know for the truth of it, and I can't tell what he will do. He is so extraordinary a man, that perhaps he will devise something extraordinary. What do you think of Madame de Genlis' last work?'

'I have not read it, ma'am.'

'Not read it?'

17.25*

WILLIAM SAY (1768–1834), engraver, after JAMES NORTHCOTE (1746–1831)
S. T. Coleridge Esq.
Undated
Mezzotint
14¼ x 10⅞ in. (36 x 27.6 cm)
Published by William Say, London
Yale Center for British Art, Paul Mellon Collection, B1977.14.9858

The poet, literary critic, and philosopher Samuel Taylor Coleridge, with his friend William Wordsworth, are regarded as founders of the Romantic movement in England. Coleridge wrote the poems 'The Rime of the Ancient Mariner' and 'Kubla Khan', as well as the major prose work *Biographia Literaria.* His criticism, especially on Shakespeare, was highly influential and he helped introduce German idealist philosophy to English-speaking culture. He was a major influence on Ralph Waldo Emerson and American Transcendentalism. Queen Charlotte's library catalogue contains a significant number of contemporary works, including those by Coleridge and Jane Austen.

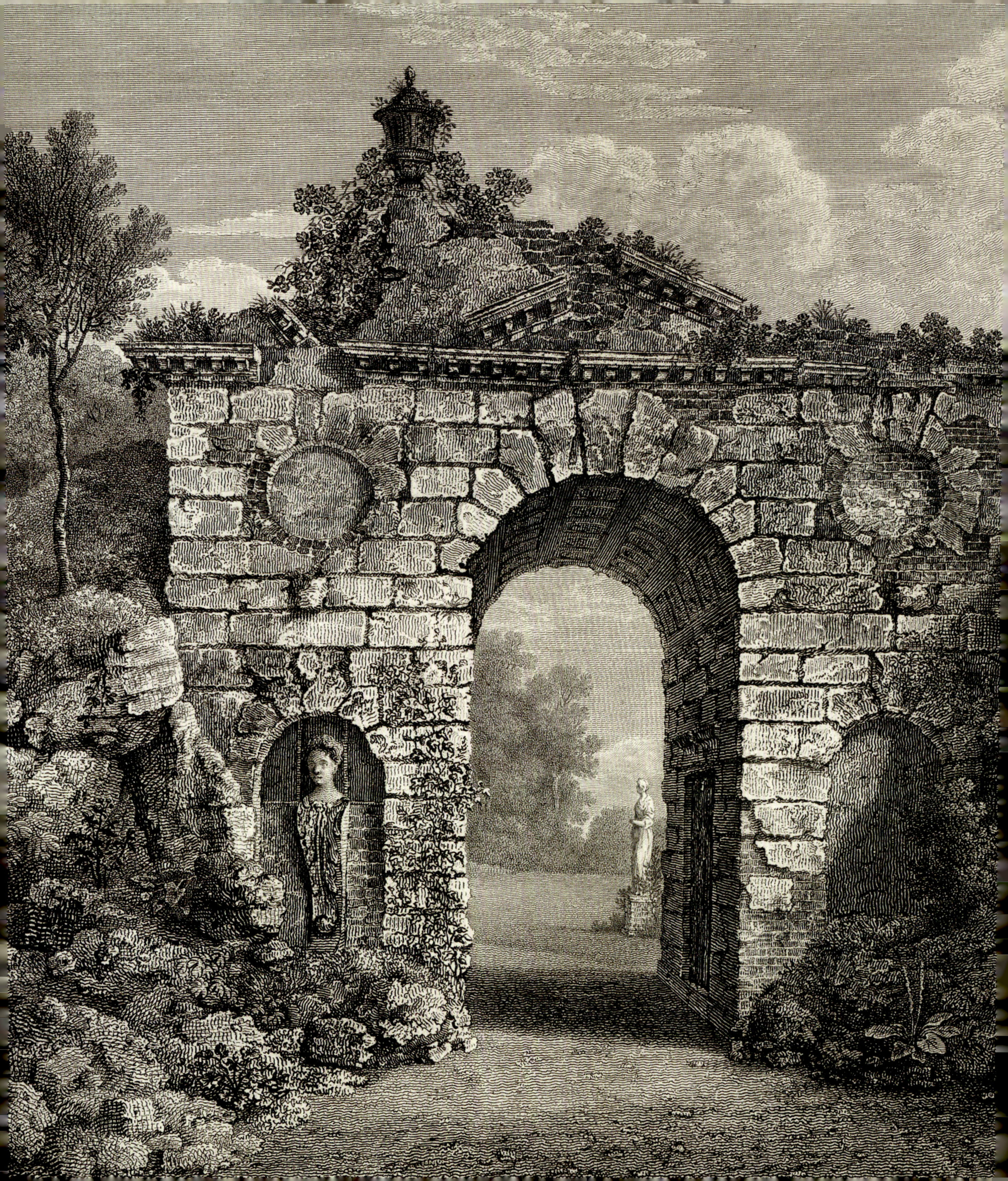

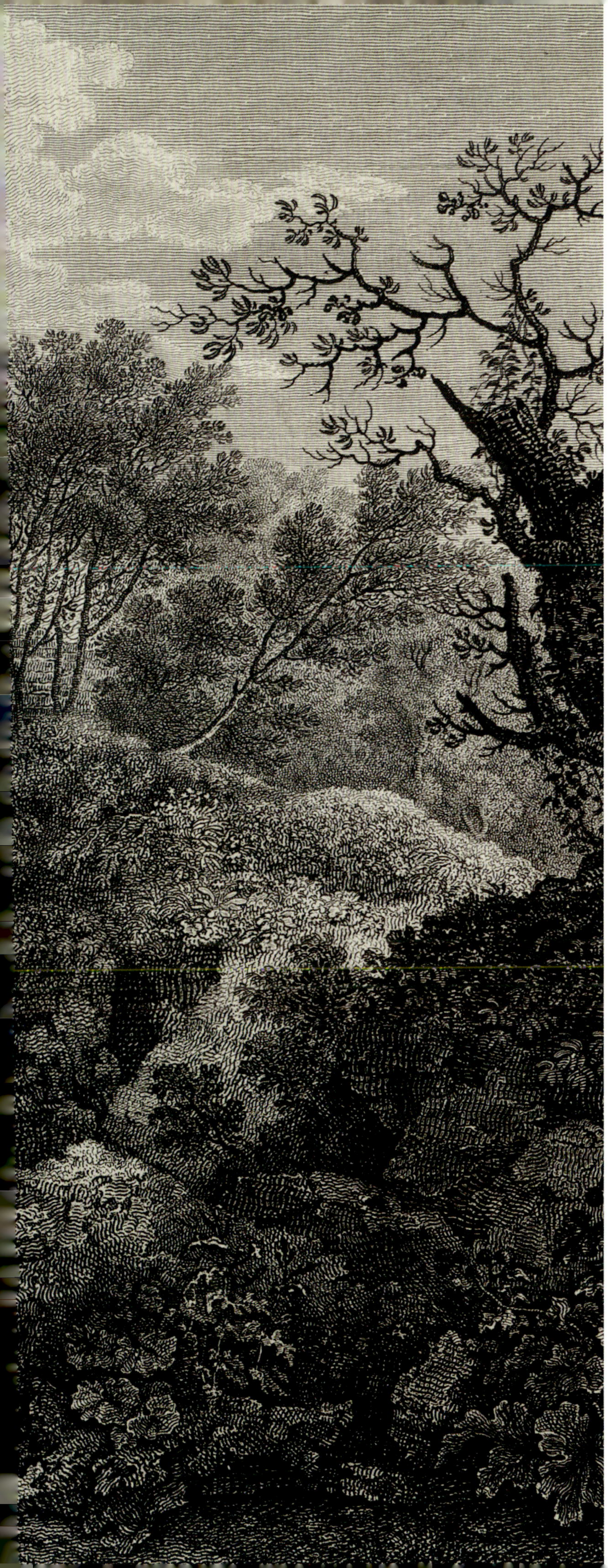

Architecture and Gardens

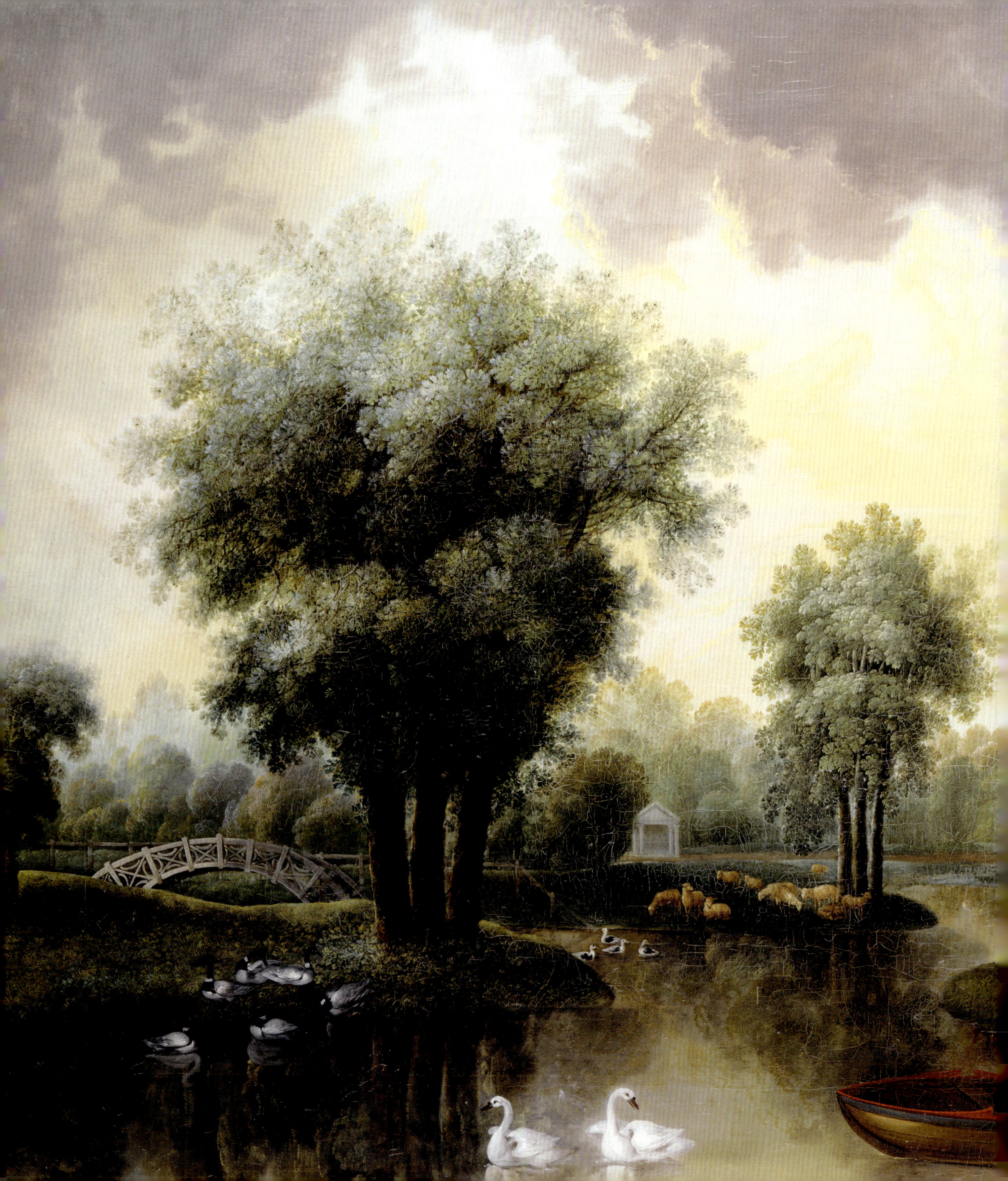

The Royal Estate

Roisin Inglesby

While not all of the Princesses' architectural projects and gardens survive, we have ample evidence of the ways in which Caroline of Ansbach, Augusta of Saxe-Gotha, and Charlotte of Mecklenburg-Strelitz separately, and collectively, created and enriched their homes, and the gardens which surrounded them, for themselves and their families. The topographical and architectural prints, drawings, and watercolours that remain do more than merely tell us what the exteriors of the Princesses' homes looked like; their very existence also provides us with clues about how their architectural ventures were perceived, the social connotations of their projects, and the balance that Caroline, Augusta, and Charlotte struck between their private and public lives. At the same time, views of the interiors of the Princesses' homes remained largely elusive. With some notable exceptions, such as the engraved diagrams by George Vertue illustrating picture-hanging arrangements at Kensington Palace, depictions of the Princesses' living quarters are rare until Charlotte's reign, their absence an indication of the limits of the public gaze and the relative domestic seclusion enjoyed by Caroline and Augusta.

In focusing on why and for whom these works were—or were not—created, we can see the Princesses' home-making in the context of the vibrant information culture of the time, and in relation to its audience—the British public. In England during the long eighteenth century there were huge developments in printmaking techniques. Images could be reproduced quickly and cheaply and entered the larger realm of print culture, lapped up by an increasingly culturally literate public. Topographical prints were especially popular, serving as illustrations for the guidebooks that proliferated from the 1730s onwards.

Caroline was particularly astute in recognizing the value of public engagement with her architectural projects. While visitors were permitted to visit her pioneering gardens at Richmond, many more encountered this rural home through widely published maps and prints. A map from 1734 by the French cartographer John Rocque provides an explanation of the Queen's lodgings in both English and French, an indication of the international appeal of views of the royal residences (fig. 18.01). Although she had used Richmond as a rural retreat since the future George II was expelled from court in 1717, Caroline's work on the site began in earnest after she received Richmond Lodge as a dower house upon her husband's accession to the throne in 1727. She employed Charles Bridgeman to remodel the grounds, while William Kent designed a series of imaginative garden structures; amongst these were the Hermitage (begun in 1730) and Merlin's Cave (1735). A popular mass-produced print, published by the London print seller Carington Bowles, depicts the exterior of the Hermitage, a rustic cottage of cyclopean masonry, nestled into the surrounding landscape (cat. 19.08). A later print published by Kent's protégé John Vardy in 1744, shows the building's interior (cat. 17.08). A typical cutaway section drawing, popular in architectural representations since Colen Campbell's *Vitruvius Britannicus* of 1715–25, Vardy's engraving reflects the architectural restraint of Kent's Palladian interior, an unexpected juxtaposition

Detail of p. 331, cat. 18.05

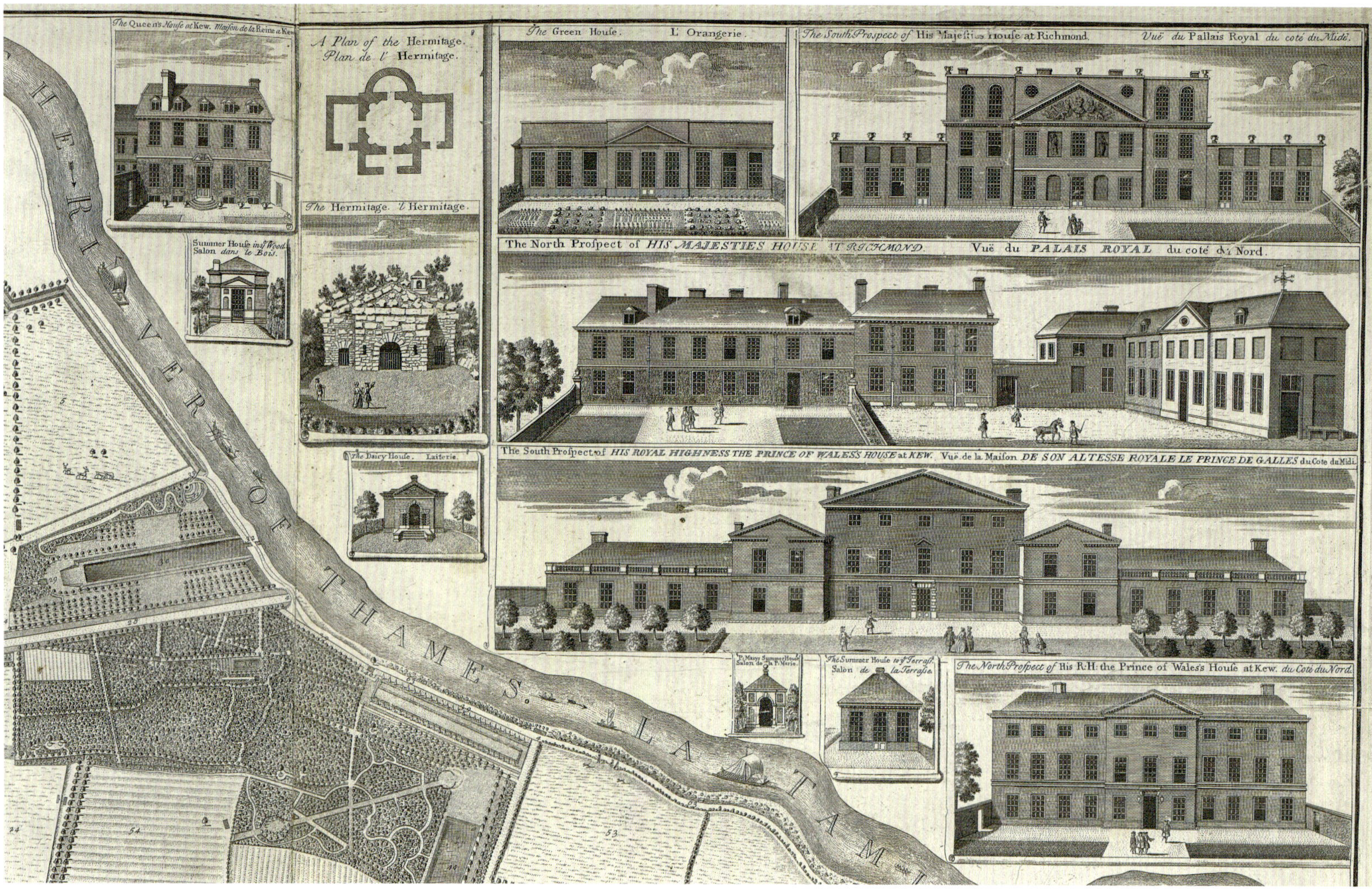

FIG. 18.01 John Rocque, *Plan of the House, Gardens, Park & Hermitage of Their Majesties at Richmond: and of Their R.H. the Prince of Wales & the Princess Royal of Kew* (detail), 1734, etching and engraving on paper, 24⅜ x 37¾ in. (62 x 96 cm). Yale Center for British Art, Paul Mellon Collection (cat. 18.02)

with the building's rustic exterior. The print shows Kent's symmetrical design, which included an elaborate canopy bed to the left, a pedimented bookcase to the right, and two classically inspired reclining couches, beneath two of the five busts of British intellectuals that Caroline had commissioned from the Italian sculptor Giovanni Battista Guelphi. In contrast to Vardy's somewhat sterile representation of the Hermitage, Kent's own more whimsical depiction of 1730–31 is designed to flatter Caroline—and by proxy Kent, her architect of choice—and shows a bust of the Queen herself, celebrated as benevolent champion of the arts. The surrounding cherubs hold a plan of the Hermitage, in a self-referential celebration of Caroline's architectural patronage.

Merlin's Cave, designed to be seen with the guidance of resident poet Stephen Duck, received comparable, though less favourable, public attention. Publications such as Edmund Curll's *The Rarities of Richmond* capitalized on and reinforced popular interest, and focused on its main attraction, the life-sized wax models depicting the mythical conjurer Merlin, Elizabeth of York, Queen Elizabeth I, and Britomart from Edmund Spenser's *Faerie Queene* (see cat. 12.07). Vardy's engraving reflects the theatrical nature of Merlin's Cave, in which the central scene is illuminated, as if by spotlights, from windows in the roof above. Kent's pioneering rustic wooden furniture can also be seen in the wings on either side.

Some commentators, including the King himself, thought Merlin's Cave an absurd exercise in whimsy, but criticisms aside, Caroline's initiatives at Richmond were hugely popular. They provided 'talking points' that contributed to her role at the centre of court culture, while the dissemination of prints allowed members of fashionable society to feel themselves part of her endeavours.[1] In turn, the elegant public were themselves depicted in popular prints, such as in an engraving by John-Baptiste-Claude

Chatelain after the Italian painter Marco Ricci, which shows a flock of visitors to he garden dressed in tastefully rustic dress.

The intense publicity surrounding Caroline's Richmond sits at odds with our expectations of the home as a private retreat, secluded from the public gaze. Yet it is not helpful to make clear distinctions between the Princesses' public roles and their domestic spheres. With the exception of Frogmore, which can more properly be understood as a retreat from the public eye, the Princesses' homes necessarily constituted a degree of public performance. The creation of admirable homes enabled Caroline, Augusta, and Charlotte to control their self-representation in a society that was increasingly conscious of being observed.[2] Additionally, their domestic pursuits, such as garden design and botany, frequently overlapped with the concerns of the nation, and could therefore never be seen as entirely private enterprises.

However, despite the public representation of the Princesses' gardens, not everything was done with an eye to an audience. It is clear that all three cultivated the pastoral ambience of Richmond and Kew, and Augusta in particular seems to have valued her gardens, especially those surrounding the White House, her country residence at Kew, as places of repose. In contrast with printed views of the gardens showing visitors jostling for space, the series of oil paintings she commissioned from the Swiss painter Johann Jakob Schalch in 1759–60 shows a deserted and serene landscape. In Schalch's views, the White House and a small Chinese-style bridge seem to be the only man-made intrusions into a bucolic pastoral scene (p. 324, cats. 18.05–18.06).

Augusta also cultivated Kew as a rural retreat for the education of her children. A print by her fourth son, Prince Henry Frederick, shows an idealized sylvan impression of Kew inhabited by country people, even while the contemporary engraving by William Woollett shows the grounds populated by fashionable ladies in elegant dresses.

Despite the natural appearance of the landscape in Schalch's views, in reality Augusta's work at Kew was every bit as far-reaching as Caroline's at Richmond. Inheriting the White House upon the death of her husband in 1751, Augusta set to work creating a nine-acre botanical garden, the foundation of today's Royal Botanic Gardens. In 1757 she employed the Anglo-Swedish architect William Chambers as resident architect and tutor to the young Prince of Wales. Chambers encouraged the Prince to see the grounds as a textbook of architectural styles, and together they populated the gardens at Kew with numerous structures which included a Chinese pagoda, a mosque, and a cathedral in the fashionable 'Gothick' style. Not merely whimsy, these buildings were serious exercises in architectural experimentation. Visitors were able to experience an array of architectural projects which included Augusta's aviary, a menagerie, and the 'Great Stove', which would heat greenhouses and ensure that delicate plants retrieved by plant hunters from tropical habitats could thrive. These constructions were recorded in watercolours which form part of a volume of views of Kew compiled by Chambers and dedicated to John Stuart, third Earl of Bute, adviser and friend to Augusta and the driving force behind much of her work on the gardens (see cats. 16.17 and 20.06). The quality of the watercolours, supplied by well-known architectural perspectivists such as Thomas Sandby, reveals the pride with which Chambers compiled this record of his work (cats. 29.37, 29.38). In a letter to Bute he wrote: 'My Lord, as I believe your Lordship has yet no plan of Kew Gardens, I have made bold to send you one bound up with the Original designs from which the plates of Kew were engraved.'[3] The 'plates of Kew' to which Chambers refers came from the printed version of the album, published in 1763 as *Plans, Elevations, Sections and Perspective Views of the Gardens and Buildings at Kew in Surry.* The publication was paid for by the King and dedicated to the Princess Dowager Augusta. The volume contains a plate showing the Great Pagoda which still stands in Kew Gardens (see cat. 19.13).

After Augusta's death in February 1772, George III and Charlotte were free to install their own family in the White House. A survey map made in about 1771 shows that the first steps had already been taken to integrate the grounds at Richmond and Kew, though the two estates were not formally joined until the early nineteenth century. Despite the fact that Richmond Lodge was deemed by many as an unsuitable royal residence, it had served as the royal family's main country retreat since 1717, and once again in the 1760s, and Charlotte clearly felt some affection for the home in which she had spent part of her honeymoon.[4] Fondness for Richmond Lodge must surely have been the impetus for Sandby's attractive watercolour depicting the young princes playing in front of the house, created as a record drawing shortly before it was demolished in 1772 (cat. 19.16). On moving her family to Kew, however, Charlotte wrote to her brother that 'it will be better and more private.'[5] Privacy, then, had become an increasing concern for the royal family, and would be jealously guarded at Charlotte's new home at Frogmore, purchased in 1792.

While there is no doubt that Caroline, Augusta, and, in particular, Charlotte, did at times seek domestic seclusion, it was understood that their homes and gardens were also powerful vehicles for public engagement. In creating spaces that drew upon contemporary fashions and philosophies of design they were able to forge common links with the aristocracy, while allowing the public a carefully controlled glimpse into an idealized form of royal domesticity. Rather than being static monuments formally set in a constructed environment, the Princesses saw their homes and gardens as evolving arenas for the cultivation of family life, adapted to current concerns and the immediate needs of the Royal Household, and had no compunction in undoing the work of their predecessors in order to achieve their own ends and express new ideas. The structures that Kent created for Caroline at Richmond, with their celebration of British history and contemporary worthies and heroes, were swept away by the work of Lancelot 'Capability' Brown in the early 1770s, while the exotic structures which populated Augusta's Kew had started to be dismantled by the 1780s. One of Charlotte's own ventures at Kew, a cottage in which she and her daughters could enjoy the fashionable pastime of drinking tea, still remains. On the whole, however, it is perhaps wise to see these buildings, which always contained an element of the stage set, as inherently transient and intended to meet immediate desires rather than being designed to endure forever. At the same time, the Princesses' gardens have left a legacy; notably Caroline's garden at Kensington celebrated in Rocque's map of 1736 and Augusta's botanical pursuits at Kew, which became increasingly formalized and scientific as the century progressed.

In their own ways, the Princesses all created homes and gardens that centred on their families and sent powerful dynastic messages. Caroline, famously interested in genealogy, cultivated elaborate pictorial family trees in the portrait gallery at Kensington and in her wax figures in Merlin's Cave. At the heart of Augusta's Kew was the education of the future George III, whose love of architecture never left him. In turn, Charlotte capitalized on her large family to create an idealized image of domesticity that could be celebrated in private and simultaneously paraded in public in the family's promenades at Windsor.

The prints and drawings made at the time are a helpful insight into the dual nature of the Princesses' homes and the gardens they established as a setting for them. At one end of the spectrum is William Kent's characteristically whimsical sketch showing a shepherdess in pastoral clothing being attended by a fawn, against the backdrop of the Hermitage at Richmond (cat. 19.07); an intimate fantasy of 'Arcadia' recorded in a private drawing. At the other extreme are the highly finished and much-reproduced prints churned out to provide the public—at home and abroad—with an insight into the detailed arrangements of the Princesses' homes and gardens. In the context of the national craze for topographical prints and developments in the English tradition of landscape painting, views such as *A East View of the Bridge and part of his Majesties Garden and Park* by Bernard Lens III (see p. 135, fig. 6.02) show how the cultivation of attractive environments enabled the Princesses to engage with fashionable culture and participate in creating an image of England that was exported to the world.[6] Ultimately, all three Princesses were able to negotiate the line between their public and private homes. They understood that the implications of their domestic lives were far-reaching, and the creation of admirable houses and gardens was consequently one of the most tangible ways in which these German Princesses successfully made England their home.

NOTES

1. Andrew Hanham, 'Caroline of Brandenburg-Ansbach and the 'Anglicisation' of the House of Hanover', in *Queenship in Europe, 1660–1815: The Role of the Consort*, ed. Clarissa Campbell Orr (Cambridge: Cambridge Univ. Press, 2004), 295.
2. John Brewer, 'This, That and the Other: Public, Social and Private in the Seventeenth and Eighteenth Centuries', in *Shifting the Boundaries: Transformation of the Languages of Public and Private in the Eighteenth Century*, ed. Dario Castiglione and Lesley Sharp (Exeter: Univ. of Exeter Press, 1995), 15–18.
3. Quoted in John Harris, *A Catalogue of British Drawings for Architecture, Decoration, Sculpture and Landscape Gardening, 1550–1900, in American Collections* (Upper Saddle River, NJ: Gregg Press, 1971), 57.
4. Jane Roberts, ed., *George III & Queen Charlotte: Patronage, Collecting and Court Taste* (London: Royal Collection, 2004), 109.
5. Roberts, ed., *George III & Queen Charlotte*, 129.
6. For the growth of London and its environs as 'consumed landscapes', see Elizabeth McKellar, *Landscapes of London: The City, the Country and the Suburbs, 1660–1840* (New Haven and London: Yale Univ. Press, 2013), 79–107.

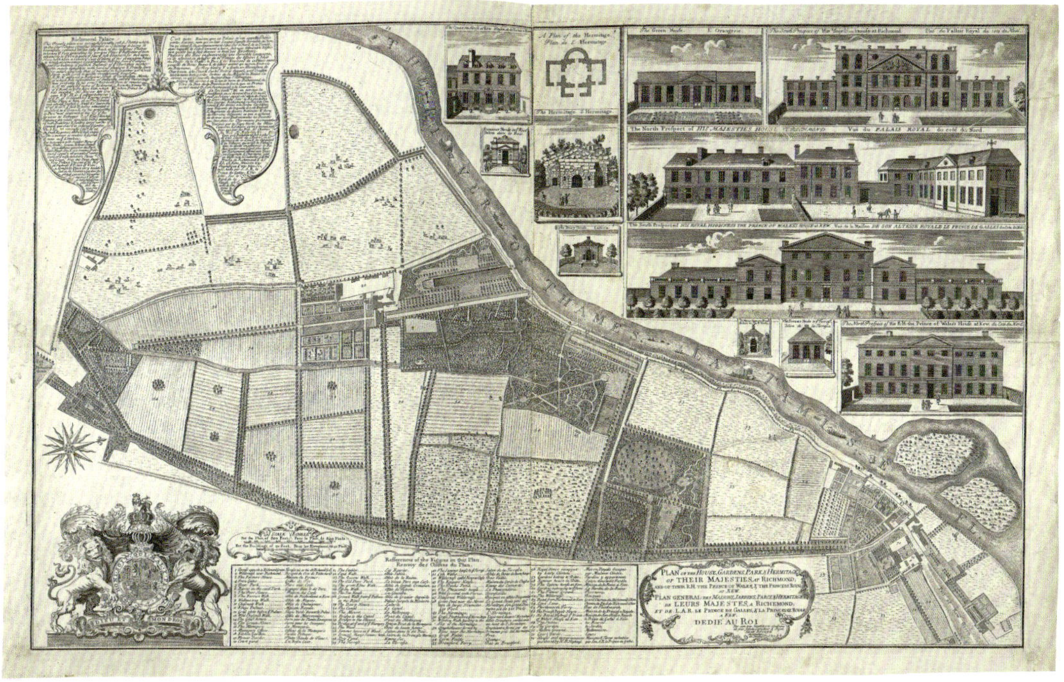

18.02

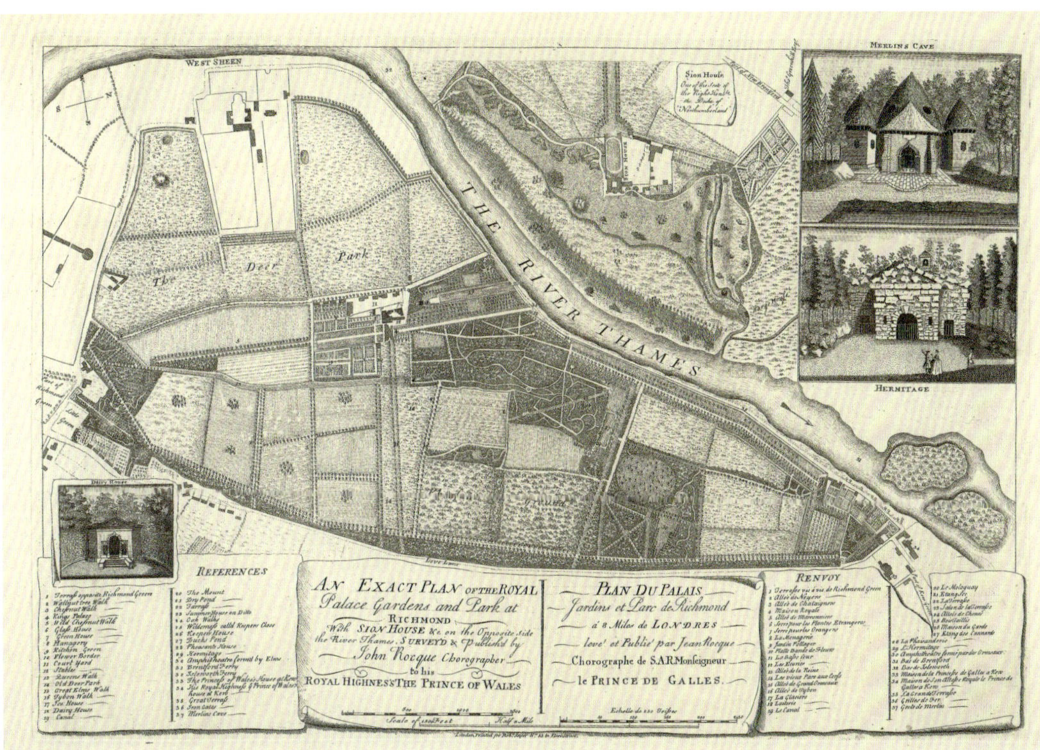

18.03

THE ROYAL ESTATES AT RICHMOND AND KEW

18.02*

JOHN ROCQUE (c. 1709–1762)
Plan of the House, Gardens, Park & Hermitage of Their Majesties at Richmond: and of Their R.H. the Prince of Wales & the Princess Royal of Kew
1734
Etching and engraving on paper
Published by J. Rocque, [London]
24⅜ x 37¾ in. (62 x 96 cm)
Yale Center for British Art, Paul Mellon Collection, Folio B G 4

18.03

JOHN ROCQUE (c. 1709–1762)
An Exact Plan of the House, Gardens, Park & Hermitage of Their Majesties at Richmond: and of Their R.H. the Prince of Wales & the Princess Royal of Kew
1754
Etching and engraving on paper
24⅜ x 37¾ in. (62 x 96 cm)
Published by Robert Sayer (1724/25–1794), London
Yale Center for British Art, Paul Mellon Collection, Folio A G 14, copy 3

The celebrated cartographer John (Jean) Rocque produced two plans of the Richmond garden. The first version in 1734 was followed by a second in 1754. In both he inserted images of the series of buildings found in the garden, together with a key to their location. The 1734 plan contains the most comprehensive list and includes a view of Richmond Lodge itself. It provides the only evidence of the appearance of the library range of the house, where Caroline's substantial collection of books was once housed.

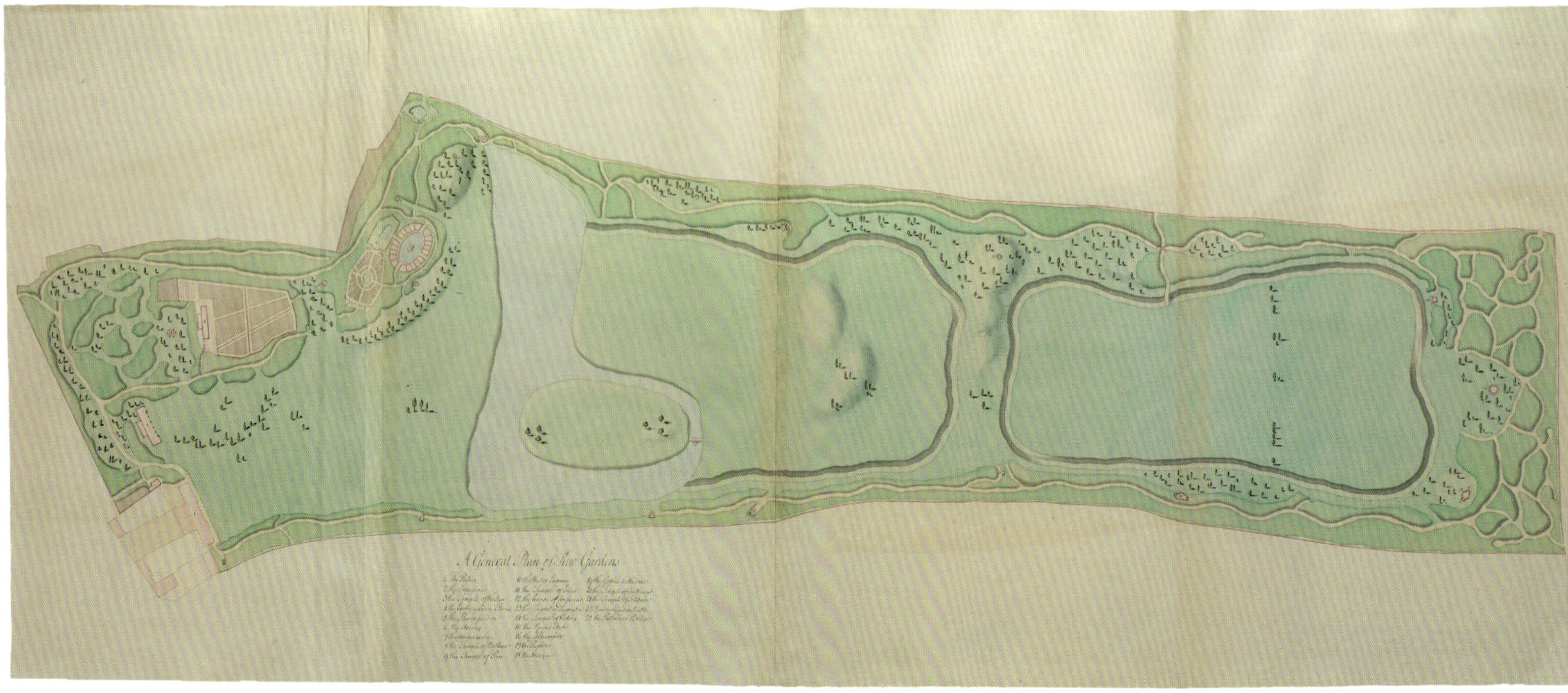

18.04

18.04*

WILLIAM CHAMBERS (1723–1796)
General Plan of the Gardens at Kew
c. 1763
Watercolour and ink on paper
20⅞ x 47⅝ in. (53 x 121 cm)
The Metropolitan Museum of Art, Harris Brisbane Dick Fund, 1925, 25.19.44

Sir William Chambers's plan of the gardens for the White House at Kew reveals that the estate was relatively long and thin, with the White House, remodelled for Frederick, Prince of Wales, which Augusta continued to use after his death, at one end. Behind this was a lawn looking into a lake, with an island so constructed to bring the sense that this was a meandering watercourse. Beyond this there were two lawns, divided by stands of mature trees, salvaged from an earlier garden layout. Around the edges of the garden was a belt of trees cut through with meandering paths. It was in this part of the garden that the aviary, menagerie, greenhouses, and forcing houses, and the 'Great Stove' for the plant collections were located, as well as the many folly temples.

TRANSFORMING 'THE DESART'

18.05†

JOHANN JAKOB SCHALCH (1723–1789)
The Gardens at Kew
1760
Oil on canvas
29¾ x 40¼ in. (75.6 x 102.1 cm)
Royal Collection Trust, RCIN 403517

18.06

JOHANN JAKOB SCHALCH (1723–1789)
The Gardens at Kew by Moonlight
1760
Oil on canvas
30 x 41 in. (76.1 x 104.2 cm)
Royal Collection Trust, RCIN 403516

Commissioned by Augusta, Princess of Wales, these are two of a set of five views of her gardens at Kew by Johann Schalch, a Swiss painter who spent a decade working in England (1754–63). The series was completed in 1759 and comprises images of different aspects of the garden, seen at different times of day. Many of the garden buildings recently completed, or still in construction to designs by William Chambers, were included, such as the Chinese Bridge, the Temple of Bellona, and the Pagoda. The meandering watercourse, 'the gentle stream . . . taught to serpentize seemingly at its leisure', is in fact an ornamental lake, cut irregularly to appear far more expansive. The paintings demonstrate how the grounds were planted to take advantage of existing trees, and how, by grading the new copses to provide eye-catchers to frame a view, it was possible to give height to the very flat site. Chambers, who was very quick to acknowledge Lord Bute's contribution to the schemes in his preface to his *Plan . . . of the Gardens and Buildings at Kew* (1763), claimed 'what was once a desart [*sic*] is now an Eden.'

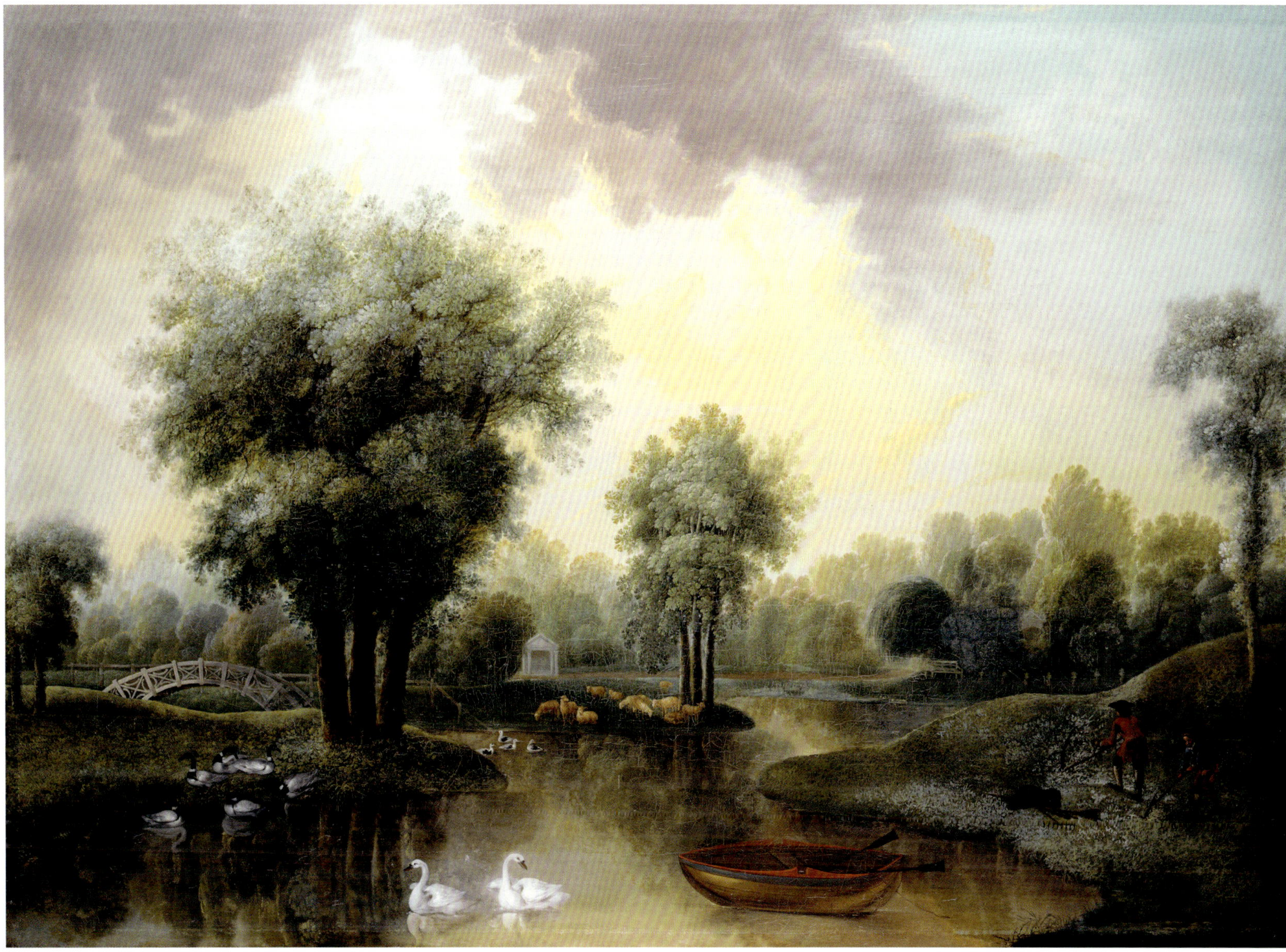

18.05

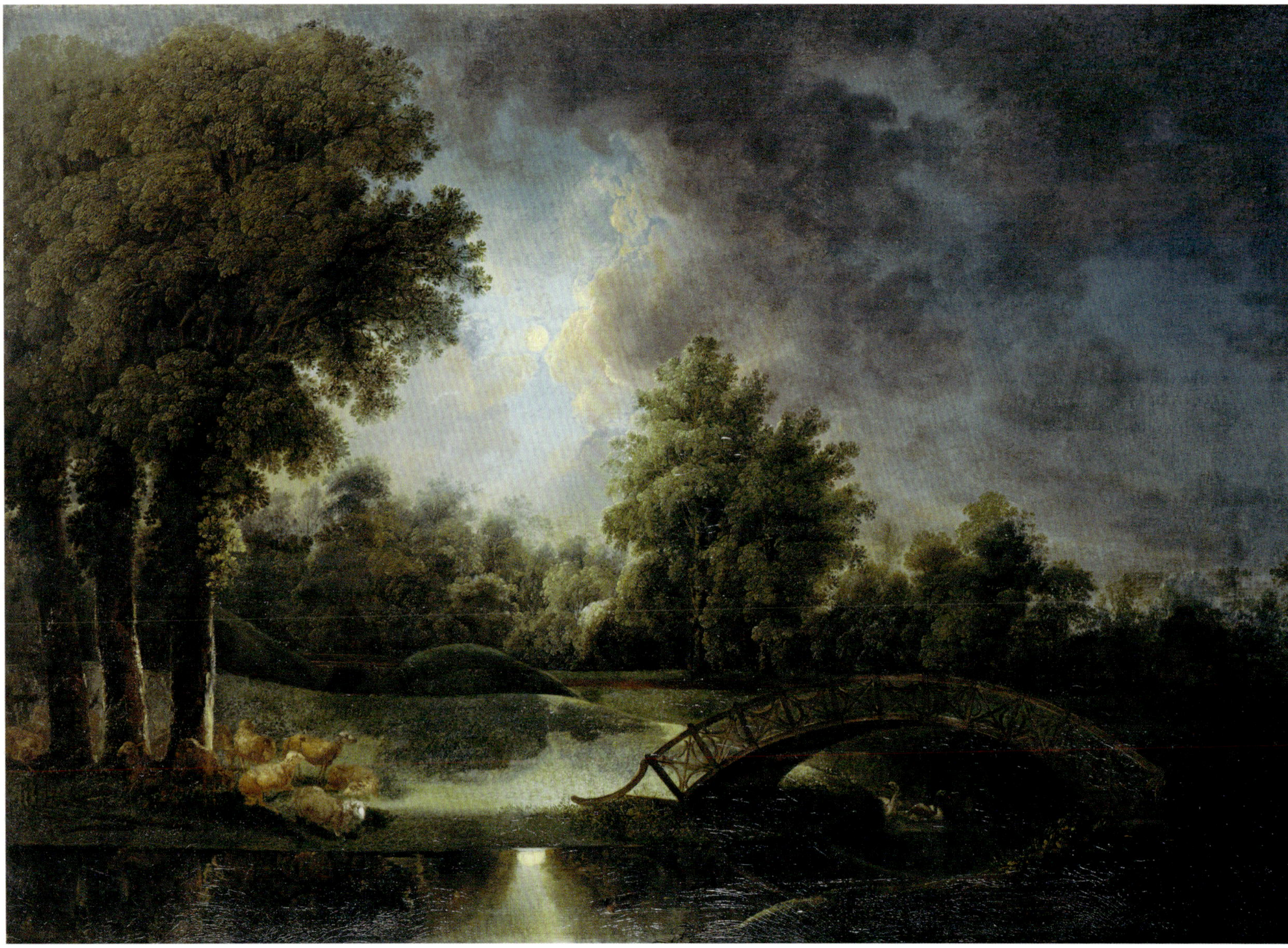

18.06

18.07*

GEORGE RITSO (active 1763)
Kew Gardens: A Poem, Humbly Inscribed to Her Royal Highness the Princess Dowager of Wales
1763
10¾ in. (27.3 cm)
Printed by M. Lewis and sold by R. and J. Dodsley, London
The Morgan Library and Museum, Miss Julia P. Wightman Bequest, PML 152030
Exhibited: Dedication (illustrated)

George Ritso was a member of Princess Augusta's household; little is known of him except the account he gave in the dedication of *Kew Gardens: A Poem*. There he described how Augusta took on his care and interests after the death of his parents, both of whom had been in her employment. The poem narrates the course of a single day at Kew, which awakes into full bloom not with the sunrise but with the arrival of its royal mistress. Ritso evoked the abundance and diversity of life—and pleasures—at Kew, and his description of various times of day recalls the series of garden views commissioned by Augusta from the Swiss painter Johann Jakob Schalch (see cats. 18.05–18.06). This copy of Ritso's poem belonged to Queen Charlotte.

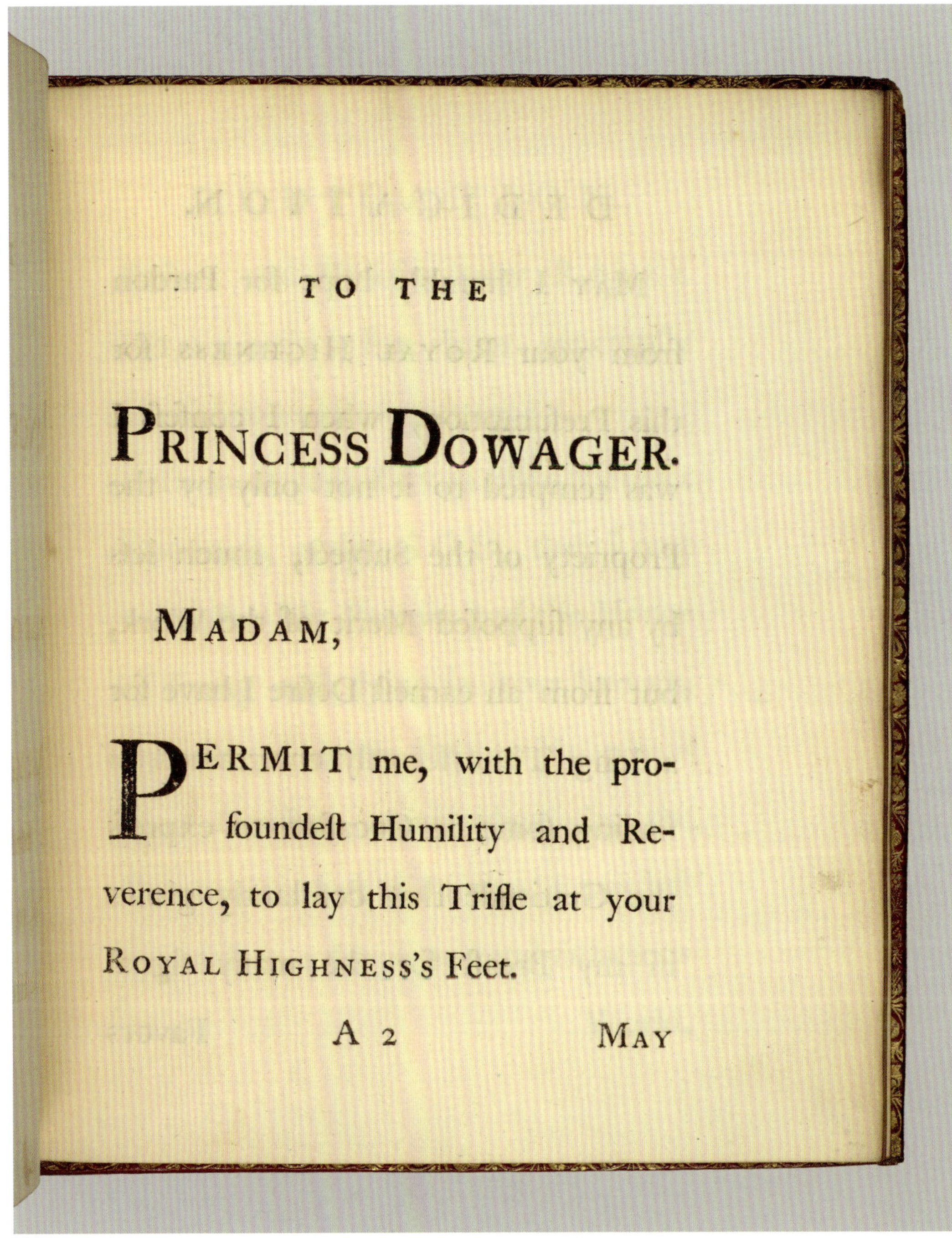
TO THE

PRINCESS DOWAGER.

MADAM,

PERMIT me, with the profoundeſt Humility and Reverence, to lay this Trifle at your ROYAL HIGHNESS's Feet.

A 2 MAY

18.07

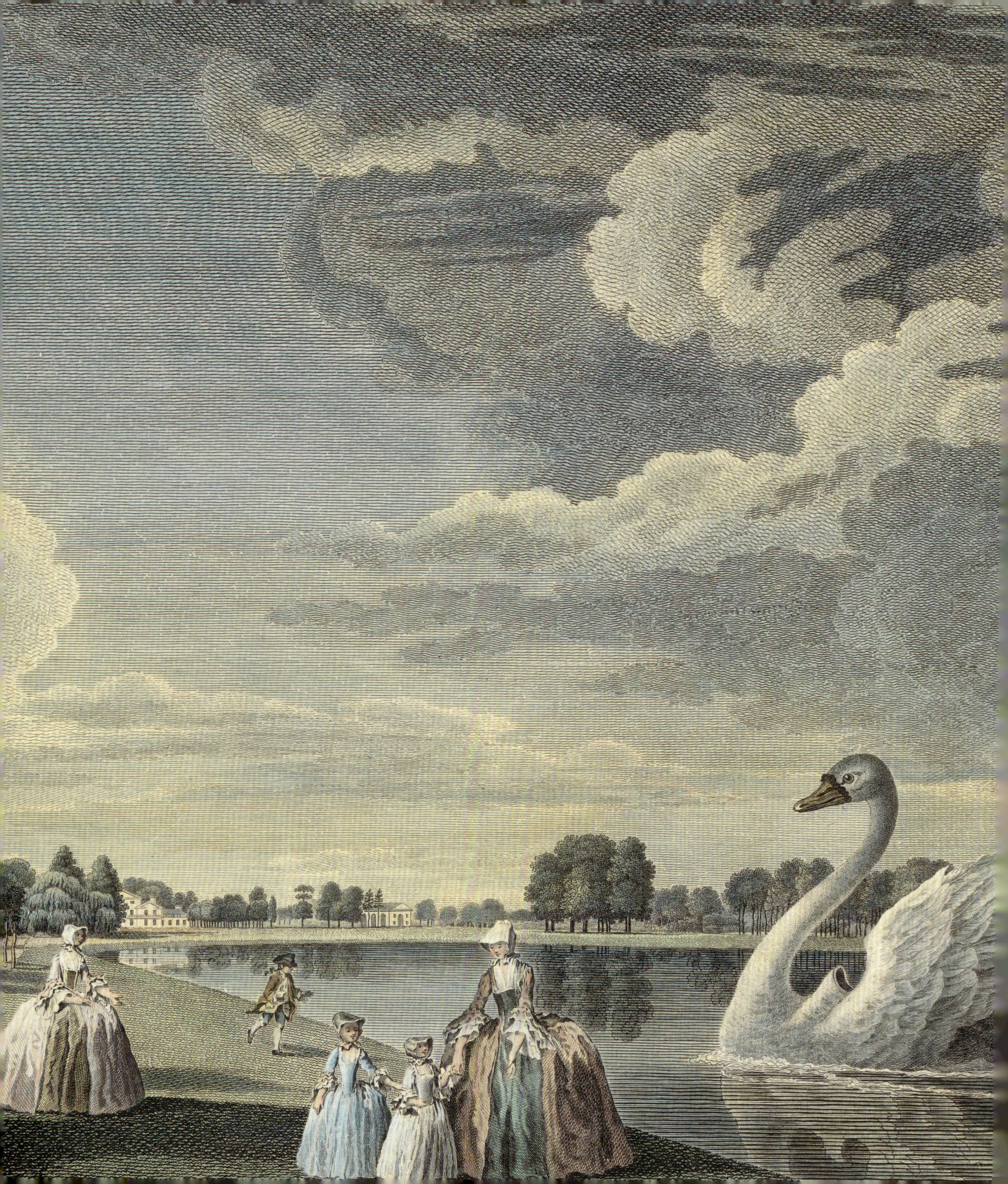

Architectural Patronage

Lee Prosser

The architectural patronage of Caroline of Ansbach, Augusta of Saxe-Gotha, and Charlotte of Mecklenburg-Strelitz could be characterized respectively as philosophical, exotic, and whimsical. This is perhaps an over-simplification, but a reflection that much of their direct architectural taste is inseparable from the other passion of the age: landscape gardening. This is also evident in projects that touch many of the royal residences. Caroline most notably sponsored the transformation of Kensington Gardens, with an ornamental pond as a centrepiece, while Augusta and her husband Frederick, Prince of Wales, were active at Carlton House. Charlotte is best known for the cultivation of her own private 'trianon' at Frogmore on the Windsor estate later in the eighteenth century, complete with its own garden follies and Gothic ruin. Nowhere encapsulates the contribution of all three Princesses more emphatically, however, than Kew Gardens, now a great botanical institution, but one which grew from two distinct but adjoining eighteenth-century royal pleasure grounds developed by Caroline, Augusta, and Charlotte. Here, a collection of buildings, some surviving, others lost or never built, expressed the talents of rising architects such as William Kent and Sir William Chambers in the furnishing of a sylvan, bucolic landscape with fantastical adornments (p. 334). Caroline's contribution has been much studied as part of a wider, more general movement, combining a political message with philosophical discourse. The contribution of Augusta is much less well understood; the earlier works in royal gardens cannot be distinguished from those of her husband Frederick, while several important buildings at Kew are usually attributed to her son George III. Most, however, including the Great Pagoda (see cat. 19.13), were paid for out of her own Privy Purse and reflect an intense burst of activity from the late 1750s, sponsored by the Princess and acknowledged as such by the architect of many of the buildings. Charlotte has been given less credit, but the Queen's Cottage, which came twenty-five years before her other great project at Frogmore, deserves greater scrutiny for its subtlety and wider influence.

Kew, or more properly Richmond Gardens, was more than just a country estate for Caroline; acquired after she and her husband were expelled from the royal palaces in 1717, it became the rural haven to which she retreated in the years following. Césare de Saussure wrote in June 1726 that 'the princess takes a great interest in the gardens, which are spacious and she has greatly embellished them.'[1] In fact it was only after her accession as Queen in 1727 and the grant of a generous parliamentary allowance of £100,000 per annum, that she was fully able to indulge her passions, extending her estate to four hundred acres and building initially a modest dairy house, a pavilion, a domed temple on a mount, and a riverside terrace.[2] Tours of other aristocratic gardens such as Cliveden, Claremont, Esher Place, and elsewhere, all in the full flow of development, inspired her to build further at Kew, and with the help of Kent she added buildings with a distinct philosophical rationale; models as well as plans remain in the Royal Collection (figs. 19.01–19.02). These buildings included the Hermitage, at an extravagant £1,140, and Merlin's Cave, in fact

Detail of p. 345, cat. 19.15

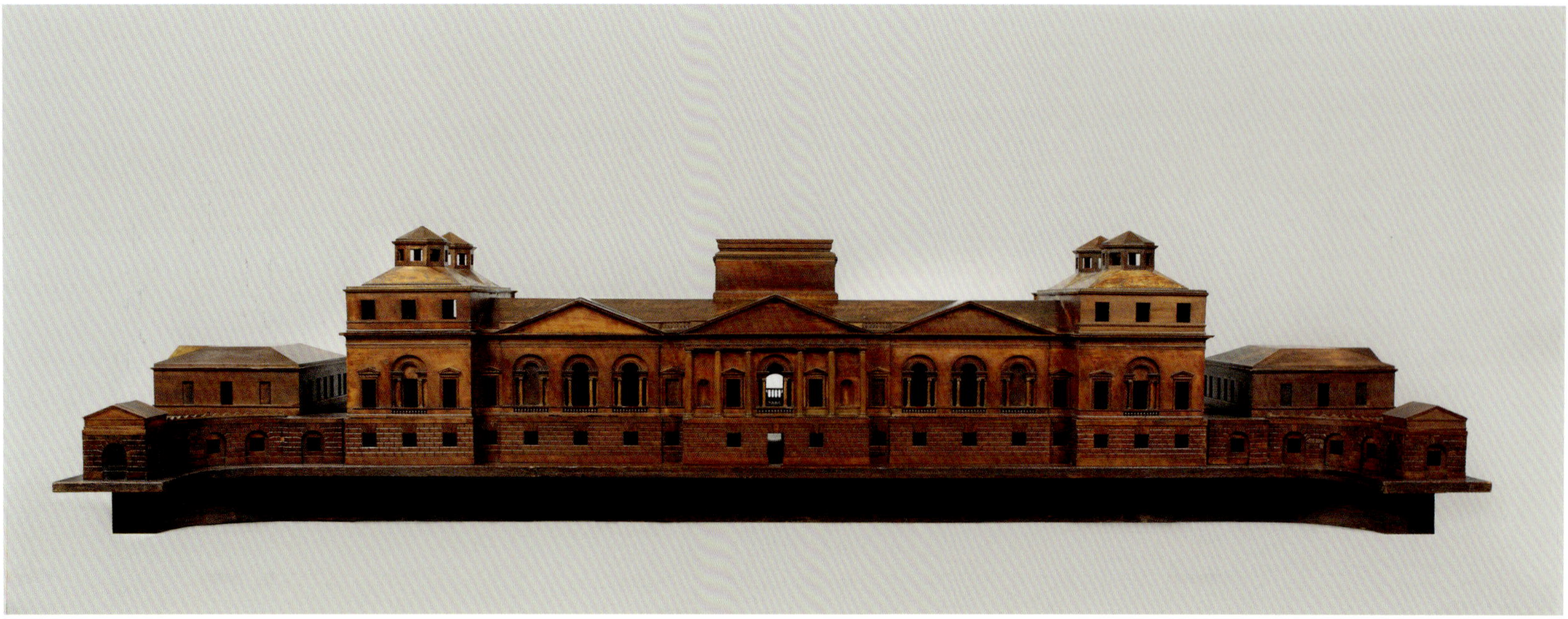

FIG. 19.01 John Marsden, for William Kent, Model for a new palace at Richmond, c. 1735, pearwood, 20 ¼ x 93 x 51⅛ in. (51.5 x 236 x 135 cm). Royal Collection Trust

a rustic cottage with extraordinary beehive-shaped thatched roofs, which formed her own temple celebrating British history, populated with full-length waxwork figures including those of Merlin, Britomart, and Queen Elizabeth I—all drawn from Edmund Spenser's *Faerie Queene* (cats. 19.10–19.11).[3] Caroline went one step further, installing the 'rustic poet' Stephen Duck as custodian and interpreter, though Horace Walpole still called it 'an unintelligible puppet show', and the King was also dismissive: 'you deserve to be abused for such childish, silly stuff', he is reported to have remarked on her endeavours.[4]

Augusta's contributions to the adjoining estate at Kew Gardens were long in the making. Her husband, Frederick, Prince of Wales, was hamstrung in his ambition by the short financial strings on which his parents kept him—less than half the allowance his father had enjoyed as Prince of Wales—and in his early years money seems to have been expended mainly on rebuilding his princely residence and the ostentatious royal barge, on horses and guns. The development of his gardens was much slower, more considered, and indeed fairly devoid of political inference, though a reference in the Duchy of Cornwall accounts mentions a dome on a mount—similar, perhaps, to the Palladian temple built in 1734 at Carlton House—and an enigmatic rustic 'witch house' may also have been erected. Ultimately, the only building of note ever completed in his lifetime was the Chinese 'House of Confucius', probably by Joseph Goupy or Chambers, of 1749.[5] It was following the Prince's death in 1751 that most attention was paid to the garden itself. Augusta directed John Haverfield the elder to complete works 'according to the plan', of which we know nothing. The rise of John Stuart, third Earl of Bute as confidante and muse for Augusta, and his appointment as finishing tutor to her son, the future King, which brought in turn the promotion of Chambers as architect to the Princess, introduced a dynamic impetus for building.

Chambers's shrewd self-promotion may have been encouraged by an unexecuted design for a grandiose mausoleum for the dead Frederick; but from then on, Kew became an extraordinary testing ground for didactic and experimental architecture, embracing different orders and styles, colours and materials into every conceivable construction, as if the lessons of the day were rapidly transformed into built form and any whim was realized (fig. 19.03). Temples and a Gothic cathedral, an orangery, Chinese bridge, menageries, pavilions, revolving seats, and the startling pagoda were all erected between 1757 and 1763 (fig. 19.04). Augusta appears to have become a relentless patron, taking up architecture with gusto. Chambers complained that

> The prince employs me three mornings in a week to teach him architecture; the building (and) other decorations at Kew fill up the remaining time. The princess has the rest of the week which is scarcely sufficient as she is forever adding new embellishments at Kew, all which I direct the execution (and) measure the work.[6]

Theatricality was a predominant theme, emphasizing the link between the gardens and architecture. Buildings such as the Theatre of Augusta, of flimsy, ephemeral materials such as wood

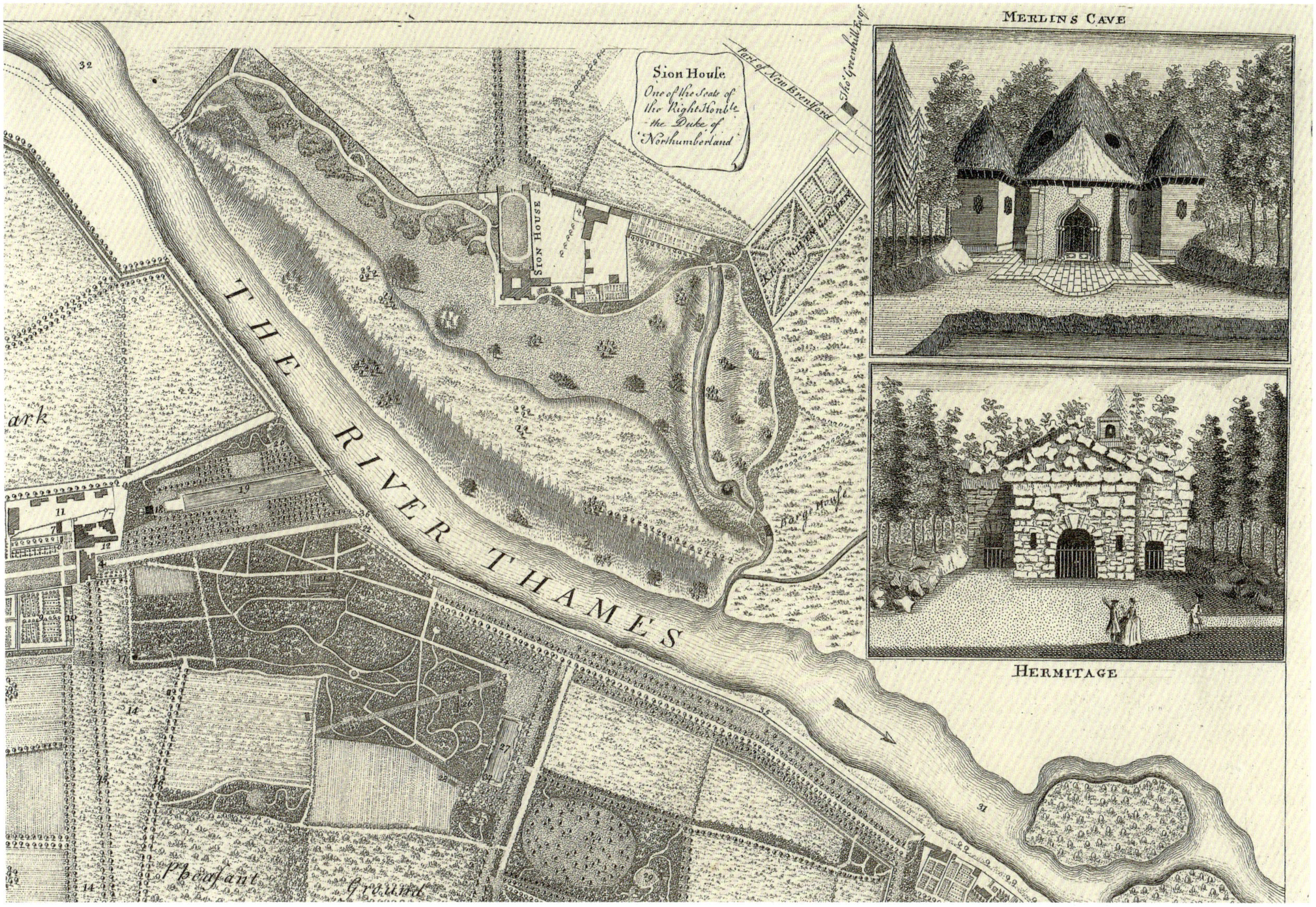

FIG. 19.02 John Rocque, *An Exact Plan of the Royal Palace Gardens and Park at Richmond* (detail), 1754, engraving, 24⅜ x 7¾ in. (62 x 96 cm). Yale Center for British Art, Paul Mellon Collection (cat. 18.03)

and canvas, may have been erected as temporary structures for outdoor performances. This reflected Frederick's and Augusta's love of theatre, and their use of the garden as a stage for performance may also be seen as an influence in the laying out of the gardens of their other residences, such as the outdoor amphitheatre at Cliveden. Before long, a string of buildings and artfully contrived ruins spread through the gardens; some were useful like the Chinese menagerie, others purely ornamental, but all could be taken in on a circuit, culminating in the picturesque and exotic triad of Moorish Alhambra, Turkish Mosque, and Chinese Pagoda (fig. 19.04). The Pagoda, erected between 1761 and 1762, caused a sensation in the press for its varnished iron roofs and green and white colouring, which has been corroborated by recent analysis. Each corner was adorned with carved dragons with an iridescent finish. Unfortunately, problems with these experimental materials led to it being stripped back to a more conventional form with the loss of the iron roofs and beasts by around 1784. When Prince Ludwig von Bentheim-Steinfurt travelled from Westphalia in Germany seeking inspiration for his own garden, and visited Kew in 1783 with his folio of engravings in hand, he found to his dismay that many of Augusta's fantasies had already been swept away.

By the time of the arrival of Charlotte in 1761, the concept of Kew Gardens was largely complete. Augusta retreated to Carlton House and no new buildings were added in the following years. The exception to this is Queen Charlotte's Cottage, a small teahouse and pavilion constructed in 1771 as a modification to, or on the site of, the old menagerie-keeper's house, supposedly to the design of the Queen herself.

The cottage is a two-storied thatched building, partly of brick, with projecting stair-turrets of nogged timber-framing, making it essentially symmetrical, and identical when viewed

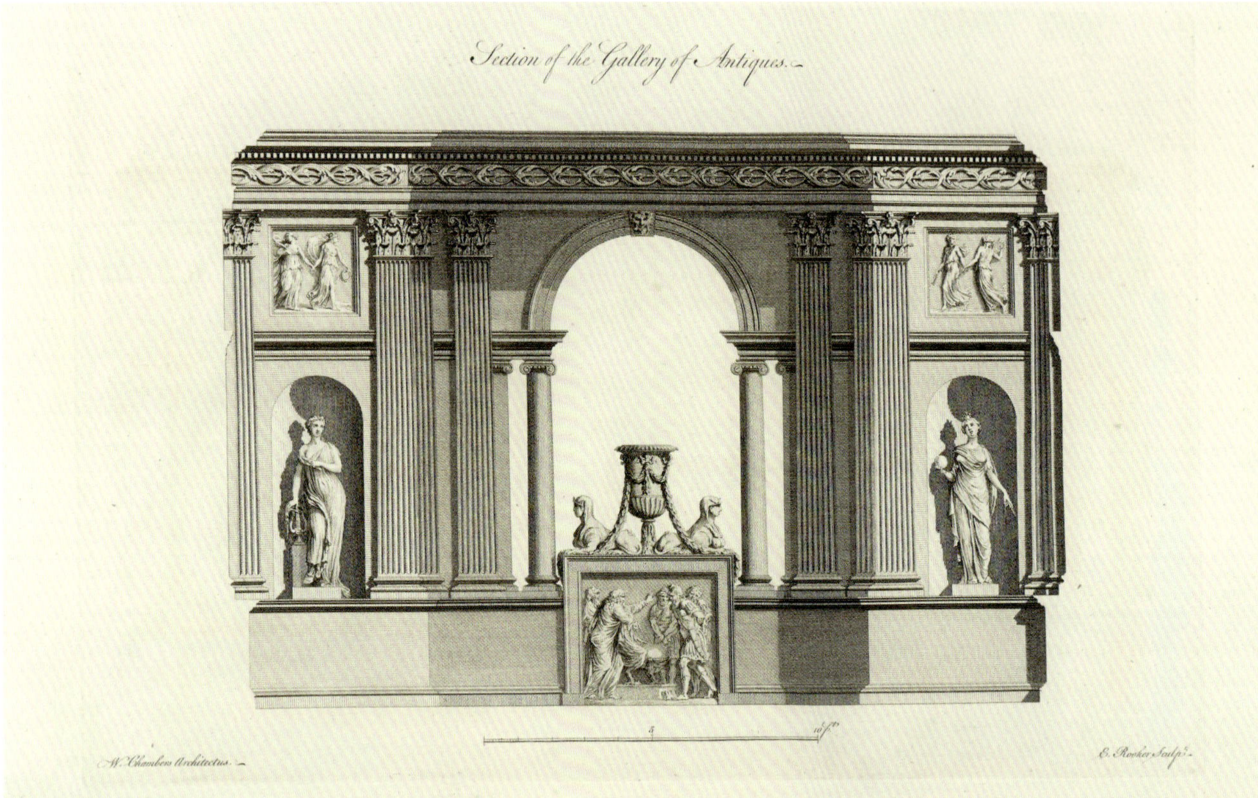

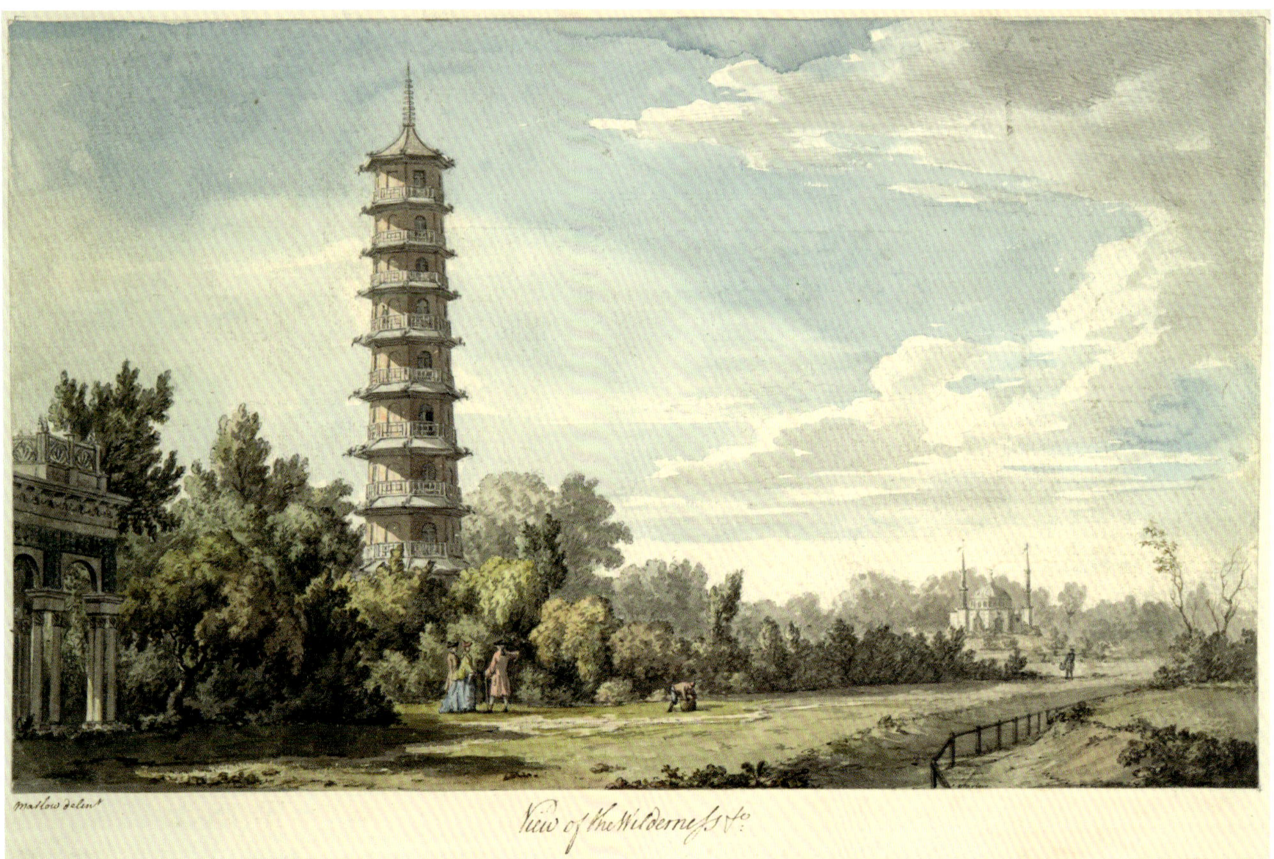

FIG. 19.03 After William Chambers, 'Gallery of Antiques', in *Plans, Elevations, Sections and Perspective Views of the Gardens and Buildings at Kew in Surry* (London: J. Haberkorn, 1763). Historic Royal Palaces

FIG. 19.04 William Marlow, *View of the Wilderness at Kew*, 1763, watercolour on paper, 11$\frac{1}{16}$ x 17$\frac{13}{16}$ in. (28.1 x 45.2 cm). The Metropolitan Museum of Art, New York, Harris Brisbane Dick Fund, 1925

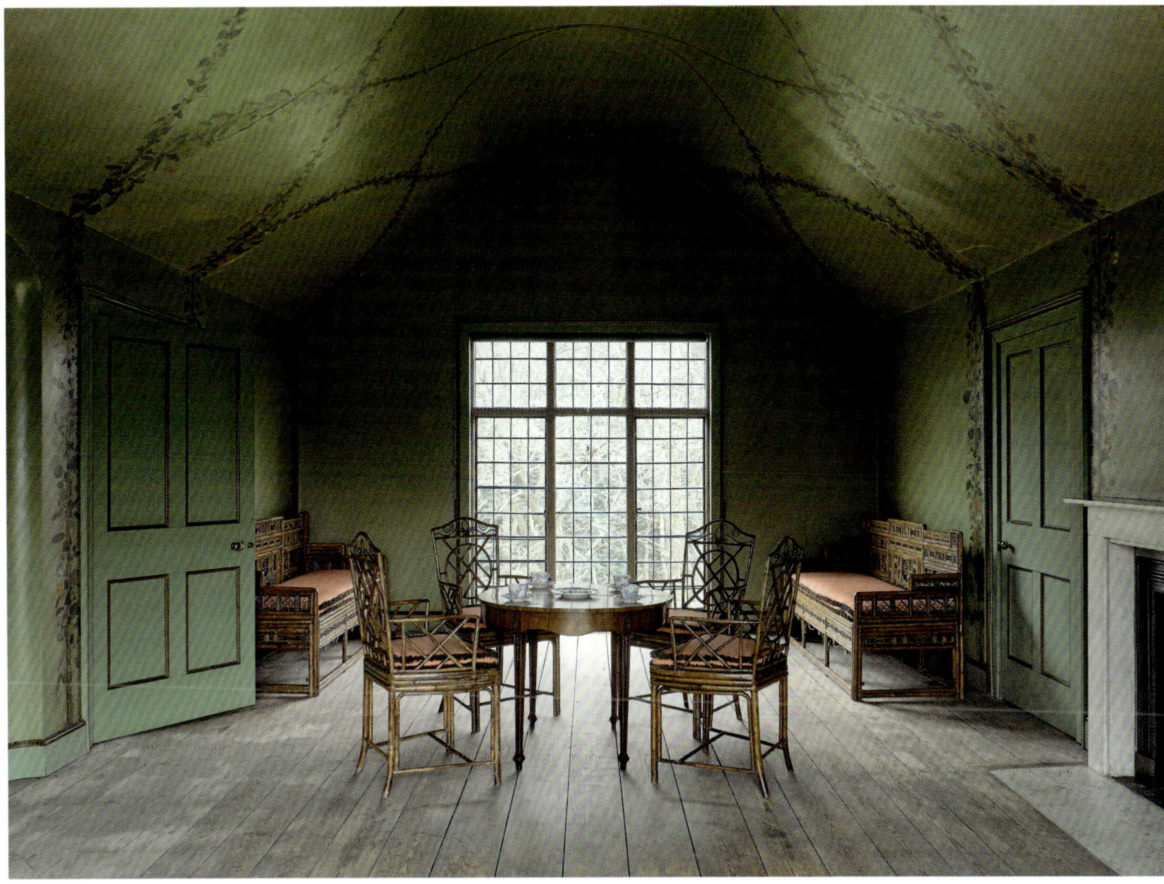

FIG. 19.05 The Picnic Room at Queen Charlotte's Cottage, Kew. Photograph by James Brittain

from both principal sides. But like Caroline's Hermitage, which had been established with an "imaginative juxtaposition" between a rustic exterior and more conventional classical interior, the rusticity of the cottage conceals generously proportioned inner rooms comprising a print room and a picnic room, flanked by sweeping, if spartan, staircases that would not disgrace a grand London town house.[7] Analysis has revealed that the windows, which are of the old-fashioned transom and mullion form, and the brickwork, are both recycled from older seventeenth-century buildings. The whole effect of exposed timber was likely to have been covered by ivy, resulting in the ultimate in studied contrivance and theatricality, while concealing within order, comfort, and conventional classicism, with fitted carpets, curtains, and elegant furniture (fig. 19.05).

Kew's buildings may represent what Thomas Sandby called 'a fresh thirst after novelty' and, indeed, the collection was celebrated by Chambers with some pride in his own publication, *Plans, Elevations, Sections and Perspective Views of the Gardens and Buildings at Kew in Surry* (see cats. 19.13, 19.16, and 29.35–29.38).[8] While some commentators criticized Kew for its 'ridiculous profusion of temples, grottoes, artificial ruins, imitative mosques and pagodas . . . in the very worst taste' and 'expensive trumpery', the *London Magazine* noted that Kew Gardens was 'the only princely thing about this country'.[9]

NOTES

1. Césare de Saussure, *A Foreign View of England in the Reigns of George I and George II*, trans. and ed. Madame van Muyden (London: John Murray, 1902), 143.
2. R. Desmond, *Kew: The History of the Royal Botanic Gardens* (London: Harvill Press with The Royal Botanic Gardens, Kew, 1995), 5.
3. Joanna Marschner, *Queen Caroline: Cultural Politics at the Early Eighteenth-Century Court* (New Haven and London: Yale Univ. Press, 2014), 52.
4. John Hervey, Baron Hervey, *Some Materials towards Memoirs of the Reign of King George II*, ed. Romney Sedgwick (London: Batsford, 1963), 109.
5. Goupy appears in Chambers's book as the author, though Harris has asserted that Chambers was the architect. Goupy's employment as the Prince's painter may imply that he merely decorated the building.
6. Quoted in Robin Middleton, 'Chambers, W., "A Treatise on Civil Architecture", London 1759', in *Sir William Chambers: Architect to George III*, ed. John Harris and Michael Snodin (New Haven and London: Yale Univ. Press, 1996), 69.
7. Marschner, *Queen Caroline*, 58.
8. David Watkin, *Sir John Soane: Enlightenment Thought and the Royal Academy Lectures* (Cambridge: Cambridge Univ. Press, 1996), 374.
9. See, respectively, John Fisher Murray, *Environs of London, Western Division* (Edinburgh and London: William Blackwood, 1842), and *London Magazine, or Gentleman's Monthly Intelligencer* 43, Aug. 1774, 359.

CAROLINE'S ARCHITECTURAL CONTRIBUTION TO RICHMOND AND KEW

19.06

BARTHOLOMEW DANDRIDGE
(c. 1691–c. 1754)
William Kent
c. 1736
Oil on canvas
35 in. x 28¼ in. (88.9 cm x 71.8 cm)
National Portrait Gallery, London, NPG 1557

William Kent, a protégé of Richard Boyle, third Earl of Burlington, was awarded his first royal commission by George I in 1722. After Sir James Thornhill had presented an expensive estimate for the decorations of the new state rooms at Kensington Palace, William Kent was pushed forward with his offer to complete the work for a more reasonable sum. It is likely that Caroline met Kent for the first time as this project was completed. Later, as her architect, her adviser in all matters artistic, and as her friend, he undertook work for her in the gardens at Richmond and Kensington, not only influencing the layout of the grounds, but also designing buildings for their ornament. He was commissioned to produce paintings, and to undertake conservation of existing works. In 1734, Caroline and Geoge II would climb the scaffolding tower 'forty foot high' to inspect Kent's work repairing and conserving the canvases by Rubens that were set into the ceiling at the Banqueting House in Whitehall. This work has long been described as a portrait of William Kent; however the identification has yet to be proved.

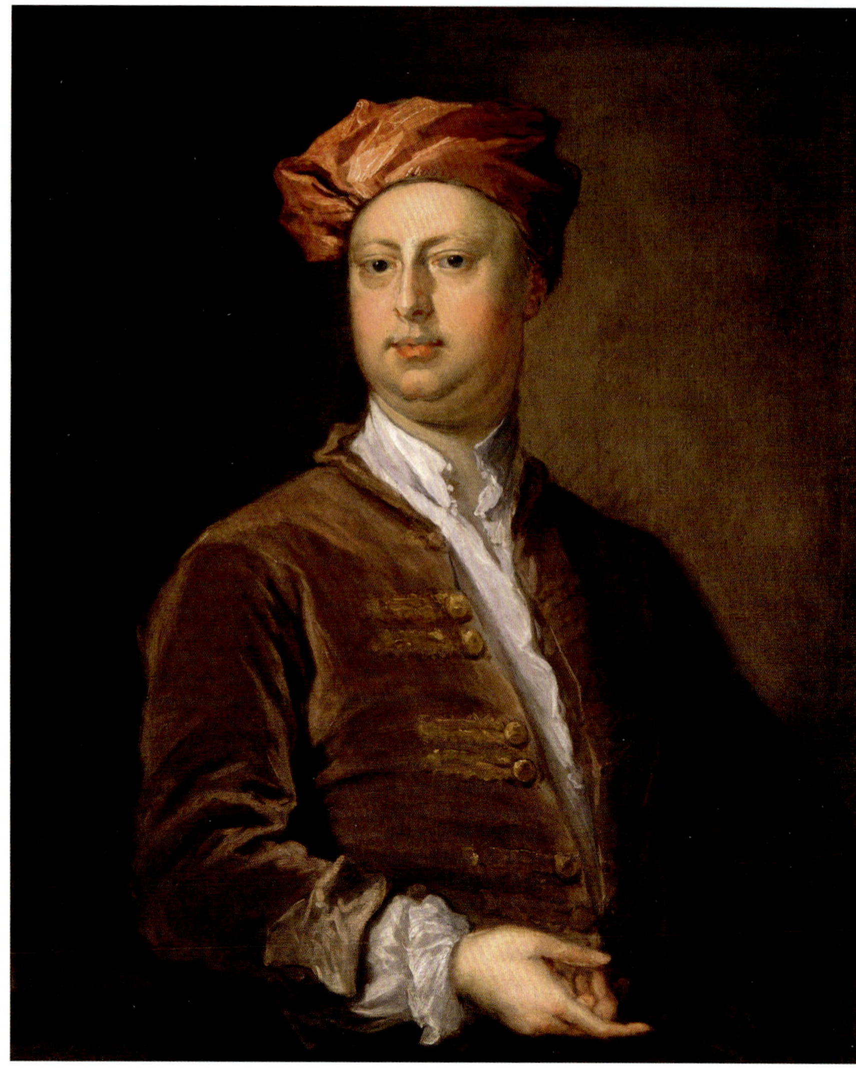

19.06

THE HERMITAGE

19.07*

WILLIAM KENT (c. 1685–1748)
Design for Queen Caroline's Hermitage in Richmond Gardens
c. 1730
Pencil, pen, and wash on paper
12 x 18¼ in. (30.4 x 46.2 cm)
Sir John Soane's Museum, London, Adam vol. 56.25

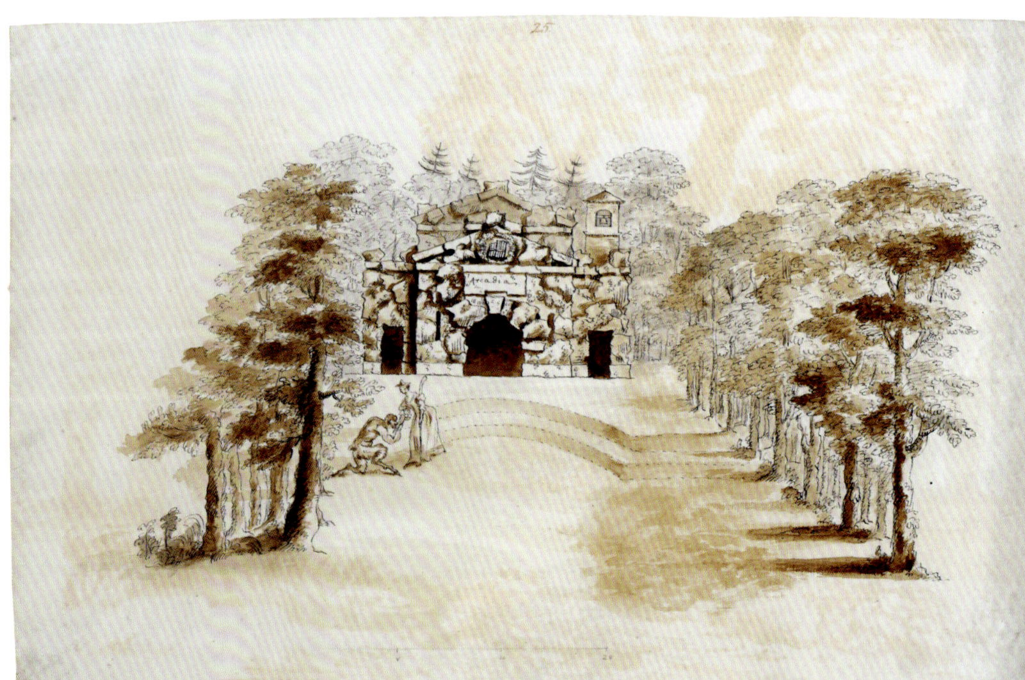

19.07

* = YCBA only
† = HRP only

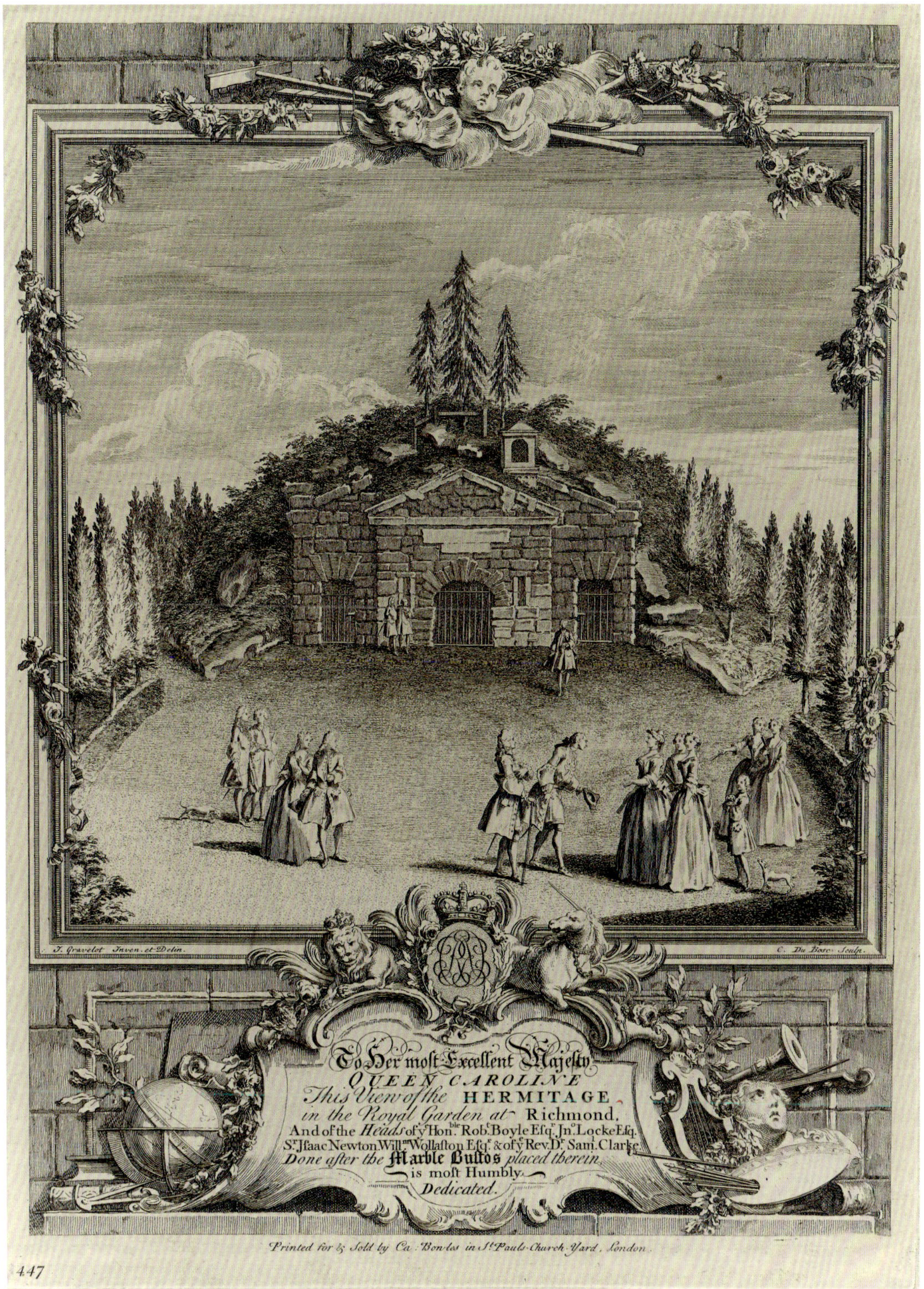

19.08

19.08

CLAUDE DU BOSC (1682–1745), engraver, after **HUBERT-FRANÇOIS GRAVELOT** (1699–1773)
View of the Hermitage in the Royal Garden at Richmond
1730–37
Etching
14½ x 10¼ in. (36.6 x 26 cm)
Published by Carington Bowles (1724–1793), London
British Museum, 1876,0708.2441

The Hermitage, situated towards the north of Richmond Gardens, was arguably William Kent's first garden building. It preceded the celebrated projects at Stowe. For the Hermitage Kent set up an imaginative juxtaposition between the rustic exterior elevation and the classically proportioned interior. The drama and surprise that resulted may have been calculated to give greater force to the messages conveyed by the display of a sculptural series celebrating contemporary or near contemporary natural philosophers and theologians. The Hermitage was completed in 1732. Caroline contributed significant funds for the project from her Privy Purse.

Kent made a series of drawings which may have been in preparation for the Hermitage commission. In one, which shows a hermitage set in a woody clearing, in front of an artificial hillock topped with pines, a fashionable woman, possibly Caroline, is depicted being serenaded by a satyr. Du Bosc's etching after Gravelot, with its dedication to Caroline, served as frontispiece to a set of mezzotints illustrating the five marble busts of contemporary scientists and philosophers by Giovanni Battista Guelphi which were set up in the Hermitage (cats. 16.11–16.14).

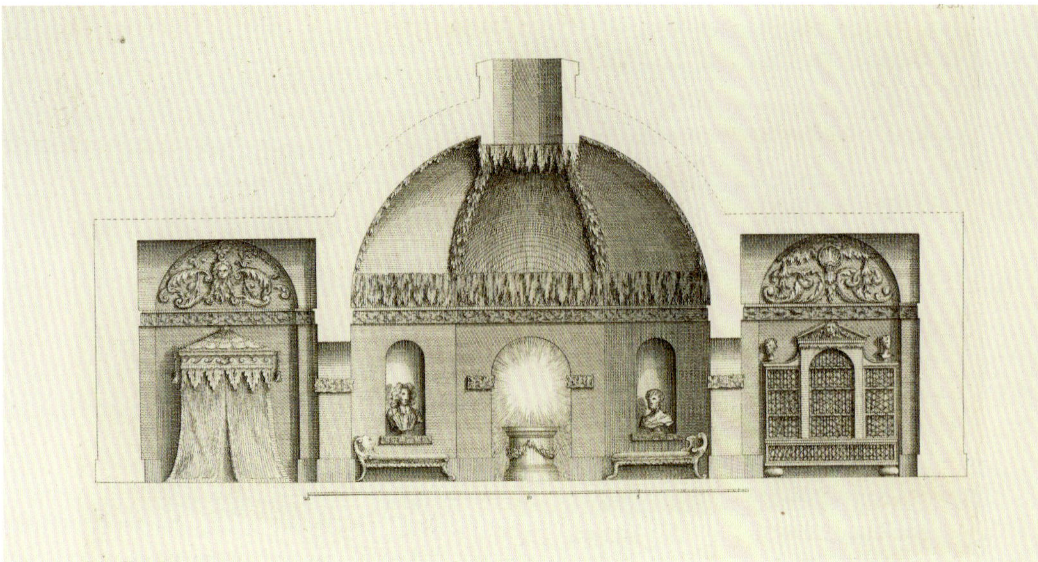

19.09

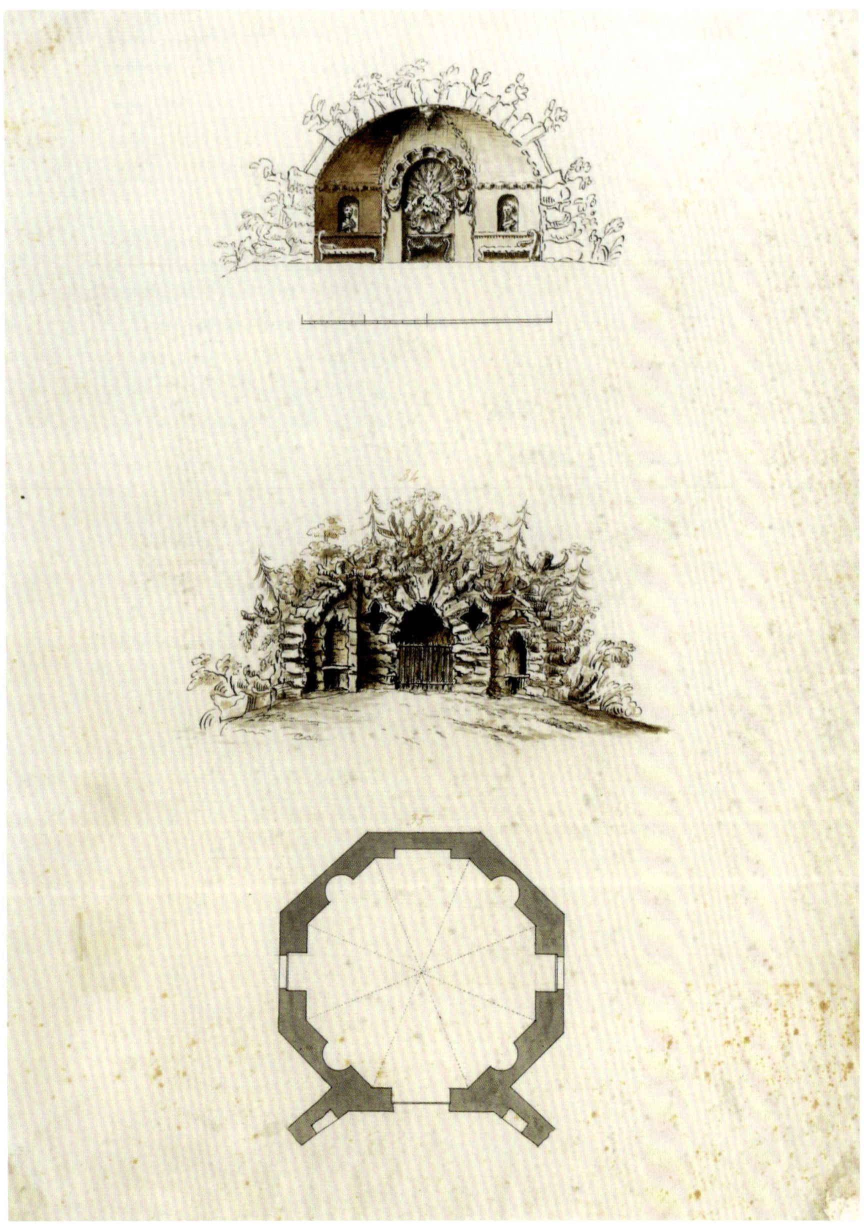

19.10

19.09†

JOHN VARDY (1718–1765), engraver, after **INIGO JONES** (1573–1652) and **WILLIAM KENT** (c. 1685–1748)
Some Designs of Mr Inigo Jones and Mr Wm Kent
1744
10½ in. (26 cm)
J. Vardy, London
Royal Collection Trust, RCIN 1150776
Exhibited: Plate 33, 'A Section through the Hermitage in the Royal Gardens at Richmond'

19.10

WILLIAM KENT (c. 1685–1748)
'Studies for a Rustic Seat or Hermitage possibly for Queen Caroline's Richmond Gardens'
c. 1730
Pen and ink, and wash on paper
Sir John Soane's Museum, London Adam 56/33–35

In 1744, John Vardy published his volume *Some Designs of Mr Inigo Jones and Mr Wm Kent*, which included illustrations, elevations, and sections of many of Kent's projects undertaken for Caroline including the Hermitage, Merlin's Cave, and other folly temples in the garden of Richmond Lodge. There are sometimes multiple images relating to the same feature, suggesting that drafts were included, but many can be linked with contemporary descriptions, or the financial accounts for the construction of the buildings. In the section concerning the Hermitage, the portrait busts commissioned by Caroline to honour her contemporary heroes may be seen arranged on plinths around the walls.

MERLIN'S CAVE

19.11*

THOMAS BOWLES (1689/90?–1767), engraver, after **WILLIAM KENT** (c. 1685–1748), architect
Merlin's Cave in the Royal Gardens at Richmond
1736
Engraving on paper
Tipped into Beinecke Rare Book & Manuscript Library, Yale University, Plays 37

By 1735 Merlin's Cave had been constructed to a design by William Kent within Richmond Gardens. It was located a short distance from the Hermitage, and visitors to the garden could take it in as they followed one of a series of meandering paths through the estate. A dancing lawn was laid out in front of it, with the intention that the Cave should form an appropriate backdrop to the enjoyment of rural pursuits. In 1715, as Princess of Wales, Caroline had presided over a rural sports day at Hampton Court, at which local boys and girls competed in running races for prizes of smocks, petticoats, and hoods. She had quickly learnt to dance traditional country dances such as the 'Hemp Dressers', which she introduced into court occasions.

The Cave was constructed by Henry Flitcroft and Andrews Jelfe. They may have brought their own contribution to the project. Both had a serious interest in contemporary archaeology and had encountered the archaeologist and antiquary, William Stukeley. The dramatic beehive-shaped roofs of the Cave may have been influenced by contemporary ideas about 'primitive' British architectural forms associated with the Druids.

19.12

SIR JOSHUA REYNOLDS (1723–1792)
Sir William Chambers
c. 1756
Oil on canvas
35 x 27 in. (88.9 x 68.6 cm)
National Portrait Gallery, London, NPG 27

William Chambers's path to establishment success was secured by his appointment as architect to Princess Augusta at Kew, and as architectural tutor to George, Prince of Wales. He noted that 'The prince employs me three mornings a week to teach him architecture; the buildings (and) other decorations at Kew fill up the remaining time.' The architectural drawings made jointly by Chambers and the prince stand as what must have been one of the most thorough courses of education in architecture received by any monarch. The variety and contrast introduced into the landscape at Kew by Chambers was presented as characteristic of Chinese gardens. The building projects were also seen as part of the prince's architectural education, for between 1757 and 1763 he was witness to the design and building of more than twenty structures, of which the Alhambra, Pagoda, and Mosque were perhaps the most famous. The decision to isolate the horticultural buildings as a designed entity within an ornamental enclosure separate from, but adjacent to, the pleasure grounds, anticipated the innovations of Humphry Repton and John Claudius Loudon.

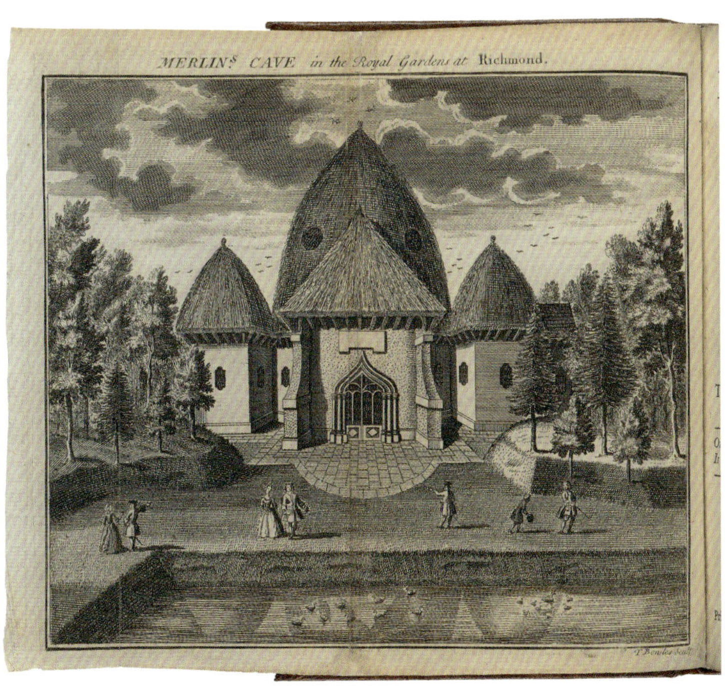

19.11

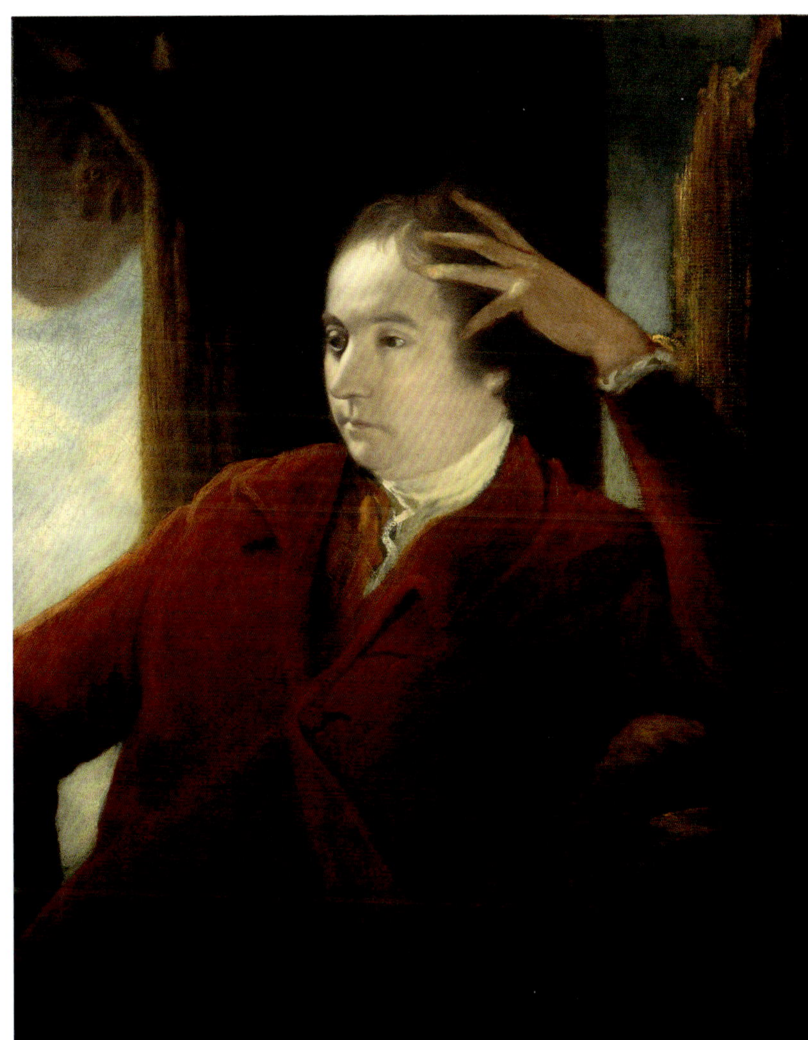

19.12

AUGUSTA'S ARCHITECTURAL CONTRIBUTIONS TO RICHMOND AND KEW

19.13

WILLIAM CHAMBERS (1723–1796)
Plans, Elevations, Sections and Perspective Views of the Gardens and Buildings at Kew in Surry
1763
20⅞ in. (53 cm)
J. Haberkorn, London
* Exhibited: T. Miller (active 1765–1775), engraver, after William Chambers (1723–1796), *Elevation of the Great Pagoda as first intended*, plate 23, 1763, engraving, 21¼ x 14¾ in. (54 x 37.5 cm). Yale Center for British Art, Paul Mellon Collection, S 419.75A (Folio A); (not illustrated)
† Exhibited: T. Miller (active 1765–1775), engraver, after William Chambers (1723–1796), *Elevation of the Great Pagoda as first intended*, plate 23, 1763, engraving, 21¼ x 14¾ in. (54 x 37.5 cm). Royal Botanic Gardens, Kew, Library, Art & Archives Collections, Ref: 9999740 (illustrated)

Chambers's Pagoda is one of the few surviving buildings from his Chinese phase. It stands at the furthest reach of Kew Gardens and was originally flanked by the Alhambra and the Mosque. It was 163 feet (49.6 metres) high, divided into ten storeys of decreasing diameter. In his 1763 folio Chambers included floor plans, a cross section, and two elevations of slightly different design. The building as realized had dragon mounts set on the roofs of each tier, bells hanging from their mouths. These would have ensured that the building made an aural contribution to the experience of the garden. Chambers's achievement at Kew Gardens is commemorated by one of the most sumptuous folios ever produced on a single garden: *Plans, Elevations, Sections and Perspective Views of the Gardens and Buildings at Kew in Surry* (1763). It was dedicated to Princess Augusta and paid for by George III. As well as including images of the many elaborate garden buildings designed and constructed through the course of the 1760s, there is a series of views and a plan which suggests how these worked into the greater landscape.

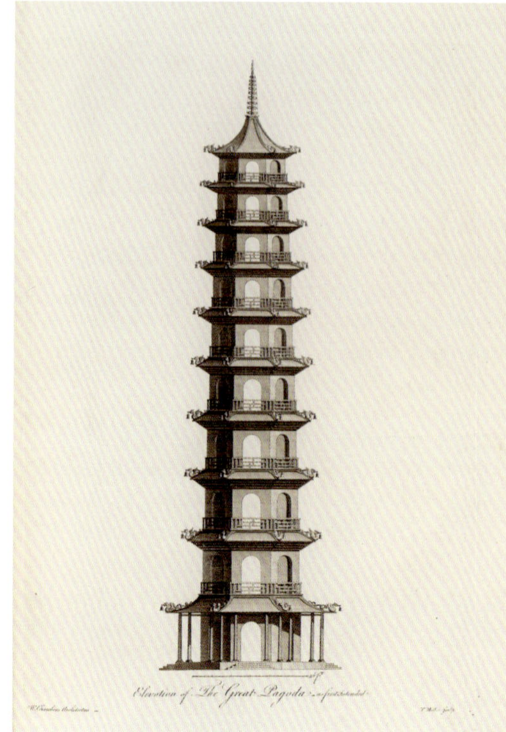

19.13

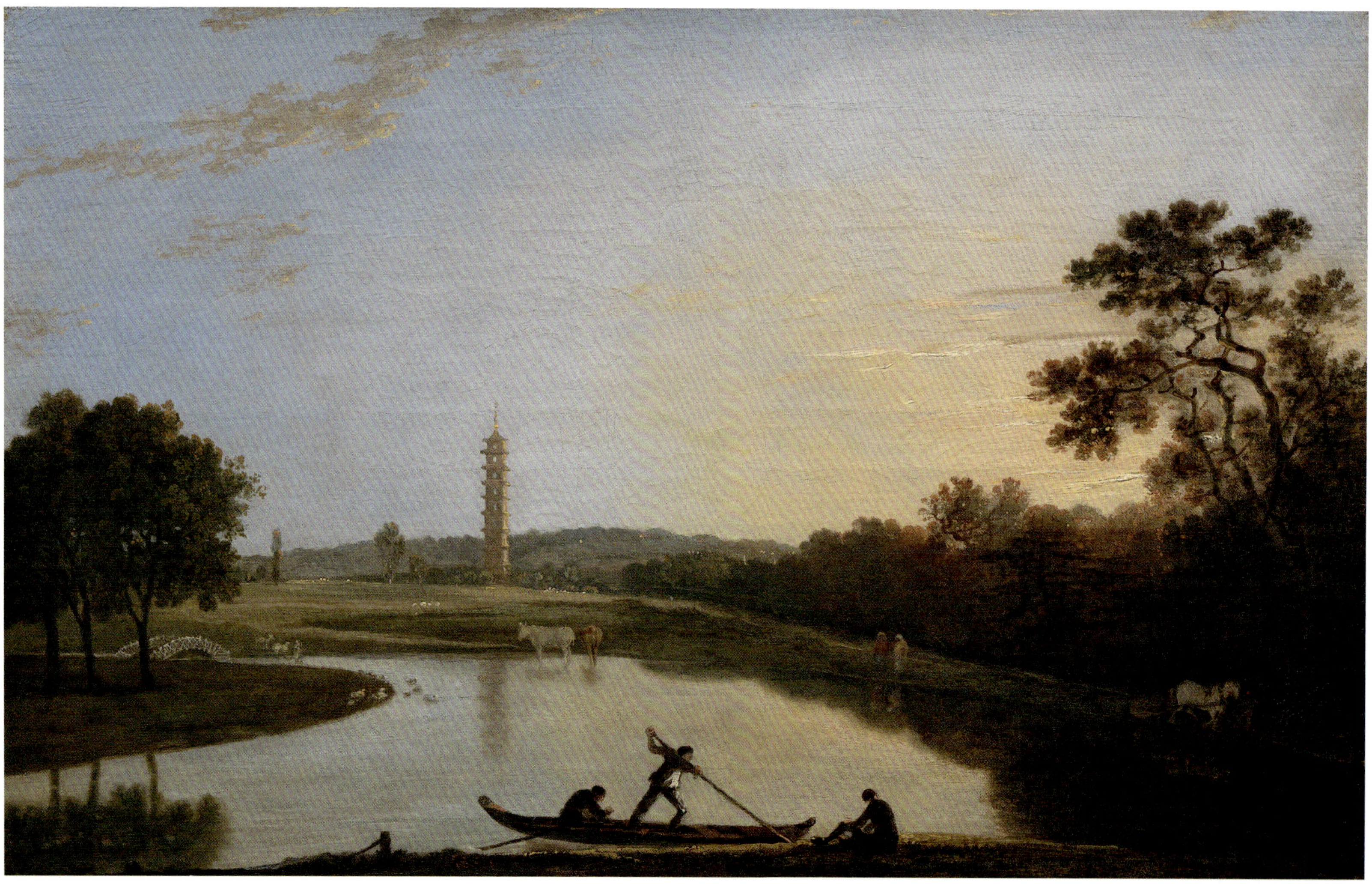

19.14

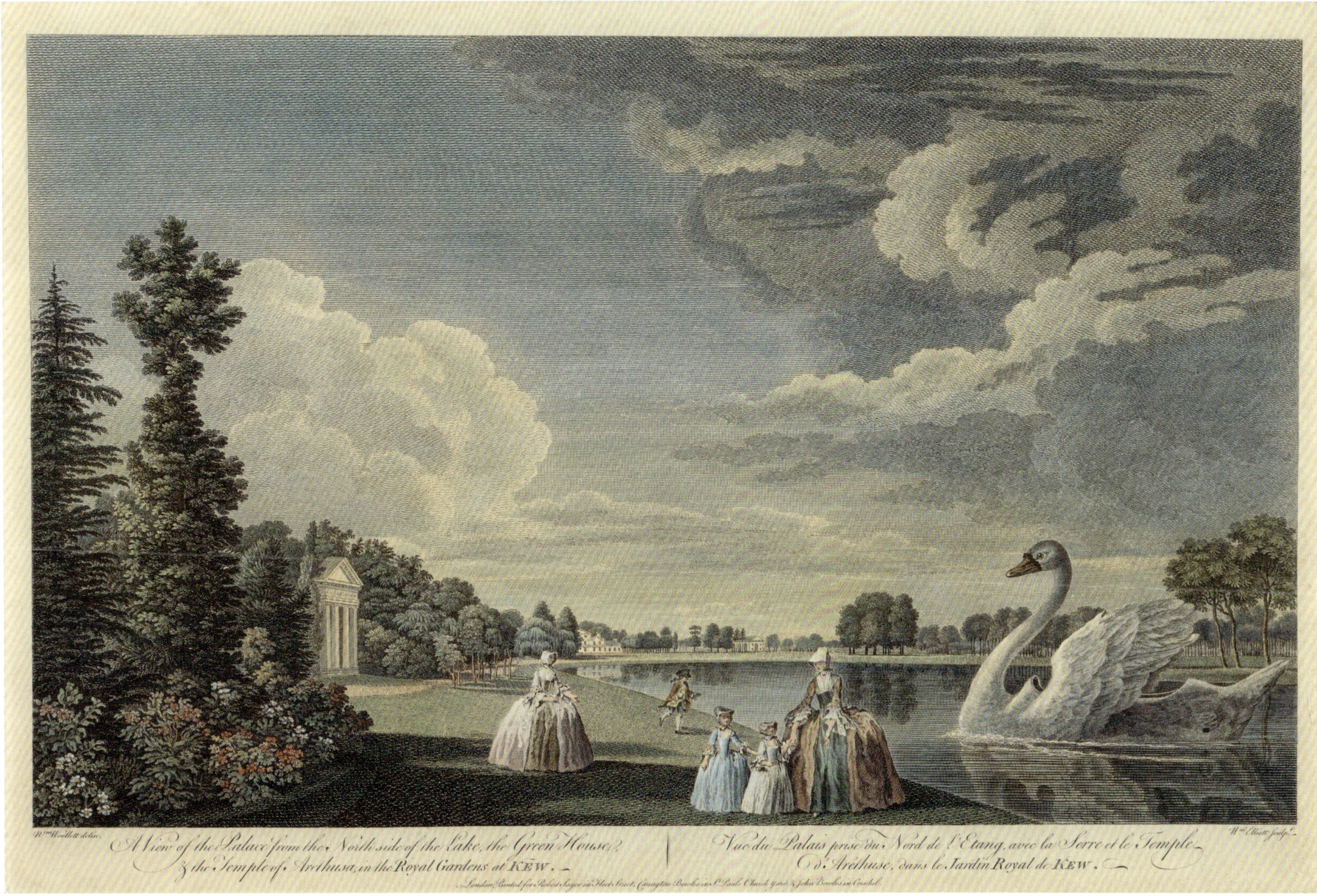

19.15

19.14*

RICHARD WILSON (c. 1714–1782)
Kew Gardens: The Pagoda and Bridge
1762
Oil on canvas
18¾ x 28¾ in. (47.6 x 73 cm)
Yale Center for British Art, Paul Mellon Collection, B1976.7.172

No single work played a greater part in defining Richard Wilson's reputation than *The Destruction of the Children of Niobe* (1759–60) which gained Wilson his only royal patron, in the person of the Duke of Cumberland. About two years later the painter tried to interest the King himself in two views of Kew Garden—*The Ruined Arch at Kew* (1761–2) and *Kew Gardens: The Pagoda and Bridge* (1761–2)—but apparently the pictures failed to please. According to Benjamin Booth, the leading later eighteenth-century collector of Wilson's works, the architect Sir William Chambers 'wish'd His Majesty to see & to purchase' the two Kew landscapes; 'They were shewn Him, but a party for Zucharelli and Flemish artists getting the better of his Judgmt He did not express that Approbation of them as was wisht nor ever employed Him.' This painting was completed before work on the Pagoda had finished, resulting in a rather impressionistic view. The painting does show, however, the skill with which the new lake had been cut around pre-existing stands of trees, to create the impression of an extensive landscape of great variety.

19.15*

WILLIAM ELLIOT (active 1774–d. 1792), engraver, after WILLIAM WOOLLETT (1735–1785)
Kew: The White House, Orangery, and Temple of Arethusa
c. 1760
Engraving, hand-coloured in watercolour
14½ x 21 in. (36.8 x 53.3 cm)
Printed for Robert Sayer (1724/5–1794), Carington Bowles (1724–1793), and John Bowles (1701?–1779), London
Yale Center for British Art, Paul Mellon Collection, B1977.14.14582

William Woollett became the most successful landscape printmaker of the mid-eighteenth century, and was eventually appointed engraver to George III. Three of Woollett's views of Kew were included in Chambers's *Plans, Elevations, Sections and Perspective Views of the Gardens and Buildings at Kew in Surry* (1763). Woollett's images were prepared later than the paintings by Johann Jakob Schalch, so it is possible to see more of Chambers's garden buildings, and these in a more finished form. The swan boat illustrated here was a seventeenth-birthday gift to George, Prince of Wales, from Augusta in 1755.

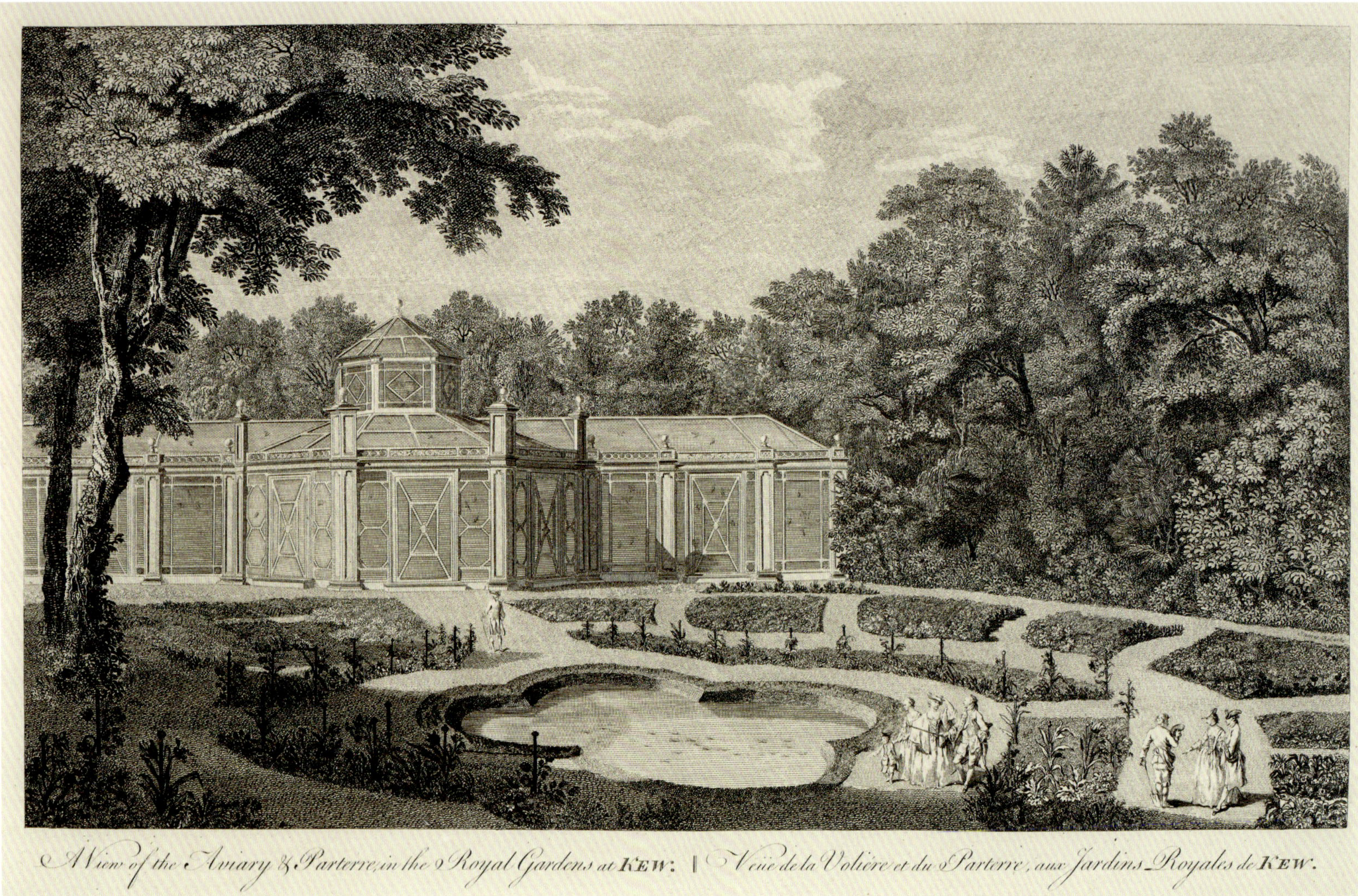

19.16

19.16†

JAMES ROBERTS THE ELDER (1725–1799), engraver, after CHARLES GRIGNION (1721–1810), etcher, after THOMAS SANDBY (1723–1798)
A View of the Aviary and Parterre, in the Royal Gardens at Kew
c. 1763
Etching
13½ x 19⅜ in. (34.3 x 49 cm)
Royal Collection Trust, RCIN 702947.l

The aviary that the architect William Chambers designed for Princess Augusta at Kew was one of several buildings inspired by his travels in China in the 1740s. On its completion in 1760 it was populated with English songbirds such as bullfinches, larks, yellowhammers, and chaffinches. The adjoining flower garden contained carnations, hyacinths, and narcissi, among other blooms, providing a pleasurable mix of scent, colour, and birdsong. This view by Thomas Sandby (see cat. 29.37) was engraved for Chambers's *Plans, Elevations, Sections and Perspective Views of the Gardens and Buildings at Kew in Surry* (1763) by Charles Grignion, a noted illustrator and engraver who trained alongside Thomas Gainsborough and worked closely with William Hogarth.

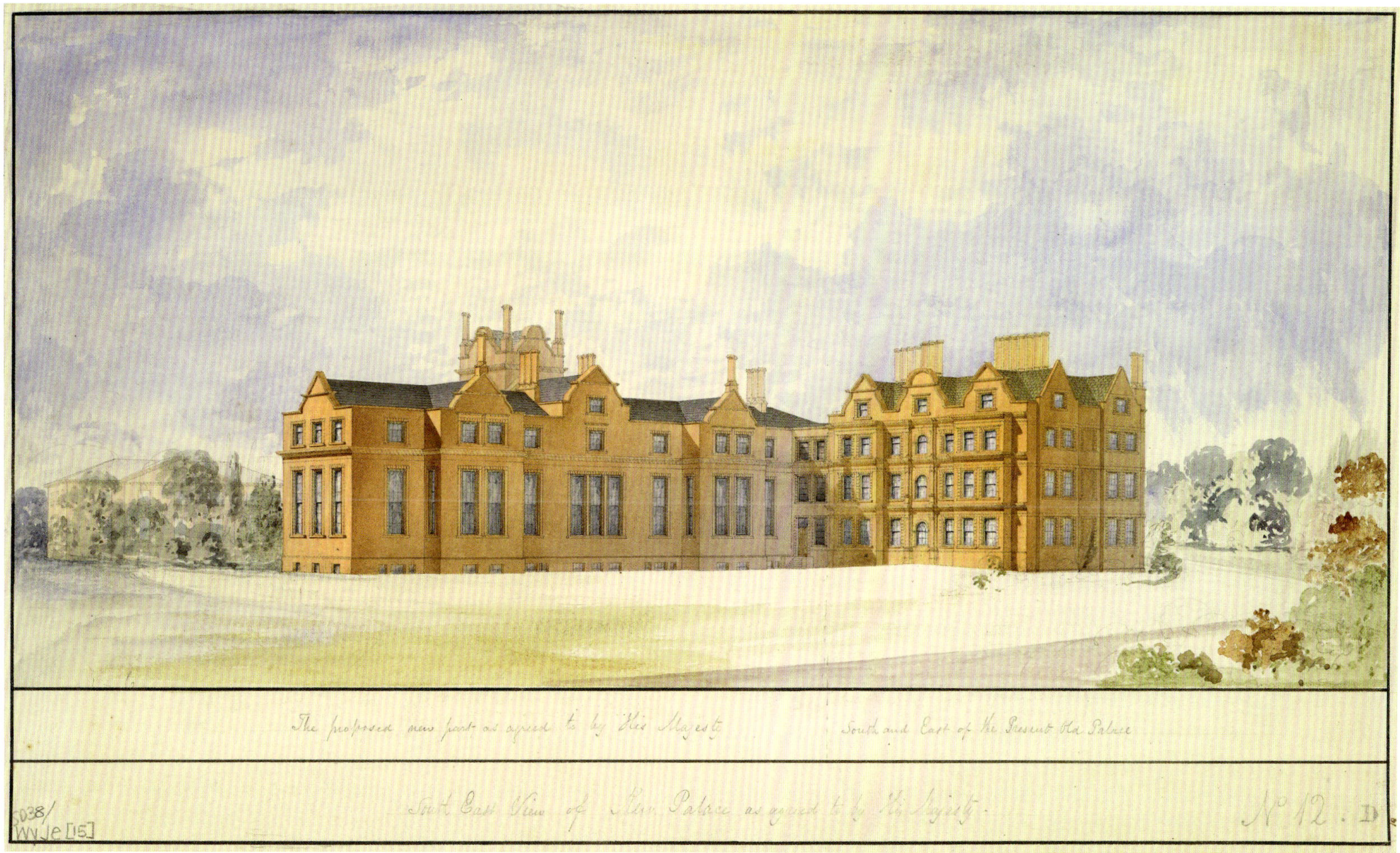

19.17

CHARLOTTE'S ARCHITECTURAL CONTRIBUTION TO KEW

19.17*

JEFFRY WYATVILLE (1766–1840)
'The proposed new part as agreed to by His Majesty . . . South and East of the Present Old Palace . . . South East View of Kew Palace as agreed to by His Majesty' [Unexecuted designs for additions to Kew Palace (The Dutch House)]
Undated
Royal Institute of British Architects, London, Library Drawings and Archives Collections
SD 38/WYJE[15]

In 1802 George III and Charlotte finally managed to negotiate the amalgamation of the gardens at Richmond Lodge with those of the White House, so the King could turn to his ambition to build a new and more spendid home. In 1794 Joseph Farington had noted 'Wyatt lately shewed hodges [*sic*] a set of designs for a Palace to be built at Kew.' The White House was demolished in 1803, in anticipation of this building project starting. The King's illness brought a halt to the building of the 'Castelated Palace' in 1806. Charlotte had always favoured the Dutch House, but though improvements were suggested to make this more comfortable, after the King's decline she rarely visited it.

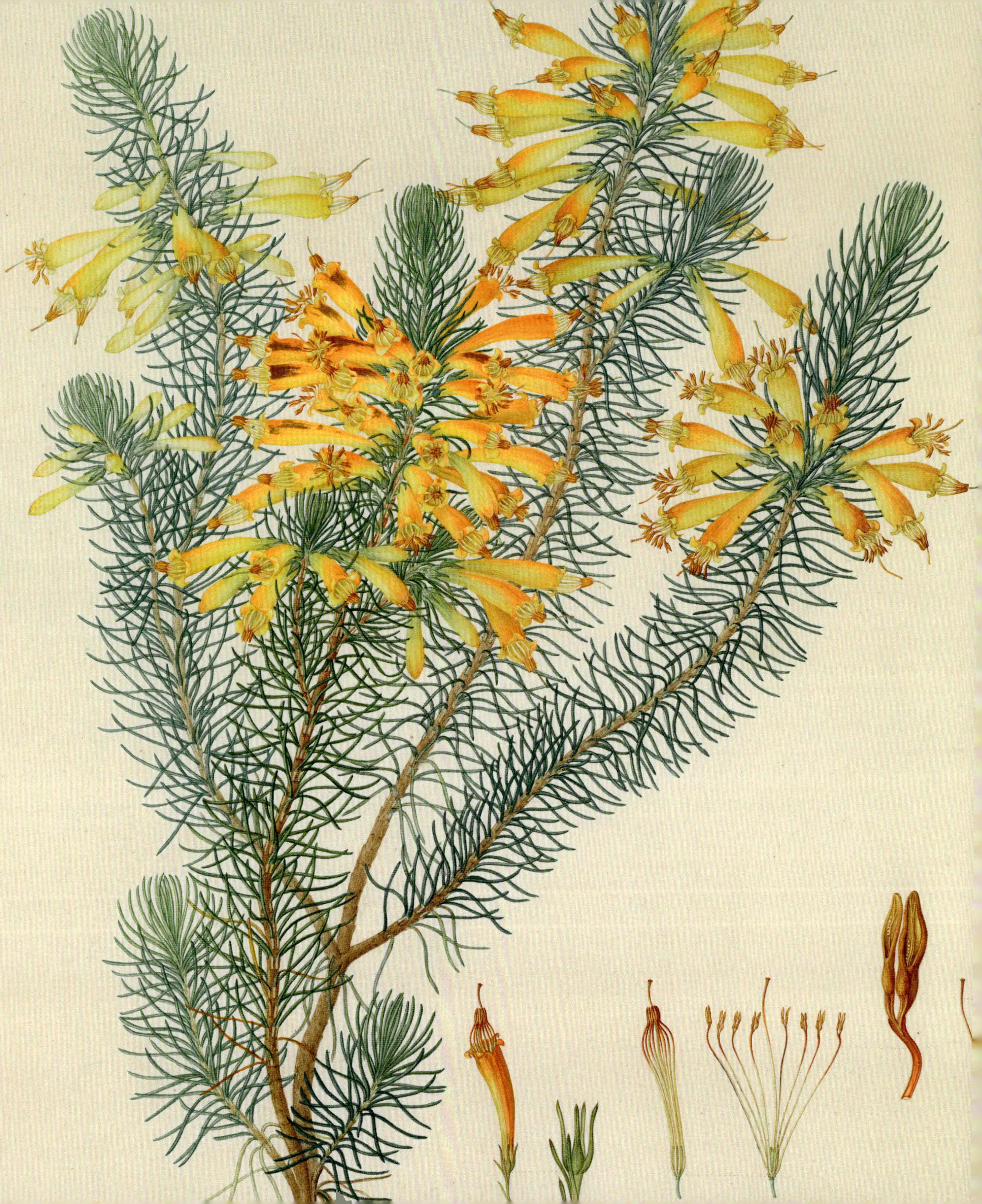

Hankering after Horticulture

Todd Longstaffe-Gowan

James Boswell remarked in his obituary of Augusta of Saxe-Gotha, Princess Dowager of Wales, in February 1772 that Her Royal Highness had a 'bounteous hand', secretly disbursing no less than £6,000 yearly to 'indigent families', and that 'the only expense in which she indulged herself was her garden at Kew, which is enriched with a variety of rare and useful plants, and is justly considered a real ornament to this country.'[1]

Despite Boswell's use of the term 'indulged', his description of the plants as 'useful' as well as 'rare' draws attention to the public function of such a garden, and so marks a subtle shift from an account of ventures into horticulture by earlier princesses. John Noorthouck relates in 1776 that in the late seventeenth century, Queen Mary II 'gave her minutes of leisure to architecture and gardening', expressing the wish that 'since it *employed many hands*, she said, *she hoped it would be forgiven her*.'[2] In contrast, Augusta is portrayed by Boswell not as requiring forgiveness for a guilty pleasure, but as fulfilling a practical, princely duty.

Augusta's implicit aim appears to have been to form a *theatrum naturae*—a highly ordered space designed to accommodate and display all species of plants—to tame and comprehend nature's productions by imposing a classificatory system that made relations between the parts and the whole intelligible and significant.[3] Her enthusiasm for gardening was initially kindled by her close friendship with her mother-in-law, Caroline of Ansbach, and in equal measure by her consort, Frederick, Prince of Wales, who from the early 1730s applied himself to a range of architectural and landscaping projects, including those at the White House at Kew and Carlton House in London.[4] John Stuart, third Earl of Bute, gave Augusta much encouragement and guidance, and recommended that her new garden should 'contain all the plants known on earth'.[5] He also acted as 'principal manager' of Kew, and supplied the impetus to see the collection of plants as an opportunity for scientific inquiry.[6] No less important were the architect Sir William Chambers and the botanist William Aiton, who from 1759 laid out and planted the new botanic garden.

Kew as a *theatrum naturae* meant that the collection and display of plants was to be reasonably methodical. However, the choice of garden buildings would be idiosyncratic.[7] Chambers observed that Augusta was 'for ever adding new embellishments at Kew'—referring, presumably, to the erection of a panoply of buildings, 'deservedly the admiration of all foreigners', including a Pagoda, an Alhambra, a Mosque, a Gothic cathedral, a Gallery of Antiquities, classical temples, a Palladian bridge, a House of Confucius, a Chinese Arch, a menagerie, an aviary, and a ruined arch (fig. 20.01).[8]

Augusta also hoped that Kew would be a place of studious retirement, much as Charlotte of Mecklenburg-Strelitz did later at Frogmore House, built on land adjacent to the Home Park at Windsor. Here the Queen, who had 'a genuine and ardent taste for the study of botany', received regular instruction from James Edward Smith, President of the Linnean Society. She also collected and studied plants, and her collection was much improved by the addition in 1788 of the herbarium of the 'scientific botanist'

Detail of p. 156, fig. 6.16

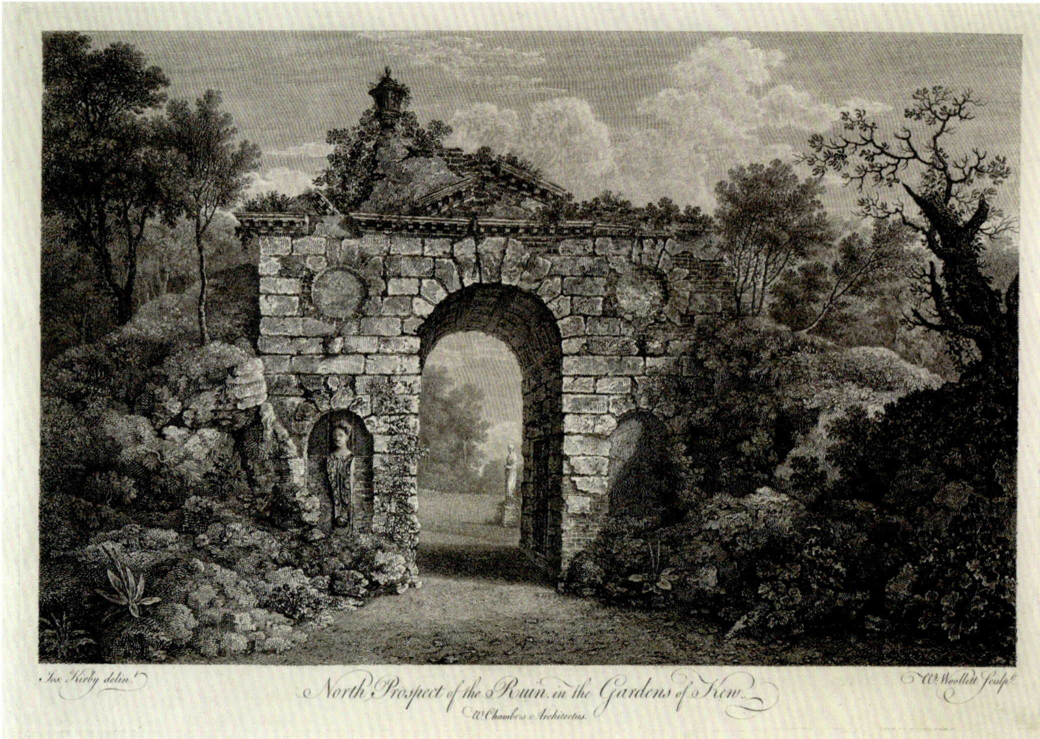

FIG. 20.01 William Woollett, after Joshua Kirby, *North Prospect of the Ruin, in the Gardens of Kew*, 1763, etching on paper, 12⅝ x 17¾ in. (32 x 45.2 cm). British Museum, London

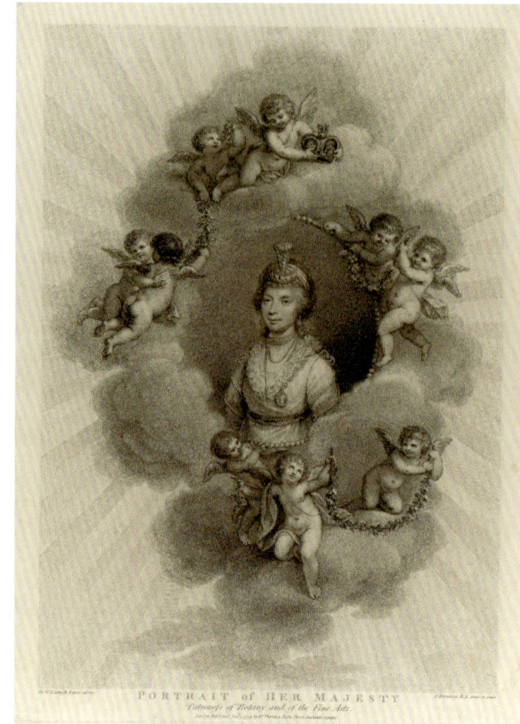

FIG. 20.02 Francesco Bartolozzi, after Sir William Beechey, 'Portrait of Her Majesty', stipple etching, in Robert Thornton, *Sexual System of Linnaeus* (London: Published for the Author, 1799), frontispiece.

John Lightfoot.[9] As Blanche Henrey observes, there is 'little doubt that Queen Charlotte, by her active interest in botany, did much to make the pursuit of this subject a fashionable pastime among "the nobility and gentry of England" of her day', as was signalled by the inclusion of her portrait as the frontispiece in Robert Thornton's *Sexual System of Linnaeus* (fig. 20.02).[10]

Although Augusta and Charlotte played a part in the layout of their respective gardens, neither was as absorbed by landscape design as Caroline, who collaborated with leading landscape theorists and practitioners. Her interest had initially been fostered by her guardian, Sophia Charlotte, Electress of Brandenburg-Prussia, and also by the latter's mother, the Dowager Electress Sophia of Hanover. Electress Sophia Charlotte had reformed the gardens at Lietzenburg (now Charlottenburg) in collaboration with Siméon Godeau, and Dowager Electress Sophia had presided over the improvement of the park and gardens at Herrenhausen. Sophia's gardens were laid out in the French Baroque taste; the gardeners who had designed them had been pupils of André Le Nôtre. Caroline, however, embraced the new English style of gardening, and her enthusiasm for it was strongly endorsed by her friends Alexander Pope, Sir Robert Walpole, and Richard Boyle, third Earl of Burlington. Her aim consisted, in her own words, 'in helping nature, not losing it in art'.[11]

She launched her efforts in September 1719 when she held a gardening conference, the aim of which was to solicit advice from 'our Men of Taste' about 'designing a princely garden'.[12] At Richmond Lodge she collaborated from 1722 to 1735 with Charles Bridgeman, Thomas Greening, and William Kent, introducing 'cultivated fields, and even morsels of a forest appearance'.[13] The scheme included Merlin's Cave, a 'subterraneous building' commissioned from Kent, inhabited by six life-size wax effigies, and possessed of a 'choice Collection of *English* books'. The poet Stephen Duck was appointed 'Cave and Library Keeper'. A contributor to *Fog's Weekly* in December 1735 declared the confection 'Hieroglyphical, Emblematical, Typical and Symbolical, conveying lessons of Policy to Princes and Ministers of State', and the Tory press ridiculed Caroline's effort to link the Hanoverian with the Tudor court by introducing the figures of Queen Elizabeth I consulting the magician Merlin. Confusion about the exact identity of the effigies was widespread and persistent. Sarah Churchill, Duchess of Marlborough, writing to her granddaughter Diana, Duchess of Bedford, in 1735, assumed 'it is taken out of Spenser's

Fairy Queen'.[14] Anne-Marie Le Page Fiquet du Boccage, the French writer, poet, letter-writer, and playwright, writing to her sister in the 1750s, by contrast described two of the figures as Anne Boleyn and Elizabeth I, 'accompanied by their nurses, persons very proper to assist at these puerile mysteries'.[15] In summing up her impression, the Duchess of Marlborough concluded:

> it is too long to give you an account of a most curious comical description which I had yesterday of what is called Merlin's Cave at Richmond, which they say is a thousand times more ridiculous than what was done for the philosophers (at the Queen's Hermitage). Merlin, it seems, was a Welsh conjuror . . . and there are a great many strange puppets dressed up in it.[16]

The Hermitage, with its sculptural programme celebrating contemporary natural philosophers and theologians, was on the other hand vaunted by some as a 'symbol of modern knowledge', had great public appeal, and inspired a rash of imitations.[17]

Caroline's improvements to the gardens at Kensington Palace, where she collaborated with Charles Bridgeman, were widely viewed as a more coherent and successful venture, achieving a happy co-mingling of art and nature, and an unaffected Englishness that was celebrated by artists as well as visitors to the Palace (fig. 20.03). She created 'well judg'd Vistos', a sprawling wilderness garden, a great basin, and a serpentine river. Her protégé, Duck, compared the gardens to Eden.[18]

The three Princesses engaged with botany and horticulture in different ways: Caroline, with her attachment to the *jardin*

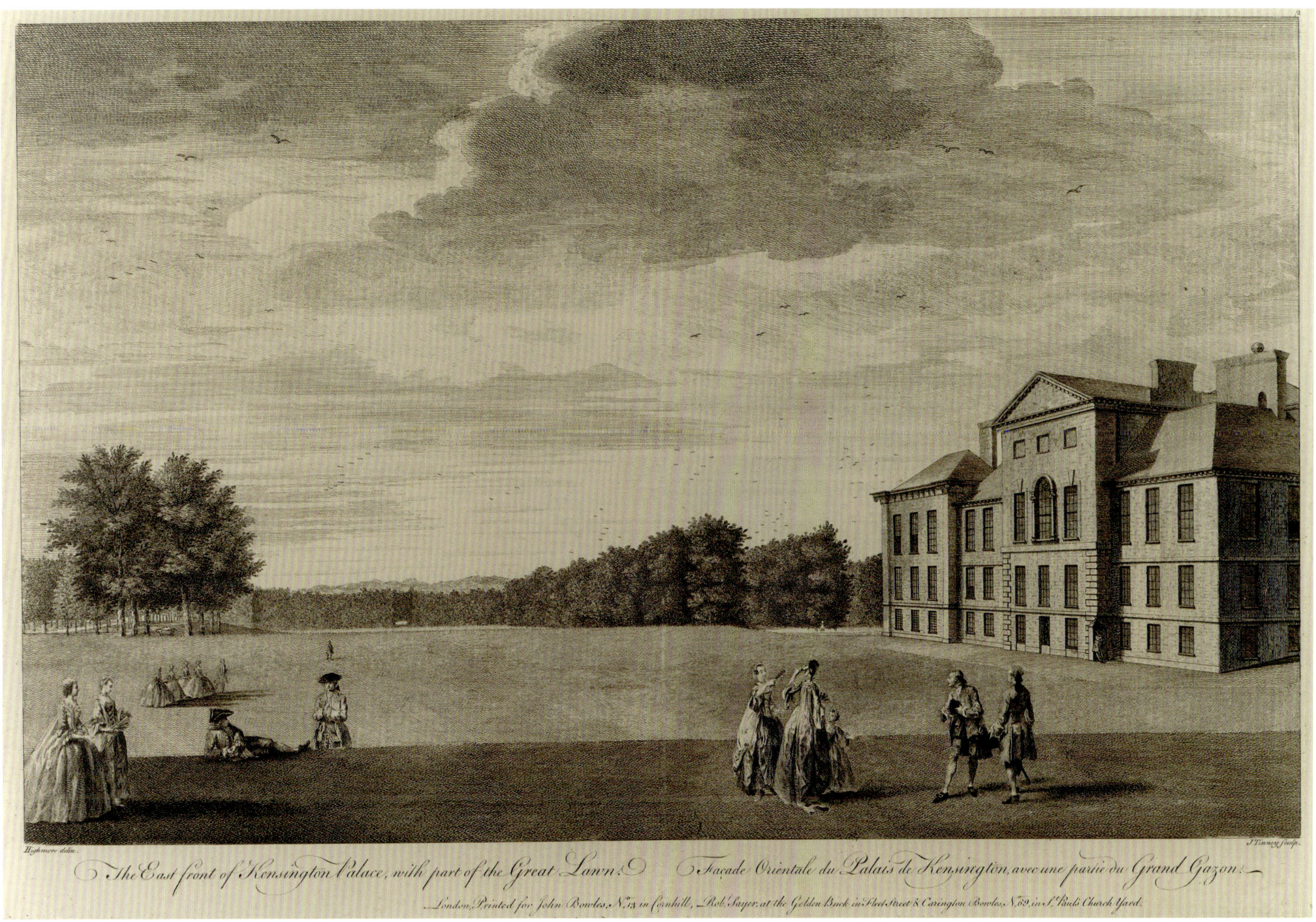

FIG. 20.03 Unknown engraver, after Anthony Highmore, *The East Front of Kensington Palace, with part of the Great Lawn*, 1744, etching and engraving on paper, 13⅝ x 19¼ in. (34.5 x 49 cm). British Museum, London

anglais, had a more personal aesthetic vision than the other two, while Charlotte had a '*violent hankering*'—as Frances Burney termed it—for the 'rational, beautiful, and enticing amusement' of pressing plants into paper and then colouring the impression.[19] However, Augusta and Charlotte did both became involved in the garden which Caroline had laid out at Richmond, adapting and enlarging it as it became possible to unite the adjoining estate belonging to the White House at Kew. This garden had a highly visible public role—not only was it on the itinerary of many foreign travellers, but it also played a central part in the exchange of plants within the British Empire, partly as a result of the efforts of Sir Joseph Banks, who became an adviser there in the early 1770s (see cat. 16.29). The fact that the Princesses' engagement with Kew never excited much surprise was perhaps because of equivocal elements both in the concept of gardening and in perceptions of the Princesses' role. Gardening could be simultaneously private and public, and princesses operated in a female sphere but one unlike that of other women. The public side of their horticultural pursuits could be identified with forms of royal femininity that never appeared to impinge disquietingly on the pursuit of economic botany in which they nonetheless participated. After all, the plant named after Charlotte was the brilliantly decorative *Strelitzia reginae*—no shrinking violet, but, on the other hand, not New Zealand spinach (cats. 20.08–20.09).[20]

NOTES

1. James Boswell, *The Scots Magazine* 34 (Feb. 1772): 109–10.
2. John Noorthouck, *An Historical and Classical Dictionary* (London: W. Strahan and T. Cadell, 1776). Original italics.
3. For a useful account of this project as it emerged in the seventeenth century, see Pamela Smith, *The Business of Alchemy: Science and Culture in the Holy Roman Empire* (Princeton, NJ: Princeton Univ. Press, 1994), 4.
4. David Coombs, 'The Garden at Carlton House of Frederick Prince of Wales and Augusta Princess Dowager of Wales', *Garden History* 25, no. 2 (Winter 1997): 154.
5. Mary Soderstrom, *Recreating Eden: A Natural History of Botanical Gardens* (Montreal: Véhicule Press, 2001), 69.
6. Undated letter to Lord Bute, reprinted in *Letters from George III to Lord Bute, 1756–1766*, ed. R. Sedgwick (London: Macmillan, 1939), 4. Alan W. Armstrong, ed., '*Forget not Mee & My Garden': Selected Letters, 1725–1768, of Peter Collinson, F.R.S.* (Philadelphia: American Philosophical Society, 2002), 229; W. J. Bean and William T. Thiselton-Dyer, *The Royal Botanic Gardens, Kew* (London and New York: Cassell and Co., 1908), 14.
7. The experience of visiting the garden is described in an extended poem by George Ritso: *Kew Gardens. A poem, humbly inscribed to Her Royal Highness the Princess Dowager of Wales* ([London:] Printed by M. Lewis and sold by R. and J. Dodsley, 1763).
8. *Gentleman's Magazine and Historical Review* 42 (1772): 400. See, in this volume, Wolf Burchard and Kathryn Jones, 'The German Princesses as British Patrons', 109–11; Roisin Inglesby, 'The Royal Estate', 325–28; Lee Prosser, 'Architectural Patronage', 335–39; and Jane Roberts, 'Frogmore and the Princesses as Artists', 375–377.
9. Abraham Rees, *The Cyclopaedia; or, Universal Dictionary of Arts, Sciences, and Literature* (London: Longman, Hurst, Rees, Orme, & Brown, 1819), under 'Lightfoot, J.'.
10. Blanche Henrey, *British Botanical and Horticultural Literature before 1800*, 3 vols. (London: Oxford Univ. Press, 1975), 2: 75.
11. John Perceval (Percival), *Manuscripts of the Earl of Egmont: Diary of Viscount Percival, afterwards first Earl of Egmont*, Historical Manuscripts Commission 63, 3 vols. (London: HMSO, 1920–3), 2: 138.
12. *The Works of Alexander Pope Esq.*, 9 vols. (London: B. Law, J. Johnson, C. Dilly, and others, 1797), 8: 227–28.
13. Roy Strong, *Royal Gardens: From Charles II to Charles, Prince of Wales* (London: Conran Octopus, 1995), 42; George Vertue, *Anecdotes of Painting in England; With Some Account of the Principal Artists; and Incidental Notes on Other Arts*, Vol. 4 (Strawberry Hill: Printed by Thomas Kirgate, 1771), 136.
14. Sarah Churchill, *Letters of a Grandmother, 1732–1735: Being the Correspondence of Sarah, Duchess of Marlborough with her Granddaughter Diana, Duchess of Bedford*, ed. G. S. Thomson (London: Jonathan Cape, 1943), 171–74.
15. Madam du Bocage [*sic*], *Letters concerning England, Holland and Italy . . . Translated from the French*, 2 vols. (London: E. and C. Dilly, 1770), 1: 50.
16. Sarah Churchill, *Letters of a Grandmother, 1732–1735: Being the Correspondence of Sarah, Duchess of Marlborough with her Granddaughter Diana, Duchess of Bedford*, ed. G. S. Thomson (London: Jonathan Cape, 1943), 171–72.
17. Stephen Knight, *Merlin: Knowledge and Power through the Ages* (Ithaca, NY: Cornell Univ. Press, 2009), 126–27.
18. 'On the Queen's Mount at Kensington', *The Gentleman's Magazine* 28 (April 1733): 206.
19. Frances Burney, *The Court Journals and Letters of Frances Burney*, Vol. 2: *1787*, ed. Stewart Cooke (Oxford: Clarendon Press, 2011), 305 notes 520, 521 (Queen Charlotte to Lord Bute, 19 March 1788).
20. New Zealand spinach and New Zealand flax were among the plants introduced to Kew by Banks.

20.04

BATTY LANGLEY (1696–1751)
New Principles of Gardening: Or, Laying out and Planting Parterres, Groves, Wildernesses, Labyrinths, Avenues, Parks, &c. After a more Grand and Rural Manner, than has been done before
1728
9⅞ in. (25 cm)
Printed for A. Bettesworth and J. Batley, London
* Oak Spring Garden Library (illustrated)
Exhibited: Plate 22, 'Shady Walk with Temple' and 'An Avenue in Perspective', engraving on paper
10¼ in. (26 cm); (not illustrated)
† Royal Botanic Gardens, Kew, Library, Art & Archives Collections, PRE-LINN-B LAN
Exhibited: Plate 22, 'Shady Walk with Temple' and 'An Avenue in Perspective', engraving on paper
10¼ in. (26 cm); (illustrated)

Batty Langley was an early popularizer of the 'naturalistic' style that revolutionized English landscape gardening in the early eighteenth century. Associated with patriotism and rustic virtue, this new approach was adopted on the estates of aristocratic patrons such as Lord Cobham at Stowe, as well as at Alexander Pope's villa at Twickenham. It rejected formal geometry, instead using winding paths, long vistas, and artfully placed groves to achieve its effects. Designs for Caroline's gardens at Richmond, by William Kent and Charles Bridgeman, used these picturesque devices to frame her Hermitage and Merlin's Cave, as well as for their political and moral associations. *New Principles of Gardening* was dedicated to George II, and Caroline's library contained a copy.

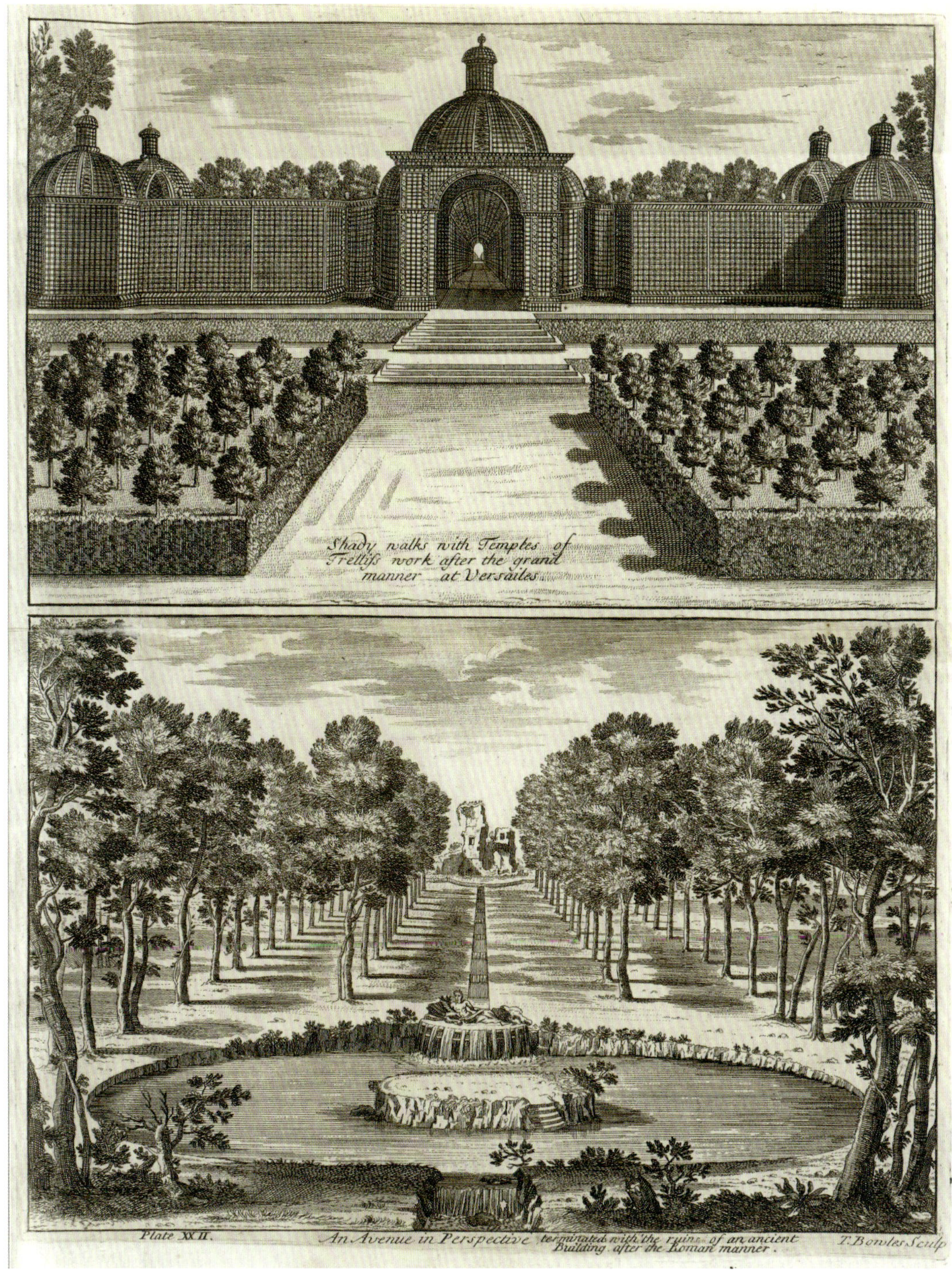

20.04

* = YCBA only
† = HRP only

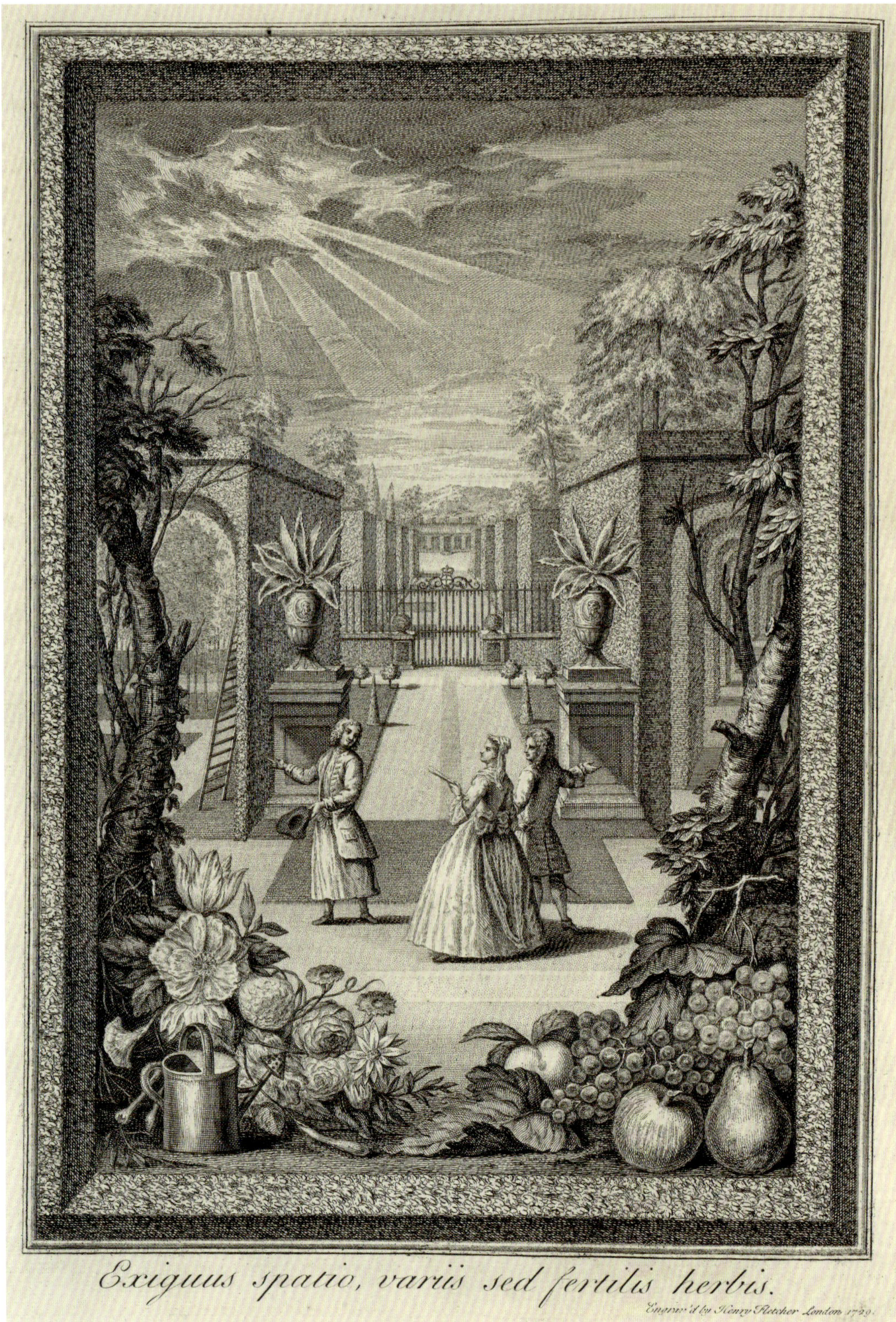

20.05

20.05*

SOCIETY OF GARDENERS, with **PHILIP MILLER** (1691–1771), editor
Catalogus plantarum, tum exoticarum tum domesticarum, quae in hortis haud procul a Londino sitis in venditionem propagantur. A Catalogue of Trees, Shrubs, Plants, and Flowers, both Exotic and Domestic, Which are Propagated for Sale in the Gardens near London
1730
18 ¼ in. (45.7 cm)
Society of Gardeners, London
Oak Spring Garden Library
Exhibited: Title page, with frontispiece, Henry Fletcher (active 1715–1744), engraver, 'Exiguus spatio, variis sed fertilis herbis' (illustrated)

The Society of Gardeners met in Newell's Coffee House in Chelsea from the later 1720s. Its *Catalogus plantarum* was published in 1730, illustrating exotic and domestic plants 'propagated for sale in the gardens near London. Divided, according to their different Degrees of Hardiness'. Finding means by which new species, especially exotics, brought into the country from markets abroad could be propagated had become a cause of concern. Philip Miller, who later would write the *Gardener's Dictionary*, was 'Clerk' of the Society and led the team of authors. It was published in English ensuring a wide circulation, even leading to some satirical skits. Caroline owned a copy of this volume.

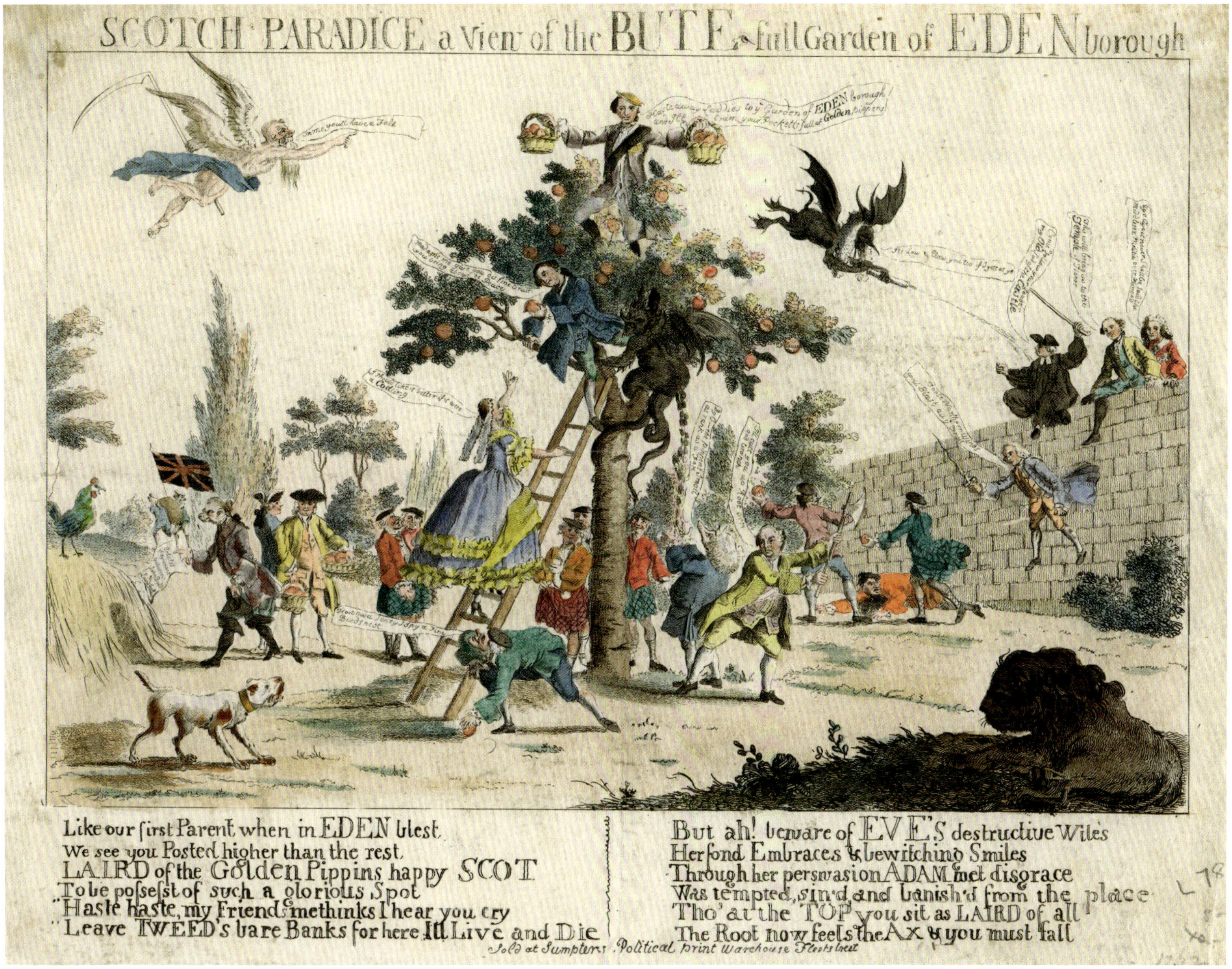

20.06

20.06*

UNKNOWN ENGRAVER
Scotch Paradice: a View of the Bute[eye]full Garden of Edenborough
1763
Hand-coloured etching on paper
9½ x 12 ¼ in. (24 x 31 cm)
Published by Edward Sumpter (active 1763–1787), London
The Lewis Walpole Library, Yale University, 763.02.00.01

The close collaboration between Princess Augusta and the Earl of Bute at Kew Gardens made it the perfect setting for satirical depictions of their relationship. Here, Augusta, Eve-like, reaches for an apple offered by a tempter, the Duc de Nivernois, the French Ambassador, bargaining for peace in the Seven Years' War. Meanwhile Bute distributes bounty, 'Golden Pippins', to his Scottish followers, one of whom peers up the Princess's skirt. A sorrowful lion, emblem of the national interest, sits in shadow in the foreground.

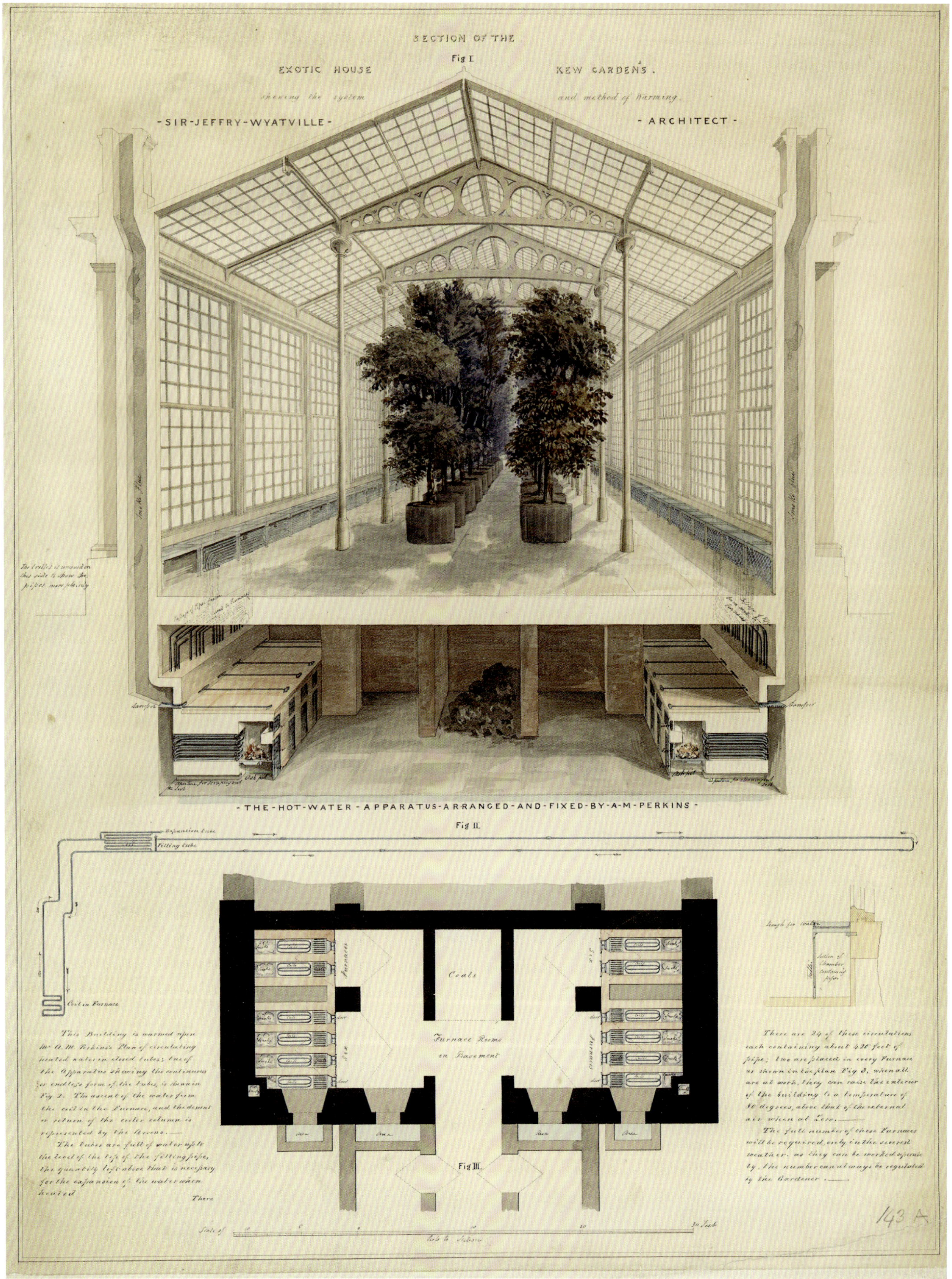
SECTION OF THE
Fig I.
EXOTIC HOUSE
KEW GARDENS.
shewing the system
and method of Warming.
-SIR-JEFFRY-WYATVILLE-
-ARCHITECT-
-THE-HOT-WATER-APPARATUS-ARRANGED-AND-FIXED-BY-A-M-PERKINS-
Fig II.
Coil in Furnace
Coals
Furnace Rooms
in Basement
Fig III.
This Building is warmed upon Mr A.M. Perkins's Plan of circulating heated water in closed tubes; one of the Apparatus shewing the continuous or endless form of the tubes, is shewn in Fig 2. The ascent of the water from the coil in the Furnace, and the descent or return of the cooler column is represented by the Arrows.—
The tubes are full of water up to the level of the top of the filling pipe, the quantity left above that is necessary for the expansion of the water when heated
There
There are 24 of these circulations each containing about 420 feet of pipe; two are placed in every Furnace as shewn in the plan Fig 3, when all are at work, they can raise the interior of the building to a temperature of 80 degrees, above that of the external air when at Zero.—
The full number of these Furnaces will be required, only in the severest weather. as they can be worked seperately. the number can always be regulated by the Gardener.—
143 A

20.07

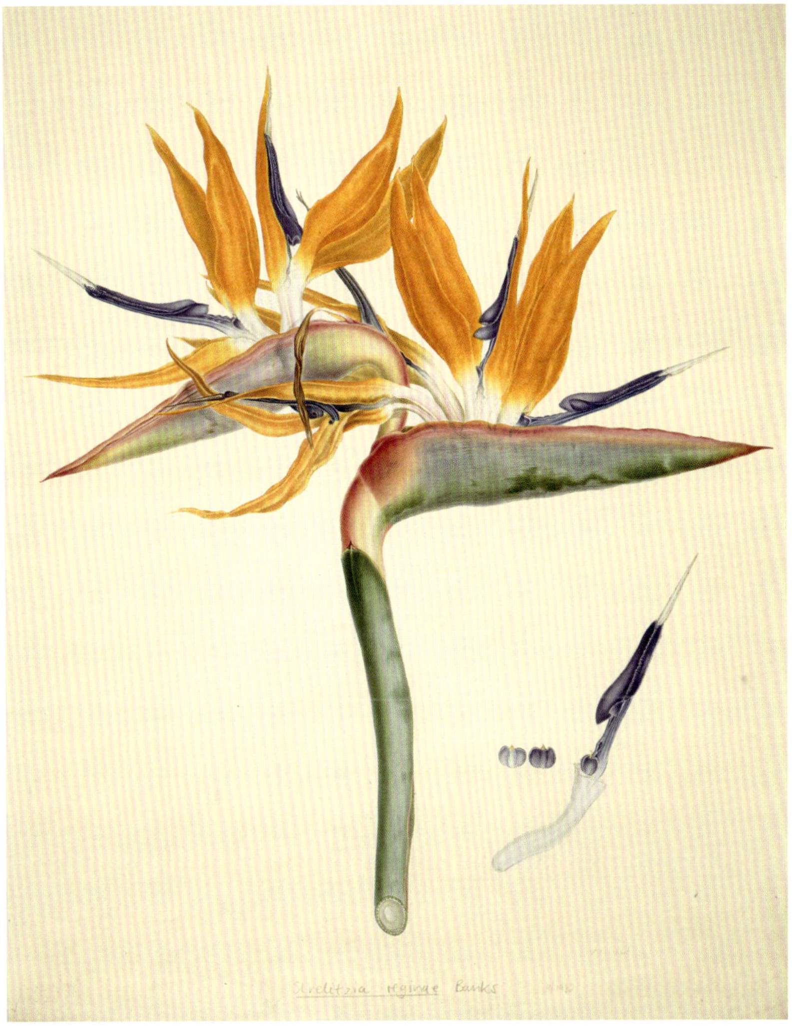

20.08

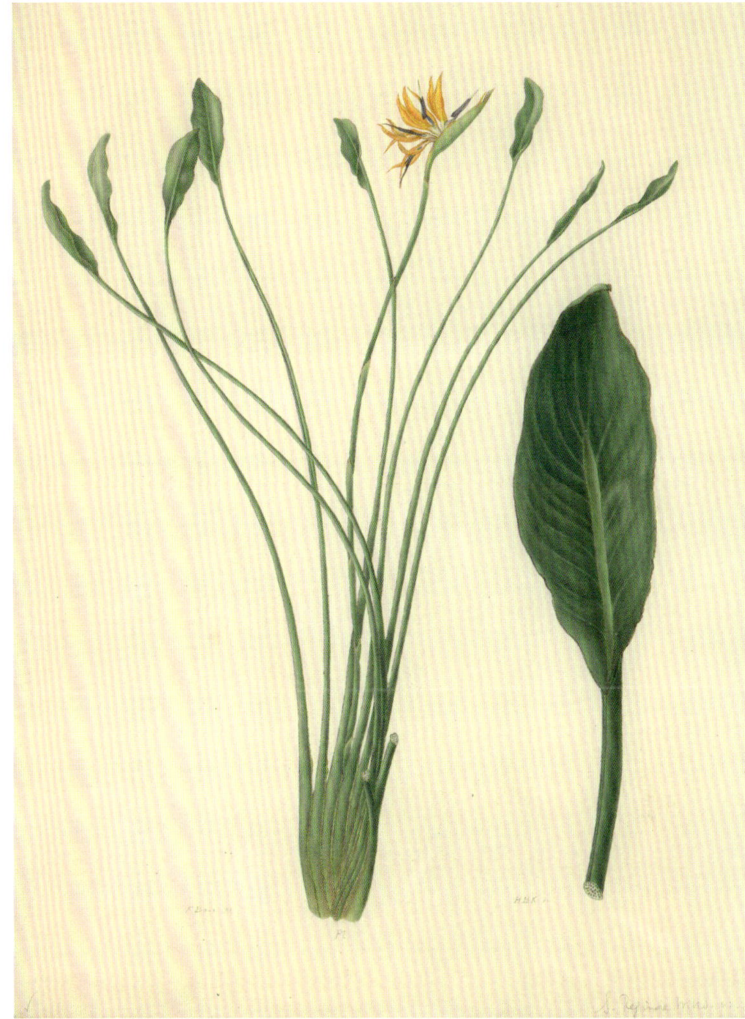

20.09

20.07*

JEFFRY WYATVILLE (1766–1840)
[Design for] 'Section of the Exotic House . . .
Kew Gardens'
1800
Pen and ink on paper
Royal Institute of British Architects, London, Library Drawings and Archives Collections, SB31/WYJE[14] (F7 [14])

This drawing shows the construction and heating system of a hothouse initially built at Kensington in 1825 and moved to Kew Gardens in 1836. The building was adapted by the architect Sir Jeffry Wyatville and included an innovative heating system that used piped hot water rather than fires whose smoke and fumes might damage delicate plants. Despite the addition of some new accommodation for exotic plants, the royal gardens at Kew were generally seen to be in decline in the 1820s and 1830s, since with the deaths of Queen Charlotte in 1818 and both George III and Sir Joseph Banks in 1820, the nexus of botanical, imperial, commercial, personal, and artistic interests operating there lost its most ardent participants and supporters.

20.08*

FRANCIS (FRANZ ANDREAS) BAUER (1758–1840)
'*Strelitzia Reginae* Banks', design for *Strelitzia Depicta*
1792–1817 [1812?]
Watercolour on paper
22⅛ x 16⅛ in. (56.2 x 41 cm)
Natural History Museum Library and Archives, London, Bauer Unit shelf C7

20.09†

FRANCIS (FRANZ ANDREAS) BAUER (1758–1840)
'*Strelitzia Reginae* Banks', design for *Strelitzia Depicta*
1792–1817 [1812?]
Watercolour on paper
22⅛ x 16⅛ in. (56.2 x 41 cm)
Natural History Museum Library and Archives, London, Bauer Unit shelf C7

This species of flower was brought from the Cape of Good Hope in 1773 by Francis Masson (1741–1805), Kew's first official plant collector, and its botanical name derives from Princess Charlotte's birthplace, Mecklenburg-Strelitz. It was not until the 1790s that the stock produced flowers in the garden at Kew. Bauer affirmed the enduring importance of the flower as a symbol of the royal family in *Strelitzia Depicta: or Coloured Figures of the known Species of the Genus Strelitzia from the Drawings in the Banksian Library*, published in 1818. The watercolours are studies for the work.

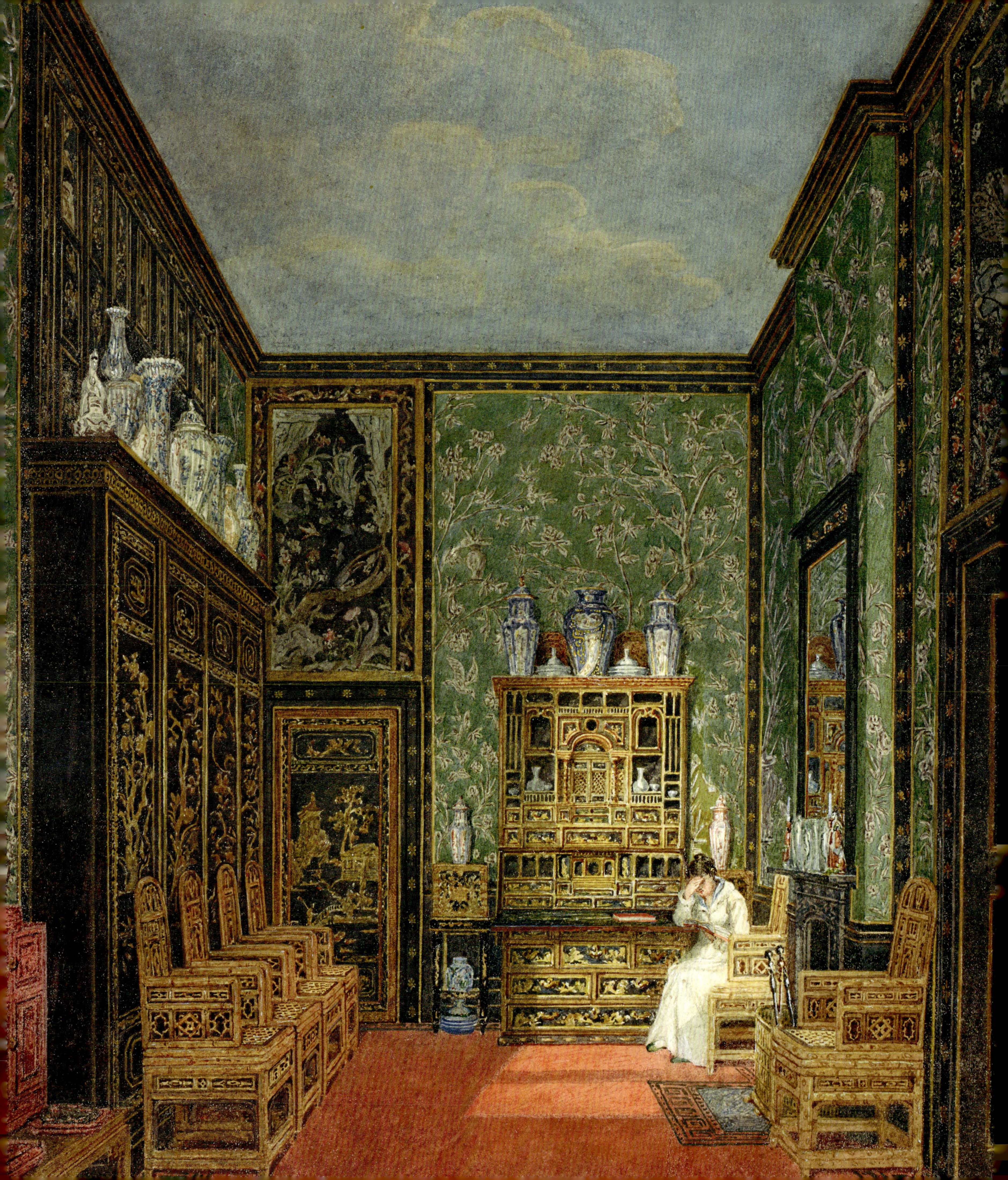

Frogmore

Samantha Howard

Away from the public gaze, the etiquette, and the intrigues of the court, Frogmore House and its gardens provided an Arcadian retreat for Charlotte of Mecklenburg-Strelitz. Frogmore was one of several royal residences, and the only house not maintained by the Office of Works, which was illustrated in William Henry Pyne's detailed three-volume publication, *The History of the Royal Residences* (1816–19), which included one hundred hand-coloured engravings of exteriors and interiors accompanied by descriptive texts.[1] Charles Wild's original watercolour illustrations of Frogmore provide us with an interesting glimpse of the surroundings in which Charlotte and her daughters pursued their literary, artistic, and 'botanising' activities.

Charlotte's Frogmore evolved from the combination of two estates, Little Frogmore and Great Frogmore, acquired in 1790 and 1792.[2] Their acquisition coincided with the King's recovery from an episode of mental illness between 1789 and 1790, after which Charlotte refused to be left alone with him. Until the early years of the nineteenth century, when staying at Windsor the royal family had used the Queen's House, also known as the Queen's Lodge, located near the castle. However, it was overlooked by a public footpath and had no garden. Located just half a mile away, the two properties at Frogmore offered the perfect opportunity for the development of a more private family retreat. With Charlotte and her daughters increasingly using the house at Little Frogmore, which was renamed Amelia Lodge, the royal architect James Wyatt presented plans for a new cottage in 1791. His design shows that he envisaged the building work to be executed to a decorative Gothic design (fig. 21.01). However, the plans for the cottage were abandoned and Amelia Lodge was demolished as part of Charlotte's scheme to develop a larger house, set within an extensive landscaped garden incorporating the adjoining estate of Great Frogmore.[3]

Although the original seventeenth-century house at Great Frogmore was sound, Charlotte was keen to make it better fit for modern royal use, and she commissioned Wyatt to undertake several alterations. One of the earliest additions, completed in 1795, were two single-storey pavilions at the north and south of the garden front, linked by an open colonnade. The north pavilion was known as the Green Pavilion. Adjacent to the southern pavilion was a room entirely decorated with flower paintings by Mary Moser, who was the leading flower painter of the age and one of the two female founder members of the Royal Academy. Wild's watercolour shows the view from the Green Pavilion looking down the colonnade to glazed doors at the southern end through which the Mary Moser room can be glimpsed (cat. 21.05).[4]

Moser was one of several artists who had taught Charlotte and her daughters to draw, especially encouraging their interest in botanical illustration.[5] A watercolour and ink drawing of a canna lily (cat. 22.06) was one of several botanical studies from a volume of drawings by Charlotte's eldest daughter, Charlotte, Princess Royal (later Queen Charlotte of Württemberg, consort of Frederick I, King of Württemberg).[6] Charlotte's enthusiasm helped to promote the study of botany as a modern female

Detail of p. 364, cat. 21.06

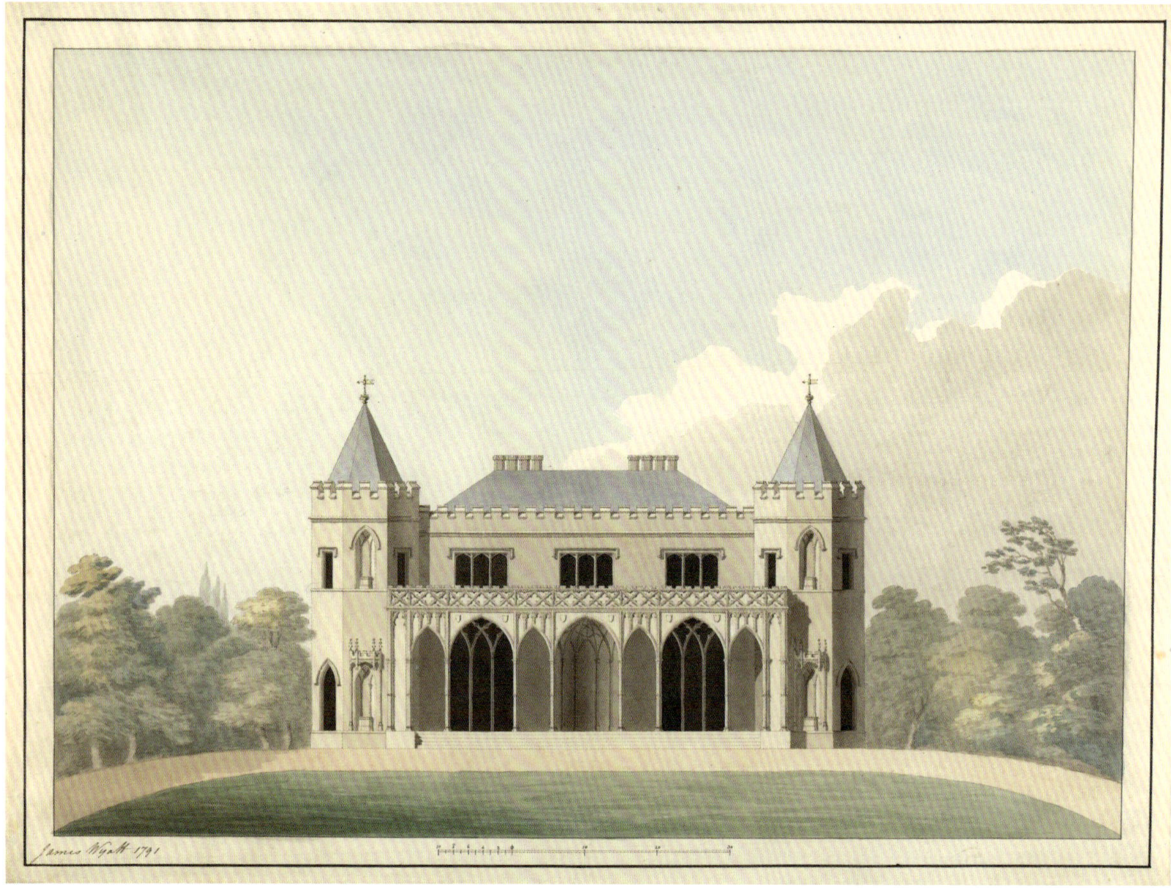

FIG. 21.01 James Wyatt, *Unexecuted Design for Little Frogmore: Elevation*, 1791, watercolour, grey wash, and pen and ink on paper, 23½ x 18 in. (59.7 x 45.7 cm). Yale Center for British Art, Paul Mellon Collection (cat. 21.03)

accomplishment—the sort of cultivated activity that was a suitable pastime for 'leisured ladies'.[7]

A love of botany would also be a contributory reason for the special friendship which grew between Charlotte and her daughters and Mary Granville, Mrs Delany (cats. 21.07–21.08). Fêted in her own day as 'a woman of fashion of all ages', Mary Delany was a prolific artist and craftswoman in a diverse variety of media.[8] She was also an active member in a number of culturally influential circles that shared a passionate interest in the natural sciences, which included Margaret Cavendish Bentinck, Duchess of Portland; John Lightfoot, the clergyman naturalist and a co-founder of the Linnean Society; and Joseph Banks and Daniel Solander, who accompanied Captain James Cook's first expedition to the South Pacific from 1768 to 1771. George III and Charlotte had been introduced to Delany at the Duchess of Portland's home Bulstrode, near Windsor, in 1776. They were both so impressed by her exquisite botanical illustrations executed in delicate cut paper—'paper mosaicks' (see cats. 21.09–21.12)—that they instructed Banks to send specimens of new foreign plants arriving at the royal gardens at Kew for her to copy, such as *Magnolia grandiflora* (Polyandria polygynia), *Sanguinaria canadensis* (bloodwort, also known as rootwort), *Polyandria monogynia* from North America, and the night-blooming *Cactus grandiflorus*, originally from the Caribbean.[9] Native plants also provided inspiration, such as *Dianthus caryophyllus*, the carnation or pink.[10]

In an atmospheric portrait by the young Cornish artist John Opie, Mrs Delany is shown wearing a locket given to her by Charlotte (the letters 'CR'—*Charlotte Regina*—are a reference to the Queen) as a token of her friendship (cat. 21.07). Commissioned in 1782 by George III, Opie imbues Delany with an air of *gravitas* by exploiting the emphasis between the tonal contrasts of his sitter's widow's weeds and the crisp whiteness of her cap, and the fichu at her throat. This feature of his work saw him acclaimed as the 'English Rembrandt' by the contemporary commentator Horace Walpole in a letter to Horace Mann: 'it is pronounced like Rembrandt, but as I told her, it does not look older than she is, but older than she does.'[11] The success of Delany's portrait no doubt enabled Opie to garner further prestigious commissions for portraits of other notable figures of the age, including Mary Wollstonecraft and Edmund Burke. Delany's connections to the

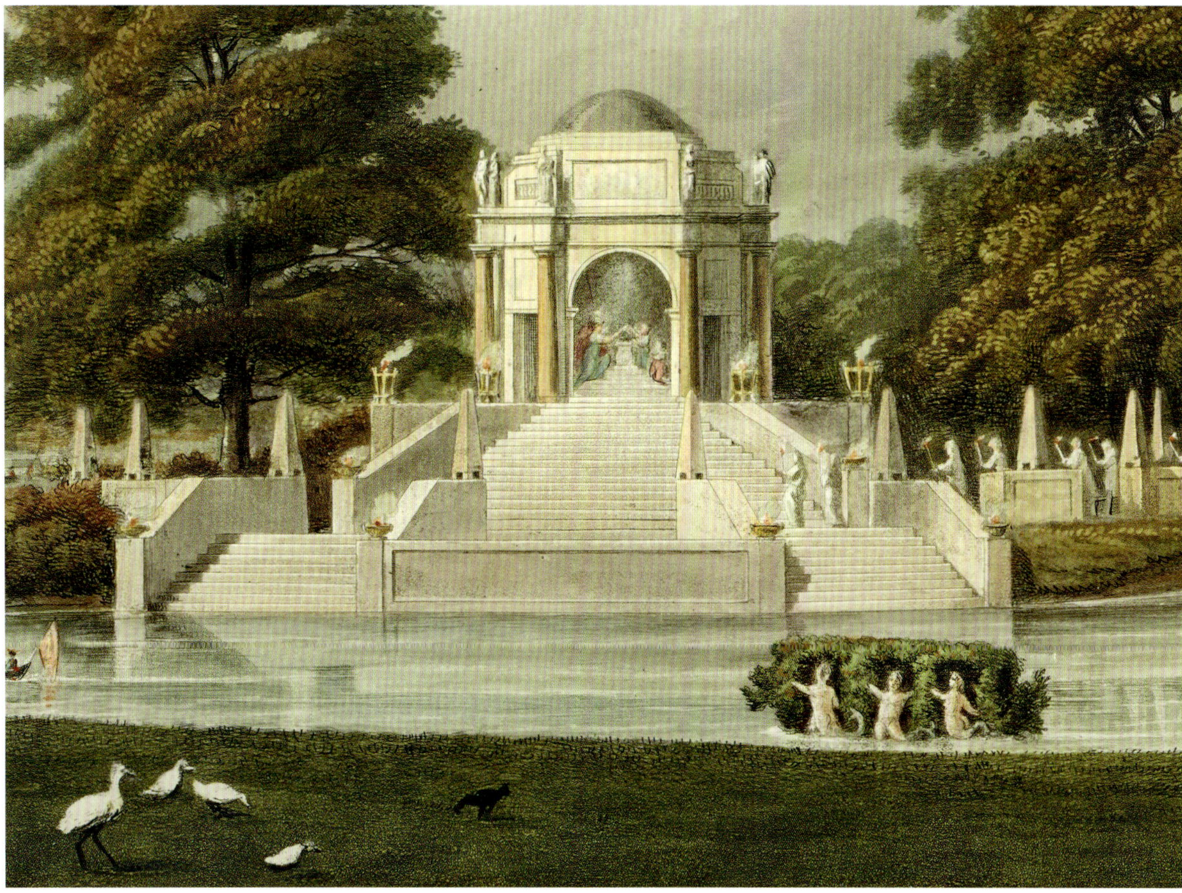

FIG. 21.02 J. Merigot, engraver, after Matthew Cotes Wyatt, *A Jubilee at Frogmore* (detail), c. 1809, etching and aquatint, 14¾ x 22 ¼ in. (37.3 x 56.3 cm). Royal Collection Trust (cat. 7.04)

royal family have also been credited as having facilitated the career of another young painter, Thomas Lawrence, after he made her portrait. The black graphite and chalk image of Delany as a rather elderly, yet spry, woman has been attributed to the artist, who worked at Windsor in the late 1780s (cat. 21.08).[12]

Even after her death in 1788, Mrs Delany remained an important influence on Charlotte and her daughters, especially with regard to their artistic and botanical activities at Frogmore. Charlotte would later produce her own botanical collages after Delany's technique, and the princesses and their Windsor friends designed and executed cut-paper garlands for the decoration of the Cross Gallery on the first floor at Frogmore in 1797.[13]

In 1804, Wyatt enlarged the two pavilions at Frogmore to make a library and anteroom at the south end of the house and balancing rooms at the north. It was in the library that Charlotte kept her botanical collections and her printing press, as well as many of her books. Wild's watercolour of the Queen's Library depicts Charlotte's painted bookcases arranged round the walls. Charlotte employed the print seller and publisher Edward Harding as her librarian and to manage the Frogmore Press. It seems that she found much solace in these activities during the distressing bouts of the King's illness. Harding produced a number of publications which included several from Charlotte's own manuscripts, such as the 'mini-card' boxed series of abridged histories—*Chronological Abridgements* (1809–17), *Translations from the German* (1812), and *Miscellaneous Poems* (1812).[14] Most of the contents of the Frogmore library was dispersed by auction after Charlotte's death in 1818. The sale catalogue includes nearly five thousand titles in German, French, and Italian as well as English.[15] Also included in the auction was the herbarium of the Reverend John Lightfoot. This had been purchased after his death in 1788 by the King as a gift for Charlotte and would later be housed in twenty-four mahogany cabinets in the Frogmore library.[16]

In the north-western corner of the house on the first floor, overlooking the garden, was the Green Closet (p. 358). With its red fitted carpet and walls hung with a heavily patterned 'Japan fabric', and other chinoiserie-inspired detail, the room provides a striking contrast to Wild's views of the classically inspired interiors which predominated within the house. The furniture depicted in Wild's view—an Indian cane cabinet and chairs, the porcelain vases on the top of the cupboards, are probably all of Asian origin. It seems likely that this room was also used to display several of the imperial gifts presented to George III by the Qianlong Emperor in 1793. These included numerous porcelain, jade, and red lacquer caskets, carved lacquer panels, and panels of silk.[17]

Outside the house, within Frogmore's spacious thirty-five-acre landscaped grounds, Charlotte commissioned Gothic ruins, a thatched hermitage, and a small octagonal temple amongst other small buildings and pavilions designed to delight the eye and intrigue the senses. The house and garden were soon established as the favourite venue for informal celebrations such as summer dances, country fêtes, and theatrical garden entertainments, especially after the King and Queen moved from the Queen's Lodge into Windsor Castle. These events were usually organized by Charlotte's third daughter, Princess Elizabeth, assisted by James Wyatt. One of Elizabeth's most ambitious outside projects was in 1809 to mark her father's Golden Jubilee. The fête was attended by nearly 1,200 people; there were fireworks and illuminations and an elaborate water pageant, celebrating the triumph of Britannia, was staged on the artificial lake especially created for the event. A highlight was the illuminated temple constructed by Wyatt after Elizabeth's own design and recorded by Wyatt's son, Matthew Cotes Wyatt. The temple was decorated with Doric columns and a dome, with a double staircase flanked by nine altars with burning incense (fig. 21.02). Lined with purple cloth, the interior displayed a tribute to the King in the form of a large transparency showing his portrait; unfortunately, owing to his increasing blindness and ill-health, the King was unable to attend the celebrations.[18] Concerts and fêtes, and other parties held indoors in the winter, continued at Frogmore until July 1817. At Charlotte's last party, two thousand people were entertained, including five hundred Eton school boys, some of whom played cricket on the lawn.[19]

NOTES

1. William Henry Pyne, *The History of the Royal Residences of Windsor Castle, St James's Palace, Carlton House, Kensington Palace, Hampton Court, Buckingham House and Frogmore*, 3 vols. (London: L. Harrison for A. Dry, 1816–19). Wild's watercolours of Frogmore in volume 1 are the same size as the image on the printed plates, which suggests that they were a guide for the artists producing the hand-painted monochrome prints; Windsor Castle and Buckingham House are featured in volumes 1 and 2 respectively.
2. Jane Roberts, *Royal Landscape: The Gardens and Parks of Windsor* (New Haven and London: Yale Univ. Press, 1997), 215–16.
3. Roberts, *Royal Landscape*, 216.
4. For Wild's watercolour, see Royal Collection (hereafter RCIN) 922121.
5. Jane Roberts, ed., *George III & Queen Charlotte: Patronage, Collecting and Court Taste* (London: Royal Collection Publications, 2004), cat. no. 137.
6. Roberts, ed., *George III & Queen Charlotte*, 75. As Charlotte's drawing of the narcissus (*Narcissus lutens polyanthus Lusitanicus*, RCIN 981137) is dated 2 Sept. 1782, she probably copied the spring flower from a plate from one of the great number of important botanical books in her mother's library at Frogmore, such as John Miller's *An Illustration of the Sexual System of Linnaeus* (1770–77).
7. Clarissa Campbell Orr, 'Queen Charlotte, "Scientific Queen"', in *Queenship in Britain, 1660–1837: Royal Patronage, Court Culture and Dynastic Politics*, ed. Clarissa Campbell Orr (Manchester: Manchester Univ. Press, 2002), 238.
8. Alicia Weisberg-Roberts, 'Introduction: Mrs Delany from Source to Subject', and Amanda Vickery, 'The Theory and Practice of Female Accomplishment', both in *Mrs Delany and her Circle*, ed. Mark Laird and Alicia Weisberg-Roberts (New Haven and London: Yale Univ. Press, 2009), 1 and 94–107 respectively.
9. British Museum Catalogue online, www.britishmuseum.org/research/collection_online/search.aspx?searchText=1897,0505.133, accessed 18 Jan. 2016.
10. For a full discussion of Delany's technique, see Kohleen Reeder, 'The "Paper Mosaick" Practice of Mrs Delany', in *Mrs Delany and her Circle*, ed. Laird and Weisberg-Roberts, 224–35.
11. Letter dated Feb. 1782, quoted in Roberts, ed., *George III & Queen Charlotte*, 191; cat. no. 165. The portrait was placed in the Queen's Bedchamber at Buckingham House, alongside that of Bishop Hurd by Thomas Gainsborough. In 1786 Fanny Burney commented: 'How worthily paired! What honour to herself, such honour to them!'
12. See Weisberg-Roberts, 'Introduction', 12–13.
13. The diarist Joseph Farington described the gallery as the 'long narrow room . . . with painted flowers, & subjects of Children &c cut in paper'. Quoted in Roberts, ed., *George III & Queen Charlotte*, 148; cat. no. 138.
14. See Jane Roberts, 'Edward Harding and Queen Charlotte', in *Burning Bright: Essays in Honour of David Bindman*, ed. Diana Dethloff, Tessa Murdoch, and Kim Sloan, with Caroline Elam (London: UCL Press, 2015), 152–53.
15. James Christie, *A Catalogue of the Genuine Library, Prints, and Books of Prints, of an Illustrious Personage, Lately Deceased* (London, 1819).
16. Roberts, ed., *George III & Queen Charlotte*, 149–50; cat. no. 140.
17. Roberts, ed., *George III & Queen Charlotte*, 150; cat. no. 141.
18. Melanie Doderer-Winkler, *Magnificent Entertainments: Temporary Architecture for Georgian Festivals* (New Haven and London: Yale Univ. Press, 2013), 74–76.
19. Roberts, ed., *George III & Queen Charlotte*, 228–29.

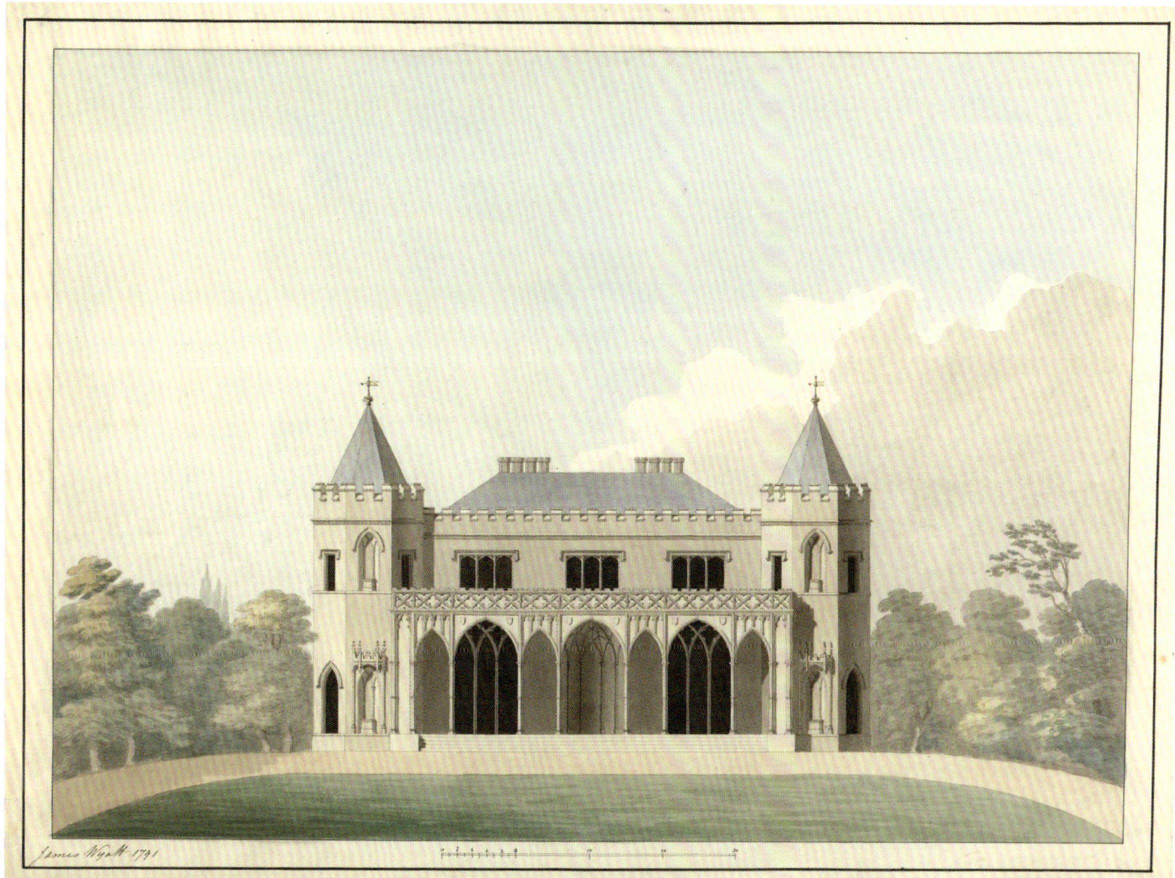

21.03

THE CREATION OF FROGMORE

21.03

JAMES WYATT (1746–1813)
Unexecuted Design for Little Frogmore: Elevation
1791
Watercolour, wash, pen and ink on paper
23½ x 18 in. (59.7 x 45.7 cm)
Yale Center for British Art, Paul Mellon Collection, B1975.2.404

In 1792 Charlotte had taken on the lease of Little Frogmore, which was promptly renamed the Queen's Cottage, or Amelia Lodge. With basic repairs completed by 1791, Princess Elizabeth would relate that 'Mama, Augusta and me would go there nearly every day. Mama sits in a very small green room which she is very fond of, reads writes and botanizes.' The house evidently gave such pleasure that later in the same year Charlotte reported 'Wyatt the Architect has made me the Prittiest plan imaginable for a gothic cottage. It consists of 4 rooms upon a floor with the Towers of which there are 4 which will make 8 closets alotted for Books, Plants, China & one for the flower pieces painted by Miss Mosert (Moser).' The project, however, was almost immediately abandoned when the possibility of acquiring the larger adjoining estate of Great Frogmore emerged in 1792. Wyatt swiftly turned his attentions to making improvements to the house there, and within a year Charlotte was celebrating her new 'sweet temple'.

21.04*

UNKNOWN ENGRAVER, UNKNOWN DESIGNER, English
'By Her Majesty's Command. Frogmore House & Gardens to be shown to [Miss Smith & her party]'
Between 1792 and 1818
Engraved admission ticket completed in manuscript and signed
3⅛ x 5⅛ in. (8 x 13 cm)
Yale Center for British Art, Paul Mellon Fund, Eph Ent 7

In 1792 George III purchased the Great Frogmore estate in Windsor Great Park for Queen Charlotte. The house was in good condition, but between 1795 and 1804 James Wyatt was employed to enlarge and modernize it to make it fit for royal inhabitants. Charlotte and her daughters took delight in making decorations, such as paper garlands, for the fêtes that took place there over the summer months. In 1795 the 'sumptuous tents of various forms and colours, and of considerable dimensions, that had formerly belonged to Tipoo Sahib, and had been presented to the Queen' were erected for such an event and enjoyed by the community. The fête for the jubilee in 1809 was perhaps the most elaborate of all. It involved a water pageant, designed by James Wyatt, and large-scale transparencies. The community in Windsor and surrounding villages were invited to the parties and to enjoy the spectacle.

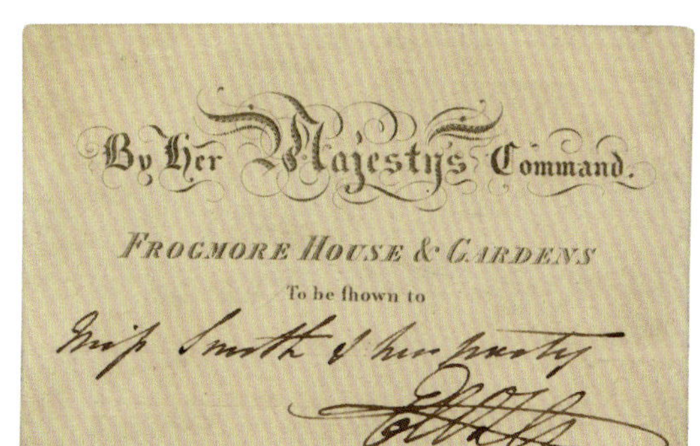

21.04

* = YCBA only
† = HRP only

21.05*

CHARLES WILD (1781–1835)
Frogmore House: The Green Pavilion
1817
Watercolour and gouache over pencil on paper
7⅞ x 10⅞ in. (20 x 27.4 cm)
Royal Collection Trust, RCIN 922121

21.06*

CHARLES WILD (1781–1835)
Frogmore House: The Green Closet
1819
Watercolour and gouache over pencil on paper
9⅞ x 7⅞ in. (25.1 x 20 cm)
Royal Collection Trust, RCIN 922123

Great Frogmore became Queen Charlotte's favourite retreat. Her earliest additions to the existing late seventeenth-century house, completed by mid-1795, were the two single-storey pavilions linked by a colonnade stretching across the garden front. The Green Pavilion was at the northern end of the colonnade; the French windows leading into the Mary Moser room are visible at the end. The Green Closet was located on the first floor at the north-western corner of the house, overlooking the garden. According to William Henry Pyne, the lacquer panelling here was 'original japan' rather than an imitation produced by Princess Elizabeth, who was skilled in the decorative technique. The cabinet and chairs were made of Indian cane and the porcelain vases on the top of the cupboards and cabinet are probably all of Far Eastern origin. This room may have housed some of the imperial gifts presented to George III by the Qianlong emperor in 1793. These included porcelain, jade, red lacquer caskets, carved lacquer panels, and silk. In the course of his visit to Frogmore in 1797, Joseph Farington noted 'some presents from the Emperor of China'. William Henry Pyne's *History of the Royal Residences* (1816–19) is a large three-volume publication with one hundred engravings of the exteriors, interiors, and furnishings of Frogmore, Windsor Castle, St James's Palace, Carlton House, Kensington Palace, and Hampton Court Palace. Charles Wild's watercolours were made to be engraved for it.

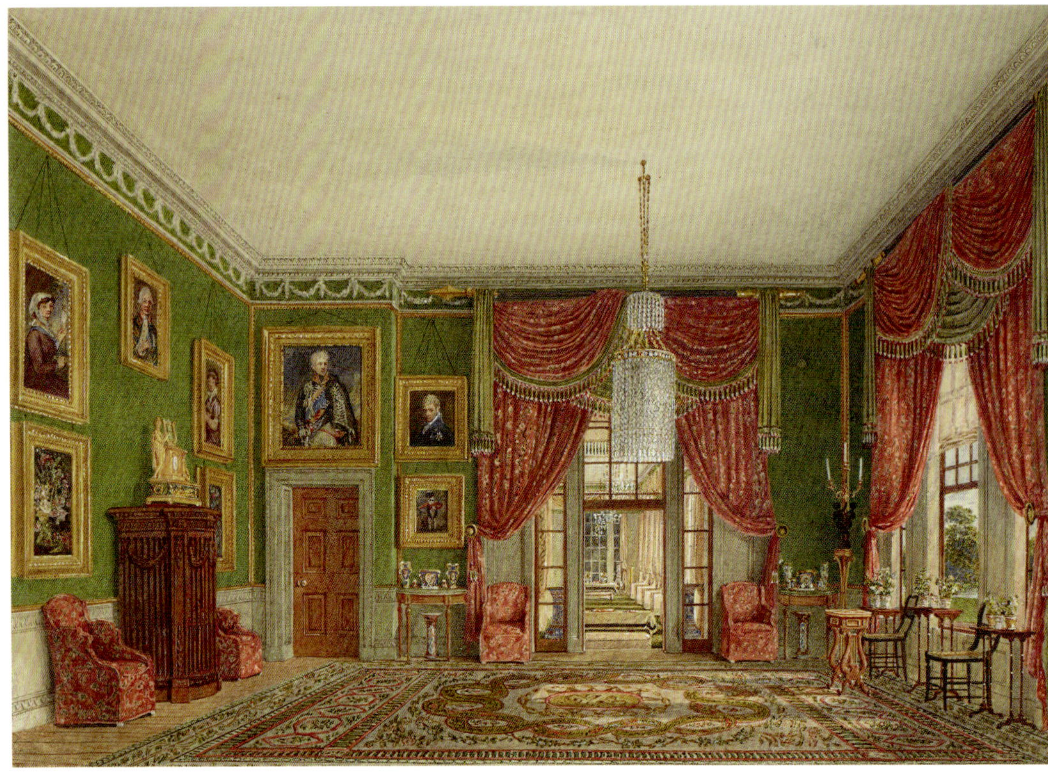

21.05

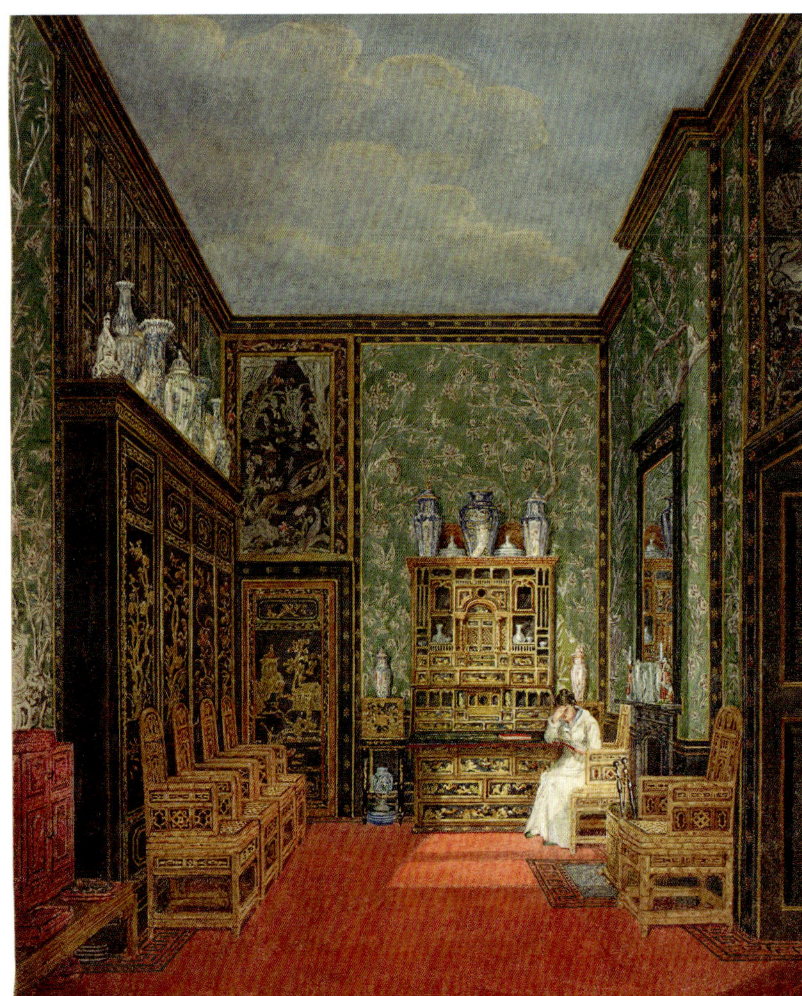

21.06

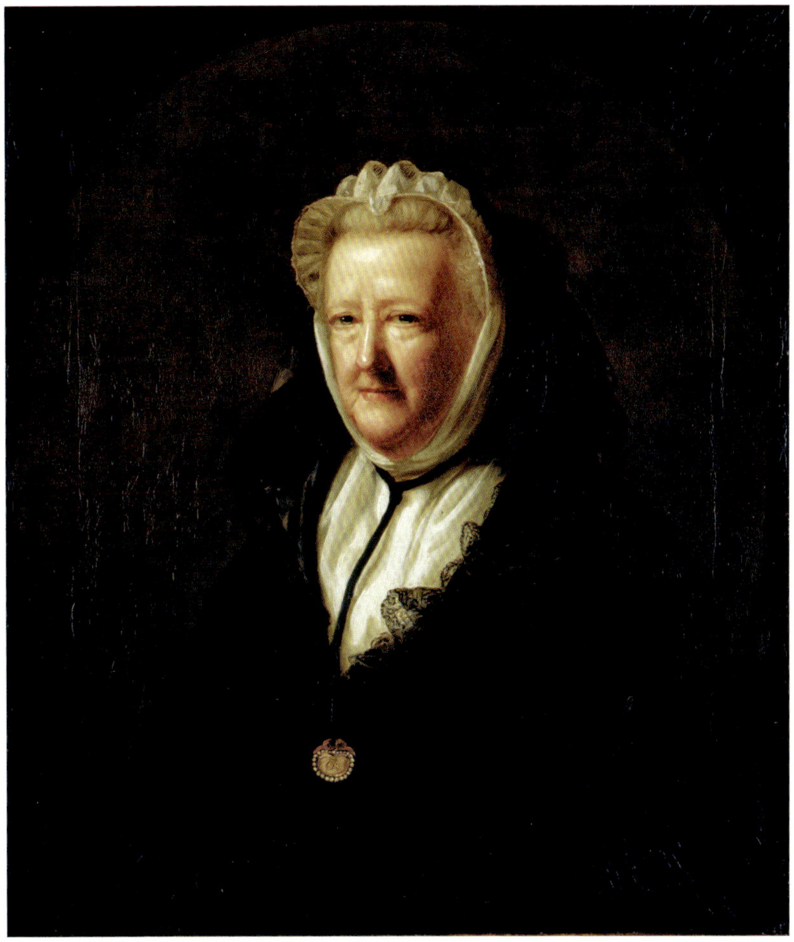

21.07

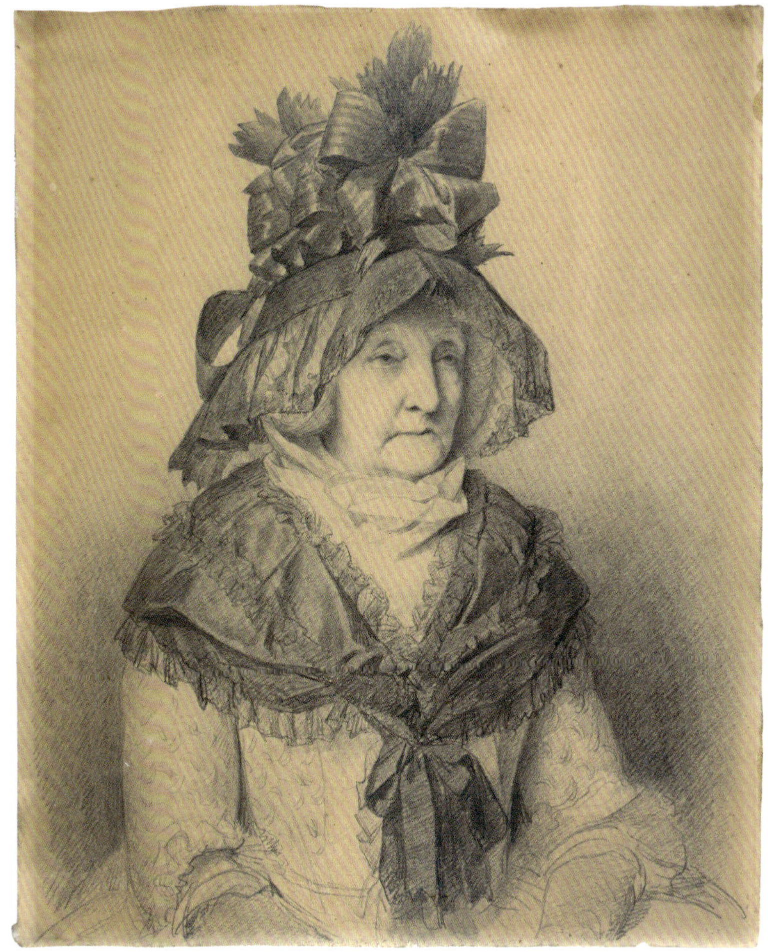

21.08

CRAFTING A PERFECT RETREAT

21.07

JOHN OPIE (1761–1807)
Mary Granville, Mrs Delany
1782
Oil on canvas
30⅛ x 25⅛ in. (76.5 x 63.6 cm)
Royal Collection Trust, RCIN 400965
This portrait was commissioned by George III and Queen Charlotte and hung in the Queen's Bedchamber at Buckingham House.

21.08*

UNATTRIBUTED, possibly by THOMAS LAWRENCE (1769–1830)
Portrait of Mrs Delany, born Granville
c. 1780–88
Pencil on brown paper
14⅜ x 11⅜ in. (36.4 x 28.8 cm)
Victoria and Albert Museum, London, Bequeathed by R. H. Stephenson, E.580–1929

Left a widow for the second time at the age of sixty-eight, after her happy marriage to Patrick Delany, Dean of Down in Ireland, Mary Granville, Mrs Delany established a pattern of staying with her friend, the dowager Duchess of Portland, for six months of each year. The two women shared a great passion for botany. Mrs Delany first met the King and Queen in 1776 at the Duchess's house, Bulstrode Park, which is thirty minutes' carriage drive to the north of Windsor. The royal visits to Bulstrode were followed by return visits to Windsor. Mrs Delany became much loved by the King, Queen, and princesses. After the Duchess of Portland's death in July 1785, Mrs Delany's small personal income proved insufficient. The King therefore arranged for her to receive an annuity of £300 and a house in Windsor, where she lived for the rest of her life.

The portrait by the Cornish artist John Opie was commissioned by George III. In it Mrs Delany wears a locket with the letters CR—a reference to Queen Charlotte, who had almost certainly given the locket to her as a token of her friendship. The portrait was soon placed in the Queen's Bedchamber at Buckingham House, alongside that of Bishop Hurd by Thomas Gainsborough. The drawing of Mrs Delany recently attributed to Thomas Lawrence, though previously associated with John Opie, John Hoppner, and Nathaniel Dance-Holland, was made at the time of her friendship with Queen Charlotte. Mrs Agnew, Mrs Delany's 'waiting woman' would write that 'Opie and Lawrence owed her much.'

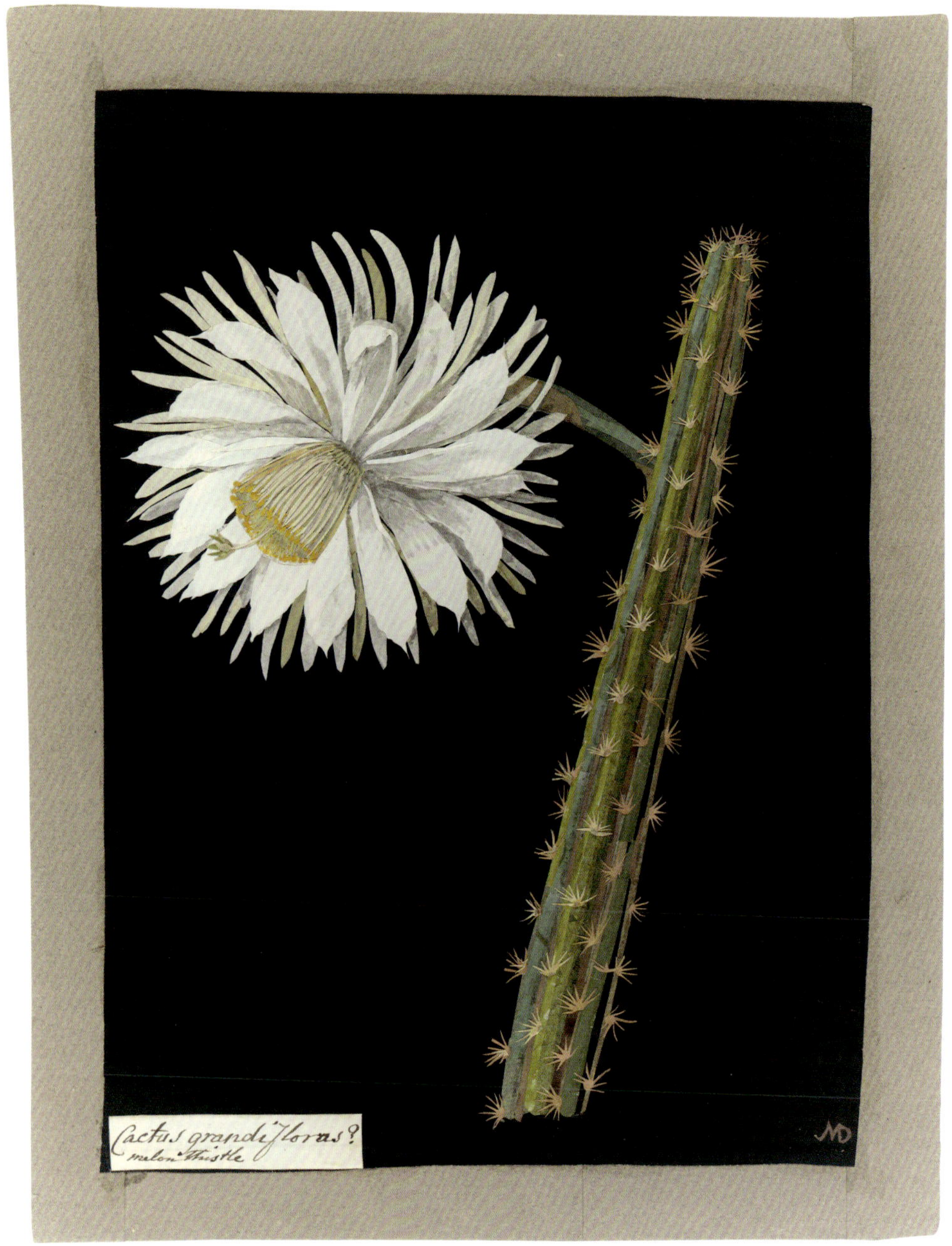

21.09

21.09

MARY DELANY (1700–1788)
Cactus Grandiflorus. Melon Thistle [Icosandria Monogynia]
1778
Collage of coloured papers, with gouache and watercolour, on black ink background
Signed with monogram on collage and inscribed on label with Latin and common name; verso inscribed with Linnaean classification, place, and date of composition: '12 augt 1778' and 'the Day the King & Queen & the Royal family were at Bulstrode'
13⅝ x 9⅝ in. (34.5 x 24.2 cm)
British Museum, 1897,0505.133

21.10

MARY DELANY (1700–1788)
Magnolia Grandiflora [Polyandria Polygynia]
1776
Collage of coloured papers, with gouache and watercolour, on black ink background
Signed with monogram in collage and inscribed on label with Latin name; verso inscribed with Linnaean classification, place, and date of composition: '[27 – crossed through] 26 Augt 1776'
13⅜ x 9¼ in. (33.8 x 23.4 cm)
British Museum, 1897,0505.557

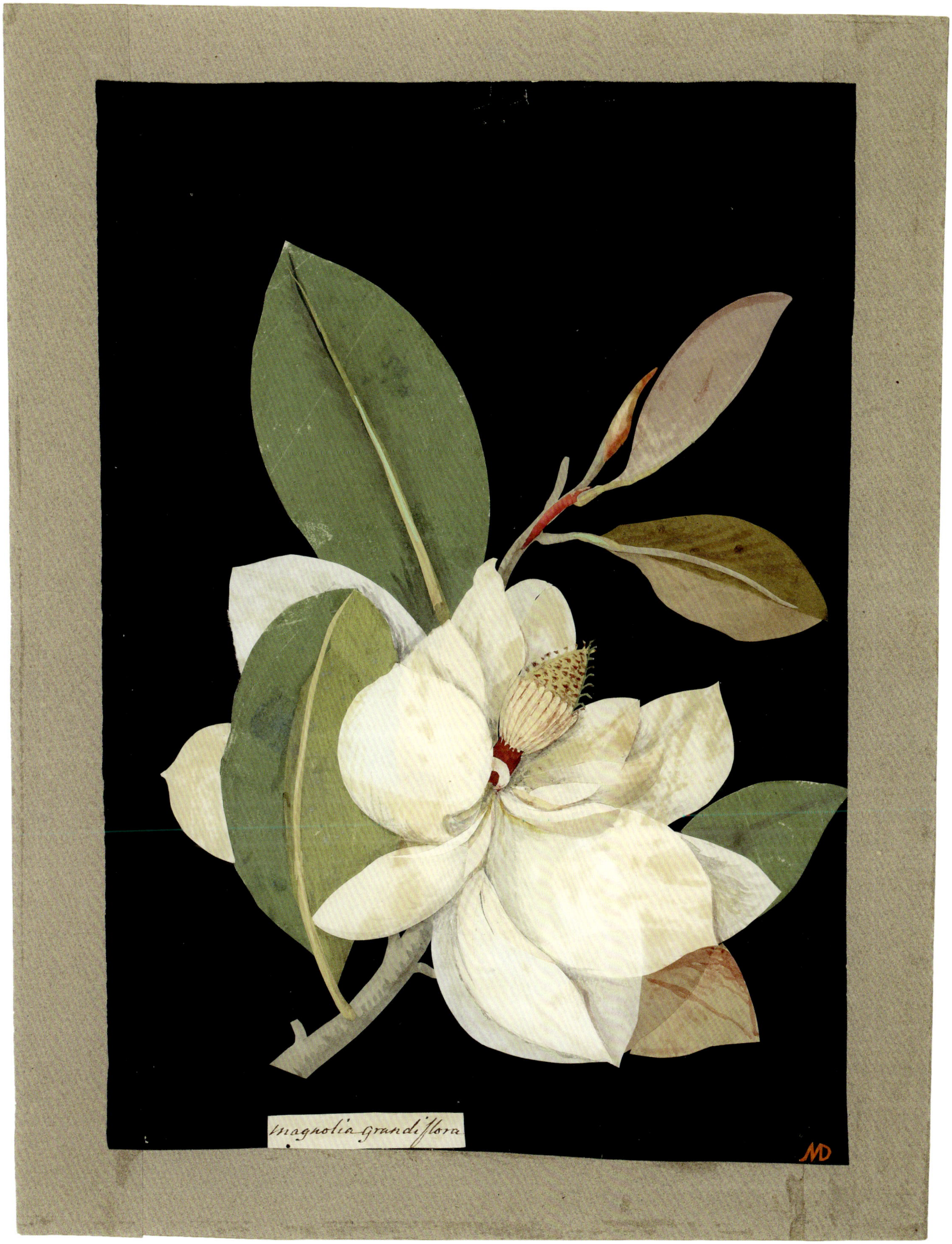

21.10

21.11*

MARY DELANY (1700–1788)
Sanguinaria Canadensis [Polyandria Monogynia]
1777
Collage of coloured papers, with gouache and watercolour, and with a leaf sample, on black ink background
Signed with monogram in collage and inscribed on label with Latin name; verso inscribed with Linnaean classification, place, and date of composition: '25 april 1777' and source of plant
9⅛ x 8¼ in. (23 x 20.7 cm)
British Museum, 1897,0505.766

21.12*

MARY DELANY (1700–1788)
Dianthus Caryophyllus. Pink or Carnation var.
1778
Collage of coloured papers, with gouache and watercolour, on black ink background
Signed with monogram in collage and inscribed on label with Latin and common name; verso inscribed with place and date of composition: '17 augt 1778' and the donor
11⅞ x 9 in. (30 x 22.7 cm)
British Museum, 1897,0505.277

Mrs Delany's interest in botany had already left a literary legacy by 1769 with her preparation of a manuscript adaption of William Hudson's *Flora Anglia*. However, it was in 1772 that she began the work for which she is best known, the *Flora Delanica*. She wrote: 'I have found a new way of imitating flowers.' This was by composing a mosaic image from hundreds of tiny pieces of carefully cut coloured paper. Over ten years she composed nearly a thousand works, which she referred to as her 'Hortus seccus', likening them to a collection of pressed and dried plants. Mrs Delany's 'flowers', unlike their botanical counterparts, have held their colour over the centuries. She taught her paper mosaic technique to Queen Charlotte and her daughters.

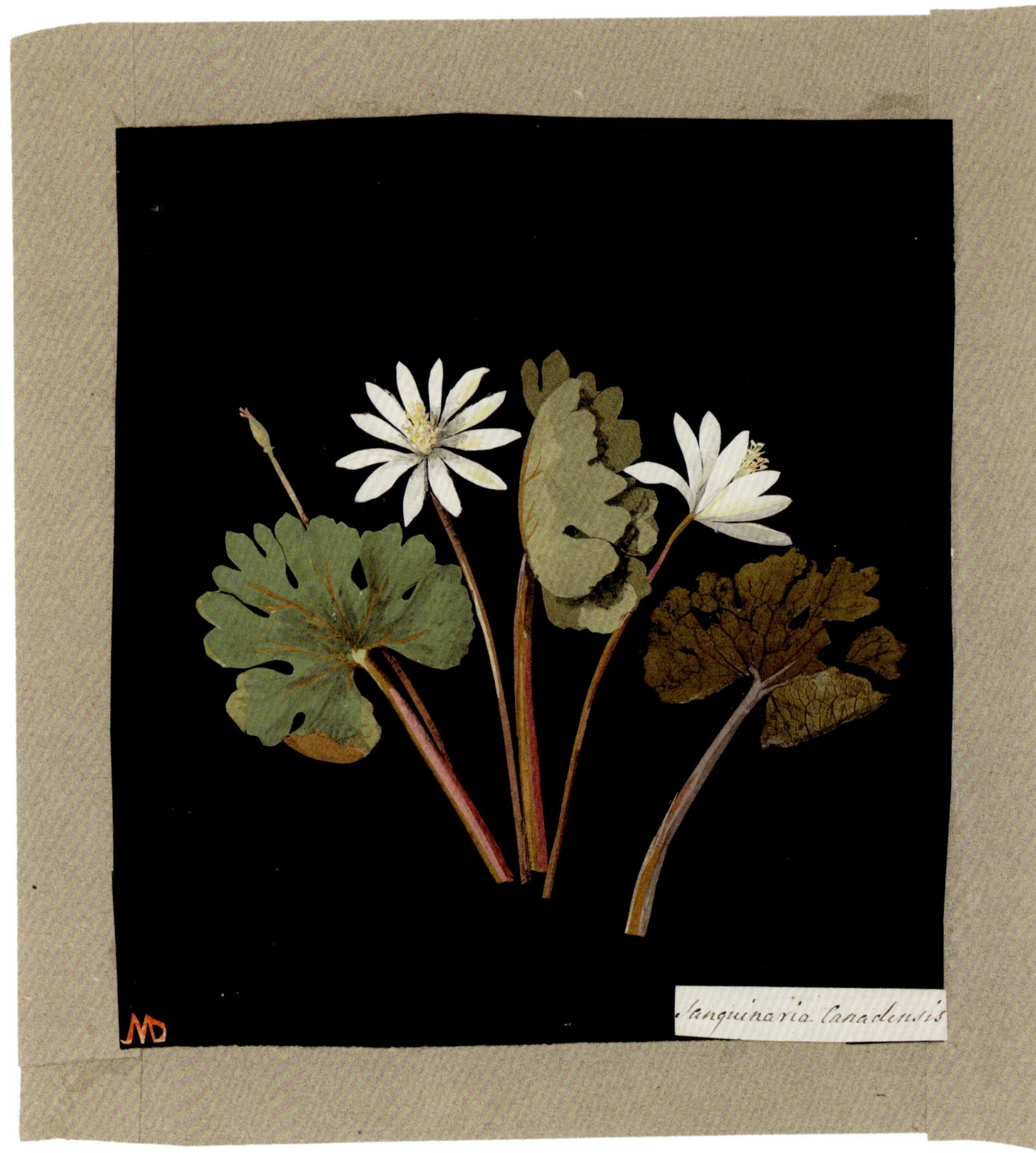

21.11

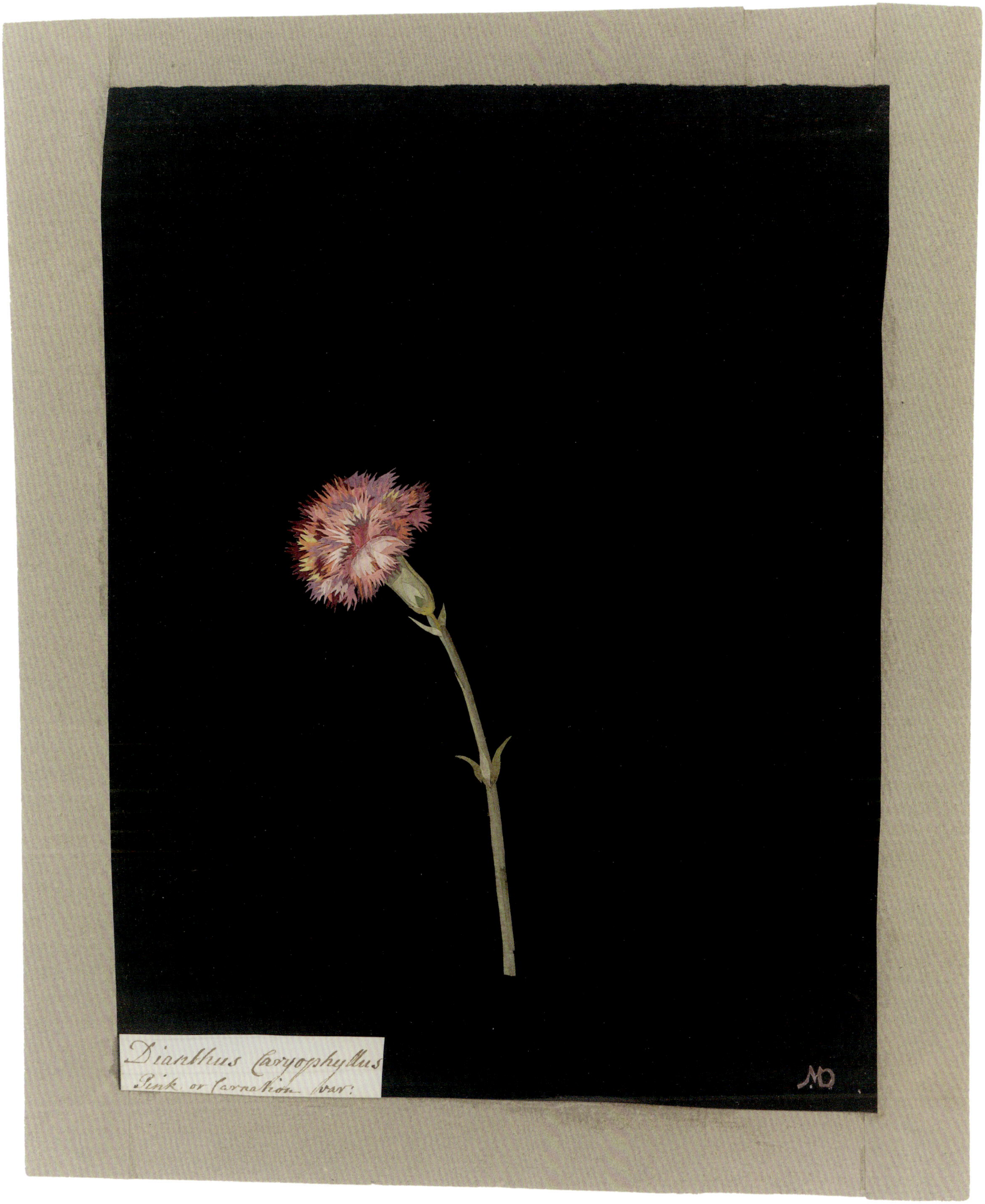

21.12

21.13

QUEEN CHARLOTTE OF MECKLENBURG-STRELITZ (1744–1818)
Needlework pocket-book
1781
Satin, coloured silks, and enamelled gold
5¾ x 3 x 1⅛ in. (14.5 x 7.5 x 2.7 cm)
Royal Collection Trust, RCIN 45126

This cream-coloured silk pocket-book hold-all, embroidered with coloured silk and metal sequins, is reputed to have been made by Queen Charlotte. It has an enamelled clasp; it is lined in pink silk and contains ten sewing implements and was a gift from her to Mary Delany in 1781. Queen Mary acquired it for the Royal Collection by 1931. The present came under cover of a letter, written at the Queen's Lodge, Windsor, on 15 December 1781, in which the Queen asked Mrs Delany to wear 'this little Pocket-Book in order to remember at times, when no dearer Person's are present, a very sincere well wisher, Friend, and affectionate Queen, Charlotte.' Queen Charlotte's skill as a needlewoman is well attested. The importance she placed on this accomplishment is underlined by her financial support of Mrs Pawsey's school for 'embroidering females' and her patronage of the unusually gifted needlewoman Mary Knowles.

21.14*

MARY LAWRANCE (d. 1830)
A Collection of Roses from Nature, frontispiece
1799
Hand-coloured etching on paper
15¾ x 12⅝ in. (40 x 32 cm)
Published by the author, London
New York Public Library, Rare Books Division, Aster, Lenox and Tilden Foundations, KW+++ (Lawrance, M. Collection of roses from nature)

Mary Lawrance Kearse was an artist, teacher of botanical drawing, and exhibitor at the Royal Academy between 1794 and 1830. Her books include *A Collection of Passion Flowers Coloured from Nature* (1799–1802), *A Collection of Roses from Nature* (1796–99) and *Sketches of Flowers from Nature* (1801). *A Collection of Roses* contains an ornamental frontispiece and ninety plates of flowers; it was intended as a manual on botanical drawing for students. Mary drew, etched, and coloured all her own illustrations. She would dedicate *A Collection of Roses* to Queen Charlotte; as a result, 'the demand for this book on roses was far in excess of the number of copies printed.' Charlotte's copy was listed among the books in her library which were dispersed at auction in 1818, and may have been referred to earlier as an aid to the art education of the Queen's daughters. Princess Charlotte and Princess Elizabeth became very proficient botanical artists.

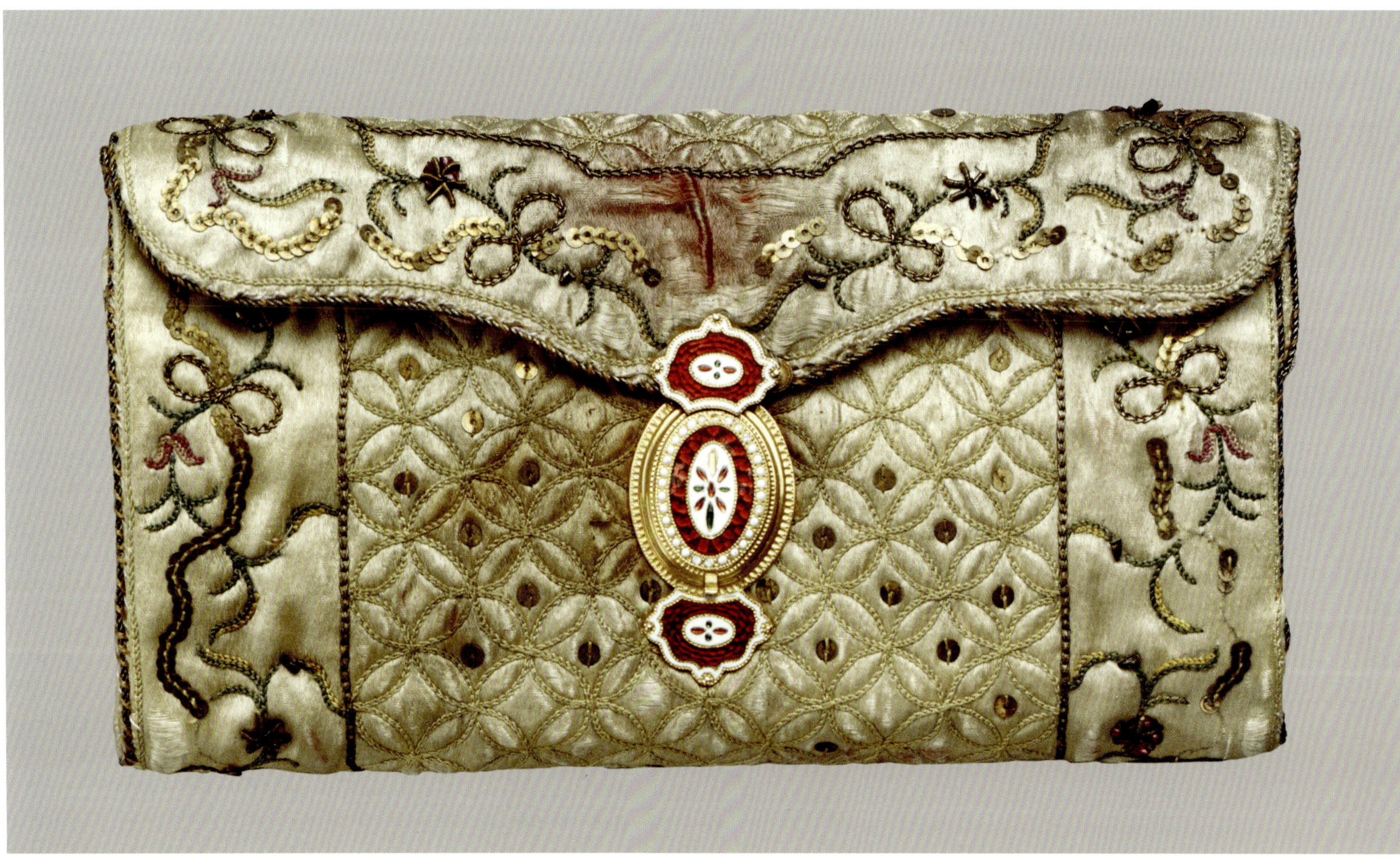

21.13

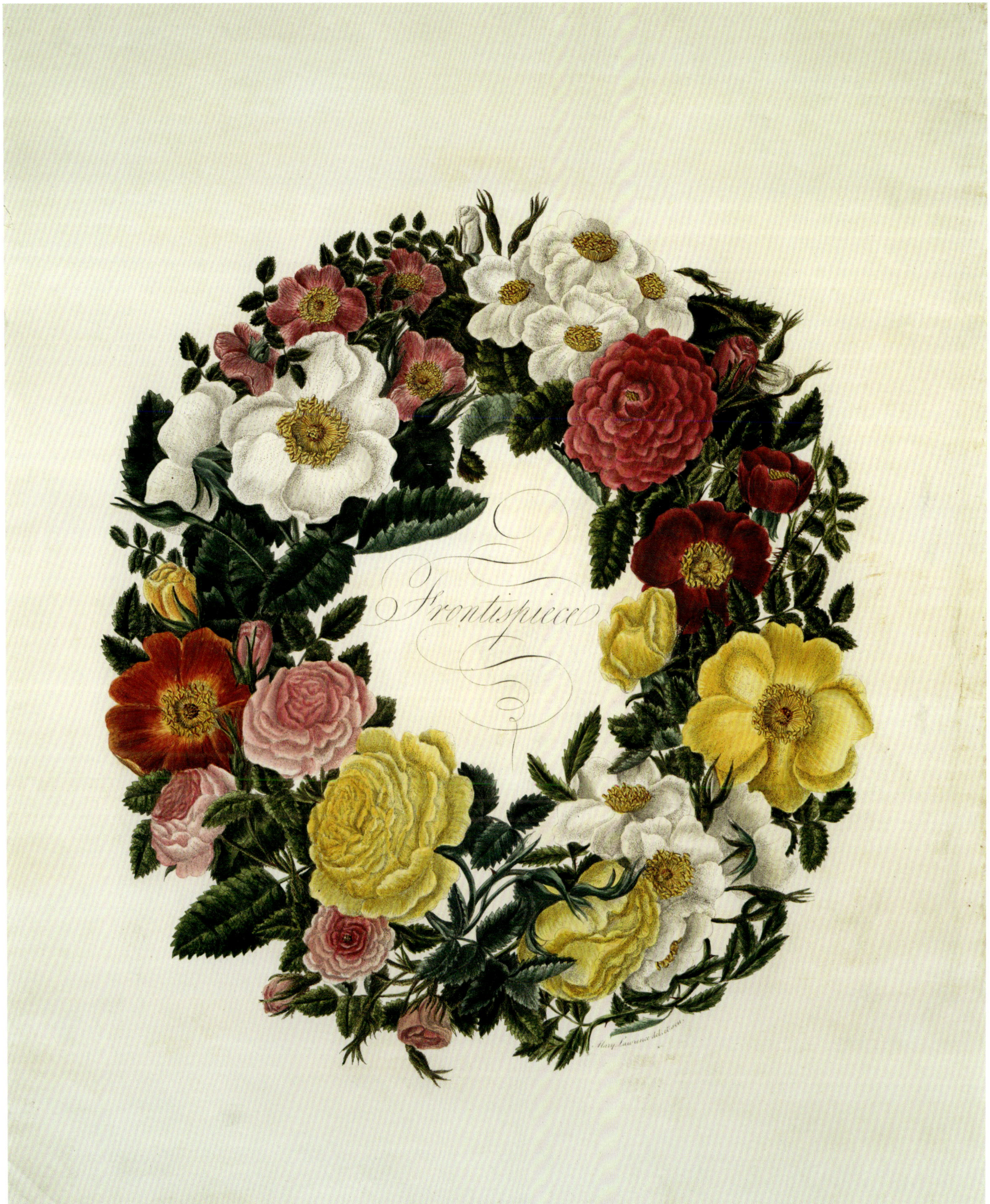

21.14

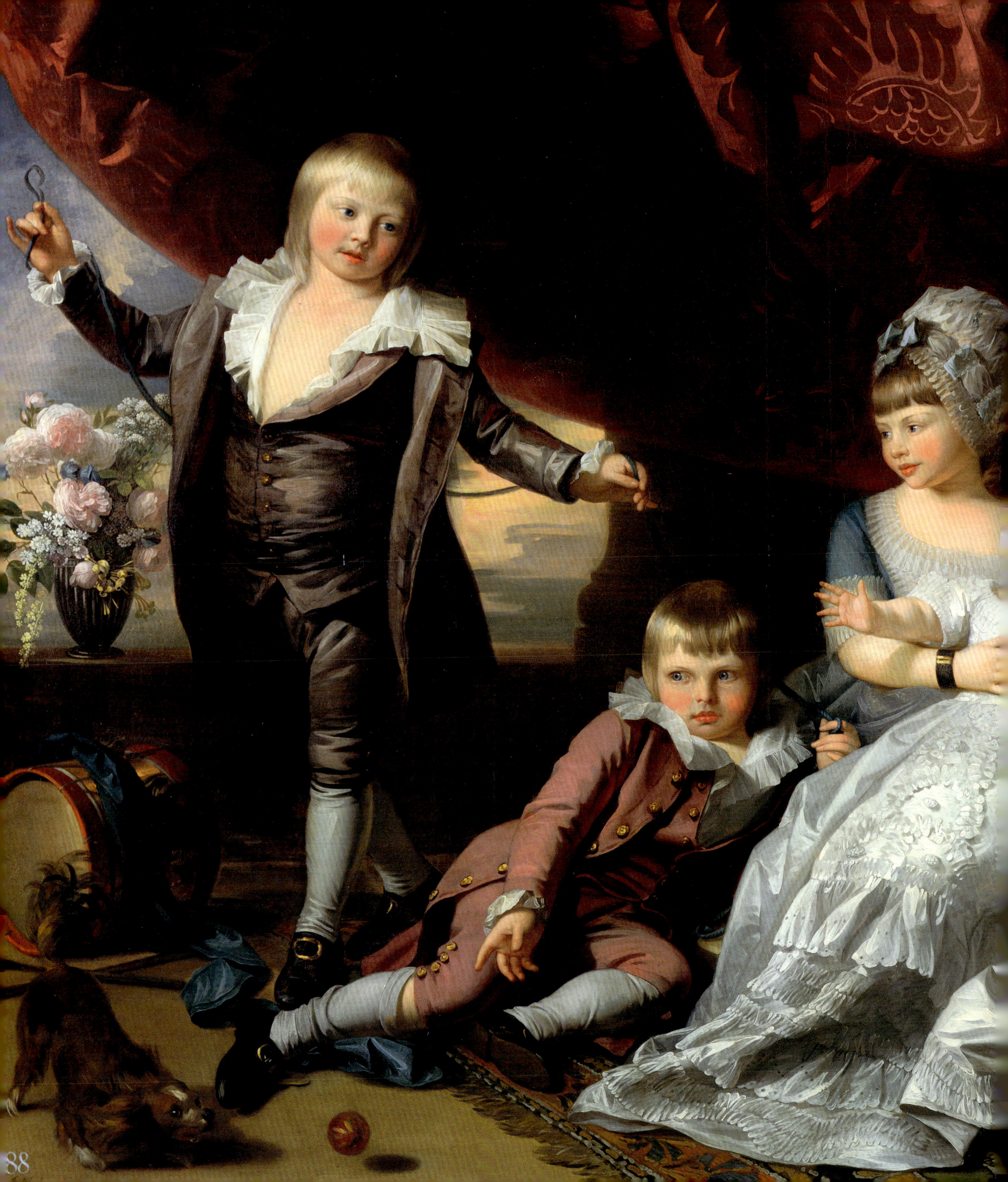

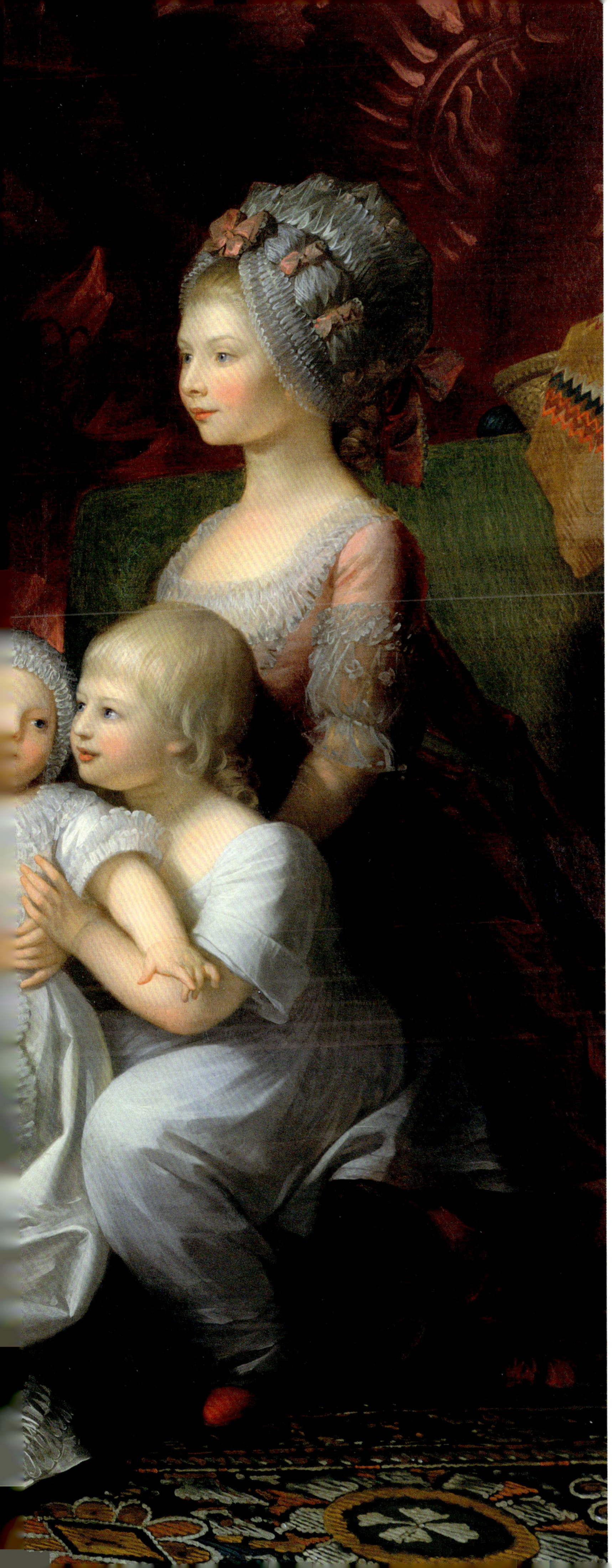

Education, Charity, and Health

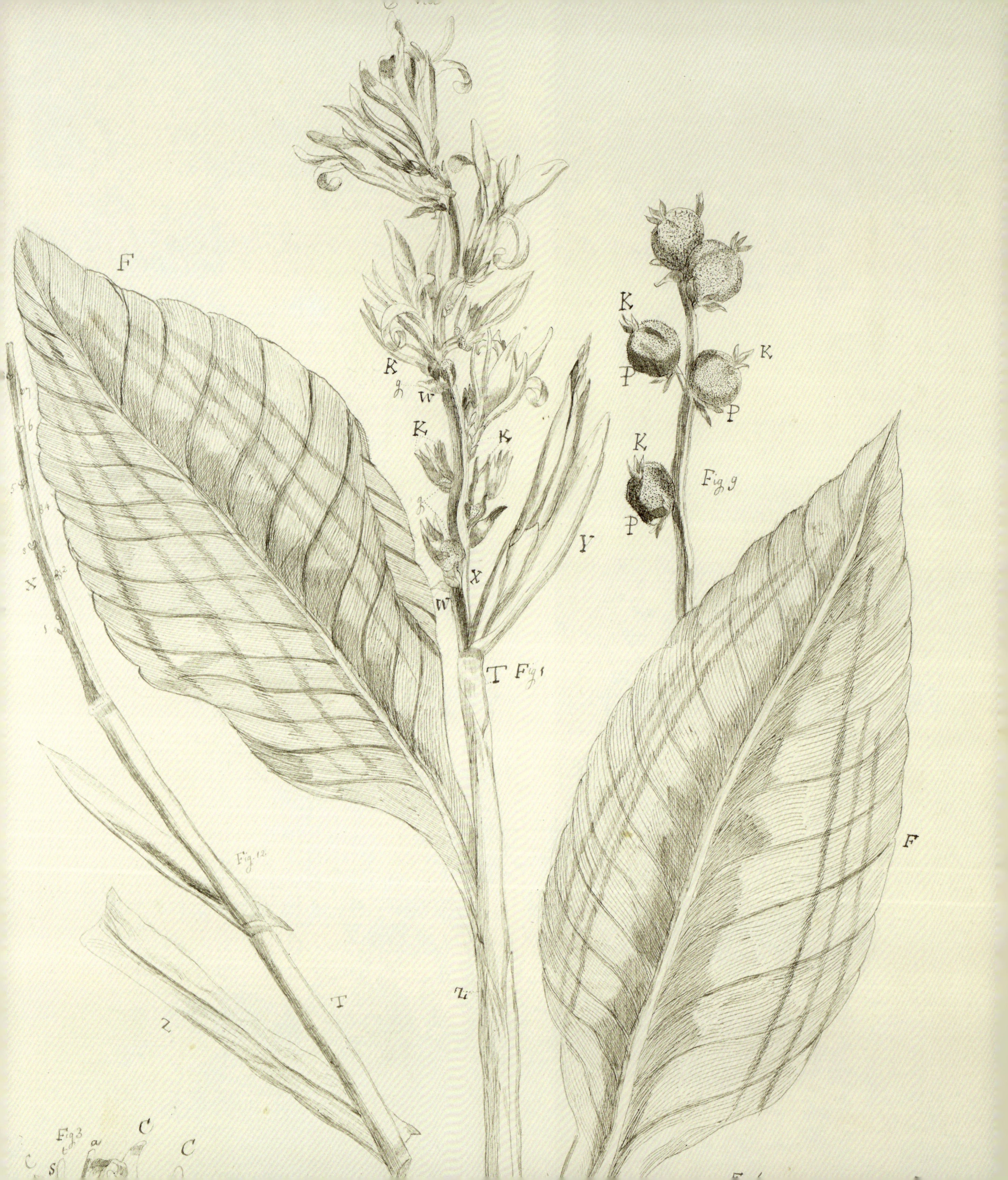
F
K
W
X
Y
T Fig 1
Fig 9
P
Fig 12
Z
Fig 3
C

Frogmore and the Princesses as Artists

Jane Roberts

In February 1790, the Queen—born Charlotte of Mecklenburg-Strelitz—asked her brother Charles (later Grand Duke of Mecklenburg-Strelitz) for 'colours to paint in gouache', requesting both 'One set of prepared colours . . . and another of powder colours to prepare here; also, if possible discover whether they use water alone to paint, or if they use gum arabic, and which gum they use . . . if they use white among their colours . . . and how they render nuances/shadows.'[1] Even without the accompanying note ('The Queen amuses herself in painting flowers, or plants, after nature'), it is clear that Charlotte was a keen and knowledgable practitioner of painting.[2] Her apartments contained a 'neat Mahogany Drawing Table on a Frame' with a drawer 'for wet and dry Colours Pencils Brushes &c' and a 'Geometrical Square for drawing in Perspective'.[3] Charlotte's diary reveals that whenever possible she joined the elder princesses for their daily morning lessons with their 'Drawing Masters'.[4] Their teachers included Benjamin West, and specialists in different techniques and subject matter, such as the flower painter Margaret Meen and the botanical artist Francis (Franz Andreas) Bauer, the result of whose lessons are clear in the works of her eldest daughter (p. 374, cat. 22.07).[5] Although very few works that are definitely by Charlotte have survived, she is the only one of the 'Enlightened Princesses' for whom documented artefacts are known (see cat. 21.13).[6]

At the time of the Queen's letter to her brother, negotiations were under way for the purchase of the lease of the first of the two estates at Frogmore, less than a mile to the south-east of Windsor Castle. Thereafter, Frogmore was to occupy a key place in the creative lives of the royal family.[7] In 1793 Charlotte described it as her 'sweet temple' and 'little paradise', adding 'I mean this place to furnish me with fresh amusements every day.'[8] Three years later she gave Frances (Fanny) Burney 'a full description of her new house at Frogmore, its fitting up, and the share of each Princess in its decoration . . . "I spend" she cried, "there almost constantly all my mornings. I rarely come home but just before dinner, merely to dress."'[9] By early 1797, Charlotte, Princess Royal (later Queen of Württemberg) had produced for Frogmore '36 drawings in pen-&-ink . . . besides ornaments on chair-covers &c, and in another room 24 of her drawings, from prints after Berghem &c'.[10] Her work at Frogmore today includes porcelain decorated by her, using blanks produced at Ludwigsburg, her married home.[11] But it was the Queen's third daughter, Princess Elizabeth (later Landgravine of Hesse-Homburg), who excelled in royal creativity (fig. 22.01). In 1793 she galvanized the female workforce of Windsor (royal and non-royals alike) to produce hundreds of yards of cut-paper garlands (of flowers and foliage) to decorate walls and tables for a fête at Frogmore.[12] By 1797 she had decorated 'a long narrow room with painted flowers, and subjects of children cut in paper, and finished by Tomkins', and in 1807 she was 'busy putting up my Japan room at Frogmore'.[13] Additions to the garden at Frogmore included an octagonal temple, a thatched hermitage, the surviving Gothic Ruin, and 'Princess Elizabeth's Thatched Barn'—a rustic ballroom used during the Queen's many entertainments at Frogmore—all erected to

Detail of p. 380, cat. 22.06

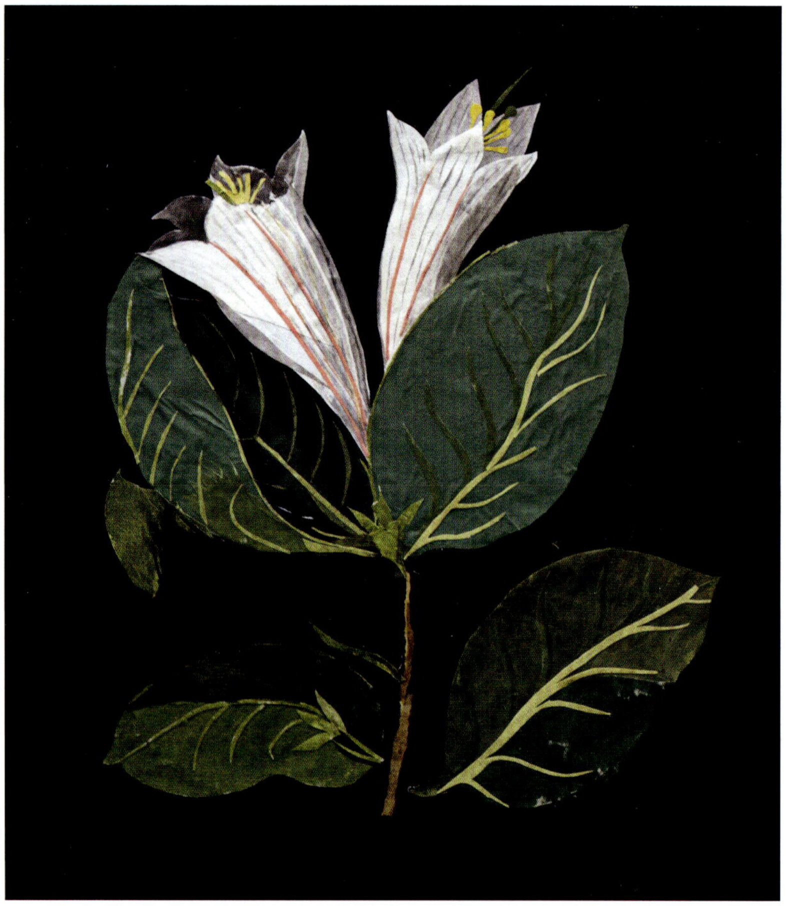

FIG. 22.01 Attributed to Princess Elizabeth, *Portlandia Grandiflora*, 1786–88, collage of coloured papers, with gouache and watercolour on paper. The Newport Garden Club, Newport, RI

Elizabeth's designs.[14] The rooms added to Frogmore after 1800 included a library where, with the help of her librarian Edward Harding, the Queen produced a number of large manuscript volumes 'in her own hand' with illustrations selected by her, as well as a small number of printed books and cards, the output of the Frogmore Press established in the house in 1809 (fig. 22.02; see also pp. 309–10, and fig. 17.06).[15] For the last three decades of Charlotte's life, Frogmore was the creative base of the royal family.

Throughout her life Charlotte was actively involved in the encouragement of useful ladies' work, knowledge of which was also part of her daughters' education. Benjamin West's portrait of the Queen and the Princess Royal in 1776, and the needlework pocket-book presented to Mary Granville, Mrs Delany in December 1781, reveal the high standards of Charlotte's own needlework (cat. 21.13).[16] For the Prince of Wales's marriage in 1795, Charlotte worked a coat and waistcoat for the King and nightdresses for the bride; 'a silken quilt . . . which she sewed with her own hands', was given to Mrs Siddons during a visit to Windsor in 1813.[17] In 1777 a visitor to the Queen's Lodge described chairs upholstered with 'knotted floss silk of different shades sewn on to imitate flowers', much of which had been worked by Charlotte.[18] Mrs Delany emerges as an important mentor—and friend—in Charlotte's artistic activities. During the late 1770s she had explained to the Queen (and later Princess Elizabeth too) her 'mosaic paper work' technique (see cats. 21.09–21.12), and during Mrs Delany's last years she also taught the Queen to use a spinning wheel.[19] The compliment was returned when the Queen showed Mrs Delany and Margaret Cavendish Bentinck, Duchess of Portland her 'frame for weaving of fringe', and explained its use (cat. 22.10).[20]

The handiwork of Charlotte and her daughters dates from a particularly productive and well-documented period, when it had become customary for well-educated people of both sexes to be accomplished artistically as well as musically. Although Caroline of Ansbach and Augusta of Saxe-Gotha would have received drawing lessons as part of their education, no works by them have survived. But their children's artefacts are useful indicators of the

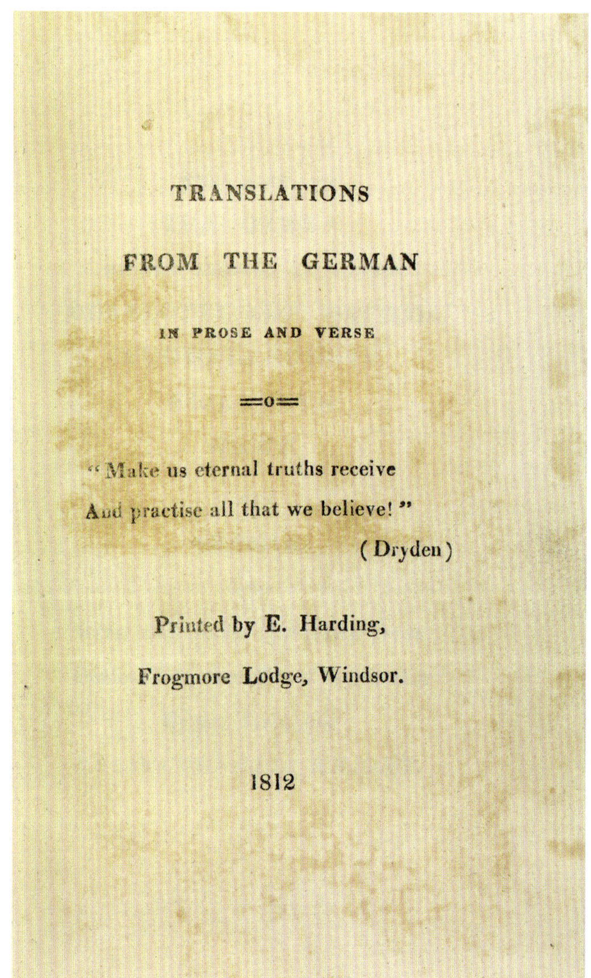

TRANSLATIONS

FROM THE GERMAN

IN PROSE AND VERSE

=o=

" Make us eternal truths receive
And practise all that we believe! "
(Dryden)

Printed by E. Harding,

Frogmore Lodge, Windsor.

1812

FIG. 22.02 Frogmore Press, *Translations from the German*, title page, 1812. Royal Collection Trust continuing importance placed on the acquisition of such skills. George III had received 'a near professional training in architecture' from William Chambers, as well as advanced tuition in perspectival drawing from Joshua Kirby.[21] For three decades from the late 1750s the King produced numerous architectural designs, some of which were doubtless intended to be carried out.[22] Earlier, he and his brother Prince Edward, Duke of York, had received more basic tuition in drawing.[23] The fact that early drawings by each of the princes (carefully preserved by Charlotte) include an identical group of buildings, indicates that these lessons focused on the production of copies.[24] A small number of etchings have survived by their younger brother Henry Frederick, later Duke of Cumberland, have survived.[25] Charlotte's children were also taught to etch—and thereby produced multiple copies of their own works for use as gifts (cat. 22.08).[26]

Many artefacts by Caroline's thoroughly educated children have also survived. In the early 1730s both the boys and the girls were taught lathe-turning and gemstone-working, while several finished oil paintings and drawings have survived by both Princess Anne (later Princess of Orange) and Princess Caroline, produced under the tutelage of Philippe Mercier.[27] In 1735 Bernard Lens III, drawing master to Caroline's younger children, issued a series of prints entitled *A New Drawing Book for* ye *Use of His Royal Highness* ye *Duke of Cumberland, their Royal Highness the Princess Mary and Princess Louisa* (cat. 23.07).[28] The miniature painted by Mary of her sister Louisa as 'Spring' indicates the high standard that could be reached by a diligent royal pupil at around the same time (cat. 23.08).[29] According to the inscription on the backing paper, it was painted 'for her Royal Mother Queen Caroline, who wore it in her Snuff Box', a reminder of the very personal nature of the artistic productions of the royal family.

NOTES

1. 'Couleurs pour peindre a Guache / ein Satz de Couleurs toute preparé . . . et un autre en poudre pour preparé ici, de même s'il est possible d'obtenir quelque information si ils font Usage de l'eau pure pour peindre ou si font Usage de Gom Wasser, et de quelle Gome ils se servent . . . si ils servent du blanc dans leur Couleurs . . . et de qu'elle manière ils font leur nuances.' Mecklenburgisches Landeshauptarchiv, Schwerin. Quoted in part in Olwen Hedley, *Queen Charlotte* (London: John Murray, 1975), 179, and in http://people.virginia.edu/~jlc5f/charlotte/letter_intro.html. (translation slightly adapted by this author).
2. Jane Roberts, *Royal Artists: From Mary Queen of Scots to the Present Day* (London: Grafton Books, 1987), 65, 66, 214 (note 37). 'La Reine s'amuse a peindre des Fleurs, ou plutôt les Plantes, d'après Nature.'
3. The National Archives (TNA) LC 9/294 (John Bradburn's account to Christmas 1767), Roberts, *Royal Artists*, 214 (note 37).
4. *Memoirs of the Court of George III*, Vol. 4: *The Diary of Queen Charlotte, 1789 and* 1795, ed. Michael Kassler (London and Vermont: Pickering & Chatto, 2015), 32–33, 36, 44–45, 50–51, 59, 60 (all for 1789).
5. Roberts, *Royal Artists*, 65, 69–73; *Court of George III*, 4: 32, 59.
6. For works by Charlotte, see Roberts, *Royal Artists*, 66–67; Jane Roberts, ed., *George III & Queen Charlotte: Patronage, Collecting and Court Taste* (London: Royal Collection, 2004), 72; and a tracing of a drawing by Marco Ricci recently on the German art market. In Princess Augusta's effects, auctioned following her death as lot 19, there is described 'a mahogany drawing box with apparatus'. Christies, *A Catalogue of the superb Jewels, Trinkets, Plates, Gold and Silver Medals, China &c. of a Person of great Rank*, 1 Dec.–3 Feb. 1773, 3. No paintings by Augusta's hand are known to have survived.
7. Jane Roberts, *Royal Landscape: The Gardens and Parks of Windsor* (New Haven and London: Yale Univ. Press, 1997), chap. 16, 211–36.
8. Roberts, *Royal Landscape*, 217, 218.
9. Quoted in Roberts, *Royal Landscape*, 219.
10. Roberts, *Royal Landscape*, 219, 560 (note 79).
11. Roberts, ed., *George III & Queen Charlotte*, 78–79.
12. Roberts, ed., *George III & Queen Charlotte*, 147–48.
13. Roberts, *Royal Landscape*, 219; Roberts, ed., *George III & Queen Charlotte*, 146–48. For illustrations of the two rooms at Frogmore decorated by Princess Elizabeth, see William Henry Pyne, *The History of the Royal Residences of Windsor Castle, St James's Palace, Carlton House, Kensington Palace, Hampton Court, Buckingham House and Frogmore*, 3 vols. (London: L. Harrison for A. Dry, 1816–19), 1 (Frogmore): opp. 17, 21. For the surviving lacquer panels, see Roberts, *Royal Artists*, 84.
14. Roberts, *Royal Landscape*, 225–26.
15. Jane Roberts, 'Edward Harding and Queen Charlotte', in *Burning Bright: Essays in Honour of David Bindman*, ed. Diana Dethloff, Tessa Murdoch, and Kim Sloan, with Caroline Elam (London: UCL Press, 2015), 146–59.
16. For Benjamin West's portrait, see Royal Collection (RCIN) 404573. On the pocketbook, see Roberts, ed., *George III & Queen Charlotte*, 71–72.
17. For the coat, waistcoat, and nightdresses, see *The Later Correspondence of George III, 1783–1810*, ed. Arthur Aspinall, 5 vols. (Cambridge: Cambridge Univ. Press, 1962–70), 1: xxix–xxx. For the 'silken quilt', see Thomas Campbell, *Life of Mrs Siddons*, 2 vols. (London: Effingham Wilson, Royal Exchange, 1834), 2: 348 (n.).
18. Roberts, *Royal Artists*, 68.
19. *Memoirs of the Court of George III*, Vol. 2: *Mary Delany (1700–1788) and the Court of King George III*, ed. Alain Kerhervé (London and Vermont: Pickering & Chatto, 2015), 19, 31. The attribution of the Newport flower mosaics to Princess Elizabeth is supported by their dates (1786–88), which fall during Mrs Delany's residence at Windsor, when she was in close contact with the Princess (see *Court of George III*, 2: 207, 211, 219). For Princess Elizabeth's decorative paper-cutting at Frogmore in the 1790s, see Hedley, *Queen Charlotte*, 139. On Mrs Delany's instruction on the spinning-wheel, see *Court of George III*, 2: 63, 67; 4: 50, 72.
20. *Court of George III*, 2: 72–73, 79.
21. David Watkin, *The Architect King: George III and the Culture of the Enlightenment* (London: Royal Collection, 2004), 55.
22. Watkin, *Architect King*; Roberts, ed., *George III & Queen Charlotte*, 93–102; Roberts, *Royal Artists*, 57–65.
23. George III, like his mother Augusta, had been taught by Joseph Goupy (1686–1770) (Roberts, *Royal Artists*, 56).
24. Compare RCIN 980199 of 1752 by Prince Edward, Duke of York, with RCIN 980271 by the future George III.
25. BM 1863,0110.110 and 1867,0309.505
26. Roberts, *Royal Artists*, 73, 76.
27. Joanna Marschner, *Queen Caroline: Cultural Politics at the Early Eighteenth-Century Court* (New Haven and London: Yale Univ. Press, 2014), 166 (lathe-turning and gem-working); 116–15.
28. Kim Sloan, *'A Noble Art': Amateur Artists and Drawing Masters, c. 1600–1800* (London: British Museum Press, 2000), 114–15.
29. RCIN 420900, for which see Richard Walker, *The Eighteenth and Early Nineteenth Century Miniatures in the Collection of Her Majesty The Queen* (Cambridge: Cambridge Univ. Press, 1992), 8–9.

22.03*

GAINSBOROUGH DUPONT (1754–1797), engraver, after **THOMAS GAINSBOROUGH** (1727–1788)
The Eldest Princesses
1793
Mezzotint on paper
Published by Gainsborough Dupont, London
24½ x 18 in. (62.2 x 45.7 cm)
Yale Center for British Art, Paul Mellon Fund, B1970.3.365

Mezzotint of Charlotte, Princess Royal, later Queen Charlotte of Württemberg; Princess Augusta Sophia; and Princess Elizabeth; the three eldest daughters of George III and Queen Charlotte. Princess Charlotte is seated to the right, Princess Augusta stands in the centre and Princess Elizabeth is standing to the left, with her left arm linked into Augusta's right. All the princesses have elaborately curled hair, and wear round-necked low-cut gowns, cut with full skirts and a high waist. Lace fichus draped round their shoulders provide an elegant modesty. The daughters of Queen Charlotte joined her at Frogmore to engage in intellectual and creative pursuits. They became able musicians, artists, and botanists.

The painting of the princesses by Thomas Gainsborough was commissioned by their brother, George IV, then Prince of Wales, at the cost of 300 guineas. It was cut down in the nineteenth century, which means Gainsborough Dupont's mezzotint is now the only record of its original form.

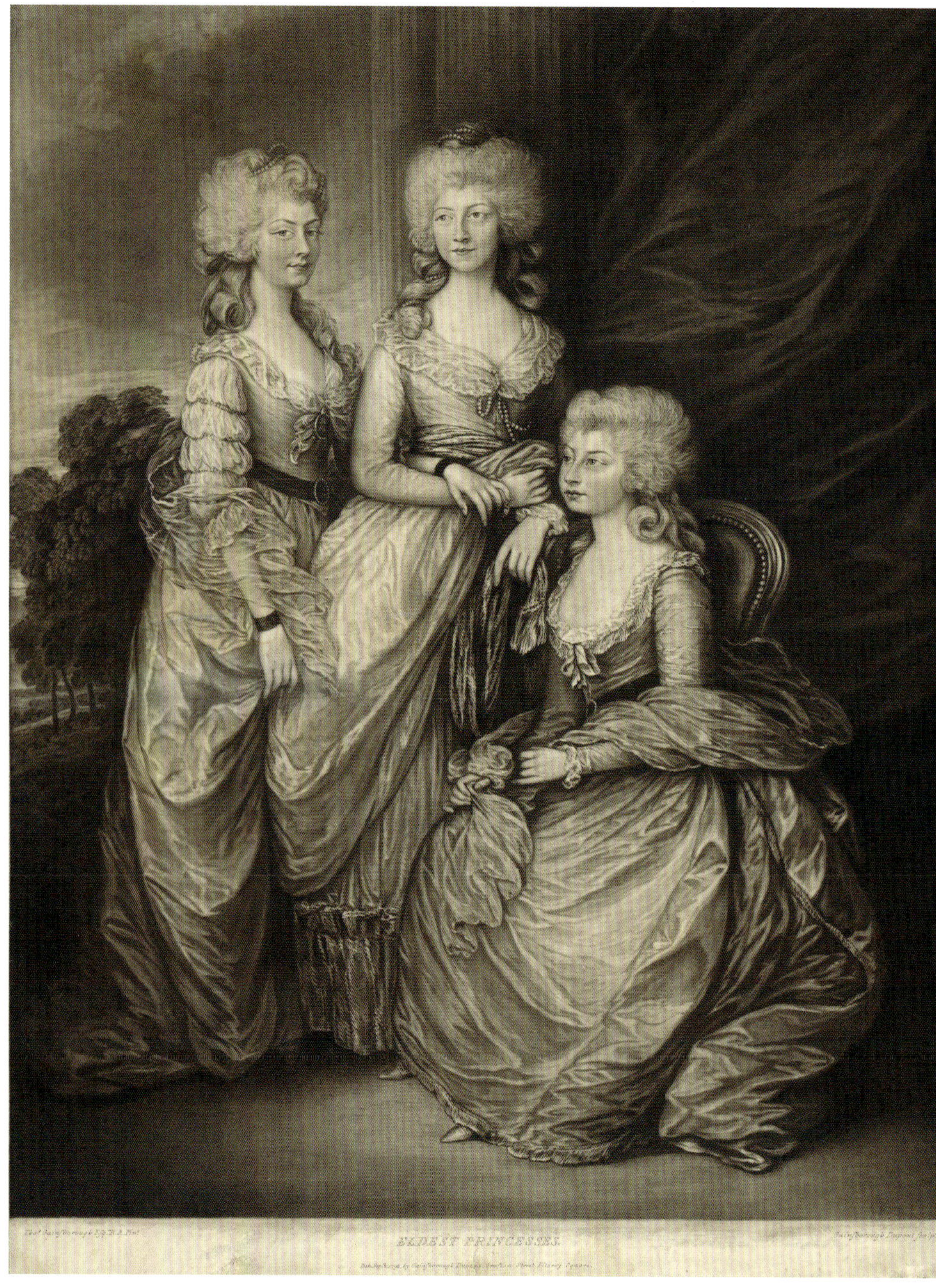

22.03

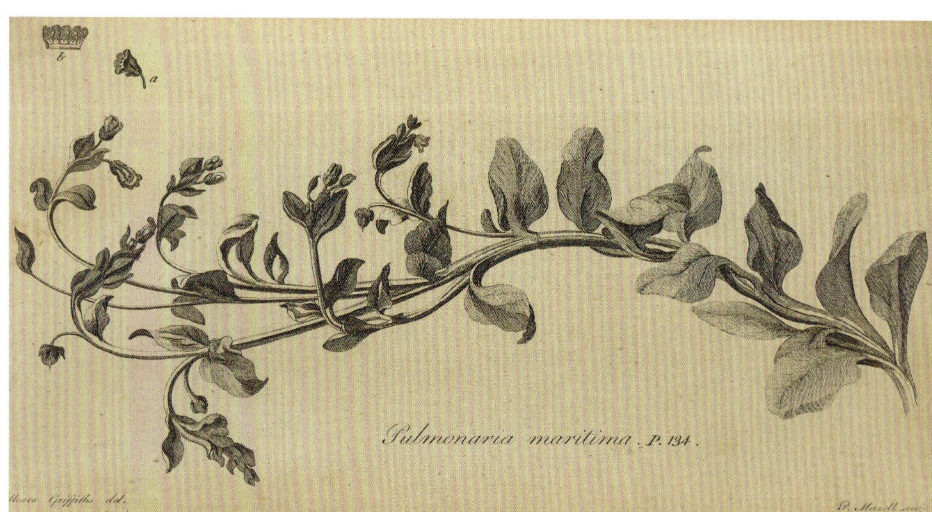

22.04

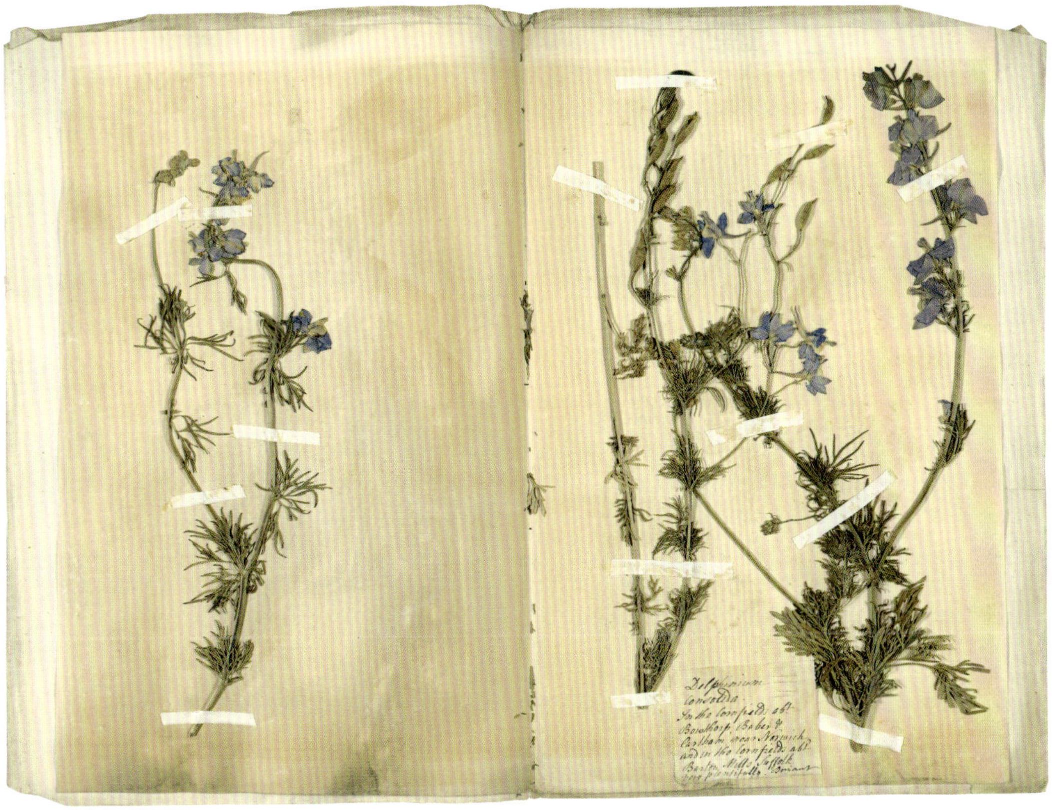

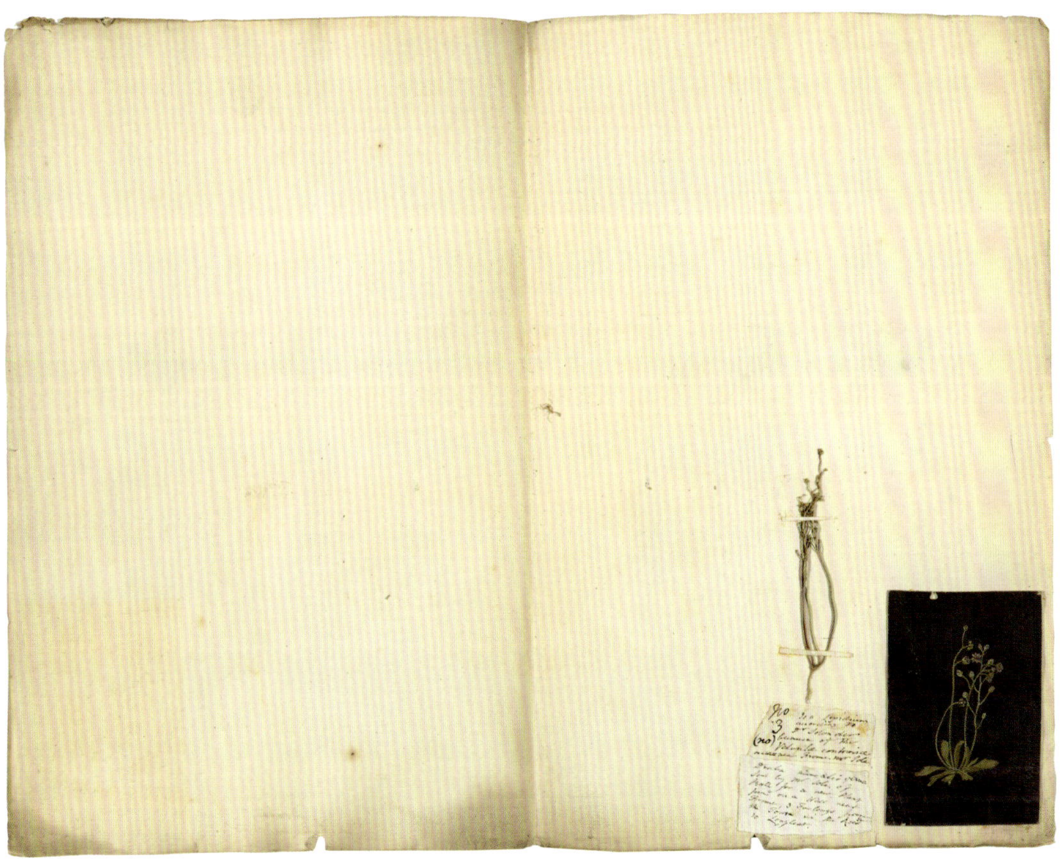

22.05

BOTANIZING AT FROGMORE

22.04*

JOHN LIGHTFOOT (1735–1788)
Flora Scotica: or, a Systematic Arrangement, in the Linnaean Method, of the Native Plants of Scotland and the Hebrides (2 vols)
1777
8¾ in. (22 cm)
Printed for B. White, London
Beinecke Rare Book & Manuscript Library, Yale University, Skde 425
Exhibited: Peter Mazell (active 1761–1797), engraver, after Moses Griffiths (1749–1819), '*Pulmonaria maritima*', engraving on paper, vol. 1, page 134 (illustrated p. 378)

22.05

JOHN LIGHTFOOT (1735–1788), collector
Herbarium sheets with *Delphinium* and *Lepidium*
Late 18th century
Delphinium: botanical specimen
16 x 12⅞ in. (40.5 x 32.7 cm)
Lepidium: botanical specimen with collage of coloured papers and drawing
16¼ x 13 in. (41.4 x 33 cm)
John Lightfoot Herbarium, Royal Botanic Gardens, Kew, Library, Art & Archives Collections

The naturalist and Church of England clergyman John Lightfoot was chaplain to the Duchess of Portland, and a friend of many in her circle, including Mrs Delany and Sir Joseph Banks. Lightfoot toured Scotland in 1772; *Pulmonaria maritima*, a flowering plant described at the opening shown here (see facing page), was native to the coast of western Scotland and the neighbouring islands, including Arran and Bute, ancestral home of the Earl of Bute. It was likely collected by him along with other native species, including *delphinium* and *lepidium* of the mustard family. After Lightfoot's death in 1788, his herbarium was purchased by George III as a gift for Queen Charlotte, who kept it in her library at Frogmore; the specimens are now housed in the herbarium at Kew. Attached to the preserved specimen of *lepidium* is a small painting of the plant worked on a black ground. This technique of botanical illustration was one with which Queen Charlotte's daughters experimented. A copy of *Flora Scotica* was listed in the catalogue of Charlotte's library.

* = YCBA only
† = HRP only

22.06†

CHARLOTTE, PRINCESS ROYAL, LATER QUEEN CHARLOTTE OF WÜRTTEMBERG, CONSORT OF FREDERICK I, KING OF WÜRTTEMBERG (1766–1828), copyist, after JOHANN SEBASTIAN MÜLLER [JOHN SEBASTIAN MILLER] (c. 1715–1792)
Canna Lilly
1783
Pen and ink over pencil on paper
18½ x 13⅛ in. (46.9 x 33.4 cm)
Royal Collection Trust, RCIN 981149

22.07

CHARLOTTE, PRINCESS ROYAL, LATER QUEEN CHARLOTTE OF WÜRTTEMBERG, CONSORT OF FREDERICK I, KING OF WÜRTTEMBERG (1766–1828)
Rose
1782–84
Watercolour, and pen and ink, over pencil
Approx. 18½ x 13⅛ in. (46.9 cm x 33.4 cm)
Shirley Sherwood Gallery of Botanical Art; Royal Botanic Gardens, Kew, Library, Art & Archives Collections, 2015/22

Queen Charlotte's daughters as well as her sons received an artistic education. Charlotte, Princess Royal, and Princess Elizabeth in particular shared their mother's interest in botany. Some of the botanical studies included in the volume of drawings made by Charlotte, Princess Royal, between 1782 and 1784 were copied from plates in books, but others were worked directly from plant specimens. According to Robert Thornton, by the end of the century 'There was not a plant in the Gardens at Kew . . . But has either been drawn by her gracious Majesty, or some of the Princesses, with a grace and skill which reflect on these personnages the highest honour.'

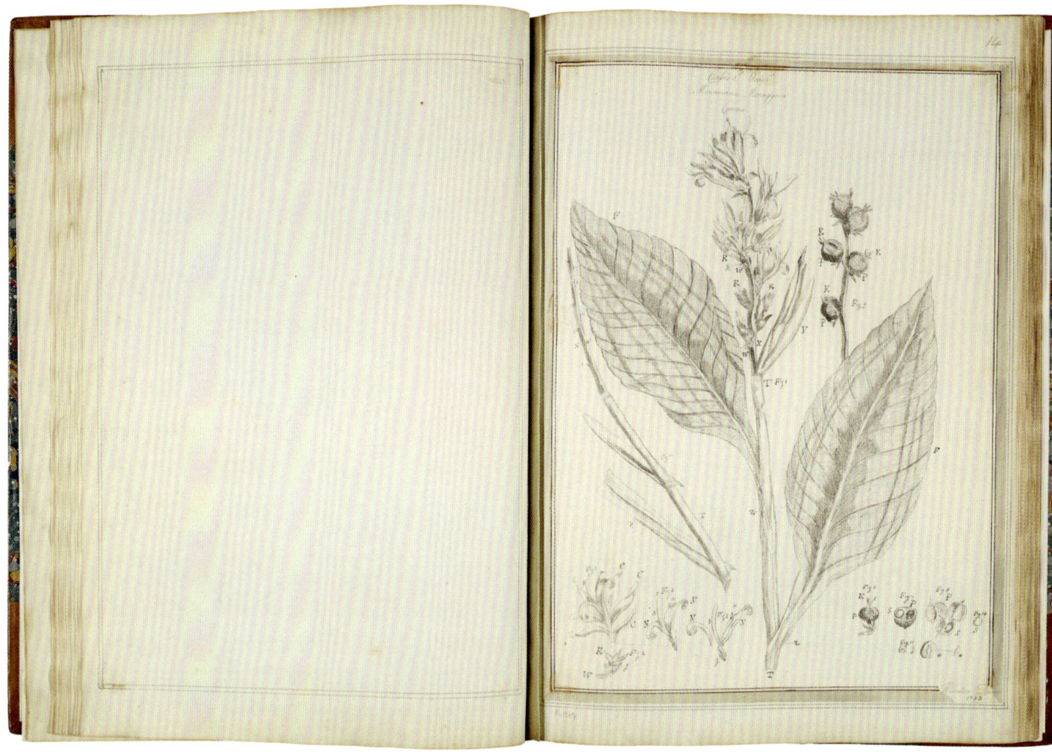

22.06

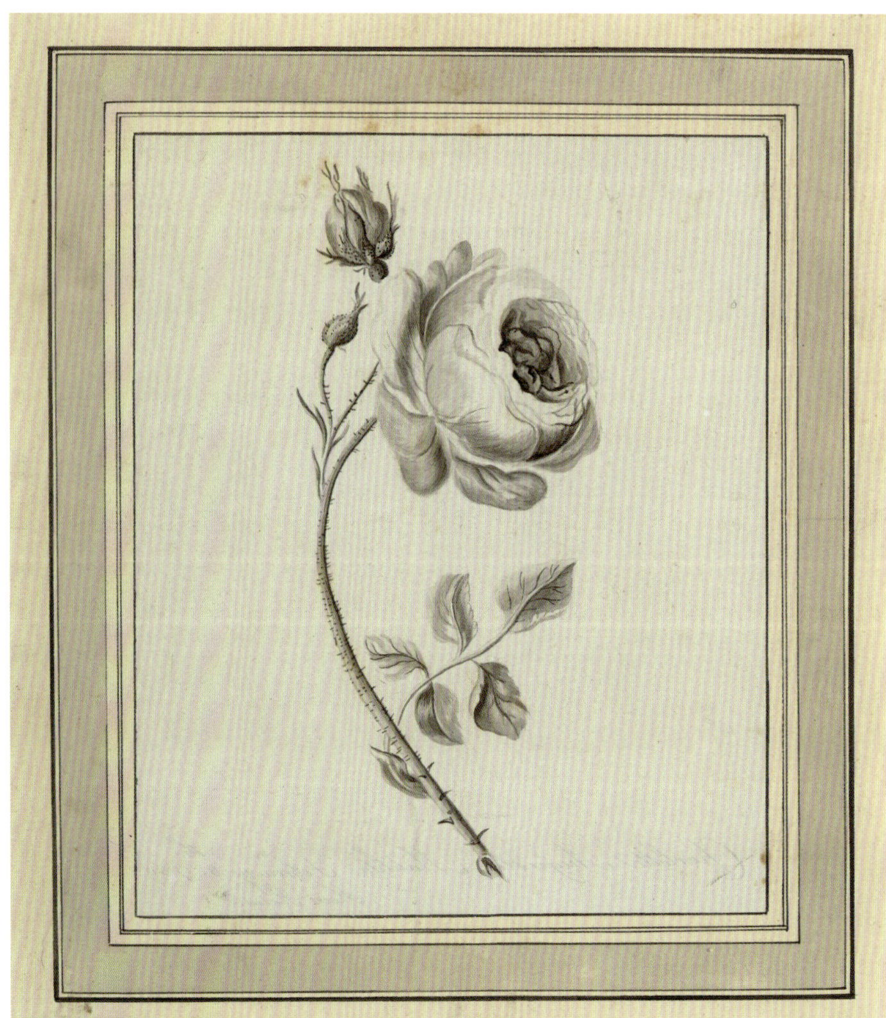

22.07

FROGMORE AND ARTIST PRINCESSES

22.08*

QUEEN CHARLOTTE OF WÜRTTEMBERG (1766–1828), copyist, after **GIULIO CLOVIO** (1498–1578)
Minerva
c. 1785
Etching with ink and wash border
11 x 8 in. (28 x 20.4 cm)
Royal Collection Trust, RCIN 816795

In November 1785, Mrs Delany wrote: 'I have been several evenings at the Queen's Lodge, with no other company but their own most lovely family. They sit round a large table, on which are books, work, pencils and paper . . . the younger part of the family are drawing, working etc, etc.' Many artists were drawn in to support the family's artistic ambition. This etching by Charlotte, Princess Royal, is after a drawing of the head of Minerva in George III's collection by the sixteenth-century artist Giulio Clovio (RCIN 990453). The Princess's etching dates from the mid-1780s, when she was receiving tuition in this process from Biagio Rebecca and Peltro William Tomkins, and the small printing press at Frogmore could possibly have been deployed.

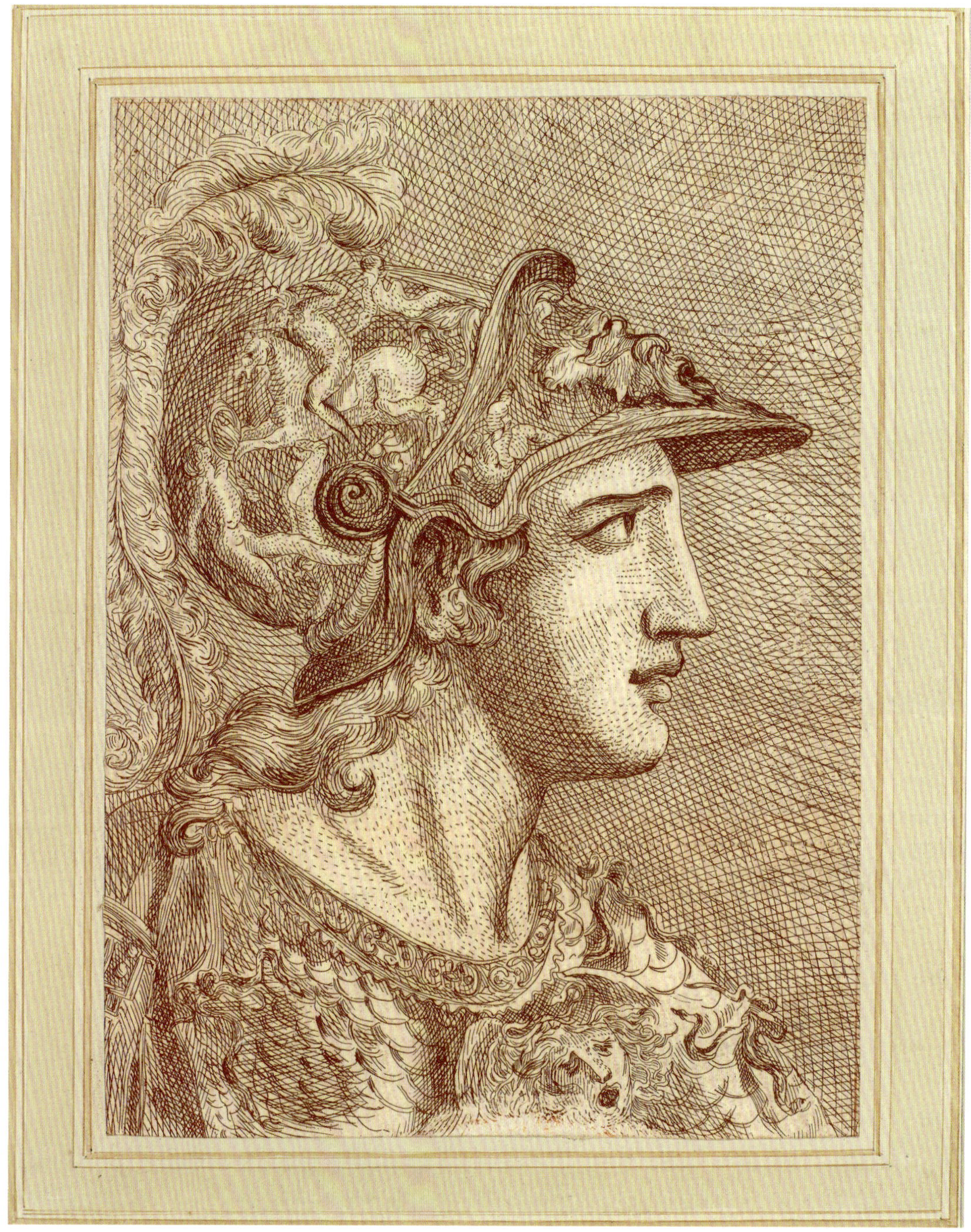

22.08

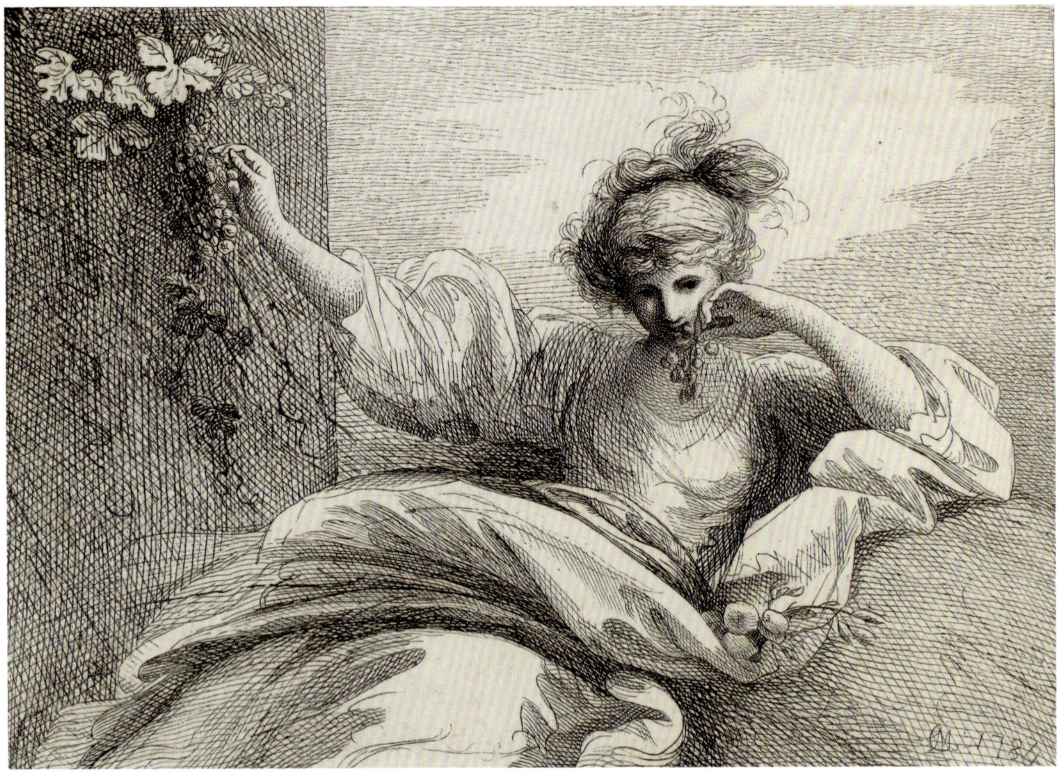

22.09

22.09*

CHARLOTTE, PRINCESS ROYAL (1766–1828), engraver, after BENJAMIN WEST (1738–1820)
The Five Senses: Taste
1784
Etching
7¼ x 10⅛ in. (18.3 x 25.7 cm)
Royal Collection Trust, RCIN 816786

The American-born artist Benjamin West provided art lessons for Charlotte, Princess Royal, and her sister Elizabeth between about 1780 and 1809. He also served as History Painter to George III and was the second president of the Royal Academy in the years 1792–1805 and 1806–20. He rented a house in Park Street, Windsor, which ensured his services could be called on whenever required. *Taste* is one of a set of allegories etched in 1784 after drawings by West; it depicts a young girl tasting a grape. The others showed *Sight*, *Sound*, *Smell* and *Touch*. West's original drawings have been dispersed subsequently but *Taste* is extant, signed and dated 1784; *Sound* and *Smell* appeared on the art market in 1968 and 1975 respectively.

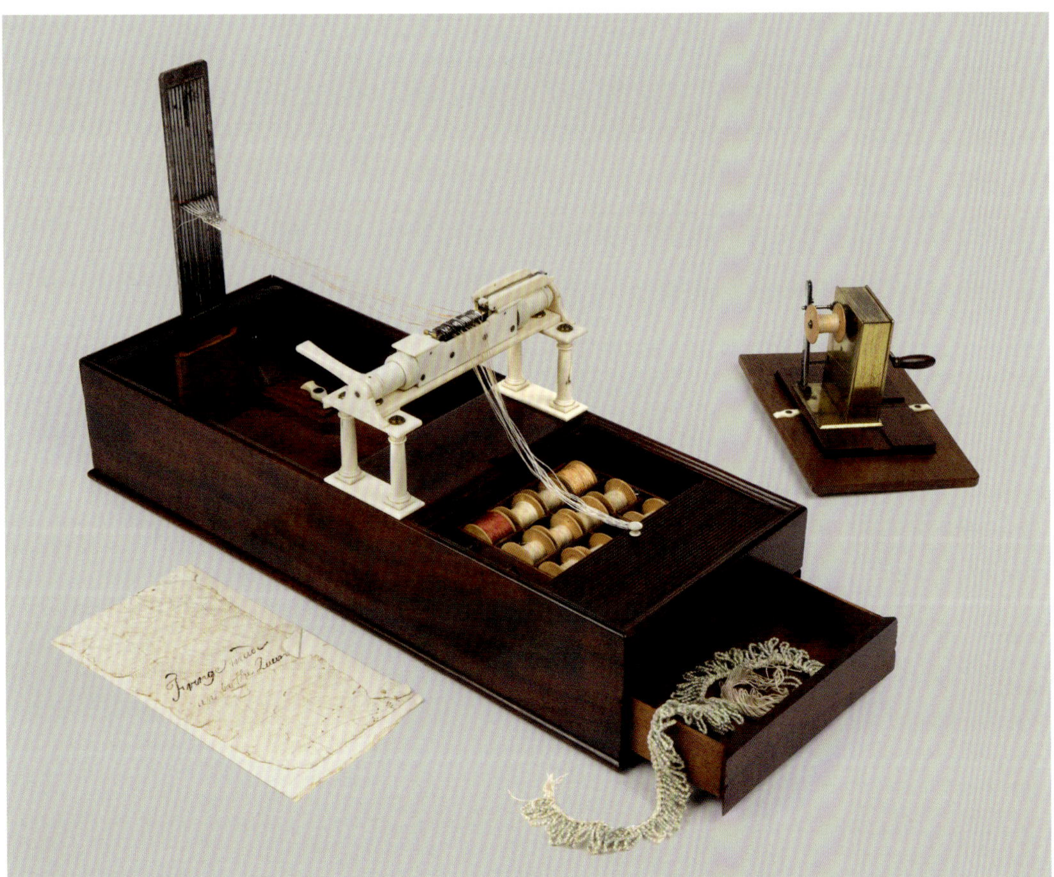

22.10

22.10

UNKNOWN MAKER
Fringe loom
c. 1790
Mahogany with rose wood, ivory, polished steel and brass
7 x 6½ x 17 in. (17.5 x 16.5cm x 43 cm)
Historic Royal Palaces, HRP 3006155

Queen Charlotte of Mecklenburg-Strelitz presented this fringe-making machine to Mrs Delany. It has a sixteen-bobbin rack for thread and a polished steel needle comb, contained within a mahogany case with concealed drawers for the accommodation of additional equipment. Fanny Burney, Madame d'Arblay, recorded Mrs Delany's delight: 'she showed me the most excellent and ingenious loom, which the Queen had made her a present of last summer at Windsor, for making fringe; and a gold knitting needle given her by the King. And she told me the whole history of their manner of presenting them, with a sort of grateful simplicity that was quite affecting.' Charlotte's skill as a needlewoman is attested by her surviving work.

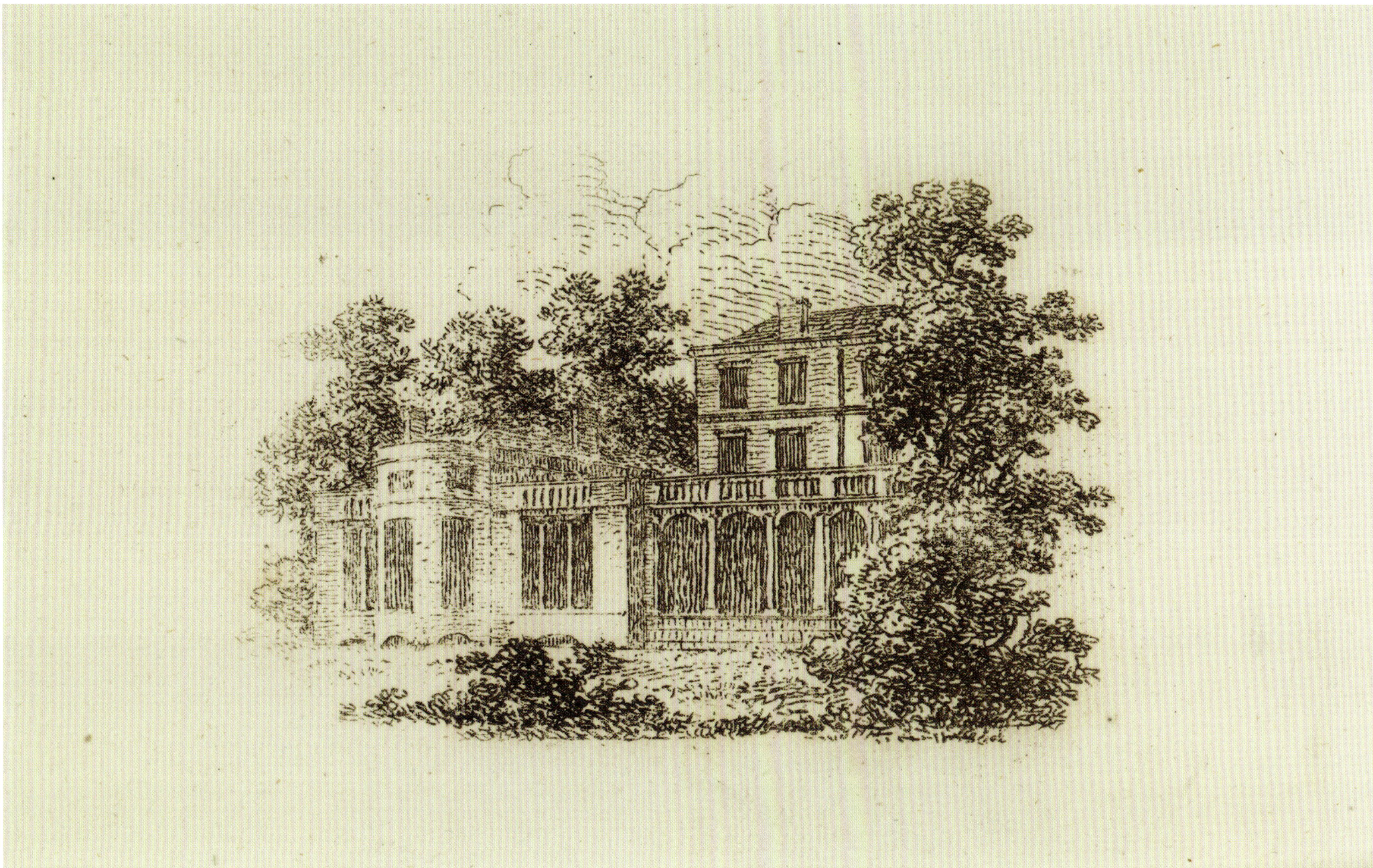

22.11

THE FROGMORE PRESS

22.11*

ELLIS CORNELIA KNIGHT (1757–1837),
author and editor
Miscellaneous Poems
1812
Printed by E. Harding, Frogmore Lodge, Windsor [England]
7⅞ in. (20 cm)
The Lewis Walpole Library, Yale University,
53 Sp4 812
Exhibited: Frontispiece, 'Frogmore' (illustrated)

When Madame de Genlis visited Windsor in 1782, she noted 'the Queen of England really enjoys reading, and at Windsor, where the princess lived in complete privacy, M. Delac was daily summoned to read for three or four hours.' Jean André Deluc (1727–1817) was French Reader to Queen Charlotte. The Queen's Library was first concentrated in the rooms to the north of the Queen's apartments in Buckingham House, but the core transferred to Frogmore House in 1804. Her collection was extensive. Books on religious subjects were most numerous, but there were a substantial number on botanical subjects. Edward Harding, her Librarian, set up a small printing press at Frogmore, known as the Frogmore Press. It produced only two titles, of which this is one. The author and artist Ellis Cornelia Knight was employed as a companion to Queen Charlotte between 1805 and 1813. The other products of the press were sets of cards of historical facts used for the teaching of children (see pp. 309–10, fig. 17.06).

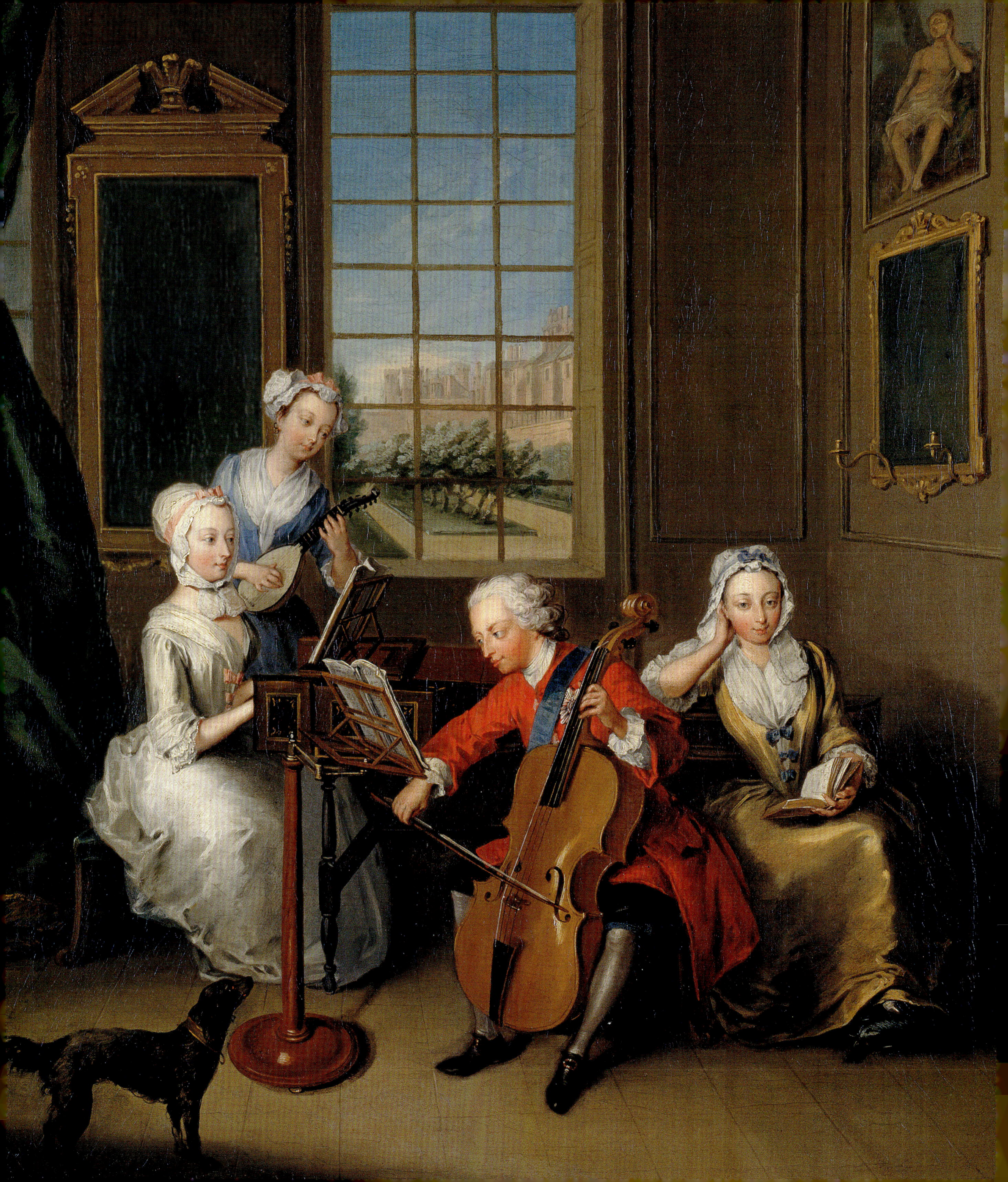

Royal Education

Florence Grant

The education provided for the Hanoverian royal children did not vary greatly from widespread prescriptions for the upbringing of other privileged children in the eighteenth century.[1] They studied history and geography, languages, mathematics, drawing, music, and dancing under the supervision of tutors and governesses. The governess appointed for Caroline's elder daughters was Jane Martha Temple (fig. 23.01), the clever daughter of Sir John Temple and second wife of William Bentick, first Earl of Portland. Some of their written exercises still survive. Boys often received military training, while girls cultivated accomplishments such as miniature painting and botanical illustration. One distinguishing factor was frequent access to tutors who were renowned practitioners in their respective fields, drawn into court circles by royal patronage. Those from whom Caroline of Ansbach, Augusta of Saxe-Gotha, and Charlotte of Mecklenburg-Strelitz commissioned paintings, musical compositions, and buildings, also instructed their children. The resulting relationships could benefit artists' commercial endeavours, they sometimes involved multiple generations on both sides, and they prepared the princes and princesses for the kind of informed and engaged patronage they were expected to perform as adults.

Philippe Mercier's *'The Music Party': Frederick, Prince of Wales with his Three Eldest Sisters* (1733) encapsulates this interconnection of patronage and education (p. 384, cat. 23.06). From 1729, Mercier held a dual appointment as principal painter and librarian to Prince Frederick, and he was also employed by George II and Caroline to provide instruction in painting to Princess Anne and her sisters Amelia and Caroline.[2] Frederick, Anne, and Caroline are shown in an informal domestic setting performing music, while the middle sister, Amelia, looks up from a volume of Milton to listen. Frederick, as the eldest son and heir to the throne, takes the role of soloist in this depiction of dynastic harmony, but Anne was the sibling most celebrated for musical accomplishments—she was trained at the keyboard by George Frideric Handel, who received patronage and support from Caroline which helped to cement his career in London.[3]

By 1733, Bernard Lens III had succeeded Mercier as drawing master to Caroline's younger children. He was limner to both George I and George II, and was one of a small group of artists who taught miniature painting to aristocratic and courtly amateurs in the eighteenth century.[4] His portrait of Princess Louisa is inspired by another miniature, *Louisa, Queen of Denmark as Spring* (c. 1735–40), painted by her sister Princess Mary under Lens's tutelage (cat. 23.08). This type of painting, executed in watercolour on tablets of ivory, would have been challenging for a beginner, but Lens also provided more approachable tasks for his pupils.

From 1735–36 Lens etched a copybook for his royal pupils entitled *A New Drawing Book for ye Use of His Royal Highness ye Duke of Cumberland, their Royal Highness the Princess Mary and Princess Louisa*. Before attempting to draw or paint from life, children would learn to copy from drawings and prints, and the appearance of a military encampment on the title page of the *New Drawing Book* confirms that for the princes especially, drawing was crucial not only to an appreciation of the fine arts, but

Detail of p. 391, cat. 23.06

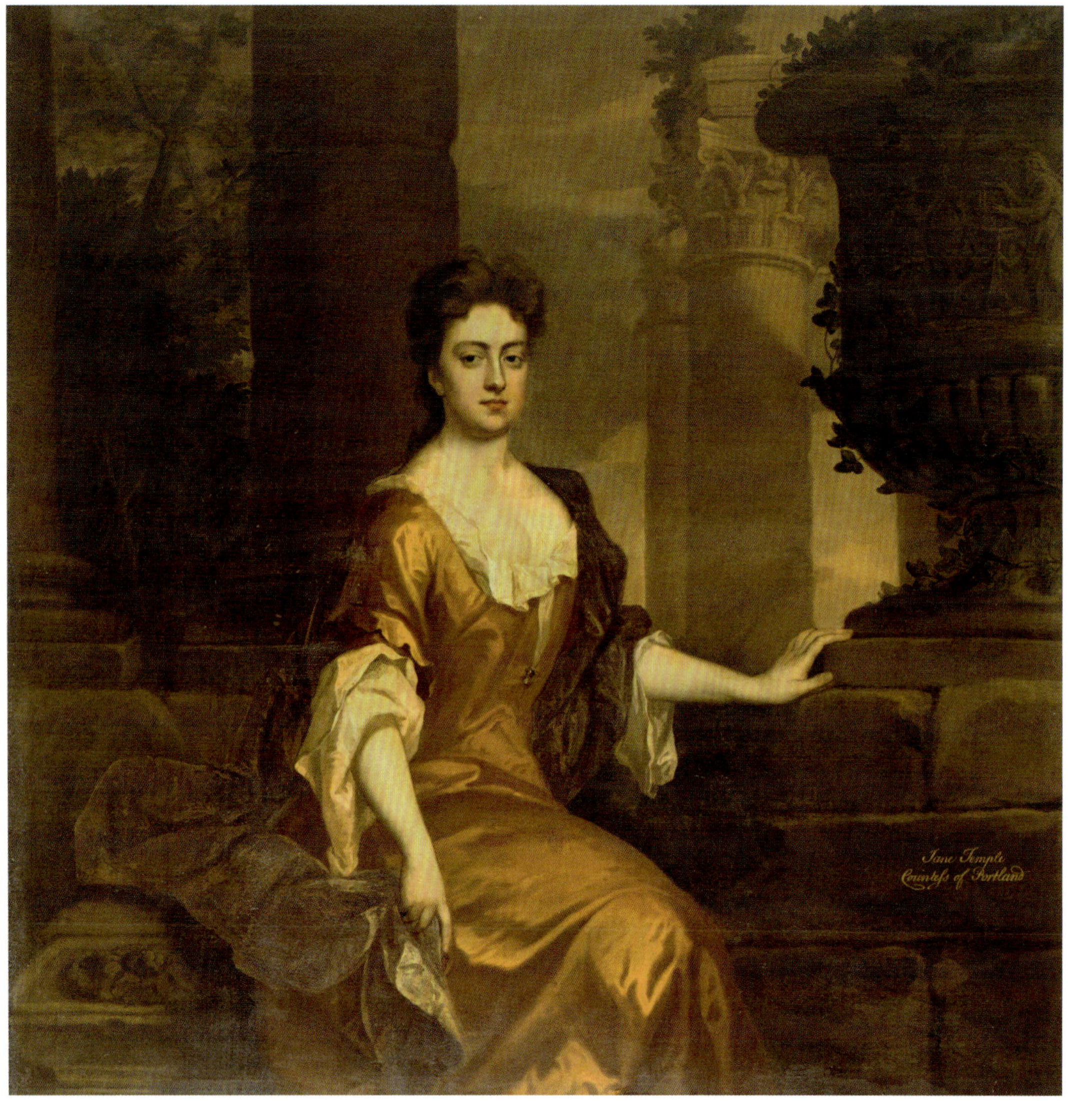

FIG. 23.01 Michael Dahl, *Jane Martha Temple*, 1690–99, oil on canvas, 65 x 61 in. (165 x 155 cm). National Trust, Petworth, West Sussex

also to the understanding of fortifications and the precise observation needed for military command.[5] Lens's inclusion of the princesses' names in the dedication suggests that in this generation it was considered appropriate for girls to be conversant in these more masculine uses of drawing—an attitude that would all but vanish by the early nineteenth century.[6]

Another courtly accomplishment that Caroline's sons and daughters practised was the art of turning various materials on a lathe. Turning had been a well-established part of European princely education since the sixteenth century, and could be used to produce simple, regular shapes, such as the turned knife handles made in amber by Princess Anne (cat. 23.10), or far more complex works such as the ivory pagoda by Princess Louisa (cat. 23.12), or a boxwood portrait medallion of the Emperor Vespasian by Prince William, Duke of Cumberland (cat. 23.11). More complex turning exercises demanded considerable understanding of mechanics and three-dimensional geometry, and could be counted as part of a comprehensive training in mathematics.[7] Interest in the art of turning would continue in this generation and the next: Augusta paid John Bradburn for a lathe in 1769, and boxes turned by George III survive in the royal collections of both Brunswick and Great Britain.

Augusta's influence over her children's education came most directly through her patronage and friendships after her husband's death in 1751. Her eldest sons' first preceptor, the Church of England clergyman Francis Ayscough, was Frederick's Chaplain and Clerk of the Closet, and the curriculum that the boys followed between his appointment in 1744 and Frederick's death was determined by the boys' father.[8] A doctrinally and politically controversial figure, Ayscough appears with his pupils Prince George and Prince Edward in a conversation piece painted in the late 1740s by Richard Wilson (cat. 23.15). After Frederick's death Ayscough was dismissed by George II and there followed years of almost constant bickering and upheaval among the men whom the King appointed to take his place. Although Augusta took a conciliatory approach with the aging George II, she appears to have been consulted very little in these appointments. She was nonetheless gravely concerned with her children's education—taking as her own priority the cultivation of piety and moral character, rather than learning or accomplishments for their own sake.[9]

Despite her frequent portrayal by historians as a loyal wife and submissive daughter-in-law, Augusta pictured herself as the head of her own household, even before the death of her husband. In 1739 she commissioned a remarkable group portrait from Jean-Baptiste van Loo (fig. 23.02). Augusta is the central, seated figure; Prince George is seated on a cushion at her right shoulder; and Prince Edward is held by Lady Archibald Hamilton, the Princess of Wales's Keeper of the Privy Purse and Mistress of the Robes. Princess Augusta stands to her mother's left, while her governess, the Hon. Arabella Herbert, stands at the far left of the picture holding Prince George's hat.[10]

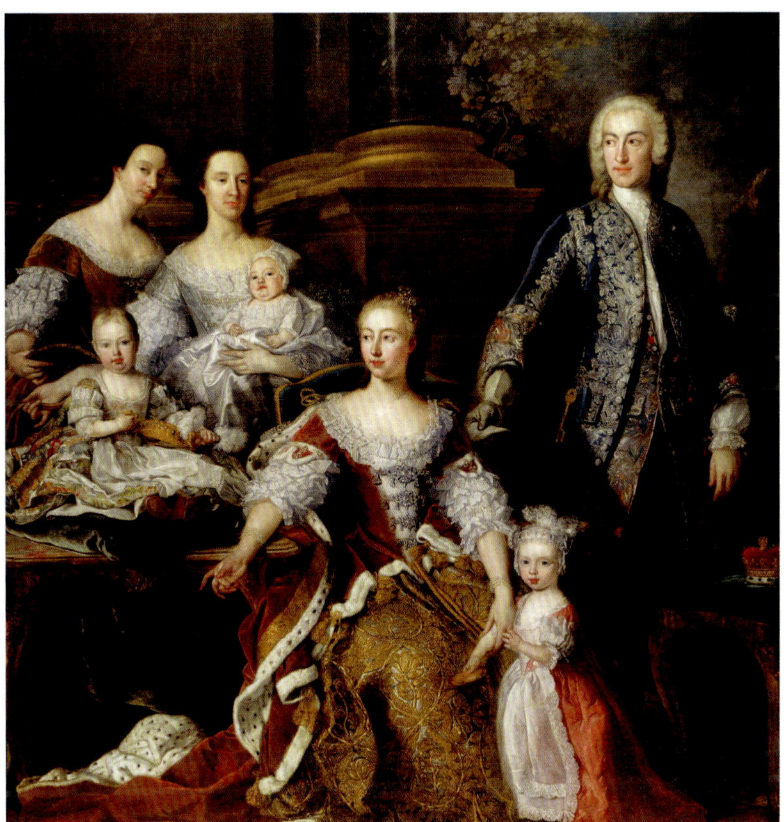

FIG. 23.02 Jean-Baptiste van Loo, *Augusta, Princess of Wales, with Members of her Family and Household*, 1739, oil on canvas, 88 x 79⅛ in. (223.4 x 201.0 cm). Royal Collection Trust

The confidence that Augusta projected in this portrait emerged in the interconnections between education and her friendships, personal interests, and patronage. By 1756 she had established her close friend John Stuart, third Earl of Bute, as Prince George's mentor and tutor in a range of subjects including history, law, and natural philosophy. The Prince, hitherto slow to learn, or at least reluctant to apply himself, subsequently thrived. The garden of the White House, the Princess's residence at Kew, was the focus of Augusta and Bute's collaborative projects in horticulture and botany; it was also the site of her intensive patronage of William Chambers, who in the late 1750s was kept busy both as architect to Augusta and tutor in architecture to Prince George. George was also taught perspective drawing by Joshua Kirby, who later worked with Chambers in his capacity as Clerk of the Works at Kew and Richmond.[11] Architectural drawing, as well as natural philosophy and astronomy, would become a lifelong interest of George III, and many of his designs for buildings—some, like the plan and elevation of a Doric Temple, delicately finished in ink wash—survive in the Royal Collection (cat. 23.16).[12] This study and Prince Edward's 1752 *Landscape Study* show that these later works built on elementary training in drawing through copying, in the manner of Lens's *New Drawing Book*.[13]

In Johan Zoffany's *Queen Charlotte with her Two Eldest Sons*, the next generation of princes appears, in costumes that may refer to François Fénelon's influential educational text, *Telemachus* of 1699 (see p. 25, fig. 1.15). They are on the boundary between learning and make-believe—still inhabiting a childhood version of the masculine, princely roles intended for them.[14] Benjamin West's *Queen Charlotte with Charlotte, Princess Royal*, one of a series of family portraits commissioned by George III between 1776 and 1778, shows the Queen and her daughter inhabiting an equally feminine role, embroidering or tatting with the material spread out between them (see p. 455, fig. 28.02, and cat. 13.20). Although the other portraits in West's series do not distinguish so strictly between the roles of sons and daughters, this pairing suggests something of the hardening lines around acceptable female accomplishments towards the end of the century.

Charlotte and her daughters energetically exploited the possibilities open to them. Charlotte, Princess Royal, became particularly skilled in etching; she produced prints after designs provided for her by West, as well as after drawings already in the royal collections (cats. 22.08–22.09).[15]

Many of the other activities Charlotte's daughters pursued, such as paper cutting and flower painting, involved high levels of skill and serious study—especially in botany—and it is remarkable that Bernard Lens III, the drawing master of their great-aunts

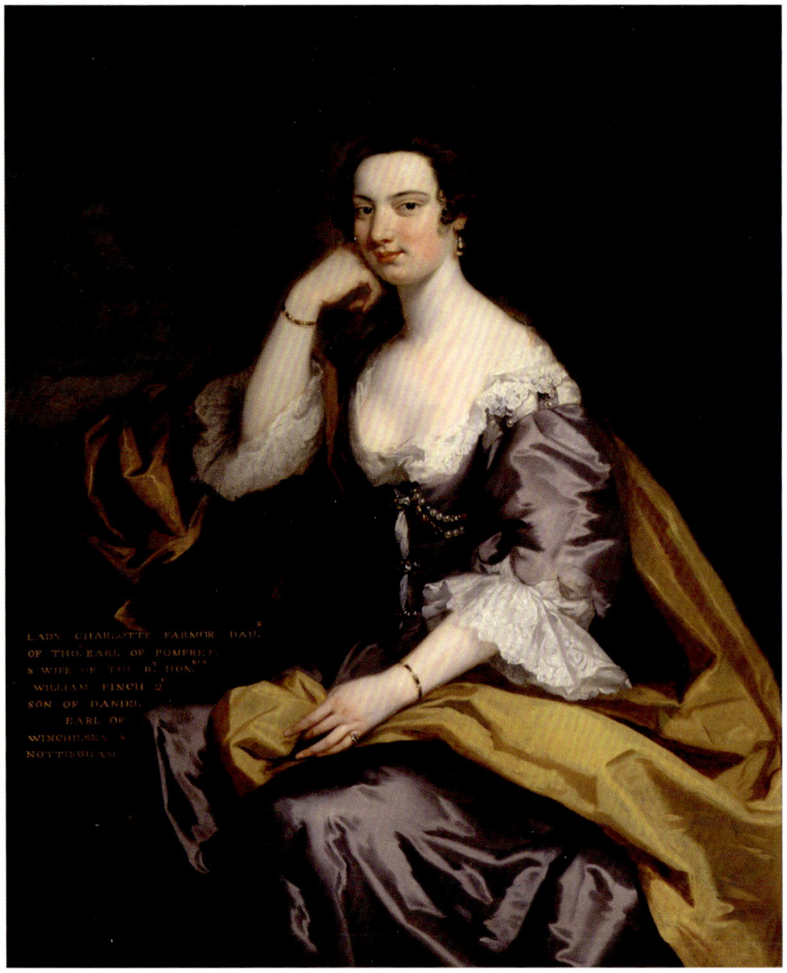

FIG. 23.03 John Robinson, *Lady Charlotte Finch*, c. 1740–45, oil on canvas, 49¾ x 39¾ in. (126.2 cm x 101 cm). National Portrait Gallery, London

Princess Mary and Princess Louisa, had also taught their older friend and companion in botanical pursuits, Mary Granville, Mrs Delany.[16] Nonetheless, the great number of flower paintings and drawings, including especially striking examples depicting a canna lily and a rose produced by Charlotte, Princess Royal (cats. 22.06–22.07), were also representative of the newly restricted range of artistic genres considered appropriate for aristocratic and royal women.[17] In these genres, flower painting among them, the Queen was able to extend her patronage to professional women artists such as Margaret Meen and Mary Moser, who instructed the princesses.[18] For the princes, in contrast, landscape drawing continued to be strongly associated with military training. *A View of St Leonard's Hill*, produced in outline by the draughtsman John Alexander Gresse and coloured by Prince Augustus, appears to follow a pedagogical method similar to the one used at the Royal Military Academy at Woolwich.[19]

The late eighteenth and early nineteenth centuries saw an increase in toys, books, and other educational materials designed for and marketed to children.[20] Teaching aids such as *The Spelling Box*, *The Grammar Box*, and *The Figure Box* represent a small part of the moral and educational works of Ellenor Fenn, whose popular book of dialogues for teaching reading, *Cobwebs to Catch Flies* (1783–84) was also in the royal nursery.[21] Lady Charlotte Finch, who served as royal governess for over thirty years, was also a progressive educator, engaging with a broad range of subjects through creative play and other innovative methods (fig. 23.03).[22] In the 1760s, she commissioned an extensive set of puzzle maps intended as aids to the study of geography (cat. 23.22). Along with such toys, this is the first generation for which a significant quantity of baby clothes, such as the shoes belonging to Princess Augusta, or the Prince of Wales's robe, survive—making the remnants of childhood appear suddenly familiar.

NOTES

1. Compare the programme laid out by John Locke in his influential *Some Thoughts Concerning Education*, published in 1693 and frequently reprinted throughout the eighteenth century.
2. Desmond Shawe-Taylor, ed., *The First Georgians: Art and Monarchy, 1714–1760* (London: Royal Collection Trust, 2014), 388; Jane Roberts, *Royal Artists: From Mary Queen of Scots to the Present Day* (London: Grafton Books, 1987), 53–55.
3. See in this volume, Donald Burrows and Berta Joncus, 'Music and Representation', 203–206.
4. Kim Sloan, *'A Noble Art': Amateur Artists and Drawing Masters, c. 1600–1800* (London: British Museum Press, 2000), 45–46, 114.
5. Sloan, *'A Noble Art'*, 114.
6. Sloan, *'A Noble Art'*, 213–14.
7. Shawe-Taylor, *First Georgians*, 299.
8. Stanley Ayling, *George the Third* (London: Collins, 1972), 29–30.
9. Ayling, *George the Third*, 32–33.
10. Oliver Millar, *The Tudor, Stuart and Early Georgian Pictures in the Collection of Her Majesty the Queen*, 2 vols. (London: Phaidon, 1963), 1: 178.
11. David Watkin, *The Architect King: George III and the Culture of the Enlightenment* (London: Royal Collection, 2004), 60, 61–65.
12. Royal Collection (RCIN) 980479.
13. Roberts, *Royal Artists*, 58.
14. Desmond Shawe-Taylor, *The Conversation Piece: Scenes of Fashionable Life* (London: Royal Collection Trust, 2009), 112.
15. Jane Roberts, ed., *George III & Queen Charlotte: Patronage, Collecting and Court Taste* (London: Royal Collection, 2004), 73–74.
16. Sloan, *'A Noble Art'*, 45, 214.
17. Album containing botanical illustrations in watercolour and pencil by Charlotte, Princess Royal, later Queen of Württemberg (1766–1828), RCIN 970325. There is a small collection of works associated with the Princess Royal in the collections of the Royal Botanic Gardens, Kew.
18. Roberts, ed., *George III & Queen Charlotte*, 75–76, 83.
19. Sloan, *'A Noble Art'*, 140–41.
20. Roy Porter, *Enlightenment: Britain and the Creation of the Modern World* (London and New York: Penguin, 2000), 268, 351–52.
21. David Stoker, 'Fenn, Ellenor, Lady Fenn (1744–1813)', *Oxford Dictionary of National Biography* (Oxford: Oxford Univ. Press, 2004; online edn, May 2005), http://www.oxforddnb.com/view/article/9279, accessed 27 April 2016.
22. Jill Shefrin, 'Finch, Lady Charlotte (1725–1813)', *Oxford Dictionary of National Biography* (Oxford: Oxford Univ. Press, 2004; online edn, Jan. 2008), http://www.oxforddnb.com/view/article/75524, accessed 27 April 2016.

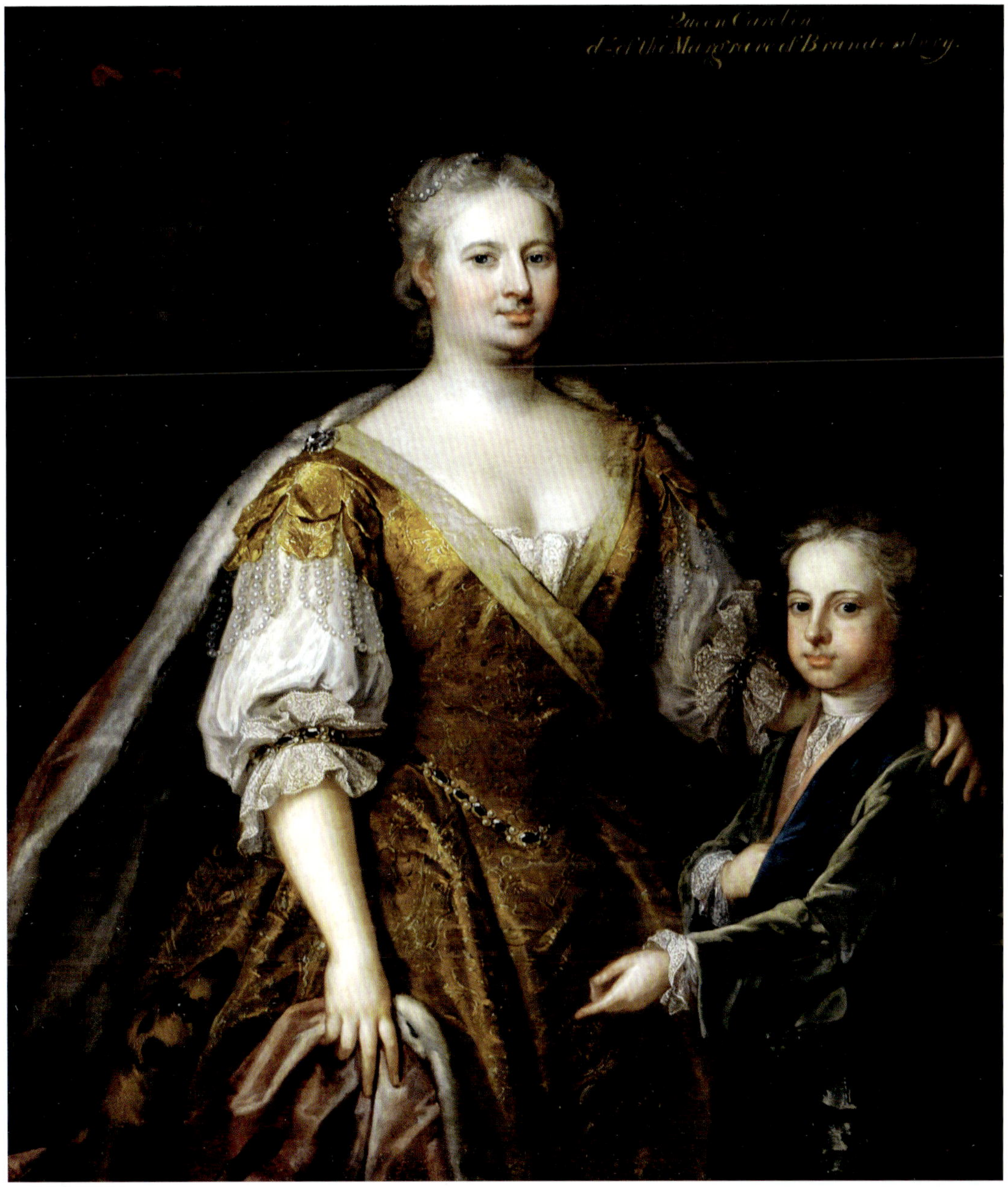

23.04

CAROLINE AND THE NURTURING OF HER FAMILY

23.04

WILLIAM AIKMAN (1682–1731)
Queen Caroline (of Brandenburg Ansbach) and her son Prince William Augustus, Duke of Cumberland
c. 1730
Oil on canvas
49 x 39 in. (124.5 x 99.1 cm)
National Trust Collections, Hardwick Hall, NT 1129123

Caroline and her youngest son, William Augustus, had a close bond, possibly because of her early separation from her eldest son Frederick, the death of a second son as a tiny infant and the stillbirth of a third. It is probable that George I had been approached by William's parents during the 1720s with a proposal that the British-born prince should inherit the British throne, while their eldest son remain in Hanover to succeed as elector. George I's death in 1727, and the diplomatic imperative to bring Frederick to London in 1728, put paid to any such ambition.

This three-quarter-length portrait by Aikman was possibly commissioned by Caroline in about 1730 and presented by her to Lady Dorothy Savile, Countess of Burlington. Lady Burlington had served her as a Lady of the Bedchamber from 1727. She was a gifted amateur artist, and her portrait of Lord Grantham hung among the dynastic collections in Caroline's Picture Closet. She would also draw Caroline on her deathbed.

23.05*

CHARLES PHILIPS (1708–1748/49)
Frederick, Prince of Wales
1731
Oil on canvas
19¼ x 14⅛ in. (48.9 x 35.9 cm)
Yale Center for British Art, Paul Mellon Collection, B1981.25.504

Frederick, Prince of Wales, arrived in London in December 1728 at the age of twenty-one, having been abruptly taken from a ball in Hanover by emissaries sent by his father, George II. His meddling in family marriage negotiations between Prussia and Brandenburg-Ansbach had resulted in political embarrassment for George II, and it was deemed safer to have him drawn more closely into the family circle. The Prince, however, would never be fully integrated.

The naive charm of Charles Philips's earliest portrait groups had attracted commissions, as noted by George Vertue, who, in 1731, recorded 'painting small figure portraits & conversations [which] has met with great encouragement amongst People of Fashion—even some of ye Royal Family'. As is characteristic, Philips's portrait of Prince Frederick is brightly coloured, showing the young man fashionably dressed, and has an uncluttered directness. In 1737 Philips would be commissioned by Frederick to make a pair of portraits of him and his new wife Augusta.

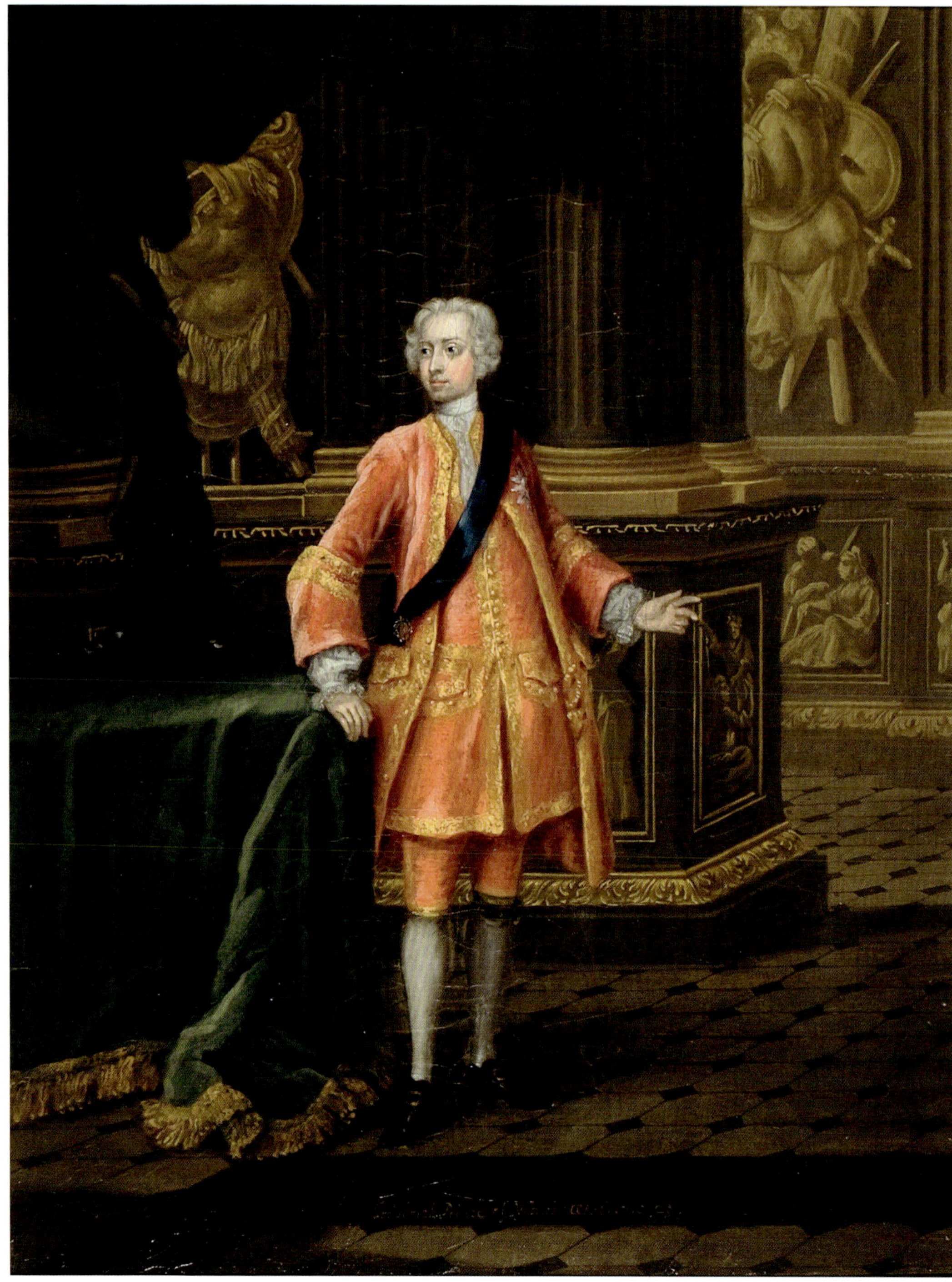

23.05

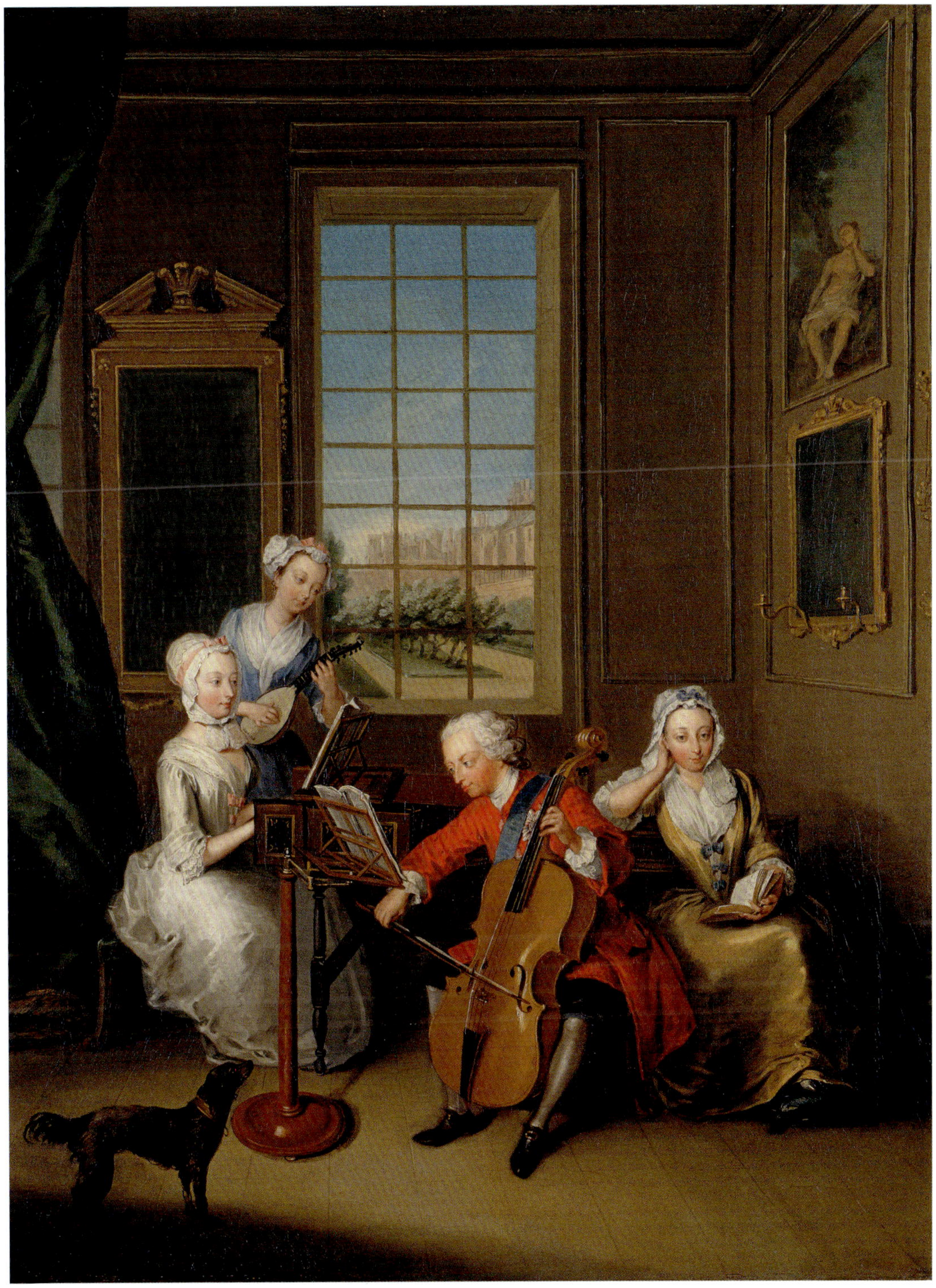

23.06

ART EDUCATION

23.06

PHILIPPE MERCIER (1689–1760)
'The Music Party': Frederick, Prince of Wales, with his Three Eldest Sisters
1733
Oil on canvas
31¼ x 22¾ in. (79.4 x 57.8 cm)
Royal Collection Trust, RCIN 402414

Philippe Mercier, a pupil of Antoine Pesne, was art tutor to the three eldest daughters of George II and Queen Caroline. This painting was probably commisioned by Frederick, Prince of Wales, the princesses' brother, in 1733. Educated in Hanover, he did not join the family until 1728. All of Caroline's children received a thorough musical education, with the eldest princesses tutored by George Frideric Handel. In the painting Prince Frederick is shown playing the cello, accompanied by Princess Anne who plays the harpsichord and Princess Caroline the mandora, a form of lute. Their sister Amelia sits reading a volume of Milton's works as they play. This is believed to be the prime version of the painting; the furniture depicted can be identified in household accounts suggesting this is an interior in one of the royal homes, though the view through the window makes precise identification difficult.

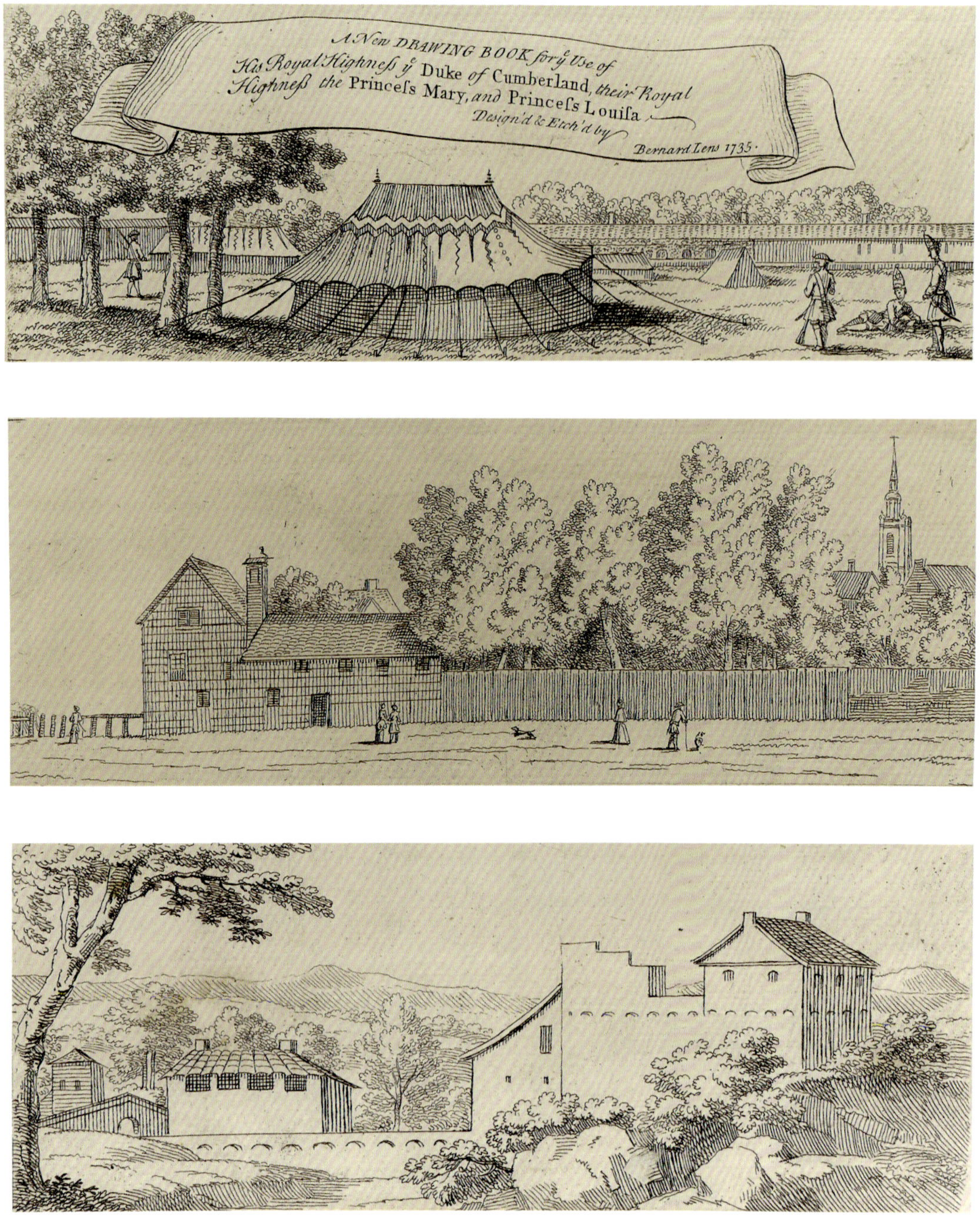

23.07

23.07*

BERNARD LENS III (1681–1740), designer and engraver
A New Drawing Book for use of His Royal Highness the Duke of Cumberland, their Royal Highness the Princess Mary and Princess Louisa
1735
Set of etchings
Each: approx. 2½ x 6⅞ in. (6.4 x 17.4 cm)
Title plate lettered with 'Design'd and etch'd by Bernard Lens 1735'
British Museum, 1872,1012.3659–3676
Exhibited: frontispiece (1872,1012.3659); 1872,1012.3660; and 1872, 1012.3662

Bernard Lens III was the deviser of this book of drawing exercises, from which models were taken for lessons for the younger children of George II and Caroline. It is possible that Lens's brother Edward may also have assisted in its compilation. He too had an interest in art tuition for the young, serving as his father had before him as Drawing Master at Christ's Hospital School. The plates indicate that Lens had selected subjects of particular importance and interest to the royal children, including landscapes, architecture (especially fortifications), and gardening, as well as military and naval scenes, some derived from drawing manuals of the period by Gerard de Lairesse and Charles-Alphonse Du Fresnoy.

Several works by the royal children survive showing how the exercises were addressed. A series of landscapes made by Princess Mary as a child, now in the Schloss Fasanerie, Fulda, Hesse in Germany are closely derived. The *Drawing Book* was still evidently in use during the arts lessons provided for Prince Frederick and Augusta's children.

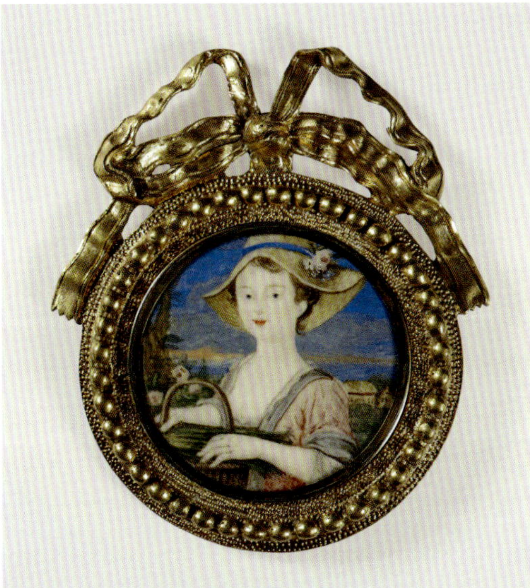

23.08

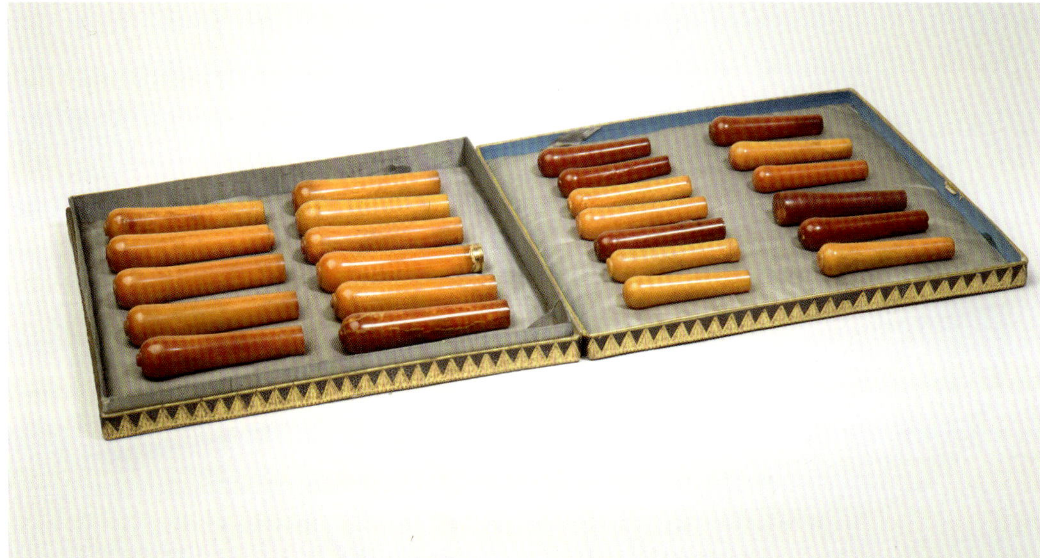

23.10

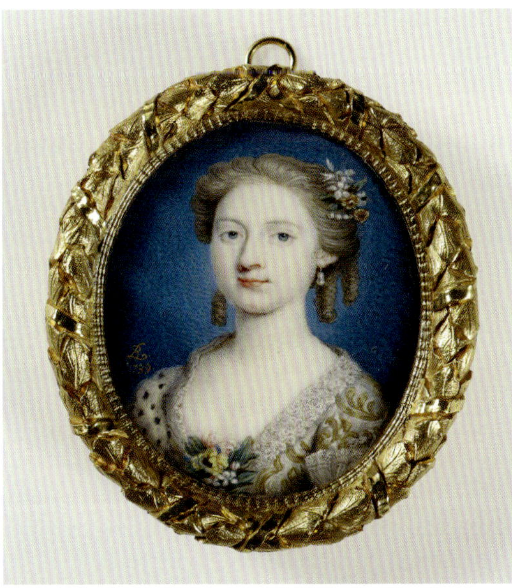

23.09

23.08†

PRINCESS MARY (1723–1772)
Louisa, Queen of Denmark (1724–1751) as Spring
1735–40
Watercolour on ivory
2½ in. (6.5 cm) diameter
Royal Collection Trust, RCIN 420900

Bernard Lens III served as drawing tutor to Caroline's younger children succeeding Philippe Mercier, who had taught Princesses Anne, Amelia, and Caroline. All the children achieved a competence as artists. Anne was perhaps the most adept. A number of her works, including a compelling self-portrait, and copies of old masters in the royal collection survive in the Royal Collection of the Netherlands. A copy of a painting by Mercier of a *Shepherd and Shepherdess*, made by Princess Caroline, is in the collection of the Earl of Wemyss at Stanway House in Gloucestershire. Lens's course of tuition for the younger children included exercises in perspective, as well as figure and nature studies. This miniature was made under his instruction and bears the inscription: 'This Portrait of Louisa, Queen of Denmark in the Character of Spring, was Drawn by her Sister Mary Princess of Hesse for her Royal Mother Queen Caroline, who wore it in her Snuff Box.'

23.09†

BERNARD LENS III (1681–1740)
Princess Louisa
1739
Watercolour on ivory
1⅞ x 1½ in. (4.7 x 3.9 cm)
Royal Collection Trust, RCIN 420180

His appointment as drawing tutor to Caroline's younger children, brought Bernard Lens III knowledge of the extensive collection of works by Isaac Oliver and Nicholas Hilliard laid out in Caroline's Picture Closet at Kensington Palace. In this miniature of Princess Louisa, Lens has adopted a deliberately anachronistic approach, reverting to the blue background characteristic of the sixteenth and seventeenth centuries. He may have been influenced specifically by Isaac Oliver's miniature of Princess Elizabeth, later Queen of Bohemia (RCIN 420031). Made to be set in a locket, this miniature would have then hung on a chain or a ribbon around the neck.

PRINCELY ARTS

23.10

PRINCESS ANNE OF HANOVER
(1709–1759)
Knife handles
1730–40
Turned amber and gold
3½ x⅞ in. (9 x 2.1 cm)
Royal Collection of the Netherlands, MU / 0346

Like all of George II and Caroline's children, Princess Anne was given instruction in lathe turning. The skill was part of the education of both boys and girls in the royal houses of Europe. An anonymous obituary when she died in 1759 noted that under the tutelage of the artist Kramer Anne 'progressed quite far in the turning of ivory and amber'. In 1740, after her marriage to William IV, King of the Netherlands, Frederick the Great of Prussia sent her some amber, writing 'Sachant que votre Altesse Royale se plait quelque fois à s'amuser des ouvrages de l'ambre, je prens la liberté de lui envoyer quelques morceaux les meilleurs j'ai pu deterre en Prusse. [Knowing that your Royal Highness is sometimes pleased to amuse herself in working amber, I take the liberty of sending her some of the best pieces I have found in Prussia.]' This set of turned amber knife handles may have been worked from Frederick the Great's gift. One has been given a gold mount, in anticipation of the knife blade being attached.

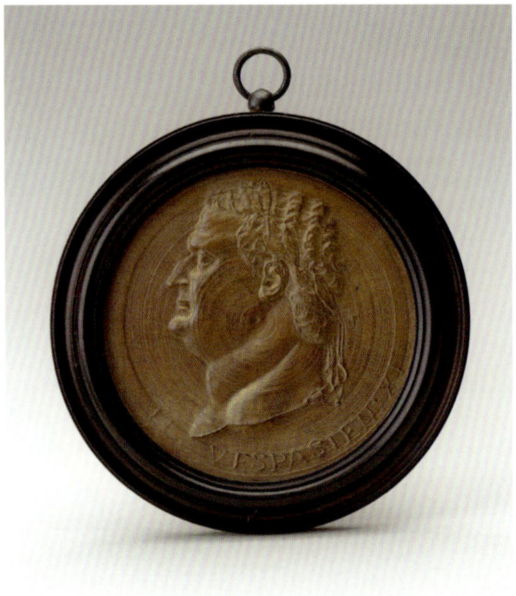

23.11

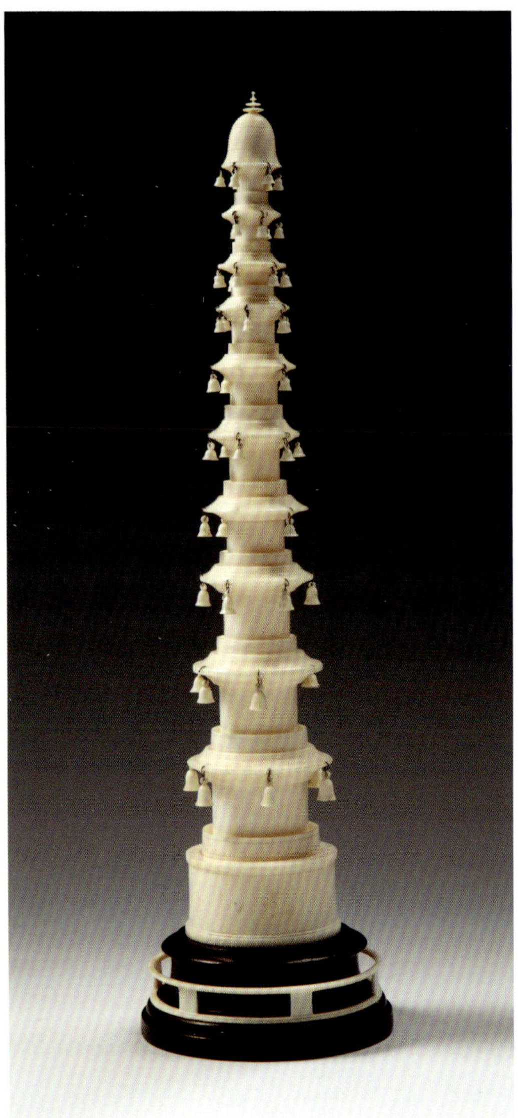

23.12

23.11

PRINCE WILLIAM AUGUSTUS, DUKE OF CUMBERLAND (1721–1765)
Emperor Vespasian
1731
Turned boxwood, with ebonized wood frame
3⅝ x 3¼ x ⅜ in. (9.1 x 8.4 x 1.0 cm)
Royal Collection Trust, RCIN 70396

According to a note on the reverse, Prince William Augustus, Duke of Cumberland, youngest son of George II and Caroline, turned this relief at Barn Elms, the home of the diplomat and courtier Stephen Poyntz, who was employed from midsummer 1731 as governor and steward to the household of the ten-year-old Duke. The plaque has been turned on a portrait-copying lathe. It is set in an ebonized frame with a suspension ring. A bill from John Bradburn records that a wheel-turning lathe was supplied to Augusta, Princess Dowager of Wales, in 1769 at a cost of £22 13s. George III certainly learned the skill, as at least two boxes by his hand survive, one in the Royal Collection (RCIN 65779) and the other in the Herzog Anton Ulrich Collection, Brunswick.

23.12

PRINCESS LOUISA OF HANOVER
(1724–1751)
Miniature pagoda
1740
Turned ivory on a stand of ebony and ivory
9½ in.; incl. stand 10½ in. (24.2 cm; incl. stand 26.6 cm)
The Royal Danish Collection, Rosenborg Castle, Chronological Collection of the Danish Kings, Copenhagen, 13-424

The pagoda was carved by Princess Louisa, daughter of George II and Caroline. Ten of the eleven storeys are adorned by carved, hanging bells, and the piece is finished with a thimble shaped cupola. Following her marriage to Frederick V of Denmark and Norway, Princess Louisa took lessons in ivory turning from Lorenz Spengler (1720–1807), who had earlier operated from a workshop in London. Three turnings attributed to her survive, and are presumed to have been made as gifts to her mother-in-law, Queen Dowager Sophie Magdalene, in whose effects this piece is listed in 1772. In 1775 it was transferred to the Royal Danish *Kunstkammer* where it was recorded in inventories made by Spengler.

BOOK LEARNING

23.13

PRINCESS ANNE OF HANOVER
(1709–1759)
'Istoria delle rivolutioni accadute nella Romana Republicia, Vol. II Lib. V'
1720–30
Manuscript, bound
9¾ in. (25 cm)
Inscribed 'Extrait Histoire de L'Empire Romain, etc . . . extrait par Anne'
Royal Collection of the Netherlands, Archive Anne of Hannover, A30-VII-470-II
Exhibited: Title page (illustrated)

23.14

PRINCE WILLIAM AUGUSTUS, DUKE OF CUMBERLAND (1721–1765)
The Original Exercise Book of HRH the Duke of Cumberland in his own Handwriting
1727–28
Manuscript, bound
Approx. 7⅞ in. (20 cm)
British Library, Add. MS. 16941
† Exhibited: Title page (not illustrated)
* Exhibited: p. 2 (Latin exercises, headed 'J. T. Phillips Guilielmo Serenissime Cumberlandia and Epistola stile simplice atque facili . . .'); (illustrated)

Six bound schoolbooks belonging to Princess Anne, the eldest daughter of George II and Caroline, are preserved in the Royal Archive of the Netherlands in The Hague. One volume contains a lengthy transcription in Italian of the history of the Roman Empire, and an addendum with grammatical exercises. Anne used her own thorough programme of instruction as a basis for the education she devised for her own children.

The exercise book belonging to six-year-old Prince William Augustus, Anne's youngest brother, is in the British Library. It contains a series of letters composed by the prince's tutor Jenkin Thomas Phillips for him to copy. From 1727 to his death in 1755, Phillips served George II as Historiographer.

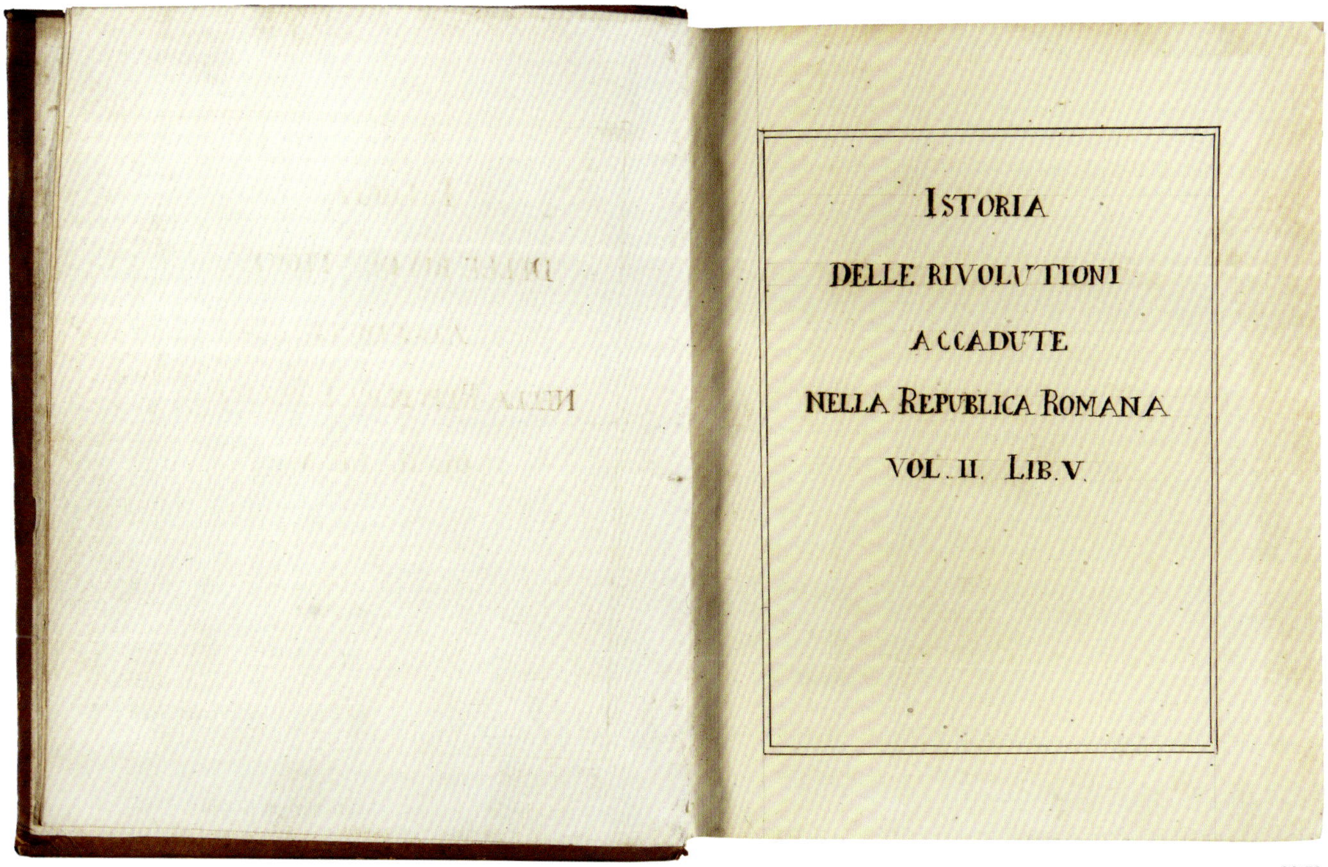
ISTORIA
DELLE RIVOLUTIONI
ACCADUTE
NELLA REPUBLICA ROMANA
VOL. II. LIB. V.

23.13

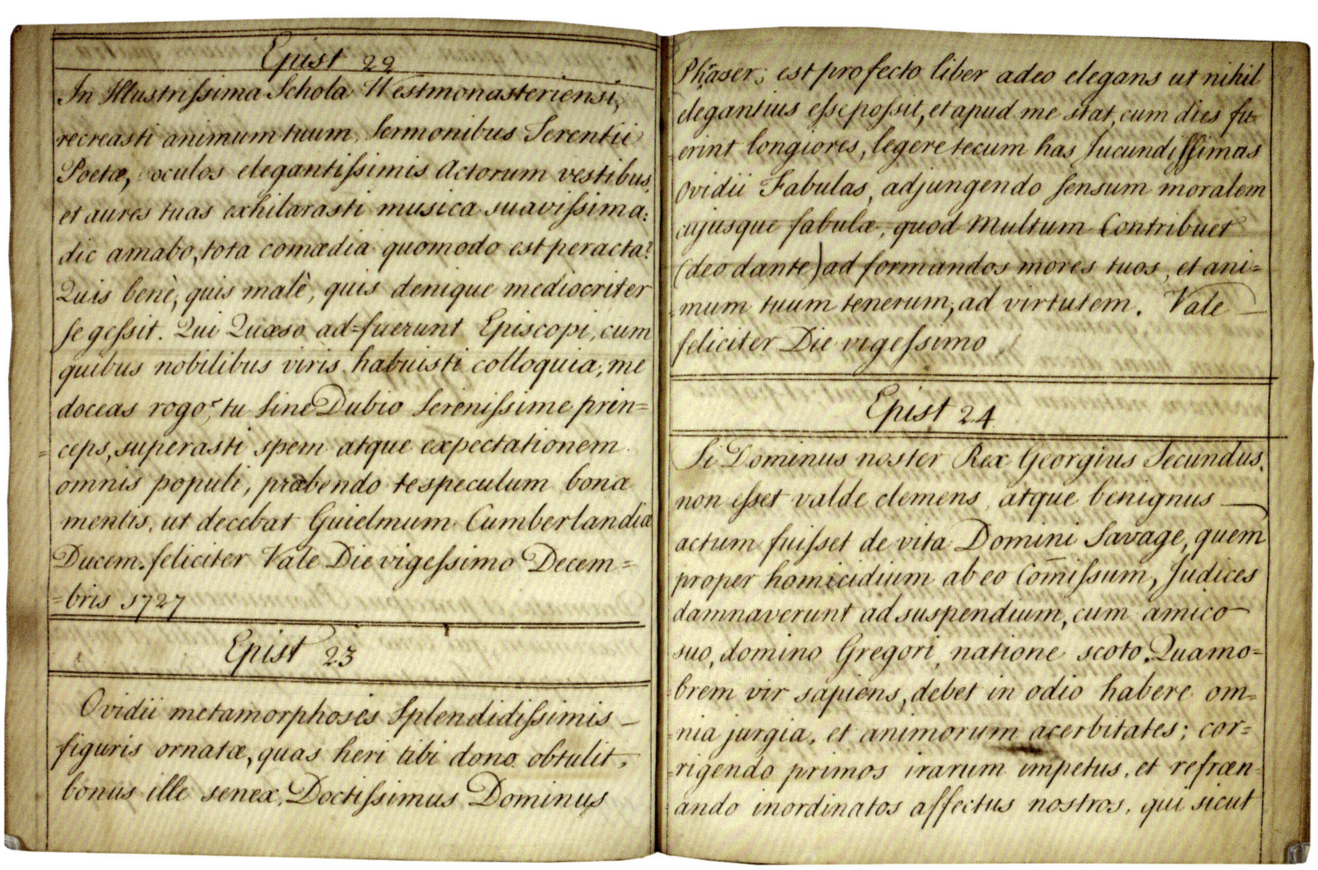
Epist 22
In Illustrissima Schola Westmonasteriensi, recreasti animum tuum Sermonibus Terentii Poetæ, oculos elegantissimis Actorum vestibus et aures tuas exhilarasti musica suavissima: dic amabo, tota comœdia quomodo est peracta? Quis benè, quis malè, quis denique mediocriter se gessit. Qui Quæso ad-fuerunt Episcopi, cum quibus nobilibus viris habuisti colloquia; me doceas rogo. tu sine Dubio Serenissime princeps, superasti spem atque expectationem omnis populi, præbendo te speculum bonæ mentis, ut decebat Guielmum Cumberlandiæ Ducem. feliciter Vale Die vigessimo Decembris 1727
Epist 23
Ovidii metamorphoses Splendidissimis figuris ornatæ, quas heri tibi dono obtulit, bonus ille senex, Doctissimus Dominus
Pkäser; est profecto liber adeo elegans ut nihil elegantius esse possit, et apud me stat, cum dies fuerint longiores, legere tecum has Jucundissimas Ovidii Fabulas, adjungendo sensum moralem cujusque fabulæ, quod Multum Contribuet (deo dante) ad formandos mores tuos, et animum tuum tenerum, ad virtutem. Vale feliciter Die vigessimo
Epist 24
Si Dominus noster Rex Georgius Secundus, non esset valde clemens atque benignus actum fuisset de vita Domini Savage, quem propter homicidium ab eo Commissum, Judices damnaverunt ad suspendium, cum amico suo, domino Gregori, natione scoto. Quamobrem vir sapiens, debet in odio habere omnia jurgia, et animorum acerbitates; corrigendo primos irarum impetus, et refrænando inordinatos affectus nostros, qui sicut

23.14

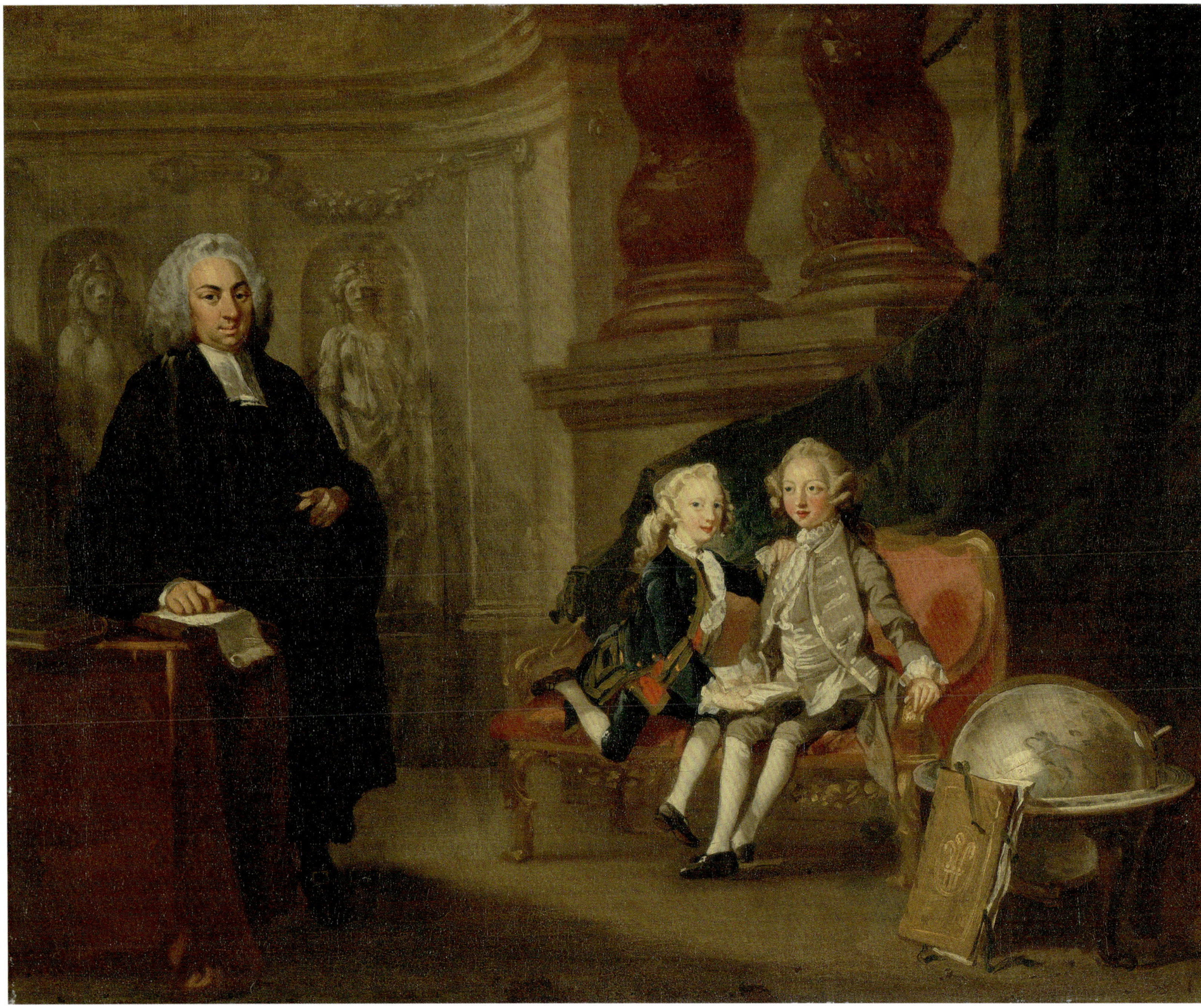

23.15

AUGUSTA AND EDUCATING THE YOUNG

23.15

RICHARD WILSON (c. 1714–1782)
Francis Ayscough with the Prince of Wales (later King George III) and Edward Augustus, Duke of York and Albany
c. 1749
Oil on canvas
25 x 30⅛ in. (63.5 x 76.5 cm)
Yale Center for British Art, Paul Mellon Collection, B1981.25.689

It was probably through his early patrons, the Lytteltons of Hagley, Worcestershire, that Richard Wilson received the commission for this painting from the Reverend Dr Francis Ayscough, then tutor to the children of Frederick, Prince of Wales. The large picture of Ayscough with Frederick's two eldest sons, Prince George and Prince Edward Augustus, dated 1748–49, is in the collection of the National Portrait Gallery, London. There are two smaller autograph versions, one at Tate Britain, London, and the other in the Yale Center for British Art. Ayscough was Clerk of the Closet to Frederick and Preceptor to his and Princess Augusta's sons, George and Edward. In 1749 George Lewis Scott, a mathematician and member of the Royal Society, was employed as an assistant to Ayscough. Through him George III became the first British monarch to have a scientific education.

23.16

George III (1738–1820)
A Doric Temple
c. 1760
Pen and ink over pencil and wash on paper
22½ x 15½ in. (57.2 x 39.4 cm)
Royal Collection Trust, RCIN 980479

In addition to receiving tuition in drawing from Joshua Kirby, it is evident that from the early 1750s George, Prince of Wales, received instruction in architectural drawing from William Chambers. Chambers in fact complained as he sought to balance his various responsibilities: 'my hands are full of work, but my pockets are not filld with money. The prince employs me three mornings a week to teach him Architecture: the buildings and other decorations at Kew fill up the remaining time.' Prince George's designs show the classical orders of architecture, and there is a series of designs for temples and another for domestic buildings.

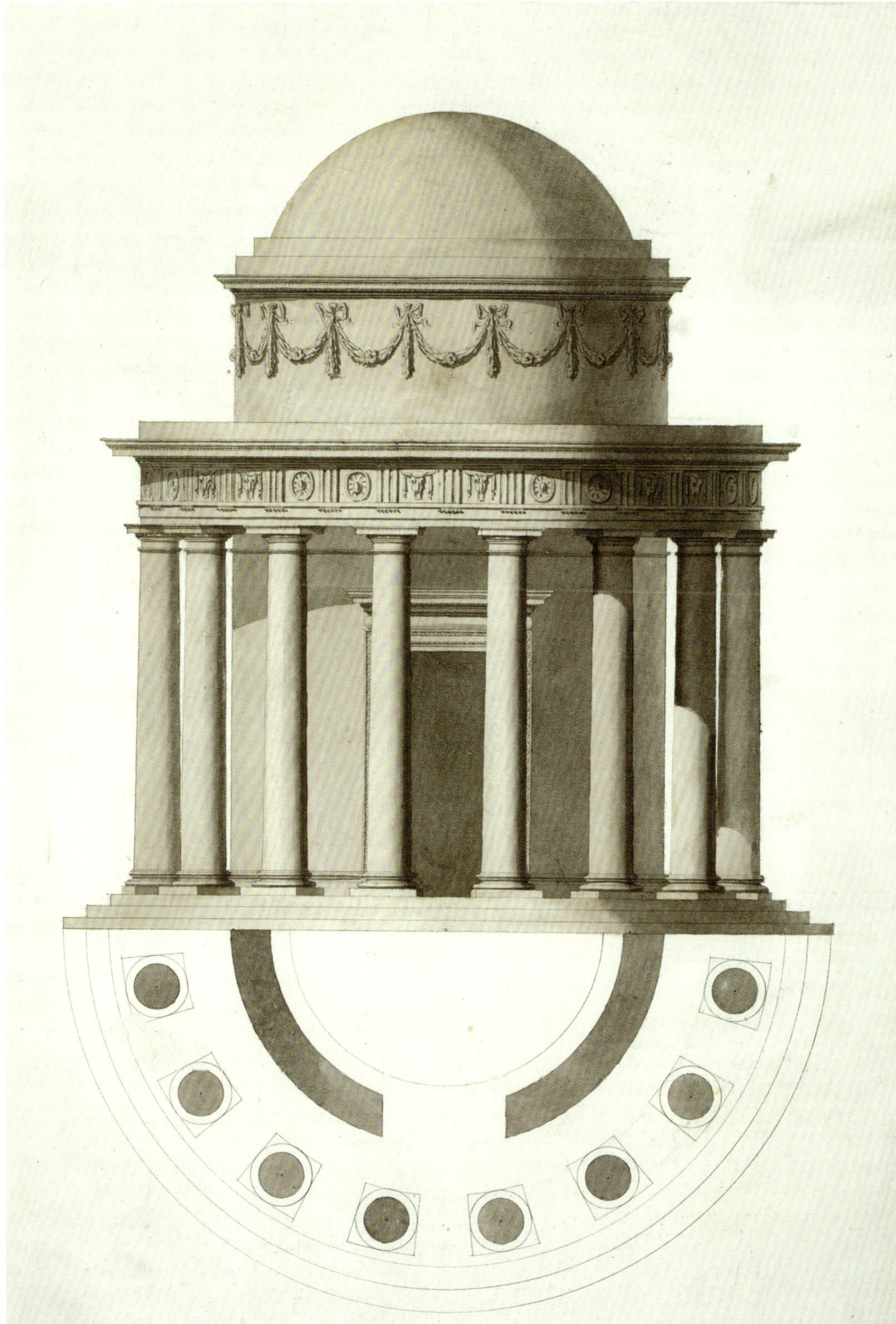

23.16

They seem to have been derived from work by Kirby and Chambers, and they are skilfully and delicately rendered. This design is for what may be George III's first architectural project. According to Henry Angelo, whose father taught fencing to the royal children, 'a small temple, erected at Kew Gardens, which is much admired, and is engraved in the folio works of Sir William Chambers, is said to have been entirely the design of the late King, whilst Prince of Wales.'

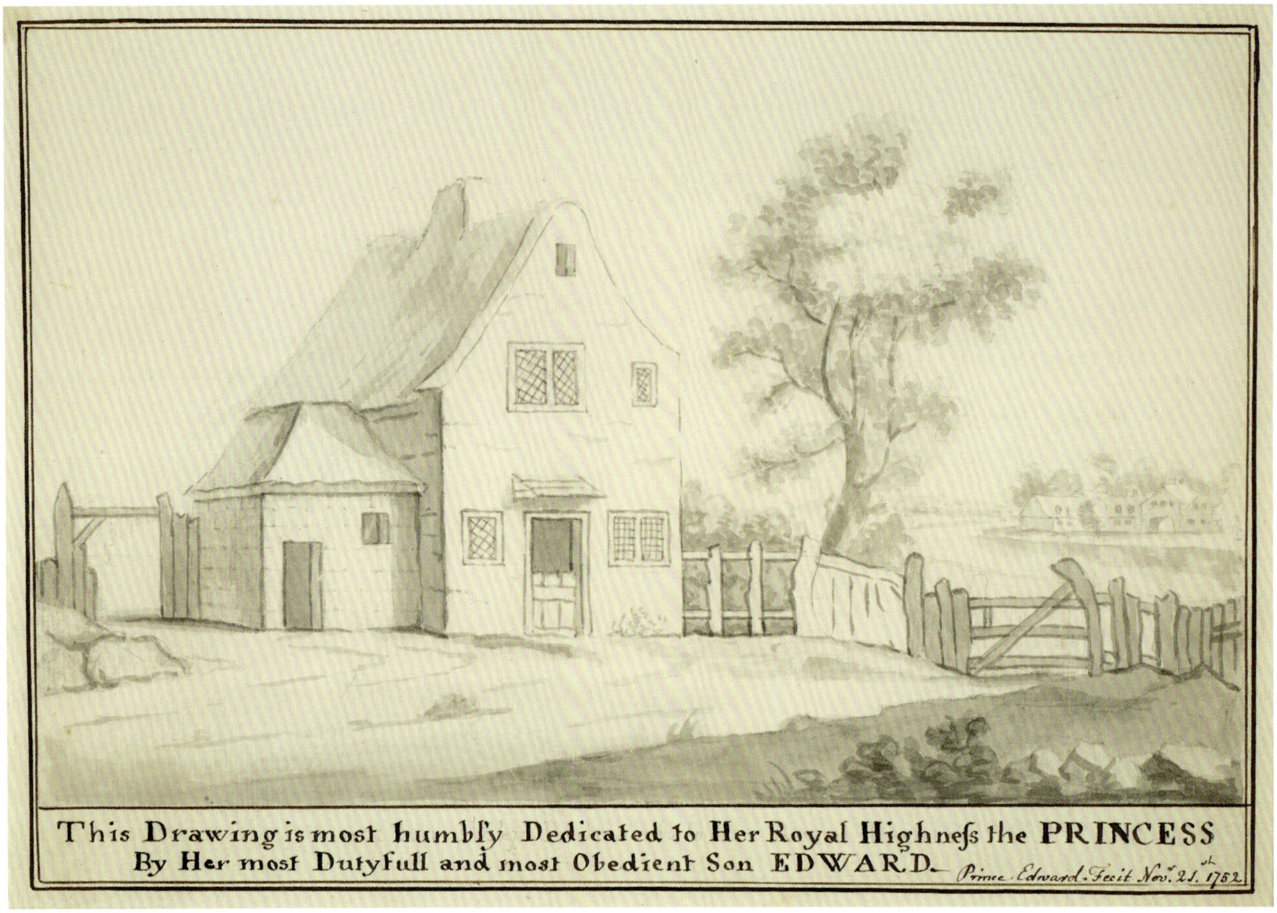

23.17

23.17

PRINCE EDWARD AUGUSTUS, DUKE OF YORK (1739–1767)
A Cottage
1752
Pencil, wash, and pen and ink
10⅝ x 13⅞ in. (26.9 x 35 cm)
Inscribed: 'This Drawing is most humbly Dedicated to Her Royal Highness the PRINCESS By Her most Dutyfull and most Obedient Son EDWARD- Prince Edward Fecit Nov.r 21.st 1752'
Royal Collection Trust, RCIN 980199

The manual prepared by Bernard Lens III for the art education of the children of George II and Caroline continued to provide useful exercises for the children of Prince Frederick and Augusta, during lessons supervised by Joseph Goupy. This drawing of a cottage made by Prince Edward is derived directly from a model in Lens's manual, and was intended to encourage the pupil to explore the principles of perspective.

CHARLOTTE AND EDUCATING THE YOUNG

23.18†

BENJAMIN WEST (1738–1820)
Princess Augusta, Princess Elizabeth, Prince Ernest, Prince Augustus, Prince Adolphus, and Princess Mary
1776
Oil on canvas
66 ¼ x 71 in. (168.2 x 180.3 cm)
Royal Collection Trust, RCIN 404574

23.19*

VALENTINE GREEN (1739–1813), engraver, after BENJAMIN WEST (1738–1820)
Children of George III and Queen Charlotte
1778
Mezzotint
23 x 33 in. (58.4 x 83.8 cm)
Historic Royal Palaces, HRP3000791

Benjamin West, described after 1772 as 'Historical Painter to the King', also provided painting lessons for the children of George III and Charlotte. Charlotte and twelve of her children would be the subject of five double or group portraits commissioned from West between 1776 and 1778 that were hung on completion in the King's Closet at St James's Palace. Every portrait is filled with action, instruction, and affection. In this scene the Princes Ernest and Augustus are pulling along a little cart in which the infant Princess Mary is supported by her elder sisters, Princesses Augusta and Elizabeth, and brother Prince Adolphus. All the children are depicted dressed informally in styles designed for comfort and ease. The two eldest boys wear open-necked shirts under their suits. The youngest boy wears a dress. The pink dress worn by Princess Augusta still has leading reins attached to the shoulders. A small spaniel plays with a ball in front of a toy drum. West was paid 300 guineas for this work which he exhibited at the Royal Academy in 1777.

The mezzotint is one of several Green made after paintings by Benjamin West in his capacity as Engraver to the King. It bears the royal coat of arms and an inscription below, including the publisher's address: 'Publish'd Septr: 29th, 1778 by John Boydell Engraver in Cheapside London'.

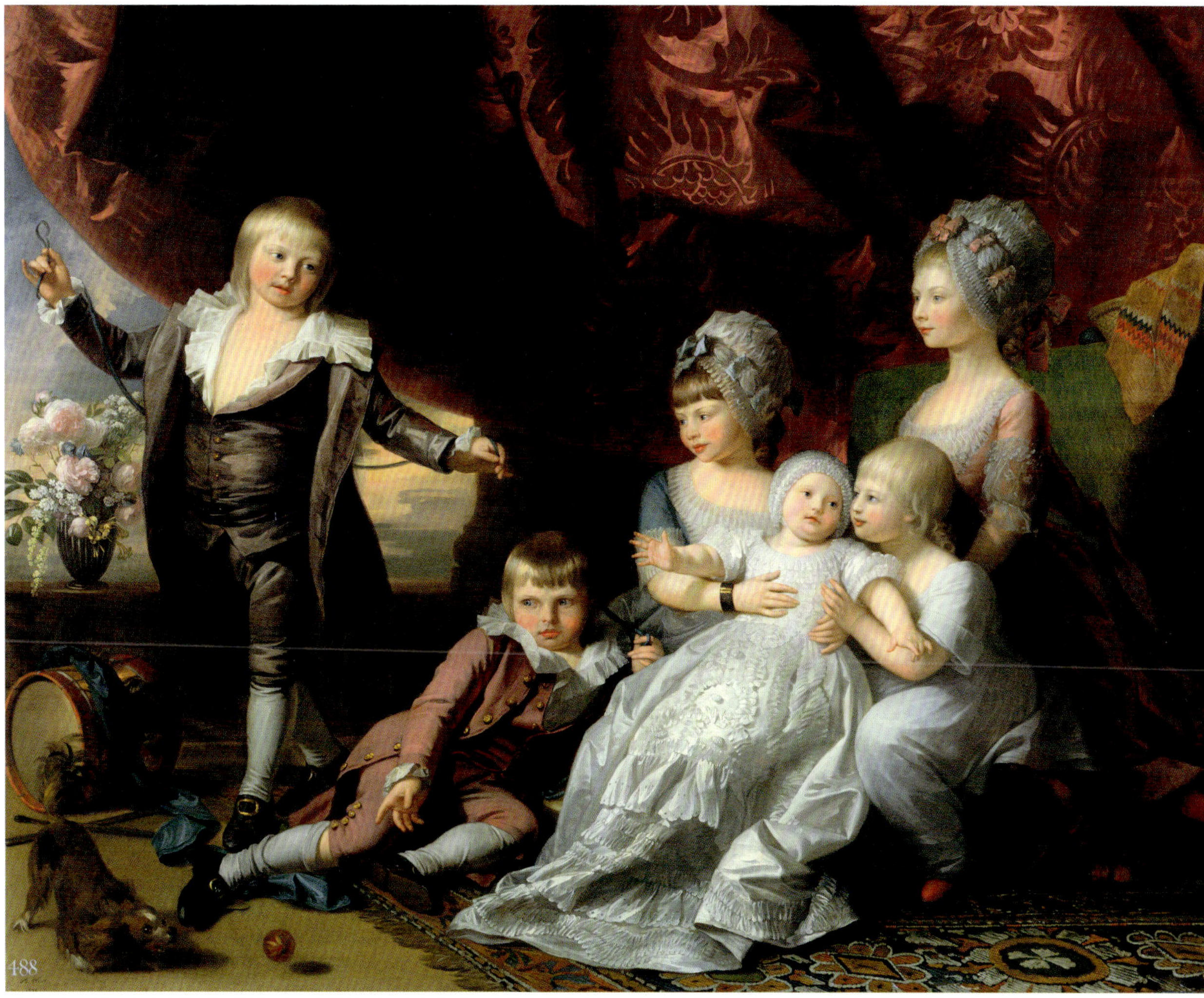

23.18

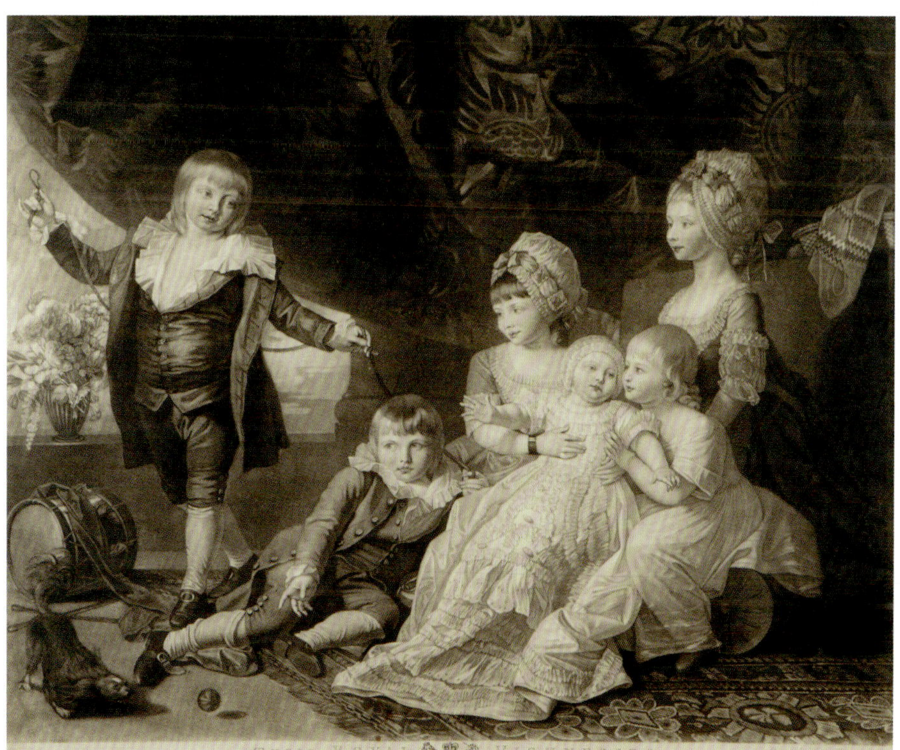

23.19

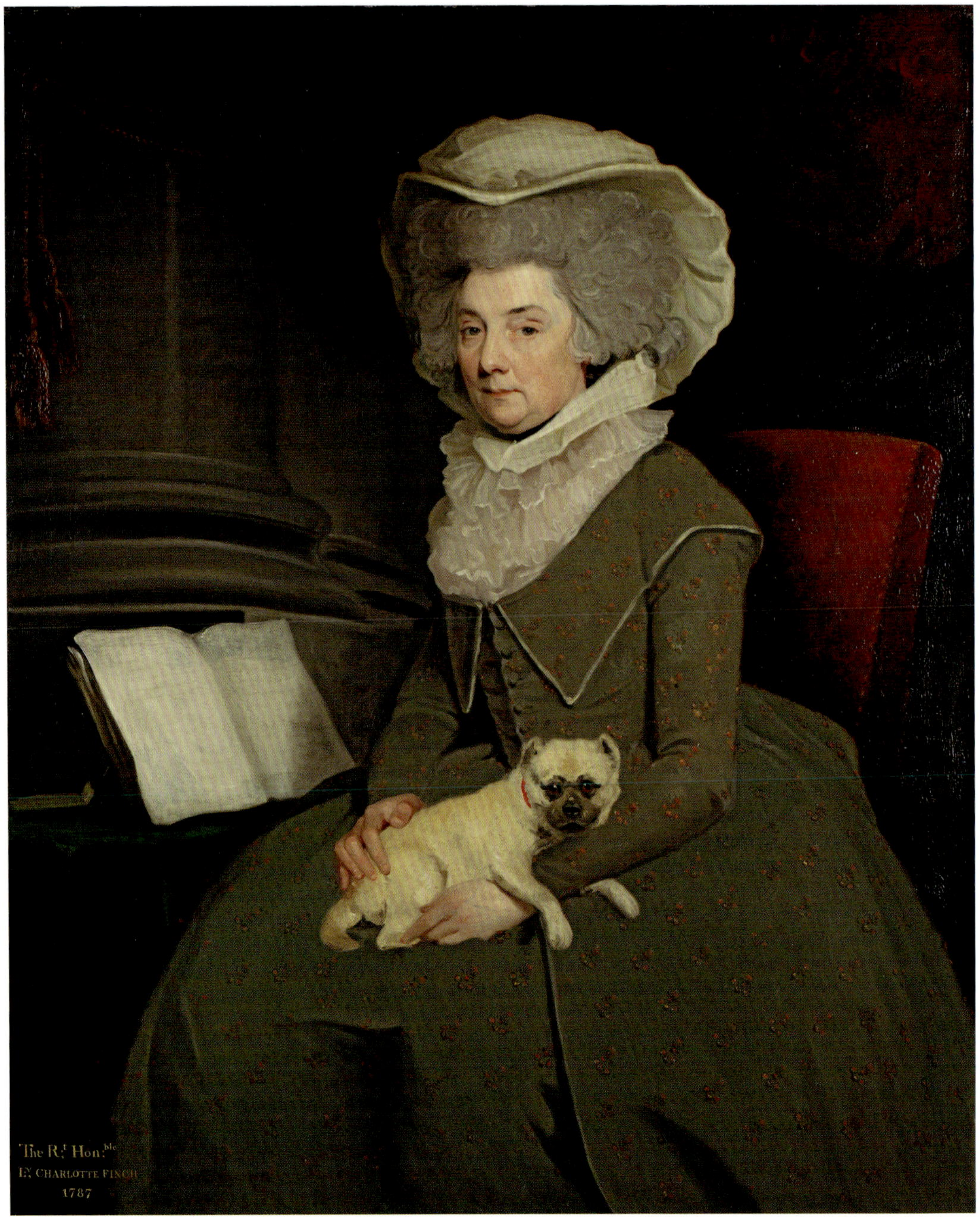

23.20

23.20*

WILLIAM HOPKINS (active 1787–1811)
Lady Charlotte Finch (née Fermor), Governess to the Children of King George III
1787
Oil on canvas
49 x 39¾ in. (124.5 x 101 cm)
UK Government Art Collection, GAC 3810

Lady Charlotte Finch (1725–1813), royal governess, was one of the daughters of the Earl and Countess of Pomfret, of Easton Neston, Northamptonshire. Both her parents had held court appointments: her father was Master of the Horse to Queen Caroline and her mother was one of Caroline's Ladies of the Bedchamber. In 1746 Charlotte married William Finch, Vice-Chamberlain of the King's Household from 1742. Charlotte and he had four daughters and a son. Lady Finch was appointed royal governess on the birth of George, Prince of Wales, in August 1762. Horace Walpole described her appointment as 'a choice so universally approved that I do not think she will be abused even in North Briton'. She would be responsible for the Prince of Wales and his seven brothers until they were set up in their own establishments with governors, and for the six princesses until they were twenty-one. She managed the nursery and supervised the staff assigned to each child, including the sub-governess, personal attendants, assistant governesses, and other teachers, some of whom were drawn from Queen Charlotte's attendants. The portrait was probably commissioned by her family and hung in the family home, Burley House, in Rutland.

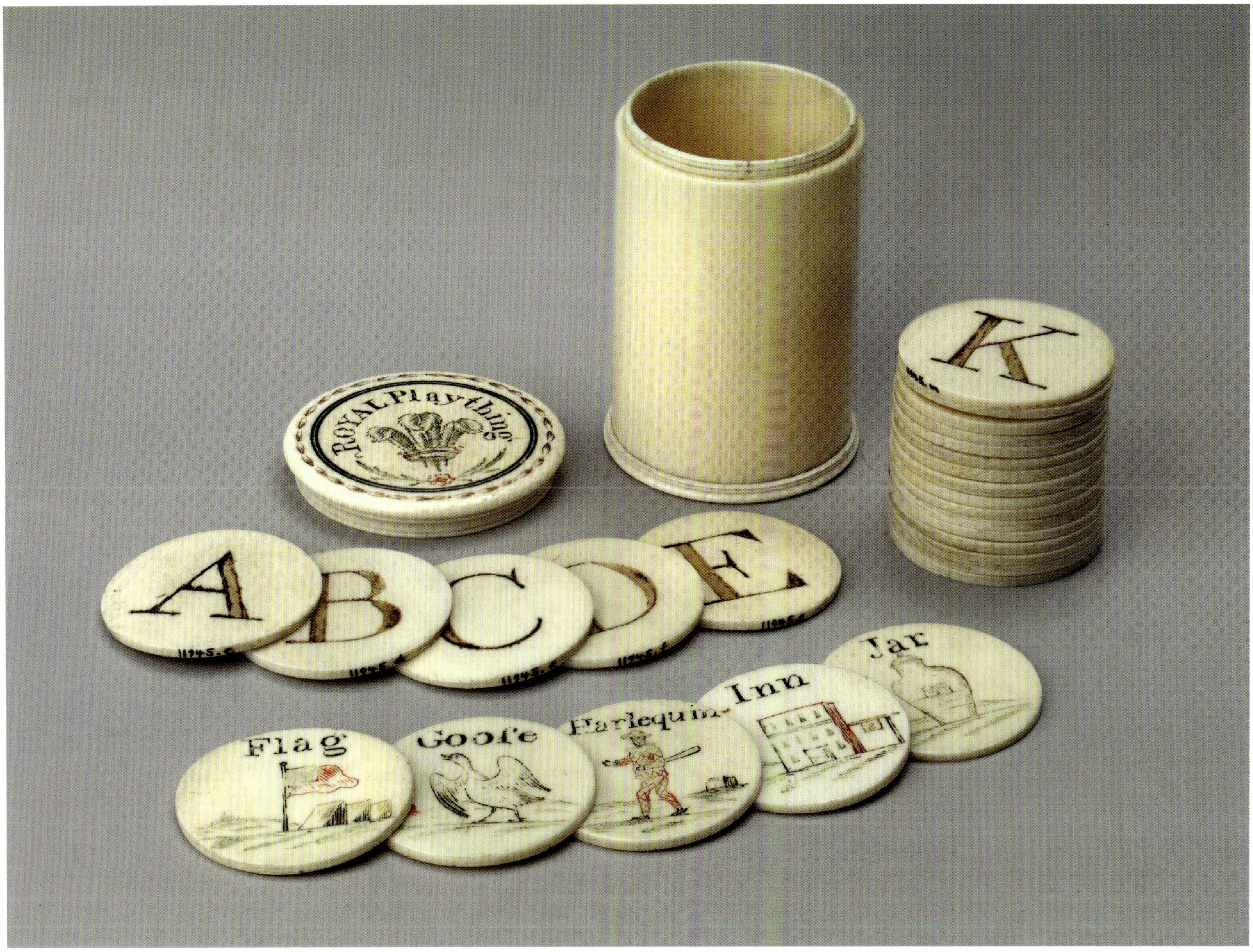

23.21

23.21†

UNKNOWN MAKER
Teaching counters
1760–1800
Twenty-six circular counters each with an incised and painted capital letter
Bone or ivory
Container: 2⅞ x 2 x 1⅝ in. (7.4 x 5.1 x 4 cm);
Counter: 1⅝ in. (4cm) diameter
Royal Collection Trust, RCIN 11945.a–ab

Lady Charlotte Finch, governess to the children of George III and Charlotte, supervised a progressive nursery focused on child-centred learning. At a time when educational philosophers encouraged aristocratic mothers to participate more actively in their children's education, her skill in teaching her own children was much praised. The royal children's instruction incorporated creative play, most of the arts and sciences, as well as the more traditional subjects and accomplishments. The court thus helped to set a fashion for new educational theories and practices. Lady Finch was an advocate of educational philosophies promoted by Ellenor Fenn, who compiled a number of games and other teaching aids. *The Art of Teaching in Sport* (1785?) was designed to accompany 'a set of toys, for enabling ladies to instil the rudiments of spelling, reading, grammar, and arithmetic, under the idea of amusement'. Lady Finch deployed flash cards to encourage skills in mental arithmetic, grammatical construction, and vocabulary in the royal nursery. This set of counters seems to have been a spelling aid.

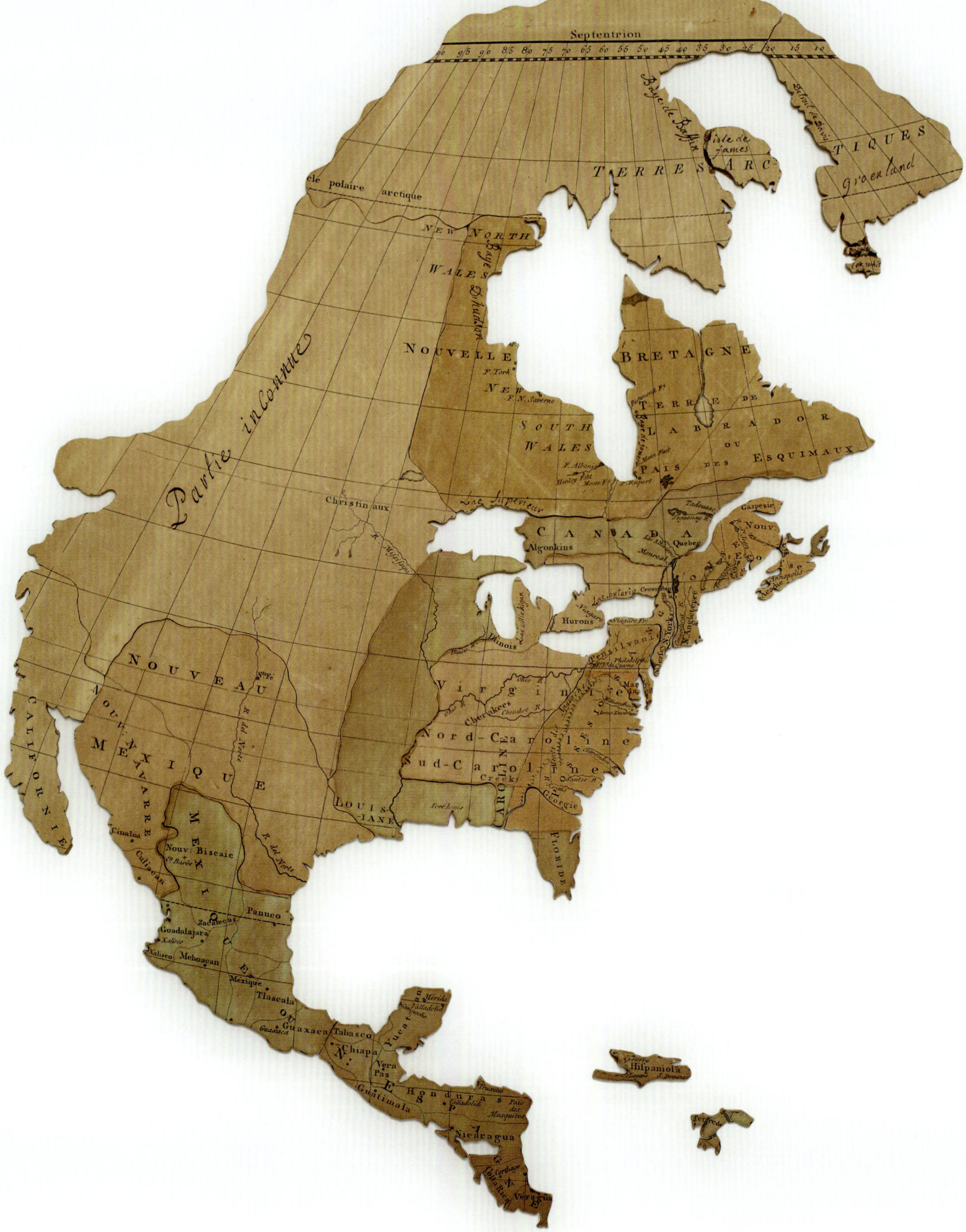
Septentrion
Partie inconnue
NOUVELLE
BRETAGNE
TERRE DE LABRADOR OU PAIS DES ESQUIMAUX
CANADA
NOUVEAU MEXIQUE
CALIFORNIE
LOUISIANE
Nord-Caroline
Sud-Caroline
FLORIDE
Hispaniola
Nicaragua
Honduras
Guatimala
Tabasco
Chiapa
Mexique
Panuco
Guadalajara
Zacatecas
Tlascala
Guaxaca
Cinaloa
Culiacan
Hurons
Algonkins
Quebec
Christinaux
Creeks
Cherokees
Chiapa
groenland
TERRES ARCTIQUES

23.22

23.22

LADY CHARLOTTE FINCH (1725–1813), designer, after **JEAN PALAIRET** (1697–1774), cartographer
Group of dissected maps (Continent of America)
c. 1755
Engraving mounted on thin mahogany, cut to shape
18½ x 14⅛ in. (47 x 36 cm)
Victoria and Albert Museum, B.6-2011/ Historic Royal Palaces, 3004942, Presented by the Art Fund jointly to the V&A Museum of Childhood and the Historic Royal Palaces

This dissected map is one of eight constructed from plates in the first edition (1755) of Jean Palairet's *Atlas méthodique composé pour l'usage de son altesse sérénissime monseigneur le prince d'Orange et de Nassau, Stadhouder des Sept Provinces-Unies* . . . This was a set of maps designed for educational purposes. The maps of the continents of Europe, Asia, Africa, and North and South America would have been cut up to Lady Charlotte Finch's direction, and the resulting jigsaws were housed in her puzzle cabinet commissioned in her capacity as Governess to the children of George III and Charlotte. The cabinet also contained two puzzles cut from maps prepared by John Spilsbury that were intended for a juvenile audience and were purposefully designed for dissections.

23.23†

GEORGE, PRINCE OF WALES, LATER GEORGE IV
Lessons and Songs (2 vols)
c. 1770
Manuscript
7⅞ in. (20 cm)
British Library, RM 23.g.20
Exhibited: Vol. 2: ff. 13v–14r ('Shepherd I have lost my love . . .'); (not illustrated) and ff. 4v–5r ('Ye shepherds give ear to my lay. . .'); (illustrated)

The first of the two volumes, which are bound as a set, is a lesson book containing scales and simple figuring exercises for the harpsichord appropriate for the musical education of George, Prince of Wales, the eldest son of George III and Queen Charlotte, as a young child. The second volume contains songs and ballads, and may have been used for family music making. The songs are in both English and Italian, and include compositions by Felice Giardini, a friend of J. C. Bach, C. F. Abel, and J. C. Fischer, who was a talented vioinist and composer.

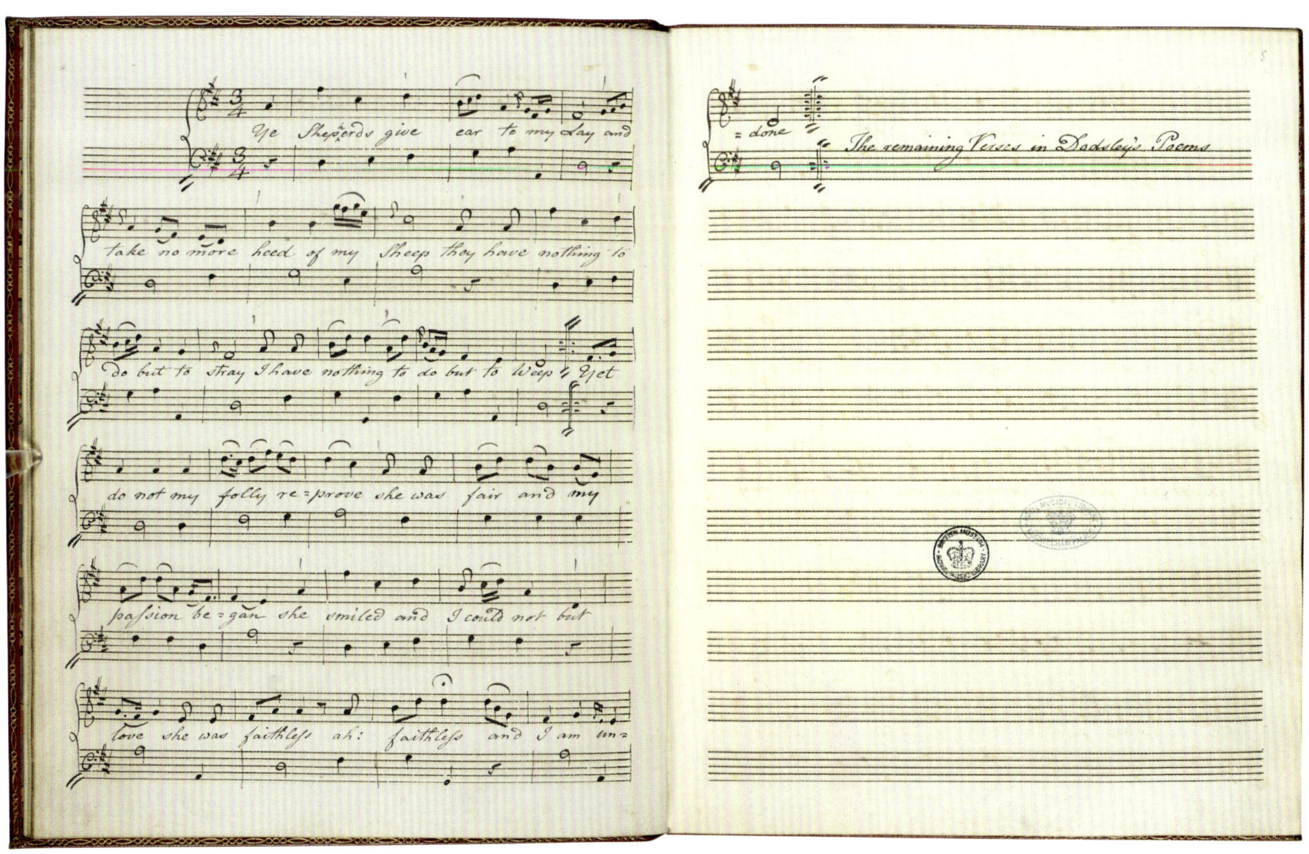

23.23

23.24

Royal children's music book
1787–89
Manuscript, bound
9½ in. (24 cm)
Beinecke Rare Book & Manuscript Library, Yale University, OSB MSS 146, box 183, folder 944
Exhibited: 'Yankey [*sic*] Doodle'

While the children of Frederick and Augusta and George III and Charlotte may not have matched the children of George II and Caroline in musical skill, they were all given a thorough musical education, including in singing. Song books surviving from the royal music libraries include fashionable contemporary works in Italian and French, but also many popular and traditional songs, such as 'Yankee Doodle'. With its origins in a middle Dutch harvest song of the fifteenth century, it had become as popular song both in Britain and America by the late eighteenth century. During the Seven Years' War it had been sung both by British forces to mock what they regarded as the disorganized 'Yankee' troops, and by the 'Yankee' troops as a song of defiance against the British, to celebrate their own military leaders.

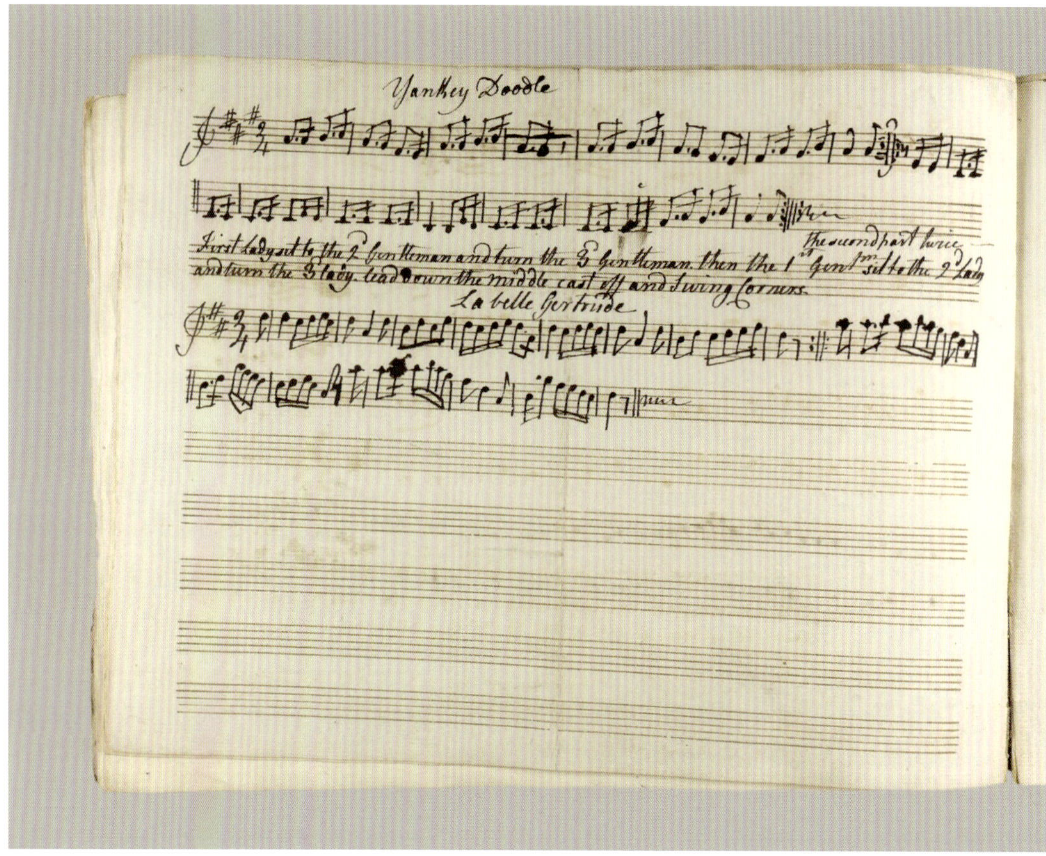

23.24

23.25

UNKNOWN MAKER
Child's shoes worn by Princess Augusta, daughter of George III and Queen Charlotte
Silk and silk ribbon (uppers), kid (lining), and leather (sole)
Each: Length: approx. 4⅜ (11.1 cm), height: approx. 1⅝ in. (4.2 cm)
Royal Collection Trust, RCIN 71938.1–2

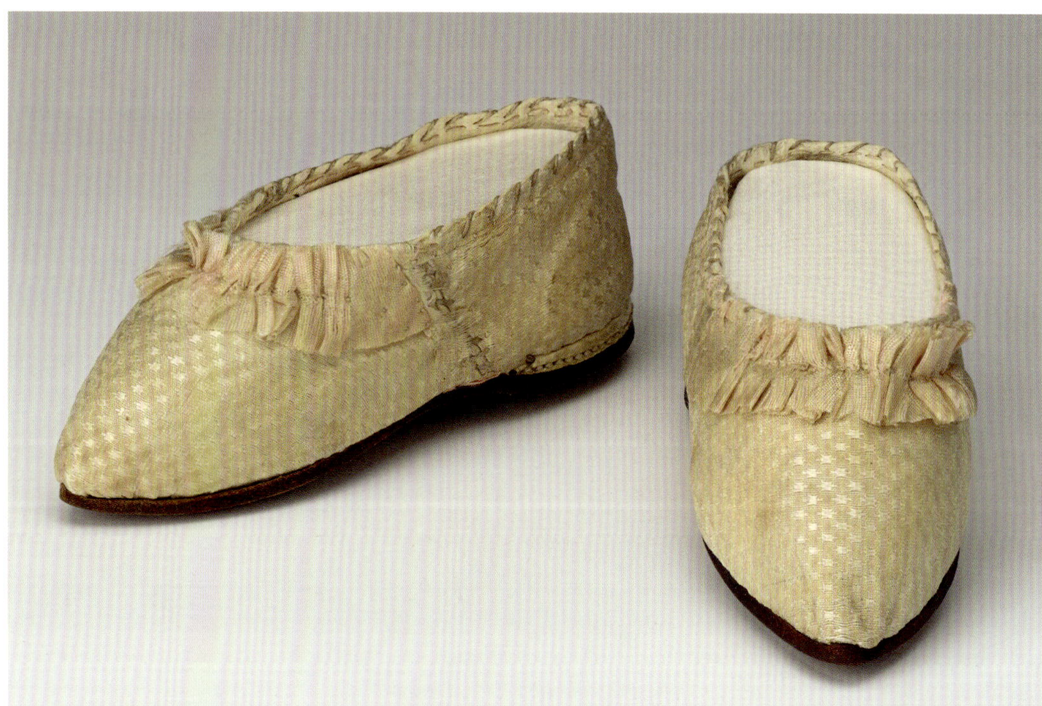

23.25

23.26

UNKNOWN MAKERS

Baby robe and bonnet belonging to George, Prince of Wales (later George IV)
Robe: Ivory silk satin and silk satin ribbon.
Bonnet: Fine cream-coloured cotton and Hollie point needle lace (not illustrated)
Robe: bust/chest: 17½ in. (44.5 cm); waist: 18½ in. (47 cm); length: 39⅞ in. (101.4 cm)
Bonnet: depth: 6½ in. (16.5 cm); circumference: 12⅝ in. (32 cm)
Historic Royal Palaces, HRP 3501059; HRP 3501055

It is not until the mid-eighteenth century that examples of dress worn by the royal children survive in any number. Mrs Scott, wet nurse to George, Prince of Wales, the eldest son of George III and Charlotte, carefully preserved pieces of dress worn by the infant prince. The robe is of simple construction, fastening behind with ribbon bows, allowing the baby to be swiftly dressed to be shown off to admiring courtiers. The tiny bonnet has a needle lace inset set into the back, worked with the design of three ostrich feathers, the Prince of Wales's heraldic signifier. A second bonnet made of fine wool trimmed with ruched silk ribbon would have been more appropriate in cold weather. There is also a baby corset, of fine linen covered with a bright red silk, and stiffened with cotton twine stitched tightly into channels. The wearing of a supportive garment like this was thought to encourage the healthy development of the baby's back. Princess Augusta's silk-covered shoes must have been made in about 1770, when she was about two years old. The princess is depicted at about this age, in her mother's arms, in Johan Zoffany's painting *George III, Queen Charlotte and their Six Eldest Children* completed in 1770 (RCIN 400501).

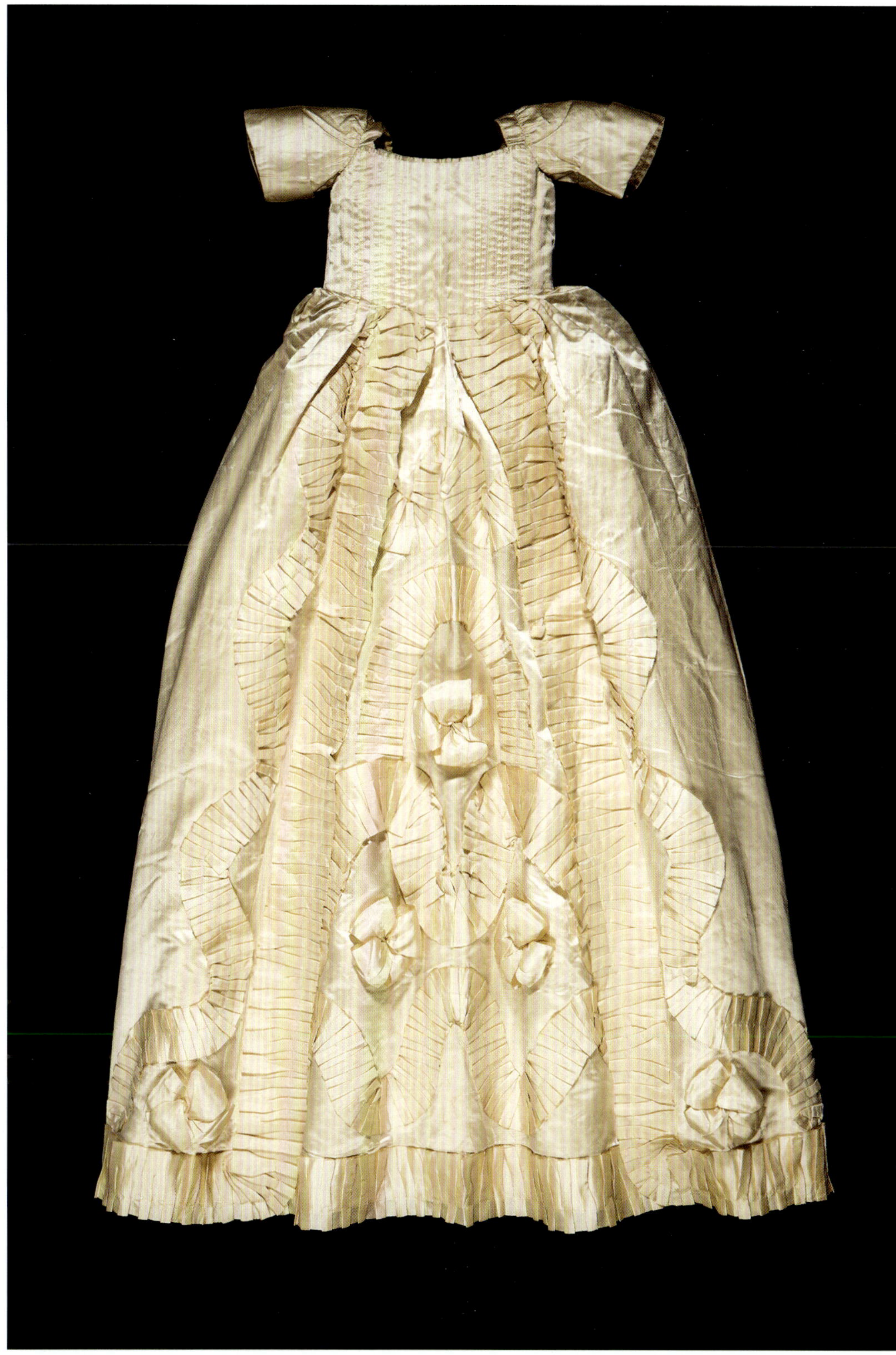

23.26

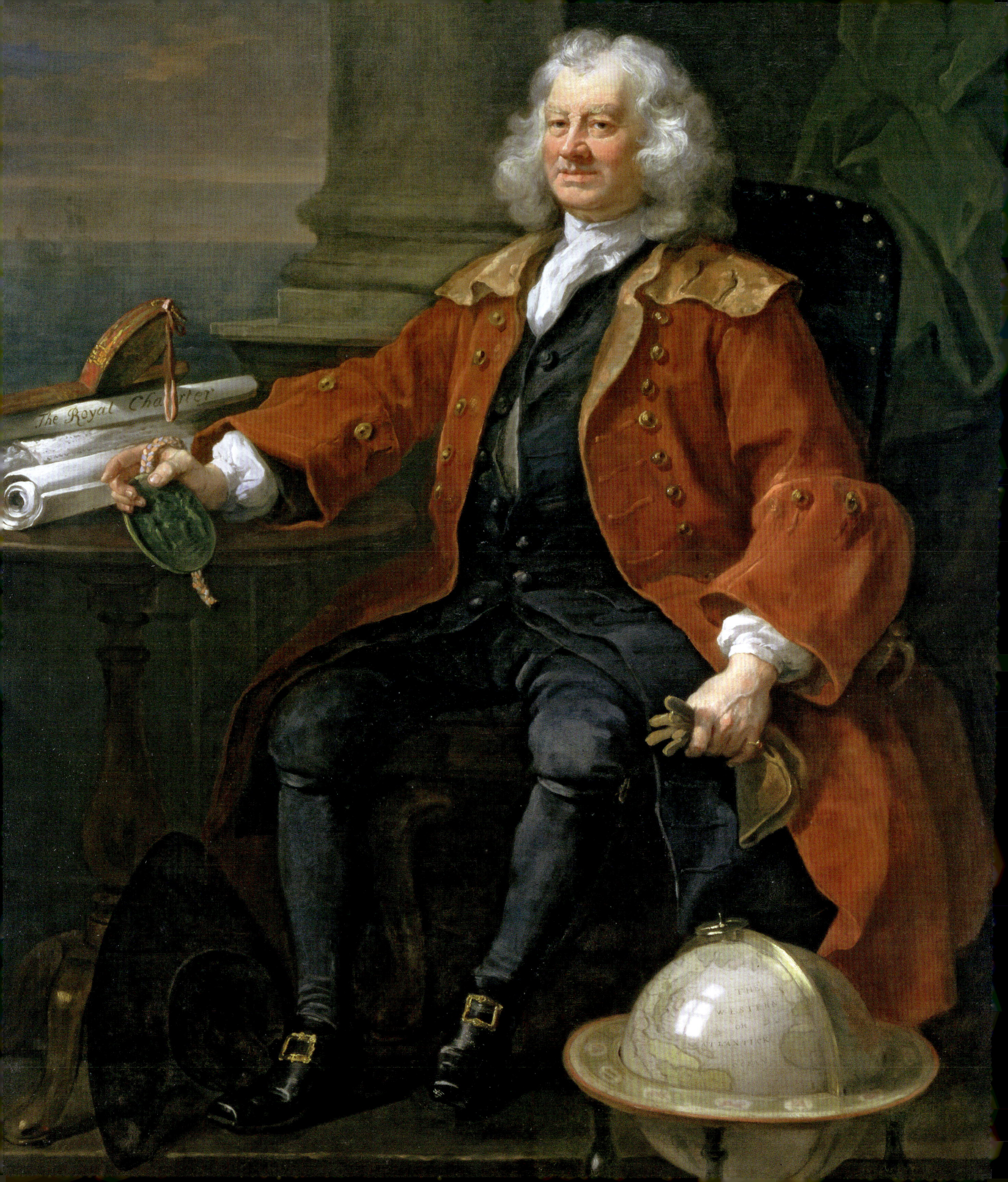
The Royal Charter

Charity and Social Work

Frank Prochaska

In the early eighteenth century English and German royalty could look back on a long, if limited, tradition of charity, which over the decades was to evolve in ways advantageous both to the British crown and to philanthropy.[1] In Protestant Germany the Pietist movement encouraged individual religious observance and the living of an active Christian life, which put a premium on good works. Such traditions fitted well with those of England, where new methods of charitable organization were under way in the eighteenth century.

English sovereigns had long distributed money and clothing to the poor, while several London hospitals, including St Bartholomew's and St Thomas's, were institutional expressions of crown patronage. The Royal Hospital and Collegiate Church of St Katharine by the Tower, located next to the Tower of London, was founded by Queen Matilda, wife of King Stephen, in 1147 in memory of two of her children, Baldwin and Matilda, who had died in infancy (fig. 24.01). Its endowment was early increased by two Queens Consort, Eleanor of Castile and Philippa of Hainault. After a dispute over its control, Queen Eleanor granted a new Charter, reserving the Foundation's patronage to the Queens of England in 1273, and this charge had been honoured by the women of the royal family over succeeding centuries. Queen Anne's reign witnessed some additional remarkable displays of royal bounty, including the establishment of a leper hospital near Bodmin.

The Hanoverian succession brought new influences to bear on such a long-established practice of charitable support within the community. One of German Pietism's expressions had been in the founding of orphanages and institutions for abandoned children, such as August Hermann Francke's home for foundlings in Halle established in 1698. Caroline of Ansbach, with her knowledge of such initiatives and empathy towards Pietist ideals, would have been receptive to the ambition of Thomas Coram to establish a Foundling Hospital in London (fig. 24.02). George Frideric Handel, whom Caroline had engaged to provide music lessons for her children, actually came from Halle, and would

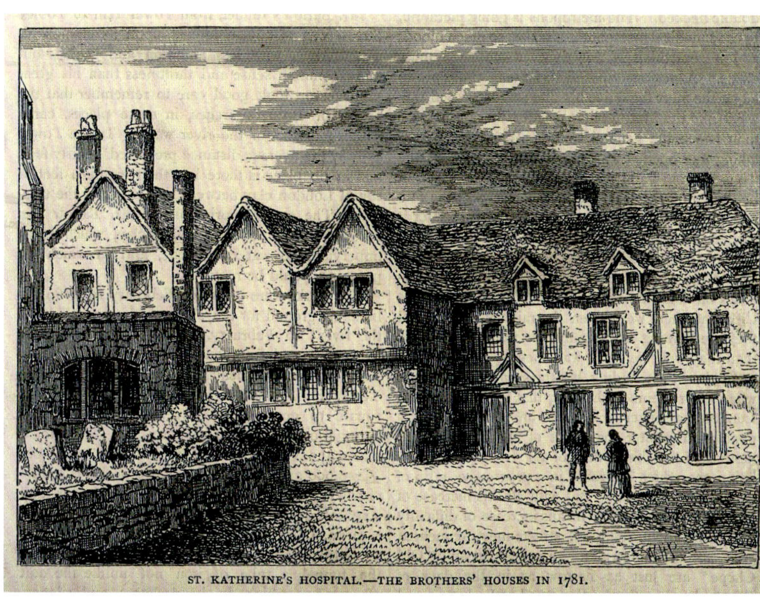

FIG. 24.01 W. H. P., *St Katharine's Hospital: The Brothers' Houses in 1781*, 1781, wood engraving, 4¼ x 5¾ in. (10.6 x 14.7 cm). Wellcome Library, London

Detail of p. 408, fig. 24.02

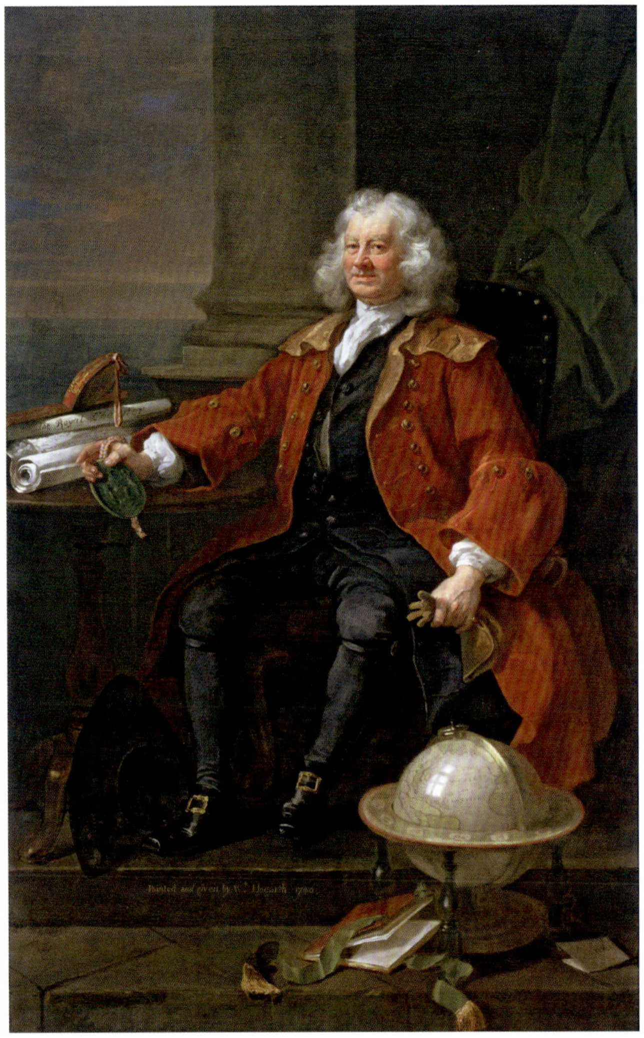

FIG. 24.02 William Hogarth, *Captain Thomas Coram*, 1740, oil on canvas, 94 x 58 in. (239 x 147.5 cm). Foundling Museum, London

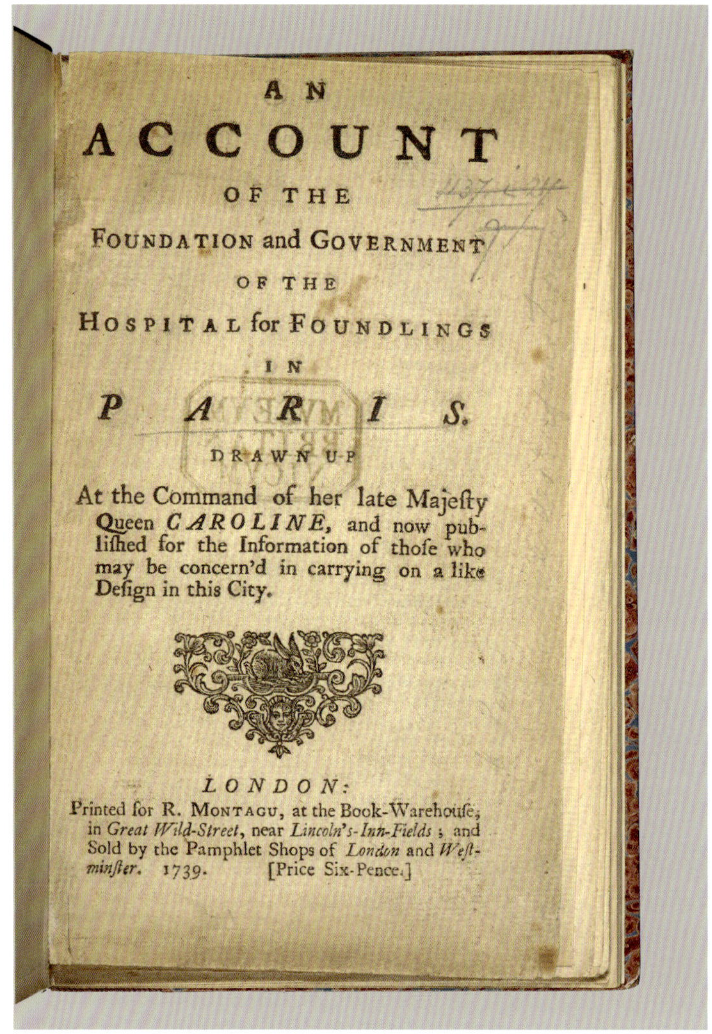

AN
ACCOUNT
OF THE
FOUNDATION and GOVERNMENT
OF THE
HOSPITAL for FOUNDLINGS
IN
PARIS.
DRAWN UP
At the Command of her late Majesty Queen CAROLINE, and now published for the Information of those who may be concern'd in carrying on a like Design in this City.

LONDON:
Printed for R. MONTAGU, at the Book-Warehouse, in *Great Wild-Street*, near *Lincoln's-Inn-Fields*; and Sold by the Pamphlet Shops of *London* and *Westminster*. 1739. [Price Six-Pence.]

FIG. 24.03 Staff of the Hôpital des Enfants-Trouvés in Paris, *An Account of the Foundation and Government of the Hospital for Foundlings in Paris—Drawn up at the Command of the Late Queen Caroline, and now published for the Information of those who may be concern'd in carrying on a like Design in this City.* (London, 1739). British Library

have known the Francke institution directly, so it is not surprising that he too gave it his whole-hearted support.

Of course, institutions that took in abandoned children were common, especially in Roman Catholic Europe. Caroline in fact commissioned the gathering of research to support Coram's plan from a distinguished European Catholic model, the Hôpital des Enfants-Trouvés in Paris (figs. 24.03). While the report presumably provided useful intelligence for the Foundling Hospital administrators, it would only be published in 1739, two years after Caroline's death.[2] Apart from her support for the Foundling Hospital, it is less clear just how generous Caroline was to the poor and benighted, in part because royal charity in her day was much less institutional and therefore less well documented.[3] Much royal beneficence turned on casual almsgiving and discreet donations, though was sometimes punctuated by grand philanthropic gestures.

For his part, George II made numerous donations to needy claimants and would arouse the expectations of hospital voluntarists by giving a royal charter to the London Foundling Hospital. However, when his daughter Princess Amelia was presented with a petition for the Hospital, it was rejected in no uncertain terms.[4] Her reaction was unusual for a Hanoverian princess.

Though the soul of discretion, Augusta of Saxe-Gotha's Privy Purse accounts in the middle decades of the eighteenth century illustrate that she gave away large sums to indigent families.[5] They also list a host of hospitals, charity schools, and societies that benefited from her largesse, including the Edinburgh Infirmary, the Westminster Infirmary, the Foundling Hospital, schools in Kew and Savoy, the Danish Mission, and the Fund (later Society) for Decay'd Musicians, now the Royal Society of Musicians.

As Augusta's institutional benefactions suggest, the structure of charity was changing. As the country became increasingly urban and industrial, poverty and distress were becoming more concentrated and more visible. Consequently, traditional forms of philanthropy such as casual almsgiving and charitable trusts, were seen to be inadequate. By the beginning of the nineteenth century the hallmark of philanthropy was collective effort, with a growing array of charitable institutions, each in need of patrons and subscribers. Increasingly, well-connected institutions sought the highest accolade—royal patronage. Such associations were of mutual benefit, for they brought cash and respectability to the societies, while advertising the royal family's beneficence. Charitable patronage created dependent clients, making royal influence visible beyond politics.

Apart from the pressures of industrial development and a growing population, the eighteenth-century evangelical revival propelled a religious activism that enveloped the royal family as it did the propertied classes. The French Revolution and the suffering caused by the prolonged war with France, which coloured the lives of a generation, had similar effects. War and revolution invigorated piety and paternalism and deepened the conservative hue of British philanthropy. Against this background, the monarchy and the charitable establishment found common cause in the reign of George III.

Both George III and Charlotte of Mecklenburg-Strelitz had shown a penchant for philanthropy well before the great troubles of the reign began. Over the years, the vicissitudes of politics and family life deepened their association with a philosophy of social betterment. Indeed, they came to head an expanding charitable movement that patriotic Britons readily linked to national purpose. The irreversible decline in the King's power following Great Britain's defeat in the American War of Independence (1775–83) also invigorated royal charity, for in the patronage of worthy causes the monarchy could identify with the burgeoning middle classes while remaining aloof from parliamentary conflict.

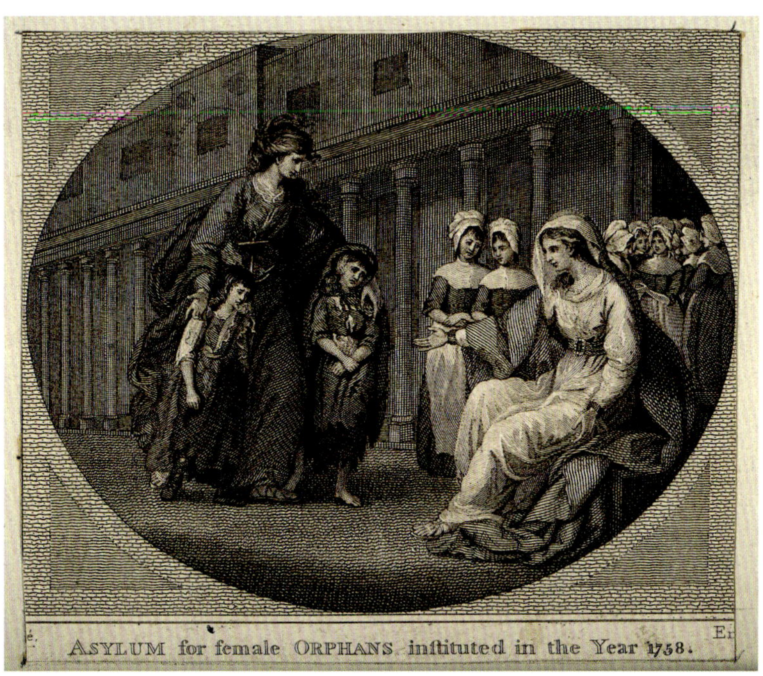

FIG. 24.04 Unknown engraver, after unknown artist, *Two Children are Brought to the Nurses of the Asylum for Female Orphans, Lambeth, London*, 1758, etching, 3½ x 4¼ in. (9.1 x 11 cm). Wellcome Library, London

Charlotte's upbringing in Germany, with its charitable traditions, had left her with a marked sense of social duty. Like other royal women, her philanthropy was primarily a pragmatic, largely unsystematic, response to immediate, individual distress. She nevertheless showed some discrimination in her social work and a preference for causes relating to women and children. Typically, she worked behind the scenes, the embodiment of those female virtues of modesty and compassion which respectable society increasingly encouraged in women of all social classes. In Windsor, she and her daughters made inquiries into local families and established a school for poor children. In one of the more curious royal charitable traditions, she dispensed money to deserving women who had given birth to triplets; a tradition that survived until 1957.[6]

Over the years, a host of begging letters from newly established institutions reached Charlotte's desk. Along with her daughters, who would also become notable for their charities, she worked on behalf of the Ladies' Society for the Education and Employment of the Female Poor and the Asylum or House of Refuge for the Reception of Friendless and Deserted Orphan Girls (24.04). She was a frequent visitor to Bailbrook Lodge in Bath, which assisted genteel widows. Her patronage included the National Society for Promoting the Education of the Poor in the Principles of the Established Church; Westminster Hospital; the Female Friendly Society; and the Lying-in Hospital, later Queen Charlotte's Hospital. On her marriage, while inheriting the patronage of the Hospital of St Katharine's, she initiated an association with the Magdalen Hospital—a refuge for prostitutes—which was propitious, for the royal seal of approval enabled the institution 'to triumph over the prejudices raised against it in the public mind'.[7]

Charlotte's last public act was a visit to the National Society for Promoting the Education of the Poor at the Egyptian Hall in the Mansion House in 1818, a grand event attended by leading churchmen and the social reformer and philanthropist Elizabeth Fry. She contributed the sum of £500. The level of the Queen's generosity had long alarmed her financial advisers, who cautioned her against 'profuse liberality'.[8] According to her treasurer and

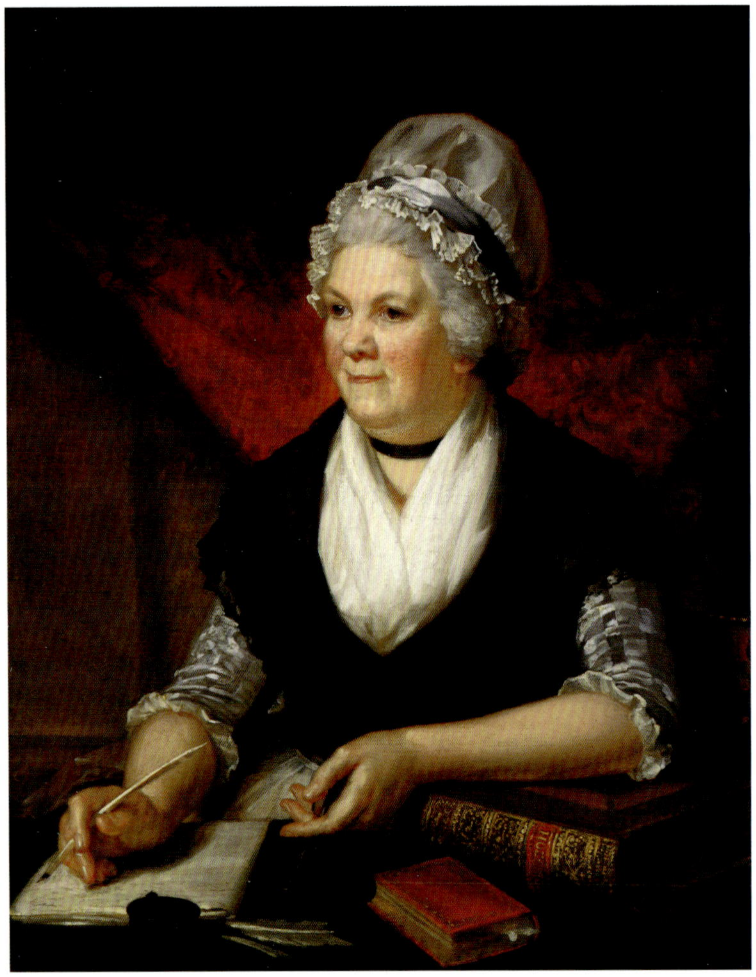

FIG. 24.05 Henry Howard, *Sarah Trimmer*, exhibited 1798, oil on canvas, 35½ x 27 in. (90.2 x 68.6 cm). National Portrait Gallery, London

later executor, her charities, estimated at £5,000 a year, were so extensive that there was 'never a pecuniary balance in her favour'.[9]

The obituaries of Charlotte said little about her privileges and luxurious surroundings. Rather, they emphasized her familial virtue, domestic religion, and charitable conduct. In the words of one churchman:

> Withstanding all the calamities which convulsed the world, and the dangers that threatened these kingdoms . . . [Queen Charlotte] was spared to foster all the charities which her prudence and benevolence had planned or preserved and to exercise her maternal regards for her own numerous family and her solicitude for the British people.[10]

Though her charities were indeed extensive, by the time of her death in 1818 loyalists had embellished them in the interest of public safety. Meanwhile, evangelicals such as Sarah Trimmer, known as a prolific author of pedagogical texts and the founding editor of the periodical *The Guardian of Education* (published 1802–6), had turned her into a model of female propriety palatable to middle-class culture and suitable for the education of young ladies (fig. 24.05).

According to her biographer Olwen Hedley, Charlotte's contribution to social history was 'an untiring interest in charities', which served as 'a substitute for the political influence which . . . she never wished to exercise'.[11] In her fashion, she represented the remarkable shift that was under way in the history of the monarchy, though, like the courtiers of her day, she little recognized it. At a time when the power of the crown had entered a downward spiral, it was gradually reaffirming its importance and popularity through good works. In allying with ascendant middle England, Charlotte and other Georgian princesses encouraged the feminization of philanthropy and the monarchy, which was to transform charity and the crown in the centuries to follow. During his reign, George III was the patron of nine charitable institutions, while Charlotte probably added as many or more. Today, Queen Elizabeth II oversees eight hundred, while the royal family's collective patronage now extends to about 3,500 organizations.[12]

NOTES

1. See Frank Prochaska, *Royal Bounty: The Making of a Welfare Monarchy* (New Haven and London: Yale Univ. Press, 1995).
2. The title page reads: *An Account of the Foundation and Government of the Hospital for Foundlings in Paris, drawn up at the Command of her late Majesty Queen Caroline, and now published for the Information of those who may be concern'd in carrying on a like Design in this City*. Much of this information has come from John Styles's research on the Foundling Museum, some of which is touched on in his *Threads of Feeling: The London Foundling Hospital's Textile Tokens, 1740–1770* (London: Foundling Museum, 2010) and Foundling Museum guidebooks.
3. Joanna Marschner, *Queen Caroline: Cultural Politics at the Early Eighteenth-Century Court* (New Haven and London: Yale Univ. Press, 2014). See also Joanna Marschner, 'Queen Caroline of Anspach and the European Princely Museum Tradition', in *Queenship in Britain, 1660–1837: Royal Patronage, Court Culture and Dynastic Politics*, ed. Clarissa Campbell Orr (Manchester: Manchester Univ. Press, 2002), 130–42.
4. R. H. Nichols and F. A. Wray, *The History of the Foundling Hospital, London* (Oxford: Oxford Univ. Press, 1935), 18.
5. John Watkins, *Memoirs of Her Most Excellent Majesty Sophia-Charlotte, Queen of Great Britain* (London: H. Colburn, 1819), 233; Christine Gerrard, 'Queens-in-waiting: Caroline of Anspach and Augusta of Saxe-Gotha as Princesses of Wales', in *Queenship in Britain*, ed. Campbell Orr, 143–59.
6. Thomas Williams, *A Brief Memoir of Her Late Majesty Queen Charlotte* (London: W. Simpkin and R. Marshall, 1819), 66. See also Ronald Allison and Sarah Riddell, eds., *The Royal Encyclopedia* (London: Macmillan, 1991), 546.
7. Watkins, *Memoirs of Her Most Excellent Majesty Sophia-Charlotte*, 189.
8. Watkins, *Memoirs of Her Most Excellent Majesty Sophia-Charlotte*, 578, 599–600.
9. Williams, *Brief Memoir of Her Late Majesty Queen Charlotte*, 69–73.
10. Rev. S. Piggott, *Female Virtue and Domestic Religion Recommended by the Example of our Late Illustrious Queen Charlotte* (London: Longman and Co., 1818), 2.
11. Olwen Hedley, *Queen Charlotte* (London: John Murray, 1975), 92.
12. Prochaska, *Royal Bounty*, 13, 275, 277.

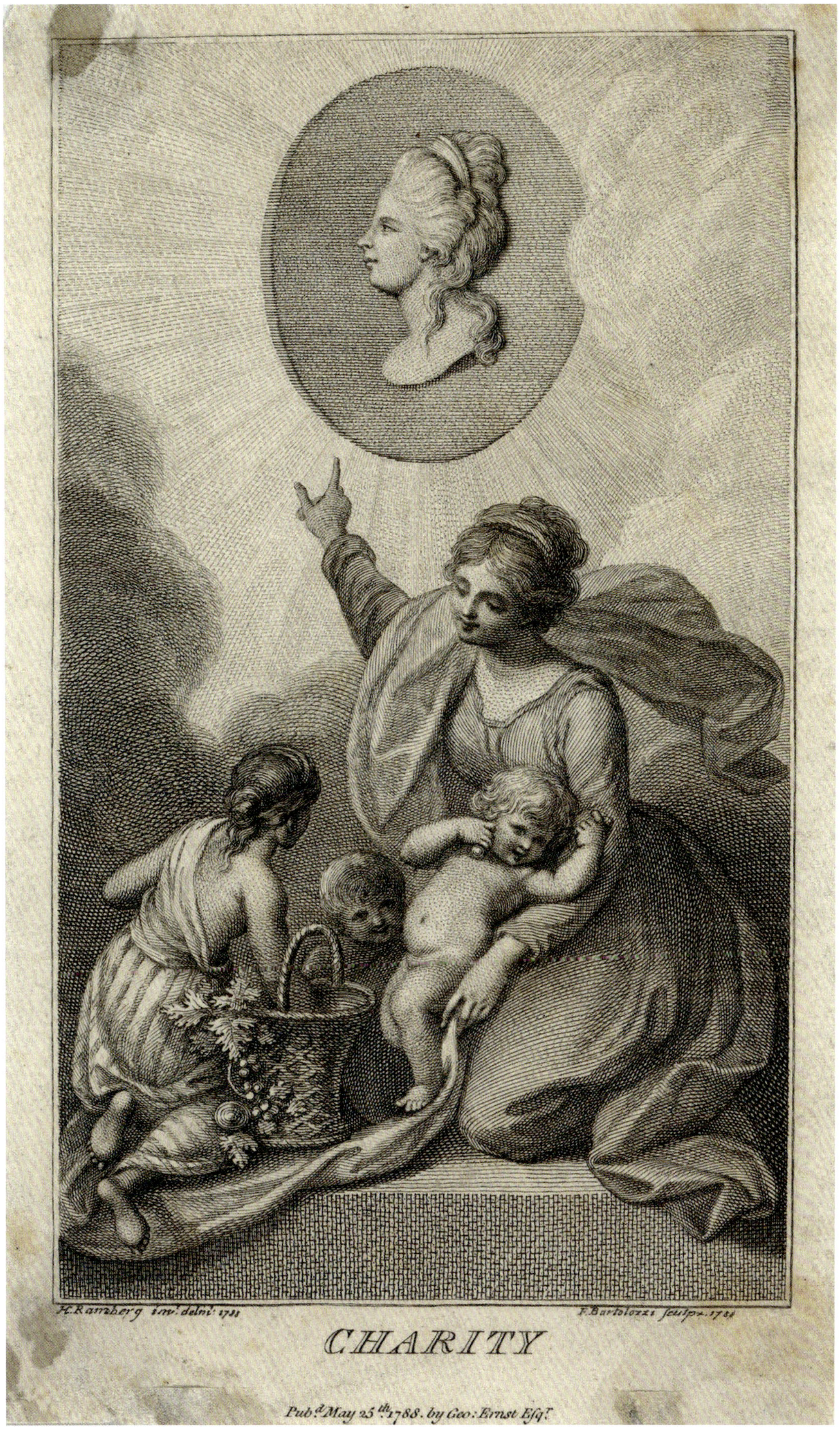

24.06

24.06*

FRANCESCO BARTOLOZZI (1728–1815), engraver, after **JOHANN HEINRICH RAMBERG** (1763–1840)
'Charity'
1788
Etching and engraving on paper
6 x 3⅝ in. (15 x 8.8 cm)
In Arnaud Berquin (1749–1791), *L'Amico di fanciulli*
Published by George Ernst, London
The Metropolitan Museum of Art, Gift of Walter Smith, 1967, 67.749.41(13)

Arnaud Berquin's most famous work, *L'Ami des Enfans* (1782–83), was first translated into English by Mary Stockdale and published in London in 1783–84 by her father. Berquin's work positions childhood reading as a familial exercise. The book includes plays that have parts for every member of the family, so that the experience of reading embraces all. Johann Heinrich Ramberg's portrait of Charlotte was incorporated into the frontispiece of the Italian edition of Berquin's work, aligning her support for causes touching the care of children with this text extolling the value of the nuclear family.

Ramberg studied with Benjamin West under the special protection of George III. He painted several portraits of members of the British royal family, including Charlotte, and the princesses Mary, Elizabeth, and Sophia. In 1788, with the help of his royal patron, Ramberg visited the Netherlands, and afterwards Germany and Italy, returning to Hanover in 1792, when he was appointed electoral court painter.

* = YCBA only
† = HRP only

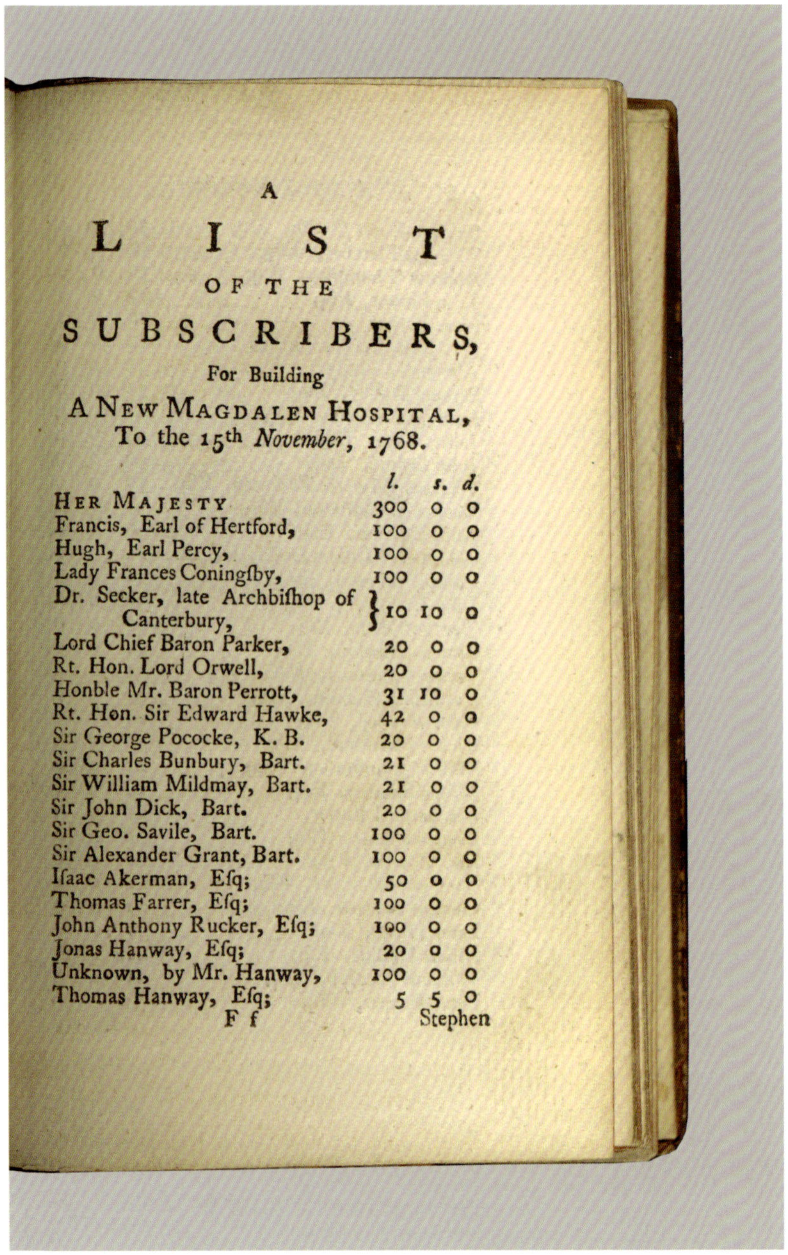

A

LIST

OF THE

SUBSCRIBERS,

For Building

A NEW MAGDALEN HOSPITAL,

To the 15th *November*, 1768.

	l.	s.	d.
HER MAJESTY	300	0	0
Francis, Earl of Hertford,	100	0	0
Hugh, Earl Percy,	100	0	0
Lady Frances Coningſby,	100	0	0
Dr. Secker, late Archbiſhop of Canterbury,	10	10	0
Lord Chief Baron Parker,	20	0	0
Rt. Hon. Lord Orwell,	20	0	0
Honble Mr. Baron Perrott,	31	10	0
Rt. Hon. Sir Edward Hawke,	42	0	0
Sir George Pococke, K. B.	20	0	0
Sir Charles Bunbury, Bart.	21	0	0
Sir William Mildmay, Bart.	21	0	0
Sir John Dick, Bart.	20	0	0
Sir Geo. Savile, Bart.	100	0	0
Sir Alexander Grant, Bart.	100	0	0
Iſaac Akerman, Eſq;	50	0	0
Thomas Farrer, Eſq;	100	0	0
John Anthony Rucker, Eſq;	100	0	0
Jonas Hanway, Eſq;	20	0	0
Unknown, by Mr. Hanway,	100	0	0
Thomas Hanway, Eſq;	5	5	0

F f Stephen

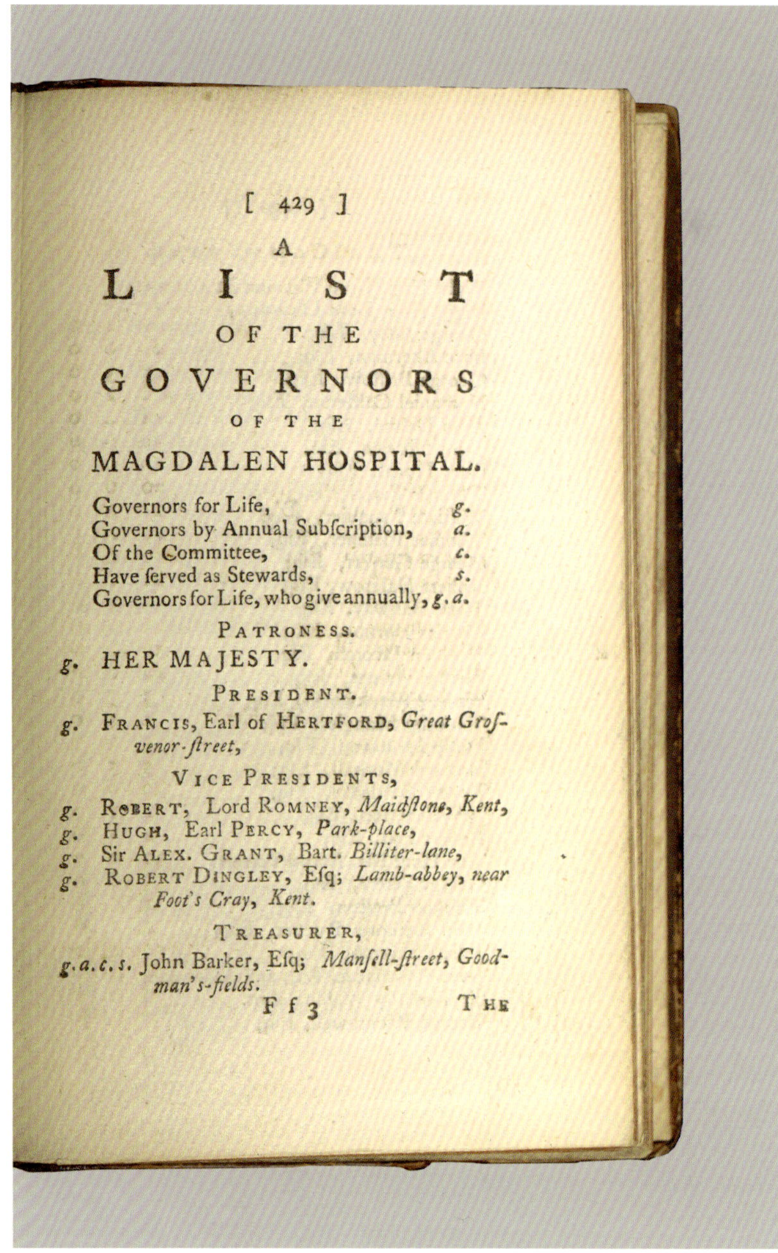

[429]

A

LIST

OF THE

GOVERNORS

OF THE

MAGDALEN HOSPITAL.

Governors for Life, *g.*
Governors by Annual Subſcription, *a.*
Of the Committee, *c.*
Have ſerved as Stewards, *s.*
Governors for Life, who give annually, *g. a.*

PATRONESS.

g. HER MAJESTY.

PRESIDENT.

g. FRANCIS, Earl of HERTFORD, *Great Groſvenor-ſtreet,*

VICE PRESIDENTS,

g. ROBERT, Lord ROMNEY, *Maidſtone, Kent,*
g. HUGH, Earl PERCY, *Park-place,*
g. Sir ALEX. GRANT, Bart. *Billiter-lane,*
g. ROBERT DINGLEY, Eſq; *Lamb-abbey, near Foot's Cray, Kent.*

TREASURER,

g. a. c. s. John Barker, Eſq; *Manſell-ſtreet, Goodman's-fields.*

F f 3 THE

24.07

24.07

WILLIAM DODD (1729–1777)
An Account of the Rise, Progress and Present State of the Magdalen Hospital, for the Reception of Penitent Prostitutes . . .
1770
7 in. (17.7 cm)
Printed by W. Faden, sold by J. Knox, etc., London
Wellcome Library, London, 34868/A
* Exhibited: pp. 424–25 (list of subscribers); (illustrated above left)
† Exhibited: pp. 428–29 (list of governors); (illustrated above right)

The first English, non-denominational Magdalen Hospital was established in (Great) Prescot Street in 1758. Six penitents were admitted on the first day; by 1769, 1,500 had passed through its doors, most having stayed for about three years. During its time in Prescot Street, ideas for a larger venue developed, and building started in 1769 at St George's Fields, Southwark—made accessible by the building of Blackfriars Bridge. The new hospital opened in 1772. Charlotte was a major subscriber, donating the substantial sum of £300 towards the new building in 1768. She remained a governor for life. Though the move away from the East End lessened its popular appeal, fundraising continued: in the octagonal chapel, a choir of inmates sang behind a screen, and at services each Sunday the mystery and beauty of music from unseen voices attracted a fashionable congregation of potential supporters.

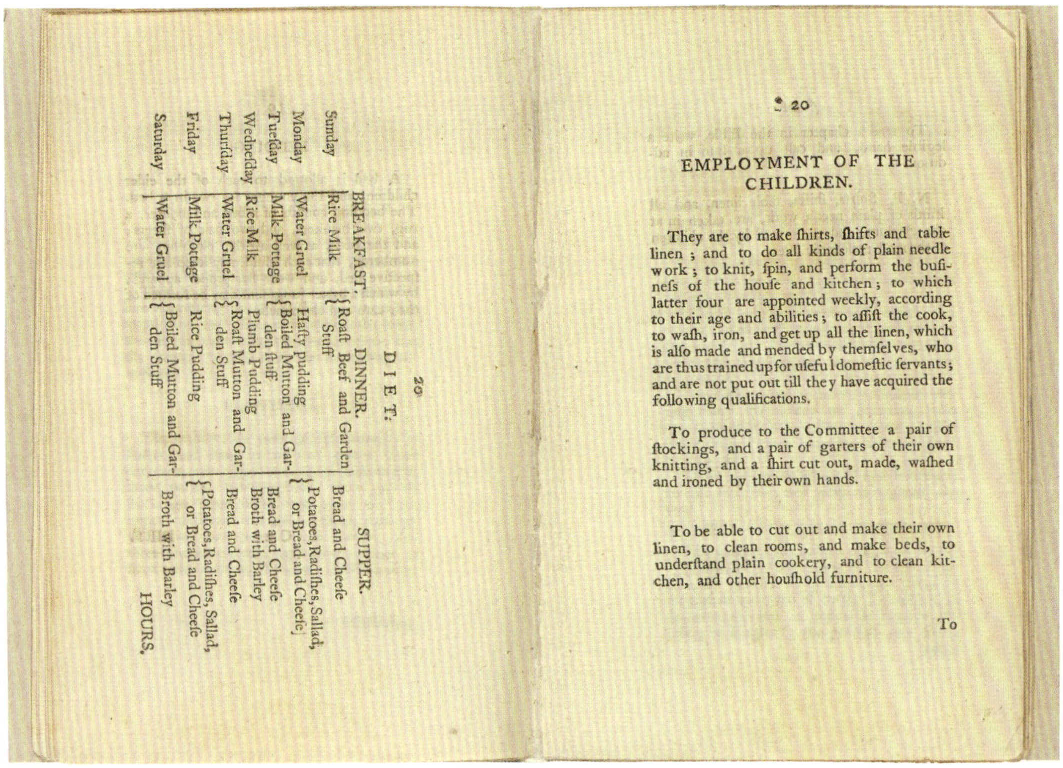

20

DIET.

	BREAKFAST.	DINNER.	SUPPER.	HOURS,
Sunday	Rice Milk	Roast Beef and Garden Stuff	Bread and Cheese	
Monday	Water Gruel	Hasty pudding	Potatoes, Radishes, Sallad, or Bread and Cheese	
Tuesday	Milk Portage	Boiled Mutton and Garden stuff	Bread and Cheese	
Wednesday	Rice Milk	Plumb Pudding	Broth with Barley	
Thursday	Water Gruel	Roast Mutton and Garden Stuff	Bread and Cheese	
Friday	Milk Portage	Rice Pudding	Potatoes, Radishes, Sallad, or Bread and Cheese	
Saturday	Water Gruel	Boiled Mutton and Garden Stuff	Broth with Barley	

20

EMPLOYMENT OF THE CHILDREN.

They are to make shirts, shifts and table linen; and to do all kinds of plain needle work; to knit, spin, and perform the business of the house and kitchen; to which latter four are appointed weekly, according to their age and abilities; to assist the cook, to wash, iron, and get up all the linen, which is also made and mended by themselves, who are thus trained up for useful domestic servants; and are not put out till they have acquired the following qualifications.

To produce to the Committee a pair of stockings, and a pair of garters of their own knitting, and a shirt cut out, made, washed and ironed by their own hands.

To be able to cut out and make their own linen, to clean rooms, and make beds, to understand plain cookery, and to clean kitchen, and other houshold furniture.

To

24.08

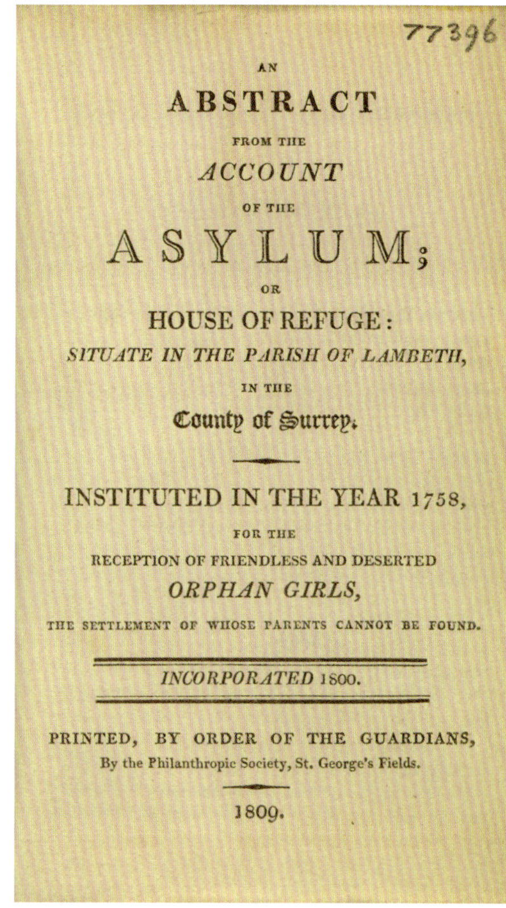

AN
ABSTRACT
FROM THE
ACCOUNT
OF THE
ASYLUM;
OR
HOUSE OF REFUGE:
SITUATE IN THE PARISH OF LAMBETH,
IN THE
County of Surrey.

INSTITUTED IN THE YEAR 1758,
FOR THE
RECEPTION OF FRIENDLESS AND DESERTED
ORPHAN GIRLS,
THE SETTLEMENT OF WHOSE PARENTS CANNOT BE FOUND.

INCORPORATED 1800.

PRINTED, BY ORDER OF THE GUARDIANS,
By the Philanthropic Society, St. George's Fields.

1809.

24.09

24.08*

An Account of the Institution, and Proceedings of the Guardians, of the Asylum or House of Refuge, Situated on the Surry side of Westminster Bridge, for the reception of orphan girls residing within the bills of mortality, whose settlements cannot be found.
1761
7⅞ in. (20 cm)
Printed by order of the Guardians, London
Beinecke Rare Book & Manuscript Library, Yale University, WC 8892
Exhibited: pp. 20–21 (illustrated)

24.09†

An Abstract from the Account of the Asylum, or House of Refuge, situate in the Parish of Lambeth, in the County of Surrey
1809
6¾ in. (17 cm)
Printed by the Philanthropic Society, London
Wellcome Library, London, 32028/P
Exhibited: Frontispiece, title page (illustrated), and pp. 14–15

The girls' asylum house is a particularly fine illustration of mid-eighteenth-century philanthropy in action. Its aim, as stated in the opening paragraph of this book, was 'to preserve friendless and deserted girls from those dangers and misfortunes, to which their distressed situation exposes them'. The list of the asylum's guardians, beginning with the Queen, included the Archbishop of Canterbury, the Earl of Macclesfield, and many other establishment figures. The London magistrate Sir John Fielding (1721–1780) had set up a number of charitable organizations for both boys and girls in the capital. Whether or not he was actually the author of this book, it is certain that he played a significant part in drafting its principles and regulations. The book is remarkably detailed, not least about the diet, employment, and education of the girls. The children were eventually to be found apprenticeships at the age of fifteen (or sooner, if qualified) 'as domestic servants in reputable families residing in England; the greatest care being taken by the guardians themselves to enquire into the characters of masters and mistresses, before the children are entrusted to them'.

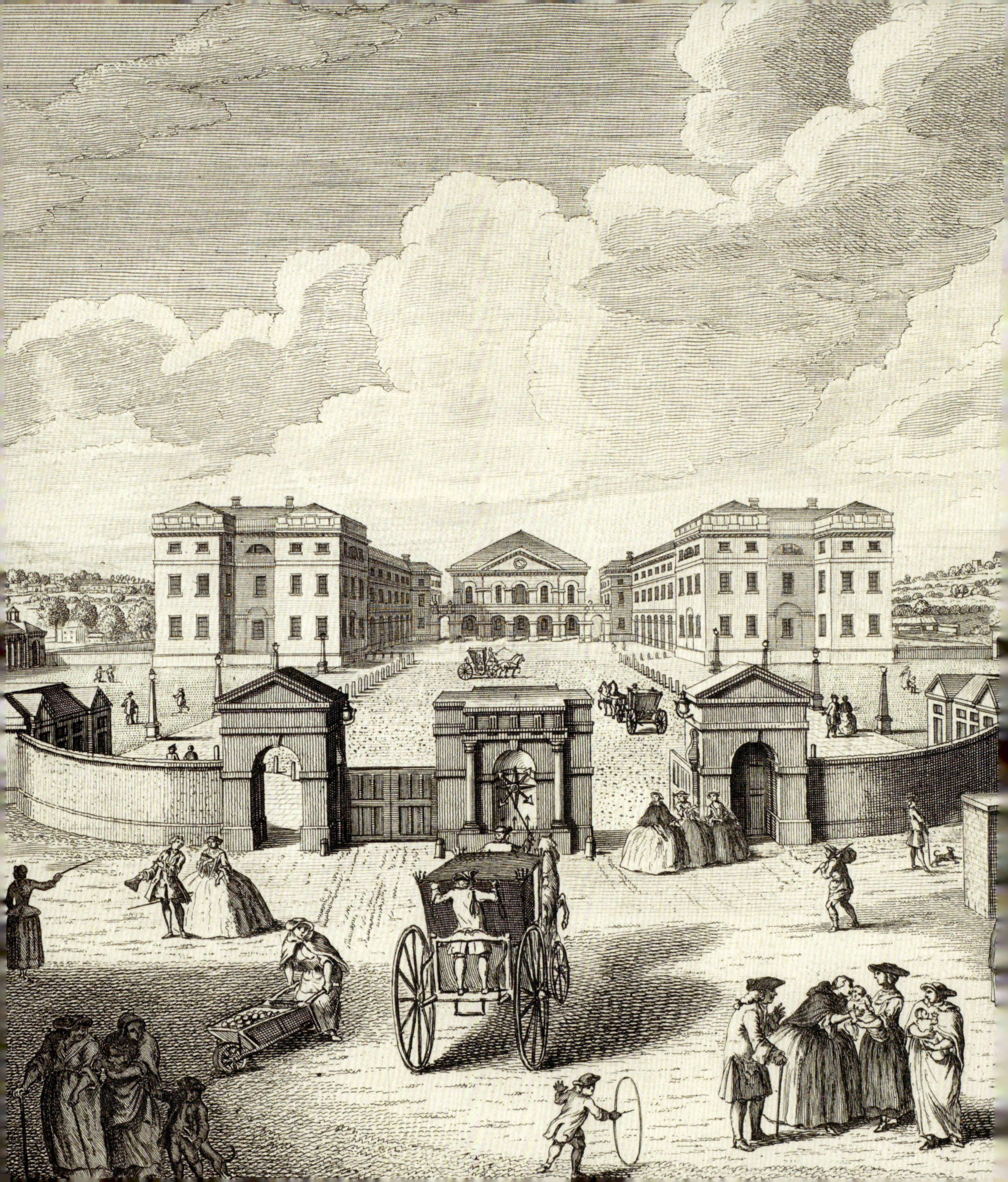

The Foundling Hospital, Women's Health, and the Welfare of Children

Tyler Griffith and Samantha Howard

For the European elite, philanthropy had long been seen not simply as a mark of power, but as a responsibility incumbent upon people of rank from both a religious and social standpoint. Donna T. Andrew summarizes this religious dynamic in her observation that 'the traditional notion of beneficence was that charity was a basic obligation entailed by the ownership of property, that as God was the sole Proprietor, and men only his stewards, charity to the poor, to God's poor, was not mercy, but justice.'[1] Equally important, 'for gentle people, the prime requisite, both psychological and economic, was keeping up appearances'; philanthropy often functioned as one of the crucial visible gestures that 'kept up appearances'. It served as a socially sanctioned and moderated activity that testified to the donor's nobility as much as dancing, music playing, or any of the other aristocratic activities that emerged from and created the notion of aristocracy itself.[2]

Even though philanthropy formed a crucial part of the *noblesse oblige*, through the nature of their royal positions, Caroline of Ansbach, Augusta of Saxe-Gotha, and Charlotte of Mecklenburg-Strelitz set themselves apart in the directions of their charitable activities. All three demonstrated a marked interest in—even enthusiasm for—promoting initiatives that helped women. This started with the act of giving birth. Childbirth had long been primarily the domain of midwives, but during the reign of George III and Queen Charlotte the theory and management of childbirth gained momentum as an important and discrete field of medical research. Charlotte helped indirectly to advance the study of obstetrics in her appointment of William Hunter, the well-known anatomist and obstetrician, as her physician in extra-ordinary in 1762, and she probably commissioned from Allan Ramsay the portrait of the doctor in 1764 to commemorate this appointment (cat. 26.09). Hunter is best known today for his monumental work on human pregnancy, *The Gravid Uterus*, which he began advertising in 1751, although the work was not published until 1774, with a dedication to the King, and well after Hunter had successfully supervised Charlotte's first birth (cat. 26.10).

We may read into Charlotte's motives not only a concern to encourage practices that would benefit both mother and child, but also a broader interest in systematizing and spreading medical practices that would benefit women's health more generally. Equally important to her was the provision of a safe, healthy environment in which women could give birth and receive support during their pregnancy. Very soon after her arrival in London, Charlotte appears to have begun informally supporting the Lying-in Hospital, originally founded on Jermyn Road in 1739 by Sir Richard Manningham, which relocated to St Marylebone in 1752. By 1809, Charlotte was its official royal patroness. Charles White, himself a student of William Hunter, co-founded the Manchester Royal Infirmary, developed the 'birthing chair', and wrote treatises on the proper care and management of pregnant women within such an institutional setting (cat. 26.11).[3] Charlotte similarly supported—with varying degrees of openness—institutions such as the House of Refuge for Orphan Girls founded in 1758 by Sir John Fielding, and the Magdalen Hospital for the

Detail of p. 421, cat. 25.05

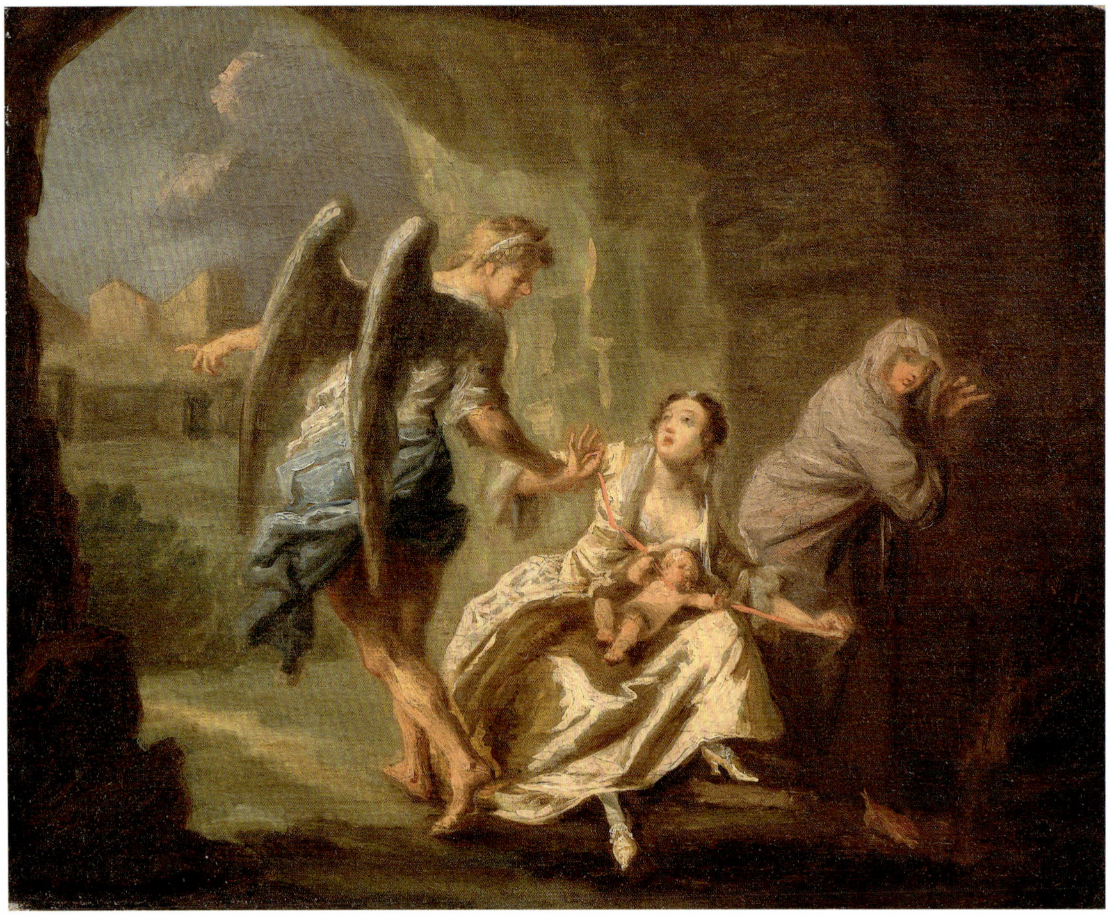

FIG. 25.01 Joseph Highmore, *The Angel of Mercy*, c. 1746, oil on canvas, 23½ x 19 in. (59.7 x 48.3 cm). Yale Center for British Art, Paul Mellon Collection (cat. 25.08)

Reception of Penitent Prostitutes, of which she became the royal patroness in 1765. She continued her involvement actively until her death in 1818.

These charitable enterprises did more than support women from a medical point of view: they also operated as organizations in which women could become leaders themselves. At the House of Refuge, numerous women—independently of their husbands—participated as Guardians of the Asylum, the formal advisory committee for the institution.[4] At the Magdalen Hospital, the head officer was not the male Warden but instead the female Head Matron.[5] Institutions such as the House of Refuge and Magdalen Hospital were not, strictly speaking, state institutions; they were associated charities that were not strongly religiously denominational, took a variety of patients, and were beyond the scope of purely state enterprises.

Charity thus functioned not simply within a religious framework, nor only as a socially acceptable way of affirming one's position. Institutions such as these became the means whereby women could exercise agency and contribute to national improvement, even while they were still being formally denied participation in 'the state' per se. The Princesses' alignment with such initiatives won them international recognition, as etchings such as *Charity* of 1798 by Francesco Bartolozzi illustrate in the case of Charlotte (cat. 24.06). In both ambition and practice, the charitable endeavours of Caroline, Augusta, and especially Charlotte, went beyond what was necessary to keep up appearances.

The idea of national improvement, as executed through charity, comes to the foreground in another domain of the Princesses' charitable activities: namely, children's welfare, and above all in their continuing interest and promotion of the Foundling Hospital in London. There was an especially strong history of philanthropy focusing on children's welfare in the German states, especially through the agency of August Hermann Francke, who founded and developed the successful 'Francke Foundations' in Halle in Saxony. These Pietist organizations shared an ideology which aligned with the Latitudinarian beliefs held by Caroline and, to a certain extent, Augusta and Charlotte too. Numerous German women—either originating in Hanover or having migrated there through marriage—involved themselves with similar programmes focused on children's welfare, such as Countess

Louise Henrietta of Nassau in the late seventeenth century and Wilhelmine Amalia of Brunswick-Lüneburg in the early eighteenth century.

The example of the Foundling Hospital in London, however, offers an interesting departure from the examples in the German tradition. Although the Royal Founding Charter presented to Thomas Coram describes essentially Protestant founding principles, Caroline is credited with commissioning a study of the Paris foundling hospital—the Hôpital des Enfants-Trouvés—a Catholic institution.[6] Certain elements of the London Foundling Hospital, such as the use of tokens to cement the relationship between mother and child, had been common practice in Europe (cats. 25.10–25.14).

Whatever its religious origins, there was a very real need in London for a foundling hospital, as the eighteenth century saw a marked demographic contraction across all levels of society due to a variety of factors such as war, disease, emigration, infanticide, and child desertion. *The Angel of Mercy* by Joseph Highmore (c. 1746), which depicts an angel directing a mother and child to the Foundling Hospital at the moment of an intended infanticide, spoke of actual practice in eighteenth-century London (fig. 25.01). Seen from the vantage point of a ruler, the high mortality rates, especially those amongst infants, put the state at risk of losing some of the most crucial participants in its society: sailors, soldiers, agricultural workers, labourers, and domestic staff above all. William Hogarth, in his undated *Study of the Foundling Hospital* was certainly aware of these motives: it is no coincidence that he places distinctly merchant-military schooners in the background, while a man works the field in the left middle ground and a builder examines a combination square on the steps of the Hospital (fig. 25.02). It was precisely these professions that Hogarth saw graduates of the Foundling Hospital potentially filling.

The Princesses' activity with the Foundling Hospital points to a decreasing confidence in the state's ability to manage society's needs. England's Poor Law provision, established by Queen Elizabeth I, had been the most comprehensive state-run welfare system in Europe, but by the eighteenth century it struggled to cover the complex needs of an increasingly urban society, and, more importantly, was being abused and discredited. The Foundling Hospital, the lying-in hospitals, and House of Refuge provided a new model for societal support and regulation, being only semi-official and relying on significant non-state support for their financial and managerial well-being. Coram's charter lists over one hundred nobles, and just as many supporters from the middling classes, as Governors and Guardians who saw to the financial support of the institution.[7] As an associated charity, there was a new requirement for, in particular, sources of income outside the funds that could be provided by the state.

The Foundling Hospital's location, first at Hatton Garden, then as now the heart of the jewellery trade in London, and subsequently in Bloomsbury, which was rapidly becoming a fashionable destination owing to the civic projects of the Dukes of

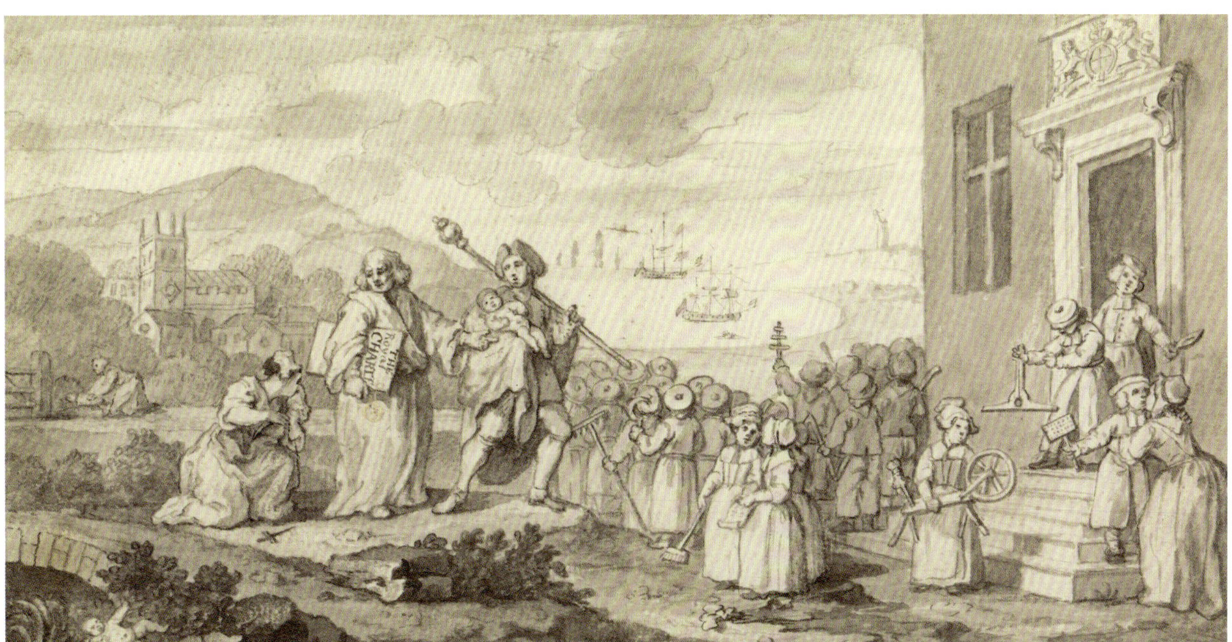

FIG. 25.02 William Hogarth, *Study for the Foundlings*, c. 1739, ink with wash on paper, 4⅜ x 8⅜ in. (11.1 x 21.3 cm). Yale Center for British Art, Paul Mellon Collection (cat. 25.09)

Bedford, contributed to its success. The Hospital was as much a public spectacle for London's well-to-do as it was a strictly charitable foundation. Consider, for example, Benjamin Cole's *View of Foundling Hospital* of around 1756, which openly positions groups of the bourgeoisie observing and discussing the foundlings, fetchingly attired in their regulation dress, as part of their fashionable round (cat. 25.06). The Hospital also became a place for performance: most famously, of course, because of its association with George Frideric Handel, one of its many notable benefactors. Handel may have been inspired by the astute creative philanthropy of his benefactor colleagues, but he also recognized that the chapel of the Hospital, built with funds largely provided by his own royal patron, had potential as a performance venue that could benefit both the charity and his own reputation. In May 1749 he approached the hospital's governors and offered to conduct a benefit concert. The programme included the first performance of Handel's 'specially written' Foundling Hospital Anthem, which opened with a text adapted from Psalm 41—'Blessed are they that consider the poor and needy . . . they deliver the poor that crieth, the fatherless'—in a clear appeal to the audience to support the work of the hospital. The music was borrowed from Handel's funeral anthem for Queen Caroline, his oratorio *Susanna*, and music by Antonio Lotti and Johann Kuhnau, the piece ending with the Hallelujah Chorus, lifted directly from *Messiah*—a work that few of his audience in London would then have known.

The benefit concert was a huge musical and financial success, and soon after, Handel donated an organ to the new chapel. The following year he returned to conduct a second benefit concert for the charity. This time he chose *Messiah*. The event was so oversubscribed—wealthy supporters had to be turned away on the night of the performance—that Handel was asked to repeat the concert two weeks later. To show its gratitude, the Hospital made Handel a governor. The organ which occupies so visible a position in the engraving after John Sanders of *The Chapel of the Foundling Hospital* of 1774 (cat. 25.07) was used to accompany the children singing at divine service, to which a fashionable congregation flocked regularly for entertainment as well as for their spiritual benefit. The Foundling Hospital is an early example of what we might now call a 'commercial charity'—a philanthropic organization dedicated to the common good, but relying heavily on marketing techniques from the commercial world to generate income and support itself financially.[8]

NOTES

1. Donna T. Andrew, *Philanthropy and Police: London Charity in the Eighteenth Century* (Princeton, NJ: Princeton Univ. Press, 1989), 277.
2. Andrew, *Philanthropy and Police*, 291.
3. Charles White, *A treatise on the management of pregnant and lying-in women, and the means of curing, but more especially of preventing the principal disorders to which they are liable. Together with some new directions concerning the delivery of the child and placenta in Natural Births. Illustrated with Cases. By Charles White, F.R.S. Member of the Corporation of Surgeons in London, and Surgeon to the Manchester Infirmary* (London: W. Johnston, 1772). In his influential treatise, White advised mothers to get out of bed as soon after delivery as possible and stressed the importance of cleanliness and ventilation; importantly, he also recognized the analogy between puerperal and surgical fevers. White dedicated his treatise to his former teacher, Hunter, and it became an international best seller, being reprinted several times in later editions, translated into both French and German, and also published in North America.
4. Amongst the female guardians, there were also the 'middling sorts' and the gentry as well as the nobility. See 'A List of the Guardians', in *An Account of the Institution, and Proceedings of the Guardians, of the Asylum or House of Refuge, Situated on the Surry side of Westminster Bridge, for the reception of orphan girls residing within the bills of mortality, whose settlements cannot be found* (London: Asylum for Orphan Girls, 1761), 27–33.
5. The duties of the Matron and her assistants are discussed in William Dodd's *An Account of the Rise, Progress, and Present state of the Magdalen Hospital, for the Reception of Penitent Prostitutes. Together with Dr. Dodd's sermons, preached before the president, vice-presidents, governors, &c.*, 4th ed. (London: W. Faden, 1770), XIII, 397–98.
6. The report, published in 1739, was dedicated to Caroline posthumously. *An Account of the Foundation and Government of the Hospital for Foundlings in Paris. Drawn up at the command of her late Majesty Queen Caroline, and now published for the Information of those who may be concern'd in carrying on a like Design in this City* (London: R. Montagu, 1739).
7. Foundling Hospital, *A List of the Present Governors and Guardians of the Hospital for the Maintenance and Education of Exposed and Deserted Young Children, incorporated by His Majesty's royal charter, bearing date the 17th day of October, 1739, and elected since to the 25th of June, 1740, inclusive* (London, 1740).
8. Foundling Hospital, *The Royal charter establishing a hospital for the maintenance and education of exposed and deserted young children* (London, 1740). The charter makes this clear in its description of the creation and establishment of 'a corporation, or body politic'; see 'Preamble'.

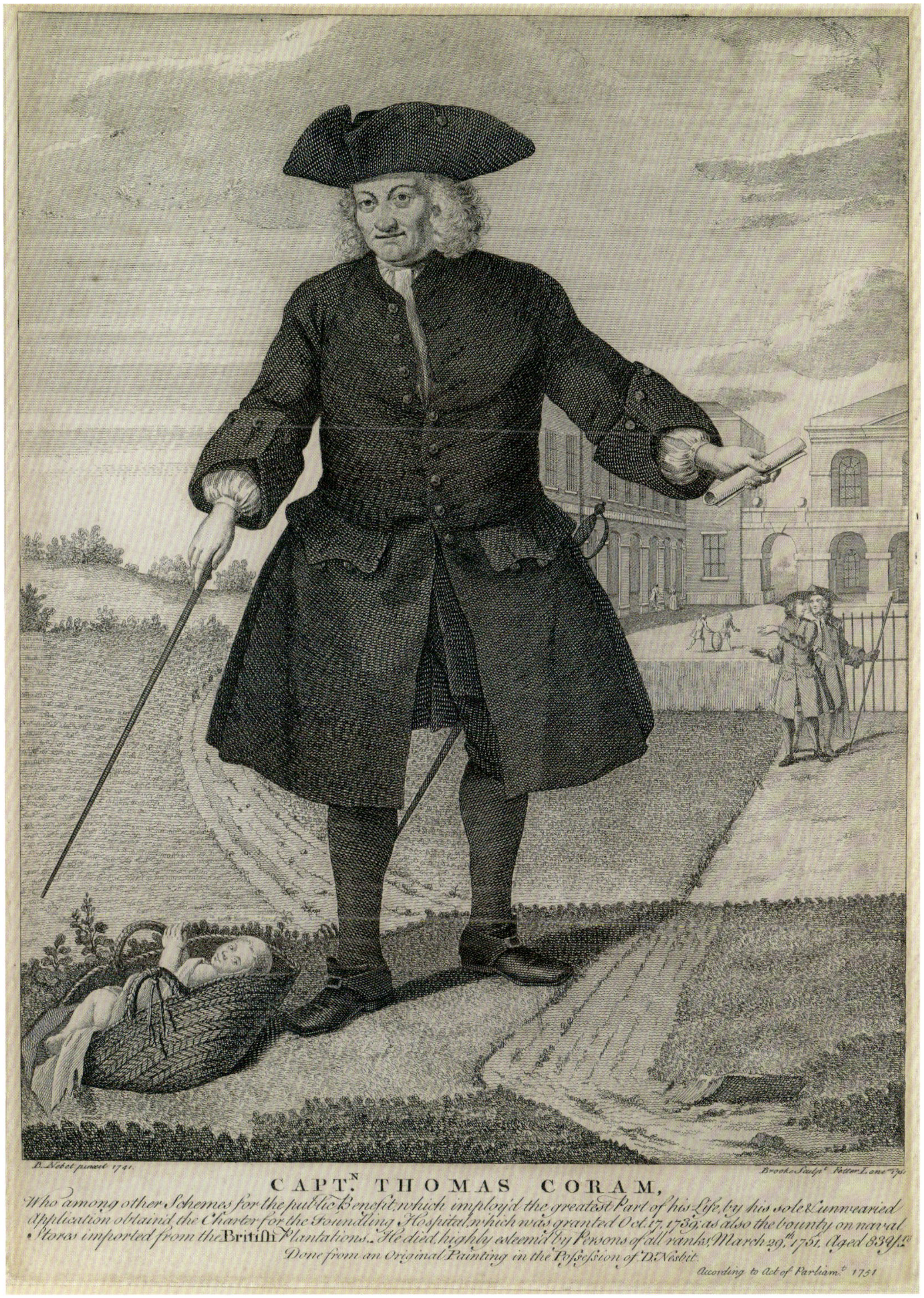

25.03

THOMAS CORAM AND THE FOUNDLING HOSPITAL

25.03[†]

JOHN BROOKE (active 1750), engraver, after **BALTHASAR NEBOT** (active 1730–1765)
Captain Thomas Coram
1751
Etching and engraving on paper
19 x 14 in. (48.4 x 35.6 cm)
Published by J. Brooke, London
Foundling Museum, London, FM638

Thomas Coram spent many years as a ship builder and merchant with business interests extending to North America. His work brought him prosperity, but as he travelled between ports he saw deprivation and its impact on children. After observing abandoned children, some dead, others dying, while living in Rotherhithe, he resolved in 1722 to create a foundling hospital inspired by examples of those in existence elsewhere in Europe. His efforts initially attracted only modest interest, as well as some opposition. Coram was poorly educated, relatively rough-mannered, and lacked a patron, as well as wealth and pedigree. The turning point in his campaign was the 'ladies petition' of 1729, signed by peeresses, and with the patronage of Caroline. Coram's petitions came before the King in council on 21 July 1737. The Foundling Hospital charter was signed by George II on 14 August 1739 and the first meeting of the governors was held at Somerset House that November.

* = YCBA only
† = HRP only

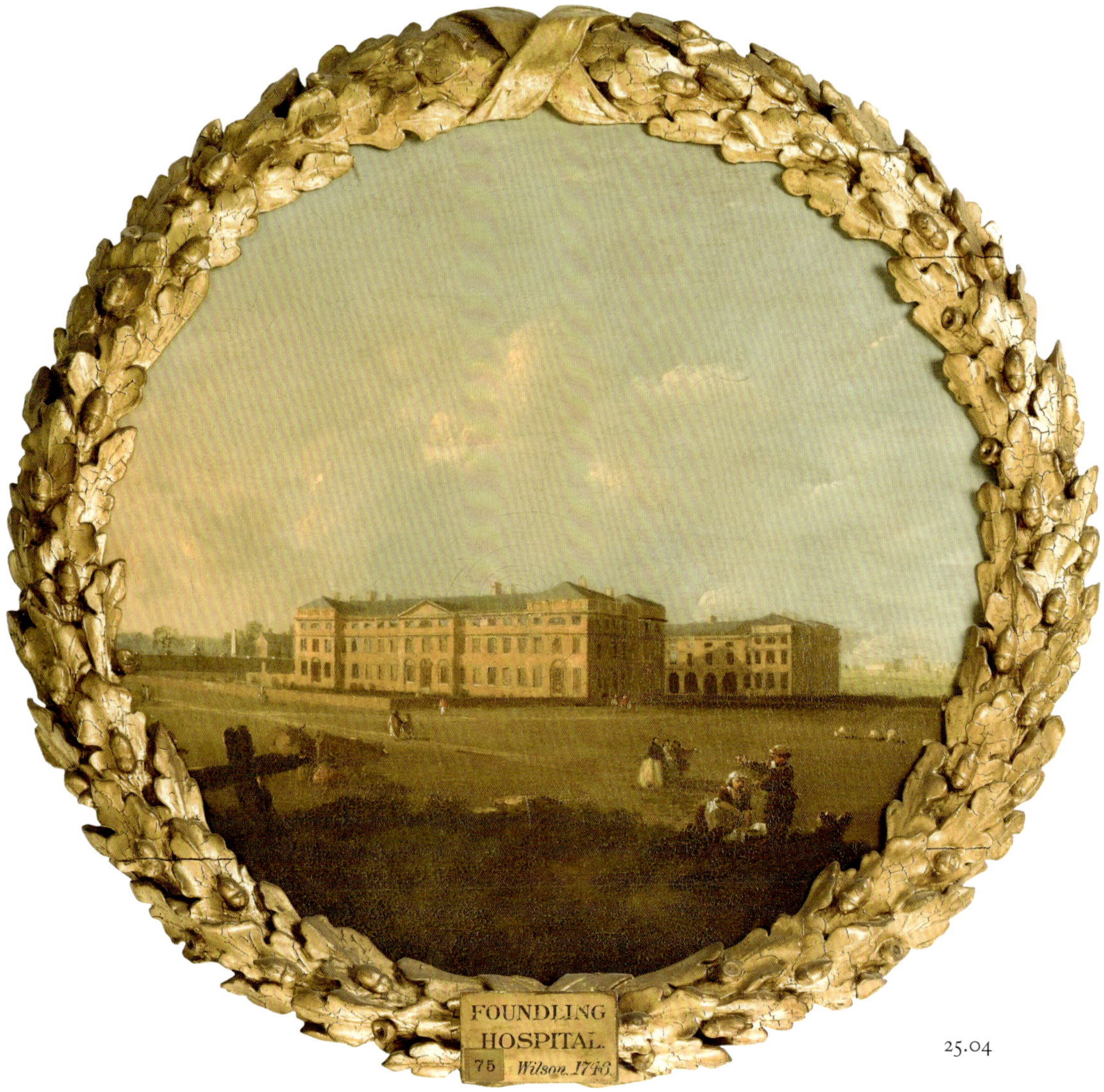

25.04

25.04†

RICHARD WILSON (c. 1714–1782)
The Foundling Hospital
1746
Oil on canvas
20⅞ x 20⅞ in. (53 x 53 cm)
Foundling Museum, London, FM65

Richard Wilson was among several artists who served as governors of the Foundling Hospital. He presented this painting to the institution upon his election. The original carved, gilded, and gessoed frame was probably the work of William Hallett, who was paid £11 4s. 'For 8 carved oval frames for pictures' on 15 November 1746. The East Wing and the Chapel were still unfinished at the date of Wilson's painting although he has omitted any evidence of this, presumably in order to present a more idealized image. The gift from Wilson to the Hospital included a second roundel: St George's Hospital is seen from the north-west, looking from Hyde Park. The pair were hung in the Court Room of the Foundling Hospital, one of the first public spaces for the display of British art. The series of roundels was completed by works by Edward Haytley of Chelsea Hospital and Bethlehem (Bedlam) Hospital; Thomas Gainsborough of Charterhouse; and Samuel Wale of Christ's Hospital and Greenwich Hospital; the group of works celebrated the network of hospitals set up in London by public subscription to improve the health of the citizenry at large.

25.05*

NATHANIEL PARR (d. 1751), engraver, after **LOUIS-PHILIPPE BOITARD** (active 1734–1760)
A View of the Foundling Hospital
1753
Etching and engraving on paper
12¾ x 18¼ in. (32.4 x 46.4 cm)
The Lewis Walpole Library, Yale University, Topos L847 no. 27+

25.06†

BENJAMIN COLE (1697–1783), engraver
'View of the Foundling Hospital'
In William Maitland, *The History of London from Its Foundation to the Present Time*
London, 1756
Etching and engraving on paper
8 ¼ in x 12 ¼ in. (20.8 x 33.1 cm)
Foundling Museum, London, FM164

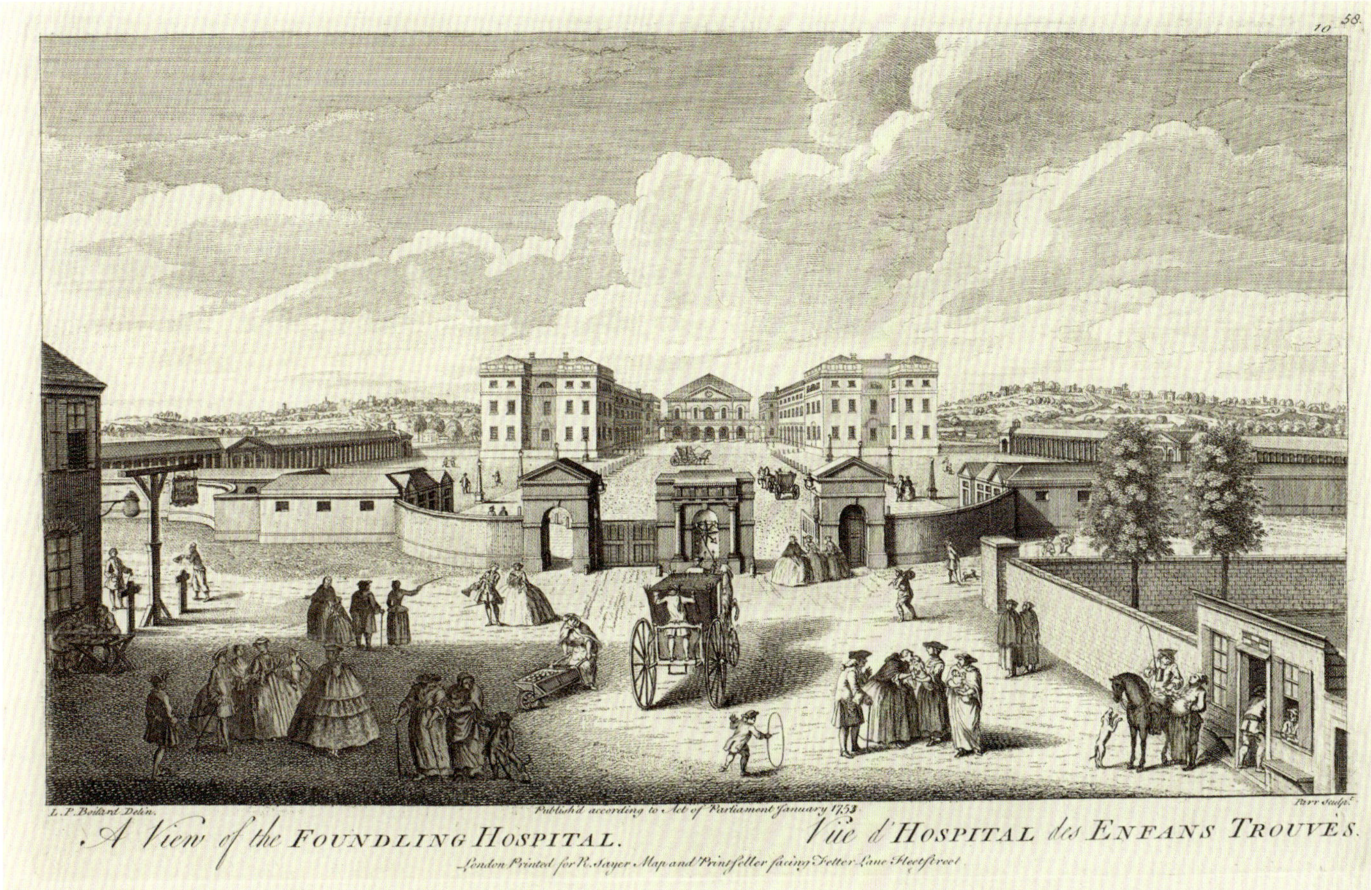
10 58
L. P. Boitard Delin.
Publish'd according to Act of Parliament January 1753
Parr Sculp.
A View of the Foundling Hospital.
Vüe d'Hospital des Enfans Trouvés.
London Printed for R. Sayer Map and Printseller facing Fetter Lane Fleetstreet

25.05

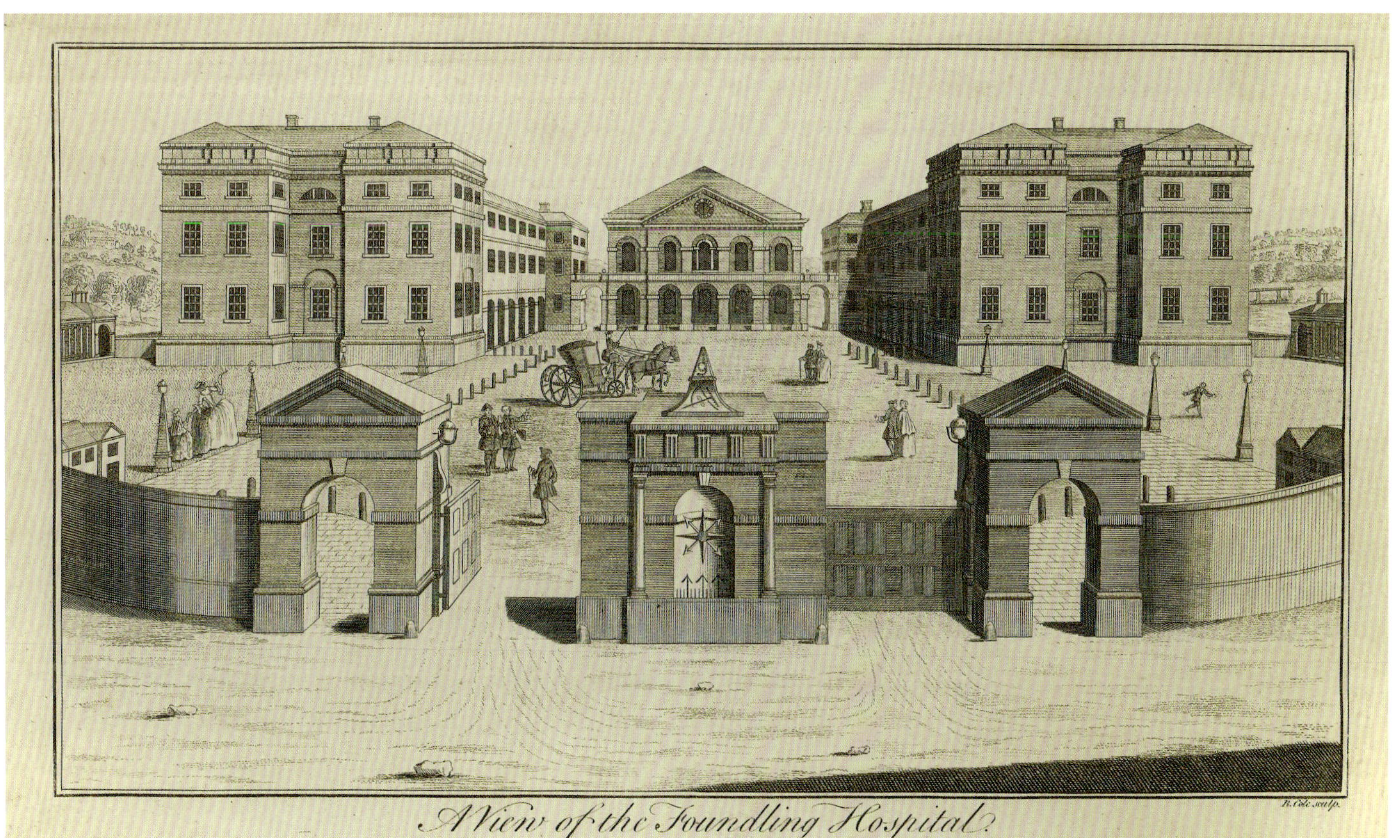
R. Cole sculp.
A View of the Foundling Hospital.

25.06

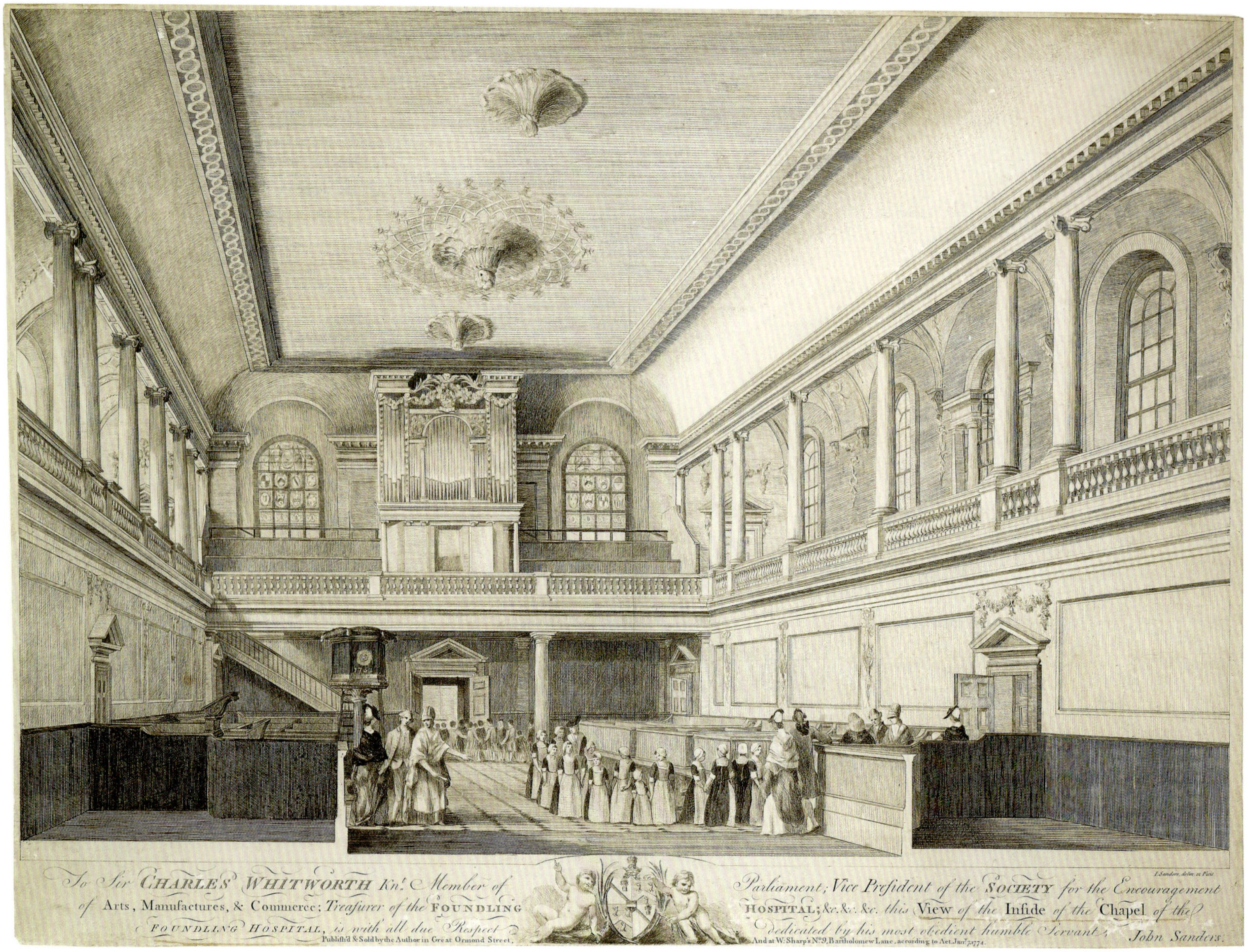

25.07

25.07

JOHN SANDERS (1768–1826), designer and engraver
The Chapel of the Foundling Hospital
1773
Engraving
18⅛ x 22⅝in. (46 x 57.5 cm)
Published by the author, London
Foundling Museum, London, FM179

The Foundling Hospital of London was incorporated by Royal Charter in 1739 for 'the maintenance and education of exposed and deserted young children'. When it first opened its doors on 25 March 1741 it was in temporary quarters in Hatton Garden. The new purpose-built hospital at Lamb's Conduit Fields began to receive children on 1 October 1745. The new building was designed by Theodore Jacobsen: a plain brick building with two wings and a chapel, built around an open courtyard. The western wing was finished in October 1745. An eastern wing was added in 1752 'in order that the girls might be kept separate from the boys'. Augusta donated money to fund the construction of the Foundling Hospital's chapel. The organ was the gift of George Frideric Handel, who served as a governor. From 1750 the annual charity performances of *Messiah* in the chapel became an enormously successful fund-raising and publicity occasion, bringing the fashionable and wealthy into the heart of the organization.

THE FOUNDLINGS

25.08*

JOSEPH HIGHMORE (1692–1780)
The Angel of Mercy
c. 1746
Oil on canvas
23½ x 19 in. (59.7 x 48.3 cm)
Yale Center for British Art, Paul Mellon Collection, B1981.25.362

The Court Room of the Foundling Hospital was located in its west wing. It would be filled with works presented by artists with an interest in the institution. *The Angel of Mercy* had been painted in anticipation that it would be presented to the Hospital for its art collection. However, the secular nature of its iconography, which shows a mother in the act of infanticide, interrupted by an angel who gestures towards the Foundling Hospital as her salvation, seems to have created an unease in the artist and he presented another work with the subject of Hagar and Ishmael, derived from a biblical source, which he seems to have considered more appropriate. *Hagar and Ishmael* by Joseph Highmore hung on the south wall.

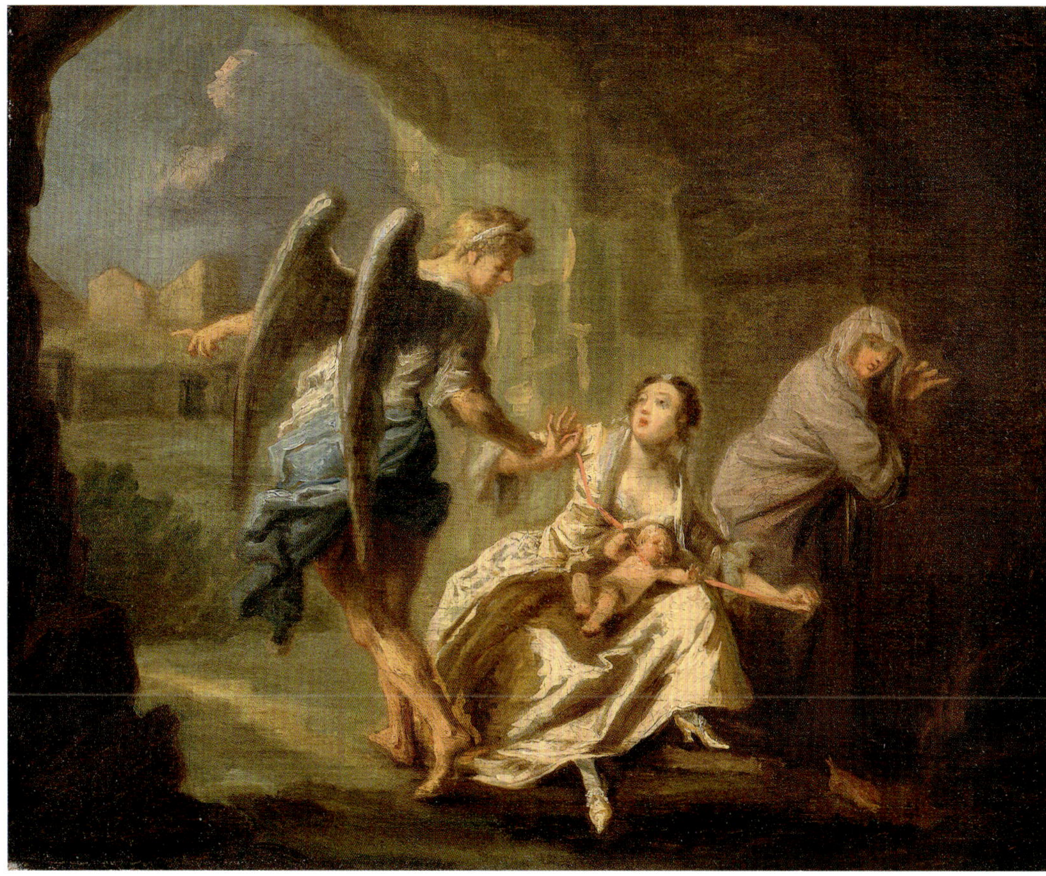

25.08

25.09*

WILLIAM HOGARTH (1697–1764)
Study for the Foundlings
c. 1739
Ink with wash on paper
4⅜ x 8⅜ in. (11.1 x 21.3 cm)
Yale Center for British Art, Paul Mellon Collection, B1975.4.1258

William Hogarth, who was childless, had a long association with the Foundling Hospital and served as a founding governor. He designed the children's uniforms and the coat of arms, and he and his wife Jane fostered foundling children. Hogarth was also instrumental in the setting up of a permanent art exhibition in the new buildings, encouraging other artists to produce work for the Hospital. He donated his painting *Moses Brought Before Pharaoh's Daughter* for the decoration of the Court Room, and *March of the Guards to Finchley* was also obtained by the Hospital after Hogarth donated lottery tickets for a sale of his works, and the Hospital drew the winning ticket.

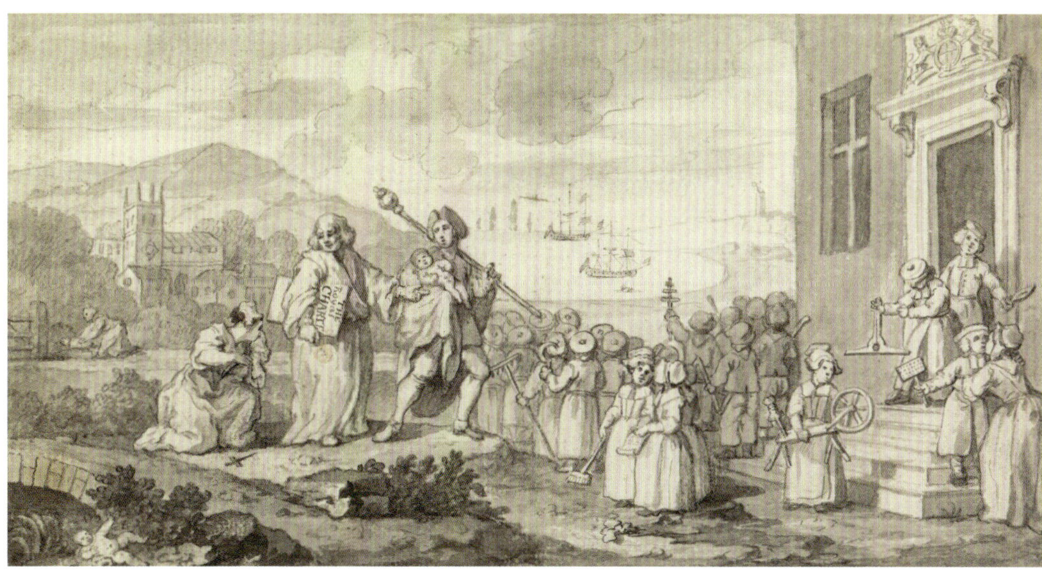

25.09

25.10

Padlock with band
Padlock: 1¼ x¾ x⅛ in. (3.2 x 2 x 0.4 cm)
Band: 2½ x 1⅜ x⅜ in. (6.2 x 3.5 x 1cm)
Foundling Museum, London, 2005.5.382
Token for child number 5706, who was admitted to the Foundling Hospital in September 1757. He was renamed James Newington. He died the following year.

25.11

Coral necklace
Approx. 9 in. (23 cm) unclasped; approx. 4⅜ in. (11 cm) clasped
Foundling Museum, London, 2005.5.156

25.12

Love token (six pence, silver, [York], reign of King William III)
1696–97
3/4 in. (2 cm) diameter
Foundling Museum, London, 2005.5.299

25.13

Metal engraved heart
9⅜ x 8¾ in. (23.8 x 22.2 cm)
Foundling Museum, London, 2005.5.62
Token for child number 291, who was admitted to the Foundling Hospital in February 1746. He was renamed Robert Sammon. He died on 19 March in the same year.

25.14

Punched and notched coin (three pence, silver, reign of King Charles II)
1680
Diameter:⅝ in. (1.7 cm)
Foundling Museum, London, 2005.5.239

The first infants were admitted to the Founding Hospital in London in 1741. To be eligible the babies were to be aged two months or less and initially numbers were limited, with annual admissions never exceeding two hundred. In 1742, to ensure fairness a ballot system was instituted, whereby the mother drew a white or black ball which would denote success or failure for her application. Successful mothers were required to 'affix on each child some particular writing, or other distinguishing mark or token, so that the children may be known thereafter if necessary.' As entry to the Hospital became a form of adoption, the child was given a new name. But should the mother's circumstances change, she had the right to retrieve her child by reference to the identifying token. The majority of the tokens left were fragments of textiles, cut probably from the clothes the infants were wearing when presented. However, mothers sometimes left a coin, piece of jewellery, or small metal trinket such as a padlock.

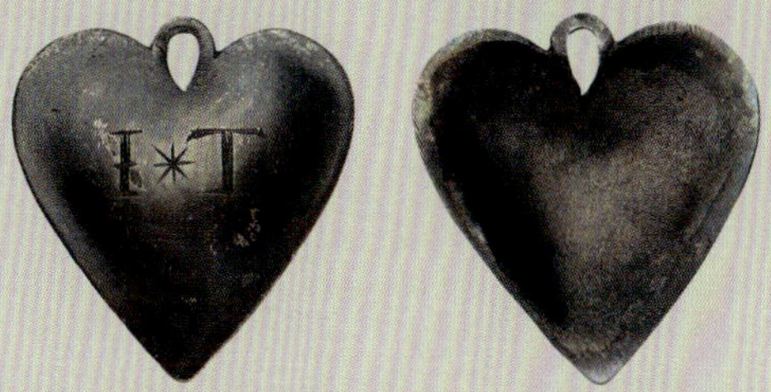

25.13

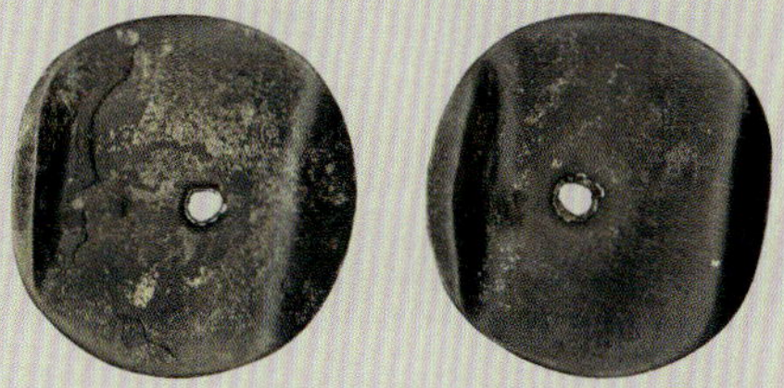

25.12

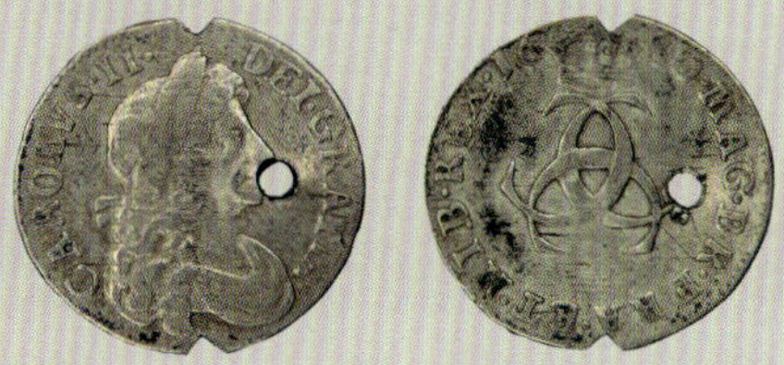

25.14

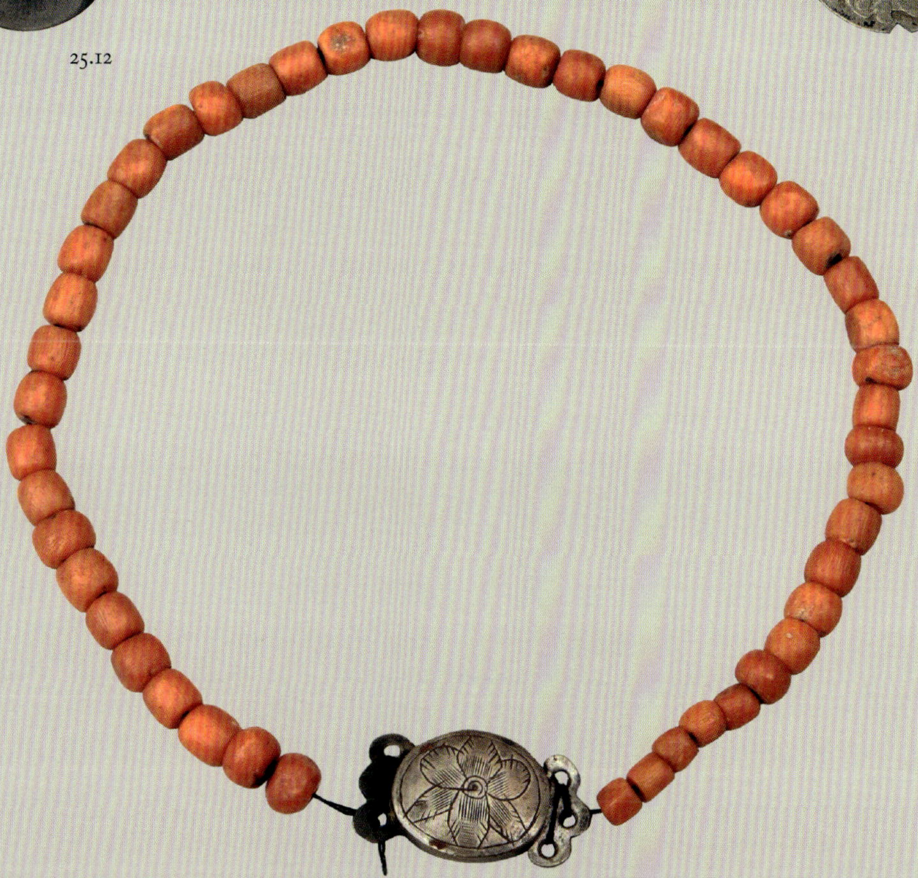

25.11

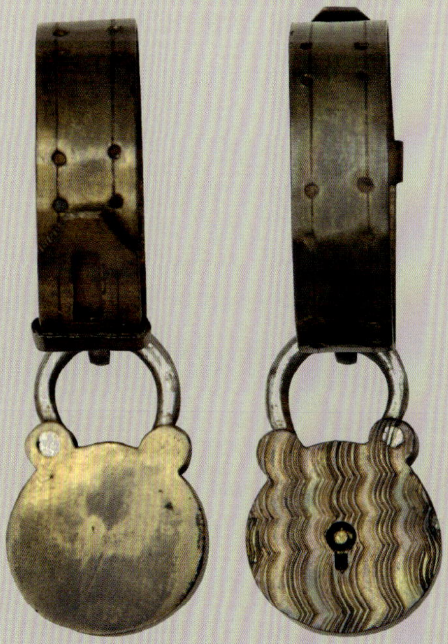

25.10

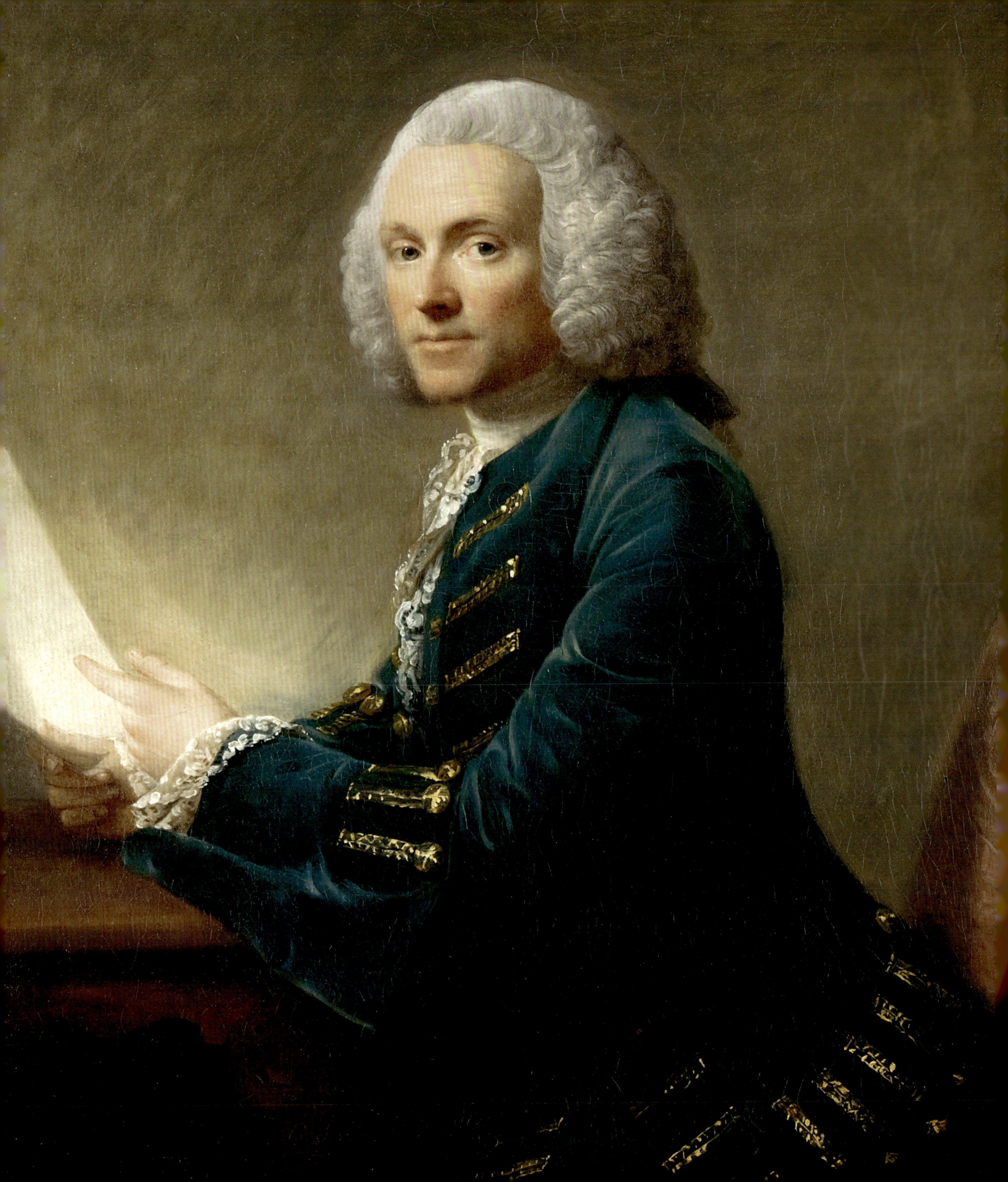

Medical Practice

Craig Ashley Hanson

This essay considers the medical patronage of Caroline of Ansbach and Charlotte of Mecklenburg-Strelitz and how they embraced and nurtured innovations in Georgian medicine. Their support of smallpox inoculation, medical Newtonianism, and advances in obstetrics emerge as crucial components of their legacies as enlightened princesses.

The Hanoverian court, especially through the three Princesses, Caroline, Augusta, and Charlotte, played an essential role in the life of eighteenth-century medicine, from the Royal Society to the appeal of quacks, from publishing projects to the collecting of objects, and from foundlings to the royal nursery. If the claim may sound like a platitude, this is not at all a common position within the history of medicine. Recent accounts of the period by John Brewer, Roy Porter, and others have emphasized the social and economic transformations of the eighteenth century, attendant upon an emerging commercial sphere, broadly under the umbrella of modernity.[1] However, the story of enlightened modernization paid relatively little attention to court or the crown. Scholars have more recently begun to emphasize the ways the court and commercial spheres were, in fact, productively bound together. If a medical marketplace meant a momentous shift away from the Royal College of Physicians, the traditional centre of medical power, the most fashionable physicians of the eighteenth century nonetheless continued to benefit from appointments to the Royal Household.[2] For too long these positions have been viewed as incidental to the careers of these doctors.

In medical terms, the dynastic shift from the Stuarts to the Hanoverians can be dated to 27 April 1714, when Queen Anne was the last monarch to touch her subjects for scrofula, known as the 'King's Evil'. The practice had carried a particular symbolic importance for her predecessors in asserting royal powers. While Georgian rulers were content to let this regal performance slip quietly into the past, the ceremony persisted in France even into the early nineteenth century; at his coronation in 1722 Louis XV touched some two thousand scrofula victims, and the practice was revived briefly with the Bourbon restoration in 1815.[3]

And yet with Anne's passing, there were already present at her deathbed two of the physicians who would play enormous roles for the next four decades: Hans Sloane and Richard Mead. Sloane, as a royal physician, was already established, whereas Mead was on the cusp, having been sent to attend to the dying Queen as a protégé of Dr John Radcliffe, who scandalously refused to go. In this sense too it was a changing-of-the-guard moment for the British medical establishment, as Radcliffe himself would die a few months later, and Mead would move into his mentor's house, inherit many of his wealthy patients, and soon enough return to the Royal Household with an ordinary appointment under both George II and Caroline.[4]

In 1736, the year before she died, Caroline presented Mead with a portrait of herself, which had been painted in 1735 by Jacopo Amigoni (see p. 85, fig. 4.11). The step of the dais includes this inscription: 'GIVEN BY HER MAJESTY QUEEN CAROLINE/ TO HER MOST HUMBLE SERVANT Dr. MEAD.

Detail of p. 435, cat. 26.09

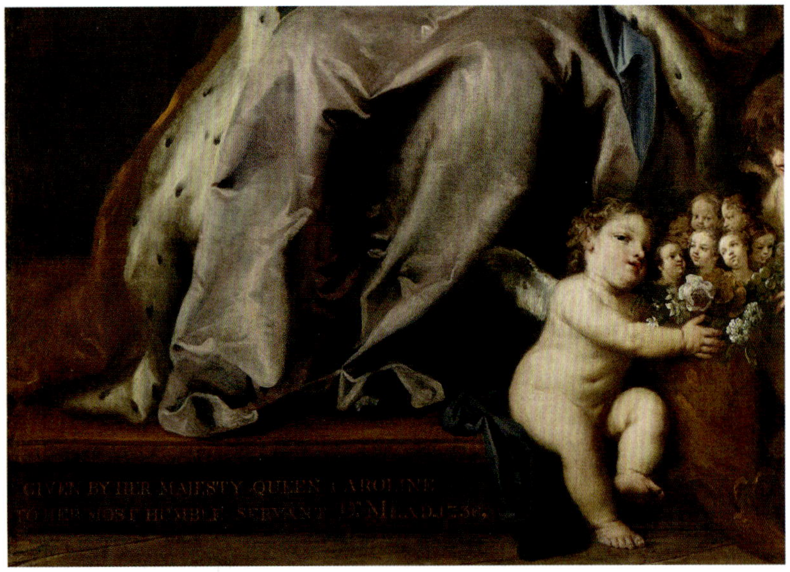

FIG. 26.01 Jacopo Amigoni, *Caroline Wilhelmina of Brandenburg-Ansbach* (detail; see p. 85, fig. 4.11), 1735, oil on canvas, 96 x 60 in. (243.9 cm x 152.4 cm). National Portrait Gallery, London

1736' (fig. 26.01). As a social document, the portrait points to the changing relationship between members of the royal family and physicians, and the inclusion in the lower right-hand corner of a cornucopia with Caroline's seven surviving children casts the picture as a royal maternal image. Understanding the nuances of Amigoni's portrait along with the relationship it marked with one of her physicians, however, requires a richer appreciation of her connections to the medical community, particularly around the issue of children's health.

At the top of the household medical hierarchy was the physician ordinary to the person of the monarch, followed by second, third, or fourth ordinary physicians. Extra-ordinary physicians held less-privileged positions, as the label denoted occasional consultations—though extra-ordinary physicians could have a greater impact, depending upon the circumstances that brought them to attend upon the monarch. There was fluidity between the positions, along with varying degrees of influence. David Hamilton, for instance, was in 1703 appointed third physician to Anne and shortly thereafter knighted. He was later bumped up to second physician, but even that does not signify the degree to which he served as a kind of confidant and counsellor. That Sir David was a 'man-midwife' or 'accoucheur' might initially account for his standing with Anne, who was so frequently, if also tragically and unsuccessfully, pregnant—except that by 1703, Anne had already faced her last miscarriage. It was also common for royal appointments to transfer from one member of the royal family to another; Hamilton for instance would go on to be appointed physician to Caroline on her arrival in London as Princess of Wales in 1714. Likewise, Dr Sloane served as physician not only to Anne but also to George I and George II, as did Sir John Shadwell.[5]

Surgeons were understood to be of inferior professional rank to physicians. They were, after all, much more prone to come into physical contact with blood and the discharges of diseased bodies. The painting of William Cheselden, one of Caroline's surgeons, *William Cheselden Giving an Anatomical Demonstration to Six Spectators in the Anatomy-theatre of the Barber-Surgeons' Company* (fig. 26.02), might mark the difference. Because of their expertise with the messier parts of healthcare, surgeons also could be called to take on vital tasks. It was John Ranby who bore the responsibility of operating on the strangulated umbilical hernia which would cause Caroline's death in 1737. It did not obstruct his career, however: he went on to become surgeon to George II.

The apothecaries also often benefited from reappointments. During the Stuart period James Chase, for instance, served James II, William, and Anne. Daniel Malthus similarly served both Anne and George I. Such continuity is also to be found within the same family across generations. Thus Malthus's brother-in-law Thomas Graham was apothecary to George I, while his son, Daniel Graham, served George II and Caroline. That they operated a profitable shop concurrently provides another instance of royal patronage and commercial enterprise working together.

As one might expect given the chain of command, apothecaries took direction from the physicians, but it is important to bear in mind the material qualities of their work. Quite apart from their role in administering medicines, it was the apothecary's job to mark both the beginning and end of the monarch's reign. They prepared both the oil to be administered by the Archbishop of Canterbury during the coronation ceremony and the embalming materials. Throughout the eighteenth century, £206 was the standard rate for the service, and was paid to Daniel Graham for the oils, which he had prepared apparently according to the same formula as in 1714, for the coronation of George II and Caroline in 1727. Incidentally, Graham's annual salary from the crown was £200.[6] Another member of the Graham family received two payments of £50 for the embalming materials for both Caroline and later Frederick, Prince of Wales, while Ernst August Jäger carried out the preparation of the Queen's body. At the death of George II, the surgeon Ranby together with the physician Edward Wilmot (Dr Mead's son-in-law) informed George III of his grandfather's final two wishes: 'that his body should be embalmed with a double quantity of perfumes, and that it should be laid close to that of the queen'.[7]

This level of detail may seem like a distraction from the enlightened interests and patronage of these royal women, but the

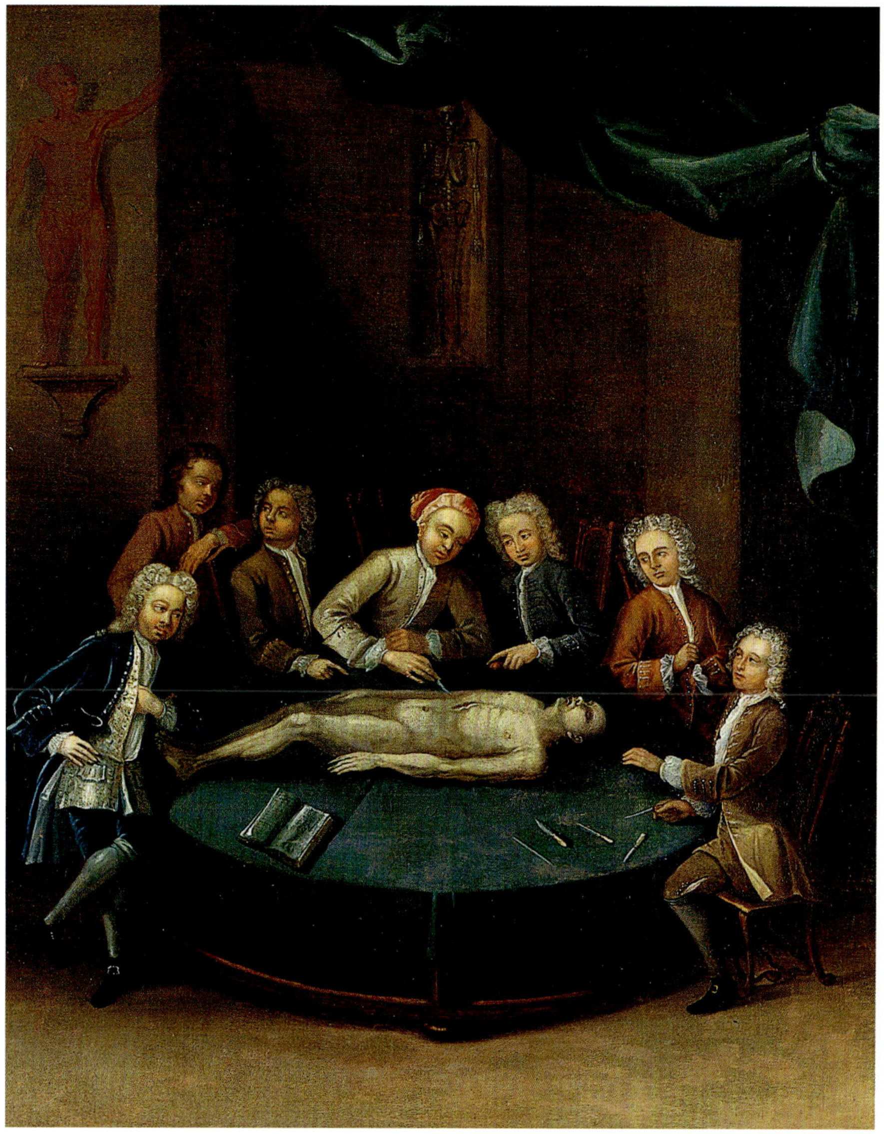

FIG. 26.02 Charles Philips, *William Cheselden Giving an Anatomical Demonstration to Six Spectators in the Anatomy-theatre of the Barber-Surgeons' Company, London*, 1730/40, oil on canvas, 31⅜ x 23¾ in. (79.7 x 60.5 cm). Wellcome Library, London

relationships between members of the royal family and medical practitioners were premised on the physical (and mental) care of a royal person. Corporeal interaction lay at the core of these relationships, with physical knowledge and trust playing an essential role.

With this sense of trust in mind, we come to the introduction of inoculation into Britain in the 1720s, culminating in the inoculation of Caroline's daughters Amelia and Caroline, aged eleven and nine years old respectively, in April 1722. The story has been told with various points of emphasis, but rarely with any interest in the Princess of Wales's involvement.[8]

In June 1714, the Royal Society of London received a letter from Emanuele Timoni on the topic of inoculation. Based in Constantinople, he had visited England in 1703, at which point he had been elected a Fellow of the Society. From this point forward, inoculation would be a recurring subject for the learned organization. In fact, as early as 1700 the Royal Society had discussed nasal inoculation (the so-called Chinese method) under the direction of royal physician Martin Lister, but it was not published, and reports of the process would only be circulated in 1720. Accounts of the history of European inoculation typically progress from Timoni's 1714 letter, published in *The Philosophical Transactions*, to the initiatives to encourage wide-scale inoculation promoted by Lady Mary Wortley Montagu. Having survived the infection in 1715, she was intrigued by inoculation as performed in Turkish and Greek medical practice, which she encountered between 1716 and 1718 while her husband served as Ambassador in Constantinople. After her son was successfully inoculated there in 1718, she would write to her father, who was a friend of Dr Samuel Garth as well as others within the court circle championing its efficacy. After her return to London, during the smallpox epidemic in 1721 she elected to have her three-year-old daughter inoculated under the auspices of Dr Charles Maitland, the Scottish surgeon who had been attached to the British Embassy in Constantinople. He agreed that the procedure might be observed by three members of the College of Surgeons. The little girl fared well, and the success of the experiment was quickly promoted by its witnesses. Sir Hans Sloane suggests that it was at this point Caroline became 'her [Lady Mary's] firm supporter and stood by her without quelling'.[9] Caroline had survived smallpox in 1707, but the disease nearly took the life of her eldest daughter, Anne, in 1720. From this point on it was the royal promotion of the practice of inoculation which would see its gradual adoption as a means of improving public health.

According to Sloane, Caroline initiated the first series of moderated inoculation experiments to take place in Britain: 'To secure her other children, and for the common good, [she] begged the lives of six condemned criminals' to assist with these.[10] On 14 June 1721, Lord Townsend wrote for legal advice regarding the practice and the legality of human experiments. Three days later came the response that it would be lawful on two grounds: firstly, if the prisoners' lives were already 'in the power of his Majesty', and secondly, if 'carrying the practice to perfection may tend to the General Benefit of Mankind.'[11]

By early August, six prisoners from Newgate prison had been selected, and on 9 August they were inoculated, with some two-dozen medical men observing.[12] Maitland carried out the procedure under the direction of Sloane and George I's German physician, Johann Steigertahl.[13] Sloane's account of the occasion was not written until the 1730s and did not appear in print until 1756.[14]

More experiments were performed and reported in the newspapers, with the royal authorization of George I and Caroline usually emphasized. In February 1722 six more people were inoculated, with reports listing where they could be found for follow-up observation; and a few weeks later, in March, the royal champions directed that five orphans from St James's Parish were to be inoculated. These cases too were reported in newspapers, and the children similarly made available for viewing afterwards. A letter from Claude Amyand, one of the royal surgeons involved, is especially informative for the orphans' experiment. He writes to Dr Sloane with concern over the poor health of the children proposed for the procedure:

> These children are very miserable, & their Royal Highnesses apprehend that the state of their boils [?] does not make them fit for an experiment. However I submit that to your better judgment. . . . The princess will be glad to know whether you think these wanting, & therefore came to wait on you on this account.[15]

While several of the words are illegible, 'princess' is perfectly clear, as is the 'Royal Highnesses' of the first sentence. Such private correspondence between court physicians thus reinforces the involvement reported on in the newspapers.

The following month Caroline had her daughters Amelia and Caroline inoculated, with Amyand performing the procedure under the watch of Sloane and Maitland. The next year, Prince Frederick, who was still residing in Hanover, was also inoculated.

We might recall just how widespread the devastating impact of smallpox was in the seventeenth and eighteenth centuries, not only for the population generally (an epidemic devastated Boston

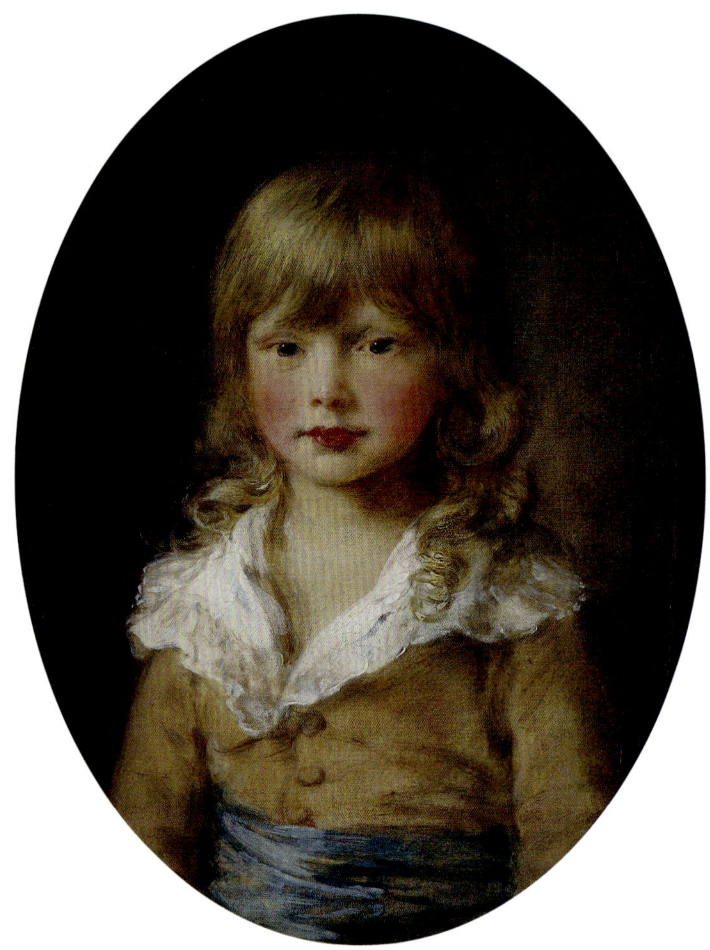

FIG. 26.03 Thomas Gainsborough, *Prince Octavius*, 1782, oil on canvas, 22¼ x 16½ in. (56.4 x 42 cm). Royal Collection Trust

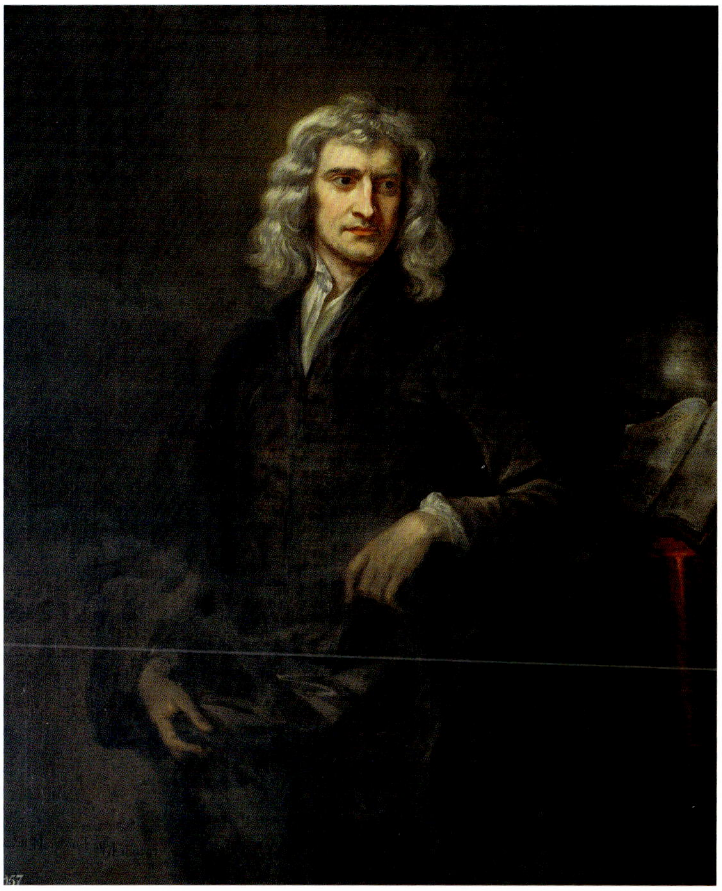

FIG. 26.04 Charles Jervas, after Sir Godfrey Kneller, *Sir Isaac Newton*, 1689, oil on canvas, 49½ x 40 in. (125.8 x 101.6 cm). Royal Collection Trust

in 1721) but also for royal families. Anne Stuart, later Queen Anne, who contracted the disease just before the wedding of her sister Mary Stuart (later Mary II) to William, Stadholder of the Netherlands (later William III), had been too ill to attend the ceremony, and two of her daughters, Anne Sophia and Mary, died as infants of smallpox. While William also contracted the disease and survived, it would cut short the life of Queen Mary in 1694.

Recent circumstances in France would have further underscored the seriousness of the threat. While the Bourbon court had long expected Louis to succeed his father, Louis XIV the Sun King, the Grand Dauphin died of smallpox in April 1711 leaving *his* son, Louis, duc de Bourgogne, the Petit Dauphin and Louis XIV's grandson, to inherit the throne. He too, however, died in February 1712 soon after his wife, and their older son likewise died just a month later—all three apparently of measles—leaving only one other son to inherit the throne. Consequently, upon the death of Louis XIV in 1715, that child, the Sun King's great-grandson, inherited the throne as a five-year-old. In short, Louis XV offered a sobering lesson. The fact that a disease could decimate a line of royal succession in a matter of months would have been perfectly clear to George and Caroline in 1722, when they began inoculating their own children.

The potential dividends, then, of inoculation were enormous, yet it brought its own risks: an estimated 2 to 3 percent of those treated would die from the procedure. Charlotte's two youngest children, Octavius and Alfred, suffered just such a fate decades later. Painted in the autumn of 1782, Thomas Gainsborough's portraits of the royal children were exhibited the following summer at the Royal Academy (fig. 26.03). While Octavius sat for the artist just before becoming ill, Alfred's picture was painted after the boy's death. Both George and Charlotte were shaken. George wrote, 'There will be no Heaven for me if Octavius is not there', and Charlotte broke down in tears in front of the portraits when they were exhibited in the Royal Academy that summer.[16]

In the 1720s, Fellows of the Royal Society compiled all records of inoculations they could assemble. James Jurin's *Account of the Success of Inoculating the Smallpox in Great Britain* of 1725 is dedicated to Caroline, and it was precisely the kind of analysis Mead thought should characterize the medical profession. He identified the mastery of classical languages and mathematics as the two crucial intellectual skills which distinguished physicians from empirics or quacks. The doctors who commanded the highest respect in Georgian London were members of the Royal Society, and they usually incorporated some version of Newtonianism into their medical theory.

Rather than depend on the German practitioners who had accompanied the Hanoverians, George and Caroline energetically embraced not only British physicians and surgeons but also Newtonianism. Caroline was eager to meet Sir Isaac Newton after her arrival in England, and by February 1716, she had entertained him, with Samuel Clarke, to hear about 'Sir Isaac's system of Philosophy'.[17] The busts Caroline commissioned from Giovanni Battista Guelphi of Newton, Clarke, Robert Boyle, John Locke, and William Wollaston visually mark her interest and engagement with the world of the contemporary scientist (see cats. 16.10–16.14). To complement the sculptural series, she may have acquired Charles Jervas's copy of Sir Godfrey Kneller's portrait of Newton for the royal collection (fig. 26.04), together with painted images of the other contemporary heroes immortalized by Guelphi. George II and Caroline aligned themselves with the foremost medical practitioners of the age, despite what might be their political leanings, in keeping with their greater strategy to present themselves as English rather than foreign monarchs. This can be seen in Caroline's commitment to Mead and Cheselden, who were devotees of Newton, at the same time as Caroline's appointment of John Freind in 1727, who despite being a medical Newtonian was also a Jacobite, implicated in the 1722 plot against the Hanoverians. He spent several months imprisoned in the Tower, and was released only at Mead's pleading. The move towards Newtonianism is all the more striking given Caroline's close connections with Newton's rival Gottfried Wilhelm Leibniz.

The Royal Society connection was one of the major shifts within the medical profession between the Stuarts and Hanoverians. Previously, it was the Royal College of Physicians that benefited from strong connections and protection from the crown. As Harold Cook has argued, that ceased to be the case after 1688, and various groups including surgeons and apothecaries played increasingly important roles, as a medical marketplace emerged.[18] One of Caroline's physicians, Alexander Stuart, exemplifies this shift. He was elected Fellow of the Royal Society in 1714, conducted inoculations in 1725, and in October 1728 was named physician to the Queen. Only the *next* month did he become a Fellow of the Royal College of Physicians, the designation coming as a seeming afterthought. Largely thanks to the lobbying of the royal surgeons Ranby and Cheselden, the surgeons broke from the Company of Barbers to form the Company of Surgeons in 1745, which became the Royal College of Surgeons in 1800.

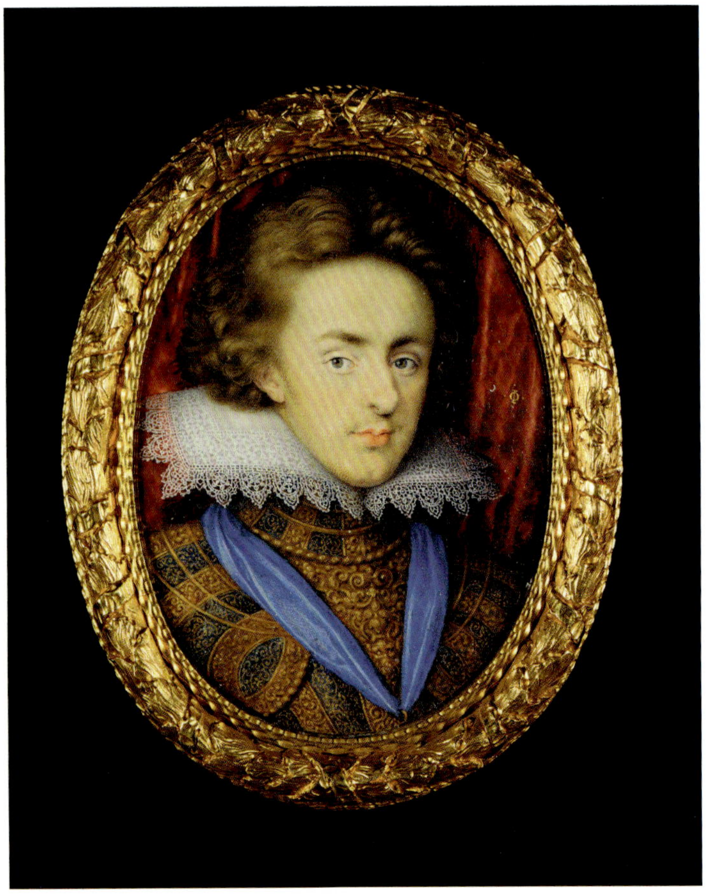

FIG. 26.05 Isaac Oliver, *Henry Frederick, Prince of Wales*, 1612, watercolour on vellum laid on playing card, 3 x 2¼ in. (7.7 x 5.6 cm). Royal Collection Trust

In such paintings as Amigoni's portrait of 1735, Caroline was not simply casting herself as a good mother, but rather using motherhood to advance her claim for popular support within the English nation.[19] Her role in promoting the practice of inoculation appears to combine the fate of the royal children with the public good, dissociating her and her husband's reputation from that of George I, who was early characterized as a foreign and disinterested monarch.

The physicians to the Georgian court were empowered by the Royal Society to set their practice of medicine in a wide-ranging sphere of scholarship and inquiry. Mead and Sloane set a standard of collecting and patronage that even rivalled that of the royal family. Mead owned one of the most important collections of art outside the nobility; Frederick, Prince of Wales, acquired miniatures from him, to the chagrin of Horace Walpole. A seventeenth-century portrait by Isaac Oliver of Henry Frederick, Prince of Wales (fig. 26.05), depicts a prince of an earlier century who would, like Prince Frederick, die before acceding to the throne. Mead's library numbered some ten thousand volumes, including a Shakespeare second folio that had belonged to Charles I and was annotated by him just before his execution.

William Hunter was the heir to Mead's legacy, combining learned medical practice and ambitious collecting.[20] At the auction of Mead's collections in 1754, Hunter acquired books and works of art, nurturing connections with his predecessor.

Hunter played a leading role in changes in obstetrics, within which Charlotte would be implicated. Hunter officially became her physician in extraordinary on 18 September 1762. While physicians had long been on hand at royal births, the delivery itself traditionally had been in the hands of midwives. Initially, this was the case for Charlotte, but with her fourth child, Charlotte, born on 29 September 1766, the Queen dismissed the royal midwife and placed herself entirely in the care of Hunter. This appears to have been the first time a royal mother went through labour with a physician delivering the child. Lady Mary Coke, in her diary for 14 September 1766, wrote: 'The Queen was to be brought to bed by Dr. Hunter instead of the old woman but that it was to be kept a secret as if the fate of the country depended on this change.'[21] Charlotte therefore deserves credit for this 'revolutionary' departure from traditional gendered expectations of the role of a physician at birth.[22] Whereas fashionable ladies had been coming round to the idea of 'man-midwives' for decades, Charlotte brought royal childbirth, a crucial function ensuring the success of a monarchy, up to date through her support of the period's most important obstetrician.[23]

Hunter's reputation today rests largely on his extraordinary anatomical publication, *The Gravid Uterus* (cat. 26.10). The doctor had been working on the project since the 1750s, though it appeared only in 1774. With thirty-four prints, the book offered by far the most precise and engagingly naturalistic images of a foetus within the uterus ever produced.[24] The full title includes Charlotte's name: *Anatomia uteri humani gravidi tabulis illustrata, auctore Gulielmo Hunter, serenissimae Reginae Charlottae medico extraordinario, in Academia Regali anatomiae professore, et Societatum, Regiae et Antiquariae, socio.*, and the book is dedicated to the King.

We should see in the royal couple's support of the publication, public affirmation for the obstetrician whose more immediate service to the family could hardly be acknowledged with such striking directness. While in the 1720s, Caroline could visibly embrace inoculation by supporting trials and inoculating her own children, in the 1760s and 1770s, Charlotte's intimate relationship with the most advanced medical knowledge of her day, while perhaps harder to bring out into the open for reasons of propriety, was just as significant.

NOTES

1. See, for instance, John Brewer, *The Pleasures of the Imagination: English Culture in the Eighteenth Century* (New York: Farrar, Straus & Giroux, 1997), and Roy Porter, *The Creation of the Modern World: The Untold Story of the British Enlightenment* (New York: Norton, 2000).
2. For the fortunes of the Royal College of Physicians in the late seventeenth century, see Harold Cook, *The Decline of the Old Medical Regime in Stuart London* (Ithaca, NY: Cornell Univ. Press, 1986), and Harold Cook, 'Living in Revolutionary Times: Medical Change under William and Mary', in *Patronage and Institutions: Science, Technology, and Medicine at the European Court, 1500–1750*, ed. Bruce T. Moran (Rochester, NY: Boydell Press, 1991), 111–35.
3. As a child, Samuel Johnson was touched by Queen Anne. For the last instance of the Queen performing the ritual, see Elizabeth Lane Furdell, *The Royal Doctors, 1485–1714: Medical Personnel at the Tudor and Stuart Courts* (Rochester, NY: Univ. of Rochester Press, 2001). For the King's Evil generally, see Roy Porter, *The Greatest Benefit to Mankind: A Medical History of Humanity from Antiquity to the Present* (London: Harper Collins, 1997), esp. 282.
4. See Craig Ashley Hanson, *The English Virtuoso: Art, Medicine, and Antiquarianism in the Age of Empiricism* (Chicago: Univ. of Chicago Press, 2009), 167–69.
5. Furdell, *Royal Doctors*, 239–40.
6. *Verses on the King coronation of . . . George II and Queen Caroline . . . spoken by the scholars of Westminster School* (London, 1761), 61–62. For the apothecaries and amounts, see Leslie Matthews, *The Royal Apothecaries* (London: Wellcome Historical Medical Library, 1967), 158.
7. Norman Moore, rev. Kaye Bagshaw, 'Sir (George) Edward Wilmot, first baronet (1693–1786)', *Oxford Dictionary of National Biography* (Oxford: Oxford Univ. Press, 2004; online ed.), http://www.oxforddnb.com/view/article/29620.
8. Genevieve Miller, *The Adoption of Inoculation for Smallpox in England and France* (Philadelphia: Univ. of Pennsylvania Press, 1957). For a brief, recent account from a medical perspective, see Robin Weiss and José Esparza, 'The Prevention and Eradication of Smallpox: A Commentary on Sloane (1755) "An Account of Inoculation"', *Philosophical Transactions of the Royal Society B* 370 (April 2015).
9. See also an acknowledgment of Caroline's contribution to the development of this medical practice, in an article by Richard Steele, published in *The Plain Dealer*, London, 3 July 1724, quoted in S. L. Kotar and J. E. Gessler, *Smallpox: A History* (Jefferson, NC, and London: McFarland, 2013), 16.
10. Hans Sloane, 'An Account of Inoculation by Sir Hans Sloane, Bart. Given to Mr. Ranby, to be Published, Anno 1736. Communicated by Thomas Birch, D. D. Secret. R. S.', *Philosophical Transactions* 49 (London, 1755–56): 517.
11. Townsend wrote to the Attorney and Solicitor General, Sir Robert Raymond and Sir Philip Yorke, asking, 'Whether His Majesty may by Law Grant his Gracious Pardon to the two Malefactors under Sentence of Death upon Condition that they will suffer to be try'd upon them the Experiment of Inoculating the Small pox'; quoted in Miller, *Adoption of Inoculation for Smallpox*, 76 (with references for the primary sources).
12. Londa Schiebinger, 'Human Experimentation in the Eighteenth Century: Natural Boundaries and Valid Testing', in *The Moral Authority of Nature*, ed. Lorraine Daston and Fernando Vidal (Chicago: Univ. of Chicago Press, 2004), 384–408.
13. As Furdell notes in *Royal Doctors*, 246–47, several German physicians arrived with George I, but these were the exception within the royal medical entourage.
14. Sloane, 'An Account of Inoculation', 516–20.
15. British Library (BL) Sloane MS 4076.
16. Jane Roberts, ed., *George III and Queen Charlotte: Patronage, Collecting and Court Taste* (London: Royal Collection, 2009), 64.
17. Lady Cowper records the meeting, as quoted in Joanna Marschner, *Queen Caroline: Cultural Politics at the Early Eighteenth-Century Court* (New Haven and London: Yale Univ. Press, 2014), 74. Marschner's chapter on 'Caroline and the Natural Philosophers', 149–69, is especially important.
18. Cook, 'Living in Revolutionary Times', 111–35.
19. For a detailed discussion of Amigoni's portrait, see, in this volume, Cassandra Albinson and Mark Hallett, 'Cornucopia: Portraiture and the Imperatives of Reproduction', 77–107. See also Christine Gerrard, 'Queens-in-Waiting: Caroline of Anspach and Augusta of Saxe-Gotha as Princesses of Wales', in *Queenship in Britain, 1660–1837: Royal Patronage, Court Culture, and Dynastic Politics*, ed. Clarissa Campbell Orr (Manchester: Manchester Univ. Press, 2002), 143–61.
20. Helen McCormack, 'A Collector of the Fine Arts in Eighteenth-Century Britain: Dr William Hunter (1718–1783)' (PhD thesis, Univ. of Glasgow, 2010).
21. Quoted in C. Helen Brock, *The Correspondence of Dr William Hunter, 1740–83*, 2 vols. (London: Routledge, 2008), 1: 243. Deborah Chetwynd was seamstress and laundress for Charlotte from 1761 to 1788.
22. Lisa Forman Cody, *Birthing the Nation: Sex, Science, and the Conception of Eighteenth-Century Britons* (Oxford: Oxford Univ. Press, 2005), 215.
23. For the acceptance of male practitioners generally, see Adrian Wilson, *The Making of Man-Midwifery: Childbirth in England, 1660–1770* (Cambridge: Harvard University Press, 1995).
24. See Caroline Grigson, '"An Universal Language": William Hunter and the Production of *The Anatomy of the Human Gravid Uterus*', in *William Hunter's World: The Art and Science of Eighteenth-Century Collecting*, ed. E. Geoffrey Hancock, Nick Pearce, and Mungo Campbell (Farnham: Ashgate Publishing, 2015), 59–80; Ludmilla Jordanova, 'Gender, Generation, and Science: William Hunter's Obstetrical Atlas', in *William Hunter and the Eighteenth-Century Medical World*, ed. W. F. Bynum and R. S. Porter (Cambridge: Cambridge Univ. Press, 1985), 385–413; and Lyle Massey, 'Pregnancy and Pathology: Picturing Childbirth in Eighteenth-Century Obstetric Atlases', *The Art Bulletin* 87 (March 2005): 73–91.

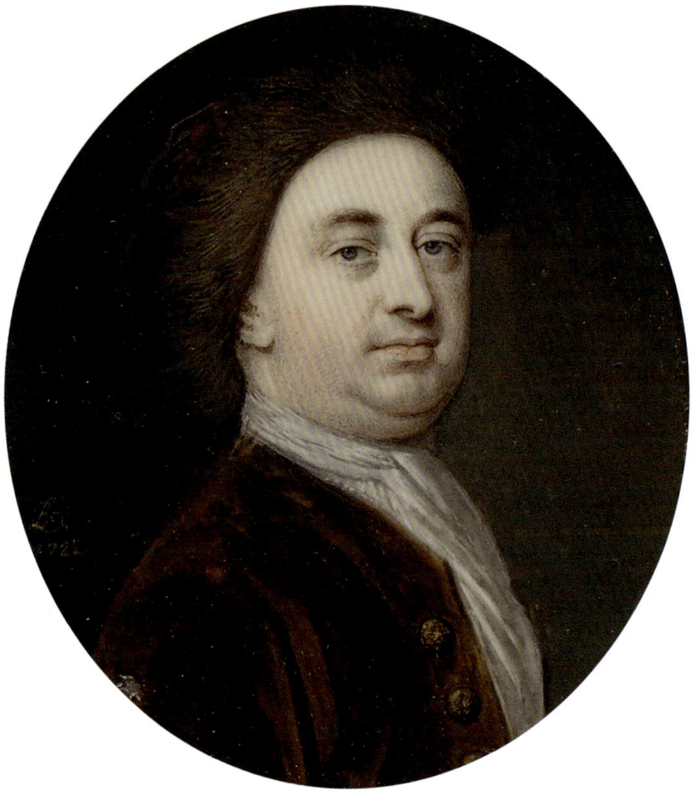

26.06

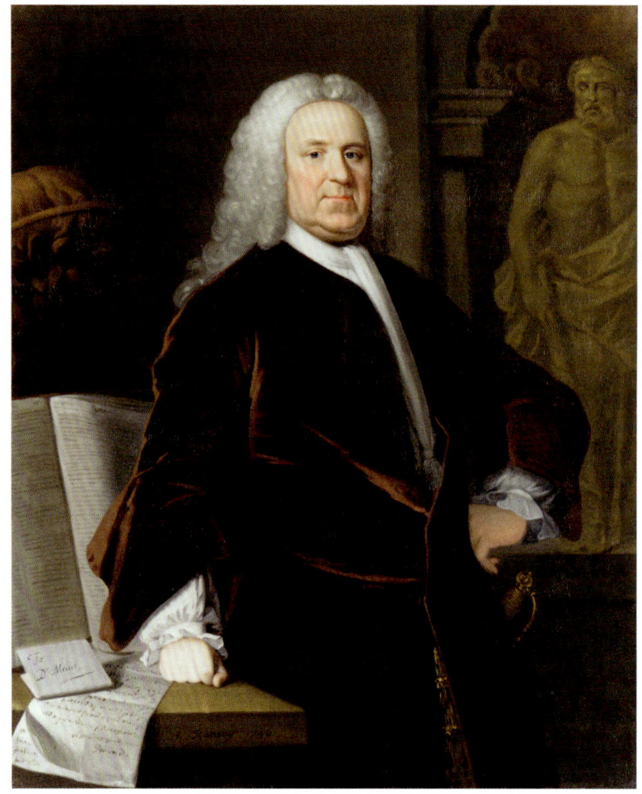

26.07

ROYAL DOCTORS

26.06*

BERNARD LENS III (1681–1740)
William Cheselden
1722
Watercolour and gouache
2⅜ x 2⅜ in. (6 x 6 cm)
Beinecke Rare Book & Manuscript Library, Yale University, James Marshall and Marie-Louise Osborn Collection, Osborn c419

In 1727 Cheselden was appointed surgeon to Queen Caroline. He had already, as an ambitious entrepreneurial anatomist, instigated a successful course of thirty-five lectures on anatomy, comparative anatomy, and animal economy, combined with indications for surgical operations. In 1711 his lecture abstract *Syllabus sive index humani corporis partium anatomicus* was published and later he joined Francis Hauksbee, a teacher of physics, to offer one of the first pre-clinical training courses in Britain. His successful practice brought him into contact with many eminent individuals. He was a close friend of Dr Richard Mead, Sir Hans Sloane, physician and collector, as well as the painter Jonathan Richardson and Alexander Pope. Bernard Lens, a member of the Rose and Crown club with Hogarth, Zincke, and Vertue, would succeed Philippe Mercier as art tutor to Caroline's children by 1773.

26.07*

ALLAN RAMSAY (1713–1784)
Richard Mead
1740
Oil on canvas
49¾ x 39¾ in. (126.4 x 101 cm)
National Portrait Gallery, London, NPG 15

Richard Mead, a successful London physician who had attended Queen Anne on her deathbed in 1714, was appointed physician to George II in 1727. In the same year he attended Sir Isaac Newton in his final illness. In 1721 Mead had joined Sloane to conduct trials of smallpox inoculation among condemned prisoners at Newgate. All recovered, and this success helped to establish the practice of inoculation, although it remained controversial.

Mead was a significant collector of books, pictures, medals, and antiquities. At his house in Great Ormond Street he instituted gatherings which attracted some of the greatest intellectual figures of the day: Alexander Pope, Martin Folkes, Edmond Halley, William Hogarth, and the young Allan Ramsay. In 1737 he requested Ramsay serve as his agent in Naples in the acquisition of a volume of drawings by Pietro Santi Bartoli, and on his return to London in 1738 he received his first commission from Mead for a portrait. In 1747 Ramsay completed a second full-length portrait which Mead would present to the Foundling Hospital.

* = YCBA only
† = HRP only

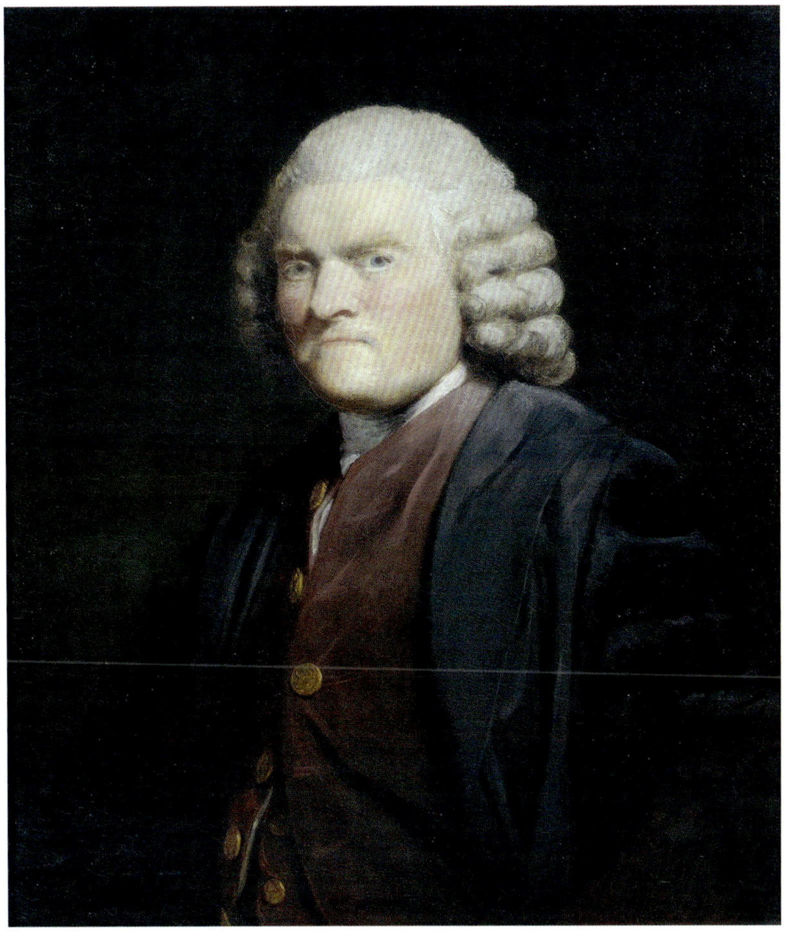

26.08

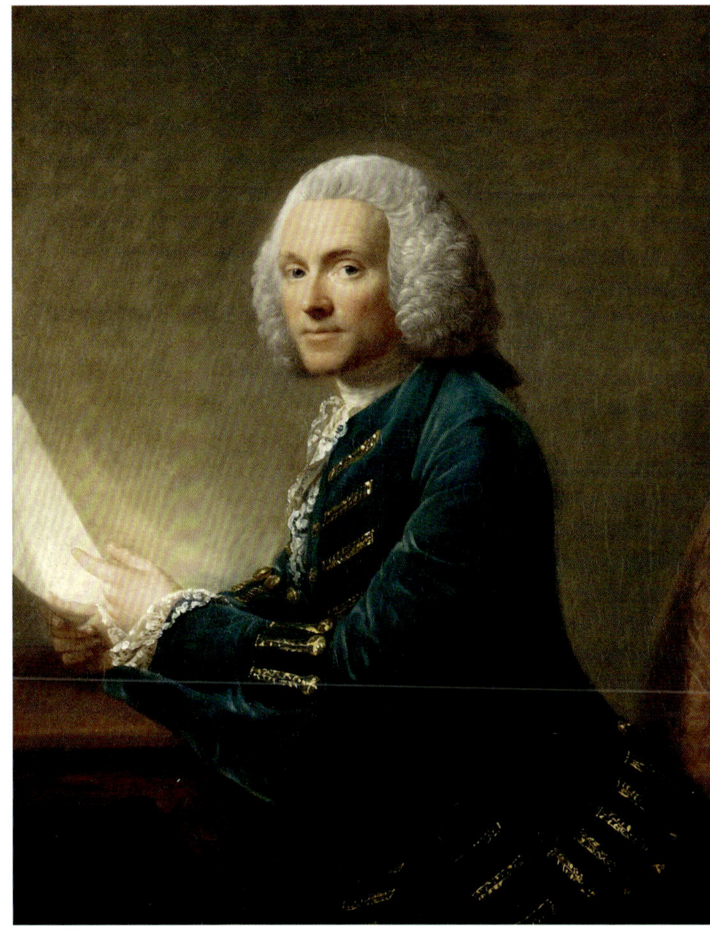

26.09

26.08*

SIR JOSHUA REYNOLDS (1723–1792)
Sir John Pringle
1774
Oil on canvas
30¼ x 25⅛ in. (76.7 x 63.7 cm)
Royal Society, London, RS.9682

The Scotsman John Pringle was an innovator in military medicine and was admitted a Fellow of the Royal Society in 1745. After settling in London in 1748 he set up in medical practice, publishing *Observations on the Nature and Cure of Hospital and Jayl Fevers* two years later, followed by , 'Experiments upon septic and antiseptic substances, with remarks relating to their use in the theory of medicine' in the *Philosophical Transactions* of the Royal Society. For the latter he received the Copley gold medal in 1752. In 1749 he became physician-in-ordinary to the Duke of Cumberland and de speciali gratia, physician to Queen Charlotte in 1761. He was gazetted physician-in-ordinary to George III in 1774. Having served as President of the Royal Society since 1772, he was forced to resign in 1778, probably as a result of an argument with George III over the design of lightning conductors. It was at this date he was also dismissed as royal physician. When the Council of the Society requested that his portrait be made in 1777, Pringle presented this work by Reynolds, privately commissioned earlier.

DR WILLIAM HUNTER AND CHILDBIRTH

26.09

ALLAN RAMSAY (1713–1784)
Dr William Hunter
1764–65
Oil on canvas
37⅞ x 29½ in. (96 x 75 cm)
Hunterian Museum and Art Gallery, University of Glasgow, GLAHA 44026

After starting his medical education in Scotland, William Hunter went to London in 1740 to learn midwifery from William Smellie. He also attended the anatomy lectures of Frank Nicholls and the Newtonian John Desaguliers's natural philosophy course. Subsequently Hunter started to build a surgical and midwifery practice, and investigated female reproductive anatomy using animals, holding appointments as man-midwife at the Middlesex Hospital and the new London Lying-in Hospital, before becoming consultant physician in 1759. George III was advised to employ Hunter to care for Queen Charlotte in her first pregnancy in 1761. The safe birth of a son in August 1762 was followed on 18 September by Hunter's appointment as Physician-Extraordinary to the Queen. He attended all her pregnancies until his death and was on friendly terms with the royal couple.

Allan Ramsay, a fellow Scot, was a few years older than William Hunter. The date of the portrait Ramsay made of his friend—the direct connection and exchange between painter and sitter is evident—is not recorded, but must be around 1765, when Hunter was completing plans for the furnishing of his new house in Windmill Street, Soho, in London.

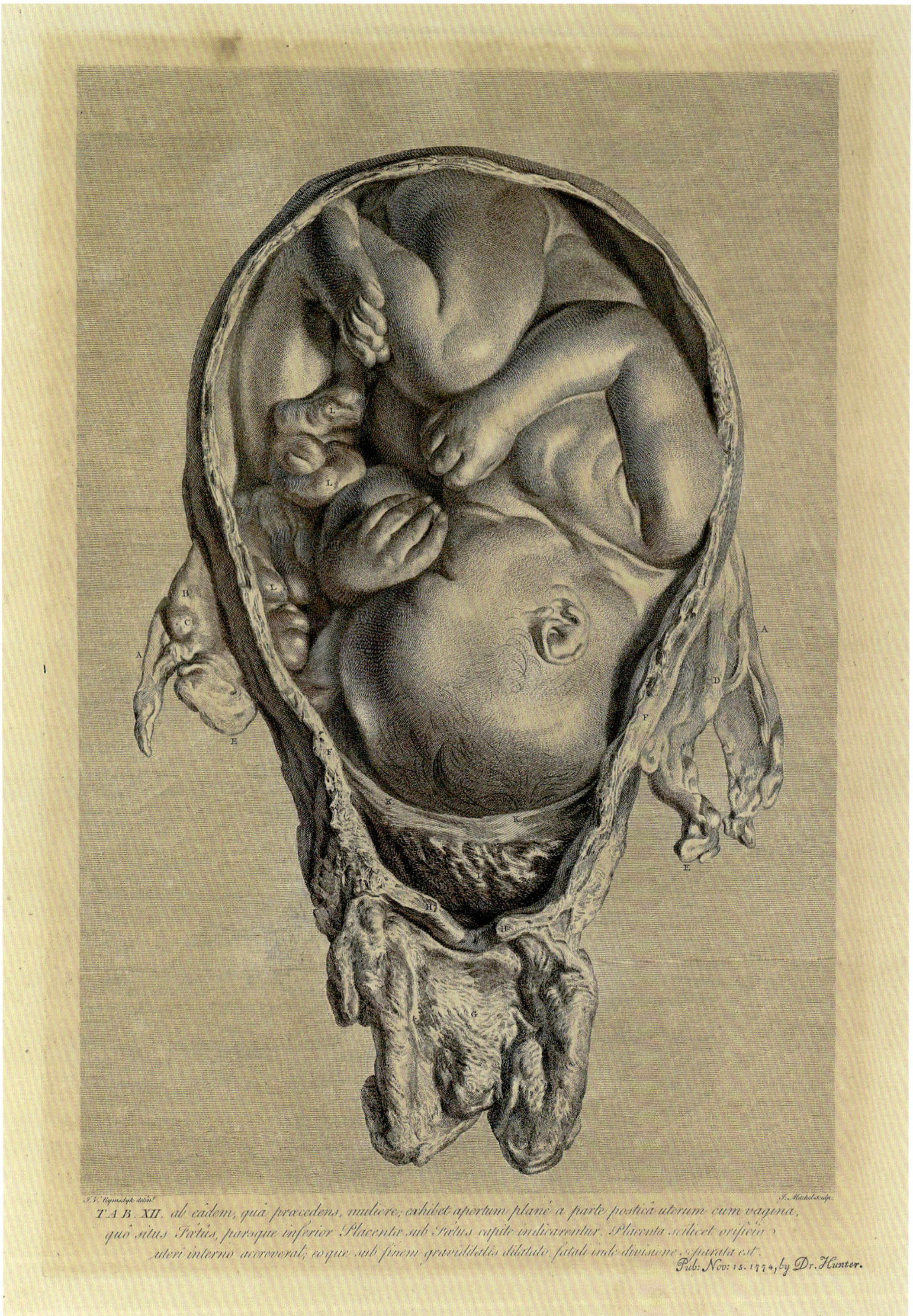

26.10

26.10

WILLIAM HUNTER (1718–1783)
Anatomia Uteri Humani Gravidi Tabulis Illustrata / The anatomy of the human gravid uterus exhibited in figures, by William Hunter, physician extraordinary to the Queen, professor of anatomy in the Royal Academy, and Fellow of the Royal and Antiquarian Societies
1774
26⅜ in. (67 cm)
Printed by J. Baskerville, Birmingham; sold in London by S. Baker and G. Leigh, T. Cadell, D. Wilson and G. Nicol, and J. Murray
* Medical History Library, Yale University, Folio Baskerville 1774h (illustrated)
† The Royal Society Library, London, RCN 45452
Exhibited: Jan van Rymsdyk (active from 1750, d. c. 1789), printmaker, and Robert Strange (1721–1792), engraver, 'Six Months', plate 12

William Hunter had his first opportunity to dissect a full-term human gravid uterus in 1750 and was able to determine the relationship between the maternal and fetal blood systems in the placenta. The dissections, beautifully illustrated by Jan van Rymsdyk, his main anatomical artist, were engraved and exhibited in 1752 to solicit subscriptions for their publication. They form the first ten plates in Hunter's *Anatomia Uteri Humani Gravidi Tabulis Illustrata*, published in 1774, which covered all stages of pregnancy. Hunter's understanding of female anatomy certainly contributed to his successful midwifery practice.

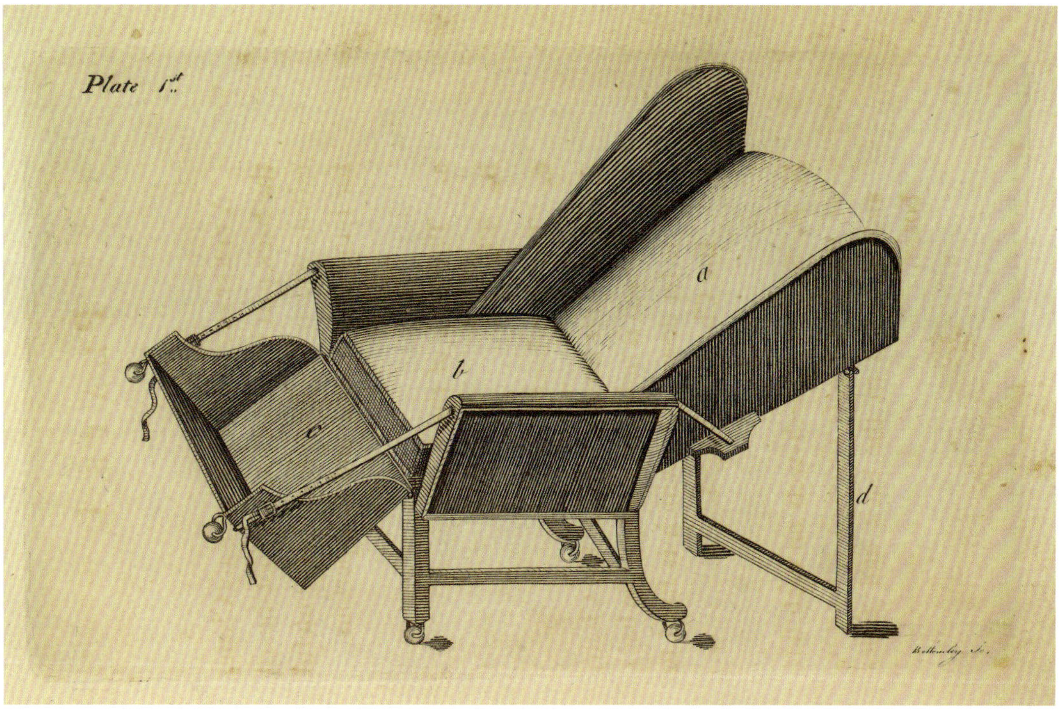

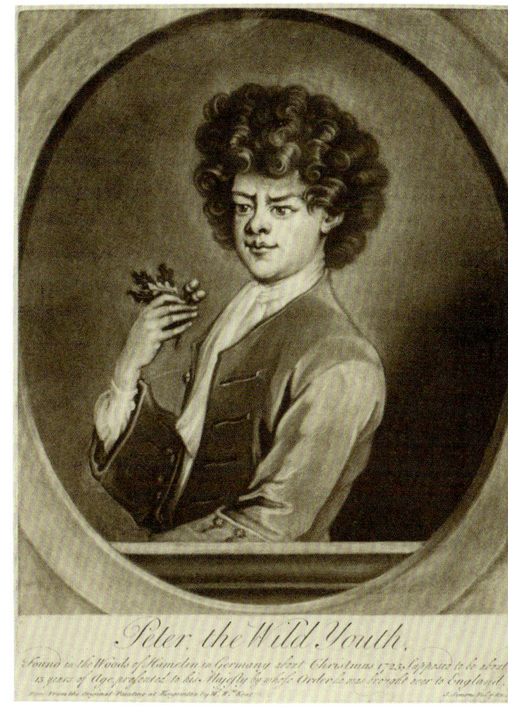

26.11

26.12

26.11†

CHARLES WHITE (1728–1813)
A Treatise on the Management of Pregnant and Lying-in Women and the Means of Curing, but more especially of preventing the principal disorders to which they are liable. Together with some new directions concerning the delivery of the child and placenta
1773
6¼ in. (16 cm)
Printed for Edward and Charles Dilly, London
Wellcome Library, London, 54806/B
Exhibited: 'Birthing Chair', p. 142

Manchester-born man-midwife Charles White was instrumental in founding the Manchester Infirmary. This charity, funded by subscriptions from the wealthy, provided medical advice and treatment for the poor. White served as honorary surgeon from 1752 until 1790. In his influential book, *Treatise on the Management of Pregnant and Lying-in Women* (1773), he advised that women should give birth 'naturally'. He advised mothers to get out of bed as soon after delivery as possible and stressed the importance of cleanliness and ventilation. The book, dedicated to his former teacher William Hunter, was translated into both French and German and also published in North America.

Queen Charlotte was patron to the Lying-in Charity established in 1757 to support and provide advice for poor women giving birth in their own homes, echoing White's medical philosophy. In 1809 Charlotte became patron to the General Lying-in Hospital in Bayswater, which was renamed Queen Charlotte's Hospital in her honour.

THE HEALTH OF THE MIND

26.12*

JOHN SIMON (c. 1675–1751), engraver, after **WILLIAM KENT** (c. 1685–1748)
Peter the Wild Youth
c. 1725
Mezzotint on paper
14 x 10⅛ in. (35.4 x 25.6 cm)
Published by J. Simon, London
Royal Collection Trust, RCIN 660069

The child known as Peter the Wild Boy was found in the forest in northern Germany, alone, unsocialized, and unable to speak. After being taken into the protection of George I, he was brought to the court in London in 1726. Caroline, then Princess of Wales, petitioned the King to allow her to assume responsibility for the boy. After this was agreed she appointed Dr John Arbuthnot as his guardian. His condition aroused widespread philosophical interest. This mezzotint by Simon reproduces the portrait of Peter in William Kent's wall paintings completed for the King's staircase in Kensington Palace.

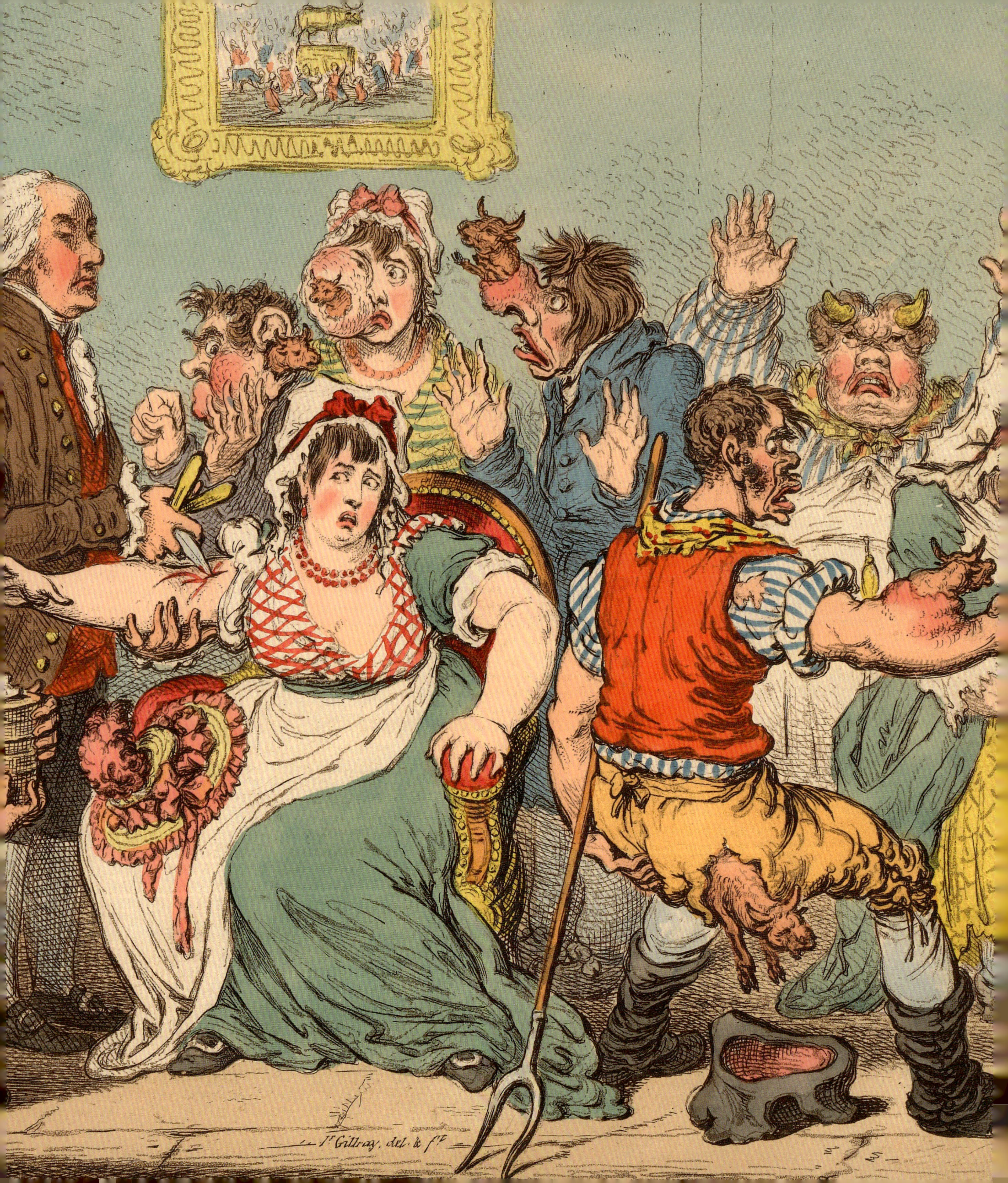
Js Gillray, del. & ft

Inoculation

Tyler Griffith and Samantha Howard

Highly contagious and frequently fatal, smallpox was acknowledged as one of the biggest killers of the eighteenth century, particularly amongst children. Symptoms began with a high fever, abdominal cramps, and vomiting, followed by the eruption of a rash which would develop into painful, round suppurating blisters (pox) on the face, arms, and legs. Those who did survive were often left with disfiguring scars.

The knowledge that if a person survived smallpox they would be rendered immune to subsequent infections with the disease had been recorded as early as the fifth century BC, but it was not until the late seventeenth and early eighteenth centuries that the practice of inoculation (also known as variolation) began to be practised more widely in western Europe.[1] Lady Mary Wortley Montagu, wife of Edward Wortley Montagu, the British Ambassador in Constantinople, had witnessed the procedure there which involved inserting pus taken from a smallpox pustule into an incision, usually on the arm of a patient. Lady Mary, having survived the disease a few years earlier, though left with horrible scars, arranged for her son, Edward, to be inoculated under the supervision of the embassy surgeon, Charles Maitland, before they returned to England in 1718. Lady Mary and her son were depicted in Turkish dress in a painting by the French artist Jean Baptiste Vanmour, who was based at the Ottoman court (fig. 27.01).

Upon her return to London, Lady Mary campaigned to promote the practice amongst her friends at court. News of this medical advance was of great interest to Caroline of Ansbach. As a survivor of the disease which had killed her father, Caroline was all too well aware of its deadly nature. After her daughter Princess Anne caught smallpox in 1720, she took the initiative to protect the rest of her children. With the encouragement of the royal physician, Hans Sloane, she persuaded George I to grant approval to Maitland to conduct inoculation experiments using prisoners and orphans—all of whom survived. In 1722, convinced of its safety, Caroline arranged for her daughters Princess Amelia and Princess Caroline to undergo the procedure; James Jurin, the Secretary to the Royal Society and a physician at Guy's Hospital, records the number of procedures undertaken for the year 1724, including Maitland's visit to Hanover to inoculate Caroline's eldest son, Prince Frederick, and his retinue, in his *Account of the Success of Inoculating the Smallpox in Great Britain* of 1725.[2]

Caroline was praised in the press for her role in helping to promote the practice, and also in Richard Holland's *Observations on the Small pox: or, an Essay to Discover a More Effectual Method of Cure* of 1728, which is dedicated to her (cat. 27.05). By the mid-century, inoculation had gained widespread acceptance. Charlotte of Mecklenburg-Strelitz followed Caroline's example and publicly advocated the procedure following the inoculation in 1768 of two of her children, Charlotte, Princess Royal, and Prince William Henry. When Omai—the first Polynesian to visit England—returned with Captain James Cook's second expedition to the South Pacific in 1774, he was inoculated against smallpox as a precaution—if not least a diplomatically prudent and humanitarian measure. The procedure was arranged by the botanist and explorer

Detail of p. 449, cat. 27.13

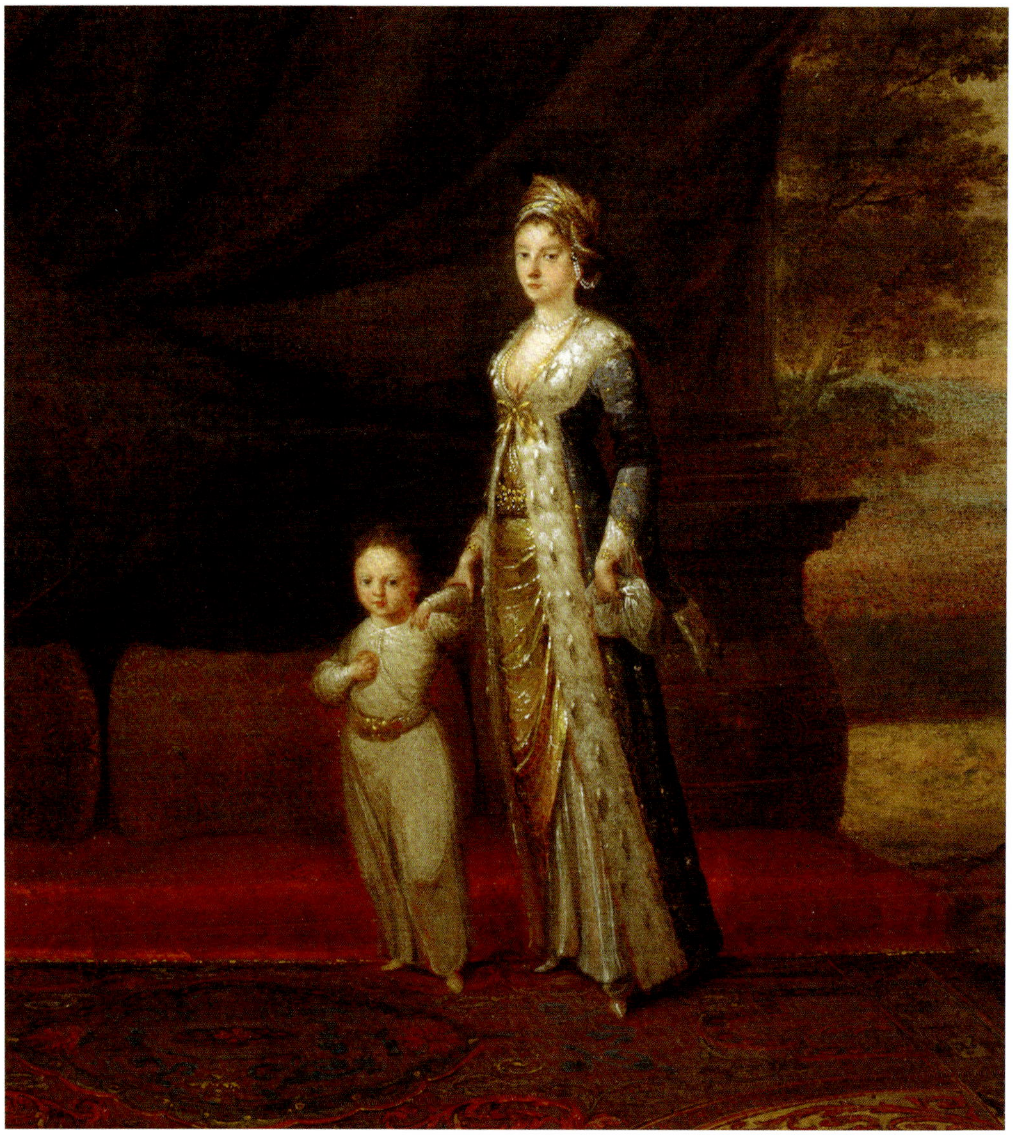

FIG. 27.01 Attributed to Jean Baptiste Vanmour (1671–1737, *Lady Mary Wortley Montagu with her Son, Edward Wortley Montagu, and Attendants* (detail), c. 1717, oil on canvas, 27¼ x 35⅜ in. (69.3 x 90.0 cm). National Portrait Gallery, London (cat. 27.03)

Sir Joseph Banks, who had accompanied Cook on his first great voyage of 1768–71. The devastating consequences of smallpox contagion had been cruelly highlighted in 1773, when Banks had arranged for a party of Inuits from Newfoundland and Labrador to visit London. On their homeward journey they presented the signs of smallpox infection and all died except for one, who recovered but carried the disease back to Labrador, thereby eradicating the whole community.[3] During the two years Omai spent in London, he was much admired for his intelligence and charm and was fêted by London society at large. Banks presented Omai at court before he returned home on Cook's third voyage to the Pacific in 1776.

Whilst the practice of inoculation proved very effective and saved many lives, it was not without its dangers. Live smallpox matter was introduced into the body in the anticipation that a mild form of the disease would result and therefore provide lifelong immunity, but there was no way of identifying the strength of the virus. There was always the risk that the patient might contract a full-blown case of the disease. George III and Charlotte were devastated by such outcomes, with the deaths of their sons, two-year-old Alfred in 1782 and four-year-old Octavius six months later, following smallpox inoculations.

Another disadvantage was that after the procedure the patient would still be contagious and, unless quarantined, could infect others. Dedicated smallpox hospitals were set up for this purpose, such as the Hospital for the Small-Pox and Inoculation based in St Pancras in London. Established in 1746, the hospital consisted of three houses: one for the purposes of inoculation, one

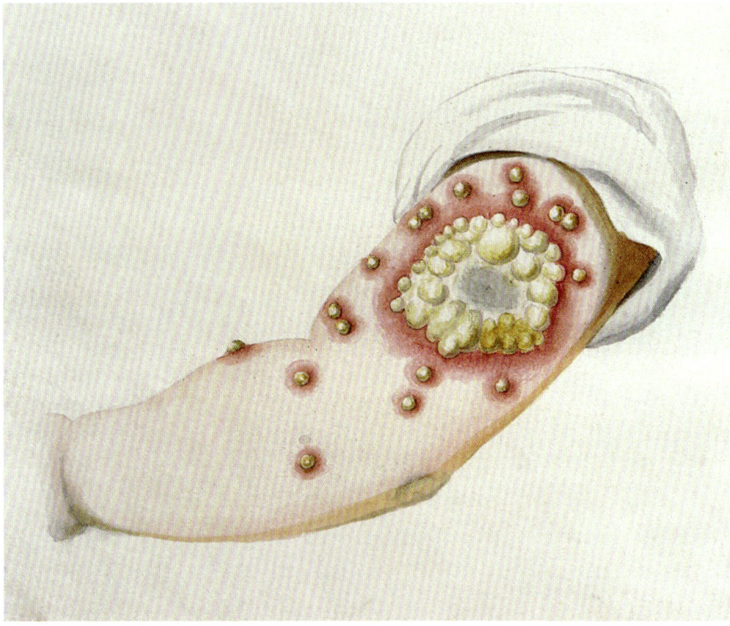

FIG. 27.02 George Kirtland, 'Fourteenth day: smallpox' [showing effects of inoculation], 1802, watercolour on paper, 11 x 8¼ in. (28 x 21 cm). Wellcome Library, London (cat. 27.11)

for quarantine, and the remaining house for patients who had contracted smallpox in the 'natural way'.[4]

It was not until towards the end of the century that a procedure known as vaccination was brought to the attention of the world by Edward Jenner. As a country doctor, Jenner became increasingly interested in a theory drawn from folklore that milkmaids who suffered the mild disease of cowpox never contracted smallpox. In 1796 he conducted an experiment on a young local boy, James Phipps, extracting pus from a cowpox pustule and inserting it into an incision he made with a lancet into the boy's arm. Jenner subsequently proved the efficacy of the procedure he termed 'vaccination' (from 'vacca', the Latin for cow) by exposing Phipps to smallpox: the boy remained healthy. Although both vaccination and inoculation work by stimulating the body's immune system to produce antibodies that combat disease, vaccination was a far safer procedure as the body was not exposed to the powerful live smallpox virus. The series of watercolour illustrations of 1802 by George Kirtland offers a striking visual representation of the considerably milder appearance of the cowpox pustule compared with that of smallpox, and chronicles the development of the respective diseases over a series of days (fig. 27.02).

Nevertheless, Jenner's efforts to promote his vaccine were at first met with hostility and ridicule, especially from fellow medical practitioners, the clergy, and the press; it was not understood how immunity to the disease could be induced in this way, and even Jenner had no explanation why it worked as the virus was invisible under microscopes of the time. These anxieties manifested themselves in several satires, such as James Gillray's 'The Cow-Pock—or—the Wonderful Effects of the New Inoculation!' of 1802, in which Jenner's patients are depicted metamorphosing into grotesque bovine creatures (cat. 27.13).

Whilst Jenner was not the first 'vaccinator'—indeed, the earliest practitioner of vaccination against smallpox in England was the West Country farmer Benjamin Jesty, who vaccinated his family with the cowpox virus in 1774—he was the first to openly conduct experimental trials, and tirelessly championed the practice of vaccination over smallpox which ultimately led to the eradication of the latter across the world.[5] In John Raphael Smith's portrait of Jenner of 1800, now in the Wellcome Library, London, the artist's depiction of a milkmaid and her cows behind his sitter commemorates the inspiration behind Jenner's legacy as the 'father of immunology' (cat. 27.09). Charlotte, who met Jenner first in 1800, would in 1814 agree to become patron to a movement, which ultimately proved unsuccessful, to draw up a testimonial honouring him. It was to be signed by mothers throughout the country whose families had been saved from illness and disfigurement by the practice of vaccination.[6]

NOTES

1. Cary P. Gross and Kent A. Sepkowitz, 'The Myth of the Medical Breakthrough: Smallpox, Vaccination, and Jenner Reconsidered', *International Journal of Infectious Diseases* 3, no. 1 (July–Sept. 1998): 54–60 (56).
2. Cited in Joanna Marschner, *Queen Caroline: Cultural Politics at the Early Eighteenth-Century Court* (New Haven and London: Yale Univ. Press, 2014), 153.
3. Patrick O'Brian, *Joseph Banks* (London: Harvill Press, 1997), 181–82.
4. Anon., *An Account of the Rise, Progress and State of the Hospital, for Relieving Poor People Afflicted with the Small-pox, and for Inoculation* (St Pancras, London: Hospital for the Small-Pox and Inoculation, 1755), 4.
5. Patrick J. Pead, 'Benjamin Jesty: New Light in the Dawn of Vaccination', *The Lancet* 362, no. 9401 (20 Dec. 2003): 2104–09.
6. John Baron, *The Life of Edward Jenner, M.D.*, 2 vols. (London: Henry Colburn, 1827–38), 1: 379; 2: 210–11.

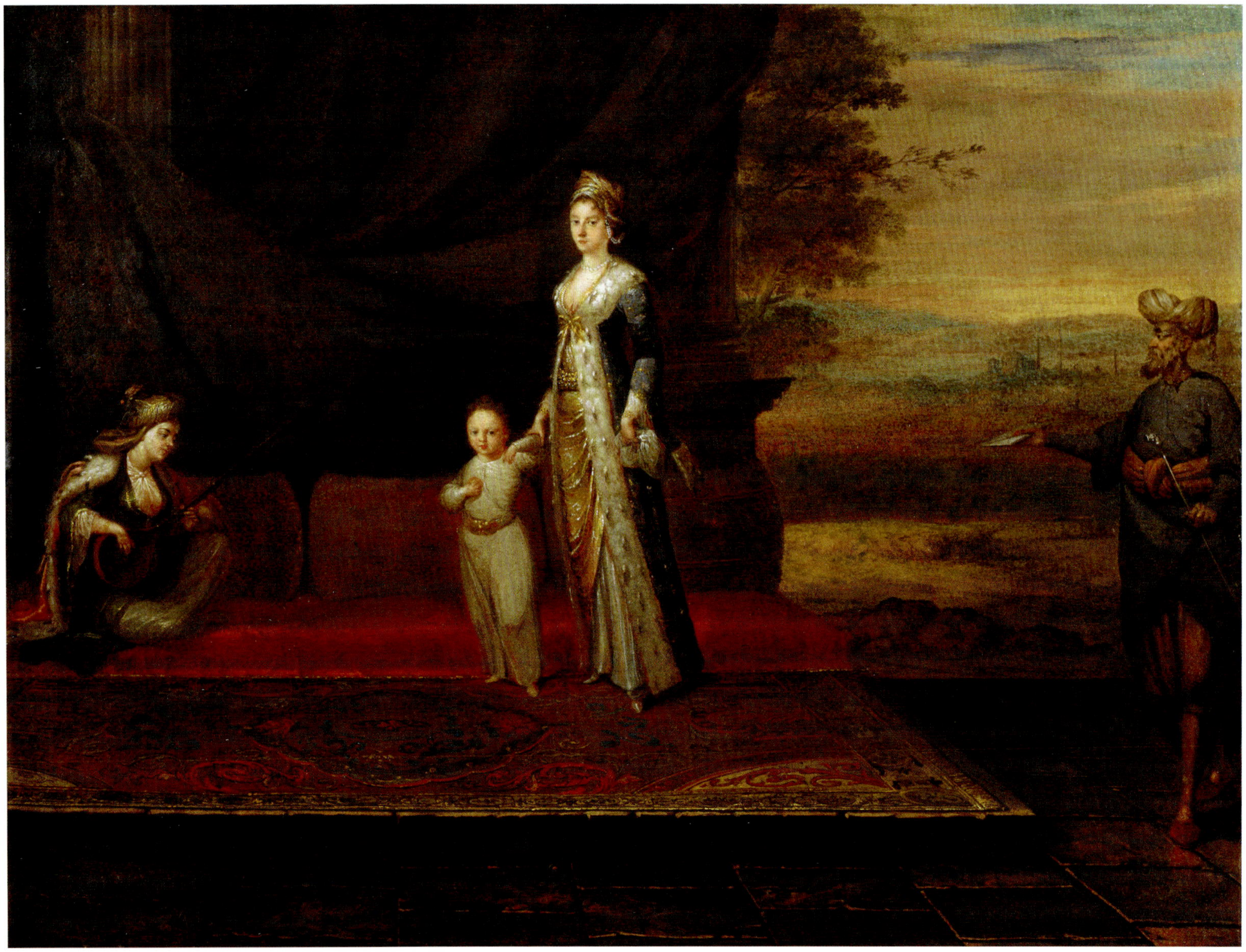

27.03

27.03

ATTRIBUTED TO JEAN BAPTISTE VANMOUR (1671–1737)
Lady Mary Wortley Montagu with her son, Edward Wortley Montagu, and attendants
c. 1717
Oil on canvas
27¼ x 35⅜ in. (69.3 x 90.0 cm)
National Portrait Gallery, London, NPG 3924

Marquis Charles de Ferriol invited Jean Baptiste Vanmour to accompany him to Constantinople when he was appointed as French Ambassador there in 1699, and commissioned him to make one hundred portraits of the local people. After De Ferriol returned to France in 1711, Vanmour stayed on, working for other diplomats. In 1716 Lady Mary Wortley Montagu accompanied her husband, Edward, on his embassy to Constantinople, and they lived there until 1718. Her second child, a daughter, was born during this time, and she had the opportunity to observe aspects of Middle Eastern medical practice, including a form of inoculation. Lady Mary was painted by Vanmour accompanied by her eldest child, Edward, who was inoculated before their return to England. He is wearing Turkish dress, and his mother described the detail of this in a letter to her sister Lady Mar: 'The first part of my dress is a pair of drawers, very full . . . over this hangs my smock, of a fine white silk gauze edged with embroidery . . . my caftan of the same stuff with my drawers, is a robe exactly fitted to my shape, and reaching to my feet . . . The curdee is a loose robe . . . being of rich brocade (mine is green and gold) either lined with ermine, or sables.'

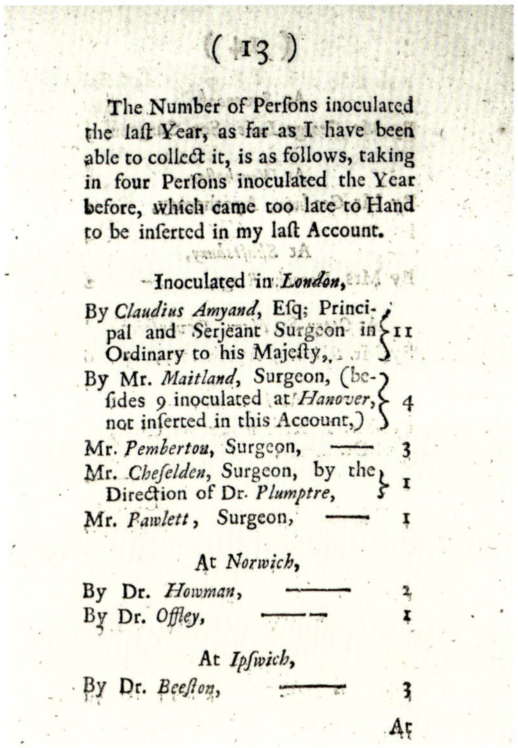

(13)

The Number of Perſons inoculated the laſt Year, as far as I have been able to collect it, is as follows, taking in four Perſons inoculated the Year before, which came too late to Hand to be inſerted in my laſt Account.

Inoculated in *London*,

By *Claudius Amyand*, Eſq; Principal and Serjeant Surgeon in Ordinary to his Majeſty,	11
By Mr. *Maitland*, Surgeon, (beſides 9 inoculated at *Hanover*, not inſerted in this Account,)	4
Mr. *Pemberton*, Surgeon, —	3
Mr. *Cheſelden*, Surgeon, by the Direction of Dr. *Plumptre*,	1
Mr. *Pawlett*, Surgeon, —	1

At *Norwich*,

By Dr. *Howman*, —	2
By Dr. *Offley*, —	1

At *Ipſwich*,

By Dr. *Beeſton*, —	3

At

27.04

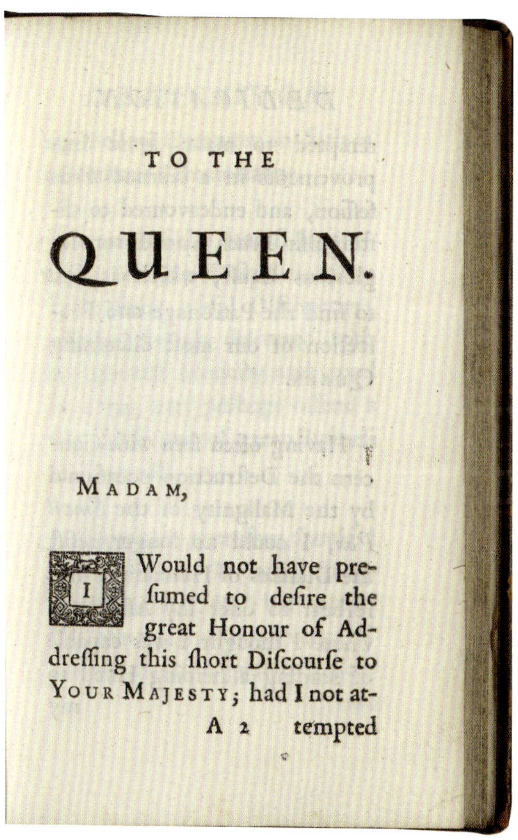

TO THE

QUEEN.

MADAM,

I Would not have preſumed to deſire the great Honour of Addreſſing this ſhort Diſcourſe to YOUR MAJESTY; had I not attempted

A 2

27.05

27.04*

JAMES JURIN (1684–1750)
An Account of the Success of Inoculating the Small-pox in Great-Britain, for the year 1726
1727
7⅞ in. (20 cm)
J. Peele, London
Medical History Library, Yale University, Inoc Vac
Exhibited: Title page

27.05

RICHARD HOLLAND (1688–1730)
Observations on the Small pox: or, An Essay to Discover a More Effectual Method of Cure
1728
Printed for J. Brindley, London
* Medical History Library, Yale University, Inoc Vac
Exhibited: Title page
† Wellcome Library, London, MST.106
Exhibited: Title page and pp. 1–2, dedication to Queen Caroline (illustrated)

As part of the debate about the efficacy of inoculation in the second decade of the eighteenth century, in which Caroline, as Princess of Wales was implicated, using his position as secretary to the Royal Society, James Jurin placed an advertisement in the *Philosophical Transactions* inviting readers to send accounts of their experiences with the practice. Over sixty individuals responded and from these accounts Jurin calculated the relative odds of dying from smallpox inoculation and natural smallpox. He published the results of his calculations in a series of annual pamphlets. Contemporaries were generally persuaded by Jurin's figures and his pamphlets were acknowledged as central to the establishment of smallpox inoculation in England.

Richard Holland was elected a Fellow of the Royal Society on 30 November 1726. He wrote *Observations on the Small Pox, or, An Essay to Discover a More Effectual Method of Cure* in 1728. It was dedicated to Caroline in recognition of her support for the medical procedure.

27.06†

JOHN HUNTER (1728–1793), surgeon and anatomist
'The skin in small-pox'
18th century
Anatomical specimen
Length approx. 6 ¼ in – 9⅞ in. (16–25 cm)
Hunterian Museum, University of Glasgow, Presented by Trustees of the Hunterian Collection in 1799, GLAHM 121115
Not illustrated

Dr John Hunter was the younger brother of Dr William Hunter, and followed his brother into medical practice in London. After appointments to posts at St Bartholemew's and St George's Hospitals he decided to devote himself to anatomy and pathology, especially of infectious diseases such as smallpox. In the course of this work he amassed over two thousand pathological preparations to aid his research. His observations using such preparations made contributions to oncology, knowledge about blood circulation, and to tissue and bone transplantation. His collection of specimens was purchased and preserved by Parliament in 1799 and the great part is now in the care of the Royal College of Surgeons.

27.07*

SIR JOSHUA REYNOLDS (1723–1792)
Omai
c. 1774
Oil on canvas
23¾ x 20¾ in. (60.3 x 52.7 cm)
Yale University Art Gallery, Gift of the Associates in Fine Arts, 1936.128

Mai (called 'Omai' in Britain), a native of the island of Raiatea, encountered Captain James Cook and Joseph Banks when they arrived on Tahiti first in 1769 and again on Cook's second voyage to the Southern Hemisphere in April 1773. After returning with the party to London in 1774, on 17 July of that year, Mai was presented by Banks to George III and Charlotte at Kew. The King recommended that Mai should be inoculated, offering to cover any costs for this, as well as providing funds to support him while he was in Britain. He was successfully inoculated by Dr Thomas Dinsdale a few weeks later. Mai's celebrity can be measured by the number of portraits of him made over the two years he spent in Britain. In 1774 he was painted by Nathaniel Dance-Holland. In 1774/75, Sir Joshua Reynolds began a series of studies, of which this is one, towards a full length portrait which would be exhibited at the Royal Academy in 1776 and engraved the following year (see cat. 27.08). William Parry painted Sir Joseph Banks with Mai and Doctor Daniel Solander in 1776 (see cat. 30.12). In the paintings Mai typically is dressed in a white tapa, the cloth worn by Tahitian chiefs; this symbol of status reflected his unofficial role as ambassador from his homeland in Polynesia.

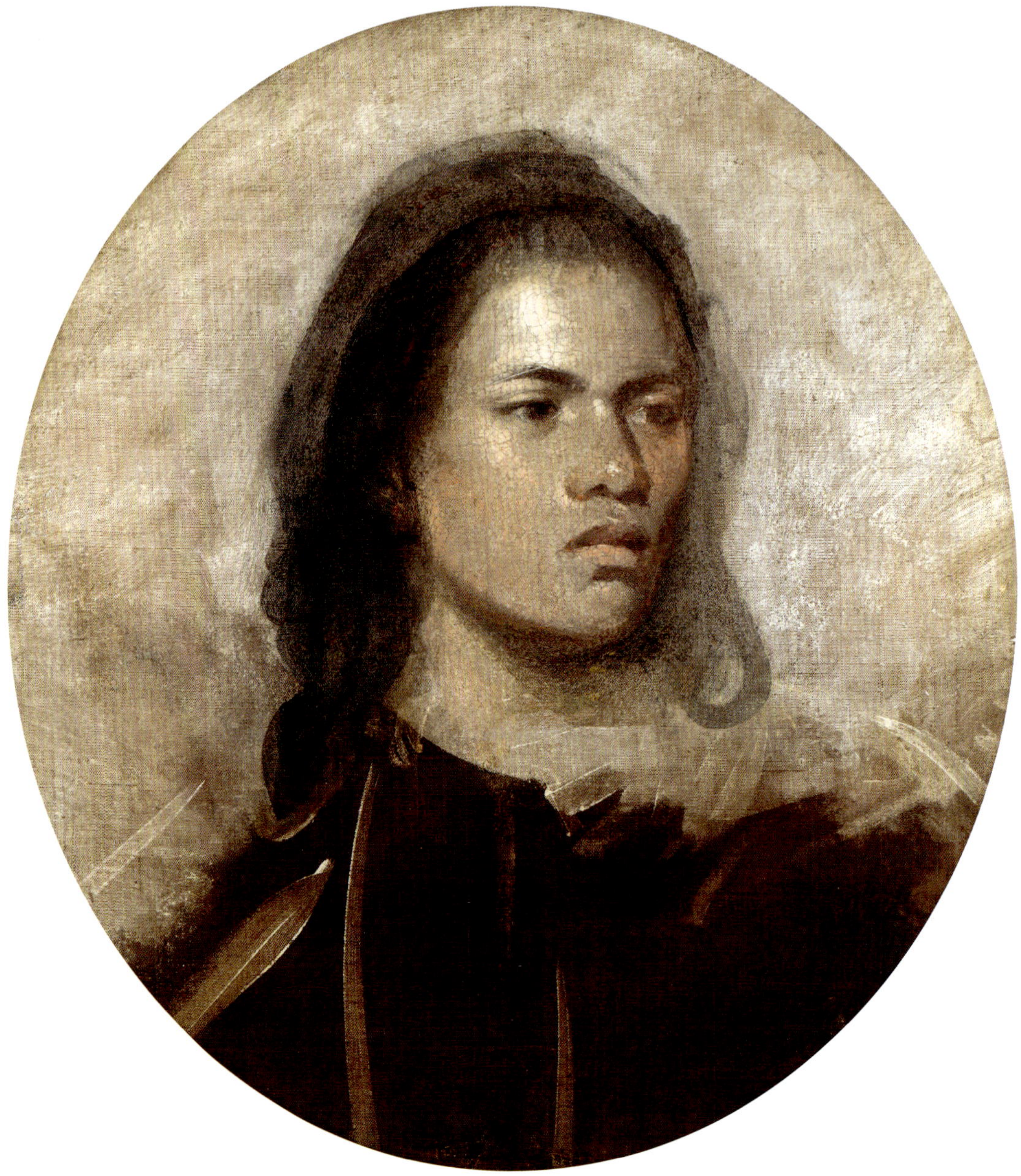

27.07

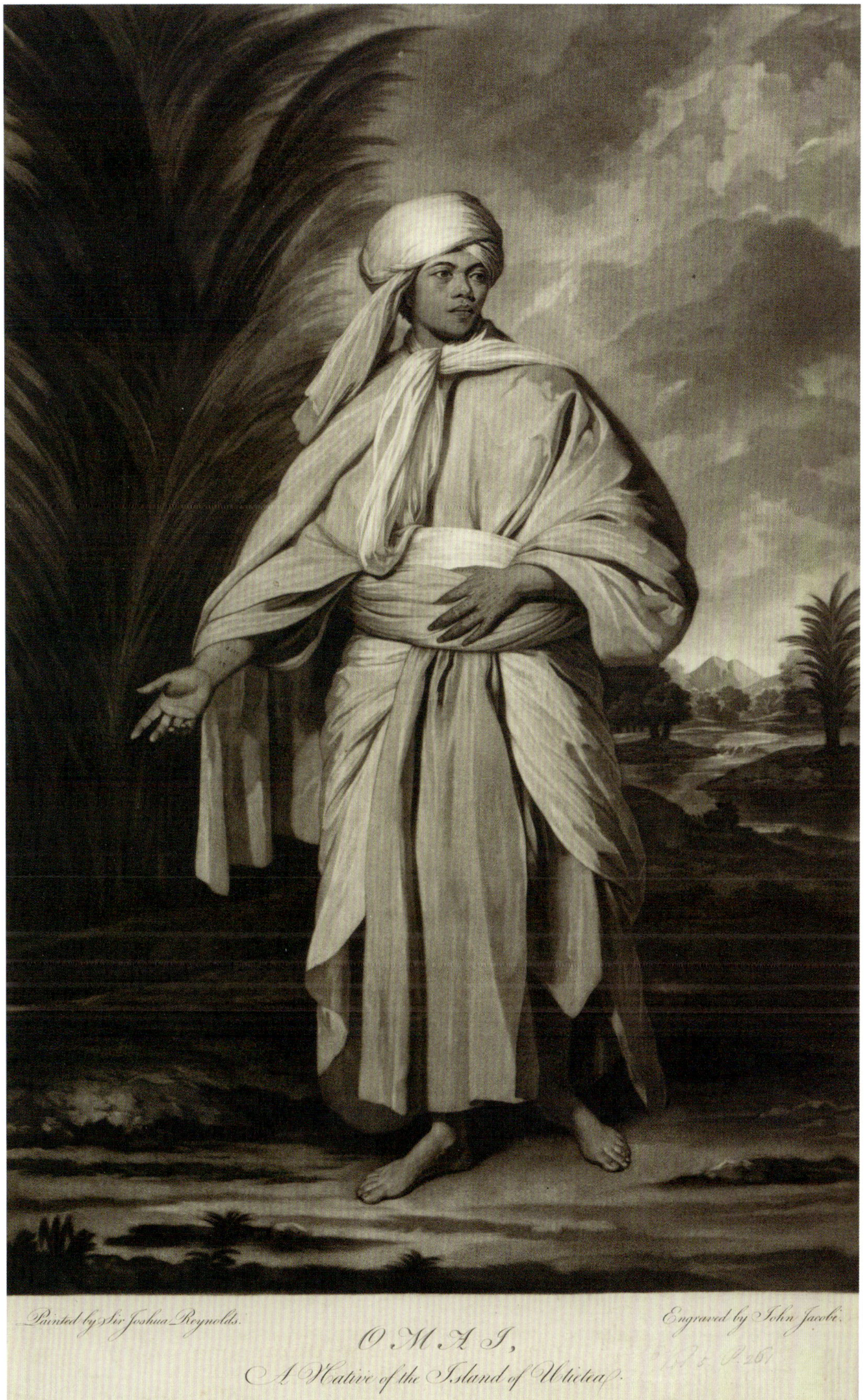

27.08

27.08*

JOHANN JACOBÉ (1733–1797), engraver, after
SIR JOSHUA REYNOLDS (1723–1792)
Omai
1777
Mezzotint
24½ x 15 in. (62.2 x 38.1 cm)
Published by John Jacobi at Mrs Sledge's, London
Inscribed 'OMAJ'; below 'A Native of the Island of Utietca'
Yale Center for British Art, Paul Mellon Collection, B1977.14.10146

Little is known of Mai's own impressions of life in London (he was called 'Omai' in Britain), but during his stay he aroused intense interest from a range of commentators and artists, including the society portraitist and President of the Royal Academy, Sir Joshua Reynolds. Reynolds exhibited his full-length portrait of Mai at the Royal Academy exhibition in 1776, and the picture was reproduced in mezzotint the following year. Despite the fact that Mai wore English clothes throughout his stay in the country, Reynolds (like William Parry) portrayed him in flowing robes and a turban with elements of traditional Tahitian dress, a costume that suggested exotic otherness.

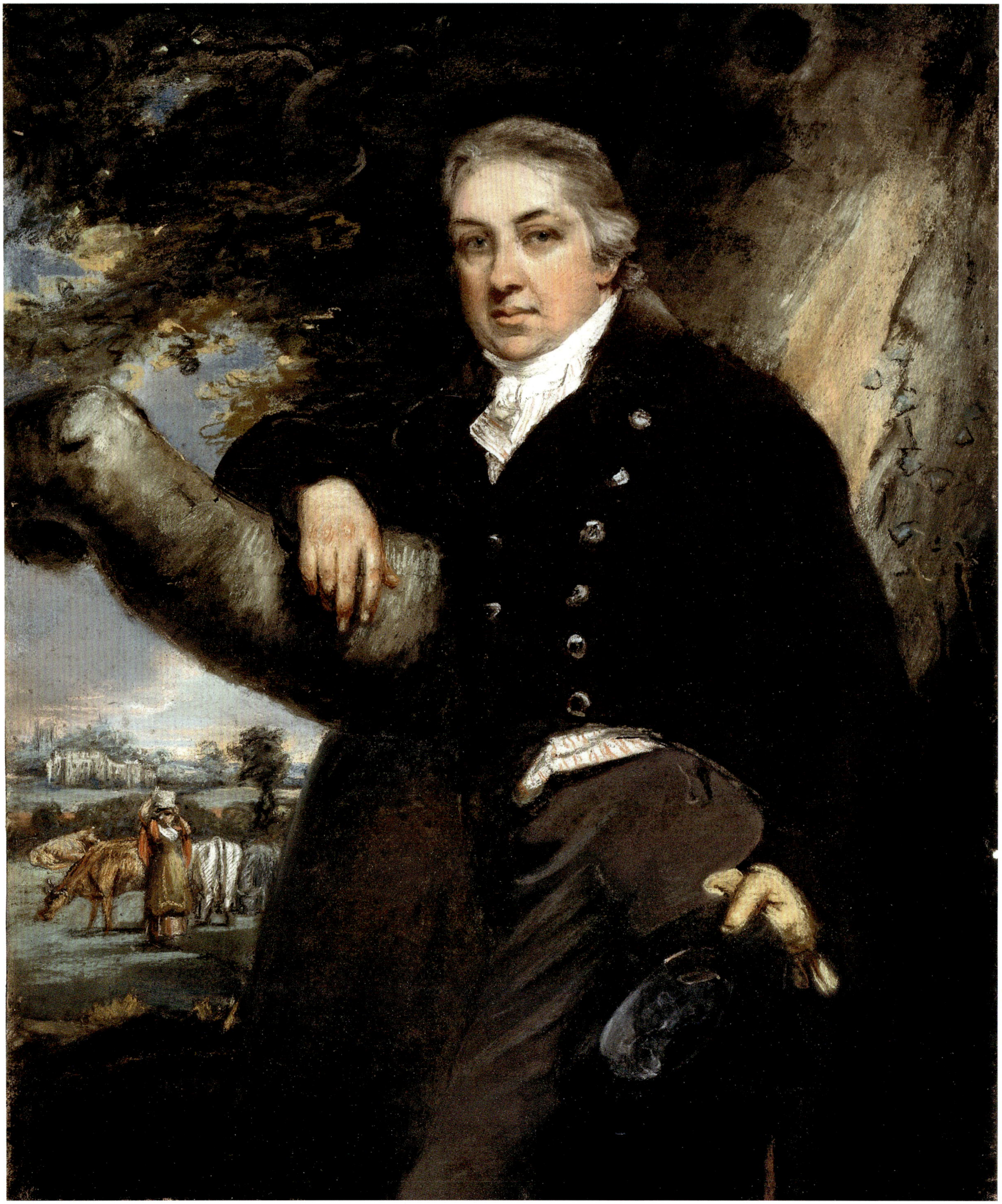

27.09

VACCINATION

27.09*

ATTRIBUTED TO JOHN RAPHAEL SMITH (1752–1812)
Edward Jenner
Undated
Oil on canvas
17¾ x 13¾ in. (45 x 35 cm)
Wellcome Library, London, 45707i

27.10

ATTRIBUTED TO SAVIGNY AND COMPANY
Lancet belonging to Edward Jenner
1790s
Steel blade with tortoiseshell handle
⅛ x 2⅜ x 2¼ in. (2 x 6 x 5.6 cm)
Science Museum, London, A600037

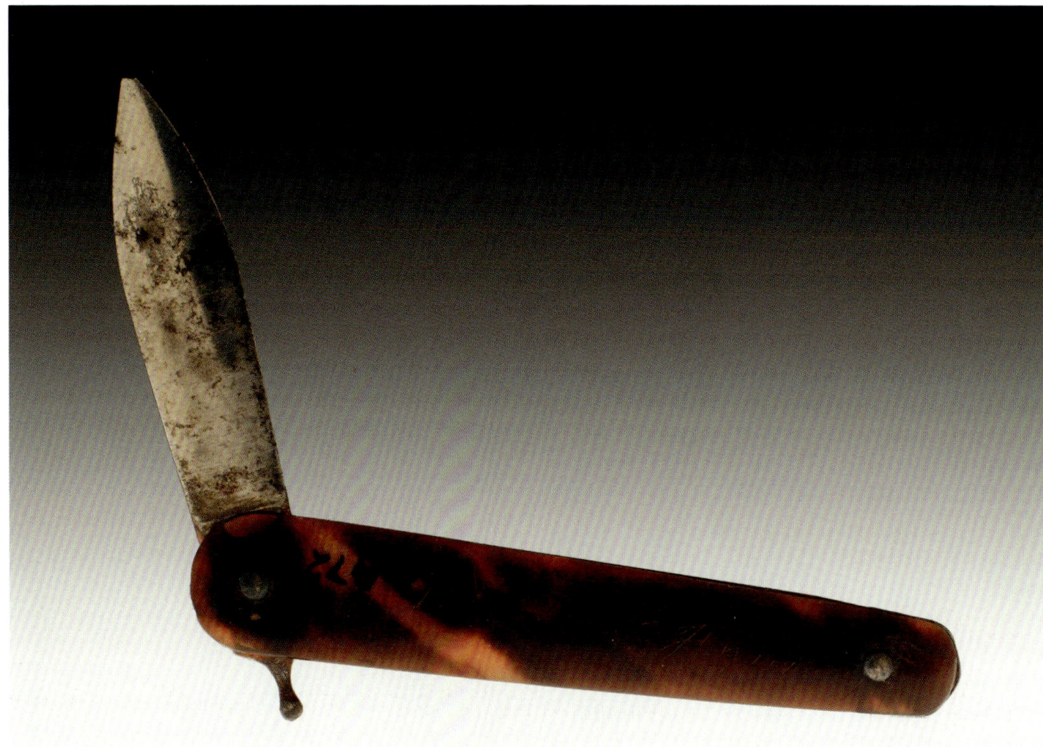

27.10

Edward Jenner, an experienced practitioner of inoculation, was aware that in rural areas those who contracted cowpox were rendered immune to natural and inoculated smallpox. By 1770 Jenner began to collect data to test this hypothesis, and he performed the first documented cowpox inoculation on 14 May 1796. This was probably not the first ever vaccination: the work of Benjamin Jesty, a Dorset farmer, only came to light after Jenner's success became known. Each experiment involved transferring pus containing the cowpox bacillus into the bloodstream of the patient with the aid of a lancet. Jenner's lancet was probably made by Savigny and Company. In the course of his vaccination crusade he corresponded with and was received by George III and Queen Charlotte, and by the Prince of Wales. Charlotte gave her support to a petition raised in the name of the mothers of Britain to honour Jenner for his contribution to the better management of this dangerous disease. He was painted by John Raphael Smith, the teacher of J. M. W. Turner. Now best known as a mezzotint engraver, Smith reproduced some forty works of Reynolds, and was appointed engraver to George, Prince of Wales, the eldest son of George III and Charlotte.

27.11

GEORGE KIRTLAND (active 1796–1815), copyist, after CHARLES GOLD (active 1801, d. 1842)
'Collection of thirty watercolour drawings of smallpox and cowpox inoculation'
1802
Bound volume of 30 sheets, watercolour on paper
11 in. (28 cm)
Wellcome Library, London, WMS 3115
Exhibited: 'Fourteenth day: smallpox [left] and cowpox [right]' (illustrated)

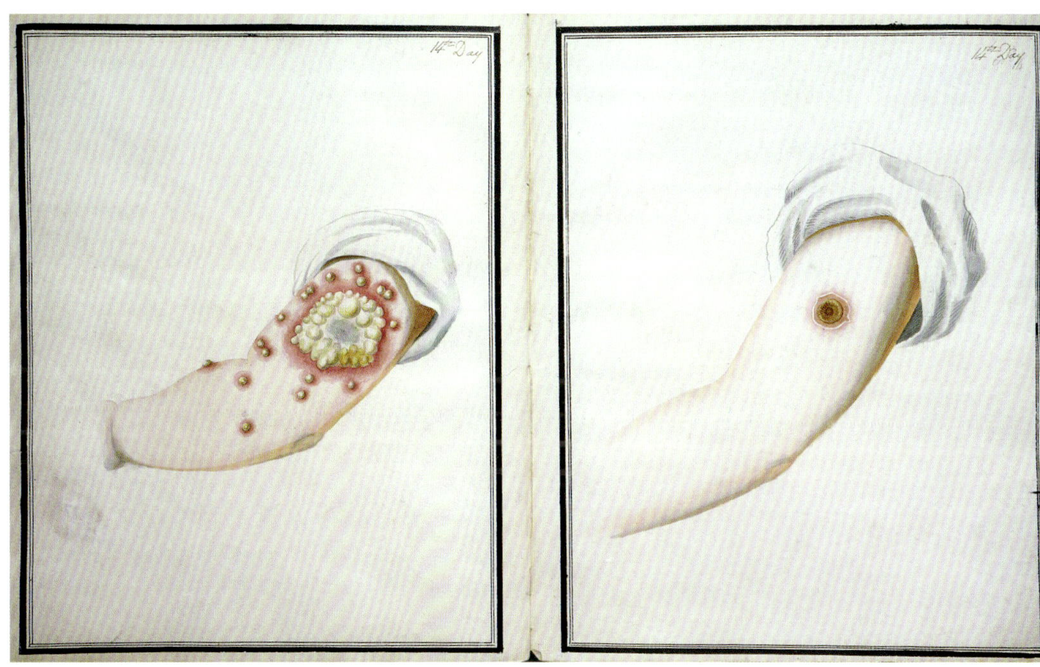

27.11

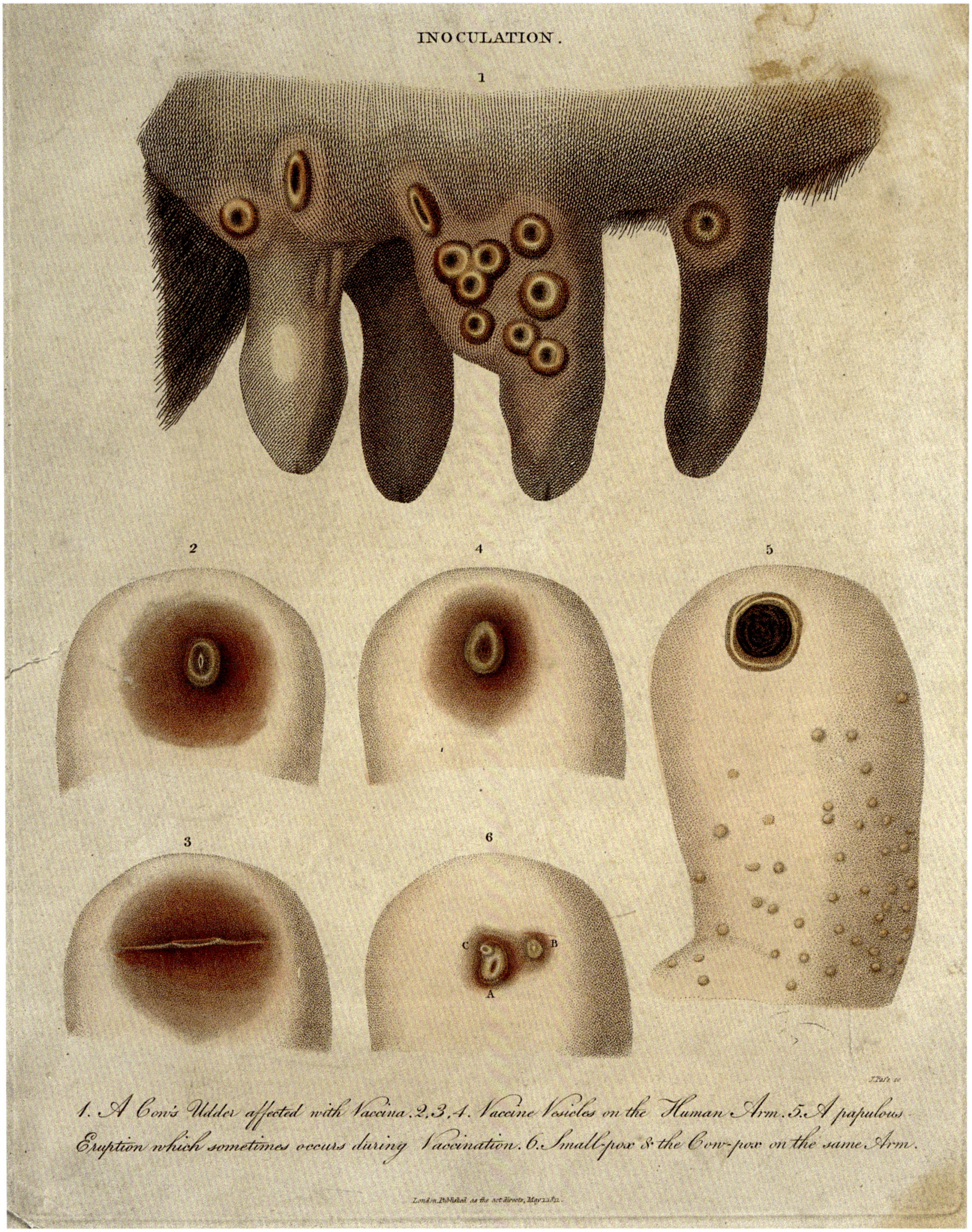
INOCULATION.
1
2
4
5
3
6
C
B
A
1. A Cow's Udder affected with Vaccina. 2, 3, 4. Vaccine Vesicles on the Human Arm. 5. A papulous Eruption which sometimes occurs during Vaccination. 6. Small-pox & the Cow-pox on the same Arm.
London, Published as the act directs

27.12

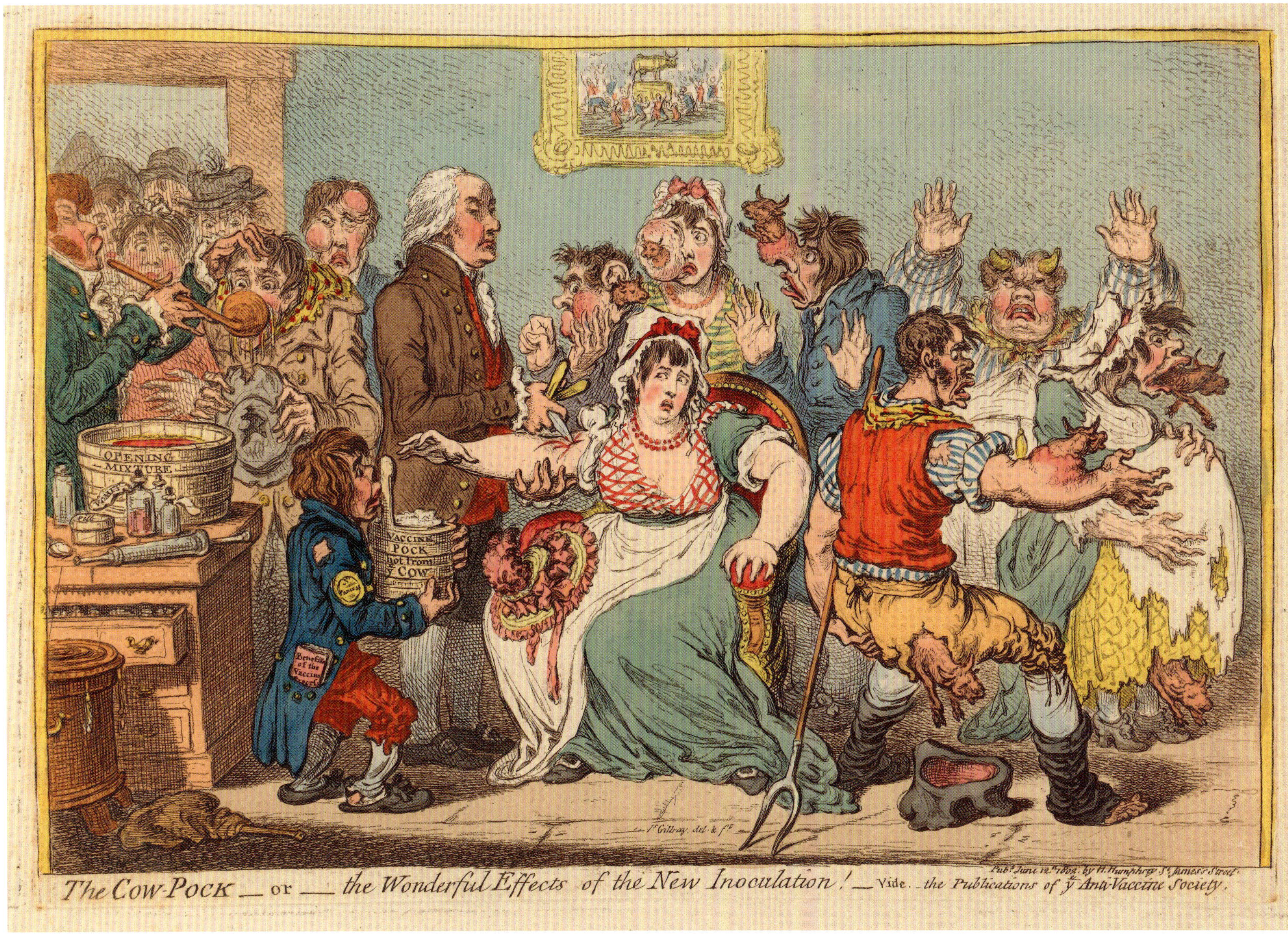

27.13

27.12*

J. PASS (active 1811–1831)
A cow's udder with vaccinia pustules and human arms exhibiting both smallpox and cowpox pustules
1811
Stipple engraving and etching, with watercolour
10⅜ x 8 in. (26.4 x 20.2 cm)
Published as the Act directs, London
Wellcome Library, London, 20186i

Edward Jenner concluded that cowpox, which could be acquired by arm-to-arm transfer, was a safe alternative to inoculation and produced lifelong immunity to smallpox. Despite his three monographs: *Inquiry into the Causes and Effects of the Variolae vaccinae a Disease . . . Known by the name of the Cow-pox* (1798), *Further Observations on the Variolae vaccinae* (1799), and *A Continuation of Facts and Observations Relative to the Variolae vaccinae or Cowpox* (1800), there was some initial confusion over the appearance of the lesions of inoculated smallpox and cowpox. This led to a series of medical illustrations made by George Kirtland, J. Pass, and others.

27.13†

JAMES GILLRAY (1757–1815)
The Cow-pock, or, The Wonderful Effects of the New Inoculation!
1802
Hand-coloured etching with aquatint on paper
11⅛ x 15¼ in. (28.4 x 38.6 cm)
Published by Hannah Humphrey, London
* Beinecke Rare Book & Manuscript Library, Yale University, Auchincloss Gillray v. 10
† Wellcome Library, London, 11752i

James Gillray's caricature shows a scene at the Smallpox and Inoculation Hospital at St Pancras, with cowpox vaccine being administered to frightened young women, and cows emerging from different parts of people's bodies. Opponents of vaccination had depicted cases of recipients developing bovine features—a possibility exaggerated by Gillray. Although the central figure is often assumed to be Edward Jenner, this may not be so. The director of the Smallpox Hospital William Woodville had originally supported Jenner, but he and his colleague George Pearson were in dispute with Jenner by the time the caricature was published. It has been suggested that the central figure represents Pearson. The only clue to identity here is the badge on the arm of the boy which confirms his connection with Woodville's hospital.

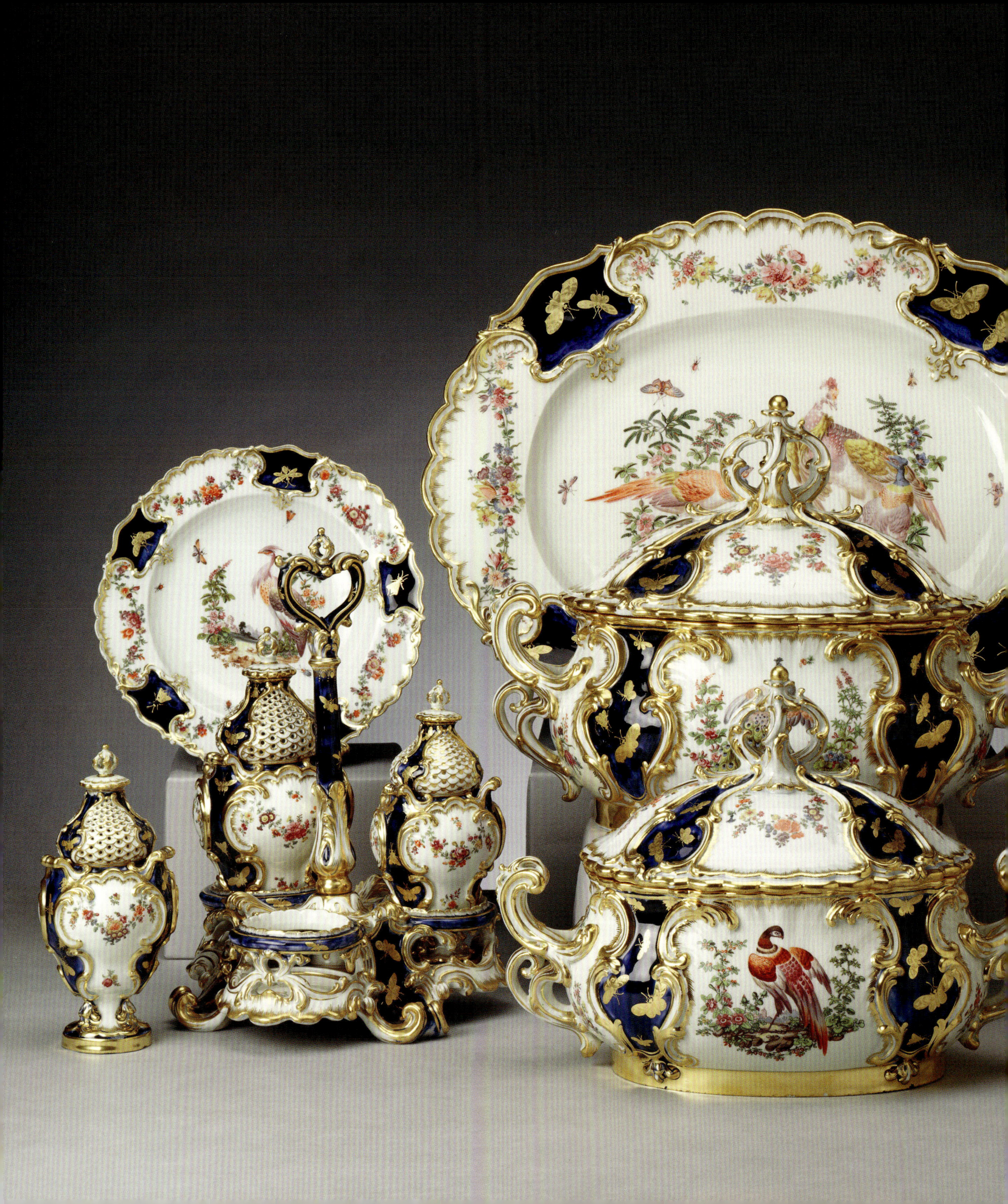

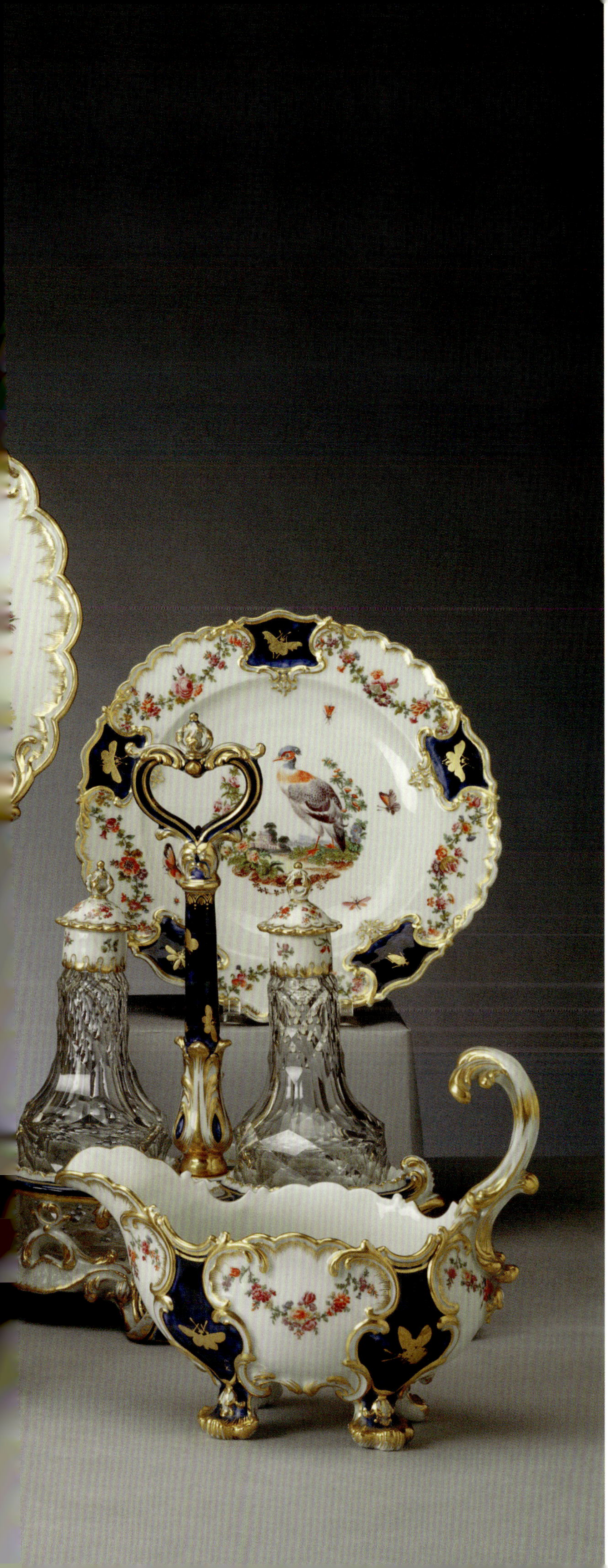

Mercantile Culture and the Wider World

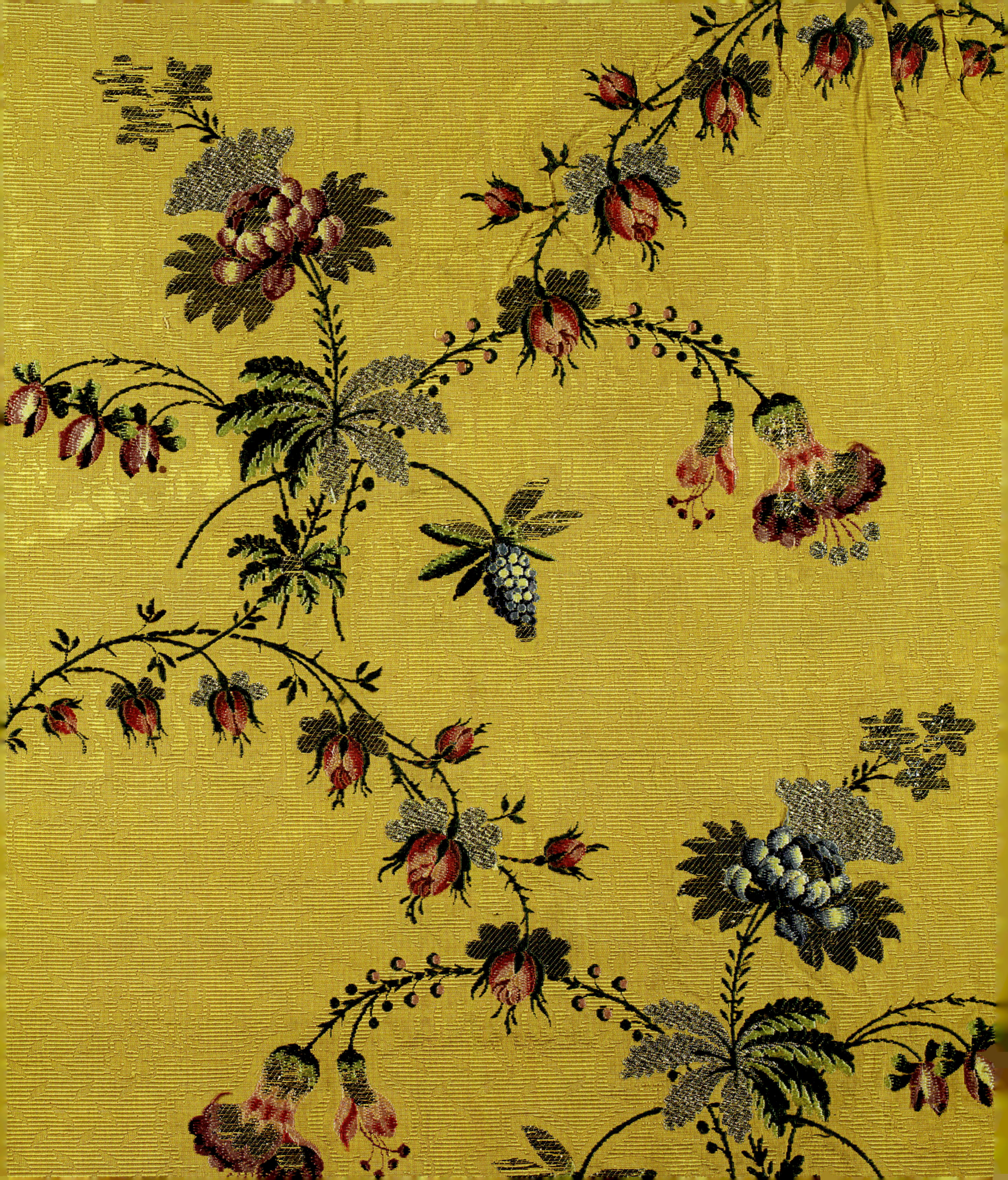

Trade, Industry, and Empire

John Styles

Eighteenth-century Britons excelled at commerce. The stereotype of the British as a nation of shopkeepers was amply justified, if by shopkeeping we mean the practice of trade. Britain enjoyed the most successful mercantile economy in eighteenth-century Europe. No other major European economy was more committed to trade, with the exception of the Netherlands, a country whose economic pre-eminence was already in the past. Britain's self-image was that of a trading nation. 'Our trade is the most considerable of the whole world', boasted the author of the *Present State of Great Britain* in 1755, 'and, indeed, Great Britain is, of all other countries, the most proper for trade.'[1]

The value of British trade increased more than fivefold between 1714 and 1818. At the start of the eighteenth century Britain traded predominantly with continental Europe, but subsequently the most dynamic growth was in its exchanges with the extra-European world, especially the Americas. Colonial imports were crucial for this expansion of trade beyond Europe. They came in the form of raw materials or exotic manufactured goods, such as furniture, the finest of the productions veneered with ivory as presented to Queen Charlotte (cat. 28.20), or metalwork—the jewelled Huma bird with which she was presented after it was taken from the throne of Tipu Sultan of Mysore, India, a supreme example (fig. 28.01)—as well as the ultra-fashionable muslins for women's dresses that came from India in the 1780s and 1790s. Yet imports of manufactured goods from the colonies, however fashionably exotic they might be, did not fit easily into the mercantilist framework that dominated British thinking about trade and empire. Ideally, colonial possessions were expected to serve as exclusive markets for goods manufactured in the mother country. In return, they were expected to furnish British manufacturers with those raw materials Britain could not provide, such as the Caribbean mahogany from which Thomas Chippendale shaped his furniture, or South Carolina indigo to dye the blue check fabric woven in Lancashire.

Markets and raw materials were crucial because Britain was the industrial powerhouse of the age. In 1777, François Lacombe, the author of a guidebook for French visitors to London, listed 'the objects that no people can furnish in such a range and quality as the English', along with the goods the English exported 'in profusion' to France, 'to the shame of the ministry at Versailles'.[2] The list was a long one, a testimony to Britain's industrial might at a time when the Industrial Revolution had barely begun. Lacombe's list included cotton, woollen, and silk textiles, the decorative metal buttons and boxes produced in Birmingham, iron railings and grates, scientific instruments, clocks, furniture from tea stands to mahogany tables, tin-glazed pottery and the new white earthenware, and even coaches. The list reflects the wealth of innovation which transformed British manufacturing across the eighteenth century. It included new kinds of things, from cut glass to creamware dinner services. It included new materials, like mahogany for furniture or Turkey red dye. It included new techniques, like transfer printing on pottery or power spinning of cotton yarn for calicoes. It included new design ideas, like those

Detail of p. 458, cat. 28.05

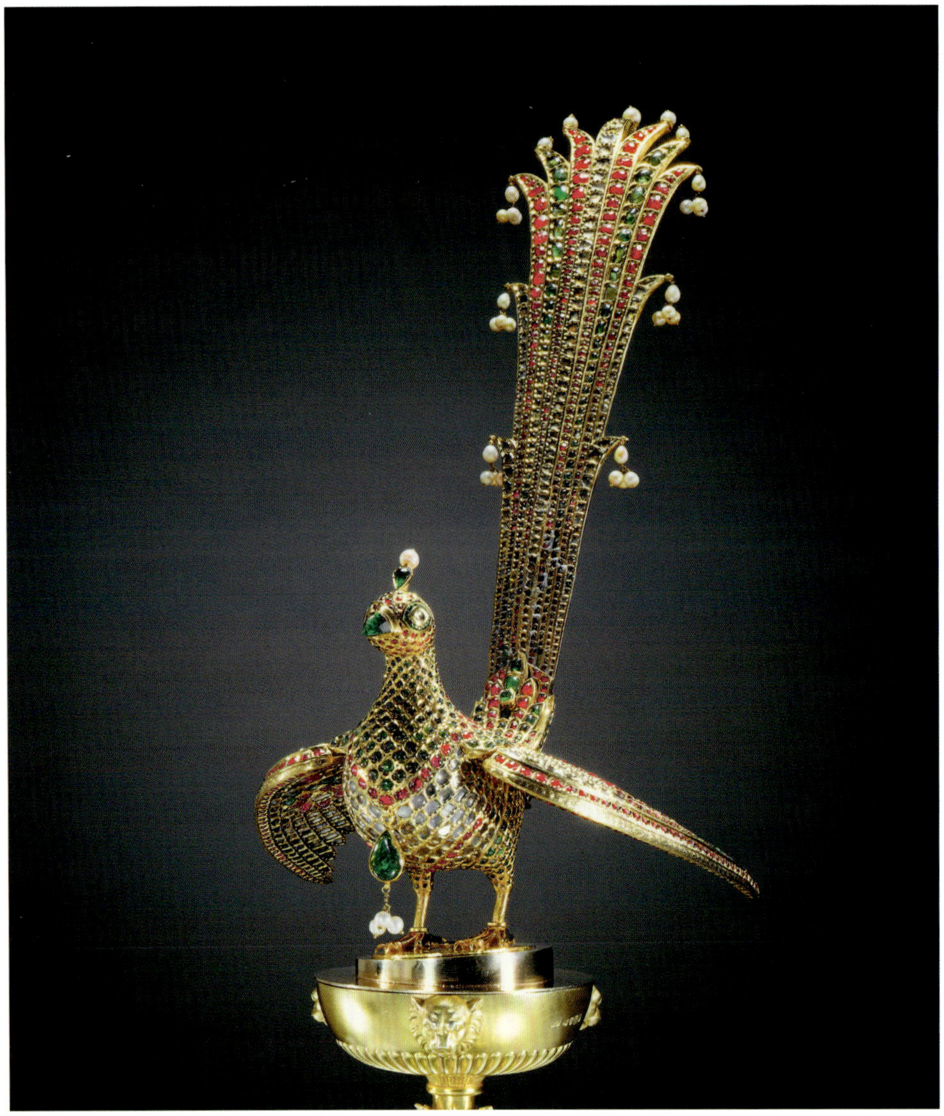

FIG. 28.01 *Huma Bird (Bird of Paradise)*, c. 1787–91, gold, rubies, emeralds, diamonds, pearls, and silver gilt, 16⅝ x 7⅞ x 11⅛ in. (42 x 20 x 28 cm). Royal Collection Trust

developed by Josiah Wedgwood in the Staffordshire potteries, carefully adapted to suit a variety of markets (see cats. 28.09–28.11). In 1776 Wedgwood wrote to his business partner Thomas Bentley about styles of decoration and colours that would appeal to overseas markets: 'this Pattern was intended chiefly for abroad, & Foreigners in general will bear higher coloring, & more forcible contrasts than the English.'[3] Above all, Lacombe's list reveals Georgian Britain's capacity for invention and improvement, underpinned by a widely held British conviction that scientific research was the key to improvements in agriculture and industry. It was a phenomenon that astonished, delighted, and sometimes, as in Lacombe's case, troubled foreigners.

Most of the goods in Lacombe's list could reach their purchasers in France only by being smuggled, because they were either banned in France or subject to huge customs duties. Like most states in eighteenth-century Europe, France espoused the mercantilist assumption that one country's economic improvement was likely to result in setbacks for the rest. States repeatedly adopted beggar-my-neighbour policies. High customs duties excluded foreign imports, while state subsidies and monopolies encouraged domestic production. France under the rule of its Sun King, Louis XIV (reigned 1643 to 1715), exemplified these policies. Impossibly high tariff barriers were erected against British manufactured goods, including staple British exports like the famous English woollen broadcloths. Concurrently, the French state established subsidized royal manufactures, with exclusive monopoly rights, to produce replacements for goods previously imported. Louis also turned his vast, brand-new royal palace at Versailles into an international showcase for the products of French industry, especially luxury products such as Gobelins tapestries,

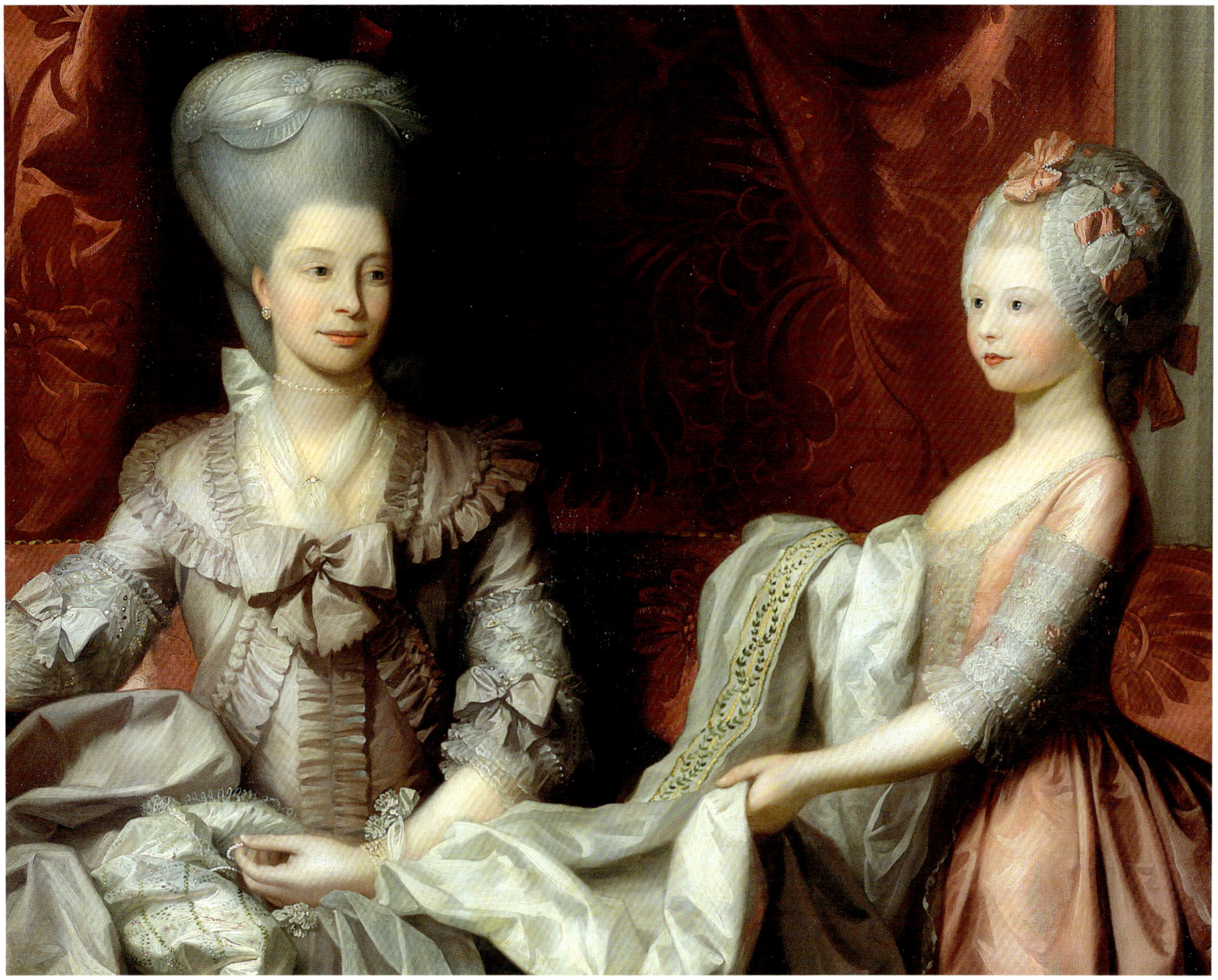

FIG. 28.02 Benjamin West, *Queen Charlotte with Charlotte, Princess Royal*, 1776, oil on canvas, 66¼ x 81 in. (168.2 x 205.7 cm). Royal Collection Trust

or the silk fabrics made at Lyons which set the fashions for the elites of Europe. It was Louis's aim that France should lead European taste and that its industries should dominate the European market.

Louis XIV's example was copied by other European rulers, including many in those parts of Germany where the princesses who married into the British royal family were raised. German rulers built royal palaces in the latest French styles and used state subsidies and monopolies to promote industry in their territories, most notably in Saxony, where the first European porcelain was made at a royal manufactory at Meissen. Eighteenth-century Britain shared many of the same mercantilist assumptions about trade, industry, and empire as its continental neighbours. Yet the capacity of the British state to implement them was limited in ways that constrained royal initiative, a legacy of seventeenth-century conflicts between crown and Parliament. As Parliament gained authority vis-à-vis the crown towards the end of the seventeenth century, royal power to grant monopolies to manufacturers was severely curtailed. In eighteenth-century Britain, a proliferation of great royal manufactories enjoying monopoly privileges, of the kind that flourished in other parts of Europe, became impossible. State-endorsed monopolies were restricted to short-term patent rights granted to more-or-less genuine innovators, and were stripped of official financial benefits. At the same time, the triumph of political and legal attitudes hostile to corporate privilege led to the end of attempts to channel state support and regulation of manufacturing through guilds. By the 1720s guilds were in steep decline across major sectors of British industry.[4]

The eclipse of industrial guilds and monopolies in Britain did not, however, mean the end of state support for manufacturing. Nor did it result in any diminution in the state's desire to secure economic advantage over Britain's international competitors. Instead, formal state support for manufacturing became ever-more narrowly focused on the regulation of international trade by means of tariffs and prohibitions. Eighteenth-century British tariffs distinguished between raw materials, semi-processed goods, and finished manufactured goods. Raw silk paid a lower customs duty than silk yarn, while imports of most silk fabrics were subjected to extremely high duties, and eventually banned. Tariffs also distinguished between different producers. French mid-market linens were effectively priced out of the English market in retaliation for French tariffs on British woollen textiles. Equivalent German mid-market linens were admitted, subject to a moderate tariff.[5] The outcome amounted to a subtle industrial policy, designed to encourage British manufacturers at the expense of their overseas rivals, while fostering the supply of raw materials, especially from British colonies where several types of manufacturing were prohibited.

As a consequence, British attachment to trade, industry, and empire posed a challenge for the three German-born royal Princesses, Caroline of Ansbach, Augusta of Saxe-Gotha, and Charlotte of Mecklenburg-Strelitz. Raised in German courts where the ruling family's intervention in support of the state's trade and industry was normal and often expected, they joined a monarchy whose initiative in this sphere was constrained. Yet their new country was one of the richest in Europe, its royal court considerably wealthier than those they left behind, its government intent on international economic supremacy, and its population convinced of its nation's commercial destiny. As members of a royal dynasty that periodically had to justify itself in the face of opposition, how were they to respond?

For all three Princesses, the answer lay in the judicious use of the resources at their disposal at court, resources that were partly a matter of purchasing power, but more importantly of prestige. All three succeeded in publicizing their support for British enterprise and empire, without compromising the image of royal

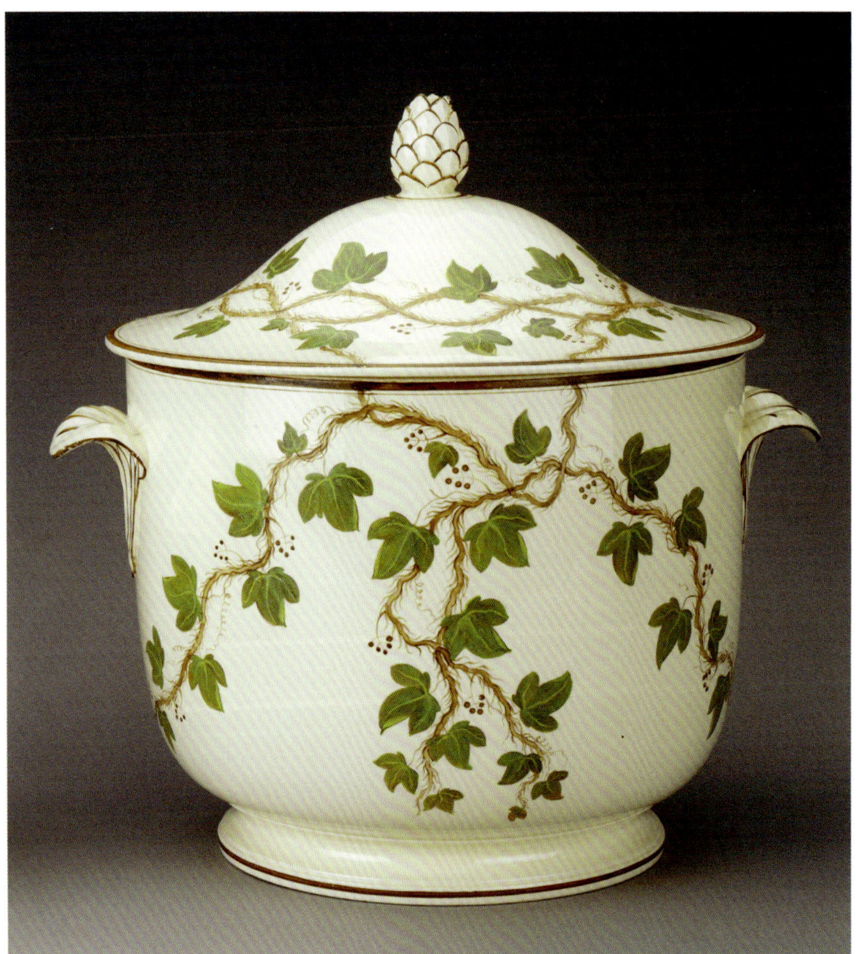

FIG. 28.03 Wedgwood, covered urn, c. 1795, creamware 'Queen's Ware' (earthenware with lead glaze), 12¾ x 12¼ x 12¼ in. (32.2 x 31.2 x 31.2 cm). Royal Collection Trust

frugality cultivated by the first three Hanoverian monarchs.[6] They did so by acquiring and endorsing British products, by supporting charitable initiatives to educate children who would become British workers, and by encouraging British science, innovation, and exploration.

British royal expenditure was frugal only by the standards of royal excess set by rulers elsewhere in Europe. The Hanoverian monarchs may not have built new palaces, but the finery on display at royal events matched any court in continental Europe. Monarchs were expected to dress magnificently. Where they led, their courtiers followed. Queen Caroline consistently promoted British-made textiles in her dress, as well as endorsing the production of raw silk in the American colonies.[7] Princess Augusta and Frederick, Prince of Wales, went further at their court in the late 1740s, insisting that all the fine clothing worn by their courtiers should be British-made. Their display of sartorial patriotism was reinforced by an ostentatious family visit to the silk weavers of the Spitalfields district of London (cats. 28.04–28.07).[8]

Queen Charlotte, too, bought Spitalfields silks, but she had no other option once imports of foreign silk fabrics were prohibited by Parliament in 1765. Benjamin West's portrait of the Queen, and her eldest daughter, of 1776, which shows mother and child working on the trimming of a fine textile, while illustrating Charlotte's interest and personal engagement in the crafts of textile manufacture, also alludes to her concern to support the textile industry in Britain more generally (fig. 28.02). More influential was her order that year of a large tea set from the Staffordshire potter Josiah Wedgwood. The tea set was not made from the porcelain that previously graced royal tables, but from Wedgwood's newly perfected cream-coloured earthenware. Its light body and brilliant, clear glaze were the product of Wedgwood's painstaking chemical experiments (fig. 28.03). He was not only a brilliant technical innovator, but an inspired commercial publicist. Tradespeople who supplied the Royal Household already displayed the royal crest in their shops and in their advertising to lure customers. With Charlotte's support, Wedgwood reinvented the practice, branding his creamware as 'Queen's Ware' and styling himself 'Potter to Her Majesty'.[9] And Wedgwood was not the only industrial innovator to benefit from Charlotte's encouragement. Tableware as well as decorative ceramics were also commissioned from the Chelsea and Worcester porcelain factories. Matthew Boulton, Wedgwood's fellow Midlands industrialist, was favoured with purchases of his ormolu clocks and vases in the 1770s (cat. 28.19).

The runaway success of Wedgwood's Queen's Ware connected Charlotte very publicly with the leading edge of technical innovation in the early years of the Industrial Revolution. Yet her approval of Wedgwood's creative manipulation of her patronage was consistent with the conduct of her two royal predecessors. Despite limited room for manoeuvre, all three German Princesses succeeded in persuading their new British subjects of their shared patriotic commitment to the nation's commercial and imperial progress.

NOTES

1. *Magnae Britanniae Notitia: Or, the Present State of Great Britain* (London, 1755), 42.
2. Un Atheronome de Berne [François Lacombe], *Observations sur Londre et ses Environs* (Paris and London, 1777), 226.
3. Josiah Wedgwood to Thomas Bentley, 15 Dec. 1776, in *Correspondence of Josiah Wedgwood*, Vol. 2: *1772–1780*, ed. Eufemia Farrar (London: Women's Printing Society, 1903), 224.
4. John Styles, 'Spinners and the Law: Regulating Yarn Standards in the English Worsted Industries, 1550–1800', *Textile History* 44, no. 2 (2013): 159.
5. Raymond L. Sickinger, 'Regulation or Ruination: Parliament's Consistent Pattern of Mercantilist Regulation of the English Textile Trade, 1660–1800', *Parliamentary History* 19, no. 2 (2000): 211–32.
6. For royal frugality, see Hannah Smith, *Georgian Monarchy: Politics and Culture, 1714–1760* (Cambridge: Cambridge Univ. Press, 2006), 116–20.
7. Joanna Marschner, 'Queen Caroline of Ansbach: Attitudes to Clothes and Cleanliness, 1727–1737', *Costume* 31, no. 1 (1997): 29.
8. Christine Gerrard, 'Queens-in-waiting: Caroline of Anspach and Augusta of Saxe-Gotha as Princesses of Wales', in *Queenship in Britain, 1660–1837: Royal Patronage, Court Culture and Dynastic Politics*, ed. Clarissa Campbell Orr (Manchester: Manchester Univ. Press, 2002), 156.
9. Neil McKendrick, 'Josiah Wedgwood: An Eighteenth-Century Entrepreneur in Salesmanship and Marketing Techniques', *Economic History Review* 12 (1960): 408–32.

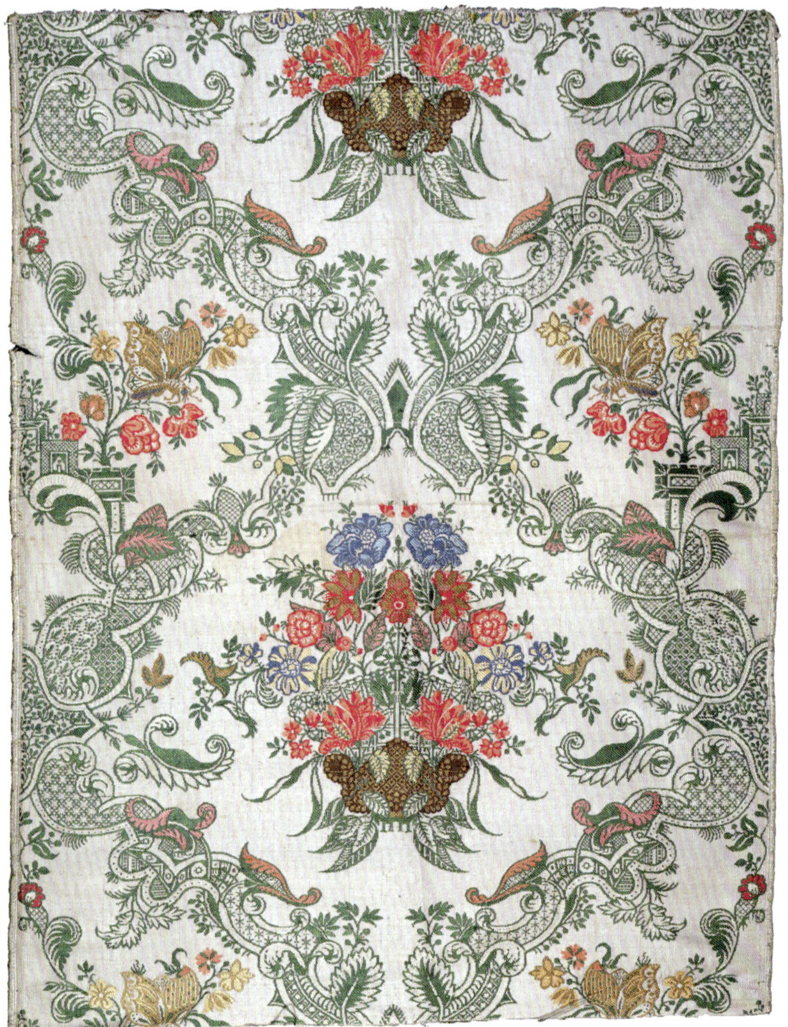

28.04

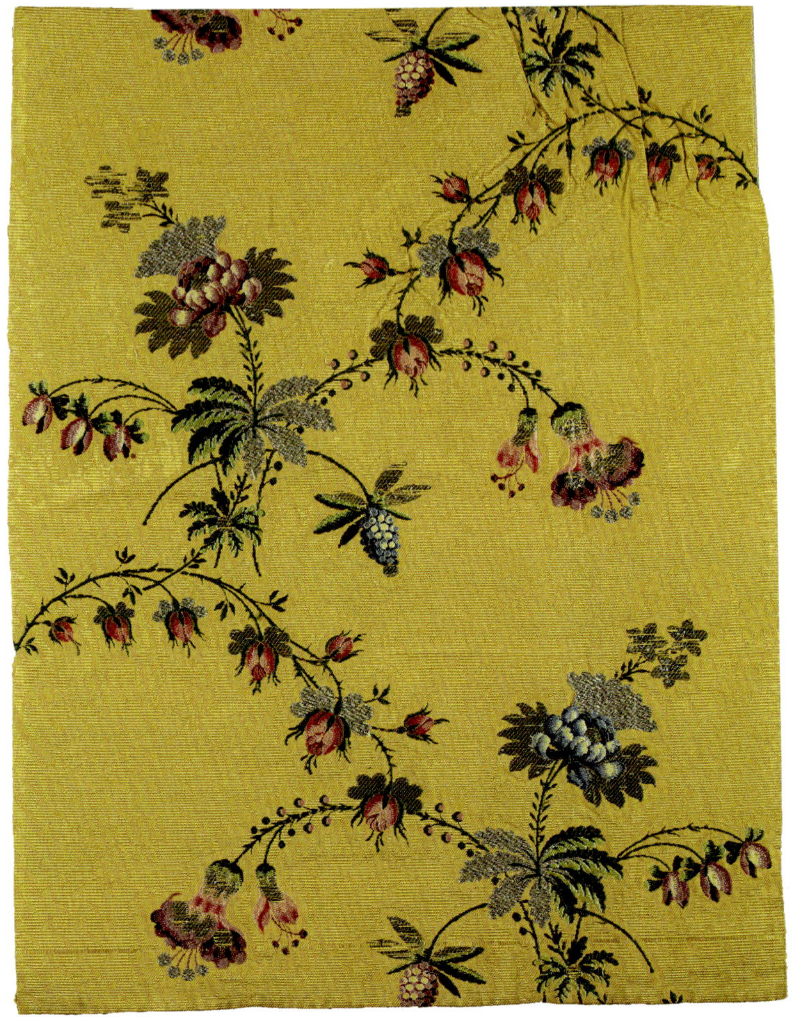

28.05

TEXTILE ART

28.04

UNKNOWN SPITALFIELDS WEAVERS, London
Dress fabric
1720–30
Brocaded silk
26⅝ x 20¼ in. (67.5 x 51.5 cm)
Victoria and Albert Museum, T.15-1951

28.05†

UNKNOWN SPITALFIELDS WEAVERS, London
Dress fabric
1750–60
Brocaded silk with silk and metal threads
40⅛ x 19¼ in. (102 x 49cm)
Victoria and Albert Museum, Given by Miss D. M. Gower, T.161&A.1959

28.06*

PROBABLY DANIEL VAUTIER (French, active c. 1749), after ANNA MARIA GARTHWAITE (1690–1763), designer
Skirt panel
1749
Silk, plain weave with supplementary patterning warps and supplementary brocading wefts
37 x 20 in. (94 x 50.8 cm)
The Metropolitan Museum of Art, Rogers Fund, 2004, 2004.416

* = YCBA only
† = HRP only

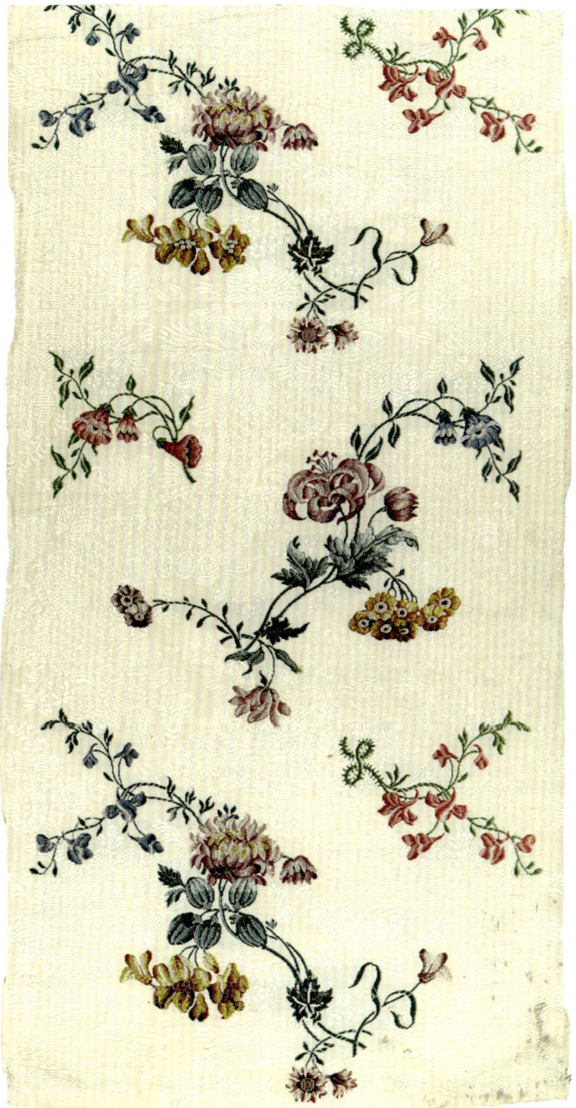

28.06

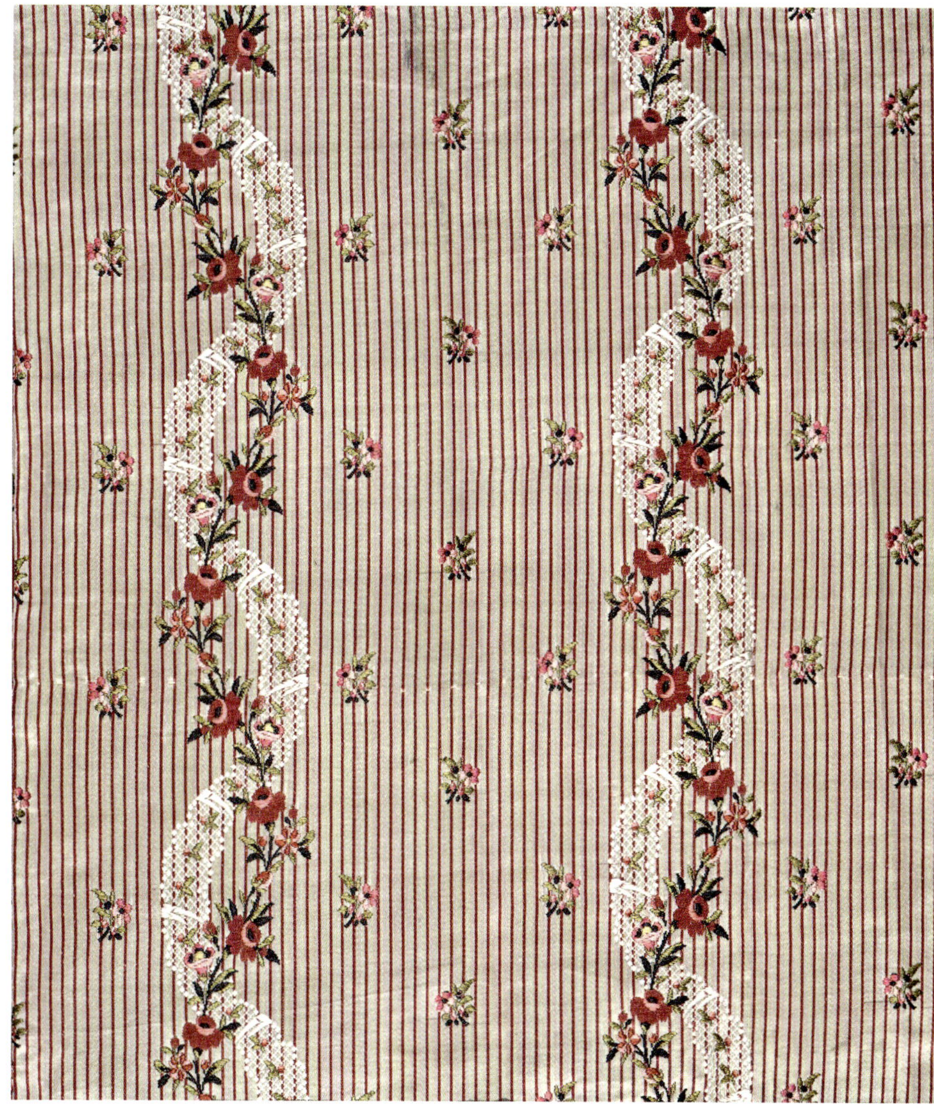

28.07

28.07

UNKNOWN SPITALFIELDS WEAVERS, London
Skirt panel
1775–80
Brocaded silk satin
60½ x 38¼ in. (154 x 97 cm)
Victoria and Albert Museum, Bequeathed by R. E. Wynne-Roberts and Mrs D. L. Wynne-Roberts, T.77-1961

Caroline, Augusta, and Charlotte, in their roles of encouragers of trade and industry, wore fabrics grown and woven both in Britain and its colonies. English silk weaving was concentrated in Spitalfields in East London. Many of the leading craftsmen were of Huguenot (French Protestant) origin. In 1735 Caroline accepted a length of silk which had been reeled by Piedmontese textile workers who had been sent to Georgia by the Trustees of the Georgia Society in an attempt to encourage an embryonic silk industry there. The thread was later woven up for the Queen at Sir Thomas Lombe's Mill in Derby. Wearing this dress of the new American silk at her Drawing Room, Caroline strongly vocalised her support of the initiative. Her taste in textiles was for stoutly woven silks in rich colours.

Frederick, Prince of Wales, and Augusta took two of their children with them for a well-publicized visit to workshops in Spitalfields in 1750, and 'expressed great satisfaction at their fine and curious manufactures'. A month later they would inspect the products of the wool-combers and weavers of woollens during their tour of the south west of England. The couple took a special pride in wearing British textiles on court occasions. For the King's birthday in 1749 it was noted that 'the Prince and Princess of Wales were observed, with great satisfaction to lovers of this country, to be richly dressed in our own manufacture.' Augusta is depicted wearing crisp light silk textiles, often trimmed with fine lace.

In 1765, challenges to the English silk weaving industry as the result of the Duke of Bedford's successful rejection of the bill to raise the import tax on Italian silks, brought demonstrations from weavers in London. The following year protectionist legislation prohibiting French imports brought the same craftsmen out on the streets in jubilation. Charlotte and her daughters found their choices scrutinized by an ever-greater and more curious public audience. Despite restricting the import of silks and embroidered lawns and muslins from India since the 1720s, these products became increasingly desirable as the century progressed. However, for court occasions the wearing of muslin was not encouraged by Charlotte; with her daughters she continued to promote English silks as the industry struggled to remain economically viable in the face of foreign competition.

28.08

UNKNOWN MAKERS, England
Queen's silk samples
1794
Silk, silk satin, silk brocade woven with metal thread, silk velvet
11¾ x 4 in. (29.8 x 10.1 cm)
Historic Royal Palaces, HRP 3501554.a-J/K/L

Within the Royal Ceremonial Dress Collection there are twenty-eight small samples of dress-weight silk brocade glued to slips of paper, with hand-written annotations indicating which were worn by the Queen and which by the princesses, sometimes with the date of the occasion. It is possible to identify twelve of the fragments directly as Spitalfields products. There are matching samples in pattern books, part of the Warner Archive in the Victoria and Albert Museum, from the weaving firm Batchelor, Ham and Perigal, which had traded from Spital Square since the late seventeenth century, and specialized in silks woven with metal threads, supplying the mercers Ibbetsons and King's. The reason for the preparation of this archive of textiles samples is unknown. The most likely answer is that it was prepared by Mrs Elizabeth Bland. A regular customer at several of the mercers supplying the Queen and her daughters, she collected a large variety of gossip and ephemera from her suppliers to send to a friend, Mrs Harriot Wrightson of Cusworth Hall in Yorkshire, whose young family prevented her from travelling to London to take part in court occasions herself. The sample sheet relating to the wedding of George, Prince of Wales, and Princess Caroline of Brunswick, which took place at the Chapel Royal in St James's Palace on 9 April 1795, reveals that the bride wore a red velvet mantle over a dress of silver tissue. Charlotte wore a petticoat of silver tissue, under a robe of white silk woven with gold. Samples of the silks, each with their small figured patterns, worn by Princesses Royal, Augusta, Elizabeth, and Sophia complete the set. One of the Queen's mercer's Harding and Howell, was granted permission to sell a style of light silk with small figured patterns as 'Queen's Silk', which Charlotte may have seen as a way of encouraging sales of such textiles.

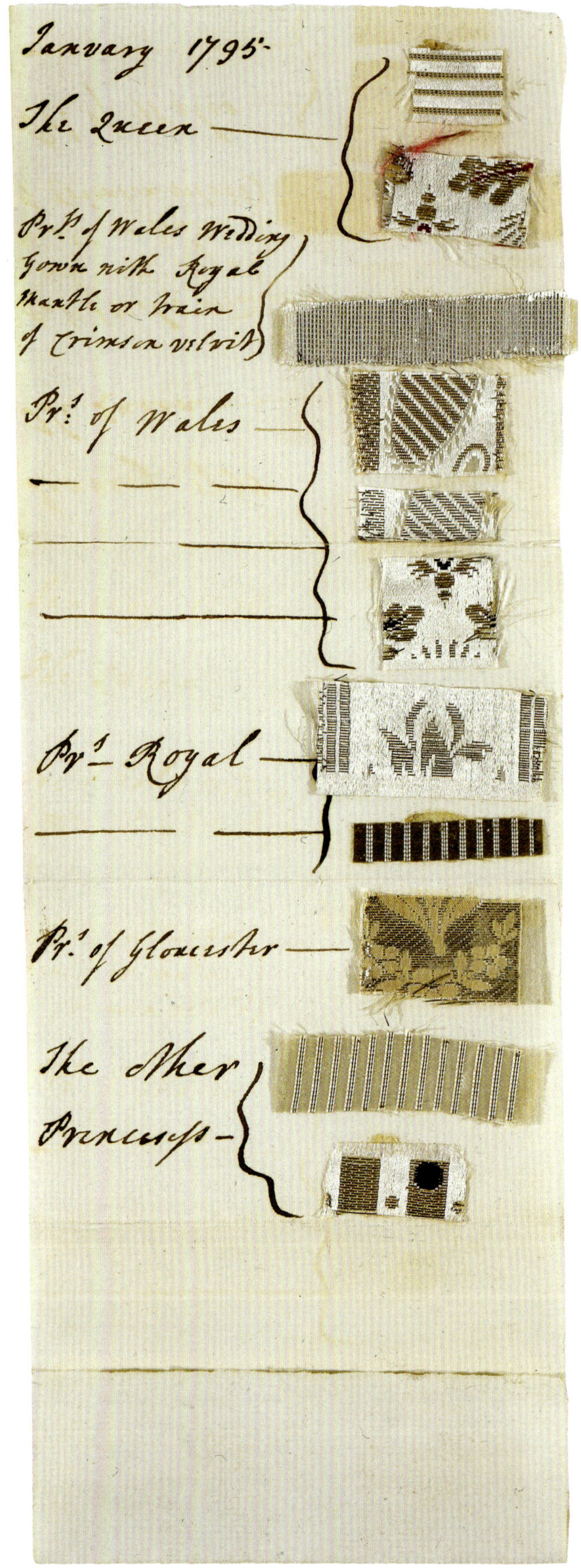

28.08

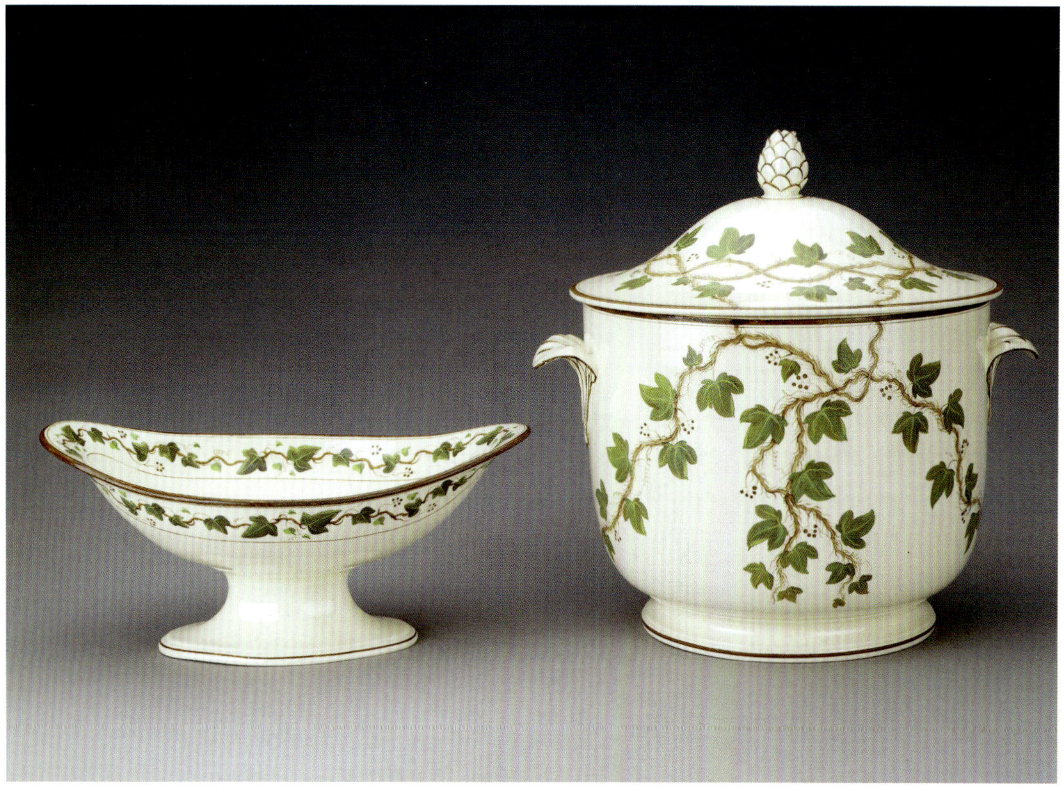

28.09

BRITISH CERAMICS

28.09

WEDGWOOD
Centre dish and covered urn
c. 1795
Creamware 'Queen's ware' (earthenware with lead glaze)
Centre dish: 5¼ x 11⅝ x 7⅝ in. (13.2 x 29.5 x 19.3 cm)
Urn: 12¾ x 12¼ x 12¼ in. (32.2 x 31.2 x 31.2 cm)
Royal Collection Trust, RCIN 53002.1 and RCIN 53050.1

In 1765 Josiah Wedgwood opened his first London showrooms in Charles Street, off Grosvenor Square, and in June he received a commission from Charlotte to make an elaborate creamware tea service 'with a gold ground & raised flowers upon it in green, in the same manner of the green flowers raised upon the *mehons* [melons]'. The following year he was officially appointed Potter to Her Majesty and his creamware was renamed 'Queen's ware', with her permission. This creamware urn and cover, one of a pair, each with an artichoke finial and handles shaped in the form of splayed stylized foliage, decorated throughout with trails of ivy, is likely to have been another commission from the Queen. The centre dish is of the same period.

28.10*

WEDGWOOD
Teapot
c. 1763
Earthenware with transfer print
5¾ x 8¼ x 5 in. (14.6 x 21 x 127 cm)
The Mint Museum, Charlotte, NC, Gift of the Delhom Service League in memory of Martha Higgins Fishburne, 1997, 1997.86A-B

Shortly after receiving his first commission from Charlotte, Josiah Wedgwood sent her 'a box of patterns . . . amongst which I intend sending two setts of Vases, Creamcolour engine turn's & printed'. In 1761 a small quantity of Wedgwood's fine creamware had been bought by John Sadler of Liverpool, probably for experiments in transfer-printed decoration; by this means moderately priced ceramic could be elaborately embellished. From 1762 Sadler and his partner Guy Green decorated increasingly large quantities, and this became a mainstay of Wedgwood's business. Commemorative wares were produced in quantity, some celebrating Charlotte, Wedgwood's patron, after which the ceramic type was named. The transfer-printed decoration of this teapot is taken from a mezzotint made of Charlotte by Thomas Frye and his pupil William Pether, in 1762. Frye had captured her likeness, together with that of George III when they visited the theatre. On the teapot the image is reversed.

28.11*

WEDGWOOD
Lobed dish and plate
1780–90
Creamware 'Queen's ware' (earthenware with lead glaze)
Dish: 1½ x 11½ x 9 in. (3.8 x 29.2 x 22.9 cm)
Plate:¾ x 8 in. (1.9 x 20.3 cm)
Winterthur Museum, Garden and Library, Wilmington, DE, 1960.0146 and 1960.0147.003

Josiah Wedgwood was fully cognizant of the benefit to be derived from royal interest in his wares. He wrote to his business partner Thomas Bentley in 1772, 'The Great People have had their vases in their Palaces long enough for them to be seen & admir'd by the Middling Class of People, which Class we know are vastly, I had almost said infinitely, superior in number, to the Great.' Colonial expansion in America presented tantalizing new markets, though with acknowledgment that there might be distinct local tastes to be satisfied. In 1766 Wedgwood wrote 'Green dessert ware is often wanted, in reality for the West India Islands. I have a few crates on hand, some gilt, some plain.' This 'Queen's ware' dish and plate, with its green decoration, is part of a service associated with a family with connections to the West Indies in the mid-eighteenth century.

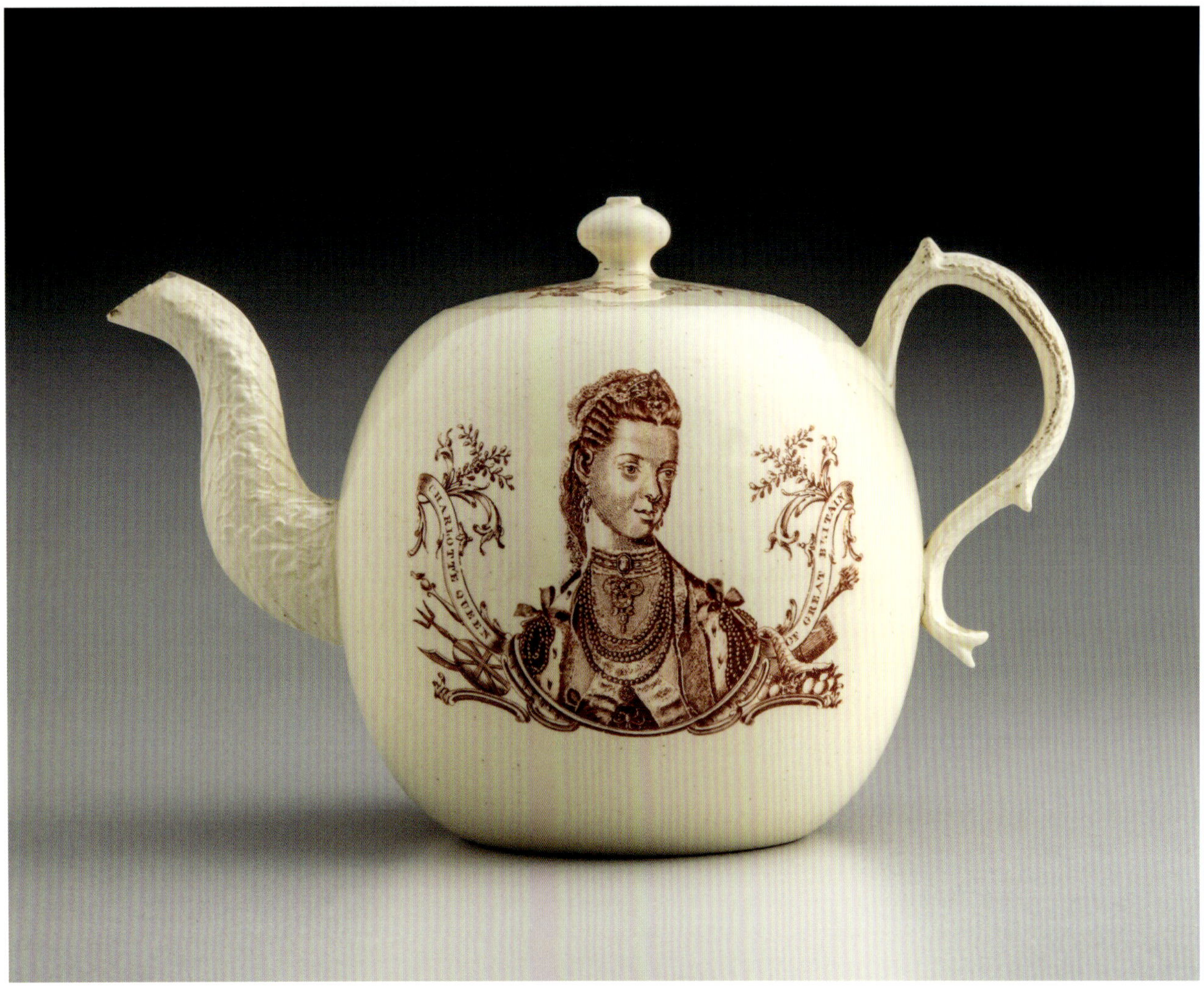

28.10

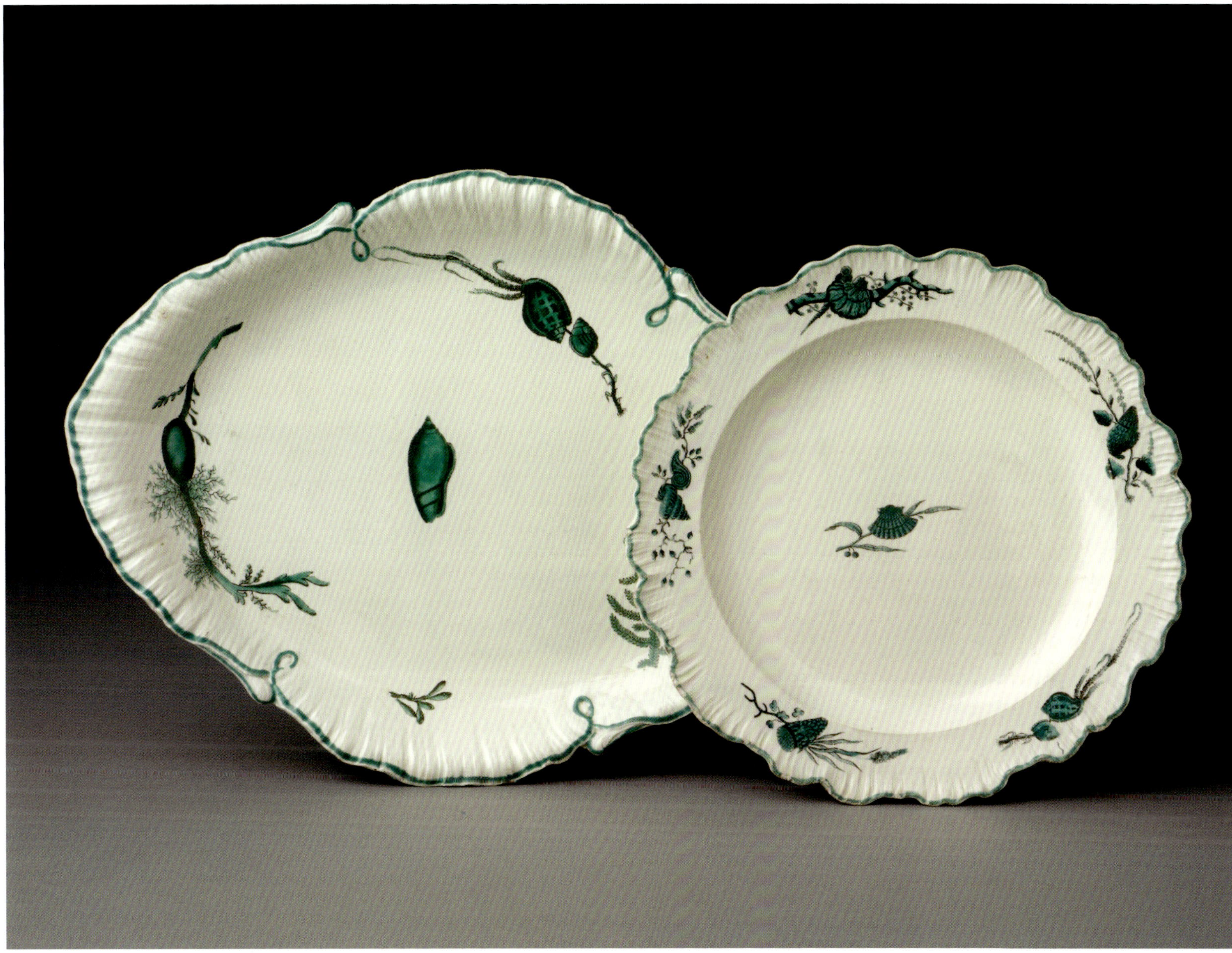

28.11

28.12*

WEDGWOOD & BENTLEY, manufacturers, after **WILLIAM HACKWOOD** (d. 1836), modeller, after **ISAAC GOSSET** (1713–1799), designer
Queen Charlotte
c. 1776
Blue and white jasperware, with contemporary ormolu frame
2 ¼ x 2½ in. (5.72 x 6.35 cm)
Yale Center for British Art, Paul Mellon Collection, B2001.2.1415

During the 1770s Josiah Wedgwood developed a form of fine stoneware which he called jasperware. Its almost jewel-like finish led to it being used most frequently for small decorative wares, inlays for furniture, cameos and jewellery. Typically the pieces were given a moulded decoration, or 'sprigging', usually in a white paste, to contrast with the coloured ground. Jasperware portrait miniatures were collected not only by the elite, but also a newly prosperous middling sort. This cameo was modelled by Wedgwood's designer William Hackwood, after a wax portrait of Charlotte by Isaac Gossett (see also cats. 13.12–13.13). Wedgwood and Bentley had taken pains to cultivate influential patrons, in the interest of furthering their commercial success, not only within the royal family, but also with ambassadors and noble families throughout Britain and Europe. These connections gave them access to historic works that could be used as models or inspiration. When rendered in jasperware, for a modest cost works could be acquired for private pleasure, for their beauty, artistic merit, and technological achievement.

28.13*

UNKNOWN MAKER after **WILLIAM HACKWOOD** (d. 1836) or **HENRY WEBBER** (1754–1826), designer, for **JOSIAH WEDGWOOD** (1730–1795)
Signet ring with inset gem carved with seal of the London Committee of the Society for Effecting the Abolition of the Slave Trade ('But Am I not a Man and a Brother')
Gold with red stone, possibly jasper
After 1787
Face: 1⅛ x¾ in. (2.8 x 2.1 cm)
Yale Center for British Art, Gift of Susan M. Yecies, Yale, B.A. 1971, Cabinet Z1 5a

This medallion, first made in 1787, became a popular icon in the British movement for the abolition of the slave trade in the late eighteenth and early nineteenth centuries. Josiah Wedgwood probably engaged sculptor Henry Webber to create the design of a kneeling slave, his hands in chains. The modeller William Hackwood then prepared the medallion for production in Wedgwood's black jasperware against a white jasperware ground to resemble cameo gemstones of antiquity. It was intended to promote the work of the Committee for the Abolition of the Slave Trade, founded in 1787 by Thomas Clarkson, who in 1786 published his 'Essay on the Slavery and Commerce of the Human Species'. Wedgwood was a member of the Committee—later known as the Society for the Abolition of the Slave Trade—so it is likely that distribution of the medallions took place through the organization, and that Wedgwood bore the costs himself. The medallion expressed in material form the growing horror at the barbarous practices of the transatlantic slave trade, and the premises upon which that trade thrived. While the sentiment expressed in the motto 'AM I NOT A MAN AND A BROTHER' which encircles the medallion, is bold, the image of the kneeling man reveals the religious, moral, and paternalistic values of the society members. George III and Charlotte demonstrated their support for the movement by restricting the consumption of sugar within the royal household. Satirists, however, construed this policy merely as evidence of the parsimony of the royal couple (see cat. 15.09).

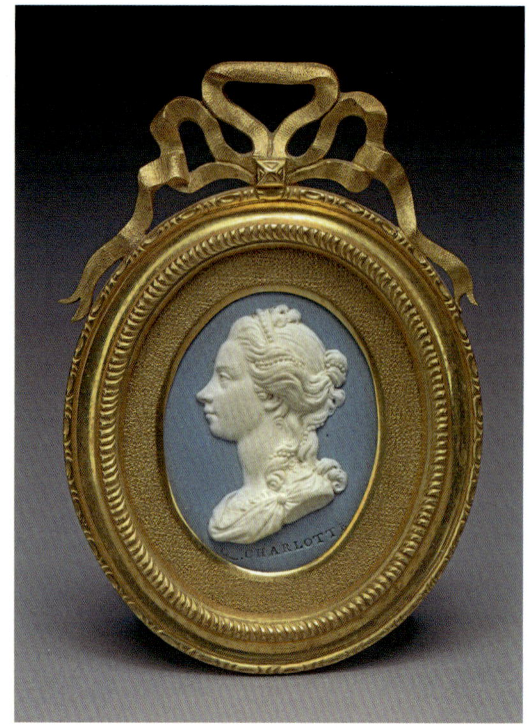

28.12

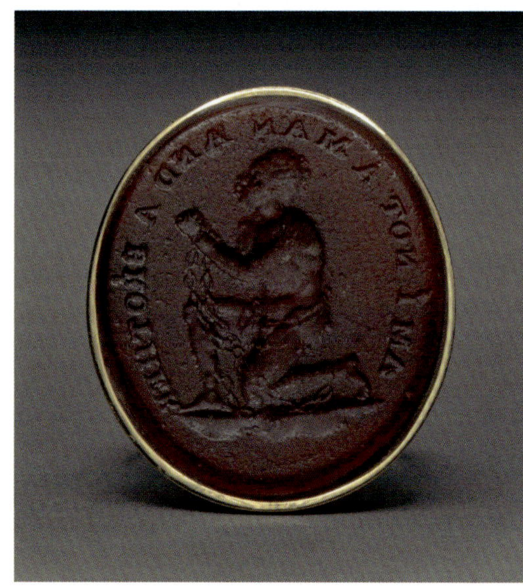

28.13

* = YCBA only
† = HRP only

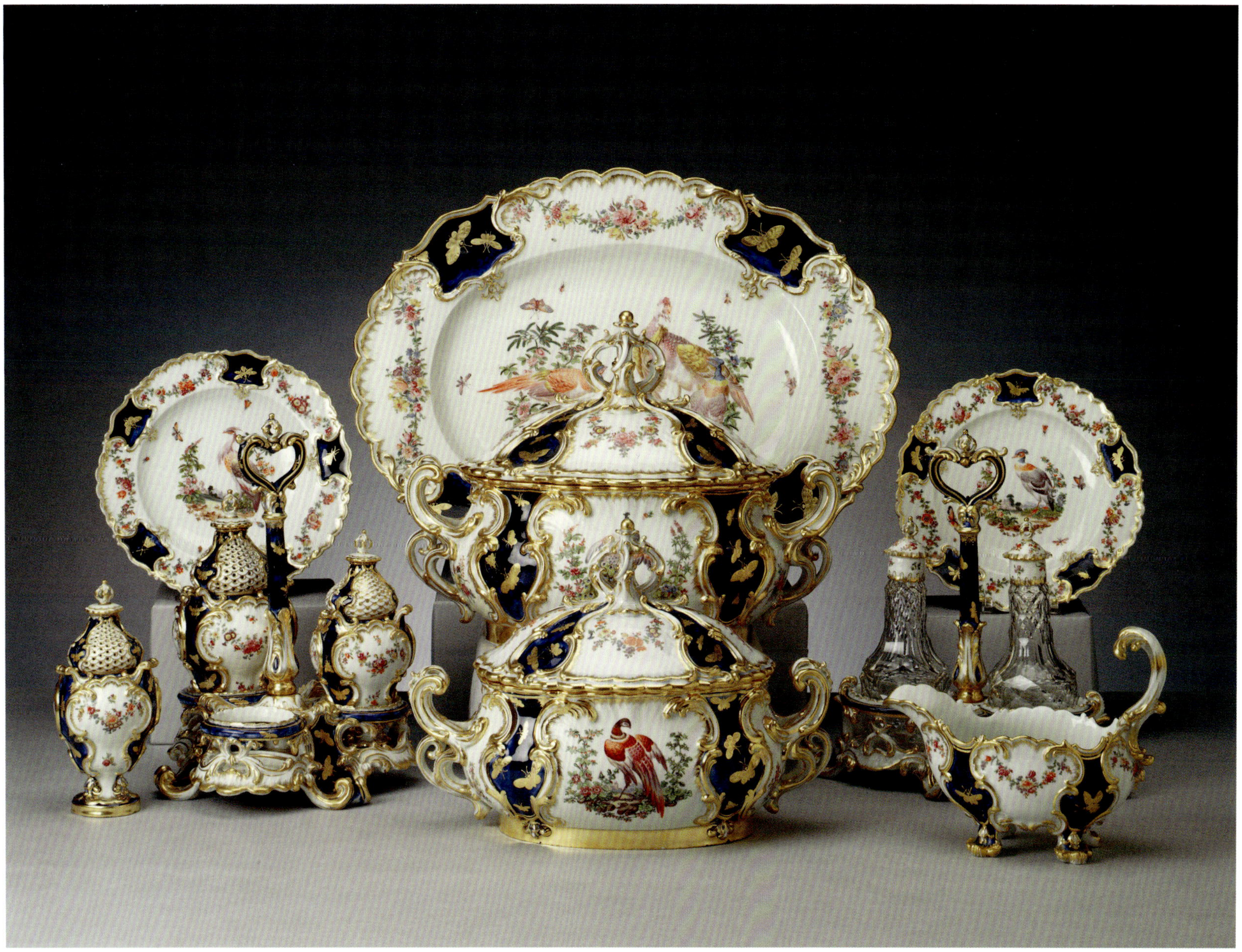

28.14

28.14

CHELSEA (LAWRENCE STREET FACTORY)
Oval platter and small tureen ('Mecklenburg service')
c. 1763
Soft paste porcelain with Mazarin blue glaze
Platter: 2¼ x 20⅛ x 16⅛ in. (5.7 x 51.0 x 41.0 cm)
Tureen with cover: 5⅝ x 10¾ x 8⅜ in. (14.2 x 27.2 x 21.2 cm)
Royal Collection Trust, RCIN 57022.1 and RCIN 57012.3 (illustrated with full set)

This service was a gift from Charlotte to her eldest brother, Adolphus Frederick IV, Duke of Mecklenberg-Strelitz in 1764. It comprised all the pieces required for the first course and dessert, as well as tea and coffee wares and candlesticks. It made a striking impression on Thomas Nugent who visited the palace at Neustrelitz in 1766: 'a complete service of Chelsea porcelane, rich and beautiful in fancy beyond expression. I really never saw any Dresden porcelane near so fine.' The Mecklenburg Service proved influential in the development of British ceramic design. The intense blue enamel, known as 'Mazarin blue', made in imitation of the dark-blue enamels pioneered at the Sèvres factory, would set the standard for blue glazes produced at this factory and elsewhere later in the century. The form of several of the pieces in the service reveals the earlier experience of Nicholas Sprimont, the Chelsea factory owner, as a silversmith. Sprimont had earlier supplied silver to Frederick, Prince of Wales.

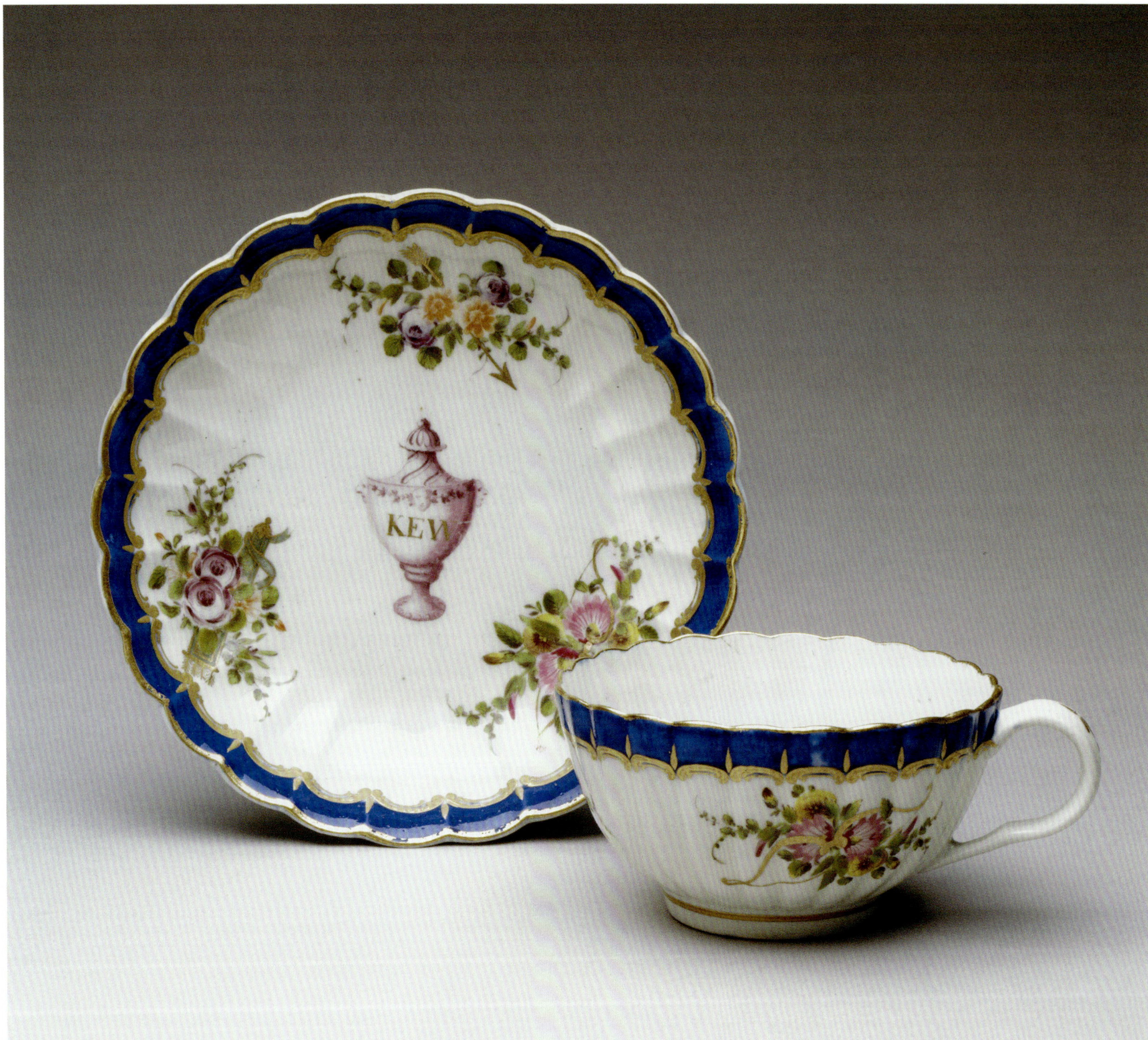

28.15

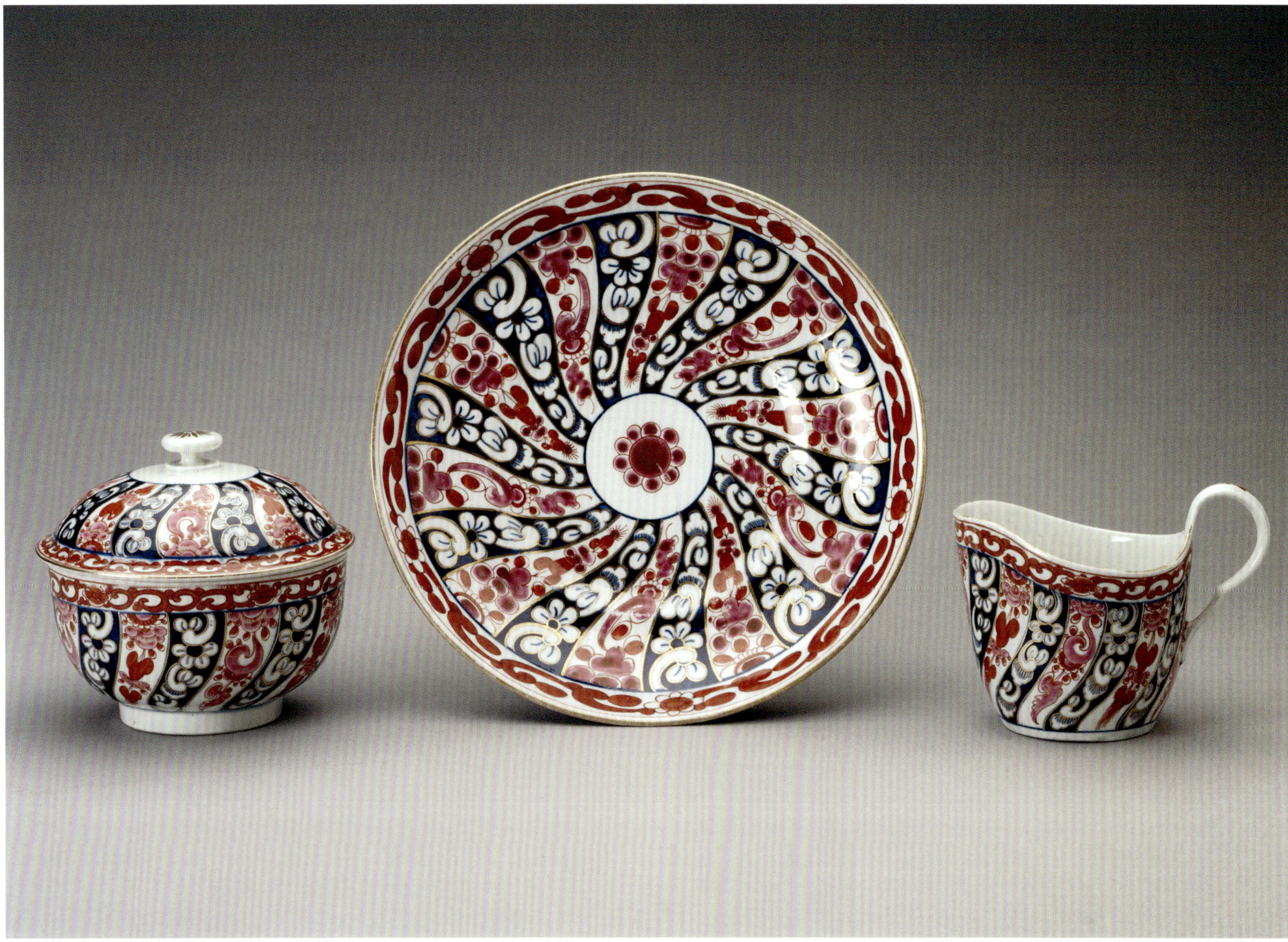

28.16

28.15

WORCESTER PORCELAIN COMPANY
Tea cup and saucer ('Kew service')
c. 1780
Porcelain, painted and gilded
Teacup: 1⅞ x 4¼ x 3⅜ in. (4.8 x 10.7 x 8.6 cm)
Saucer: 5⅜ in. (13.8 cm) diameter
Royal Collection Trust, RCIN 73111

28.16

WORCESTER PORCELAIN COMPANY
Saucer dish, sugar bowl and cover, milk jug (Queen Charlotte's pattern)
c. 1780
Porcelain
Saucer dish: 8½ in. (21.7 cm) diameter
Sugar bowl: 4⅜ x 4⅞ in. (11.3 x 12.3 cm)
Milk jug: 3¾ x 4¾ in. (9.5 x 12.2 cm)
Royal Collection Trust, RCIN 73297, RCIN 73294 a–b, RCIN 73295

In 1788 George III and Charlotte visited the Worcester Porcelain factory. It was subsequently granted a royal warrant, and thereafter was known as the 'Royal Porcelain Works'.

The decoration of the teacup and saucer with its floral sprays pierced with bows, arrows, and quivers, was known as the 'Marriage' pattern. However, the insertion of the urn, inscribed with the word 'Kew', suggests that this tea service was an early purchase by the royal family for use at the White House, Kew, or in one of the folly temples that adorned its garden.

The decoration of the saucer dish, sugar bowl, and milk jug is derived from patterns found on wares imported from Japan, and had been a staple of the Worcester factory from the late 1760s. It was renamed the 'Queen Charlotte' pattern, possibly to commemorate the visit of George III and Charlotte to the factory in 1788.

28.17

DUESBURY & CO., manufacturers, after **JOHN BACON** (1740–1799), modeller, after **RICHARD EARLOM** (1743–1822), engraver, after **JOHAN ZOFFANY** (1733–1810)
Queen Charlotte with Princess Augusta and Princess Charlotte
c. 1773
Biscuit porcelain and porcelain with turquoise glaze and gilding
9 ¼ x 7½ x 6⅞ in. (23.5 x 19.0 x 17.5 cm)
Royal Collection Trust, RCIN 37022

This figure group, showing Charlotte with her daughters, the Princess Royal and Princess Augusta, is one of a set of three which combine to reproduce in porcelain the royal family as depicted by Johan Zoffany in his portrait of 1770, showing George III, Queen Charlotte and their six eldest children. Made in a biscuit porcelain, it was produced by 'Mr Dusbury's Derby and Chelsea Manufactury of Porcelaines' in about 1773. In a brochure advertising his productions on sale, the figure groups were first in the list of 'the Principal Additions made in this Year'. It is likely that the modelling of the group was achieved by John Bacon, who worked after the engraving of Zoffany's painting by Richard Earlom. The potential for mass production of such a piece brought an opportunity for a broad constituency to acquire a work both inspired by a painting in the royal collection and commemorating the royal family.

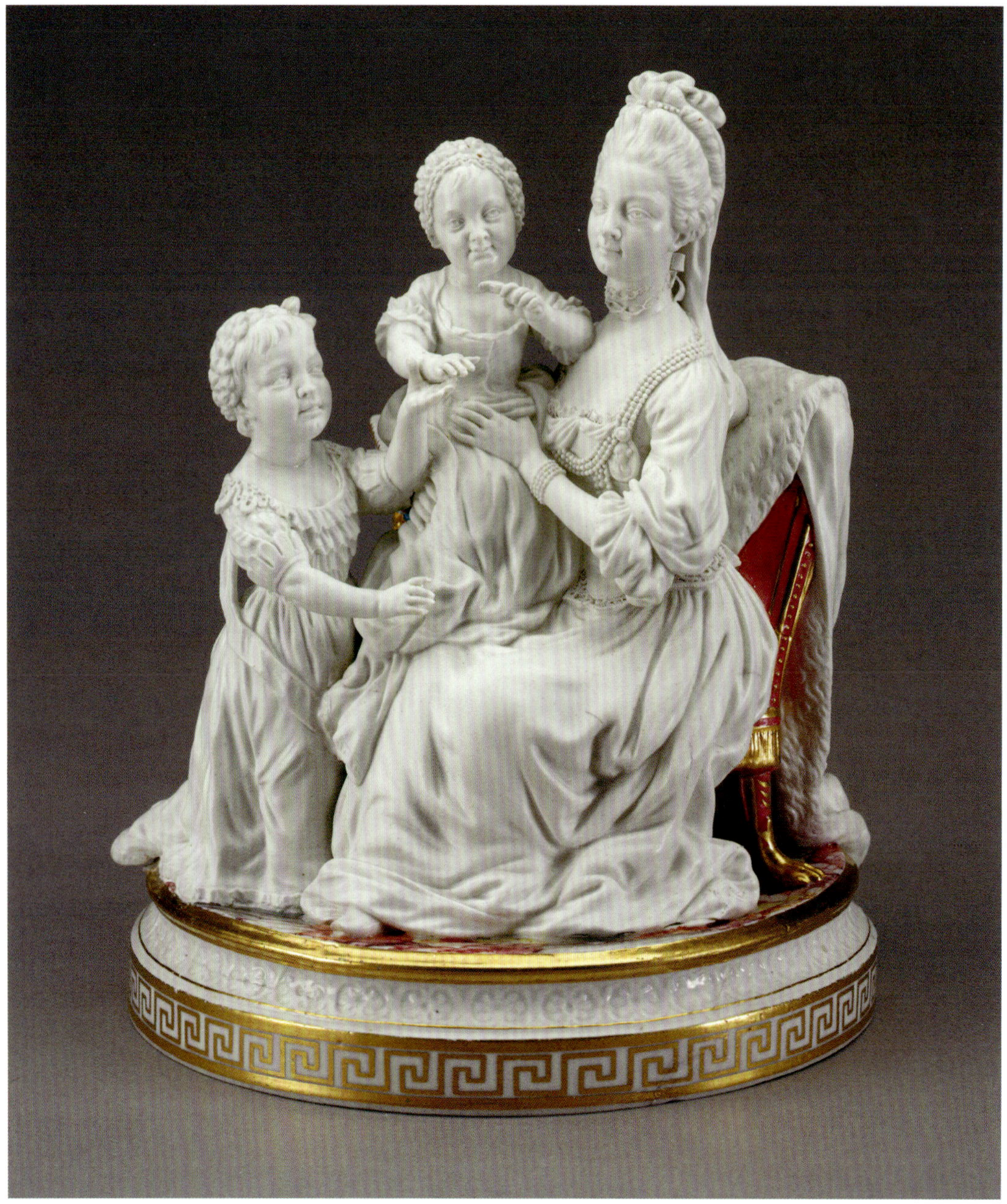

28.17

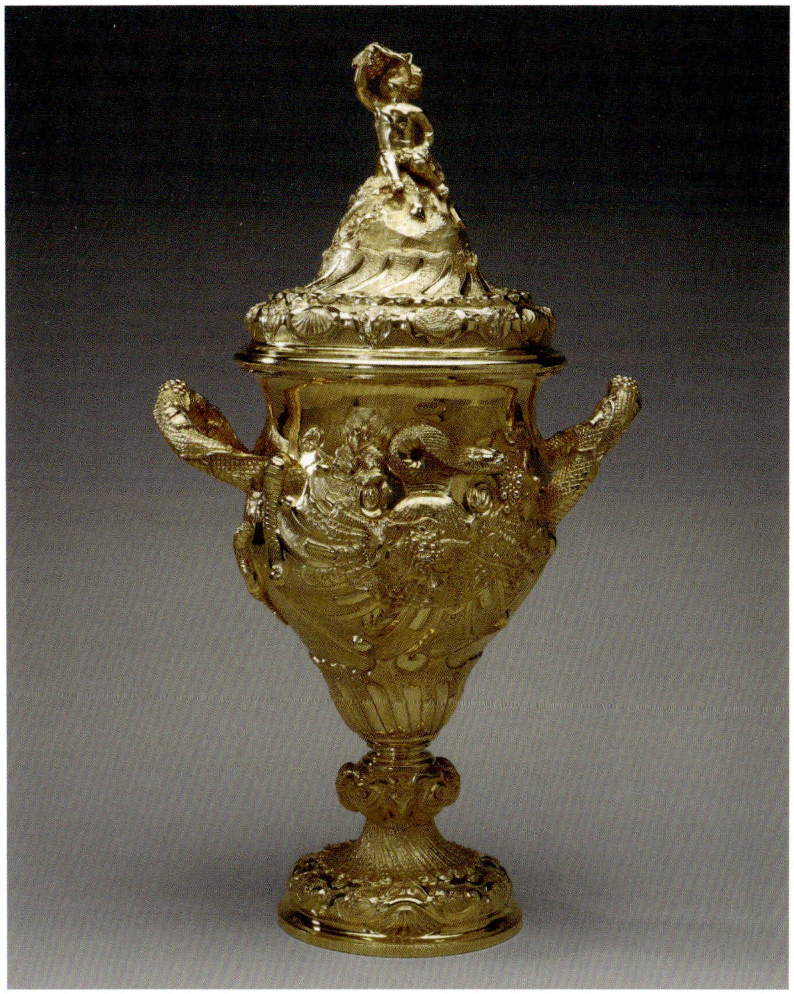

28.18

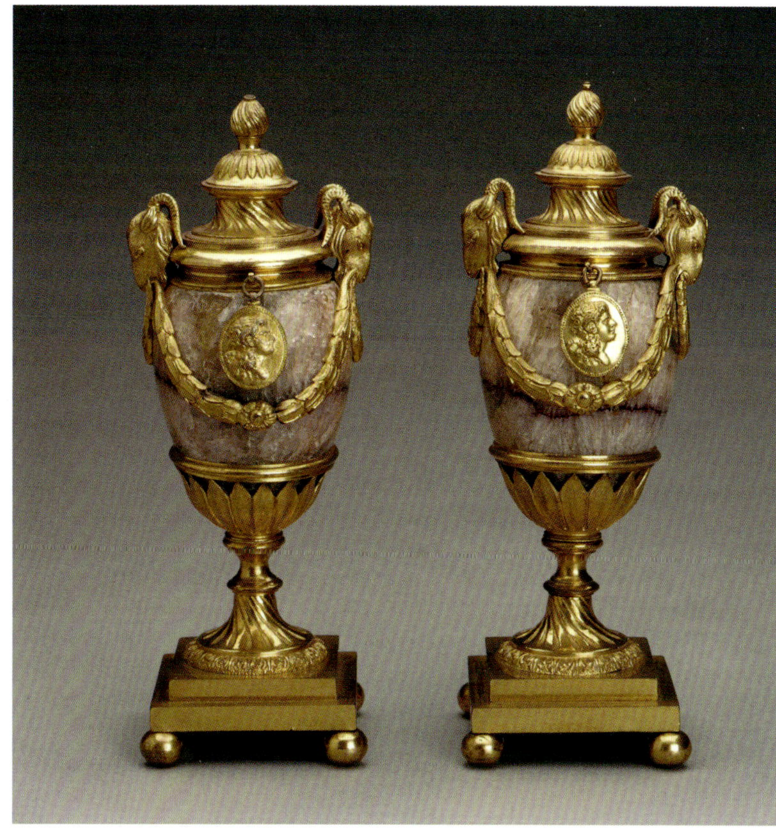

28.19

BRITISH METALWORK

28.18*

ATTRIBUTED TO PAUL JACQUES DE LAMERIE (1688–1751)
Cup and cover
c. 1740
Silver gilt, cast, embossed, and chased
17¾ x 9½ x 6⅜ in. (45 x 24 x 16.2 cm)
Royal Collection Trust, RCIN 51016

This cup was acquired by Frederick, Prince of Wales, in January 1740. In his ledger it is listed as 'fine cheased gilt cup and cover', and it cost him £61 3s. Although unmarked it bears many close similarities to works by Paul de Lamerie, including the Bath Cup, presented by Frederick to the city of Bath in 1738, following the visit he made there with Augusta to promote the trades and industries of the West Country. This piece demonstrates a highly developed use of Rococo ornament and asymmetrical design, and its modelling is of exceptional quality. It has been suggested that it has design characteristics which relate to a particular craftsman in the Lamerie workshop, traditionally known as the 'Maynard modeller' after the Maynard sideboard dish. This modeller remains unidentified but may be related to the Kändler family who were important in the development of the Meissen factory in Dresden. It is likely that Frederick, Prince of Wales, gave the cup to Augusta as a present. An inventory made in 1832 of plate in the royal collection recorded specifically that it had belonged to the Princess.

28.19*

MATTHEW BOULTON (1728–1809)
Pair of candle vases
1770–71
Blue john and gilt bronze
8 x 3 x 3 in. (20.5 x 7.7 x 7.7 cm)
Royal Collection Trust, RCIN 6828 1–2

Matthew Boulton was an entrepreneur and manufacturer of 'toys' or small-scale novelties in Birmingham. A member of the Lunar Society, he was a friend of Benjamin Franklin, James Watt, and Dr Erasmus Darwin. The potter Josiah Wedgwood was another member of the society, and the relationship between science and industry is well illustrated in his plan to develop a research establishment in the Staffordshire Potteries. Wedgwood and Boulton collaborated over the production of cameo jewellery set in cut-steel mounts.

In 1768 Josiah Wedgwood reported to Thomas Bentley: 'Mr Boulton is picking up Vases, & going to make then in Bronze. You know how old China bowles, Jarrs &c are mounted in Metal, he proposes an alliance betwixt the Pottery and Metal branches, Viz, that we should make such things as will be suitable for mounting, & not have a pott look. & he will finish them with mounts.' Boulton's essays, as well as incorporating ceramics, would also take in semiprecious stones such as blue john. In 1770 Boulton's 'scheming to be sent for by his Majesty' was successful. It is likely that this pair of blue john vases, of the pattern known as 'Goat's Head', was acquired by George III and Charlotte following this visit.

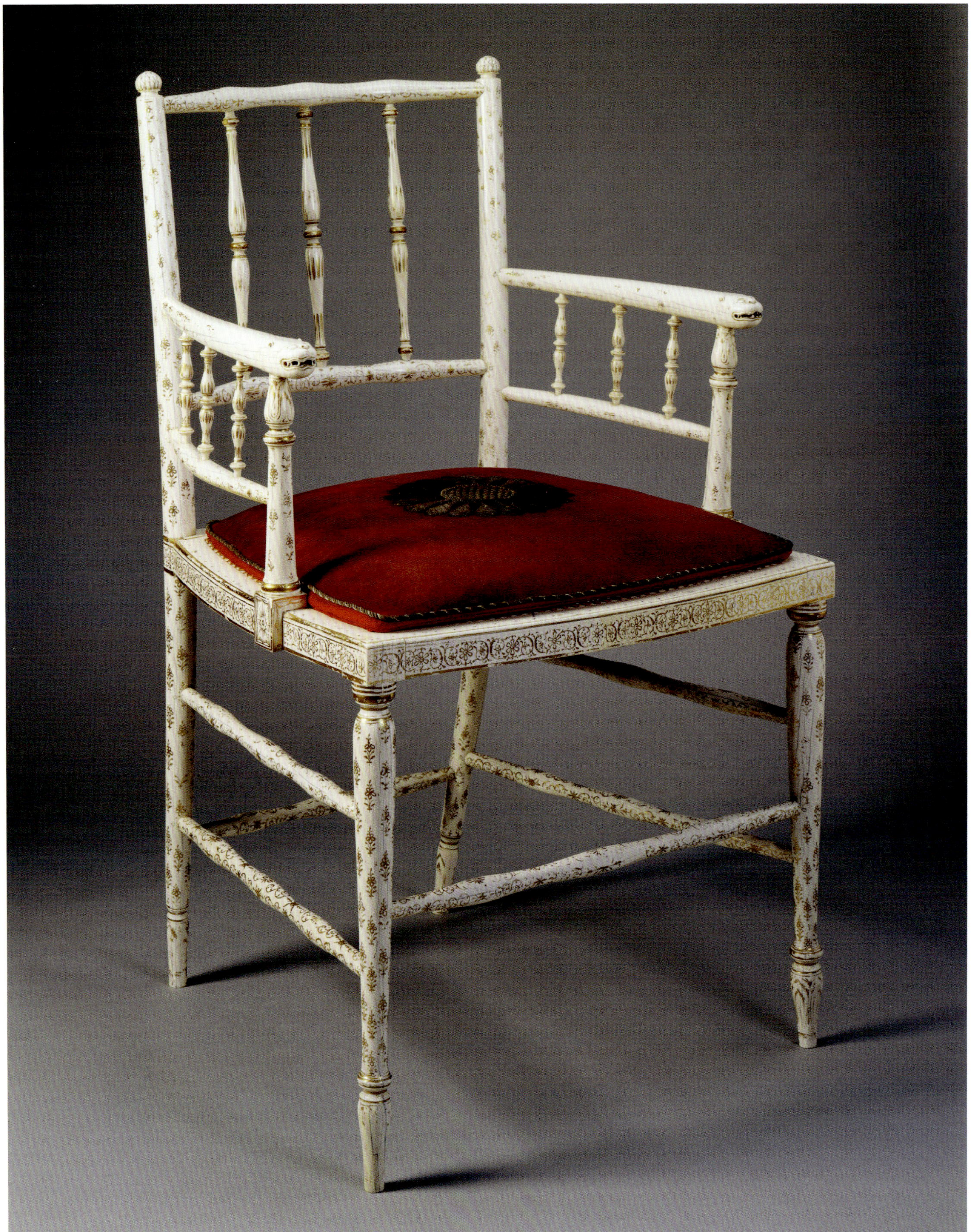

28.20

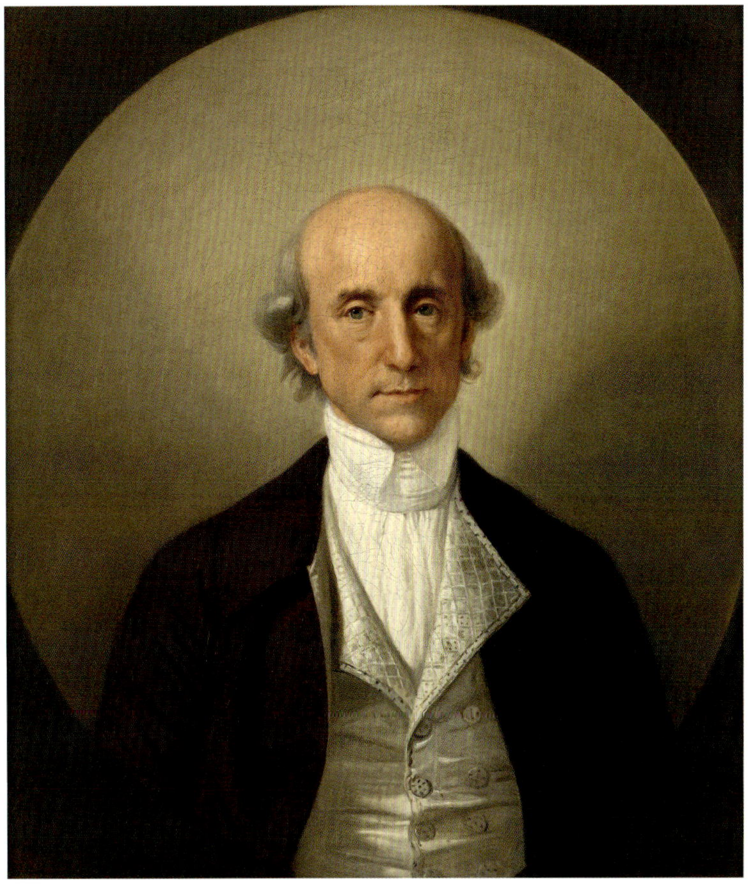

28.21

Warren Hastings was the first Governor of the Presidency of Fort William (Bengal), the head of the Supreme Council of Bengal, and thereby the first de facto Governor-General of India from 1773 to 1784, during which time he helped extend and regularize the nascent Raj created by Clive of India. Hastings resigned and returned to England in 1785. In 1787 he was accused of corruption and impeached, but after a long trial he was acquitted in 1795. He was made a Privy Counsellor in 1814.

Hastings served as intermediary for many gifts presented to George III and Charlotte from rulers of the Indian states. After the substantial gift of many pieces of ivory veneered furniture in 1784, in 1785 Charlotte received 'A state bed of rich and very curious workmanship [which] was carried to the Queen's Palace as a present from Lady [*sic*] Hastings [Marian Hastings, Hasting's second wife], brought from India, which exceeds anything in grandeur seen in this country.' The delivery of diamonds, the gift of Nizam Ali Kahn in 1786, would lead to popular accusations that Hastings, by that date discredited, was trying to buy royal favour.

From 1783 Johan Zoffany spent six years travelling and working in India. This painting was made shortly after the artist's arrival in Calcutta, where he was introduced to Warren Hastings, who became his chief benefactor. It is an informal image showing Hastings modestly dressed, and was probably commissioned as a gift for his friend, Sir Elijah Impey, Chief Justice in Calcutta.

INDIA AND EMPIRE

28.20†

UNKNOWN MAKER, Murshidabad, India
Armchair
c. 1790
Ivory, partly painted and gilt, velvet with silver embroidery
35 x 22⅛ x 21¼ in. (88.8 x 56 x 54 cm)
Royal Collection Trust, RCIN 11197

This chair is one of a set of four that probably formed part of a group acquired by Warren Hastings, Governor-General of Bengal, as a gift from Mani Begum, widow of Nawab Mir Jafar. The chairs were presented to Queen Charlotte in about 1784 by Hastings's wife, Marian. This type of furniture, of which the Queen's 1819 sale catalogues include twenty pieces, was made in Murshidabad from the third quarter of the eighteenth century. In 1781 George III bought Charlotte a bureau of the same manufacture from the sale of Alexander Wynch, Governor of Madras from 1773 to 1775. Charlotte continued to acquire items for her 'Indian Room' at Frogmore.

After the Battle of Plassey in 1757, ivory carvers in Murshidabad increasingly sought to make goods that were desirable to Anglo-Indian consumers. The circulation of print sources, such as Thomas Chippendale's *The Gentleman and Cabinet-maker's Director* (1754) and George Hepplewhite's *The Cabinet Maker and Upholsterer's Guide* (1788) facilitated this process. Between 1760 and 1780, ivory carvers in Vizagapatam introduced new techniques involving ivory veneer, a method that was soon adopted elsewhere. A thin layer of ivory was applied to the wooden carcass by means of glue and rivets. The surface was then decorated by engraving the ivory and filling the design with black varnish.

28.21

JOHAN ZOFFANY (1733–1810)
Warren Hastings
1783–84
Oil on canvas
28½ x 24¼ in. (72.4 x 61.6 cm)
Yale Center for British Art, Paul Mellon Fund, B2012.17

28.22*

UNKNOWN ENGRAVER
Queen of Hearts cover'd with diamonds
1786
Etching on paper
7¾ x 5½ in. (19.6 x 13.9 cm)
Unknown publisher
Library of Congress, British Cartoon Prints Collection, PC1-6978

While flattering at first glance, this satire by an unknown artist connected Queen Charlotte's fondness for jewels with accusations of personal corruption against Warren Hastings, the Governor of Bengal. Hastings had presented a large diamond to George III and the stone, which was publicly construed as a bribe, made its way into Charlotte's hands. The Queen is depicted wearing a large fashionable turban headdress, an allusion to Hastings's Indian associations, and sports an enormous quantity of diamond jewellery.

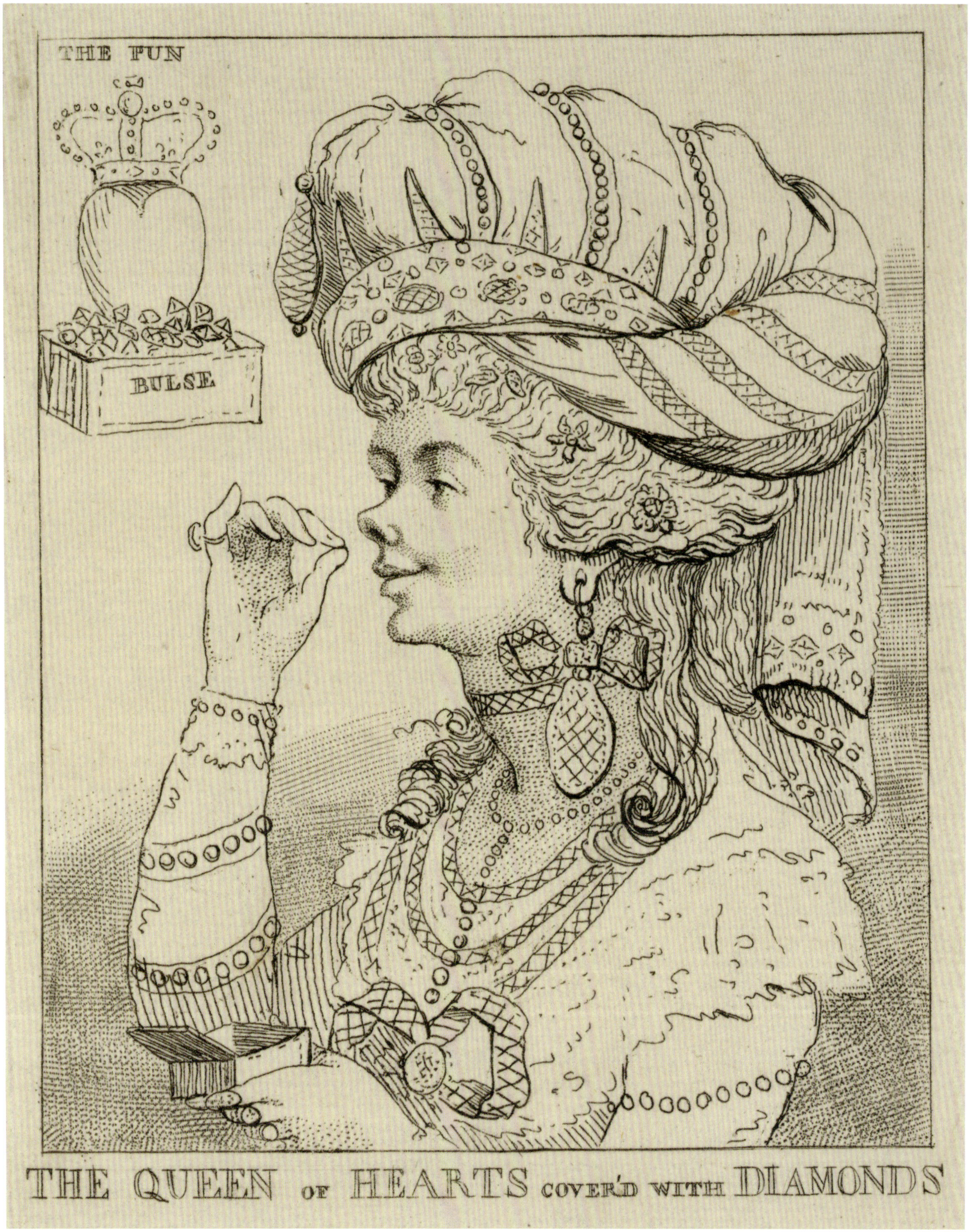

28.22

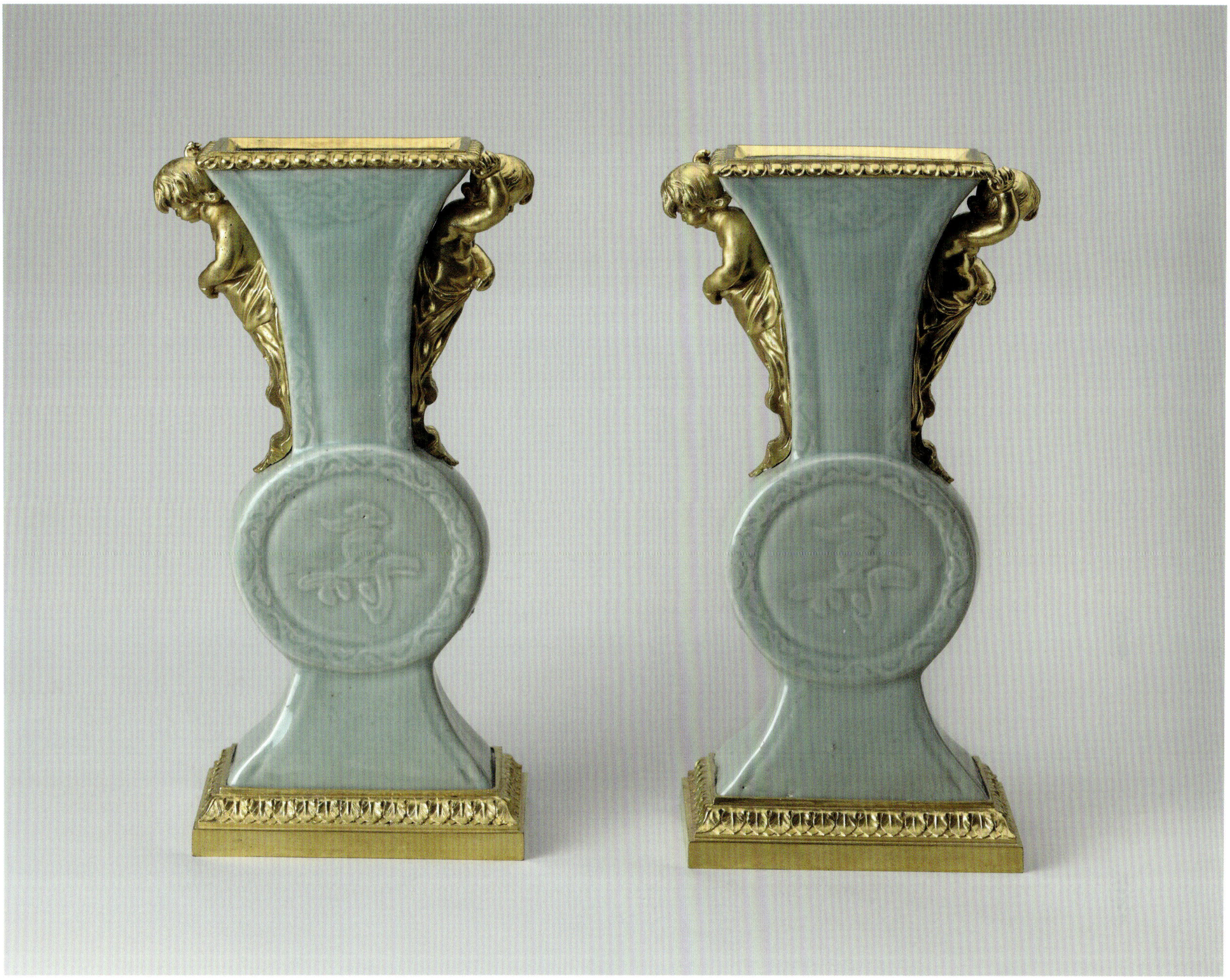

28.23

28.23

UNKNOWN MAKERS, China and France
Pair of porcelain beakers
Vases: 1725–50; mounts: third quarter of the 18th century
Porcelain with celadon glaze, Jingdezhen Jiangxi province, China; mounted in gilt metal, France
11⅞ x 6⅛ x 3¾ in. (30 x 15.5 x 9.5 cm)
Royal Collection Trust, RCIN 2310

This pair of Chinese celadon beakers was a gift for Charlotte from the Qianlong Emperor. The gilt bronze mounts were fitted in London. The vases are probably those described in the Queen's sale catalogues of 1819 as 'A pair of sea-green Mandarin beakers of flat shape, with Chinese characters indented, very richly mounted with or-moulu, in handsome old French taste'. They were bought for George, Prince of Wales, and later installed in the Royal Pavilion, Brighton.

The Macartney Embassy (1793) was the first British diplomatic mission to China. Adviser to the delegation was Sir Joseph Banks, President of the Royal Society, who had ambitions to gather valuable plants from all over the world to be studied at the Botanic Gardens at Kew and the newly established Calcutta Botanical Garden in Bengal. He insisted that gardeners and artists be present on the expedition. To further these ends, Macartney met the Qianlong Emperor on 14 September. Several thousand guests were present to witness a formal exchange of gifts as well as a letter from George III to the Emperor.

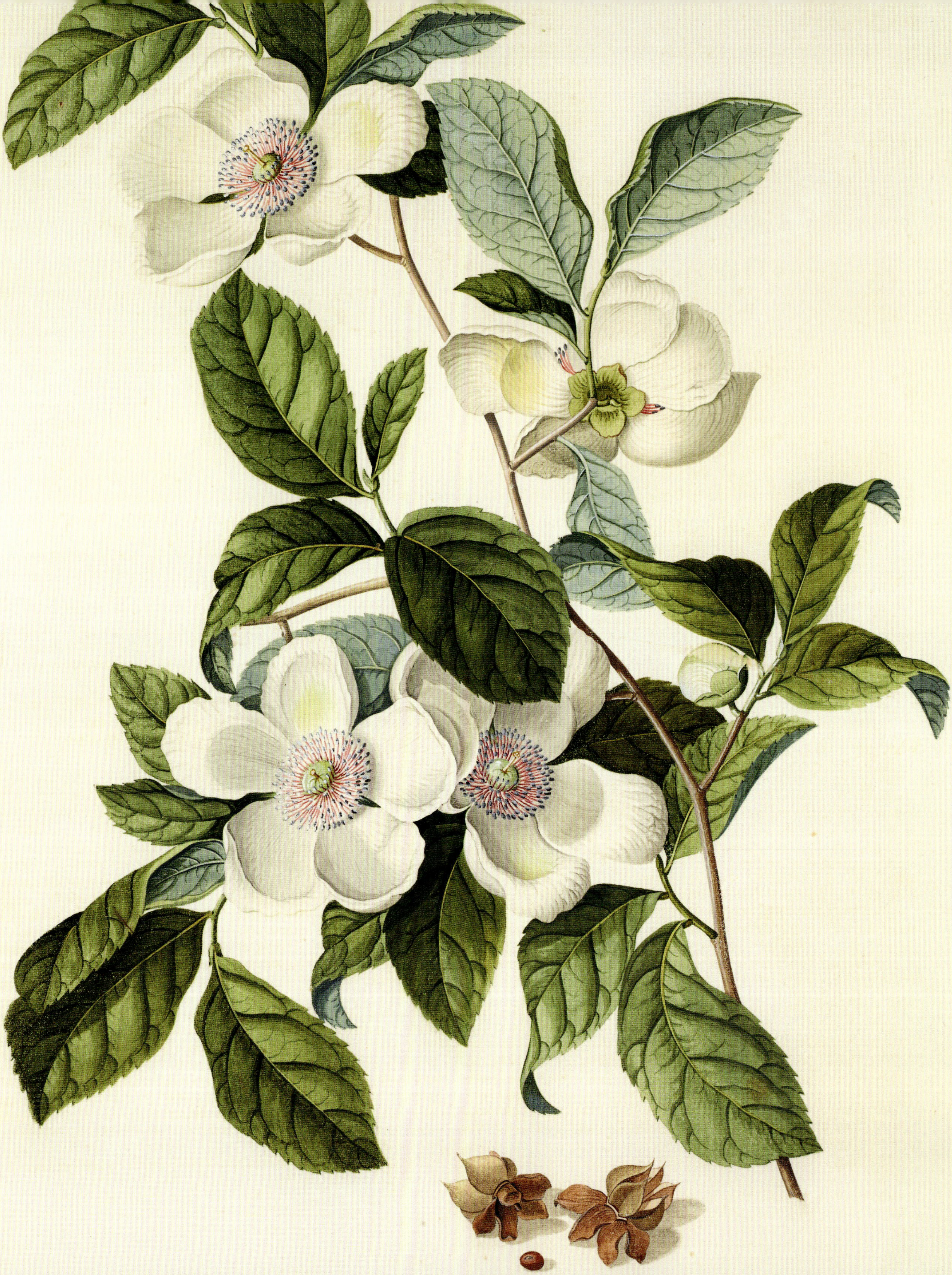

The Natures of Britain and Empire

Tyler Griffith and Amy Meyers

THE FIRST EMPIRE, ECONOMIC BOTANY, AND NETWORKS OF EXCHANGE

At the time of Caroline's coronation in 1727, Britain found itself in a period of relative peace, between the War of the Spanish Succession and the Seven Years' War. The Peace of Utrecht of 1713 had ensured the continued possession of Britain's holdings on the Indian subcontinent; enlarged the empire to incorporate French domains in North America and Spanish territories in the Mediterranean; and granted to Britain a monopoly on the slave trade that provided enforced human labour to fuel the empire's agrarian prosperity. The period between Caroline of Ansbach's death in 1737 and Charlotte of Mecklenburg-Strelitz's coronation in 1761, when Augusta of Saxe-Gotha was Princess of Wales and subsequently Princess Dowager, witnessed violent uprisings in the territories controlled by the East India Company, and increasingly aggressive confrontations with colonists in North America. Charlotte's tenure lasted until 1818, a fifty-seven-year period that was marked by substantial losses in North America through the War of Independence and the foundation of the United States of America—but also by significant territorial expansion across Canada, the Caribbean, Africa, the Middle East, South Asia, China, and Australia. Britain's pursuit of a global empire created a tightly knit, interconnected world that stimulated unprecedented levels of exchange: the circulation of natural resources, people, and ideas characterized Britain's First Empire.[1] William Kent's allegorical drawing, *Queen Caroline, as Britannia, receiving the Gifts of the Earth* (c. 1727–40), adapts a conventional image of the enthroned monarch—in this case, a female consort—receiving tribute, by situating the scene not in the traditional palace or court of justice, but on the edge of a quay (fig. 29.01). On the right, a ship arrives into port carrying 'gifts of the earth' from Britain's expanding dominions, giving the picture a political specificity beyond the symbolic portrayal of a queen's power. Kent's drawing may or may not have been known to other artists of the time, but an image of Caroline as the steward of imperial production and exchange occurs in other imagery of the period. For example, the design of 1736 by Hubert-François Gravelot for Jernegan's Lottery Medal, situates Caroline among stylized palm trees with the message, 'Growing Arts Adorn Empire'.[2] Additionally, *in exergue* George is shown 'reigning', while Caroline is shown 'protecting'—a concise account of the critical role played by the consort in the development of the First Empire (see cats. 13.06–13.07).

Caroline received the gifts of the earth not only in the form of new plants and animals from the colonies, but also in the form of textual and pictorial representations created by naturalists who sought to define the flora and fauna of Britain's imperial domains, and to demonstrate their utility to the nation. In 1729, when the naturalist Mark Catesby began to publish the first overarching natural history of Britain's colonies, *The Natural History of Carolina, Florida and the Bahama Islands*, he was granted an audience with Caroline to present her with the first part, consisting of twenty plates and their associated texts.[3] When he completed the first volume in 1731, he dedicated it to the Queen. This was particularly

Detail of p. 490, cat. 29.18

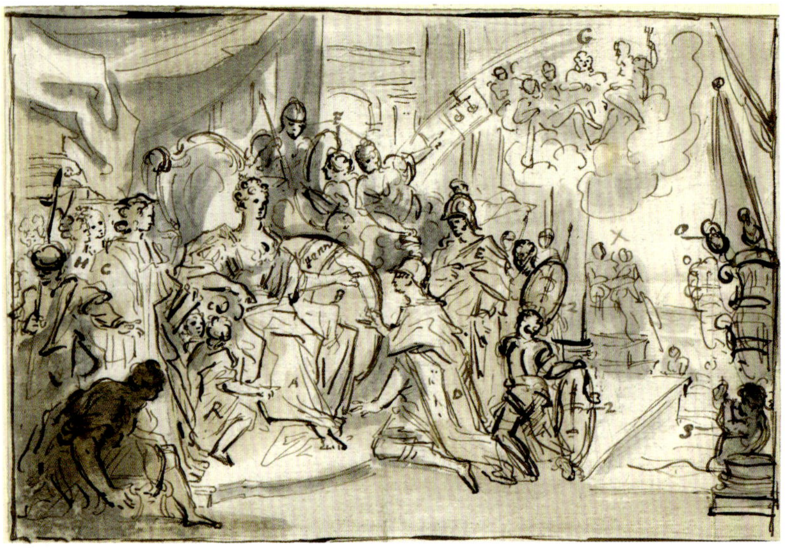

FIG. 29.01 William Kent, *Queen Caroline, as Britannia, Receiving the Gifts of the Earth*, 1730–37, pen and ink and wash over pencil on paper, 5¾ x 7⅞ in. (14.5 x 20.9 cm). British Museum, London (cat. 13.03)

fitting, since Caroline had a strong interest in botany and natural history in general. She commissioned Charles Bridgeman to design extensive gardens for Richmond Lodge and then Kensington Palace, and kept menageries at Richmond and the Tower of London. When Catesby completed the second volume of his *Natural History* in 1743, he dedicated it to Caroline's daughter-in-law, Augusta, Princess of Wales, since the Queen had died in 1737. Augusta followed in Caroline's footsteps as an avid student of natural history, partnering with her husband, Frederick, Prince of Wales, in the development of the grounds of the White House at Kew, encouraged by Frederick's close associate John Stuart, third Earl of Bute, who also was a supporter of Catesby.

Catesby's *Natural History* provides readers not simply with a visual and textual catalogue of the flora and fauna of the southern colonies of North America and the Bahamas, but also with a carefully articulated analysis of the interplay he observed between animals and plants, as well as between human beings and their environments, which he attributed in part to the processes of empire.[4] Though a horticulturalist by profession, Catesby also developed an interest in the animals of North America and the Caribbean, giving particular attention to birds because of their close association with the plants that were the special focus of his work. Many of the one hundred plates that constitute Volume I of the *Natural History* pair birds with the plants with which Catesby understood them to be related. The one hundred and twenty plates that make up Volume II and its Appendix often depict reptiles, mammals, and even fish, in association with plants and other natural productions from their habitats. Although to some extent following the precedent of Sir Hans Sloane's *A Voyage to the Islands Madera, Barbados, Nieves, S. Christophers and Jamaica* (Vol. I, 1707; Vol. II, 1725; see cat. 29.15), and Maria Sibylla Merian's *Metamorphosis Insectorum Surinamensium* (1705), Catesby presented a radical approach to natural history in the period, asserting that Britain not only possessed the plants and animals of its colonial holdings, but was responsible for the stewardship of the complex natural and human relationships that the colonizing mission was both creating and disrupting.

The original gouaches from which Catesby made his prints were purchased by George III in 1768, and undoubtedly were known to Charlotte, whose deep interest in natural history followed that of Caroline and Augusta, and was reflected in her prodigious activities in extending the gardens at Kew as well as at Frogmore, and in creating a menagerie containing animals from across the empire. Many of these original drawings are complete compositions by Catesby himself, which he etched almost exactly; however, some depict a single organism that he then combined with drawings of other animals and plants, to produce a composite image from multiple gouaches. There also are instances when two gouaches depict the organisms that appear in three plates or more. For example, the two gouaches depicting the *Spinus tristis* (American goldfinch), the *Passerina ciris* (painted finch, also called the painted bunting), and the *Passerina cyanea* (blue linnet, also called the indigo bunting), became three plates in the printed version of the *Natural History* (figs. 29.02–29.06). Augusta actually received the birds represented in these gouaches when she was visited at Richmond by the South Carolina planter Eliza Pinckney during the summer of 1753.[5] Undoubtedly, Pinckney was familiar with Catesby's publication and chose these songbirds at least in part because they appear sequentially in plates 43, 44, and 45 in Volume I of the *Natural History*, a publication that she must have known was closely associated with the Princess to whom she was making her gift. The story of Pinckney's gift illustrates the complex physical and intellectual exchanges that enabled the study of the natural world and gave it symbolic value during the First Empire.

Catesby had practical, even openly commercial, applications in mind for much of his work. This can be seen especially clearly in his posthumous *Hortus Britanno-Americanus* of 1763, which was marketed to a continental readership under the title of *Hortus Europae Americanus* (fig. 29.07). The *Hortus* details Catesby's experience in adapting American species to the 'climates and soils' of the British Isles and Europe more generally. He concerns himself specifically with practical questions: methods for shipping seeds and live specimens successfully across the Atlantic, how to store seeds, what will grow where, and what conditions need to be

TOP LEFT

FIG. 29.02 Mark Catesby, *American Goldfinch, the Acacia, and the water locust*, 1722–26, watercolour and gouache over pen and ink on paper, 14⅞ x 10⅝ in. (37.8 x 26.8 cm) Royal Collection Trust (cat. 30.09)

TOP RIGHT

FIG. 29.03 Mark Catesby, *The Painted Finch and the Loblolly Bay*, 1722–26, watercolour and gouache over pencil on paper, 14¾ x 10⅝ in. (37.4 x 26.8 cm). Royal Collection Trust (cat. 30.08)

BOTTOM LEFT

FIG. 29.04 Mark Catesby, artist and engraver, '*Carduelis Americanus*: The American Goldfinch, [and] *Acacia Abruae foliis*: Acacia', 1729–30, hand-coloured etching, in Mark Catesby, *The Natural History of Carolina, Florida, and the Bahama Islands . . .* 2 vols. (London: Published by the Author, 1731–43, Appendix, 1747), vol. 1, plate 43. Royal Botanic Gardens, Kew, Library, Art & Archives Collections

BOTTOM RIGHT

FIG. 29.05 Mark Catesby, artist and engraver, '*Fringilla Tricolor*: The Painted Finch, [and] *Alcea Floridana quinque capsularis*: the Loblolly Tree', 1729–30, hand-coloured etching, in Mark Catesby, *The Natural History of Carolina, Florida, and the Bahama Islands . . .* 2 vols. (London: Published by the Author, 1731–43, Appendix, 1747), vol. 1, plate 44. Royal Botanic Gardens, Kew, Library, Art & Archives Collections

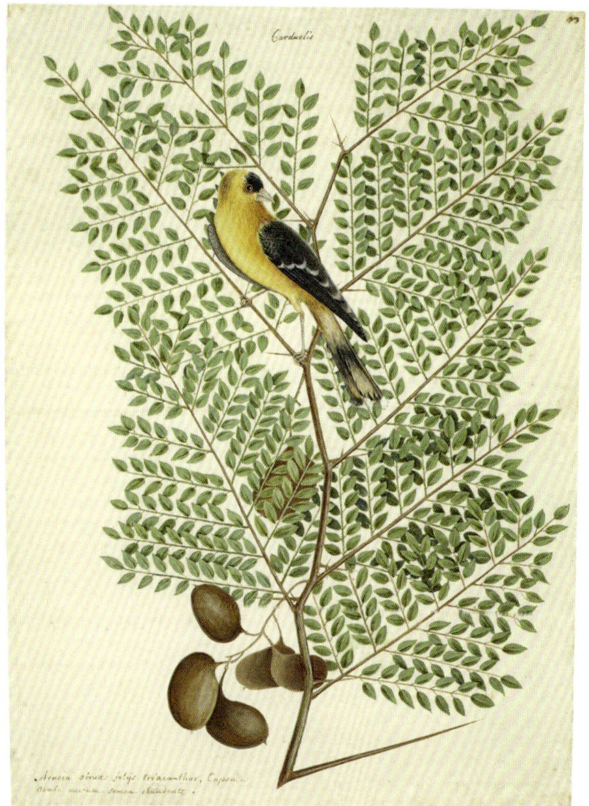

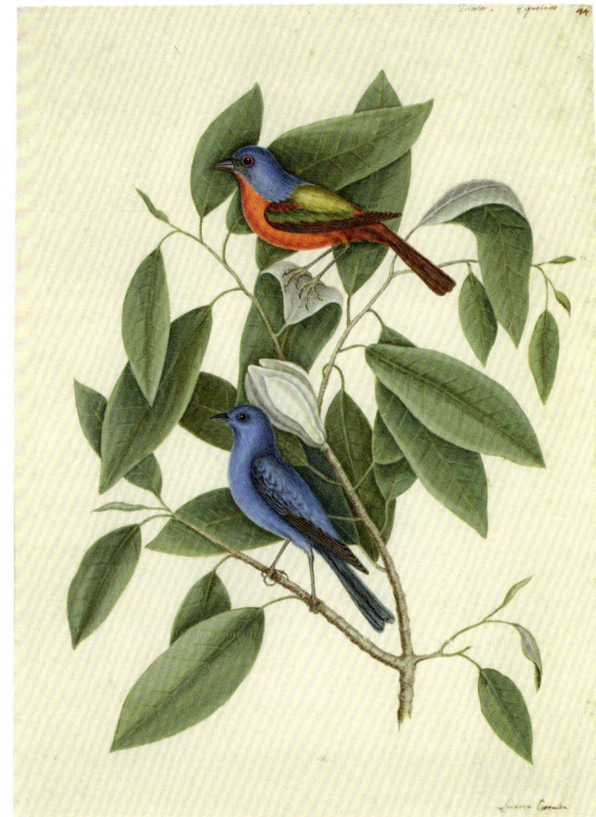

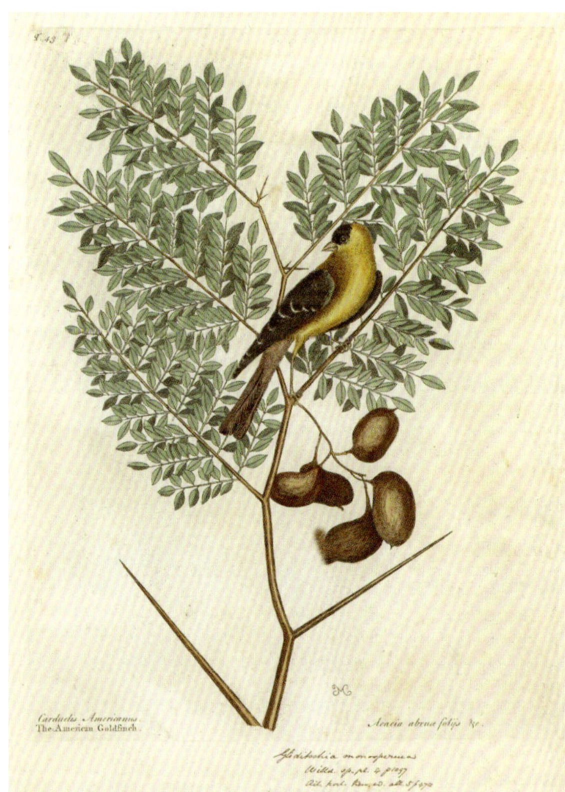

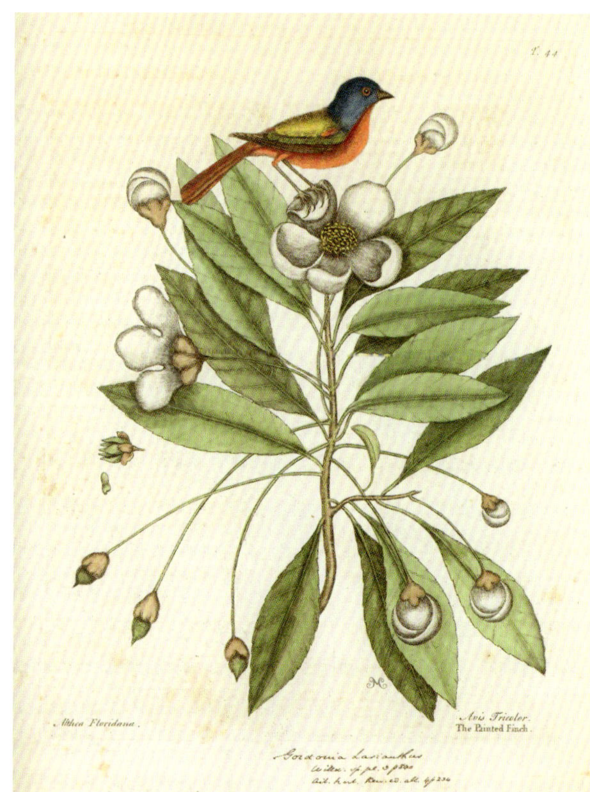

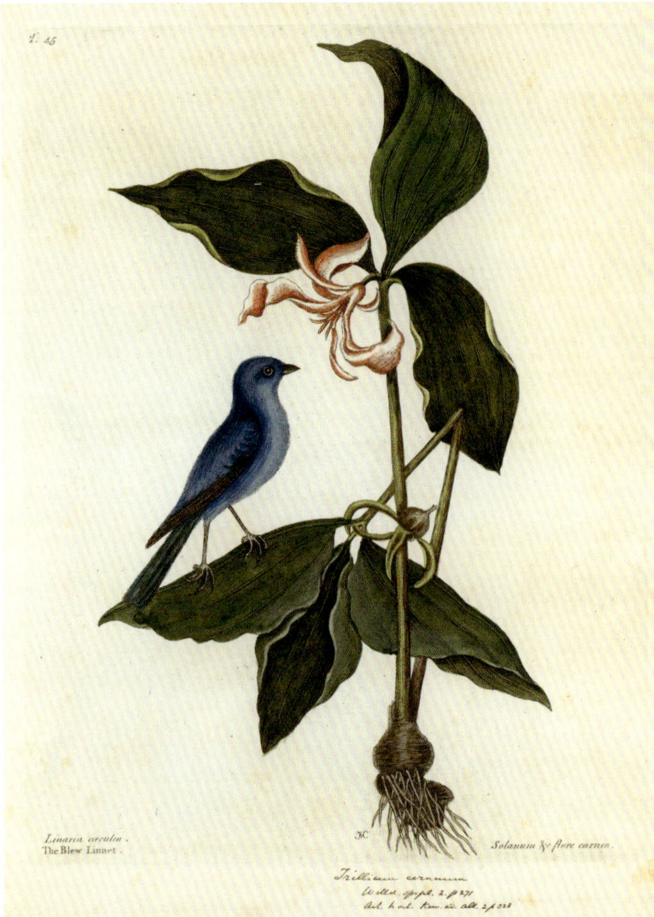

FIG. 29.06 Mark Catesby, artist and engraver, '*Linaria Caerulea*: The Blew Linnet, [and] *Solanum triphyllon* [Catesby's Trillium]', 1729–30, hand-coloured etching, in Mark Catesby, *The Natural History of Carolina, Florida, and the Bahama Islands* . . . 2 vols. (London: Published by the Author, 1731-43, Appendix, 1747), vol. 1, plate 45. Royal Botanic Gardens, Kew, Library, Art & Archives Collections

produced to ensure successful cultivation.[6] His knowledge of methods of transplantation was garnered after his return from his second trip to the Americas in 1726, while he was producing the *Natural History* and working as a nurseryman to supply royal, aristocratic, gentry and merchant and other gardens with plants from the Americas. As the *Hortus* demonstrates, Catesby's publication was aimed at more than the advancement of pure knowledge of the natural world; it had economic utility and reflected the ideology of imperialism in the gardens and menageries of such agents of empire as Caroline, Augusta, and Charlotte, and those in their circles.

Just as Caroline and Augusta had served as Catesby's patrons, so Charlotte served as the patron of a horticulturist of North American flora, though one of far less distinction. The Philadelphia plantsman William Young had been born in Germany, but emigrated with his family to the British colonies of North America at the age of two.[7] In 1763, his family settled on a farm in Kingsessing, Pennsylvania, near the renowned naturalist John Bartram who became King's Botanist for North America in 1765 and was awarded a pension of £50 a year for his incomparable activities as a plant collector until the War of Independence. Young had begun his career as a nurseryman and collector of North American plants by 1761, tutored in part by Bartram. He sailed for England in 1764 to promote his own business, and by 1765, he had been received by George III, with whom he spoke about the natural productions of North America. Charlotte bestowed on him the title of Queen's Botanist, with the generous annual stipend of £300. The royal favour shown Young, in part because of his German birth and because he spoke the language fluently, made Bartram uneasy, although he was assured by the London merchant Peter Collinson who was responsible for the transport of most of the animals and plants that came to Britain from across the Empire in the period, that Young had become a ridiculous fop. Indeed, within the year, Young's extravagance and inappropriate behaviour at court earned him dismissal and ultimately imprisonment for debt. Though Young was sent home and told never to return to court, he continued to be retained by Charlotte at a reduced stipend equivalent to Bartram's. When he arrived back in Philadelphia, Young flaunted his title, and continued to send botanical materials to the Queen. Although Charlotte distanced herself from him, Young sailed to London several more times with large cargoes of plants. While he suffered socially from a dubious reputation, he carried on an active business with British nurseries and plant collectors until the War of Independence interrupted trade. His careful collecting of seeds and plants, their successful propagation, and his meticulous packing methods gained him the respect of important British collectors, and several of his later introductions made their way into royal gardens. An important set of his pressed plant specimens and associated botanical watercolours, executed in a unique style reminiscent of Pennsylvania German *fraktur*, are now in the Natural History Museum in London (fig. 29.08). Young revived his transatlantic business after the war's end, but died prematurely from an accident while botanizing in 1785.

TRANSITION TO THE SECOND EMPIRE

After the thirteen colonies along the eastern seaboard of North America won independence from Britain in 1783, imperial focus shifted to long-established, but increasingly significant, arenas of colonisation across much of the rest of the globe. Despite its presence in India for nearly two centuries, the British East India Company did not solidify its domination of the Indian subcontinent until the 1750s, when its military and political support of Mir Jafar at the Battle of Plassey and the following coup installed him

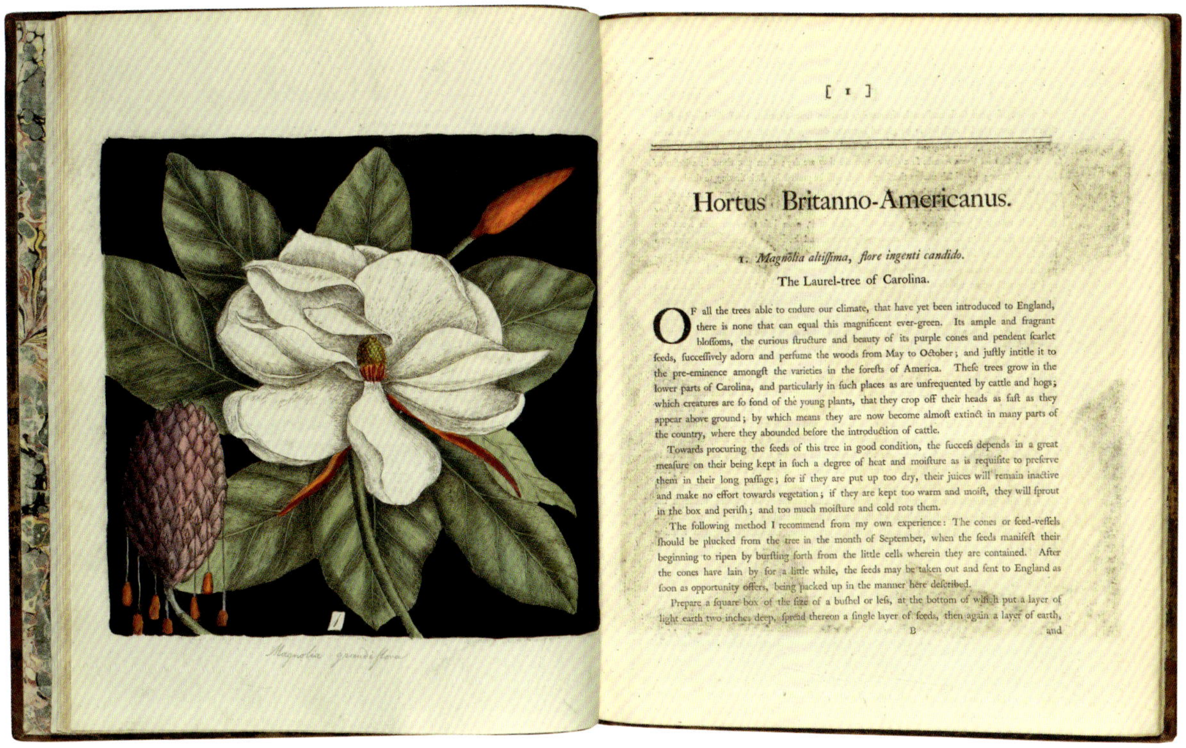

FIG. 29.07 Mark Catesby, artist and engraver, '*Magnolia Grandiflora*', hand-coloured engraving, in Mark Catesby, *Hortus Europae Americanus: or, A collection of 85 curious trees and shrubs, the produce of North America; adapted to the climates and soils of Great-Britain, Ireland, and most parts of Europe, &c.* . . . (London: Printed by W. Richardson and S. Clark for J. Ryall, 1767), plate 1. Oak Spring Garden Library, Upperville, VA (cat. 29.22)

as Nawab in 1757. The inroads at Bengal—motivated almost entirely by commercial interest—precipitated the creation of the British Governorship of Bengal and, ultimately, the establishment of the British Raj, itself a curious confluence of British and Indian culture.[8] A set of four ivory armchairs from Murshidabad (see cat. 28.20) was probably a gift from Mani Begum, the widow of Mir Jafar, to Warren Hastings at or near the time he became head of the Supreme Council of Bengal and *de facto* Governor-General of British India.[9] The chairs, created by Indian ivory carvers, blend Western form with Bengali ornament, reflecting the growing interest in combining aesthetic elements from both British and Indian culture, and distilling the vast cultural shifts that were taking place simultaneously across the political and economic landscapes of the region. The complicated history of the chairs tells an equally important story about the significance of formalized exchange at the end of the eighteenth century: commissioned by Mani Begum, they were presented to Hastings as a gift; Hastings gave them to his wife, Marian; Marian—herself a German noblewoman who arrived in London at the encouragement of Charlotte's Mistress of the Robes—subsequently gave them to Queen Charlotte; and Charlotte probably gave them to her son, the future George IV.[10]

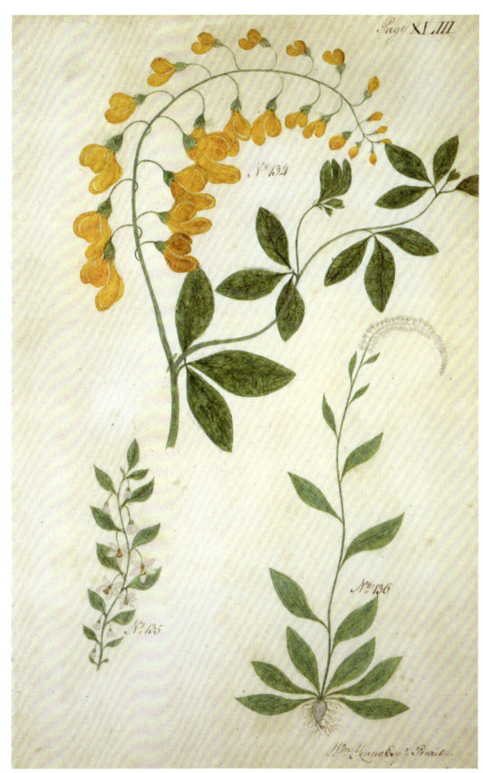

FIG. 29.08 William Young, artist, '134. *Indiofera*, 135. *Andromeda*, 136. *Tetramdynamia*,' c. 1767, watercolour on paper, 15 x 9½ in. (38 x 24 cm), in 'A Natural History of Plants Containing the Productions of North and South Carolina of upwards of 300 different species which have been carefully collected, drawn, and coloured from the life in the year 1767', 1767, p. 43. Natural History Museum, London (cat. 29.23)

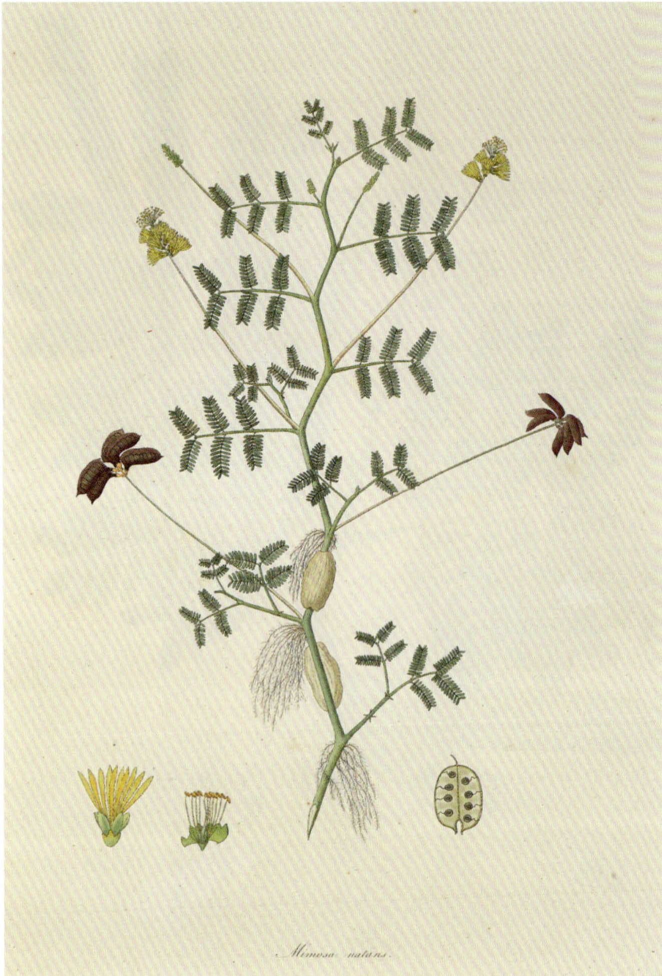

FIG. 29.09 Daniel MacKenzie, engraver, after unidentified artist, '*Neptunia oleracea* Lour. [as *Mimosa natans* L.f.]', hand-coloured etching, in William Roxburgh, *Plants of the Coast of Coromandel: Selected from Drawings and Descriptions presented to the Hon. Court of Directors of the East India Company*, 2 vols. (London: Printed by W. Bulmer and Co. for G. Nicol, 1795–1819), vol. 2, plate 119. The New York Botanical Garden, The LuEsther T. Mertz Library (see cat. 29.24)

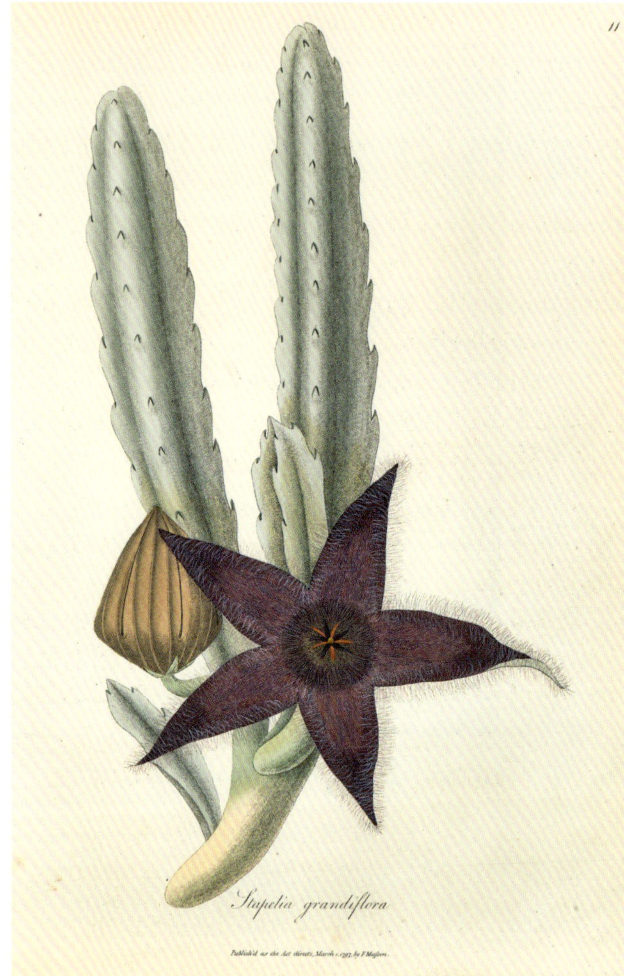

FIG. 29.10 Daniel MacKenzie, engraver, after Francis Masson, 'Stapelia Grandiflora', hand-coloured etching, in Francis Masson, *Stapeliae novae, or, A Collection of Several New Species of that Genus, Discovered in the Interior Parts of Africa* (London: George Nicol, 1796), plate 11. The New York Botanical Garden, The LuEsther T. Mertz Library (see cat. 29.30)

The East India Company had also been carrying on limited trade with China for almost two hundred years, but it was not until the last decades of the eighteenth century, when George III and Charlotte came to the throne that a permanent—if unsuccessful—diplomatic post was established in Peking (now Beijing).[11] Many Chinese artefacts, such as a pair of porcelain beakers from 1725–50 (cat. 28.23), entered the royal collection as diplomatic gifts from the Emperor and his family, and Charlotte displayed these objects prominently. Visitors to Frogmore noted 'presents from the Emperor of China' in the 1790s, and Charles Wild's watercolour of the Green Closet at Frogmore (cat. 21.06) shows Chinese vases among other treasures.[12] It is unknown exactly where Charlotte displayed this particular pair of beakers, although evidence suggests that they were in the Yellow Anteroom of the Royal Pavilion in Brighton during her lifetime, and in the Queen's Corridor at Buckingham Palace sometime after her death.[13] As well as being simply decorative, the display of objects from regions under British influence and control served to demonstrate the power of colonial expansion back home by representing its global dominion.

But bringing home the empire involved the living stuff of the natural world as well. The study of the plants native to the global range of colonial expansion continued into the nineteenth century, notably through the work of William Roxburgh and Francis Masson, who were closely associated with the development of the gardens at Kew by George III and Charlotte. Roxburgh's knowledge of Indian flora was so comprehensive that he was called 'The Linnaeus of India' and 'The Father of Indian Botany'.[14] His three-volume *Plants of the Coast of Coromandel* of 1795–1819 (fig. 29.09) contains 300 prints made from drawings that were in his personal collection of over 2,500 botanical studies, assembled during his stay in India from 1776 to 1793 with the support of the naturalist, Sir Joseph Banks. Roxburgh became Natural Historian of the East India Company and oversaw the Botanic Garden at Calcutta as its superintendent. Like many East India Company employees studying botany, he almost always employed Indian artists to make life-sized drawings of his specimens, adding to the corpus of his own botanical studies.[15] Masson, the first plant collector to be sent abroad upon the appointment of Banks as director of Kew in 1797, sailed with James Cook to Africa and subsequently to the Caribbean, the Iberian Coast, New England, and Canada. His *Stapeliae novae* of 1796 (fig. 29.10), which he compiled during his trip to Africa, proved to be his most influential work. He documented a host of species from the interior of the continent that were previously unknown to European botanists, setting the standard for African botany until well into the nineteenth century.[16]

Animals were also transported to Britain from the far reaches of the globe. One of the most famous was Queen Charlotte's African zebra, which would be satirized as the 'Queen's Ass' (see cats. 15.05–15.07).[17] As one of the first living specimens of its kind to reach Britain, the female zebra was a gift from Sir Thomas Adams, Commander of *H.M.S. Terpsichore*, who acquired the animal in South Africa in 1762. Immediately upon her arrival in Britain, she was added to the Royal Menagerie at Buckingham Palace.[18] George Stubbs portrayed her in 1763 (fig. 29.11) in a series of paintings of exotic animals, including the *Nilgai* painted in 1769–70 (cat. 29.51) to which he probably gained access through Charlotte's physician William Hunter, with whom he was engaged in the study of comparative anatomy.

However, Stubbs's depiction of Charlotte's zebra is far more than a scientific examination of the animal; its power has endeared it to many as one of the greatest animal paintings of all time. Standing quietly in profile, the zebra looks uncomfortably exposed and out of place, her black-and-white striped coat contrasting starkly with the dark backdrop of the English wood in which she has been placed, rather than blending protectively into the environment. Her slightly forlorn and quizzical expression is disturbingly affecting, conveying the sense that she is puzzling over how she has arrived at such a strange and lonely place.

Human beings were also brought to Britain from the farthest reaches of the empire, either as slaves or as visitors, whose

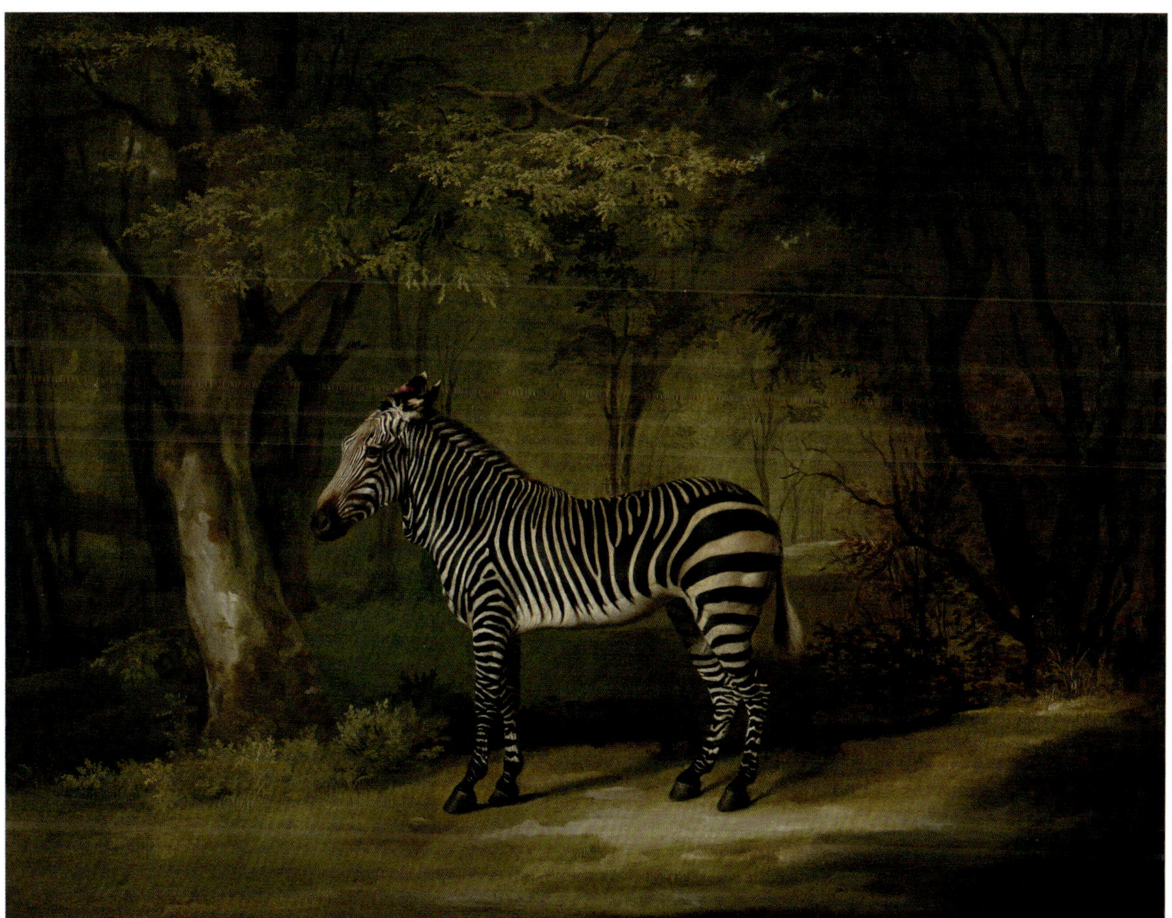

FIG. 29.11 George Stubbs, *Zebra*, 1763, oil on canvas, 40½ x 50¼ in. (102.9 x 127.6 cm). Yale Center for British Art, Paul Mellon Collection (cat. 29.50)

perceived exoticism attracted widespread attention. Omai became the second native Pacific Islander to visit Europe when he arrived aboard the H.M.S. *Adventure* in London in 1774 on the return of Captain James Cook's second Pacific voyage.[19] Sir Joseph Banks introduced him to aristocratic and scientific circles, and he quickly became a celebrity both at the salons of the elite and at meetings of learned institutions such as the Royal Society.[20] He also visited the royal family numerous times, having audiences with both George III and Queen Charlotte. William Parry, in his group portrait of 1775–76 of Omai, Banks, and Daniel Solander, the botanical artist who accompanied Banks on the voyage (fig. 29.12), cleverly manipulated the conventions of the conversation piece to capture the excitement of Omai's visit to Britain. Omai gazes out towards the viewer in a heroic pose, while Banks adopts a classically rhetorical stance to highlight the encounter with Solander, who records Omai as he would a botanical specimen.[21] The most famous depiction of Omai is the grand-manner portrait by Sir Joshua Reynolds, finished around 1776 (fig. 29.13), for which there is a preparatory sketch from 1774 (see cat. 27.07). Reynolds's imposing portrait was widely circulated in prints sold to collectors in search of the exotic through a mezzotint from 1777 by Johann Jacobé (see cat. 27.08).[22]

The success of the empire always depended on more than raw military force and political strategy; it was propelled forward by complex and often insidious networks of economic exchange involving natural productions and people, as well as a more nuanced, but equally potent, transfer of philosophical concepts and cultural practices. The objects discussed in this short essay speak to more than exploration and exploitation. They also speak to shifting relationships among peoples, animals, plants, and a global ecosystem in constant transformation, resulting from the intricate processes of imperialism and colonization. By the time of Charlotte's death in 1818, Britain and the world were at the beginning of the 'Imperial Century'—a period characterized by British military, political, and economic dominance, but also by dramatic, dynamic, and accelerated cultural and intellectual interchanges that continue to affect the world today in almost every way.

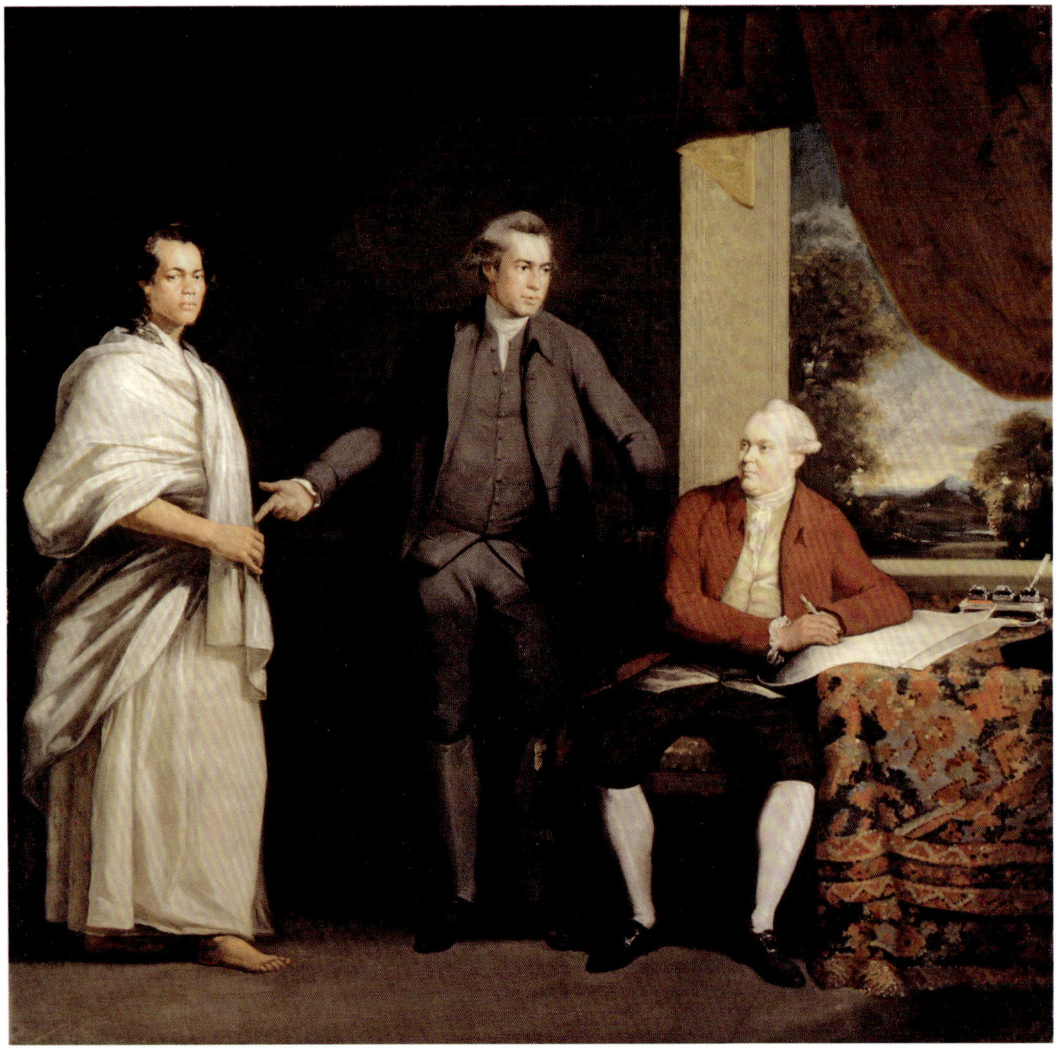

FIG. 29.12 William Parry, *Omai (Mai), Sir Joseph Banks, and Daniel Charles Solander*, 1775–76, oil on canvas, 60 x 60 in. (152.4 x 152.4 cm). National Portrait Gallery, London, Purchased jointly with the Captain Cook Memorial Museum, Whitby, and the National Museums & Galleries of Wales, with help from a private benefactor and local trusts, the Art Fund, the Garfield Weston Foundation, Flora Fraser and Peter Soros, Sir Christopher Ondaatje, Linda L. Brownrigg, Randolph and Lara Lerner, Jon and Lillian Lovelace, the Clore Duffield Foundation, Sir Harry Djanogly, Hans and Märit Rausing, Lawrence Banks, the Swan Trust, Amanda Sebestyen, Sir David Attenborough, Lord Plymouth, and Lord Windsor, and many other donations, 2003 (cat. 30.11)

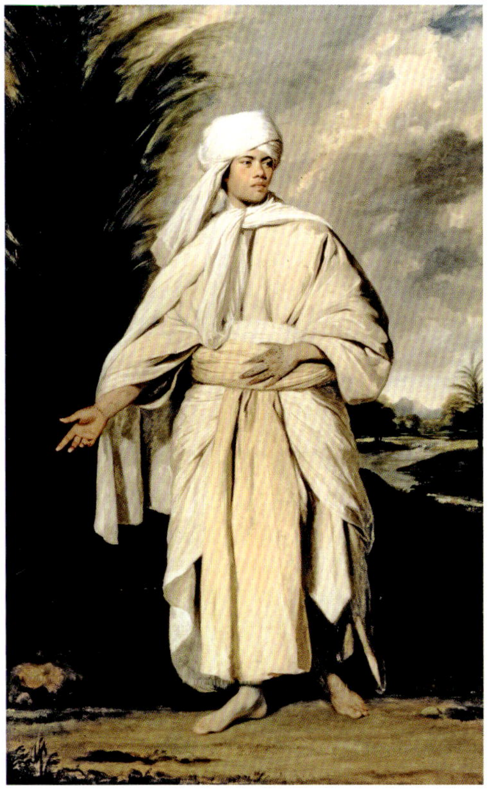

FIG. 29.13 Sir Joshua Reynolds, *Portrait of Omai*, c. 1776, 92.9 x 57.3 in. (236 x 145.5 cm) oil on canvas. Private Collection (see also cat. 27.08)

NOTES

1. The exchange of goods and ideas as a formalized network of trust and benefit formed a critical facet of the notion of empire. See Natasha Glaisyer, 'Networking: Trade and Exchange in the Eighteenth-Century British Empire', *Historical Journal* 47, no. 2 (2004): 451–76.
2. Roughly 30,000 medals were sold as tickets to a lottery for a giant silver cistern now in the State Hermitage Museum in St Petersburg, Russia. Desmond Shawe-Taylor, ed., *The First Georgians: Art and Monarchy, 1714–1760* (London: Royal Collection Trust, 2014), 317.
3. On Catesby and his *Natural History*, see George Frederick Frick and Raymond Phineas Stearns, *Mark Catesby: The Colonial Audubon* (Urbana, IL: Univ. of Illinois Press, 1961); Henrietta McBurney, *Mark Catesby's Natural History of America: The Watercolors from the Royal Library, Windsor Castle* (London: Merrell Holberton, 1997); Amy R. W. Meyers and Margaret Beck Pritchard, eds., *Empire's Nature: Mark Catesby's New World Vision* (Chapel Hill, NC: Univ. of North Carolina Press, 1998); and E. Charles Nelson and David J. Elliott, eds., *The Curious Mister Catesby: A 'Truly Ingenious' Naturalist Explores New Worlds* (Athens, GA: Univ. of Georgia Press, 2015). See also, in this volume, Mark Laird, 'An August Domain: Gardening and Collecting at Kew', 135–59, and Ben Marsh, 'Visitor from South Carolina: Mrs Eliza Pinckney', 515–19.
4. On Catesby's environmental interpretation of the natural world, see Amy R. W. Meyers, 'Sketches from the Wilderness: Changing Conceptions of Nature in American Natural History Illustration, 1680–1880' (PhD diss., Yale University, 1985), 46–112; Meyers, '"The Perfecting of Natural History": Mark Catesby's Drawing of American Flora and Fauna in the Royal Library, Windsor Castle', in McBurney, *Mark Catesby's Natural History of America*, 11–25; and Meyers, 'Picturing a World in Flux: Mark Catesby's Response to Environmental Interchange and Colonial Expansion', in *Empire's Nature*, ed. Meyers and Pritchard, 228–61. On Catesby and the politics of the visual culture of natural history, see Meyers and Pritchard, eds., *Empire's Nature*.
5. See, in this volume, Ben Marsh, 'Visitor from South Carolina: Mrs Eliza Pinckney', 515–19.
6. For a discussion of the transatlantic transport of specimens, see Christopher M. Parsons and Kathleen S. Murphy, 'Ecosystems under Sail: Specimen Transport in the Eighteenth-Century French and British Atlantics', *Early American Studies* 10, no. 3 (2012): 503–39.
7. On William Young's life and career, see John W. Hashberger, 'William Young, Jr., of Philadelphia, Queen's Botanist', *Torreya* 17, no. 6 (June 1917): 91–99.
8. Robert Skelton, *The Indian Heritage: Court Life and Arts under Mughal Rule* (London: Victoria and Albert Museum, 1982), 168 ff.
9. Jane Roberts, *George III & Queen Charlotte: Patronage, Collecting and Court Taste* (London: Royal Collection, 2004), 381; P. J. Marshall, 'The Personal Fortune of Warren Hastings', *Economic History Review* 17, no. 2 (1964): 284–300.
10. K. Ghose, 'Note on Ivory Chair & Teapot in Victoria Memorial Collection', *Bulletin of the Victoria Memorial* 10 (1976): 53–55; K. Ghose, 'Additional Note on Ivory Chair in Victoria Memorial Collection', *Bulletin of the Victoria Memorial* 12 (1979): 25–27.
11. Roberts, *George III & Queen Charlotte*, 382.
12. Roberts, *George III & Queen Charlotte*, 383.
13. Royal Collection (RCIN) 2310.1–2, Royal Collection Trust, Queen Charlotte Sale 1819, possibly 'A pair of sea-green Mandarin beakers of flat shape, with Chinese characters indented, very richly mounted with or-moulu, in handsome old French taste' (Christie's Third Day's Sale Catalogue, 28, no. 41). Recorded in the Yellow Anteroom of the Royal Pavilion, Brighton as 'A pair of sea green Beakers, flat shape circular centres, with Chinese Characters impressed, mounted with ormolu mouldings at the top and bottom & Infant handles, twelve inches' (RCT inventory 1829B, 56); sent to Buckingham Palace in March 1847 (RCT inventory 1829A, 28) and in 1914 noted in the Private Apartments. Noted in the Queen's Corridor, Buckingham Palace in 1921 (RCT inventory 1829B, 56). Internal Reference 56; Royal Collection Trust, Internal Reference, 1829A, 28.
14. On Roxburgh in general, see Tim Robinson, *William Roxburgh: The Founding Father of Indian Botany* (Chichester: Phillimore, 2008). It is worth pointing out that scholars continued to use Roxburgh's work as a standard reference for Indian botany through the close of the twentieth century. See, for example, L. L. Forman, 'Notes Concerning the Typification of Names of William Roxburgh's Species of Phanerogams', *Kew Bulletin* 52, no. 3 (1997): 513–34.
15. See David Platzer, 'Botanical Watercolours from Calcutta', *British Art Journal* 7, no. 2 (2006): 117.
16. C. Lyte, *The Plant Hunters* (London: Orbis, 1983), 23–35; Mike Fraser and Liz Fraser, *The Smallest Kingdom: Plants and Plant Collectors at the Cape of Good Hope* (Richmond: Kew Publishing, 2011), passim.
17. Christopher Plumb, *The Georgian Menagerie: Exotic Animals in Eighteenth-Century London* (London: I. B. Tauris, 2015), 185–93.
18. Douglas Fordham, 'George Stubbs's *The Zebra* and the Spectacle of Fine Art at the End of the Seven Years' War', in *The Culture of the Seven Years' War: Empire, Identity, and the Arts in the Eighteenth-Century Atlantic World*, ed. Frans De Bruyn and Shaun Regan (Toronto: Univ. of Toronto Press, 2013).
19. Louis Antoine de Bougainville had brought the first Pacific Islander, Ahu-toru, to Paris several years before in 1768. Much like Omai, Ahu-toru was received by the French crown and became a celebrity figure during his stay in Paris. On Omai, see E. H. McCormick, *Omai: Pacific Envoy* (Oxford: Oxford Univ. Press, 1978).
20. Jocelyn Hackforth-Jones, 'Mai/Omai in London and the South Pacific: Performativity, Cultural Entanglement, and Indigenous Appropriation', in *Material Identities*, ed. Joanna Sofaer (Oxford: Wiley-Blackwell, 2007), 13–30.
21. Jos Hackforth-Jones and others, *Between Worlds: Voyagers to Britain, 1700–1850* (London: National Portrait Gallery, 2007), 52.
22. For the British fascination with 'exotic' visitors from the Pacific, see Kate Fullager, '"Savages that are Among Us": Mai, Bennelong, and British Imperial Culture, 1774–1795', *Eighteenth Century* 49, no. 3 (2008): 211–37.

Sir Hans Sloane (1660–1753)

29.14*

STEPHEN SLAUGHTER (1697–1765)
Sir Hans Sloane, Bt.
1736
Oil on canvas
49½ x 39¾ in. (125.7 x 101 cm)
National Portrait Gallery, London, NPG 569

In 1727 George II renewed the appointment of Sir Hans Sloane as Physician-in-Ordinary, Sloane 'having been before constantly employ'd about the whole Royal Family, & always honour'd with the Esteem & favour of the Queen Consort'. He had previously served as Physician Extraordinary to Queen Anne and George I. Caroline's involvement with the promotion of inoculation against smallpox had drawn her to his attention. He would later contribute books to her library. They exchanged letters about botany, and in 1734 while George II was in Hanover, she took her children to view Sloane's collection. The first elements of this collection were obtained on his youthful botanizing expeditions, and his voyage to Jamaica in 1687, as physician to the governor of the colony, the second Duke of Albemarle, resulted in a substantial increase both in the size of the collection and in the perception of its value. He bought the collection of seeds and fruits formed by the plant anatomist and physiologist Nehemiah Grew, and he acquired collections of plants developed by many involved in horticulture, agronomy, and botany. Specimens (including shells as well as plants) were contributed by naturalists and others involved in global expeditions for the purposes of trade and empire, including the explorer William Dampier. Sloane was 'always ready, on proper Notice to admit the Curious to the sight of his Museum', and it became one of the sights of London for visiting connoisseurs and scientists, most notably the Swedish naturalist Carl Linnaeus, who came in August 1736 bearing a letter of introduction from the Dutch botanist and chemist Herman Boerhaave. In this portrait by Stephen Slaughter, Sloane holds a copy of the illustration by the Dutch artist Everhard Kick of the Jamaican plant known as lace bark, published in Sloane's *A voyage to the Islands Madera, Barbados, Nieves, S. Christophers and Jamaica . . .* (1707–25). Slaughter was a London-born painter who studied at Sir Godfrey Kneller's academy and later became keeper of the king's pictures.

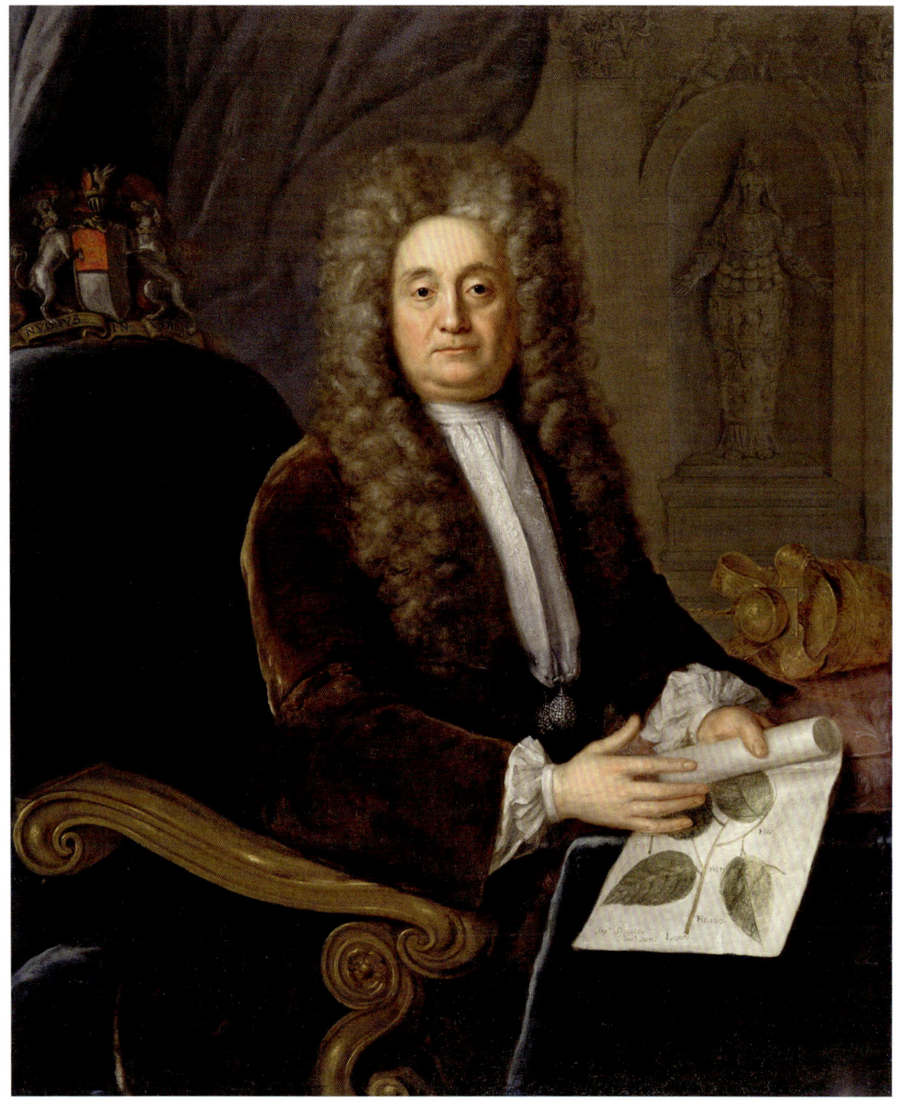

29.14

29.15

SIR HANS SLOANE (1660-1753)
A voyage to the Islands Madera, Barbados, Nieves, S. Christophers and Jamaica ...,
Printed by B. M. for the author, London, 1707–25
14 ¼ in. (36 cm)
* Yale Center for British Art, Paul Mellon Collection, QH109.J5 S66 1707+ Oversize vol. 1
1707
Exhibited: [Transcription of Afro-Caribbean Music], pp. l–li
† Yale Center for British Art, Paul Mellon Collection, QH109.J5 S66 1707+ Oversize vol. 2
1725
Exhibited: Attributed to Michael Vandergucht (1660–1725), engraver, [Stringed Instruments from South Asia, Jamaica, and West Africa], c. 1707, engraving, plate 3

In 1687, the physician and collector Hans Sloane accepted an appointment in the household of the governor of Jamaica. At this time the island's economic focus was shifting from piracy to sugar production, which depended on the labour of enslaved Africans. In 1688 Sloane visited a sugar plantation where he and another guest, a French musician named Baptiste, witnessed a performance of African music and dance. Sloane published Baptiste's transcription of the music in the account of his travels to the West Indies, along with an engraving of stringed instruments from South Asia, Jamaica, and West Africa. Sloane would later serve as physician to George II and he was connected with Caroline through their many shared interests, including botany, book collecting, and the promotion of inoculation against smallpox. Although in most copies of Sloane's *Voyage* the musical notation and the illustration of instruments both appear in volume one, the owner of this copy bound the latter into volume two.

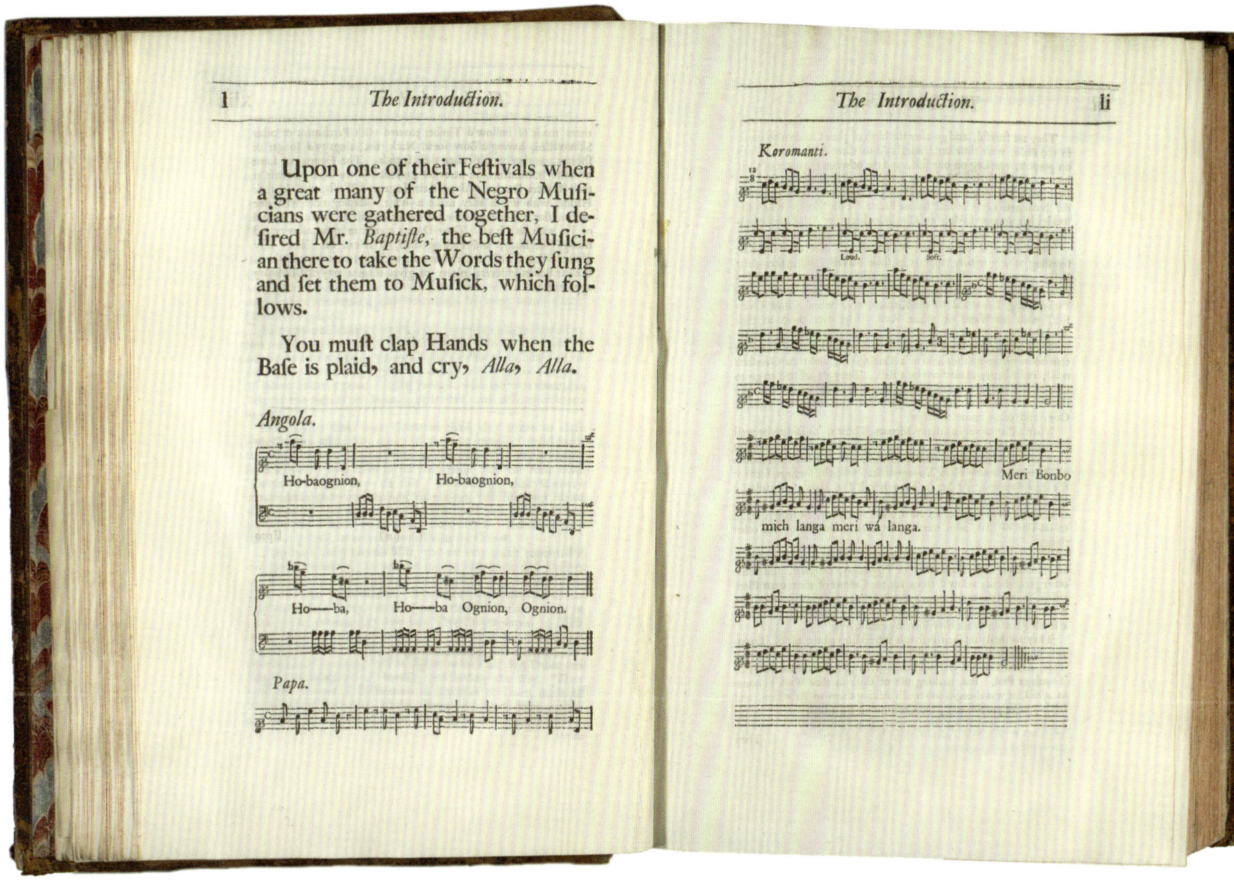
l
The Introduction.
Upon one of their Festivals when a great many of the Negro Musicians were gathered together, I desired Mr. Baptiste, the best Musician there to take the Words they sung and set them to Musick, which follows.
You must clap Hands when the Base is plaid, and cry, Alla, Alla.
Angola.
Ho-baognion, Ho-baognion,
Ho—ba, Ho—ba Ognion, Ognion.
Papa.
The Introduction.
li
Koromanti.
Meri Bonbo
mich langa meri wa langa.

29.15a

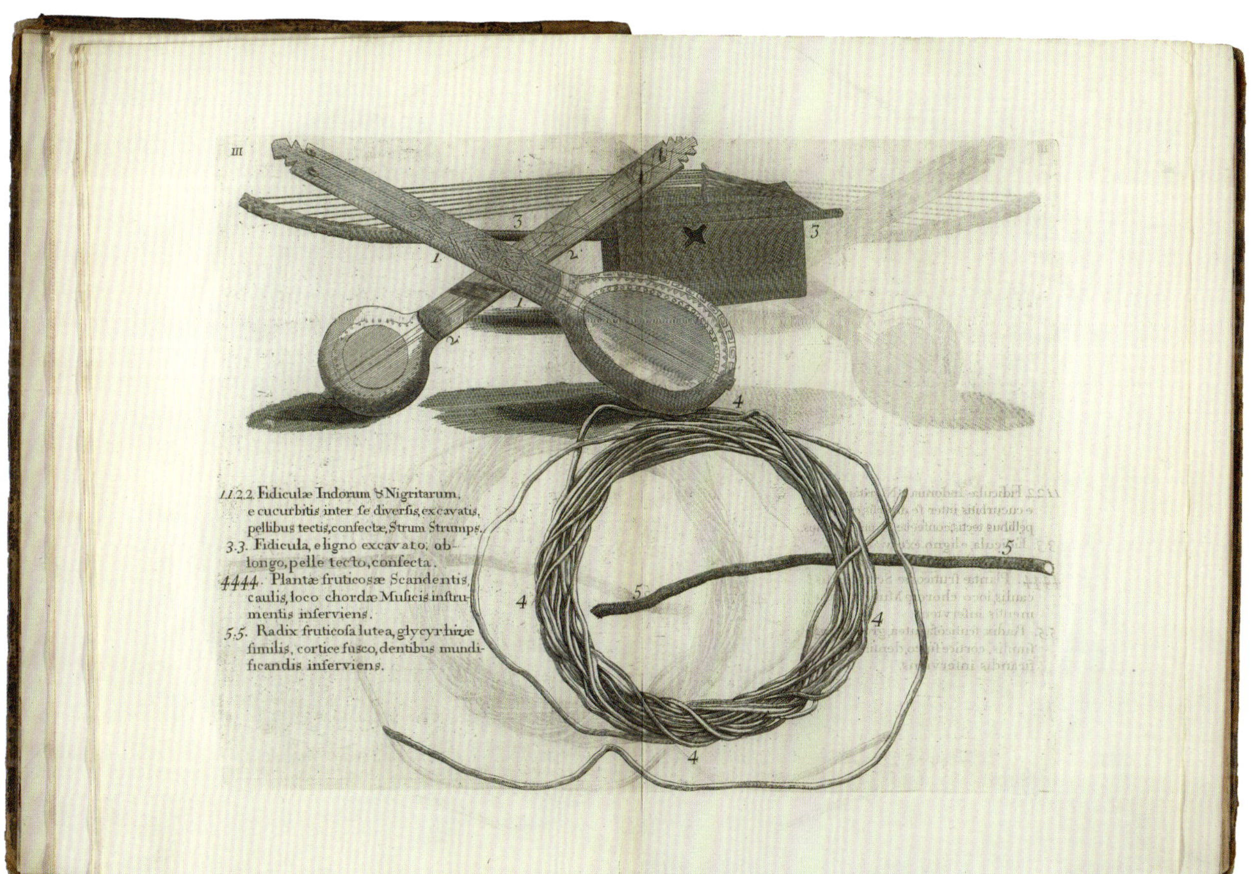
1.1.2.2. Fidiculæ Indorum & Nigritarum, e cucurbitis inter se diversis, excavatis, pellibus tectis, confectæ, Strum Strumps.
3.3. Fidicula, e ligno excavato, oblongo, pelle tecto, confecta.
4444. Plantæ fruticosæ Scandentis, caulis, loco chordæ Musicis instrumentis inserviens.
5.5. Radix fruticosa lutea, glycyrhizæ similis, cortice fusco, dentibus mundificandis inserviens.

29.15b

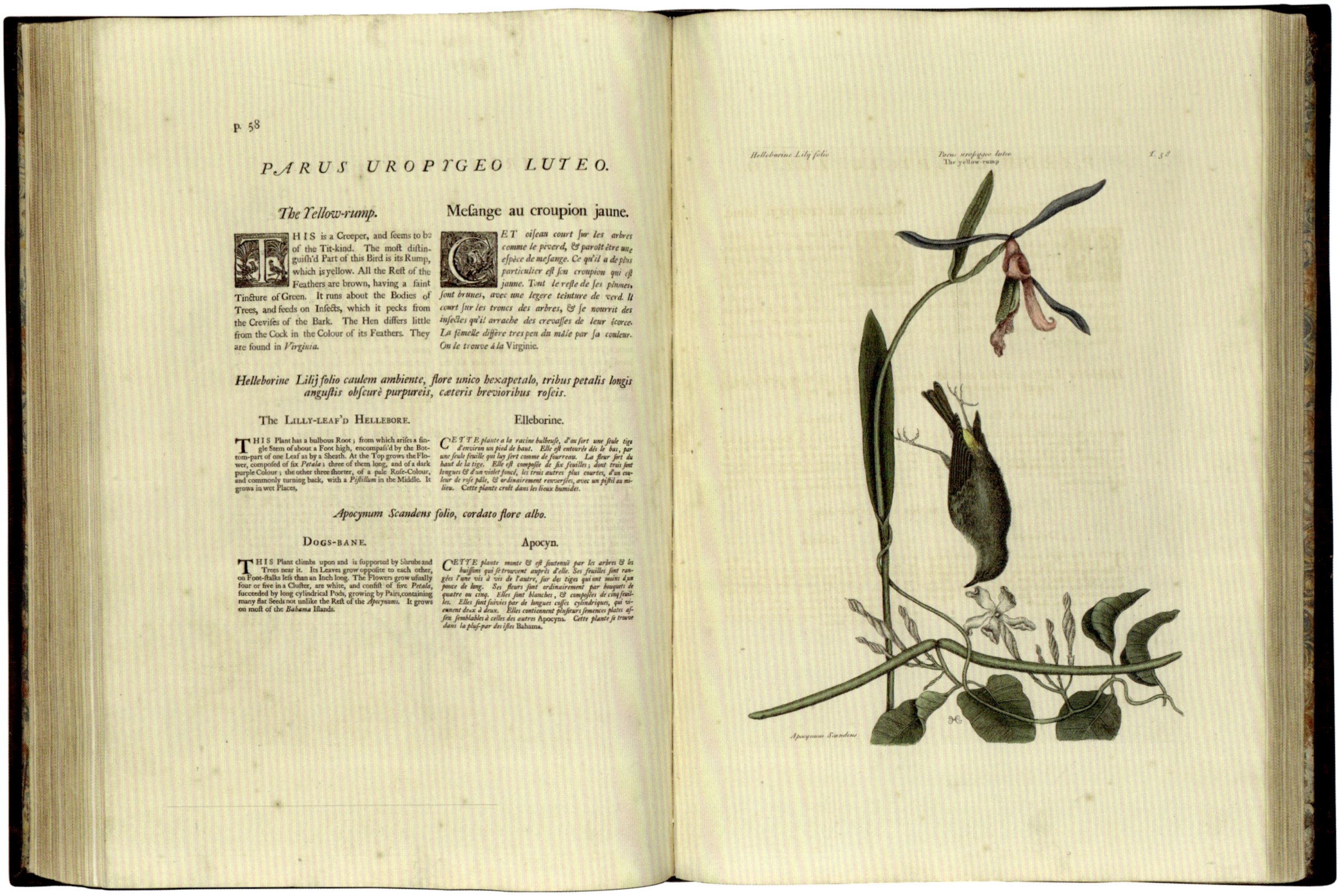

p. 58

PARUS UROPYGEO LUTEO.

The Yellow-rump.

THIS is a Creeper, and ſeems to be of the Tit-kind. The moſt diſtinguiſh'd Part of this Bird is its Rump, which is yellow. All the Reſt of the Feathers are brown, having a faint Tincture of Green. It runs about the Bodies of Trees, and feeds on Inſects, which it pecks from the Creviſes of the Bark. The Hen differs little from the Cock in the Colour of its Feathers. They are found in *Virginia*.

Meſange au croupion jaune.

CET oiſeau court ſur les arbres comme le piverd, & paroît être une eſpèce de meſange. Ce qu'il a de plus particulier eſt ſon croupion qui eſt jaune. Tout le reſte de ſes plumes, ſont brunes, avec une legere teinture de verd. Il court ſur les troncs des arbres, & ſe nourrit des inſectes qu'il arrache des crevaſſes de leur écorce. La femelle diffère tres peu du mâle par ſa couleur. On le trouve à la Virginie.

Helleborine Lilij folio caulem ambiente, flore unico hexapetalo, tribus petalis longis anguſtis obſcurè purpureis, cæteris brevioribus roſeis.

The Lilly-leaf'd Hellebore.

THIS Plant has a bulbous Root; from which ariſes a ſingle Stem of about a Foot high, encompaſs'd by the Bottom-part of one Leaf as by a Sheath. At the Top grows the Flower, compoſed of ſix *Petala*; three of them long, and of a dark purple Colour; the other three ſhorter, of a pale Roſe-Colour, and commonly turning back, with a *Piſtillum* in the Middle. It grows in wet Places,

Elleborine.

CETTE plante a la racine bulbeuſe, d'ou ſort une ſeule tige d'environ un pied de haut. Elle eſt entourée dès le bas, par une ſeule feuille qui luy ſert comme de fourreau. La fleur ſort du haut de la tige. Elle eſt compoſée de ſix feuilles; dont trois ſont longues & d'un violet foncé, les trois autres plus courtes, d'un couleur de roſe pâle, & ordinairement renverſées, avec un piſtil au milieu. Cette plante croît dans les lieux humides.

Apocynum Scandens folio, cordato flore albo.

Dogs-bane.

THIS Plant climbs upon and is ſupported by Shrubs and Trees near it. Its Leaves grow oppoſite to each other, on Foot-ſtalks leſs than an Inch long. The Flowers grow uſually four or five in a Cluſter, are white, and conſiſt of five *Petala*, ſucceeded by long cylindrical Pods, growing by Pairs, containing many flat Seeds not unlike the Reſt of the *Apocynums*. It grows on moſt of the *Bahama* Islands.

Apocyn.

CETTE plante monte & eſt ſoutenuë par les arbres & les buiſſons qui ſe trouvent auprès d'elle. Ses feuilles ſont rangées l'une vis à vis de l'autre, ſur des tiges qui ont moins d'un pouce de long. Ses fleurs ſont ordinairement par bouquets de quatre ou cinq. Elles ſont blanches, & compoſées de cinq feuilles. Elles ſont ſuivies par de longues coſſes cylindriques, qui viennent deux à deux. Elles contiennent pluſieurs ſemences plates aſſez ſemblables à celles des autres Apocyns. *Cette plante ſe trouve dans la pluſ-par des iſles* Bahama.

29.16a

29.16

MARK CATESBY (1683–1749)
The Natural History of Carolina, Florida, and the Bahama Islands . . .
2 vols., 1731–43; Appendix, 1747
20 in. (52 cm)
Published by the author, London
*Oak Spring Garden Library
Exhibited: Mark Catesby, artist and engraver, 'Yellow-rump, The Lilly-leaf'd Hellebore, Dogs-Bane', 1729–30, hand-coloured etching, vol. 1, plate 58 (illustrated above)
†Royal Botanic Gardens, Kew, Library, Art & Archives Collections, qF13.4
Exhibited: Mark Catesby, artist and engraver, 'The Flamingo', 1730–31, hand-coloured etching, vol. 1, plate 73 (illustrated p. 487)

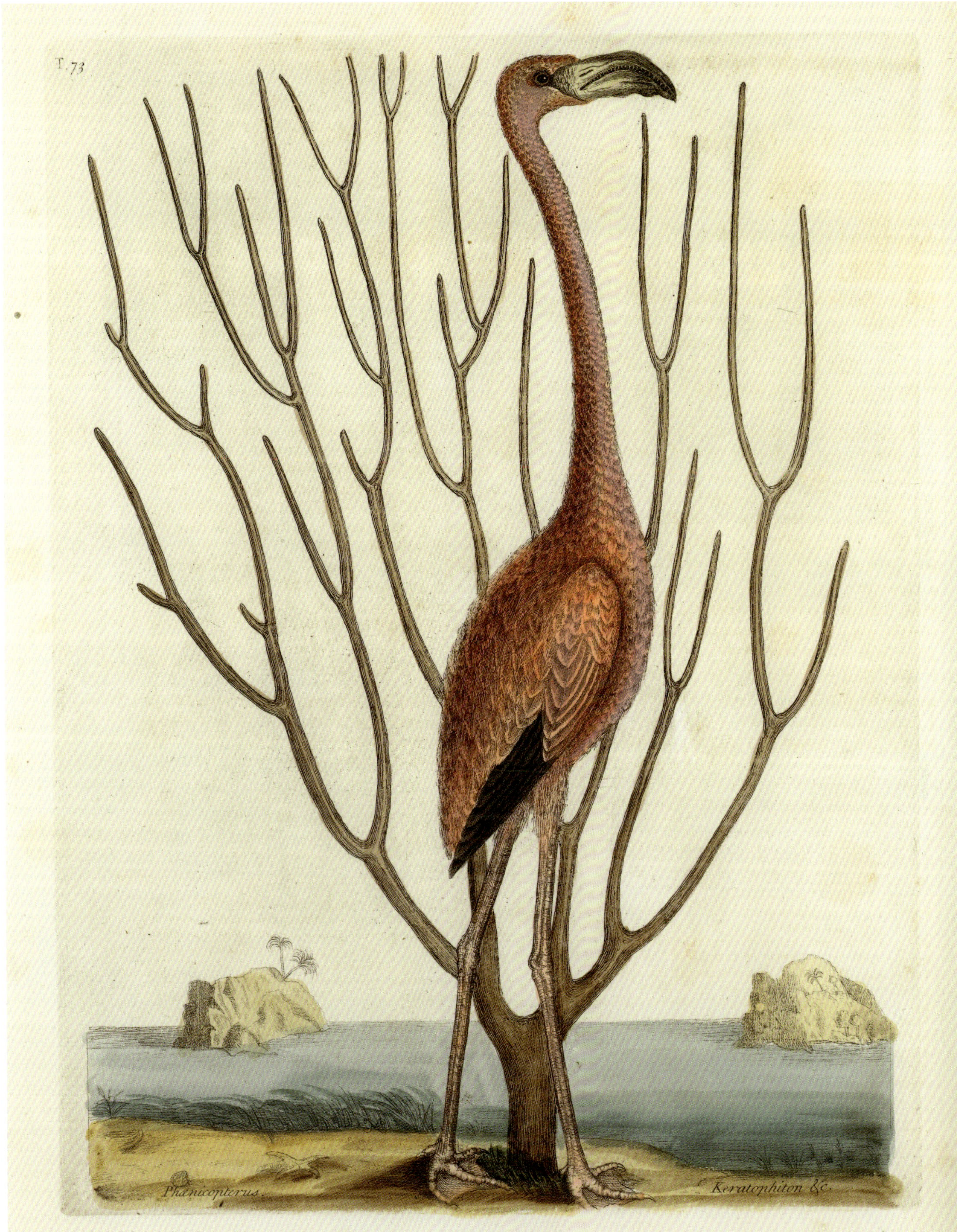

29.16b

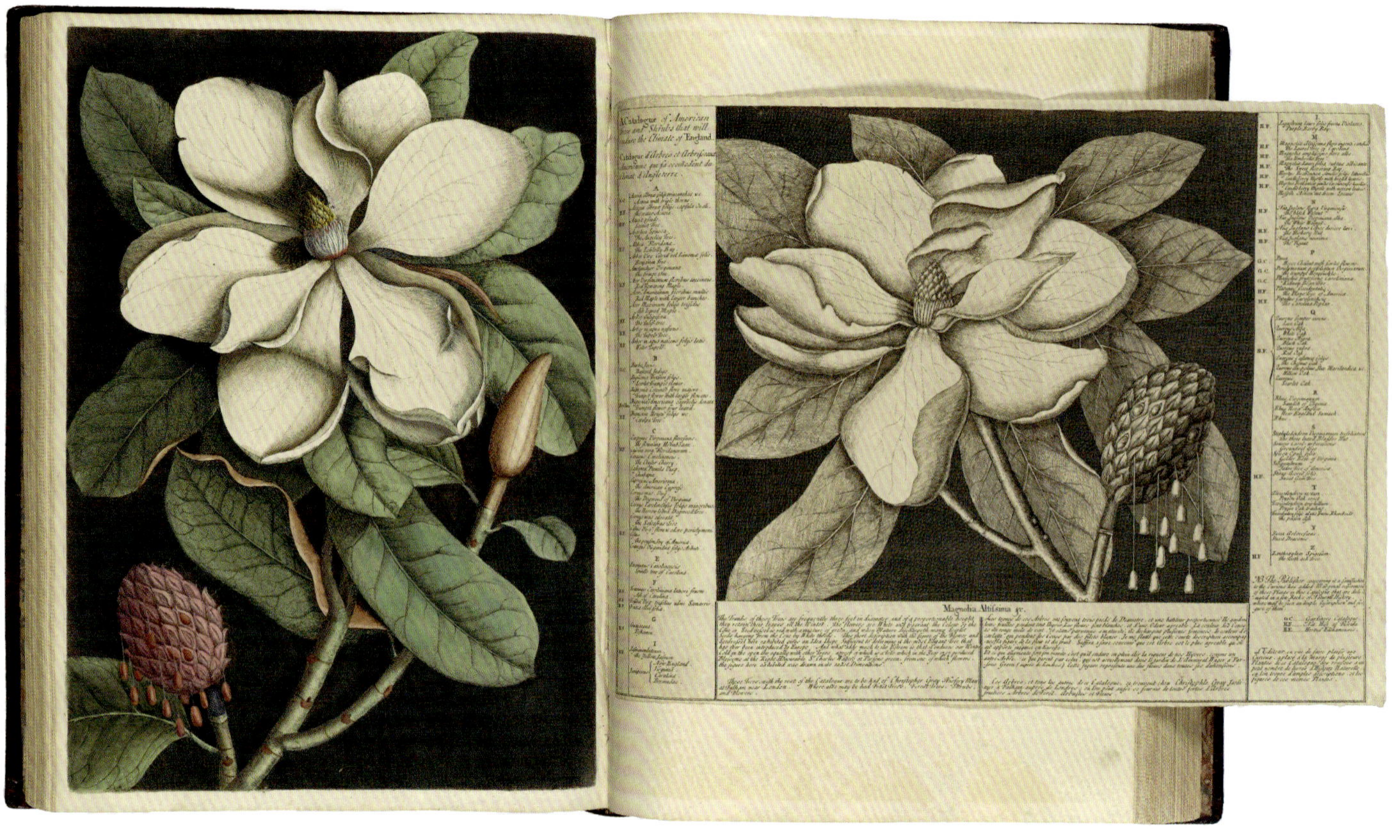

29.17a

29.17

MARK CATESBY (1683–1749)
The Natural History of Carolina, Florida, and the Bahama Islands . . .
2 vols., 1731–43; Appendix 1747
20½ in. (52 cm)
Published by the author, London
*Oak Spring Garden Library
Exhibited: left, Georg Dionysius Ehret (1708–1770), artist and engraver, 'Magnolia altissima', 1737–38, hand-coloured etching, vol. 2, plate 61; right, Mark Catesby, artist and engraver, *A Catalogue of American Trees and Shrubs that will endure the Climate of England*, c. 1737, etching, published by Christopher Gray, Fulham (illustrated above)
†Royal Botanic Gardens, Kew, Library, Art & Archives Collections, qF13.4
Exhibited: Mark Catesby, artist and engraver (squirrel after Everhard Kick [1636–after 1701]), 'The Flying Squirrel', before 1738, hand-coloured etching, vol. 2, plate 76 (not illustrated); and Mark Catesby, artist and engraver (squirrel after Everhard Kick [1636–after 1701]), 'The Flying-Squirrel, Its Posture and Manner of Flying', [and] *Viscum Caryophylloides*, before 1738, hand-coloured etching, vol. 2, plate 77 (illustrated p. 489)

Mark Catesby's *Natural History of Carolina, Florida, and the Bahama Islands . . .* was the first major illustrated natural history of the British colonies in North America, and revolutionized the understanding of plant and animal life there. During his travels to North American and the West Indies between 1712 and 1719 and again from 1722 to 1726, Catesby accumulated hundreds of specimens and preliminary sketches that formed the basis for this important work. Catesby dedicated the first volume to Queen Caroline in honour of her significant support for the study of botany in the development of her gardens, particularly at Richmond. He presented the initial part of his first volume to Caroline at court in May 1729. Following the Queen's death in 1737, the second volume found its ideal patron in Princess Augusta, whose efforts would profoundly shape gardening and horticulture in Britain. On his return to London, Catesby's work as a horticulturalist, first in association with Thomas Fairchild and then with Christopher Gray, was crucial to the introduction of American plant species in Britain.

While many of Catesby's plates combined plants and animals that occurred together in nature, addressing the relationships between species in specific environments, others appear as assemblies of disparate species, often creatively composed from a strange and slightly amusing perspective. In figure 29.16a (p. 486), the yellow-rumped warbler from Virginia hangs lifeless from the stalk of hellebore, found in the southern colonies of North America, while the dogs-bane, a native of the Bahamas, twines below.

Figure 29.17a represents a collaboration between Mark Catesby and the great botanical illustrator Georg Dionysius Ehret, who provided ten drawings and two etchings for *The Natural History of Carolina, Florida, and the Bahama Islands . . .* Ehret, a native of Heidelberg, settled in London in 1735 and became the pre-eminent botanical artist of his generation. The precision and beauty of his paintings won him the patronage of the great Swedish taxonomist, Carl Linnaeus, and, in England, of Sir Hans Sloane and Richard Mead, among many others. The magnolia so captivated English gardeners and naturalists that its flowering was a notable event, and both Ehret and Catesby made drawings of the one that bloomed in the garden of Sir Charles Wager at Parson's Green in 1737. The facing sheet, which was bound into this copy by the book's owner, is a catalogue of American plants sold by Catesby's associate, the Fulham nurseryman Christopher Gray. The central image, a magnolia in full bloom, was drawn and etched by Catesby, in emulation of the magnificent plate Ehret had produced for the second volume of Catesby's *Natural History*.

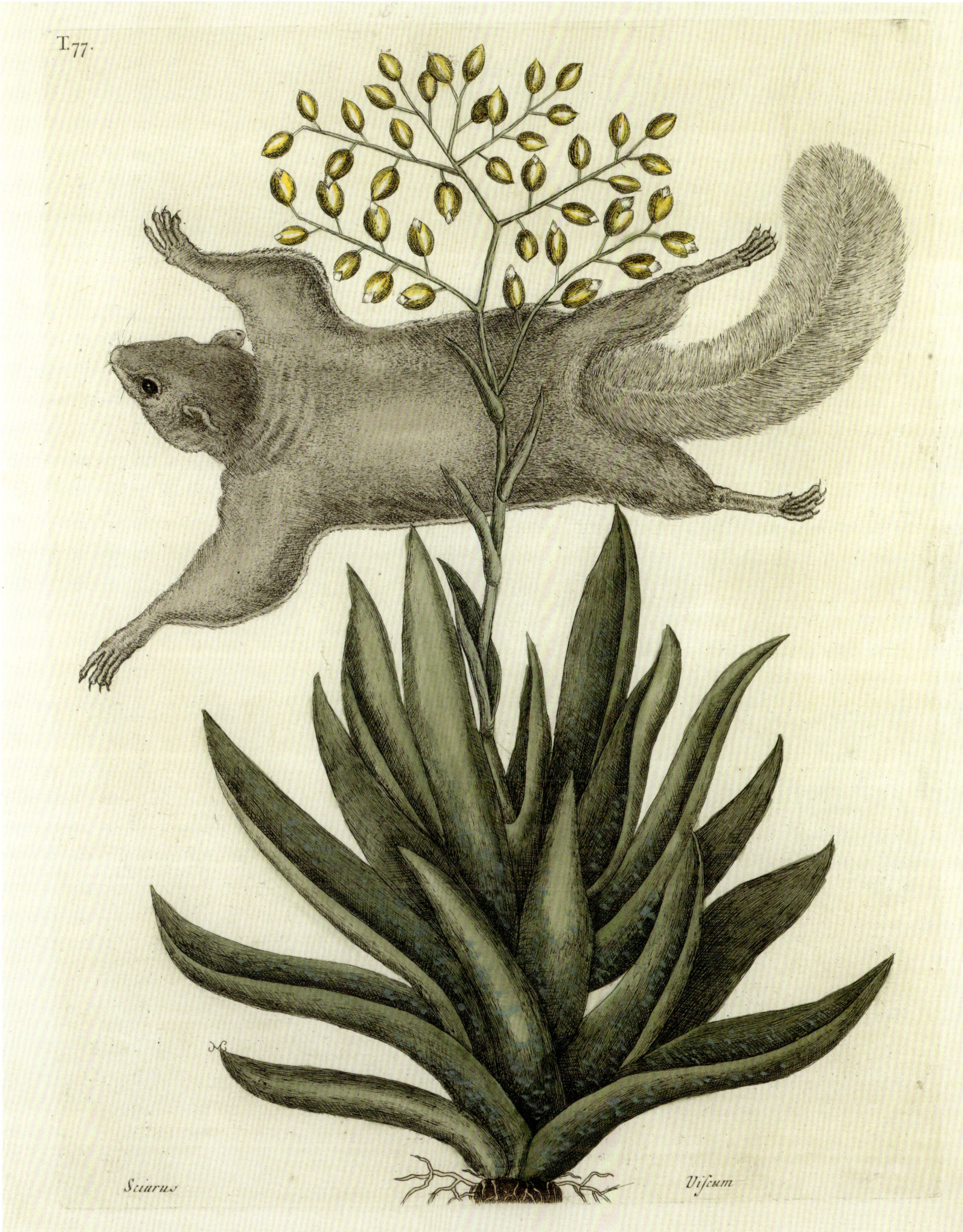

29.17b

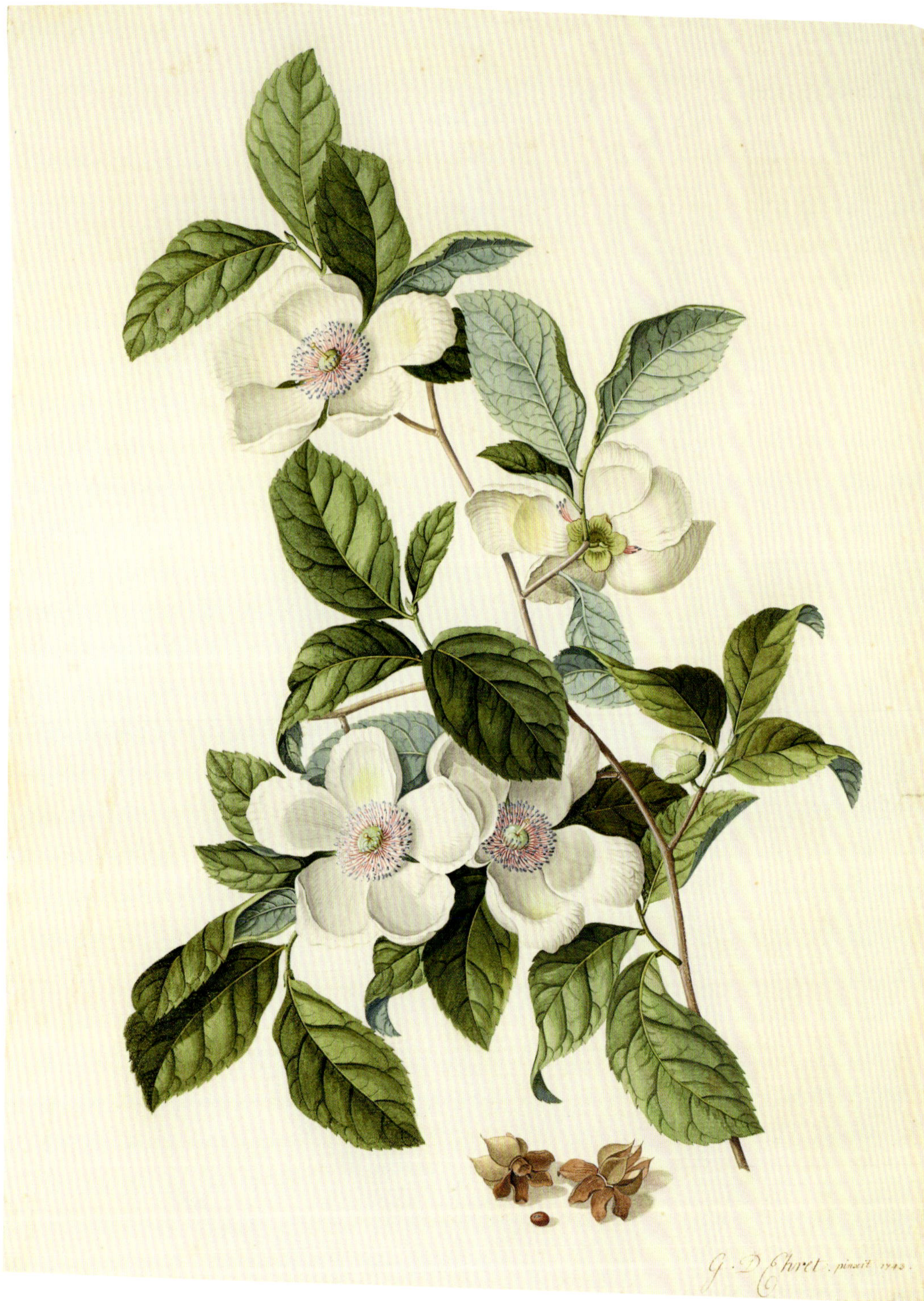

29.18

29.18*

GEORG DIONYSIUS EHRET (1708–1770)
Silky Camellia
c. 1742
Watercolour and gouache over pencil on paper
16⅜ x 12 in. (41.5 x 30.3 cm)
Royal Collection Trust, RCIN 926082

Georg Dionysius Ehret was celebrated during his lifetime as England's leading botanical illustrator, and he acted for a time as tutor in botanical drawing to the Duchess of Portland's daughters. He provided this preparatory drawing for plate thirteen in the appendix to volume two of Mark Catesby's *Natural History of Carolina, Florida, and the Bahama Islands . . .* (published in 1743 and dedicated to Augusta, Princess of Wales; Appendix, 1747). The silky camellia, or *Stewartia malacodendron*, belongs to a genus named by Linnaeus in honour of John Stuart, third Earl of Bute, a close friend of Augusta as well as her adviser and collaborator in the development of the botanical gardens at Kew. The plant depicted here was sent to Catesby in 1742 by a correspondent in Virginia, and it flowered in his Fulham garden three months later.

29.19

MARK CATESBY (1683–1749)
Bison Americanus and Pseudo Acacia hispida floribus noseis
1738–47
Watercolour and gouache over pen and ink on paper
14¾ x 10⅝ in. (37.2 x 27 cm)
Royal Collection Trust, RCIN 926090

When the pioneering naturalist and artist Mark Catesby went 'out a Buffello hunting' in 1723, he not only observed those 'awful creature[s]', but also collected specimens of the pseudo-acacia, which he took to be part of their diet. Despite Catesby's insistence on observation from life, this bison was copied from a drawing made by the Dutch artist Everhard Kick for the royal physician Sir Hans Sloane, for whom Catesby also made drawings, and who was a subscriber to the naturalist's *Natural History*. The twig bearing the plant's leaves and flowers is greatly magnified to make its identifying features visible. The resulting picture is rich with information but rather strange and humorous in appearance, as is often the case with Catesby's images. The majority of Catesby's preparatory drawings for his *Natural History of Carolina, Florida, and the Bahama Islands . . .* were acquired by George III in 1768. The bison and pseudo-acacia appeared as plate twenty of the appendix to volume two, which appeared in 1747.

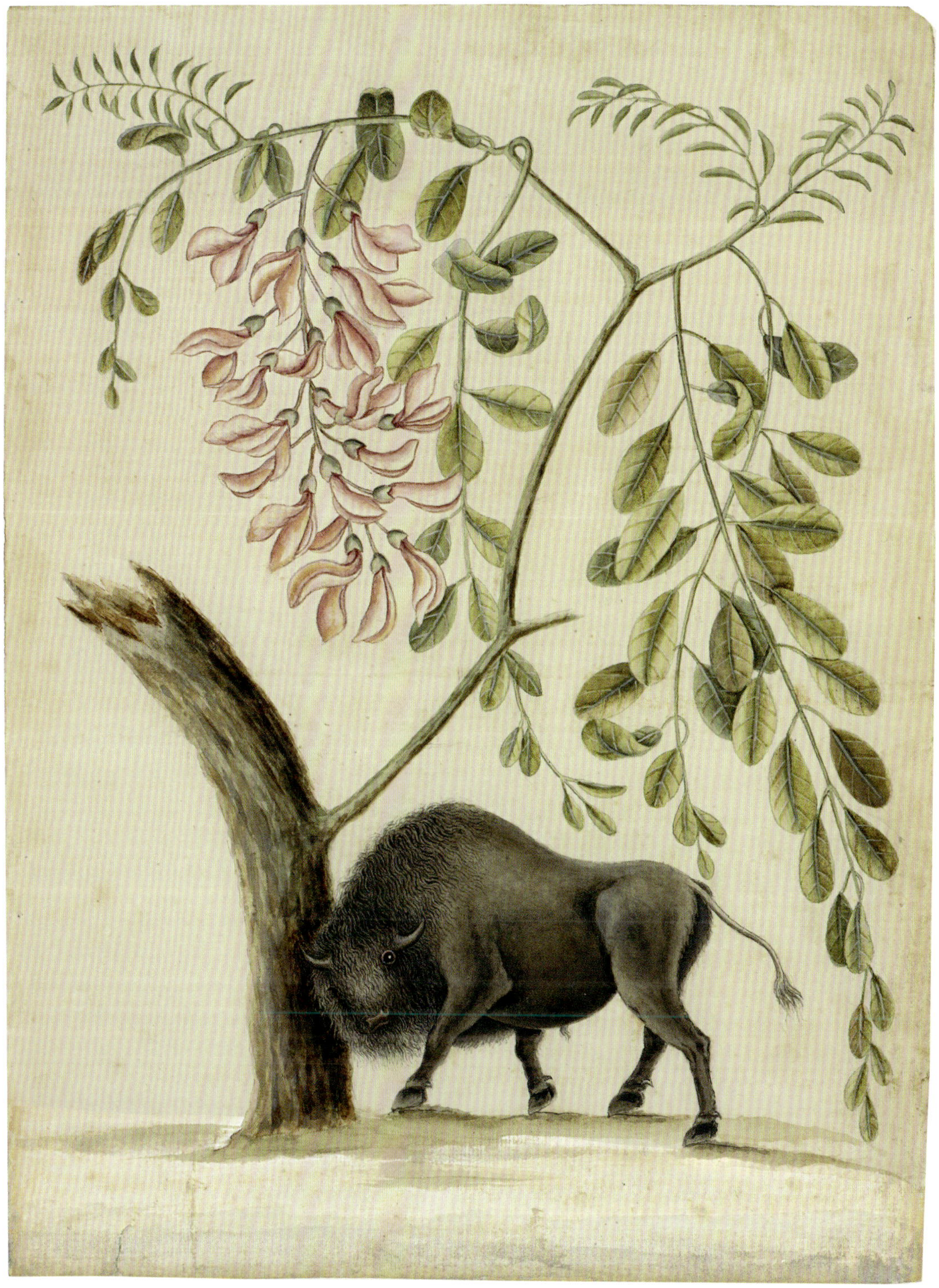

29.19

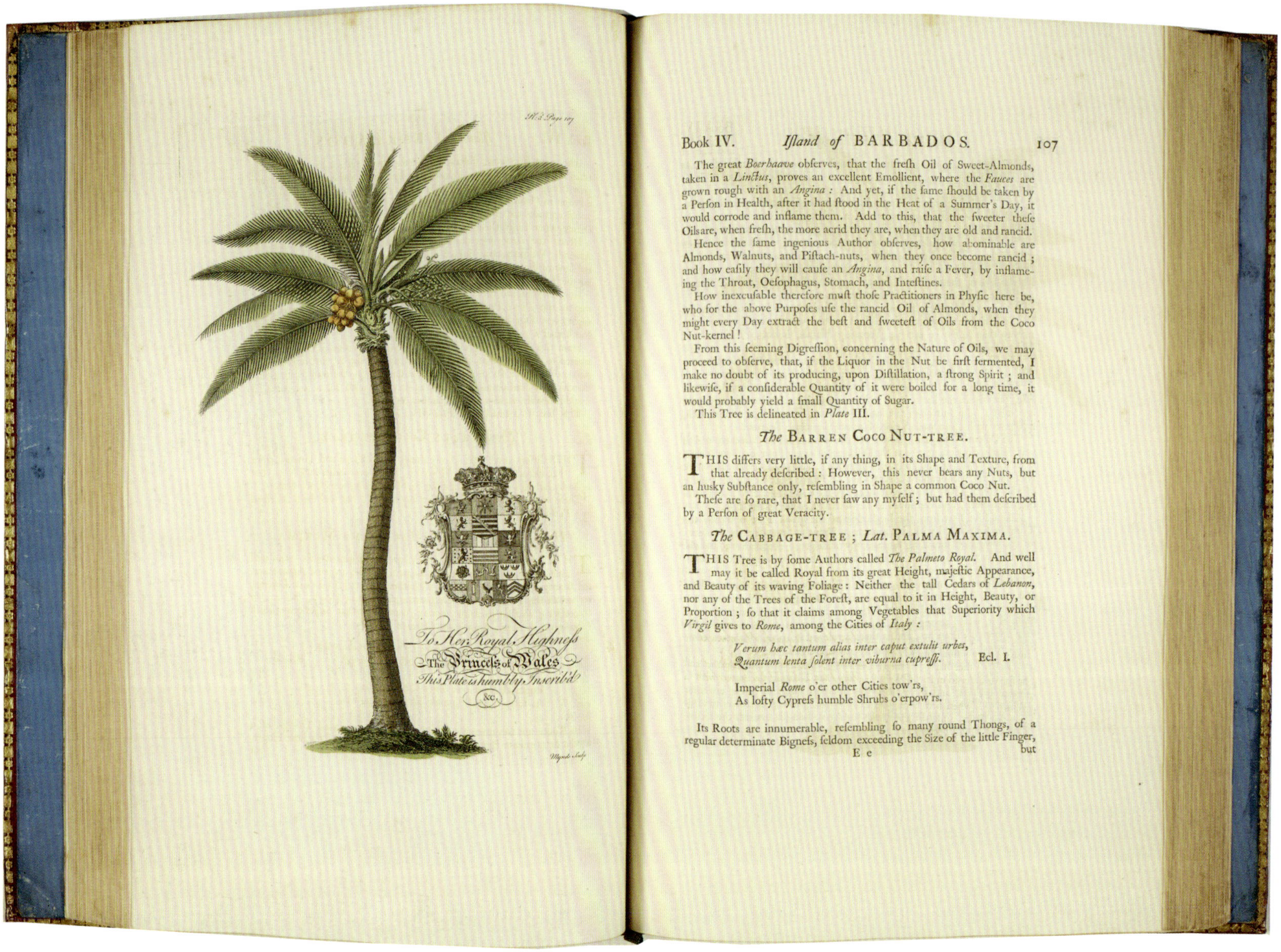
To Her Royal Highness The Princess of Wales This Plate is humbly Inscrib'd &c.

Book IV. *Iſland of* BARBADOS. 107

The great *Boerhaave* obſerves, that the freſh Oil of Sweet-Almonds, taken in a *Linctus*, proves an excellent Emollient, where the *Fauces* are grown rough with an *Angina* : And yet, if the ſame ſhould be taken by a Perſon in Health, after it had ſtood in the Heat of a Summer's Day, it would corrode and inflame them. Add to this, that the ſweeter theſe Oils are, when freſh, the more acrid they are, when they are old and rancid.

Hence the ſame ingenious Author obſerves, how abominable are Almonds, Walnuts, and Piſtach-nuts, when they once become rancid ; and how eaſily they will cauſe an *Angina*, and raiſe a Fever, by inflameing the Throat, Oeſophagus, Stomach, and Inteſtines.

How inexcuſable therefore muſt thoſe Practitioners in Phyſic here be, who for the above Purpoſes uſe the rancid Oil of Almonds, when they might every Day extract the beſt and ſweeteſt of Oils from the Coco Nut-kernel !

From this ſeeming Digreſſion, concerning the Nature of Oils, we may proceed to obſerve, that, if the Liquor in the Nut be firſt fermented, I make no doubt of its producing, upon Diſtillation, a ſtrong Spirit ; and likewiſe, if a conſiderable Quantity of it were boiled for a long time, it would probably yield a ſmall Quantity of Sugar.

This Tree is delineated in *Plate* III.

The BARREN COCO NUT-TREE.

THIS differs very little, if any thing, in its Shape and Texture, from that already deſcribed : However, this never bears any Nuts, but an husky Subſtance only, reſembling in Shape a common Coco Nut.

Theſe are ſo rare, that I never ſaw any myſelf ; but had them deſcribed by a Perſon of great Veracity.

The CABBAGE-TREE ; *Lat.* PALMA MAXIMA.

THIS Tree is by ſome Authors called *The Palmeto Royal.* And well may it be called Royal from its great Height, majeſtic Appearance, and Beauty of its waving Foliage : Neither the tall Cedars of *Lebanon*, nor any of the Trees of the Foreſt, are equal to it in Height, Beauty, or Proportion ; ſo that it claims among Vegetables that Superiority which *Virgil* gives to *Rome*, among the Cities of *Italy* :

Verum hæc tantum alias inter caput extulit urbes,
Quantum lenta ſolent inter viburna cupreſſi. Ecl. I.

Imperial *Rome* o'er other Cities tow'rs,
As lofty Cypreſs humble Shrubs o'erpow'rs.

Its Roots are innumerable, reſembling ſo many round Thongs, of a regular determinate Bigneſs, ſeldom exceeding the Size of the little Finger, but

E e

29.20

THE LEGACY OF SLOANE AND CATESBY

29.20

GRIFFITH HUGHES (1707–1758)
History of Barbados in Ten Books . . .
1750
16¾ in. (42.5 cm)
Printed for the author, London
*Oak Spring Garden Library
Exhibited: James Mynde (active 1740–1770), after unknown artist (possibly Georg Dionysius Ehret, 1708–1770), 'The Coco Nut-tree; Lat. Palma Nucifera', c. 1750, etching, plate 3 (not illustrated)
†Royal Collection Trust, RCIN 1081286
Exhibited: James Mynde (active 1740–1770), engraver, after unidentified artist (possibly Georg Dionysius Ehret, 1708–1770), 'The Coco Nut-tree; Lat. Palma Nucifera', c. 1750, hand-coloured etching, plate 3 (illustrated)

The Church of England clergyman and naturalist Griffith Hughes was rector of St Lucy's, Barbados, from 1736 to around 1748. George Dionysius Ehret prepared some plates for his book, a lavish production with almost one thousand subscribers, including George II, Princess Augusta (to whom this plate is dedicated), and her husband, Frederick, Prince of Wales. The author used his description of the coconut tree, shown here, to debunk the 'chimerical' qualities previously attributed to it, including the claim that the plant could produce, in his words, 'Bread, Water, Wine, Vinegar, Brandy, Oil, Honey, Cups, Spoons, Besoms [brooms], Masts, Nails, Needles, and Covering for Houses'. This copy of Hughes's book was bound for Frederick and was later in George III's library at Cumberland Lodge.

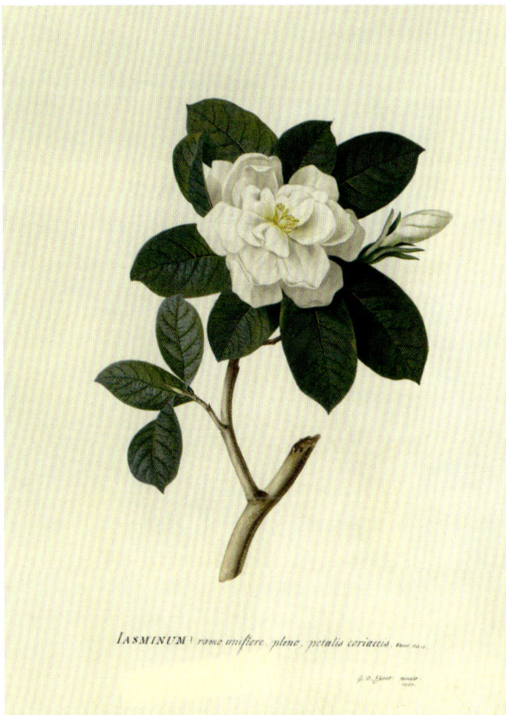

29.21

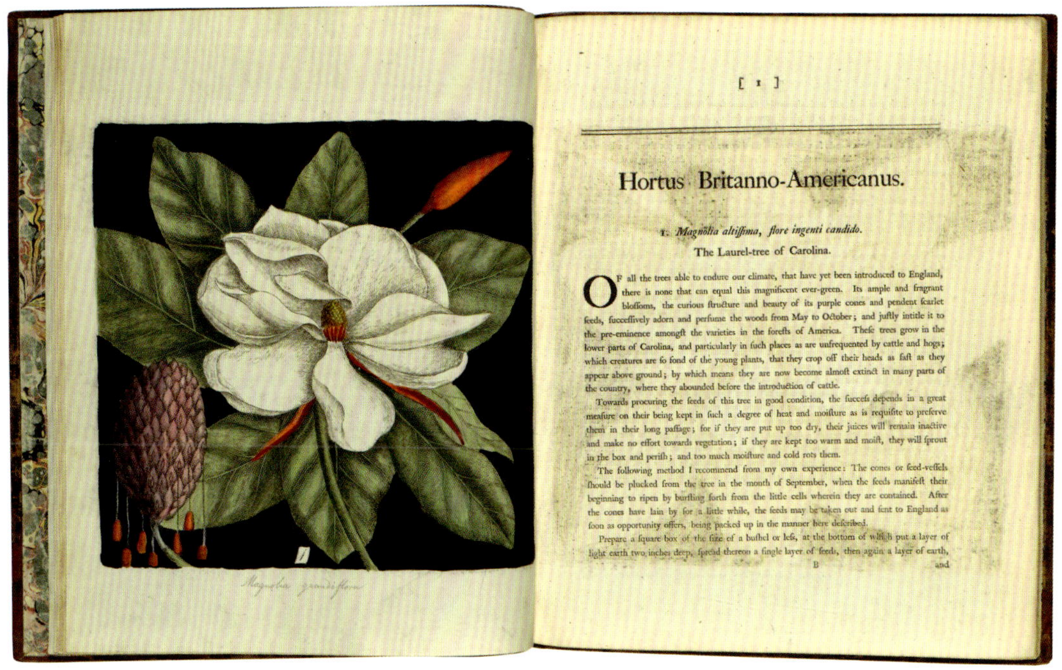

29.22

29.21

GEORG DIONYSIUS EHRET (1708–1770)
Jasminum ramo uniflore pleno petalis coriaieis
[Cape Jasmine]
1760
Gouache on vellum
19⅝ x 13¾ in. (50 x 35 cm)
Royal Botanic Gardens, Kew, Library, Art & Archives Collections, EHR.11

The Cape Jasmine is native to southern China, but by the eighteenth century it had been carried to South Africa by Dutch traders. In 1754 a captain of the British East India Company sent a live example back to England. When it finally bloomed in 1758, the plant's beauty and fragrance drew prominent botanists and artists such as John Ellis, Peter Collinson, and Georg Dionysius Ehret to the hothouse in Essex where it grew. Ellis unsuccessfully sought Carl Linnaeus's approval for naming first its genus and then its species in honour of Princess Augusta. The two men eventually agreed to call the plant *Gardenia jasminoides*, paying tribute instead to Ellis's friend the naturalist and physician Alexander Garden, who lived in Charleston, South Carolina.

29.22*

MARK CATESBY (1683–1749)
Hortus Europae Americanus: or, A collection of 85 curious trees and shrubs, the produce of North America; adapted to the climates and soils of Great-Britain, Ireland, and most parts of Europe, &c. . .
1767
15 in. (38 cm)
Printed by W. Richardson and S. Clark for J. Ryall, London
Oak Spring Garden Library
Exhibited: Mark Catesby, artist and engraver, 'Magnolia grandiflora', struck c. 1767 from a plate etched before 1749, hand-coloured etching, plate 1

Best known for his ground-breaking two-volume *Natural History of Carolina, Florida, and the Bahama Islands . . .*, Mark Catesby also contributed to the transformation of natural history and horticulture in Britain by sending back viable seeds of American plants to England. When he finally returned to London in 1726 from his two expeditions to North America and the Caribbean, he was associated first with the London nurseryman Thomas Fairchild, and then with Christopher Gray in Fulham, studying how best to adapt and care for these new arrivals in the local soil and climate. These investigations were enormously influential, and Catesby has been called 'one of the most important experimental horticulturalists of his generation'. His posthumously published *Hortus Britanno-Americanus* (1763) gave readers the benefit of his practical experience cultivating American transplants, most of which were his own introductions, and provided comprehensive details about their requirements, development, care, and uses. To extend the book's market appeal, the second edition was issued under the alternate title *Hortus Europae Americanus*. Catesby designed his etchings for the book after the plates he originally produced for his *Natural History*.

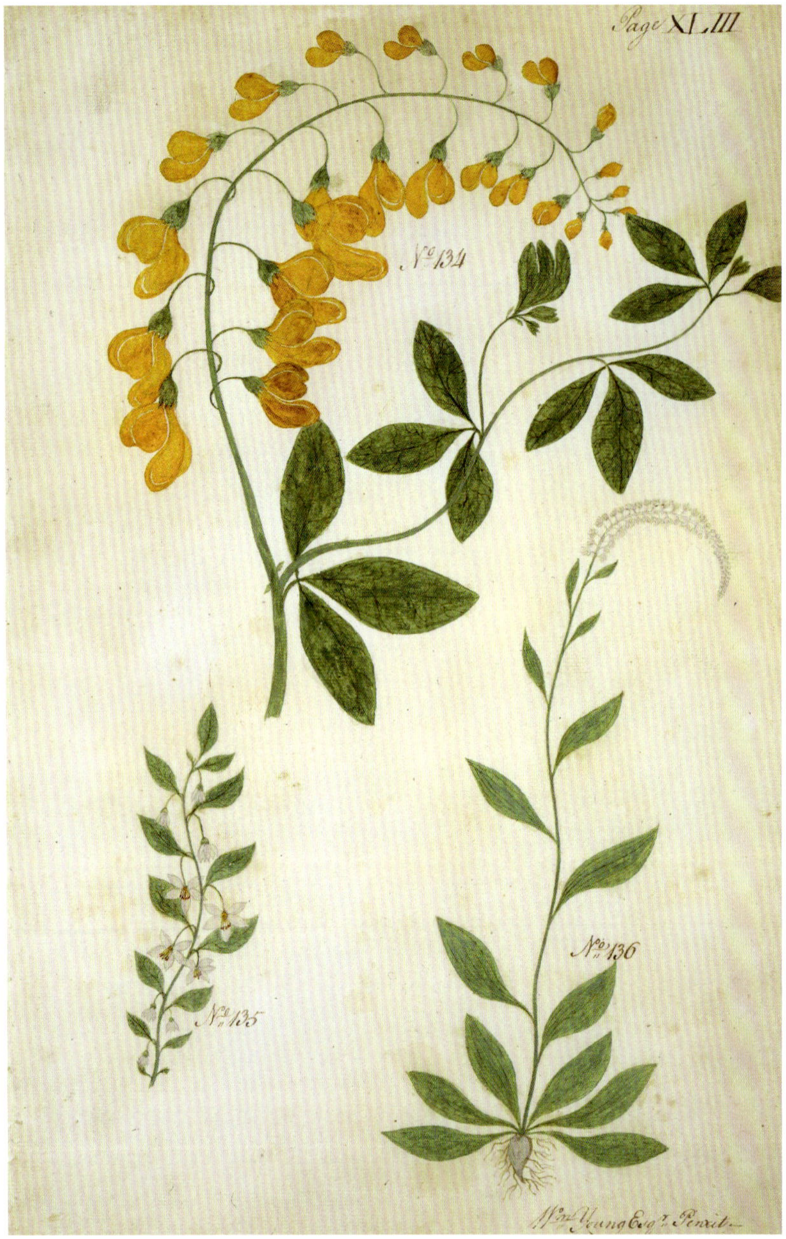

29.23a

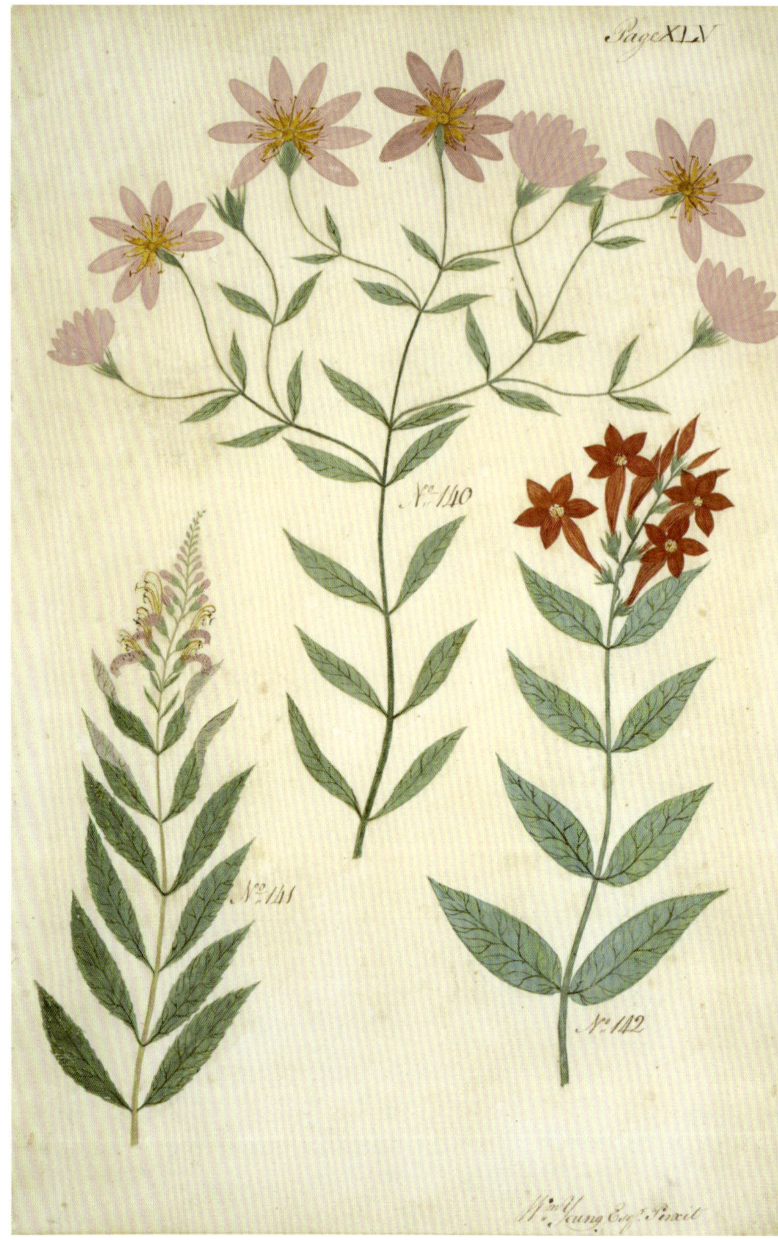

29.23b

29.23*

WILLIAM YOUNG (1742–1785)
'A Natural History of Plants Containing the Productions of North and South Carolina of upwards of 300 different species which have been carefully collected, drawn, and coloured from the life in the year 1767'
1767
15 in.
Natural History Museum, London, 999801824917002081 (sheet 43) and 999801824917202081 (sheet 45)
Exhibited: William Young, artist, '134. *Indiofera*, 135. *Andromeda*, 136. *Tetramdynamia*,' c. 1767, watercolour on paper, 15 x 9½ in. (38 x 24 cm), p. 43 (29.23a); and William Young, artist, '140. *Centaria*, 141. *Dydynamia*, 142. Pink', c. 1767, watercolour on paper, 15 x 9½ in. (38 x 24 cm), p. 45 (29.23b)

Around 1764, the Pennsylvania botanist William Young, the son of German immigrants, sent an unsolicited shipment of plants to Queen Charlotte, who responded by summoning him to London and appointing him Botanist to the Queen. Young, despite his inexperience, managed to bring the first live example of a Venus flytrap—a plant that both entertained and puzzled contemporaries with its apparent sentience—from America to England. Leading London naturalists, such as Peter Collinson and John Fothergill, cautiously welcomed Young when he arrived in London in 1765, and John Hill, assistant to the Earl of Bute at Kew, served as his tutor. Despite this promising beginning, Young kept unfortunate company and ended up in a debtors' prison. Queen Charlotte paid his debts on the condition that he depart immediately for America; she nonetheless allowed him to keep his title and a reduced salary of £50 a year. In 1767, Young produced an album of 302 watercolours of plants he had studied on an expedition through North and South Carolina, and he dedicated the work to the Queen and King. The watercolours illustrated here are examples from that album.

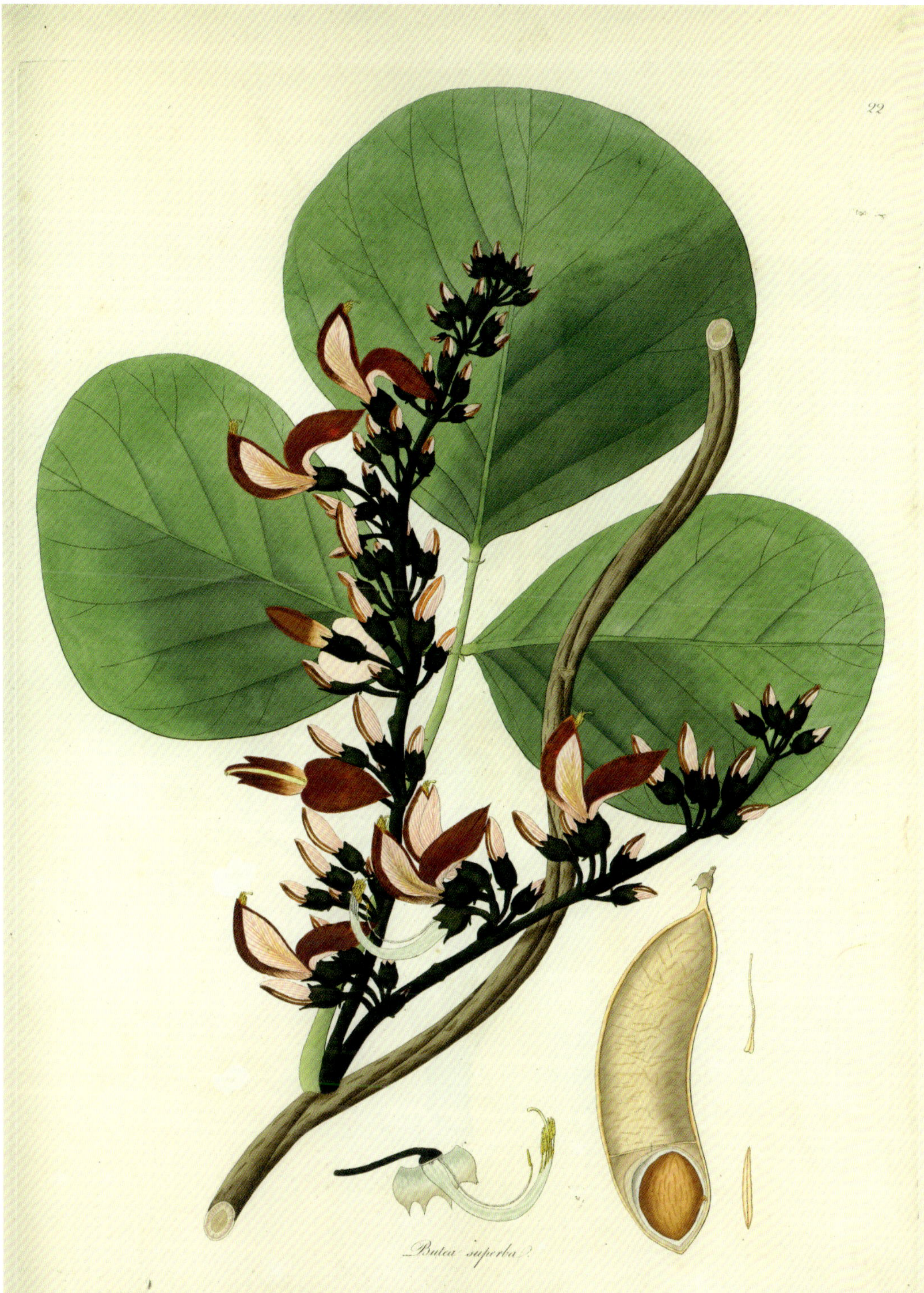

29.24

29.24

WILLIAM ROXBURGH (1751–1815)
Plants of the coast of Coromandel: selected from drawings and descriptions presented to the Hon. Court of Directors of the East India Company (3 vols)
1795–1819
23¼ x 19¾ in. (59 x 50 cm)
Printed by W. Bulmer and Co. for G. Nicol, London
*New York Botanical Garden, Mertz Library, lf QK 349.8 .R6
Exhibited: Unknown engraver, after unidentified artist, India, '*Neptunia oleracea* Lour. [as *Mimosa natans* L.f.]', c. 1798, hand-coloured etching, vol. 2, plate 119 (illustrated p. 480, fig. 29.09)
†Royal Botanic Gardens, Kew, Library, Art & Archives Collections, qF5.2
Exhibited: Unknown engraver, after unidentified artist, India, '*Roxburghia gloriosoides*', c. 1795, hand-coloured engraving, vol. 1, plate 32 (not illustrated); and unknown engraver, after unknown artist, India, '*Butea superba*', c. 1795, hand-coloured engraving, vol. 1, plate 22 (illustrated)

The botanist and surgeon William Roxburgh was appointed official naturalist in Madras (Chennai) by the East India Company in 1790. Two Indian artists provided him with drawings of the plants in his collection, which he sent, with descriptions, in consignments to Sir Joseph Banks, who was then directing the activities of the royal gardens at Kew. By 1794 Banks had received around five hundred of these drawings, and began to arrange for their publication as *Plants of the Coast of Coromandel*. Its contents included plants of the genus *Roxburghia* (see fig. 29.09), named for the author by Jonas Dryander; and of the genus *Butea*, named in honour of the third Earl of Bute, who had helped Princess Augusta to develop her gardens at Kew and continued as a supporter of Charlotte's botanical activities. Roxburgh observed that the *Mimosa natans* was a 'sensitive' plant whose leaves moved in response to the touch. In the eighteenth century, sensitive and carnivorous plants (such as the Venus flytrap, introduced from America by William Young, botanist to Queen Charlotte) challenged established ways of thinking about the natural world because they seemed to blur the boundary between the plant and animal kingdoms. A copy of Roxburgh's book was in Queen Charlotte's extensive botanical library.

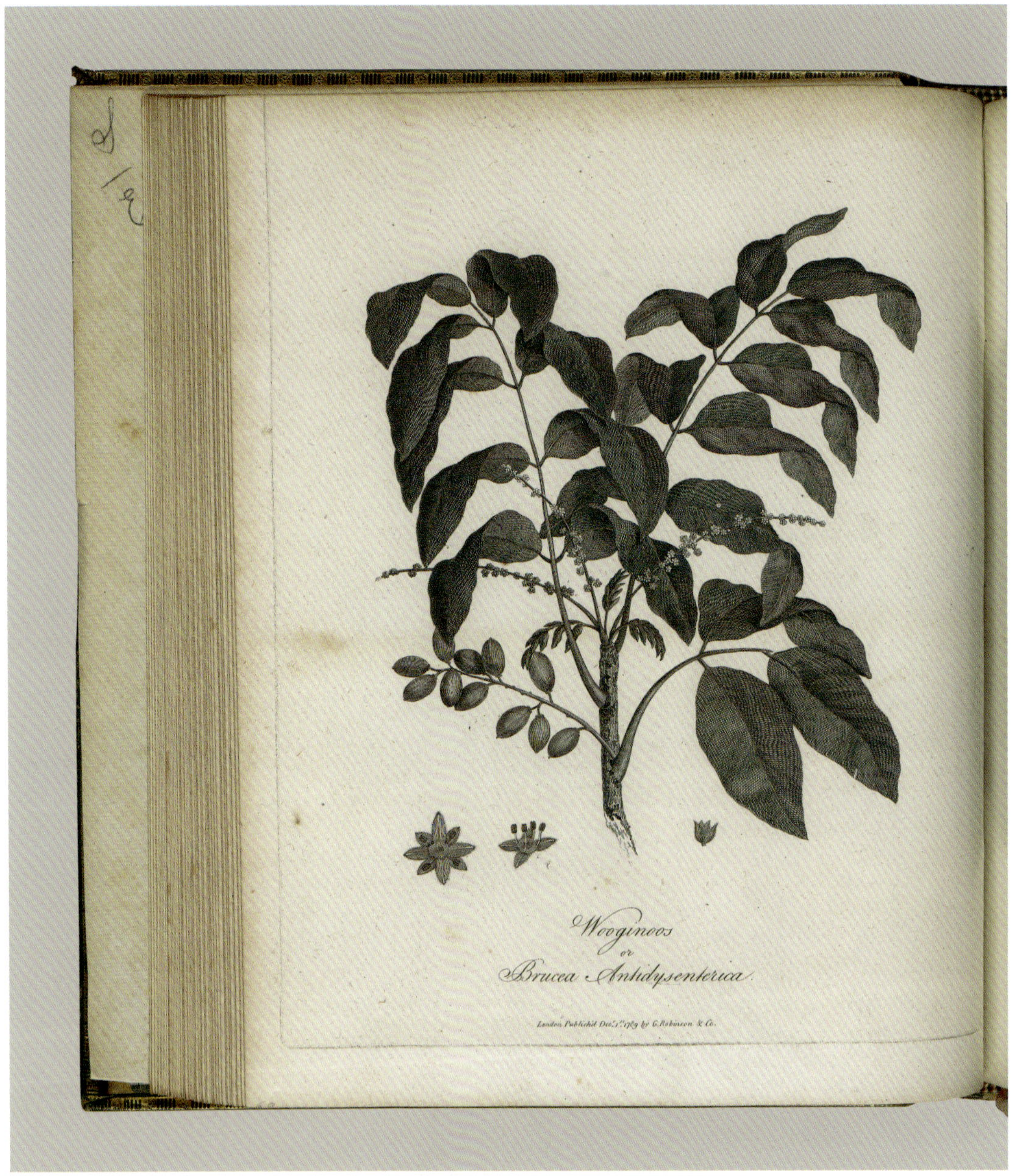

29.25

ILLUSTRATIONS FOR BRUCE'S *TRAVELS*

29.25*

JAMES BRUCE (1730–1794)
Travels to Discover the Source of the Nile . . . (5 vols)
1790
12.75 in. (32.4 cm)
Printed by J. Ruthven, for G.G.J. and J. Robinson, Edinburgh
Yale Center for British Art, Paul Mellon Collection, DT377 B88 1790+ Oversize, copy 2
Exhibited: Probably James Heath (1757–1834), engraver, after James Bruce, 'Brucea antidysenterica', 1789, line engraving and etching, vol. 5, facing p. 69

29.26*

LUIGI BALUGANI (1737–1771)
Brucea antidysenterica
1769–71
Pen and ink on paper
12 x 7⅞ in. (30.5 x 20 cm)
Yale Center for British Art, Paul Mellon Collection, B1977.14.9047 verso
Not illustrated

29.27*

LUIGI BALUGANI (1737–1771)
Brucea antidysenterica
1769–71
Pencil with pen and ink on paper
10⅛ x 8 in. (25.7 x 20.3 cm)
Yale Center for British Art, Paul Mellon Collection, B1977.14.9031 verso

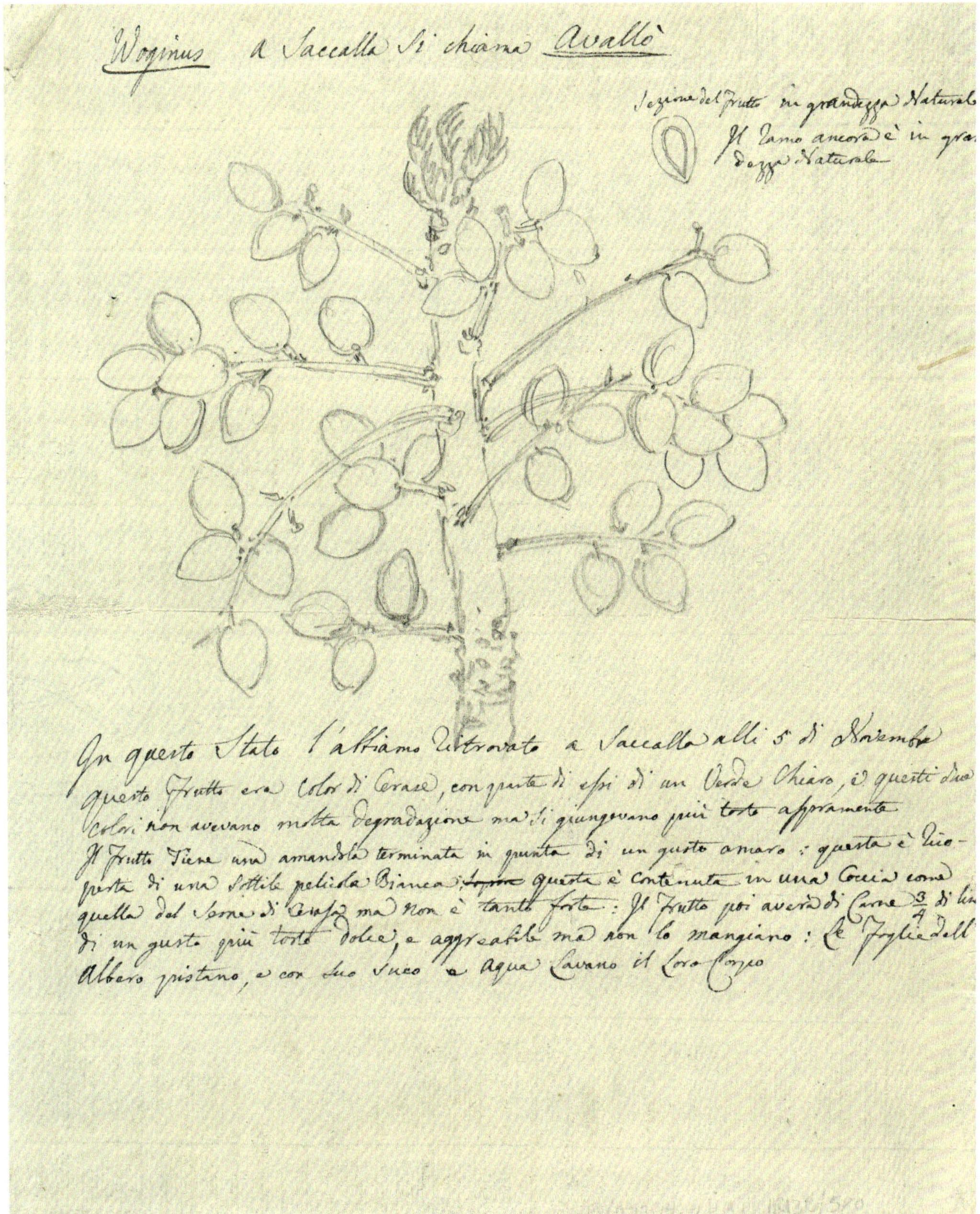

29.27

29.28*

ATTRIBUTED TO JAMES BRUCE
(1730–1794)
Brucea antidysenterica J. Miller (James Bruce's Tree): finished drawing with stem of fruits added
1769–73
Watercolour, gouache, and pencil on paper
14⅜ x 10½ in. (36.5 x 26.7 cm)
Yale Center for British Art, Paul Mellon Collection, B1977.14.8918
Not illustrated

29.29

FREDERICK POLYDORE NODDER
(active 1773–1800)
Brucea antidysenterica J. Miller (James Bruce's Tree): finished drawing of part of stem with flowering shoots and leafy shoots
1777
Watercolour, gouache, and pencil on vellum
18^13^⁄16 x 14^1^⁄16 in. (47.8 x 35.7 cm)
Yale Center for British Art, Paul Mellon Collection, B1977.14.8882

Luigi Balugani, an Italian architect and draughtsman, accompanied the Scottish adventurer James Bruce on his journeys through Africa and the Middle East from 1765 until Balugani's death in Ethiopia in 1771. He was initially hired to produce drawings of antiquities commissioned from Bruce by George III, but he also made hundreds of drawings and sketches of the plants and animals they encountered along the way. After visiting Algiers, Syria, and Egypt, Bruce was persuaded to go in search of the source of the Blue Nile. He and Balugani reached the spring of Gish Abay in Ethiopia in 1770, although they were not the first Europeans to do so. In Ethiopia, Bruce was treated for dysentery with the powdered roots of a shrub known locally as 'Wooginoos'. He sent seeds of the plant back to Kew, where the plant was successfully cultivated, and Joseph Banks named it *Brucea antidysenterica* in his honour. Bruce returned to England in 1774, but his account of his journey was received with widespread disbelief; discouraged, he waited sixteen years before publishing his five-volume *Travels to Discover the Source of the Nile . . .* (1790; 29.25).

On his return to England, Bruce attempted to claim authorship of all Balugani's drawings, but soon met with scepticism. When he presented the views of Palmyra to George III and Queen Charlotte, Charlotte pointedly asked him if he 'had not had help'. Bruce conceded that in this case he had, but never relinquished his claim to the botanical drawings. The hands of both men are apparent in this group, in the figures as well as the inscriptions. The engraving of *Brucea antidysenterica* that Bruce published in volume five of his *Travels* was based on a drawing (29.28) attributed to him, but more likely the work of Balugani. The finished quality of this drawing suggests that it was made in the relative comfort of a camp, working from field sketches. The rough sketches themselves (29.26, 29.27) show different stages of the plant's seasonal development, and the drawing combines the flowers and fruit to give a more complete picture of its life cycle.

The botanical draughtsman and engraver Frederick Polydore Nodder is thought to have had a German background, but no details of his early life are known. He was described in Royal Academy exhibition catalogues of the late 1780s as 'Draughtsman in Natural History' and 'Botanic Painter to her Majesty', Queen Charlotte. As well as producing illustrations for works such as Erasmus Darwin's *Botanic Garden* (1789), he finished many of the drawings by Luigi Balugani that James Bruce brought back to England after the artist's death in Ethiopia. Nodder made the watercolour illustrated here (cat. 29.29) at Kew, from one of the plants cultivated from seeds that Bruce sent back during his expedition.

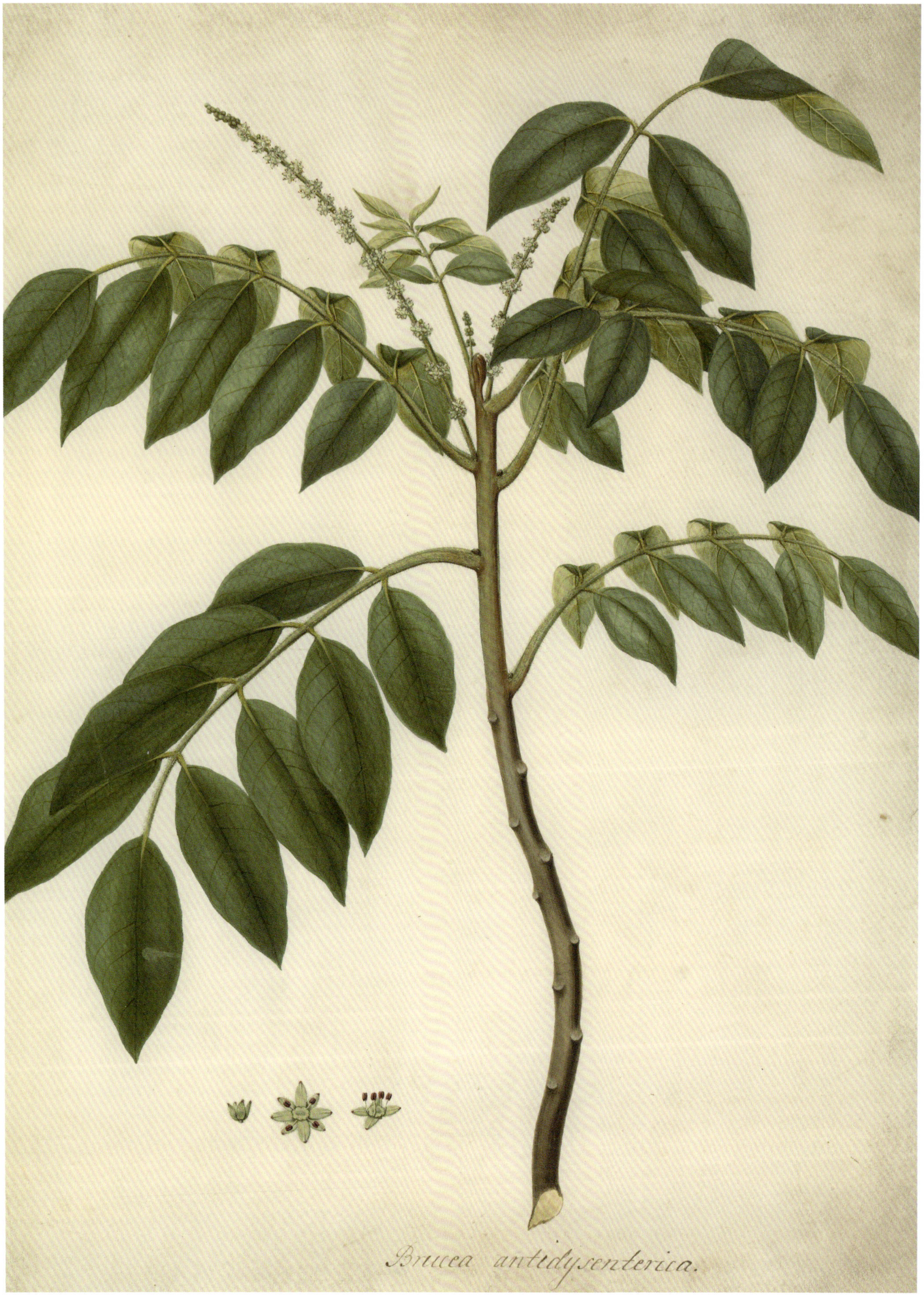

29.29

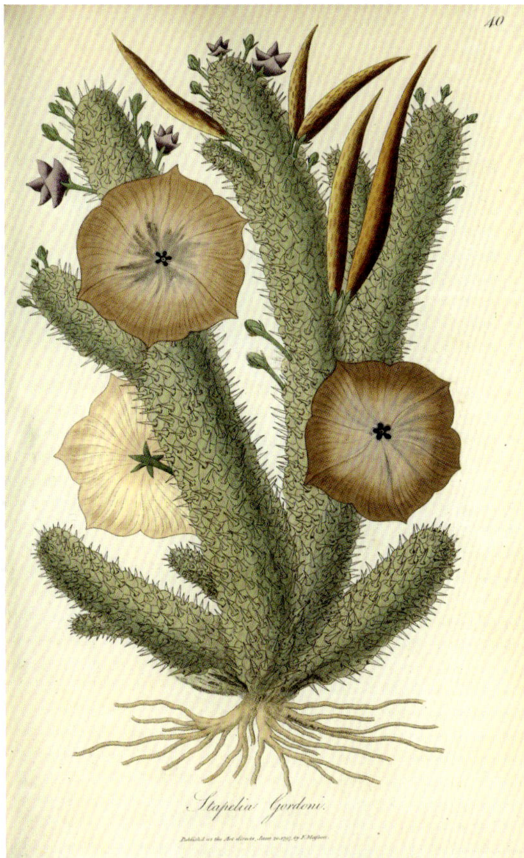

29.30

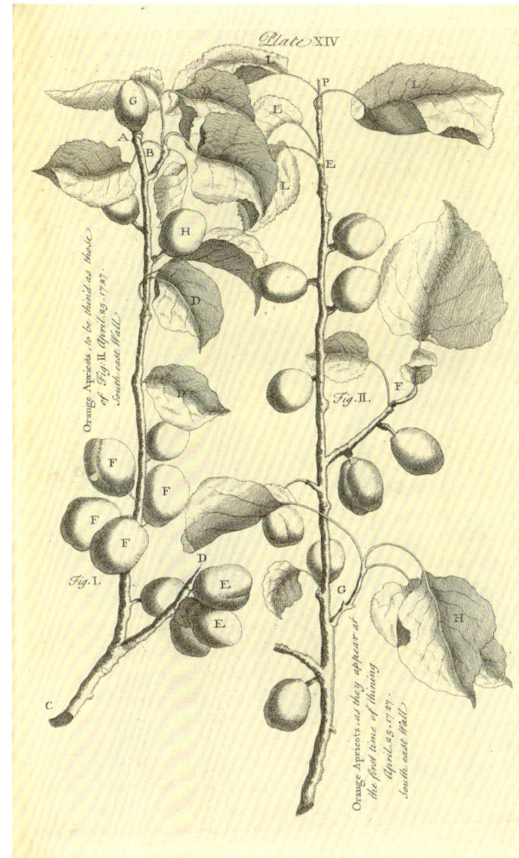

29.31

29.30

FRANCIS MASSON (1741–1805)
Stapeliae novae, or, A Collection of Several New Species of that Genus, Discovered in the Interior Parts of Africa
1796
15 in. (38 cm)
George Nicol, London
* The New York Botanical Garden, The LuEsther T. Mertz Library, f QL 278 .S8 M37
Exhibited: Daniel Mackenzie (active 1770–1795) engraver, after Francis Masson, '*Stapelia grandiflora*', c. 1796–97, hand-coloured etching and stipple engraving, plate 11 (illustrated p. 480, fig. 29.10)
† Royal Botanic Gardens, Kew, Library, Art & Archives Collections, q107 STA
Exhibited: Daniel Mackenzie (active 1770–1795), engraver, after Francis Masson, '*Stapelia grandiflora*', 1797, hand-coloured etching and engraving, plate 11 (not illustrated); and Daniel Mackenzie (active 1770–1795), engraver, after Robert Jacob Gordon (1743–1795), 'Stapelia gordoni', 1796[–97], hand-coloured etching and stipple engraving, plate 40 (illustrated)

In 1771 or 1772, William Aiton, the head gardener at Kew, chose Francis Masson to travel to South Africa as the garden's first official plant collector. Masson arrived in 1772 and made three collecting trips inland before leaving for the Canaries, Azores, and West Indies in 1776. He returned to England, then departed again for the Cape in 1786. From his travels in South Africa he introduced over one thousand new plant species to Britain. These included forty-two species of the genus *Stapelia*, which Masson described and illustrated in *Stapeliae novae*, a work owned by Queen Charlotte. In his preface, Masson credited Colonel Robert Jacob Gordon, commander of the Dutch garrison at the Cape, with the discovery of several other species of this genus, including *Stapelia gordoni*, which he named for the colonel. The *Stapelia grandiflora* is also known as the 'carrion plant', for the smell with which it attracts the flies that pollinate its flowers.

ROYAL GARDENS AT RICHMOND AND KEW

29.31*

BATTY LANGLEY (1696–1751)
Pomona: or, The Fruit-Garden Illustrated. Containing Sure Methods for Improving All the Best Kinds of Fruits Now Extant in England
1729
15⅜ in. (39 cm)
Published by G. Strahan, London
Beinecke Rare Book & Manuscript Library, Yale University, 1987 +121
Exhibited: Thomas Bowles (1712?–1762), engraver, after Batty Langley, [Method of Thinning Apricots], engraving, plate 14

Batty Langley's work on fruit trees, published shortly after his *New Principles of Gardening* (1728), was dedicated to Queen Caroline, and the dedication reprised themes of fruitfulness and nourishment of both family and nation often associated with queens consort. Langley recommended that fruit trees be included among 'shady Walks, Avenues, Groves, Wildernesses, Grotto's, Fountains, Canals, &c.' in the new, 'natural' style of gardening advocated by him and implemented for Caroline at Richmond. *Pomona* contains practical discussions of grafting, thinning fruit (as in this plate, showing apricots), and pruning branches; it nonetheless makes frequent reference to Stephen Hales's *Vegetable Staticks* (1727), a more theoretical work on the movement of fluids in plants. Hales became Princess Augusta's clergyman and also encouraged her botanical studies.

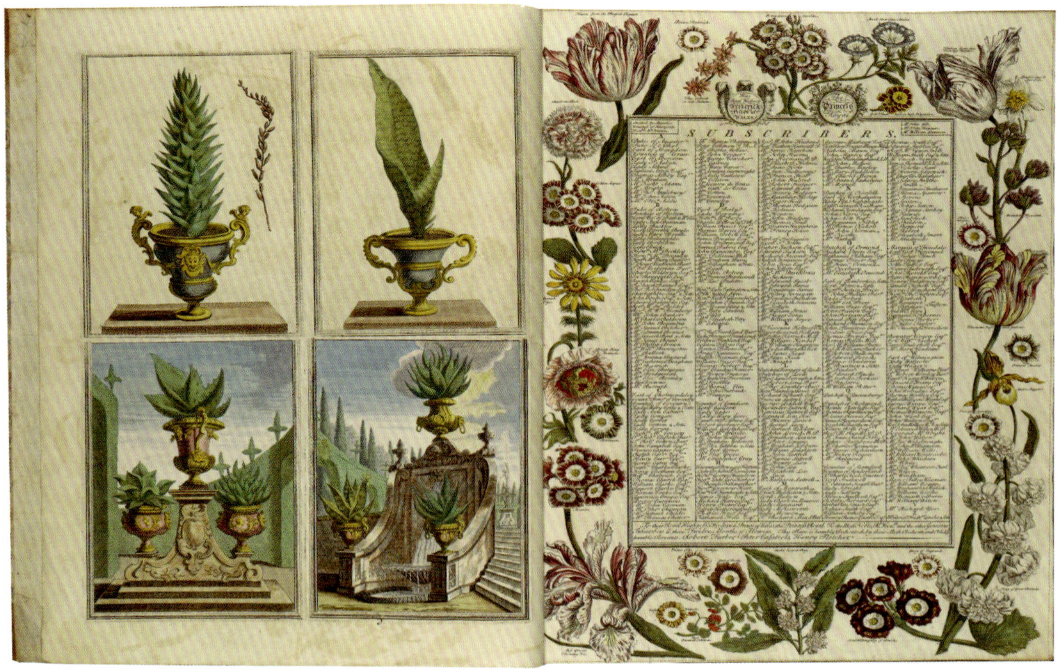

29.32

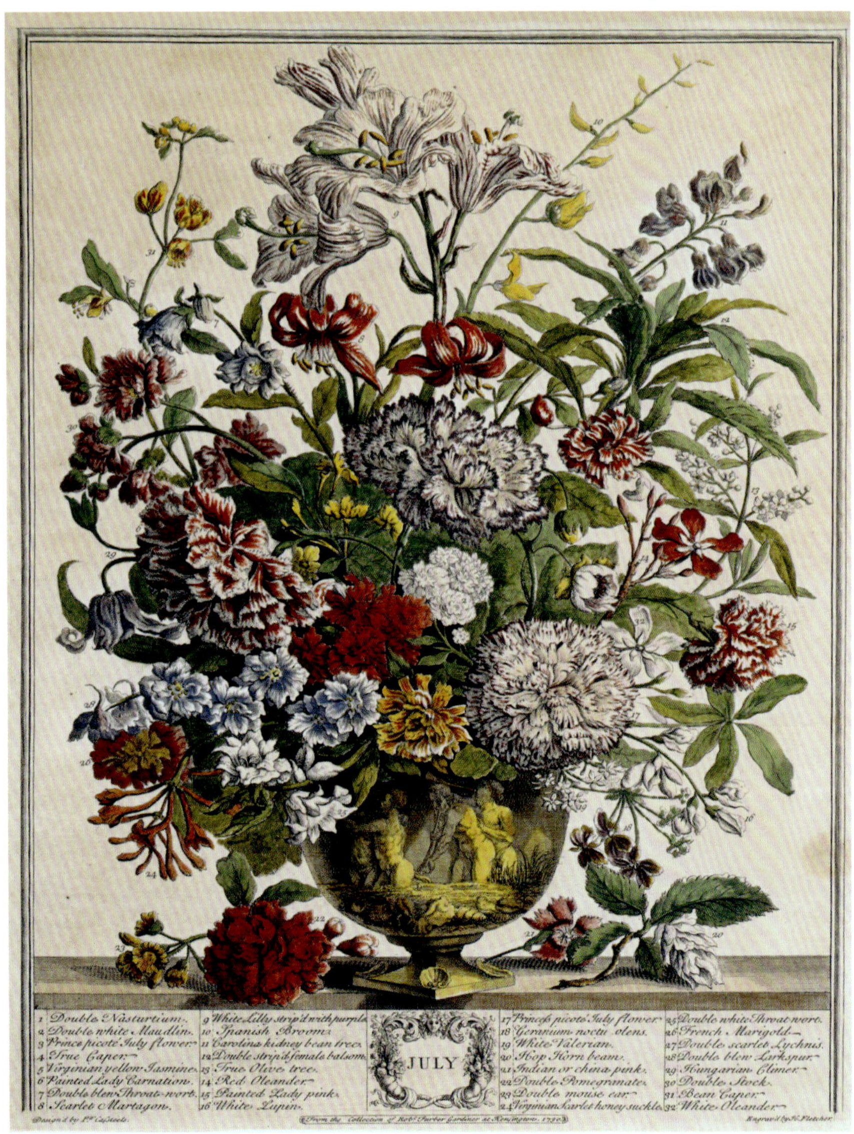

29.33

ROBERT FURBER (c. 1674–1756)

29.32

HENRY FLETCHER (active 1715–1744), engraver, after **PIETER CASTEELS** (1684–1749)
Twelve Months of Flowers
1730
Hand-coloured etching on paper
23⅝ in x 18⅞ in. (60 x 48 cm)
Published by Robert Furber, London
* Oak Spring Garden Library
Exhibited: left, leaf with four engravings of succulents in pots; right, Henry Fletcher, engraver, after Pieter Casteels, *List of Subscribers*, 1732 (illustrated)
†Natural History Museum Library and Archives, London, 44NHM ALMA
Exhibited: Henry Fletcher, engraver, after Pieter Casteels, *List of Subscribers*, 1732 (not illustrated)

29.33*

HENRY FLETCHER (active 1715–1744), engraver, after **PIETER CASTEELS** (1684–1749)
Twelve Months of Flowers: July
1730
Hand-coloured etching on paper
16⅛ x 12⅛ in. (41 x 30.8 cm)
Published by Robert Furber, London
The Chipstone Foundation, 1959.10.08

29.34*

HENRY FLETCHER (active 1715–1744), engraver, after **PIETER CASTEELS** (1684–1749)
Twelve Months of Fruits: June
1732
Hand-coloured etching and engraving on paper
21⅝ x 17¼ in. (54.9 x 43.8 cm)
Published by Robert Furber, London
Oak Spring Garden Library

Robert Furber, a member of the Society of Gardeners, established a plant nursery in Kensington around 1700. His business was integral to the networks of commerce and exploration that linked the royal gardens with the wider world. He was active in the acclimatization and distribution of North American plant species, including tulip trees and magnolias, and received shipments of plants and seeds from Mark Catesby. Furber produced multiple engraved catalogues, which functioned as decorative prints as well as advertising the richness and variety of his stock.

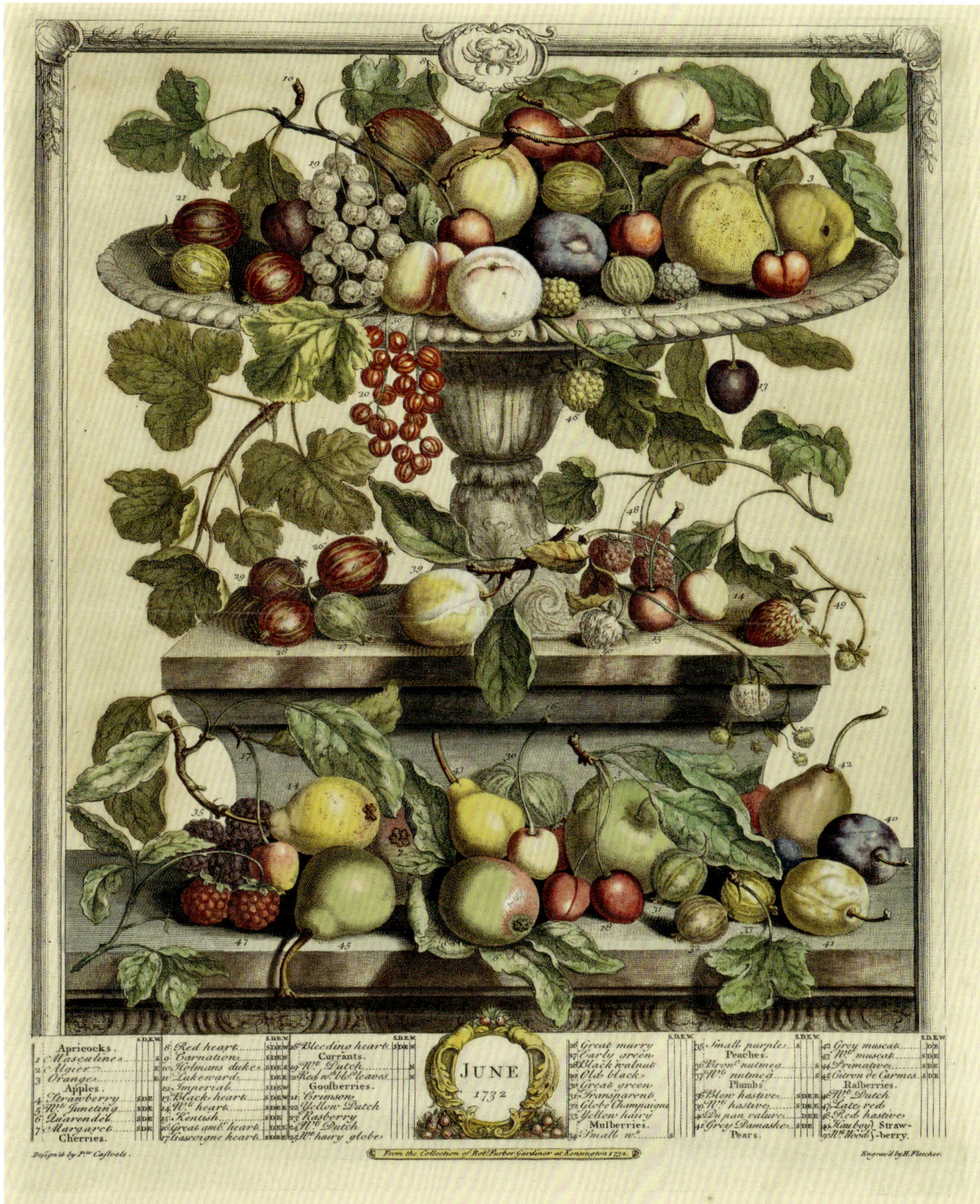

29.34

Twelve Months of Flowers, a set of a dozen leaves published in 1730, acted as a visual catalogue for Furber's business. The plates were engraved by Henry Fletcher (known for his natural history illustration) after the Dutch artist Pieter Casteels, who established himself in London in 1708 and became a renowned painter of flowers and exotic birds. Their commercial success was a testament to the popularity of floriculture, and women constituted a remarkably high proportion of the 345 individuals named in the list of subscribers. Furber dedicated the work to Queen Caroline (the border of the subscribers list features a primrose auricula named in her honour), her son Frederick, Prince of Wales, and her eldest daughter Anne, the Princess Royal (29.32).

The sheets of *Twelve Months of Flowers* were sold individually, and some examples of the series, like the copy from the Oak Spring Garden Library, were bound by their owners (in this case, along with other botanical engravings). Furber's selection of blooms for *July* includes Virginian yellow jasmine, which Mark Catesby depicted in the first volume of his *Natural History of Carolina, Florida, and the Bahama Islands* (1731), as well as the Carolina bean tree, a type of wisteria introduced to England by Catesby, and the Virginian scarlet honeysuckle, which the Philadelphia botanist John Bartram supplied to English patrons (29.33). The success of *Twelve Months of Flowers* led Furber to publish *Twelve Months of Fruits* two years later, with plates designed by Casteels and engraved by Fletcher (29.34).

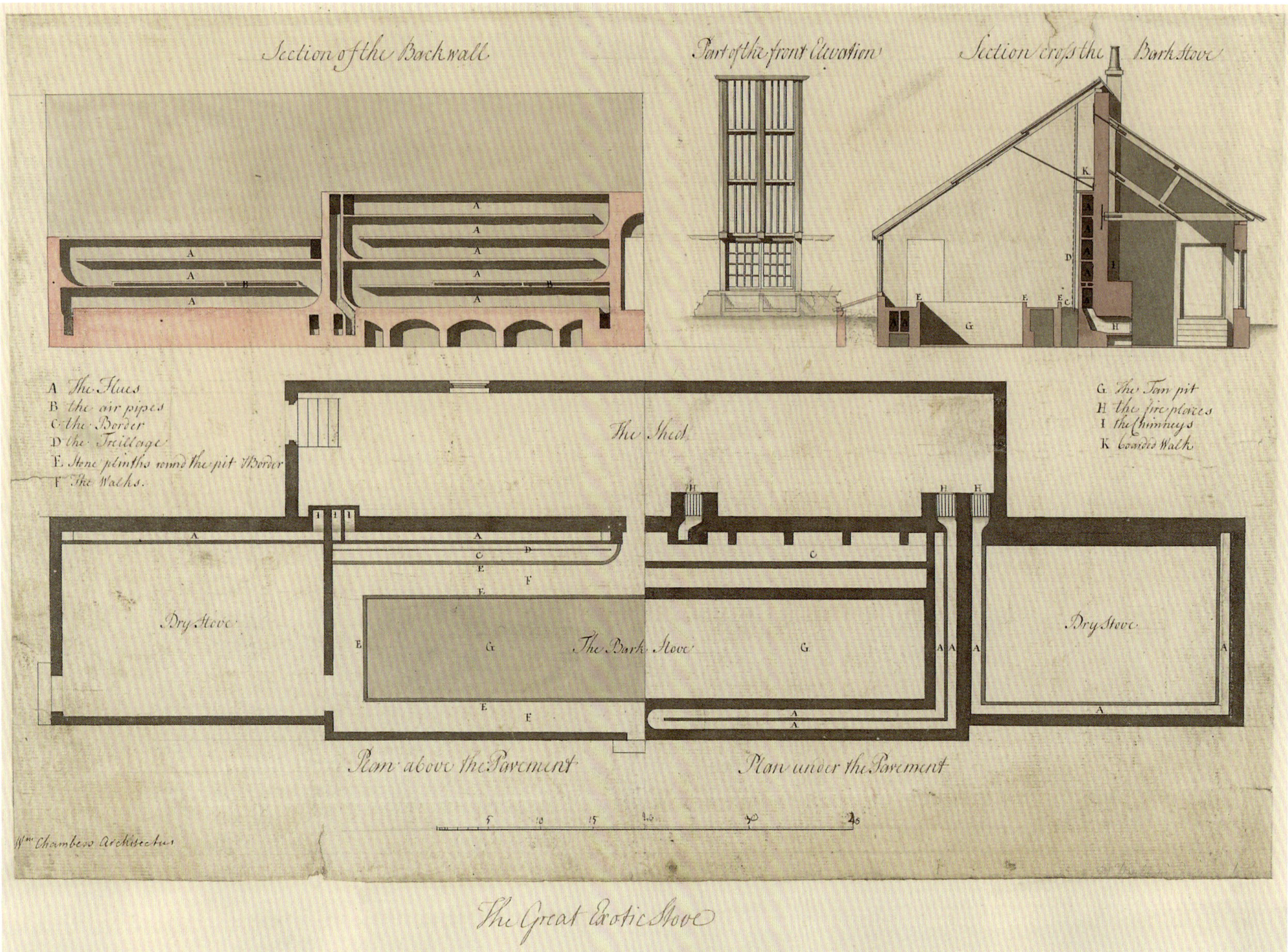
Section of the Backwall
Part of the front Elevation
Section cross the Bark Stove
A The Flues
B the air pipes
C the Border
D the Trillage
E Stone plinths round the pit & Border
F The Walks
G The Tan pit
H the fire places
I the Chimneys
K Cowdes Walk
The Shed
Dry Stove
The Bark Stove
Dry Stove
Plan above the Pavement
Plan under the Pavement
Wm Chambers Architectus
The Great Exotic Stove

29.35

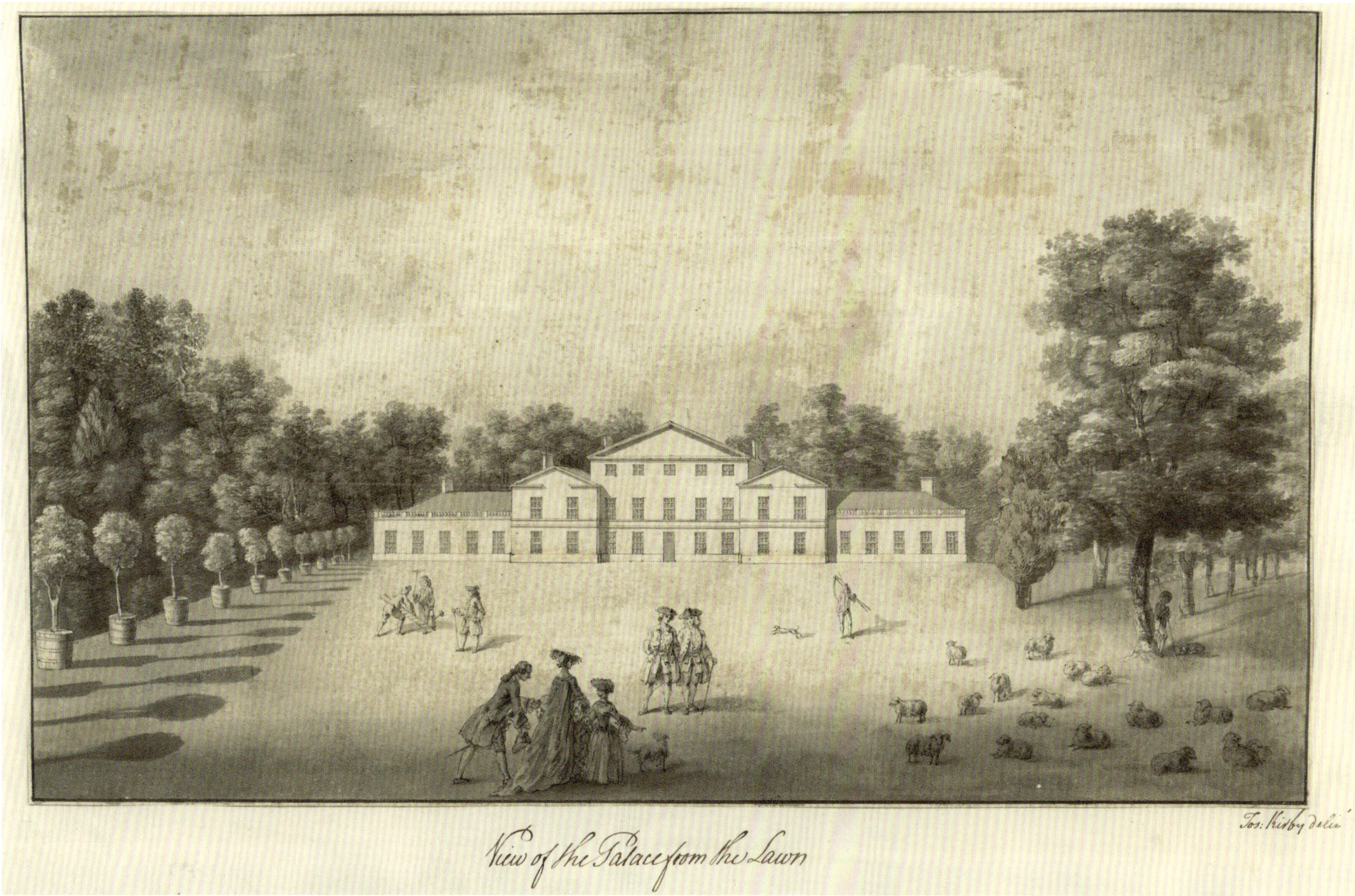

29.36

WILLIAM CHAMBERS (1722–1796) AND THE GARDENS AT KEW

29.35*

WILLIAM CHAMBERS (1723–1796)
Plans, Elevations, Sections, and Perspective Views of the Gardens and Buildings at Kew
1763
Drawings in watercolour, ink, and gouache on paper, bound
21⅛ in. (53.7 cm)
The Metropolitan Museum of Art, New York, Harris Brisbane Dick Fund, 1925, 25.19(1-44)
Exhibited: William Chambers, 'The Great Exotic Stove at Kew', 1763, ink, wash, and watercolour on paper, 25.19.7

29.36*

WILLIAM CHAMBERS (1723–1796)
Plans, Elevations, Sections, and Perspective Views of the Gardens and Buildings at Kew
1763
Drawings in watercolour, ink, and gouache on paper, bound
21⅛ in. (53.7 cm)
The Metropolitan Museum of Art, New York, Harris Brisbane Dick Fund, 1925, 25.19(1-44)
Exhibited: Joshua Kirby (1716–1774), 'View of the Palace from the lawn at Kew', 1763, ink and wash on paper, 25.19.36

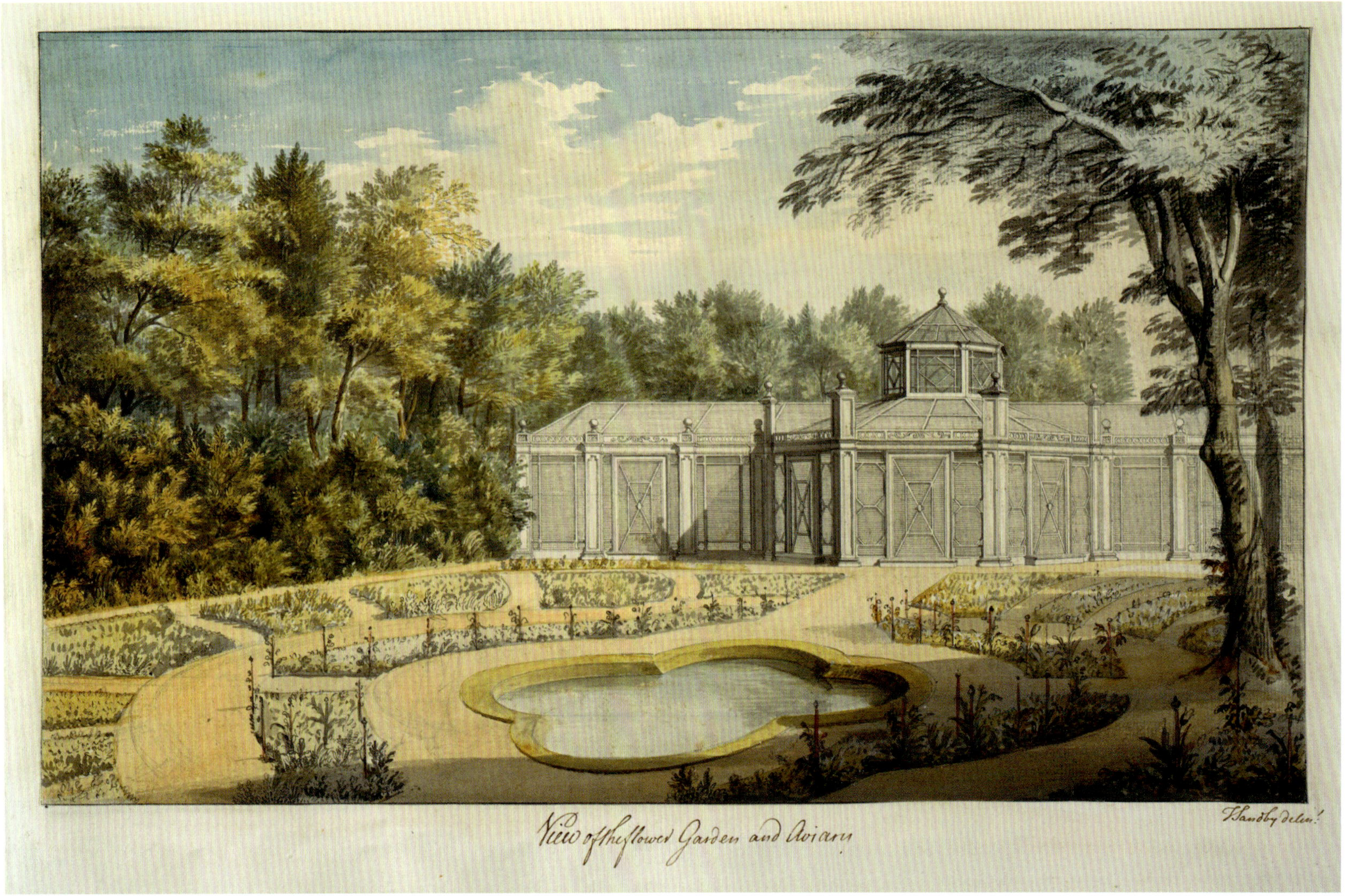

29.37

29.37*

WILLIAM CHAMBERS (1723–1796)
Plans, Elevations, Sections, and Perspective Views of the Gardens and Buildings at Kew
1763
Drawings in watercolour, ink, and gouache on paper, bound
21⅛ in. (53.7 cm)
The Metropolitan Museum of Art, New York, Harris Brisbane Dick Fund, 1925, 25.19(1-44)
Exhibited: Thomas Sandby (1723–1798), 'View of Flower Garden and Aviary at Kew', 1763, watercolour on paper, 25.19.39

29.38*

WILLIAM CHAMBERS (1723–1796)
Plans, Elevations, Sections, and Perspective Views of the Gardens and Buildings at Kew
1763
Drawings in watercolour, ink, and gouache on paper, bound
21⅛ in. (53.7 cm)
The Metropolitan Museum of Art, New York, Harris Brisbane Dick Fund, 1925, 25.19(1-44)
Exhibited: Thomas Sandby (1723–1798), 'A View of the Menagerie, and its Pavillion, at Kew', 1763, watercolour on paper, 25.19.40

Augusta's husband, Frederick, Prince of Wales, arrived in London from Germany in December 1728, less than a year after his mother, Queen Caroline, had taken the lease on the Dutch House at Kew. When Frederick's parents had themselves moved to London from Germany on the accession of George I, Frederick had been left behind as a representative of the family in Hanover. Now aged twenty-one, he knew little of his sisters and possibly he wished for a family rapprochement. He soon took a long lease on the Capel House at Kew and in 1731 also purchased its contents from its owner, Elizabeth St André. From 1732, Frederick set about remodelling the house with the assistance of William Kent. Kent would design a new great staircase, chimneypieces, furniture and painted ceilings. Isaac Mansfield decorated the gallery with plasterwork of flowers, foliage, and feathers. The house became known as the White House because of its new white render. The architect and draughtsman Joshua Kirby, who produced the drawing of Kew Palace for Chambers illustrated here (29.36), acted as tutor in perspective drawing to Frederick and Augusta's son George, Prince of Wales, in the late 1750s. After succeeding to the throne in 1760, George III appointed

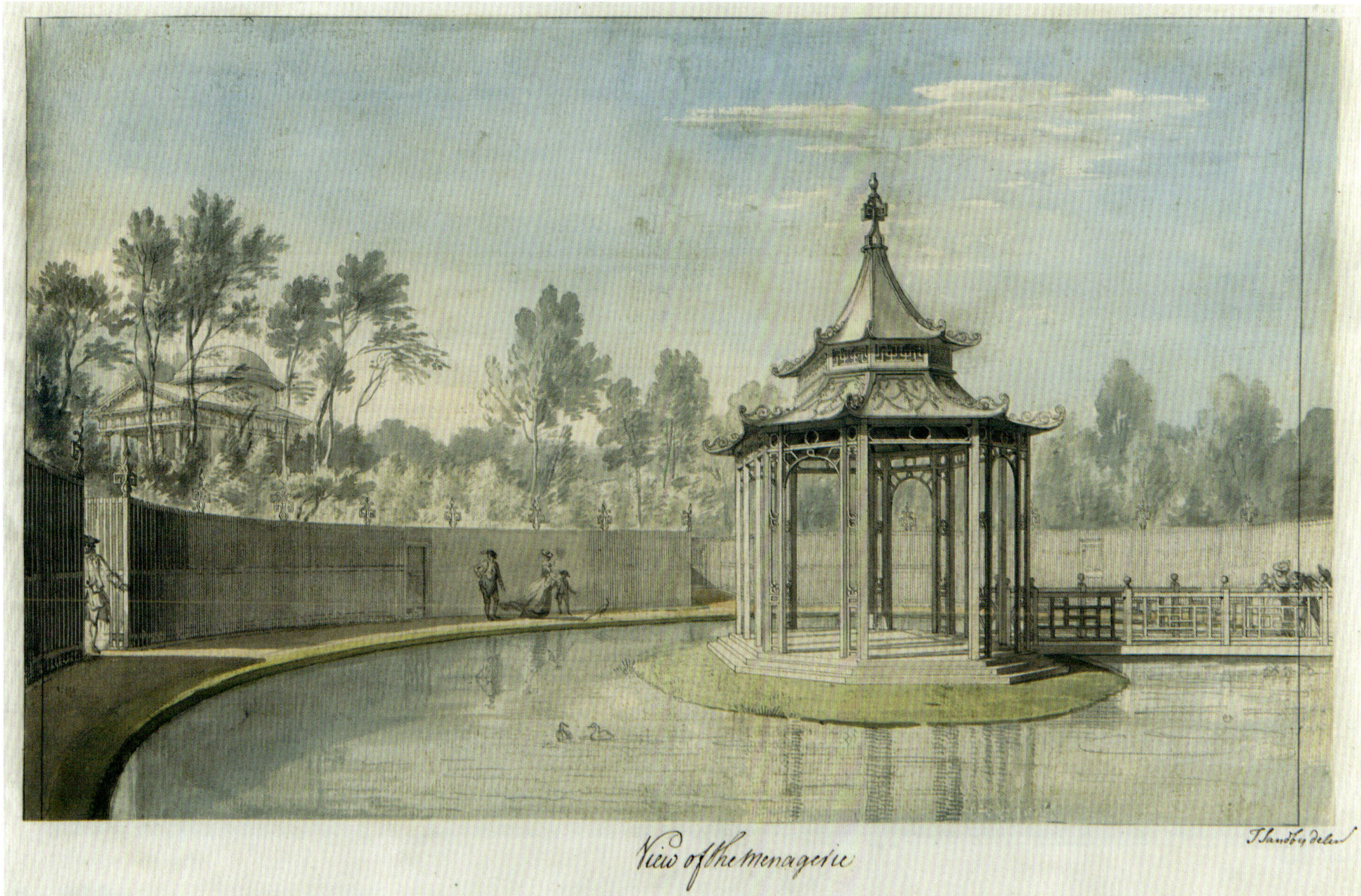

29.38

Kirby and his son as joint clerks of the works at Richmond and Kew.

In 1757, Princess Augusta appointed William Chambers her architect at Kew, and architectural tutor to her son George, Prince of Wales. Chambers took the prince through an exceptionally thorough training in architectural drawing and design, and, at Kew, provided buildings in a dazzling array of styles for Augusta. Chambers had spent several years in China in the 1740s, in the service of the Swedish East India Company. These travels provided inspiration for the Chinese-style buildings he executed for Princess Augusta at Kew, such as the menagerie pavilion (29.38) and the Aviary (29.37; see also cat. 19.16). The menagerie housed large birds such as pheasants; its pool was stocked with goldfish. Many other drawings for Chambers's richly illustrated *Plans, Elevations, Sections, and Perspective Views of the Gardens and Buildings at Kew* were made by the architect and draughtsman Thomas Sandby, a founding member of the Royal Academy who had long been in the service of Augusta's brother-in-law, the Duke of Cumberland. The book, published in 1763, was dedicated to Princess Augusta. Chambers also presented the preparatory drawings for his publication, bound into the album illustrated here, to John Stuart, third Earl of Bute, in recognition of his contributions to Augusta's vision for Kew.

The unpublished plan included in Lord Bute's album indicates how Chambers's buildings worked into the greater landscape (cat. 18.04). The estate was relatively long and thin, with the White House, remodelled for Frederick, Prince of Wales, and which Augusta continued to use after his death, at one end. Behind this was a lawn looking over a lake, with an island so constructed to create the sense that the lake was a meandering watercourse. Beyond were two lawns, divided by stands of mature trees, salvaged from an earlier garden layout. Around the edges of the garden was a belt of trees cut through with meandering paths. It was in this part of the garden that the aviary, menagerie, greenhouses and forcing houses, and the 'Great Stove' for the plant collections were located, as well as the many folly temples.

The 'stove' or hothouse designed by Chambers (29.35), possibly with involvement from George, Prince of Wales, was supplied through the Earl of Bute's extensive collections of plants from Asia, Africa, the Americas, and Europe. As networks of botanical exchange expanded across the globe, such stoves housed an ever-increasing variety of species, while knowledge about acclimatization of more delicate exotics also advanced.

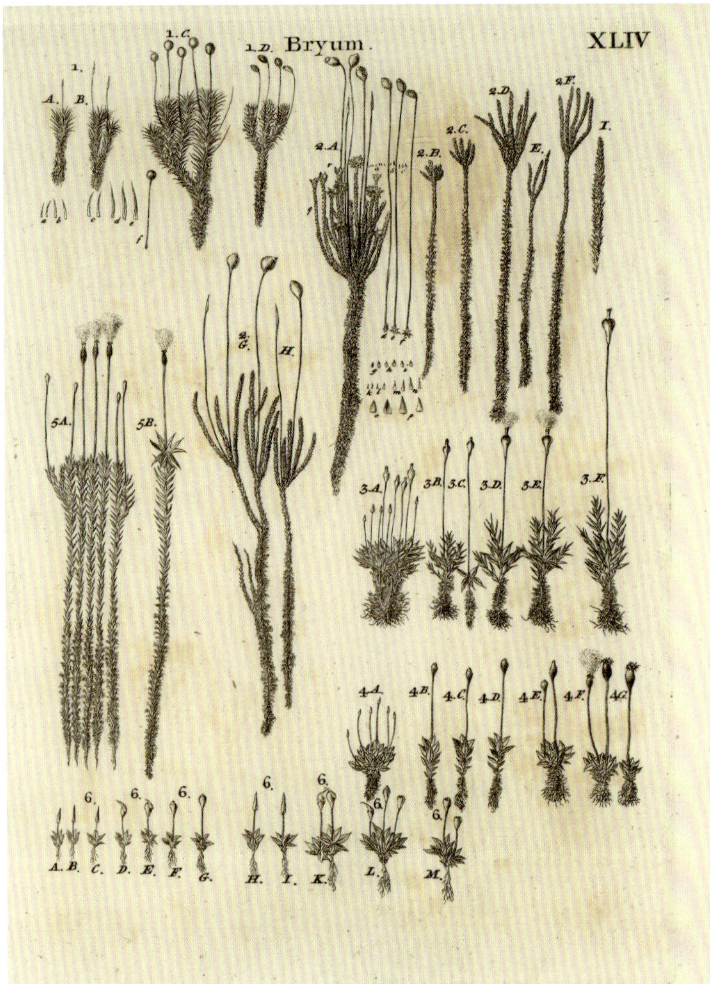

29.39

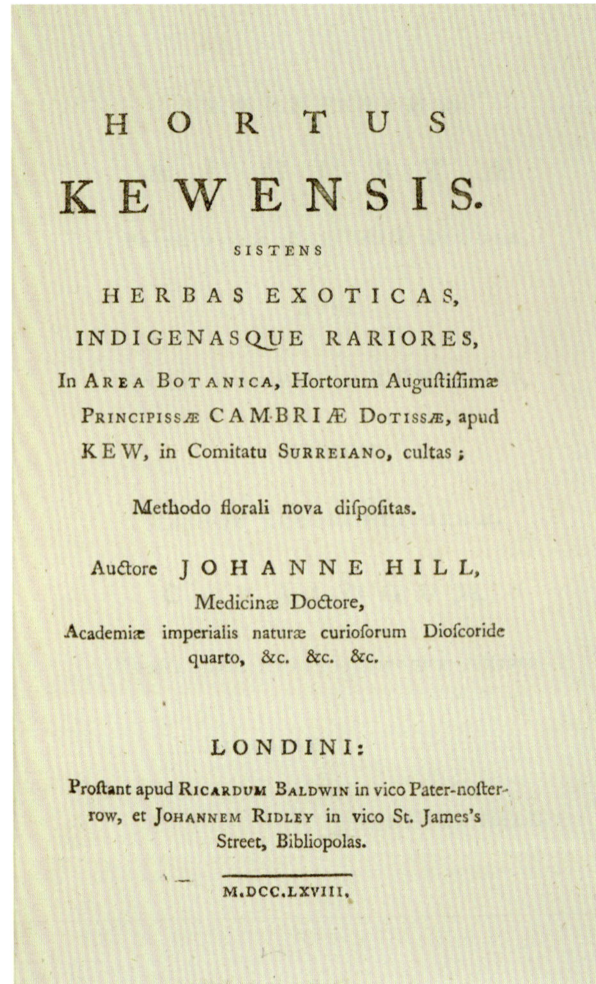

HORTUS
KEWENSIS.
SISTENS
HERBAS EXOTICAS,
INDIGENASQUE RARIORES,
In AREA BOTANICA, Hortorum Auguſtiſſimæ
PRINCIPISSÆ CAMBRIÆ DOTISSÆ, apud
KEW, in Comitatu SURREIANO, cultas;
Methodo florali nova diſpoſitas.
Auctore JOHANNE HILL,
Medicinæ Doctore,
Academiæ imperialis naturæ curioſorum Dioſcoride
quarto, &c. &c. &c.
LONDINI:
Proſtant apud RICARDUM BALDWIN in vico Pater-noſter-
row, et JOHANNEM RIDLEY in vico St. James's
Street, Bibliopolas.
M.DCC.LXVIII.

29.40

BOTANICAL PUBLICATIONS AND THE GARDENS AT KEW

29.39*

JOHANN JAKOB DILLENIUS (1684–1747)
Historia muscorum: a general history of land and water, &c., mosses and corals, containing all the known species
1768
11½ in. (29 cm)
J. Millan, London
Royal Botanic Gardens, Kew, Library, Art & Archives Collections
Exhibited: Johann Jakob Dillenius, artist and engraver, '*Bryum*', c. 1741, engraving, plate 44 (illustrated)

The German-born botanist Johann Jakob Dillenius was appointed the first professor of botany at the University of Oxford in 1734. Mosses and corals were sources of mystery for naturalists in the eighteenth century; the reproductive mechanisms of the former were unknown, and the latter's basic identity as animal or plant was under question. This copy of *Historia muscorum*, considered Dillenius's most significant work, belonged to Queen Charlotte and was later in the library of the nineteenth-century botanist William Orde Newnham. It is annotated 'W. O. Newnham, Sanct. Johann. Coll. E Libris Reginae Charlotte'.

29.40

JOHN HILL (c. 1714–1775)
Hortus Kewensis, Sistens Herbas Exoticas, Indigenasque Rariores, in Area Botanica Hortorum Augustissimae Principissae Cambriae Dotissae apud Kew in Comitatu Surreiano Cultas
1768
8⅞ in. (22.5 cm)
Richard Baldwin, London
Royal Collection Trust, RCIN 1057340
Exhibited: Title page

The botanist, apothecary, gossip columnist, and occasional actor John Hill remains an enigmatic yet important figure in the history of the development of the gardens at Kew during Augusta's time. The Princess appointed him to assist his patron, John Stuart, third Earl of Bute, in the management of her botanic gardens, but his precise role is undocumented. Hill's *Hortus Kewensis* (1768) was the first catalogue of the plants being grown in the botanic section of Kew. In all, it recorded 3,400 species, demonstrating the richness and global reach of the garden near the end of Augusta's lifetime—this and other works by Hill published in the late 1760s and early 1770s listed plants from North America, the West Indies, China, South Africa, Italy, and Portugal.

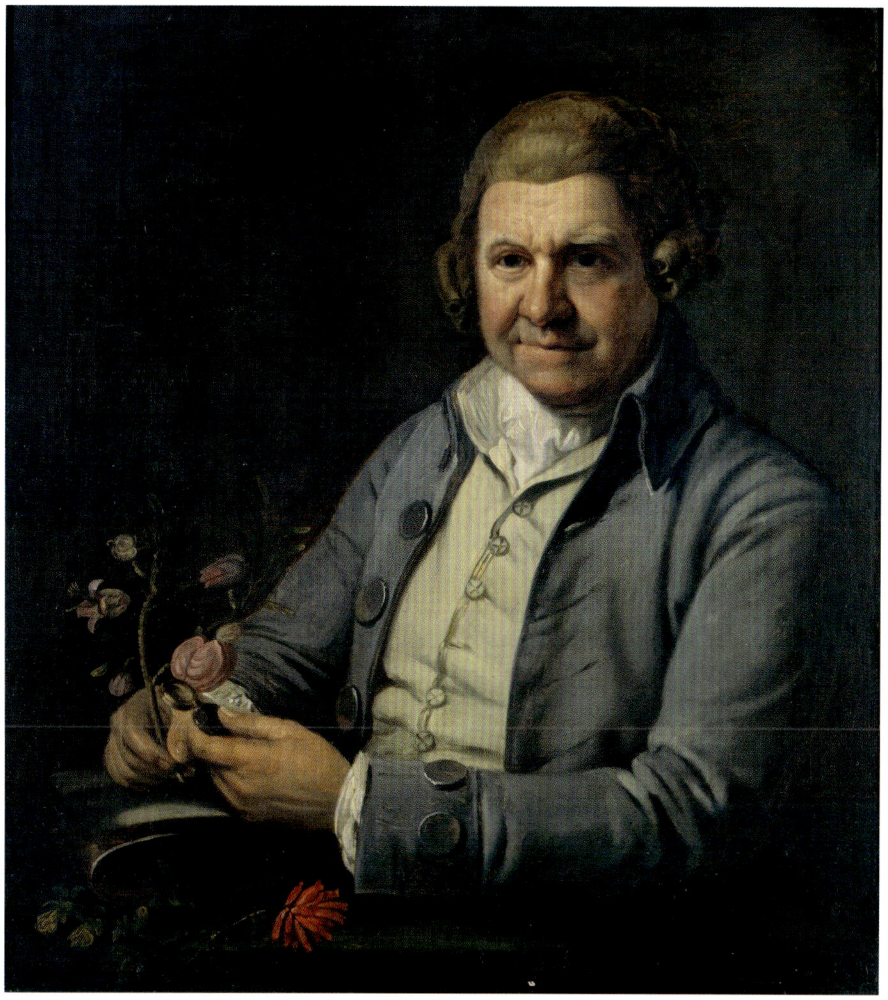

29.41

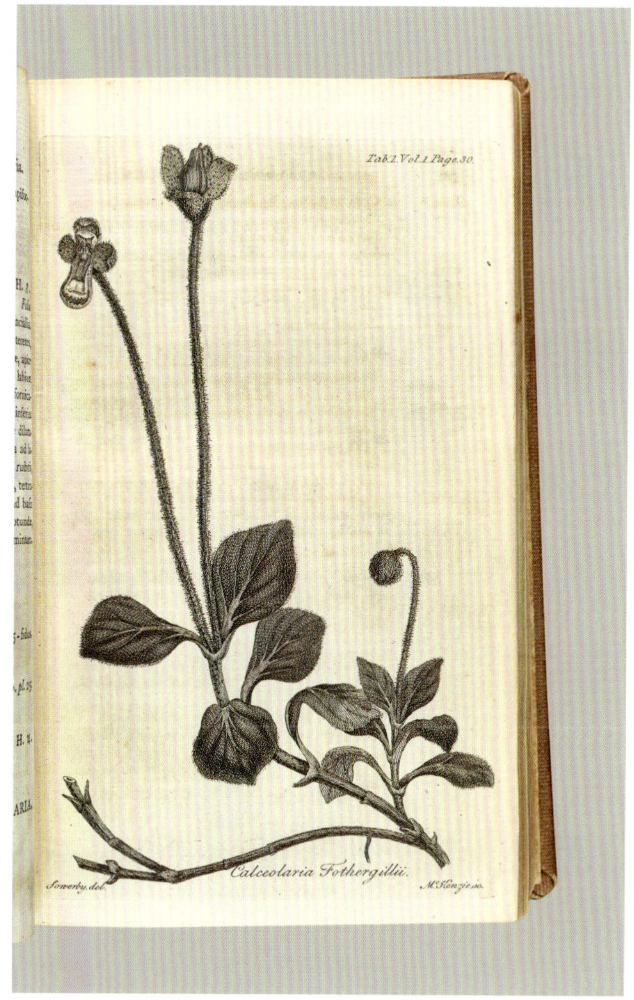

29.42

WILLIAM AITON (1731–1793)

29.41

EDMUND BRISTOW (1787–1876)
William Aiton
Undated
Oil on canvas
15⅜ x 13⅜ in. (39 x 34 cm)
Royal Botanic Gardens, Kew, Library, Art & Archives Collections, HO 005

William Aiton was one of the most important British horticulturalists of the eighteenth century, and the first gardener to oversee the entire royal estate at Kew. He began his career as assistant to Philip Miller at the Chelsea Physic Garden and in 1759 was initially hired at Kew to manage Princess Augusta's garden of exotic plants. Until his death in 1793, Aiton worked closely with Sir Joseph Banks, the botanist, explorer, and close friend of George III and Queen Charlotte, to expand the reach of Kew's collections and to systematize the acquisition of new plant species. This portrait shows Aiton holding what appears to be a sprig of Chinese lantern, a South African species named *Aitonia capensis* in his honour (it was later renamed). Edmund Bristow, a painter of portraits, still lifes, and genre scenes, lived his entire life in the vicinity of Windsor and Eton. He had no formal training, but early in his career enjoyed the patronage of Princess Elizabeth.

29.42*

WILLIAM AITON (1731–1793)
Hortus Kewensis; or a Catalogue of Plants Cultivated in the Royal Botanic Garden at Kew (3 vols)
1789
9⅛ in. (23 cm)
Published by George Nicol, London
The New York Botanical Garden, The LEsther T. Mertz Library, QK 73 .K4 A35
Exhibited: Unknown engraver, after James Sowerby (1757–1822), '*Calceolaria Fothergillii*', c. 1789, etching and line engraving, vol. 1, plate 1 (illustrated)

In 1773 William Aiton, along with Sir Joseph Banks's curator-librarians Daniel Solander and Jonas Dryander (both pupils of Linnaeus), began compiling a new *Hortus Kewensis*. It described nearly 5,600 species from specimens now in the Natural History Museum, London. By the time of its publication, Francis Masson's expeditions across South Africa had added hundreds of new species, along with other notable introductions from Australia, South America, China, and Siberia. The plant shown here, *Calceolaria fothergillii*, is native to the Falkland Islands and was introduced to Britain by the Quaker physician and naturalist John Fothergill in 1777. The artist and publisher James Sowerby, who provided the preparatory drawing for this plate, illustrated many works of botany and natural history and became an associate of Sir Joseph Banks through his membership in the Linnaean Society. Aiton's *Hortus Kewensis* was dedicated to George III, and a copy was listed in Queen Charlotte's library catalogue.

29.43

JOHN STUART, 3RD EARL OF BUTE
(1713–1792)
Botanical Tables: containing the different familys of British plants (9 vols)
1785
11⅞ in. (30.2 cm)
Printed privately for the author
Royal Collection Trust, RCIN 1123772
*Exhibited: Johann Sebastian Müller (1715–1792), artist and engraver, 'Pink: *Dianthus caryophyllus*', c. 1785, hand-coloured engraving, vol. 6, facing p. 405 (illustrated)
†Exhibited: Frontispiece, vol. 1 (not illustrated)

29.44

UNKNOWN MAKER
Cabinet containing Lord Bute's *Botanical Tables*
1784
Painted satinwood and gilt bronze
13¾ x 17 x 8½ in. (35 x 43.3 x 21.5 cm)
Royal Collection Trust, RCIN 37017

As the royal gardens at Kew gathered species of plants from around the world, botanists also turned to the study of native British plants with renewed interest. The Earl of Bute's *Botanical Tables*, like John Lightfoot's *Flora Scotica*, epitomized this trend. Along with his sometime assistant John Hill, Bute was critical of the Linnaean system of classification and organized his species according to alternative principles. Only twelve copies of his botanical tables were printed, at Bute's expense. In the dedication to Queen Charlotte he wrote that the work was intended for a female audience; he presented seven of the copies to women prominent in the field, including Charlotte; the Duchess of Portland; and Catherine II, Empress of Russia. The botanical plates were drawn and engraved by Johann Sebastian Müller, a native of Nuremberg, Germany, who settled in London in 1744. Müller became distinguished as an engraver of portraits and historical prints, but his principal interest lay in botanical illustration. The *Dianthus caryophyllus*, or 'pink', shown here, is a familiar species of carnation.

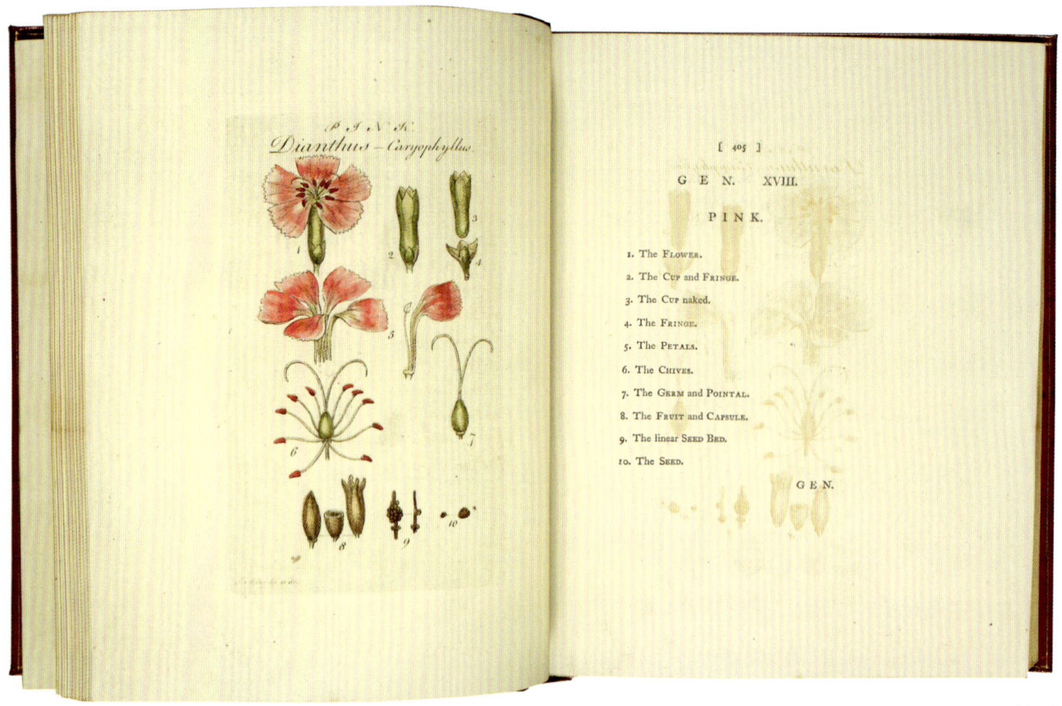

29.43

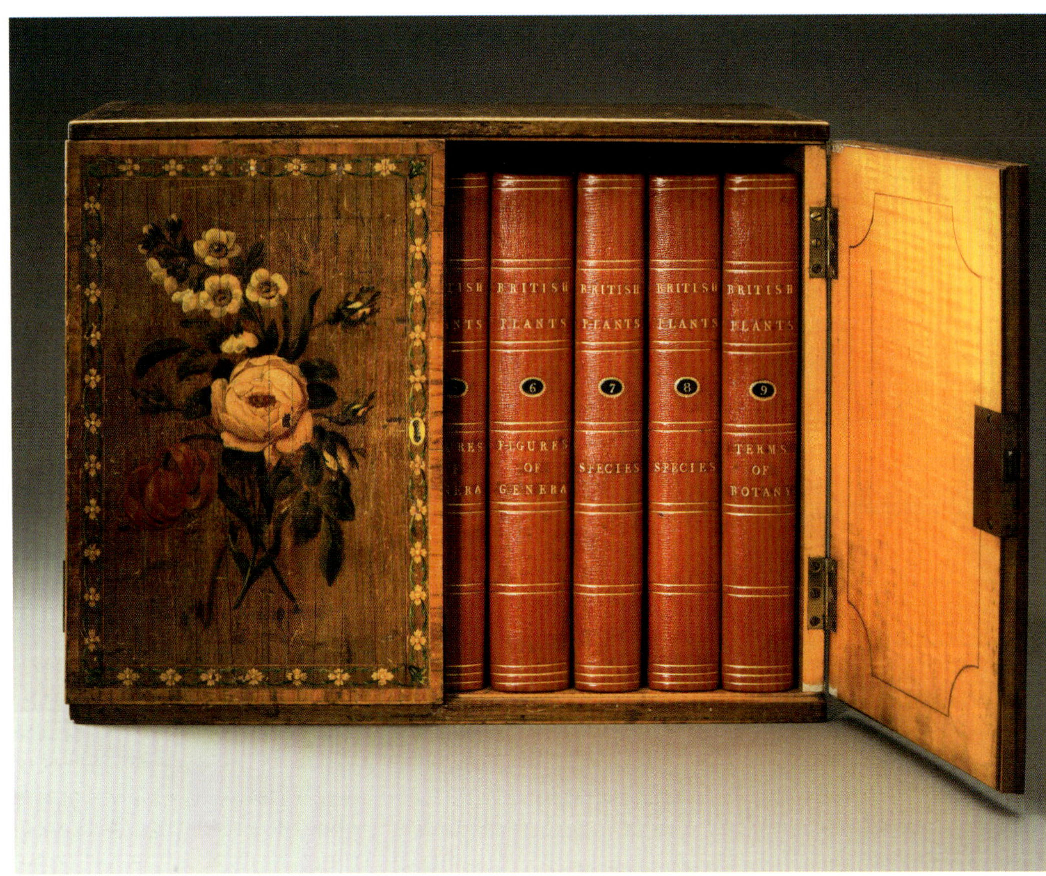

29.44

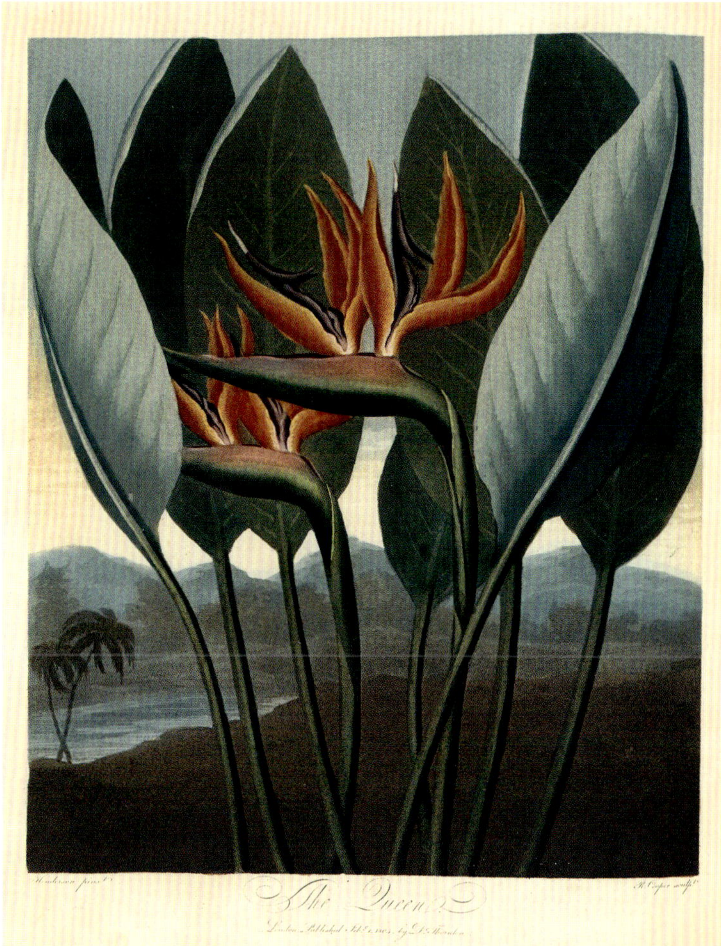

29.45

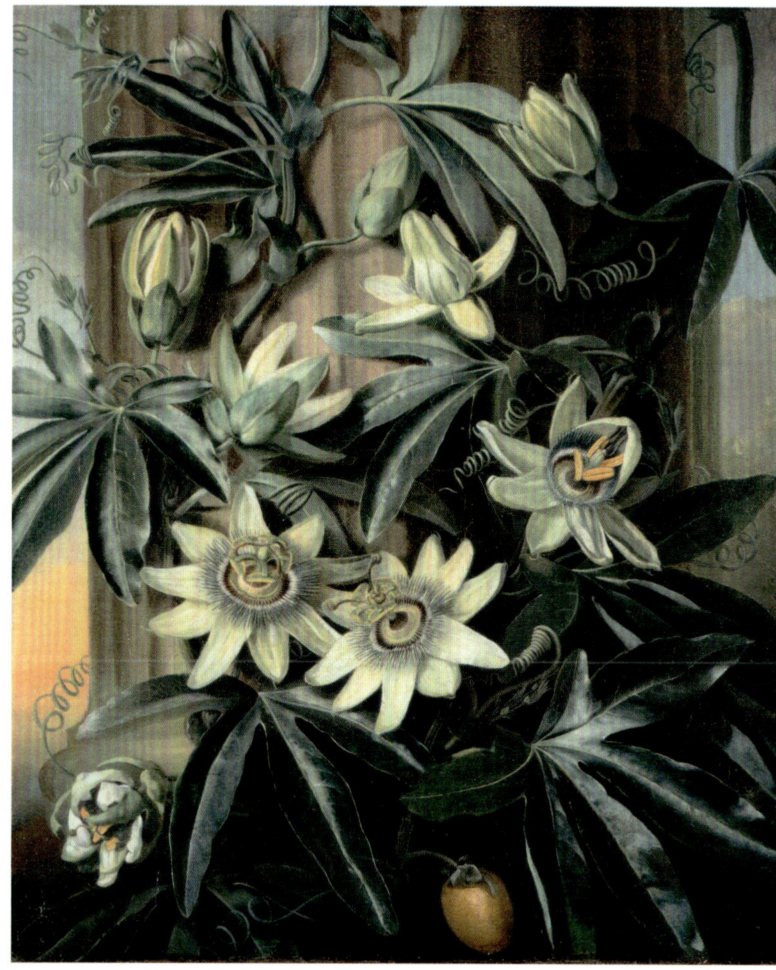

29.46

ROBERT JOHN THORNTON (1768–1837) AND THE 'TEMPLE OF FLORA'

29.45[#]

ROBERT JOHN THORNTON (1768–1837)
New Illustration of the Sexual System of Carolus von Linnaeus . . .
1807
23¼ in. (59 cm)
Printed for the publisher [Thornton] by T. Bensley, London
Beinecke Rare Book & Manuscript Library, Yale University, SSi7x 807T
Exhibited: R[obert?] Cooper (active 1800–1836?), engraver, after Peter Charles Henderson (active 1799, d. 1829), 'The Queen [*Strelitzia regina*]', 1804, hand-coloured stipple and line engraving, in Robert Thornton, *The Temple of Flora*, plate 11

29.46[*]

PHILIP REINAGLE (1748–1833)
'Blue Passion Flower' for *The 'Temple of Flora'* by *Robert Thornton*
1800
Oil on canvas
17⅞ x 14¾ in. (45.4 x 37.5 cm)
Yale Center for British Art, Paul Mellon Collection, B1981.25.516

The physician and botanist Robert John Thornton was appointed lecturer in medical botany at Guy's and St Thomas's hospitals in London in 1796. In the following year he began to advertise his *New Illustration of the Sexual System of Carolus von Linnaeus . . .*, which he issued in parts starting in 1799. The book's main sections include an introduction to the reproductive cycles of plants, an explanation of the Linnaean system of classification, and a collection of spectacular coloured prints known collectively as *The Temple of Flora*. Thornton's ambitious vision for this work reveals how botany at the turn of the nineteenth century could combine systematic science, aesthetic feeling, and poetic imagination and expression.

Strelitzia regina (see cats. 20.08–20.09), native to South Africa, was introduced to England by Francis Masson, Kew's first official plant collector, who sent hundreds of specimens back to Kew during his expeditions there in the 1770s and 1780s. The plant was successfully cultivated, and Sir Joseph Banks named it in honour of Queen Charlotte. The frontispiece of Thornton's publication also featured a dedicatory portrait of the Queen as 'Patroness of Botany, and of the Fine Arts' (p. 350, fig. 20.02). The plates for Thornton's *Temple of Flora* were complex and technically innovative, each one produced, after a specially commissioned painting, with a combination of aquatint, mezzotint, stipple engraving, and line engraving.

The painter Philip Reinagle served his apprenticeship in the studio of Allan Ramsay, where he was engaged in copying Ramsay's state portraits of George III and Charlotte. He created *Blue Passion Flower* and several other paintings to be engraved for the *Temple of Flora*. The blue passion flower, a native of Brazil and Peru, had been cultivated in Europe since the late seventeenth century; it took its place in Thornton's publication among familiar plants such as roses and tulips, as well as more recent arrivals, such as the bird of paradise flower (*Strelitzia regina*; cat. 29.45).

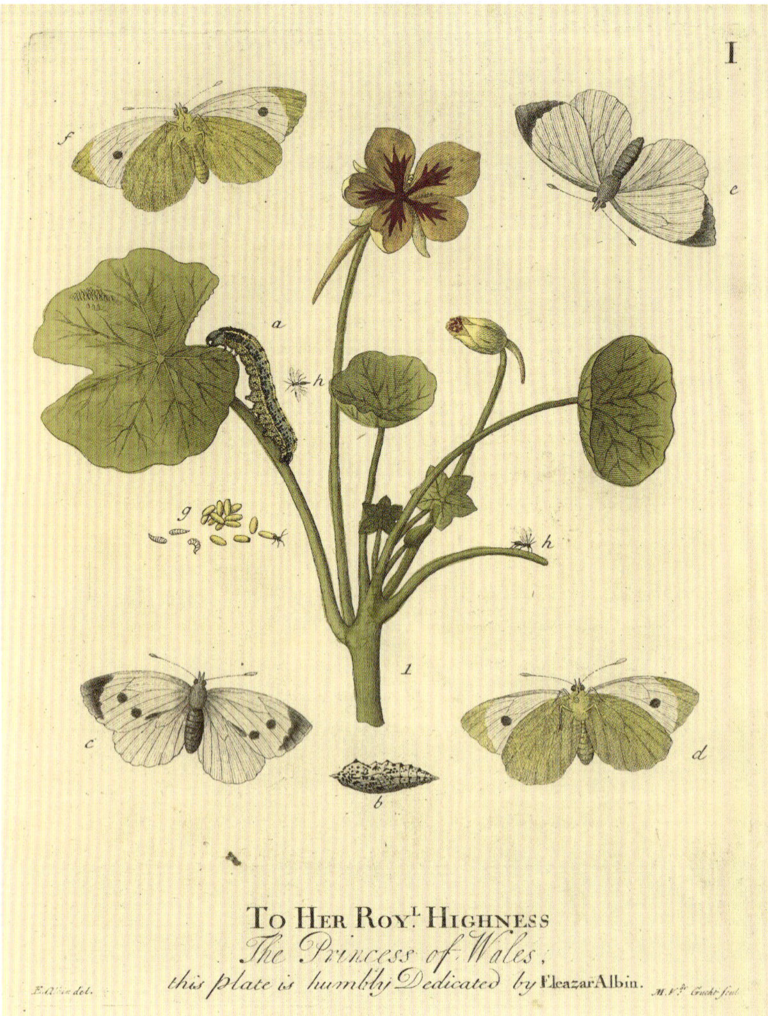

29.47

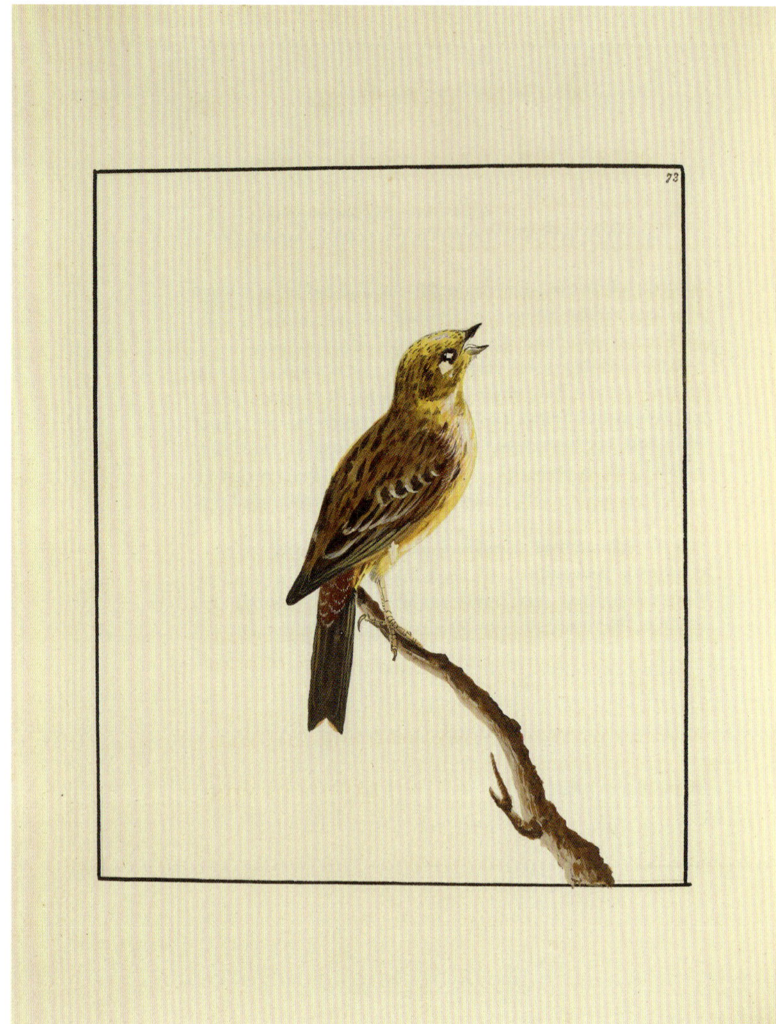

29.48

THE STUDY OF ANIMALS IN THE FIELD AND MENAGERIES

29.47*

ELEAZAR ALBIN (active 1713–1759)
Natural History of English Insects . . .
1720
11⅜ in. (29 cm)
Printed for the author, and sold by William and John Innys at the West End of St Pauls, London
Beinecke Rare Book & Manuscript Library, Yale University, ST11 G5 +720a
Exhibited: Unknown engraver (possibly Gerard Vandergucht, 1696–1776), after Eleazar Albin, artist, '*Papilio albus vulgaris major*: the Great White Butterfly', 1720, hand-coloured engraving, plate 1 (illustrated)

The naturalist and watercolour painter Eleazar Albin undertook his *Natural History of English Insects* at the behest of Mary, Duchess Dowager of Beaufort. Albin dedicated the entire publication, as well as the first plate, to Caroline, as Princess of Wales, who was among the book's subscribers. The great white butterfly, or cabbage butterfly, shown here, was considered a pest. Albin's plate documents the insect's full life cycle and shows a caterpillar feeding on one of its favourite plants, the nasturtium. In doing so, it presents a picture of the interconnectedness of species in particular environments, a mode of thinking about nature drawn from works such as those of Maria Sibylla Merian and developed further by Mark Catesby in his *Natural History of Carolina, Florida, and the Bahama Islands . . .* (1731–43, Appendix 1747).

29.48*

WILLIAM LEWIN (1747–1795)
The Birds of Great-Britain: With Their Eggs, Accurately Figured (7 vols)
1789–94
13 in. (33 cm)
For the author, London
Beinecke Rare Book & Manuscript Library, Yale University, Sy13 G5 +789L
Exhibited: William Lewin, 'Yellow Bunting', 1790, watercolour on paper, vol. 2, plate 73 (illustrated)

Princess Augusta's household accounts show that she purchased buntings and other songbirds for her aviary at Kew. The birds she kept there represented predominantly British species, such as those depicted by the artist and naturalist William Lewin. Lewin's seven-volume work enjoyed the patronage of the Duchess of Portland, a close friend of George III and Charlotte, as well as Mary Delany. The Duchess allowed Lewin to draw the eggs included in her renowned collection of natural history specimens. The first edition of *The Birds of Great-Britain* was published by subscription and limited to only sixty copies, since Lewin himself painted every illustration by hand.

29.49*

GEORGE EDWARDS (1694–1773)
A Natural History of Uncommon Birds: and of some other rare and undescribed animals, quadrupedes, reptiles, fishes, insects, &c . . . (4 vols)
[1743]–51
11⅞ in. (30 cm)
Printed for the author, at the College of Physicians, London
Beinecke Rare Book & Manuscript Library, Yale University, Sy4 +743E
Exhibited: George Edwards, artist and engraver, 'Hummingbird', 1742, hand-coloured etching, vol. 1, plate 33 (illustrated)

The ornithologist George Edwards learned to etch from the naturalist-artist Mark Catesby, and both men were part of the networks of natural-historical correspondence and exchange that included the royal physician and collector Sir Hans Sloane, Carl Linnaeus in Sweden, and the Bartram family in Philadelphia. The hummingbird shown in this plate was native to Jamaica, and Edwards produced this picture from a preserved specimen brought back to England by a 'Capt. Chandler at Stepney'. Augusta probably turned to this, among other books by Edwards, Mark Catesby, and Eleazar Albin, for information about the birds in her aviary at Kew; the Earl of Bute acquired a copy for her in 1758.

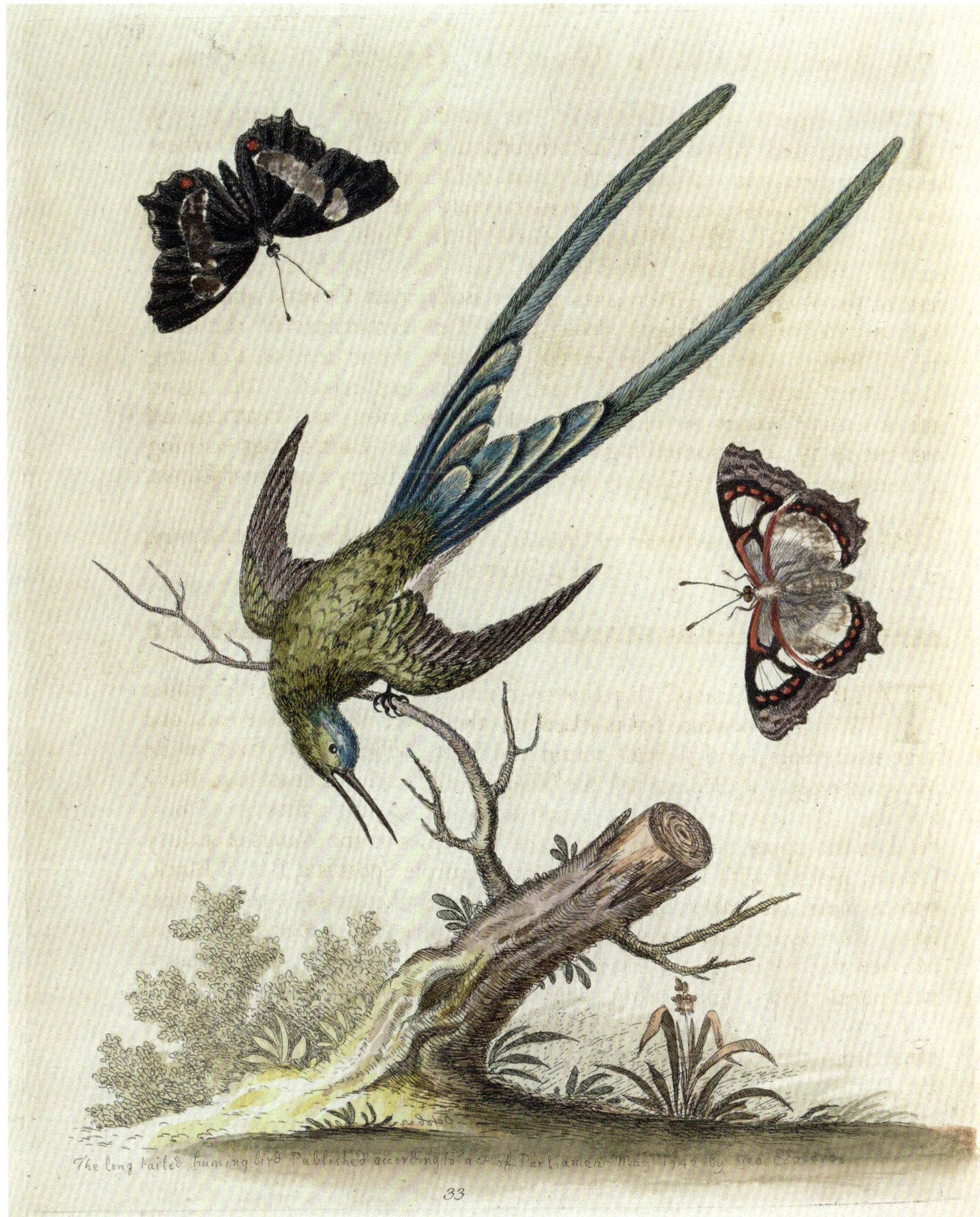

29.49

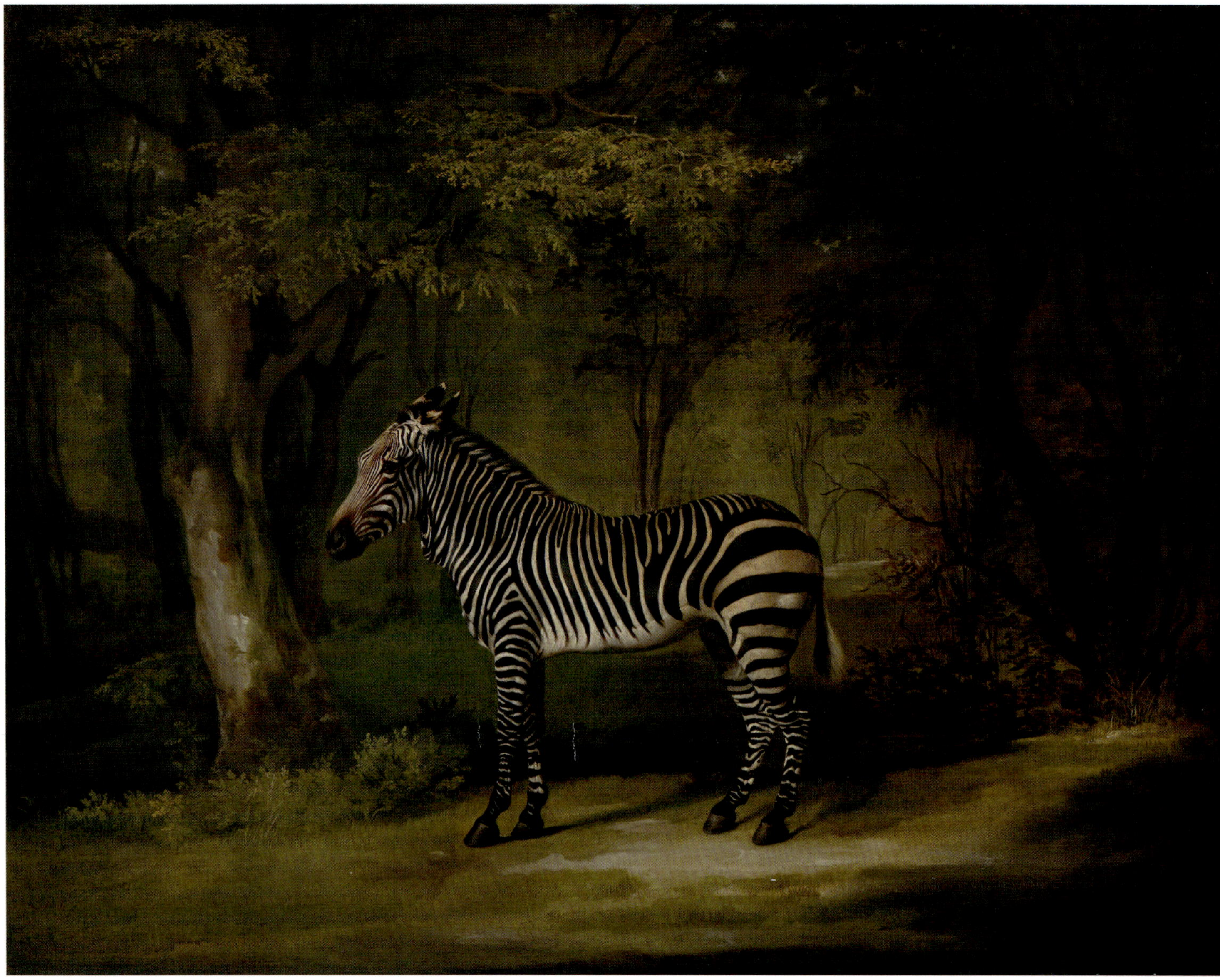

29.50

29.50*

GEORGE STUBBS (1724–1806)
Zebra
1763
Oil on canvas
40½ x 50¼ in. (102.9 x 127.6 cm)
Yale Center for British Art, Paul Mellon Collection, B1981.25.617

In 1762 the colonial governor of South Africa sent a pair of zebras to London as a gift for the young Queen Charlotte. The female of the pair survived the voyage and was installed in the menagerie at Buckingham House, where she became a focus of public attention. It is likely that the celebrated anatomist and painter George Stubbs gained access to the creature through his friendship with William Hunter, the physician who attended Charlotte in childbirth, and for whom Stubbs would come to paint animals by 1769 (Hunter also served as Professor of Anatomy at the Royal Academy of Arts in London). The intended recipient of Stubbs's painting is unknown, since it remained in his studio at the time of his death. But the zebra's image gained a life of its own in the popular press, appearing in rude songs and caricatures for decades after as 'the queen's ass' (see cat. 15.07).

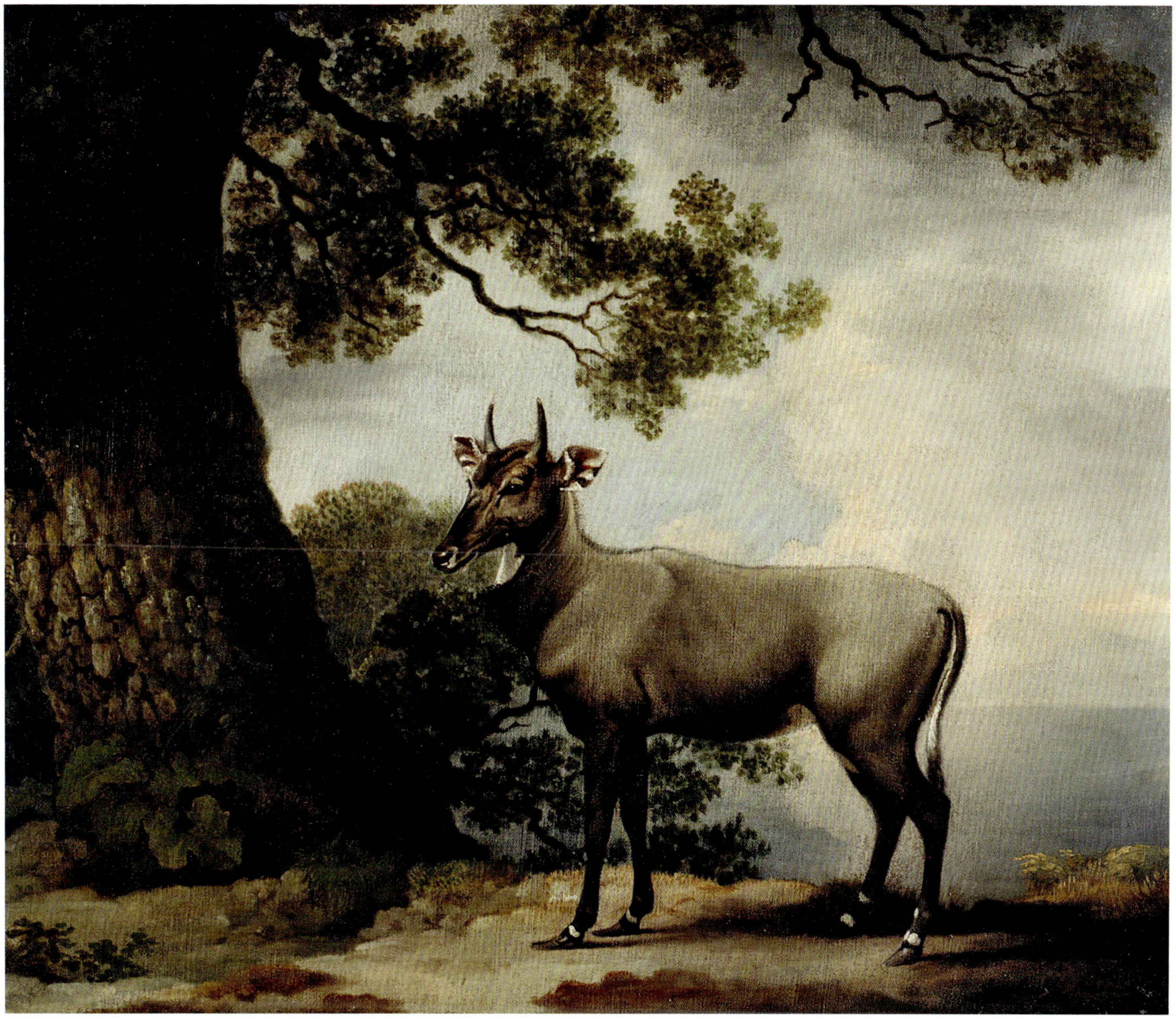

29.51

29.51†

GEORGE STUBBS (1724–1806)
The Nilgai
1769
Oil on canvas
24⅝ x 28¾ in. (62.3 x 73 cm)
Hunterian Museum and Art Gallery, University of Glasgow, GLAHA 43821

Queen Charlotte kept a pair of Nilgai (or Nylghau)—Asian antelope—in her menagerie at Kew. Her physician William Hunter, with whom she shared an interest in exotic species, borrowed the animals in order to observe them and have one of them painted by his friend George Stubbs. This, the first of many such paintings commissioned from Stubbs by William Hunter and his brother, the surgeon and anatomist John Hunter, was intended to illustrate a paper about the animal that William delivered to the Royal Society in 1771. In his paper, Hunter expressed his gratitude to the Queen for allowing him to keep and study the animals, emphasizing 'her Majesty's desire to encourage every useful or curious inquiry in natural knowledge'.

29.52†

WILLIAM HUNTER (1718–1783), anatomist and collector
Nylghau eyeball ('Eye of Neel ghaw')
18th century
Zoological specimen
6 ¼ x 9⅞ in. (10 x 25 cm)
Hunterian Museum Zoology Collection, University of Glasgow, GLAHM:119953
Not illustrated

When one of Queen Charlotte's Nilgai (or Nylghau) died at Kew, she donated the animal's body to her physician William Hunter for dissection. Hunter followed in the footsteps of other royal physicians who had been distinguished collectors, such as Hans Sloane and Richard Mead. He preserved the creature's eye in his collection of anatomical specimens, which he bequeathed, along with a vast collection spanning natural history, coins and medals, art and antiquities, and rare books and manuscripts to the University of Glasgow.

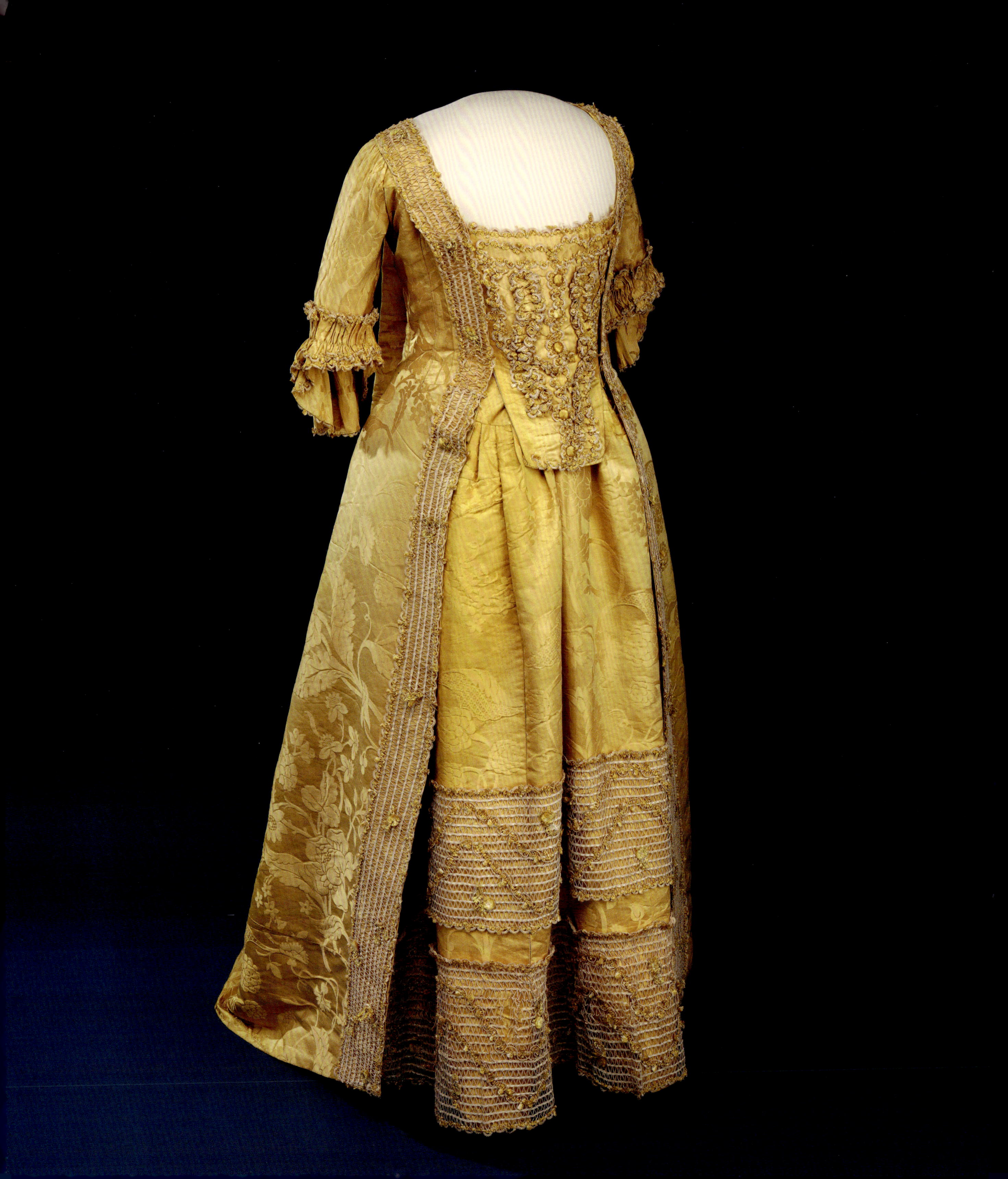

Visitor from South Carolina: Mrs Eliza Pinckney

Ben Marsh

It must have been with great apprehension and excitement that the Pinckney family, plantation owners and slaveholders from South Carolina awaited their crowning moment of imperial exposure: an audience at the White House at Kew with the royal family in the summer of 1753. Gaining permission to meet Augusta of Saxe-Gotha, then Princess Dowager of Wales, and some of her children—aged between two and sixteen—and choreographing the visit had required entreaties, strong connections, and a rigorous checking of credentials. The Pinckneys' London contacts reminded them 'how great a favour' they did them, how 'it was a thing very rarely permit[ted]', and informed them that it was only thanks to the Pinckneys' high standing in colonial society, demonstrated by Charles Pinckney's membership of the twelve-man royal council of South Carolina, that they had secured the opportunity. The Pinckneys undertook to make no requests, and merely to 'show the affection' they had for 'the Family'. What perhaps cemented the rapid approval of their royal audience, was that they brought with them 'some birds from Carolina'; an 'Indigo bird, a Nonparreil, and a Yellow bird': that is, a *Passerina cyanea* (blue linnet or indigo bunting), a *Passerina ciris* (painted finch or painted bunting), and a *Spinus tristis* (American goldfinch)—presumably males, whose vibrant summer plumage brought splashes of vivid blue, scarlet, green, and gold, with exotic vocalizations to match (cats. 30.08–30.09).[1]

The colonial visitors were disappointed on their first attempt to gain access, for—perhaps because of the challenges of coordinating children and animals—they were late, 'and found the Princess gone a airing with the Princess Augusta' (her sixteen-year-old daughter). Fobbed off by an apologetic servant, they reluctantly left their birdcages with a gracious note, penned by (or on behalf of) their nearly five-year-old daughter, Harriott Pinckney, and returned home in their coach 'lamenting as we went [the] uneasy situation of those who . . . were dependance [*sic*] on a Court'. But a reprieve came later that evening, with a warm message from Augusta addressed to young Miss Pinckney inviting her to attend again the next day. The following afternoon, the colonial Americans arrived more punctually, 'in full dress', Eliza perhaps wearing the 'bricked' salmon-coloured gown currently in the collection of the Charleston Museum, in the same sack-back style as her three-piece gold silk damask gown, now in the Smithsonian Institution, Washington (p. 514, figs. 30.01–30.02 and cat. 30.06).[2] Their dependant subject status was emphasized by their being made to wait for some time in the parlour as Augusta got dressed at her leisure, before ordering them to be escorted to her dressing room. At that point Eliza Lucas Pinckney, the South Carolina planter, wife, and mother, reported that much of the formality and mystique evaporated, and nerves calmed. The two mothers and their families enjoyed an enlightening social occasion that would leave deep impressions and material remains. As Pinckney reflected in a letter relating the details of the encounter to an unidentified correspondent, now in the collection of The National Society of The Colonial Dames of America in the State of South Carolina, it all seemed 'pretty extraordinary to an american' (see Appendix, pp. 533–37).

See also p. 522, cat. 30.06

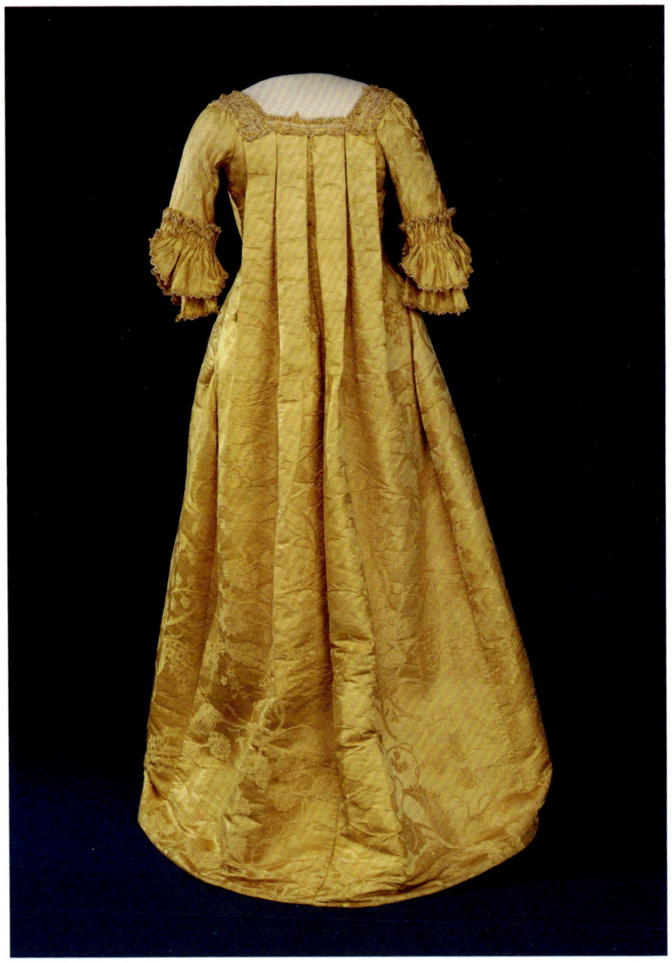

FIG. 30.01 Unknown makers, South Carolina and London, Sack-back gown made for Eliza Lucas Pinckney (rear view), c. 1753, silk damask. Smithsonian Institution, National Museum of American History, Behring Center, Washington DC. Presented to the Smithsonian by the Descendants of Eliza Pinckney, Division of Home and Community Life (cat. 30.06)

Since the dawn of English Atlantic imperialism, royal authority and sponsorship had fostered colonial development; by the eighteenth century, systems of colonial governance and imperial administration allowed for plenty of direct and indirect crown political patronage. But royal patronage was also significant when it came to target imperial objectives of an economic or sociocultural nature. Certain commodities and projects appealed particularly to members of the royal family keen to bolster their credibility as patrons of knowledge, commerce, manufacturing, and the arts. Although the American colonies were held to be culturally backward, they presented sites for the patronage of natural history, ethnography, agricultural experimentation or 'improvement', Christian proselytizing, and imperial philanthropy. As settlers and explorers, naturalists and missionaries accrued empirical experience and built up transatlantic networks in these different fields, the prospects of royal patronage increased, and the recipients were keen to express their gratitude in the form of dedications and appropriate gifts.

Mark Catesby dedicated the two volumes of his magnificently illustrated *Natural History of Carolina, Florida and the Bahama Islands*, based on expeditions in 1712–19 and 1722–26, to Caroline of Ansbach and Augusta respectively, recognizing their beneficence and support of imperial botany (cats. 3.08–3.09). Caroline acquiesced to royal patronage on reviewing his drawings; she was a perfect choice because of her well-known interest in gardening and landscapes. Given her death in 1737 before the completion of the second volume, Catesby sensibly dedicated this to Augusta, who was beginning to establish herself as a patron of imperial botany. That year, she and her husband Frederick, Prince of Wales, took on Caroline's landscape architect, William Kent, to redesign their residence at Kew, and it was there that Augusta and her children met in the flesh the indigo and painted buntings and the American goldfinch which were illustrated in Catesby's pages, 'one of them . . . a favourite of hers'.[3] Catesby's many original images in gouache made for the two volumes were later purchased by George III for the royal collection in 1768.

The 1730s witnessed a burst of activity that tied the royal family to innovative colonial projects in the southern American colonies. The founding of Georgia, designed to shore up the southern frontier against Spanish threats from Florida, was justified by its Trustees, amongst them James Oglethorpe, Thomas Coram, and John Perceval (Percival), first Earl of Egmont, because of its capacity to take in the worthy London poor and persecuted German Protestants and to yield new natural resources that would furnish materials for British manufacturers and merchants.[4] These ideas resonated with philanthropic and scientific interests in Britain, including those championed by the Princesses Caroline, Augusta, and Charlotte of Mecklenburg-Strelitz: supporting orphans, encouraging moral imperialism, integrating German, English, and Scottish peoples, and rivalling leading European industries in scientific botany, fashion, and dress. Publicized moments involving visits and gifts were ways for the Princesses to highlight their intellectual investments, drawing on their figurative claim to be mothers, not just to their children, but also to the nation and the wider empire.

Reverend Samuel Wesley celebrated Caroline in this way in a poem of March 1736, for wearing an early sample of Georgia-reeled silk to King George II's Birthday Levee. 'Admiring Strangers . . . Shall own was never so Majestick seen, / As in her home-wrought Silks, the *British* QUEEN.'[5] A specimen of the silk, processed in Thomas Lombe's famous mill the previous spring, had impressed Caroline, who directed it to be worked up into a 'beautiful pattern' of her choice by Daniel Booth, a weaver in Little Moorfields. The Queen would later take Lord Egmont to task over claims that this silk was actually Italian, though she accepted his answer that 'we debauched two Italians, Piedmontese, away, and sent them to Georgia to cultivate the silk.'[6]

Two years earlier, Oglethorpe and the Trustees had engineered another publicity coup at court, around the diplomatic visit and reception of a party of native Americans of the Yamacraw nation in 1734. William Verelst's painting of 1734 or 1735 of this exotic embassy was hastily commissioned by the Office of the Georgia Trustees, though it took years to complete, as he scrupulously filled out the individual features of its subjects, each of whom contributed to his fee. In its layout, the bewigged gentlemen appear ennobled and enlightened in contrast to the partially clad visitors, whose backdrop is darker and includes wild animals. The group portrait conveyed the Europeans' conviction of their own superiority and their sense of imperial mission: they believed that the curiosities of the wider world (whether peoples, animals, or plants) could stimulate enlightenment when brought to London (fig. 30.03). Here they could be studied, tamed, harmonized, and improved, and provide a new understanding of the natural productions of the wild Americas.[7] Though the Yamacraw chiefs were depicted in their deerskins for the portrait, the Trustees reclothed them for their court visit; the *mico* (leader) and his wife, Senauki, were styled as a king and queen, 'dress'd in Scarlet, trimm'd with Gold'.

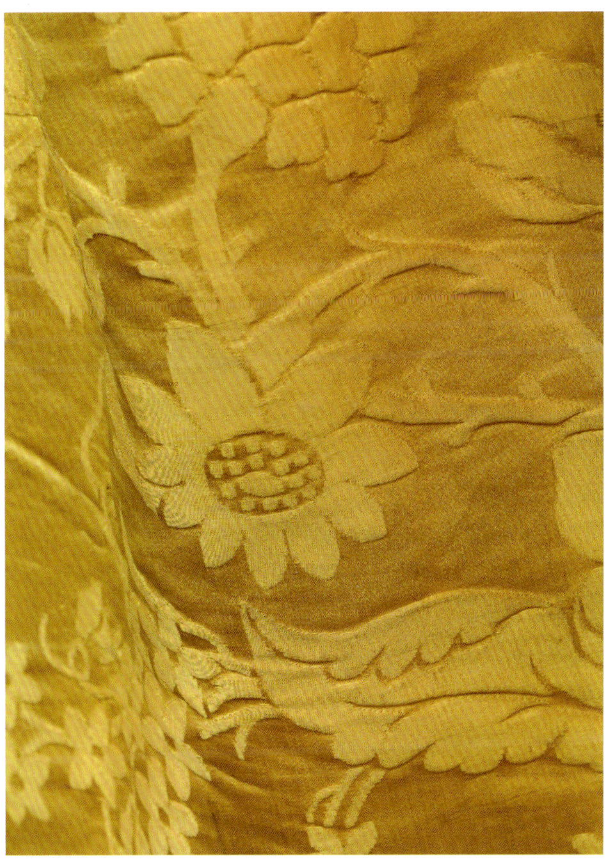

FIG. 30.02 Detail of fig. 30.01 (cat. 30.06)

The popular press reported favourably on the embassy. The *Grub-Street Journal* announced that the elderly Yamacraw leader Tomochichi had demonstrated 'the most natural eloquence' in his speeches before the royals. Caroline reportedly 'stroked' the face of Tomochichi's nephew and heir, Toonahowi, and introduced him to her own children.[8] Following an exchange of gifts, the delegation later enjoyed trips to theatres, hospitals, schools, and Lord Mayor's day; an audience with the Archbishop of Canterbury; and visits to Hampton Court and Kensington Palace.[9]

The event, as with the case of the American silk, involved some sleight of hand, for the Yamacraws were only a minor band of the Muskogee people, settled along the Savannah River, ravaged by smallpox and surviving in the wastelands caused by British slave trading and wars with native American peoples in recent decades. Little wonder that their priorities for their trip were securing mutual defence and fair commerce with the new settlements. Contrary to the public munificence, Egmont uncharitably described 'Queen' Senauki in his private diary as 'an old, ugly creature, who dresses their meat'. But the romanticizing of American connections worked both ways: the Hanoverian dynasty, by their widely reported involvement in and support of international ventures, enhanced its profile in the eyes of its subject peoples, offering a unifying dynamic that helped mitigate against the fragmenting impulses of expansion. In a way, this was a larger imperial variant of what the observant Tomochichi noticed on his tour through Kensington Palace: 'They carry'd him thro a great many houses (by which he meant rooms) to make him believe the King's Palace consisted of many, but he was Surprised to find he return'd by the Same Stairs he went up, by which he found it was Still but One house.'[10] Caroline, Augusta, and Charlotte acted as matriarchs of the many chambers of empire and industry, self-consciously performing the part of 'the common Mother and Protectress of us and all our Children', in Tomochichi's phrase to Caroline in August 1734.[11]

In the two hours shared by Augusta and Eliza Pinckney in the dressing room at Kew Palace in 1753, by which time the Georgia Trustees had given up on their idealistic experiment, the public dimensions of royal patronage had little part to play. The gifts were personal and the exchanges unaffected. These two women of privilege in their respective worlds questioned, discussed, and admired one another's accomplishments and lifestyles—giving shape to their preconceptions about empire, environment, and family. Augusta and Eliza inhabited gendered cages, like all eighteenth-century women. But because both had enjoyed a good education and could exercise considerable agency and independence, they were alert to each other's interests and experiences. The conversation went far beyond the superficial protocol of diplomatic embassies or formal visits, and, in the absence of attendants, was inevitably punctuated by the presence of children. In Pinckney's telling, the conversation shed light on the ways in which humour, discipline, gender, and novelty were mobilized. Certain topics, particularly shared lifecycles and imperial reference points, connected a member of the royal family to their subjects in distant colonies.

Both women were inquisitive about the challenges of motherhood and each other's attitudes to parenting, dwelling on behaviour and education in order to find common ground between themselves. Augusta introduced four of her children: her eldest daughter, Augusta, Elizabeth (aged twelve), William (nine), and Henry (seven). After half an hour 'poor little' Harriott Pinckney broke into embarrassed tears, at which point the Pinckney parents were minded to withdraw. However the Princess Dowager 'stoopd upon her knee to her . . . then took her in her lap', and called for her youngest children to provide distraction, telling Caroline (around two), Frederick (three), and 'poorly' Louisa (four) matter-of-factly when they arrived, that 'Miss P[inckney] from So[uth] Carolina was come to see them and to go and kiss her.' Eliza was impressed at Augusta's tactility, attentiveness, and 'her familiar engaging manner', which included a good deal of humour and informality. When Augusta playfully asked her two-year-old, 'have you ever been in the Corner my Queen[?]' and Caroline responded 'no Madm . . . never in the corner', Augusta corrected her: 'I am afraid you have.' She threatened to punish Frederick at one point for teasing Harriott, and when he retorted he would go to Carolina, she shrugged, 'Well then, good by to you.' Eliza Pinckney proved increasingly forthright as her confidence grew, opining during a discussion of breastfeeding 'that it appeard very strange to me to hear of people puting their children out to Nurse . . . we had no such practises in Carolina.' Augusta seemed 'vastly pleased' by this, deeming putting out to be 'unnatural', a verdict that placed her in line with advanced medical authorities such as William Cadogan, who urged that maternal colostrum was beneficial, and celebrated new philanthropic institutions such as the Foundling Hospital which promised further advances.[12] Augusta was intrigued to learn about the racial dimensions of wet-nursing in South Carolina, where slaves were often used to suckle infants within the household: 'surprized at the suckling blacks . . . the princess stroakd H[arriott']s cheek [and] sayd it made no alteration in the complexion.'[13]

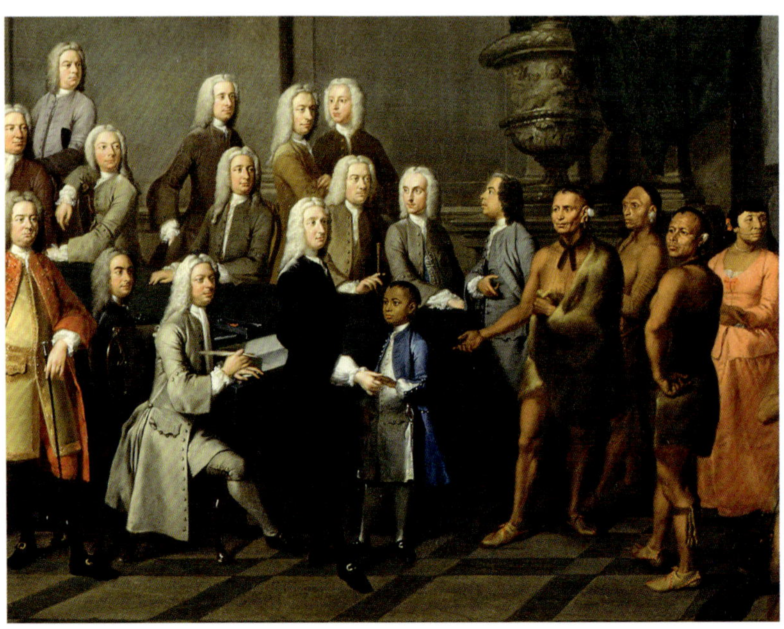

FIG. 30.03 William Verelst, *Audience Given by the Trustees of Georgia to a Delegation of Creek Indians* (detail), 1734–35, oil on canvas. Winterthur Museum, Art Gallery and Garden, Gift of Henry Francis du Pont (cat. 30.04)

No less than issues of family, health, and education—what Eliza characterized as 'many little domestick questions'—the empire featured prominently as a framework for shared identity, bridging the disparate accents and geographies that each of them understood. Eliza discussed Atlantic journeys, noted the consumer goods used by Augusta's family (such as the 'great [deal] of China upon two Cabinets', chairs and carpets), and informed her of the names of colonial governors and rivers. Augusta introduced her seven-year-old, Harry, intentionally as 'a sailour', reminding him that 'may be you may go there [to Carolina] If there should be a war, wch: I hope there will not for we have had Enough of that'; while Charles Pinckney told Prince William that he hoped 'to have another [son] for the Sea'. The subjects of conversation remained generally elitist—fine wines, how best to eat turtle—and included Eliza's summary of the progress of silk production in South Carolina, in which Augusta showed particular interest.

For all of the mutual respect, common imperial ground, and shared interests, Eliza Pinckney's account of the South Carolinians' visit to Kew also exposed some of the colonial divergences and 'american' sensibilities developing by the third quarter of the century. When Augusta laughed at young Harriott Pinckney's response that she 'likd England . . . not so well as Carolina', and the Princess reflected that 'it was very natural for such a little woman as she was to love her own Country best', Augusta perhaps drew on her own emotional childhood memories of relocation from Saxe-Gotha, and an assumption that maturity would bring a more inclusive viewpoint. The British American colonies at this point were expanding at an unprecedented pace, becoming a vital consumer base for British manufactures and increasingly prominent in global production and exchange. South Carolina's main contributions were its slave-produced rice and indigo, the latter a dyestuff that Eliza Pinckney had been instrumental in bringing from the Caribbean to the Lower South.[14] A subsequent gift that integrated the colony, the indigo dye, the bird, and the silk was a 'Piece of Silk Damask of the Growth and Product of [their own] Plantation . . . and dyed a fine Blue with CAROLINA INDICO', delivered by the Pinckneys to Augusta in 1755. In response, the Princess declared 'her Satisfaction in seeing such Improvements made in the Produce of our Colonies', promising to honour it (like Caroline before her) 'with her own Wearing'.[15]

As an octogenarian, Harriott Pinckney Horry, the one-time tearful toddler perched on the Princess's knee, recollected that the quantity of Eliza's silk from which this damask was wrought up had been enough to 'produce many yards of very rich wearing apparel'. A later descendant claimed in 1896 that Augusta's damask was one of 'three beautiful dresses woven of it' while Eliza was in London, a second purportedly gifted to Philip Stanhope, fourth Earl of Chesterfield (a supporter of colonial interests and agricultural experimentation). Harriott still possessed the third home-grown silk gown in 1829, when she described it as 'one of the ancient dresses' of her mother's. She cut off a breadth of the surviving gold damask to send to two girls to encourage them to take up sericulture, noting that 'the Antiqutity [*sic*] of its make' was 'scarcely less a curiosity than its fabric' (see fig. 30.02).[16] But though the silk had survived, the larger British Atlantic family had not. When imperial push came to republican shove in the 1770s, both Harriott Pinckney's elder brothers would become Continental Army officers. It was a mark of Eliza Pinckney's social standing in the newly United States that, at his own request, George Washington acted as her pallbearer when she succumbed to breast cancer in May 1793. Like Augusta, she had lived longer a widow than a wife, but also like her, she had invested in as many projects as the gendered proscriptions of her society allowed.[17]

NOTES

1. Throughout this essay, quotations from this letter are from 'Eliza Lucas Pinckney to Unidentified person [1753]', in *The Papers of Eliza Lucas Pinckney and Harriott Pinckney Horry Digital Edition*, ed. Constance Schulz (Charlottesville, VA: Univ. of Virginia Press, Rotunda, 2012), http://rotunda.upress.virginia.edu/PinckneyHorry/ELP1015, accessed 25 Feb. 2015, transcribed here, pp. 533–37. For a wider contextualization of the Pinckneys and their London visit, see Constance B. Schulz, 'Eliza Lucas Pinckney and Harriott Pinckney Horry: A South Carolina Revolutionary-Era Mother and Daughter', in *South Carolina Women: Their Lives and Times*, ed. Marjorie J. Spruill, Joan M. Johnson, and Valinda W. Littlefield, 3 vols. (Athens, GA: Univ. of Georgia Press, 2009–12), 1: 79–108. According to Carl Linnaeus's 1758 enlarged tenth edition of his *Systema Naturae*, the indigo bunting was described as genus *Tanagra* and the painted bunting as genus *Emberiza*.
2. The pink dress is currently being conserved and can be viewed at the Charleston Museum's website: http://charlestonmuseum.tumblr.com/post/15618476027. The gold dress was gifted to the costume collection of the Smithsonian's National Museum of American History by Pinckney's direct descendants, having been on loan there since 1912.
3. On natural history, botany, and imperial patronage, see Walter Edgar, *South Carolina: A History* (Columbia, SC: Univ. of South Carolina Press, 1998), 177–79; Amy R. W. Meyers and Margaret Beck Pritchard, eds., *Empire's Nature: Mark Catesby's New World Vision* (Chapel Hill, NC: Univ. of North Carolina Press, 1998), 15–19; and Susan Scott Parrish, *American Curiosity: Cultures of Natural History in the Colonial British Atlantic World* (Chapel Hill, NC: Univ. of North Carolina Press, 2006).
4. For examples, see Benjamin Martyn, *Reasons for Establishing the Colony of Georgia, With Regard to the Trade of Great Britain, the Increase of our People, and the Employment and Support it will afford to great Numbers of our own Poor, as well as foreign persecuted Protestants* (London: W. Meadows, 1733); and [James Edward Oglethorpe], *A New and Accurate Account of the Provinces of South-Carolina and Georgia* (London: Printed for J. Worrall, 1732). For a wider discussion, see Milton L. Ready, 'Philanthropy and the Origins of Georgia', in *Forty Years of Diversity: Essays on Colonial Georgia*, ed. Harvey H. Jackson and Phinizy Spalding (Athens, GA: Univ. of Georgia Press, 1984).
5. John C. Stephens, ed., *Georgia, and Two Other Occasional Poems on the Founding of the Colony, 1736* (Atlanta, GA: Emory Univ. Publications, 1950), 327, 336.
6. Allen D. Candler and others, eds., *The Colonial Records of the State of Georgia* (Atlanta and Athens, GA, 1904–), Vol. 3: 130; Vol. 20: 69, 152, 211–12.
7. Phinizy Spalding and Harvey H. Jackson, eds., *Oglethorpe in Perspective: Georgia's Founder after Two Hundred Years* (Tuscaloosa, AL: Univ. of Alabama Press, 1989), 153–56; Kumkum Chatterjee and Clement Hawes, eds., *Europe Observed: Multiple Gazes in Early Modern Encounters* (Lewisburg, PA: Bucknell Univ. Press, 2008), 178–87.
8. Robert G. McPherson, ed., *The Journal of the Earl of Egmont: Abstract of the Trustees Proceedings for Establishing the Colony of Georgia, 1732–1738* (Athens, GA: Univ. of Georgia Press, 1962), 61.
9. *Grub-Street Journal* (London), 8 Aug. 1734; *The Gentleman's Magazine* 4 (1734), 449, 571. McPherson, ed., *The Journal of the Earl of Egmont*, 60–61; John Percival, Earl of Egmont, *Diary of the First Earl of Egmont (Viscount Percival)*, 3 vols. (London: Historic Manuscripts Commission, 1920–23), 2: 112–33; Timothy J. Shannon, '"This Wretched Scene of British Curiosity and Savage Debauchery": Performing Indian Kingship in Eighteenth-Century Britain', in *Native Acts: Indian Performance, 1603–1832*, ed. Joshua D. Bellin and Laura L. Mielke (Lincoln, NE: Univ. of Nebraska Press, 2011), 221–33; Nancy Shoemaker, *A Strange Likeness: Becoming Red and White in Eighteenth-Century North America* (Oxford: Oxford Univ. Press, 2004), 35–39.
10. McPherson, ed., The *Journal of the Earl of Egmont*, 61; Viscount Percival, *Diary*, 2: 114, 117.
11. *The Gentleman's Magazine* 4 (1734), 449.
12. William Cadogan, *An Essay upon Nursing, and the Management of Children, from Their Birth to Three Years of Age* (London: J. Roberts, 1748), 13–16.
13. Paula A. Treckel, 'Breastfeeding and Maternal Sexuality in Colonial America', *Journal of Interdisciplinary History* 20, no. 1 (Summer 1989): 25–51, esp. 49–50 on Pinckney; Nora Doyle, '"The Highest Pleasure of Which Woman's Nature Is Capable": Breast-Feeding and the Sentimental Maternal Ideal in America, 1750–1860', *Journal of American History* 97, no. 4 (2011), 958–73.
14. Barbara L. Bellows, 'Eliza Lucas Pinckney: The Evolution of an Icon', *South Carolina Historical Magazine* 106, no. 2 (2005): 147–65; David L. Coon, 'Eliza Lucas Pinckney and the Reintroduction of Indigo Culture in South Carolina', *Journal of Southern History* 42, no. 1 (1976): 61–76.
15. *South Carolina Gazette*, 3–10 April 1755; Ben Marsh, 'Silk Hopes in Colonial South Carolina', *Journal of Southern History* 78, no. 4 (2012): 807–54.
16. Harriott Pinckney Horry to Mrs. [Margaret Anne] Waties, c. 1829, Harriott Horry Papers (South Caroliniana Library); Harriott Horry Ravenel, *Eliza Pinckney* (New York: Scribner's, 1896), 131.
17. Schulz, 'Eliza Lucas Pinckney and Harriott Pinckney Horry', 101.

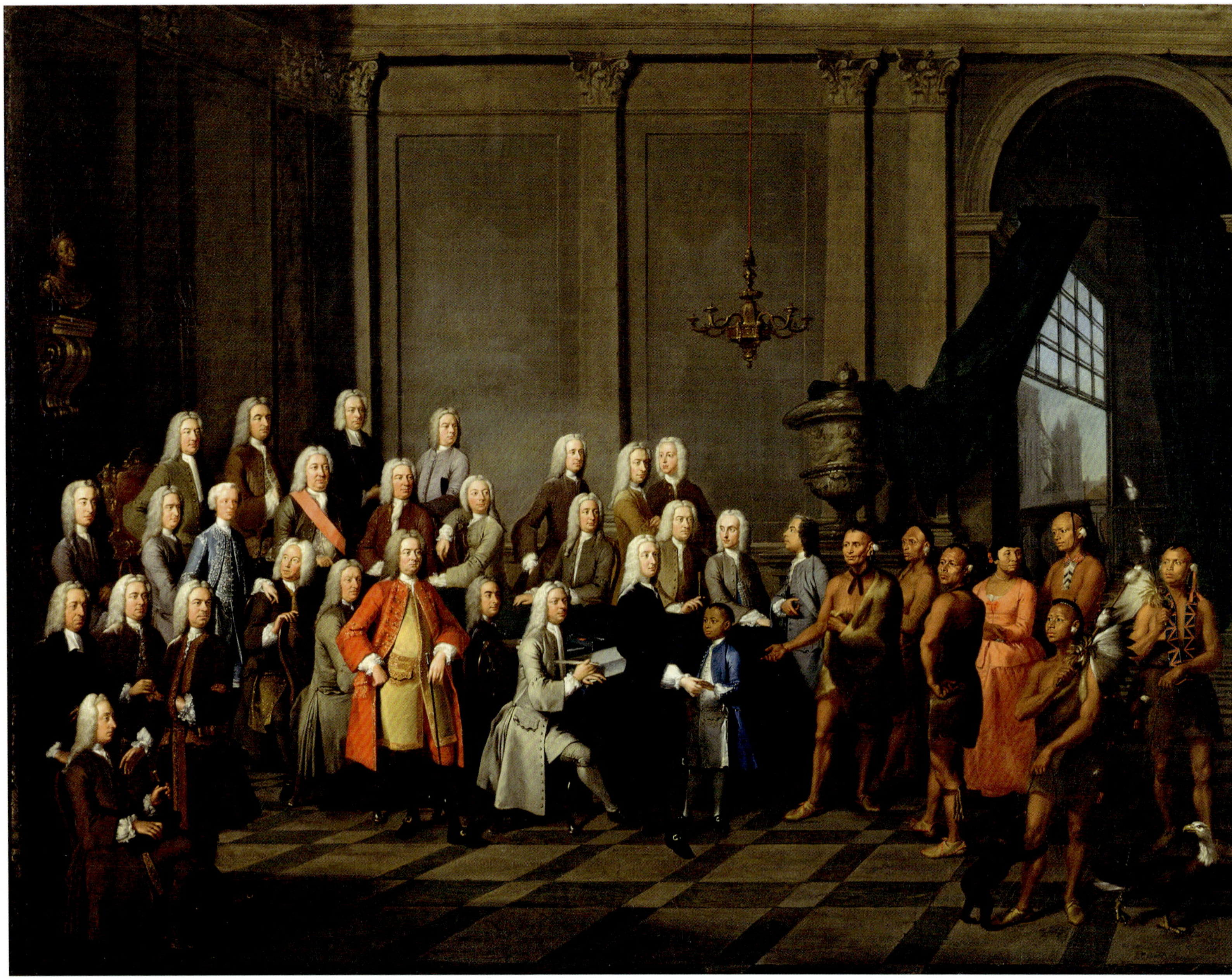

30.04

30.04

WILLIAM VERELST (1704–1752)
Audience Given by the Trustees of Georgia to a Delegation of Creek Indians
1734–35
Oil on canvas
48½ x 61⅜ in. (123.2 x 155.9 cm)
Winterthur Museum, Art Gallery and Garden, Gift of Henry Francis du Pont, 1956.0567 A

The colony of Georgia was established in 1732, and in 1734 a delegation from the Yamacraw, a Creek tribe in the region, accompanied the colony's founder, James Oglethorpe, to London for a series of negotiations with the Georgia trustees. This painting depicts an encounter between the two groups and was commissioned by the trustees to commemorate the visit. At Kensington, George II, Queen Caroline, and their children received the delegation, including the chief, Tomochichi, his wife, Senauki, and his nephew and heir, Toonahowi. Tomochichi and George II exchanged assurances of friendship between their nations, and Caroline took particular interest in the young Toonahowi, to whom Prince William presented a gun and a gold watch.

* = YCBA only
† = HRP only

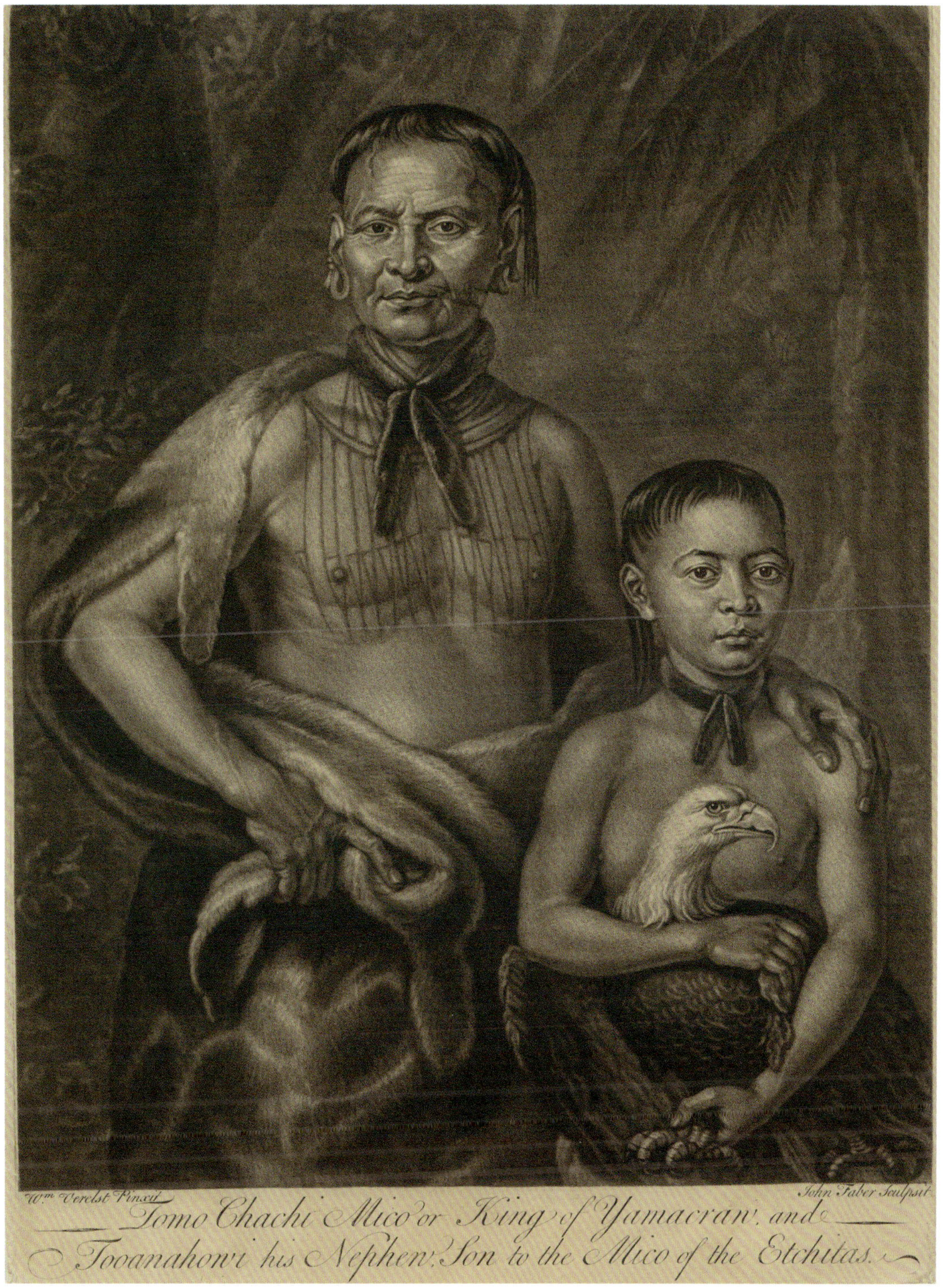

30.05

30.05*

JOHN FABER THE YOUNGER (1695–1756), engraver, after **WILLIAM VERELST** (1704–1752)
Tomo Chachi Mico, or King of Yamacraw, and Tooanahowi [sic], His Nephew
c. 1734
Mezzotint
14⅛ x 10⅛ in. (35.8 x 25.7 cm)
Yale Center for British Art, Paul Mellon Fund, B1970.3.674
Inscribed: 'Tomo Chachi Mico or King of Yamacran, and / Tooanahowi his Nephew, Son to the Mico of the Etchitas'

Tomochichi was the leader of the delegation of Yamacraws who travelled to London with James Oglethorpe, the founder of the Georgia colony, in 1734. In negotiations with the Georgia trustees he sought assurances that commerce between his people and British settlers would be regulated and that instruction 'in the Knowledge of the English' be made available to the children of the Creek nation. His nephew, then a boy of fifteen, accompanied him on the voyage to London. Although both sitters acquired specially made English clothes during their stay in London (Tomochichi had made for him 'a fine scarlet banian . . . edged everywhere with rabbit's fur, white, and . . . a gold galloon lace down the edges'), this portrait shows them in an imagined 'American' setting against a woodland background—Tomochichi wears a deerskin robe and his nephew and heir, Toonahowi, holds a bald eagle. During their audience with the royal family, Tomochichi presented eagle feathers to George II as an emblem of peace. Deer skin was a major export of the southern colonies.

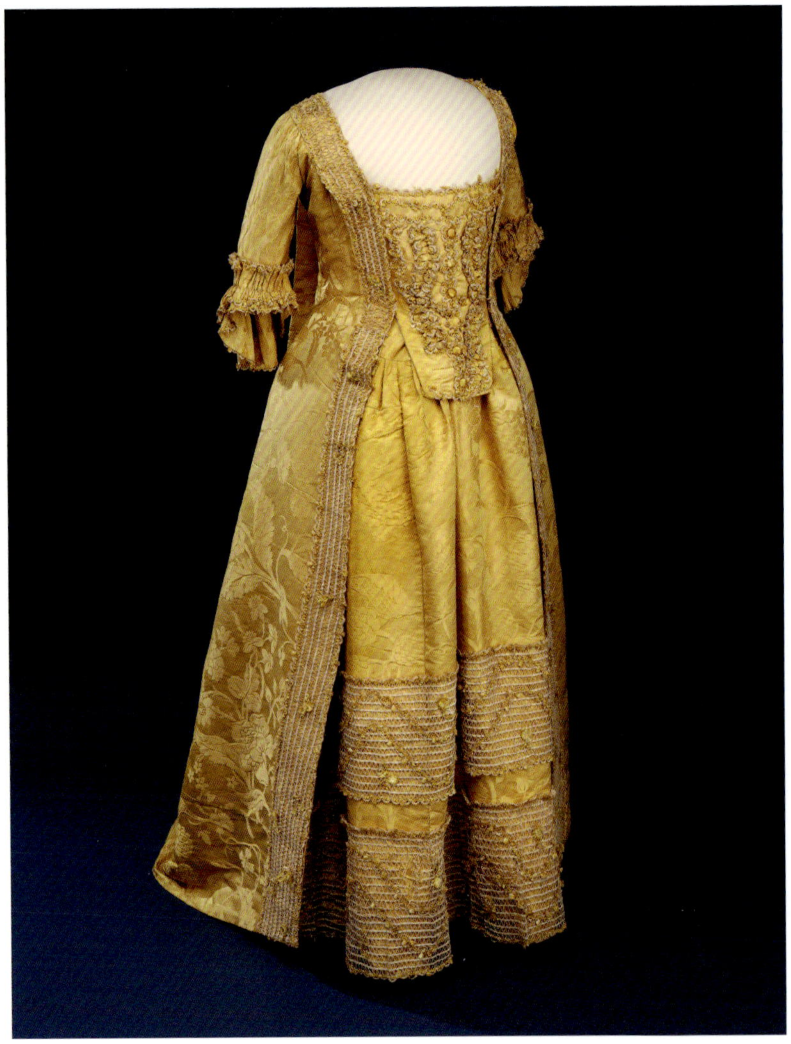

30.06

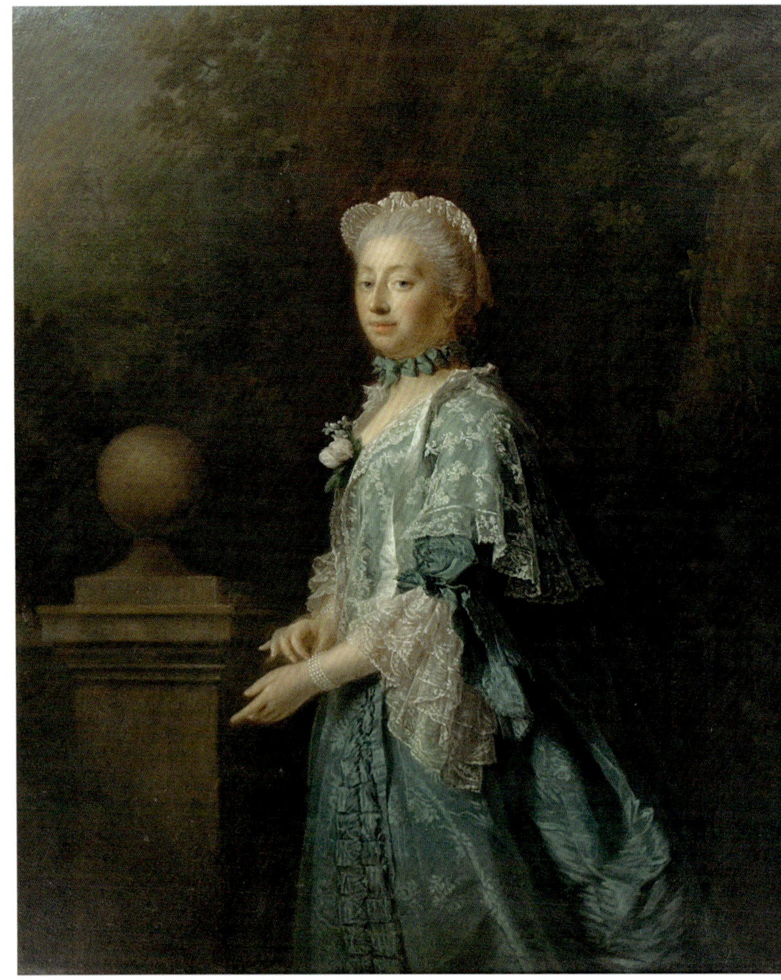

30.07

30.06*

UNKNOWN MAKERS, South Carolina and London
Sack-back gown comprising robe, petticoat and stomacher, made for Eliza Lucas Pinckney
c. 1753
Silk damask trimmed with lace, silk tufted braid and buttons
Robe: Nape to hem, centre back approx. 57in. Petticoat: hem diameter approx. 66in. Stomacher: 12¾ x 9½
Smithsonian Institution, National Museum of American History, Behring Center, Washington, DC. Presented to the Smithsonian by the Descendants of Eliza Pinckney, Division of Home and Community Life

Eliza Pinckney took over the management of her father's South Carolina plantation at the age of sixteen. Her experiments with cultivating indigo in the 1740s, based on the knowledge of her African slaves, were successful, leading to its widespread adoption as an export crop in the colony. After her marriage she took on the challenge of silk production. The silk for this gown was raised by slaves on her plantation, woven in London, and made up into three dresses, one of which she wore for her audience with Princess Augusta at the White House, Kew, in 1753. This dress is comprised of a robe, a petticoat, and a stomacher. The robe is of sack-back construction: the fullness of the back of the robe is caught into the neckband and hangs in deep folds to the slightly trained hem. The gathered petticoat, stomacher, and the robings and double-tiered sleeve ruffles of the robe are trimmed with deep bands of lace, edged and trimmed with yellow tufted silk braid matching the colour of the silk damask. The stomacher is decorated with applied silk braid with buttons at the centre front.

30.07

ALLAN RAMSAY (1713–1784)
Augusta, Dowager Princess of Wales
1769
Oil on canvas
55½ x 45 in. (141 x 114.2 cm)
Collection of H.R.H. Heriditary Prince Ernst August of Hanover, Duke of Brunswick and Lüneberg

After 1767, when Ramsay was appointed Principal Painter in Ordinary he undertook a series of commissions for members of the royal family. Most depicted members of the family of George III and Charlotte, and they included a posthumous portrait of Princess Elizabeth Albertina, Charlotte's mother, after a model by Daniel Woge. She had died shortly before her daughter had left for England. Augusta paid £52 for her portrait, which was commissioned as a gift for the court in Brunswick. There is a preparatory study in red and white chalk in the collection of the National Gallery of Scotland. It was

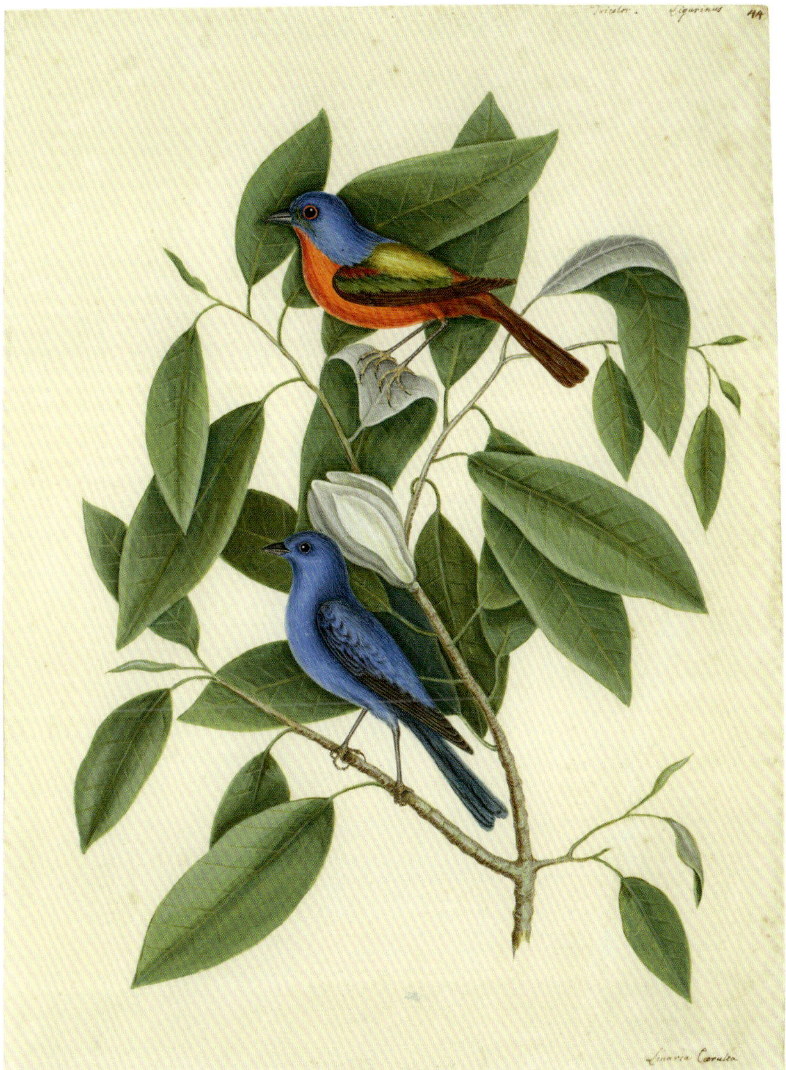

30.08

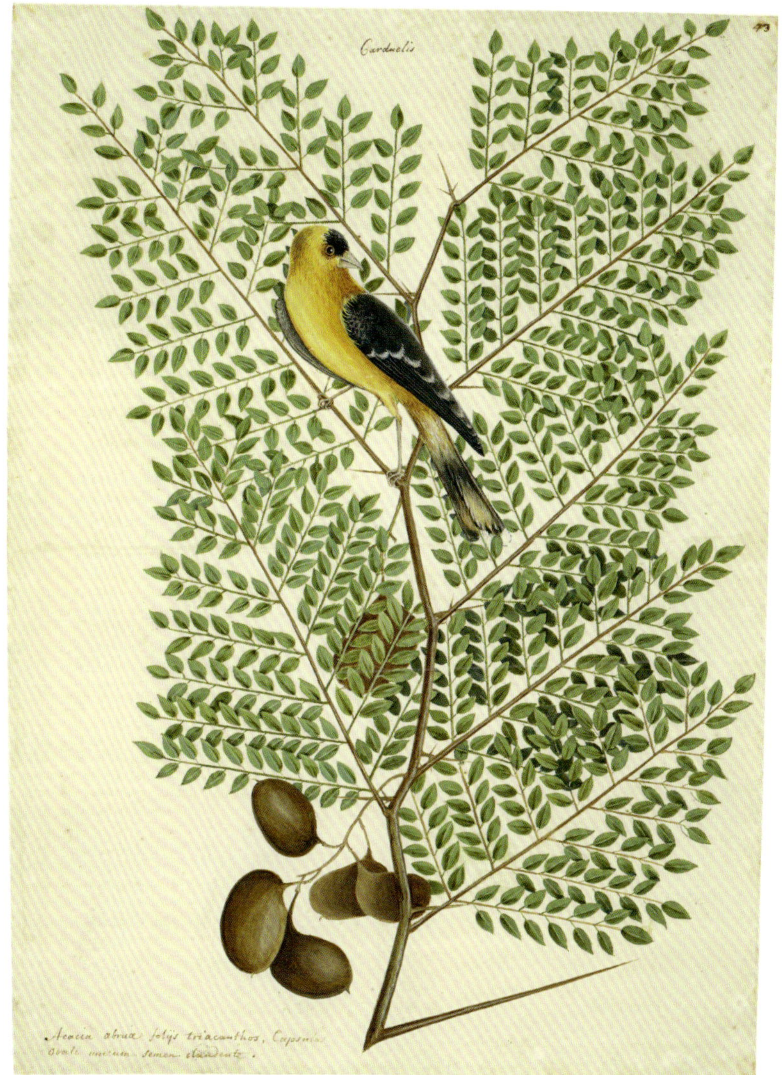

30.09

not the first time that she had sat to the artist. In 1764 Ramsay had painted a full-length portrait for John Stuart, third Earl of Bute (see cat. 13.11). Both paintings of the Princess depict her in a garden, probably that at the White House, Kew, which she completed after the death of her husband, Frederick, Prince of Wales, in 1751, with the help and advice of Ramsay's fellow Scotsman and patron, Lord Bute. In this painting she wears a sack-back gown, similar to the one made for Mrs Pinckney and now at the Smithsonian (cat. 30.06).

30.08

MARK CATESBY (1683–1749)
The Painted Finch and the Loblolly Bay
1722–26
Watercolour and gouache over pencil on paper
14¾ x 10⅝ in. (37.4 x 26.8 cm)
Royal Collection Trust, RCIN 925875

Mark Catesby's *Natural History of Carolina, Florida, and the Bahama Islands . . .* was the first publication to both illustrate and describe North American birds, which he considered more spectacular but less melodious than English ones. He devoted most of the first volume to birds, in consideration of their beauty and their close relation to the plants on which they fed and nested. Painted finches were prized as caged birds in the American colonies because of their vibrant colouring, and Catesby recounted seeing four or five of them at the home of the governor of South Carolina. Princess Augusta's fondness for keeping birds was well known; both a painted finch and a blue linnet were among those given to her as gifts by the Pinckney family of South Carolina in 1753.

30.09

MARK CATESBY (1683–1749)
The American Goldfinch, the Acacia and the water locust
1722–26
Watercolour and gouache over pen and ink on paper
14⅞ x 10⅝ in. (37.8 x 26.8 cm)
Royal Collection Trust, RCIN 924857

Like the painted finch, the American goldfinch was often kept as a caged bird because of its bright colouring. Catesby described goldfinches as more common in New York than in Virginia or Carolina, but the Pinckney family of South Carolina nonetheless managed to acquire one as a gift for Augusta when they visited her at the White House in 1753. Some years after their visit, William Chambers would complete the Chinese-style aviary for Augusta at Kew, where she kept a chorus of English songbirds. She would have been familiar with the American goldfinch, the painted finch, and the blue linnet from Catesby's *Natural History of Carolina, Florida, and the Bahama Islands . . .*, the second volume of which was dedicated to her in 1743.

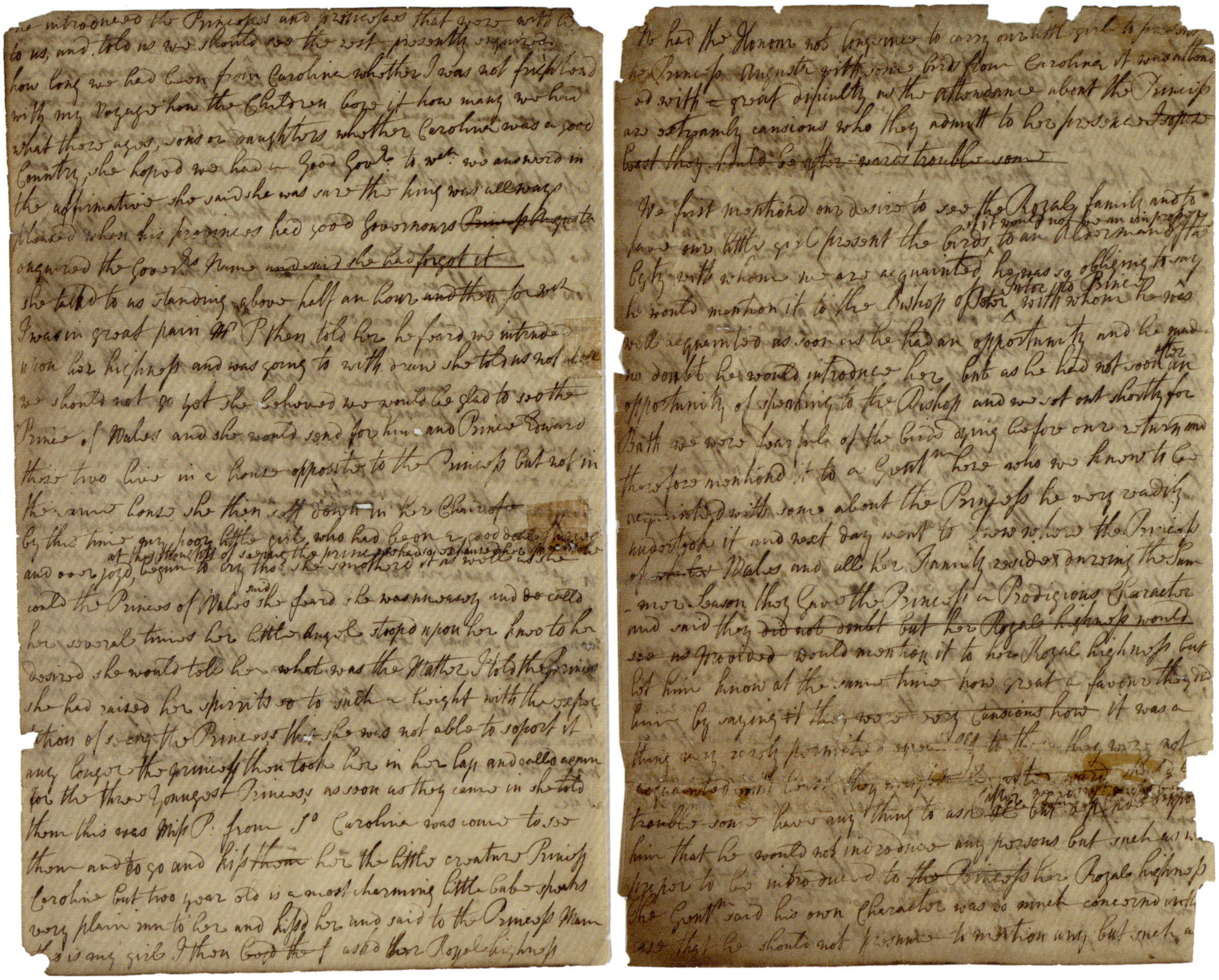

30.10

30.10

ELIZA LUCAS PINCKNEY (1722–1793)
Letter to an unidentified person
c. 1753
Ink on paper
11¾ x 14¼ in. (29.8 x 36.2 cm)
The National Society of The Colonial Dames of America in the State of South Carolina, ELP1015
Exhibited: pp. 4, 1 (sheet 1, illustrated); pp. 5–6 (sheet 2); pp. 10–11 (sheet 3)
See facsimile and transcription, pp. 533–37

This letter gives an account of Eliza Pinckney's remarkable and candid discussion with Princess Augusta, then Princess Dowager of Wales, during her visit with her husband, Charles, and daughter Harriott to the White House at Kew in the summer of 1753. Mrs Pinckney was impressed with the kind and informal welcome they received, and recorded in detail the affectionate banter between Augusta and her children as well as the conversation among the adults, which touched on the Pinckneys' voyage to England; the American birds they had brought as a gift to the Princess; the government, indigenous peoples, geography, and architecture of South Carolina; hurricanes and earthquakes; and 'many little domestick questions', including breast-feeding and the 'manner of dressing turtle'.

MAI'S VISIT TO LONDON

30.11

WILLIAM PARRY (1743–1791)
Omai (Mai), Sir Joseph Banks and Daniel Charles Solander
1775–76
Oil on canvas
60 x 60 in. (152.4 x 152.4 cm)
National Portrait Gallery, London, Purchased jointly with the Captain Cook Memorial Museum, Whitby and the National Museums & Galleries of Wales, with help from a private benefactor and local trusts, the Art Fund, the Garfield Weston Foundation, Flora Fraser and Peter Soros, Sir Christopher Ondaatje, Linda L. Brownrigg, Randolph and Lara Lerner, Jon and Lillian Lovelace, the Clore Duffield Foundation, Sir Harry Djanogly, Hans and Märit Rausing, Lawrence Banks, the Swan Trust, Amanda Sebestyen, Sir David Attenborough, Lord Plymouth and Lord Windsor and many other donations, 2003, NPG 6652

30.12

UNKNOWN ENGRAVER, after an **UNKNOWN ARTIST**
'Mr Banks presenting Omaih the native of Otaheite to her Majesty'
In *The Ladies most Elegant and Convenient Pocket Book, for the year 1775*
1774
Engraving
5 ¼ x 6 ¼ in. (14 x 16 cm.)
Printed for J. Wheble; and sold by J. Bew, in Pater-Noster Row, London
Historic Royal Palaces, HRP 3005001

The Swedish naturalist Daniel Solander and the botanist and explorer Joseph Banks accompanied Captain James Cook on his first great voyage across the Pacific Ocean, observing the transit of Venus in Tahiti in 1769. Mai (called 'Omai' in Britain) was the first Tahitian to visit Britain, travelling to London with Cook's second expedition in 1774. Banks introduced Mai to London society, ensuring that he was inoculated against smallpox beforehand. Banks's close relationship with the royal family ensured Mai's welcome at court, and Banks and Solander presented him to George III and Queen Charlotte at Kew shortly after his arrival. This portrait was probably commissioned by Banks to commemorate the three sitters' friendship before Mai's return to Tahiti in 1776. The reception was also commemorated in an engraving which was inserted into several periodical journals, ensuring it had a wide circulation.

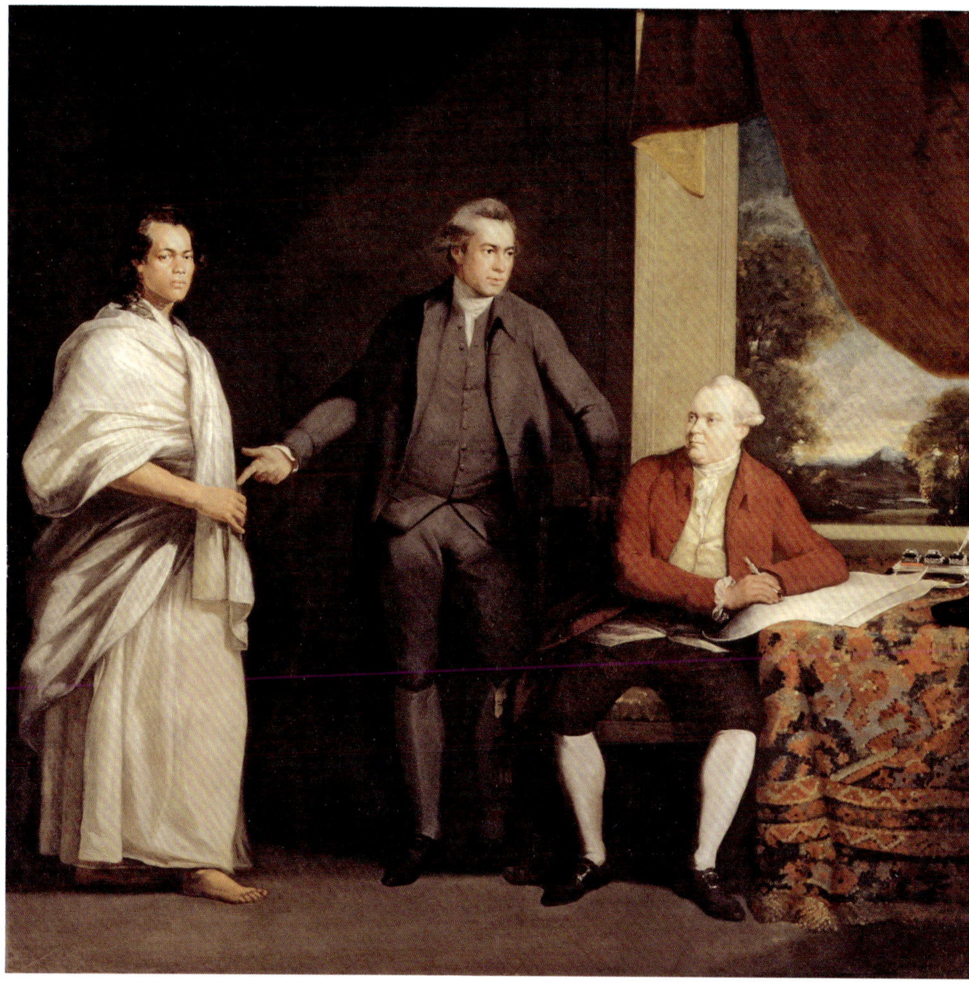

30.11

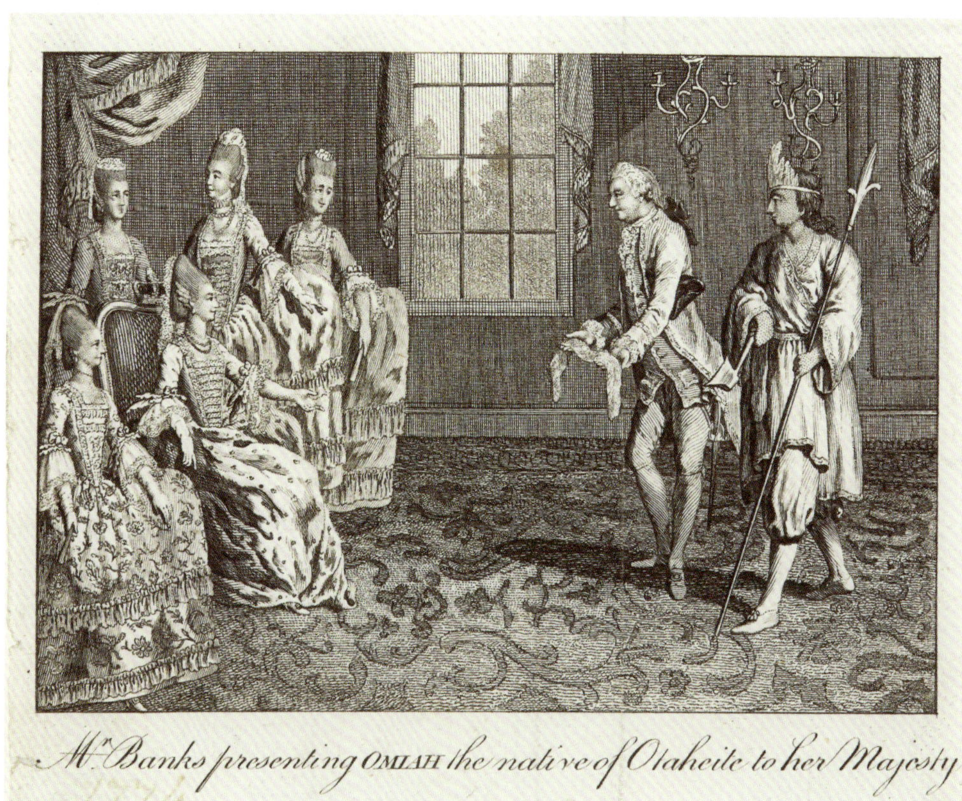

30.12

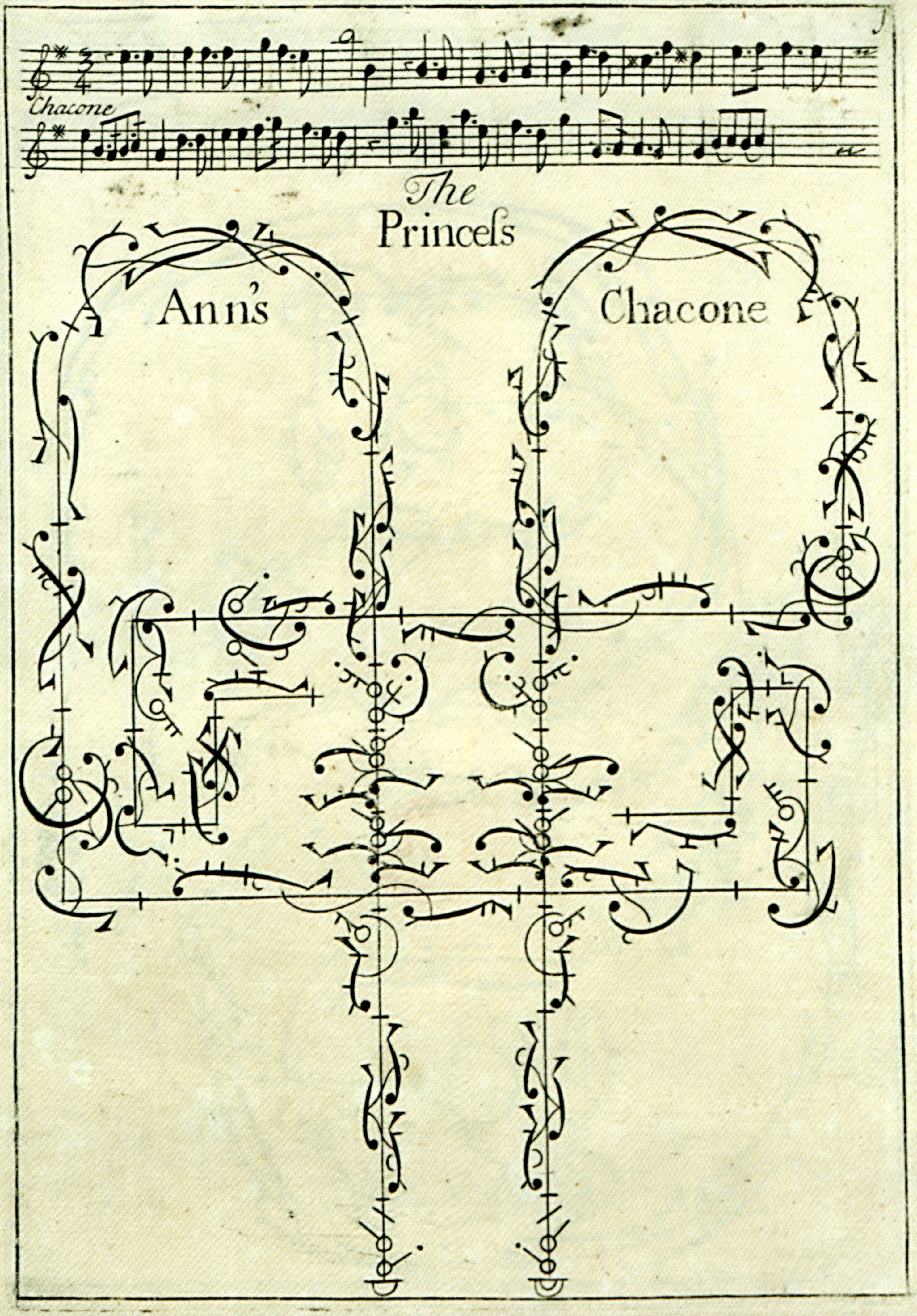
Chacone
The Princess Ann's Chacone

Curating the Intangible

Glenn Adamson

For the curatorial team that created the exhibition *Enlightened Princesses: Caroline, Augusta, Charlotte, and the Shaping of the Modern World*, the problem of the ephemeral has been particularly crucial. Many of the signal achievements of these three women, and the court culture that they helped to build, were fleeting in their character. The music that sounded in Kensington Palace; the plantings that shaped Kew Gardens; the transitory impressions afforded by magic lanterns and other period technology; the conversations that constituted the very medium in which the Enlightenment unfolded: all are lost to us. Nonetheless, at an early stage, the project's lead curator Joanna Marschner, and Amy Meyers, the organizing curator at the Yale Center for British Art, determined that such intangibles should have a prominent place in the project. The question was how this might be done.

This postscript details three key initiatives undertaken by the exhibition team, still in progress at the time of writing, each of which addresses an intangible, sometimes ephemeral cultural form of the eighteenth century. As we will see, not all these intangibles are created equal. One of our case studies concerns the refraction of light, an optical phenomenon that can be observed in the same way today as it was three centuries ago. The second addresses courtly dance, which was recorded in the form of notation and so can be restaged in the present, albeit only by highly specialized performers. Our final initiative arises from a single conversation, fragments of which were written down in a letter. None of these three historical experiences can be reproduced exactly; our goal was never to recreate the past. Instead, we sought to create experiences for the museum visitor that are, in their own way, as evocative as those which animated the Enlightenment in its own day.

NEWTON'S OPTICKS

As Joanna Marschner describes in her essay for this volume,[1] Caroline of Ansbach committed herself to the patronage of Isaac Newton's research upon her arrival in England, a politically astute decision that involved downplaying her previous connection with his rival Gottfried Wilhelm Leibniz. Through the association with Newton, she was seen to be quite literally fostering the leading light of British science, at the culmination of his long career. He was seventy-three years of age when he demonstrated, in her apartments at Kensington Palace, how sunlight could be refracted into rainbows and then reconstituted into white light through the use of lenses and prisms. We have a good idea of the nature of these experiments, because similar ones had been executed by the Royal Society's appointed demonstrator John Theophilus Desaguliers three years earlier, in 1713. Desaguliers (whose work Caroline and George II eagerly followed) kept careful notes of his demonstrations, explicating Newton's discoveries in a way that was more easily understood than the great scientist's groundbreaking publication of his own research in 1704 (fig. 31.01).[2] Desaguliers described in detail how he used the features of the demonstration room to create the desired effects, for example:

Detail of p. 209, cat. 11.06

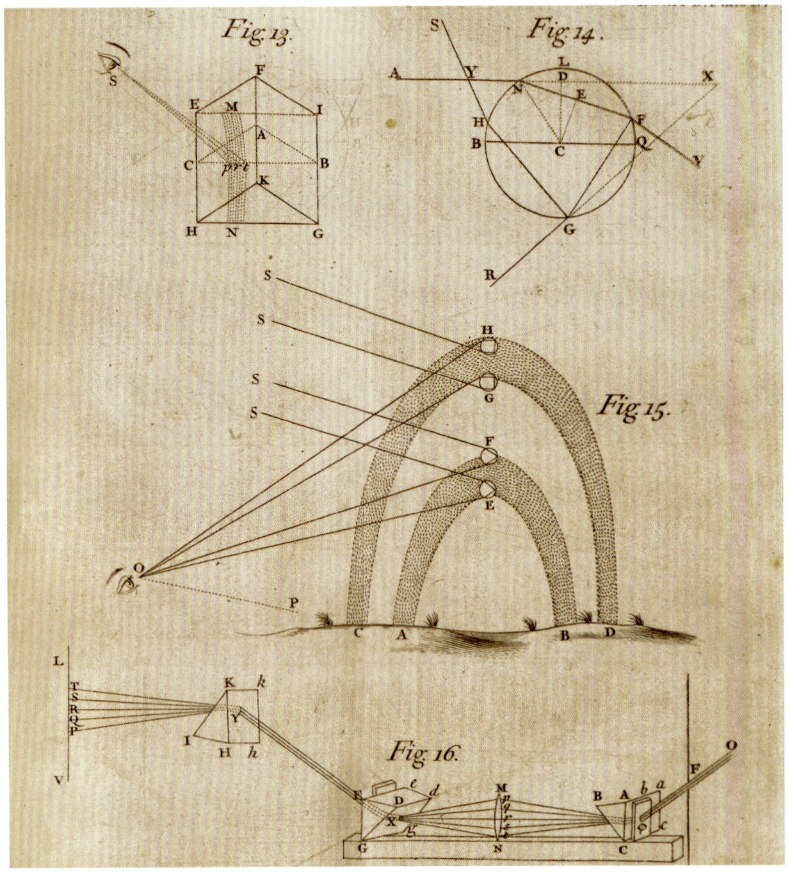

FIG. 31.01 Unknown engraver after unknown artist, "Experiments on the Nature of Colour: Figs. 15 and 16", in Sir Isaac Newton, *Opticks: or, A Treatise of the Reflexions, Refractions, Inflexions and Colours of Light. Also Two Treatises of the Species and Magnitude of Curvilinear Figures* (London: Printed for S. Smith, and B. Walford, 1704), plate IV. Science Museum, London

> the Sun shining, I made two Holes in the Window Shut of a darkened Room, thro' which letting the Sun's Beams pass, by means of two prisms (one near each Hole) I open'd the Rays coming from the Sun into the two colour'd Spectra . . . about 20 Feet from the Window where the Sun's Light came in, I caus'd a Piece of white Paper ¾ inch broad and 5 inches long, to be held within the refracted Rays . . . which produc'd these Colours in such manner, that by turning the Prisms round their Axes, I cou'd make the red Ray of the *Spectrum*, made by the one Prism fall upon one half of the Paper, and the purple Ray of the spectrum made by the other Prism fall on the other Half.[3]

This was exactly the sort of procedure that Newton and Caroline would have conducted at court, and in turn, the sort of effect we wanted to convey to the contemporary viewer. What could be a more appropriate subject for an exhibition devoted to an examination of the Enlightenment, than the discovery of optics, and the powerful metaphorical and aesthetic implications of experimentation with light itself?

Fittingly, our exploration has refracted into two different paths. First, Tyler Griffith, postdoctoral research associate at the YCBA, took the lead on a collaborative project with faculty and students from the Yale Center for Engineering Innovation and Design, in the context of a course offered to first-year undergraduates. Even at a university of Yale's size and quality, this class stands out for its innovative combination of theoretical engineering, hands-on training, and creativity. During the semester when the course is offered each fall, clients like ourselves with a problem to solve are matched with groups of students, who learn on the job. The diversity of topics is impressive—other clients for the course have included a football coach seeking technical support for his team's pass coverage drills, and our colleagues at the Yale University Art Gallery who are preparing an exhibition about the modernist kinetic artist Thomas Wilfred. (As it happens, that latter project also involves creating didactic materials helping to explain lighting effects to the museum visitor.) The team working on *Enlightened Princesses* has devised didactic materials that can be used to explicate the nature of light. We are asking them, in effect, to step into the role of Newton himself, showing visitors how light can be analyzed into its parts.

As part of our consideration of Newtonian optics, we have also had a lively intellectual exchange with the architects Anna Dyson and Octavio Perez, of the Center for Architecture Science and Ecology, Rensselaer Polytechnic Institute; and Michelle Addington, of the Yale School of Architecture. Addington has tailored her graduate seminar, entitled 'Studies in Light and Materials', to prepare students for the semester in which *Enlightened Princesses* will be on display at the YCBA. These experts are extending optical studies of the kind undertaken in the eighteenth century into the present day. They seek to shape the fall of natural light within architectural environments, using prisms, mirrors, and other equipment to direct solar illumination where it is needed, thus reducing dependence on artificial light and bringing about a lower carbon footprint. Dyson and Perez also pose questions about the effects of light on wellbeing, drawing from recent cognitive science that emphasizes physiological effects of spectrum shift, and even non-visible wavelengths.

One primary difference between this ongoing research and that of Newton's time has to do with the scale of the equipment involved. Today's manufacturing technologies allow for the creation of thin plastics that are embedded with innumerable microscopic diffraction elements. Such applications must be subtly calibrated to take into account shifts in natural light, over the course of the day and across seasonal changes, but when deployed properly can focus or split light to produce startlingly beautiful effects. Optical effects seem particularly well suited to the Yale Center for British Art, conceived by its architect, Louis Kahn, with light as a principal design consideration. Inspired in part by their work, we will be creating rainbows in the YCBA's grand entrance court, an undertaking that might best be considered an architectural experiment rather than an art installation. By projecting a large-scale spectrum in Kahn's vaulting space, we provide an appropriately monumental, yet still intangible, tribute to Newton's discoveries and Caroline's enlightened patronage.

THE PRINCESS'S PASSEPIED

As Joseph Roach explains in his contribution to this volume, the rhetoric of royalty was about much more than conspicuous consumption; it was a subtle and complex messaging system, which forwarded and reinforced political and social objectives.[4] What Walter Bagehot (cited by Roach) calls the 'visible form' of government, was nowhere more explicitly foregrounded than in dance. It was part of a passion for music that the Hanoverians brought to Britain—Caroline's patronage of George Frideric Handel was just as noteworthy as her support for Newton—and was explicitly theatrical, in that it involved the combined display of bodily skill, fashionable dress, and cultural taste.[5] Dance as stagecraft could function explicitly as propaganda, as when two of Caroline's daughters danced at court to demonstrate their good health following their inoculation against smallpox (then an experimental and risky procedure).[6]

Our route into this particular ephemeral practice was also one of the most visually unexpected objects in the *Enlightened Princesses* exhibition—a book of dance notation, which includes several pieces composed specially for court (see cats 11.05, 11.06). The graphic treatments of the choreography—complex interlacings of abstract symbols, each denoting a particular step, gesture, or attitude of posture—would primarily have been used by dancing masters to teach to their charges, who of course included members of the royal family. The transcription of dance is a difficult challenge, and remains so today; the system used in this case, which derived from French precedents, is incomprehensible not only to the average contemporary viewer but even to today's professional dancers.

Fortunately, we were able to turn to two specialists in baroque music, the dancers Ricardo Barros, artistic director of Mercurius Company, and Edith Lalonger, director of the troupe Plaisirs des Nations. Barros and Lalonger are not only able to read period notation, but also execute these dances with consummate grace. On the strength of their expertise, we conceived a plan: we would ask them to execute 'The Princess's Passepied', composed for Caroline in 1715, in the state apartments at Kensington Palace.[7] For the occasion, we also called upon musicians who played on period-appropriate instruments, and

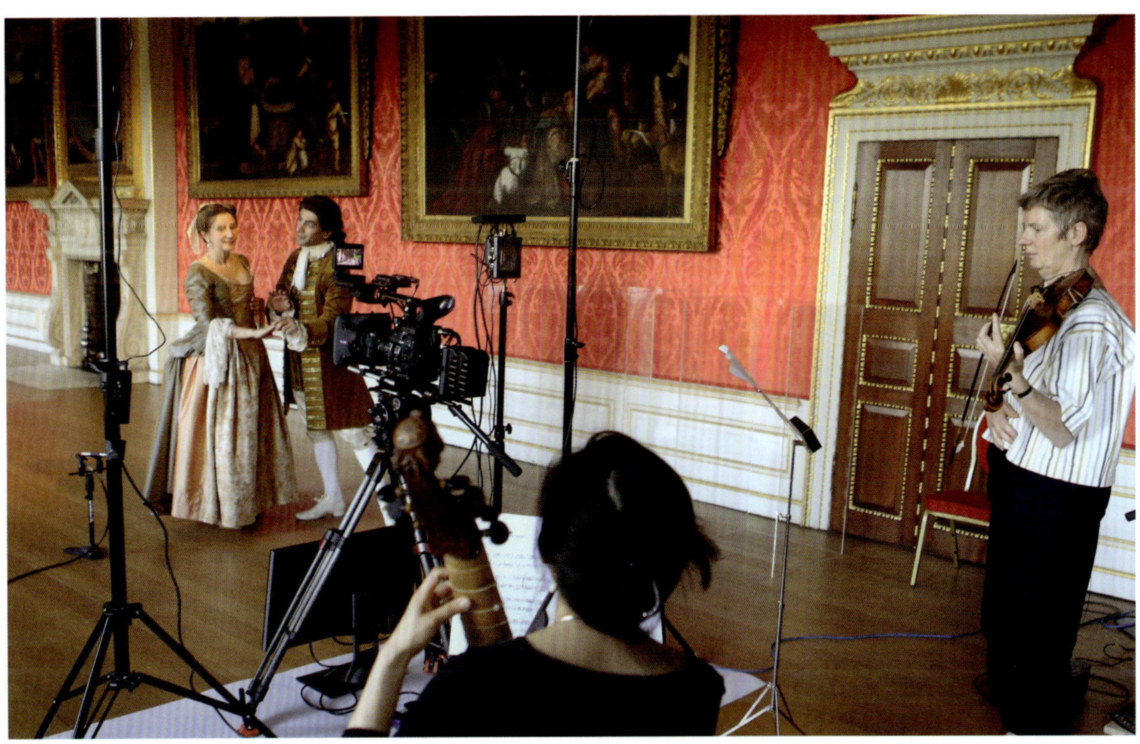

FIG. 31.02 Dancers and Musicians performing 'The Princess's Passepied' at Kensington Palace, 2016, photograph by Fred Fabre

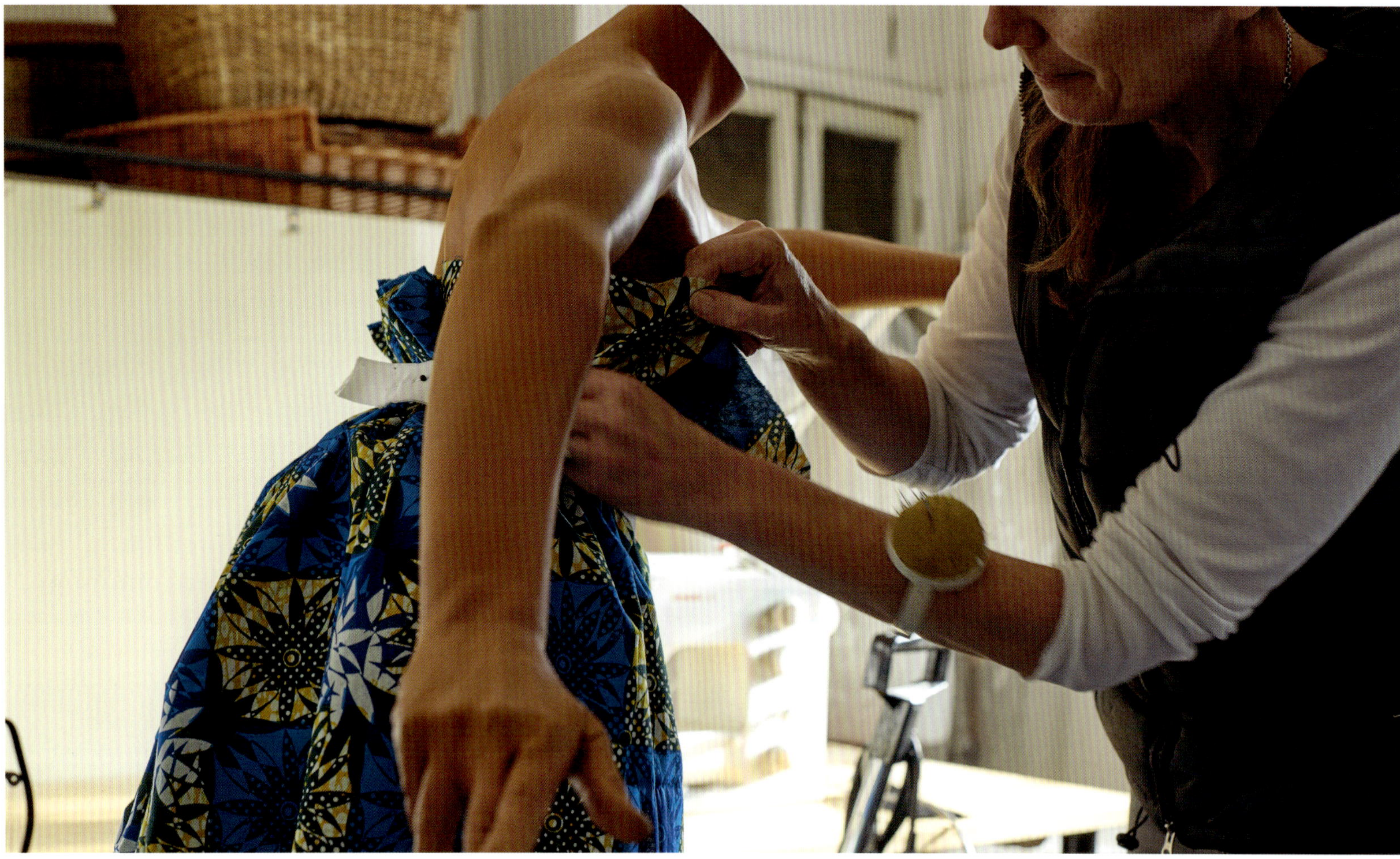

FIG. 31.03 Yinka Shonibare, *Mrs Pinckney and the Emancipated Birds of South Carolina*. Costumier Dee Sheehan dressing the figure, work in progress, December 2016. Photograph by Jonathan Bassett

commissioned filmmaker Fred Fabre to capture the performance in high definition digital video. This part of the process was completed in July 2016, and the results were amazing to behold: the couple created elegant symmetries, in period costume and in the exact location where such dances were staged in the eighteenth century (fig. 31.02). The sound was a particularly powerful element, as the music was accompanied by the gentle swish of Lalonger's skirt against the ground, and the light taps of both dancers' heels in the echoing hall.

To enhance further the visitor's experience, we asked Wendall K. Harrington and Ruppert Bohle at the Yale School of Drama to assist us in bringing the dance notation itself to life. Harrington and Bohle, veteran stage designers with particular expertise in projected imagery, are creating an animation of the abstracted movement that can be projected on the gallery or palace floor, in perfect time with the music. At the Center, the film of the dance, meanwhile, can be projected in the gallery at life-sized scale, nearly floor-to-ceiling, giving the illusion of a window into Kensington Palace, a portal to the past. Our hope is that visitors will imaginatively enter this historic space and the experience of the dance. It will certainly be a lesson in the formidable skills that court attendants were obliged to learn. We hope to suggest to visitors that dance was a key art form for eighteenth-century Britain: as aesthetically ambitious as painting and sculpture, and as vital a form of embodied knowledge as Enlightenment science.

THE ART OF CONVERSATION

Of all the ephemeral phenomena relevant to the story of the Enlightenment, perhaps the most significant was the art of conversation. It was through fostering discourse, live and in person, that Caroline, Augusta, and Charlotte had their greatest influence.[8] Though most of these exchanges are lost to us, eighteenth-century epistolary practices often did include transcriptions of conversations, particularly those involving royal persons. So it is that we have the substance of an exchange between Princess Augusta and Eliza Lucas Pinckney, which occurred in 1753 at the

White House in Kew. As recounted by Ben Marsh, Pinckney came to court with the express intention of showcasing her plantation's productivity to Augusta, who she believed could be enrolled as a support given her championing of British textile industries.[9] Her commercial success in growing indigo, and her cultivation of silk on an experimental basis, were perfectly aligned to the extractive economics that underpinned British mercantile strategy in the Americas. Equally well suited to courtly imperatives was her ability to satisfy Augusta's curiosity concerning the norms and customs of the Carolinas.

As exciting as the preserved traces of this conversation are for the historian, they are also shadowed by the most reprehensible aspect of eighteenth-century British Atlantic culture. Pinckney owned slaves and depended upon their labour and expertise in cultivating indigo, a common crop in West Africa. Pinckney's decision to wear a silk gown woven of fibres from her own plantation literally introduced the fabric of her slave economy to the court. She also gave Augusta a length of indigo-dyed silk as a present, supplementing the other gifts she had brought, including three live birds. Particularly unsettling for the modern reader is a passage in the conversation where Augusta expresses surprise that Pinckney's children are no darker in complexion for having been suckled by African-American wet-nurses.

To convey the full cultural freight of Pinckney and Augusta's encounter, we knew that we would need to do more than write an explanatory text. As with the less problematical topics of optics and dance, we wanted to materialize the conversation; but in this case, we also wanted to stress that issues of agency were at stake. We wanted to capture not just Pinckney's and Augusta's roles in the encounter, but also that of the slave labourers an ocean away from the court. And so we turned to an artist who has given much thought to such questions over the course of his career: Yinka Shonibare MBE (RA).

Known internationally for his sculptures incorporating a wide palette of historical referents, Shonibare has a close relationship with the Yale Center for British Art, having mounted an exhibition of works relating to Lord Nelson in the galleries in autumn 2016. He often works with the contrast between African and European visual culture, creating hybrids (for example) in which British costume is executed in patterned kente cloths. Narrative is ever present in his work, and is used critically; he invents fantastical departures from 'real' history to pose questions about the underlying shape, motives, and present legacy of that history.

Shonibare's commissioned sculpture for *Enlightened Princesses* is a quintessential example of his artistic methodology. He has imagined a female figure that could be seen as a stand-in for Eliza Pinckney or Augusta, but also functions as an allegory of Britain's transatlantic empire (fig. 31.03). The figure stands atop a globe of the world (as it appeared when the conversation between the two women occurred), a position that bespeaks both power and precariousness. The figure is garbed in a dress which reflects Mrs Pinckney's original garment, but in a cloth that evokes the textile heritage of the African slaves who were crucial to the success of her business enterprises. A culminating gesture is the replacement of the sculpture's head with a cage, with its door flung open. A flock of birds take flight around the figure, as if in escape—a literal reference to the buntings and goldfinch that Pinckney brought to court, but metaphorically, a representation of the slaves who lived and died under her control, and to Mrs Pinckney herself, a colonial subject of the Princess, who would only be given her liberty by the War of Independence in which her sons fought for the American cause.

Shonibare's extraordinary work indicates both the limits and the possibilities of curating the intangible. It is not in any sense a replacement for the artefacts or ideas of the African Americans who toiled on Pinckney's plantation. Nothing could provide that, nor should we forget: this erasure is itself a tragedy and an outrage. What Shonibare does provide, however, is an imaginative overlay. His art engages the discourse of the past, inventively and sensitively. He succeeds brilliantly at thinking through the problem of the intangible, in ways that parallel our efforts in representing science and movement. History is never entirely preserved. But when we see that a crucial narrative is missing, we should not fall silent. Rather, we should allow our imaginations to take flight.

NOTES

1. See, in this volume, Joanna Marschner, 'Enlightened Princesses: Caroline, Augusta, and Charlotte and the Shaping of the Modern World', 18–20.
2. Sir Isaac Newton, *Opticks: Or, a Treatise of the Reflexions, Refractions, Inflexions and Colours of Light* (London, 1704).
3. John Theophilus Desaguliers, 'An Account of Some Experiments of Light and Colours, Formerly Made by Sir Issac Newton, and Mention'd in His Opticks, Lately Repeated Before the Royal Society', *Philosophical Transactions* 29 (1 Jan. 1714): 437–38.
4. See, in this volume, Joseph Roach, 'Public and Private Performance', 163–66.
5. See, in this volume, Donald Burrows and Berta Joncus, 'Music and Representation', 203–6.
6. See, in this volume, Tyler Griffith and Samantha Howard, 'Inoculation', 431–33.
7. John Essex, *For the Further Improvement of Dancing*, 2nd ed (1715). The volume is dedicated to Caroline, then Princess of Wales.
8. See, in this volume, Clarissa Campbell Orr, 'Conversations, Knowledge, and Influence', 279–83.
9. See, in this volume, Ben Marsh, 'Visitor from South Carolina: Mrs Eliza Pinckney', 515–19.

We first mentiond our desire to see the Royal family an
and our little girl present the birds [it would not be an imp] to an Alderman
ity with whom we are acquainted he was so obliging t
e would mention it to the Bishop of Peter [Tutor to the Prince] with whom he
ell acquainted as soon as he had an opportunity and he
o doubt he would introduce her but as he had not soon
pportunity of speaking to the Bishop and we set out shortly
ath we were fearful of the birds dying before our retu
e mentiond it to a Gentm here who we knew to
erefore mentiond
cquainted with some about the Princess he very read
undertook it and next day went to Kew where the Pr
of ~~[illegible]~~ Wales and all her Family reside during the
mer Season they Gave the Princess a Prodigious Charac
nd said they ~~did not doubt but her Royal highness wo~~
~~e us Graciously~~ would mention it to her Royal highness
et him know at the same time how great a favour t
im by saying ~~it they were very Cautious how~~ it wa
thing very rarely permitted [it] ... all to those they were
[illegible] ... 18 ... afterward ...

Eliza Lucas Pinckney to Unidentified Person (1753)

PAGE I

We had the Honour not long since to carry our little girl to present the Princess Augusta with some birds from Carolina. It was attended with great difficulty as the attendance about the Princ[ess] are extreamly causious who they admit to her presence ~~I so p<. . .> least they should be after wards trouble some~~.

We first mentioned our desire to see the Royal family and t[o] have our little girl present the birds ^if it would not be an improp<. . .> to an Alderman of [the] Cyty with whom we are acquainted he was so obligin to s[ay] he would mention it to the Bishop of Peter ^[Tutor to Prince?] with whom he was well acquainted as soon as he had an opportunity and he ma[de] no doubt he would introduce her but as he had not soon after an<. . .> opportunity of speaking to the Bishop and we set out shortly for Bath we were fearful of the birds dying before our return <. . .> therefore mentioned it to a Gentm here who we know to be acquainted with some about the Princess he very readily undertook it and next day went to Kew where the Prince[ss] of ~~whales~~ Wales and all her Family reside during the Summer Season; they Gave the Princess a Prodigious Character and said they ~~did not doubt but her Royal highness would so <. . .> provided~~ would mention it to her Royal Highness [but] let him know at the same time how great a favour the[y did] him, by saying ~~it they were very Consious how~~ it was a thing very rarely permit[ted, especially to those they were not] acquainted with least they ~~might be after ward~~ should ~~b[e] trouble some~~ have anything to ask after wards w[ch]: might be [troublesome to the Princess] <. . .> but depended upon him that he would not introduce any persons but such as w[ere] proper to be presented to ~~the Princess~~ her Royal highne[ss.]

The Gentm said his own Character was so much concernd in t[he] case, that he should not presume to mention any but such a[s he]

PAGE 2

knew to be persons of Character and Distinction in the Country from whence they came as he knew this Gentm was, that he was one of his Majestys Council of So. Carolina and ~~a Gentm that his Majesty~~ had nothing to ask but ^was desirous to show the affection he had to his Majesty and all his Royal House, and his inclination to see the Family, that his Majesty had himself ^taken particular notice of him honoured him with a conference since his arrival upon wch: this last ~~last~~ especially they said they would let the Princess know they returnd and said the Princess would see us and we were desired to go at Eleven o'Clock any day the next week wch in a few days we did, we exceeded our time a little and found the Princess gone ~~out~~ a airing with the Princess Augusta ^and it was uncertain when she would return we carried the birds in the Coach with us and wrote a Card to give the Child in her hand in case we should not go in with her. The card was this Miss Harriotte Pinckney, daughter to to CP Esqr, one of his Majestys Council of So Carolina pays her Duty to her highness and humbly begs leave to present her with ~~a pres~~ an Indigo bird, a Non-

Detail of p. 524, cat. 30.10

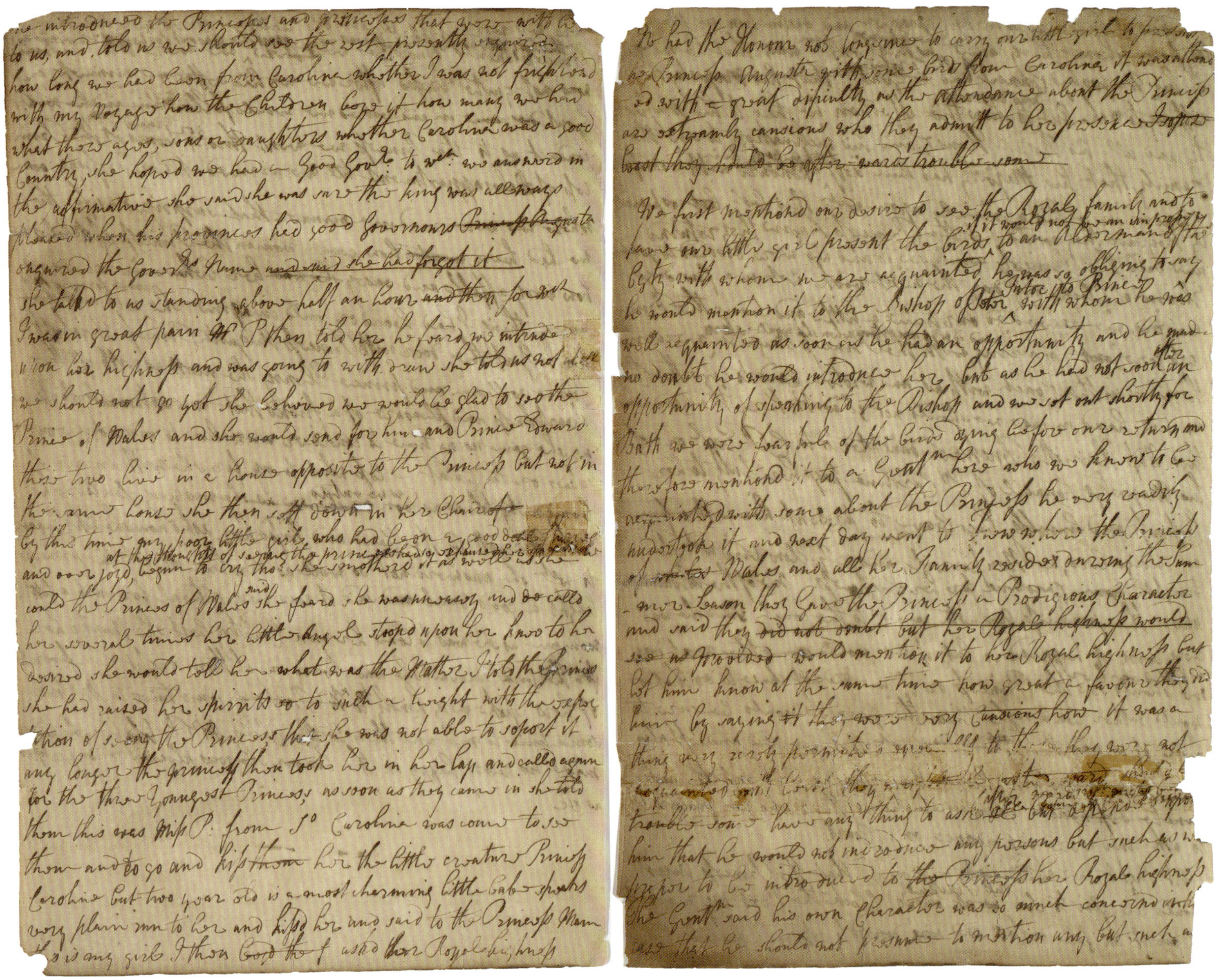

parreil, and a Yellow bird wch she has brought from Carolina for her highness. One of the attendance upon the princess Augusta came to the coach said she was very sorry it happend so but if we would come the next day a little earlier we should see the Princess or if we did not chuse to do that or would rather leave the birds the princess would be sure to [hear of us, and to have them; wch: last we did, and left the Card. alsoe, and returned] home, lamenting as we went ~~the unfor[tu]nate dependece~~ uneasy situation of those who had favours to ask or were dependance on a Court.

At Night we ~~were <. . . .>~~ had a message that the Princess Augusta [w]ould be glad to See ~~[us?]~~ Miss Pinckney at one o'Clock the next day we acc[ordingly] went in full dress and was desired to sit in a parlour [below?]

PAGE 3

[where we were recd by an old lady, a foreigner, till the] Princess should know we were there ^This Lady told us the Princess was very sorry [she was out yesterday afternoon] we were told the Princess was not quite dressd after we ~~we~~ had sett some minuets a Gent ~~[do ?]~~ came in and desired we would follow him and he would conduct us to the Princess we went thro: 3 or 4 grand rooms of the Princess of Wales appartment till we arrived at her dressing room where we were recd in a manner that surprized us for tho' we had heard ~~much~~ how good a woman the Princess of Whales w[as] how very affable and easey ~~she was~~ her behaveour exceeded every thing I had heard or could imagine she ^came forward and recd us ~~at the~~ her self at the door with Princess Augusta, Princess Elizabeth, Prince william, and prince Henry she mett us with all the chearfulness and pleasure of a friend that was extreamly glad to see us she gave us no time to consider ~~or w~~ how to introduce our selves or to be at a loss what to say for she with an air of benignity told us as soon as we enterd ~~as she saw us~~ she was very glad to see us took Harriott by the hand and kissd her askd how she likd England to wch: she answerd not so well as Carolina at wch: the princess laughd a good deal and said it was very natural for such a little woman as she was to love her own Country best she thankd her for the birds and said she was afeard one of them might be a favourite of hers spoke very kindly to us ~~to Mr Pinckney~~. sometimes to Mr Pinckney, some times to me and then to the Child. Mr P: told her she had made us very happy in the honour she was pleased to confer upon us &c.

PAGE 4

She introduced the Princes and princesses that were with h[er] to us, and told us we should see the rest presently enquired how long we had been from Carolina whether I was not frightend with my Voyage how the Children bore it how many we had what their ages, sons or daughters whether Carolina was a good Country she hoped we had a Good Govr: to wch: we answerd in the affirmative she said she was sure the king was allways pleased when his provinces had good Governours ~~Princess Augusta~~ enquired the Goverrs Name and said she had forgot it.

she talkd to us standing about half an hour ~~and then~~ for wch I was in great pain Mr P then told her he feard we intruded upon her highness and was going to with draw she told us not at all we should not go yet she believed we would be glad to see the Prince of Wales and she would send for him and Prince Edward these two live in a house just opposite to the Princess but not in the same house she then sett down in her Chair ~~of a~~ by this time my poor little girl who had been a good deal [flurried] and overjoyd, at the thoughts of seeing the princesses had so exhaused her <. . .> she, began to cry tho: she smotherd it as well as she could the Princess of Wales said she feard she was uneasey and ~~de~~ calld her several times her little Angel, stoopd upon her knee to her and desired she would tell her what was the Matter I told the Princes[s] she had raised her spirrits ~~so~~ to such a height with the expectation of seeing the Princes's that she was not able to soport it any longer the princess then took her in her lap and called again for the three youngest Princes's as soon as they came in she told them this was Miss P: from So Carolina was come to see them and to go and kiss ~~them~~ her the little creature Princess Caroline but two year old is a most charming little babe speaks very plain run to her and kissd her and said to the Princess [Marm?] this is my girl I then ~~begd the f <. . .>~~ asked her Royal Highness

PAGE 5

[if she would permit me to kiss the little one, she replyd, pray do,] and ordered prince Frederick 3 years old, to come ~~to me~~ and ask me if he was not a good pretty little foot boy.

I should observe that as soon as we were introduced the attendance all withdrew and the princess shut the door and when the Princess ordered the little ones in ~~Pr~~ there was ~~no Servant~~ none of the attendance calld nor when she sent for the Prince of ~~wh~~ Wales but the Princess Augusta went out of the room herself on these Messages to some one without wch: was 4 times while we stayd, there was in the room a great [deal] of China upon two Cabinets the Princess got up herself and reachd one of the figures to please Harriott and another time desired the Princess Augusta to get one wch: was out of her reach she got a chair and stood up and reachd it she then calld for a little chair for one of the little ones who I fancy was not well for, 'tis not usual for any ~~body~~ to sit in her presence. wch: Princess Augusta brought herself.

this, you imagine must seem pretty extraordinary to an american.

the two youngest set themselves down upon the Carpet at her feet, I told her highness for by this time I could converse with as much ease with her as with almost any of my acquaintance such was her condescension and her familiar engaging manner.

I told her ~~the little~~ princess Caroline (the youngest of all) was very humble she sayd she was a pre[tty] good Girl; then addressd herself to her have you ever been in the Corner my Queen no Madm says the pretty creature never ~~not~~ in the corner I am afraid you have replyd the Princess upon wch Prince Frederick sayd No Madm she was never in the Cor[ner] but that Sister has pointing to Harriott who he ~~saw~~ had seen ~~been~~ cry[ing] her Mamma puts her in the Corner sometimes the Princess held up her finger at him told him twas he that would be put in the Corner then Ill go to Ca[rolina, says he. Well then, good by to you; replyd the Princess.]

PAGE 6

[S]he then bid H set down by her in the chair Princess Emelia had rose out ~~of wh~~ I told her I could not suffer her to set in her presence puh-puh says the Princess she knows nothing of all that and sat her down and told her she had no pretty things here for her but when she went to London she would get something that was pretty ~~for~~ and send her by this time the little ones were called to dinner I observed that tho: they were quite easy in their behaviour and seem'd to be under no restraint Young as they were they

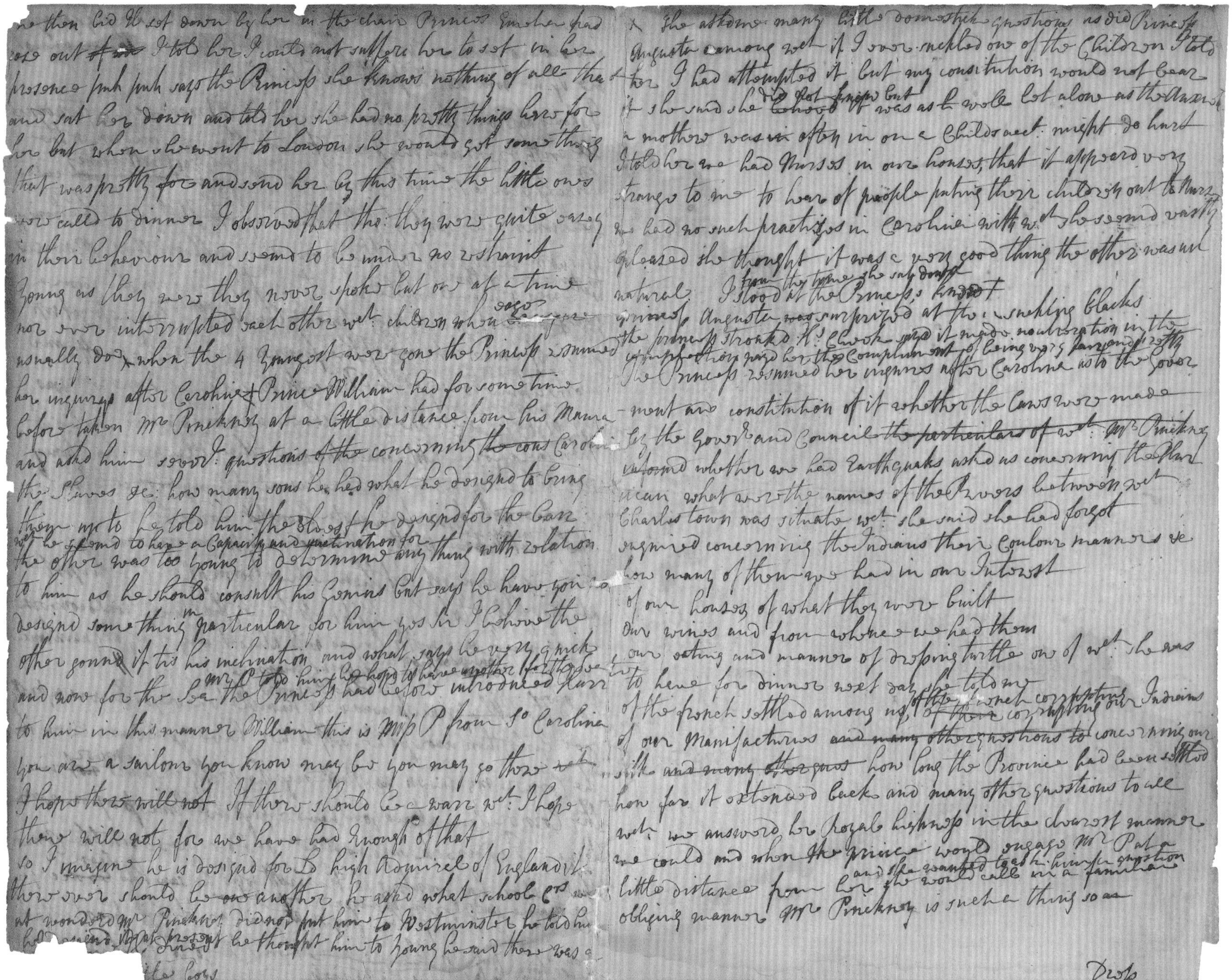
ne then bid the set down by her in the chair Princess Emelia had
rose out of I told her I could not suffer her to set in her
presence pish pish says the Princess she knows nothing of all that
and sat her down and told her she had no pretty things here for
her but when she went to London she would get some thing
that was pretty for and send her by this time the little ones
were called to dinner I observed that tho' they were quite easy
in their behaviour and seemd to be under no restraint
young as they were they never spoke but one at a time
nor ever interrupted each other wch. children when eager
usually do when the 4 youngest were gone the Princess resumed
her inquirys after Carolina Prince William had for some time
before taken Mr Pinckney at a little distance from his Mama
and asked him several questions of the concerning the cons Carolina
the Slaves &c: how many sons he had what he designed to bring
them up to he told him the eldest he designed for the Barr
wch he seemd to have a Capacity and inclination for
the other was too young to determine any thing with relation
to him as he should consult his Genius but says he have you no
designd some thing in particular for him yes sir I believe the
other young. If tis his inclination and what says he very quick
and now for the Sea the Princess had before introduced Harr
to him in this manner William this is Mrs P from So Carolina
you are a sailor you know may be you may go there
I hope there will not If there should be a warr wch. I hope
there will not for we have had enough of that
so I imagine he is designd for Lo high Admiral of England it
there ever should be one another he asked what school &c
at wondered Mr Pinckney did not put him to Westminster he told him
... he thought him to young he said there was a
... boys

x She asked me many little domestick questions as did Princess
Augusta among wch if I ever suckled one of the Children I told
her I had attempted it but my constitution would not bear
it she said she did not know but it was as well let alone as the anxiety
a mother was in after in on a Childs acct. might do hurt
I told her we had Nurses in our houses, that it appeard very
strange to me to hear of people puting their children out to Nurse
we had no such practices in Carolina with wch. she seemd vastly
pleased she thought it was a very good thing the other was unnatural
I stood at the Princess's knee from the time she sat down
Princess Augusta was surprized at the suckling blacks
the Princess stroked Hs Cheek ... it made no alteration in the
complection ... the Compliment of being very fair and pretty
The Princess resumed her inquirys after Carolina as to the Gover-
-ment and constitution of it whether the Laws were made
by the Govr. and Council the particulars of wch. Mr Pinckney
informd whether we had Earthquaks asked us concerning the Harbour
... what were the names of the Rivers between wch
Charles town was situate wch. she said she had forgot
enquired concerning the Indians their Colour manners &c
how many of them we had in our Interest
of our houses of what they were built
Our wines and from whence we had them
our eating and manner of dressing turtle one of wch. she was
to have for dinner next day she told me
of the french settled among us, of the french corrupting our Indians
of our Manufactures and many other questions to concerning our
silk and many other ques how long the Province had been settled
how far it extended back and many other questions to all
wch. we answered her Royal highness in the clearest manner
we could and when the Prince would engage Mr P at a
little distance from her and she wanted to ask him a question she would call in a familiar
obliging manner Mr Pinckney is such a thing so or

Drop

never spoke but one at a time nor ever interrupted each other wch: children when eager ~~<. . .>~~ usually do when the 4 youngest were gone the Princess resumed her inquiries after Carolina + Prince William had for sometime before taken Mr Pinckney at a little distance from his Mama and asked him sevral: questions ~~of the~~ concerning ~~the cons~~ Caro[lina] the Slaves &c: how many sons he had and what he designd to bring them up to he told him the eldest he designed for the barr wch he seemd to have capacity and inclination for the other was too young to determine any thing with relation to him as he should consult his Genius but says he have you [not] designd something in particular for him yes Sir I believe the other gown if tis his inclination and what says he very quick and none for the Sea Mr P told him he hopd to have another for the Sea the Princess had before introduced Harr to him in this manner William this is Miss P from So Carolina you are a sailour you know may be you may go there ~~wch I hope there will not~~ If there should be a warr, wch: I hope there will not for we have had Enough of that.

so I imagine he is designd for Ld High Admiral of England if there ever should be ~~one~~ another he askd what school Crs [was] at wondered Mr Pinckney did not put him to Westminster he told him he [design]nd it, but at the present time he thought him to young he said there was [a] <. . .> [for little boys.]

PAGE 7

X She askd me many little domestiek questions as did Princess Augusta among wch if I ever suckled one of the Children I told her I had attempted it but my constitution would not bear it she said she did not know but ~~believd~~ it was as ~~[l?]~~ well let alone as the anxie[ty] a mother was ~~in~~ often in on a Child's acct: might do hurt I told her we had Nurses in our houses, that it appeard very strange to me to hear of people puting their children out to Nurse we had no such practizes in Carolina with wch she seemed vastly pleased she thought it was a very good thing, the other was unnatural. I from the time she sat down stood at the Princess's knee+

princess Augusta ~~was~~ surprized at the suckling blacks the princess stroakd Hs: cheek sayd it made no alreration in the complexion payd her the compliment of being very fair and pretty The Princess resumed her inquires after Carolina as to the Government and constitution of it whether the Laws were made by the Goverr and Council ~~the particulars of wch: Mr Pinckney in formd~~

whether we had Earthquaks askd us concerning the Hurrican

what were the names of the Rivers between wch Charles town was situate wch: she said she had forgot

enquired concerning the Indians their colour manners &c how many of them we had in our Interest of our houses, of what they were built

our wines and from whence we had them

our eating and manner of dressing turtle one of wch she was to have for dinner next day she told me of the french settled among us, of the french corrupting ~~of them~~ corrupting our Indians

[of] our Manifactures ~~and many other questions to~~ concerning our Silk ~~and many other ques~~ how long the Province had been settled

how far it extended back and many other questions to all wch. we answerd her Royal highness in the clearest manner we could and when the Prince would engage Mr P. at a little distance from her ^and she wanted to ask him a question she would call in a familiar obliging manner Mr Pinckney is such a thing so ~~an~~

[Dress?]

PAGE 8 [BLANK]

PAGE 9

[who live in a house opposite to her, so that we saw all nine children to gether, and the Princess in the midst, and a most lovely family it was We kissed after we had been there two hours her Royal Higness hand and then withdrew she ordered Prince Edward to see us to the Door.

I hope you will pardon this intruding on your time I know there are many Chit chat negligent things wch: have a tolerable air in conversation that make but a poor appearance when one comes to write them down and subscribe to them in a formal manner but when I begin to write to my friends in Carolina I dont know how to conclude and this desire of conversing with them may make me a very troublesome Corrispondant tho: I hope it will at the same time show how much I am]

[Dr Madm
Yr affectionate and obedt Servt]
[E. P]

REPRINTED FROM:

Eliza Lucas Pinckney to Unidentified person, [1753], in *The Papers of Eliza Lucas Pinckney and Harriott Pinckney Horry Digital Edition*, ed. Constance Schulz. Charlottesville: University of Virginia Press, Rotunda, 2012. http://rotunda.upress.virginia.edu/PinckneyHorry/ELP1015

The Papers of Eliza Lucas Pinckney and Harriott Pinckney Horry Digital Edition
The University of Virginia Press
© 2012–2016 by the Rector and Visitors of the University of Virginia

Bibliography

ARCHIVAL SOURCES

DUCHY OF CORNWALL OFFICE

Frederick Prince of Wales, household accounts and vouchers, 1729–51.

THE ROYAL ARCHIVES (RA)

Papers and correspondence. Queen Caroline.
RA Add. MS. 17/75
RA Add. MS. 52760–52766
RA Add. MS. 52773
RA Add. MS. 52824
RA Add. MS. 52895–52900
RA Add. MS. 52936
RA Geo Add. MS. 1/49
RA Geo Add. MS. 2
RA Geo Add. MS. 16
RA Geo Add. MS. 17/65/2
RA Geo Add. MS. 28/2–18
RA Geo Add. MS. 54000
RA Geo Add. MS. 55616
Accounts and bills. Queen Caroline.
RA Add. MS. 53993–54005. May 1730–March 1731.
RA Add. MS. 54006–54014. May 1731–Sept. 1731.
RA Add. MS. 54015–54023. Sept. 1731–33
'A Catalogue of the Royal Library of Her late Majesty, Queen Caroline, distributed into faculties', 1743. RCIN 1028932.
Papers and Correspondence. Princess Augusta.
RA. Add. MS. 54237–54250. 1751–71.
Accounts and bills. Princess Augusta.
RA Add. MS. 54969–55205. 1749–51.
RA Add. MS. 55114–55661. 1736–71
RA Add. MS. 55206–55421. 1749–51.
RA Add. MS. 55432–55651. 1755–71.
RA Add. MS. 73956–74334. 1736–61.
RA Geo Add. MS. 17/65/1.
Papers and Correspondence. Queen Charlotte.
RA Add. MS. 43/4–22.
RA Add. MS. 45/1–3 [Diary, 1789.]
RA Add. MS. 36357B [To Mme de la Fite, 1770s?].
RA Add. MS. 36413–36774.
Accounts and bills. Queen Charlotte.
RC Add. MS. 36836–36948.
RA Add. MS. 36950.
Buckingham House, 'Queens Apartments, No.3b. [wall plan]', c. 1774. RCIN 926333
Buckingham House, 'The Queen's Dressing Room (later the Blue Drawing Room) . . . Hanging plan', c. 1774. RCIN 926331
RA Establishment books 21–30.

THE SURVEYOR'S OFFICE, ROYAL COLLECTION TRUST

Barnard, F. A. Windsor Library Catalogue, 'A Catalogue of His Majesty's library at Windsor 1780', rev. 1812. RCIN 1028949.
A Catalogue of Pictures Belonging to King James the Second (Copied from a ms. in the Library of the Earl of Oxford): To which is Appended a Catalogue of the Pictures, Drawings, Limnings, Enamels, Models in Wax and the Ivory Carvings etc. at Kensington in Queen Caroline's Closet, next the State Bedchamber, London, 1758, with ms. notes by Horace Walpole. RCIN 1112557.
'Catalogue of the Pictures in the Palace at Kensington', 1818. RCIN 11125430.
'A Catalogue of his Royal Highness the Prince of Wales's Pictures in Leicester House', Sept. 1749. RA Geo Add. MS. 1/54.
'Catalogue of the Pictures at the late Queen's house, St James's Park. With: Catalogue of the pictures at St James's Palace', 1819. RCIN 1112572.
'The Collection of Pictures Paintings &c at Kensington, Hampton-court and in the Castle of Windsor 1750', 3 vols. RCIN 1112905–1112907.
Combe, Charles. *Catalogue of the Several Series of Modern Medals & Coins in His Majesty's Collection*, 1771, rev. 1814.
Garrard, R. & S. 'Descriptive Inventory of the Various Services of Plate, etc, at Windsor Castle and Buckingham Palace, belonging to the Crown', 3 vols. RCIN 1048090–1048092.
—. *Descriptive Inventory of the Various Services of Plate, &c. belonging to the Crown*, London, 1872. RCIN 1114696.
George I Inventory, transcript of BL Stowe MS. 567, 'Royal Pictures at Kensington, Hampton Court, Windsor & St. James's', [before 1723?]. RCIN 1112573.
[Glover, John]. *A Catalogue of the Miniatures in the Possession of Her Majesty the Queen*, London, 1851.
Inventory A, 'A catalogue of the drawings and prints as they are arranged in book cases', c. 1770–1840. RCIN 1155585.

'An Inventory of the household furniture at Buckingham Palace taken May 2nd 1825'. RCIN 1114747.
'Inventory of Miniatures in the Royal Library, Windsor Castle', March 1870. RCIN 1112528.
Inventory Taken of Sundry Jewels etc. at Windsor Castle 16 Sept. 1830, and the Following Days by Messrs. Bridge in the Presence of Major General Wheatley, Major General Stephenson and Sir Frederick Watson, 1830–38. RCIN 1114749
Lowman, Henry, and others. 'A Catalogue taken of the Pictures, which are in their Majesties Publick, and Private Lodgings, in the Palace at Kensington', c. 1736. RCIN 1112553.
Jutsham, R. D. 'An account of Furniture &c. received and deliver'd by Benjamin Jutsham on account of His Royal Highness The Prince of Wales at Carlton House', 1806–20. RCIN 1112484.
—. 'Ledger of furniture &c received by Benjamin Jutsham', vol. 2, 23 June 1816–17 Dec. 1829. RCIN 1112775.
—. 'A Catalogue of arms: the property of HRH The Princes of Wales at Carlton House, compiled by Benjamin Jutsham', 7 vols., late eighteenth century–1827, RCIN 1113358–1113365.
Kew House Inventory, 'Catalogue of His Majesty's pictures at Kew House', c. 1800–1805. RCIN 1112543-4
Kew Library Catalogue. 'A Catalogue of His Majesty's library at Kew', 1785. RCIN 1028953.
Library Catalogue. 'A Catalogue of His Majesty's library at Kew', 1780. RCIN 1028954.
'A List of Her Majesties Pictures in Kensington, Hampton Court and Windsor Castle', c. 1705–10. RCIN 1112574.
Royal Collection Inventory. *VR Inv. 24 Feb. 1866*, 1866.
Rundell, Bridge, and Co. 'Inventory of sundry articles of jewellery & in the strong closet in Windsor Castle belonging to Her Majesty', 1837. RCIN 1114746.
Skerrett, Marianne, and John Glover. List bound into 'Catalogue of Enamel Miniatures in Room 243', by P. Fraser Tythers, [1845–50?]. RCIN 1112529b.
Vertue, George. *The Collections of Pictures, Paintings etc at Kensington, Hampton Court and in the Castle of Windsor*, 1750. RCIN 1112785.
—. *A Catalogue and Description of Charles the First's Capital Collection of Pictures, Limnings, Statues etc.*, London, 1756.
Vulliamy, B. L. 'List of clocks belonging to His Majesty taken by command of His Majesty William the 4th by B. L. Vulliamy', 1835. RCIN 1114929.

ALTHORP HOUSE, NORTHAMPTONSHIRE

MS. B8

BODLEIAN LIBRARY, OXFORD

Bodleian Maps 40

BRITISH LIBRARY, DEPARTMENTS OF MANUSCRIPTS AND MAPS (BL)

BL Add. MS. 4308–4309. Thomas Birch Papers, Stephen Hales to Thomas Birch.
BL Add. MS. 11511
BL Add. MS. 15752
BL Add. MS. 16941
BL Add. MS. 20013, ff. 1–13. 'A List of Pictures at Kensington House', c. 1710.
BL Add. MS. 20101
BL Add. MS. 22256. Stafford Papers.
BL Add. MS. 22626. Suffolk Papers, Lord Chesterfield to Mrs Howard (afterwards Lady Suffolk), 13 Aug. 1728.
BL Add. MS. 32684. Newcastle Papers, royal letters, 1732–68.
BL Add. MS. 41354. Martin Papers, vol. 9.
BL Add. MS. 47067. Egmont Papers, memorandum book, Sept. 1736–May 1737.
BL Add. MS. 47068. Egmont Papers, journal, May 1737–July 1738.
BL Add. MS. 47069. Egmont Papers, journal, 12 July 1738–[?].
BL Add. MS. 47070. Egmont Papers, journal, 22 Aug. 1739–25 Dec. 1740.
BL Add. MS. 47071. Egmont Papers, journal, Dec. 1740–Aug. 1747.
BL Add. MS. 51396
BL Add. MS. 61860. North (Sheffield Park) Papers, vol. 1, 1737–1828.
BL Add. MS. 81088. Dodington Papers, diary, March 1749–Feb. 1761.
BL C120.h.6 (1–7)
BL Egerton MS. 1717
BL Sloane MS. 567
BL Sloane MS. 1968
BL Sloane MS. 3516
BL Sloane MS. 4034
BL Sloane MS. 4068
BL Sloane MS. 4076
BL Stowe MS. 567. 'Inventory of Pictures at Kensington, Hampton Court, Windsor and St James's', c. 1720.
BL Maps H. 258
BL Maps K. Top. 16

BRITISH LIBRARY, ROYAL MUSIC COLLECTION

BL HS.74/1042. 'An Anthem for the use of the Magdalen Chapel', Magdalen Hospital. London, 1761.
BL MS. 1.9.d.11, f. 161r. [George Frideric Handel], lessons composed for Princess Louisa.
BL RM 21.a.9. J. C. Bach, Harpsichord lessons, fingerings, etc.
BL RM 22.a.17. J. C. Bach, 'Ode on the auspicious Arrival and Nuptials of Her Present Most Gracious Majesty, Queen Charlotte by John Costerman'.
BL RM 22.9.16. J. C. Bach, 'Happy Morn, Auspicious Rise', 1761.
BL RM 22.A.17. [J. C. Bach], 'Ode on the auspicious arrival and Nuptials of her present most gracious Majesty Queen Charlotte, by John Lockman. Set to music by Mr Bach'.

CHATSWORTH HOUSE, DERBYSHIRE

Chatsworth. Kent/Burlington House 26/69/76
Chiswick & Chatsworth & Miscellaneous Drawings, 46 and 50
MS. 127/6
MS. 212/2
MS. 230/5

KONINKLIJK HUISARCHIEF, THE HAGUE, THE NETHERLANDS

KH A17/404
KH A17/412
KH A17/430
KH A17/470
KH A30.430. Letters from Caroline of Hanover; George I; George II; Frederick, Prince of Wales; William, Duke of Cumberland; Augusta of Saxe-Gotha; Princess Mary; and Princess Amelia.
KH A.30.430.A. Letters from Frederick William of Prussia.
KH A30.470. School books of Anne of Hanover.
KH A30.480. Mourning regulations for Caroline of Ansbach, 1737–38.

THE LEWIS WALPOLE LIBRARY, FARMINGTON, CT

LWL 49/1609. 'A Collection of the most remarkable tracts published in the reign of King George III'.
LWL MS. 10. Augusta, Princess of Wales, accounts.
LWL Vol. 74. Dr Samuel Clarke, letters.
Queen Charlotte's Collection of Hogarth Works.

LINNEAN SOCIETY, LONDON

GB-110/JES/COR/6/45. Aylmer Bourke Lambert to James Edward Smith, 1 Oct. 1805.

MOUNT STUART, BUTE ARCHIVE

BU 98/2
BU 98/3
BU 98/4
BU 98/5
BU 98/6
BU 98/7
BU 98/8
BU 98/9
BU 98/10
BU 108/97
BU 109 NiL
BU 114 NiL
BU 115 NiL
BU 118/2
BU 118/9

THE NATIONAL ARCHIVES, KEW (TNA)

Audit Office Papers
AO1, AO3

Lord Chamberlain's Papers
LC2, LC5, LC9, LC11

Lord Steward's Papers
LS1, LS13, LS58

Map Room
MR 696

Shaw, William Arthur, *Calendar of Treasury Books and Papers: Preserved in Her Majesty's Record Office*, London, 1897–1903.

Treasury Papers
T1, T27, T29, T52, T54, T56

Works Papers
Works 1, Works 4, Works 5, Works 6, Works 32

NATURAL HISTORY MUSEUM, LONDON

Young, William. 'A Natural History of plants containing the productions of North and South Carolina of upwards of three hundred different species which have been carefully collected, drawn, and coloured from Life in the year 1767. To which will be added a description of the soil and Nature of their growth, as likewise remarks of a great many of their medicinal qualities, with a general description of the province' [96 sheets with watercolour on paper, and 5 pages ms.], 1767, Philadelphia.

PREUSSISCHES STAATSARCHIV, HANOVER

HStA Hanover Dep. 103, XXIV, Nr. 2487 (Marianburg). 'Specification d'Argenterie de S. A. Madame l'Electrice. Inventer 1709'.

THE ROYAL BOTANIC GARDENS, KEW

Rocque, John. *An Exact Plan of the Royal Palace Gardens and Park at Richmond*, [London], 1734.
—. *An Exact Plan of the Royal Palace Gardens and Park at Richmond*, London, 1754.

ROYAL INSTITUTE OF BRITISH ARCHITECTS, LONDON

Adam, Robert. Sketchbook, c. 1750.

SIR JOHN SOANE'S MUSEUM, LONDON

Adam vol. 56/25, 33–34
Adam vol. 147/192–198

SOUTH CAROLINIANA LIBRARY, UNIVERSITY OF SOUTH CAROLINA, COLUMBIA

Manuscripts fiche M.7. Pinckney family papers, 1703–1847.
The Papers of Eliza Lucas Pinckney and Harriott Pinckney Horry Digital Edition, ed. Constance B. Schulz, Charlottesville, VA, 2012, http://rotunda.upress.virginia.edu/PinckneyHorry/default.xqy.

VICTORIA AND ALBERT MUSEUM, LONDON

V&A E.871-1928–E.885-1928. William Kent, preparatory drawings for *The Faerie Queene*.

CONTEMPORARY NEWSPAPERS AND JOURNALS

Daily Courant
Daily Gazetteer
Daily Journal
Daily Post
Dawks's News Letter
Free Briton
General Evening Post
Gentleman's Magazine
London Chronicle
London Daily Post and General Advertiser
London Journal
London Magazine and Monthly Chronologer
The Political State of Great Britain
The Spectator

PUBLISHED SOURCES: BOOKS

Adamson, J., ed. *The Princely Courts of Europe: Ritual, Politics and Culture under the Ancien Régime, 1500–1750*, London, 1998.
Addison, Joseph. *To Her Royal Highness the Princess of Wales, with the Tragedy of Cato, Nov. 1714, To Sir Godfrey Kneller, on his Picture of the King*, London, 1716.
Aiton, William. *Hortus Kewensis: or a catalogue of the plants cultivated in the Royal Botanic Garden at Kew*, London, 1789.
Aiton, W. T., and Franz Bauer. *Delineations of Exotick Plants Cultivated in the Royal Garden at Kew*, London, 1796.
Albin, Eleazar. *A Natural History of English insects*, London, 1749.
Alexander, David. *Richard Newton and English Caricature in the 1790s*, Manchester, 1998.
Alexander, H. G., ed. *The Leibniz-Clarke Correspondence, together with extracts from Newton's Principia and Opticks*, Manchester, 1956.
Alexander, Michael. *Omai, Noble Savage*, London, 1977.
Allen, Brian. *Francis Hayman*, New Haven and London, 1987.
Allison, Ronald, and Sarah Riddell, eds. *The Royal Encyclopedia*, London, 1991.

Alvensleben, U. von, and H. Reuther. *Herrenhausen: Die Sommerresidenz des Welfen*, Hanover, 1966.
André, Bernard. *The Life of Henry VII*, trans. and intro. Daniel Hobbins, New York, 2011.
Andrew, Donna T. *Philanthropy and Police: London Charity in the Eighteenth Century*, Princeton, 1989.
Angelo, Henry. *Reminiscences of Henry Angelo, with memoirs of his late father and friends, including numerous original anecdotes and curious traits of the most celebrated characters that have flourished during the last eighty years*, 2 vols., London, 1828.
Anglo, Sydney. *Images of Tudor Kingship*, London, 1992.
Anon. *An Account of the Foundation and Government of the Hospital for Foundlings in Paris*, London, 1739.
Anon. *An account of the rise, progress and state of the hospital, for relieving poor people afflicted with the small-pox, and for inoculation*, London, 1755.
Anon. *The Conjugal Lovers*, London, 1724.
Anon. *An Enquiry how the Wild Youth, Lately taken in the Woods near Hanover (and now brought over to England) could be there Left, and by what Creature he could be Suckled, Nursed and Brought Up*, London, 1726.
Anon. *The English Poems collected from the Oxford and Cambridge verses on the Death of his Royal Highness Frederick Prince of Wales*, Edinburgh, 1752.
Anon. *Verses on the King coronation of . . . George II and Queen Caroline, . . . spoken by the scholars of Westminster School*, London, 1761.
Anon. [Swift, Jonathan?]. *It never Rains but it Pours: Or London Strow'd with Rarities*, 2 vols., London, 1726.
Arbuthnot, J. *Mr Maitland's Account of Inoculating the Small Pox Vindicated*, London, 1722.
Arbuthnot, Mrs. *Journal of Mrs. Arbuthnot*, ed. F. Bamford and the Duke of Wellington, 2 vols., London, 1950.
Arciszewska, Barbara. *The Hanoverian Court and the Triumph of Palladio: The Palladian Revival in Hanover and England, c. 1700*, Warsaw, 2002.
Arkell, R. L. *Caroline of Ansbach: George the Second's Queen*, Oxford, 1939.
Armstrong, Alan W., ed. *'Forget not Mee & My Garden': Selected Letters, 1725–1768, of Peter Collinson, F.R.S.* Philadelphia, 2002.
Aspinall, A., ed. *The Later Correspondence of George III, 1783–1810*, 5 vols., Cambridge, 1962–70.
Attenborough, D., S. Owens, M. Clayton, and R. Alexandratos. *Amazing Rare Things: The Art of Natural History in the Age of Discovery*, London, 2007.
Ayling, Stanley. *George the Third*, London, 1972.
Baily, F. E. *Sophia of Hanover and her Times*, London, 1936.
Baker, Christopher, William Haupman, and others, eds. *Jean-Étienne Liotard, 1702–1789*, London, 2015.
Baker, Kenneth. *George III: A Life in Caricature*, New York, 2007.
Baker-Smith, V. P. M. *A Life of Anne of Hanover, Princess Royal*, Leiden, 1995.
—. *Royal Discord: The Family of George II*, London, 2008.
Balfour-Paul, Jenny. *Indigo*. London, 1998.
Banks, Audrey. 'Inventing Omai: Britain's Reception of the Tahitian Visitor Mai in the Last Quarter of the Eighteenth Century', BA Thesis, Princeton University, 2007.
The Banks Letters. A Calendar of the Manuscript Correspondence of Sir Joseph Banks preserved in the British Museum (Natural History), ed. W. R. Dawson, London, 1958.
Baron, John. *The Life of Edward Jenner, M.D.*, 2 vols., London, 1827–38.
Barringer, Tim, Gillian Forrester, and Barbaro Martinez Ruiz, eds. *Art and Emancipation in Jamaica: Isaac Mendes Belisario and His Worlds*, New Haven and London, 2009.
Bartoschek, Gerd, Veronica Biermann, and others. *Sophie Charlotte und ihr Schloss: Ein Musenhof des Barock in Brandenburg-Preussen*, Munich, 1999.
Batey, M. *Alexander Pope: The Poet and the Landscape*, London, 1999.
Bayley, Peter, ed. *Spenser, The Faerie Queene: A Casebook*, London, 1977.
Beach, M.C., and E. Koch. *The King of the World: The Padshahnama, an Imperial Mughal Manuscript from the Royal Library, Windsor Castle*, London and Washington, DC, 1997.
Beaglehole, J. C., ed. *The Journals of Captain Cook on his Voyages of Discovery*, vol. 2: *The Voyage of the Resolution and Adventure, 1772–1775*, Cambridge, 1961.
Bean, W. J., and William T. Thiselton-Dyer. *The Royal Botanic Gardens, Kew*, London and New York, 1908.
Beattie, John M. *The English Court in the Reign of George I*, Cambridge, 1967.
Bellaigue, Geoffrey de, gen. ed. *The James A. De Rothschild Bequest at Waddesdon Manor: Catalogue of Drawings for Architecture, Design and Ornament*, 2 vols., Waddesdon Manor, 2006.
Bellaigue, Geoffrey de, and Oliver Millar. *George III, Collector and Patron*, London, 1974.
Bellin, Joshua D., and Laura L. Mielke, eds. *Native Acts: Indian Performance, 1603–1832*, Lincoln, NE, 2011.
Belon, Pierre. *L'Histoire de la nature des oyseaux*, Paris, 1555.
Bemrose, W. *Bow, Chelsea and Derby Porcelain*, London, 1898.
Bencard, M. *The Hen in the Egg*, Copenhagen, 1999.
Bending, Stephen. *Green Retreats: Women, Gardens and Eighteenth-Century Culture*, Cambridge, 2013.
Bentley, R. *A Proposal for Building a Royal Library, and Establishing it by Act of Parliament*, London, 1697.
—. *The Correspondence of Richard Bentley*, 2 vols., London, 1842.
Bickham, G. *Apelles Britannicus, Book I: The Royal Palace of Hampton Court; Book II: The Royal Palace of Kensington*, London, 1741.
—. *Deliciae Britannicae: Or the Curiosities of Hampton Court and Windsor Castle*, London, 1742.
Bindman, D. *Hogarth and his Times: Serious Comedy*, London, 1997.
Binns, R. W. *A Century of Potting in the City of Worcester, being the History of the Royal Porcelain Works from 1751–1851*, 2nd ed., London and Worcester, 1877.
Biographia Britannica: Or, the Lives of the Most Eminent Persons who have Flourished in Great Britain and Ireland from the Earliest Ages, down to the Present Times, Collected from the Best Authorities . . ., 6 vols., London, 1747–66.
Birch, T. *Heads and Characters of Illustrious Persons of Great Britain with their Portraits Engraven by Mr. Houbraken and Mr. Vertue: With their Lives and Characters*, 2 vols., London, 1743.
Black, Jeremy. *Natural and Necessary Enemies: Anglo-French Relations in the Eighteenth Century*, London, 1986.
Black, Peter, ed. *'My Highest Pleasures': William Hunter's Art Collection*, London, 2007.
Blake, Robin. *George Stubbs and the Wide Creation: Animals, People and Places in the Life of George Stubbs, 1724–1806*, London, 2005.
Blanning, T. C. W. *The Culture of Power and the Power of Culture: Old Regime Europe, 1660–1789*, Oxford, 2002.
Boalch, D. H. *Makers of the Harpsichord and Clavichord, 1440–1840*, 3rd ed., Oxford, 1995.
Bocage, Madam du [*sic*]. *Letters concerning England, Holland and Italy . . . Translated from the French*, 2 vols., London, 1770.
Bold, J. *Greenwich: An Architectural History of the Royal Hospital for Seamen and the Queen's House*, New Haven and London, 2000.
Bolton, A. T. *The Architecture of Robert and James Adam*, 2 vols., London, 1922.
Boorsch, S. *Fireworks! Five Centuries of Pyrotechnics in Print and Drawings*, New York, 2000.
Borman, T. *King's Mistress, Queen's Servant: The Life and Times of Henrietta Howard*, London, 2007.
Borth, Helmut. *Herzoghaus Mecklenburg-Strelitz: Von gekrönten Häuptern, blaublütigen Kuckuckskindern und der Mirower Fürstengruft*, Friedland, 2011.
Boswell, J. *The Life of Samuel Johnson, LLD*, 4 vols., London, 1799.
Bowden, Delia K. *Leibniz as a Librarian: And, Eighteenth-Century Libraries in Germany*, London, 1969.

Bowden, Jean K. *John Lightfoot, his Work and Travels: With a Biographical Introduction and a Catalogue of the Lightfoot Herbarium*, Kew and Pittsburgh, PA, 1989.
Bowett, A. *Early Georgian Furniture, 1715–1740*, Woodbridge, 2009.
Brandsch, Juliane Ricarda, ed. *Ernst der Fromme (1601–1675): Bauherr und Sammler*, Gotha, 2001.
Bredekamp, H. *Leibniz und die Revolution der Gartenkunst: Herrenhausen, Versailles und die Philosophie der Blatter*, Berlin, 2013.
—. *The Lure of Antiquity and the Cult of the Machine: The Kunstkammer and the Evolution of Nature, Art and Technology*, trans. Allison Brown, Princeton, 1995.
Brett, R. L. *The Third Earl of Shaftesbury: A Study in Eighteenth-Century Literary Theory*, London, 1951.
Brewer, John. *The Pleasures of the Imagination: English Culture in the Eighteenth Century*, London and New York, 1997.
Brewer, John, and Roy Porter, eds. *Consumption and the World of Goods*, London, 1993.
Brewer, John, and Susan Staves. *Early Modern Concepts of Property*, London, 1995.
Brinkley, Roberta Florence. *Arthurian Legend in the Seventeenth Century*, Baltimore, MD, and London, 1932.
Britt, A. S., Jr., and A. R. Dees, eds. *Selected Eighteenth Century Manuscripts, Georgia Historical Society Collections* 20, Savannah, 1980.
Brock, C. Helen, ed. *The Correspondence of Dr William Hunter, 1740–83*, 2 vols., London, 2008.
Brooke, J. *King George III*, London, 1972.
Brown, L. A. *A Catalogue of British Historical Medals, 1760–1960*, vol. 1: *The Accession of George III to the Death of William IV*, London, 1980.
Brown, Stuart, ed. *British Philosophy and the Age of Enlightenment*, London, 1996.
Brownell, M. R. *Alexander Pope and the Arts of Georgian England*, Oxford, 1978.
Browning, J. D., ed. *Education in the 18th Century*, New York and London, 1979.
Bruyn, Frans de, and Shaun Regan, eds. *The Culture of the Seven Years' War: Empire, Identity, and the Arts in the Eighteenth-Century Atlantic World*, Toronto, 2013.
Bucholz, R. O. *The Augustan Court: Queen Anne and the Decline of Court Culture*, Stanford, CA, 1993.
Burney, Charles. *An Account of the Musical Performances . . . in Commemoration of Handel*, London, 1785.
—. *The State of Music in France and Italy*, London, 1773.
Butterfield, H. *George III and the Historians*, London, 1957.
—. *The Whig Interpretation of History*, London, 1931.
Bynum, W. F., and R. S. Porter, eds. *William Hunter and the Eighteenth-Century Medical World*, Cambridge, 1985.
Cadogan, William. *An Essay upon Nursing, and the Management of Children, from Their Birth to Three Years of Age*, London, 1748.
Calmann, Gerta. *Ehret: Flower Painter Extraordinary*, Boston, MA, 1977.
Campbell, C. *Vitruvius Britannicus or the British Architect*, London, 1715.
Campbell Orr, Clarissa, ed. *Queenship in Britain, 1660–1837: Royal Patronage, Court Culture and Dynastic Politics*, Manchester, 2002.
—. *Queenship in Europe, 1660–1815: The Role of the Consort*, Cambridge, 2004.
Candler, Allen D., and others, eds. *The Colonial Records of the State of Georgia*, Atlanta and Athens, GA, 1904–16.
Casley, D. *Catalogue of the Manuscripts of the King's Library: An appendix to the Catalogue of the Cottonian Library, together with an account of the books burnt or damaged by a late fire . . .*, London, 1734.
Cassirer, Ernst. *The Philosophy of the Enlightenment*, trans. Fritz C. A. Koelin and James P. Pettegrove, Princeton, 1951.
Castiglione, Dario, and Lesley Sharp, eds. *Shifting the Boundaries: Transformation of the Languages of Public and Private in the Eighteenth Century*, Exeter, 1995.
Catesby, Mark, *The Natural History of Carolina, Florida and the Bahama Islands: containing the figures of birds, beasts, fishes, serpents, insects, and plants: particularly, the forest-trees, shrubs, and other plants, not hitherto described, or very incorrectly figured by authors . . .*, 2 vols., London, 1729–43; appendix 1747.
Chaloner Smith, J. *British Mezzotint Portraits*, 5 vols., London, 1878–83.
Chambers, William. *Plans, Elevations, Sections, and Perspective Views of the Gardens and Buildings at Kew in Surry . . .*, London, 1763.
Chatterjee, Kumkum, and Clement Hawes, eds. *Europe Observed: Multiple Gazes in Early Modern Encounters*, Lewisburg, PA, 2008.
Chenevix Trench, C., *George II*, London, 1973.
Chesterfield, Philip Dormer Stanhope, Earl of. *Lord Chesterfield's Letters*, ed. and intro. D. Roberts, Oxford, 1992.
—. *The Letters of Philip Dormer Stanhope, Earl of Chesterfield: Including Numerous Letters now first Published*, ed. 5 vols., London, 1845–53.
Christies [London]. *A Catalogue of the superb Jewels, Trinkets, Plates, Gold and Silver Medals, China &c. of a Person of great Rank, 1 Dec.–3 Feb. 1773*, London, 1773.
Christie, Manson and Wood. *A Catalogue of the Genuine Library, Prints, and Books of Prints, of an Illustrious Personage, Lately Deceased*, London, 1819.
—. *A Catalogue of a Superbe Assemblage of Jewels, Trinkets . . . which will be sold by Mr Christie, 17–18 May, 1819*, London, 1819.
—. *A Catalogue of the Remaining Part of a Valuable Collection of Curiosities, which will be sold by Auction by Mr Christie, Monday May 24, 1819 and the following days*, London, 1819.
Churchill, Sarah. *Letters of a Grandmother, 1732–1735: Being the Correspondence of Sarah, Duchess of Marlborough with her Granddaughter Diana, Duchess of Bedford*, ed. G. S. Thomson, London, 1943.
Clark, Anna. *Scandal: The Sexual Politics of the British Constitution*, Princeton and Oxford, 2004.
Clarke, Alured. *An Essay towards the Character of Her Late Majesty Caroline, Queen Consort of Great Britain*, London, 1738.
Clarke, G. B., ed. *Descriptions of Lord Cobham's Gardens at Stowe, 1700–1750*, Aylesbury, 1990.
Clarke, G., J. Marsden, R. Wheeler, M. Bevington, and T. Knox. *Stowe Landscape Gardens*, London, 1997.
Clarke, Dr Samuel, ed. *A collection of papers, which passed between the late learned Mr. Leibnitz, and Dr. Clarke, in the years 1715 and 1716: relating to the principles of natural philosophy and religion*, London, 1717.
Clayton, Martin. *Holbein to Hockney: Drawings from the Royal Collection*, London, 2004.
—. *Poussin: Works on Paper: Drawings from the Collection of Her Majesty Queen Elizabeth II*, London, 1995.
Clifton, G. *Directory of British Scientific Instrument Makers, 1550–1851*, London, 1996.
Cloake, J. *Palaces and Parks of Richmond and Kew:* vol. 2, *Richmond Lodge and the Kew Palaces*, Chichester, 1996.
Cody, Lisa Forman. *Birthing the Nation: Sex, Science, and the Conception of Eighteenth-Century Britons*, Oxford, 2005.
Coke, Lady Mary. *The Letters and Journals of Lady Mary Coke*, 4 vols., Edinburgh, 1889–96.
Colley, Linda. *Britons: Forging the Nation, 1707–1837*, New Haven and London, 1992.
Collins Baker, C. H. *Principal Pictures in the Royal Collection at Windsor Castle*, London 1937.
Colvin, H. M. *The History of the King's Works*, 6 vols., London, 1963–82.
—. *Royal Buildings*, London, 1968.
Connaughton, Richard. *Omai: The Prince Who Never Was*, London, 2005.
Constable, W. G. *Richard Wilson*, London, 1953.
Cook, Harold. *The Decline of the Old Medical Regime in Stuart London*, Ithaca, NY, 1986.
Cook, James. *An account of the voyages undertaken by the order of His present Majesty for making discoveries in the southern hemisphere. And successively performed by Commodore Byron, Captain Wallis, Captain Carteret, and Captain Cook, in the Dolphin, the Swallow, and the Endeavour: drawn up from the journals which were kept by the several commanders, and from the papers of Sir Joseph Banks, bart.*, London, 1785.

Cook, James. *A voyage to the Pacific Ocean: undertaken, by the command of His Majesty, for making discoveries in the Northern Hemisphere: to determine the position and extent of the west side of North America, its distance from Asia, and the practicability of a northern passage to Europe; performed under the direction of Captains Cook, Clerke, and Gore, in his Majesty's ships the Resolution and Discovery, in the years 1776, 1777, 1778, 1779, and 1780*, Dublin, 1784.
Cooper, Anthony Ashley, 3rd Earl of Shaftesbury. *Characteristicks of Men, Manners, Opinions, Times* . . . , 3 vols., London, 1711.
—. 3rd Earl of Shaftesbury. *Second Characters: Or the Language of Forms*, 3 vols., London, 1778.
Cooper, Paul Martyn. *Images of Nature: The Bauer Brothers*, London, 2015.
Cornforth, John. *Early Georgian Interiors*, New Haven and London, 2004.
Corp, E. *A Court in Exile: The Stuarts in France, 1689–1718*, Manchester and New York, 2004.
—. *The King over the Water: Portraits of the Stuarts in Exile after 1689*, Edinburgh, 2001.
Cowper, F. T. de Grey, Earl Cowper. *The Manuscripts of the Earl Cowper, KG, Preserved at Melbourne Hall, Derbyshire* . . . , 12th Report Royal Commission on Historical Manuscripts, appendix pt. 1–3, 1888–89.
Cowper, Mary, Countess. *Diary of Mary, Countess Cowper: Lady of the Bedchamber to the Princess of Wales, 1714–1720*, ed. Hon. C. S. Cowper, London, 1864.
Coxe, W. *Anecdotes of George Frederick Handel, and John Christopher Smith*, London, 1799.
—. *Memoirs of the Life and Administration of Sir Robert Walpole, Earl of Orford*, 4 vols., London, 1816.
Curll, E. *The Rarities of Richmond: Being Exact Descriptions of the Royal Hermitage and Merlin's Cave, with his Life and Prophesies*, London, 1736.
Dale, W. *Tschudi the Harpsichord Maker*, London, 1913.
D'Arblay, Madame. *Diary and Letters of Madame D'Arblay (1778–1840)*, ed. Charlotte Barrett and A. Dobson, 6 vols., London, 1904–5.
Dann, T. *Die Koniglichen Prunkappartements im hannoverschen Leineschlob: Untersuchungen zu Raumfolgen in der 1. Halfte des 19. Jahrhunderts*, Hanover, 2000.
Dann, Uriel. *Hanover and Great Britain, 1740–1760: Diplomacy and Survival*, Leicester, 1991.
Daston, Lorraine, and Fernando Vidal, eds. *The Moral Authority of Nature*, Chicago, 2004.
Davenport, C. *Royal English Bookbindings*, London, 1896.
Davis, Mary Rose. *Stephen Duck: The Thresher Poet*, Orono, ME, 1927.
Dean, Christopher. *Arthur of England: English Attitudes to King Arthur and the Knights of the Round Table in the Middle Ages and the Renaissance*, Toronto, 1987.
Defoe, Daniel. *The History of the Life and Adventures of Mr Duncan Campbell: A gentleman, who, tho' Deaf and Dumb, Writes Down any Stranger's Name at first Sight* . . . , London, 1728.
Desmond, R. *Kew: The History of the Royal Botanic Gardens*, London, 1995.
Dethloff, Diana, Tessa Murdoch, and Kim Sloan, with Caroline Elam, eds. *Burning Bright: Essays in Honour of David Bindman*, London, 2015.
Deutsch, O. E. *Handel: A Documentary Biography*, New York, 1995.
Dezallier d'Argentville, Antoine Joseph. *The Theory and Practice of Gardening*, trans. John James, Farnborough, 1969.
Doderer-Winkler, M. *Magnificent Entertainments: Temporary Architecture for Georgian Festivals*, New Haven and London, 2013.
Dodington, George Bubb. *The Diary of the Late George Bubb Dodington, Baron of Melcombe Regis . . . now first published by Henry Pennruddocke Wyndham*, London, 1784.
—. *The Political Journal of George Bubb Dodington*, ed. J. Carswell and L. A. Dralle, Oxford, 1965.
Dodsley, Robert. *London and its Environs Described*, 6 vols., London, 1761.
Dolan, Brian. *Josiah Wedgwood: Entrepreneur to the Enlightenment*, London, 2004.
Donald, Diana. *The Age of Caricature: Satirical Prints in the Reign of George III*, New Haven and London, 1996.
Douglas Scott, J. *Sir Godfrey Kneller and the English Baroque Portrait*, Oxford, 1983.
Doyle, Kathleen, and Scot McKendrick, eds. *1000 Years of Royal Books and Manuscripts*, London, 2013.
Drinkuth, Friederike. *Queen Charlotte: A Princess from Mecklenburg-Strelitz Ascends the Throne of England*, Schwerin, 2011.
Dryden, J. *Merlin: Or, the British Inchanter. And King Arthur the British Worthy*, London, 1736.
Duindam, Jeroen. *Dynasties: A Global History of Power, 1300–1800*, Cambridge, 2015.
Duncker, Max, and others, ed., *Politische Correspondenz Friedrichs des Grossen*, 47 vols., Berlin, 1893.
Edgar, Walter. *South Carolina: A History*, Columbia, SC, 1998.
Edgcumbe, R. *The Art of the Gold Chaser*, London, 2000.
Edgeworth, Maria. *Letters for Literary Ladies*, ed. Claire Connolly, London, 1993.
Edwards. G. *A Natural History of Birds*, London, 1750.
—. *Some Gleanings of Natural History*, 3 vols., London, 1758–64.
—. *Some Memoirs of the Life and Works of George Edwards, Fellow of the Royal and Antiquarian Societies*, London, 1776.
Edwards, R., and M. Jourdain. *Georgian Cabinet-Makers*, 1st edn., London, 1944.
—. *Georgian Cabinet-makers, c. 1700–1800*, 3rd ed., London, 1955.
Eger, E., and L. Peltz. *Brilliant Women: 18th-Century Bluestockings*, London, 2008.
Egerton, Judy. *George Stubbs: Anatomist and Animal Painter*, London, 1976.
Egmont, John Percival [Perceval], Earl of. *Diary of the First Earl of Egmont (Viscount Percival)*, vol. 2, London, 1923.
—. *Manuscripts of the Earl of Egmont: Diary of Viscount Percival. Afterwards First Earl of Egmont*, 3 vols., London, 1920–23.
Einberg, E., and Jones, R., eds. *Manners & Morals: Hogarth and British Painting, 1700–1760*, London, 1987.
Eisler, William. *Les Médailles des Dassiers de Genève*, Milan, 2010.
Elstob, Elizabeth. *Rudiments of Grammar for the English-Saxon Tongue*, London, 1715.
Erffa, Helmut von, and Allen Staley. *The Paintings of Benjamin West*, New Haven and London, 1986.
Esdaile, K. A. *The Art of Rysbrack in Terracotta*, London, 1932.
Eustace, Katharine. *Michael Rysbrack: Sculptor, 1694–1770*, Bristol, 1982.
Evans, E. D. *A History of Wales, 1660–1815*, Cardiff, 1976.
Evans, Joan. *A History of the Society of Antiquaries*, Oxford, 1956.
Evans, M., ed. *Princes as Patrons: The Art Collections of the Princes of Wales from the Renaissance to the Present Day*, London, 1999.
Evans, Mark, Clare Browne, and Arnold Nesselrath, eds. *Raphael: Cartoons and Tapestries for the Sistine Chapel*, London, 2010.
Evelyn, John. *The Diary of John Evelyn*, ed. Austin Dobson, 3 vols., London and New York, 1906.
Fara, Patricia. *Pandora's Breeches: Women, Science and Power in the Enlightenment*, London, 2004.
Farington, Joseph, R.A. *The Diaries of Joseph Farington*, ed. K. Garlick, A. Macintyre, and K. Cave, 16 vols., London and New Haven, 1978–84 (index vol., 1998).
—. *The Farington Diary*, ed. James Greig, 8 vols., London, [1922–28].
—. 'Memoir of the life of Sir Joshua Reynolds', in E. Malone, *The Literary Works of Sir Joshua Reynolds*, 3 vols., London, 1819.
Fauchier-Magnan, Adrien. *The Small German Courts in the Eighteenth Century*, trans. Mervyn Savill, London, 1958.
Feuillet, Raoul Auger. [*Recueil de contredances*. English]. *For the further improvement of dancing . . . translated from the French . . . With a collection of country dances, and a new French dance call'd the Princess's Passpied compos'd and writt in characters by John Essex*, London, c. 1715.
Field, Ophelia. *The Kit-Cat Club: Friends who Imagined a Nation*, London, 2008.
Findlen, Paula. *Possessing Nature: Museums, Collecting and Scientific Culture in Early Modern Italy*, Berkeley, CA, and London, 1994.

Fisher Murray, John. *Environs of London, Western Division*, Edinburgh and London, 1842.
Fittschen, K. Die *Bildnisgalerie in Herrenhausen bei Hannover: Zur Rezeptions- und Sammlungsgeschichte antiker Portrats*, Gottingen, 2006.
Fitzmaurice, Edmond George Petty-Baron Fitzmaurice. *Life of William, Earl of Shelburne, afterwards first Marquess of Lansdowne, with extracts from his papers and correspondence*, 2 vols., London, 1912.
Foister, Susan. *Holbein in England*, London, 2006.
Ford, Mr. *George Vertue: Sale of Pictures, Coins and Drawings*, 17–19 March, 1759.
—. *George Vertue: Sale of Library, 18–19 March, 1757*, London, 1757.
Forgan, Sophie. *Captain Cook's Endeavor*, Whitby, 1997.
Fortescue, John, ed. *The Correspondence of King George III from 1760 to December 1783*, 6 vols., London, 1927–28.
Fox, C. *The Arts of Industry in the Age of Enlightenment*, New Haven and London, 2010.
Fox, John Howard, Mark H. Waddicor, and Derek A. Watts, eds. *Studies in Eighteenth Century French Literature presented to Robert Niklaus*. Exeter, 1975.
Franklin, Colin. *Lord Chesterfield: His Character and Characters*, Aldershot, 1993.
Franklin, Julian H. *John Locke and the Theory of Sovereignty: Mixed Monarchy and the Right of Resistance in the Political Thought of the English Revolution*, Cambridge, 1978.
Fraser, Flora. *Unruly Queen: The Life of Queen Caroline*, London, 1996.
Fraser, Mike, and Liz Fraser. *The Smallest Kingdom: Plants and Plant Collectors at the Cape of Good Hope*, Richmond, Surrey, 2011.
Fresnoy, C. A. Du., *The Art of Painting*, trans. J. Dryden, London, 1716.
Frick, George Frederick, and Raymond Phineas Stearns. *Mark Catesby: The Colonial Audubon*, Urbana, IL, 1961.
Furdell, Elizabeth Lane. *The Royal Doctors, 1485–1714: Medical Personnel at the Tudor and Stuart Courts*, Rochester, NY, 2001.
Gartner, Heinz. *J. C. Bach: Mozart's Mentor and Friend*, London, 1994
Gascoigne, John. *Joseph Banks and the English Enlightenment: Useful Knowledge and Polite Culture*, Cambridge, 1994.
Gatrell, Vic. *City of Laughter: Sex and Satire in Eighteenth-Century London*, London, 2006.
Gay, J. *John Gay: Dramatic Works*, ed. John Fuller, 2 vols., Oxford, 1983.
—. *The Letters of John Gay*, ed. C. F. Burgess, Oxford, 1966.
—. *Poems on Several Occasions*, 2 vols., London, 1762.
George III. *Letters from George III to Lord Bute, 1756–1766*, ed. Romney Sedgwick, London, 1939.
Gerbino, A., and S. Johnston. *Compass and Rule: Architecture as Mathematical Practice in England, 1550–1750*, New Haven and London, 2009.
Gerrard, C. *The Patriot Opposition to Walpole: Politics, Poetry, and National Myth, 1725–1742*, Oxford, 1994.
Gerrard, C., ed. *A Companion to Eighteenth-Century Poetry*, Oxford, 2006.
Godfrey, Richard. *James Gillray: The Art of Caricature*, London, 2001.
Goff, M., J. Goldfinch, and others. *Georgians Revealed: Life, Style and the Making of Modern Britain*, London, 2013.
Goodison, N. *Matthew Boulton: Ormolu*, London, 2002.
Goodman, Dena, *The Republic of Letters: A Cultural History of the French Enlightenment*, Ithaca, NY, and London, 1994.
Goodman, Katherine R. *Amazons and Apprentices: Women and the German Parnassus in the Early Enlightenment*, Rochester, NY, and Woodbridge, 1999.
Gordon, Peter, and Denis Lawton. *Royal Education: Past, Present and Future*, London, 1999.
Gowing, Laura. *Gender Relations in Early Modern England*, Harlow, 2012.
Granger, Rev. J. *A Biographical History of England*, 5th ed., 6 vols., London, 1825.
Granville, Mary, Mrs Delany. *The Autobiography and Correspondence of Mary Granville, Mrs Delany*, ed. Lady Llanover, series 1, 3 vols., London, 1861; series 2, 3 vols., London, 1862.
Gray, J. M. *James and William Tassie: A Biography and Critical Sketch with a Catalogue of their Portrait Medallions*, Edinburgh, 1894.
Griffiths, Antony, ed. *Landmarks in Print Collecting: Connoisseurs and Donors at the British Museum since 1753*, London, 1996.
—. *The Print in Stuart Britain, 1603–1689*, London, 1998.
Grimwade, A. *London Goldsmiths, 1697–1837: Their Marks and Lives*, London, 1976.
—. *Rococo Silver, 1727–1765*, London, 1974.
Groom, S., and L. Prosser. *Kew Palace: The Official Illustrated History*, London and New York, 2006.
Grundy, Isabel. *Lady Mary Wortley Montagu: Comet of the Enlightenment*, Oxford, 1999.
[Guardians of the House of Asylum]. *An account of the institution and proceedings of the Guardians of the Asylum or House of Refuge, situate in the parish of Lambeth, in the County of Surry: for the reception of orphan girls, whose settlements cannot be found*, London, 1786.
Gwynn, Stephen. *The Life and Friendships of Dean Swift*, London, 1933.
Hackforth-Jones, Jocelyn, David Bindman, and Stephanie Pratt, eds. *Between Worlds: Voyagers to Britain, 1700–1850*, London, 2007.
Hales, Stephen. *A description of ventilators: whereby great quantities of fresh air may with ease be conveyed into mines, gaols, hospitals*, London, 1743.
—. *Statical Essays, containing Vegetable Staticks*, London, 1728.
Hallett, Mark. *Reynolds: Portraiture in Action*, New Haven and London, 2014.
—. *The Spectacle of Difference: Graphic Satire in the Age of Hogarth*, New Haven and London, 1999.
Hallett, Mark, and C. Riding, eds. *Hogarth*, London, 2006.
Halsband, Robert. *Lord Hervey: Eighteenth-Century Courtier*, Oxford, 1973.
—. *The Life of Lady Mary Wortley Montagu*, London, 1956.
Hanham, A. A. 'Trustees for establishing the colony of Georgia in America', *ODNB*, Oxford, 2006 online edn, May 2016, http://www.oxforddnb.com/view/theme/95206, accessed 4 Nov 2016.
Hanson, Craig Ashley. *The English Virtuoso: Art, Medicine, and Antiquarianism in the Age of Empiricism*, Chicago, 2009.
Harcourt, Edward William, ed. *The Harcourt Papers*, 14 vols., Oxford, 1880–95.
Harding, N. *Hanover and the British Empire, 1700–1837*, Woodbridge, 2007.
Harmsen, T. H. B. M. *Antiquarianism in the Augustan Age: Thomas Hearne, 1678–1735*, Oxford, 2000.
Harris, F. *A Passion for Government: The Life of Sarah, Duchess of Marlborough*, Oxford, 1991.
Harris, J. *The British Hero: Or, a Discourse . . .* , London, 1715.
Harris, John. *A Catalogue of British Drawings for Architecture, Decoration, Sculpture and Landscape Gardening, 1550–1900, in American Collections*, Upper Saddle River, NJ, 1971.
—. *The Palladian Revival: Lord Burlington, his Villa and Garden at Chiswick*, London, New Haven, 1994.
—. *Sir William Chambers, Knight of the Polar Star*, London, 1970.
—. *William Kent, 1685–1748: A Poet on Paper*, London, 1998.
Harris, John, Geoffrey de Bellaigue, and Oliver Millar. *Buckingham Palace*, London, 1968.
Harris, John, and Michael Snodin, eds. *Sir William Chambers: Architect to George III*, New Haven and London, 1996.
Hartshorne, A., ed. *Memoirs of a Royal Chaplain, 1729–1763: The Correspondence of Edmund Pyle, D.D. Chaplain in ordinary to George II, with Samuel Kerrich, D.D. vicar of Dersingham*, London, 1905.
Haspels, J. *Automatic Musical Instruments, their Mechanics and their Music, 1580–1820*, Koedijk, 1987
Hatton, Ragnhild. *George I: Elector and King*, London, 1978.
Hawkins, E., ed. *Medallic Illustrations of the History of Great Britain and Ireland to the Death of George II*, London, 1885.
Hayden, R. *Mrs. Delany and Her Flower Collages*, 2nd ed., London, 1992.
Haywood, Ian. *Romanticism and Caricature*, Cambridge, 2013.
Heard, Kate, and Lucy Whitaker. *The Northern Renaissance: Dürer to Holbein*, London, 2011.

Hedley, Olwen. *Queen Charlotte*, London, 1975.
Henderson, A. *The Life of William Augustus, Duke of Cumberland (1766)*, ed. R. Macpherson, London, 2010.
Henrey, Blanche. *British Botanical and Horticultural Literature before 1800*, 3 vols., London, 1975.
Hervey, John, Baron Hervey. *Letter-Books of John Hervey, First Earl of Bristol*, 3 vols., Wells, 1894.
—. *Lord Hervey and his Friends, 1726–1738: Based on Letters from Holland House, Melbury and Ickworth*, ed. Earl of Ilchester, London, 1950.
—. *Lord Hervey's Memoirs: Edited from a Copy of the Original Manuscripts in the Royal Archives at Windsor Castle*, ed. Romney Sedgwick, London, 1952.
—. *Memoirs of the Reign of George the Second: From his Accession to the Death of Queen Caroline*, ed. J. W. Croker, 3 vols., London, 1848.
—. *Some Materials towards Memoirs of the Reign of King George II*, ed. Romney Sedgwick, London, 1931.
Hervey, Mary, Lady Hervey. *Letters of Mary Lepel, Lady Hervey: With a Memoir, and Illustrative Notes*, ed. J. W. Croker, London, 1821.
Hesse, Carla. *The Other Enlightenment: How French Women Became Modern*, Princeton, NJ, 2001.
Hetherington, Michelle. *Cook & Omai: The Cult of the South Seas*, Canberra, 2001,
Hibbert, Christopher. *George III: A Personal History*, London, 1998.
Hill, John. *The British herbal: an history of plants and trees, natives of Britain, cultivated for use, or raised for beauty*, London, 1756.
Hind, A. M. *Engraving in England in the Sixteenth and Seventeenth Centuries*, 3 vols., Cambridge, 1952–64.
Hoffmann, Nancy, and John C. Van Horne, eds. *America's Curious Botanist: A Tercentennial Reappraisal of John Bartram (1699–1777)*, Philadelphia, 2004.
Hogwood, C. *Handel*, London, 1984.
Holenstein, André, Hubert Steinke, and Martin Stuber, eds. *Scholars in Action: The Practise of Knowledge and the Figure of the Savant in the 18th Century*, Leiden, 2013.
Hopkirk, Mary. *Queen Adelaide*, London, 1946.
Howard, Henrietta, Countess of Suffolk, afterwards Berkeley, Hon. Mrs George. *Letters to and from Henrietta, Countess of Suffolk, and her Second Husband, the Hon. George Berkeley, from 1712 to 1767*, ed. J. W. Croker, 2 vols., London, 1824.
Huddleston, R., ed. *A New Edition of Toland's History of the Druids*, Montrose, 1814.
Hughes, Griffith. *The Natural History of Barbados, in ten books*, 10 vols., London, 1750.
Huish, R. *The Public and Private life of His Late Excellent and Most Gracious Majesty George the Third*, London, 1905.
Hunt, John Dixon, *William Kent: Landscape Garden Designer*, London, 1987.
Hunt, John Dixon, and P. Willis, eds., *The Genius of the Place: The English Landscape Garden, 1620–1820*, London, 1975.
Hunt, John Dixon., ed. *The Oxford Book of Garden Verse*, London, 1993.
Hyde, Melissa, and Jennifer Milam, eds. *Women, Art and the Politics of Identity in Eighteenth-Century Europe*, Aldershot, 2003.
Impey, E. *Kensington Palace: The Official Illustrated History*, London and New York, 2003.
Impey, O., and A. MacGregor, eds. *The Origins of Museums: The Cabinet of Curiosities in Sixteenth- and Seventeenth-Century Europe*, Oxford, 1980.
Jackson, Harvey H., and Phinizy Spalding Ready, eds. *Forty Years of Diversity: Essays on Colonial Georgia*, Athens, GA, 1984.
Jackson-Stops, Gervase, ed. *The Fashioning and Furnishing of the British Country House*, Washington, DC, Hanover, and London, 1989.
—. *The Treasure Houses of Britain: Five Hundred Years of Private Patronage and Art Collecting*, London and New Haven, 1985.
Jacob, Margaret C. *The Newtonians and the English Revolution, 1689–1720*, Hassocks, 1976.
Jacobsen, Roswitha, and Hans-Jörg Ruge, eds. *Ernst der Fromme (1601–1675): Staatsmann und Reformer*, Chicago, 2002 .
Jaffer, A. *Furniture from British India and Ceylon: A Catalogue of the Collections in the Victoria and Albert Museum and the Peabody Essex Museum*, London, 2001.
Jagger, C. *Royal Clocks*, London, 1983.
Jardine, Lisa. *Ingenious Pursuits: Building the Scientific Revolution*, London, 1999.
Jefferson, T. *A Summary View of the Rights of British America*, London, 1774.
Jenkins, G. H. *The Foundations of Modern Wales*, Oxford, 1987.
Jennings, Judith. *Religion and Radicalism in the Long Eighteenth Century: The Ingenious Quaker and her Connections*, Aldershot, 2006.
Joeres, R. E. B., and M. J. Maynes. *German Women in the Eighteenth and Nineteenth Centuries: A Social and Literary History*, Bloomington, 1986.
Johnson, Edward Mead. *Francis Cotes*, Oxford, 1976.
Jones, Inigo. *The Designs of Inigo Jones: Consisting of Plans and Elevations for Publick and Private Buildings: Published by William Kent with Some Additional Designs*, 2 vols., London, 1770.
Kassler, Michael, ed. *Memoirs of the Court of George III*, vol. 2: *Mary Delany (1700–1788) and the Court of King George III*, ed. Alain Kerhervé; vol. 4: *The Diary of Queen Charlotte, 1789 and 1795*, ed. Michael Kassler, London and Brookfield, VT, 2015.
Kelley, Donald R., and David H. Sacks, eds. *The Historical Imagination in Early Modern Britain: History, Rhetoric and Fiction, 1500–1800*, Cambridge, 1997.
Kelly, J. M. *The Society of Dilettanti: Archaeology and Identity in the British Enlightenment*, New Haven and London, 2009.
Kerslake, J. *National Portrait Gallery: Early Georgian Portraits*, 2 vols., London, 1977.
Keynes, M. *The Iconography of Sir Isaac Newton to 1800*, Woodbridge, 2005.
Kidd, Colin. *British Identities before Nationalism: Ethnicity and Nationhood in the Atlantic World, 1600–1800*, Cambridge, 1999.
King, J. N. *Spenser's Poetry and the Reformation Tradition*, Princeton, NJ, 1990.
King, R. *Royal Kew*, London, 1985.
King, W. *Chelsea Porcelain*, London, 1922.
Kliger, S. *The Goths in England: A Study in Seventeenth and Eighteenth Century Thought*, Cambridge, MA, 1952.
Knight, Stephen. *Merlin: Knowledge and Power through the Ages*, Ithaca, NY, 2009.
Krause, Friedhilde, ed. *Handbuch der Historischen Buchbestände in Deutschland*, vol. 19: *Thüringen A–G*, Hildesheim, 1998.
Kremers, Anorthe, and Elisabeth Reich, eds. *Loyal Subversion: Caricatures from the Personal Union between England and Hanover (1714–1837)*, Göttingen and Bristol, CT, 2014.
Kroll, Frank-Lothar, and Martin Muncke, eds. *Hannover-Coburg Gotha-Windsor: Problems and Perspectives of a Comparative Anglo German Dynastic History from the 18th to the 20th Century*, Berlin, 2015.
Kroll, Maria. *Sophie, Electress of Hanover: A Personal Portrait*, London, 1973.
L'Abbe, Anthony. *A new collection of dances. To be sold at Mr. Barreau's; and at Mr. Roussau's*, London, c. 1725.
—. *The Prince of Wales, a new dance for the year 1727. Writt by Mr Pemberton: and sold by him*, London, 1727.
—. *The Princess Amelia a new dance . . . 1718. Writ by Mr: Pemberton and sold by him*, London, 1718.
—. *The Princess Anna a new dance . . . 1716. Writ by Mr: Pemberton and sold by him*, London, 1716.
—. *The Princess Ann's chacone a new dance . . . 1719. Writ by Mr: Pemberton, and sold by him*, London, 1719.
—. *The Princess Royale a new dance . . . 1715, writ by Mr: Pemberton and sold by him*, London, 1715.
—. *Queen Caroline a new dance . . . 1728. Writt by Mr Pemberton: and sold by him*, London, 1728.
[Lacombe, François] Un Atheronome de Berne. *Observations sur Londre et ses Environs*, Paris and London, 1777.
Laird, Mark. *A Natural History of English Gardening, 1650–1800*, New Haven and London, 2015.

Laird, Mark, and Alicia Weisberg Roberts, eds. *Mrs. Delany and her Circle*, New Haven and London, 2009.

Laking, G. F. *The Furniture of Windsor Castle*, London, 1905.

Langley, Batty. *Gothic Architecture Improved, by Rules and Proportions in Many Grand Designs . . .* , London, 1747.

—. *New Principles of Gardening: Or, The Laying out and Planting Parterres . . .* [London, 1728], Farnborough, 1971.

Lawrence, Cynthia, ed. *Women and Art in Early Modern Europe: Patrons, Collectors and Connoisseurs*, University Park, PA, 1997.

Lees-Milne, J. *Earls of Creation: Five Great Patrons of Eighteenth-Century Art*, London, 1962.

Lesemann, Silke, and Annette von Stieglitz, eds. *Stand und Repräsentation, Kultur- und Sozialgeschichte des hannoverschen Adels vom siebzehnten bis zum neunzehnten Jahrhundert*, Hanover, 2004.

Levey, Michael. *The Later Italian Pictures in the Collection of Her Majesty the Queen*, Cambridge, 1964.

Levine, J. M. *Humanism and History: Origins of Modern English Historiography*, Ithaca, NY, and London, 1987.

Lipking, Laurence. *The Ordering of the Arts in Eighteenth-Century England*, Princeton, NJ, 1970.

Lloyd, C. *Gainsborough and Reynolds: Contrasts in Royal Patronage*, London, 1994.

—. *The Queen's Pictures: Old Masters from the Royal Collection*, London, 1994.

—. *The Queen's Pictures: Royal Collectors through the Centuries*, London, 1991.

—. *The Quest for Albion: Painting and Patronage and the British Monarch*, London, 1998.

Lloyd, C., and V. Remington. *Masterpieces in Little: Portrait Miniatures from the Collection of Her Majesty the Queen*, London, 1997.

Loche, R., and M. Roethlisberger. *L'Opera completa di Liotard*, Milan, 1979.

Lockman, John. *Verses most humbly addressed to the Princess Dowager of Wales. On Her Majesty's Birthday Nov. 30, 1757, by John Lockman, Secretary to the Society of the Free British Fishery. Presented to Her Royal Highness in Leicester House*, London, 1756.

Lodge, P., ed. *Leibniz and his Correspondents*, Cambridge, 2004.

Lonsdale, R. *Dr Charles Burney: A Literary Biography*, Oxford, 1965.

Lybbe Powys, Mrs Philip. *Passages from the Diary of Mrs. Philip Lybbe Powys, 1756–1808*, ed. E. J. Climenson, London, 1899.

Lyte, C. *The Plant Hunters*, London, 1983.

Macalpine, I., and R. Hunter. *George III and the Mad-Business*, London, 1969.

MacGregor, A., ed. *Sir Hans Sloane: Collector, Scientist, Antiquary, Founding Father of the British Museum*, London, 1994.

—. *The Late King's Goods: Collections, Possessions and Patronage of Charles I in the Light of the Commonwealth Sale Inventories*, London and Oxford, 1989.

Mackintosh, I. *The Georgian Playhouse*, London, 1975.

MacLeod, Catharine, and Julia Marciari Alexander, eds. *Painted Ladies: Women at the Court of Charles II*, London, 2001.

Marples, Morris. *Poor Fred and the Butcher: Sons of George II*, London, 1970.

Marschner, Joanna. 'Caroline of Ansbach: The Queen, Collecting and Connoisseurship at the Early Georgian Court', PhD diss., University College London, 2007.

—. *Queen Caroline: Cultural Politics at the Early Eighteenth-Century Court*, New Haven and London, 2014.

Marsden, Jonathan, ed. *The Wisdom of George the Third*, London, 2005.

Marshall, P. J. *The Impeachment of Warren Hastings*, Oxford, 1965.

Martyn, Benjamin. *Reasons for Establishing the Colony of Georgia, With Regard to the Trade of Great Britain, the Increase of our People, and the Employment and Support it will afford to great Numbers of our own Poor, as well as foreign persecuted Protestants*, London, 1733.

Mason, William. *An Heroic Epistle to Sir William Chambers*, 2nd ed., London, 1773.

Matthews, Leslie. *The Royal Apothecaries*, London, 1967.

McBurney, Henrietta. *Mark Catesby's Natural History of America: The Watercolors from the Royal Library, Windsor Castle*, London, 1997.

McCarthy, Michael. *The Origins of the Gothic Revival*, New Haven and London, 1987.

McCormack, Helen. 'A Collector of the Fine Arts in Eighteenth-Century Britain: Dr William Hunter (1718–1783)', PhD diss., University of Glasgow, 2010.

McCormick, E. H. *Omai: Pacific Envoy*, Oxford, 1978.

McKechnie, S. *British Silhouete Artists and their Work, 1760–1860*, London, 1978.

McKellar, Elizabeth. *Landscapes of London: The City, the Country and the Suburbs, 1660–1840*, New Haven and London, 2013.

McPherson, Robert G., ed. *The Journal of the Earl of Egmont: Abstract of the Trustees Proceedings for Establishing the Colony of Georgia, 1732–1738*, Athens, GA, 1962.

Melville, Lewis. *Lady Suffolk and her Circle*, London, 1924.

Merriman, James Douglas. *The Flower of Kings: A Study of the Arthurian Legend in England between 1485 and 1835*, Lawrence, KS, 1972.

Meyers, Amy R. W. 'Sketches from the Wilderness: Changing Conceptions of Nature in American Natural History Illustration, 1680–1880', PhD diss., Yale University, 1985.

Meyers, Amy R. W., ed. *Knowing Nature: Art and Science in Philadelphia, 1740–1840*, New Haven and London, 2014.

Meyers, Amy R. W., and Margaret Beck Pritchard, eds. *Empire's Nature: Mark Catesby's New World Vision*, Chapel Hill, NC, 1998.

Millar, Oliver. *The Tudor, Stuart and Early Georgian Pictures in the Collection of Her Majesty The Queen*, 2 vols., London, 1963.

—. *The Later Georgian Pictures in the Collection of Her Majesty the Queen*, 2 vols., London, 1969.

—. *The Queen's Pictures*, London, 1977.

—. *Zoffany and his Tribuna*, London, 1967.

Miller, Genevieve. *The Adoption of Inoculation for Smallpox in England and France*, Philadelphia, 1957.

Miller, John. *Illustratio Systematis Sexualis Linnaei. An Illustration of the Sexual System of Linnaeus*, London, [1770]–77.

Milton, John. *The Poetical Works of John Milton*, ed. J. Mitford, 2 vols., London, 1851.

Montagu, Elizabeth. *Elizabeth Montagu, the Queen of the Blue-Stockings: Her Correspondence from 1720 to 1761*, ed. E. J. Climenson, 2 vols., London, 1906.

—. *The Letters of Mrs Elizabeth Montagu*, ed. M. Montagu, 4 vols., London, 1809–13.

Montagu, Lady Mary Wortley. *The Complete Letters of Lady Mary Wortley Montagu*, ed. Robert Halsband, 3 vols., Oxford, 1965–67.

—. *Essays and Poems; and Simplicity: A Comedy*, ed. R. Halsband and I. Grundy, Oxford, 1977.

—. *The Letters and Works of Lady Mary Wortley Montagu*, ed. Lord Wharncliffe, 2 vols., London, 1837.

Moore, A. W., ed. *Houghton Hall: The Prime Minister, the Empress and the Heritage*, Norwich, 1996.

—. *Norfolk and the Grand Tour: Eighteenth-Century Travellers Abroad and their Souvenirs*, Norwich, 1985.

Moore, Lucy. *Amphibious Thing: The Life of Lord Hervey*, London, 2000.

Mora, Bruce T., ed. *Patronage and Institutions: Science, Technology, and Medicine at the European Court, 1500–1750*, Rochester, NY, 1991.

More, Hannah. *Hints towards forming the Character of a Young Princess*, 2 vols., London, 1805.

—. *Slavery, A Poem*, London, 1788.

Morris, Robert. *An Essay in Defence of Ancient Architecture: Or, a Parallel of the Ancient Buildings with the Modern . . .* , London, 1728.

Morton, Alan Q., and Jane A. Wess. *Public and Private Science: The King George III Collection*, Oxford, 1993.

Mowl, T. *Gentlemen and Players: Gardeners of the English Landscape*, Stroud, 2000.
—. *William Kent: Architect, Designer, Opportunist*, London, 2006.
Mowl, T., and B. Earnshaw. *John Wood: Architect of Obsession*, London, 1988.
Mozart, W. A. *The Letters of Mozart and his Family*, ed. E. Anderson, London, 1938.
Murdoch, Alexander. *British History, 1660–1832: National Identity and Local Culture*, London, 1998.
Myrone, Martin. *George Stubbs*, London, 2002.
Myrone, M., and L. Peltz, eds. *Producing the Past: Aspects of Antiquarian Culture and Practice, 1700–1850*, Aldershot, 199.
Naber, Johanna W. A. *Onze Vorstinnen uit het Huis van Orange-Nassau in het Stad Stadhuderlijk Tijdperle*, 2 vols., Haarlem, 1911.
The Naturalist's Pocket Magazine or, Compleat Cabinet of Nature, London, 1799–1802.
Neal, Daniel. *A Narrative of the Method and Success of Inoculating the Small-Pox in New England. By Mr Benj. Colman. With a Reply to the Objections Made against it from Principles of Conscience. In a letter from a Minister at Boston [Rev Wm Cooper] to which is now Prefixed an Historical Introduction*, London, 1722.
Nelson, E. Charles, and David J. Elliott, eds. *The Curious Mister Catesby: A 'Truly Ingenious' Naturalist Explores New Worlds*, Athens, GA, 2015.
Newman, Gerald. *The Rise of English Nationalism: A Cultural History, 1740–1830*, Basingstoke, 1997.
Newton, M. *Savage Girls and Wild Boys: A History of Feral Children*, London, 2002.
Newton, Isaac. *Opticks: Or, a Treatise of the Reflexions, Refractions, Inflexions and Colours of Light*, 2 vols., London, 1704–25.
Nichols, R. H., and F. A. Wray. *The History of the Foundling Hospital, London*, Oxford, 1935.
Nixon, H. M., and M. Foot. *The History of Decorated Bookbinding in England*, London, 1992.
Nokes, D. *John Gay: A Profession of Friendship*, Oxford, 1995.
Noorthouck, John. *An Historical and Classical Dictionary*, London, 1776.
Northumberland, Duchess of. *The Diaries of a Duchess: Extracts from the Diaries of the First Duchess of Northumberland (1716–1776)*, ed. James Greig, London, 1926.
Novellis, Mark de. *Pallas Unveil'd: The Life and Art of Dorothy Savile, Countess of Burlington (1699–1758)*, London, 1999.
Nugent, T. *Travels through Germany . . . with a Particular Account of the County of Mecklenburg*, 2 vols., London, 1768.
Nussbaum, Felicity A. *Torrid Zones: Maternity, Sexuality, and Empire in Eighteenth-Century English Narratives*, Baltimore, MD, 1995.
O'Brian, Patrick. *Joseph Banks*, London, 1997.
O'Brien, Karen. *Women and Enlightenment in Eighteenth-Century Britain*, Cambridge, 2009.
Ogborn, Miles. *Global Lives: Britain and the World, 1550–1800*, Cambridge, 2008.
[Oglethorpe, James Edward]. *A New and Accurate Account of the Provinces of South-Carolina and Georgia*, London, 1732.
Oppé, A. P. *English Drawings, Stuart and Georgian Periods, in the Collection of His Majesty The King at Windsor Castle*, London, 1950.
Oresko, Robert, G. C. Gibbs, and Hamish Scott, eds. *Royal and Republican Sovereignty in Early Modern Europe: Essays in Memory of Ragnhild Hatton*, Cambridge, 1997.
Orléans, Charlotte-Elisabeth, Duchesse d'. *Briefe der Herzogin Elisabeth Charlotte von Orléans*, ed. W. L. Holland, 7 vols., Stuttgart, 1843–81.
—. *Correspondance complète de Madame, Duchesse d'Orléans*, trans. M. G. Brunet, 2 vols., Paris, 1857.
—. *Correspondance de Madame, Duchesse d'Orléans*, trans. E. Jaeglé, 3 vols., Paris, 1890.
—. *Fragmens de Lettres Originales de Madame Charlotte-Elizabeth de Bavière, Veuve de Monsieur, Frère unique de Louis XIV*, Hambourg, 1788.
—. *Letters from Liselotte: Elisabeth-Charlotte, Princess Palatine and Duchess of Orleans, 'Madame', 1652–1722*, trans. and ed. Maria Kroll, London, 1970.
—. *The Letters of Madame: The Correspondence of Elisabeth Charlotte of Bavaria, Princess Palatine, Duchess of Orleans, 1661–1708*, trans. and ed. G. S. Stevenson, 3 vols., New York, 1924.
—. *Secret Memoirs of the Court of Louis XIV and of the Regency*, London, 1895.
Osborn, Sarah. *Political and Social Letters of a Lady of the Eighteenth Century, 1721–1771*, ed. E. F. D. Osborn, London, 1890.
Oulton, W. C. *The Authentic and Impartial Memoirs of her Late Majesty Charlotte, Queen of Great Britain and Ireland*, London, 1819.
Outram, Dorinda. *Panorama of the Enlightenment*, London, 2006.
Owen, A. L. *The Famous Druids: A Survey of Three Centuries of Literature on the Druids*, Oxford, 1962.
Owen, David. *English Philanthropy, 1660–1960*, Cambridge, MA, 1965.
Paas, Sigrun, ed. *Liselotte von der Pfalz: Madame am Hofe der Sonnenkönigs*, Heidelberg, 1996.
Pagden, Anthony. *European Encounters with the New World*, New Haven and London, 1993.
Paintin, E. *The King's Library*, London, 1989.
Papendiek, C. L. H. *Court and Private Life in the Time of Queen Charlotte: Being the Journals of Mrs. Papendiek, Assistant-Keeper of the Wardrobe and Reader to Her Majesty*, ed. Mrs. Vernon Delves Broughton, 2 vols., London, 1887.
Pares, R. *King George III and the Politicians*, Oxford, 1953.
Parker, K. T. *The Drawings of Hans Holbein in the Collection of His Majesty the King at Windsor Castle*, London, New York, and Oxford, 1945.
Parrish, Susan Scott. *American Curiosity: Cultures of Natural History in the Colonial British Atlantic World*, Chapel Hill, NC, 2006.
Pennington, M., ed. *A series of letters between Mrs. Elizabeth Carter and Miss Catherine Talbot*, 4 vols., London, 1809.
Phillips, Nicholas, and Quentin Skinner, eds. *Political Discourse in Early Modern Britain*, Cambridge, 1993.
Philpot, Stephen. *An Essay on the Advantage of a Polite Education Joined with a Learned One*, London, 1747.
Piacenti, Kirsten Aschengreen, and John Boardman. *Ancient and Modern Gems and Jewels in the Collection of Her Majesty The Queen,* London, 2008.
Piggott, S. *William Stukeley: An Eighteenth Century Antiquary*, Oxford, 1985.
—. *Ancient Britons and the Antiquarian Imagination: Ideas from the Renaissance to the Regency*, London, 1989.
Piggott, Rev. S. *Female Virtue and Domestic Religion Recommended by the Example of our Late Illustrious Queen Charlotte*, London, 1818.
Pinckney, Elise, ed. *The Letterbook of Eliza Lucas Pinckney, 1739–1762*, Columbia, SC, 1972.
Pittock, Murray G. H. *Celtic Identities and the British Image*, Manchester, 1999.
Plumb, Christopher. *The Georgian Menagerie: Exotic Animals in Eighteenth-Century London*, London, 2015.
Plumb, J. H. *Sir Robert Walpole: The King's Minister*, 2 vols., London, 1972.
Pocock, J. G. A. *Virtue, Commerce and History: Essays on Political Thought and History, Chiefly in the Eighteenth Century*, Cambridge, 1985.
The Polite Preceptor or a Collection of Entertaining and Instructive Essays, London, 1776.
Pöllnitz, Carl Ludwig von [La Saxe Galante]. *The Amorous Adventures of Augustus of Saxony . . . Translated from the French by a Gentleman of Oxford*, London, 1929.
—. *Memoirs of Charles-Lewis, Baron de Pöllnitz*, 5 vols., London, 1745.
Pointon, Marcia. *Brilliant Effects: A Cultural History of Gem Stones and Jewellery*, New Haven and London, 2009.
—. *Hanging the Head: Portraiture and Social Formation in Eighteenth-Century England*, New Haven and London, 1993.
Pope, Alexander. *The Correspondence of Alexander Pope*, ed. G. Sherburn, 5 vols., Oxford, 1956.
—. *The Dunciad in Four Books*, ed. Valerie Rumbold, 2nd ed., Harlow, 2009.

—. *Selected Poetry and Prose*, ed. W. K. Wimsatt, New York, 1972.
—. *The Works of Alexander Pope*, ed. J. W. Croker, intro. and notes W. Elwin and W. J. Courthope, 10 vols., London, 1871–89.
—. *The Works of Alexander Pope Esq.*, 9 vols., London, 1797.
Porter, Roy. *The Creation of the Modern World: The Untold Story of the British Enlightenment*, New York, 2000.
—. *English Society in the Eighteenth Century*, London, 1982.
—. *Enlightenment: Britain and the Creation of the Modern World*, London and New York, 2000.
—. *The Greatest Benefit to Mankind: A Medical History of Humanity from Antiquity to the Present*, London, 1997.
Postle, Martin, ed. *Johan Zoffany RA: Society Observed*, New Haven and London, 2011.
Preuss, J. D. E., ed. *Œuvres de Frédéric le Grand*, 31 vols., Berlin, 1846–57.
Prochaska, Frank. *Royal Bounty: The Making of a Welfare Monarchy*, New Haven and London, 1995.
Pyke, E. J. *A Biographical Dictionary of Wax Modellers*, Oxford, 1973.
Pyne, William Henry. *The History of the Royal Residences of Windsor Castle, St James's Palace, Carlton House, Kensington Palace, Hampton Court, Buckingham House and Frogmore*, 3 vols., London, 1816–19.
Quennell, Peter. *Caroline of England: An Augustan Portrait*, London, 1939.
Rameau, Pierre. *Le Maître a danser*, Paris, 1725.
Rapin de Thoyras, M. *The History of England . . . translated into English with additional notes . . . by N. Tindall*, 21 vols., London, 1757–63.
Rauser, Amelia. *Character Unmasked: Irony, Authenticity, and Individualism in Eighteenth-Century English Prints*, Newark, NJ, 2008.
Raven, James, ed. *Lost Libraries: The Destruction of Great Book Collections Since Antiquity*, Basingstoke, 2004.
Rees, Abraham. *The Cyclopaedia; or, Universal Dictionary of Arts, Sciences, and Literature*. London, 1819.
Reid, Margaret J. C. *The Arthurian Legend: Comparison of Treatment in Modern and Medieval Literature: A Study in the Literary Value of Myth and Legend*, London, 1938.
Reilly, R. *Wedgwood*, 2 vols., London, 1989.
Reilly, R., and G. Savage. *Wedgwood: The Portrait Medallions*, London, 1973.
Reynolds, Graham. *The Sixteenth- and Seventeenth-Century Miniatures in the Collection of Her Majesty the Queen*, London, 1999.
Reynolds, Joshua. *Discourses on Art*, ed. Robert R. Wark, San Marino, CA, 1959.
Ribeiro, A. *The Art of Dress: Fashion in England and France 1750 to 1820*, New Haven and London, 1995.
Richardson, T. *The Arcadian Friends: Inventing the English Landscape Garden*, London, 2007.
Rieckhof Nackas, Jörn Tillmann. *Schloss Friedenstein in Gotha: Entwürfe für die Residenz Ernsts des Frommen (1601–1675)*, Berlin, 2010.
Ritso, George. *Kew Gardens. A poem, humbly inscribed to Her Royal Highness the Princess Dowager of Wales*, [London], 1763.
Roberts, Jane. *Drawings by Holbein from the Court of Henry VIII*, New York, 1987.
—. 'Edward Harding and Queen Charlotte', in *Burning Bright: Essays in Honour of David Bindman*, ed. Diana Dethloff, Tessa Murdoch, and Kim Sloan, with Caroline Elam, London, 2015, 146–59.
—. *Holbein and the Court of Henry VIII: Drawings and Miniatures from the Royal Library, Windsor Castle*, Edinburgh, 1993.
—. *Royal Artists: From Mary Queen of Scots to the Present Day*, London, 1987.
—. *Royal Landscape: The Gardens and Parks of Windsor*, New Haven and London, 1997.
—. *Royal Treasures: A Golden Jubilee Celebration*, London, 2002.
Roberts, Jane, ed. *George III & Queen Charlotte: Patronage, Collecting and Court Taste*, London, 2004.
Robin, George. *Catalogue of the Classic Contents of Strawberry Hill Collected by Horace Walpole, 25 April–21 May 1842*, London, 1842.
Robins, Jane. *Rebel Queen: The Trial of Queen Caroline*, London, 2006.
Robinson, John Martin. *Temples of Delight: Stowe Landscape Gardens*, London, 1994.
Robinson, Tim. *William Roxburgh: The Founding Father of Indian Botany*, Chichester, 2008.
Roche, Sophie von la. *Sophie in London: Being the Diary of Sophie von la Roche*, trans. C. Williams, London, 1933.
Rorschach, Kimerly. *The Early Georgian Landscape Garden*, ex. cat., New Haven, 1983.
Rosenthal, Angela. *Angelica Kauffman: Art and Sensibility*, New Haven and London, 2006.
Rosenthal, Michael. *The Art of Thomas Gainsborough: 'A Little Business for the Eye'*, New Haven and London, 1999.
Ross, Angus, ed. *The Correspondence of Dr John Arbuthnot*, Munich, 2006.
Ross, Charles, ed. *The Correspondence of Charles, First Marquis Cornwallis*, 3 vols., London, 1859.
Rott, Herbert W., Werner Busch, Tim Clayton, and others, eds. *George Stubbs, 1724–1806: Science into Art*, Munich, 2012.
Rouquet, André. *The Present State of the Arts in England*, London, 1755.
Royal Collection. *Holbein and the Court of Henry VIII*, London, 1978.
—. *A King's Purchase: King George III and the Collection of Consul Smith*, London, 1993.
Royal Society of Musicians. *List of benefactors and subscribers to a fund for the support of decayed musicians and their families*, London, 1790.
Ruggles, Dede Fairchild. *Sound and Scent in the Garden*, Washington, DC, 2017.
Russell, F. *John, 3rd Earl of Bute: Patron and Collector*, London, 2004.
Ryan, Thomas. *The History of Queen Charlotte's Lying-in Hospital, from its Foundation in 1752 to the Present Time*, London, 1885.
Ryder, D. *The Diary of Dudley Ryder, 1715–16*, ed. William Matthews, London, 1939.
Sambrook, James. *The Eighteenth Century: The Intellectual History and Cultural Context of English Literature, 1700–1789*, London, 1993.
Sanders, Mary F., *The Life and Times of Queen Adelaide*, London, 1915.
Sandon, H. *Flight and Barr Worcester Porcelain, 1783–1840*, Woodbridge, 1978.
Sandon, J. *The Dictionary of Worcester Porcelain*, vol. 1: *1751–1851*, Woodbridge, 1993.
Saussure, César de. *A Foreign View of England in the Reigns of George I and George II*, ed. and trans. Madame Van Muyden, London, 1902.
—. *Letters from London, 1725–1730*, trans. P. Scott, Newnham, 2006.
Savage, Richard. *The Poetical Works of Richard Savage*, ed. Clarence Tracy, Cambridge, 1962.
Schaich, Michael, ed. *Monarchy and Religion: The Transformation of Royal Culture in Eighteenth-Century Europe*, Oxford, 2007.
Schnath, Georg. *Das Leineschloss: Kloster, Fürstensitz, Landtagsgebäude*, Hanover, 1962.
Schuhmann, Gunther. *Die Markgrafen von Brandenburg-Ansbach:Eine Bilddokumentation zur Geschichte der Hohenzollern in Franken*, Ansbach, 1980.
[Scott, Mrs G. L.] *The History of Mecklenburgh; from the First Settlement of the Vandals in that Country, to the Present Time*, London, 1762.
Scott, Hamish, ed. *The European Nobilities in the Seventeenth and Eighteenth Centuries*, 2 vols., London, 1995.
Scott, K. *St. James's Palace: A History*, London, 2010.
Seymour, Frances, Duchess of Somerset. *Correspondence between Frances Countess of Hertford (afterwards Duchess of Somerset) and Henrietta Louisa, Countess Pomfret between the years 1738 and 1741*, ed. W. Bingley, 2nd ed., 2 vols., London, 1805.
Sharp, Tony. *Pleasure and Ambition: The Life, Loves and Wars of Augustus the Strong, 1670–1707*, London, 2001.
Shawe-Taylor, Desmond. *The Conversation Piece: Scenes of Fashionable Life*, London, 2009.
—. *The Georgians: Eighteenth-Century Portraiture and Society*, London, 1990.

Shawe-Taylor, Desmond, ed. *The First Georgians: Art and Monarchy, 1714–1760*, London, 2014.
Shearman, John. *The Early Italian Pictures in the Collection of Her Majesty The Queen*, Cambridge, 1983.
——. *Raphael's Cartoons in the Collection of Her Majesty The Queen, and the Tapestries for the Sistine Chapel*, London, 1972.
Shefrin, J. *Such Constant Affectionate Care: Lady Charlotte Finch, Royal Governess, and the Children of George III*, Los Angeles, 2003.
Shoemaker, Nancy. *A Strange Likeness: Becoming Red and White in Eighteenth-Century North America*, Oxford, 2004.
Simms, B. *Three Victories and a Defeat: The Rise and Fall of the First British Empire, 1714–1785*, London, 2007.
Simms, Brendan, and Torsten Riotte, eds. *The Hanoverian Dimension in British History, 1714–1837*, Cambridge, 2007.
Simon, Edith. *The Making of Frederick the Great*, London, 1963.
Simon, J., ed. *Handel: A Celebration of his Life and Times, 1685–1759*, London, 1985.
Singer, A. *The Lion and the Dragon: The Story of the First British Embassy to the Emperor Qianlong in Peking, 1792–1794*, London, 1992.
Siris, P. *The Princess Anna*, London, 1715.
Skelton, R. 'Europe and India', in *Europa und die Kunst des Islam 15 bis 18 Jahrhundert*, ed. Elisabeth Liskar, Vienna, 1985, 33–42.
——. *The Indian Heritage: Court Life and Arts under Mughal Rule*, London, 1982.
Sloan, Kim. *'A Noble Art': Amateur Artists and Drawing Masters, c. 1600–1800*, London, 2000.
Sloan, Kim, and Andrew Burnett, eds. *Enlightenment: Discovering the World in the Eighteenth Century*, London, 2003.
Smart, A. *Allan Ramsay: Painter, Essayist and Man of Enlightenment*, New Haven and London, 1992.
Smart, Alastair, and John Ingamells, eds. *Allan Ramsay: A Complete Catalogue of his Paintings*, New Haven and London, 1999.
Smiles, Sam. *The Image of Antiquity: Ancient Britain and the Romantic Imagination*, New Haven and London, 1994.
Smith, Bernard. *European Vision in the South Pacific*, New Haven and London, 1985.
Smith, H. Clifford. *Buckingham Palace: Its Furniture, Decoration and History*, London, 1930.
Smith, Hannah. *Georgian Monarchy: Politics and Culture, 1714–1760*, Cambridge, 2006.
Smith, Pamela. *The Business of Alchemy: Science and Culture in the Holy Roman Empire*, Princeton, NJ, 1994.
Snodin, Michael, ed. *Horace Walpole's Strawberry Hill*, New Haven and London, 2009.
Society of Antiquaries of London. *A Description of Nine Historical Prints, Representing Kings, Queens, Princes, etc. of the Tudor Family, Selected, Drawn and Engraved from the Original Paintings, by George Vertue, late Engraver to the Society of Antiquaries of London*, London, 1776.
Society of Gardeners. *A Catalogue of Trees, Shrubs, Plants, and Flowers, both Exotic and Domestic, which are propagated for sale, in the Gardens near London*, London, 1730.
Soderstrom, Mary. *Recreating Eden: A Natural History of Botanical Gardens*, Montreal, 2001.
Sofaer, Joanna, ed. *Material Identities*, Oxford, 2007
Solkin, David H. *Art in Britain, 1660–1815*, New Haven and London, 2015.
——. *Richard Wilson: The Landscape of Reaction*, London, 1982.
Sophia of Hanover. *Memoirs of Sophia: Electress of Hanover, 1630–1680*, trans. H. Forester, London, 1888.
Sotheby's. *The Royal House of Hanover/ Das Konigshaus von Hannover, Schloss Marienburg, 5–15 October 2005*, 3 vols., London, 2005.
Spalding, Phinizy, and Harvey H. Jackson, eds. *Oglethorpe in Perspective: Georgia's Founder after Two Hundred Years*, Tuscaloosa, AL, 1989.
Spence, J. *Anecdotes, Observations and Characters of Books and Men: Collected from the Conversation of Mr Pope . . .*, ed. S. W. Singer, London, 1820.
Spencer, S. I. *French Women and the Age of Enlightenment*, Bloomington, 1984.
Spenser, Edmund. *The Fairy Queen* [first published 1590–96], London, 1996.
Spruill, Marjorie J., Joan M. Johnson, and Valinda W. Littlefield, eds. *South Carolina Women: Their Lives and Times*, 3 vols., Athens, GA, 2009–12.
St Clair, William. *The Reading Nation in the Romantic Period*, Cambridge, 2004.
St John Hope, W. H. *Windsor Castle*, 2 vols., London, 1913.
Stephens, John C., ed. *Georgia, and Two Other Occasional Poems on the Founding of the Colony, 1736*, Atlanta, GA, 1950.
Stewart, J. Douglas. *Sir Godfrey Kneller and the English Baroque Portrait*, Oxford, 1983.
Stirling, A. M. W. *Coke of Norfolk and his Friends*, London, 1908.
Stobel, Heidi A. *The Artistic Marriage of Queen Charlotte (1744–1818): How a Queen Promoted both Art and Female Artists in English Society*, Lewiston, NY, 2011.
Stone, B., trans. and intro. *King Arthur's Death: Alliterative Morte Arthure and Stanzaic le morte Arthur*, London, 1988.
String, T. C., and M. Bull, eds. *Tudorism: Historical Imagination and the Appropriation of the Sixteenth Century*, Oxford, 2011.
Strong, Roy. *The Artist and the Garden*, New Haven and London, 2000.
——. *Henry, Prince of Wales and England's Lost Renaissance*, London, 1986.
——. *Royal Gardens: From Charles II to Charles, Prince of Wales*, London, 1995.
Stuart, D. M. *The Daughters of George III*, London, 1939.
Stukeley, William. *Itinerarium Curiosum: Or an Account of the Antiquitys and Remarkable Curiositys in Nature or Art, Observ'd in Travels thro' Great Britain*, London, 1724.
——. *Palaeographia Sacra: or Discourses on Sacred Subjects*, London, 1763.
Styles, John. *Threads of Feeling: The London Foundling Hospital's Textile Tokens, 1740–1770*, London, 2010.
Sundon, Viscountess. *Memoirs of Viscountess Sundon: Mistress of the Robes to Queen Caroline, Consort of George II*, ed. A. T. Thomson, 2 vols., London, 1847.
Sweet, Rosemary. *Antiquaries: The Discovery of the Past in Eighteenth-Century Britain*, London, 2004.
Swift, Jonathan. *The Correspondence of Jonathan Swift, D.D.*, ed. F. Elrington Ball, 6 vols., London, 1910–14.
Switzer, S. *Ichnographia Rustica: Or, the Nobleman, Gentleman, and Gardener's Recreation*, London, 1718.
Tatton-Brown, Tim, and Richard Mortimer, eds. *Westminster Abbey: The Lady Chapel of Henry VII*, Woodbridge and Rochester, NY, 2003.
Taylor, Kristina, and Robert Peel. *Passion, Plants and Patronage: 300 Years of the Bute Family Landscapes*, London, 2012.
Taylor, S., R. Connors, and C. Jones, eds. *Hanoverian Britain and Empire: Essays in Memory of Philip Lawson*, Woodbridge, 1998.
Tennant, C. M. *Peter the Wild Boy*, London, 1938.
Thicknesse, Philip. *Man-Midwifery Analysed: And the Tendency of That Practice Detected and Exposed*, London, 1763.
Thomas, P. D. G. *Politics in Eighteenth-Century Wales*, Cardiff, 1998.
Thompson, A. C. *Britain, Hanover and the Protestant Interest, 1688–1756*, Woodbridge, 2006.
——. *George II: King and Elector*, New Haven and London, 2011.
Thomson, James, and David Mallet. *Alfred, an Opera. Altered from the Play written by Mr Thomson and Mr Mallet, in honour of the birthday of Her Royal Highness, the young Princess Augusta*, London, 1740.
Thornhill, James. *An Explanation of the Painting in the Royal Hospital at Greenwich*, London, 1730.
Thornton, Robert. *New Illustration of the Sexual System of Carolus von Linnaeus: Comprising An Elucidation of the Several Parts of the Fructification; A Prize Dissertation on the Sexes of Plants; A Full Explanation of the Classes, and Orders, of the Sexual System; and the Temple of Flora, or Garden of Nature, Being Picturesque, Botanical, Coloured Plates of Select Plants Illustrative of the Same, with Descriptions*, 2 vols., London, 1807.

Thrush, Coll-Peter. *Indigenous London: Native Travelers at the Heart of Empire*, New Haven and London, 2016.
Tobin, Beth Fowkes. *The Duchess's Shells: Natural History Collecting in the Age of Cook's Voyages*, New Haven and London, 2014.
Toland, Mr. *An Account of the Courts of Prussia and Hanover*, London, 1705.
Tomlinson, Kellom. *The art of dancing*, London, 1735.
Trimmer, Sarah. *The Oeconomy of Charity: or, an Address to Ladies concerning Sunday Schools; the Establishment of Schools of Industry under Female Inspection; and the Distribution of Voluntary Benefactions: to which is added an Appendix containing an Account of the Sunday-Schools in Old Brentford*, London, 1787.
Uffenbach, Zacharias Conrad von. *London in 1710: From the Travels of Z. C. von Uffenbach*, ed. and trans. W. H. Quarrell and Margaret Mare, London, 1934.
Urbach, Karina, ed. *Royal Kinship: Anglo-German Family Networks, 1815–1914*, Munich, 2008.
Vardy, John. *Some Designs of Mr Inigo Jones and Mr Wm Kent*, London, 1744.
Varnhagen von Ense, August. *Leben der Königin von Preußen Sophie Charlotte*, Berlin, 1837.
Vergil, Polydore, ed. and trans. Denys Hay, *The Anglica Historia of A.D. 1485–1537*, London, 1950.
Vivian, Frances. *A Life of Frederick, Prince of Wales, 1707–1751: A Connoisseur of the Arts*, ed. Roger White, Lewiston, NY, and Lampeter, 2006.
Voltaire [pseud. for François-Marie Arouet]. *Letters Concerning the English Nation*, intro. Charles Whibley, London, New York, and Toronto, 1926.
Voltaire, *Lettres Philosophiques*, ed. R. Pomeau, Paris and Montreal, 1964.
Wagstaffe, William. *A Letter to Dr Friend, showing the Danger and Uncertainty of Inoculating the Small Pox*, London, 1722.
Waley, M. I. *Islamic Manuscripts in the Royal Collection: A Concise Catalogue*, vol. 6, *Manuscripts of the Middle East*, Leiden, 1992.
Walker, Richard. *The Eighteenth and Early Nineteenth Century Miniatures in the Collection of Her Majesty The Queen*, Cambridge, 1992.
Walpole, Horace, Earl of Orford. *A Selection of the Letters of Horace Walpole*, ed. W. S. Lewis, London, 1926.
—. *Aedes Walpolianae: Or, a Description of the Collection of Pictures at Houghton Hall in Norfolk, the Seat of the Right Honourable Sir Robert Walpole, Earl of Orford*, London, 1752.
—. *Anecdotes of Painting in England: With some Account of the Principal Artists*, ed. Rev. J. Dallaway, 3 vols., London, 1848.
—. *The Duchess of Portland's Museum*, intro. W. S. Lewis, New York, 1936.
—. *Horace Walpole: Memoirs and Portraits*, ed. M. Hodgart, London, 1963.
—. *Letters of Horace Walpole*, selected and ed. C. D. Yonge, 2 vols., London, 1891.
—. *Memoirs of King George the Second*, ed. John Brooke, 3 vols., New Haven and London, 1985.
—. *Memoirs of the Reign of King George the Second: Collected from the original MSS. with a Preface and Notes*, ed. Lord Holland, 3 vols., London, 1846.
—. *Reminiscences Written by Mr Horace Walpole in 1788 for the Amusement of Miss Mary and Miss Agnes Berry*, ed. Paget Toynbee, Oxford, 1924.
—. *The Yale Edition of Horace Walpole's Correspondence*, ed. W. S. Lewis, Charles Bennett, and Andrew G. Hoover, 48 vols., New Haven and London, 1937–83.
Walters, John. *The Royal Griffin: Frederick, Prince of Wales, 1707–51*, London, 1972.
Ware, I. *Designs of Mr Inigo Jones and Others*, London, 1735.
—. *The Plans, Elevations, and Sections, Chimney Pieces, and Ceilings of Houghton in Norfolk, the Seat of Sir R, Walpole*, engraved Foudrinier, London, 1735.
Watkin, David. *The Architect King: George III and the Culture of the Enlightenment*, London, 2004.
—. *The Royal Interiors of Regency England*, London and Melbourne, 1984.
—. *Sir John Soane: Enlightenment Thought and the Royal Academy Lectures*, Cambridge, 1996.
Watkins, John. *Memoirs of Her Most Excellent Majesty Sophia-Charlotte, Queen of Great Britain*, London, 1819.
Watkins, John. *Representing Elizabeth in Stuart England: Literature, History, Sovereignty*, Cambridge, 2002.
Watson, J. Steven. *The Reign of George III, 1760–1815*, Oxford, 1960.
Weaver, Jace. *The Red Atlantic: American Indigenes and the Making of the Modern World, 1000–1927*, Chapel Hill, NC, 2014.
Webb, M. I. *Michael Rysbrack: Sculptor*, London, 1954.
Weber, Susan, ed. *William Kent: Designing Georgian Britain*, New Haven and London, 2013.
Webster, Mary. *Johan Zoffany, 1733–1810*, New Haven and London, 2011.
Wedgwood, Josiah. *Correspondence of Josiah Wedgwood, 1781–1794: with an appendix containing some letters on canals and Bentley's pamphlet on inland navigation*, London, 1906.
—. *Letters of Josiah Wedgwood, 1762–1780*, 2 vols., London, 1903.
—. *The Selected Letters of Josiah Wedgwood*, ed. A. Finer and G. Savage, London, 1965.
Whately, M., *Observations on Modern Gardening*, Dublin, 1770.
Whinney, M. *Sculpture in Britain, 1530–1830*, Harmondsworth, 1964.
Whitaker, L., and M. Clayton. *The Art of Italy in the Royal Collection: Renaissance and Baroque*, London, 2007.
Whitaker, Muriel. *The Legends of King Arthur in Art*, Cambridge, 1990.
Whitworth, R. *William Augustus, Duke of Cumberland: A Life*, London, 1992.
Wiesner, Merry E. *Women and Gender in Early Modern Europe*, Cambridge, 2000.
Wilkins, W. H. *Caroline the Illustrious*, 2 vols., London, 1901.
Williams, Basil. *The Whig Supremacy, 1714–1760*, Oxford, 1952.
Williams, Kate. *Becoming Queen*, London, 2008.
Williams, Thomas. *A Brief Memoir of Her Late Majesty Queen Charlotte*, London, 1819.
Williamson, G. C. *Catalogue of a Collection of Miniatures: The Property of His Royal Highness Prince Ernest Augustus William Adolphus George Frederick, Duke of Cumberland, Duke of Brunswick-Lüneburg, K.G., G.C.H.*, London, 1914.
Willis, Peter. *Charles Bridgeman and the English Landscape Garden*, Newcastle upon Tyne, 2002.
Wilson, M. I. *William Kent: Architect, Designer, Painter, Gardener, 1685–1748*, London, 1984.
Wilson, Peter H. *Absolutism in Central Europe*, London, 2000.
Wittkower, R. *Palladio and English Palladianism*, London, 1974.
Wollaston, H. *British Official Medals for Coronation and Jubilees*, Long Eaton, 1978.
Woolf, D. R. *Reading History in Early Modern England*, Cambridge, 2000.
Worsley, L. *Courtiers: The Secret History of Kensington Palace*, London, 2010.
Young, Sir George. *Poor Fred: The People's Prince*, Oxford, 1937.
Young, Percy M. *The Bachs, 1500–1850*, London, 1970.

PUBLISHED SOURCES: PERIODICALS

Addison, Joseph. 'The Pleasures of the Imagination', *The Spectator*, nos. 411–21, 1712.
Alexander, D. 'George Vertue as an Engraver', *Walpole Society* 70 (2008): 207–517.
Arciszewska, B. 'A Villa fit for a King: The Role of Palladian Architecture in the Ascendancy of the House of Hanover under George I', *Canadian Art Review* 19 (1992): 41–112.
Ariew, R. 'Leibniz on the Unicorn and various other Curiosities', *Early Science and Medicine* 3, no. 4 (1998): 267–88.
Barber, P. 'Royal Geography: The Development and Destiny of King George III's Geographical Collections', in preparation.
Barthélemy E. de. 'Inventaire du Mobilier de la Duchesse d'Orléans, Mère du Régent, après son Decès en 1722', *Bulletin du Comité des Travaux Historiques*, 1883.
Beard, G. 'Three Eighteenth-Century Cabinet-makers: Moore, Goodison and Vile', *The Burlington Magazine* 119 (July 1977): 479–86.
Bellaigue, G. de. 'Huzza the King is Well!', *The Burlington Magazine* 126 (June 1984): 325–31.

Bellows, Barbara L. 'Eliza Lucas Pinckney: The Evolution of an Icon', *South Carolina Historical Magazine* 106, no. 2 (2005): 147–65 .
Bertolini Meli, D. 'Caroline, Leibniz and Locke', *Journal of the History of Ideas* 60, no. 3 (July 1999): 469–86.
Bignamini, Ilaria. 'George Vertue, Art Historian and the Art Institutions of London', *Walpole Society* 54 (1991): 1–48.
Binney, M. 'Schloss Favorite, Ludwigsburg', *Country Life*, 6 July 1989: 104–9.
Bird, Rufus. 'The Furniture and Furnishing of St James's Palace, 1714–1715', *Furniture History* 50 (2014): 147–203.
Bowett, Adam. 'George I's Furniture at Kensington Palace', *Apollo* 162 (Nov. 2005): 37–46.
Braun, Heinz. 'Ansbacher Spätbarock: Ein Beitrag zur Stilgeschichte der Baukunst im Fürstentum Brandenburg-Onolzbach, 1695–1791', *Jahrbuch für Fränkische Landesforschung* 16 (1956): 455–91.
Brooke, John. 'The Library of King George III', *Yale University Library Gazette* 52, no. 1 (1977): 33–45.
Brown, Gregory. 'Leibniz's Endgame and the Ladies of the Court', *Journal of the History of Ideas* 65, no. 1 (Jan. 2004): 75–100.
Brown, L. A. 'Queen Charlotte's Medal for the Recovery of George III', *British Numismatic Journal* 72 (2002): 183–84.
Bryant, Jenny A., Linda M. Irvine, and Emma Ruffle. 'Insights into the Life and Work of Rev. John Lightfoot (1735–1788), with Particular Reference to his Algal Herbarium and its Conservation', *The Linnean* 28, no. 1 (April 2012): 26–43.
Bryant, Julius. 'Queen Caroline's Richmond Lodge by William Kent: An Architectural Model Unlocked', *The Burlington Magazine* 157 (May 2015): 325–30.
—. 'William Kent's Whimsical Marginalia', *Master Drawings* 53, no. 1 (2015): 99–108.
Bullion, John L. 'The Origins and Significance of Gossip about Princess Augusta and Lord Bute, 1755–1756', *Studies in Eighteenth-Century Culture* 21 (1992): 245–65.
—. 'The Prince's Mentor: A New Perspective on the Friendship between George III and Lord Bute during the 1750s', *Albion: A Quarterly Journal Concerned with British Studies* 21, no. 1 (Spring 1989): 34–55.
Burchard, Wolf. 'Face to Face with the First Georgians', *The Georgian* 1 (2014): 4–8.
—. 'St James's Palace: George II's and Queen Caroline's Principal London Residence', *The Court Historian* 16, no. 2 (Dec. 2011): 177–203.
Campbell Orr, Clarissa. 'New Perspectives on Hanoverian History', *The Historical Journal* 52, no. 2 (2009): 513–29 .
—. 'Queen Charlotte as Patron: Some Intellectual and Social Contexts', *The Court Historian* 6, no. 3 (2001): 183–212.
Carré, J. 'Lord Burlington's Garden at Chiswick', *Garden History* 1, no. 3 (1973): 23–30.
Clarke, G. 'Grecian Taste and Gothic Virtue: Lord Cobham's Gardening Programme and its Iconography', *Apollo* 97 (1973): 566–71.
Clement, Edward Henry, and Samuel Eliot Morison, eds. 'Letters of Thomas Coram', *Proceedings of the Massachusetts Historical Society* (third series) 56 (Oct. 1922–June 1923): 1–68.
Clifford, T. 'The Chelsea-Derby Royal Family Groups', *The Burlington Magazine* 127 (supplement, September 1985): 12–15.
—. 'J. J. Spangler: A Virtuoso Swiss Modeller at Derby', *Connoisseur* (June 1978): 146–55.
—. 'Vulliamy Clocks and British Sculpture', *Apollo* 132 (Oct. 1990): 226–37.
Clinton, Katherine B. 'Femme and Philosophe: Enlightenment Origins of Feminism', *Eighteenth-Century Studies* 8, no. 3 (Spring 1975): 283–99.
Colton, J. 'Kent's Hermitage for Queen Caroline at Richmond', *Architectura* 2 (1974): 181–91.
—. 'Merlin's Cave and Queen Caroline: Garden Art as Political Propaganda', *Eighteenth-Century Studies* 10, no. 1 (Autumn 1976): 1–20.
Coombs, David. 'The Garden at Carlton House of Frederick Prince of Wales and Augusta Princess Dowager of Wales', *Garden History* 25, no. 2 (Winter 1997): 153–77 .
Coon, David L. 'Eliza Lucas Pinckney and the Reintroduction of Indigo Culture in South Carolina', *Journal of Southern History* 42, no. 1 (1976): 61–76.
Corder, Jim. 'Spenser and the Eighteenth-Century Informal Garden', *Notes and Queries* 204 (1959): 19–21.
Corner, Betsy Copping. 'Dr. Ibis and the Artists: A Sidelight upon Hunter's Atlas, The Gravid Uterus', *Journal of the History of Medicine and Allied Sciences* 6 (Winter 1951): 1–21 .
Cornforth, John. 'Houghton Hall, Norfolk I', *Country Life*, 30 April 1987, 124–29.
—. 'Houghton Hall, Norfolk II', *Country Life*, 7 May 1987, 104–8.
—. 'The Growth of an Idea: The Making of Houghton's Landscape', *Country Life*, 14 May 1987, 162–68.
—. 'Kensington Palace, London', *Country Life*, 5 Jan. 1995, 32–37.
Crawforth, M. A. 'Instrument Makers in the London Guilds', *Annals of Science* 44 (1987): 319–37.
Dawson, A. 'Creamware, Cameos and Coffee Cans', *Country Life*, 5 Sept. 1985, 638–40.
Desaguliers, J. T. 'An Account of an Experiment to Prove an Interspers'd Vacuum; Or to Shew that all Places are not Equally Full', *Philosophical Transactions of the Royal Society* 30 (1717–19): 717–20.
Doort, Abraham van der. 'Abraham van der Doort's Catalogue of the Collections of Charles I', ed. Oliver Millar, *Walpole Society* 37 (1960).
Doyle, Nora. '"The Highest Pleasure of Which Woman's Nature Is Capable": Breast-Feeding and the Sentimental Maternal Ideal in America, 1750–1860', *Journal of American History* 97, no. 4 (2011): 958–73.
Drinkuth, Friedrike. 'Zur Identifikation und Bedeutung einer Aussteuerliste von Herzogin Elisabeth Albertine zu Mecklenburg-Strelitz.' *Mecklenburgische Jahrbücher* 129 (2014): 265–82.
Edwards, Phyllis. 'Sir Joseph Banks and the Botany of Captain Cook's Three Voyages of Exploration', *Pacific Studies* 2, no. 1 (1978): 20–43.
Eichholz, Jeffrey P. 'William Kent's Career as Literary Illustrator', *Bulletin of the New York Public Library* 70 (1966): 620–46.
Eustace, Katharine. 'The Politics of the Past: Stowe and the Development of the Historical Portrait Bust', *Apollo* 148 (July 1998): 31–40.
Fabianski, M. 'Iconography of the Architecture of the Ideal: *Musaea* in the Fifteenth to Sixteenth Centuries', *History of Collections* 2, no. 2 (1990): 95–134.
Findlen, P. 'The Museum: Its Classical Etymology and Renaissance Genealogy', *Journal of the History of Collections* 1, no. 1 (1989): 59–78.
Forman, L. L. 'Notes Concerning the Typification of Names of William Roxburgh's Species of Phanerogams', *Kew Bulletin* 52, no. 3 (1997): 513–34.
Fountain, R. B. 'George Stubbs (1724–1806) as an Anatomist', *Section of the History of Medicine* 61 (1968): 639–45.
Fullager, Kate. '"Savages that are Among Us": Mai, Bennelong, and British Imperial Culture, 1774–1795', *Eighteenth Century* 49, no. 3 (2008): 211–37.
Gerrard, C. 'The Castle of Indolence and the Opposition to Walpole', *The Review of English Studies* (new series) 41, no. 161 (Feb. 1990): 45–64.
Ghose, K. 'Additional Note on Ivory Chair in Victoria Memorial Collection', *Bulletin of the Victoria Memorial* 12 (1979): 25–27.
—. 'Note on Ivory Chair & Teapot in Victoria Memorial Collection', *Bulletin of the Victoria Memorial* 10 (1976): 53–55.
Giometti, C. 'Giovanni Battista Guelfi: New Discoveries', *Sculpture Journal* 3 (1999): 26–43.
Glaisyer, Natasha. 'Networking: Trade and Exchange in the Eighteenth-Century British Empire', *Historical Journal* 47, no. 2 (2004): 451–76.
Goff, Moira. 'Desnoyer: Charmer of the Georgian Age', *Historical Dance* 4, no. 2 (2012): 3–10.
Gosset, M. H. 'A Family of Modellers in Wax', *Proceedings of the Huguenot Society of London* 3 (1888–91): 540–68.

Gutfleish, B., and J. Menzhausen. '"How a Kunstkammer Should Be Formed": Gabriel Kaltemarckt's Advice to Christian I of Saxony on the Formation of an Art Collection', *Journal of the History of Collections* 1, no. 1 (1989): 3–32.

Gwynn, D. 'The Charles Clay Organ Clock in the Royal Collection (Windsor Castle) and its Connection with Georg Frideric Handel', *British Institute of Organ Studies Reporter* 21, no. 2 (Jan. 1997).

Harcourt, Mary, Lady. 'Mrs Harcourt's Diary of the Court of King George III', *Miscellanies of the Philobiblon Society* 13 (1871–72).

Harris, Frances. '"The Honourable Sisterhood": Queen Anne's Maids of Honour', *The British Library Journal* 19, no. 2 (Autumn 1993): 181–98.

Harris, John. 'The Bicentenary of Kew Gardens I: Fate of Royal Buildings', *Country Life*, 28 May 1959, 1182.

—. 'Esher Place, Surrey', *Country Life*, 2 April 1987, 94–97.

Hashberger, John W. 'William Young, Jr., of Philadelphia, Queen's Botanist', *Torreya* 17, no. 6 (June 1917): 91–99.

Haskell, F. 'The Apotheosis of Newton in Art', *Texas Quarterly* 10, no. 3 (Autumn 1967): 218–37.

Hessler-Cable, Mabel. 'The Idea of a Patriot King in the Propaganda of the Opposition to Walpole 1735–1739', *Philosophical Quarterly* 18 (1939): 124–30.

Hinderaker, Eric. 'The "Four Indian Kings" and the Imaginative Construction of the British Empire', *William and Mary Quarterly* (3rd series) 53 (1996): 487–526.

Hodson, Y. 'Prince William, Royal Map Collector', *The Map Collector* 44 (1988): 2–12.

Hunt, John Dixon. 'A Silent and Solitary Hermitage', *Annual Report of the York Georgian Society* (1970): 47–61.

Jackson-Stops, Gervase. '"A Noble Simplicity": Pyne's Views of Buckingham House', *Apollo* 138 (Aug. 1993): 112–24.

Jaffer, A. 'Tipu Sultan, Warren Hastings and Queen Charlotte: The Mythology and Typology of Anglo-Indian Ivory Furniture', *The Burlington Magazine* 141 (1999): 271–81.

Jansen, D. J. 'Antiquarian Drawings and Prints as Collectors' Items', *Journal of the History of Collections* 6, no. 2 (1994): 181–88.

Jarvis, Charles E., Ashley DuVal, and Peter R. Crane, 'Gardenia Jasinoides, a Traditional Chinese Dye Plant becomes a Garden Ornamental in Europe', *Curtis's Botanical Magazine* 31, no. 1 (2014): 80–98.

Jay, Emma. 'Court Patronage Reconsidered: The English Literature in Queen Caroline's Library', *Library and Information History* 30, no. 2 (2014): 75–89.

—. 'Queen Caroline's Library and its European Contexts', *Book History* 9 (2006): 31–55.

Jenkins, S. 'Queen Caroline's Taste: The Furnishing and Functioning of the Queen's Private Apartments at Hampton Court', *Apollo* 142 (May 1996): 20–24.

Keynes, Simon. 'The Cult of King Alfred', *Anglo-Saxon England* 28 (1999): 225–356.

King, G. 'Two New Letters from Princess Amelia', *Handel Jarhbuch* 40/41 (1994–95): 169–71.

King, Richard C. 'On Princess Anne's Lessons with Handel', *Newsletter of the American Handel Society* 7 no. 2 (Aug. 1992): 4–5.

Klein, L. 'Coffeehouse Civility, 1660–1714: An Aspect of Post-Courtly Culture in England', *Huntington Library Quarterly* 59, no. 1 (1996): 30–52.

—. 'Politeness and the Interpretation of the British Eighteenth Century', *The Historical Journal* 45, no. 4 (2002): 869–98.

Kremeier, Jarl. 'George I and George II as Patrons of Architecture in Hanover', *The Georgian Group Journal* 22 (2014): 1–20.

Lister, Jenny. 'Twenty Three Samples of Silk: Silks Worn by Queen Charlotte and the Princesses at Birthday Balls, 1791–1794', *Costume* 37 (2003): 51–65.

—. 'Twenty Three Samples of Spitalfields Silk (II): The Wedding Clothes at the Nuptials of HRH the Prince of Wales with Princess Caroline of Brunswick', unpublished essay, 2002.

Löwenstein, Uta. 'So Hält der Engel Hand die Cron aus Engelland. Feierlichkeiten bei der Hochzeit des Landgrafen Friedrich von Hessen-Kassel und der Prinzessin Maria von Grossbritannien', *Hessische Heimat* 44, no. 4 (1994): 135–39.

—. 'Apartment of the Countess Dorothea Friederike von Hanau-Lichtenberg at her Death in the Year 1731', unpublished article.

Maier, Josef. 'Residenzschloss Ansbach: Gestalt und Ausstattung im Wandel der Zeit', *Jahrbuch des Historischen Vereins für Mittelfranken* 100 (2005): 58–60.

Mallet, J. V. G. 'Rococo English Porcelain, a Study in Style', *Apollo* 90 (Aug. 1969): 100–13.

Marks, Arthur. 'An Anatomical Drawing by Alexander Cozens', *Journal of the Warburg and Courtauld Institutes* 30 (1967): 434–38.

Marschner, J. M. 'Baths and Bathing at the Early Georgian Court', *Furniture History* 31 (1995): 23–28.

—. 'Queen Caroline of Ansbach: Attitudes to Clothes and Cleanliness, 1727–1737', *Costume* 31, no. 1 (1997): 28–37.

Marsh, Ben. 'Silk Hopes in Colonial South Carolina', *Journal of Southern History* 78, no. 4 (2012): 807–54.

Marshall, P. J. 'The Personal Fortune of Warren Hastings', *Economic History Review* 17, no. 2 (1964): 284–300.

Martin, Eliza Layne. 'Eliza Lucas Pinckney: Production and Consumption in the Atlantic World', *World History Connected* 7, 1 (Feb. 2010): http://worldhistoryconnected.press.illinois.edu/7.1/martin.html.

Marx, Hans Joachim. '"A love of music to distraction": Musik im Leben der englischen Königen Charlotte (1744–1818)', *Archiv für Muskwissenschaft* 71, no. 1 (2014): 1–20.

McGeary, T. 'Frederick, Prince of Wales as Print Collector', *Print Quarterly* 19, no. 3 (Aug. 2002): 254–60.

McKendrick, Neil. 'Josiah Wedgwood: An Eighteenth-Century Entrepreneur in Salesmanship and Marketing Techniques', *Economic History Review* 12 (1960): 408–32.

McMillan, Patrick D., and Amy Hackney Blackwell. 'The Vascular Plants collected by Mark Catesby in South Carolina: Combining the Sloane and Oxford Herbaria' (2013): www.phytoneuron.net/2013Phytoneuron/73PhytoN-CatesbySloaneOxford.pdf.

Manfredi, M. 'Jacopo Amigoni: A Venetian painter in Georgian London', *The Burlington Magazine* 147 (Oct. 2005): 676–79.

Murdoch, T. 'The King's Cabinet-Maker: The Giltwood Furniture of James Moore the Elder', *The Burlington Magazine* 145 (June 2003): 408–20.

Nash, R. C. 'South Carolina Indigo, European Textiles, and the British Atlantic Economy in the Eighteenth Century', *Economic History Review* 63, no. 2 (2010): 362–92.

Oman, C. 'Silver Designs by William Kent', *Apollo* 95 (Jan. 1972): 22–24.

Owen, F. 'Joshua Kirby (1716–74): A Biographical Sketch', *Gainsborough's House Review* (1995/96): 61–75.

Parsons, Christopher M., and Kathleen S. Murphy. 'Ecosystems under Sail: Specimen Transport in the Eighteenth-Century French and British Atlantics', *Early American Studies* 10, no. 3 (2012): 503–39.

Peace, Mary. 'The Magdalen Hospital and the Fortunes of Whiggish Sentimentality in Mid-Eighteenth Century Britain: "Well-Grounded" Exemplarity vs. "Romantic" Exceptionality', *The Eighteenth Century* 48, no. 2 (Summer 2007): 125–48.

Pead, Patrick J. 'Benjamin Jesty: New Light in the Dawn of Vaccination', *The Lancet* 362, no. 9401 (20 Dec. 2003): 2104–109.

Percy, Victoria, and Gervase Jackson-Stops. '"Exquisite taste and tawdry ornament": The Travel Journals of the 1st Duchess of Northumberland—II', *Country Life*, 7 Feb. 1974, 250–52.

Platzer, David. 'Botanical Watercolours from Calcutta', *British Art Journal* 7, no. 2 (2006): 117.

Pointon, M. '"Surrounded with Brilliants": Miniature Portraits in Eighteenth-Century England', *The Art Bulletin* 83, no. 1 (2001): 48–71.
Pollard, G. 'Matthew Boulton and Conrad Heinrich Küchler', *Numismatic Chronicle* (7th series) 10 (1970): 259–318.
Prance, Ghillean T. 'A Brief History of Conservation at the Royal Botanic Gardens, Kew', *Kew Bulletin* 65, no. 4 (2010): 501–8.
Ramsden, C. 'Bookbinders to George III and his Immediate Descendants and Collaterals', *The Library* (5th series) 13 (1958): 186–93.
Roberts, Hugh. 'Metamorphosis in Wood: Royal Library Furniture in the Eighteenth and Nineteenth Centuries', *Apollo* 131 (June 1990): 383–90.
——. 'So Beautiful a Style', *Country Life*, 23 Nov. 1995, 58–59.
Von Rohr, A. 'Bildnisse der Sophie von der Pfalz und des Kurfursten Ernst August von Hannover', *Niederdeutsche Beitrage zur Kunstgeschichte* 20 (1981): 127–50.
Rorschach, Kimerly. 'Frederick Prince of Wales (1707–1751) as Collector and Patron', *Walpole Society* 55 (1988/90): 1–76.
Russell, F. 'King George III's Picture Hang at Buckingham House', *The Burlington Magazine* 129 (1987): 524–31.
Sanders, Julie. 'Caroline Salon Culture and Female Agency: The Countess of Carlisle, Henrietta Maria and Public Theatre', *Theatre Journal* 52, no. 4 (Dec. 2000): 449–64.
Schuhmann, Günther. 'Ansbacher Bibliotheken vom Mittelalter bis 1806', *Schriften des Instituts für Fränkischen Landesforschung und der Universität Erlangen* 8 (1961): 90–96.
Schulz, E. 'Notes on the History of Collecting and of Museums in the Light of Selected Literature of the Sixteenth to the Eighteenth Century', *Journal of the History of Collections* 2, no. 2 (1990): 205–18.
Schutte, R. A. 'The Kunst- und Naturalienkammer of Duke Anton Ulrich of Brunswick-Luneburg at Scloss Salzdahlum: Cabinet Collections, Literature and Science in the First Part of the Eighteenth Century', *Journal of the History of Collections* 9, no. 1 (1997): 79–115.
Sherlock, Peter. 'The Monuments of Elizabeth Tudor and Mary Stuart: King James and the Manipulation of Memory', *Journal of British Studies* 46, no. 2 (April 2007): 263–89.
Sicca, C. '"Like a shallow cave by nature made": William Kent's "Natural" Architecture at Richmond', *Architectura* 16 (1986): 68–82.
Sickinger, Raymond L. 'Regulation or Ruination: Parliament's Consistent Pattern of Mercantilist Regulation of the English Textile Trade, 1660–1800', *Parliamentary History* 19, no. 2 (2000): 211–32.
Simon, Jacob. 'Frame Studies II: Allan Ramsay and Picture Frames', *The Burlington Magazine* 136 (July 1994): 444–55.
Steguweit, Wolfgang, and Martin Hoernes. 'Gothas Gold und Silber: Der Münzschatz Herzog Friedrichs II. von Sachsen-Gotha-Altenburg kehrt zurück nach Friedenstein', *Arsprototo* 4 (2011): 38–41.
St John, Harold. 'The First Collection of Hawaiian Plants by David Nelson in 1779: Hawaiian Plant Studies 55', *Pacific Science* 32, no. 3 (1978): 315–24.
Styles, John. 'Spinners and the Law: Regulating Yarn Standards in the English Worsted Industries, 1550–1800', *Textile History* 44, no. 2 (2013): 145–70.
Sweet, Julie Anne. 'Bearing the Feathers of the Eagle: Tomochichihi's Trip to England', *Georgia Historical Quarterly* 86, no. 3 (Autumn 2002): 339–71.
Symes, M. 'The Landscaping of Esher Place', *Journal of Garden History* 8, no. 4 (1988): 63–96.
Thorp, J. 'Anthony L'Abbé and the Court of St James's: Dances to Honour the Hanoverian Royal Family', *The Court Historian* 15, no. 2 (2010): 189–206.
Treckel, Paula A. 'Breastfeeding and Maternal Sexuality in Colonial America', *Journal of Interdisciplinary History* 20, no. 1 (Summer 1989): 25–51.
Turner, G. L'E. 'The Auction Sales of the Earl of Bute's Instruments, 1793', *Annals of Science* 23, no. 3 (1967): 213–14.
——. 'Queen Charlotte's Protractor and Joshua Kirby', *Bulletin of the Scientific Instrument Society* 35 (Dec. 1992): 23–24.
Ulbricht, Otto. 'The Debate about Foundling Hospitals in Enlightenment Germany: Infanticide, Illegitimacy and Infant Mortality Rates', *Central European History* 18 (1 Sept. 1985): 211–56.
Vertue, George. 'Vertue Note Books I–VI', *Walpole Society* 18, 20, 22, 24, 26, 30 (1930–50).
Walpole, Horace. 'Walpole's Journals of Visits to Country Seats', ed. Paget Toynbee, *Walpole Society* 16 (1928).
Walton, K.-M. 'Queen Charlotte's Dressing Table', *Furniture History* 11 (1975): 112–83.
Webb, M. I. 'Busts of Sir Isaac Newton', *Country Life*, 25 Jan. 1952, 216–18.
Weiss, Robin, and José Esparza. 'The Prevention and Eradication of Smallpox: A Commentary on Sloane (1755) "An Account of Inoculation"', *Philosophical Transactions of the Royal Society* B 370 (April 2015).
Wilson, M. I. 'A Taste for the Strange: William Kent's Book Illustrations', *Country Life*, 7 Nov. 1985, 1396–98.
Wilson, Peter H. 'Prussia's Relations with the Holy Roman Empire, 1740–1786', *Historical Journal* 51, no. 2 (June 2008): 337–71.
Winterbottom, M. '"Such Massy Pieces of Plate": Silver Furnishings in the English Royal Palaces 1660–1702', *Apollo* 160 (2002): 19–26.
Wittkower, R. 'Lord Burlington and William Kent', *Archaeological Journal* 102 (1946): 760–61.
Woodbridge, Kenneth. 'William Kent as Landscape-Gardener: A Re-appraisal', *Apollo* 100 (Aug. 1974): 126–37.
Woodward, J. 'Amigoni as a Portrait Painter in England', *The Burlington Magazine* 99 (1957): 20–21.
Worsley, G. 'Houghton Hall, Norfolk', *Country Life*, 4 March 1993, 50–53.

Index

NOTE: Page numbers in italics refer to an illustration. Page numbers followed by 'n' to information in a note. Page numbers in bold italics refer to the catalogue entry for a work or item by an artist or maker. Artists' works appear at the end of their index entry. Works by unknown artists are entered by title.

Photography Credits

Every effort has been made to credit the photographers and sources of all illustrations in this volume; if there are any errors or omissions, please contact Yale University Press so that corrections can be made in any subsequent edition.

Beinecke Rare Book & Manuscript Library, Yale University: pages 30, 438; figs. 2.01, 2.02, 6.03, 11.10, 16.06, 16.26, 17.03, 17.12, 17.14, 17.16, 17.21, 17.24, 19.11, 22.04, 23.24, 24.08, 26.06, 27.13, 29.31, 29.45, 29.47, 29.48, 29.49

© British Library Board: pages 202, 526; figs. 5.13, 6.10, 11.05, 11.06, 11.08, 11.11, 11.12, 12.05, 12.07, 17.05, 17.18, 23.14, 23.23, 24.03

By permission of the Richmond Borough Art Collection at Orleans House Gallery: fig. 6.02

Château de Versailles, France / Bridgeman Images: fig. 2.04

Collection of H.R.H. Hereditary Prince Ernst August of Hanover, Duke of Brunswick and Lüneburg/Photograph: Städtisches Museum Braunschweig, Jakob Adolphi: fig. 30.07

© Coram in the care of the Foundling Museum, London: figs. 25.03, 25.04, 25.06, 25.07, 25.10, 25.11, 25.12, 25.13, 25.14

© Coram in the care of the Foundling Museum, London/Bridgeman Images: page 406; fig. 24.02

Fine Arts Museums of San Francisco, Achenbach Foundation for Graphic Arts: fig. 4.12

Fine Arts Museum of San Francisco, Palace of the Legion of Honor, Mildred Anna Williams Collection: fig. 8.02

© Fitzwilliam Museum, Cambridge: figs. 6.14, 11.01

Folger Shakespeare Library, Washington, DC: cover detail (middle row, right); page viii; fig. 2.03

Photograph by Fred Fabre: fig. 31.02

The Trustees of the Goodwood Collection / Bridgeman Images: fig. 4.04

Historic Royal Palaces: figs. 13.24, 15.02, 15.04, 17.02, 17.08, 19.03, 19.05, 22.10, 23.19, 23.26, 28.08, 30.12

Reproduced by kind permission of the House of Orange-Nassau, Historic Collections Trust. Koninklijk Huisarchief, The Hague: figs. 23.10, 23.13

© Courtesy of the Huntington Art Collections, San Marino, California: fig. 11.09

Library of Congress, British Cartoon Prints Collection: fig. 28.22

Library of Congress, Washington, DC, Photochrom Print Collection: fig. 5.17

Lady Baker, née Muir (Lord and Lady Baker Collection): figs. 13.12, 13.13

Photograph by Jonathan Bassett: fig. 31.03

Mount Stuart Trust, Mount Stuart: figs. 13.11, 16.19, 16.20

Museum of the History of Science, University of Oxford: fig. 16.25

© National Portrait Gallery, London: figs. 4.10, 4.11, 8.01, 11.07, 11.16, 16.02, 16.11, 16.21, 16.27, 16.29, 17.15, 19.06, 19.12, 23.03, 24.05, 26.01, 26.07, 27.01, 27.03, 29.12, 29.14

© National Portrait Gallery, London / National Museum Cardiff / Captain Cook Memorial Museum, Whitby: fig. 30.11

©National Maritime Museum, Greenwich, London: fig. 6.11

© National Trust Images: cover detail (middle row, middle); figs. 2.06, 13.10, 23.01, 23.04

National Trust, Uppark House and Garden, West Sussex, UK / Bridgeman Images: figs. 3.09, 8.03

The New York Public Library, Rare Books Division, Astor, Lenox and Tilden Foundations: fig. 21.14

Oak Spring Garden Library, Upperville, VA: back cover; figs. 6.04, 20.04, 20.05, 20.06, 29.07, 29.16, 29.17, 29.22, 29.32, 29.34

© Philip Mould, Historical Portraits/Bridgeman Images: figs. 3.07, 3.08

Private collection: pages 160–161; fig. 7.02

Private collection: fig. 29.13

© Royal Albert Memorial Museum & Art Gallery, Exeter City Council: page xiv; figs. 1.14, 13.14,

Royal Botanic Gardens, Kew, Library, Art & Archives Collections: figs. 3.06, 6.01, 6.09, 6.15a, 6.15b, 15.01, 19.13, 22.05, 22.07, 29.04, 29.05, 29.06, 29.16, 29.17, 29.21, 29.24, 29.30, 29.39, 29.41

Royal Collection Trust/© Her Majesty Queen Elizabeth II 2016: cover detail (top row, left); cover detail (bottom row, left); cover detail (bottom row, right); frontispiece; pages vi, x, 108, 176, 184, 216–217, 218, 228, 304, 324, 358, 372–373, 374, 384, 450–451, 474; figs. 1.01, 1.02, 1.03, 1.04, 1.05, 1.06, 1.08, 1.09, 1.15, 3.01, 3.03, 4.03, 4.05, 4.06, 4.07, 4.08, 4.09, 4.13, 4.14, 4.15, 4.16, 4.18, 4.19, 4.20, 4.21, 4.22, 4.23, 4.24, 4.26, 4.27, 5.01, 5.02, 5.03, 5.04, 5.05, 5.06, 5.07, 5.08, 5.09, 5.10, 5.11, 5.12, 5.15, 5.16, 5.18, 5.19, 5.21, 5.22, 5.23, 6.05, 6.06, 7.01, 7.04, 8.04, 9.01, 9.02, 9.03, 9.04, 9.05, 9.06, 9.07, 9.08, 9.09, 9.10, 9.11, 10.01, 10.02, 10.04, 10.05, 10.06, 10.07, 10.08, 10.09, 10.10, 10.11, 10.12, 10.13, 10.14, 10.15, 10.16, 10.17, 10.18, 10.19, 10.20, 10.21, 10.22, 10.23, 10.24, 10.25, 10.26, 10.27, 10.28, 10.29, 10.30, 10.31, 10.32, 11.02, 11.03, 11.04, 11.14, 12.01, 12.02, 12.03, 12.04, 12.06, 12.08, 12.09, 12.10, 13.02, 13.05, 13.18, 13.21, 13.23, 13.27, 16.01, 16.05, 16.10, 16.12, 16.13, 16.14, 16.18, 17.01, 17.04, 17.06, 17.07, 17.10, 17.17, 17.19, 17.20, 18.05, 18.06, 19.01, 19.09, 19.16, 21.02, 21.05, 21.06, 21.07, 21.13, 22.02, 22.06, 22.08, 22.09, 23.02, 23.06, 23.08, 23.09, 23.11, 23.16, 23.17, 23.18, 23.21, 23.25, 26.03, 26.04, 26.05, 26.12, 28.01, 28.02, 28.03, 28.09, 28.14, 28.15, 28.16, 28.17, 28.18, 28.19, 28.20, 28.23, 29.02, 29.03, 29.18, 29.19, 29.20, 29.40, 29.43, 29.44, 30.08, 30.09

Royal Institute of British Architects, Library Drawings & Archives Collections: figs. 19.17, 20.07

© Science Museum/Science & Society Picture Library: cover detail (top row, middle); figs. 1.10, 7.03, 16.09, 16.24, 27.10, 31.01
Scottish National Gallery. On loan from the Scottish Government, 2011: fig 16.17
Scottish National Gallery/Antonia Reeve: fig. 13.16
Smithsonian Institution, National Museum of American History, Division of Home and Community Life, Washington, DC: page 514; figs. 30.01, 30.02, 30.06
The Chipstone Foundation, photo by Gavin Ashworth: fig. 29.33
© The Hunterian Museum and Art Gallery, University of Glasgow 2016: page 426; figs. 16.03, 26.09, 29.51
The Lewis Walpole Library, Yale University: pages 264, 414; figs. 2.07, 2.08, 2.09, 10.03, 11.15, 14.01, 14.09, 14.13, 15.03, 15.06, 15.07, 15.09, 16.15, 20.06, 22.11, 25.05
The Metropolitan Museum of Art, Morris K. Jesup Fund, 1955/Art Resource, NY: fig. 16.31
The Metropolitan Museum of Art, New York, Gift of Walter Smith, 1967: fig. 24.06
The Metropolitan Museum of Art, New York, Harris Brisbane Dick Fund, 1925: figs. 18.04, 19.04, 29.35, 29.36, 29.37, 29.38,
The Metropolitan Museum of Art, Rogers Fund, 2004: fig. 28.06
The Mint Museum, Charlotte, NC, Gift of the Delhom Service League in memory of Martha Higgins Fishburne, 1997: fig. 28.10
The Morgan Library and Museum, New York, Miss Julia P. Wightman Bequest: fig. 18.07
The National Society of The Colonial Dames of America in the State of South Carolina: cover detail (bottom row, middle), pages 532, 534, 536; fig. 30.10
The Newport Garden Club, Newport, RI: figs. 6.12, 22.01
The New York Botanical Garden, The LuEsther T. Mertz Library: figs. 29.09, 29.10, 29.42
The Provost and Fellows of The Queen's College, Oxford: fig. 13.17
The Royal Danish Collection, Rosenborg Castle: fig. 23.12
© The Royal Society: figs. 16.04, 26.08
The State Hermitage Museum, St. Petersburg, Russia/Bridgeman Images: page 52; fig. 3.10
The Stuart Collection: fig. 16.32
© The Trustees of the Natural History Museum, London: page 348; figs. 6.07, 6.08, 6.13, 6.16, 20.09, 29.08, 29.23, 29.23
© The Trustees of the Natural History Museum, London / Bridgeman Images: fig. 20.08
© Timm Allrich / Staatliche Schlösser und Gärten Mecklenburg-Vorpommern: fig. 5.20
Trustees of Sir John Soanes's Museum: pages 276-277; figs. 5.14, 17.09, 19.07, 19.10
© Trustees of the British Museum: cover detail (middle row, left); pages 76, 132, 162, 252, 322-323; figs. 1.12, 1.13, 3.02, 4.01, 4.02, 4.17, 7.07, 13.03, 14.02, 14.03, 14.04, 14.05, 14.06, 14.07, 14.08, 14.10, 14.12, 15.05, 15.08, 16.07, 19.08, 20.01, 20.02, 20.03, 21.09, 21.10, 21.11, 21.12, 23.07, 29.01
© UK Government Art Collection: figs. 4.25, 13.01, 13.20, 23.20
University of Aberdeen: figs. 3.04, 16.28
© Victoria and Albert Museum, London: pages 170, 452; figs. 1.11, 8.06, 16.16, 16.33, 16.34, 21.08, 23.22, 28.04, 28.05, 28.07
Courtesy of The Victoria Art Gallery, Bath and North East Somerset Council: fig. 13.09
Wellcome Library, London: figs. 24.01, 24.04, 24.07, 24.09, 26.02, 26.11, 27.02, 27.04, 27.09, 27.11, 27.12
Copyright: Dean and Chapter of Westminster: fig. 3.05
Winterthur Museum, Art Gallery and Garden: fig. 28.11
Winterthur Museum, Art Gallery and Garden, Gift of Henry Francis du Pont: figs. 30.03, 30.04
Yale Center for British Art, Gift of David Bindman/Richard Caspole: fig. 7.06
Yale Center for British Art, Gift of Margaret Wimsatt in memory of William Kurtz Wimsatt Jr/Richard Caspole: fig. 17.13
Yale Center for British Art, In Memory of John V. McCarthy/Richard Caspole: fig. 17.11
Yale Center for British Art, Paul Mellon Collection/ Richard Caspole: cover detail (top row, right), page 334; figs. 7.05, 8.05, 13.06, 13.07, 13.15, 13.19, 13.25, 13.26, 13.28, 14.11, 15.10, 16.23, 16.30, 17.22, 17.23, 17.25, 18.01, 18.02, 18.03, 19.02, 19.14, 19.15, 21.01, 21.03, 21.04, 22.03, 23.05, 23.15, 25.01, 25.02, 25.08, 25.09, 27.08, 28.12, 29.11, 29.15, 29.25, 29.26, 29.27, 29.28, 29.29, 29.46, 29.50, 30.05,
Yale Center for British Art, Paul Mellon Collection, Gift of Susan M. Yecies, Yale, B.A. 1971/Richard Caspole: fig. 28.13
Yale Center for British Art, Paul Mellon Fund/ Richard Caspole: fig. 28.21
Yale University Art Gallery, Gift of the Associates in Fine Arts: fig. 27.07
Yale University Art Gallery, Transfer from the Yale University Library, Numismatic Collection, 2001: figs. 1.07, 13.04, 13.08
Yale University Art Gallery, Transfer from the Yale University Library, Numismatic Collection, 2001, Gift of Irving Dillaye-Vann, B.A. 1897: fig. 13.22
Yale University Library: fig. 2.05
Yale University, Irving S. Gilmore Music Library: fig. 11.13
Yale University, Medical History Library: page 278; figs. 16.08, 16.22, 26.10, 27.05

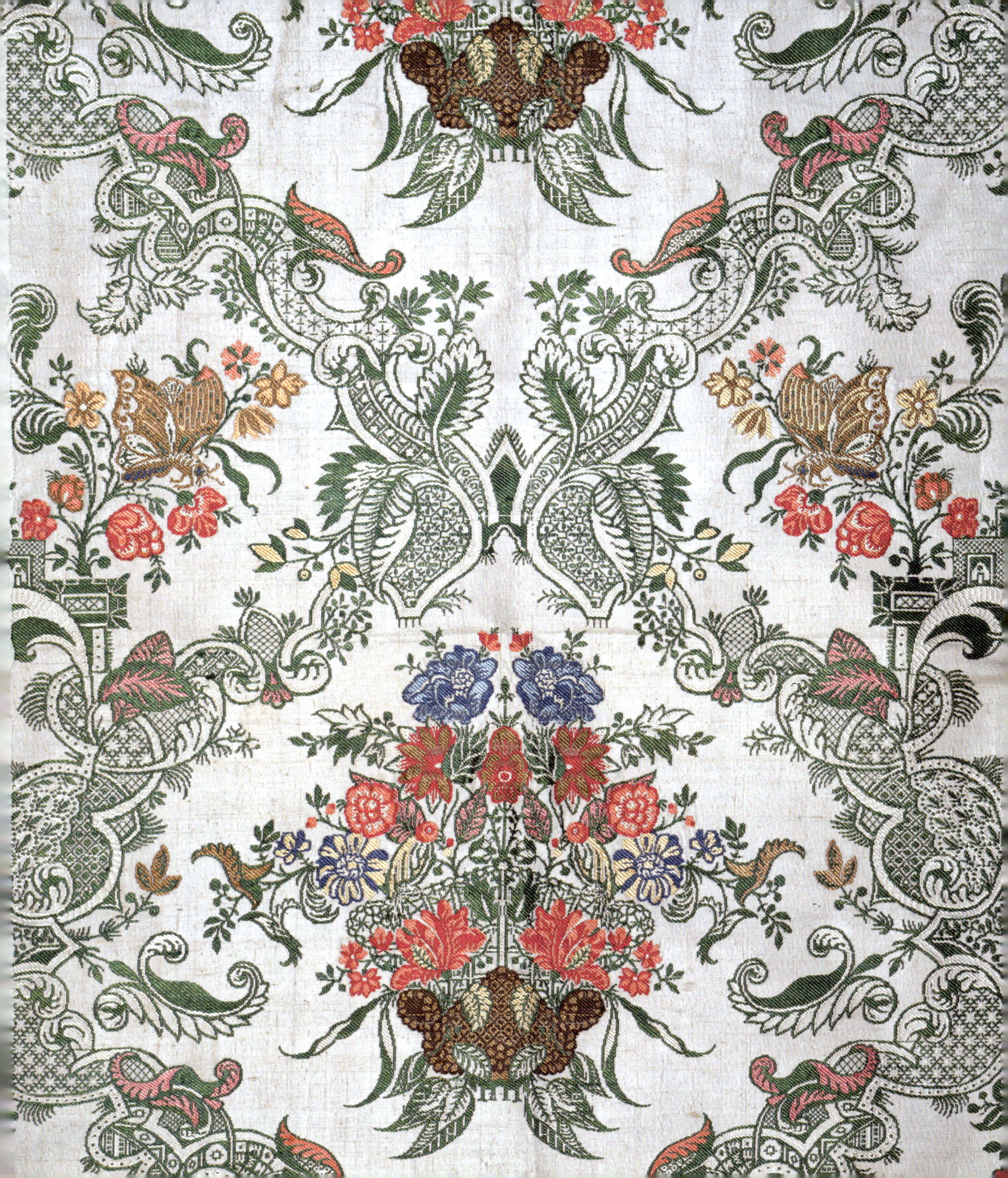

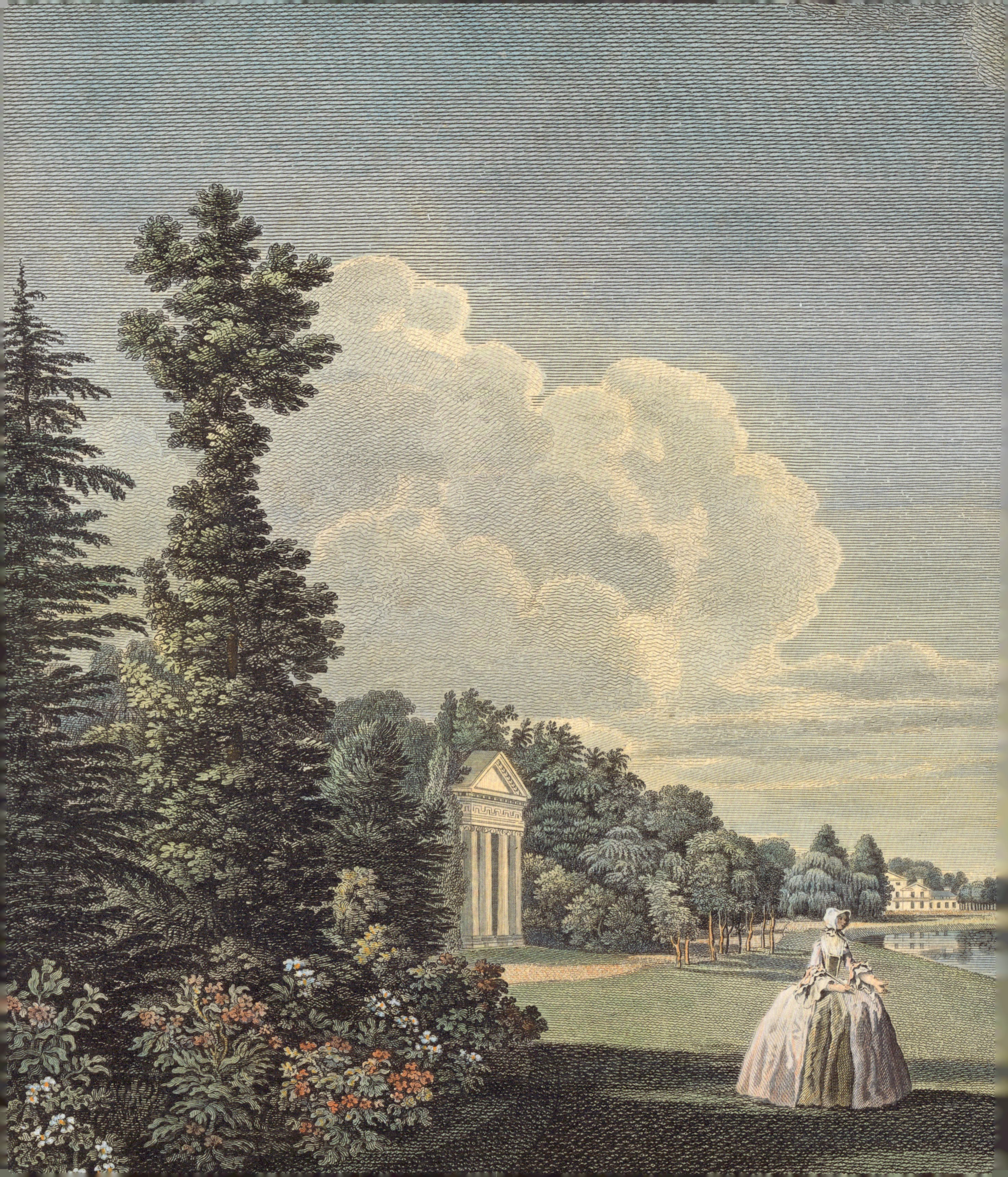